Encyclopedia of Algorithms

Ming-Yang Kao

Editor

Encyclopedia of Algorithms

Second Edition

Volume 3

Q–Z

With 379 Figures and 51 Tables

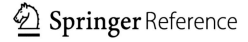

Editor
Ming-Yang Kao
Department of Electrical Engineering
and Computer Science
Northwestern University
Evanston, IL, USA

ISBN 978-1-4939-2863-7   ISBN 978-1-4939-2864-4   (eBook)
ISBN 978-1-4939-2865-1   (print and electronic bundle)
DOI 10.1007/ 978-1-4939-2864-4

Library of Congress Control Number: 2015958521

Printed on acid-free paper

This Springer imprint is published by SpringerNature

The registered company is Springer Science+Business Media LLC New York

Preface

The Encyclopedia of Algorithms provides researchers, students, and practitioners of algorithmic research with a mechanism to efficiently and accurately find the names, definitions, and key results of important algorithmic problems. It also provides further readings on those problems.

This *encyclopedia* covers a broad range of algorithmic areas; each area is summarized by a collection of entries. The entries are written in a clear and concise structure so that they can be readily absorbed by the readers and easily updated by the authors. A typical encyclopedia entry is an in-depth mini-survey of an algorithmic problem written by an expert in the field. The entries for an algorithmic area are compiled by area editors to survey the representative results in that area and can form the core materials of a course in the area.

This 2nd edition of the encyclopedia contains a wide array of important new research results. Highlights include works in tile self-assembly (nanotechnology), bioinformatics, game theory, Internet algorithms, and social networks. Overall, more than 70 % of the entries in this edition and new entries are updated.

This reference work will continue to be updated on a regular basis via a live site to allow timely updates and fast search. Knowledge accumulation is an ongoing community project. Please take ownership of this body of work. If you have feedback regarding a particular entry, please feel free to communicate directly with the author or the area editor of that entry. If you are interested in authoring a future entry, please contact a suitable area editor. If you have suggestions on how to improve the Encyclopedia as a whole, please contact me at kao@northwestern.edu. The credit of this Encyclopedia goes to the area editors, the entry authors, the entry reviewers, and the project editors at Springer, including Melissa Fearon, Michael Hermann, and Sylvia Blago.

About the Editor

Ming-Yang Kao is a Professor of Computer Science in the Department of Electrical Engineering and Computer Science at Northwestern University. He has published extensively in the design, analysis, and applications of algorithms. His current interests include discrete optimization, bioinformatics, computational economics, computational finance, and nanotechnology. He serves as the Editor-in-Chief of Algorithmica.

He obtained a B.S. in Mathematics from National Taiwan University in 1978 and a Ph.D. in Computer Science from Yale University in 1986. He previously taught at Indiana University at Bloomington, Duke University, Yale University, and Tufts University. At Northwestern University, he has served as the Department Chair of Computer Science. He has also cofounded the Program in Computational Biology and Bioinformatics and served as its Director. He currently serves as the Head of the EECS Division of Computing, Algorithms, and Applications and is a Member of the Theoretical Computer Science Group.

For more information, please see www.cs.northwestern.edu/~kao

Area Editors

Algorithm Engineering

Giuseppe F. Italiano* Department of Computer and Systems Science, University of Rome, Rome, Italy

Department of Information and Computer Systems, University of Rome, Rome, Italy

Rajeev Raman* Department of Computer Science, University of Leicester, Leicester, UK

Algorithms for Modern Computers

Alejandro López-Ortiz David R. Cheriton School of Computer Science, University of Waterloo, Waterloo, ON, Canada

Algorithmic Aspects of Distributed Sensor Networks

Sotiris Nikoletseas Computer Engineering and Informatics Department, University of Patras, Patras, Greece

Computer Technology Institute and Press "Diophantus", Patras, Greece

Approximation Algorithms

Susanne Albers* Technical University of Munich, Munich, Germany

Chandra Chekuri* Department of Computer Science, University of Illinois, Urbana-Champaign, Urbana, IL, USA

Department of Mathematics and Computer Science, The Open University of Israel, Raanana, Israel

Ming-Yang Kao Department of Electrical Engineering and Computer Science, Northwestern University, Evanston, IL, USA

Sanjeev Khanna* University of Pennsylvania, Philadelphia, PA, USA

Samir Khuller* Computer Science Department, University of Maryland, College Park, MD, USA

---

*Acknowledgment for first edition contribution

Average Case Analysis

Paul (Pavlos) Spirakis* Computer Engineering and Informatics, Research and Academic Computer Technology Institute, Patras University, Patras, Greece

Computer Science, University of Liverpool, Liverpool, UK

Computer Technology Institute (CTI), Patras, Greece

Bin Packing

Leah Epstein Department of Mathematics, University of Haifa, Haifa, Israel

Bioinformatics

Miklós Csűrös Department of Computer Science, University of Montréal, Montréal, QC, Canada

Certified Reconstruction and Mesh Generation

Siu-Wing Cheng Department of Computer Science and Engineering, Hong Kong University of Science and Technology, Hong Kong, China

Tamal Krishna Dey Department of Computer Science and Engineering, The Ohio State University, Columbus, OH, USA

Coding Algorithms

Venkatesan Guruswami* Department of Computer Science and Engineering, University of Washington, Seattle, WA, USA

Combinatorial Group Testing

Ding-Zhu Du Computer Science, University of Minnesota, Minneapolis, MN, USA

Department of Computer Science, The University of Texas at Dallas, Richardson, TX, USA

Combinatorial Optimization

Samir Khuller* Computer Science Department, University of Maryland, College Park, MD, USA

Compressed Text Indexing

Tak-Wah Lam Department of Computer Science, University of Hong Kong, Hong Kong, China

Compression of Text and Data Structures

Gonzalo Navarro Department of Computer Science, University of Chile, Santiago, Chile

Computational Biology

Bhaskar DasGupta Department of Computer Science, University of Illinois, Chicago, IL, USA

Tak-Wah Lam Department of Computer Science, University of Hong Kong, Hong Kong, China

Computational Counting

Xi Chen Computer Science Department, Columbia University, New York, NY, USA

Computer Science and Technology, Tsinghua University, Beijing, China

Computational Economics

Xiaotie Deng AIMS Laboratory (Algorithms-Agents-Data on Internet, Market, and Social Networks), Department of Computer Science and Engineering, Shanghai Jiao Tong University, Shanghai, China

Department of Computer Science, City University of Hong Kong, Hong Kong, China

Computational Geometry

Sándor Fekete Department of Computer Science, Technical University Braunschweig, Braunschweig, Germany

Computational Learning Theory

Rocco A. Servedio Computer Science, Columbia University, New York, NY, USA

Data Compression

Paolo Ferragina* Department of Computer Science, University of Pisa, Pisa, Italy

Differential Privacy

Aaron Roth Department of Computer and Information Sciences, University of Pennsylvania, Levine Hall, PA, USA

Distributed Algorithms

Sergio Rajsbaum Instituto de Matemáticas, Universidad Nacional Autónoma de México (UNAM) México City, México

Dynamic Graph Algorithms

Giuseppe F. Italiano* Department of Computer and Systems Science, University of Rome, Rome, Italy

Department of Information and Computer Systems, University of Rome, Rome, Italy

Enumeration Algorithms

Takeaki Uno National Institute of Informatics, Chiyoda, Tokyo, Japan

Exact Exponential Algorithms

Fedor V. Fomin Department of Informatics, University of Bergen, Bergen, Norway

External Memory Algorithms

Herman Haverkort Department of Computer Science, Eindhoven University of Technology, Eindhoven, The Netherlands

Game Theory

Mohammad Taghi Hajiaghayi Department of Computer Science, University of Maryland, College Park, MD, USA

Geometric Networks

Andrzej Lingas Department of Computer Science, Lund University, Lund, Sweden

Graph Algorithms

Samir Khuller* Computer Science Department, University of Maryland, College Park, MD, USA

Seth Pettie Electrical Engineering and Computer Science (EECS) Department, University of Michigan, Ann Arbor, MI, USA

Vijaya Ramachandran* Computer Science, University of Texas, Austin, TX, USA

Liam Roditty Department of Computer Science, Bar-Ilan University, Ramat-Gan, Israel

Dimitrios Thilikos AlGCo Project-Team, CNRS, LIRMM, France

Department of Mathematics, National and Kapodistrian University of Athens, Athens, Greece

Graph Drawing

Seokhee Hong School of Information Technologies, University of Sydney, Sydney, NSW, Australia

Internet Algorithms

Edith Cohen Tel Aviv University, Tel Aviv, Israel

Stanford University, Stanford, CA, USA

I/O-Efficient Algorithms

Herman Haverkort Department of Computer Science, Eindhoven University of Technology, Eindhoven, The Netherlands

Kernels and Compressions

Gregory Gutin Department of Computer Science, Royal Holloway, University of London, Egham, UK

Massive Data Algorithms

Herman Haverkort Department of Computer Science, Eindhoven University of Technology, Eindhoven, The Netherlands

Mathematical Optimization

Ding-Zhu Du Computer Science, University of Minnesota, Minneapolis, MN, USA

Department of Computer Science, The University of Texas at Dallas, Richardson, TX, USA

Mechanism Design

Yossi Azar* Tel-Aviv University, Tel Aviv, Israel

Mobile Computing

Xiang-Yang Li* Department of Computer Science, Illinois Institute of Technology, Chicago, IL, USA

Modern Learning Theory

Maria-Florina Balcan Department of Machine Learning, Carnegie Mellon University, Pittsburgh, PA, USA

Online Algorithms

Susanne Albers* Technical University of Munich, Munich, Germany

Yossi Azar* Tel-Aviv University, Tel Aviv, Israel

Marek Chrobak Computer Science, University of California, Riverside, CA, USA

Alejandro López-Ortiz David R. Cheriton School of Computer Science, University of Waterloo, Waterloo, ON, Canada

Parameterized Algorithms

Dimitrios Thilikos AlGCo Project-Team, CNRS, LIRMM, France

Department of Mathematics, National and Kapodistrian University of Athens, Athens, Greece

Parameterized Algorithms and Complexity

Saket Saurabh Institute of Mathematical Sciences, Chennai, India

University of Bergen, Bergen, Norway

Parameterized and Exact Algorithms

Rolf Niedermeier* Department of Mathematics and Computer Science, University of Jena, Jena, Germany

Institut für Softwaretechnik und Theoretische Informatik, Technische Universität Berlin, Berlin, Germany

Price of Anarchy

Yossi Azar* Tel-Aviv University, Tel Aviv, Israel

Probabilistic Algorithms

Sotiris Nikoletseas Computer Engineering and Informatics Department, University of Patras, Patras, Greece

Computer Technology Institute and Press "Diophantus", Patras, Greece

Paul (Pavlos) Spirakis* Computer Engineering and Informatics, Research and Academic Computer Technology Institute, Patras University, Patras, Greece

Computer Science, University of Liverpool, Liverpool, UK

Computer Technology Institute (CTI), Patras, Greece

Quantum Computing

Andris Ambainis Faculty of Computing, University of Latvia, Riga, Latvia

Radio Networks

Marek Chrobak Computer Science, University of California, Riverside, CA, USA

Scheduling

Leah Epstein Department of Mathematics, University of Haifa, Haifa, Israel

Scheduling Algorithms

Viswanath Nagarajan University of Michigan, Ann Arbor, MI, USA

Kirk Pruhs* Department of Computer Science, University of Pittsburgh, Pittsburgh, PA, USA

Social Networks

Mohammad Taghi Hajiaghayi Department of Computer Science, University of Maryland, College Park, MD, USA

Contributors

Karen Aardal Centrum Wiskunde & Informatica (CWI), Amsterdam, The Netherlands

Department of Mathematics and Computer Science, Eindhoven University of Technology, Eindhoven, The Netherlands

Ittai Abraham Microsoft Research, Silicon Valley, Palo Alto, CA, USA

Adi Akavia Department of Electrical Engineering and Computer Science, MIT, Cambridge, MA, USA

Réka Albert Department of Biology and Department of Physics, Pennsylvania State University, University Park, PA, USA

Mansoor Alicherry Bell Laboratories, Alcatel-Lucent, Murray Hill, NJ, USA

Noga Alon Department of Mathematics and Computer Science, Tel-Aviv University, Tel-Aviv, Israel

Srinivas Aluru Department of Electrical and Computer Engineering, Iowa State University, Ames, IA, USA

Andris Ambainis Faculty of Computing, University of Latvia, Riga, Latvia

Christoph Ambühl Department of Computer Science, University of Liverpool, Liverpool, UK

Nina Amenta Department of Computer Science, University of California, Davis, CA, USA

Amihood Amir Department of Computer Science, Bar-Ilan University, Ramat-Gan, Israel

Department of Computer Science, Johns Hopkins University, Baltimore, MD, USA

Spyros Angelopoulos Sorbonne Universités, L'Université Pierre et Marie Curie (UPMC), Université Paris 06, Paris, France

Anurag Anshu Center for Quantum Technologies, National University of Singapore, Singapore, Singapore

Alberto Apostolico College of Computing, Georgia Institute of Technology, Atlanta, GA, USA

Vera Asodi Center for the Mathematics of Information, California Institute of Technology, Pasadena, CA, USA

Peter Auer Chair for Information Technology, Montanuniversitaet Leoben, Leoben, Austria

Pranjal Awasthi Department of Computer Science, Princeton University, Princeton, NJ, USA

Department of Electrical Engineering, Indian Institute of Technology Madras, Chennai, Tamilnadu, India

Adnan Aziz Department of Electrical and Computer Engineering, University of Texas, Austin, TX, USA

Moshe Babaioff Microsoft Research, Herzliya, Israel

David A. Bader College of Computing, Georgia Institute of Technology, Atlanta, GA, USA

Michael Bader Department of Informatics, Technical University of Munich, Garching, Germany

Maria-Florina Balcan Department of Machine Learning, Carnegie Mellon University, Pittsburgh, PA, USA

Hideo Bannai Department of Informatics, Kyushu University, Fukuoka, Japan

Nikhil Bansal Eindhoven University of Technology, Eindhoven, The Netherlands

Jérémy Barbay Department of Computer Science (DCC), University of Chile, Santiago, Chile

Sanjoy K. Baruah Department of Computer Science, The University of North Carolina, Chapel Hill, NC, USA

Surender Baswana Department of Computer Science and Engineering, Indian Institute of Technology (IIT), Kanpur, Kanpur, India

MohammadHossein Bateni Google Inc., New York, NY, USA

Luca Becchetti Department of Information and Computer Systems, University of Rome, Rome, Italy

Xiaohui Bei Division of Mathematical Sciences, School of Physical and Mathematical Sciences, Nanyang Technological University, Singapore, Singapore

József Békési Department of Computer Science, Juhász Gyula Teachers Training College, Szeged, Hungary

Djamal Belazzougui Department of Computer Science, Helsinki Institute for Information Technology (HIIT), University of Helsinki, Helsinki, Finland

Aleksandrs Belovs Computer Science and Artificial Intelligence Laboratory, MIT, Cambridge, MA, USA

Aaron Bernstein Department of Computer Science, Columbia University, New York, NY, USA

Vincent Berry Institut de Biologie Computationnelle, Montpellier, France

Randeep Bhatia Bell Laboratories, Alcatel-Lucent, Murray Hill, NJ, USA

Andreas Björklund Department of Computer Science, Lund University, Lund, Sweden

Eric Blais University of Waterloo, Waterloo, ON, Canada

Mathieu Blanchette Department of Computer Science, McGill University, Montreal, QC, Canada

Markus Bläser Department of Computer Science, Saarland University, Saarbrücken, Germany

Avrim Blum School of Computer Science, Carnegie Mellon University, Pittsburgh, PA, USA

Hans L. Bodlaender Department of Computer Science, Utrecht University, Utrecht, The Netherlands

Sergio Boixo Quantum A.I. Laboratory, Google, Venice, CA, USA

Paolo Boldi Dipartimento di Informatica, Università degli Studi di Milano, Milano, Italy

Glencora Borradaile Department of Computer Science, Brown University, Providence, RI, USA

School of Electrical Engineering and Computer Science, Oregon State University, Corvallis, OR, USA

Ulrik Brandes Department of Computer and Information Science, University of Konstanz, Konstanz, Germany

Andreas Brandstädt Computer Science Department, University of Rostock, Rostock, Germany

Department of Informatics, University of Rostock, Rostock, Germany

Gilles Brassard Université de Montréal, Montréal, QC, Canada

Vladimir Braverman Department of Computer Science, Johns Hopkins University, Baltimore, MD, USA

Tian-Ming Bu Software Engineering Institute, East China Normal University, Shanghai, China

Adam L. Buchsbaum Madison, NJ, USA

Costas Busch Department of Computer Science, Lousiana State University, Baton Rouge, LA, USA

Jaroslaw Byrka Centrum Wiskunde & Informatica (CWI), Amsterdam, The Netherlands

Department of Mathematics and Computer Science, Eindhoven University of Technology, Eindhoven, The Netherlands

Jin-Yi Cai Beijing University, Beijing, China

Computer Sciences Department, University of Wisconsin–Madison, Madison, WI, USA

Mao-cheng Cai Chinese Academy of Sciences, Institute of Systems Science, Beijing, China

Yang Cai Computer Science, McGill University, Montreal, QC, Canada

Gruia Calinescu Department of Computer Science, Illinois Institute of Technology, Chicago, IL, USA

Colin Campbell Department of Physics, Pennsylvania State University, University Park, PA, USA

Luca Castelli Aleardi Laboratoire d'Informatique (LIX), École Polytechnique, Bâtiment Alan Turing, Palaiseau, France

Katarína Cechlárová Faculty of Science, Institute of Mathematics, P. J. Šafárik University, Košice, Slovakia

Nicolò Cesa-Bianchi Dipartimento di Informatica, Università degli Studi di Milano, Milano, Italy

Amit Chakrabarti Department of Computer Science, Dartmouth College, Hanover, NH, USA

Deeparnab Chakrabarty Microsoft Research, Bangalore, Karnataka, India

Erin W. Chambers Department of Computer Science and Mathematics, Saint Louis University, St. Louis, MO, USA

Chee Yong Chan National University of Singapore, Singapore, Singapore

Mee Yee Chan Department of Computer Science, University of Hong Kong, Hong Kong, China

Wun-Tat Chan College of International Education, Hong Kong Baptist University, Hong Kong, China

Tushar Deepak Chandra IBM Watson Research Center, Yorktown Heights, NY, USA

Kun-Mao Chao Department of Computer Science and Information Engineering, National Taiwan University, Taipei, Taiwan

Bernadette Charron-Bost Laboratory for Informatics, The Polytechnic School, Palaiseau, France

Ioannis Chatzigiannakis Department of Computer Engineering and Informatics, University of Patras and Computer Technology Institute, Patras, Greece

Shuchi Chawla Department of Computer Science, University of Wisconsin–Madison, Madison, WI, USA

Shiri Chechik Department of Computer Science, Tel Aviv University, Tel Aviv, Israel

Chandra Chekuri Department of Computer Science, University of Illinois, Urbana-Champaign, Urbana, IL, USA

Department of Mathematics and Computer Science, The Open University of Israel, Raanana, Israel

Danny Z. Chen Department of Computer Science and Engineering, University of Notre Dame, Notre Dame, IN, USA

Ho-Lin Chen Department of Electrical Engineering, National Taiwan University, Taipei, Taiwan

Jianer Chen Department of Computer Science, Texas A&M University, College Station, TX, USA

Ning Chen Division of Mathematical Sciences, School of Physical and Mathematical Sciences, Nanyang Technological University, Singapore, Singapore

Xi Chen Computer Science Department, Columbia University, New York, NY, USA

Computer Science and Technology, Tsinghua University, Beijing, China

Siu-Wing Cheng Department of Computer Science and Engineering, Hong Kong University of Science and Technology, Hong Kong, China

Xiuzhen Cheng Department of Computer Science, George Washington University, Washington, DC, USA

Huang Chien-Chung Chalmers University of Technology and University of Gothenburg, Gothenburg, Sweden

Markus Chimani Faculty of Mathematics/Computer, Theoretical Computer Science, Osnabrück University, Osnabrück, Germany

Francis Y.L. Chin Department of Computer Science, University of Hong Kong, Hong Kong, China

Rajesh Chitnis Department of Computer Science, University of Maryland, College Park, MD, USA

Minsik Cho IBM T. J. Watson Research Center, Yorktown Heights, NY, USA

Rezaul A. Chowdhury Department of Computer Sciences, University of Texas, Austin, TX, USA

Stony Brook University (SUNY), Stony Brook, NY, USA

George Christodoulou University of Liverpool, Liverpool, UK

Marek Chrobak Computer Science, University of California, Riverside, CA, USA

Chris Chu Department of Electrical and Computer Engineering, Iowa State University, Ames, IA, USA

Xiaowen Chu Department of Computer Science, Hong Kong Baptist University, Hong Kong, China

Julia Chuzhoy Toyota Technological Institute, Chicago, IL, USA

Edith Cohen Tel Aviv University, Tel Aviv, Israel

Stanford University, Stanford, CA, USA

Jason Cong Department of Computer Science, UCLA, Los Angeles, CA, USA

Graham Cormode Department of Computer Science, University of Warwick, Coventry, UK

Derek G. Corneil Department of Computer Science, University of Toronto, Toronto, ON, Canada

Bruno Courcelle Laboratoire Bordelais de Recherche en Informatique (LaBRI), CNRS, Bordeaux University, Talence, France

Lenore J. Cowen Department of Computer Science, Tufts University, Medford, MA, USA

Nello Cristianini Department of Engineering Mathematics, and Computer Science, University of Bristol, Bristol, UK

Maxime Crochemore Department of Computer Science, King's College London, London, UK

Laboratory of Computer Science, University of Paris-East, Paris, France

Université de Marne-la-Vallée, Champs-sur-Marne, France

Miklós Csürös Department of Computer Science, University of Montréal, Montréal, QC, Canada

Fabio Cunial Department of Computer Science, Helsinki Institute for Information Technology (HIIT), University of Helsinki, Helsinki, Finland

Marek Cygan Institute of Informatics, University of Warsaw, Warsaw, Poland

Artur Czumaj Department of Computer Science, Centre for Discrete Mathematics and Its Applications, University of Warwick, Coventry, UK

Bhaskar DasGupta Department of Computer Science, University of Illinois, Chicago, IL, USA

Constantinos Daskalakis EECS, Massachusetts Institute of Technology, Cambridge, MA, USA

Mark de Berg Department of Mathematics and Computer Science, TU Eindhoven, Eindhoven, The Netherlands

Xavier Défago School of Information Science, Japan Advanced Institute of Science and Technology (JAIST), Ishikawa, Japan

Daniel Delling Microsoft, Silicon Valley, CA, USA

Erik D. Demaine MIT Computer Science and Artificial Intelligence Laboratory, Cambridge, MA, USA

Camil Demetrescu Department of Computer and Systems Science, University of Rome, Rome, Italy

Department of Information and Computer Systems, University of Rome, Rome, Italy

Ping Deng Department of Computer Science, The University of Texas at Dallas, Richardson, TX, USA

Xiaotie Deng AIMS Laboratory (Algorithms-Agents-Data on Internet, Market, and Social Networks), Department of Computer Science and Engineering, Shanghai Jiao Tong University, Shanghai, China

Department of Computer Science, City University of Hong Kong, Hong Kong, China

Vamsi Krishna Devabathini Center for Quantum Technologies, National University of Singapore, Singapore, Singapore

Olivier Devillers Inria Nancy – Grand-Est, Villers-lès-Nancy, France

Tamal Krishna Dey Department of Computer Science and Engineering, The Ohio State University, Columbus, OH, USA

Robert P. Dick Department of Electrical Engineering and Computer Science, University of Michigan, Ann Arbor, MI, USA

Walter Didimo Department of Engineering, University of Perugia, Perugia, Italy

Ling Ding Institute of Technology, University of Washington Tacoma, Tacoma, WA, USA

Yuzheng Ding Xilinx Inc., Longmont, CO, USA

Michael Dom Department of Mathematics and Computer Science, University of Jena, Jena, Germany

Riccardo Dondi Università degli Studi di Bergamo, Bergamo, Italy

Gyorgy Dosa University of Pannonia, Veszprém, Hungary

David Doty Computing and Mathematical Sciences, California Institute of Technology, Pasadena, CA, USA

Ding-Zhu Du Computer Science, University of Minnesota, Minneapolis, MN, USA

Department of Computer Science, The University of Texas at Dallas, Richardson, TX, USA

Hongwei Du Department of Computer Science and Technology, Shenzhen Graduate School, Harbin Institute of Technology, Shenzhen, China

Ran Duan Institute for Interdisciplinary Information Sciences, Tsinghua University, Beijing, China

Devdatt Dubhashi Department of Computer Science, Chalmers University of Technology, Gothenburg, Sweden

Gothenburg University, Gothenburg, Sweden

Adrian Dumitrescu Computer Science, University of Wisconsin–Milwaukee, Milwaukee, WI, USA

Iréne Durand Laboratoire Bordelais de Recherche en Informatique (LaBRI), CNRS, Bordeaux University, Talence, France

Stephane Durocher University of Manitoba, Winnipeg, MB, Canada

Pavlos Efraimidis Department of Electrical and Computer Engineering, Democritus University of Thrace, Xanthi, Greece

Charilaos Efthymiou Department of Computer Engineering and Informatics, University of Patras, Patras, Greece

Michael Elkin Department of Computer Science, Ben-Gurion University, Beer-Sheva, Israel

Matthias Englert Department of Computer Science, University of Warwick, Coventry, UK

David Eppstein Donald Bren School of Information and Computer Sciences, Computer Science Department, University of California, Irvine, CA, USA

Leah Epstein Department of Mathematics, University of Haifa, Haifa, Israel

Jeff Erickson Department of Computer Science, University of Illinois, Urbana, IL, USA

Constantine G. Evans Division of Biology and Bioengineering, California Institute of Technology, Pasadena, CA, USA

Eyal Even-Dar Google, New York, NY, USA

Rolf Fagerberg Department of Mathematics and Computer Science, University of Southern Denmark, Odense, Denmark

Jittat Fakcharoenphol Department of Computer Engineering, Kasetsart University, Bangkok, Thailand

Piotr Faliszewski AGH University of Science and Technology, Krakow, Poland

Lidan Fan Department of Computer Science, The University of Texas, Tyler, TX, USA

Qizhi Fang School of Mathematical Sciences, Ocean University of China, Qingdao, Shandong Province, China

Martín Farach-Colton Department of Computer Science, Rutgers University, Piscataway, NJ, USA

Panagiota Fatourou Department of Computer Science, University of Ioannina, Ioannina, Greece

Jonathan Feldman Google, Inc., New York, NY, USA

Vitaly Feldman IBM Research – Almaden, San Jose, CA, USA

Henning Fernau Fachbereich 4, Abteilung Informatikwissenschaften, Universität Trier, Trier, Germany

Institute for Computer Science, University of Trier, Trier, Germany

Paolo Ferragina Department of Computer Science, University of Pisa, Pisa, Italy

Johannes Fischer Technical University Dortmund, Dortmund, Germany

Nathan Fisher Department of Computer Science, Wayne State University, Detroit, MI, USA

Abraham Flaxman Theory Group, Microsoft Research, Redmond, WA, USA

Paola Flocchini School of Electrical Engineering and Computer Science, University of Ottawa, Ottawa, ON, Canada

Fedor V. Fomin Department of Informatics, University of Bergen, Bergen, Norway

Dimitris Fotakis Department of Information and Communication Systems Engineering, University of the Aegean, Samos, Greece

Kyle Fox Institute for Computational and Experimental Research in Mathematics, Brown University, Providence, RI, USA

Pierre Fraigniaud Laboratoire d'Informatique Algorithmique: Fondements et Applications, CNRS and University Paris Diderot, Paris, France

Fabrizio Frati School of Information Technologies, The University of Sydney, Sydney, NSW, Australia

Engineering Department, Roma Tre University, Rome, Italy

Ophir Frieder Department of Computer Science, Illinois Institute of Technology, Chicago, IL, USA

Hiroshi Fujiwara Shinshu University, Nagano, Japan

Stanley P.Y. Fung Department of Computer Science, University of Leicester, Leicester, UK

Stefan Funke Department of Computer Science, Universität Stuttgart, Stuttgart, Germany

Martin Fürer Department of Computer Science and Engineering, The Pennsylvania State University, University Park, PA, USA

Travis Gagie Department of Computer Science, University of Eastern Piedmont, Alessandria, Italy

Department of Computer Science, University of Helsinki, Helsinki, Finland

Gábor Galambos Department of Computer Science, Juhász Gyula Teachers Training College, Szeged, Hungary

Jianjiong Gao Computational Biology Center, Memorial Sloan-Kettering Cancer Center, New York, NY, USA

Jie Gao Department of Computer Science, Stony Brook University, Stony Brook, NY, USA

Xiaofeng Gao Department of Computer Science, Shanghai Jiao Tong University, Shanghai, China

Juan Garay Bell Laboratories, Murray Hill, NJ, USA

Minos Garofalakis Technical University of Crete, Chania, Greece

Olivier Gascuel Institut de Biologie Computationnelle, Laboratoire d'Informatique, de Robotique et de Microélectronique de Montpellier (LIRMM), CNRS and Université de Montpellier, Montpellier cedex 5, France

Leszek Gąsieniec University of Liverpool, Liverpool, UK

Serge Gaspers Optimisation Research Group, National ICT Australia (NICTA), Sydney, NSW, Australia

School of Computer Science and Engineering, University of New SouthWales (UNSW), Sydney, NSW, Australia

Maciej Gazda Department of Mathematics and Computer Science, Eindhoven University of Technology, Eindhoven, The Netherlands

Raffaele Giancarlo Department of Mathematics and Applications, University of Palermo, Palermo, Italy

Gagan Goel Google Inc., New York, NY, USA

Andrew V. Goldberg Microsoft Research – Silicon Valley, Mountain View, CA, USA

Oded Goldreich Department of Computer Science, Weizmann Institute of Science, Rehovot, Israel

Jens Gramm WSI Institute of Theoretical Computer Science, Tübingen University, Tübingen, Germany

Fabrizio Grandoni IDSIA, USI-SUPSI, University of Lugano, Lugano, Switzerland

Roberto Grossi Dipartimento di Informatica, Università di Pisa, Pisa, Italy

Lov K. Grover Bell Laboratories, Alcatel-Lucent, Murray Hill, NJ, USA

Xianfeng David Gu Department of Computer Science, Stony Brook University, Stony Brook, NY, USA

Joachim Gudmundsson DMiST, National ICT Australia Ltd, Alexandria, Australia

School of Information Technologies, University of Sydney, Sydney, NSW, Australia

Rachid Guerraoui School of Computer and Communication Sciences, EPFL, Lausanne, Switzerland

Heng Guo Computer Sciences Department, University of Wisconsin–Madison, Madison, WI, USA

Jiong Guo Department of Mathematics and Computer Science, University of Jena, Jena, Germany

Manoj Gupta Indian Institute of Technology (IIT) Delhi, Hauz Khas, New Delhi, India

Venkatesan Guruswami Department of Computer Science and Engineering, University of Washington, Seattle, WA, USA

Gregory Gutin Department of Computer Science, Royal Holloway, University of London, Egham, UK

Michel Habib LIAFA, Université Paris Diderot, Paris Cedex 13, France

Mohammad Taghi Hajiaghayi Department of Computer Science, University of Maryland, College Park, MD, USA

Sean Hallgren Department of Computer Science and Engineering, The Pennsylvania State University, University Park, State College, PA, USA

Dan Halperin School of Computer Science, Tel-Aviv University, Tel Aviv, Israel

Moritz Hardt IBM Research – Almaden, San Jose, CA, USA

Ramesh Hariharan Strand Life Sciences, Bangalore, India

Aram W. Harrow Department of Physics, Massachusetts Institute of Technology, Cambridge, MA, USA

Prahladh Harsha Tata Institute of Fundamental Research, Mumbai, Maharashtra, India

Herman Haverkort Department of Computer Science, Eindhoven University of Technology, Eindhoven, The Netherlands

Meng He School of Computer Science, University of Waterloo, Waterloo, ON, Canada

Xin He Department of Computer Science and Engineering, The State University of New York, Buffalo, NY, USA

Lisa Hellerstein Department of Computer Science and Engineering, NYU Polytechnic School of Engineering, Brooklyn, NY, USA

Michael Hemmer Department of Computer Science, TU Braunschweig, Braunschweig, Germany

Danny Hendler Department of Computer Science, Ben-Gurion University of the Negev, Beer-Sheva, Israel

Monika Henzinger University of Vienna, Vienna, Austria

Maurice Herlihy Department of Computer Science, Brown University, Providence, RI, USA

Ted Herman Department of Computer Science, University of Iowa, Iowa City, IA, USA

John Hershberger Mentor Graphics Corporation, Wilsonville, OR, USA

Timon Hertli Department of Computer Science, ETH Zürich, Zürich, Switzerland

Edward A. Hirsch Laboratory of Mathematical Logic, Steklov Institute of Mathematics, St. Petersburg, Russia

Wing-Kai Hon Department of Computer Science, National Tsing Hua University, Hsin Chu, Taiwan

Seokhee Hong School of Information Technologies, University of Sydney, Sydney, NSW, Australia

Paul G. Howard Akamai Technologies, Cambridge, MA, USA

Peter Høyer University of Calgary, Calgary, AB, Canada

Li-Sha Huang Department of Computer Science and Technology, Tsinghua University, Beijing, China

Yaocun Huang Department of Computer Science, The University of Texas at Dallas, Richardson, TX, USA

Zhiyi Huang Department of Computer Science, The University of Hong Kong, Hong Kong, Hong Kong

Falk Hüffner Department of Math and Computer Science, University of Jena, Jena, Germany

Thore Husfeldt Department of Computer Science, Lund University, Lund, Sweden

Lucian Ilie Department of Computer Science, University of Western Ontario, London, ON, Canada

Sungjin Im Electrical Engineering and Computer Sciences (EECS), University of California, Merced, CA, USA

Csanad Imreh Institute of Informatics, University of Szeged, Szeged, Hungary

Robert W. Irving School of Computing Science, University of Glasgow, Glasgow, UK

Alon Itai Technion, Haifa, Israel

Giuseppe F. Italiano Department of Computer and Systems Science, University of Rome, Rome, Italy

Department of Information and Computer Systems, University of Rome, Rome, Italy

Kazuo Iwama Computer Engineering, Kyoto University, Sakyo, Kyoto, Japan

School of Informatics, Kyoto University, Sakyo, Kyoto, Japan

Jeffrey C. Jackson Department of Mathematics and Computer Science, Duquesne University, Pittsburgh, PA, USA

Ronald Jackups Department of Pediatrics, Washington University, St. Louis, MO, USA

Riko Jacob Institute of Computer Science, Technical University of Munich, Munich, Germany

IT University of Copenhagen, Copenhagen, Denmark

Rahul Jain Department of Computer Science, Center for Quantum Technologies, National University of Singapore, Singapore, Singapore

Klaus Jansen Department of Computer Science, University of Kiel, Kiel, Germany

Jesper Jansson Laboratory of Mathematical Bioinformatics, Institute for Chemical Research, Kyoto University, Gokasho, Uji, Kyoto, Japan

Stacey Jeffery David R. Cheriton School of Computer Science, University of Waterloo, Waterloo, ON, Canada

Madhav Jha Sandia National Laboratories, Livermore, CA, USA

Zenefits, San Francisco, CA, USA

David S. Johnson Department of Computer Science, Columbia University, New York, NY, USA

AT&T Laboratories, Algorithms and Optimization Research Department, Florham Park, NJ, USA

Mark Jones Department of Computer Science, Royal Holloway, University of London, Egham, UK

Tomasz Jurdziński Institute of Computer Science, University of Wrocław, Wrocław, Poland

Yoji Kajitani Department of Information and Media Sciences, The University of Kitakyushu, Kitakyushu, Japan

Shahin Kamali David R. Cheriton School of Computer Science, University of Waterloo, Waterloo, ON, Canada

Andrew Kane David R. Cheriton School of Computer Science, University of Waterloo, Waterloo, ON, Canada

Mamadou Moustapha Kanté Clermont-Université, Université Blaise Pascal, LIMOS, CNRS, Aubière, France

Ming-Yang Kao Department of Electrical Engineering and Computer Science, Northwestern University, Evanston, IL, USA

Alexis Kaporis Department of Information and Communication Systems Engineering, University of the Aegean, Karlovasi, Samos, Greece

George Karakostas Department of Computing and Software, McMaster University, Hamilton, ON, Canada

Juha Kärkkäinen Department of Computer Science, University of Helsinki, Helsinki, Finland

Petteri Kaski Department of Computer Science, School of Science, Aalto University, Helsinki, Finland

Helsinki Institute for Information Technology (HIIT), Helsinki, Finland

Hans Kellerer Department of Statistics and Operations Research, University of Graz, Graz, Austria

Andrew A. Kennings Department of Electrical and Computer Engineering, University of Waterloo, Waterloo, ON, Canada

Kurt Keutzer Department of Electrical Engineering and Computer Science, University of California, Berkeley, CA, USA

Mohammad Reza Khani University of Maryland, College Park, MD, USA

Samir Khuller Computer Science Department, University of Maryland, College Park, MD, USA

Donghyun Kim Department of Mathematics and Physics, North Carolina Central University, Durham, NC, USA

Jin Wook Kim HM Research, Seoul, Korea

Yoo-Ah Kim Computer Science and Engineering Department, University of Connecticut, Storrs, CT, USA

Valerie King Department of Computer Science, University of Victoria, Victoria, BC, Canada

Zoltán Király Department of Computer Science, Eötvös Loránd University, Budapest, Hungary

Egerváry Research Group (MTA-ELTE), Eötvös Loránd University, Budapest, Hungary

Lefteris Kirousis Department of Computer Engineering and Informatics, University of Patras, Patras, Greece

Jyrki Kivinen Department of Computer Science, University of Helsinki, Helsinki, Finland

Masashi Kiyomi International College of Arts and Sciences, Yokohama City University, Yokohama, Kanagawa, Japan

Kim-Manuel Klein University Kiel, Kiel, Germany

Rolf Klein Institute for Computer Science, University of Bonn, Bonn, Germany

Adam Klivans Department of Computer Science, University of Texas, Austin, TX, USA

Koji M. Kobayashi National Institute of Informatics, Chiyoda-ku, Tokyo, Japan

Stephen Kobourov Department of Computer Science, University of Arizona, Tucson, AZ, USA

Kirill Kogan IMDEA Networks, Madrid, Spain

Christian Komusiewicz Institute of Software Engineering and Theoretical Computer Science, Technical University of Berlin, Berlin, Germany

Goran Konjevod Department of Computer Science and Engineering, Arizona State University, Tempe, AZ, USA

Spyros Kontogiannis Department of Computer Science, University of Ioannina, Ioannina, Greece

Matias Korman Graduate School of Information Sciences, Tohoku University, Miyagi, Japan

Guy Kortsarz Department of Computer Science, Rutgers University, Camden, NJ, USA

Nitish Korula Google Research, New York, NY, USA

Robin Kothari Center for Theoretical Physics, Massachusetts Institute of Technology, Cambridge, MA, USA

David R. Cheriton School of Computer Science, Institute for Quantum Computing, University of Waterloo, Waterloo, ON, Canada

Ioannis Koutis Computer Science Department, University of Puerto Rico-Rio Piedras, San Juan, PR, USA

Dariusz R. Kowalski Department of Computer Science, University of Liverpool, Liverpool, UK

Evangelos Kranakis Department of Computer Science, Carleton, Ottawa, ON, Canada

Dieter Kratsch UFM MIM – LITA, Université de Lorraine, Metz, France

Stefan Kratsch Department of Software Engineering and Theoretical Computer Science, Technical University Berlin, Berlin, Germany

Robert Krauthgamer Weizmann Institute of Science, Rehovot, Israel

IBM Almaden Research Center, San Jose, CA, USA

Stephan Kreutzer Chair for Logic and Semantics, Technical University, Berlin, Germany

Sebastian Krinninger Faculty of Computer Science, University of Vienna, Vienna, Austria

Ravishankar Krishnaswamy Computer Science Department, Princeton University, Princeton, NJ, USA

Danny Krizanc Department of Computer Science, Wesleyan University, Middletown, CT, USA

Piotr Krysta Department of Computer Science, University of Liverpool, Liverpool, UK

Gregory Kucherov CNRS/LIGM, Université Paris-Est, Marne-la-Vallée, France

Fabian Kuhn Department of Computer Science, ETH Zurich, Zurich, Switzerland

V.S. Anil Kumar Virginia Bioinformatics Institute, Virginia Tech, Blacksburg, VA, USA

Tak-Wah Lam Department of Computer Science, University of Hong Kong, Hong Kong, China

Giuseppe Lancia Department of Mathematics and Computer Science, University of Udine, Udine, Italy

Gad M. Landau Department of Computer Science, University of Haifa, Haifa, Israel

Zeph Landau Department of Computer Science, University of California, Berkelely, CA, USA

Michael Langberg Department of Electrical Engineering, The State University of New York, Buffalo, NY, USA

Department of Mathematics and Computer Science, The Open University of Israel, Raanana, Israel

Elmar Langetepe Department of Computer Science, University of Bonn, Bonn, Germany

Ron Lavi Faculty of Industrial Engineering and Management, Technion, Haifa, Israel

Thierry Lecroq Computer Science Department and LITIS Faculty of Science, Université de Rouen, Rouen, France

James R. Lee Department of Computer Science and Engineering, University of Washington, Seattle, WA, USA

Stefano Leonardi Department of Information and Computer Systems, University of Rome, Rome, Italy

Pierre Leone Informatics Department, University of Geneva, Geneva, Switzerland

Henry Leung Department of Computer Science, The University of Hong Kong, Hong Kong, China

Christos Levcopoulos Department of Computer Science, Lund University, Lund, Sweden

Asaf Levin Faculty of Industrial Engineering and Management, The Technion, Haifa, Israel

Moshe Lewenstein Department of Computer Science, Bar-Ilan University, Ramat-Gan, Israel

Li (Erran) Li Bell Laboratories, Alcatel-Lucent, Murray Hill, NJ, USA

Mengling Li Division of Mathematical Sciences, Nanyang Technological University, Singapore, Singapore

Ming Li David R. Cheriton School of Computer Science, University of Waterloo, Waterloo, ON, Canada

Ming Min Li Computer Science and Technology, Tsinghua University, Beijing, China

Xiang-Yang Li Department of Computer Science, Illinois Institute of Technology, Chicago, IL, USA

Vahid Liaghat Department of Computer Science, University of Maryland, College Park, MD, USA

Jie Liang Department of Bioengineering, University of Illinois, Chicago, IL, USA

Andrzej Lingas Department of Computer Science, Lund University, Lund, Sweden

Maarten Löffler Department of Information and Computing Sciences, Utrecht University, Utrecht, The Netherlands

Daniel Lokshtanov Department of Informatics, University of Bergen, Bergen, Norway

Alejandro López-Ortiz David R. Cheriton School of Computer Science, University of Waterloo, Waterloo, ON, Canada

Chin Lung Lu Institute of Bioinformatics and Department of Biological Science and Technology, National Chiao Tung University, Hsinchu, Taiwan

Pinyan Lu Microsoft Research Asia, Shanghai, China

Zaixin Lu Department of Mathematics and Computer Science, Marywood University, Scranton, PA, USA

Feng Luo Department of Mathematics, Rutgers University, Piscataway, NJ, USA

Haiming Luo Department of Computer Science and Technology, Shenzhen Graduate School, Harbin Institute of Technology, Shenzhen, China

Rune B. Lyngsø Department of Statistics, Oxford University, Oxford, UK

Winton Capital Management, Oxford, UK

Bin Ma David R. Cheriton School of Computer Science, University of Waterloo, Waterloo, ON, Canada

Department of Computer Science, University of Western Ontario, London, ON, Canada

Mohammad Mahdian Yahoo! Research, Santa Clara, CA, USA

Hamid Mahini Department of Computer Science, University of Maryland, College Park, MD, USA

Veli Mäkinen Department of Computer Science, Helsinki Institute for Information Technology (HIIT), University of Helsinki, Helsinki, Finland

Dahlia Malkhi Microsoft, Silicon Valley Campus, Mountain View, CA, USA

Mark S. Manasse Microsoft Research, Mountain View, CA, USA

David F. Manlove School of Computing Science, University of Glasgow, Glasgow, UK

Giovanni Manzini Department of Computer Science, University of Eastern Piedmont, Alessandria, Italy

Department of Science and Technological Innovation, University of Piemonte Orientale, Alessandria, Italy

Madha V. Marathe IBM T.J. Watson Research Center, Hawthorne, NY, USA

Alberto Marchetti-Spaccamela Department of Information and Computer Systems, University of Rome, Rome, Italy

Igor L. Markov Department of Electrical Engineering and Computer Science, University of Michigan, Ann Arbor, MI, USA

Alexander Matveev Computer Science and Artificial Intelligence Laboratory, MIT, Cambridge, MA, USA

Eric McDermid Cedar Park, TX, USA

Catherine C. McGeoch Department of Mathematics and Computer Science, Amherst College, Amherst, MA, USA

Lyle A. McGeoch Department of Mathematics and Computer Science, Amherst College, Amherst, MA, USA

Andrew McGregor School of Computer Science, University of Massachusetts, Amherst, MA, USA

Brendan D. McKay Department of Computer Science, Australian National University, Canberra, ACT, Australia

Nicole Megow Institut für Mathematik, Technische Universität Berlin, Berlin, Germany

Manor Mendel Department of Mathematics and Computer Science, The Open University of Israel, Raanana, Israel

George B. Mertzios School of Engineering and Computing Sciences, Durham University, Durham, UK

Julián Mestre Department of Computer Science, University of Maryland, College Park, MD, USA

School of Information Technologies, The University of Sydney, Sydney, NSW, Australia

Pierre-Étienne Meunier Le Laboratoire d'Informatique Fondamentale de Marseille (LIF), Aix-Marseille Université, Marseille, France

Ulrich Meyer Department of Computer Science, Goethe University Fankfurt am Main, Frankfurt, Germany

Daniele Micciancio Department of Computer Science, University of California, San Diego, La Jolla, CA, USA

István Miklós Department of Plant Taxonomy and Ecology, Eötvös Loránd University, Budapest, Hungary

Shin-ichi Minato Graduate School of Information Science and Technology, Hokkaido University, Sapporo, Japan

Vahab S. Mirrokni Theory Group, Microsoft Research, Redmond, WA, USA

Neeldhara Misra Department of Computer Science and Automation, Indian Institute of Science, Bangalore, India

Joseph S.B. Mitchell Department of Applied Mathematics and Statistics, Stony Brook University, Stony Brook, NY, USA

Shuichi Miyazaki Academic Center for Computing and Media Studies, Kyoto University, Kyoto, Japan

Alistair Moffat Department of Computing and Information Systems, The University of Melbourne, Melbourne, VIC, Australia

Mark Moir Sun Microsystems Laboratories, Burlington, MA, USA

Ashley Montanaro Department of Computer Science, University of Bristol, Bristol, UK

Tal Mor Department of Computer Science, Technion – Israel Institute of Technology, Haifa, Israel

Michele Mosca Canadian Institute for Advanced Research, Toronto, ON, Canada

Combinatorics and Optimization/Institute for Quantum Computing, University of Waterloo, Waterloo, ON, Canada

Perimeter Institute for Theoretical Physics, Waterloo, ON, Canada

Thomas Moscibroda Systems and Networking Research Group, Microsoft Research, Redmond, WA, USA

Yoram Moses Department of Electrical Engineering, Technion – Israel Institute of Technology, Haifa, Israel

Shay Mozes Efi Arazi School of Computer Science, The Interdisciplinary Center (IDC), Herzliya, Israel

Marcin Mucha Faculty of Mathematics, Informatics and Mechanics, Institute of Informatics, Warsaw, Poland

Priyanka Mukhopadhyay Center for Quantum Technologies, National University of Singapore, Singapore, Singapore

Kamesh Munagala Levine Science Research Center, Duke University, Durham, NC, USA

J. Ian Munro David R. Cheriton School of Computer Science, University of Waterloo, Waterloo, ON, Canada

Joong Chae Na Department of Computer Science and Engineering, Sejong University, Seoul, Korea

Viswanath Nagarajan University of Michigan, Ann Arbor, MI, USA

Shin-ichi Nakano Department of Computer Science, Gunma University, Kiryu, Japan

Danupon Nanongkai School of Computer Science and Communication, KTH Royal Institute of Technology, Stockholm, Sweden

Giri Narasimhan Department of Computer Science, Florida International University, Miami, FL, USA

School of Computing and Information Sciences, Florida International University, Miami, FL, USA

Gonzalo Navarro Department of Computer Science, University of Chile, Santiago, Chile

Ashwin Nayak Department of Combinatorics and Optimization, and Institute for Quantum Computing, University of Waterloo, Waterloo, ON, Canada

Amir Nayyeri Department of Electrical Engineering and Computer Science, Oregon State University, Corvallis, OR, USA

Jesper Nederlof Technical University of Eindhoven, Eindhoven, The Netherlands

Ofer Neiman Department of Computer Science, Ben-Gurion University of the Negev, Beer Sheva, Israel

Yakov Nekrich David R. Cheriton School of Computer Science, University of Waterloo, Waterloo, ON, Canada

Jelani Nelson Harvard John A. Paulson School of Engineering and Applied Sciences, Cambridge, MA, USA

Ragnar Nevries Computer Science Department, University of Rostock, Rostock, Germany

Alantha Newman CNRS-Université Grenoble Alpes and G-SCOP, Grenoble, France

Hung Q. Ngo Computer Science and Engineering, The State University of New York, Buffalo, NY, USA

Patrick K. Nicholson Department D1: Algorithms and Complexity, Max Planck Institut für Informatik, Saarbrücken, Germany

Rolf Niedermeier Department of Mathematics and Computer Science, University of Jena, Jena, Germany

Institut für Softwaretechnik und Theoretische Informatik, Technische Universität Berlin, Berlin, Germany

Sergey I. Nikolenko Laboratory of Mathematical Logic, Steklov Institute of Mathematics, St. Petersburg, Russia

Sotiris Nikoletseas Computer Engineering and Informatics Department, University of Patras, Patras, Greece

Computer Technology Institute and Press "Diophantus", Patras, Greece

Aleksandar Nikolov Department of Computer Science, Rutgers University, Piscataway, NJ, USA

Nikola S. Nikolov Department of Computer Science and Information Systems, University of Limerick, Limerick, Republic of Ireland

Kobbi Nisim Department of Computer Science, Ben-Gurion University, Beer Sheva, Israel

Lhouari Nourine Clermont-Université, Université Blaise Pascal, LIMOS, CNRS, Aubière, France

Yoshio Okamoto Department of Information and Computer Sciences, Toyohashi University of Technology, Toyohashi, Japan

Michael Okun Weizmann Institute of Science, Rehovot, Israel

Rasmus Pagh Theoretical Computer Science, IT University of Copenhagen, Copenhagen, Denmark

David Z. Pan Department of Electrical and Computer Engineering, University of Texas, Austin, TX, USA

Peichen Pan Xilinx, Inc., San Jose, CA, USA

Debmalya Panigrahi Department of Computer Science, Duke University, Durham, NC, USA

Fahad Panolan Institute of Mathematical Sciences, Chennai, India

Vicky Papadopoulou Department of Computer Science, University of Cyprus, Nicosia, Cyprus

Fabio Pardi Institut de Biologie Computationnelle, Laboratoire d'Informatique, de Robotique et de Microélectronique de Montpellier (LIRMM), CNRS and Université de Montpellier, Montpellier cedex 5, France

Kunsoo Park School of Computer Science and Engineering, Seoul National University, Seoul, Korea

Srinivasan Parthasarathy IBM T.J. Watson Research Center, Hawthorne, NY, USA

Apoorva D. Patel Centre for High Energy Physics, Indian Institute of Science, Bangalore, India

Matthew J. Patitz Department of Computer Science and Computer Engineering, University of Arkansas, Fayetteville, AR, USA

Mihai Pătraşcu Computer Science and Artificial Intelligence Laboratory (CSAIL), Massachusetts Institute of Technology (MIT), Cambridge, MA, USA

Maurizio Patrignani Engineering Department, Roma Tre University, Rome, Italy

Boaz Patt-Shamir Department of Electrical Engineering, Tel-Aviv University, Tel-Aviv, Israel

Ramamohan Paturi Department of Computer Science and Engineering, University of California at San Diego, San Diego, CA, USA

Christophe Paul CNRS, Laboratoire d'Informatique Robotique et Microélectronique de Montpellier, Université Montpellier 2, Montpellier, France

Andrzej Pelc Department of Computer Science, University of Québec-Ottawa, Gatineau, QC, Canada

Jean-Marc Petit Université de Lyon, CNRS, INSA Lyon, LIRIS, Lyon, France

Seth Pettie Electrical Engineering and Computer Science (EECS) Department, University of Michigan, Ann Arbor, MI, USA

Marcin Pilipczuk Institute of Informatics, University of Bergen, Bergen, Norway

Institute of Informatics, University of Warsaw, Warsaw, Poland

Michał Pilipczuk Institute of Informatics, University of Warsaw, Warsaw, Poland

Institute of Informatics, University of Bergen, Bergen, Norway

Yuri Pirola Università degli Studi di Milano-Bicocca, Milan, Italy

Olivier Powell Informatics Department, University of Geneva, Geneva, Switzerland

Amit Prakash Microsoft, MSN, Redmond, WA, USA

Eric Price Department of Computer Science, The University of Texas, Austin, TX, USA

Kirk Pruhs Department of Computer Science, University of Pittsburgh, Pittsburgh, PA, USA

Teresa M. Przytycka Computational Biology Branch, NCBI, NIH, Bethesda, MD, USA

Pavel Pudlák Academy of Science of the Czech Republic, Mathematical Institute, Prague, Czech Republic

Simon J. Puglisi Department of Computer Science, University of Helsinki, Helsinki, Finland

Balaji Raghavachari Computer Science Department, The University of Texas at Dallas, Richardson, TX, USA

Md. Saidur Rahman Department of Computer Science and Engineering, Bangladesh University of Engineering and Technology, Dhaka, Bangladesh

Naila Rahman University of Hertfordshire, Hertfordshire, UK

Rajmohan Rajaraman Department of Computer Science, Northeastern University, Boston, MA, USA

Sergio Rajsbaum Instituto de Matemáticas, Universidad Nacional Autónoma de México (UNAM), México City, México

Vijaya Ramachandran Computer Science, University of Texas, Austin, TX, USA

Rajeev Raman Department of Computer Science, University of Leicester, Leicester, UK

M.S. Ramanujan Department of Informatics, University of Bergen, Bergen, Norway

Edgar Ramos School of Mathematics, National University of Colombia, Medellín, Colombia

Satish Rao Department of Computer Science, University of California, Berkeley, CA, USA

Christoforos L. Raptopoulos Computer Science Department, University of Geneva, Geneva, Switzerland

Computer Technology Institute and Press "Diophantus", Patras, Greece

Research Academic Computer Technology Institute, Greece and Computer Engineering and Informatics Department, University of Patras, Patras, Greece

Sofya Raskhodnikova Computer Science and Engineering Department, Pennsylvania State University, University Park, PA, USA

Rajeev Rastogi Amazon, Seattle, WA, USA

Joel Ratsaby Department of Electrical and Electronics Engineering, Ariel University of Samaria, Ariel, Israel

Kaushik Ravindran National Instruments, Berkeley, CA, USA

Michel Raynal Institut Universitaire de France and IRISA, Université de Rennes, Rennes, France

Ben W. Reichardt Electrical Engineering Department, University of Southern California (USC), Los Angeles, CA, USA

Renato Renner Institute for Theoretical Physics, Zurich, Switzerland

Elisa Ricci Department of Electronic and Information Engineering, University of Perugia, Perugia, Italy

Andréa W. Richa School of Computing, Informatics, and Decision Systems Engineering, Ira A. Fulton Schools of Engineering, Arizona State University, Tempe, AZ, USA

Peter C. Richter Department of Combinatorics and Optimization, and Institute for Quantum Computing, University of Waterloo, Waterloo, ON, Canada

Department of Computer Science, Rutgers, The State University of New Jersey, New Brunswick, NJ, USA

Liam Roditty Department of Computer Science, Bar-Ilan University, Ramat-Gan, Israel

Marcel Roeloffzen Graduate School of Information Sciences, Tohoku University, Sendai, Japan

Martin Roetteler Microsoft Research, Redmond, WA, USA

Heiko Röglin Department of Computer Science, University of Bonn, Bonn, Germany

José Rolim Informatics Department, University of Geneva, Geneva, Switzerland

Dana Ron School of Electrical Engineering, Tel-Aviv University, Ramat-Aviv, Israel

Frances Rosamond Parameterized Complexity Research Unit, University of Newcastle, Callaghan, NSW, Australia

Jarek Rossignac Georgia Institute of Technology, Atlanta, GA, USA

Matthieu Roy Laboratory of Analysis and Architecture of Systems (LAAS), Centre National de la Recherche Scientifique (CNRS), Université Toulouse, Toulouse, France

Ronitt Rubinfeld Massachusetts Institute of Technology (MIT), Cambridge, MA, USA

Tel Aviv University, Tel Aviv-Yafo, Israel

Atri Rudra Department of Computer Science and Engineering, State University of New York, Buffalo, NY, USA

Eric Ruppert Department of Computer Science and Engineering, York University, Toronto, ON, Canada

Frank Ruskey Department of Computer Science, University of Victoria, Victoria, BC, Canada

Luís M.S. Russo Departamento de Informática, Instituto Superior Técnico, Universidade de Lisboa, Lisboa, Portugal

INESC-ID, Lisboa, Portugal

Wojciech Rytter Institute of Informatics, Warsaw University, Warsaw, Poland

Kunihiko Sadakane Graduate School of Information Science and Technology, The University of Tokyo, Tokyo, Japan

S. Cenk Sahinalp Laboratory for Computational Biology, Simon Fraser University, Burnaby, BC, USA

Michael Saks Department of Mathematics, Rutgers, State University of New Jersey, Piscataway, NJ, USA

Alejandro Salinger Department of Computer Science, Saarland University, Saarbücken, Germany

Sachin S. Sapatnekar Department of Electrical and Computer Engineering, University of Minnesota, Minneapolis, MN, USA

Shubhangi Saraf Department of Mathematics and Department of Computer Science, Rutgers University, Piscataway, NJ, USA

Srinivasa Rao Satti Department of Computer Science and Engineering, Seoul National University, Seoul, South Korea

Saket Saurabh Institute of Mathematical Sciences, Chennai, India

University of Bergen, Bergen, Norway

Guido Schäfer Institute for Mathematics and Computer Science, Technical University of Berlin, Berlin, Germany

Dominik Scheder Institute for Interdisciplinary Information Sciences, Tsinghua University, Beijing, China

Institute for Computer Science, Shanghai Jiaotong University, Shanghai, China

Christian Scheideler Department of Computer Science, University of Paderborn, Paderborn, Germany

André Schiper EPFL, Lausanne, Switzerland

Christiane Schmidt The Selim and Rachel Benin School of Computer Science and Engineering, The Hebrew University of Jerusalem, Jerusalem, Israel

Markus Schmidt Institute for Computer Science, University of Freiburg, Freiburg, Germany

Dominik Schultes Institute for Computer Science, University of Karlsruhe, Karlsruhe, Germany

Robert Schweller Department of Computer Science, University of Texas Rio Grande Valley, Edinburg, TX, USA

Shinnosuke Seki Department of Computer Science, Helsinki Institute for Information Technology (HIIT), Aalto University, Aalto, Finland

Pranab Sen School of Technology and Computer Science, Tata Institute of Fundamental Research, Mumbai, India

Sandeep Sen Indian Institute of Technology (IIT) Delhi, Hauz Khas, New Delhi, India

Maria Serna Department of Language and System Information, Technical University of Catalonia, Barcelona, Spain

Rocco A. Servedio Computer Science, Columbia University, New York, NY, USA

Comandur Seshadhri Sandia National Laboratories, Livermore, CA, USA

Department of Computer Science, University of California, Santa Cruz, CA, USA

Jay Sethuraman Industrial Engineering and Operations Research, Columbia University, New York, NY, USA

Jiří Sgall Computer Science Institute, Charles University, Prague, Czech Republic

Rahul Shah Department of Computer Science, Louisiana State University, Baton Rouge, LA, USA

Shai Shalev-Shwartz School of Computer Science and Engineering, The Hebrew University, Jerusalem, Israel

Vikram Sharma Department of Computer Science, New York University, New York, NY, USA

Nir Shavit Computer Science and Artificial Intelligence Laboratory, MIT, Cambridge, MA, USA

School of Computer Science, Tel-Aviv University, Tel-Aviv, Israel

Yaoyun Shi Department of Electrical Engineering and Computer Science, University of Michigan, Ann Arbor, MI, USA

Ayumi Shinohara Graduate School of Information Sciences, Tohoku University, Sendai, Japan

Eugene Shragowitz Department of Computer Science and Engineering, University of Minnesota, Minneapolis, MN, USA

René A. Sitters Department of Econometrics and Operations Research, VU University, Amsterdam, The Netherlands

Balasubramanian Sivan Microsoft Research, Redmond, WA, USA

Daniel Sleator Department of Computer Science, Carnegie Mellon University, Pittsburgh, PA, USA

Michiel Smid School of Computer Science, Carleton University, Ottawa, ON, Canada

Adam Smith Computer Science and Engineering Department, Pennsylvania State University, University Park, State College, PA, USA

Dina Sokol Department of Computer and Information Science, Brooklyn College of CUNY, Brooklyn, NY, USA

Rolando D. Somma Theoretical Division, Los Alamos National Laboratory, Los Alamos, NM, USA

Wen-Zhan Song School of Engineering and Computer Science, Washington State University, Vancouver, WA, USA

Bettina Speckmann Department of Mathematics and Computer Science, Technical University of Eindhoven, Eindhoven, The Netherlands

Paul (Pavlos) Spirakis Computer Engineering and Informatics, Research and Academic Computer Technology Institute, Patras University, Patras, Greece

Computer Science, University of Liverpool, Liverpool, UK

Computer Technology Institute (CTI), Patras, Greece

Aravind Srinivasan Department of Computer Science, University of Maryland, College Park, MD, USA

Venkatesh Srinivasan Department of Computer Science, University of Victoria, Victoria, BC, Canada

Gerth Stølting Department of Computer Science, University of Aarhus, Århus, Denmark

Jens Stoye Faculty of Technology, Genome Informatics, Bielefeld University, Bielefeld, Germany

Scott M. Summers Department of Computer Science, University of Wisconsin – Oshkosh, Oshkosh, WI, USA

Aries Wei Sun Department of Computer Science, City University of Hong Kong, Hong Kong, China

Vijay Sundararajan Broadcom Corp, Fremont, CA, USA

Wing-Kin Sung Department of Computer Science, National University of Singapore, Singapore, Singapore

Mario Szegedy Department of Combinatorics and Optimization, and Institute for Quantum Computing, University of Waterloo, Waterloo, ON, Canada

Stefan Szeider Department of Computer Science, Durham University, Durham, UK

Tadao Takaoka Department of Computer Science and Software Engineering, University of Canterbury, Christchurch, New Zealand

Masayuki Takeda Department of Informatics, Kyushu University, Fukuoka, Japan

Kunal Talwar Microsoft Research, Silicon Valley Campus, Mountain View, CA, USA

Christino Tamon Department of Computer Science, Clarkson University, Potsdam, NY, USA

Akihisa Tamura Department of Mathematics, Keio University, Yokohama, Japan

Tiow-Seng Tan School of Computing, National University of Singapore, Singapore, Singapore

Shin-ichi Tanigawa Research Institute for Mathematical Sciences (RIMS), Kyoto University, Kyoto, Japan

Eric Tannier LBBE Biometry and Evolutionary Biology, INRIA Grenoble Rhône-Alpes, University of Lyon, Lyon, France

Alain Tapp Université de Montréal, Montréal, QC, Canada

Stephen R. Tate Department of Computer Science, University of North Carolina, Greensboro, NC, USA

Gadi Taubenfeld Department of Computer Science, Interdiciplinary Center Herzlia, Herzliya, Israel

Kavitha Telikepalli CSA Department, Indian Institute of Science, Bangalore, India

Barbara M. Terhal JARA Institute for Quantum Information, RWTH Aachen University, Aachen, Germany

Alexandre Termier IRISA, University of Rennes, 1, Rennes, France

My T. Thai Department of Computer and Information Science and Engineering, University of Florida, Gainesville, FL, USA

Abhradeep Thakurta Department of Computer Science, Stanford University, Stanford, CA, USA

Microsoft Research, CA, USA

Justin Thaler Yahoo! Labs, New York, NY, USA

Sharma V. Thankachan School of CSE, Georgia Institute of Technology, Atlanta, USA

Dimitrios Thilikos AlGCo Project-Team, CNRS, LIRMM, France

Department of Mathematics, National and Kapodistrian University of Athens, Athens, Greece

Haitong Tian Department of Electrical and Computer Engineering, University of Illinois at Urbana-Champaign, Urbana, IL, USA

Ioan Todinca INSA Centre Val de Loire, Universite d'Orleans, Orléans, France

Alade O. Tokuta Department of Mathematics and Physics, North Carolina Central University, Durham, NC, USA

Laura Toma Department of Computer Science, Bowdoin College, Brunswick, ME, USA

Etsuji Tomita The Advanced Algorithms Research Laboratory, The University of Electro-Communications, Chofu, Tokyo, Japan

Csaba D. Tóth Department of Computer Science, Tufts University, Medford, MA, USA

Department of Mathematics, California State University Northridge, Los Angeles, CA, USA

Luca Trevisan Department of Computer Science, University of California, Berkeley, CA, USA

John Tromp CWI, Amsterdam, The Netherlands

Nicolas Trotignon Laboratoire de l'Informatique du Parallélisme (LIP), CNRS, ENS de Lyon, Lyon, France

Jakub Truszkowski Cancer Research UK Cambridge Institute, University of Cambridge, Cambridge, UK

European Molecular Biology Laboratory, European Bioinformatics Institute (EMBL-EBI), Wellcome Trust Genome Campus, Hinxton, Cambridge, UK

Esko Ukkonen Department of Computer Science, Helsinki Institute for Information Technology (HIIT), University of Helsinki, Helsinki, Finland

Jonathan Ullman Department of Computer Science, Columbia University, New York, NY, USA

Takeaki Uno National Institute of Informatics, Chiyoda, Tokyo, Japan

Ruth Urner Department of Machine Learning, Carnegie Mellon University, Pittsburgh, USA

Jan Vahrenhold Department of Computer Science, Westfälische Wilhelms-Universität Münster, Münster, Germany

Daniel Valenzuela Department of Computer Science, Helsinki Institute for Information Technology (HIIT), University of Helsinki, Helsinki, Finland

Marc van Kreveld Department of Information and Computing Sciences, Utrecht University, Utrecht, The Netherlands

Rob van Stee University of Leicester, Leicester, UK

Stefano Varricchio Department of Computer Science, University of Roma, Rome, Italy

José Verschae Departamento de Matemáticas and Departamento de Ingeniería Industrial y de Sistemas, Pontificia Universidad Católica de Chile, Santiago, Chile

Stéphane Vialette IGM-LabInfo, University of Paris-East, Descartes, France

Sebastiano Vigna Dipartimento di Informatica, Università degli Studi di Milano, Milano, Italy

Yngve Villanger Department of Informatics, University of Bergen, Bergen, Norway

Paul Vitányi Centrum Wiskunde & Informatica (CWI), Amsterdam, The Netherlands

Jeffrey Scott Vitter University of Kansas, Lawrence, KS, USA

Berthold Vöcking Department of Computer Science, RWTH Aachen University, Aachen, Germany

Tjark Vredeveld Department of Quantitative Economics, Maastricht University, Maastricht, The Netherlands

Magnus Wahlström Department of Computer Science, Royal Holloway, University of London, Egham, UK

Peng-Jun Wan Department of Computer Science, Illinois Institute of Technology, Chicago, IL, USA

Chengwen Chris Wang Department of Computer Science, Carnegie Mellon University, Pittsburgh, PA, USA

Feng Wang Mathematical Science and Applied Computing, Arizona State University at the West Campus, Phoenix, AZ, USA

Huijuan Wang Shandong University, Jinan, China

Joshua R. Wang Department of Computer Science, Stanford University, Stanford, CA, USA

Lusheng Wang Department of Computer Science, City University of Hong Kong, Hong Kong, Hong Kong

Wei Wang School of Mathematics and Statistics, Xi'an Jiaotong University, Xi'an, Shaanxi, China

Weizhao Wang Google Inc., Irvine, CA, USA

Yu Wang Department of Computer Science, University of North Carolina, Charlotte, NC, USA

Takashi Washio The Institute of Scientific and Industrial Research, Osaka University, Ibaraki, Osaka, Japan

Matthew Weinberg Computer Science, Princeton University, Princeton, NJ, USA

Tobias Weinzierl School of Engineering and Computing Sciences, Durham University, Durham, UK

Renato F. Werneck Microsoft Research Silicon Valley, La Avenida, CA, USA

Matthias Westermann Department of Computer Science, TU Dortmund University, Dortmund, Germany

Tim A.C. Willemse Department of Mathematics and Computer Science, Eindhoven University of Technology, Eindhoven, The Netherlands

Ryan Williams Department of Computer Science, Stanford University, Stanford, CA, USA

Tyson Williams Computer Sciences Department, University of Wisconsin–Madison, Madison, WI, USA

Andrew Winslow Department of Computer Science, Tufts University, Medford, MA, USA

Paul Wollan Department of Computer Science, University of Rome La Sapienza, Rome, Italy

Martin D.F. Wong Department of Electrical and Computer Engineering, University of Illinois at Urbana-Champaign, Urbana, IL, USA

Prudence W.H. Wong University of Liverpool, Liverpool, UK

David R. Wood School of Mathematical Sciences, Monash University, Melbourne, VIC, Australia

Damien Woods Computer Science, California Institute of Technology, Pasadena, CA, USA

Lidong Wu Department of Computer Science, The University of Texas, Tyler, TX, USA

Weili Wu College of Computer Science and Technology, Taiyuan University of Technology, Taiyuan, Shanxi Province, China

Department of Computer Science, California State University, Los Angeles, CA, USA

Department of Computer Science, The University of Texas at Dallas, Richardson, TX, USA

Christian Wulff-Nilsen Department of Computer Science, University of Copenhagen, Copenhagen, Denmark

Mingji Xia The State Key Laboratory of Computer Science, Chinese Academy of Sciences, Beijing, China

David Xiao CNRS, Université Paris 7, Paris, France

Dong Xu Bond Life Sciences Center, University of Missouri, Columbia, MO, USA

Wen Xu Department of Computer Science, The University of Texas at Dallas, Richardson, TX, USA

Katsuhisa Yamanaka Department of Electrical Engineering and Computer Science, Iwate University, Iwate, Japan

Hiroki Yanagisawa IBM Research – Tokyo, Tokyo, Japan

Honghua Hannah Yang Strategic CAD Laboratories, Intel Corporation, Hillsboro, OR, USA

Qiuming Yao University of Missouri, Columbia, MO, USA

Chee K. Yap Department of Computer Science, New York University, New York, NY, USA

Yinyu Ye Department of Management Science and Engineering, Stanford University, Stanford, CA, USA

Anders Yeo Engineering Systems and Design, Singapore University of Technology and Design, Singapore, Singapore

Department of Mathematics, University of Johannesburg, Auckland Park, South Africa

Chih-Wei Yi Department of Computer Science, National Chiao Tung University, Hsinchu City, Taiwan

Ke Yi Hong Kong University of Science and Technology, Hong Kong, China

Yitong Yin Nanjing University, Jiangsu, Nanjing, Gulou, China

S.M. Yiu Department of Computer Science, University of Hong Kong, Hong Kong, China

Makoto Yokoo Department of Information Science and Electrical Engineering, Kyushu University, Nishi-ku, Fukuoka, Japan

Evangeline F.Y. Young Department of Computer Science and Engineering, The Chinese University of Hong Kong, Hong Kong, China

Neal E. Young Department of Computer Science and Engineering, University of California, Riverside, CA, USA

Bei Yu Department of Electrical and Computer Engineering, University of Texas, Austin, TX, USA

Yaoliang Yu Machine Learning Department, Carnegie Mellon University, Pittsburgh, PA, USA

Raphael Yuster Department of Mathematics, University of Haifa, Haifa, Israel

Morteza Zadimoghaddam Google Research, New York, NY, USA

Francis Zane Lucent Technologies, Bell Laboraties, Murray Hill, NJ, USA

Christos Zaroliagis Department of Computer Engineering and Informatics, University of Patras, Patras, Greece

Norbert Zeh Faculty of Computer Science, Dalhousie University, Halifax, NS, Canada

Li Zhang Microsoft Research, Mountain View, CA, USA

Louxin Zhang Department of Mathematics, National University of Singapore, Singapore, Singapore

Shengyu Zhang The Chinese University of Hong Kong, Hong Kong, China

Zhang Zhao College of Mathematics Physics and Information Engineering, Zhejiang Normal University, Zhejiang, Jinhua, China

Hai Zhou Electrical Engineering and Computer Science (EECS) Department, Northwestern University, Evanston, IL, USA

Yuqing Zhu Department of Computer Science, California State University, Los Angeles, CA, USA

Department of Computer Science, The University of Texas at Dallas, Richardson, TX, USA

Sandra Zilles Department of Computer Science, University of Regina, Regina, SK, Canada

Aaron Zollinger Department of Electrical Engineering and Computer Science, University of California, Berkeley, CA, USA

Uri Zwick Department of Mathematics and Computer Science, Tel-Aviv University, Tel-Aviv, Israel

Q

Quadtrees and Morton Indexing

Herman Haverkort[1] and Laura Toma[2]
[1]Department of Computer Science, Eindhoven
University of Technology, Eindhoven, The
Netherlands
[2]Department of Computer Science, Bowdoin
College, Brunswick, ME, USA

Keywords

IO-efficient algorithms; Segment intersection;
Space decomposition; Space-filling curve

Years and Authors of Summarized Original Work

2002; Hjaltason, Samet
2006; Agarwal, Arge, Danner
2010; de Berg, Haverkort, Thite, Toma
2013; McGranaghan, Haverkort, Toma

Problem Definition

The quadtree describes a class of data structures
for geometric objects. A quadtree partitions space
hierarchically using a stopping rule that decides
when a region is small enough so that it does not
need to be subdivided further. If the space is d
dimensional, a quadtree recursively divides a d-
dimensional hypercube containing the input data
into 2^d hypercubes until each region satisfies the
given stopping rule. In 2D, the hypercubes are
squares. Three-dimensional quadtrees are also
known as *octrees*. Quadtrees have been used for
many types of data, such as points, line segments,
polygons, rectangles, curves, and images, and for
many types of applications. For a detailed presentation, we refer to the book by Samet [10]. While
their worst-case behavior is good only in some
simple cases, quadtrees perform well empirically
in many applications.

A quadtree can be stored as a tree that corresponds to the hierarchical subdivision of the input
region. A region that is subdivided further is then
represented by a node with four children, one for
each quadrant; the cells that are not subdivided
further constitute the leaves of the tree and represent a subdivision of the input region. A quadtree
with m leaves has exactly $(m - 1)/3$ internal
nodes and $4m/3 - 1/3$ nodes in total. Hence it
can be described by a sequence of $4m/3 - 1/3$
bits representing the nodes of the tree in preorder,
where each internal node is represented by a 1
(meaning that the next bit encodes its first child)
and each leaf is represented by a 0. However,
for efficient navigation one would typically use
a pointer-based data structure. Alternatively, one
may store only the leaves of the tree, ordered
along a space-filling curve. This variant of the
quadtree is called the *linear quadtree* and was
introduced by Gargantini [3]. The linear quadtree
has smaller memory requirements as it does not
store the tree structure but only the data in the

© Springer Science+Business Media New York 2016
M.-Y. Kao (ed.), *Encyclopedia of Algorithms*,
DOI 10.1007/978-1-4939-2864-4

leaves. This makes it particularly useful when dealing with large data.

In this entry we focus on quadtree construction algorithms that are efficient on very large data. To analyze these algorithms, we use the ▶ I/O-Model and the ▶ Cache-Oblivious Model. We'll use the terms linear quadtree and quadtree subdivision interchangeably. We define the size of a subdivision as the number of cells it contains and the size of a cell is the size of the data (points and edges) it contains/intersects.

The Complexity of Quadtrees for Points in the Plane

Let P be a set of n points in the plane and assume, for simplicity, that the points lie in the unit square. A quadtree for P corresponds to a recursive subdivision of the unit square into four equal regions, called canonical squares, quadrants, or cells, until each square contains at most one point. Following customary terminology in the computational geometry literature (and in deviation from Samet [10]), we refer to this generically as a *point quadtree*.

In the worst case, the size of a quadtree subdivision on P cannot be bounded by a function of n. If ϵ is the distance between the two closest points in P, the worst-case complexity is $\Theta(n \lg \frac{1}{\epsilon})$, and the corresponding tree may have a large number of empty nodes. A *compressed quadtree* is a quadtree where paths of nodes that each have three empty children are merged into a single node along with their empty children; the region corresponding to the merged node is called a *donut* and represents the difference between two canonical squares. A compressed quadtree for a set of n points in the plane such that each cell contains at most one point has size $\Theta(n)$ and height $\Theta(n)$ in the worst case.

The Complexity of Quadtrees for Line Segments in the Plane

Let $\mathcal{E}$ be a set of n non-intersecting line segments in the plane – for example, the edges of a planar subdivision – and assume, as above, that the edges lie in the unit square. We refer to a quadtree for $\mathcal{E}$ generically as an *edge quadtree* and assume that each edge is stored in all the cells

that it intersects. The simplest way to define an edge quadtree may be to take a point quadtree on the endpoints of the edges and then store each edge with the leaves that correspond to the quadtree cells intersected by the edge. We denote by l the number of intersections between $\mathcal{E}$ and the cells in the subdivision. Even if we use a compressed quadtree, in the worst case, there can be $\Theta(n)$ cells that each intersects $\Theta(n)$ edges, so $l = \Theta(n^2)$, and the quadtree will have size $\Theta(n + l) = \Theta(n^2)$. Other edge quadtrees can be defined by formulating stopping criteria that allow subdividing cells further in order to limit the number of edges that intersect each cell; this will result in a subdivision with more cells but smaller number of edges per cell. However, obtaining a good trade-off between the size of a cell (number of points and edges inside or intersecting it) and the number of cells in the subdivision is not possible in the worst case. Note that an edge quadtree that splits a region until it intersects a single edge will result in a subdivision of unbounded size since the distance between two edges can be arbitrarily small.

Quadtrees and Morton Indexing

Quadtrees are often used in conjunction with a z-order space-filling curve. A z-order, or Morton order, can be understood as a mapping from two-dimensional (in general multidimensional) data to one dimension. We use a z-order curve that visits the four quadrants of the initial square, recursively, in the order top left, top right, bottom left, and bottom right. This order gives a well-defined ordering between any two canonical squares in the subdivision. If we define canonical squares to be closed on the top and left side and open on the bottom and right side, the z-order also gives a well-defined ordering between any two points in the input region. Let $p = (p_x, p_y)$ be a point in the unit square $[0, 1)^2$, with the x-axis oriented from left to right and the y-axis oriented from top to bottom. We define the z-index $Z(p)$ of p to be the value in the range $[0, 1)$ obtained by interleaving the bits in the fractional parts of p_x and p_y, starting with a bit from p_y. The value $Z(p)$ is sometimes called the *Morton block index* of p. The z-order of two points in the unit square is the order of their z-indices. A crucial property is that the z-indices of all points in a canonical square σ form an interval $[z_1, z_2)$ of $[0, 1)$, where z_1 is the z-index of the top left corner of σ. A donut cell is the difference between two canonical squares $[z_1, z_2)$ and $[z_3, z_4)$, and thus, it is the union of two intervals $[z_1, z_3)$ and $[z_4, z_2)$.

With this notation, a (compressed) quadtree subdivision corresponds to a subdivision $\mathcal{Q}$ of the z-order curve and can be viewed as a set of consecutive, adjacent, nonoverlapping intervals, covering $[0, 1)$, in z-order: $\mathcal{Q} = \{[z_1 = 0, z_2), [z_2, z_3), \ldots\}$. Each interval $[z_i, z_{i+1})$ corresponds to a cell σ_i, which is either a canonical square or a part of a donut. We note that this representation does not make any assumptions on the stopping criterion used to generate the quadtree subdivision and thus works on any quadtree subdivision, no matter how many points are in a region and whether it is compressed or not. A linear edge quadtree can therefore be represented as a set of key-edge pairs, where each intersection of an edge e with a quadtree cell σ corresponding to an interval $[z_1, z_2]$ is represented by storing edge e with key z_1; thus each cell stores all edges that intersect it [2,4].

Key Results

Point Quadtrees
Agarwal et al. [1] described an algorithm for constructing a quadtree on a set of n points in the plane such that each cell contains $O(k)$ points; the algorithm runs in $O(\frac{n}{B} \frac{h}{\log M/B})$ I/O's, where h is the height of the quadtree. Effectively, this is $O(\text{sort}(n))$ I/O's only when $h = O(\log n)$, which is true when the points are nicely distributed. A bound on the size of the quadtree is not given, and the quadtree is not compressed, which means the quadtree size can be unbounded in the worst case. The algorithms were implemented and tested as part of an application to interpolate LIDAR datasets, which are nicely distributed and unlikely to cause worst-case behavior.

De Berg et al. [2] described an algorithm to construct a compressed quadtree subdivision with at most one point per cell in $O(\text{sort}(n))$ I/O's, as a step in the construction of their Guard-quadtree for edges which is discussed below. Haverkort et al. [4] describe a simple generalization of this algorithm which constructs a compressed quadtree subdivision of $O(n/k)$ cells with at most k points per cell in the same I/O-bound. Thus, compared to the algorithm by Agarwal et al., a stronger bound on the I/O-complexity is obtained, along with an upper bound on the number of cells in the subdivision.

PM Quadtrees
A variety of edge quadtrees were described by Samet and various co-authors [5–7,9,11,12]. All of these solutions are aimed at subdividing the cells that intersect too many edges, while also limiting the total size of the quadtree and being able to construct it I/O-efficiently.

The *PM quadtree* [11] allows a region to contain more than one edge if the edges meet at a vertex inside the region; otherwise it keeps subdividing it. Variants of PM quadtrees differ in how to handle regions that contain no vertices (only edges). The *segment quadtree* [12] is a linear quadtree in which a leaf cell is either empty, contains one edge and no vertices, or contains precisely one vertex and its incident edges. The most versatile structure within the PM family is the *PMR quadtree* [9], a linear quadtree where each region may have a variable number of segments and regions are split if they contain more than a predetermined threshold of edges. The tree is built incrementally, by inserting each segment into all the regions that it intersects. When a region contains more segments than a predetermined splitting threshold, the region is split, once, into four quadrants. Improved algorithms for the construction (or *bulk loading*) of the PMR quadtree were described in [5–7]. These algorithms are developed and optimized with massive data in mind and use I/O-efficient sorting as one of the steps. It is reported that in many cases (although not in the worst case), the I/O-cost of the bulk-loading algorithm is the same as that of external sorting [5]. The algorithms are reported to perform well in practice, but there are several disadvantages: the the resulting quadtree depends on the insertion order; complexity is analyzed

Q

in terms of various parameters that depend on the data; and the performance is not worst-case optimal. On the plus side, the algorithms can handle insertions and work in situations where the data is dynamic.

Star-Quadtrees

The Star-quadtree by De Berg et al. [2] is designed for fat triangulations (a triangulation is fat if every angle of every triangle is larger than some fixed positive constant δ). A Star-quadtree is a linear, uncompressed edge quadtree that splits a region until all edges intersecting a region are incident on one common endpoint (similar to the PM quadtree by Samet and Webber [11]). The Star-quadtree can be built on any set of edges in the plane, but, when the input is a fat triangulation, it can be shown that this stopping rule creates (1) a quadtree of $\Theta(n)$ size and (2) each leaf cell in the quadtree (each cell in the subdivision) intersecting $\Theta(1)$ edges. The height of the quadtree can still be $\Theta(n)$, which makes a top-down construction, such as that used by Agarwal et al. [1], height dependent and not optimal. The authors of the Star-quadtree describe a completely different algorithm for its construction that crucially exploits the stopping criterion and runs in $O(\text{sort}(n))$ I/O's if the input is a fat triangulation.

Guard-Quadtrees

The Guard-quadtree by De Berg et al. [2] is designed for sets of non-intersecting edges of low density – a set of edges has density λ if any disk D is intersected by at most λ edges whose length is at least the diameter of D. For a given set of n edges, the authors define a set of at most $4n$ guards, namely, the vertices of the minimum axis-parallel bounding rectangles of the individual edges. The Guard-quadtree is a linear, compressed edge quadtree that splits a region until it contains at most one guard. As the set of guards is a superset of the endpoints of the edges, this leads to a subdivision that is more refined than a quadtree built only on the endpoints of edges. The stopping rule, together with compression, leads to a quadtree subdivision that has $O(n)$ cells and each cell intersects $O(1)$

edges, provided the set of edges to be stored has low density. Furthermore, the quadtree can be constructed in $O(\text{sort}(n))$ I/O's in this case.

K-Quadtrees

Combining ideas from De Berg et al. [2] with packing more vertices in a region, Haverkort et al. [4] described an I/O-efficient edge quadtree referred to as a K-quadtree. For any $k \geq 1$, the K-quadtree is a compressed, linear quadtree built on the endpoints of the edges, with $O(n/k)$ cells in total and such that each cell contains $O(k)$ vertices (and such that each edge is stored in all cells that they intersect). Each cell in the subdivision can intersect $O(n)$ edges in the worst case. For $k = 1$, a K-quadtree is a linear, compressed edge quadtree with $O(n)$ cells and at most one vertex per cell. Larger values of k can be chosen to trade off between the number of cells $O(n/k)$ and the number of vertices in a cell $O(k)$.

The algorithm for building a K-quadtree has two steps: First it builds, in $O(\text{sort}(n))$ I/O's, a linear, compressed quadtree subdivision on the endpoints of $\mathcal{E}$ with $O(n/k)$ cells in total and such that each cell contains $O(k)$ vertices. This step is a simple generalization of the algorithm for building Guard-quadtrees from [2]. In the second step, the K-quadtree construction algorithm computes the intersections between the edges and the subdivision in $O(\text{sort}(n + l))$ I/O's, where $l = O(n^2/k)$ is the total number of intersections.

The main idea of the second step of the algorithm is to split the set of edges into edges of positive slope $\mathcal{E}_+$ and edges of negative slope $\mathcal{E}_-$ and compute the intersections of each set separately. The intersections of $\mathcal{E}_+$ with the subdivision are computed by *time-forward processing*, as follows. The cells of the quadtree subdivision are scanned in z-order. At any point during this scan, there is a *frontier*: an xy-monotone curve that constitutes the boundary between the cells that have already been scanned and the cells that are still to be scanned. The algorithm relies on the property that an edge of positive slope intersects the cells in the subdivision in z-order (a similar property holds for the edges of negative slope and a reflected version of the z-order). During the scan, each edge of $\mathcal{E}_+$ is passed on, from each intersected quadtree cell to the next, through a supporting data structure that stores the edges intersecting the frontier.

Unlike standard instantiations of time-forward processing, the supporting data structure is not a priority queue, but it is a list, implemented as two stacks, containing the edges that intersect the frontier, in order along the frontier. At each point in time, the list starts at the bottom of one stack and goes up to the top and then down the other stack. The cutting point between the two stacks corresponds to the current scanning position in the list; scanning backward or forward in the list for lookups and updates is implemented by moving elements from one stack to another. The key to I/O-efficiency is that the total amount of scanning that is needed to maintain the supporting data structure is linear in the output size, incurring only $O(\text{scan}(l))$ I/Os.

As the algorithm relies only on the basic building blocks of I/O-efficient sorting, scanning, and stacks, it is also easy to implement cache obliviously.

Compared to a quadtree that employs a stopping criterion that aims to bound the number of edges intersecting a cell (like PMR, Star- and Guard-quadtrees), the simpler K-quadtree has a couple of advantages: (1) the resulting subdivision size is smaller; (2) the total size of the quadtree (the number of intersections between edges and the subdivisions) is also smaller since the size of the subdivision is smaller; and (3) the quadtree can be built in $O(\text{sort}(n + l))$ I/O's, without making any assumptions about the input.

Datasets

Common test datasets for 2D quadtrees are triangulated terrains and USA TIGER data. They represent relatively simple classes of inputs; however they arise frequently in practice and have been used extensively as test beds for spatial index structures. The TIGER dataset consists of 50 datasets, one for each state, containing roads, railways, boundaries, and hydrography in the state. The size of a dataset ranges from 115,626 edges (Delaware) to 40.4

million edges (Texas). The TIGER datasets can be downloaded from http://www.census.gov/cgi-bin/geo/shapefiles2013/main.

Experimental Results

Since many of the quadtree algorithms perform much better in practice than their theoretical worst-case bounds, experimental analysis is an important way to assess their merits. Some of the early experimental analysis of quadtrees performance on massive data was by Hjaltason and Samet [6]. They describe ample results concerning practical performance of PMR quadtrees in terms of construction time, insertions and bulk insertions, comparison with R-tree bulk-loading, and, as an application, performance of spatial join using quadtrees to store the datasets. Their test data consists of TIGER datasets ranging from 40K lines to approximately 260K edges on a machine with 64MB RAM.

Agarwal et al. [1] implemented and tested their I/O-efficient point quadtree part of an application to interpolate LIDAR datasets, where it was used specifically for batched neighbor finding (finding the points in all neighbor leaves for each quadtree leaf). The algorithms are scalable up to at least 500 million points (20GB raw data) (their platform was an Intel 3.4GHz machine with 1GB RAM running Linux).

Haverkort et al. [4] described an experimental analysis of K-quadtrees reporting on the construction time and size of the quadtree (number of cells and number of edge-cell intersections) for various values of k, as well as computing a spatial join using K-quadtrees. The K-quadtree construction algorithm is efficient and scalable, with the running time getting faster as more points are packed into a leaf. Even though the number of edges intersecting a cell may be large, the average size of a cell stays low and the total size of the quadtree is linear. Their tests use TIGER data with the largest bundle, corresponding to the entire USA, having approximately 427 million edges, on a machine with 512MB RAM. A comparison with the PMR quadtree results of

Hjaltason and Samet [6] is difficult because of the difference in platforms.

Extensions

A series of recent results have shown that compressed quadtrees and Delaunay triangulations are equivalent structures, in the sense that a compressed quadtree of a set of points P in the plane can be computed in linear time given the Delaunay triangulation $DT(P)$; and the other way around, the Delaunay triangulation $DT(P)$ can be computed in linear time given a compressed quadtree of P; see, for example, Löffler and Mulzer [8]. In the I/O-model, both problems can be solved in $O(\text{sort}(n))$ I/O's. This naturally brings the question of whether one can be computed from the other in $O(\text{scan}(n))$ I/O's.

Recommended Reading

1. Agarwal PK, Arge L, Danner A (2006) From point cloud to grid DEM: a scalable approach. In: Proceedings of the 12th symposium on spatial data handling, Vienna, pp 771–788
2. de Berg M, Haverkort H, Thite S, Toma L (2010) Star-quadtrees and guard-quadtrees: I/O-efficient indexes for fat triangulations and low-density planar subdivisions. Comput Geom 43(5):493–513
3. Gargantini I (1982) An effective way to represent quadtrees. Commun ACM 25(12):905–910
4. Haverkort H, Toma L, Wei BP (2013) An edge quadtree for external memory. In: Proceedings of the 12th international symposium on experimental algorithms, Rome, pp 115–126
5. Hjaltason G, Samet H (1999) Improved bulk-loading algorithms for quadtrees. In: Proceedings of the ACM international symposium on advances in GIS, Kansas City, pp 110–115
6. Hjaltason GR, Samet H (2002) Speeding up construction of PMR quadtree-based spatial indexes. VLDB J 11:190–137
7. Hjaltason G, Samet H, Sussmann Y (1997) Speeding up bulk-loading of quadtrees. In: Proceedings of the ACM international symposium on advances in GIS, Las Vegas
8. Löffler M, Mulzer W (2011) Triangulating the square and squaring the triangle: quadtrees and delaunay triangulations are equivalent. In: Proceedings of the 22nd ACM-SIAM symposium on discrete algorithms (SODA), San Francisco, pp 1759–1777
9. Nelson R, Samet H (1987) A population analysis for hierarchical data structures. In: Proceeding of the SIGMOD, San Francisco, pp 270–277
10. Samet H (2006) Foundations of multidimensional and metric data structures. Morgan-Kaufmann, San Francisco
11. Samet H, Webber R (1985) Storing a collection of polygons using quadtrees. ACM Trans Graph 4(3):182–222
12. Samet H, Shaffer C, Webber R (1986) The segment quadtree: a linear quadtree-based representation for linear features. In: Data structures for raster graphics pp 91–123

Quantification of Regulation in Networks with Positive and Negative Interaction Weights

Colin Campbell[1] and Réka Albert[2]
[1]Department of Physics, Pennsylvania State University, University Park, PA, USA
[2]Department of Biology and Department of Physics, Pennsylvania State University, University Park, PA, USA

Keywords

Complex systems; Edge weights; Graph theory; Interaction strength; Network communicability; Regulation; Signal transduction; Systems biology

Years and Authors of Summarized Original Work

2011; Campbell, Thakar, Albert

Problem Definition

A network representation of a complex system comprises nodes, which represent system elements, and edges, which represent interactions between the elements. Networks may be described in terms of their topology; for instance, some nodes may be connected to an atypically large number of other nodes, and some may act as bridge nodes that participate in paths between

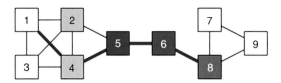

Quantification of Regulation in Networks with Positive and Negative Interaction Weights, Fig. 1 Common network measures applied to a sample 9-node network with symmetric interactions. Darker nodes have higher betweenness centrality (i.e., they tend to act as a bridge between other pairs of nodes); note that even nodes with low degree (i.e., few connections) may have high betweenness centrality. Highlighted edges show a shortest path (length 4) between nodes 1 and 8

many other pairs of nodes (Fig. 1). For a review of topological network measures, see [1–3].

In some contexts, this topological structure serves as a basis for a dynamical description, where nodes are characterized by a dynamic variable that is regulated by the node's interactions. For instance, in the Boolean framework, nodes are either ON or OFF (1 or 0, respectively) [4]. In biological regulatory networks, where interactions between system elements can represent both upregulation and downregulation, one common dynamic scheme is summative [5]:

$$x_i\,(t+\tau) = \mathrm{sgn}\left(\sum_j E_{j,i} x_j\,(t)\right),$$

where $E_{j,i}$ is the weight of the interaction from node j to node i and absent interactions have a weight of 0 by definition. In such a framework, the state change of a node can propagate to the node(s) it directly regulates, then to the node(s) they regulate, and so on. This information flow across a network is sometimes referred to as network communicability [6]. A topological analysis of the network should, in principle, give insight into its dynamical structure and address questions such as, "Which nodes yield a strong influence on many other nodes?," "Which nodes are regulated in a complex way by many other nodes?," and "Which nodes seem to have a peripheral impact on the dynamics of the network?"

However, while networks have been used to explicate the structure and function of a large and diverse array of complex systems, most network measures consider the most general properties of networks and are therefore ill-suited for application to specialized networks. The positive and negative edge weights typically used in biological regulatory networks are one such specialization: standard network measures do not consider edge weights of opposite sign and are therefore ill-equipped to fully capture the dynamical implications of their topology.

Here, we address this shortcoming by developing a suite of topological measures that address the regulatory relationship between nodes that are connected by edges with both positive and negative edge weights. We first consider node-node interactions and then summarize those measures to quantify both the regulatory impact of a node on the entire network and of the entire network on a node.

Key Results

To first consider node-node relationships, we introduce two complementary measures. The **weighted node-node path count** from node i to node j considers both the number of paths from node i to node j and their length:

$$\omega_{ij} \equiv \sum_{l=1}^{l_{\max}} \frac{p_{lij}^+ + p_{lij}^-}{l}$$

Here, p_{lij}^+ and p_{lij}^- respectively indicate the number of positive and negative paths from node i to node j of length l. While we here consider a path to be positive if it contains 0 or an even number of negative edges, note that this measure effectively ignores the sign of the paths. To take this into consideration, we introduce the **node-node path influence**:

$$\pi_{ij} \equiv \sum_{l=1}^{l_{\max}} \frac{p_{lij}^+ - p_{lij}^-}{l}$$

π_{ij} is therefore bounded by the range $[-\omega_{ij},\,\omega_{ij}]$. ω_{ij} indicates the regulatory strength insofar as it is large when there are many short paths between the nodes and decreases when the paths are few and/or long; π_{ij} indicates the overall regulatory nature of those interactions. Values close to

0 relative to $|\omega_{ij}|$ indicate mixed (complex) regulation, while values close to $|\omega_{ij}|$ indicate overall positive or negative regulation.

Node-network relationships may be assessed by cumulating these measures with a fixed source or target node. The **node path influence**, ι_i, and **node path susceptibility**, σ_j, take this into account for a fixed source and target node, respectively:

$$\iota_i \equiv \sum_j \pi_{ij} \omega_{ij}$$

$$\sigma_j \equiv \sum_i \pi_{ij} \omega_{ij}.$$

The summative product results in large absolute values for these measures only when the regulation is both strong and consistent in sign. Nodes receiving low values are regulated weakly and/or in a complex way.

In cases where edge weights take on values other than ± 1, the p_{ijl} values may readily be modified to, for instance, the sum of the mean interaction weights of the pertinent paths. This modification reduces to the above definition when edge weights are restricted to ± 1. In both cases, however, the above measures are characterized by the parameter l_{max}, which represents the longest path considered by the algorithm. Counting all paths of arbitrary length for all but the simplest networks is computationally intractable, and so in practice l_{max} must generally be a low number. We therefore introduce a complementary measure, **strength of connection**, which considers paths of arbitrary length through network erosion. The measure is determined for any two nodes i and j via a procedure that assigns every node a characteristic value. In the below pseudo-algorithm, these values are stored in a dictionary d.

```
N = number of nodes in graph
d = {}
for node in graph: d[node] = infinity

if i != j:
  while a path exists between i and j:
    SP = [nodes on the shortest path between i and j]
    SPL = length of shortest path between i and j
    if SPL == 1:
      delete edge between i and j
      d[i] = d[j] = SPL
    else:
      for node in SP:
        if d[node] == infinity: d[node] = SPL
        if (node != i and node != j): remove node from graph
  return sum(1/(values in d))/(N/2)
else:
    while a cycle containing i exists:
      SP = [nodes on the shortest cycle containing node i]
      SPL = length of shortest cycle containing node i
      if SPL == 1:
        delete self-edge
        d[i] = SPL
      else:
        for node in SP:
          if d[node] == infinity: d[node] = SPL
          if node != i: remove node from graph
    return sum(1/(values in d))/((N+2)/2)
```

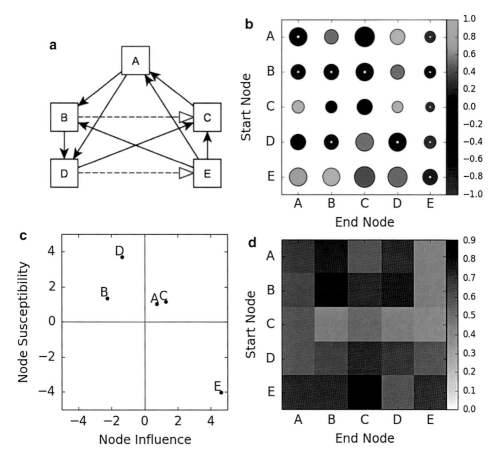

Quantification of Regulation in Networks with Positive and Negative Interaction Weights, Fig. 2 Adapted from Figs. 1 and 2 of [7]. (a) A fully connected 5-node network. *Solid black arrows* indicate positive regulation, while *dashed, red arrows* indicate negative regulation. (b) A circle at position i, j has a size proportional to ω_{ij} $(\max(\omega_{ij}) = 2.75)$ and color determined by π_{ij}/ω_{ij}, with positive, neutral, and negative sign corresponding to *green, black*, or *red* coloring, respectively. Circles are additionally identified with a small *white concentric circle* if $\pi_{ij}/\omega_{ij} \leq -0.2$. (c) A scatter plot of node path influence, ι, and node path susceptibility, σ. (d) The strength of connection measure indicates which node pairs remain well connected under network erosion; the values vary significantly despite each node having equal degree and the network being strongly connected

The normalization factors force the returned value to be bounded by 1 [7]. While this algorithm does not consider the sign of paths, it is straightforward to modify it to, e.g., include only those paths that are of the specified sign. Such a modification would then yield both a positive strength of connection and a negative strength of connection. We demonstrate the above-defined measures for a simple network in Fig. 2.

Applications

The analytical measures introduced above may be applied to any network with both positive and negative edge weights. Biological regulatory networks are a prime example of complex systems that are often modeled in this way. For example, the measures have been applied to explicate the regulatory cross talk of a network of the immune response responding to both respiratory bacteria and allergen [7]. The measures stand to inform the dynamical regulation between nodes from a strictly topological perspective and thereby (1) provide insight into systems where the dynamic behavior is poorly understood and (2) complement dynamic analysis in systems where the regulatory behavior is understood.

Open Problems

The methodology discussed here considers the topology of a network with weighted positive and negative interactions. However, network analysis often involves an investigation of network dynamics, where the details of the interactions encoded in the network topology play a pivotal role. The role of network topology in constraining network dynamics is an active area of study (see, e.g., [8]).

Recommended Reading

1. Albert R, Barabási A-L (2002) Statistical mechanics of complex networks. Rev Mod Phys 74:47–97. doi:10.1103/RevModPhys.74.47
2. Newman MEJ (2010) Networks: an introduction. Oxford University Press, Oxford/New York
3. Newman MEJ (2012) Communities, modules and large-scale structure in networks. Nat Phys 8:25–31. doi:10.1038/nphys2162
4. Wang R-S, Saadatpour A, Albert R (2012) Boolean modeling in systems biology: an overview of methodology and applications. Phys Biol 9:055001. doi:10.1088/1478-3975/9/5/055001
5. Campbell C, Yang S, Albert R, Shea K (2011) A network model for plant–pollinator community assembly. Proc Natl Acad Sci 108:197–202. doi:10.1073/pnas.1008204108
6. Estrada E, Hatano N, Benzi M (2012) The physics of communicability in complex networks. Phys Rep 514:89–119. doi:10.1016/j.physrep.2012.01.006
7. Campbell C, Thakar J, Albert R (2011) Network analysis reveals cross-links of the immune pathways activated by bacteria and allergen. Phys Rev E 84:031929. doi:10.1103/PhysRevE.84.031929
8. Zañudo JGT, Albert R (2013) An effective network reduction approach to find the dynamical repertoire of discrete dynamic networks. Chaos Interdiscip J Nonlinear Sci 23:025111. doi:10.1063/1.4809777

Quantum Algorithm for Element Distinctness

Andris Ambainis
Faculty of Computing, University of Latvia, Riga, Latvia

Keywords

Element distinctness; Quantum algorithms; Quantum search; Quantum walks

Years and Authors of Summarized Original Work

2004; Ambainis

Problem Definition

In the *element distinctness* problem, one is given a list of N elements $x_1, \ldots, x_N \in \{1, \ldots, m\}$ and one must determine if the list contains two equal elements. Access to the list is granted by submitting queries to a black box, and there are two possible types of query.

Value Queries. In this type of query, the input to the black box is an index i. The black box outputs x_i as the answer. In the quantum version of this model, the input is a quantum state that may be entangled with the workspace of the algorithm. The joint state of the query, the answer register, and the workspace may be represented as $\sum_{i,y,z} a_{i,y,z} |i, y, z\rangle$, with y being an extra register which will contain the answer to the query and z being the workspace of the algorithm. The black box transforms this state into $\sum_{i,y,z} a_{i,y,z} |i, (y + x_i) \mod m, z\rangle$. The simplest particular case is if the input to the black box is of the form $\sum_i a_i |i, 0\rangle$. Then, the black box outputs $\sum_i a_i |i, x_i\rangle$. That is, a quantum state consisting of the index i is transformed into a quantum state, each component of which contains x_i together with the corresponding index i.

Comparison Queries. In this type of query, the input to the black box consists of two indices i, j. The black box gives one of the three possible answers: "$x_i > x_j$", "$x_i < x_j$," or "$x_i = x_j$." In the quantum version, the input is a quantum state consisting of basis states $|i, y, z\rangle$, with i, j being two indices and z being algorithm's workspace.

There are several reasons why the element distinctness problem is interesting to study. First of all, it is related to sorting. Being able to sort $x_1, \ldots, x_N$ enables one to solve the element distinctness by first sorting $x_1, \ldots, x_N$ in increasing order. If there are two equal elements $x_i = x_j$, then they will be the next one to another in the sorted list. Therefore, after one has sorted $x_1, \ldots, x_N$, one must only check the sorted list to see if each element is different from the next one. Because of this relation, the element distinctness problem captures some of the same difficulty as sorting. This has led to a long line of research on classical lower bounds for the element distinctness problem (cf. [5, 11, 21] and many other papers).

Second, the central concept of the algorithms for the element distinctness problem is the notion of a collision. This notion can be generalized in different ways, and its generalizations are useful for building quantum algorithms for various graph-theoretic problems (e.g., triangle finding [18]) and matrix problems (e.g., checking matrix identities [12]).

A generalization of element distinctness is element k-distinctness [3], in which one must determine if there exist k different indices $i_1, \ldots, i_k \in \{1, \ldots, N\}$ such that $x_{i1} = x_{i2} = \ldots = x_{ik}$. A further generalization is the k-subset finding problem [14], in which one is given a function $f(y_1, \ldots, y_k)$ and must determine whether there exist $i_1, \ldots, i_k \in \{1, \ldots, N\}$ such that $f(x_{i1}, x_{i2}, \ldots, x_{ik}) = 1$ (where $x_1, \ldots, x_N$ are the input data).

Key Results

Element Distinctness: Summary of Results

In the classical (non-quantum) context, the natural solution to the element distinctness problem is done by sorting, as described in the previous section. This uses $O(N)$ value queries (or $O(N \log N)$ comparison queries) and $O(N \log N)$ time. Any classical algorithm requires $\Omega(N)$ value or $\Omega(N \log N)$ comparison queries. If the algorithm is restricted to $o(N)$ space, stronger lower bounds are known [21].

In the quantum context, Buhrman et al. [13] gave the first nontrivial quantum algorithm, using $O(N^{3/4})$ queries. Ambainis [3] then designed a new algorithm, based on a novel idea using quantum walks. Ambainis' algorithm uses $O(N^{2/3})$ queries and is known to be optimal: Aaronson and Shi [1, 2, 15] have shown that any quantum algorithm for element distinctness must use $\Omega(N^{2/3})$ queries.

For quantum algorithms that are restricted to storing r values x_i (where $r < N^{2/3}$), the best algorithm runs in $O(N/\sqrt{r})$ time.

All of these results are for value queries. They can be adapted to the comparison query model, with a log N factor increase in the complexity. The time complexity is within a polylogarithmic $O(\log^c N)$ factor of the query complexity, as long as the computational model is sufficiently general [3]. (Random access quantum memory is necessary for implementing any of the known quantum algorithms.)

Using the quantum walk methods, one can also solve the k-distinctness problem [3]. This gives a quantum algorithm for k-distinctness (and k-subset finding) that uses $O(N^{k/(k+1)})$ value queries and $O(N^{k/(k+1)})$ memory. For the case when the memory is restricted to $r < N^{k/(k+1)}$ values of x_i, it suffices to use $O(r + (N^{k/2})/(r^{(k-1)/2}))$ value queries. The results generalize to comparison queries and time complexity, with a polylogarithmic factor increase in the time complexity (similarly to the element distinctness problem). For the k-subset finding problem, Belovs and Rosmanis [8] have shown that there is a function $f(y_1, \ldots, y_k)$ for which $\Omega(N^{k/(k+1)})$ queries are also necessary.

For the k-distinctness problem, a better quantum algorithm has been recently developed by Belovs [6], using the learning graph approach. It solves 3-distinctness using $O(N^{5/7})$ value queries and k-distinctness using $O\left(N^{1-\frac{2^k-2}{2^k-1}}\right)$ value queries. The algorithm for 3-distinctness can be implemented so that it runs in time $O(N^{5/7}\log^c N)$ [9]. It is an open problem to construct a time-efficient implementation for $k > 3$.

Element Distinctness: The Methods

Ambainis' algorithm has the following structure. Its state space is spanned by basic states $|T\rangle$, for all sets of indices $T \subseteq \{1,\ldots,N\}$ with $|T| = r$. The algorithm starts in a uniform superposition of all $|T\rangle$ and repeatedly applies a sequence of two transformations:

1. Conditional phase flip: $|T\rangle \to -|T\rangle$ for all T such that T contains i, j with $x_i = x_j$, and $|T\rangle \to |T\rangle$ for all other T;
2. Quantum walk: perform $O(\sqrt{r})$ steps of quantum walk, as defined in [3]. Each step is a transformation that maps each $|T\rangle$ to a combination of basis states $|T'\rangle$ for T' that differ from T in one element.

The algorithm maintains another quantum register, which stores all the values of $x_i, i \in T$. This register is updated with every step of the quantum walk.

If there are two elements i, j such that $x_i = x_j$, repeating these two transformations $O(N/r)$ times increases the amplitudes of $|T\rangle$ containing i, j. Measuring the state of the algorithm at that point with high probability produces a set T containing i, j. Then, from the set T, we can find i and j.

The basic structure of [3] is similar to Grover's quantum search, but with one substantial difference. In Grover's algorithm, instead of using a quantum walk, one would use Grover's *diffusion transformation*. Implementing Grover's diffusion requires $\Omega(r)$ updates to the register that stores $x_i, i \in T$. In contrast to Grover's diffusion, each step of quantum walk changes T by one element, requiring just one update to the list of $x_i, i \in T$. Thus, $O(\sqrt{r})$ steps of quantum walk can be performed with $O(\sqrt{r})$ updates, quadratically better than Grover's diffusion. And, as shown in [3],

the quantum walk provides a sufficiently good approximation of diffusion for the algorithm to work correctly.

This was one of the first uses of quantum walks to construct quantum algorithms. Ambainis, Kempe, and Rivosh [4] then generalized it to handle searching on grids (described in another entry of this encyclopedia). Their algorithm is based on the same mathematical ideas, but has a slightly different structure. Instead of alternating quantum walk steps with phase flips, it performs a quantum walk with two different walk rules – the normal walk rule and the "perturbed" one. (The normal rule corresponds to a walk without a phase flip and the "perturbed" rule corresponds to a combination of the walk with a phase flip.)

Generalization to Arbitrary Markov Chains

Szegedy [20] and Magniez et al. [19] have generalized the algorithms of [4] and [3], respectively, to speed up the search of an arbitrary Markov chain. The main result of [19] is as follows.

Let P be an irreducible Markov chain with state space X. Assume that some states in the state space of P are *marked*. Our goal is to find a marked state. This can be done by a classical algorithm that runs the Markov chain P until it reaches a marked state (Algorithm 1).

There are three costs that contribute to the complexity of Algorithm 1:

1. **Setup cost** S: the cost to sample the initial state x from the initial distribution.
2. **Update cost** U: the cost to simulate one step of a random walk.
3. **Checking cost** C: the cost to check if the current state x is marked.

The overall complexity of the classical algorithm is then $S + t_2(t_1 U + C)$. The required t_1 and t_2 can be calculated from the characteristics of the Markov chain P. Namely,

Proposition 1 ([19]) *Let P be an ergodic, yet symmetric Markov chain. Let $\delta > 0$ be the eigenvalue gap of P, and assume that whenever*

Algorithm 1: Search by a classical random walk

---

1. Initialize x to a state sampled from some initial distribution over the states of P.
2. t_2 times repeat:
   (a) If the current stage y is marked, output y and stop;
   (b) Simulate t_1 steps of random walk, starting with the current state y.
3. If the algorithm has not terminated, output "no marked state."

---

the set of marked states M is nonempty, we have $|M|/|X| \geq \epsilon$. Then there are $t_1 = O(1/\delta)$ and $t_2 = O(1/\epsilon)$ such that Algorithm 1 finds a marked element with high probability.

Thus, the cost of finding a marked element classically is $O(S + 1/\epsilon(1/\delta U + C))$. Magniez et al. [19] construct a quantum algorithm that finds a marked element in $O(S' + 1/\epsilon(1/\sqrt{\delta}U' + C'))$ steps, with S', U', and C' being quantum versions of the setup, update, and checking costs (in most of applications, these are of the same order as S, U, and C). This achieves a quadratic improvement in the dependence on both ϵ and δ.

The element distinctness problem is solved by a particular case of this algorithm: a search on the Johnson graph. The Johnson graph is the graph whose vertices v_T correspond to subsets $T \subseteq \{1, \ldots, N\}$ of size $|T| = r$. A vertex v_T is connected to a vertex v'_T, if the subsets T and T' differ in exactly one element. A vertex v_T is marked if T contains indices i, j with $x_i = x_j$.

Consider the following Markov chain on the Johnson graph. The starting probability distribution s is the uniform distribution over the vertices of the Johnson graph. In each step, the Markov chain chooses the next vertex v'_T from all vertices that are adjacent to the current vertex v_T, uniformly at random. While running the Markov chain, one maintains a list of all $x_i, i \in T$. This means that the costs of the classical Markov chain are as follows:

- Setup cost of $S = r$ queries (to query all $x_i, i \in T$ where v_T is the starting state).

- Update cost of $U = 1$ query (to query the value $x_i, i \in T' - T$, where v_T is the vertex before the step and v'_T is the new vertex).
- Checking cost of $C = 0$ queries (the values $x_i, i \in T$ are already known to the algorithm, and no further queries are needed).

The quantum costs S', U', and C' are of the same order as S, U, and C.

For this Markov chain, it can be shown that the eigenvalue gap is $\delta = O(1/r)$ and the fraction of marked states is $\epsilon = O((r^2)/(N^2))$. Thus, the quantum algorithm runs in time

$$O\left(S' + \frac{1}{\sqrt{\epsilon}}\left(\frac{1}{\sqrt{\delta}}U' + C'\right)\right)$$

$$= O\left(S' + \sqrt{r}\left(\frac{N}{r}U' + C'\right)\right)$$

$$= O\left(r + \frac{N}{\sqrt{r}}\right).$$

Learning Graphs

Another framework that generalizes the element distinctness is the learning graphs by Belovs [6]. A learning graph is a structure that describes algorithm's information about the input data. Using this approach, many quantum algorithms can be described as sequences of high-level instructions (which can be compiled into a standard quantum query algorithm). For example, the element distinctness algorithm corresponds to a sequence of three operations:

1. Load $O(N^{2/3})$ values x_i for randomly chosen $i \in \{1, 2, \ldots, N\}$.
2. Load one of the two equal elements x_i.
3. Load the other equal element x_j.

Belovs [6] describes rules for determining the complexity of each step. In the algorithm above, the complexities are $O(N^{2/3})$, $O(\sqrt{N})$, and $O(N^{2/3})$, respectively. This results in the same overall complexity of $O(N^{2/3})$.

The learning graph approach has been used to construct new quantum algorithms for k-distinctness [7], triangle-finding [6], and other tasks.

Applications

Magniez et al. [19] showed how to use the ideas from the element distinctness algorithm as a subroutine to solve the *triangle problem*. In the triangle problem, one is given a graph G on n vertices, accessible by queries to an oracle, and they must determine whether the graph contains a triangle (three vertices v_1, v_2, v_3 with $v_1 v_2$, $v_1 v_3$, and $v_2 v_3$ all being edges). This problem requires $\Omega(n^2)$ queries classically. Magniez et al. [19] showed that it can be solved using $O(n^{1.3} \log^c n)$ quantum queries, with a modification of the element distinctness algorithm as a subroutine. This has been improved by several authors. Currently, the best quantum algorithm for triangle finding is by Le Gall [17] which uses $O(n^{1.25} \log^c n)$ queries. It is also based on quantum walks but uses them in a much more complex way.

The methods of Szegedy [20] and Magniez et al. [19] can be used as subroutines for quantum algorithms for checking matrix identities [12,18].

Bernstein et al. [10] have used the element distinctness algorithm to design a quantum algorithm for the subset sum problem, by combining the element distinctness algorithm with ideas from classical algorithms for subset sum. The resulting algorithm solves the subset sum problem for n numbers in $2^{(0.241+o(1))n}$ time steps, under some heuristic assumptions that are similar to the ones that are assumed for classical subset sum algorithms. The best classical algorithm uses $2^{(0.291+o(1))n}$ time steps.

Open Problems

1. How many queries are necessary to solve the element distinctness problem if the memory accessible to the algorithm is limited to r items, $r < N^{2/3}$? The algorithm of [3] gives $O(N/\sqrt{r})$ queries, and the best lower bound is $\Omega(N^{2/3})$ queries.
2. Consider the following problem:

    Graph collision [18]. The problem is specified by a graph G (which is arbitrary but known in advance) and variables $x_1, \ldots, x_N \in \{0, 1\}$, accessible by queries to an oracle. The task is to determine if G contains an edge uv such that $x_u = x_v = 1$. How many queries are necessary to solve this problem?

    The element distinctness algorithm can be adapted to solve this problem with $O(N^{2/3})$ queries [18], but there is no matching lower bound. Is there a better algorithm? A better algorithm for the graph collision problem would immediately imply a better algorithm for the triangle problem.

Cross-References

▶ Quantum Analogues of Markov Chains
▶ Quantum Algorithm for Finding Triangles
▶ Quantum Algorithm for Search on Grids
▶ Quantum Algorithms for Matrix Multiplication and Product Verification
▶ Quantum Search

Recommended Reading

1. Aaronson S, Shi Y (2004) Quantum lower bounds for the collision and the element distinctness problems. J ACM 51(4):595–605
2. Ambainis A (2005) Polynomial degree and lower bounds in quantum complexity: collision and element distinctness with small range. Theor Comput 1:37–46
3. Ambainis A (2007) Quantum walk algorithm for element distinctness. SIAM J Comput 37(1):210–239
4. Ambainis A, Kempe J, Rivosh A (2006) Coins make quantum walks faster. In: Proceedings of the ACM/SIAM symposium on discrete algorithms (SODA'06), Miami, pp 1099–1108
5. Beame P, Saks M, Sun X, Vee E (2003) Time-space trade-off lower bounds for randomized computation of decision problems. J ACM 50(2):154–195
6. Belovs A (2012) Span programs for functions with constant-sized 1-certificates: extended abstract. In: Proceedings of ACM symposium on theory of computing (STOC'12), New York, pp 77–84
7. Belovs A (2012) Learning-graph-based quantum algorithm for k-distinctness. In: Proceedings of IEEE conference on foundations of computer science (FOCS'12), New Brunswick, pp 207–216
8. Belovs A, Rosmanis A (2013) On the power of non-adaptive learning graphs. In: IEEE conference on computational complexity, Palo Alto, pp 44–55
9. Belovs A, Childs A, Jeffery S, Kothari R, Magniez F (2013) Time-efficient quantum walks for 3-distinctness. In: Proceedings of international colloquium on automata, languages and programming (ICALP'13), Riga, vol 1, pp 105–122

10. Bernstein D, Jeffery S, Lange T, Meurer A (2013) Quantum algorithms for the subset-sum problem. In: Proceedings of international workshop on post-quantum cryptography (PQCrypto'13), Limoges, pp 16–33
11. Borodin A, Fischer M, Kirkpatrick D, Lynch N (1981) A time-space tradeoff for sorting on non-oblivious machines. J Comput Syst Sci 22:351–364
12. Buhrman H, Spalek R (2006) Quantum verification of matrix products. In: Proceedings of the ACM/SIAM symposium on discrete algorithms (SODA'06), Miami, pp 880–889
13. Buhrman H, Durr C, Heiligman M, Høyer P, Magniez F, Santha M, de Wolf R (2005) Quantum algorithms for element distinctness. SIAM J Comput 34(6):1324–1330
14. Childs AM, Eisenberg JM (2005) Quantum algorithms for subset finding. Quantum Inf Comput 5:593
15. Kutin S (2005) Quantum lower bound for the collision problem with small range. Theor Comput 1:29–36
16. Le Gall F (2014, to appear) Improved quantum algorithm for triangle finding via combinatorial arguments. In: Proceedings of the IEEE symposium on foundations of computer science (FOCS 2014), Philadelphia
17. Magniez F, Nayak A (2005) Quantum complexity of testing group commutativity. In: Proceedings of the international colloquium automata, languages and programming (ICALP'05), Lisbon, pp 1312–1324
18. Magniez F, Santha M, Szegedy M (2007) Quantum algorithms for the triangle problem. SIAM J Comput 37(2):413–424
19. Magniez F, Nayak A, Roland J, Santha M (2007) Search by quantum walk. In: Proceedings of the ACM symposium on the theory of computing (STOC'07), San Diego, pp 575–584
20. Szegedy M (2004) Quantum speed-up of Markov Chain based algorithms. In: Proceedings of the IEEE conference on foundations of computer science (FOCS'04), Rome, pp 32–41
21. Yao A (1994) Near-optimal time-space tradeoff for element distinctness. SIAM J Comput 23(5):966–975

Quantum Algorithm for Factoring

Sean Hallgren
Department of Computer Science and Engineering, The Pennsylvania State University, University Park, State College, PA, USA

Years and Authors of Summarized Original Work

1994; Shor

Problem Definition

Every positive integer n has a unique decomposition as a product of primes $n = p_1^{e_1} \cdots p_k^{e_k}$, for prime number p_i, and positive integer exponent e_i. Computing the decomposition $p_1, e_1, \ldots, p_k, e_k$ from n is the factoring problem.

Factoring has been studied for many hundreds of years, and exponential time algorithms for it were found to include trial division, Lehman's method, Pollard's ρ method, and Shank's class group method [1]. With the invention of the RSA public-key cryptosystem in the late 1970s, the problem became practically important and started receiving much more attention. The security of RSA is closely related to the complexity of factoring, and in particular, it is only secured if factoring does not have an efficient algorithm. The first subexponential-time algorithm is due to Morrison and Brillhard [4] using a continued fraction algorithm. This was succeeded by the quadratic sieve method of Pomerance and the elliptic curve method of Lenstra [5]. The number field sieve [2, 3], found in 1989, is the best-known classical algorithm for factoring and runs in time $\exp(c(\log n)^{1/3}(\log \log n)^{2/3})$ for some constant c. Shor's result is a polynomial-time quantum algorithm for factoring.

Key Results

Theorem 1 ([2, 3]) *There is a subexponential-time classical algorithm that factors the integer n in time $\exp(c(\log n)^{1/3}(\log \log n)^{2/3})$.*

Theorem 2 ([6]) *There is a polynomial-time quantum algorithm that factors integers. The algorithm factor n in time $O((\log n)^2(\log n \log n)(\log \log \log n))$ plus polynomial in $\log n$ post-processing which can be done classically.*

Applications

Computationally hard number theoretic problems are useful for public-key cryptosystems.

The RSA public-key cryptosystem, as well as others, requires that factoring not to have an efficient algorithm. The best-known classical algorithms for factoring can help determine how secure the cryptosystem is and what key sizes to choose. Shor's quantum algorithm for factoring can break these systems in polynomial time using a quantum computer.

Open Problems

It is open whether there is a polynomial-time classical algorithm for factoring.

Cross-References

▶ Quantum Algorithm for Solving Pell's Equation
▶ Quantum Algorithm for the Discrete Logarithm Problem
▶ Quantum Algorithms for Class Group of a Number Field

Recommended Reading

1. Cohen H (1993) A course in computational algebraic number theory. Graduate texts in mathematics, vol 138. Springer, Berlin/Heidelberg/ New York
2. Lenstra A, Lenstra H (eds) (1993) The development of the number field sieve. Lecture notes in mathematics, vol 1544. Springer, Berlin
3. Lenstra AK, Lenstra HW Jr, Manasse MS, Pollard JM (1990) The number field sieve. In: Proceedings of the twenty second annual ACM symposium on theory of computing, Baltimore, 14–16 May 1990, pp 564–572
4. Morrison M, Brillhart J A method of factoring and the factorization of F7
5. Pomerance C Factoring. In: Pomerance C (ed) Cryptology and computational number theory. Proceedings of symposia in applied mathematics, vol 42. American Mathematical Society, Providence, p 27
6. Shor PW (1997) Polynomial-time algorithms for prime factorization and discrete logarithms on a quantum computer. SIAM J Comput 26:1484–1509

Quantum Algorithm for Finding Triangles

Stacey Jeffery[1] and Peter C. Richter[2,3]
[1]David R. Cheriton School of Computer Science, University of Waterloo, Waterloo, ON, Canada
[2]Department of Combinatorics and Optimization, and Institute for Quantum Computing, University of Waterloo, Waterloo, ON, Canada
[3]Department of Computer Science, Rutgers, The State University of New Jersey, New Brunswick, NJ, USA

Keywords

Quantum query complexity; Triangle finding

Years and Authors of Summarized Original Work

2014; Le Gall

Problem Definition

A *triangle* is a clique of size three in an undirected graph. Triangle finding has been the subject of extensive study as a basic search problem whose quantum query complexity is still open, in contrast to unstructured search [6] and element distinctness [1].

This survey concerns quantum query algorithms for triangle finding. A *quantum query algorithm* for a search problem $P = \{M_f\}_f$ is a sequence of unitary operators $Q_f = U_k O_f U_{k-1} O_f U_1 O_f U_0$ such that if $M_f \neq \emptyset$, measuring $|\psi_f\rangle = Q_f|0\rangle$ yields a member of M_f, the set of objects associated with input f, with probability $\geq 2/3$. The operators O_f are *oracle queries*, $O_f : |x\rangle|a\rangle \mapsto |x\rangle|a \oplus f(x)\rangle$, which yield information about f, whereas the U_j are independent of f. The quantum query complexity of P is the minimum number of oracle queries required by a quantum query algorithm for P.

In the context of triangle finding, the function f is the adjacency matrix of an undirected graph on vertices $[n]$, $G \subseteq [n]^2$, with $m = |G|$ edges, where $(a, b) \in G \Rightarrow (b, a) \in G$, by convention. The associated set, M_G, is the set of triangles in G.

Problem 1 (Triangle finding)

INPUT: The adjacency matrix f of a graph G on n vertices.

OUTPUT: A triangle: $(a, b, c) \in [n]^3$ such that $(a, b), (b, c), (a, c) \in G$, if one exists.

A lower bound of $\Omega(n)$ on the quantum query complexity of the triangle finding problem follows from a reduction from search [5]. It is easy to see that the randomized query complexity of the triangle finding problem is $\Theta(n^2)$.

Key Results

Progress on the quantum query complexity of triangle finding has closely followed the development of quantum algorithmic techniques for search problems. The first upper bounds were based on increasingly clever use of the structure of the problem, combined with amplitude amplification [4]. The first bound to go beyond the amplitude amplification framework, achieving $\tilde{O}(n^{13/10})$ [10], was one of the first applications of the quantum walk search technique introduced by Ambainis in his element distinctness algorithm [1] and extended in [13] and [11]. The next bound of $O(n^{35/27})$ was the first application of a new quantum algorithmic technique, the learning graph framework [3]. This finding led to the development of extensions to the quantum walk search technique to give a $\tilde{O}(n^{35/27})$ quantum walk algorithm for triangle finding [7]. The next improvement to $O(n^{9/7})$ also used the learning graph framework [9], whereas the most recent upper bound of $O(n^{5/4})$ uses, once again, a quantum walk search algorithm [8].

An $O(n + \sqrt{nm})$ Algorithm Using Amplitude Amplification

A trivial application of Grover's quantum search algorithm solves the triangle finding problem with $O(n^{3/2})$ quantum queries by searching over $[n]^3$. Buhrman et al. [5] improved this upper bound in the special case where G is sparse (i.e., $m = o(n^2)$).

The algorithm searches for an edge $(a, b) \in G$ in $O(\sqrt{|[n]^2|/m}) = O(n/\sqrt{m})$ quantum queries and then for $c \in [n]$ such that (a, b, c) is a triangle in $O(\sqrt{n})$ quantum queries. The second step succeeds when (a, b) is a triangle edge, which happens with probability at least $1/m$ when G contains a triangle, so applying amplitude amplification to this procedure gives a $O(\sqrt{m}(n/\sqrt{m} + \sqrt{n})) = O(n + \sqrt{nm})$ upper bound:

Theorem 1 (Buhrman et al. [5]) *Using quantum amplitude amplification, the triangle finding problem can be solved in $O(n + \sqrt{nm})$ quantum queries.*

An $\tilde{O}(n^{10/7})$ Algorithm Using Amplitude Amplification

The algorithm of Szegedy et al. [10, 12] is also based on amplitude amplification; however, it exploits additional combinatorial structure in the triangle finding problem.

For $A \subseteq [n]$ and $w \in [n]$, define $\Delta_G(A, w) := \{(u, v) \in A^2 : (u, w), (v, w) \in G\}$. Choose a random subset $X \subseteq [n]$ of size $n^\chi \log n$, for $\chi = 3/7$. Query $X \times [n]$ and search for an edge in $E_X := \bigcup_{w \in X} \Delta_G([n], w)$, which can be determined from $G \cap (X \times [n])$, using $O(|X|n + \sqrt{|E_X|}) = \tilde{O}(n^{1+\chi})$ queries. Either a triangle is found, or $E_X \cap G = \emptyset$.

Let $G' := [n]^2 \setminus E_X$. If a triangle is not found in the first step, then $G \subseteq G'$. Fix $\alpha = \beta = 1/7$. Szegedy et al. show that for most X, G' can be partitioned into (T, E), such that T has $O(n^{3-\alpha})$ triangles and $|E \cap G| = O(n^{2-\beta} + n^{2-\chi+\alpha+\beta})$, in $\tilde{O}(n^{1+\alpha+\beta})$ queries (or a triangle is found in the process). If $G \subseteq G'$, any triangle in G either lies in T, in which case it can be found in $O(\sqrt{n^{3-\alpha}})$ queries using quantum search, or intersects E, in which case it can be found in

$O(n + \sqrt{n|G \cap E|})$ queries using the algorithm of Buhrman et al. This gives the following:

Theorem 2 (Szegedy; Magniez, Santha and Szegedy [10, 12]) *Using amplitude amplification, the triangle finding problem can be solved in $\tilde{O}(n^{10/7})$ quantum queries.*

An $\tilde{O}(n^{13/10})$ Algorithm Using Quantum Walks

A more efficient algorithm for the triangle finding problem was obtained by Magniez et al. [10], using the quantum walk search technique introduced by Ambainis [1].

Given oracle access to a function defining a relation $M \subseteq [n]^k$, Ambainis' quantum walk search procedure finds $(a_1, \ldots, a_k) \in M$ if $M \neq \emptyset$. The algorithm walks on sets $A \subseteq [n]$ of size n^α, keeping track of some data structure $D(A)$ for the current state A and transitioning, in superposition, from A to A' for $A' \subseteq [n]$ of size n^α such that $|A \setminus A'| = 1$. Assume access to a quantum procedure Φ that determines if $A^k \cap M \neq \emptyset$ using $D(A)$, with *checking cost* C queries. Suppose $D(A)$ can be constructed from scratch at *setup cost* S queries and modified from $D(A)$ to $D(A')$ when $|A \setminus A'| = 1$ at an *update cost* U. Then the procedure finds an element of M in $O(\mathsf{S} + (\frac{n}{n^\alpha})^{k/2}(\sqrt{n^\alpha}\mathsf{U} + \mathsf{C}))$ quantum queries. (For details, see the encyclopedia entry on element distinctness.)

For a fixed graph $G \subseteq [n]^2$, consider the *graph collision* problem on G, where an input f defines the binary relation $M_f \subseteq [n]^2$ satisfying $(u, u') \in M_f$ if $f(u) = f(u') = 1$ and $(u, u') \in G$. Setting $k = 2$, it is a simple exercise to see that a quantum walk search algorithm solves this problem with $O(n^\alpha + (\frac{n}{n^\alpha})(\sqrt{n^\alpha} \cdot 1 + 0)) = O(n^\alpha + n^{1-\alpha/2})$ queries. Setting $\alpha = 2/3$ gives an upper bound of $O(n^{2/3})$ quantum queries for graph collision.

Magniez et al. [10] solve triangle finding using a quantum walk algorithm whose checking subroutine is based on graph collision. Let M be the set of triangle edges. Define $D(A) = G \cap A^2$. Then $\mathsf{S} = n^{2\alpha}$ initial queries are needed to set up $D(A)$, and $\mathsf{U} = n^\alpha$ new queries are needed to update $D(A)$, where α is now $3/5$. The check-

ing step consists of an algorithm that, given a known subgraph $H = G \cap A^2$ on n^α vertices, decides if H contains a triangle edge using $\mathsf{C} = \tilde{O}(\sqrt{n}n^{2/3\alpha})$ queries, as follows. For any $v \in [n]$, define f_v on A by $f_v(u) = 1$ if $(u, v) \in G$. An edge $(a, b) \in A^2$ is a graph collision in f_v on $G \cap A^2$ if and only if (a, b, v) is a triangle, so searching for $v \in [n]$ for which f_v has a graph collision, using $O(\sqrt{n}(n^\alpha)^{2/3})$ quantum queries, is equivalent to deciding if $G \cap A^2$ contains a triangle edge. Repeat $\Theta(\log n)$ times, to decrease the error to $1/n^{\Theta(1)}$, since the subroutine is called many times. This gives the following:

Theorem 3 (Magniez, Santha, and Szegedy [10]) *Using a quantum walk search procedure, the triangle finding problem can be solved in $\tilde{O}(n^{13/10})$ quantum queries.*

An $O(n^{35/27})$ Algorithm Using Learning Graphs

The learning graph framework, introduced by Belovs [3], allows for the construction of a quantum algorithm from a particular type of edge-weighted graph called a *learning graph*. For further details, refer to [3]. The first application of this framework was a new upper bound on the quantum query complexity of triangle finding.

A learning graph may be constructed in stages, corresponding to searching for more and more specialized structures, which will eventually contain a 1-certificate for the problem being solved. In Belovs' application to triangle finding, the first part of the learning graph corresponds to searching for an n^α-vertex subset of $[n]$, A, containing two triangle vertices a and b. The next two stages correspond to searching for an $n^{2\alpha-\sigma}$-edge graph on A, H, which contains the triangle edge $\{a, b\}$. The final stages correspond to the graph collision subroutine used in [10] to decide if any edge of the queried subgraph H is a triangle edge. Using $\alpha = 2/3$ and $\sigma = 1/27$ gives the following:

Theorem 4 (Belovs [3]) *Using a learning graph algorithm, the triangle finding problem can be solved in $O(n^{35/27}) = O(n^{1.2963})$ quantum queries.*

Additionally, a quantum walk search algorithm based on this learning graph construction solves triangle finding in $\tilde{O}(n^{35/27})$ queries [7].

An $O(n^{9/7})$ Algorithm Using Learning Graphs

The next upper bound on the quantum query complexity of triangle finding, due to Lee et al. [9], also uses a learning graph. The first part of their learning graph corresponds to searching for an n^α-vertex subset $A \subseteq [n]$, containing a triangle vertex a. The next part corresponds to searching for an n^β-vertex subset $B \subseteq [n]$, containing a vertex, b, from the same triangle as a. The final part corresponds to the graph collision subroutine used in [10], but optimized for an unbalanced bipartite graph, used to decide if any edge of $G \cap (A \times B)$ is a triangle edge. Using $\alpha = 4/7$ and $\beta = 5/7$ gives the following:

Theorem 5 (Lee, Magniez and Santha [9]) *Using a learning graph algorithm, the triangle finding problem can be solved in $O(n^{9/7}) = O(n^{1.2858})$ quantum queries.*

As with the previous algorithm, there exists a quantum walk search algorithm based on this learning graph construction that solves triangle finding in $\tilde{O}(n^{9/7})$ queries [7].

An $\tilde{O}(n^{5/4})$ Algorithm Using Quantum Walks

The best known upper bound on the quantum query complexity of triangle finding is an algorithm by Le Gall [8]. Le Gall's algorithm uses the quantum walk search technique, as in the $\tilde{O}(n^{13/10})$-query algorithm, combined with a more clever utilization of the combinatorial structure of triangle finding, similar to that of the $\tilde{O}(n^{10/7})$-query algorithm, and a quantum search algorithm of Ambainis that finds an x such that $\Phi(x) = 1$ in cost $O(\sqrt{\sum_x C(x)^2})$, where $C(x)$ is the cost to compute $\Phi(x)$ [2].

The algorithm begins, like the $\tilde{O}(n^{10/7})$ algorithm, by choosing a random $X \subseteq [n]$ of size $n^\chi \log n$ and searching for a triangle in $X \times [n]^2$. This is done by quantum search on $X \times [n]^2$, using $O(\sqrt{|X \times [n]^2|}) = \tilde{O}(n^{1+\chi/2})$ quantum queries. If no triangle is found, as in the $\tilde{O}(n^{10/7})$

algorithm, the rest of the algorithm will make use of the fact that $E_X \cap G = \emptyset$, although in this case, since $X \times [n]$ is not queried, E_X is not known.

The rest of the algorithm consists of the following four levels of recursion:

1. Using a quantum walk search algorithm, search for a set $A \subseteq [n]$ of size n^α such that A^2 contains a triangle edge. Maintain a data structure, $D(A)$, encoding $G \cap (A \times X)$.
2. For any $A \subseteq [n]$, to check if A^2 contains a triangle edge, search for a vertex $c \in [n]$ such that $A^2 \times \{c\}$ contains a triangle.
3. For any $A \subseteq [n]$ and $c \in [n]$, to check if $A^2 \times \{c\}$ contains a triangle, use a quantum walk search algorithm to search for a set $B \subseteq A$ of size n^β such that $B^2 \times \{c\}$ contains a triangle. Maintain a data structure, $D^c(B)$, encoding $G \cap (B \times \{c\})$.
4. For any $B \subseteq [n]$ and $c \in [n]$, to check if $B^2 \times \{c\}$ contains a triangle, search for an edge in $\Delta_G(B,c) \setminus E_X$. Here the algorithm exploits the fact that there is no edge in E_X. The set $E_X \cap B^2$ can be determined from $G \cap (A \times X)$.

Constructing $D(A)$ costs $\mathsf{S} = |A \times X| = \tilde{O}(n^{\alpha+\chi})$ queries. Mapping $D(A)$ to $D(A')$ costs $\mathsf{U} = 2|X| = \tilde{O}(n^\chi)$ queries. Let $\Phi(A) = 1$ if A^2 has a triangle edge. Then if C is the quantum query complexity of computing $\Phi(A)$, the quantum query complexity of finding a triangle in $G \setminus E_X$ is $O\left(\mathsf{S} + \frac{n}{n^\alpha}\left(\sqrt{n^\alpha}\mathsf{U} + \mathsf{C}\right)\right) = \tilde{O}\left(n^{\alpha+\chi} + n^{1+\chi-\alpha/2} + n^{1-\alpha}\mathsf{C}\right).$

Let $\Phi_A(c) = 1$ if $A^2 \times \{c\}$ contains a triangle. To compute $\Phi(A)$, search for $c \in [n]$ such that $\Phi_A(c) = 1$. Let $\mathsf{C}'(c)$ be the cost of computing $\Phi_A(c)$, which will vary in c. Then by [2], $\mathsf{C} = O\left(\sqrt{\sum_{c\in[n]} \mathsf{C}'(c)}\right).$

Let $\Phi^c(D^c(B)) = 1$ if $B^2 \times \{c\}$ has a triangle. To compute $\Phi_A(c)$, search for $B \subseteq A$ such that $\Phi^c(D^c(B)) = 1$. Creating $D^c(B)$ costs $\mathsf{S}'' = |B \times \{c\}| = n^\beta$ queries. Mapping $D^c(B)$ to $D^c(B')$ costs $\mathsf{U}'' = 2$ queries. If computing $\Phi^c(D^c(B))$ costs $\mathsf{C}''(c)$, computing $\Phi_A(c)$ costs $\mathsf{C}'(c) = \mathsf{S}'' + \frac{n^\alpha}{n^\beta}\left(\sqrt{n^\beta}\mathsf{U}'' + \mathsf{C}''(c)\right) = O\left(n^\beta + n^{\alpha-\beta/2} + n^{\alpha-\beta}\mathsf{C}''(c)\right)$ queries.

Observe that $\Phi^c(D^c(B)) = 1$ if and only if $\Delta_G(B, c)$ contains an edge. Since $G \cap E_X = \emptyset$, one need only search $\Delta_G(B, c) \setminus E_X$ for an edge. The set $\Delta_G(B, c)$ can be determined from $D^c(B)$, and $E_X \cap A^2$ can be determined from $D(A)$, so $\Delta_G(B, c) \setminus E_X$ is known. Thus, $\mathsf{C}''(c) = O(\sqrt{|\Delta_G(B, c) \setminus E_X|})$.

Using combinatorial arguments, Le Gall proves an upper bound on $|\Delta_G(B, c) \setminus E_X|$ relative to $|\Delta_G(A, c) \setminus E_X|$ for *most* B, allowing him to use further combinatorial arguments to show an upper bound on $\mathsf{C} = \sqrt{\sum_{c \in [n]} \mathsf{C}'(c)^2}$ of $O(n^{1/2+\chi} + n^{1/2+\beta} + n^{1/2+\alpha-\beta/2} + n^{1/2+\alpha-\chi/2})$. Setting $\chi = \beta = 1/2$ and $\alpha = 3/4$ then gives the following:

Theorem 6 (Le Gall [8]) *Using a quantum walk search algorithm, the triangle finding problem can be solved in $\tilde{O}(n^{5/4})$ quantum queries.*

The quantum query complexity of triangle finding is still open, as the best known lower bound is $\Omega(n)$.

Cross-References

▶ Quantum Algorithm for Element Distinctness
▶ Quantum Algorithm for the Collision Problem
▶ Quantum Analogues of Markov Chains

Recommended Reading

1. Ambainis A (2007) Quantum walk algorithm for element distinctness. SIAM J Comput 37(1):210–239
2. Ambainis A (2010) Quantum search with variable times. Theory Comput Syst 47(3):786–807
3. Belovs A (2012) Span programs for functions with constant-sized 1-certificates. In: Proceeding of STOC, New York, pp 77–84
4. Brassard G, Høyer P, Mosca M, Tapp A (2002) Quantum amplitude amplification and estimation. In: Quantum computation and quantum information: a millennium volume. AMS contemporary mathematics series millennium volume, vol 305. American Mathematical Society, Providence, pp 53–74
5. Buhrman H, Dürr C, Heiligman M, Høyer P, Santha M, Magniez F, de Wolf R (2005) Quantum algorithms for element distinctness. SIAM J Comput 34(6):1324–1330
6. Grover LK (1996) A fast quantum mechanical algorithm for database search. In: Proceeding of STOC, Philadelphia, pp 212–219
7. Jeffery S, Kothari R, Magniez F (2013) Nested quantum walks with quantum data structures. In: Proceeding of SODA, New Orleans, pp 1474–1485
8. Le Gall F (2014) Improved quantum algorithm for triangle finding via combinatorial arguments. In: Proceeding of FOCS, Philpelphia, pp 216–225. quant-ph/1407.0085
9. Lee T, Magniez F, Santha M (2013) Improved quantum query algorithms for triangle finding and associativity testing. In: Proceeding of SODA, New Orleans, pp 1486–1502
10. Magniez F, Santha M, Szegedy M (2007) Quantum algorithms for the triangle problem. SIAM J Comput 37(2):413–424
11. Magniez F, Nayak A, Roland J, Santha M (2011) Search via quantum walk. SIAM J Comput 40(1):142–164. quant-ph/0608026
12. Szegedy M (2003) On the quantum query complexity of detecting triangles in graphs. quant-ph/0310107
13. Szegedy M (2004) Quantum speed-up of Markov chain based algorithms. In: Proceeding of FOCS, Rome, pp 32–41

Quantum Algorithm for Search on Grids

Andris Ambainis
Faculty of Computing, University of Latvia, Riga, Latvia

Keywords

Spatial search

Years and Authors of Summarized Original Work

2005; Ambainis, Kempe, Rivosh

Problem Definition

Consider an $\sqrt{N} \times \sqrt{N}$ grid, with each location storing a bit that is 0 or 1. The locations on the grid are indexed by (i, j), where $i, j \in$

$\{0, 1, \ldots, \sqrt{N} - 1\} \cdot a_{i,j}$ denotes the value stored at the location (i, j).

The task is to find a location storing $a_{i,j} = 1$. This problem is as an abstract model for search in a two-dimensional database, with each location storing a variable $x_{i,j}$ with more than two values. The goal is to find $x_{i,j}$ that satisfies certain constraints. One can then define new variables $a_{i,j}$ with $a_{i,j} = 1$ if $x_{i,j}$ satisfies the constraints and search for i, j satisfying $a_{i,j} = 1$.

The grid is searched by a "robot," which at any moment of time is at one location i, j. In one time unit, the robot can either examine the current location or move one step in one of the four directions (left, right, up, or down).

In a probabilistic version of this model, the robot is probabilistic. It makes its decisions (querying the current location or moving) randomly according to prespecified probability distributions. At any moment of time, such a robot is at a probability distribution over the locations of the grid. In the quantum case, one has a "quantum robot" [5] which can be in a quantum superposition of locations (i, j) and is allowed to perform transformations that move it at most one step at a time.

There are several ways to make this model of a "quantum robot" precise [1] and they all lead to similar results.

The simplest to define is the Z-local model of [1]. In this model, the robot's state space is spanned by states $|i, j, a\rangle$ with i, j representing the current location and a being the internal memory of the robot. The robot's state $|\psi\rangle$ can be any quantum superposition of those: $|\psi\rangle = \sum_{i,j,a} \alpha_{i,j,a} |i, j, a\rangle$, where $\alpha_{i,j,a}$ are complex numbers such that $\sum_{i,j,a} |\alpha_{i,j,a}|^2 = 1$. In one step, the robot can either perform a query of the value at the current location or a Z-local transformation.

A query is a transformation that leaves i, j parts of a state $|i, j, a\rangle$ unchanged and modifies the a part in a way that depends only on the value $a_{i,j}$. A Z-local transformation is a transformation that maps any state $|i, j, a\rangle$ to a superposition that involves only states with robot being either at the same location or at one of the four adjacent

locations ($|i, j, b\rangle, |i-1, j, b\rangle, |i+1, j, b\rangle, |i, j-1, b\rangle$ or $|i, j+1, b\rangle$ where the content of the robot's memory b is arbitrary).

The problem generalizes naturally to d-dimensional grid of size $N^{1/d} \times N^{1/d} \times \ldots \times N^{1/d}$, with robot being allowed to query or move one step in one of the d directions in one unit of time.

Key Results

Early Results

This problem was first studied by Benioff [5] who considered the use of the usual quantum search algorithm by Grover [9] in this setting. Grover's algorithm allows to search a collection of N items $a_{i,j}$ with $O(\sqrt{N})$ queries. However, it does not respect the structure of a grid. Between any two queries, it performs a transformation that may require the robot to move from any location (i, j) to any other locations (i', j'). In the robot model, where the robot in only allowed to move one step in one time unit, such transformation requires $O(\sqrt{N})$ steps to perform. Implementing Grover's algorithm, which requires $O(\sqrt{N})$ such transformations, therefore, takes $O(\sqrt{N}) \times O(\sqrt{N}) = O(N)$ time, providing no advantage over the naive classical algorithm.

The first algorithm improving over the naive use of Grover's search was proposed by Aaronson and Ambainis [1] who achieved the following results:

- Search on $\sqrt{N} \times \sqrt{N}$ grid, if it is known that the grid contains exactly one $a_{i,j} = 1$ in $O(\sqrt{N} \log^{3/2} N)$ steps.
- Search on $\sqrt{N} \times \sqrt{N}$ grid, if the grid may contain an arbitrary number of $a_{i,j} = 1$ in $O(\sqrt{N} \log^{5/2} N)$ steps.
- Search on $N^{1/d} \times N^{1/d} \times \ldots \times N^{1/d}$ grid, for $d \geq 3$, in $O(\sqrt{N})$ steps.

They also considered a generalization of the problem, search on a graph G, in which the robot moves on the vertices v of the graph G and searches for a variable $a_v = 1$. In one step, the robot can examine the variable a_v corresponding to the current vertex v or move to another vertex w adjacent to v. Aaronson and Ambainis [1] gave an algorithm for searching an arbitrary graph with grid-like expansion properties in $O(N^{1/2+o(1)})$ steps. The main technique in those algorithms was the use of Grover's search and its generalization, amplitude amplification [6], in combination with "divide-and-conquer" methods recursively breaking up a grid into smaller parts.

Quantum Walks

The next algorithms were based on quantum walks [3,7,8]. Ambainis, Kempe, and Rivosh [3] presented an algorithm, based on a discrete time quantum walk, which searches the two-dimensional $\sqrt{N} \times \sqrt{N}$ in $O(\sqrt{N} \log N)$ steps, if the grid is known to contain exactly one $a_{i,j} = 1$ and in $O(\sqrt{N} \log^2 N)$ steps in the general case. Childs and Goldstone [8] achieved a similar performance, using continuous time quantum walk. Curiously, it turned out that the performance of the walk crucially depended on the particular choice of the quantum walk, both in the discrete and continuous time, and some very natural choices of quantum walk (e.g., one in [7]) failed.

Besides providing an almost optimal quantum speedup, the quantum walk algorithms also have an additional advantage: their simplicity. The discrete quantum walk algorithm of [3] uses just two bits of quantum memory. Its basis states are $|i, j, d\rangle$, where (i, j) is a location on the grid and d is one of the four directions: $\leftarrow$, $\rightarrow$, $\uparrow$, and $\downarrow$. The basic algorithm consists of the following simple steps:

1. Generate the state $\sum_{i,j,d} \frac{1}{2\sqrt{N}} |i, j, d\rangle$.

2. $O(\sqrt{N} \log N)$ times repeat
   1. Perform the transformation

$$C_0 = \begin{pmatrix} -\frac{1}{2} & \frac{1}{2} & \frac{1}{2} & \frac{1}{2} \\ \frac{1}{2} & -\frac{1}{2} & \frac{1}{2} & \frac{1}{2} \\ \frac{1}{2} & \frac{1}{2} & -\frac{1}{2} & \frac{1}{2} \\ \frac{1}{2} & \frac{1}{2} & \frac{1}{2} & -\frac{1}{2} \end{pmatrix}$$

2. On the states $|i, j, \leftarrow\rangle$, $|i, j, \rightarrow\rangle$, $|i, j, \uparrow\rangle$, $|i, j, \downarrow\rangle$, if $a_{i,j} = 0$ and the transformation $C_1 = -I$ on the same four states if $a_{i,j} = 1$.

3. Move one step according to the direction register and reverse the direction:

$$|i, j, \rightarrow\rangle \to |i + 1, j, \leftarrow\rangle,$$

$$|i, j, \leftarrow\rangle \to |i - 1, j, \rightarrow\rangle,$$

$$|i, j, \uparrow\rangle \to |i, j - 1, \downarrow\rangle,$$

$$|i, j, \downarrow\rangle \to |i, j + 1, \uparrow\rangle.$$

In case, if $a_{i,j} = 1$ for one location (i, j), a significant part of the algorithm's final state will consist of the four states $|i, j, d\rangle$ for the location (i, j) with $a_{i,j} = 1$. This can be used to detect the presence of such location. More precisely, if we run the algorithm for $O(\sqrt{N \log N})$ steps and measure the state, we obtain one of the four states $|i, j, d\rangle$ with probability $\Theta(1/\log N)$.

We can increase the probability of algorithm finding the right location (i, j) by either repeating the algorithm or using quantum amplitude amplification. Quantum amplitude amplification [6] takes a quantum algorithm that succeeds with a small probability ε and increases the success probability to 3/4, by repeating the quantum algorithm $O(1/\sqrt{\epsilon})$ times. In our case, $\epsilon = \Theta(1/\log N)$ which means that it suffices to repeat the basic algorithm $O(\sqrt{\log N})$ times. This increases the running time from $O(\sqrt{N \log N})$ for the basic algorithm to $O(\sqrt{N} \log N)$.

A quantum algorithm for search on a grid can be also derived by designing a classical algorithm that finds $a_i, j = 1$ by performing a random walk on the grid and then applying Szegedy's general translation of classical random walks to quantum random chains, with a quadratic speedup over the classical random walk algorithm [15]. The resulting algorithm is similar to the algorithm of [3] described above and has the same running time.

For an overview on related quantum algorithms using similar methods, see [2, 10].

Further Developments

The running time of the algorithm has been improved to $O(\sqrt{N \log N})$ time steps if the grid is known to contain exactly one (i, j) with $a_{i,j} = 1$ and $O(\sqrt{N} \log^{1.5} N)$ steps in the general case. This can be achieved in two different ways. First, Tulsi [16] showed how to modify the quantum walk so that, after $O(\sqrt{N \log N})$ steps, it finds the right (i, j) with a constant probability. This eliminates the need to use amplitude amplification.

Second, Ambainis et al. [4] showed that the same result can be achieved without modifying the quantum walk, by a simple classical postprocessing. That is, even if the quantum walk does not find the right (i, j), its final state is much more likely to be (i', j') that is close to (i, j). One can then run the quantum walk for $O(\sqrt{N \log N})$ steps once, measure the result, obtain a location (i', j'), and search the nearby locations for (i, j) with $a_{i,j} = 1$.

Search algorithms similar to the original 2D search algorithm have been analyzed for a number of other graphs (e.g., for hierarchical networks [12]).

Applications

The quantum algorithm for search on the grid by Ambainis, Kempe, and Rivosh [3] has been generalized by Szegedy [15], obtaining a general procedure for speeding up classical Markov chains (described in more detail in the article on Quantization of Markov Chains). Szegedy's generalization concerns a class of algorithms called *Search by Random Walk* in which one performs a random walk on some search space until finding an element with a certain property. Szegedy [15] showed that if a classical random walk finds a marked element in T steps (on average), there is a quantum algorithm that detects the existence of a marked element in $O(\sqrt{T})$ steps.

It is an open problem to extend Szegedy's algorithm so that it not only detects the existence of an element with the desired property but also finds it in $O(\sqrt{T})$ time steps. (This is known as the "finding problem".) A step in this direction was made by Magniez et al. [11] who generalized Tulsi's algorithm for search on the grid [16] to solve the finding problem in $O(\sqrt{T})$ steps whenever the classical random walk is vertex transitive and the search space has a unique element with the desired property.

Quantum algorithms for spatial search are also useful for designing quantum communication protocols for the set disjointness problem. In the set disjointness problem, one has two parties holding inputs $x \in \{0, 1\}^N$ and $y \in \{0, 1\}^N$ and they have to determine if there is $i \in \{1, \ldots, N\}$ for which $x_i = y_i = 1$. (One can think of x and y as representing subsets $X, Y \subseteq \{1, \ldots, N\}$ with $x_i = 1 (y_i = 1)$ if $i \in X(i \in Y)$. Then, determining if $x_i = y_i = 1$ for some i is equivalent to determining if $X \cap Y \neq \emptyset$.)

The goal is to solve the problem, communicating as few bits between the two parties as possible. Classically, $\Omega(N)$ bits of communication are required [13]. The optimal quantum protocol [1] uses $O(\sqrt{N})$ quantum bits of communication and its main idea is to reduce the problem to spatial search. As shown by the $\Omega(\sqrt{N})$ lower bound of [14], this algorithm is optimal.

Cross-References

▶ Quantum Algorithm for Element Distinctness
▶ Quantum Analogues of Markov Chains
▶ Quantum Search

Recommended Reading

1. Aaronson S, Ambainis A (2003) Quantum search of spatial regions. In: Proceedings of the 44th annual IEEE symposium on foundations of computer science (FOCS), Cambridge, pp 200–209
2. Ambainis A (2003) Quantum walks and their algorithmic applications. Int J Quantum Inf 1:507–518
3. Ambainis A, Kempe J, Rivosh A (2005) Coins make quantum walks faster. In: Proceedings of SODA'05, Vancouver, pp 1099–1108
4. Ambainis A, Backurs A, Nahimovs N, Ozols R, Rivosh A (2012) Search by quantum walks on two-dimensional grid without amplitude amplification. In: Proceedings of TQC'2012, Tokyo, pp 87–97

Q

5. Benioff P (2002) Space searches with a quantum robot. In: Quantum computation and information, Washington, DC, 2000. Contemporary mathematics, vol 305. American Mathematical Society, Providence, pp 1–12

6. Brassard G, Høyer P, Mosca M, Tapp A (2002) Quantum amplitude amplification and estimation. In: Quantum computation and information, Washington, DC, 2000. Contemporary mathematics, vol 305. American Mathematical Society, Providence, pp 53–74

7. Childs AM, Goldstone J (2004) Spatial search by quantum walk. Phys Rev A 70:022314

8. Childs AM, Goldstone J (2004) Spatial search and the Dirac equation. Phys Rev A 70:042312

9. Grover L (1996) A fast quantum mechanical algorithm for database search. In: Proceedings of the 28th STOC, Philadelphia. ACM, New York, pp 212–219

10. Kempe J (2003) Quantum random walks – an introductory overview. Contemp Phys 44(4):302–327

11. Magniez F, Nayak A, Richter P, Santha M (2012) On the hitting times of quantum versus random walks. Algorithmica 63(1–2):91–116

12. Marquezino F, Portugal R, Boettcher S (2011) Quantum search algorithms on hierarchical networks. In: IEEE information theory workshop (ITW), Paraty, pp 247–251

13. Razborov A (1992) On the distributional complexity of disjointness. Theor Comput Sci 106(2):385–390

14. Razborov AA (2002) Quantum communication complexity of symmetric predicates. Izv Russ Acad Sci Math 67:145–159

15. Szegedy M (2004) Quantum speed-up of Markov Chain based algorithms. In: Proceedings of FOCS'04, Rome, pp 32–41

16. Tulsi A (2008) Faster quantum-walk algorithm for the two-dimensional spatial search. Phys Rev A 78:012310

Quantum Algorithm for Solving Pell's Equation

Sean Hallgren
Department of Computer Science and Engineering, The Pennsylvania State University, University Park, State College, PA, USA

Years and Authors of Summarized Original Work

2002; Hallgren

Problem Definition

Pell's equation is one of the oldest studied problem in number theory. For a positive square-free integer d, Pell's equation is $x^2 - dy^2 = 1$, and the problem is to compute integer solutions x, y of the equation [8, 10]. The earliest algorithm for it uses the continued fraction expansion of $\sqrt{d}$ and dates back to 1000 a.d. by Indian mathematicians. Lagrange showed that there are an infinite number of solutions of Pell's equation. All solutions are of the form $x_n + y_n\sqrt{d} = (x_1 + y_1\sqrt{d})^n$, where the smallest solution, (x_1, y_1), is called the fundamental solution. The solution (x_1, y_1) may have exponentially many bits in general in terms of the input size, which is $\log d$, and so cannot be written down in polynomial time. To resolve this difficulty, the computational problem is recast as computing the integer closest to the regulator $R = \ln(x_1 + y_1\sqrt{d})$. In this representation, solutions of Pell's equation are positive integer multiples of R.

Solving Pell's equation is a special case of computing the unit group of number field. For a positive non-square integer Δ congruent to 0 or 1 mod 4, $K = \mathbb{Q}(\sqrt{\Delta})$ is a real quadratic number field. Its subring $\mathcal{O} = \mathbb{Z}\left[\frac{\Delta+\sqrt{\Delta}}{2}\right] \subseteq Q(\sqrt{\Delta})$ is called the quadratic order of discriminant Δ. The unit group is the set of invertible elements of $\mathcal{O}$. Units have the form $\pm\varepsilon^k$, where $k \in \mathbb{Z}$, for some $\varepsilon > 1$ called the fundamental unit. The fundamental unit ε can have exponentially many bits, so an approximation of the regulator $R = \ln\varepsilon$ is computed. In this representation the unit group consists of integer multiples of R. Given the integer closest to R there are classical polynomial-time algorithms to compute R to any precision. There are also efficient algorithms to test if a given number is a good approximation to an integer multiple of a unit or to compute the least significant digits of $\varepsilon = e^R$ [1,3].

Two related and potentially more difficult problems are the principal ideal problem and computing the class group of a number field. In the principal ideal problem, a number field and an ideal I of $\mathcal{O}$ are given, and the problem is to decide if the ideal is principal, i.e., whether there exists α such that $I = \alpha\mathcal{O}$. If it is principal, then one can ask for an approximation of $\ln\alpha$. There are efficient classical algorithms to verify that a number is close to $\ln\alpha$ [1, 3]. The class group of a number field is the finite abelian group defined by taking the set of fractional ideals modulo the principal fractional ideals. The class number is the size of the class group. Computing the unit group, computing the class group, and solving the principal ideal problems are three of the main problems of computational algebraic number theory [3]. Assuming the GRH, they are in NP ∩ CoNP [9].

Key Results

The best known classical algorithms for the problems defined in the last section take subexponential time, but there are polynomial-time quantum algorithms for them [5, 7].

Theorem 1 *Given a quadratic discriminant Δ, there is a classical algorithm that computes an integer multiple of the regulator to within one. Assuming the GRH, this algorithm computes the regulator to within one and runs in expected time* $\exp\left(\sqrt{(\log\Delta)\log\log\Delta}\right)^{O(1)}$.

Theorem 2 *There is a polynomial-time quantum algorithm that, given a quadratic discriminant Δ, approximates the regulator to within δ of the associated order $\mathcal{O}$ in time polynomial in $\log\Delta$ and $\log\delta$ with probability exponentially close to one.*

Corollary 1 *There is a polynomial-time quantum algorithm that solves Pell's equation.*

The quantum algorithm for Pell's equation uses the existence of a periodic function on the reals which has period R and is one-to-one within each period [5, 7]. There is a discrete version of this function that can be computed efficiently. This function does not have the same periodic property since it cannot be evaluated at arbitrary real numbers such as R, but it does approximate the situation well enough for the quantum algorithm. In particular, computing the approximate

period of this function gives R to the closest integer or, in other words, computes a generator for the unit group.

Theorem 3 *There is a polynomial-time quantum algorithm that solves the principal ideal problem in real quadratic number fields.*

Corollary 2 *There is a polynomial-time quantum algorithm that can break the Buchmann-Williams key-exchange protocol in real quadratic number fields.*

Theorem 4 *The class group and class number of a real quadratic number field can be computed in quantum polynomial time assuming the GRH.*

In general, one can ask to find the unit group of an arbitrary degree number field $\mathbb{Q}(\theta)$, where θ is the root of a polynomial with rational coefficients. There are two parameters associated with this problem. The first is the discriminant, which generalizes parameter above. The second is the degree n of the number field as a vector space over the rational numbers. In the above example the degree is fixed at 2. The unit group of an arbitrary degree number can also be computed efficiently by a quantum algorithm.

Theorem 5 ([4]) *The unit group of a number field can be computed by a quantum algorithm in time polynomial in $\log$ the discriminant, and the degree n.*

This last result uses a major generalization of the hidden subgroup problem to continuous functions. A new method is used to compute the function that is polynomial time in the degree of the number field and solves the hidden subgroup problem for continuous groups.

Applications

Computationally hard number theoretic problems are useful for public key cryptosystems. There are reductions from factoring to Pell's equation and Pell's equation to the principal ideal problem, but no reductions are known in the opposite direction. The principal ideal problem forms the basis of the Buchmann-Williams key-exchange protocol

[2]. Identification schemes based on this problem have been proposed by Hamdy and Maurer [6]. The classical exponential-time algorithms help determine which parameters to choose for the cryptosystem. The best known algorithm for Pell's equation is exponentially slower than the best factoring algorithm. Systems based on these harder problems were proposed as alternatives in case factoring turns out to be polynomial time solvable. The efficient quantum algorithms can break these cryptosystems.

Open Problems

Lattice-based cryptography is the leading class of candidates for primitives secure against quantum computers. Recent systems have used lattices from number fields in order to make them more efficient. It is an open question whether lattice-based systems are secure against quantum computers, given that quantum computers have an exponential advantage over classical computers for some problems in number fields.

Cross-References

▶ Quantum Algorithm for Factoring
▶ Quantum Algorithms for Class Group of a Number Field

Recommended Reading

1. Buchmann J, Thiel C, Williams HC (1995) Short representation of quadratic integers. In: Bosma W, van der Poorten AJ (eds) Computational algebra and number theory, Sydney 1992. Mathematics and its applications, vol 325. Kluwer Academic, Dordrecht, pp 159–185
2. Buchmann JA, Williams HC (1989) A key exchange system based on real quadratic fields (extended abstract). In: Brassard G (ed) Advances in cryptology-CRYPTO '89, 20–24 Aug 1990. Lecture notes in computer science, vol 435. Springer, Berlin, pp 335–343
3. Cohen H (1993) A course in computational algebraic number theory. Graduate texts in mathematics, vol 138. Springer, Berlin/Heidelberg
4. Eisentraeger K, Hallgren S, Kitaev A, Song F (2014) A quantum algorithm for computing the unit group of an arbitrary degree number field. In: Proceedings of the 46th ACM symposium on theory of computing
5. Hallgren S (2007) Polynomial-time quantum algorithms for Pell's equation and the principal ideal problem. J ACM 54(1):1–19
6. Hamdy S, Maurer M (1999) Feige-fiat-shamir identification based on real quadratic fields. Techincal report TI-23/99. Technische Universität Darmstadt, Fachbereich Informatik. http://www.informatik.tu-darmstadt.de/TI/Veroeffentlichung/TR/
7. Jozsa R (2003) Notes on Hallgren's efficient quantum algorithm for solving Pell's equation. Techincal report, quant-ph/0302134
8. Lenstra HW Jr (2002) Solving the Pell equation. Not Am Math Soc 49:182–192
9. Thiel C (1995) On the complexity of some problems in algorithmic algebraic number theory. Ph.D. thesis, Universität des Saarlandes, Saarbrücken
10. Williams HC (2002) Solving the Pell equation. In: Proceedings of the millennial conference on number theory, pp 397–435

Quantum Algorithm for the Collision Problem

Gilles Brassard[1], Peter Høyer[2], and Alain Tapp[1]
[1]Université de Montréal, Montréal, QC, Canada
[2]University of Calgary, Calgary, AB, Canada

Keywords

Collision; Grover's algorithm; Quantum algorithm

Years and Authors of Summarized Original Work

1998; Brassard, Høyer, Tapp

Problem Definition

A function F is said to be *r-to-one* if every element in its image has exactly r distinct preimages.

Input : an r-to-one function F.

Output : x_1 and x_2 such that $F(x_1) = F(x_2)$.

Key Results

The algorithm presented here finds collisions in arbitrary r-to-one functions F after only $O(\sqrt[3]{N/r})$ expected evaluations of F. The algorithm uses the function as a black box, that is, the only thing the algorithm requires is the capacity to evaluate the function. Again assuming the function is given by a black box, the algorithm is optimal [1], and it is more efficient than the best *possible* classical algorithm which has query complexity $\Omega(\sqrt{N/r})$. The result is stated precisely in the following theorem and corollary.

Theorem 1 *Given an r-to-one function $F : X \to Y$ with $r \geq 2$ and an integer $1 \leq k \leq N = |X|$, algorithm* **Collision**(F, k) *returns a collision after an expected number of $O(k + \sqrt{N/(rk)})$ evaluations of F and uses space $\Theta(k)$. In particular, when $k = \sqrt[3]{N/r}$, then* **Collision**(F, k) *uses an expected number of $O(\sqrt[3]{N/r})$ evaluations of F and space $\Theta(\sqrt[3]{N/r})$.*

Corollary 1 *There exists a quantum algorithm that can find a collision in an arbitrary r-to-one function $F : X \to Y$, for any $r \geq 2$, using space S and an expected number of $O(T)$ evaluations of F for every $1 \leq S \leq T$ subject to $ST^2 \geq |F(X)|$ where $F(X)$ denotes the image of F.*

The algorithm uses as a procedure a version of Grover's search algorithm. Given a function H with domain size n and a target y, **Grover**(H, y) returns an x such that $H(x) = y$ in expected $O(\sqrt{n})$ evaluations of H.

Collision(F, k):

1. Pick an arbitrary subset $K \subseteq X$ of cardinality k. Construct a table L of size k where each item in L holds a distinct pair $(x, F(x))$ with $x \in K$.

2. Sort L according to the second entry in each item of L.

3. Check if L contains a collision, that is, check if there exist distinct elements $(x_0, F(x_0))$, $(x_1, F(x_1)) \in L$ for which $F(x_0) = F(x_1)$. If so, go to step 6.

4. Compute $x_1 = $ **Grover**$(H, 1)$ where $H : X \to \{0, 1\}$ denotes the function defined by $H(x) = 1$ if and only if there exists $x_0 \in K$ so that $(x_0, F(x)) \in L$ but $x \neq x_0$. (Note that x_0 is unique if it exists since we already checked that there are no collisions in L.)

5. Find $(x_0, F(x_1)) \in L$.

6. Output the collision $\{x_0, x_1\}$.

Applications

This problem is of particular interest for cryptology because some functions known as *hash functions* are used in various cryptographic protocols. The security of these protocols crucially depends on the presumed difficulty of finding collisions in such functions.

Recommended Reading

1. Aaronson S, Shi Y (2004) Quantum lower bounds for the collision and the element distinctness problems. J ACM (JACM) 51(4):595–605
2. Brassard G, Høyer P, Tapp A (1998) Quantum algorithm for the collision problem. In: 3rd Latin American theoretical informatics symposium (LATIN'98). LNCS, vol 1380. Springer, Berlin, pp 163–169
3. Brassard G, Høyer P, Mosca M, Tapp A (2002) Quantum amplitude amplification and estimation. In: Lomonaco SJ (ed) Quantum computation & quantum information science. AMS contemporary mathematics series millennium volume, vol 305, pp 53–74
4. Boyer M, Brassard G, Høyer P, Tapp A (1996) Tight bounds on quantum searching. In: Proceedings of the fourth workshop on physics of computation, Boston, pp 36–43
5. Carter JL, Wegman MN (1979) Universal classes of hash functions. J Comput Syst Sci 18(2):143–154
6. Grover LK (1996) A fast quantum mechanical algorithm for database search. In: Proceedings of the 28th annual ACM symposium on theory of computing, Philadelphia, pp 212–219

7. Nielsen MA, Chuang IL (2000) Quantum computation and quantum information. Cambridge University Press, Cambridge
8. Stinson DR (1995) Cryptography: theory and practice. CRC, Boca Raton

Quantum Algorithm for the Discrete Logarithm Problem

Pranab Sen
School of Technology and Computer Science,
Tata Institute of Fundamental Research,
Mumbai, India

Keywords

Logarithms in groups

Years and Authors of Summarized Original Work

1994; Shor

Problem Definition

Given positive real numbers $a \neq 1, b$, the logarithm of b to base a is the unique real number s such that $b = a^s$. The notion of the *discrete logarithm* is an extension of this concept to general groups.

Problem 1 (Discrete logarithm)

INPUT: Group $G, a, b \in G$ such that $b = a^s$ for some positive integer s.

OUTPUT: The smallest positive integer s satisfying $b = a^s$, also known as the *discrete logarithm* of b to the base a in G.

The usual logarithm corresponds to the discrete logarithm problem over the group of positive reals under multiplication. The most common case of the discrete logarithm problem is when the group $G = \mathbb{Z}_p^*$, the multiplicative group of integers between 1 and $p - 1$ modulo p, where p is a prime. Another important case is when the group G is the group of points of an elliptic curve over a finite field.

Key Results

The discrete logarithm problem in $\mathbb{Z}_p^*$, where p is a prime, as well as in the group of points of an elliptic curve over a finite field is believed to be intractable for randomized classical computers. That is, any, possibly randomized, algorithm for the problem running on a classical computer will take time that is superpolynomial in the number of bits required to describe an input to the problem. The best classical algorithm for finding discrete logarithms in $\mathbb{Z}_p^*$, where p is a prime, is Gordon's [4] adaptation of the number field sieve which runs in time $\exp(O((\log p)^{1/3}(\log \log p)^{2/3}))$.

In a breakthrough result, Shor [9] gave an efficient quantum algorithm for the discrete logarithm problem in any group G; his algorithm runs in time that is polynomial in the bit size of the input.

Result 1 ([9]) There is a quantum algorithm solving the discrete logarithm problem in any group G on n-bit inputs in time $O(n^3)$ with probability at least 3/4.

Description of the Discrete Logarithm Algorithm

Shor's algorithm [9] for the discrete logarithm problem makes essential use of an efficient quantum procedure for implementing a unitary transformation known as the *quantum Fourier transform*. His original algorithm gave an efficient procedure for performing the quantum Fourier transform only over groups of the form $\mathbb{Z}_r$, where r is a "smooth" integer, but nevertheless, he showed that this itself sufficed to solve the discrete logarithm in the general case. In this article, however, a more modern description of Shor's algorithm is given. In particular, a result by Hales and Hallgren [5] is used which shows that the quantum Fourier transform over any finite cyclic group $\mathbb{Z}_r$ can be efficiently approximated to inverse-exponential precision.

A description of the algorithm is given below. A general familiarity with quantum notation on the part of the reader is assumed. A good introduction to quantum computing can be found in the book by Nielsen and Chuang [8]. Let $(G, a, b, \bar{r})$ be an instance of the discrete logarithm problem, where $\bar{r}$ is a supplied upper bound on the order of a in G. That is, there exists a positive integer $r \leq \bar{r}$ such that $a^r = 1$. By using an efficient quantum algorithm for order finding also discovered by Shor [9], one can assume that the order of a in G is known, that is, the smallest positive integer r satisfying $a^r = 1$. Shor's order-finding algorithm runs in time $O((\log \bar{r})^3)$. Let $\epsilon > 0$. The discrete logarithm algorithm works on three registers, of which the first two are each t qubits long, where $t := O(\log r + \log(1/\epsilon))$, and the third register is big enough to store an element of G. Let U denote the unitary transformation

$$U : |x\rangle|y\rangle|z\rangle \mapsto |x\rangle|y\rangle|z \otimes (b^x a^y)\rangle,$$

where $\oplus$ denotes bitwise XOR. Given access to a reversible oracle for group operations in G, U can be implemented reversibly in time $O(t^3)$ by repeated squaring.

Let $\mathbb{C}[\mathbb{Z}_r]$ denote the Hilbert space of functions from $\mathbb{Z}_r$ to complex numbers. The computational basis of $\mathbb{C}[\mathbb{Z}_r]$ consists of the delta functions $\{|l\rangle\}_{0 \leq l \leq r-1}$, where $|\rangle$ is the function that sends the element l to 1 and the other elements of $\mathbb{Z}_r$ to 0. Let $\text{QFT}_{\mathbb{Z}_r}$ denote the *quantum Fourier transform* over the cyclic group $\mathbb{Z}_r$ defined as the following unitary operator on $\mathbb{C}[\mathbb{Z}_r]$:

$$\text{QFT}_{\mathbb{Z}_r} : |x\rangle \mapsto r^{-1/2} \sum_{y \in \mathbb{Z}_r} e^{-2\pi i xy/r} |y\rangle.$$

It can be implemented in quantum time $O(t \log(t/\epsilon) + \log^2(1/\epsilon))$ up to an error of ϵ using one t-qubit register [5]. Note that for any $k \in \mathbb{Z}_r, \text{QFT}_{\mathbb{Z}_r}$ transforms the state $r^{-1/2} \sum_{x \in \mathbb{Z}_r} e^{-2\pi i kx/r} |x\rangle$ to the state $|k\rangle$. For any

integer $l, 0 \leq l \leq r - 1$, define

$$|\hat{l}\rangle := r^{-1/2} \sum_{k=0}^{r-1} e^{-2\pi i lk/r} |a^k\rangle. \quad (1)$$

Observe that $\{|\hat{l}\rangle\}_{0 \leq l \leq r-1}$ forms an orthonormal basis of $\mathbb{C}[\langle a \rangle]$, where $\langle a \rangle$ is the subgroup generated by a in G and is isomorphic to $\mathbb{Z}_r$, and $\mathbb{C}[\langle a \rangle]$ denotes the Hilbert space of functions from $\langle a \rangle$ to complex numbers.

Algorithm 1 (Discrete logarithm)

INPUT: Elements $a, b \in G$, a quantum circuit for U, the order r of a in G.

OUTPUT: With constant probability, the discrete logarithm s of b to the base a in G.

RUNTIME: A total of $O(t^3)$ basic gate operations, including four invocations of $\text{QFT}_{\mathbb{Z}_r}$ and one of U.

PROCEDURE:

1. Repeat Steps (a)–(e) twice, obtaining $(sl_1 \bmod r, l_1)$ and $(sl_2 \bmod r, l_2)$.
   (a) $|0\rangle|0\rangle|0\rangle$
   (b) $\mapsto r^{-1} \sum_{x,y \in \mathbb{Z}_r} |x\rangle|y\rangle|0\rangle$
       Apply $\text{QFT}_{\mathbb{Z}_r}$ to the first two registers:
   (c) $\mapsto r^{-1} \sum_{x,y \in \mathbb{Z}_r} |x\rangle|y\rangle|b^x a^y\rangle$
       Apply U
   (d) $\mapsto r^{-1/2} \sum_{l=0}^{r-1} |sl \bmod r\rangle|l\rangle|\hat{l}\rangle$
       Apply $\text{QFT}_{\mathbb{Z}_r}$ to the first two registers:
   (e) $\mapsto (sl \bmod r, l)$
       Measure the first two registers:
2. If l_1 is not coprime to l_2, abort.
3. Let k_1, k_2 be integers such that $k_1 l_1 + k_2 l_2 = 1$. Then, output $s = k_1(sl_1) + k_2(sl_2) \bmod r$.

The working of the algorithm is explained below. From Eq. (1), it is easy to see that

$$|b^x a^y\rangle = r^{-1/2} \sum_{l=0}^{r-1} e^{2\pi i l(sx+y)/r} |\hat{l}\rangle.$$

Thus, the state in Step 1(c) of the above algorithm can be written as

$$r^{-1} \sum_{x,y \in \mathbb{Z}_r} |x\rangle |y\rangle |b^x a^y\rangle$$

$$= r^{-3/2} \sum_{l=0}^{r-1} \sum_{x,y \in \mathbb{Z}_r} e^{2\pi i l(sx+y)/r} |x\rangle |y\rangle |\hat{l}\rangle$$

$$= r^{-3/2} \sum_{l=0}^{r-1} \left[\sum_{x \in \mathbb{Z}_r} e^{2\pi i s l x/r} |x\rangle \right] \cdot \left[\sum_{y \in \mathbb{Z}_r} e^{2\pi i l y/r} |y\rangle \right] |\hat{l}\rangle.$$

Now, applying $\text{QFT}_{\mathbb{Z}_r}$ to the first two registers gives the state in Step 1(d) of the above algorithm. Measuring the first two registers gives (slmod r, l) for a uniformly distributed $l, 0 \le l \le r - 1$ in Step 1(e). By elementary number theory, it can be shown that if integers l_1, l_2 are uniformly and independently chosen between 0 and $l - 1$, they will be coprime with constant probability. In that case, there will be integers k_1, k_2 such that $k_1 l_1 + k_2 l_2 = 1$, leading to the discovery of the discrete logarithm s in Step 3 of the algorithm with constant probability. Since actually only an ϵ-approximate version of $\text{QFT}_{\mathbb{Z}_r}$ can be applied, ϵ can be set to be a sufficiently small constant, and this will still give the correct discrete logarithm s in Step 3 of the algorithm with constant probability. The success probability of Shor's algorithm for the discrete logarithm problem can be boosted to at least 3/4 by repeating it a constant number of times.

Generalizations of the Discrete Logarithm Algorithm

The discrete logarithm problem is a special case of a more general problem called the *hidden subgroup problem* [8]. The ideas behind Shor's algorithm for the discrete logarithm problem can be generalized in order to yield an efficient quantum algorithm for hidden subgroups in Abelian groups (see [1] for a brief sketch). It turns out that finding the discrete logarithm of b to the base a in G reduces to the hidden subgroup problem in the group $\mathbb{Z}_r \times \mathbb{Z}_r$ where r is the order of a in G. Besides the discrete logarithm problem, other cryptographically important functions like integer factoring, finding the order of permutations, as well as finding self-shift-equivalent polynomials over finite fields can be reduced to instances of a hidden subgroup in Abelian groups.

Applications

The assumed intractability of the discrete logarithm problem lies at the heart of several cryptographic algorithms and protocols. The first example of public-key cryptography, namely, the Diffie-Hellman key exchange [2], uses discrete logarithms, usually in the group $\mathbb{Z}_p^*$ for a prime p. The security of the US national standard Digital Signature Algorithm (see [7] for details and more references) depends on the assumed intractability of discrete logarithms in $\mathbb{Z}_p^*$, where p is a prime. The ElGamal public-key cryptosystem [3] and its derivatives use discrete logarithms in appropriately chosen subgroups of $\mathbb{Z}_p^*$, where p is a prime. More recent applications include those in elliptic curve cryptography [6], where the group consists of the group of points of an elliptic curve over a finite field.

Cross-References

▶ Abelian Hidden Subgroup Problem
▶ Quantum Algorithm for Factoring

Recommended Reading

1. Brassard G, Høyer P (1997) An exact quantum polynomial-time algorithm for Simon's problem. In: Proceedings of the 5th Israeli symposium on theory of computing and systems, Ramat-Gan, 17–19 June 1997, pp 12–23

2. Diffie W, Hellman M (1976) New directions in cryptography. IEEE Trans Inf Theor 22:644–654
3. ElGamal T (1985) A public-key cryptosystem and a signature scheme based on discrete logarithms. IEEE Trans Inf Theor 31(4):469–472
4. Gordon D (1993) Discrete logarithms in GF(p) using the number field sieve. SIAM J Discret Math 6(1):124–139
5. Hales L, Hallgren S (2000) An improved quantum Fourier transform algorithm and applications. In: Proceedings of the 41st annual IEEE symposium on foundations of computer science, Redondo Beach, pp 515–525
6. Hankerson D, Menezes A, Vanstone S (2004) Guide to elliptic curve cryptography. Springer, New York
7. Menezes A, van Oorschot P, Vanstone S (1997) Handbook of applied cryptography. CRC Press, Boca Raton
8. Nielsen M, Chuang I (2000) Quantum computation and quantum information. Cambridge University Press, Cambridge
9. Shor P (1997) Polynomial-time algorithms for prime factorization and discrete logarithms on a quantum computer. SIAM J Comput 26(5):1484–1509

Quantum Algorithm for the Parity Problem

Yaoyun Shi
Department of Electrical Engineering and Computer Science, University of Michigan, Ann Arbor, MI, USA

Keywords

Deutsch algorithm; Deutsch-Jozsa algorithm; Parity

Years and Authors of Summarized Original Work

1985; Deutsch

Problem Definition

The *parity* of n bits $x_0, x_1, \cdots, x_{n-1} \in \{0, 1\}$ is

$$x_0 \oplus x_1 \oplus \cdots \oplus x_{n-1} = \sum_{i=0}^{n-1} x_i \quad \mathrm{mod}\ 2.$$

As an elementary Boolean function, parity is important not only as a building block of digital logic but also for its instrumental roles in several areas such as error correction, hashing, discrete Fourier analysis, pseudorandomness, communication complexity, and circuit complexity. The feature of parity that underlies its many applications is its maximum sensitivity to the input: flipping any bit in the input changes the output. The computation of parity from its input bits is quite straightforward in most computation models. However, two settings deserve attention.

The first is the circuit complexity of parity when the gates are restricted to AND, OR, and NOT gates. It is known that parity cannot be computed by such a circuit of a polynomial size and a constant depth, a groundbreaking result proved independently by Furst, Saxe, and Sipser [7] and Ajtai [1] and improved by several subsequent works.

The second, and the focus of this article, is in the decision tree model (also called the query model or the black-box model), where the input bits $x = x_0 x_1 \cdots x_{n-1} \in \{0, 1\}^n$ are known to an oracle only, and the algorithm needs to ask questions of the type "$x_i =$?" to access the input. The complexity is measured by the number of queries. Specifically, a quantum query is the application of the following query gate:

$$O_x : |i, b\rangle \mapsto |i, b \oplus x_i\rangle,$$

$$i \in \{0, \cdots, n-1\}, b \in \{0, 1\}.$$

Key Results

Proposition 1 *There is a quantum query algorithm computing the parity of 2 bits with probability 1 using 1 query.*

Proof Denote by $|\pm\rangle = \frac{1}{\sqrt{2}}(|0\rangle \pm |1\rangle)$. The initial state of the algorithm is

$$\frac{1}{\sqrt{2}}(|0\rangle + |1\rangle) \otimes |-\rangle.$$

Apply a query gate, using the first register for the index slot and the second register for the answer slot. The resulting state is

$$\frac{1}{\sqrt{2}}((-1)^{x_0}|0\rangle + (-1)^{x_1}|1\rangle) \otimes |-\rangle.$$

Applying a Hadamard gate $H = |+\rangle\langle 0| + |-\rangle\langle 1|$ on the first register brings the state to

$$(-1)^{x_0}|x_0 + x_1\rangle \otimes |-\rangle.$$

Thus measuring the first register gives $x_0 + x_1$ with certainty.

Corollary 1 *There is a quantum query algorithm computing the parity of n bits with probability 1 using $\lceil n/2 \rceil$ queries.*

The above quantum upper bound for parity is tight, even if the algorithm is allowed to err with a probability bounded away from 1/2 [6]. In contrast, any classical randomized algorithm with bounded error probability requires n queries. This follows from the fact that on a random input, any classical algorithm not knowing all the input bits is correct with precisely 1/2 probability.

Applications

The quantum speedup for computing parity was first observed by Deutsch [4]. His algorithm uses $|0\rangle$ in the answer slot, instead of $|-\rangle$. After one query, the algorithm has 3/4 chance of computing the parity, better than any classical algorithm (1/2 chance). The presented algorithm is actually a special case of the Deutsch-Jozsa Algorithm, which solves the following problem now referred to as the Deutsch-Jozsa Problem.

Problem 1 (Deutsch-Jozsa Problem) Let $n \geq 1$ be an integer. Given an oracle function $f : \{0, 1\}^n \to \{0, 1\}$ that satisfies either (a) $f(x)$ is constant on all $x \in \{0, 1\}^n$ or (b) $|\{x : f(x) = 1\}| = |\{x : f(x) = 0\}| = 2^{n-1}$, determine which case it is.

When $n = 1$, the above problem is precisely parity of 2 bits. For a general n, the Deutsch-Jozsa Algorithm solves the problem using only once the following query gate:

$$O_f : |x, b\rangle \mapsto |x, f(x) \oplus b\rangle, \quad x \in \{0, 1\}^n, \ b \in \{0, 1\}.$$

The algorithm starts with

$$|0^n\rangle \otimes |-\rangle.$$

It applies $H^{\otimes n}$ on the index register (the first n qubits), changing the state to

$$\frac{1}{2^{n/2}} \sum_{x \in \{0,1\}^n} |x\rangle \otimes |-\rangle.$$

The oracle gate is then applied, resulting in

$$\frac{1}{2^{n/2}} \sum_{x \in \{0,1\}^n} (-1)^{f(x)} |x\rangle \otimes |-\rangle.$$

For the second time, $H^{\otimes n}$ is applied on the index register, bringing the state to

$$\sum_{y \in \{0,1\}^n} \left(\frac{1}{2^n} \sum_{x \in \{0,1\}^n} (-1)^{f(x)+x \cdot y} \right) |y\rangle \otimes |-\rangle. \tag{1}$$

Finally, the index register is measured in the computational basis. The Algorithm returns "Case (a)" if 0^n is observed, otherwise returns "Case (b)."

By direct inspection, the amplitude of $|0^n\rangle$ is 1 in Case (a) and 0 in Case (b). Thus the algorithm is correct with probability 1. It is easy to see that any deterministic algorithm requires $n/2 + 1$ queries in the worst case; thus the algorithm provides the first exponential quantum versus deterministic speedup.

Note that $O(1)$ expected a number of queries are sufficient for randomized algorithms to solve the Deutsch-Jozsa Problem with a constant success probability arbitrarily close to 1. Thus the Deutsch-Jozsa Algorithm does not have much advantage compared with error-bounded randomized algorithms. One might also feel that the saving of one query for computing the parity of 2 bits by Deutsch-Jozsa Algorithm is due to the artificial definition of one quantum query. Thus the significance of the Deutsch-Jozsa Algorithm is not in solving a practical problem, but in its pioneering use of quantum Fourier transform (QFT), of which $H^{\otimes n}$ is one, in the pattern

$$QFT \to Query \to QFT.$$

The same pattern appears in many subsequent quantum algorithms, including those found by Bernstein and Vazirani [2], Simon [9], and Shor [8].

The Deutsch-Jozsa Algorithm is also referred to as Deutsch Algorithm. The algorithm as presented above is actually the result of the improvement by Cleve, Ekert, Macchiavello, and Mosca [3] and independently by Tapp (unpublished) on the algorithm in [5].

Cross-References

▶ Greedy Set-Cover Algorithms

Recommended Reading

1. Ajtai M (1983) $\sum_{1}^{1}$-formulae on finite structures. Ann Pure Appl Log 24(1):1–48
2. Bernstein E, Vazirani U (1997) Quantum complexity theory. SIAM J Comput 26(5):1411–1473
3. Cleve R, Ekert A, Macchiavello C, Mosca M (1998) Quantum algorithms revisited. Proc R Soc Lond A454:339–354
4. Deutsch D (1985) Quantum theory, the Church-Turing principle and the universal quantum computer. Proc R Soc Lond A400:97–117
5. Deutsch D, Jozsa R (1992) Rapid solution of problems by quantum computation. Proc R Soc Lond A439:553–558
6. Farhi E, Goldstone J, Gutmann S, Sipser M (1998) A limit on the speed of quantum computation in determining parity. Phys Rev Lett 81:5442–5444
7. Furst M, Saxe J, Sipser M (1984) Parity, circuits, and the polynomial time hierarchy. Math Syst Theor 17(1):13–27
8. Shor PW (1997) Polynomial-time algorithms for prime factorization and discrete logarithms on a quantum computer, SIAM J Comput 26(5):1484–1509
9. Simon DR (1997) On the power of quantum computation. SIAM J Comput 26(5):1474–1483

Quantum Algorithms for Class Group of a Number Field

Sean Hallgren
Department of Computer Science and Engineering, The Pennsylvania State University, University Park, State College, PA, USA

Years and Authors of Summarized Original Work

2005; Hallgren

Problem Definition

Associated with each number field is a finite abelian group called the class group. The order of the class group is called the class number. Computing the class number and the structure of the class group of a number field is among the main tasks in computational algebraic number theory [4].

A number field F can be defined as a subfield of the complex numbers $\mathbb{C}$ which is generated over the rational numbers $\mathbb{Q}$ by an algebraic number, i.e., $F = \mathbb{Q}(\theta)$ where θ is the root of a polynomial with rational coefficients. The ring of integers $\mathcal{O}$ of F is the subset consisting of all elements that are roots of monic polynomials with integer coefficients. The ring $\mathcal{O} \subseteq F$ can be thought of as a generalization of $\mathbb{Z}$, the ring of integers in $\mathbb{Q}$. In particular, one can ask whether $\mathcal{O}$ is a principal ideal domain and whether elements in $\mathcal{O}$ have unique factorization. Another interesting problem is computing the unit group $\mathcal{O}^*$, which is the set of invertible algebraic integers inside F, that is, elements $\alpha \in \mathcal{O}$ such that α^{-1} is also in $\mathcal{O}$.

Ever since the class group was discovered by Gauss in 1798, it has been an interesting object of study. The class group of F is the set of equivalence classes of fractional ideals of F, where two ideals I and J are equivalent if there exists $\alpha \in F^*$ such that $J = \alpha I$. Multiplication of two ideals I and J is defined as the ideal generated by all products ab, where $a \in I$ and $b \in J$. Much is still unknown about number fields, such as whether there exist infinitely many number fields with trivial class group. The question of the class group being trivial is equivalent to asking whether the elements in the ring of integers $\mathcal{O}$ of the number field have unique factorization.

In addition to computing the class number and the structure of the class group, computing the unit group and determining whether given ideals are principal, called the principal ideal problem, are also central problems in computational algebraic number theory.

Key Results

The best known classical algorithms for the class group take subexponential time [1, 2, 4]. Assuming the GRH, computing the class group, the unit group, and solving the principal ideal problem are in NP ∩ CoNP [10].

The following theorems state that the three problems defined above have efficient quantum algorithms [7, 9].

Theorem 1 *There is a polynomial-time quantum algorithm that computes the unit group of a constant degree number field.*

Theorem 2 *There is a polynomial-time quantum algorithm that solves the principal ideal problem in constant degree number fields.*

Theorem 3 *The class group and class number of a constant degree number field can be computed in quantum polynomial time assuming the GRH.*

Computing the class group means computing the structure of a finite abelian group given a set of generators for it. When it is possible to efficiently multiply group elements (including computing large powers of elements) and efficiently

compute unique representations of each group element, then this problem reduces to the standard hidden subgroup problem over the integers and therefore has an efficient quantum algorithm. Ideal multiplication is efficient in number fields. For imaginary number fields, there are efficient classical algorithms for computing group elements with a unique representation, and therefore there is an efficient quantum algorithm for computing the class group.

For real number fields, there is no known way to efficiently compute unique representations of class group elements. As a result, the classical algorithms typically compute the unit group and class group at the same time. A quantum algorithm [7] is able to efficiently compute the unit group of a number field and then use the principal ideal algorithm to compute a unique quantum representation of each class group element. Then the standard quantum algorithm can be applied to compute the class group structure and class number.

Applications

There are factoring algorithms based on computing the class group of an imaginary number field. One is exponential time and the other is subexponential time [4].

Computationally hard number theoretic problems are useful for public key cryptosystems. Pell's equation reduces to the principal ideal problem, which forms the basis of the Buchmann-Williams key-exchange protocol [3]. Identification schemes have also been based on this problem by Hamdy and Maurer [8]. The classical exponential-time algorithms help determine which parameters to choose for the cryptosystem. Factoring reduces to Pell's equation, and the best known algorithm for it is exponentially slower than the best factoring algorithm. Systems based on these harder problems were proposed as alternatives in case factoring turns out to be polynomial time solvable. The efficient quantum algorithms can break these cryptosystems.

Open Problems

The unit group of an arbitrary degree number field has an efficient quantum algorithm [6], and computing the class group and solving the principal ideal problem are related to this problem. One open problem is to compute certain towers of number fields with special properties, such as an infinite family with constant root discriminant [5].

Cross-References

▶ Quantum Algorithm for Factoring
▶ Quantum Algorithm for Solving Pell's Equation

Recommended Reading

1. Biasse JF, Fieker C Subexponential class group and unit group computation in large degree number fields. LMS J Comput Math 17:385–403
2. Buchmann J (1990) A subexponential algorithm for the determination of class groups and regulators of algebraic number fields. In: Goldstein C (ed) Séminaire de Théorie des Nombres, Paris 1988–1989. Progress in mathematics, vol 91. Birkhäuser, Boston, pp 27–41
3. Buchmann JA, Williams HC (1990) A key exchange system based on real quadratic fields (extended abstract). In: Brassard G (ed) Advances in cryptology-CRYPTO '89, 20–24 Aug 1989. Lecture notes in computer science, vol 435. Springer, Berlin, pp 335–343
4. Cohen H (1993) A course in computational algebraic number theory. Graduate texts in mathematics, vol 138. Springer, Berlin/Heidelberg
5. Eisentraeger K, Hallgren, S (2010) Algorithms for ray class groups and Hilbert class fields. In: Proceedings of the 21st ACM-SIAM symposium on discrete algorithms (SODA)
6. Eisentraeger K, Hallgren S, Kitaev A, Song F (2014) A quantum algorithm for computing the unit group of an arbitrary degree number field. In: Proceedings of the 46th ACM symposium on theory of computing
7. Hallgren S (2005) Fast quantum algorithms for computing the unit group and class group of a number field. In: Proceedings of the 37th ACM symposium on theory of computing
8. Hamdy S, Maurer M (1999) Feige-fiat-shamir identification based on real quadratic fields. Techincal report TI-23/99. Technische Universität Darmstadt,
Fachbereich Informatik. http://www.informatik.tu-darmstadt.de/TI/Veroeffentlichung/TR/
9. Schmidt A, Vollmer U (2005) Polynomial time quantum algorithm for the computation of the unit group of a number field. In: Proceedings of the 37th ACM symposium on theory of computing
10. Thiel C (1995) On the complexity of some problems in algorithmic algebraic number theory. Ph.D. thesis, Universität des Saarlandes, Saarbrücken

---

Quantum Algorithms for Graph Connectivity

Aleksandrs Belovs
Computer Science and Artificial Intelligence Laboratory, MIT, Cambridge, MA, USA

Keywords

Finding small subgraphs within large graphs; UPATH; USTCON

Years and Authors of Summarized Original Work

2012; Belovs, Reichardt

Problem Definition

The input is an undirected simple graph G on n vertices. The graph is given by its adjacency matrix: For any two vertices u and v, one can query whether u and v are connected by an edge. (Note that classical algorithms usually have access to G in the form of incidence lists. However, specification of the input graph in the form of adjacency matrix is standard in quantum algorithms.) Two special vertices of the graph, s and t, are selected. The task is to detect whether s and t lie in the same connected component of G. Quantum algorithms for this problem are described.

Classically, this problem can be solved in quadratic time by a variety of algorithms. It is easy to see that this is optimal. Also, the st-connectivity problem is a canonical example of a problem in RL (the class of problems solvable in randomized logspace) [1]. Later, it was shown to be in L (deterministic logspace) [8].

Key Results

Previous Algorithm

Dürr et al. [5] gave a quantum algorithm with the following properties. Its query complexity is $O(n^{3/2})$, and its time complexity is the same up to logarithmic factors. The algorithm repeatedly executes a quantum subroutine that uses $O(\log n)$ qubits and requires quantum read-only access to $O(n \log n)$ classical bits. This memory is changed between the runs of the quantum subroutine.

The algorithm is based on Borůvka's algorithm [4]. It solves a more general problem of finding a minimum spanning tree of G. In particular, the algorithm outputs a list of the connected components of G.

Main Algorithm

Theorem 1 ([3]) *Consider the st-connectivity problem on an n-vertex graph G with the additional promise: Either s and t lie in different components of G, or they are connected by a path of length at most d. The above problem can be solved by a quantum algorithm in $\tilde{O}(n\sqrt{d})$ time, $O(n\sqrt{d})$ queries, and $O(\log n)$ space.*

Thus, in the worst case of $d = n - 1$, the complexity of the algorithm is the same as of the algorithm by Dürr et al. But if d is small, this algorithm performs better. This promise appears quite naturally in practice.

The algorithm is based on the quantum algorithm for evaluating span programs [6, 7].

Applications

The st-connectivity algorithm or its modifications can be used as a quantum version of dynamic programming. In general, quantum algorithms provide no advantage in implementing dynamic programming. The algorithm of Theorem 1, although it does not have the full power of dynamic programming, attains a quadratic speedup (for small values of d). In [3], this algorithm is combined with the color-coding approach [2] to solve the problem of finding small subgraphs.

For example, consider the problem of detecting the presence of a k-path in an input graph G given by its adjacency matrix. (We assume that $k = O(1)$.) Color each vertex of G in a color from $\{0, 1, \ldots k\}$ independently and uniformly at random. Leave only those edges of G that connect vertices whose colors differ by exactly 1. Add two new vertices s and t, connect s to all vertices of color 0, and connect t to all vertices of color k. Denote the resulting graph by G'.

We say that a k-path in G is colored correctly, if the colors of its vertices go from 0 to k starting with one of its end points. Thus, for any k-path of G, the probability it is colored correctly is $\Omega(1)$.

Execute the algorithm of Theorem 1 on G' with $d = k + 2$. If G contains a correctly colored k-path, then G' has a path of length $k + 2$ from s to t; hence, the algorithm accepts. On the other hand, if s and t are connected by a path in G', then G contains a k-path (not necessary correctly colored). Hence, if there is no k-path in G, the algorithm rejects for any coloring of G. By repeating the algorithm constant a number of times with different colorings, it is possible to distinguish these two cases.

Classically, color coding is capable of finding a subgraph H in the input graph, if H is an arbitrary fixed tree. In the quantum case, the class of graphs is narrower.

Problem 1 (Subgraph/not-a-minor promise problem) Let H be a fixed simple graph. The input is a graph G given by its adjacency matrix. The task is to distinguish two cases:

- The graph G contains H as a subgraph.
- The graph G does not contain H as a minor.

Classically, this problem requires $\Omega(n^2)$ queries even if H is a single edge. The quantum query lower bound is $\Omega(n)$.

Theorem 2 ([3]) *Assume that H is a triangle or an edge-subdivision of a star. The subgraph/not-a-minor promise problem for H on an n-vertex input graph can be solved by a quantum algorithm in $\tilde{O}(n)$ time. The algorithm uses $O(n)$ queries. If H is an edge-subdivision of a star, the algorithm uses $O(\log n)$ space.*

Corollary 1 *Assume that H is a path or an edge-subdivision of a claw (a 3-star). There exists a quantum algorithm that detects whether an n-vertex input graph contains H as a subgraph in $\tilde{O}(n)$ time. The algorithm uses $O(n)$ queries and $O(\log n)$ space.*

Cross-References

▶ Color Coding
▶ Single-Source Shortest Paths

Recommended Reading

1. Aleliunas R, Karp RM, Lipton RJ, Lovasz L, Rackoff C (1979) Random walks, universal traversal sequences, and the complexity of maze problems. In: Proceedings of 20th IEEE FOCS, San Juan, pp 218–223
2. Alon N, Yuster R, Zwick U (1995) Color-coding. J ACM 42:844–856
3. Belovs A, Reichardt BW (2012) Span programs and quantum algorithms for st-connectivity and claw detection. In: Proceedings of 20th ESA, Ljubljana. LNCS, vol 7501, pp 193–204
4. Borůvka O (1926) O jistém problému minimálním (About a certain minimal problem). Práce mor Přírodověd spol v Brně (Acta Societ Scient Natur Moravicae) 3:37–58 (In Czech)
5. Dürr C, Heiligman M, Høyer P, Mhalla M (2004) Quantum query complexity of some graph problems. In: Proceedings of 31st ICALP, Turku. LNCS, vol 3142, pp 481–493. Springer
6. Reichardt BW (2009) Span programs and quantum query complexity: the general adversary bound is nearly tight for every boolean function. arXiv:0904.2759
7. Reichardt BW (2011) Reflections for quantum query algorithms. In: Proceedings of 22nd ACM-SIAM SODA, San Francisco, pp 560–569
8. Reingold O (2008) Undirected connectivity in logspace. J ACM 55(4):17

Quantum Algorithms for Matrix Multiplication and Product Verification

Robin Kothari[1,2] and Ashwin Nayak[3]
[1]Center for Theoretical Physics, Massachusetts Institute of Technology, Cambridge, MA, USA
[2]David R. Cheriton School of Computer Science, Institute for Quantum Computing, University of Waterloo, Waterloo, ON, Canada
[3]Department of Combinatorics and Optimization, and Institute for Quantum Computing, University of Waterloo, Waterloo, ON, Canada

Keywords

Boolean matrix multiplication; Matrix product verification; Quantum algorithms

Years and Authors of Summarized Original Work

2006; Buhrman, Špalek
2012; Jeffery, Kothari, Magniez

Problem Definition

Let S be any algebraic structure over which matrix multiplication is defined, such as a field (e.g., real numbers), a ring (e.g., integers), or a semiring (e.g., the Boolean semiring). If we use $+$ and $\cdot$ to denote the addition and multiplication operations over S, then the matrix product C of two $n \times n$ matrices A and B is defined as $C_{ij} := \sum_{k=1}^{n} A_{ik} \cdot B_{kj}$ for all $i, j \in \{1, 2, \ldots, n\}$. Over the Boolean semiring, the addition and multiplication operations are the logical OR and logical AND operations, respectively, and thus, the matrix product C is defined as $C_{ij} := \bigvee_{k=1}^{n}(A_{ik} \wedge B_{kj})$. In this article we consider the following problems.

Problem 1 (Matrix multiplication)

INPUT: Two $n \times n$ matrices A and B with entries from S.
OUTPUT: The matrix $C := AB$.

Problem 2 (Matrix product verification)

INPUT: Three $n \times n$ matrices A, B, and C with entries from S.

OUTPUT: A bit indicating whether or not $C = AB$.

The matrix multiplication problem is a well-studied problem in classical computer science. The straightforward algorithm for matrix multiplication that computes each entry separately using its definition uses $O(n^3)$ operations. In 1969, Strassen [17] presented an algorithm that multiplies matrices over any ring using only $O(n^{2.807})$ operations, showing that the straightforward approach was suboptimal. Since then there have been many improvements and the complexity of matrix multiplication remains an area of active research.

Surprisingly, the matrix product verification problem can be solved faster. In 1979, Freivalds [6] presented an optimal $O(n^2)$ time bounded-error probabilistic algorithm to solve the matrix product verification problem over any ring using a randomized fingerprinting technique, which has found numerous other applications in theoretical computer science (see, e.g., Ref. [15]).

In the quantum setting, these problems are traditionally studied in the model of quantum query complexity, where we assume the entries of the input matrices are provided by a black box or an oracle. The query complexity of an algorithm is the number of queries made to the oracle. The bounded-error quantum query complexity of a problem is the minimum query complexity of any quantum algorithm that solves the problem with bounded error, i.e., it outputs the correct answer with probability greater than (say) $2/3$. The time complexity of an algorithm refers to the time required to implement the remaining non-query operations. In this article we only consider bounded-error quantum algorithms.

Key Results

It is not known if quantum algorithms can improve the time complexity of the general matrix multiplication problem compared to classical algorithms. Improvements are possible for matrix product verification and special cases of the matrix multiplication problem, as described below.

Matrix Product Verification over Rings

According to Buhrman and Špalek [3], matrix product verification was first studied (in an unpublished paper) by Ambainis, Buhrman, Høyer, Karpinski, and Kurur. Using a recursive application of Grover's algorithm [7], they gave an $O(n^{7/4})$ query algorithm for the problem. The first published work on the topic is due to Buhrman and Špalek [3], who gave an $O(n^{5/3})$ query algorithm for matrix product verification over any ring using a generalization of Ambainis' element distinctness algorithm [1]. This algorithm also achieves the same query complexity over semirings and more general algebraic structures. The algorithm can easily be cast in the quantum walk search framework of Magniez, Nayak, Roland, and Santha [14] as explained in the survey by Santha [16]. More interestingly, they presented an algorithm with time complexity $\tilde{O}(n^{5/3})$ for the problem over fields and integral domains. Their algorithm uses the same technique used by Freivalds [6] and is therefore also time efficient over arbitrary rings. Buhrman and Špalek also proved a lower bound showing that any bounded-error quantum algorithm must make at least $\Omega(n^{3/2})$ queries to solve the problem over the field $\mathbb{F}_2$. This lower bound can be extended to all rings [10].

Theorem 1 (Matrix product verification over rings) *The matrix product verification problem over any ring can be solved by a quantum algorithm with query complexity $O(n^{5/3})$ and time complexity $\tilde{O}(n^{5/3})$. Furthermore, any quantum algorithm must make $\Omega(n^{3/2})$ queries to solve the problem over a ring.*

Buhrman and Špalek also studied the relationship between the complexity of their algorithm and the number of incorrect entries in the purported product, C, and showed that their algorithm performs better when C has a large number of incorrect entries [3].

Matrix Multiplication over Rings

The quantum query complexity of multiplying two $n \times n$ matrices is easy to characterize in terms of the input size. Clearly the query complexity is upper bounded by the input size, $O(n^2)$. On the other hand, if A equals the identity matrix, then $C = B$ and in this case the matrix multiplication problem is equivalent to learning all the bits of an input of size n^2, which requires $\Omega(n^2)$ queries. This follows, for example, from the fact that computing the parity of n^2 bits requires $\Omega(n^2)$ queries [2, 5]. This shows that the quantum query complexity of matrix multiplication is $\Theta(n^2)$, which is the same as the classical query complexity. Similarly, no quantum algorithm is known to improve the time complexity of matrix multiplication over rings compared to classical algorithms.

Buhrman and Špalek [3] studied the matrix multiplication problem in terms of n and an additional parameter ℓ, the number of nonzero entries in the output matrix C, and showed the following result.

Theorem 2 *The matrix multiplication problem over any ring can be solved by a quantum algorithm with query and time complexity upper bounded by*

$$\tilde{O}(n^{5/3}\ell^{2/3}) \text{ when } 1 \leq \ell \leq \sqrt{n},$$
$$\tilde{O}(n^{3/2}\ell) \quad \text{when } \sqrt{n} \leq \ell \leq n, \text{ and}$$
$$\tilde{O}(n^2\sqrt{\ell}) \quad \text{when } n \leq \ell \leq n^2,$$

where ℓ is the number of nonzero entries in the output matrix C.

When ℓ is small, this algorithm achieves subquadratic time complexity and when ℓ approaches n^2, its time complexity is close to $O(n^3)$, which is trivial and slower than known classical algorithms. A detailed comparison of this quantum algorithm with classical algorithms may be found in Ref. [3].

Boolean Matrix Product Verification

Buhrman and Špalek [3] also studied the matrix product verification problem over the Boolean semiring and showed that the problem can be solved with query and time complexity $O(n^{3/2})$.

On the other hand, the best known lower bound is only $\Omega(n^{1.055})$ queries due to Childs, Kimmel, and Kothari [4].

Theorem 3 (Boolean matrix product verification) *The Boolean matrix product verification problem can be solved by a quantum algorithm with query complexity $O(n^{3/2})$ and time complexity $\tilde{O}(n^{3/2})$. Furthermore, any quantum algorithm must make $\Omega(n^{1.055})$ queries to solve the problem.*

Boolean Matrix Multiplication

As before, the quantum query complexity of multiplying two $n \times n$ Boolean matrices is $\Theta(n^2)$, since it is at least as hard as learning n^2 input bits. The time complexity of Boolean matrix multiplication can be improved to $\tilde{O}(n^{2.5})$ by observing that the inner product of two Boolean vectors of length n can be computed with $O(\sqrt{n})$ queries using Grover's algorithm [7]. This observation also speeds up matrix multiplication over some other semirings.

Similar to the matrix multiplication problem over rings, Boolean matrix multiplication can be studied in terms of an additional parameter ℓ, the number of nonzero entries in the output matrix. Indeed, the problem has been extensively studied in this setting.

Buhrman and Špalek [3] observed that two Boolean matrices can be multiplied with query complexity $O(n^{3/2}\sqrt{\ell})$. This upper bound was improved by Vassilevska Williams and Williams [18], who presented an algorithm with query complexity $\tilde{O}(\min\{n^{1.3}\ell^{17/30}, n^2 + n^{13/15}\ell^{47/60}\})$, which was then improved by Le Gall [11]. Finally, Jeffery, Kothari, and Magniez [8] presented a quantum algorithm for Boolean matrix multiplication that makes $\tilde{O}(n\sqrt{\ell})$ queries. These upper bounds are depicted in Fig. 1. The log factors present in their algorithm were later removed to yield an algorithm with query complexity $O(n\sqrt{\ell})$ [9]. Jeffery, Kothari, and Magniez [8] also proved a matching lower bound of $\Omega(n\sqrt{\ell})$ when $\ell \leq \epsilon n^2$ for any constant $\epsilon < 1$. Their algorithm can also be modified to achieve time complexity $\tilde{O}(n\sqrt{\ell} + \ell\sqrt{n})$ [12].

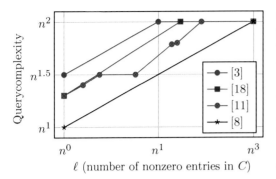

Quantum Algorithms for Matrix Multiplication and Product Verification, Fig. 1 Upper bounds on the quantum query complexity of Boolean matrix multiplication

Theorem 4 (Boolean matrix multiplication)
The Boolean matrix multiplication problem can be solved by a quantum algorithm with query complexity $O(n\sqrt{\ell})$. Furthermore, any quantum algorithm that solves the problem must make $\Omega(n\sqrt{\ell})$ queries when $\ell \leq \epsilon n^2$ for any constant $\epsilon < 1$. Boolean matrix multiplication can be solved in time $\tilde{O}(n\sqrt{\ell} + \ell\sqrt{n})$.

Recently the problem has also been studied in terms of the sparsity of the input matrix. Le Gall and Nishimura [13] present algorithms with improved time complexity in this case. Their algorithm's time complexity is a complicated function of the parameters and the reader is referred to Ref. [13] for details.

Matrix Multiplication over Other Semirings
Le Gall and Nishimura [13] recently initiated the study of matrix multiplication over semirings other than the Boolean semiring and presented algorithms with improved time complexity for the $(\max, \min)$-semiring and related semirings.

Open Problems

Several open problems remain in the time and query complexity settings. In the time complexity setting, a major open problem is whether quantum algorithms can solve the matrix multiplication problem faster than classical algorithms over any ring. In the query complexity setting, the complexity of matrix product verification over rings and the Boolean semiring remains open. The best upper and lower bounds are presented in Theorems 1 and 3. A more comprehensive survey of the quantum query complexity of matrix multiplication and its relation to other problems studied in quantum query complexity such as triangle finding and graph collision can be found in the first author's PhD thesis [10], which also contains additional open problems.

Cross-References

▶ Quantum Algorithm for Element Distinctness
▶ Quantum Analogues of Markov Chains
▶ Quantum Search

Recommended Reading

1. Ambainis A (2007) Quantum walk algorithm for element distinctness. SIAM J Comput 37(1):210–239
2. Beals R, Buhrman H, Cleve R, Mosca M, de Wolf R (2001) Quantum lower bounds by polynomials. J ACM 48(4):778–797
3. Buhrman H, Špalek R (2006) Quantum verification of matrix products. In: Proceedings of 17th ACM-SIAM symposium on discrete algorithms, Miami, pp 880–889
4. Childs AM, Kimmel S, Kothari R (2012) The quantum query complexity of read-many formulas. In: Algorithms – ESA 2012. Volume 7501 of lecture notes in computer science. Springer, Heidelberg, pp 337–348
5. Farhi E, Goldstone J, Gutmann S, Sipser M (1998) Limit on the speed of quantum computation in determining parity. Phys Rev Lett 81(24):5442–5444
6. Freivalds R (1979) Fast probabilistic algorithms. In: Mathematical foundations of computer science. Volume 74 of lecture notes in computer science. Springer, Berlin, pp 57–69
7. Grover LK (1996) A fast quantum mechanical algorithm for database search. In: Proceedings of the 28th ACM symposium on theory of computing (STOC 1996), Philadelphia, pp 212–219
8. Jeffery S, Kothari R, Magniez F (2012) Improving quantum query complexity of boolean matrix multiplication using graph collision. In: Automata, languages, and programming. Volume 7391 of lecture notes in computer science. Springer, Berlin/Heidelberg, pp 522–532
9. Kothari R (2014) An optimal quantum algorithm for the oracle identification problem. In: Proceedings

of the 31st international symposium on theoretical aspects of computer science (STACS 2014), Lyon. Volume 25 of Leibniz international proceedings in informatics (LIPIcs), pp 482–493

10. Kothari R (2014) Efficient algorithms in quantum query complexity. PhD thesis, University of Waterloo

11. Le Gall F (2012) Improved output-sensitive quantum algorithms for Boolean matrix multiplication. In: Proceedings of the 23rd ACM-SIAM symposium on discrete algorithms (SODA 2012), Kyoto, pp 1464–1476

12. Le Gall F (2012) A time-efficient output-sensitive quantum algorithm for Boolean matrix multiplication. In: Algorithms and computation. Volume 7676 of lecture notes in computer science. Springer, Berlin, pp 639–648

13. Le Gall F, Nishimura H (2014) Quantum algorithms for matrix products over semirings. In: Algorithm theory – SWAT 2014. Volume 8503 of lecture notes in computer science. Springer, Berlin, pp 331–343

14. Magniez F, Nayak A, Roland J, Santha M (2011) Search via quantum walk. SIAM J Comput 40(1):142–164

15. Motwani R, Raghavan P (1995) Randomized algorithms. Cambridge University Press, New York

16. Santha M (2008) Quantum walk based search algorithms. In: Theory and applications of models of computation. Volume 4978 of lecture notes in computer science. Springer, New York, pp 31–46

17. Strassen V (1969) Gaussian elimination is not optimal. Numerische Mathematik 13:354–356

18. Williams V V, Williams R (2010) Subcubic equivalences between path, matrix and triangle problems. In: Proceedings of the 51st IEEE symposium on foundations of computer science (FOCS 2010), Las Vegas, pp 645–654

Quantum Algorithms for Simulated Annealing

Sergio Boixo[1] and Rolando D. Somma[2]
[1]Quantum A.I. Laboratory, Google, Venice, CA, USA
[2]Theoretical Division, Los Alamos National Laboratory, Los Alamos, NM, USA

Keywords

Adiabatic quantum state transformations; Combinatorial optimization; Quantum algorithms; Simulated annealing

Years and Authors of Summarized Original Work

2008; Somma, Boixo, Barnum, Knill
2009; Boixo, Knill, Somma
2014; Chiang, Xu, Somma

Problem Definition

This problem is concerned with the development of quantum methods to speed up classical algorithms based on simulated annealing (SA).

SA is a well-known and powerful strategy to solve discrete combinatorial optimization problems [1]. The search space $\Sigma = \{\sigma_0, \ldots, \sigma_{d-1}\}$ consists of d configurations σ_i, and the goal is to find the (optimal) configuration that corresponds to the global minimum of a given cost function $E : \Sigma \to \mathbf{R}$. Monte Carlo implementations of SA generate a stochastic sequence of configurations via a sequence of Markov processes that converges to the low-temperature Gibbs (probability) distribution, $\pi_{\beta_m}(\Sigma) \propto \exp(-\beta_m E(\Sigma))$. If β_m is sufficiently large, sampling from the Gibbs distribution outputs an optimal configuration with large probability, thus solving the combinatorial optimization problem. The annealing process depends on the choice of an annealing schedule, which consists of a sequence of $d \times d$ stochastic matrices (transition rules) $S(\beta_1), S(\beta_2), \ldots, S(\beta_m)$. Such matrices are determined, e.g., by using Metropolis-Hastings [2]. The real parameters β_j denote a sequence of "inverse temperatures." The implementation complexity of SA is given by m, the number of times that transition rules must be applied to converge to the desired Gibbs distribution (within arbitrary precision). Commonly, the stochastic matrices are sparse, and each list of nonzero conditional probabilities and corresponding configurations, $\{\Pr_\beta(\sigma_j|\sigma_i), j : \Pr_\beta(\sigma_j|\sigma_i) > 0\}$, can be efficiently computed on input (i, β). This implies an efficient Monte Carlo implementation of each Markov process. When a lower bound on the spectral gap of the stochastic matrices (i.e., the difference between the two largest eigenvalues) is known and given by $\Delta > 0$, one can choose $(\beta_{k+1} - \beta_k) \propto \Delta/E_{\max}$ and $\beta_0 = 0$, $\beta_m \propto \log \sqrt{d}$. $E_{\max}$ is an upper bound on $\max_\sigma |E(\sigma)|$. The constants of proportionality depend on the error probability ϵ, which is the probability of not finding an optimal solution after the transition rules have been applied. These choices result in a complexity $m \propto E_{\max} \log \sqrt{d}/\Delta$ for SA [3].

Quantum computers can theoretically solve some problems, such as integer factorization, more efficiently than classical computers [4]. This work addresses the question of whether quantum computers could also solve combinatorial optimization problems more efficiently or not. The answer is satisfactory in terms of Δ (Section "Key Results"). The complexity of a quantum algorithm is determined by the number of elementary steps needed to prepare a quantum state that allows one to sample from the Gibbs distribution after measurement. Similar to SA, such a complexity is given by the number of times a unitary corresponding to the stochastic matrix is used. For simplicity, we assume that the stochastic matrices are sparse and disregard the cost of computing each list of nonzero conditional probabilities and configurations, as well as the cost of computing $E(\sigma)$. We also assume $d = 2^n$ and the space of configurations Σ is represented by n-bit strings. Some assumptions can be relaxed.

Problem

INPUT: *An objective function* $E : \Sigma \to \mathbf{R}$, *sparse stochastic matrices* $S(\beta)$ *satisfying the detailed balance condition, a lower bound* $\Delta > 0$ *on the spectral gap of* $S(\beta)$, *an error probability* $\epsilon > 0$.

OUTPUT: *A random configuration* $\sigma_i \in \Sigma$ *such that* $\Pr(\sigma_i \in S_0) \geq 1 - \epsilon$, *where* S_0 *is the set of optimal configurations that minimize* E.

Key Results

The main result is a quantum algorithm, referred to as quantum simulated annealing (QSA), that solves a combinatorial optimization problem with high probability using $m_Q \propto E_{\max} \log \sqrt{d}/\sqrt{\Delta}$

unitaries corresponding to the stochastic matrices [5]. The quantum speedup is in the spectral gap, as $1/\sqrt{\Delta} \ll 1/\Delta$ when $\Delta \ll 1$.

Computationally hard combinatorial optimization problems are typically manifest in a spectral gap that decreases exponentially fast in $\log d$, the problem size. The quadratic improvement in the gap is then most significant in hard instances. The QSA algorithm is based on ideas and techniques from quantum walks and the quantum Zeno effect. The quantum Zeno effect can be implemented by evolution randomization [6]. Nevertheless, recent results on "spectral gap amplification" allow for other quantum algorithms that result in a similar complexity scaling [7].

Quantum Walks for QSA

A quantization of the classical random walk is obtained by first defining a $d^2 \times d^2$ unitary matrix that satisfies [8–10]

$$X|\sigma_i\rangle|\mathbf{0}\rangle = \sum_{j=0}^{d-1} \sqrt{\Pr_\beta(\sigma_j|\sigma_i)}|\sigma_i\rangle|\sigma_j\rangle . \quad (1)$$

The configuration $\mathbf{0}$ represents a simple configuration, e.g., $\mathbf{0} \equiv \sigma_0 = 0 \ldots 0$ (the n-bit string), and $\Pr_\beta(\sigma_j|\sigma_i)$ are the entries of the stochastic matrix $S(\beta)$. The other $d^2 \times d^2$ unitary matrices used by QSA are P, the permutation (swap) operator that transforms $|\sigma_i\rangle|\sigma_j\rangle$ into $|\sigma_j\rangle|\sigma_i\rangle$, and $R = \mathbb{1} - 2|\mathbf{0}\rangle\langle\mathbf{0}|$, the reflection operator over $|\mathbf{0}\rangle$.

The quantum walk is $W = X^\dagger PXPRPX^\dagger PXR$, and the detailed balance condition implies [5]

$$W \sum_{i=0}^{d-1} \sqrt{\pi_\beta(\sigma_i)}|\sigma_i\rangle|\mathbf{0}\rangle = \sum_{i=0}^{d-1} \sqrt{\pi_\beta(\sigma_i)}|\sigma_i\rangle|\mathbf{0}\rangle ,$$

$$\quad (2)$$

where $\pi_\beta(\sigma_i)$ are the probabilities given by the Gibbs distribution. (X, $X^\dagger$, and W also depend on β.) The goal of QSA is to prepare the corresponding eigenstate of W in Eq. 2, within certain precision $\epsilon > 0$, and for inverse temperature $\beta_m \propto \log d$. A projective quantum measurement of $|\sigma_i\rangle$ on such a state outputs an optimal solution in the set S_0 with probability $\Pr(S_0) \geq 1 - \epsilon$.

Evolution Randomization and QSA Implementation

The QSA is based on the idea of adiabatic state transformations [6, 11]. For $\beta = 0$, the initial eigenstate of W is $\sum_{i=0}^{d-1} |\sigma_i\rangle|0\rangle/\sqrt{d}$, which can be prepared easily on a quantum computer. The purpose of QSA is then to drive this initial state towards the eigenstate of W for inverse temperature β_m, within given precision. This is achieved by applying the sequence of unitary operations $[W(\beta_m)]^{t_m} \ldots [W(\beta_2)]^{t_2}[W(\beta_1)]^{t_1}$ to the initial state (Fig. 1). In contrast to SA, $(\beta_{k+1} - \beta_k) \propto 1/E_{\max}$ [11], but the initial and final inverse temperatures are also $\beta_0 = 0$ and $\beta_m \propto \log\sqrt{d}$. This implies that the number of different inverse temperatures in QSA is $m \propto E_{\max}\log\sqrt{d}$, where the constant of proportionality depends on ϵ. The nonnegative integers t_k can be sampled randomly according to several distributions [6]. One way is to obtain t_k after sampling multiple (but constant) times from a uniform distribution on integers between 0 and $Q - 1$, where $Q = \lceil 2\pi/\sqrt{\Delta}\rceil$. The average cost of QSA is then $m\langle t_k\rangle \propto E_{\max}\log\sqrt{d}/\sqrt{\Delta}$. One can use Markov's inequality to avoid those (improbable) instances where the cost is significantly greater than the average cost. The QSA and the values of the constants are given in detail in Fig. 1.

Analytical Properties of W

The quantum walk W has eigenvalues $e^{\pm i\phi_j}$, for $j = 0, \ldots, d - 1$, in the relevant subspace. In particular, $\phi_0 = 0 < \phi_1 \leq \ldots \leq \phi_{d-1}$ and $\phi_1 \geq \sqrt{\Delta}$ [5, 7–9]. This implies that the relevant spectral gap for methods based on quantum adiabatic state transformations is of order $\sqrt{\Delta}$. The quantum speedup follows from the fact that the complexity of such methods, recently discussed in [6, 11–13], depends on the inverse of the relevant gap.

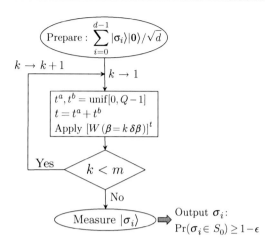

Quantum Algorithms for Simulated Annealing, Fig. 1
Flow diagram for the QSA. Under the assumptions, the input state can be easily prepared on a quantum computer by applying a sequence of n Hadamard gates on n qubits. unif$[0, Q - 1]$ is the uniform distribution on nonnegative integers in that range and $Q = \lceil 2\pi/\sqrt{\Delta}\rceil$. $\delta\beta = \beta_{k+1} - \beta_k = \epsilon/(2E_{\max})$ and $m = \lceil 2\beta_m E_{\max}/\epsilon\rceil$. Like SA, the final inverse temperature is $\beta_m = (\gamma/2)\log(2\sqrt{d}/\epsilon)$, where γ is the gap of E, that is, $\gamma = \min_{\sigma \notin S_0} E(\sigma) - E(S_0)$. The average cost of the QSA is then $mQ = \lceil 2\pi\gamma E_{\max}\log(2\sqrt{d}/\epsilon)/(\epsilon\sqrt{\Delta})\rceil$, and dependence on ϵ can be made fully logarithmic by repeated executions of the algorithm. A quantum computer implementation of W can be efficiently done by using the algorithm that computes the nonzero conditional probabilities of the stochastic matrix $S(\beta)$

Applications

Like SA, QSA can be applied to solve general discrete combinatorial optimization problems [14]. QSA is often more efficient than exhaustive search in finding the optimal configuration. Examples of problems where QSA can be powerful include the simulation of equilibrium states of Ising spin glasses or Potts models, solving satisfiability problems or solving the traveling salesman problem.

Open Problems

Some (classical) Monte Carlo implementations do not require varying an inverse temperature and apply the same (time-independent) transition rule

S to converge to the Gibbs distribution. The number of times the transition rule must be applied is the so-called mixing time, which depends on the inverse spectral gap of S [15]. The development of quantum algorithms to speed up this type of Monte Carlo algorithms remains open. Also, the technique of spectral gap amplification outputs a Hamiltonian $H(\beta)$ on input $S(\beta)$. The relevant eigenvalue of such a Hamiltonian is zero, and the remaining eigenvalues are $\pm\sqrt{\lambda_i}$, where $\lambda_i \geq \Delta$. This opens the door to a quantum adiabatic version of the QSA, in which $H(\beta)$ is changed slowly and the quantum system remains in an "excited" eigenstate of eigenvalue zero at all times. The speedup is also due to the increase in the eigenvalue gap. Nevertheless, finding a different Hamiltonian path with the same gap, where the adiabatic evolution occurs within the lowest energy eigenstates of the Hamiltonians, is an open problem.

Cross-References

▶ Quantum Algorithms for Simulated Annealing

Recommended Reading

1. Kirkpatrick S, Gelett CD, Vecchi MP (1983) Optimization by simulated annealing. Science 220:671
2. Hastings WK (1970) Monte Carlo sampling methods using Markov Chains and their applications. Biometrika 57(1):97–109
3. Aldous DJ (1982) Some inequalities for reversible Markov Chains. J Lond Math Soc s2–25:564
4. Shor P (1994) Proceedings of the 35th annual symposium on foundations of computer science, Santa Fe
5. Somma R, Boixo S, Barnum H, Knill E (2008) Quantum simulations of classical annealing processes. Phys Rev Lett 101:130504
6. Boixo S, Knill E, Somma R (2009) Eigenpath traversal by phase randomization. Quantum Inf Comput 9:0833
7. Somma R, Boixo S (2013) Spectral gap amplification. SIAM J Comput 42:593
8. Ambainis A (2004) Proceedings of the 45th symposium on foundations of computer science, Rome
9. Szegedy M (2004) Proceedings of the 45th IEEE symposium on foundations of computer science, Rome
10. Magniez F, Nayak A, Roland J, Santha M (2007) Proceedings of the 39th annual ACM symposium on theory of computing, San Diego
11. Chiang HT, Xu G, Somma R (2014) Improved bounds for eigenpath traversal. Phys Rev A 89:012314
12. Wocjan P, Abeyensinghe (2008) Speedup via quantum sampling. Phys Rev A 78:042336
13. Boixo S, Knill E, Somma R (2010). arXiv:1005.3034
14. Cook WJ, Cunningham WH, Pulleyblank WR (1998) Combinatorial optimization. Wiley, New York
15. Levin DA, Peres Y, Wilmer EL, Markov Chains and mixing times. Available at: http://research.microsoft.com/en-us/um/people/peres/markovmixing.pdf

Quantum Algorithms for Systems of Linear Equations

Aram W. Harrow
Department of Physics, Massachusetts Institute of Technology, Cambridge, MA, USA

Keywords

Filtering; Hamiltonian simulation; Linear algebra; Matrix inversion; Phase estimation

Years and Authors of Summarized Original Work

2009; Harrow, Hassidim, Lloyd
2012; Ambainis

Problem Definition

The problem is to find a vector $x \in \mathbb{C}^N$ such that $Ax = b$, for some given inputs $A \in \mathbb{C}^{N \times N}$ and $b \in \mathbb{C}^N$. Several variants are also possible, such as rectangular matrices A, including overdetermined and underdetermined systems of equations.

Unlike in the classical case, the output of this algorithm is a quantum state on $\log(N)$ qubits whose amplitudes are proportional to the entries of x, along with a classical estimate of $\|x\| :=$

$\sqrt{\sum_i |x_i|^2}$. Similarly, the input b is given as a quantum state. The matrix A is specified implicitly as a row-computable matrix. Specifying the input and output in this way makes it possible to find x in time sublinear, or even polylogarithmic, in N. The next section has more discussion of the relation of this algorithm to classical linear systems solvers.

Key Results

Suppose that:

- $A \in \mathbb{C}^{N \times N}$ is Hermitian, has all eigenvalues in the range $[-1, -1/\kappa] \cup [1/\kappa, 1]$ for some known $\kappa \geq 1$, and has $\leq s$ nonzero entries per row. The parameter κ is called the *condition number* (defined more generally to be the ratio of the largest to the smallest singular value) and s is the *sparsity*.
- There is a quantum algorithm running in time T_A that takes an input $i \in [N]$ and outputs the nonzero entries of the ith row, together with their location.
- Assume that $\|b\| = 1$ and that there is a corresponding quantum state to produce the state $|b\rangle$ that runs in time T_B.

Define $x' := A^{-1} |b\rangle$ and $x = \frac{x'}{\|x'\|}$.
We use the notation x to refer to the vector as a mathematical object and $|x\rangle$ to refer to the corresponding quantum state on $\log(N)$ qubits. For a variable T, let $\tilde{O}(T)$ denote a quantity upper bounded by $T \cdot \text{poly}\log(T)$. The norm of a vector $\|x\|$ is the usual Euclidean norm $\sqrt{\sum_i |x_i|^2}$, while for a matrix $\|A\|$ is the operator norm $\max_{\|x\|=1} \|Ax\|$, or equivalently the largest singular value of A.

Quantum Algorithm for Linear Systems
The main result is that $|x\rangle$ and $\|x'\|$ can be produced, both up to error ϵ, in time $\text{poly}(\kappa, s, \epsilon^{-1}, \log(N), T_A, T_B)$. More precisely, the following run-times are known:

$$\tilde{O}(\kappa T_B + \log(N)s^2 \kappa^2 T_A / \epsilon) \quad [5] \quad (1a)$$

$$\tilde{O}(\kappa T_B + \log(N)s^2 \kappa T_A / \epsilon^3) \quad [1] \quad (1b)$$

A key subroutine is Hamiltonian simulation, and the run-times in (1) are based on the recent improvements in this component due to [3].

Hardness Results and Comparison to Classical Algorithms

These algorithms are analogous to classical algorithms for solving linear systems of equations, but do not achieve exactly the same thing. Most classical algorithms output the entire vector x as a list of N numbers, while the quantum algorithms output the state $|x\rangle$, i.e., a superposition on $\log(N)$ qubits whose N amplitudes equal x. This allows potentially faster algorithms but for some tasks will be weaker. This resembles the difference between the Quantum Fourier Transform and the classical Fast Fourier Transform.

To compare the classical and quantum complexities for this problem, we should consider classical tasks (with classical output) that can be solved with the help of quantum linear equations algorithms. One can show that better classical algorithms for such tasks exist only if *all* quantum algorithms could be simulated more quickly by classical algorithms. This is because the linear systems problem is BQP-complete, i.e., solving large sparse well-conditioned linear systems of equations is equivalent in power to general purpose quantum computing.

To make this precise, define LinearSystemSample$(N, \kappa, \epsilon, T_A)$ to be the problem of producing a sample $i \in [N]$ from a distribution p satisfying $\sum_{i=1}^{N} |p_i - |x_i|^2| \leq \epsilon$, where $x = x'/\|x'\|$, $x' = A^{-1}b$, and $b = e_1$ (i.e., one in the first entry and zero elsewhere). Additionally the eigenvalues of A should have absolute value between $1/\kappa$ and 1, and there should exist a classical algorithm for computing the entries of a row of A that runs in time T_A. This problem differs slightly from the version described above, but only in ways that make it easier, so that it still makes sense to talk about a matching hardness result.

Theorem 1 *Consider a quantum circuit on n qubits that applies two-qubit unitary gates $U_1, \ldots, U_T$ to the $|0\rangle^{\otimes n}$ state and concludes by outputting the result of measuring the first qubit. It is possible to simulate this measurement outcome up to error ϵ by reducing to* LinearSystemSample($N, \kappa, \epsilon/2, T_A$) *with $N = O(2^n T/\epsilon)$, $\kappa = O(T/\epsilon)$, and $T_A =$ poly $\log(N)$.*

In other words, LinearSystemSample is at least as hard to solve as any quantum computation of the appropriate size. This result is nearly tight. In other words, when combined with the algorithm of [1], the relation between N, κ (for linear system solving) and n, T (for quantum circuits) is known to be nearly optimal, while the correct ϵ dependence is known up to a polynomial factor.

Theorem 1 can also rule out classical algorithms for LinearSystemSample(N, κ, ϵ, T_A). Known algorithms for the problem (assuming for simplicity that A is s-sparse) run in time poly(N) poly $\log(\kappa/\epsilon) + N T_A$ (direct solvers), N poly(κ) poly $\log(1/\epsilon) T_A$ (iterative methods), or even $s^{\kappa \ln(1/\epsilon)}$ poly $\log(N)$ (direct expansion of $x \approx \sum_{n \leq \kappa \ln(1/\epsilon)} (I - A)^n b$, assuming A is positive semidefinite). Depending on the parameters N, κ, ϵ, s, a different one of these may be optimal. And from Theorem 1 it follows (a) that any nontrivial improvement in these algorithms would imply a general improvement in the ability of classical computers to simulate quantum mechanics and (b) that such improvement is impossible for algorithms that use the function describing A in a black-box manner (i.e., as an oracle).

Applications and Extensions

Linear system solving is usually a subroutine in a larger algorithm, and the following algorithms apply it to a variety of settings. Complexity analyses can be found in the cited papers, but since hardness results are not known for them, we cannot say definitively whether they outperform all possible classical algorithms.

Machine Learning

A widely used application of linear systems of equations is to performing least-squares estimation of a model [6]. In this problem, we are given a matrix $A \in \mathbb{R}^{n \times p}$ with $n \geq p$ (for an overdetermined model) along with a vector $b \in \mathbb{R}^n$, and we wish to compute $\arg \min_{x \in \mathbb{R}^p} \|Ax - b\|$. If A is well conditioned, sparse, and implicitly specified, then the state $|x\rangle$ can be found quickly [6], and from this features of x can be extracted by measurement.

Differential Equations

Consider the differential equation [2]

$$\dot{x}(t) = A(t)x(t) + b(t) \qquad x(t) \in \mathbb{R}^N. \quad (2)$$

One of the simplest ways to solve this is to discretize time to take values $t_1 < \ldots < t_m$ and approximate

$$x(t_{i+1}) \approx x(t_i) + (A(t_i)x(t_i) + b(t_i))(t_{i+1} - t_i). \quad (3)$$

By treating $(x(t_1), \ldots, x(t_m))$ as a single vector of size Nm, we can find this vector as a solution of the linear system of equations specified by (3). More sophisticated higher-order solvers can also be made quantum; see [2] for details.

Boundary-Value Problems

The solution to PDEs can also be expressed in terms of the solution to a linear system of equations [4]. For example, in Poisson's equation we are given a function $Q : \mathbb{R}^3 \to \mathbb{R}$ and want to find $u : \mathbb{R}^3 \to \mathbb{R}$ such that $-\nabla^2 u = Q$. By defining x and b to be discretized versions of u, Q, this PDE becomes an equation of the form $Ax = b$. One challenge is that if A is the finite-difference operator (i.e., discretized second derivative) for an $L \times L \times L$ box, then its condition number will scale as L^2. Since the total number of points is $O(L^3)$, this means the quantum algorithm cannot achieve a substantial speedup. Classically this condition number is typically reduced by using preconditioners. A method for using preconditioners with the quantum linear system solver was presented in [4], along with an application to an electromagnetic scattering problem. The resulting complexity is still not known.

Cross-References

▶ Quantum Analogues of Markov Chains

Recommended Reading

1. Ambainis A (2012) Variable time amplitude amplification and quantum algorithms for linear algebra problems. In: STACS, Paris, vol 14, pp 636–647
2. Berry DW (2014) High-order quantum algorithm for solving linear differential equations. J Phys A 47(10):105301
3. Berry DW, Childs AM, Cleve R, Kothari R, Somma RD (2014) Exponential improvement in precision for simulating sparse hamiltonians. In: Proceedings of STOC 2014, New York, pp 283–292
4. Clader BD, Jacobs BC, Sprouse CR (2013) Preconditioned quantum linear system algorithm. Phys Rev Lett 110:250504
5. Harrow AW, Hassidim A, Lloyd S (2009) Quantum algorithm for solving linear systems of equations. Phys Rev Lett 15(103):150502
6. Wiebe N, Braun D, Lloyd S (2012) Quantum algorithm for data fitting. Phys Rev Lett 109:050505

Quantum Analogues of Markov Chains

Ashwin Nayak[1], Peter C. Richter[1,2], and Mario Szegedy[1]
[1]Department of Combinatorics and Optimization, and Institute for Quantum Computing, University of Waterloo, Waterloo, ON, Canada
[2]Department of Computer Science, Rutgers, The State University of New Jersey, New Brunswick, NJ, USA

Keywords

Element distinctness; Hitting time; Markov chains; Quantum algorithms; Quantum search; Quantum walks; Search problem; Spatial search; Triangle finding

Years and Authors of Summarized Original Work

2004; Szegedy

Problem Definition

Spatial Search and Walk Processes

Spatial search by *quantum walk* is database search with the additional constraint that one is required to move through the search space that obeys some locality structure. For example, the data items may be stored at the vertices of a two-dimensional grid. The requirement of moves along the edges of the grid captures the cost of accessing different items starting from some fixed position in the database.

One of possible ways of carrying out spatial search is by performing a random walk on the search space or its quantum analog, a quantum walk. The complexity of spatial search by quantum walk is strongly tied to the *quantum hitting time* [19] of the walk.

Let S, with $|S| = n$, be a finite set of *states*. Assume that a subset $M \subseteq S$ of states are *marked*. We are given a procedure C that, on input $x \in S$ and an associated data structure $d(x)$, checks whether the state x is marked. The goal is either to find a marked state when promised that $M \neq \emptyset$ (*search version*) or to determine whether M is nonempty (*decision version*).

The algorithm progresses in stages. In the *setup stage*, we access some state of S (usually a random state). In the walk stage we move from state to state, performing a spatial walk as described below. The moves are called *updates*. In addition, in the walk stage we perform *checks* to see if the current state is marked at steps selected by the algorithm.

In the classical setting, the *transition probabilities* of the spatial walk are described by a stochastic matrix $P = (p_{x,y})_{x,y \in S}$. This makes the walk a *Markov chain*. In every move the algorithm *must perform* a random transition according to P. The possible $x \rightarrow y$ moves, i.e., those with $p_{x,y} \neq 0$, form the edges of a (directed) graph G, and we say that the Markov chain P has *locality structure G*.

We define the search problem in the classical setting, which carries over to the quantum case with little modification:

INPUT: Markov chain P on set S, marked subset $M \subseteq S$ that is implicitly specified by a checking procedure $\mathcal{C}$, and the associated costs:

Cost type	Setup	Update	Checking
Notation	S	U	C

OUTPUT: a marked state if one exists (search version) or a Boolean return value that indicates whether M is empty or not (decision version).

The algorithm is required to be correct with probability at least $2/3$ in either case, the search or the decision problem. The significance of the setup cost, which is incurred only once, will be clearer when we see some applications. Often we can choose between several competing walks, and we would like to design the one with minimum total cost.

In the *quantum* case, the random process P is replaced by a *quantum walk* W_P that has the same locality structure as P. The costs S, U, C reflect the costs of quantum operations.

The Quantum Walk Algorithm

Designing a quantum analog of P is not so straightforward, since stochastic matrices have no immediate unitary equivalents. One either needs to abandon the discrete-time nature of the walk [15] or define the walk operator on a space other than $\mathbb{C}^S$. Here we take the second route.

We say that a Markov chain P is *irreducible* if its underlying digraph is strongly connected. Let P be an irreducible Markov chain, let π be its unique stationary distribution, and let P^* (with $P^* = (p^*_{x,y})$) denote the *time-reversed Markov chain*, where $p^*_{x,y} := \pi_y p_{y,x}/\pi_x$. Define the following vectors in the vector space $\mathbb{C}^S$:

$$|p_x\rangle := \sum_{y \in X} \sqrt{p_{x,y}} |y\rangle \quad \text{and}$$

$$|p^*_y\rangle := \sum_{x \in X} \sqrt{p^*_{y,x}} |x\rangle \ .$$

Define the unitary operator $W_P := R_1 R_2$ on $\mathbb{C}^{S \times S}$ as the product of the two reflections $R_2 := \sum_{x \in S} |x\rangle\langle x| \otimes (2|p_x\rangle\langle p_x| - I)$ and $R_1 := \sum_{y \in S} (2|p^*_y\rangle\langle p^*_y| - I) \otimes |y\rangle\langle y|$. The operator W_P is called the *quantum analog* of P, or the *discrete-time quantum walk operator* arising from P, and may be viewed as a walk on the *edges* of the underlying graph G. We define a "checking" operator on $\mathbb{C}^S$, based on whether or not the current state is marked: $O_M := \sum_{x \notin M} |x\rangle\langle x| - \sum_{x \in M} |x\rangle\langle x|$.

In the above description, we have suppressed the data structure associated with a state in the Markov chain for the sake of simplicity. The precise description of the operators can be derived via the isometry $|x\rangle \mapsto |x\rangle|d(x)\rangle$ between the appropriate spaces (see, e.g., Refs. [28, 29]). The data structure becomes especially significant in the context of the complexity of the operators.

A search algorithm by quantum walk is described by a quantum circuit that acts on "registers" or "wires" which are associated with the space $\mathbb{C}^S \otimes \mathbb{C}^S \otimes \mathbb{C}^k$, for some $k \geq 0$. We again suppress the registers carrying the data structure. The first two registers hold the current edge, and the last register holds auxiliary information, or work space, that drives the quantum walk. The quantum circuit implements the composition $X := X_t X_{t-1} \cdots X_1$, where each X_i is either W_P or O_M acting on the edge registers, possibly controlled by the auxiliary register, or a unitary operator independent of P and M acting on any of the registers. The circuit X is applied to a suitably constructed initial state $|\phi_0\rangle$.

We associate a cost with each operator as a measure of its complexity, with respect to a resource of interest. The resource could be circuit size or in the query model (which is the more typical application) the number of queries. We denote the cost of implementing W_P as a quantum circuit in the units of the resource of interest by U (*update cost*), the cost of constructing O_M by C (*checking cost*), and the cost of preparing the initial state, $|\phi_0\rangle$, of the algorithm by S (*setup cost*). Every time an operator is used, we incur the cost associated with it. This abstraction, implicit in Ref. [3] and made explicit in Ref. [28], allows W_P and O_M to be treated as black-box operators and provides a convenient way to capture *time complexity* or, in the quantum query model, *query complexity*. The cost of the sequence $X_t X_{t-1} \cdots X_1$ is the sum of the costs of the individual operators. The *observation probability* is the probability that we observe an element of M on measuring the first register of the final state, $|\phi_t\rangle := X|\phi_0\rangle$, in the standard basis $(|x\rangle)_{x \in S}$. In the decision version of the problem, we measure a fixed single qubit of the auxiliary register in the standard basis to obtain the output of the algorithm.

Key Results

Walk Definitions
Quantum walks were first introduced by David Meyer and John Watrous to study quantum cellular automata and quantum logspace, respectively. Discrete-time quantum walks were investigated for their own sake by Ambainis, Bach, Nayak, Vishwanath, and Watrous [4, 32] and Aharonov, Ambainis, Kempe, and Vazirani [2] on the infinite line and the n-cycle, respectively. The central issues in the early development of quantum walks included the definition of the walk operator, notions of mixing and hitting times, and the speedup achievable compared to the classical setting.

Hitting Time
Exponential quantum speedup of the hitting time between antipodes of the hypercube was shown by Kempe [19]. Childs, Cleve, Deotto, Farhi, Gutmann, and Spielman [13] presented the first oracle problem solvable exponentially faster by a quantum walk-based algorithm than by any (not necessarily walk-based) classical algorithm.

The first systematic studies of quantum hitting time on the hypercube and the d-dimensional torus were conducted by Shenvi, Kempe, and Whaley [34] and Ambainis, Kempe, and Rivosh [5]. Improving upon the Grover search-based spatial search algorithm of Aaronson and Ambainis, Ambainis et al. [5] showed that the d-dimensional torus with n nodes can be searched by quantum walk in $\sqrt{n}$ steps with observation probability $\Omega(1)$ for $d \geq 3$ and in $\sqrt{n \log n}$ in steps and observation probability $\Omega(1/\log n)$ for $d = 2$ (see also Ref. [11]). Combining the algorithm for $d = 2$ with amplitude amplification [9], we get an algorithm with observation probability $\Omega(1)$, at a cost that is a multiplicative factor of $\sqrt{\log n}$ larger.

In the results in Refs. [13, 19], the algorithm has implicit knowledge of the target state, as the walk starts from a state whose location is "related" to that of the target. It is not known if we can achieve an exponential speedup when the walk starts in a state that is independent of the target.

Element Distinctness
The first result that used a quantum walk to solve a natural algorithmic problem, the so-called *element distinctness problem*, was due to Ambainis [3]. The problem is to find out if among the set of s elements of a database, two are identical. Ambainis constructed a walk on the *Johnson graph* $J(r, s)$ whose vertices are the r-size subsets of a universe of size s (in his case the universe corresponds to the set of all database elements), with two subsets connected iff their symmetric difference has size two. A subset is marked, i.e., it is an element of M, if it captures two identical database elements. In the quantum (but also the classical) query model, the setup cost is r, which stands for the cost of downloading r (random) database elements. Update incurs a constant cost, as it requires reading a new database element and forgetting an old one. Furthermore, since we are in the query model, the checking cost is zero, since whether a state is marked can be deduced from the currently held database elements without any further download. Ambainis ingeniously balanced the costs of S and U finding that in the quantum case, the optimum choice for r is $s^{2/3}$, leading to a query complexity of $s^{2/3}$ (this is a nontrivial balance: in the classical case, the same walk gives no speedup).

In contrast, the Grover algorithm, the inspiration behind Ambainis' work, has no balancing option: its setup and update costs are zero in the query model. (The Grover search may be viewed as a quantum walk on the complete graph.) It turns out that the above walk-based quantum query algorithm with complexity $O(s^{2/3})$ matches the lower bound due to Aaronson and Shi [1].

General Markov Chains

Ambainis's result is based on the quantum hitting time of $J(r, s)$ for a marked set of relative size $\left(\frac{r}{s}\right)^2$. In Ref. [35], Szegedy investigates the hitting time of quantum walks arising from general Markov chains. His definitions (walk operator, hitting time) are abstracted directly from Ref. [3] and are consistent with prior literature, although slightly different in presentation.

For a Markov chain P, the (classical) *average hitting time* of M can be expressed in terms of the *leaking walk matrix* P_M, which is obtained from P by deleting all rows and columns indexed by states of M. Let $v_1, \ldots, v_{n-m}$, be the normalized eigenvectors of P_M, and let $\lambda_1, \ldots, \lambda_{n-m}$ be the associated eigenvalues, where $m = |M|$. Let $h(x, M)$ denote the expected time to reach M from x. Let $\mu : S \to \mathbb{R}^+$ be the initial distribution from which we start and μ' its restriction to $S \setminus M$. Denote the vector $(\sqrt{\mu'(x)})_{x \in S \setminus M}$ by u. Then the average hitting time of M is $h := \sum_{x \in S} \mu(x) h(x, M) = \sum_{k=1}^{n-m} \frac{|(v_k, u)|^2}{1 - \lambda_k}$. Although the leaking walk matrix P_M is not stochastic, one can consider the *absorbing walk* matrix $P' = \begin{bmatrix} P_M & P'' \\ 0 & I \end{bmatrix}$, where P'' is the matrix obtained from P by deleting the rows indexed by M and the columns indexed by $S \setminus M$. The walk P' behaves like P but is absorbed by the first marked state it hits. Consider the quantum analog $W_{P'}$ of P' and $|\phi_0\rangle := \sum_{x \in S} \sqrt{\pi(x)} |x\rangle |p_x\rangle$, where π is the stationary distribution of P. The state $|\phi_0\rangle$ is stationary for W_P, i.e., an eigenvector with eigenvalue 1. Define the *quantum hitting time*, H, of set M to be the smallest t for which $\|W_{P'}^t |\phi_0\rangle - |\phi_0\rangle\| \geq 0.1$. Note that the cost of $W_{P'}$ is proportional to $\mathsf{U} + \mathsf{C}$.

The motivation behind this definition of quantum hitting time is the following. The classical hitting time measures the number of iterations of the absorbing walk P' required to noticeably skew the uniform starting distribution. Similarly, the quantum hitting time bounds the number of iterations of the following quantum algorithm for detecting whether M is nonempty: At each step, apply operator $W_{P'}$. If M is empty, then $P' = P$ and the starting state is left invariant. If M is nonempty, then the angle between $W_{P'}^t |\phi_0\rangle$ and $W_P^t |\phi_0\rangle$ gradually increases (for t not too large). Using an additional *control register* to apply either $W_{P'}$ or W_P with quantum control, the divergence of these two states (should M be nonempty) can be detected. The required number of iterations is characterized by H.

It remains to compute H. When P is symmetric and *ergodic*, the expression for the classical hitting time has a quantum analog [35] (we assume $m \leq n/2$ for technical reasons):

$$H \leq \sum_{k=1}^{n-m} \frac{v_k^2}{\sqrt{1 - \lambda_k}}, \tag{1}$$

where $v_k = (v_k, u)$. Note that $u = \frac{1}{\sqrt{n}}(1, \ldots, 1)$, since P is symmetric, so v_k sum of the coordinates of v_k divided by $1/\sqrt{n}$. From (1) and the expression for h, one can derive an amazing connection between the classical and quantum hitting times:

Theorem 1 (Szegedy [35]) *Let P be symmetric and ergodic, and let h be the classical hitting time for marked set M and uniform starting distribution. Then the quantum hitting time of M is at most $\sqrt{h}$. Therefore, the cost of solving the decision version of the problem is of order $\mathsf{S} + \sqrt{h}(\mathsf{U} + \mathsf{C})$.*

One can further show:

Theorem 2 (Szegedy [35]) *If P is state-transitive and $|M| = 1$, then the marked state is observed with probability at least n/h with cost $O(\mathsf{S} + \sqrt{h}(\mathsf{U} + \mathsf{C}))$.*

The observation probability n/h can be increased to $\Theta(1)$ with $\sqrt{h/n}$ iterations of the algorithm from Theorem 2, using amplitude amplification [9]. Theorems 1 and 2 imply most quantum hitting time results of the previous section *directly*, relying only on estimates of the corresponding classical hitting times. Expression (1) is based on a fundamental connection between the eigenvalues and eigenvectors of P and W_P. Notice that $p_{y,x}^* = p_{y,x}$ for symmetric P, so $|p_y^*\rangle = |p_y\rangle$. So R_1 and R_2 are reflections through the subspaces generated by $\{|p_x\rangle \otimes |x\rangle|\ x \in S\}$ and $\{|x\rangle \otimes |p_x\rangle|\ x \in S\}$, respectively. The eigenvalues of $R_1 R_2$ can be expressed in terms of the eigenvalues of the mutual Gram matrix $D(P)$ of these systems. This matrix $D(P)$, the *discriminant matrix* of P, equals P when P is symmetric. The formula remains fairly simple even when P is not symmetric. In particular, the absorbing walk P' has discriminant matrix $\begin{bmatrix} P_M & 0 \\ 0 & I \end{bmatrix}$. Finally, the relation between $D(P)$ and the spectral decomposition of W_P is given by:

Theorem 3 (Szegedy [35]) *Let P be an arbitrary Markov chain on a finite state space S and let $\cos\theta_1 \geq \cdots \geq \cos\theta_l$ be those singular values of $D(P)$ lying in the open interval $(0,1)$, with associated singular vector pairs v_j, w_j for $1 \leq j \leq l$. Then the nontrivial eigenvalues of W_P (namely, those other than 1 and -1) and their corresponding eigenvectors are $(\mathrm{e}^{-2\mathrm{i}\theta_j}, R_1 w_j - \mathrm{e}^{-\mathrm{i}\theta_j} R_2 v_j)$ and $(\mathrm{e}^{2\mathrm{i}\theta_j}, R_1 w_j - \mathrm{e}^{\mathrm{i}\theta_j} R_2 v_j)$ for $1 \leq j \leq l$.*

Subsequent Developments

Magniez, Nayak, Roland, and Santha [29] used the Szegedy quantum analog W_P of an ergodic walk P, rather than that of its absorbing version P', to develop a *search* algorithm in the style of Ambainis [3].

Theorem 4 (Magniez, Nayak, Roland, Santha [29]) *Let P be reversible and ergodic with spectral gap $\delta > 0$. Let M have probability either zero or $\varepsilon > 0$ under the stationary distribution of P. There is a quantum algorithm solving the search problem with cost $\mathsf{S} + \frac{1}{\sqrt{\varepsilon}}(\frac{1}{\sqrt{\delta}}\mathsf{U} + \mathsf{C})$.*

The main idea here is to apply quantum phase estimation [14, 21] to the quantum walk W_P in order to implement an approximate reflection operator about the initial state. This operator is then used along with the checking operator O_M in an amplitude amplification scheme to get the final algorithm.

The average classical hitting time h may be bounded by $1/\delta\varepsilon$ (with δ, M, ε as in Theorem 4), and this bound is tight for most known applications. In these applications, the above algorithm *finds* marked elements with complexity at most that of the Szegedy algorithm. In other applications, for instance, Triangle Finding [28], where the checking cost C is much larger than the update cost U, the complexity of the algorithm in Theorem 4 is asymptotically smaller.

In the case of the two-dimensional square grid with n vertices, the average classical hitting time h is $n \log n$. This is asymptotically lesser than $1/\delta\varepsilon$ when there is a single marked element. (In this case, $1/\delta\varepsilon = n^2$.) Algorithms due to Ambainis et al. [5] and Szegedy [35] find a unique marked state with $\mathrm{O}(\sqrt{n}\log n)$ steps of quantum walk, a $\sqrt{\log n}$ factor larger than $\sqrt{h}$. Tulsi [36] showed how we may find a unique marked element in $\mathrm{O}(\sqrt{h})$ steps. Magniez, Nayak, Richter, and Santha [30] extended this result to show that for any state-transitive Markov chain, a unique marked state can be found in $\mathrm{O}(\sqrt{h})$ steps. They also devised a detection algorithm that solves the decision version of the problem for any reversible Markov chain and any number of marked elements, in $\mathrm{O}(\sqrt{h})$ steps (thus extending Theorem 1).

Krovi, Magniez, Ozols, and Roland [23] presented a different quantum algorithm for finding multiple marked elements in any *reversible* Markov chain. They introduced a notion of interpolation between any reversible chain P and its absorbing counterpart P' and used the quantum analog of the interpolated walk. In the case of a unique marked element, the resulting algorithm solves the search version of the problem with cost $\mathsf{S} + \sqrt{h}(\mathsf{U} + \mathsf{C})$. The precise relationship between the number of steps of the quantum walk taken by the algorithm in the case of more than one marked element and the corresponding classical hitting time remains open. It is known that for certain choices of P and M, the former may be asymptotically larger than $\sqrt{h}$.

The schema due to Magniez et al. [29] described above has been extended in different ways. Jeffery, Kothari, and Magniez [17] use a *quantum state* as the data structure $d(x)$ associated with a state $x \in S$ in quantum algorithms with nested walks. In this manner, they avoid the repeated overhead of setup cost in the inner quantum walks used for checking marked states. They solve several problems, including Triangle Finding, with query as well as time complexity matching, up to polylogarithmic factors, the performance of algorithms previously derived from *learning graphs* [7, 26]. Childs, Jeffery, Kothari, and Magniez [8] introduced the use of a data structure that depends on the state transition in the walk. Using this, they develop quantum algorithms with nested walks, where the recursion occurs in the update operation. The cost incurred is essentially what we would expect from Theorem 4. This extension leads to algorithms that are as efficient in *time* as in query complexity, for applications such as 3-Distinctness. Independently, Belovs designed a different quantum walk algorithm [8], which leads to a similar result for 3-Distinctness.

Applications

We list some quantum walk-based results for search problems that represent speedups over Grover search-based solutions. All are inspired by Ambainis' algorithm for element distinctness.

Triangle Finding

Suppose we are given the adjacency matrix A of a graph on n vertices and are required to determine if the graph contains a triangle (i.e., a clique of size 3), using as few queries as possible to the entries of A. The classical query complexity of this problem is $\Theta(n^2)$. Magniez, Santha, and Szegedy [28] gave an $\tilde{O}(n^{1.3})$ algorithm. This upper bound has been improved by a sequence of results [7, 25, 26, 29] (see also Ref. [17]) to $\tilde{O}(n^{5/4})$. Several of these algorithms, including

the current best algorithm due to Le Gall [25], are based on the quantum walk search framework.

Matrix Product Verification and Matrix Multiplication

Suppose we are given three $n \times n$ matrices A, B, C over a ring and are required to determine if $AB \neq C$, i.e., if there exist i, j such that $\sum_k A_{ik} B_{kj} \neq C_{ij}$. We would like to make as few queries as possible to the entries of A, B, and C. This problem has classical query complexity $\Theta(n^2)$. Buhrman and Špalek [10] gave an $O(n^{5/3})$ quantum query algorithm. They also observed that two Boolean matrices can be multiplied with query complexity $O(n^{3/2}\sqrt{\ell})$, where ℓ is the number of nonzero entries in the product. This has since been improved in a sequence of results [16, 24, 37] to $O(n\sqrt{\ell})$. The algorithm due to Le Gall [24] builds upon quantum walk algorithms. We refer the reader to Ref. [22] for further work on this topic.

Group Commutativity Testing

Suppose we are presented with a black-box group specified by its k generators and are required to determine if the group commutes using as few queries as possible to the group product operation (i.e., queries of the form "What is the product of elements g and h?"). The classical query complexity is $\Theta(k)$ group operations. Magniez and Nayak [27] gave an (essentially optimal) $\tilde{O}(k^{2/3})$ quantum query algorithm for this problem. The algorithm involves a quantum walk on the product of two graphs whose vertices are ordered l-tuples of distinct generators.

Forbidden Subgraph Property

A property of graphs is called *minor closed* when the following condition holds: if a graph has the property, then all its minors also possess the property. A graph property (which need not be minor closed) is called a *forbidden subgraph property* (FSP) if it can be described by a finite set of forbidden subgraphs. Suppose we are given the adjacency matrix A of a graph on n vertices and are required to determine if the graph has a minor closed property Π, using as few queries as possible to the entries of A. Childs and Kothari [12] show that if Π is nontrivial and is *not* FSP, then it has query complexity in $\Theta(n^{3/2})$. They complement this with a more efficient algorithm for any minor closed property Π that *is* FSP. The algorithm has query complexity $O(n^\alpha)$ for some $\alpha < 3/2$ and is based on the quantum walk search framework.

3-Distinctness

This is a generalization of the element distinctness problem. Suppose we are given elements $x_1, \ldots, x_m \in \{1, \ldots, m\}$ and are asked if there exist *three* distinct indices i, j, k such that $x_i = x_j = x_k$. The Ambainis quantum walk algorithm achieves query and time complexity $O(m^{3/4})$. The *query* complexity was improved to $O(m^{5/7})$ by Belovs [6] using a new technique – learning graphs, while the best time complexity remained unchanged. Childs et al. [8] later designed *time*-efficient query algorithms with complexity $\tilde{O}(m^{5/7})$, using extensions of the quantum walk search framework.

Open Problems

Many issues regarding quantum analogs of Markov chains remain unresolved, both for the search problem and the closely related mixing problem.

Search Problem

Can the quadratic quantum speedup of hitting time for the decision version of the problem be extended from all reversible Markov chains to all *ergodic* ones? Can quantum walks also *find* marked elements quadratically faster than classical walks, in the case of reversible Markov chains with *multiple* marked states? What other algorithmic applications of search by quantum walk can be found?

Sampling Problem

Another wide use of Markov chains in classical algorithms is in generating samples from certain probability distributions. In particular, *Markov chain Monte Carlo* algorithms work by running a carefully designed ergodic Markov chain. After

a number of steps given by the *mixing time* of P, the distribution over states is guaranteed to be ϵ-close to its stationary distribution π. Such algorithms form the basis of most randomized algorithms for approximating #P-complete problems (see, e.g., Ref. [18]). The sampling problem may be formalized as follows:

INPUT: Markov chain P, tolerance $\epsilon \in (0, 1)$.
OUTPUT: A sample from a distribution that is ϵ-close to π in total variation distance.

Notions of quantum mixing time were first proposed and analyzed on the line, the cycle, and the hypercube [2, 4, 31, 32]. Kendon and Tregenna [20] and Richter [33] have investigated the use of decoherence in improving mixing of quantum walks. Two fundamental questions about quantum mixing time remain open: What is the "most natural" definition? And when is there a quantum speedup over the classical mixing time?

Cross-References

▶ Quantum Algorithm for Element Distinctness
▶ Quantum Algorithm for Finding Triangle

Recommended Reading

1. Aaronson S, Shi Y (2004) Quantum lower bounds for the collision and the element distinctness problems. J ACM 51(4):595–605
2. Aharonov D, Ambainis A, Kempe J, Vazirani U (2001) Quantum walks on graphs. In: Proceedings of the thirty-third annual ACM Symposium on Theory of Computing, STOC '01, New York. ACM, pp 50–59
3. Ambainis A (2007) Quantum walk algorithm for Element Distinctness. SIAM J Comput 37(1):210–239
4. Ambainis A, Bach E, Nayak A, Vishwanath A, Watrous J (2001) One-dimensional quantum walks. In: Proceedings of the thirty-third annual ACM Symposium on Theory of Computing, STOC '01, New York. ACM, pp 37–49
5. Ambainis A, Kempe J, Rivosh A (2005) Coins make quantum walks faster. In: Proceedings of the sixteenth annual ACM-SIAM Symposium on Discrete Algorithms, SODA '05, Philadelphia. Society for Industrial and Applied Mathematics, pp 1099–1108

6. Belovs A (2012) Learning-graph-based quantum algorithm for k-Distinctness. In: Proceedings of the 53rd annual IEEE Symposium on Foundations of Computer Science. IEEE Computer Society, Los Alamitos, pp 207–216

7. Belovs A (2012) Span programs for functions with constant-sized 1-certificates: extended abstract. In: Proceedings of the forty-fourth annual ACM Symposium on Theory of Computing, STOC '12, New York. ACM, pp 77–84

8. Belovs A, Childs AM, Jeffery S, Kothari R, Magniez F (2013) Time-efficient quantum walks for 3-Distinctness. In: Fomin FV, Freivalds R, Kwiatkowska M, Peleg D (eds) Automata, Languages, and Programming. Volume 7965 of Lecture Notes in Computer Science. Springer, Berlin/Heidelberg, pp 105–122

9. Brassard G, Høyer P, Mosca M, Tapp A (2002) Quantum amplitude amplification and estimation. In: Quantum Computation and Information (Washington, DC, 2000). Volume 305 of Contemporary Mathematics. American Mathematical Society, Providence, pp 53–74

10. Buhrman H, Špalek R (2006) Quantum verification of matrix products. In: Proceedings of the seventeenth annual ACM-SIAM Symposium on Discrete Algorithms, SODA '06, Philadelphia. Society for Industrial and Applied Mathematics, pp 880–889

11. Childs A, Goldstone J (2004) Spatial search by quantum walk. Phys Rev A 70:022314

12. Childs AM, Kothari R (2012) Quantum query complexity of minor-closed graph properties. SIAM J Comput 41(6):1426–1450

13. Childs AM, Cleve R, Deotto E, Farhi E, Gutmann S, Spielman DA (2003) Exponential algorithmic speedup by a quantum walk. In: Proceedings of the thirty-fifth annual ACM Symposium on Theory of Computing, STOC '03, New York. ACM, pp 59–68

14. Cleve R, Ekert A, Macchiavello C, Mosca M (1998) Quantum algorithms revisited. Proc R Soc A Math Phys Eng Sci 454(1969):339–354

15. Farhi E, Gutmann S (1998) Quantum computation and decision trees. Phys Rev A 58:915–928

16. Jeffery S, Kothari R, Magniez F (2012) Improving quantum query complexity of Boolean Matrix Multiplication using Graph Collision. In: Czumaj A, Mehlhorn K, Pitts A, Wattenhofer R (eds) Automata, Languages, and Programming. Volume 7391 of Lecture Notes in Computer Science. Springer, Berlin/Heidelberg, pp 522–532

17. Jeffery S, Kothari R, Magniez F (2013) Nested quantum walks with quantum data structures. In: Proceedings of the twenty-fourth annual ACM-SIAM Symposium on Discrete Algorithms, SODA '13. SIAM, Philadelphia, pp 1474–1485

18. Jerrum M, Sinclair A (1997) The Markov Chain Monte Carlo method: an approach to approximate counting and integration. In: Hochbaum DS (ed) Approximation algorithms for NP-hard problems. PWS Publishing Co., Boston, pp 482–520

19. Kempe J (2005) Discrete quantum walks hit exponentially faster. Probab Theory Relat Fields 133(2):215–235

20. Kendon V, Tregenna B (2003) Decoherence can be useful in quantum walks. Phys Rev A 67:042315

21. Kitaev A (1995) Quantum measurements and the Abelian stabilizer problem. Technical report. quant-ph/9511026, arXiv.org

22. Kothari R (2014) Efficient algorithms in quantum query complexity. PhD thesis, University of Waterloo, Waterloo

23. Krovi H, Magniez F, Ozols M, Roland J (2014) Quantum walks can find a marked element on any graph. Technical report. arXiv:1002.2419v2, arXiv.org

24. Le Gall F (2012) Improved output-sensitive quantum algorithms for boolean matrix multiplication. In: Proceedings of the twenty-third annual ACM-SIAM Symposium on Discrete Algorithms, SODA '12. SIAM, Philadelphia, pp 1464–1476

25. Le Gall F (2014) Improved quantum algorithm for triangle finding via combinatorial arguments. In: Proceedings of the 55th annual IEEE Symposium on Foundations of Computer Science, Los Alamitos, 18–21 Oct 2014. IEEE Computer Society Press, pp 216–225

26. Lee T, Magniez F, Santha M (2013) Improved quantum query algorithms for triangle finding and associativity testing. In: Proceedings of the twenty-fourth annual ACM-SIAM Symposium on Discrete Algorithms, SODA '13. SIAM, Philadelphia, pp 1486–1502

27. Magniez F, Nayak A (2007) Quantum complexity of testing group commutativity. Algorithmica 48(3):221–232

28. Magniez F, Santha M, Szegedy M (2007) Quantum algorithms for the triangle problem. SIAM J Comput 37(2):413–424

29. Magniez F, Nayak A, Roland J, Santha M (2011) Search via quantum walk. SIAM J Comput 40:142–164

30. Magniez F, Nayak A, Richter PC, Santha M (2012) On the hitting times of quantum versus random walks. Algorithmica 63(1):91–116

31. Moore C, Russell A (2002) Quantum walks on the hypercube. In: Rolim JDP, Vadhan S (eds) Randomization and Approximation Techniques in Computer Science. Volume 2483 of Lecture Notes in Computer Science. Springer, Berlin/Heidelberg, pp 164–178

32. Nayak A, Vishwanath A (2000) Quantum walk on the line. Technical report. quant-ph/0010117, arXiv.org.

33. Richter P (2007) Quantum speedup of classical mixing processes. Phys Rev A 76:042306

34. Shenvi N, Kempe J, Birgitta Whaley K (2003) Quantum random-walk search algorithm. Phys Rev A 67:052307

35. Szegedy M (2004) Quantum speed-up of markov chain based algorithms. In: Proceedings of the 45th annual IEEE Symposium on Foundations of Computer Science. IEEE Computer Society, Los Alamitos, pp 32–41

36. Tulsi A (2008) Faster quantum walk algorithm for the two dimensional spatial search. Phys Rev A 78:012310
37. Williams VV, Williams R (2010) Subcubic equivalences between path, matrix and triangle problems. In: Proceedings of the 51st annual IEEE Symposium on Foundations of Computer Science, FOCS '10, Washington. IEEE Computer Society, pp 645–654

Quantum Approximation of the Jones Polynomial

Zeph Landau
Department of Computer Science, University of California, Berkelely, CA, USA

Keywords

AJL algorithm

Years and Authors of Summarized Original Work

2005; Aharonov, Jones, Landau

Problem Definition

A knot invariant is a function on knots (or links – i.e., circles embedded in R^3) which is invariant under isotopy of the knot, i.e., it does not change under stretching, moving, tangling, etc. (cutting the knot is not allowed). In low dimensional topology, the discovery and use of *knot invariants* is of central importance. In 1984, Jones [12] discovered a new knot invariant, now called the Jones polynomial $V_L(t)$, which is a Laurent polynomial in $\sqrt{t}$ with integer coefficients and which is an invariant of the link L. In addition to the important role it has played in low dimensional topology, the Jones polynomial has found applications in numerous fields, from *DNA* recombination [16] to statistical physics [20].

From the moment of the discovery of the Jones polynomial, the question of how hard it is to compute became important. There is a very simple inductive algorithm (essentially due to Conway [5]) to compute it by changing crossings in a link diagram, but, naively applied, this takes exponential time in the number of crossings. It was shown [11] that the computation of $V_L(t)$ is #P-hard for all but a few values of t where $V_L(t)$ has an elementary interpretation. Thus, a polynomial time algorithm for computing $V_L(t)$ for any value of t other than those elementary ones is unlikely. Of course, the #P-hardness of the problem does not rule out the possibility of good approximations. Still, the best classical algorithms to approximate the Jones polynomial at all but trivial values are exponential. Simply stated, the problem becomes:

Problem 1 For what values of t and for what level of approximation can the Jones polynomial $V_L(t)$ be approximated in time polynomial in the number of crossings and links of the link L?

Key Results

As mentioned above, exact computation of the Jones polynomial for most t is #P-hard, and the best known classical algorithms to approximate the Jones polynomial are exponential. The key results described here consider the above problem in the context of quantum rather than classical computation.

The results concern the approximation of links that are given as closures of braids. (All links can be described this way.) Briefly, a braid of n strands and m crossings is described pictorially by n strands hanging alongside each other, with m crossings, each of two adjacent strands. A braid B may be "closed" to form a link by tying its ends together in a variety of ways, two of which are the *trace closure* (denoted by B^{tr}) which joins the ith strand from the top right to the ith strand from the bottom right (for each i) and the *plat closure* (denoted by B^{pl}) which is defined only for braids with an even number of strands by connecting pairs of adjacent strands (beginning at the rightmost strand) on both the top and bottom.

Quantum Approximation of the Jones Polynomial, Fig. 1
The trace closure (*left*) and plat closure (*right*) of the same 4-strand braid

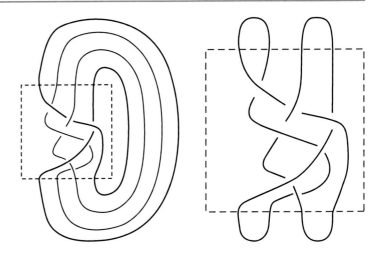

Examples of the trace and plat closure of the same 4-strand braid are given in Fig. 1.

For such braids, the following results have been shown by Aharonov, Jones, and Landau:

Theorem 1 ([2]) *For a given braid B in B_n with m crossings and a given integer k, there is a quantum algorithm which with probability $1 - c^{\Omega(n+m+k)}$ outputs a complex number r with $|r - V_{B^{tr}}\left(e^{2\pi i/k}\right)| < \epsilon d^{n-1}$ where $d = 2\cos(\pi/k)$ and ϵ is inverse polynomial in n, k, m, using time that is polynomial in n, m, k.*

Theorem 2 ([2]) *For a given braid B in B_n with m crossings and a given integer k, there is a quantum algorithm which with probability $1 - c^{\Omega(n+m+k)}$ outputs a complex number r with $|r - V_{B^{pl}}\left(e^{2\pi i/k}\right)| < \epsilon d^{n-1}$ where $d = 2\cos(\pi/k)$ and ϵ is inverse polynomial in n, k, m, using time that is polynomial in n, m, k.*

The original connection between quantum computation and the Jones polynomial was made earlier in the series of papers [6–9]. A model of quantum computation based on Topological Quantum Field Theory (*TQFT*) and Chern-Simons theory was defined in [6, 9], and Kitaev, Larsen, Freedman, and Wang showed that this model is polynomially equivalent in computational power to the standard quantum computation model in [7, 8]. These results, combined with a deep connection between *TQFT* and the value of the Jones polynomial

at particular roots of unity discovered by Witten 13 years earlier [18], implicitly implied (without explicitly formulating) an efficient quantum algorithm for the approximation of the Jones polynomial at the value $e^{2\pi i/5}$.

The approximation given by the above algorithms are additive, namely, the result lies in a given window, whose size is independent of the actual value being approximated. The formulation of this kind of additive approximation was given in [4]; this is much weaker than a multiplicative approximation, which is what one might desire (again, see discussion in [4]). One might wonder if under such weak requirements, the problem remains meaningful at all. It turns out that, in fact, this additive approximation problem is hard for quantum computation, a result originally shown by Freedman, Kitaev, and Wang:

Theorem 3 (Adapted from [8]) *The problem of approximating the Jones polynomial of the plat closure of a braid at $e^{2\pi i/k}$ for constant k, to within the accuracy given in Theorem 2, is BQP-hard.*

A different proof of this result was given in [19], and the result was strengthened by Aharonov and Arad [1] to any k which is polynomial in the size of the input, namely, for all the plat closure cases for which the algorithm is polynomial in the size of the braid.

Understanding the Algorithm

The structure of the solution described by Theorems 1 and 2 consists of four steps:

1. *Mapping the Jones polynomial computation to a computation in the Temperley-Lieb algebra.* There exists a homomorphism of the braid group inside the so-called Temperley-Lieb algebra (this homomorphism was the connection that led to the original discovery of the Jones polynomial in [12]). Using this homomorphism, the computation of the Jones polynomial of either the plat or trace closure of a braid can be mapped to the computation of a particular linear functional (called the Markov trace) of the image of the braid in the Temperley-Lieb algebra (for an essential understanding of a geometrical picture of the Temperley-Lieb algebra, see [14]).

2. *Mapping the Temperley-Lieb algebra calculation into a linear algebra calculation.* Using a representation of the Temperley-Lieb algebra, called the path model representation, the computation in step 1 is shown to be equal to a particular weighted trace of the matrix corresponding to the Temperley-Lieb algebra element coming from the original braid.

3. *Choosing the parameter t corresponding to unitary matrices.* The matrix in step 2 is a product of basic matrices corresponding to individual crossings in the braid group; an important characteristic of these basic matrices is that they have a local structure. In addition, by choosing the values of t as in Theorems 1 and 2, the matrices corresponding to individual crossings become unitary. The result is that the original problem has been turned into a weighted trace calculation of a matrix formed from a product of local unitary matrices – a problem well suited to a quantum computer.

4. *Implementing the quantum algorithm.* Finally the weighted trace calculation of a matrix described in step 3 is formally encoded into a calculation involving local unitary matrices and qubits.

A nice exposition of the algorithm is given in [15].

Applications

Since the publication [2], a number of interesting results have ensued investigating the possibility of quantum algorithms for other combinatorial/topological questions. Quantum algorithms have been developed for the case of the HOMFLY-PT two-variable polynomial of the trace closure of a braid at certain pairs of values [19]. (This entry also extends the results of [2] to a class of more generalized braid closures; it is recommended reading as a complement to [2] or [15] as it gives the representation theory of the Jones-Wenzl representations, thus putting the path model representation of the Temperley-Lieb algebra in a more general context.) A quantum algorithm for the colored Jones polynomial is given in [10].

Significant progress was made on the question of approximating the partition function of the Tutte polynomial of a graph [3]. This polynomial, at various parameters, captures important combinatorial features of the graph. Intimately associated to the Tutte polynomial is the Potts model, a model originating in statistical physics as a generalization of the Ising model to more than 2 states [17, 20]; approximating the partition function of the Tutte polynomial of a graph is a very important question in statistical physics. The work of [3] develops a quantum algorithm for additive approximation of the Tutte polynomial for all planar graphs at all points in the Tutte plane and shows that for a significant set of these points (though not those corresponding to the Potts model) the problem of approximating is a complete problem for a quantum computer. Unlike previous results, these results use non-unitary representations.

Open Problems

There remain many unanswered questions related to the computation of the Jones polynomial from both a classical and quantum computational point of view.

From a classical computation point of view, the originally stated Problem 1 remains wide open for all but trivial choices of t. A result as strong as Theorem 2 for a classical computer seems unlikely since it would imply (via Theorem 3) that classical computation is as strong as quantum computation. A result by Jordan and Shor [13] shows that the approximation given in Theorem 1 solves a complete problem for a presumed (but not proven) weaker quantum model called the one-clean-qubit model. Since this model seems weaker than the full quantum computation model, a classical result as strong as Theorem 1 for the trace closure of a braid is perhaps in the realm of possibility.

From a quantum computational point of view, various open directions seem worthy of pursuit. Most of the quantum algorithms known as of the writing of this entry are based on the quantum Fourier transform and solve problems which are algebraic and number theoretical in nature. Arguably, the greatest challenge in the field of quantum computation (together with the physical realization of large scale quantum computers) is the design of new quantum algorithms based on substantially different techniques. The quantum algorithm to approximate the Jones polynomial is significantly different from the known quantum algorithms in that it solves a problem which is combinatorial in nature, and it does so without using the Fourier transform. These observations suggest investigating the possibility of quantum algorithms for other combinatorial/topological questions. Indeed, the results described in the applications section above address questions of this type. Of particular interest would be progress beyond [3] in the direction of the Potts model, specifically either showing that the approximation given in [3] is non-trivial or providing a different non-trivial algorithm.

Cross-References

▶ Fault-Tolerant Quantum Computation
▶ Quantum Error Correction

Recommended Reading

1. Aharonov D, Arad I (2006) The BQP-hardness of approximating the Jones Polynomial. Arxiv: quant-ph/0605181
2. Aharonov D, Jones V, Landau Z (2006) A polynomial quantum algorithm for approximating the Jones polynomial. In: Proceedings of the 38th ACM symposium on theory of computing (STOC), Seattle. Arxiv:quant-ph/0511096
3. Aharonov D, Arad I, Eban E, Landau Z (2007) Polynomial quantum algorithms for additive approximations of the Potts model and other points of the Tutte plane. Arxiv:quant-ph/0702008
4. Bordewich M, Freedman M, Lovasz L, Welsh D (2005) Approximate counting and quantum computation. Comb Prob Comput 14(5–6):737–754
5. Conway JH (1970) An enumeration of knots and links, and some of their algebraic properties. In: Computational problems in abstract algebra (Proc. Conf., Oxford, 1967). Pergamon Press, New York, pp 329–358
6. Freedman M (1998) P/NP and the quantum field computer. Proc Natl Acad Sci USA 95:98–101
7. Freedman MH, Kitaev A, Wang Z (2002) Simulation of topological field theories by quantum computers. Commun Math Phys 227:587–603
8. Freedman MH, Kitaev A, Wang Z (2002) A modular Functor which is universal for quantum computation. Commun Math Phys 227(3):605–622
9. Freedman M, Kitaev A, Larsen M, Wang Z (2003) Topological quantum computation. Mathematical challenges of the 21st century. (Los Angeles, CA, 2000). Bull Am Math Soc (NS) 40(1):31–38
10. Garnerone S, Marzuoli A, Rasetti M (2006) An efficient quantum algorithm for colored Jones polynomials. ArXiv.org:quant-ph/0606167
11. Jaeger F, Vertigan D, Welsh D (1990) On the computational complexity of the Jones and Tutte polynomials. Math Proc Camb Philos Soc 108(1):35–53
12. Jones VFR (1985) A polynomial invariant for knots via von Neumann algebras. Bull Am Math Soc 12(1):103–111
13. Jordan S, Shor P (2007) Estimating Jones polynomials is a complete problem for one clean qubit. http://arxiv.org/abs/0707.2831
14. Kauffman L (1987) State models and the Jones polynomial. Topology 26:395–407
15. Kauffman L, Lomonaco S (2006) Topological quantum computing and the Jones polynomial. ArXiv.org:quant-ph/0605004
16. Podtelezhnikov A, Cozzarelli N, Vologodskii A (1999) Equilibrium distributions of topological states in circular DNA: interplay of supercoiling and knotting (English. English summary). Proc Natl Acad Sci USA 96(23):12974–129

17. Potts R (1952) Some generalized order – disorder transformations. Proc Camb Philos Soc 48:106–109
18. Witten E (1989) Quantum field theory and the Jones polynomial. Commun Math Phys 121(3):351–399
19. Wocjan P, Yard J (2008) The Jones polynomial: quantum algorithms and applications in quantum complexity theory. Quantum Inf Comput 8(1 & 2):147–180. ArXiv.org:quant-ph/0603069 (2006)
20. Wu FY (1992) Knot theory and statistical mechanics. Rev Mod Phys 64(4):1099–1131

$$|\Psi_{00}\rangle = \frac{1}{\sqrt{2}}(|00\rangle + |11\rangle),$$

$$|\Psi_{10}\rangle = \frac{1}{\sqrt{2}}(|00\rangle + |11\rangle),$$

$$|\Psi_{01}\rangle = \frac{1}{\sqrt{2}}(|01\rangle + |10\rangle),$$

$$|\Psi_{11}\rangle = \frac{1}{\sqrt{2}}(|01\rangle + |10\rangle).$$

Quantum Dense Coding

Barbara M. Terhal
JARA Institute for Quantum Information,
RWTH Aachen University, Aachen, Germany

Keywords

Dense coding; Superdense coding

Years and Authors of Summarized Original Work

1992; Bennett, Wiesner

Problem Definition

Quantum information theory distinguishes classical bits from quantum bits or qubits. The quantum state of n qubits is represented by a complex vector in $(\mathbb{C}^2)^{\otimes n}$, where $(\mathbb{C}^2)^{\otimes n}$ is the tensor product of n 2-dimensional complex vector spaces. Classical n-bit strings form a basis for the vector space $(\mathbb{C}^2)^{\otimes n}$. Column vectors in $(\mathbb{C}^2)^{\otimes n}$ are denoted as $|\psi\rangle$ and row vectors are denoted as $|\psi\rangle^{\dagger} = |\psi\rangle^{*T} \equiv \langle\psi|$. The complex inner product between vectors $|\psi\rangle$ and $|\phi\rangle$ is conveniently written as $\langle\psi|\phi\rangle$.

Entangled quantum states $|\psi\rangle \in (\mathbb{C}^2)^{\otimes n}$ are those quantum states that cannot be written as a product of some vectors $|\psi_i\rangle \in \mathbb{C}^2$, that is, $|\psi\rangle \neq \bigotimes_i |\psi_i\rangle$. The Bell states are four orthogonal (maximally) entangled states defined as

The Pauli matrices X, Y, and Z are three unitary, Hermitian 2×2 matrices. They are defined as $X = |0\rangle\langle 1| + |1\rangle\langle 0|$, $Z = |0\rangle\langle 0| - |1\rangle\langle 1|$ and $Y = iXZ$.

Quantum states can evolve dynamically under inner product preserving unitary operations $U(U^{-1} = U^{\dagger})$. Quantum information can be mapped onto observable classical information through the formalism of quantum measurements. In a quantum measurement on a state $|\psi\rangle$ in $(\mathbb{C}^2)^{\otimes n}$, a basis $\{|x\rangle\}$ in $(\mathbb{C}^2)^{\otimes n}$ is chosen. This basis is made observable through an interaction of the qubits with a macroscopic measurement system. A basis vector x is thus observed with probability $\mathbb{P}(x) = |\langle x|\psi\rangle|^2$.

Quantum information theory or more narrowly quantum Shannon theory is concerned with protocols which enable distant parties to efficiently transmit quantum or classical information, possibly aided by the sharing of quantum entanglement between the parties. For a detailed introduction to quantum information theory, see the book by Nielsen and Chuang [12].

Key Results

Superdense coding [3] is the protocol in which two classical bits of information are sent from sender Alice to receiver Bob. This is accomplished by sharing a Bell state $|\Psi_{00}\rangle_{AB}$ between Alice and Bob and the transmission of one qubit. The protocol is illustrated in Fig. 1. Given two bits b_1 and b_2, Alice performs the following unitary transformation on her half of the Bell state:

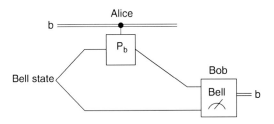

Quantum Dense Coding, Fig. 1 Dense coding. Alice and Bob use a shared Bell state to transmit two classical bits $b = (b_1, b_2)$ by sending one qubit. *Double lines* are classical bits and *single lines* represent quantum bits

$$P_{b_1 b_2} \otimes I_B |\Psi_{00}\rangle = |\Psi_{b_1 b_2}\rangle, \qquad (1)$$

i.e., one of the four Bell states. Here $P_{00} = I$, $P_{01} = X$, $P_{10} = Z$, and $P_{11} = XZ = -iY$. Alice then sends her qubit to Bob. This allows Bob to do a measurement in the Bell basis. He distinguishes the four states $|\Psi_{b1b2}\rangle$ and learns the value of the two bits b_1 and b_2.

The protocol demonstrates the interplay between classical information and quantum information. No information can be communicated by merely sharing an entangled state such as $|\Psi_{00}\rangle$ without the actual transmission of physical information carriers. On the other hand, it is a consequence of Holevo's theorem [10] that one qubit can encode at most one classical bit of information. The protocol of dense coding shows that the two resources of entanglement and qubit transmission *combined* give rise to a *superdense coding* of classical information. Dense coding is thus captured by the following resource inequality:

$$1 \text{ ebit} + 1 \text{ qubit} \geq 2 \text{ cbits}. \qquad (2)$$

In words, one bit of quantum entanglement (one ebit) in combination with the transmission of one qubit is sufficient for the transmission of two classical bits or cbits.

Dense coding can be generalized to the encoding of continuous variables, namely, the encoding of quadrature variables (x, p) of an electromagnetic field into one half of a two-mode squeezed state [2]. Such a two-mode squeezed state approximates the two-mode EPR state – in which

both quadrature variables are perfectly correlated, i.e., $x_1 = x_2$ and $p_1 = -p_2$ – in the limit of large squeezing. The authors in [2] show that the information transmission capacity through the EPR state is, in the limit of large squeezing, twice that of a direct encoding using a single transmitted mode. The scheme thus exemplifies the notion of dense coding through the use of quantum entanglement.

Quantum teleportation [4] is a protocol that is dual to dense coding. In quantum teleportation, 1 ebit (a Bell state) is used in conjunction with the transmission of two classical bits to send one qubit from Alice to Bob. Thus, the resource relation for quantum teleportation is

$$1 \text{ ebit} + 2 \text{ cbits} \geq 1 \text{ qubit}. \qquad (3)$$

The relation with quantum teleportation allows one to argue that dense coding is optimal. It is not possible to encode $2k$ classical bits in less than $m < k$ quantum bits even in the presence of shared quantum entanglement. Let us assume the opposite and obtain a contradiction. One uses quantum teleportation to convert the transmission of k quantum bits into the transmission of $2k$ classical bits. Then one can use the assumed superdense coding scheme to encode these $2k$ bits into $m < k$ qubits. As a result one can send k quantum bits by effectively transmitting $m < k$ quantum bits (and sharing quantum entanglement) which is known to be impossible.

Applications

Harrow [8] has introduced the notion of a coherent bit or cobit. The notion of a cobit is useful in understanding resource relations and trade-offs between quantum and classical information. The noiseless transmission of a qubit from Alice to Bob can be viewed as the linear map $S_q : |x\rangle_A \rightarrow |x\rangle_B$ for a set of basis states $\{|x\rangle\}$. The transmission of a classical bit can be viewed as the linear map $S_c : |x\rangle_A \rightarrow |x\rangle_B |x\rangle_E$ where E stands for the environment Eve. Eve's copy of every basis state $|x\rangle$ can be viewed as the output of a quantum measurement, and thus, Bob's state is classical. The transmission of a cobit corresponds to the linear map $S_{co} : |x\rangle_A \rightarrow |x\rangle_A |x\rangle_B$. Since Alice keeps a copy of the transmitted data, Bob's state is classical. On the other hand, the cobit can also be used to generate a Bell state between Alice and Bob. Since no qubit can be transmitted via a cobit, a cobit is weaker than a qubit. A cobit is stronger than a classical bit since entanglement can be generated using a cobit.

One can define a *coherent* version of superdense coding and quantum teleportation in which measurements are replaced by unitary operations. In this version of dense coding, Bob replaces his Bell measurement by a rotation of the states $|\Psi_{b_1 b_2}\rangle$ to the states $|b_1 b_2\rangle_B$. Since Alice keeps her input bits, the coherent protocol implements the map $|x_1 x_2\rangle_A \rightarrow |x_1 x_2\rangle_A |x_1 x_2\rangle_B$. Thus, one can strengthen the dense coding resource relation to

$$1 \text{ ebit} + 1 \text{ qubit} \geq 2 \text{ cobits.} \qquad (4)$$

Similarly, the coherent execution of quantum teleportation gives rise to the modified relation $2 \text{ cobits} + 1 \text{ ebit} \geq 1 \text{ qubit} + 2 \text{ ebits}$. One can omit 1 ebit on both sides of the inequality by using ebits catalytically, i.e., they can be borrowed and returned at the end of the protocol. One can then combine both coherent resource inequalities and obtain a resource *equality*:

$$2 \text{ cobits} = 1 \text{ qubit} + 1 \text{ ebit.} \qquad (5)$$

A different extension of dense coding is the notion of superdense coding of quantum states proposed in [9]. Instead of dense coding classical bits, the authors in [9] propose to code quantum bits *whose quantum states are known to the sender Alice*. This last restriction is usually referred to as the remote preparation of qubits, in contrast to the transmission of qubits whose states are unknown to the sender. In remote preparation of qubits, the sender Alice can use the additional knowledge about her states in the choice of encoding. In [9] it is shown that one can obtain the asymptotic resource relation

$$1 \text{ ebit} + 1 \text{ qubit} \geq 2 \text{ remotely prepared qubit(s).} \qquad (6)$$

Such relation would be impossible if the r.h.s. were replaced by 2 qubits. In that case the inequality could be used repeatedly to obtain that 1 qubit suffices for the transmission of an arbitrary number of qubits which is impossible.

The "non-oblivious" superdense coding of quantum states should be compared with the non-oblivious and asymptotic variant of quantum teleportation which was introduced in [5]. In this protocol, referred to as remote state preparation (using classical bits), the quantum teleportation inequality, Eq. (3), is tightened to

$$1 \text{ ebit} + 1 \text{ cbit} \geq 1 \text{ remotely prepared qubit(s).} \qquad (7)$$

These various resource (in)equalities and their underlying protocols can be viewed as the first in a comprehensive theory of resources inequalities. The goal of such theory [6] is to provide a unified and simplified approach to quantum Shannon theory.

Experimental Results

In [11] a partial realization of dense coding was given using polarization states of photons as qubits. The Bell state $|\Psi_{01}\rangle$ can be produced by parametric down-conversion; this state was used in the experiment as the shared entanglement between Alice and Bob. With current experimental techniques, it is not possible to carry out a low-noise measurement in the Bell basis which uniquely distinguishes the four Bell states. Thus, in [11] one of three messages, *a trit*, is encoded into the four Bell states. Using two-particle interferometry, Bob learns the value of the trit by distinguishing two of the four Bell states uniquely and obtaining a third measurement signal for the two other Bell states.

In perfect dense coding, the channel capacity is 2 bits. For the trit-scheme of [11], the ideal channel capacity is $\log 3 \approx 1.58$. Due to the noise in the operations and measurements, the authors of [11] estimate the experimentally achieved capacity as 1.13 bits. In [1] it is shown how the presence of additional entanglement of the polarized photons in their orbital momentum degree of freedom (hyperentanglement) can assist in distinguishing all 4 Bell states in a modified Bell state analyzer. A capacity of 1.63 bits is reported.

In [13] the complete protocol of dense coding was carried out using two $^9Be^+$ ions confined to an electromagnetic trap. A qubit is formed by two internal hyperfine levels of the $^9Be^+$ ion. Single-qubit and two-qubit operations are carried out using two polarized laser beams. A single qubit measurement is performed by observing a weak/strong fluorescence of $|0\rangle$ and $|1\rangle$. The authors estimate that the noise in the unitary transformations and measurements leads to an overall error rate on the transmission of the bits b of 15%. This results in an effective channel capacity of 1.16 bits.

In [7] dense coding was carried out using NMR spectroscopy. The two qubits were formed by the nuclear spins of ^1H and ^{13}C of chloroform molecules $^{13}CHCL_3$ in liquid solution at room temperature. The full dense coding protocol was implemented using the technique of temporal averaging and the application of coherent RF pulses; see [12] for details. The authors estimate an overall error rate on the transmission of the bits b of less than 10%.

Cross-References

▶ Teleportation of Quantum States

Recommended Reading

1. Barreiro J, Wei Tzu-Chieh, Kwiat P (2008) Beating the channel capacity limit for linear photonic superdense coding. Nat Phys 4:282–286
2. Braunstein S, Kimble HJ (2000) Dense coding for continuous variables. Phys Rev A 61(4):042302
3. Bennett CH, Wiesner SJ (1992) Communication via one- and two-particle operators on Einstein-Podolsky-Rosen states. Phys Rev Lett 69:2881–2884
4. Bennett CH, Brassard G, Crepeau C, Jozsa R, Peres A, Wootters WK (1993) Teleporting an unknown quantum state via dual classical and Einstein-Podolsky-Rosen channels. Phys Rev Lett 70:1895–1899
5. Bennett CH, DiVincenzo DP, Smolin JA, Terhal BM, Wootters WK (2001) Remote state preparation. Phys Rev Lett 87:077902
6. Devetak I, Harrow A, Winter A (2008) A resource framework for quantum Shannon theory. IEEE Trans Inf Theory 54(10), 4587–4618
7. Fang X, Zhu X, Feng M, Mao X, Du F (2000) Experimental implementation of dense coding using nuclear magnetic resonance. Phys Rev A 61:022307
8. Harrow AW (2004) Coherent communication of classical messages. Phys Rev Lett 92:097902
9. Harrow A, Hayden P, Leung D (2004) Superdense coding of quantum states. Phys Rev Lett 92:187901
10. Holevo AS (1973) Bounds for the quantity of information transmitted by a quantum communication channel. Problemy Peredachi Informatsii 9:3–11. English translation in: Probl Inf Transm 9:177–183 (1973)
11. Mattle K, Weinfurter H, Kwiat PG, Zeilinger A (1996) Dense coding in experimental quantum communication. Phys Rev Lett 76:4656–4659
12. Nielsen MA, Chuang IL (2000) Quantum computation and quantum information. Cambridge University Press, Cambridge
13. Schaetz T, Barrett MD, Leibfried D, Chiaverini J, Britton J, Itano WM, Jost JD, Langer C, Wineland DJ (2004) Quantum dense coding with atomic qubits. Phys Rev Lett 93:040505

Quantum Error Correction

Martin Roetteler
Microsoft Research, Redmond, WA, USA

Keywords

Quantum codes; Quantum error-correcting codes; Stabilizer codes

Years and Authors of Summarized Original Work

1995; Shor

Problem Definition

A quantum system can never be seen as being completely isolated from its environment, thereby permanently causing disturbance to the state of the system. The resulting noise problem threatens quantum computers and their great promise, namely, to provide a computational advantage over classical computers for certain problems (see also the cross-references in the section "Cross-References"). Quantum noise is usually modeled by the notion of a *quantum channel* which generalizes the classical case and, in particular, includes scenarios for communication (space) and storage (time) of quantum information. For more information about quantum channels and quantum information in general, see [19]. A basic channel is the quantum mechanical analog of the classical binary symmetric channel [17]. This quantum channel is called the *depolarizing channel* and depends on a real parameter $p \in [0, 1]$. Its effect is to randomly apply one of the Pauli spin matrices X, Y, and Z to the state of the system, mapping a quantum state ρ of one qubit to $(1 - p)\rho + p/3(X\rho X + Y\rho Y + Z\rho Z)$. It should be noted that it is always possible to map any quantum channel to a depolarizing channel by twirling operations. The basic problem of quantum error correction is to devise a mechanism that allows to recover quantum information that has been sent through a quantum channel, in particular the depolarizing channel.

Key Results

For a long time, it was not known whether it would be possible to protect quantum information against noise. Even some indication in the form of the no-cloning theorem was put forward to support the view that it might be impossible. The no-cloning theorem essentially says that an unknown quantum state cannot be copied perfectly. This dashes hopes that, similar to the classical case, a simple triple-replication and majority voting mechanism may be used in the quantum case as well. Therefore, it came as a surprise when Shor [20] found a quantum code which encodes one qubit into nine qubits in such a way that the resulting state has the ability to be protected against arbitrary single-qubit errors on each of these nine qubits. The idea is to use a concatenation of two threefold repetition codes.

One of them protects against bit-flip errors while the other protects against phase-flip errors. The quantum code is a two-dimensional subspace of the 2^9 dimensional Hibert space $(\mathbb{C}^2)^{\otimes 9}$. Two orthogonal basis vectors of this space are identified with the logical 0 and 1 states, respectively, usually called $|0\rangle$ and $|1\rangle$. Explicitly, the code is given by

$$|\underline{0}\rangle = \frac{1}{2\sqrt{2}}(|000\rangle + |111\rangle) \otimes (|000\rangle + |111\rangle)$$
$$\otimes (|000\rangle + |111\rangle),$$

$$|\underline{1}\rangle = \frac{1}{2\sqrt{2}}(|000\rangle + |111\rangle) \otimes (|000\rangle - |111\rangle)$$
$$\otimes (|000\rangle + |111\rangle).$$

The state $\alpha|0\rangle + \beta|1\rangle$ of one qubit is encoded to the state $\alpha|\underline{0}\rangle + \beta|\underline{1}\rangle$ of the nine-qubit system. The reason why this code can correct one arbitrary quantum error is as follows.

First, suppose that a bit-flip error has happened, which in quantum mechanical notation is given by the operator X. Then a majority vote of each block of three qubits $1-3, 4-6$, and $7-9$ can be computed and the bit flip can be corrected. To correct against phase-flip errors, which are given by the operator Z, the fact is used that the code can be written as $|\underline{0}\rangle = |+ ++\rangle + |- --\rangle$, $|\underline{1}\rangle = |+ ++\rangle - |- --\rangle$, where $|\pm\rangle = \frac{1}{\sqrt{2}}(|000\rangle + |111\rangle)$. By measuring each block of three in the basis $\{|+\rangle, |-\rangle\}$, the majority of the phase flips can be detected and one phase-flip error can be corrected. Similarly, it can be shown that Y, which is a combination of a bit flip and a phase flip, can be corrected.

Discretization of Noise

Even though the above procedure seemingly only takes care of bit-flips and phase-flip errors, it actually is true that an *arbitrary* error affecting a single qubit out of the nine qubits can be corrected. In particular, and perhaps surprisingly, this is also the case if one of the nine qubits is completely destroyed. The linearity of quantum mechanics allows this method to work. Linearity implies that whenever operators A and B can be corrected, so can their sum $A + B$ [8, 20, 22]. Since the (finite) set $\{1_2, X, Y, Z\}$ forms a vector space basis for the (continuous) set of all one-qubit errors, the nine-qubit code can correct an arbitrary single-qubit error.

Syndrome Decoding and the Need for Fresh Ancillas

A way to do the majority vote quantum-mechanically is to introduce two new qubits (also called ancillas) that are initialized in $|0\rangle$. Then, the results of the two parity checks for the repetition code of length three can be computed into these two ancillas. This syndrome computation for the repetition code can be done using the so-called controlled not (CNOT) gates [19] and Hadamard gates. After this, the qubits holding the syndrome will factor out (i.e., they have no influence on future superpositions or interferences of the computational qubits) and can be discarded. Quantum error correction demands a large supply of fresh qubits for the syndrome computations which have to be initialized in a state $|0\rangle$. The preparation of many such states is required to fuel active quantum error-correcting cycles, in which syndrome measurements have to be applied repeatedly. This poses great challenges to any concrete physical realization of quantum error-correcting codes.

Conditions for General Quantum Codes

Soon after the discovery of the first quantum code, general conditions required for the existence of codes, which protect quantum systems against noise, were sought after. Here the noise is modeled by a general quantum channel, given by a set of error operators E_i. The Knill-Laflamme conditions [13] yield such a characterization. Let C be the code subspace and let P_C be an orthogonal projector onto C. Then the existence of a recovery operation for the channel with error operators E_i is equivalent to the equation

$$P_C E_i^\dagger E_j P_C = \lambda_{i,j} P_C,$$

for all i and j, where $\lambda_{i,j}$ are some complex constants. This recently has been extended to the more general framework of subsystem codes

(also called operator quantum error-correcting codes) [16].

Constructing Quantum Codes

The problem of deriving general constructions of quantum codes was addressed in a series of groundbreaking papers by several research groups in the mid-1990s. Techniques were developed which allow classical coding theory to be imported to an extent that is enough to provide many families of quantum codes with excellent error correction properties.

The IBM group [3] investigated quantum channels, placed bounds on the quantum channels' capacities, and showed that for some channels, it is possible to compute the capacity (such as for the quantum erasure channel). Furthermore, they showed the existence of a five-qubit quantum code that can correct an arbitrary error, thereby being much more efficient than Shor's code. Around the same time, Calderbank and Shor [4] and Steane [21] found a construction of quantum codes from any pair C_1, C_2 of classical linear codes satisfying $C_2^\perp \subseteq C_1$. Named after their inventors, these codes are known as CSS codes.

The AT&T group [5] found a general way of defining a quantum code. Whenever a classical code over the finite field $\mathbb{F}_4$ exists that is additively closed and self-orthogonal with respect to the Hermitian inner product, they were able to find even more examples of codes. Independently, D. Gottesman [8,9] developed the theory of stabilizer codes. These are defined as the simultaneous eigenspaces of an abelian subgroup of the group of tensor products of Pauli matrices on several qubits. Soon after this, it was realized that the two constructions are equivalent.

A stabilizer code which encodes k qubits into n qubits and has distance d is denoted by $[n, k, d]$. It can correct up to $\lfloor (d - 1)/2 \rfloor$ errors of the n qubits. The rate of the code is defined as $r = k/n$. Similar to classical codes, bounds on quantum error-correcting codes are known, i.e., the Hamming, Singleton, and linear programming bounds.

Asymptotically Good Codes

Matching the developments in classical algebraic coding theory, an interesting question deals with the existence of asymptotically good codes, i.e., families of quantum codes with parameters $[[n_i, k_i, d_i]]$, where $i \geq 0$, which have asymptotically nonvanishing rate $\lim_{i \to \infty} k_i/n_i > 0$ and nonvanishing relative distance $\lim_{i \to \infty} d_i/n_i > 0$. In [4], the existence of asymptotically good codes was established using random codes. Using algebraic geometry (Goppa) codes, it was later shown by Ashikhmin, Litsyn, and Tsfasman that there are also explicit families of asymptotically good quantum codes [2]. Currently, most constructions of quantum codes are from the abovementioned stabilizer/additive code construction, with notable exception of a few nonadditive codes and some codes which do not fit into the framework of Pauli error bases.

Applications

Besides their canonical application to protect quantum information against noise, quantum error-correcting codes have been used for other purposes as well. The Preskill/Shor proof of the security of the quantum key distribution scheme BB84 relies on an entanglement purification protocol, which in turn uses CSS codes [19]. Furthermore, quantum codes have been used for quantum secret sharing, quantum message authentication, and secure multiparty quantum computations. Properties of stabilizer codes are also germane for the theory of fault-tolerant quantum computation.

Open Problems

The literature of quantum error correction is fast growing, and the list of open problems is certainly too vast to be surveyed here in detail. The following short list is highly influenced by the preference of the author.

1. It is desirable to find quantum codes for which all stabilizer generators have low weight and which at the same time allow for efficient fault-tolerant quantum computation with the encoded data. These requirements correspond to a quantum equivalent to low-density parity check (LDPC) codes. So far only a few constructions are known, but recent progress was made by Gottesman [10] who used quantum LDPC codes to show that universal fault-tolerant quantum computing with constant overhead is possible. See also [11, 15] for recent progress on quantum LDPC codes.

2. It is an open problem to find new families of quantum codes that improve on the currently best known estimates for the threshold for fault-tolerant quantum computing, in particular for codes that can be implemented on a two-dimensional fabric of qubits. An advantage might be had by using subsystem codes since they allow for simple error correction circuits. For more information about noise thresholds, see also the entry on ▸ Fault-Tolerant Quantum Computation.

3. Many quantum codes are designed for the depolarizing channel, where – roughly speaking – the error probability is improved from p to $p^{d/2}$ for a distance d code. The independence assumption underlying this model might not always be justified, and therefore, it seems imperative to consider other channels, e.g., non-Markovian local error models. Under some assumptions on the decay of the interaction strengths, threshold results for such channels have been shown [1]. However, it remains open to find constructions of good codes for non-Markovian noise and in general for noise models that are more realistic than the depolarizing channel.

Experimental Results

Active quantum error-correcting codes, such as those codes which require syndrome measurements and correction operations, as well as passive codes (i.e., codes in which the system stays in a simultaneous invariant subspace of all error operators for certain types of noise), have been demonstrated for various physical systems. First, this was shown in nuclear magnetic resonance (NMR) experiments [14]. The three-qubit repetition code, which protects one qubit against phase-flip error Z, was then demonstrated in an ion trap for beryllium ion qubits [6].

Subsequently, architectures have been proposed [18] that would in principle allow to construct scalable quantum computers based on ion traps and concatenated coding, e.g., based on the $[1, 3, 7]$ Steane code. In superconducting qubit systems, using an architecture that supports nine physical qubits, high gate fidelities have been reported [12]. This suggests that it might be possible in this architecture to achieve error rates that are below the threshold for the surface code, which is known to be around 1 % [7].

Data Sets

Markus Grassl maintains http://www.codetables. de, which contains tables of the best known quantum codes, some entries of which extend ([5], Table III). It also contains bounds on the minimum distance of quantum codes for given lengths and dimensions and contains information about the construction of the codes. In principle, this can be used to get explicit generator matrices (see also the following section "URL to Code").

URL to Code

The computer algebra system Magma (http:// magma.maths.usyd.edu.au/magma/) has functions and data structures for defining and analyzing quantum codes. Several quantum codes are already defined in a database of quantum codes. For instance, the command *BestKnownQuantumCode(F, n, k)* returns the best known quantum code (i.e., one of the highest known minimum weight) over the field F, of length n, and dimension k. It allows the user to define new quantum codes and to study its properties such as the weight distribution, automorphism, and several predefined methods for obtaining new codes from a set of given ones.

Cross-References

▶ Abelian Hidden Subgroup Problem
▶ Fault-Tolerant Quantum Computation
▶ Quantum Algorithm for Element Distinctness
▶ Quantum Algorithm for Factoring
▶ Quantum Algorithm for Finding Triangles
▶ Quantum Algorithm for Solving Pell's Equation
▶ Quantum Analogues of Markov Chains
▶ Quantum Key Distribution
▶ Teleportation of Quantum States

Recommended Reading

1. Aharonov D, Kitaev A, Preskill J (2006) Fault-tolerant quantum computation with long-range correlated noise. Phys Rev Lett 96:050504
2. Ashikhmin A, Litsyn S, Tsfasman MA (2001) Asymptotically good quantum codes. Phys Rev A 63:032311
3. Bennett CH, DiVincenzo DP, Smolin JA, Wootters WK (1996) Mixed-state entanglement and quantum error correction. Phys Rev A 54:3824–3851
4. Calderbank AR, Shor PW (1996) Good quantum error-correcting codes exist. Phys Rev A 54:1098–1105
5. Calderbank AR, Rains EM, Shor PW, Sloane NJA (1998) Quantum error correction via codes over GF(4). IEEE Trans Inf Theory 44:1369–1387
6. Chiaverini J, Leibfried D, Schaetz T, Barrett MD, Blakestad RB, Britton J, Itano WM, Jost JD, Knill E, Langer C, Ozeri R, Wineland DJ (2004) Realization of quantum error correction. Nature 432:602–605
7. Fowler AG, Mariantoni M, Martinis JM, Cleland AN (2012) Surface codes: towards practical large-scale quantum computation. Phys Rev A 86:032324
8. Gottesman D (1996) Class of quantum error-correcting codes saturating the quantum Hamming bound. Phys Rev A 54:1862–1868
9. Gottesman D (1997) Stabilizer codes and quantum error correction, Ph.D. thesis, Caltech. See also: arXiv preprint quant-ph/9705052
10. Gottesman D (2013) Fault-tolerant quantum computation with constant overhead. arXiv.org preprint, arXiv:1310.2984
11. Hastings MB (2014) Decoding in hyperbolic spaces: quantum LDPC codes with linear rate and efficient error correction. Quantum Inf Comput 14:1187–1202
12. Kelly J, Barends R, Fowler AG, Megrant A, Jeffrey E, White TC, Sank D, Mutus JY, Campbell B, Chen Y, Chen Z, Chiaro B, Dunsworth A, Hoi I-C, Neill C, O'Malley PJJ, Quintana C, Roushan P, Vainsencher A, Wenner J, Cleland AN, Martinis JM (2014) State preservation by repetitive error detection in a superconducting quantum circuit. Nature 519:66–69

13. Knill E, Laflamme R (1997) Theory of quantum error-correcting codes. Phys Rev A 55:900–911
14. Knill E, Laflamme R, Martinez R, Negrevergne C (2001) Benchmarking quantum computers: the five-qubit error correcting code. Phys Rev Lett 86:5811–5814
15. Kovalev AA, Pryadko LP (2013) Fault-tolerance of "bad" quantum low-density parity check codes. Phys Rev A 87:020304(R)
16. Kribs D, Laflamme R, Poulin D (2005) Unified and generalized approach to quantum error correction. Phys Rev Lett 94(4):180501
17. MacWilliams FJ, Sloane NJA (1977) The theory of error-correcting codes. North-Holland, Amsterdam
18. Monroe C, Raussendorf R, Ruthven A, Brown KR, Maunz P, Duan L-M, Kim J (2014) Large scale modular quantum computer architecture with atomic memory and photonic interconnects. Phys Rev A 89:022317
19. Nielsen M, Chuang I (2000) Quantum computation and quantum information. Cambridge University Press, Cambridge
20. Shor PW (1995) Scheme for reducing decoherence in quantum computer memory. Phys Rev A 52:R2493–R2496
21. Steane A (1996) Error correcting codes in quantum theory. Phys Rev Lett 77:793–797
22. Steane A (1996) Multiple-particle interference and quantum error correction. Proc R Soc Lond A 452:2551–2577

Quantum Key Distribution

Renato Renner
Institute for Theoretical Physics, Zurich, Switzerland

Keywords

Quantum cryptography

Years and Authors of Summarized Original Work

1984; CH Bennett, G Brassard
1991; A Ekert

Problem Definition

Secret keys, i.e., random bitstrings not known to an adversary, are a vital resource in cryptography (they can be used, e.g., for message encryption or authentication). The *distribution* of secret keys among distant parties, possibly only connected by insecure communication channels, is thus a fundamental cryptographic problem. *Quantum key distribution (QKD)* is a method to solve this problem using quantum communication. It relies on the fact that any attempt of an adversary to wiretap the communication would, by the laws of quantum mechanics, inevitably introduce disturbances which can be detected.

For the technical definition, consider a setting consisting of two honest parties, called *Alice* and *Bob*, as well as an adversary, *Eve*. Alice and Bob are connected by a quantum channel Q which might be coupled to a (quantum) system E controlled by Eve (see Fig. 1). In addition, it is assumed that Alice and Bob have some means to exchange classical messages *authentically*, that is, they can make sure that Eve is unable to (un-)detectably) alter classical messages during transmission. If only insecure communication channels are available, Alice and Bob can achieve this using an *authentication scheme* [19]. The scheme requires that Alice and Bob have a short initial key or at least some initial common randomness that is not entirely known to Eve [17]. This is why QKD is sometimes called *quantum key growing*.

A *QKD protocol* $\pi = (\pi_A, \pi_B)$ is a pair of algorithms for Alice and Bob, producing classical outputs S_A and S_B, respectively. S_A and S_B take values in $S \cup \{\perp\}$ where S is called *key space* and $\perp$ is a symbol (not contained in S) indicating that no key can be generated. A QKD protocol π with key space S is said to be *perfectly secure on a channel Q* if, after its execution using communication over Q, the following holds:

- $S_A = S_B$;
- if $S_A \neq \perp$, then S_A and S_B are uniformly distributed on S and independent of the state of E.

More generally, π is said to be *ε-secure on Q* if it satisfies the above conditions except with probability (at most) ε. Furthermore, π is

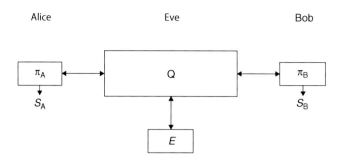

Alice   Eve   Bob

Quantum Key Distribution, Fig. 1 A QKD protocol π consists of algorithms π_A and π_B for Alice and Bob, respectively. The algorithms communicate over a quantum channel $\mathcal{Q}$ that might be coupled to a system E controlled by an adversary. The goal is to generate identical keys S_A and S_B which are independent of E

said to be ε-*robust on* $\mathcal{Q}$ if the probability that $S_A = \perp$ is at most ε. These definitions may be extended to sets of channels $\mathcal{Q}$, i.e., one demands that the conditions hold for any member of the set.

In the standard literature on QKD, protocols are typically parametrized by some positive number k quantifying certain resources needed for its execution (e.g., the amount of communication). A protocol $\pi = (\pi_k)_{k \in \mathbb{N}}$ is said to be *secure* (*robust*) on a set of channels if there exists a sequence $(\varepsilon_k)_{k \in \mathbb{N}}$ which approaches zero exponentially fast such that π_k is ε_k-secure (ε_k-robust) on this set for any $k \in \mathbb{N}$. Moreover, if the key space of π_k is denoted by $\mathcal{S}_k$, the *key rate* of $\pi = (\pi_k)_{k \in \mathbb{N}}$ is defined by $r = \lim\limits_{k \to \infty} \frac{l_k}{k}$ where $l_k := \log_2 |\mathcal{S}_k|$ is the key length.

The ultimate goal is to construct QKD protocols π which are secure against general attacks, i.e., secure on the set of *all* possible channels $\mathcal{Q}$. This ensures that an adversary cannot get any information on the generated key even if she fully controls the communication between Alice and Bob. At the same time, a protocol π should be robust on a set of realistic channels, corresponding to a situation where the noise of the channel is below a given threshold and no adversary is present. Note that, in contrast to security, robustness cannot be guaranteed on the set of all possible channels. Indeed, an adversary could, for instance, interrupt the entire communication between Alice and Bob (in which case key generation is obviously impossible).

Key Results

Protocols

On the basis of the pioneering work of Wiesner [20], Bennett and Brassard, in 1984, invented QKD and proposed a first protocol, known today as the *BB84 protocol* [3]. In 1991, Ekert invented entanglement-based QKD. His protocol is commonly referred to as E91 [8] and provides an additional level of security, termed *device independence* [1, 9]. Later, in an attempt to increase the efficiency and practicability of QKD, various extensions to the BB84 and E91 protocols as well as alternative schemes have been proposed.

QKD protocols can generally be subdivided into (at least) two subprotocols. The purpose of the first, called *distribution protocol*, is to generate a *raw key pair*, i.e., a pair of correlated classical values X and Y known to Alice and Bob, respectively. In many protocols (including BB84), Alice chooses $X = (X_1, \ldots, X_k)$ at random, encodes each of the X_i into the state of a quantum particle, and then sends the k particles over the quantum channel to Bob. Upon receiving the particles, Bob applies a measurement to each of them, resulting in $Y = (Y_1, \ldots, Y_k)$. The crucial idea now is that, by virtue of the laws of quantum mechanics, the secrecy of the raw key is a function of the strength of the correlation between X and Y; in other words, the more information about the (raw) key an adversary tries to acquire, the more disturbances she introduces.

Quantum Key Distribution, Fig. 2 Key rate of an extended version of the BB84 QKD protocol depending on the maximum tolerated channel noise (measured in terms of the bit-flip probability e) [14]

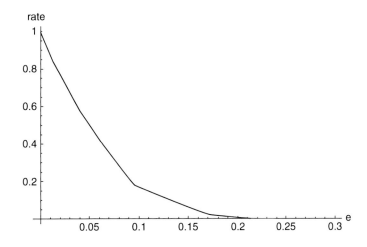

This is exploited in the second subprotocol, called *distillation protocol*. Roughly speaking, Alice and Bob estimate the statistics of the raw key pair (X, Y). If the correlation between their respective parts is sufficiently strong, they use classical techniques such as *information reconciliation* (error correction) and *privacy amplification* (see [4] for the case of a classical adversary which is relevant for the analysis of security against individual attacks and [14, 16] for the quantum-mechanical case which is relevant in the context of collective and general attacks; cf. the characterization below) to turn (X, Y) into a pair (S_A, S_B) of identical and secret keys.

Key Rate as a Function of Robustness and Security

The performance (in terms of the key rate) of a QKD protocol strongly depends on the desired level of security and robustness, as illustrated in Fig. 2. (The robustness is typically measured in terms of the *maximum tolerated channel noise*, i.e., the maximum noise of a channel Q such that the protocol is still robust on Q.) The results summarized below apply to protocols of the form described above where, for the analysis of robustness, it is assumed that the quantum channel Q connecting Alice and Bob is *memoryless* and *time invariant*, i.e., each transmission is subject to the same type of disturbances. Formally, such channels are denoted by $Q = \bar{Q}^{\otimes k}$ where $\bar{Q}$ describes the action of the channel in a single transmission.

Security Against Individual Attacks

A QKD protocol π is said to be *secure against individual attacks* if it is secure on the set of channels Q of the form $\bar{Q}^{\otimes k}$ under the constraint that the coupling to E is purely classical. Note that this notion of security is relatively weak. Essentially, it only captures attacks where the adversary applies identical and independent measurements to each of the particles sent over the channel.

The following statement can be derived from a classical argument due to Csiszár and Körner [6]. Let τ be a distribution subprotocol as described above, i.e., τ generates a raw key pair (X, Y). Moreover, let $\mathcal{S}$ be a set of quantum channels $\bar{Q}$ suitable for τ. Then there exists a QKD protocol π (parametrized by k) consisting of k executions of the subprotocol τ followed by an appropriate distillation subprotocol such that the following holds: π is robust on $Q = \bar{Q}^{\otimes k}$ for any $\bar{Q} \in \mathcal{S}$, is secure against individual attacks, and has key rate at least

$$r \geq \min_{\bar{Q} \in \mathcal{S}} H(X|Z) - H(X|Y), \qquad (1)$$

where the conditional Shannon entropies on the r.h.s. are evaluated for the joint distribution $P_{XYZ}^{\bar{Q}}$ of the raw key (X, Y) and the (classical) state Z of Eve's system E after one execution of τ on the channel Q. Evaluating the right hand side for the BB84 protocol on a channel with bit-flip probability e shows that the rate is non-negative if $e \leq 14.6\%$ [10].

Security Against Collective Attacks

A QKD protocol π is said to be *secure against collective attacks* if it is secure on the set of channels Q of the form $\bar{Q}^{\otimes k}$ with arbitrary coupling to E. This notion of security is strictly stronger than security against individual attacks, but it still relies on the assumption that an adversary does not apply joint operations to the particles sent over the channel.

As shown by Devetak and Winter [7], the above statement for individual attacks extends to collective attacks when replacing inequality (1) by

$$r \geq \min_{\bar{Q} \in \mathcal{S}} S(X|E) - H(X|Y), \qquad (2)$$

where $S(X|E)$ is the conditional von Neumann entropy evaluated for the classical value X and the quantum state of E after one execution of τ on $\bar{Q}$. For the standard BB84 protocol, the rate is positive as long as the bit-flip probability e of the channel satisfies $e \leq 11.0\%$ [18] (see Fig. 2 for a graph of the performance of an extended version of the protocol).

Security Against General Attacks

A QKD protocol π is said to be *secure against general attacks* if it is secure on the set of all channels Q. This type of security is sometimes also called *full* or *unconditional security* as it does not rely on any assumptions on the type of attacks (as long as they are constrained to the communication channel) or the resources needed by an adversary.

The first QKD protocol to be proved secure against general attacks was the BB84 protocol. The original argument by Mayers [13] was followed by various alternative proofs. Most notably, based on a connection to the problem of entanglement purification [5] established by Lo and Chau [12], Shor and Preskill [18] presented a general argument which applies to various versions of the BB84 protocol.

Later it has been shown that, for virtually *any* QKD protocol, security against collective attacks implies security against general attacks [14, 15]. In particular, the above statement about the security of QKD protocols against collective attacks, including formula 2 for the key rate, extends to security against general attacks.

Applications

Because the notion of security described above is *composable* [16] (see [2, 14] for a general discussion of composability of QKD), the key generated by a secure QKD protocol can in principle be used within any application that requires a secret key (such as one-time pad encryption). More precisely, let $\mathcal{A}$ be a scheme which, when using a *perfect* key S (i.e., a uniformly distributed bitstring which is independent of the adversary's knowledge), has some failure probability δ (according to some arbitrary failure criterion). Then, if the perfect key S is replaced by the key generated by an ε-secure QKD protocol, the failure probability of $\mathcal{A}$ is bounded by $\delta + \varepsilon$ [14].

Experimental Results

Most known QKD protocols (including BB84 and E91) only require relatively simple quantum operations on Alice and Bob's side (e.g., preparing a two-level quantum system in a given state or measuring the state of such a system). This makes it possible to realize them with today's technology. Experimental implementations of QKD protocols usually use photons as carriers of quantum information, because they can easily be transmitted (e.g., through optical fibers or free space). A main limitation, however, is noise in the transmission, which, with increasing distance between Alice and Bob, reduces the performance of the protocol (see Fig. 2). We refer to [11] for an overview on quantum cryptography with a focus on experimental aspects.

Cross-References

▶ Quantum Error Correction
▶ Teleportation of Quantum States

Recommended Reading

1. Acín A et al (2007) Device-independent security of quantum cryptography against collective attacks. Phys Rev Lett 98:230501
2. Ben-Or M, Horodecki M, Leung DW, Mayers D, Oppenheim J (2005) The universal composable security of quantum key distribution. In: Second theory of cryptography conference TCC. Lecture notes in computer science, vol 3378. Springer, Berlin, pp 386–406. Also available at http://arxiv.org/abs/quant-ph/0409078
3. Bennett CH, Brassard G (1984) Quantum cryptography: public-key distribution and coin tossing. In: Proceedings of IEEE international conference on computers, systems and signal processing. IEEE Computer Society, Los Alamitos, pp 175–179
4. Bennett CH, Brassard G, Crépeau C, Maurer U (1995) Generalized privacy amplification. IEEE Trans Inf Theory 41(6):1915–1923
5. Bennett CH, Brassard G, Popescu S, Schumacher B, Smolin J, Wootters W (1996) Purification of noisy entanglement and faithful teleportation via noisy channels. Phys Rev Lett 76:722–726
6. Csiszár I, Körner J (1978) Broadcast channels with confidential messages. IEEE Trans Inf Theory 24:339–348
7. Devetak I, Winter A (2005) Distillation of secret key and entanglement from quantum states. Proc R Soc Lond A 461:207–235
8. Ekert AK (1991) Quantum cryptography based on Bell's theorem. Phys Rev Lett 67:661–663
9. Ekert A, Renner R (2014) The ultimate physical limits of privacy. Nature 507:443–447
10. Fuchs CA, Gisin N, Griffiths RB, Niu C, Peres A (1997) Optimal eavesdropping in quantum cryptography, I. Information bound and optimal strategy. Phys Rev A 56:1163–1172
11. Gisin N, Ribordy G, Tittel W, Zbinden H (2002) Quantum cryptography. Rev Mod Phys 74:145–195
12. Lo H-K, Chau HF (1999) Unconditional security of quantum key distribution over arbitrarily long distances. Science 283:2050–2056
13. Mayers D (1996) Quantum key distribution and string oblivious transfer in noisy channels. In: Advances in cryptology – CRYPTO '96. Lecture notes in computer science, vol 1109. Springer, Berlin, pp 343–357
14. Renner R (2005) Security of quantum key distribution. Ph.D. thesis, Swiss Federal Institute of Technology (ETH) Zurich. Also available at http://arxiv.org/abs/quant-ph/0512258
15. Renner R (2007) Symmetry of large physical systems implies independence of subsystems. Nat Phys 3:645–649
16. Renner R, König R (2005) Universally composable privacy amplification against quantum adversaries. In: Second theory of cryptography conference TCC. Lecture notes in computer science, vol 3378. Springer, Berlin, pp 407–425. Also available at http://arxiv.org/abs/quant-ph/0403133
17. Renner R, Wolf S (2003) Unconditional authenticity and privacy from an arbitrarily weak secret. In: Proceedings of Crypto 2003. Lecture notes in computers science, vol 2729. Springer, Berlin, pp 78–95
18. Shor PW, Preskill J (2000) Simple proof of security of the BB84 quantum key distribution protocol. Phys Rev Lett 85:441
19. Wegman MN, Carter JL (1981) New hash functions and their use in authentication and set equality. J Comput Syst Sci 22:265–279
20. Wiesner S (1983) Conjugate coding. Sigact News 15(1):78–88

Quantum Search

Apoorva D. Patel[1] and Lov K. Grover[2]
[1]Centre for High Energy Physics, Indian Institute of Science, Bangalore, India
[2]Bell Laboratories, Alcatel-Lucent, Murray Hill, NJ, USA

Keywords

Amplitude amplification; Grover diffusion; Quantum walk; Unsorted database; Unstructured search

Years and Authors of Summarized Original Work

1996; Grover

Problem Definition

Brief Description

The search problem can be described informally as finding an item possessing a specific property, in a given set of N items. Each item either does or does not possess the specified property, and that can be checked by a binary query. The complexity of the problem is the number of such queries required to find the desired item (also called the *target* item). The items are often collected in a database and sorted to simplify the subsequent searches. When they are not sorted, there is no shortcut to the brute force method of checking each item one by one until the desired item is found. A familiar example of a database is a telephone directory. Its entries are sorted according to the names of persons but not according to the telephone numbers. Hence, it is easy to find the telephone number of a particular person, but difficult to find the name of a person to whom a particular telephone number belongs (i.e., a lookup is difficult when it is not in the same order in which the database is sorted).

The $\Omega(N)$ lower bound on search speed, based on inspection of one item at a time, is correct only for classical computers. Quantum computers can be in a superposition of multiple states, however, and so can inspect multiple items at the same time. There is no obvious lower bound on how fast a quantum search can be, nor is there an obvious technique faster than the brute force search. It turns out, though, that there is an efficient and optimal quantum search algorithm that requires only $O(\sqrt{N})$ queries [15].

This quantum algorithm is very different from the search on a classical computer [8]. The optimal classical strategy is to check the items one at a time in a random order, avoiding in later trials the items that have already been checked earlier. After η items have been checked from a uniform distribution, the probability that the search hasn't yet succeeded is $(1 - 1/N)(1 - 1/(N - 1)) \cdots (1 - 1/(N - \eta + 1))$. For $\eta \ll N$, the success probability is therefore roughly $1 - (1 - 1/N)^\eta \approx \eta/N$. Increasing this success probability to $\Theta(1)$ requires the number of items checked, η, to be $\Theta(N)$.

In contrast to classical computation, quantum computation is formulated in terms of wavelike complex amplitudes, whose interference can be used to cancel undesirable components and boost the desired component. Quantum search is then analogous to the design of a multi-element antenna array, where a careful choice of phases can boost the radiation in a particular direction. The analysis of such structures is carried out using the algebra of unitary transformations, and absolute-value squares of the amplitudes give the ob-

servation probabilities. Unitary transformations include rotations and reflections (about various directions) in the space of amplitudes, as well as local phase shifts. Rotations and reflections redistribute the amplitudes and are similar to classical transformations. On the other hand, the phase shifts are uniquely quantum; they do not alter probabilities of individual components, but affect their subsequent interference pattern. The challenge of quantum computation is to find a sequence of elementary unitary operations (i.e., quantum logic gates) that solve the given computer science problem, while ensuring that the input and the output of the quantum algorithm have clear classical interpretations.

The quantum search algorithm steadily increases the amplitude of the desired item through a series of quantum operations. Starting with an initial amplitude $1/\sqrt{(N)}$, in η steps, the amplitude increases to roughly $\eta/\sqrt{N}$, and hence, the success probability (on observation of the state) increases to η^2/N. Boosting this to $\Theta(1)$ requires only $O(\sqrt{N})$ steps, approximately the square root of the number of steps required by the best classical algorithm.

The quantum search algorithm is of wide interest because of its versatility; it can be adapted to different settings in a variety of fields, giving a new class of quantum algorithms extending well beyond the search problems. Since its discovery, it has been incorporated in solutions of many quantum problems – several of them are mentioned later in this article. Even now, two decades after the algorithm's discovery, new applications and extensions keep on appearing regularly.

Formal Construction

Let the items in the set be labeled by an index $i = 1, 2, \ldots, N$. Let the binary query be represented by an oracle $f(i)$, such that $f(i) = 1$ when i represents a desired item and $f(i) = 0$ otherwise. The quantum algorithm works in an N-dimensional vector space with complex coordinates, known as the Hilbert space. We use Dirac's notation, which is standard in the literature of quantum mechanics and quantum computation. Then the items are mapped to the N orthogonal basis vectors $|i\rangle$ of the Hilbert space, and the binary query is mapped to the selective phase-shift operator defined by $U_f |i\rangle = (-1)^{f(i)} |i\rangle$. Given the binary query $f(i)$, it is easy to construct the operator U_f using an ancilla qubit.

The problem is to start from a specific initial state $|s\rangle$ and evolve to a target state satisfying $U_f |t\rangle = -|t\rangle$, by applying a sequence of unitary operations. The number of times U_f is used in the algorithm is its query complexity. This search problem is *unstructured* because nothing is known about the solution, except the information available from the oracle that can tell whether or not a specific state is the target state.

NP-complete problems can be represented as exhaustive search problems. For example, let ϕ be a 3-SAT formula on n Boolean variables. Then the search problem is to find an assignment for the variables, $i \in \{1, 2, \ldots, N = 2^n\}$, that satisfy ϕ. This example does not involve a database and so bypasses concerns regarding how the items are stored in a physical memory device and the spatial relationship among them.

Key Results

Grover [15] showed that there indeed exists a quantum search algorithm that provides a square-root speedup over the optimal classical randomized algorithm. The algorithm has its simplest form when there is only one target item. Then the algorithm starts with an unbiased uniform superposition state $|s\rangle = (1/\sqrt{N}) \sum_{i=1}^{N} |i\rangle$ and performs $Q = O(\sqrt{N})$ iterations to evolve to the state $(U_D U_f)^Q |s\rangle$. Each iteration consists of two reflection operations:

1. $U_f = 1 - 2|t\rangle\langle t|$ is a reflection along $|t\rangle$. It uses the binary query to flip the sign of the amplitude of the target state.
2. $U_D = 2|s\rangle\langle s| - 1$ is reflection about $|s\rangle$. It can be carried out without any information about the target state. Since the action of $|s\rangle\langle s|$ gives the average amplitude state, U_D amounts to inversion about the average or overrelaxation.

At the end, the final state is measured in the $\{|i\rangle\}$ basis that encodes the item labels, and i is output.

There are several ways of analyzing the algorithm, and the geometric picture is perhaps the simplest. We observe that throughout the evolution, the quantum state stays in the two-dimensional subspace (of the Hilbert space) spanned by $|s\rangle$ and $|t\rangle$. Initially, the amplitude of the state along $|t\rangle$ is $\langle t|s\rangle = 1/\sqrt{N}$, and the angle between $|t\rangle$ and $|s\rangle$ is $\pi/2 - \theta$ with $\sin\theta = 1/\sqrt{N}$. It is a general property of linear transformations in two dimensions that a pair of reflections about two distinct axes produces a rotation, and the amount of rotation is twice the angle between the two axes. The quantum search algorithm is an alternating sequence of reflections about two different axes. Each application of the operator $U_D U_f$ rotates the quantum state from $|s\rangle$ toward $|t\rangle$ by angle 2θ. The number of iterations $\tilde{Q}$ required to exactly reach the target state is therefore given by $(2\tilde{Q} + 1)\theta = \pi/2$. In practice, we have to truncate to integer $Q = \lfloor \tilde{Q} + 0.5 \rfloor$, introducing a small error. The success probability still remains at least $\cos^2\theta = 1 - 1/N$. Asymptotically, $Q = (\pi/4)\sqrt{N}$.

The reflection about the uniform superposition state, U_D, is known as the Grover diffusion operation. When the indices are represented in binary notation, with $N = 2^n$, we have $|s\rangle = H^{\otimes n} |0\rangle^{\otimes n}$ in terms of the Hadamard operator $H = \frac{1}{\sqrt{2}} \begin{pmatrix} 1 & 1 \\ 1 & -1 \end{pmatrix}$. Then $U_D = H^{\otimes n} U_0 H^{\otimes n}$, with $U_0 = 2|0\rangle^{\otimes n}\langle 0|^{\otimes n} - 1$ being the reflection about the $|0\rangle^{\otimes n}$ state, and it can be implemented using $O(n)$ qubit-level operations. In this case, the full quantum search algorithm evolves the state $|0\rangle^{\otimes n}$ to the state $(H^{\otimes n} U_0 H^{\otimes n} U_f)^Q H^{\otimes n} |0\rangle^{\otimes n}$.

When there are M target items, instead of just one, all that is required is to replace $|t\rangle$ by $\sum_{j=1}^{M} |t_j\rangle / \sqrt{M}$ in the algorithm. The final measurement then yields one of the target items after $O\left(\sqrt{N/M}\right)$ queries. Thus, we have [15]:

Theorem 1 (Grover search) *There is a quantum black-box unstructured search algorithm* *with success probability* $\Theta(1)$, *which finds any one of the* M *target items in a set of* N *items, using* $O\left(\sqrt{N/M}\right)$ *queries.*

This algorithm has several noteworthy properties:

- The algorithm is optimal. It saturates the $\Omega\left(\sqrt{N}\right)$ lower bound [5] on the number of queries required for an unstructured quantum search. The evolution from $|s\rangle$ to $|t\rangle$ follows the shortest geodesic route in the Hilbert space at constant speed. A variational analysis shows that the algorithm cannot be improved by even a single query [43].
- The best classical search algorithm has to walk randomly through all the items, while the quantum search algorithm performs a directed walk in the Hilbert space. The square-root speedup of the quantum search algorithm can therefore be understood as the well-known result that directed walk provides a square-root speedup over random walk while covering the same distance.
- The algorithm can be looked upon as evolution of the quantum state from $|s\rangle$ to $|t\rangle$, governed by a Hamiltonian containing two terms, $|t\rangle\langle t|$ and $|s\rangle\langle s|$. The former represents a potential energy attracting the state toward $|t\rangle$, and the latter represents a kinetic energy diffusing the state throughout the Hilbert space. The algorithm is then the discrete Trotter's formula, generated by exponentiating the two terms in the Hamiltonian [17].
- Grover search does not require the full power of quantum dynamics and can be implemented using any system that obeys the superposition principle. Explicit examples using classical waves in the form of coupled oscillators have been constructed [20, 29]. In these mechanical systems, the role of the uniform superposition state is played by the center-of-mass mode, and the search problem becomes the energy focusing problem. The classical wave implementation requires the same number of queries as the quantum algorithm. The difference is that to represent N items, we

need N wave modes but only $n = \log_2 N$ qubits.
- Grover search finds with certainty a single target state out of four possibilities using a single binary query, i.e., $Q = 1$ for $N = 4$. The best classical Boolean algorithm can distinguish only two items with a single binary query, and so it needs two binary queries to carry out the same task. When the query can be factored into subqueries, e.g., the item label is searched for one digit at a time and not as a whole, the best quantum (or wave) arrangement for a database is a quaternary tree. Then every subquery reduces the search space by a factor of 4, which is a factor-of-2 advantage over the classically optimal binary tree. An additional advantage following from commutativity of superposition is that, unlike the classical case, the quantum tree does not require sorting of the database [28].
- The quantum search algorithm is robust against changes in the initial state and the operators, in sharp contrast to many other quantum processes that are highly sensitive to errors. The initial state and the diffusion operator are related in Grover search, but can be separated in a more general context [2, 40]. Let $\tilde{U}_D$ be the modified diffusion operator with $|s\rangle$ as an eigenstate with eigenvalue 1, i.e., the diffusion is translationally invariant. The algorithm then succeeds with $\Theta(1)$ probability, provided $\alpha = |\langle t|s\rangle|$ as well as the angular spectral gap of $\tilde{U}_D$ in the vicinity of identity (say $\tilde{\theta}$) are bounded away from zero. The number of queries required is $O(B^3/\alpha)$, where B^2 is related to the second moment of the eigenvalue distribution of $\tilde{U}_D$ and obeys $B^2 < 1 + (4/\tilde{\theta}^2)$, in contrast to the classical result $O(1/\alpha^2)$. Grover search, therefore, can be generalized to an entire class of algorithms that use different diffusion operators. This flexibility is one of the reasons why Grover search ideas appear frequently in quantum algorithms.
- Quantum search can be implemented so as to be robust also against faulty queries, a problem known as *bounded-error search*. When the query has a bounded coherent

error, say $\tilde{U}_f |i\rangle |1\rangle = \sqrt{p_i}(-1)^{f(i)}|i\rangle |1\rangle + \sqrt{1-p_i}|\tilde{i}\rangle |0\rangle$ with each $p_i \geq 0.9$ ($|\tilde{i}\rangle$ are arbitrary states and the last qubit is a witness for the fault), quantum search can still be implemented using $O\left(\sqrt{N}\right)$ queries [22]. The deterioration of the search algorithm depends on p_i, but that is only in the scaling constant for the number of queries.

A useful generalization of the quantum search algorithm is the *amplitude amplification* technique [10, 16], which can be applied on top of nearly any quantum algorithm for any problem. It says that given a quantum algorithm that solves a problem with a small success probability ϵ, the success probability can be increased to roughly $m^2 \epsilon$ using $O(m)$ calls to that algorithm. (Classically, the success probability can be increased to only about $m\epsilon$.) For the standard search problem, the simple algorithm that picks a random item has success probability $\epsilon = 1/N$, which the quantum search algorithm increases to $\Theta(1)$.

More formally, let V be the unitary operator corresponding to an algorithm that evolves the initial state $|s\rangle$ to $V|s\rangle$. Its success probability is $\epsilon = |\langle t|V|s\rangle|^2 = |V_{ts}|^2$. The algorithm obtained by replacing $|s\rangle$ by $V|s\rangle$ in Grover search, i.e., replacing U_D by $VU_DV^\dagger$, then increases the success probability to $\Theta(1)$ in $O\left(1/\sqrt{|V_{ts}|}\right)$ iterations. In particular, this algorithm evolves the quantum state in the two-dimensional subspace spanned by $|t\rangle$ and $V|s\rangle$, rotating it by angle $2\sin^{-1}|V_{ts}|$ at every iteration. In order to implement U_f, the algorithm needs a witness for the correctness of the output. Thus, we have [10, 16]:

Theorem 2 (Amplitude amplification) *Let $\mathcal{A}$ be a quantum algorithm that outputs a correct answer with witness, with known probability $\epsilon = \sin^2 \theta$. Furthermore, let $m = \lfloor \pi/(4\theta) \rfloor$. Then there is an algorithm $\mathcal{A}'$ that uses $2m + 1$ calls to $\mathcal{A}$ and $\mathcal{A}^{-1}$ and outputs a correct answer with probability $\epsilon' \geq 1 - \epsilon$.*

Depending on the actual implementation, it is possible to vary the quantum search algorithm somewhat from the preceding deterministic

and optimal approach and obtain small improvements:

- The algorithm needs $O\left(\sqrt{N}\log N\right)$ qubit-level operations in order to implement $H^{\otimes n}$ and U_0. The log N factor in this count can be suppressed by adding a small number of queries to the algorithm. A simple scheme divides the n qubits into k sets of n/k qubits each and uses the Grover diffusion operators $U_D{}^{(i)} = H^{\otimes(n/k)}U_0{}^{(i)}H^{\otimes(n/k)}$ that act only on one set at a time leaving the other sets unchanged [18]. Sequentially going through all the sets generates the transformation $V = \prod_{i=1}^{k}(U_D^{(i)}U_f)H^{\otimes n}$, which is then used for amplitude amplification with the initial state $|0\rangle^{\otimes n}$. Overall, the number of qubit-level operations reduce by a factor $\Theta(k)$, while the number of queries go up by a factor $1 + \Theta(kN^{-1/k})$, provided $kN^{-1/k} = o(1)$. The choice $k = \Theta(n/\log n)$ reduces the qubit-level operations to $O\left(\sqrt{N}\log n\right)$, at the cost of increasing the queries by a factor $1 + \Theta(1/\log n)$.

- Consider the partial search problem where the items are separated into N/b blocks of size b each, and only the block containing the desired item is to be located using the same U_f. In that case, the number of queries can be reduced by $0.34\sqrt{b}$ for large b [25]. The procedure first uses Grover search to make the amplitudes of nontarget blocks sufficiently small, then applies Grover search in parallel within each block to make the amplitudes of the target block sufficiently negative (amplitudes of nontarget blocks remain unchanged in this step), and then executes a final U_D operation to reduce the amplitudes of nontarget blocks to zero.

- Though Grover search proceeds from $|s\rangle$ to $|t\rangle$ with uniform speed in the Hilbert space, it slows down in terms of the success probability as it nears the target state. So one can reduce the expected number of queries, by stopping the algorithm before reaching the target state and then looking for the desired item probabilistically. That amounts to minimizing $(Q + 1)/p$, with the success probability $p = \sin^2\left((2Q + 1)\sin^{-1}(1/\sqrt{N})\right)$. This probabilistic search reduces the required number of queries asymptotically to $0.6900\sqrt{N}$ [14].

• Consider the search problem where the times required for querying different items are not the same. This can happen when the query is an algorithm $\mathcal{A}$ acting on different input states $|i\rangle$. When the query for the ith item takes time t_i to execute, unstructured quantum search can be accomplished in time

$$O\left(T = \left(\sum_{i=1}^{N} t_i^2\right)^{1/2}\right) \quad \text{when } t_i \text{ are known}$$

apriori [3]. The strategy is to divide the items into multiple groups so that items in every group have query times within a constant factor, apply Grover search within each group, and then query the groups sequentially. The number of groups needed is $O(\log N)$, and the result improves upon the global Grover search bound $O\left(\sqrt{N}t_{\text{max}}\right)$. Amplitude amplification can be used when $\mathcal{A}$ is probabilistic, and a polylogarithmic overhead in T is required when t_i are not known in advance.

Applications

NP-Complete Problems
Even though NP-complete problems have some structure, there are few known algorithms that exploit this structure to solve them, and often the only recourse left is to solve them as exhaustive search problems. Since quantum search does not assume any structure or pattern in the input data, it provides a square-root speedup in such cases.

Quantum Counting
The counting problem is to find the number of items in a set that satisfy the given query. Its quantum solution is based on the fact that the iterative evolution in Grover search is periodic, with angular frequency $\omega = 2\sin^{-1}\left(\sqrt{M/N}\right)$. The phase estimation procedure (based on quantum Fourier transform) [24] can therefore deter-

mine M approximately, up to error $\sqrt{M}$, using $O\left(\sqrt{N}\right)$ queries. Then, using the property that ω differs by $1/\sqrt{M(N - M)}$ between adjacent values of M, M can be determined exactly using $O/\sqrt{M(N - M)}$ queries [27]. For $M = o(N)$, this quantum result is a power-law improvement over the classical result of $\Theta(N)$ queries, although not as good as a square-root speedup.

Element Distinctness
An early application of Grover search was to find collisions, i.e., given oracle access to a 2-to-1 function f, find distinct arguments x, y such that $f(x) = f(y)$. The quantum *collision problem* has an $O(N^{1/3})$ algorithm [9]. The more general *element distinctness problem* is to find distinct x, y such that $f(x) = f(y)$, for an unknown function f that can be accessed only by an oracle. Ambainis discovered an optimal $O(N^{2/3})$ quantum algorithm for this problem [2]. It searches a suitably constructed graph, with the Grover diffusion operation replaced by a certain quantum walk. The vertices of the graph correspond to various subsets of items $S_j \subseteq \{1, 2, \ldots, N\}$, each of size $N^{2/3}$, two vertices are connected by an edge when the corresponding subsets differ by only one item, and the target vertices are the subsets S_j that solve the element distinctness problem.

Distributed Search
Grover search is also useful in improving communication complexity. For example, a straightforward distributed implementation of the quantum search algorithm solves the *set intersection problem* or the *appointment problem*. The result is, when A and B have respective data strings $x, y \in \{0, 1\}^N$, and they want to find an index i such that $x_i = y_i = 1$, only $O\left(\sqrt{N}\log N\right)$ qubits of communication is necessary [11]. This result has led to an exponential classical/quantum separation in the memory required to evaluate a certain total function with a streaming input [26].

Fixed-Point Search
The iterative evolution in Grover's search algorithm is cyclic, and knowledge of N is necessary

to stop it at the right time to find the target item. In contrast, fixed-point search algorithms converge monotonically to the target state. For a long time, a fixed-point quantum algorithm was considered unlikely, since any iterative unitary evolution is periodic. Surprisingly, a way out was provided by recursive unitary evolution. When the reflection operations in the amplitude amplification algorithm are replaced by selective phase shifts of $\pi/3$ (e.g., $R_i = 1 + (e^{i\pi/3} - 1)|i\rangle\langle i|$ for the state $|i\rangle$), $|\langle t|V|s\rangle|^2 = 1 - \epsilon$ implies $|\langle t|VR_s V^\dagger R_t V|s\rangle|^2 = 1 - \epsilon^3$ [19]. So each recursive substitution of the operator V by the operator $VR_s V^\dagger R_t V$ reduces the deviation of the final state from the target state to the cube of what it was before. (The corresponding best classical reduction is $O(\epsilon^2)$, e.g., by majority rule selection after three trials.) This technique does not give a square-root speedup for search, but it is useful when ϵ is small, for instance, in error correction. It has been used to design composite pulse sequences for reducing systematic errors [35]. An iterative quantum search algorithm with similar properties has been obtained combining reflection operations with non-unitary projective measurements [41]. Another recent construction is a bounded-error quantum search algorithm (i.e., success probability $p \geq 1 - \delta$) that varies the phase shifts between π and $\pi/3$ as a function of the iteration number [42]. It exhibits square-root speedup as well as convergence to the target state, provided that both δ and $|\langle t|s\rangle|$ are bounded away from zero.

Spatial Search

This is the search problem where the items belonging to a database are spread over distinct physical locations, say a d-dimensional lattice, and there is a restriction that one can proceed from any location to only its neighbors while searching for the target item. Its quantum solution replaces the global Grover diffusion operator by a local quantum walk, and Grover search becomes the $d \to \infty$ limit. The required number of queries has to obey the double lower bound $\Omega\left(dN^{1/d}, \sqrt{N}\right)$; the former arises from the finite speed of movement on the lattice and the

latter from the optimality of Grover search. The best algorithms are found in the framework of relativistic quantum mechanics. They use $O\left(\sqrt{N}\right)$ queries for $d > 2$, with the scaling constant approaching $\pi/4$ from above as $d \to \infty$ [1,33]. In the critical dimension $d = 2$, the algorithms are slowed down by logarithmic factors arising from the infrared divergence, and the best known algorithm requires $O\left(\sqrt{N \log N}\right)$ queries [39]. For non-integer values of d, the scaling behavior of the algorithm has been verified using numerical simulations on fractal lattices [32].

Markov Chain Evolution

Generic stationary stochastic processes (e.g., random walks) are defined in terms of transition matrices that encode the possible evolutionary changes at each step. Many properties of the resulting evolution (e.g., hitting time, detection, mixing, escape time) scale as negative powers of the spectral gap of the transition matrix. For Markovian evolution on bipartite graphs, the transition matrix can be separated into two disjoint parts, say $\{x\} \to \{y\}$ and $\{y\} \to \{x\}$. Szegedy constructed two reflection operators from these parts and defined a quantum evolution operator as their product [38] (classical Markov chain evolution does not allow such reflection operators). The spectral gap of this quantum evolution operator scales as the square root of the spectral gap of the original transition matrix and so speeds up the evolution the same way as Grover search does.

Recursive Search

Game-tree evaluation, which is a recursive search problem, is an extension of unstructured search. Classically, using the alpha-beta pruning technique, the value of a balanced binary AND-OR tree can be computed with $o(1)$ error in expected time $O\left(N^{\log_2[(1+\sqrt{33})/4]}\right) = O(N^{0.754})$ [36]. This is optimal even for bounded-error algorithms [37]. By applying quantum search recursively, a depth-d regular AND-OR tree can be evaluated with constant error in time $\sqrt{N} \cdot O(\log N)^{d-1}$. The log factors come from amplifying the success probability of inner searches to be close to one. Bounded-error quantum search eliminates these log factors, reducing the time to $O\left(\sqrt{N} \cdot c^d\right)$ for some constant c. Recently, an $O(N^{0.5+o(1)})$ time algorithm has been discovered for evaluating an arbitrary AND-OR tree on N variables [4, 13].

Open Problems

In several applications of the quantum search algorithm, only the leading asymptotic behavior of the query complexity is known, and attempts to suppress logarithmic corrections (when they appear) and reduce the scaling constants continue [34]. In this section, we point out some other offbeat applications.

Hamiltonian Evolution
Many conventional algorithms for simulations of quantum systems with sparse Hamiltonians use the Trotter formula with a small step size. They have a power-law dependence of the computational complexity on the simulation error and hence are not efficient. In contrast, Grover search amounts to a Trotter formula with the largest possible step size, given the projection operator nature of the terms in the Hamiltonian. A recent exciting realization is that this feature leads to only logarithmic dependence of the computational complexity on the simulation error, which is an exponential improvement. The general strategy is to decompose the sparse Hamiltonian as a sum of projection operators, formulate the evolution problem as a multi-query search problem, and then use a large step size Trotter formula to simulate it [7, 31]. This framework can also readily benefit classical simulations of quantum systems.

Molecular Biology
Many molecular processes of metabolism occur at scales, nanometer and picosecond, where quantum dynamics is relevant. They frequently involve unstructured search and transport, in the sense that correct ingredients for the processes have to be found from the mixture of molecules

floating around. Evolution over billions of years has certainly produced complex machinery to carry out these searches efficiently, although we do not fully comprehend their optimization criteria. Attempts to understand some of these processes suggest that Grover search may have played a role in their design, quite likely exploiting coherent coupled vibrational modes and not quantum superposition. An intriguing example is that the universal genetic language uses an alphabet of four letters, while a binary alphabet would be sufficient and simpler to construct during evolution [30]. Coherent vibrational dynamics of molecules also contributes to efficient energy transport during photosynthesis and to the detection of smell [23].

Ordered Search
A sequentially ordered database can be easily searched by factoring $f(i)$ into subqueries for individual digits of i. An alternative is to use a different oracle $g(i)$, such that $g(i) = 0$ when i represents items before the desired item and $g(i) = 1$ otherwise. Classically, binary search is the optimal algorithm given either $f(i)$ or $g(i)$ and requires $\lceil \log_2 N \rceil$ queries. The optimal quantum algorithm for $f(i)$ is quaternary search with $0.5\lceil \log_2 N \rceil$ queries, but surprisingly a quantum algorithm using $g(i)$ can do better. In case of $g(i)$, though the optimal solution is unknown, the query complexity for an exact algorithm has a lower bound of $0.221 \log_2 N$ [21] and a known solution of $0.433 \log_2 N$ [12] (there also exists a quantum stochastic Las Vegas algorithm with $0.32 \log_2 N$ expected queries and $o(1)$ error [6]).

Search with Additional Structure
It may be possible to speed up a search process beyond the square-root speedup of Grover search, when the problem has extra structure beyond the minimal information provided by the oracle $f(i)$. The details of the algorithm and the extent of speedup would then depend on the extra structure, and the possibilities are open to explorations. Some examples are symmetries among the items, associative memory recall with connections, and patterns in the Boolean function

to be evaluated. Another problem of interest is determination of the complete path (with certain properties) from the initial to the target state instead of locations of just the end points.

Perspective

In a lecture at the Bell Labs in 1985, Richard Feynman made an interesting observation. In the 1940s when airplanes were being developed, aeronautical engineers had proved bounds and theorems about why planes would never be able to fly faster than the speed of sound. For several years, this speed was regarded as fundamentally a bound for flights as the speed of light is for communications. However, gradually just by using intelligent design, it was discovered that airplanes could indeed fly faster than the speed of sound – only the rules of design in the new regime were very different. The question is whether the bounds on quantum computation (specifically the $\Omega\left(\sqrt{N}\right)$ bound for search) will continue to hold, or by making the rules of design very different, just as in the case of supersonic airplanes, someone will find a way around these bounds. No one has found any loophole in the arguments in the 20 years since the lower bound for quantum search was discovered, despite numerous scientists from different fields having tried their hand at it. On the other hand, even though this bound has been derived over and over again using different methods, no one has come up with a simple and short physical explanation for it, which would give one the assurance that one really understood it.

Cross-References

▶ Quantum Algorithm for Element Distinctness
▶ Routing

Recommended Reading

1. Aaronson S, Ambainis A (2005) Quantum search of spatial regions. Theory Comput 1:47–79
2. Ambainis A (2007) Quantum walk algorithm for element distinctness. SIAM J Comput 37(1):210–239
3. Ambainis A (2010) Quantum search with variable times. Theory Comput Syst 47(3): 786–807
4. Ambainis A, Childs AM, Reichardt BW, Špalek R, Zhang S (2010) Any AND-OR formula of size N can be evaluated in time $N^{1/2 + o(1)}$ on a quantum computer. SIAM J Comput 39(6):2513–2530
5. Bennett CH, Bernstein E, Brassard G, Vazirani U (1997) Strengths and weaknesses of quantum computing. SIAM J Comput 26(5):1510–1523
6. Ben-Or M, Hassidim A (2007) Quantum search in an ordered list via adaptive learning. arXiv:quant-ph/0703231
7. Berry DW, Childs AM, Cleve R, Kothari R, Somma RD (2014) Exponential improvement in precision for simulating sparse Hamiltonians. In: Proceedings of the 46th ACM symposium on theory of computing (STOC'14), New York, 31 May–3 June 2014, pp 283–292
8. Brassard G (1997) Searching a quantum phone book. Science 275(5300):627–628
9. Brassard G, Høyer P, Tapp A (1998) Quantum cryptanalysis of hash and claw-free functions. In: Proceedings of the 3rd Latin American theoretical informatics symposium (LATIN'98). Lecture notes in computer science, vol 1380. Springer, Berlin/Heidelberg, pp 163–169
10. Brassard G, Høyer P, Tapp A (1998) Quantum Counting. In: Proceedings of the 25th international colloquium on automata, languages and programming (ICALP'98). Lecture notes in computer science, vol 1443. Springer, Berlin/Heidelberg, pp 820–831
11. Buhrman H, Cleve R, Wigderson A (1998) Quantum vs. classical communication and computation. In: Proceedings of the 30th ACM symposium on theory of computing (STOC'98), Dallas, 24–26 May 1998, pp 63–68
12. Childs AM, Landahl AJ, Parrilo PA (2007) Improved quantum algorithms for the ordered search problem via semidefinite programming. Phys Rev A 75(3):032335
13. Farhi E, Goldstone J, Gutmann S (2008) A quantum algorithm for the Hamiltonian NAND tree. Theory Comput 4:169–190
14. Gingrich RM, Williams CP, Cerf NJ (2000) Generalized quantum search with parallelism. Phys Rev A 61(5):052313
15. Grover LK (1996) A fast quantum mechanical algorithm for database search. In: Proceedings of the 28th ACM symposium on theory of computing (STOC'96), Philadelphia, 22–24 May 1996, pp 212–219
16. Grover LK (1998) A framework for fast quantum mechanical algorithms. In: Proceedings of the 30th ACM symposium on theory of computing (STOC'98), Dallas, 24–26 May 1998, pp 53–62
17. Grover LK (2001) From Schrödinger's equation to the quantum search algorithm. Pramana 56:333–348
18. Grover LK (2002) Trade-offs in the quantum search algorithm. Phys Rev A 66(5):052314

Q

19. Grover LK (2005) Fixed-point quantum search. Phys Rev Lett 95:150501

20. Grover LK, Sengupta AM (2002) From coupled pendulums to quantum search. In: Brylinski RK, Chen G (eds) Mathematics of quantum computation. CRC, Boca Raton, pp 119–134

21. Høyer P, Neerbak J, Shi Y (2002) Quantum complexities of ordered searching, sorting and element distinctness. Algorithmica 34(4):429–448

22. Høyer P, Mosca M, de Wolf R (2003) Quantum search on bounded-error inputs. In: Proceedings of the 30th international colloquium on automata, languages and programming (ICALP 2003). Lecture notes in computer science, vol 2719. Springer, Berlin/Heidelberg, pp 291–299

23. Huelga SF, Plenio MB (2013) Vibrations, quanta and biology. Contemp Phys 54(4):181–207

24. Kitaev AYu, Shen AH, Vyalyi MN (2002) Classical and quantum computation. Graduate studies in mathematics, vol 47. American Mathematical Society, Providence. Section 13.5

25. Korepin VE, Grover LK (2006) Simple algorithm for partial quantum search. Quantum Inf Process 5(1):5–10

26. Le Gall F (2006) Exponential separation of quantum and classical online space complexity. In: Proceedings of the 18th annual ACM symposium on parallelism in algorithms and architectures (SPAA 2006), Cambridge, 30 July–2 Aug 2006, pp 67–73

27. Mosca M (2001) Counting by quantum eigenvalue estimation. Theor Comput Sci 264:139–153

28. Patel A (2001) Quantum database search can do without sorting. Phys Rev A 64(3):034303

29. Patel A (2006) Optimal database search: waves and catalysis. Int J Quantum Inf 4(5):815–825; Erratum: ibid. 5(3):437 (2007)

30. Patel A (2008) Towards understanding the origin of genetic languages. In: Abbott D, Davies PCW, Pati AK (eds) Quantum aspects of life. Imperial College Press, London, pp 187–219

31. Patel A (2014) Optimisation of quantum Hamiltonian evolution. In: Proceedings of the 32nd international symposium on lattice field theory, New York, 23–28 June 2014. PoS(LATTICE2014)324

32. Patel A, Raghunathan KS (2012) Search on a fractal lattice using a quantum random walk. Phys Rev A 86(1):012332

33. Patel A, Rahaman MdA (2010) Search on a hypercubic lattice using a quantum random walk: I. d>2. Phys Rev A 82(3):032330

34. Portugal R (2013) Quantum walks and search algorithms. Springer, New York

35. Reichardt BW, Grover LK (2005) Quantum error correction of systematic errors using a quantum search framework. Phys Rev A 72:042326

36. Saks M, Wigderson A (1986) Probabilistic Boolean decision trees and the complexity of evaluating game trees. In: Proceedings of the 27th annual IEEE symposium on foundations of computer science (FOCS), Toronto, 27–29 Oct 1986, pp 29–38

37. Santha M (1995) On the Monte Carlo decision tree complexity of read-once formulae. Random Struct Algorithms 6(1):75–87

38. Szegedy M (2004) Quantum speed-up of Markov chain based algorithms. In: Proceedings of the 45th annual IEEE symposium on foundations of computer science (FOCS'04), Rome, 17–19 Oct 2004, pp 32–41

39. Tulsi A (2008) Faster quantum-walk algorithm for the two-dimensional spatial search. Phys Rev A 78(1):012310

40. Tulsi A (2012) General framework for quantum search algorithms. Phys Rev A 86(4):042331

41. Tulsi T, Grover LK, Patel A (2006) A new algorithm for fixed point quantum search. Quantum Inf Comput 6(6):483–494

42. Yoder TJ, Low GH, Chuang IL (2014) Optimal fixed-point quantum amplitude amplification using Chebyshev polynomials. Phys Rev Lett 113:210501

43. Zalka C (1999) Grover's quantum searching algorithm is optimal. Phys Rev A 60(4):2746–2751

Query Release via Online Learning

Jonathan Ullman
Department of Computer Science, Columbia University, New York, NY, USA

Keywords

Computational learning theory; Differential privacy; Statistical query model

Years and Authors of Summarized Original Work

2010; Hardt, Rothblum
2012; Hardt, Rothblum, Servedio
2012; Thaler, Ullman, Vadhan
2013; Ullman
2014; Chandrasekaran, Thaler, Ullman, Wan
2014; Bun, Ullman, Vadhan

Problem Definition

Our goal is to design differentially private algorithms to answer *statistical queries* on a sensitive database. We model the database $D = (x_1, \ldots, x_n) \in (\{0, 1\}^d)^n$ as a collection of n records – one per individual – each consisting of d binary attributes. A differentially private algorithm is a randomized algorithm whose output distribution does not depend "significantly" on any one record of the database. The formal definition is as follows:

Definition 1 ([8]) An algorithm $\mathcal{A}:(\{0,1\}^d)^n \to \mathcal{R}$ is (ε, δ)-*differentially private* if for every pair of databases $D, D' \in (\{0,1\}^d)^n$ that differ on at most one row and every $S \subseteq \mathcal{R}$,

$$\mathbb{P}[\mathcal{A}(D) \in S] \le e^\varepsilon \mathbb{P}[\mathcal{A}(D') \in S] + \delta.$$

Henceforth, we will say $\mathcal{A}$ is *differentially private* if it satisfies $(1, 1/n^2)$-differential privacy. (The choice of $\varepsilon = 1$ and $\delta = 1/n^2$ is arbitrary and can be replaced with $\varepsilon = c$ and $\delta = 1/n^{1+c'}$ for any constants $c, c' > 0$ without affecting any stated results).

A *statistical query* (henceforth, simply *query*) is specified by a Boolean predicate $q : \{0,1\}^d \to \{0,1\}$. The answer to a query is the expected value of the predicate over records in the database. Abusing notation, we write

$$q(D) = \frac{1}{n} \sum_{i=1}^{n} q(x_i).$$

We wish to design a differentially private algorithm $\mathcal{A}$ that takes a database and a set of statistical queries and outputs an approximate answer to each query.

Definition 2 An algorithm $\mathcal{A}$ is α-*accurate* for a query q if $\mathcal{A}(D)$ outputs $a \in [0,1]$ such that $|a - q(D)| \le \alpha$, with probability at least $99/100$. An algorithm $\mathcal{A}$ is α-*accurate* for a set of queries $Q = \{q_1, q_2, \dots\}$ if $\mathcal{A}(D)$ outputs $(a_q)_{q \in Q}$ such that for every $q \in Q$, $|a_q - q(D)| \le \alpha$ with probability at least $99/100$.

The goal is to design differentially private algorithms that are α-accurate for sets of queries Q as large as possible. As privacy is easier to achieve when the number of records n is large, we will seek to obtain privacy and accuracy for n as small as possible. Lastly, we seek to make the algorithms as computationally efficient as possible.

Key Results

As a baseline, we will consider simple *additive perturbation* [2, 6–8], which answers each query by independently perturbing the answer with noise from a suitable distribution.

Theorem 1 *There is a differentially private algorithm $\mathcal{A}$ that takes a database $D \in (\{0,1\}^d)^n$ and a set of queries $Q = \{q_1, \dots, q_k\}$ as input, runs in time $\mathrm{poly}(n, d, |q_1| + \cdots + |q_k|)$, and is α-accurate for Q so long as $n \ge \tilde{O}(|Q|^{1/2}/\alpha)$.*

Here, $|q|$ represents the time complexity of evaluating the predicate on a single row of the database. Typically, this it is assumed to be $\mathrm{poly}(d)$.

Additive perturbation is differentially private and computationally efficient, but requires that the size of the database be polynomial in the number of queries, and thus is restricted to answering at most about n^2 queries. As we will see, it is possible to accurately answer *exponentially* more queries under differential privacy.

Answering Many Queries via No-Regret Learning

The first algorithm that improved on additive perturbation for answering arbitrary queries was given by Blum, Ligett, and Roth [3]. Surprisingly, they showed for the first time that it was possible to answer *exponentially many* queries under differential privacy. Subsequent to their work, there were several improvements in the computational efficiency, functionality, and quantitative guarantees of their algorithm. This work led to the *private multiplicative weights algorithm* of Hardt and Rothblum [10]. We summarize the capabilities of this algorithm in the following theorem.

Theorem 2 ([10]) *There is a differentially private algorithm $\mathcal{A}$ that takes a database $D \in (\{0,1\}^d)^n$ and a set of queries $Q = \{q_1, \dots, q_k\}$ as input, runs in time $\mathrm{poly}(n, 2^d, |q_1| + \cdots + |q_k|)$, and is α-accurate for $|Q|$ so long as $n \ge \tilde{O}(\sqrt{d} \log |Q|/\alpha^2)$.*

The private multiplicative weights algorithm is based on the following surprisingly simple framework: Begin with a "crude approximation" of the database D^1. Then, for $t = 1, \ldots, T$, find (in a differentially private manner) a query $q^t \in Q$ such that the approximation D^t does not give an accurate answer. That is, $|q^t(D) - q^t(D^t)| > \alpha$. Use q^t to "update" D^t into a better approximation D^{t+1}. Finally, output the answers to Q given by D^T.

Remarkably, it is possible to find a query $q^t \in Q$ such that D^t is inaccurate (or conclude that none exists) using much less data than would be required to simply answer all the queries in Q using additive perturbation. Perhaps even more surprisingly, it can be shown that if the updates are performed using the *multiplicative weights update rule*, then after $T = O(d/\alpha^2)$ iterations (independent of $|Q|!$), the database D^T will give an accurate answer to *every* $q \in Q$. This argument makes use of the guarantee that multiplicative weights update rule is a "no-regret learning algorithm" (cf. the survey of Arora, Hazan, and Kale [1] for more information about the multiplicative weights update rule). This fast convergence makes it possible to argue that the algorithm can give accurate and differentially private answers with much less data than would be required by simple additive perturbation.

Computational Complexity and Optimality

When $|Q|$ is large, the private multiplicative weights algorithm requires many fewer records n than additive perturbation (when $|Q| \gg d/\alpha^2$). One might ask whether even fewer records suffices. Bun, Ullman, and Vadhan [4] gave a negative answer to this question, and showed that the private multiplicative weights algorithm uses essentially the fewer records possible.

Theorem 3 ([4]) *There is no (even computationally inefficient) differentially private algorithm $\mathcal{A}$ that takes an arbitrary set of queries $Q = \{q_1, \ldots, q_k\}$ with $k \gg d/\alpha^2$ and a database $D \in (\{0, 1\}^d)^n$ with $n \leq \tilde{\Omega}(\sqrt{d} \log |Q|/\alpha^2)$ as input and is α-accurate for the set of queries Q.*

A drawback of the private multiplicative weights algorithm (and all known algorithms with similar properties), when compared to additive perturbation, is computational complexity.

Even when answering a polynomial number of efficiently computable queries, the running time of private multiplicative weights is dominated by the factor of 2^d, which is exponential in the number of attributes in the database. Ullman [13] showed that this is inherent, and (under a widely believed cryptographic assumption) improving on additive perturbation requires exponential running time.

Theorem 4 ([13]) *Assuming the existence of one-way functions, there is no differentially private algorithm $\mathcal{A}$ that takes an arbitrary set of queries $Q = \{q_1, \ldots, q_k\}$ database $D \in (\{0, 1\}^d)^n$ with $n \leq \tilde{\Omega}(|Q|^{1/2})$ as input, runs in time $\text{poly}(n, d, |q_1| + |q_k|)$, and is $1/3$-accurate for the set of queries Q.*

Together, these negative results show the private multiplicative weights algorithm is nearly optimal for answering large sets of arbitrary statistical queries under differential privacy.

Faster Algorithms for Marginal Queries via Efficient Learning

Given the hardness of answering arbitrary queries, there has been a significant effort to design faster differentially private algorithms that improve on additive noise for natural restricted set of queries. One such set of queries is k-*way marginals*. These queries are specified by a subset of attributes $S \subseteq [d]$ of size at most k and a pattern $t \in \{0, 1\}^{|S|}$ and asks for the fraction of records in D that have each attribute $j \in S$ set to t_j. Note that there are $\text{poly}(d^k)$ such queries, and thus, additive perturbation would require running time $\text{poly}(d^k)$ and $n \geq \text{poly}(d^k)$. On the other hand, private multiplicative weights would require running time $\text{poly}(2^d)$, but $n \geq \tilde{O}(k\sqrt{d})$ would suffice.

Most of the more effective algorithms for answering k-way marginal queries are based on the following technique, introduced by Gupta et al. [9]: View the database D as specifying a function $f_D(q) = q(D)$ that maps a query to its answer on D, and then attempt to "learn" a differentially private approximation $g_D \approx f_D$. Intuitively, the value of this approach is that learning algorithms see the evaluation of f_D on a small number of queries and then are able to predict the value of f_D on new queries. Since the learning algorithm only needs a small number of examples, it is easier to ensure differential privacy. If the queries q are "simple," then good learning algorithms may exist for the function f_D. In the case of k-way marginal queries, it turns out that f is in fact (an average of) *conjunctions*, and there are learning algorithms for this class of functions that satisfy various interesting parameter trade-offs. This technique underlies the following two results:

Theorem 5 ([12], building on [11]) *For every $k \in \mathbb{N}$, there is a differentially private algorithm $\mathcal{A}$ that takes a database $D \in (\{0,1\}^d)^n$ as input and runs in time $\mathrm{poly}(n, d^{\sqrt{k}})$, and if $n \geq \mathrm{poly}(d^{\sqrt{k}})$, $\mathcal{A}$ outputs a summary of the database that yields $1/100$-accurate answers to every k-way marginal query. That is, for every k-way marginal query q, one can obtain a $1/100$-accurate answer to q in time $\mathrm{poly}(n, d^{\sqrt{k}})$.*

Theorem 6 ([5]) *For every $k \in \mathbb{N}$, there is a differentially private algorithm $\mathcal{A}$ that takes a database $D \in (\{0,1\}^d)^n$ as input and runs in time $\mathrm{poly}(n, 2^{d^{1-1/100\sqrt{k}}})$, and if $n \geq kd^{.51}$, $\mathcal{A}$ outputs $1/100$-accurate answers to every k-way marginal query.*

We remark that there are many other algorithms for answering k-way marginal queries based on this learning approach, each achieving different parameter trade-offs and guarantees of accuracy. At the time of writing, improving these algorithms and extending these techniques to richer classes of queries remains an active area of research.

Recommended Reading

1. Arora S, Hazan E, Kale S (2012) The multiplicative weights update method: a meta-algorithm and applications. Theory Comput 8(1):121–164
2. Blum A, Dwork C, McSherry F, Nissim K (2005) Practical privacy: the SuLQ framework. In: PODS. ACM, Baltimore MD, pp 128–138
3. Blum A, Ligett K, Roth A (2013) A learning theory approach to noninteractive database privacy. J ACM 60(2):12
4. Bun M, Ullman J, Vadhan SP (2014) Fingerprinting codes and the price of approximate differential privacy. In: STOC. ACM, New York, NY, pp 1–10
5. Chandrasekaran K, Thaler J, Ullman J, Wan A (2014) Faster private release of marginals on small databases. In: ITCS. ACM, Princeton, NJ, pp 387–402
6. Dinur I, Nissim K (2003) Revealing information while preserving privacy. In: PODS. ACM, San Diego, CA, pp 202–210
7. Dwork C, Nissim K (2004) Privacy-preserving datamining on vertically partitioned databases. In: CRYPTO, Santa Barbara, CA, pp 528–544
8. Dwork C, McSherry F, Nissim K, Smith A (2006) Calibrating noise to sensitivity in private data analysis. In: Halevi S, Rabin T (eds) TCC. Lecture notes in computer science, vol 3876. Springer, New York, NY, pp 265–284
9. Gupta A, Hardt M, Roth A, Ullman J (2013) Privately releasing conjunctions and the statistical query barrier. SIAM J Comput 42(4):1494–1520
10. Hardt M, Rothblum G (2010) A multiplicative weights mechanism for privacy-preserving data analysis. In: Proceedings of the 51st foundations of computer science (FOCS). IEEE, Las Vegas, NV, pp 61–70
11. Hardt M, Rothblum GN, Servedio RA (2012) Private data release via learning thresholds. In: SODA. SIAM, Kyoto, Japan, pp 168–187
12. Thaler J, Ullman J, Vadhan SP (2012) Faster algorithms for privately releasing marginals. In: ICALP (1). Springer, Warwick, UK, pp 810–821
13. Ullman J (2013) Answering $n^{2+o(1)}$ counting queries with differential privacy is hard. In: STOC. ACM, Palo Alto, CA, pp 361–370

Quorums

Dahlia Malkhi
Microsoft, Silicon Valley Campus, Mountain View, CA, USA

Keywords

Coteries; Quorum systems; Voting systems

Years and Authors of Summarized Original Work

1985; Garcia-Molina, Barbara

Problem Definition

Quorum systems are tools for increasing the availability and efficiency of replicated services. A *quorum system* for a universe of servers is a collection of subsets of servers, each pair of which intersect. Intuitively, each quorum can operate on behalf of the system, thus increasing its availability and performance, while the intersection property guarantees that operations done on distinct quorums preserve consistency.

The motivation for quorum systems stems from the need to make critical missions performed by machines that are reliable. The only way to increase the reliability of a service, aside from using intrinsically more robust hardware, is via replication. To make a service robust, it can be installed on multiple identical servers, each one of which holds a copy of the service state and performs read/write operations on it. This allows the system to provide information and perform operations even if some machines fail or communication links go down. Unfortunately, replication incurs a cost in the need to maintain the servers consistent. To enhance the availability and performance of a replicated service, Gifford and Thomas introduced in 1979 [3, 14] the usage of *votes* assigned to each server, such that a majority of the sum of votes is sufficient to perform operations. More generally, quorum systems are defined formally as follows:

Quorum system: Assume a *universe U* of servers, $|U| = n$, and an arbitrary number of clients. A *quorum system* $Q \subseteq 2^U$ is a set of subsets of U, every pair of which intersect. Each $Q \in Q$ is called a *quorum*.

Access Protocol

To demonstrate the usability of quorum systems in constructing replicated services, quorums are used here to implement a multi-writer multi-reader atomic shared variable. Quorums have also been used in various *mutual exclusion* protocols, to achieve Consensus, and in commit protocols.

In the application, clients perform read and write operations on a variable x that is replicated at each server in the universe U. A copy of the variable x is stored at each server, along with a timestamp value t. Timestamps are assigned by a client to each replica of the variable when the client writes the replica. Different clients choose distinct timestamps, e.g., by choosing integers appended with the name of c in the low-order bits. The read and write operations are implemented as follows.

Write: For a client c to write the value v, it queries each server in some quorum Q to obtain a set of value/timestamp pairs $A = \{\langle v_u, t_u \rangle\}_{u \in Q}$; chooses a timestamp $t \in T_c$ greater than the highest timestamp value in A; and updates x and the associated timestamp at each server in Q to v and t, respectively.

Read: For a client to read x, it queries each server in some quorum Q to obtain a set of value/timestamp pairs $A = \{\langle v_u, t_u \rangle\}_{u \in Q}$. The client then chooses the pair $\langle v, t \rangle$ with the highest timestamp in A to obtain the result of the read operation. It writes back $\langle v, t \rangle$ to each server in some quorum Q'.

In both read and write operations, each server updates its local variable and timestamp to the received values $\langle v, t \rangle$ only if t is greater than the timestamp currently associated with the variable. The above protocol correctly implements the semantics of a multi-writer multi-reader atomic variable (see ▶ Linearizability).

Key Results

Perhaps the two most obvious quorum systems are the singleton, and the set of majorities, or more generally, weighted majorities suggested by Gifford [3].

Singleton: The set system $Q = \{\{u\}\}$ for some $u \in U$ is the singleton quorum system.

Weighted Majorities: Assume that every server s in the universe U is assigned a number of votes w_s. Then, the set system $Q = \{Q \subseteq U : \sum_{q \in Q} w_q > (\sum_{q \in U} w_q)/2\}$ is a quorum system called Weighted Majorities. When all the weights are the same, simply call this the system of Majorities.

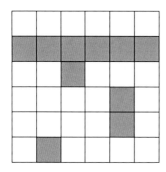

Quorums, Fig. 1 The Grid quorum system of 6×6, with one quorum shaded

An example of a quorum system that cannot be defined by voting is the following Grid construction:

Grid: Suppose that the universe of servers is of size $n = k^2$ for some integer k. Arrange the universe into a $\sqrt{n} \times \sqrt{n}$ grid, as shown in Fig. 1. A quorum is the union of a full row and one element from each row below the full row. This yields the Grid quorum system, whose quorums are of size $O(\sqrt{n})$.

Maekawa suggests in [6] a quorum system that has several desirable symmetry properties, and in particular, that every pair of quorums intersect in exactly one element:

FPP: Suppose that the universe of servers is of size $n = q^2 + q + 1$, where $q = p^r$ for a prime p. It is known that a finite projective plane exists for n, with $q + 1$ pairwise intersecting subsets, each subset of size $q + 1$, and where each element is contained in $q + 1$ subsets. Then the set of finite projective plane subsets forms a quorum system.

Voting and Related Notions

Since generally it would be senseless to access a large quorum if a subset of it is a quorum, a good definition may avoid such anomalies. Garcia-Molina and Barbara [2] call such well-formed systems *coteries*, defined as follows:

Coterie: A *coterie* $\mathcal{Q} \subseteq 2^U$ is a quorum system such that for any $Q, Q' \in \mathcal{Q} : Q \nsubseteq Q'$.

Of special interest are quorum systems that cannot be reduced in size (i.e., that no quorum in the system can be reduced in size). Garcia-Molina

and Barbara [2] use the term "dominates" to mean that one quorum system is always superior to another, as follows:

Domination: Suppose that $\mathcal{Q}, \mathcal{Q}'$ are two coteries, $\mathcal{Q} \neq \mathcal{Q}'$, such that for every $Q' \in \mathcal{Q}'$, there exists a $Q \in \mathcal{Q}$ such that $Q \subseteq Q'$. Then $\mathcal{Q}$ *dominates* $\mathcal{Q}'$. $\mathcal{Q}'$ is *dominated* if there exists a coterie $\mathcal{Q}$ that dominates it, and is *non-dominated* if no such coterie exists.

Voting was mentioned above as an intuitive way of thinking about quorum techniques. As it turns out, vote assignments and quorums are not equivalent. Garcia-Molina and Barbara [2] show that quorum systems are strictly more general than voting, i.e., each vote assignment has some corresponding quorum system but not the other way around. In fact, for a system with n servers, there is a double-exponential ($2^{2^{cn}}$) number of non-dominated coteries, and only $O(2^{n^2})$ different vote assignments, though for $n \le 5$, voting and non-dominated coteries are identical.

Measures

Several measures of quality have been identified to address the question of which quorum system works best for a given set of servers; among these, *load* and *availability* are elaborated on here.

Load

A measure of the inherent performance of a quorum system is its *load*. Naor and Wool define in [10] the load of a quorum system as the probability of accessing the busiest server in the *best* case. More precisely, given a quorum system $\mathcal{Q}$, an *access strategy* w is a probability distribution on the elements of $\mathcal{Q}$; i.e., $\sum_{Q \in \mathcal{Q}} w(Q) = 1$. $w(Q)$ is the probability that quorum Q will be chosen when the service is accessed. Load is then defined as follows:

Load: Let a strategy w be given for a quorum system $\mathcal{Q} = \{Q_1, \ldots, Q_m\}$ over a universe U. For an element $u \in U$, the load induced by w on u is $l_w(u) = \sum_{Q_i \ni u} w(Q_i)$. The load induced by a strategy w on a quorum system $\mathcal{Q}$ is

$$L_w(\mathcal{Q}) = \max_{u \in U} \{l_w(u)\}.$$

The *system load* (or just *load*) on a quorum system Q is

$$L(Q) = \min_w \{L_w(Q)\},$$

where the minimum is taken over all strategies.

The load is a best-case definition, and will be achieved only if an optimal access strategy is used, and only in the case that no failures occur. A strength of this definition is that load is a property of a quorum system, and not of the protocol using it.

The following theorem was proved in [10] for all quorum systems.

Theorem 1 *Let Q be a quorum system over a universe of n elements. Denote by $c(Q)$ the size of the smallest quorum of Q. Then $L(Q) \geq \max\{\frac{1}{c(Q)}, \frac{c(Q)}{n}\}$. Consequently, $L(Q) \geq \frac{1}{\sqrt{n}}$.*

Availability
The resilience f of a quorum system provides one measure of how many crash failures a quorum system is *guaranteed* to survive.

Resilience: The *resilience f* of a quorum system Q is the largest k such that for every set $K \subseteq U$, $|K| = k$, there exists $Q \in Q$ such that $K \cap Q = \emptyset$.

Note that, the resilience f is at most $c(Q) - 1$, since by disabling the members of the smallest quorum every quorum is hit. It is possible, however, that an f-resilient quorum system, though vulnerable to a few failure configurations of $f + 1$ failures, can survive many configurations of more than f failures. One way to measure this property of a quorum system is to assume that each server crashes independently with probability p and then to determine the probability F_p that no quorum remains completely alive. This is known as *failure probability* and is formally defined as follows:

Failure probability: Assume that each server in the system crashes independently with probability p. For every quorum $Q \in Q$ let $\mathcal{E}_Q$ be the event that Q is *hit*, i.e., at least one element $i \in Q$ has crashed. Let crash (Q) be the event that all the quorums $Q \in Q$ were hit, i.e.,

crash $(Q) = \bigwedge_{Q \in Q} \mathcal{E}_Q$. Then the system failure probability is $F_p(Q) = \Pr(\text{crash}(Q))$.

Peleg and Wool study the availability of quorum systems in [11]. A good failure probability $F_p(Q)$ for a quorum system Q has $\lim_{n \to \infty} F_p(Q) = 0$ when $p < \frac{1}{2}$. Note that, the failure probability of any quorum system whose resilience is f is at least $e^{-\Omega(f)}$. Majorities has the best availability when $p < \frac{1}{2}$; for $p = \frac{1}{2}$, there exist quorum constructions with $F_p(Q) = \frac{1}{2}$; for $p > \frac{1}{2}$, the singleton has the best failure probability $F_p(Q) = p$, but for most quorum systems, $F_p(Q)$ tends to 1.

The Load and Availability of Quorum Systems
Quorum constructions can be compared by analyzing their behavior according to the above measures. The singleton has a load of 1, resilience 0, and failure probability $F_p = p$. This system has the best failure probability when $p > \frac{1}{2}$, but otherwise performs poorly in both availability and load.

The system of Majorities has a load of $\lceil \frac{n+1}{2n} \rceil \approx \frac{1}{2}$. It is resilient to $\lfloor \frac{n-1}{2} \rfloor$ failures, and its failure probability is $e^{-\Omega(n)}$. This system has the highest possible resilience and asymptotically optimal failure probability, but poor load.

Grid's load is $O(\frac{1}{\sqrt{n}})$, which is within a constant factor from optimal. However, its resilience is only $\sqrt{n} - 1$ and it has poor failure probability which tends to 1 as n grows.

The resilience of a FPP quorum system is $q \approx \sqrt{n}$. The load of FPP was analyzed in [10] and shown to be $L(\text{FPP}) = \frac{q+1}{n} \approx 1/\sqrt{n}$, which is optimal. However, its failure probability tends to 1 as n grows.

As demonstrated by these systems, there is a tradeoff between load and fault tolerance in quorum systems, where the resilience f of a quorum system Q satisfies $f \leq nL(Q)$. Thus, improving one must come at the expense of the other, and it is in fact impossible to simultaneously achieve both optimally. One might conclude that good load conflicts with low failure probability, which is not necessarily the case. In fact, there exist quorum systems such as the

Paths system of Naor and Wool [10] and the Triangle Lattice of Bazzi [1] that achieve asymptotically optimal load of $O(1/\sqrt{n})$ and have close to optimal failure probability for their quorum sizes. Another construction is the CWlog system of Peleg and Wool [12], which has unusually small quorum sizes of $\log n - \log \log n$, and for systems with quorums of this size, has optimal load, $L(\text{CWlog}) = O(1/\log n)$, and optimal failure probability.

Byzantine Quorum Systems

For the most part, quorum systems were studied in environments where failures may simply cause servers to become unavailable (benign failures). But what if a server may exhibit arbitrary, possibly malicious behavior? Malkhi and Reiter [7] carried out a study of quorum systems in environments prone to arbitrary (Byzantine) behavior of servers. Intuitively, a quorum system tolerant of Byzantine failures is a collection of subsets of servers, each pair of which intersect in a set containing sufficiently many *correct* servers to mask out the behavior of faulty servers. More precisely, Byzantine quorum systems are defined as follows:

Masking quorum system

A quorum system $\mathcal{Q}$ is a *b-masking quorum system* if it has resilience $f \geq b$, and each pair of quorums intersect in at least $2b + 1$ elements.

The masking quorum system requirements enable a client to obtain the correct answer from the service despite up to b Byzantine server failures. More precisely, a write operation remains as before; to obtain the correct value of x from a read operation, the client reads a set of value/timestamp pairs from a quorum Q and sorts them into clusters of identical pairs. It then chooses a value/timestamp pair that is returned from at least $b + 1$ servers, and therefore must contain at least one correct server. The properties of masking quorum systems guarantee that at least one such cluster exists. If more than one such cluster exists, the client chooses the one with the highest timestamp. It is easy to see that any value so obtained was written before, and moreover, that the most recently

written value is obtained. Thus, the semantics of a multi-writer multi-reader safe variable are obtained (see ▸ Linearizability) in a Byzantine environment.

For a b-masking quorum system, the following lower bound on the load holds:

Theorem 2 *Let $\mathcal{Q}$ be a b-masking quorum system. Then $L(\mathcal{Q}) \geq \max\{\frac{2b+1}{c(\mathcal{Q})}, \frac{c(\mathcal{Q})}{n}\}$, and consequently $L(\mathcal{Q}) \geq \sqrt{\frac{2b+1}{n}}$.*

This bound is tight, and masking quorum constructions meeting it were shown.

Malkhi and Reiter explore in [7] two variations of masking quorum systems. The first, called *dissemination quorum systems*, is suited for services that receive and distribute *self-verifying* information from correct clients (e.g., digitally signed values) that faulty servers can fail to redistribute but cannot undetectably alter. The second variation, called *opaque masking quorum systems*, is similar to regular masking quorums in that it makes no assumption of self-verifying data, but it differs in that clients do not need to know the failure scenarios for which the service was designed. This somewhat simplifies the client protocol and, in the case that the failures are maliciously induced, reveals less information to clients that could guide an attack attempting to compromise the system. It is also shown in [7] how to deal with faulty clients in addition to faulty servers.

Probabilistic Quorum Systems

The resilience of any quorum system is bounded by half of the number of servers. Moreover, as mentioned above, there is an inherent tradeoff between low load and good resilience, so that it is in fact impossible to simultaneously achieve both optimally. In particular, quorum systems over n servers that achieve the optimal load of $\frac{1}{\sqrt{n}}$ can tolerate at most $\sqrt{n}$ faults.

To break these limitations, Malkhi et al. propose in [8] to relax the intersection property of a quorum system so that "quorums" chosen according to a specified strategy intersect only with very high probability. They accordingly name these *probabilistic quorum systems*. These systems admit the possibility, albeit small, that two operations will be performed at non-intersecting quorums, in which case consistency of the system may suffer. However, even a small relaxation of consistency can yield dramatic improvements in the resilience and failure probability of the system, while the load remains essentially unchanged. Probabilistic quorum systems are thus most suitable for use when availability of operations despite the presence of faults is more important than certain consistency. This might be the case if the cost of inconsistent operations is high but not irrecoverable, or if obtaining the most up-to-date information is desirable but not critical, while having no information may have heavier penalties.

The family of constructions suggested in [8] is as follows:

$W(n, \ell)$ Let U be a universe of size n. $W(n, \ell)$, $\ell \geq 1$, is the system $\langle Q, w \rangle$ where Q is the set system $Q = \{Q \subseteq U : |Q| = \ell\sqrt{n}\}$; w is an access strategy w defined by $\forall Q \in Q$, $w(Q) = \frac{1}{|Q|}$.

The probability of choosing according to w two quorums that do not intersect is less than $e^{-\ell^2}$, and can be made sufficiently small by appropriate choice of ℓ. Since every element is in $\binom{n-1}{\ell\sqrt{n}-1}$ quorums, the load $L(W(n, \ell))$ is $\frac{\ell}{\sqrt{n}} = O(\frac{1}{\sqrt{n}})$. Because only $\ell\sqrt{n}$ servers need be available in order for some quorum to be available, $W(n, \ell)$ is resilient to $n - \ell\sqrt{n}$ crashes. The failure probability of $W(n, \ell)$ is less than $e^{-\Omega(n)}$ for all $p \leq 1 - \frac{\ell}{\sqrt{n}}$, which is asymptotically optimal. Moreover, if $\frac{1}{2} \leq p \leq 1 - \frac{\ell}{\sqrt{n}}$, this probability is provably better than any (non-probabilistic) quorum system.

Relaxing consistency can also provide dramatic improvements in environments that may experience Byzantine failures. More details can be found in [8].

Applications

Just about any fault tolerant distributed protocol, such as Paxos [5] or consensus [1] implicitly builds on quorums, typically majorities. More concretely, scalable data repositories were built, such as Fleet [9], Rambo [4], and Rosebud [13].

Cross-References

▶ Concurrent Programming, Mutual Exclusion

Recommended Reading

1. Dwork C, Lynch N, Stockmeyer L (1988) Consensus in the presence of partial synchrony. J Assoc Comput Mach 35:288–323
2. Garcia-Molina H, Barbara D (1985) How to assign votes in a distributed system. J ACM 32:841–860
3. Gifford DK (1979) Weighted voting for replicated data. In: Proceedings of the 7th ACM symposium on operating systems principles, pp 150–162
4. Gilbert S, Lynch N, Shvartsman A (2003) Rambo ii: rapidly reconfigurable atomic memory for dynamic networks. In: Proceedings if the IEEE 2003 international conference on dependable systems and networks (DNS), San Francisco, pp 259–268
5. Lamport L (1998) The part-time parliament. ACM Trans Comput Syst 16:133–169
6. Maekawa M (1985) A $\sqrt{n}$ algorithm for mutual exclusion in decentralized systems. ACM Trans Comput Syst 3(2):145–159
7. Malkhi D, Reiter M (1998) Byzantine quorum systems. Distr Comput 11:203–213
8. Malkhi D, Reiter M, Wool A, Wright R (2001) Probabilistic quorum systems. Inf Comput J 170:184–206
9. Malkhi D, Reiter MK (2000) An architecture for survivable coordination in large-scale systems. IEEE Trans Knowl Data Eng 12:187–202
10. Naor M, Wool A (1998) The load, capacity and availability of quorum systems. SIAM J Comput 27:423–447
11. Peleg D, Wool A (1995) The availability of quorum systems. Inf Comput 123:210–223
12. Peleg D, Wool A (1997) Crumbling walls: a class of practical and efficient quorumsystems. Distrib Comput 10:87–98
13. Rodrigues R, Liskov B (2003) Rosebud: a scalable Byzantine-fault tolerant storage architecture. In: Proceedings of the 18th ACM symposium on operating system principles, San Francisco
14. Thomas RH (1979) A majority consensus approach to concurrency control for multiple copy databases. ACM Trans Database Syst 4:180–209

R

Radiocoloring in Planar Graphs

Vicky Papadopoulou
Department of Computer Science, University of
Cyprus, Nicosia, Cyprus

Keywords

λ-coloring; k-coloring; Coloring the square of the
graph; Distance-2 coloring

Years and Authors of Summarized Original Work

2005; Fotakis, Nikoletseas, Papadopoulou,
Spirakis

Problem Definition

Consider a graph $G(V, E)$. For any two vertices
$u, v \in V$, $d(u, v)$ denotes the distance of u, v in
G. The general problem concerns a coloring of
the graph G and it is defined as follows:

Definition 1 (k-coloring problem)
INPUT: A graph $G(V, E)$.
OUTPUT: A function $\phi : V \to \{1, \ldots, \infty\}$,
called k -coloring of G such that $\forall u, v \in V$,
$x \in \{0, 1, \ldots, k\}$: if $d(u, v) \geq k - x + 1$ then
$|\phi(u) - \phi(v)| = x$.

OBJECTIVE: Let $|\phi(V)| = \lambda_\phi$. Then λ_ϕ is the
number of colors that φ actually uses (it is
usually called *order* of G under φ). The number
$\nu_\phi = max_{v \in V}\phi(v) - min_{u \in V}\phi(u) + 1$ is usually
called the *span* of G under φ. The function φ
satisfies one of the following objectives:

- minimum span: λ_ϕ is the minimum possible
  over all possible functions φ of G;
- minimum order: ν_ϕ is the minimum possible
  over all possible functions φ of G;
- Min span order: obtains a minimum span
  and moreover, from all minimum span assignments, φ obtains a minimum order.
- Min order span: obtains a minimum order and
  moreover, from all minimum order assignments, φ obtains a minimum span.

Note that the case $k = 1$ corresponds to the well
known problem of *vertex graph coloring*. Thus,
k-coloring problem (with k as an input) is $\mathcal{NP}$-
complete [4]. The case of k-coloring problem
where $k = 2$, is called the *Radiocoloring problem*.

Definition 2 (Radiocoloring Problem (RCP) [7])
INPUT: A graph $G(V, E)$.
OUTPUT: A function $\Phi : V \to N^*$ such that
$|\Phi(u) - \Phi(v)| \geq 2$ if $d(u, v) = 1$ and
$|\Phi(u) - \Phi(v)| \geq 1$ if $d(u, v) = 2$.
OBJECTIVE: The least possible number
(order) needed to radiocolor G is denoted

© Springer Science+Business Media New York 2016
M.-Y. Kao (ed.), *Encyclopedia of Algorithms*,
DOI 10.1007/978-1-4939-2864-4

by $X_{\text{order}}(G)$. The least possible number $\max_{v \in V} \Phi(v) - \min_{u \in V} \Phi(u) + 1$ (span) needed for the radiocoloring of G is denoted as $X_{\text{span}}(G)$. Function Φ satisfies one of the followings:

- Min span RCP: Φ obtains a minimum span, i.e., $\lambda_\Phi = X_{span}(G)$;
- Min order RCP: Φ obtains a minimum order $\nu_\Phi = X_{\text{order}}(G)$;
- Min span order RCP: obtains a minimum span and moreover, from all minimum span assignments, Φ obtains a minimum order.
- Min order span RCP: obtains a minimum order and moreover, from all minimum order assignments, Φ obtains a minimum span.

A related to the RCP problem concerns to the square of a graph G, which is defined as follows:

Definition 3 Given a graph $G(V, E)$, G^2 is the graph having the same vertex set V and an edge set $E' : \{u, v\} \in E'$ iff $d(u, v) \leq 2$ in G.

The related problem is to color the square of a graph G, G^2 so that no two neighbor vertices (in G^2) get the same color. The objective is to use a minimum number of colors, denoted as $\chi(G^2)$ and called *chromatic number of the square of the graph G*. Fotakis et al. [5, 6] first observed that for any graph G, $X_{\text{order}}(G)$ is the same as the (vertex) chromatic number of G^2, i.e., $X_{\text{order}}(G) = \chi(G^2)$.

Key Results

Fotakis et al. [5, 6] studied *min span order, min order* and *min span* RCP in *planar* graph G. A planar graph, is a graph for which its edges can be embedded in the plane without crossings. The following results are obtained:

- It is first shown that the number of colors used in the *min span order RCP* of graph G is different from the chromatic number of the square of the graph, $\chi(G^2)$. In particular, it may be greater than $\chi(G^2)$.

- It is then proved that the radiocoloring problem for general graphs is hard to approximate (unless $\mathcal{NP} = ZPP$, the class of problems with polynomial time zero-error randomized algorithms) within a factor of $n^{1/2-\epsilon}$ (for any $\epsilon > 0$), where n is the number of vertices of the graph. However, when restricted to some special cases of graphs, the problem becomes easier.

  It is shown that the *min span RCP* and *min span order RCP* are $\mathcal{NP}$-complete for planar graphs. Note that few combinatorial problems remain hard for *planar* graphs and their proofs of hardness are not easy since they have to use planar gadgets which are difficult to find and understand.

- It presents a $O(n\Delta(G))$ time algorithm that *approximates* the min order of RCP, X_{order}, of a planar graph G *by a constant ratio which tends to 2 as the maximum degree $\Delta(G)$ of G increases.*

  The algorithm presented is motivated by a constructive coloring theorem of Heuvel and McGuiness [9]. The construction of [9] can lead (as shown) to an $O(n^2)$ technique assuming that a planar embedding of G is given. Fotakis et al. [5, 6] improves the time complexity of the approximation, and presents a much more simple algorithm to verify and implement. The algorithm does not need any planar embedding as input.

- Finally, the work considers the problem of *estimating the number of different radiocolorings* of a planar graph G. This is a #$\mathcal{P}$-complete problem (as can be easily seen from the completeness reduction presented there that can be done parsimonious). They authors employ here standard techniques of rapidly mixing Markov Chains and the *new method of coupling* for purposes of proving *rapid convergence* (see e.g., [10]) and present *a fully polynomial randomized approximation scheme* for estimating the number of radiocolorings with λ colors for a planar graph G, when $\lambda \geq 4\Delta(G) + 50$.

In [8] and [7] it has been proved that the problem of min span RCP is $\mathcal{NP}$-complete, even for graphs of diameter 2. The reductions use highly non-planar graphs. In [11] it is proved that the problem of coloring the square of a general graph is $\mathcal{NP}$-complete.

Another variation of RCP for planar graphs, called *distance-2-coloring* is studied in [12]. This is the problem of coloring a given graph G with the minimum number of colors so that the vertices of distance *at most* two get different colors. Note that this problem is equivalent to coloring the square of the graph G, G^2. In [12] it is proved that the distance-2-coloring problem for planar graphs is $\mathcal{NP}$-complete. As it is shown in [5, 6], this problem is different from the min span order RCP. Thus, the $\mathcal{NP}$-completeness proof in [12] certainly does not imply the $\mathcal{NP}$-completeness of min span order RCP proved in [5, 6]. In [12] a 9-approximation algorithm for the distance-2-coloring of planar graphs is also provided.

Independently and in parallel, Agnarsson and Halldórsson in [1] presented approximations for the chromatic number of square and power graphs (G^k). In particular they presented an 1.8-approximation algorithm for coloring the square of a planar graph of large degree ($\Delta(G) \geq 749$). Their method utilizes the notion of *inductiveness* of the square of a planar graph.

Bodlaender et al. in [2] proved also independently and and in parallel that the min span RCP, called λ-labeling there, is $\mathcal{NP}$-complete for planar graphs, using a similar to the approach used in [5, 6]. In the same work the authors presented approximations for the problem for some interesting families of graphs: outerplanar graphs, graphs of bounded treewidth, permutation and split graphs.

Applications

The Frequency Assignment Problem (FAP) in radio networks is a well-studied, interesting problem, aiming at assigning frequencies to transmitters exploiting frequency reuse while keeping signal interference to acceptable levels. The interference between transmitters are modeled by an interference graph $G(V, E)$, where V ($|V| = n$) corresponds to the set of transmitters and E represents distance constraints (e.g., if two neighbor nodes in G get the same or close frequencies then this causes unacceptable levels of interference). In most real life cases the network topology formed has some special properties, e.g., G is a lattice network or a planar graph. Planar graphs are mainly the object of study in [5, 6].

The FAP is usually modeled by variations of the graph coloring problem. The set of colors represents the available frequencies. In addition, each color in a particular assignment gets an integer value which has to satisfy certain inequalities compared to the values of colors of nearby nodes in G (frequency-distance constraints). A discrete version of FAP is the k-coloring problem, of which a particular instance, for $k = 2$, is investigated in [5, 6].

Real networks reserve bandwidth (range of frequencies) rather than distinct frequencies. In this case, an assignment seeks to use as small range of frequencies as possible. It is sometimes desirable to use as few distinct frequencies of a given bandwidth (span) as possible, since the unused frequencies are available for other use. However, there are cases where the primary objective is to minimize the number of frequencies used and the span is a secondary objective, since we wish to avoid reserving unnecessary large span. These realistic scenaria directed researchers to consider optimization versions of the RCP, where one aims in minimizing the span (bandwidth) or the order (distinct frequencies used) of the assignment. Such optimization problems are investigated in [5, 6].

Cross-References

▶ Channel Assignment and Routing in Multiradio Wireless Mesh Networks
▶ Graph Coloring

Recommended Reading

1. Agnarsson G, Halldórsson MM (2000) Coloring powers of planar graphs. In: Proceedings of the 11th annual ACM-SIAM symposium on discrete algorithms, pp 654–662

2. Bodlaender HL, Kloks T, Tan RB, van Leeuwen J (2000) Approximations for λ-coloring of graphs. In: Proceedings of the 17th annual symposium on theoretical aspects of computer science. Lecture notes in computer science, vol 1770. Springer, pp 395–406

3. Hale WK (1980) Frequency assignment: theory and applications. Proc IEEE 68(12):1497–1514

4. Garey MR, Johnson DS (1979) Computers and intractability: a guide to the theory of NP-completeness. W.H. Freeman

5. Fotakis D, Nikoletseas S, Papadopoulou V, Spirakis P (2000) NP completeness results and efficient approximations for radiocoloring in planar graphs. In: Proceedings of the 25th international symposium on mathematical foundations of computer science. Lecture notes of computer science, vol 1893. Springer, pp 363–372

6. Fotakis D, Nikoletseas S, Papadopoulou VG, Spirakis PG (2005) Radiocoloring in planar graphs: complexity and approximations. Theor Comput Sci Elsevier 340:514–538

7. Fotakis D, Pantziou G, Pentaris G, Spirakis P (1999) Frequency assignment in mobile and radio networks. In: Networks in distributed computing. DIMACS series in discrete mathematics and theoretical computer science, vol 45, pp 73–90

8. Griggs J, Liu D (1998) Minimum span channel assignments. In: Recent advances in radio channel assignments. Invited Minisymposium, Discrete Mathematics

9. van d Heuvel J, McGuiness S (1999) Colouring the square of a planar graph. CDAM research report series, July 1999

10. Jerrum M (1994) A very simple algorithm for estimating the number of k-colourings of a low degree graph. Random Struct Algorithm 7:157–165

11. Lin YL, Skiena S (1995) Algorithms for square roots of graphs. SIAM J Discret Math 8:99–118

12. Ramanathan S, Loyd ER (1992) The complexity of distance 2- coloring. In: Proceedings of the 4th international conference of computing and information, pp 71–74

Random Planted 3-SAT

Abraham Flaxman
Theory Group, Microsoft Research, Redmond, WA, USA

Keywords

Constraint satisfaction

Years and Authors of Summarized Original Work

2003; Flaxman

Problem Definition

This classic problem in complexity theory is concerned with efficiently finding a satisfying assignment to a propositional formula. The input is a formula with n Boolean variables which is expressed as an AND of ORs with 3 variables in each OR clause (a *3-CNF formula*). The goal is to (1) find an assignment of variables to TRUE and FALSE so that the formula has value TRUE or (2) prove that no such assignment exists. Historically, recognizing satisfiable 3-CNF formulas was the first "natural" example of an NP-complete problem, and, because it is NP-complete, no polynomial-time algorithm can succeed on all 3-CNF formulas unless P = NP [4, 10]. Because of the numerous practical applications of 3-SAT, and also due to its position as the canonical NP-complete problem, many heuristic algorithms have been developed for solving 3-SAT, and some of these algorithms have been analyzed rigorously on random instances.

Notation

A 3-CNF formula over variables $x_1, x_2, \ldots, x_n$ is the conjunction of m clauses $C_1 \wedge C_2 \wedge \ldots \wedge C_m$, where each clause is the disjunction of 3 literals, $C_i = \ell_{i_1} \vee \ell_{i_2} \vee \ell_{i_3}$, and each literal ℓ_{i_j} is either a variable or the negation of a variable (the negation of the variable x is denoted by $\bar{x}$). A 3-CNF formula is *satisfiable* if and only if there is an assignment of variables to truth values so that every clause contains at least one true literal. Here, all asymptotic analysis is in terms of n, the number of variables in the 3-CNF formula, and a sequence of events $\{\mathcal{E}_n\}$ is said to hold *with high probability* (abbreviated **whp**) if $\lim_{n \to \infty} \Pr[\mathcal{E}_n] = 1$.

Distributions

There are many distributions over 3-CNF formulas which are interesting to consider, and this chapter focuses on dense satisfiable instances. Dense satisfiable instances can be formed by conditioning on the event $\{I_{n,m}$ is satisfiable$\}$, but this conditional distribution is difficult to sample from and to analyze. This has led to research in "planted" random instances of 3-SAT, which are formed by first choosing a truth assignment φ uniformly at random and then selecting each clause independently from the triples of literals where at least one literal is set to TRUE by the assignment φ. The clauses can be included with equal probabilities in analogy to the $\mathbb{I}_{n,p}$ or $\mathbb{I}_{n,m}$ distributions above [8, 9], or different probabilities can be assigned to the clauses with one, two, or three literals set to TRUE by φ, in an effort to better hide the satisfying assignment [2,7].

Problem 1 (3-SAT)

INPUT: *3-CNF Boolean formula* $F = C_1 \wedge C_2 \wedge \cdots \wedge C_m$, *where each clause* C_i *is of the form* $C_i = \ell_{i_1} \vee \ell_{i_2} \vee \ell_{i_3}$ *and each literal* ℓ_{i_j} *is either a variable or the negation of a variable.*

OUTPUT: *A truth assignment of variables to Boolean values which makes at least one literal in each clause TRUE or a certificate that no such assignment exists.*

Key Results

A line of basic research dedicated to identifying hard search and decision problems, as well as the potential cryptographic applications of planted instances of 3-SAT, has motivated the development of algorithms for 3-SAT which are known to work on planted random instances.

Majority Vote Heuristic: If every clause consistent with the planted assignment is included with the same probability, then there is a bias towards including the literal satisfied by the planted assignment more frequently than its negation. This is the motivation behind the majority vote heuristic, which assigns each variable to the truth value which will satisfy the majority of the clauses in which it appears. Despite its simplicity, this heuristic has been proven successful **whp** for sufficiently dense planted instances [8].

Theorem 1 *When c is a sufficiently large constant and* $I \sim \mathbb{I}_{n,cn \log n}^{\phi}$, **whp** *the majority vote heuristic finds the planted assignment* φ.

When the density of the planted random instance is lower than $c \log n$, *then the majority vote heuristic will fail, and if the relative probability of the clauses satisfied by one, two, and three literals is adjusted appropriately, then it will fail miserably. But there are alternative approaches.*

For planted instances where the density is a sufficiently large constant, the majority vote heuristic provides a good starting assignment, and then the k-OPT heuristic can finish the job. The k-OPT heuristic of [6] is defined as follows: Initialize the assignment by majority vote. Initialize k to 1. While there exists a set of k variables for which flipping the values of the assignment will (1) make false clauses true and (2) will not make true clauses false, flip the values of the assignment on these variables. If this reaches a local optimum that is not a satisfying assignment, increase k and continue.

Theorem 2 *When c is a sufficiently large constant and* $I \sim I_{n,cn}^{\phi}$, *the k-OPT heuristic finds a satisfying assignment in polynomial time* **whp**. *The same is true even in the semi-random case, where an adversary is allowed to add clauses to I that have all three literals set to TRUE by* φ *before giving the instance to the k-OPT heuristic.*

A related algorithm has been shown to run in expected polynomial time in [9], and a rigorous analysis of *warning propagation (WP)*, a message passing algorithm related to survey propagation, has shown that WP is successful **whp** on planted satisfying assignments, provided that the clause density exceeds a sufficiently large constant [5].

When the relative probabilities of clauses containing one, two, and three literals are adjusted carefully, it is possible to make the majority vote assignment very different from the planted assignment. A way of setting these relative probabilities that is predicted to be difficult is discussed in [2]. If the density of these instances is high enough (and the relative probabilities are anything besides the case of "Gaussian elimination with noise"), then a spectral heuristic provides a starting assignment close to the planted assignment and local reassignment operations are sufficient to recover a satisfying assignment [7].

More formally, consider instance $I = I_{n,p1,p2,p3}$, formed by choosing a truth assignment φ on n variables uniformly at random and including in I each clause with exactly i literals satisfied by φ independently with probability p_i. By setting $p_1 = p_2 = p_3$, this reduces to the distribution mentioned above.

Setting $p_1 = p_2$ and $p_3 = 0$ yields a natural distribution on 3CNFs with a planted not-all-equal assignment, a situation where the greedy variable assignment rule generates a random assignment. Setting $p_2 = p_3 = 0$ gives 3CNFs with a planted exactly-one-true assignment (which succumb to the greedy algorithm followed by the nonspectral steps below). Also, correctly adjusting the ratios of p_1, p_2, and p_3 can obtain a variety of (slightly less natural) instance distributions which thwart the greedy algorithm. Carefully selected values of p_1, p_2, and p_3 are considered in [2], where it is conjectured that no algorithm running in polynomial time can solve $I_{n,p1,p2,p3}$ **whp** when $p_i = c_i \alpha/n^2$ and

$$0.007 < c_3 < 0.25 \qquad c_2 = (1 - 4c_3)/6$$

$$c_1 = (1 + 2c_3)/6 \quad \alpha > \frac{4.25}{7}.$$

The *spectral heuristic* modeled after the coloring algorithms of [1, 3] was developed for such planted distributions in [7]. This polynomial time algorithm which returns a satisfying assignment to $I_{n,p1,p2,p3}$ **whp** when $p_1 = d/n^2$, $p_2 = \eta_2 d/n^2$, and $p_3 = \eta_3 d/n^2$, for $0 \leq \eta_2, \eta_3 \leq 1$, and $d \geq d_{\min}$, where $d_{\min}$ is a function of η_2, η_3. The algorithm is structured as follows:

1. Construct a graph G from the 3CNF.
2. Find the most negative eigenvalue of a matrix related to the adjacency matrix of G.

3. Assign a value to each variable based on the signs of the eigenvector corresponding to the most negative eigenvalue.
4. Iteratively improve the assignment.
5. Perfect the assignment by exhaustive search over a small set containing all the incorrect variables.

A more elaborate description of each step is the following:

Step (1): Given 3CNF $I = I_{n,p1,p2,p3}$, where $p_1 = d/n^2$, $p_2 = \eta_2 d/n^2$, and $p_3 = \eta_3 d/n^2$, the graph in step (1), $G = (V, E)$, has $2n$ vertices, corresponding to the literals in I, and labeled $\{x_1, \bar{x}_1, \ldots x_n, \bar{x}_n\}$. G has an edge between vertices ℓ_i and ℓ_j if I includes a clause with both ℓ_i and ℓ_j (and G does not have multiple edges).

Step (2): Consider $G' = (V, E')$, formed by deleting all the edges incident to vertices with degree greater than $180d$. Let A be the adjacency matrix of G'. Let λ be the most negative eigenvalue of A and $\mathbf{v}$ be the corresponding eigenvector.

Step (3): There are two assignments to consider, π_+, which is defined by

$$\pi_+(x_i) = \begin{cases} T, & \text{if } \mathbf{v}_i \geq 0; \\ F, & \text{otherwise}; \end{cases}$$

and π_-, which is defined by

$$\pi_-(x) = \neg \pi_+(x).$$

Let π_0 be the better of π_+ and π_- (i.e., the assignment which satisfies more clauses). It can be shown that π_0 agrees with φ on at least $(1 - C/d)n$ variables for some absolute constant C.

Step (4): For $i = 1, \ldots, \log n$, do the following: for each variable x, if x appears in $5\varepsilon d$ clauses unsatisfied by π_{i-1}, then set $\pi_i(x) = \neg \pi_{i-1}(x)$, where ε is an appropriately chosen constant (taking $\varepsilon = 0.1$ works); otherwise set $\pi_i(x) = \pi_{i-1}(x)$.

Step (5): Let $\pi_0' = \pi_{\log n}$ denote the final assignment generated in step (4). Let $\mathcal{A}_4^{\pi_0'}$ be the set of variables which do not appear in $(3 \pm 4\varepsilon)d$ clauses as the only true literal with respect to assignment π_0', and let $\mathcal{B}$ be the set of variables which do not appear in $(\mu_D \pm \varepsilon)d$ clauses, where $\mu_D d = (3+6)d + (6+3)\eta_2 d + 3\eta_3 d + \mathcal{O}(1/n)$ is the expected number of clauses containing variable x. Form partial assignment π_1' by unassigning all variables in $\mathcal{A}_4^{\pi_0'}$ and $\mathcal{B}$. Now, for $i \geq 1$, if there is a variable x_i which appears in less than $(\mu_D - 2\varepsilon)d$ clauses consisting of variables that are all assigned by π_i', then let π_{i+1}' be the partial assignment formed by unassigning x_i in π_i'. Let π' be the partial assignment when this process terminates. Consider the graph Γ with a vertex for each variable that is unassigned in π' and an edge between two variables if they appear in a clause together. If any connected component in Γ is larger than $\log n$, then fail. Otherwise, find a satisfying assignment for I by performing an exhaustive search on the variables in each connected component of Γ.

Theorem 3 *For any constants* $0 \leq \eta_2, \eta_3 \leq 1$, *except* $(\eta_2, \eta_3) = (0,1)$, *there exists a constant* $d_{\min}$ *such that for any* $d \geq d_{\min}$, *if* $p_1 = d/n^2$, $p_2 = \eta_2 d/n^2$, *and* $p_3 = \eta_3 d/n^2$, *then this polynomial-time algorithm produces a satisfying assignment for random instances drawn from* $I_{n,p1,p2,p3}$ **whp**.

Applications

3-SAT is a universal problem, and due to its simplicity, it has potential applications in many areas, including proof theory and program checking, planning, cryptanalysis, machine learning, and modeling biological networks.

Open Problems

An important direction is to develop alternative models of random distributions which more accurately reflect the type of instances that occur in the real world.

Data Sets

Sample instances of satisfiability and 3-SAT are available on the web at http://www.satlib.org/.

URL to Code

Solvers and information on the annual satisfiability solving competition are available on the web at http://www.satlive.org/.

Recommended Reading

1. Alon N, Kahale N (1997) A spectral technique for coloring random 3-colorable graphs. SIAM J Comput 26(6):1733–1748
2. Barthel W, Hartmann AK, Leone M, Ricci-Tersenghi F, Weigt M, Zecchina R (2002) Hiding solutions in random satisfiability problems: a statistical mechanics approach. Phys Rev Lett 88:188701
3. Chen H, Frieze AM (1996) Coloring bipartite hypergraphs. In: Cunningham HC, McCormick ST, Queyranne M (eds) Integer programming and combinatorial optimization, 5th international IPCO conference, Vancouver, 3–5 June 1996. Lecture notes in computer science, vol 1084. Springer, pp 345–358
4. Cook S (1971) The complexity of theorem-proving procedures. In: Proceedings of the 3rd annual symposium on theory of computing, Shaker Heights, 3–5 May, pp 151–158
5. Feige U, Mossel E, Vilenchik D (2006) Complete convergence of message passing algorithms for some satisfiability problems. In: Díaz J, Jansen K, Rolim JDP, Zwick U (eds) Approximation, randomization, and combinatorial optimization. Algorithms and techniques, 9th international workshop on approximation algorithms for combinatorial optimization problems, APPROX 2006 and 10th International Workshop on Randomization and Computation, RANDOM 2006, Barcelona, 28–30 Aug 2006. Lecture notes in computer science, vol 4110. Springer, pp 339–350
6. Feige U, Vilenchik D (2004) A local search algorithm for 3-SAT. Technical report, The Weizmann Institute, Rehovat
7. Flaxman AD (2003) A spectral technique for random satisfiable 3CNF formulas. In: Proceedings of the fourteenth annual ACM-SIAM symposium on discrete algorithms, Baltimore. ACM, New York, pp 357–363
8. Koutsoupias E, Papadimitriou CH (1992) On the greedy algorithm for satisfiability. Inf Process Lett 43(1):53–55

R

9. Krivelevich M, Vilenchik D (2006) Solving random satisfiable 3CNF formulas in expected polynomial time. In: Proceedings of the 17th annual ACM-SIAM symposium on discrete algorithm (SODA '06), Miami. ACM
10. Levin LA (1973) Universal enumeration problems. Probl Pereda Inf 9(3):115–116

Randomization in Distributed Computing

Tushar Deepak Chandra
IBM Watson Research Center, Yorktown Heights, NY, USA

Keywords

Agreement; Byzantine agreement

Years and Authors of Summarized Original Work

1996; Chandra

Problem Definition

This problem is concerned with using the multi-writer multi-reader register primitive in the shared memory model to design a fast, wait-free implementation of consensus. Below are detailed descriptions of each of these terms.

Consensus Problems

There are n processors and the goal is to design distributed algorithms to solve the following two consensus problems for these processors.

Problem 1 (Binary consensus)
INPUT: Processor i has input bit b_i.
OUTPUT: Each processor i has output bit b_i' such that: (1) all the output bits b_i' equal the same value v; and (2) $v = b_i$ for some processor i.

Problem 2 (Id consensus)
INPUT: Processor i has a unique id u_i.
OUTPUT: Each processor i has output value u_i' such that: (1) all the output values u_i' equal the same value u; and (2) $u = u_i$ for some processor i.

Wait-Free

This result builds on extensive previous work on the shared memory model of parallel computing. Shared object types include data structures such as read/write registers and synchronization primitives such as "test and set". A shared object is said to be *wait-free* if it ensures that every invocation on the object is guaranteed a response in finite time even if some or all of the other processors in the system crash. In this problem, the existence of wait-free registers is assumed and the goal is to create a fast wait-free algorithm to solve the consensus problem. In the rest of this summary, "wait-free implementations" will be referred to simply as "implementations" i.e., the term wait-free will be omitted.

Multi-writer Multi-reader Register

Many past results on solving consensus in the shared memory model assume the existence of a single writer multi-reader register. For such a register, there is a single writer client and multiple reader clients. Unfortunately, it is easy to show that the per processor step complexity of any implementation of consensus from single writer multi-reader registers will be at least linear in the number of processors. Thus, to achieve a time efficient implementation of consensus, the more powerful primitive of a multi-writer multi-reader register must be assumed. A multi-writer multi-reader register assumes the clients of the register are multiple writers and multiple readers. It is well known that it is possible to implement such a register in the shared memory model.

The Adversary

Solving the above problems is complicated by the fact that the programmer has little control over the rate at which individual processors execute. To model this fact, it is assumed that the schedule at which processors run is picked by an adversary.

It is well-known that there is no deterministic algorithm that can solve either Binary consensus or ID consensus in this adversarial model if the number of processors is greater than 1 [6, 7]. Thus, researchers have turned to the use of randomized algorithms to solve this problem [1]. These algorithms have access to random coin flips. Three types of adversaries are considered for randomized algorithms. The *strong adversary* is assumed to know the outcome of a coin flip immediately after the coin is flipped and to be able to modify its schedule accordingly. The *oblivious adversary* has to fix the schedule before any of the coins are flipped. The *intermediate adversary* is not permitted to see the outcome of a coin flip until some process makes a choice based on that coin flip. In particular, a process can flip a coin and write the result in a global register, but the intermediate adversary does not know the outcome of the coin flip until some process reads the value written in the register.

Key Results

Theorem 1 *Assuming the existence of multi-writer multi-reader registers, there exists a randomized algorithm to solve binary consensus against an intermediate adversary with $O(1)$ expected steps per processor.*

Theorem 2 *Assuming the existence of multi-writer multi-reader registers, there exists a randomized algorithm to solve id-consensus against an intermediate adversary with $O(\log^2 n)$ expected steps per processor.*

Both of these results assume that every processor has a unique identifier. Prior to this result, the fastest known randomized algorithm for binary consensus made use of single writer multiple reader registers, was robust against a strong adversary, and required $O(n \log^2 n)$ steps per processor [2]. Thus, the above improvements are obtained at the cost of weakening the adversary and strengthening the system model when compared to [2].

Applications

Binary consensus is one of the most fundamental problems in distributed computing. An example of its importance is the following result shown by Herlihy [8]: If an abstract data type X together with shared memory is powerful enough to implement wait-free consensus, then X together with shared memory is powerful enough to implement in a wait-free manner any other data structure Y. Thus, using this result, a wait-free version of any data structure can be created using only wait-free multi-writer multi-reader registers as a building block.

Binary consensus has practical applications in many areas including: database management, multiprocessor computation, fault diagnosis, and mission-critical systems such as flight control. Lynch contains an extensive discussion of some of these application areas [9].

Open Problems

This result leaves open several problems. First, it leaves open a gap on the number of steps per process required to perform randomized consensus using multi-writer multi-reader registers against the *strong* adversary. A recent result by Attiya and Censor shows an $\Omega(n^2)$ lower bound on the total number of steps for all processors with multi-writer multi-reader registers (implying $\Omega(n)$ steps per process) [3]. They also show a matching upper bound of $O(n^2)$ on the total number of steps. However, closing the gap on the per-process number of steps is still open.

Another open problem is whether there is a randomized implementation of id consensus using multi-reader multi-writer registers that is robust to the intermediate adversary and whose expected number of steps per processor is better than $O(\log^2 n)$. In particular, is a constant run time possible? Aumann in follow up work to this result was able to improve the expected run time per process to $O(\log n)$ [4]. However, to the best of the reviewer's knowledge, there have been no further improvements.

A third open problem is to close the gap on the time required to solve binary consensus against the strong adversary with a single writer multiple reader register. The fastest known randomized algorithm in this scenario requires $O(n \log^2 n)$ steps per processor [2]. A trivial lower bound on the number of steps per processor when single-writer registers are used is $\Omega(n)$. However, to the best of this reviewers knowledge, a $O(\log^2 n)$ gap still remains open.

A final open problem is to close the gap on the total work required to solve consensus with single-reader single-writer registers against an oblivious adversary. Aumann and Kapah-Levy describe algorithms for this scenario that require $O(n \log n \exp(2\sqrt{\ln n \ln(c \log n \log^* n)})$ expected total work for some constant c [5]. In particular, the total work is less than $O(n^{1+\epsilon})$ for any $\epsilon > 0$. A trivial lower bound on total work is $\Omega(n)$, but a gap remains open.

Cross-References

▶ Asynchronous Consensus Impossibility
▶ Atomic Broadcast
▶ Byzantine Agreement
▶ Implementing Shared Registers in Asynchronous Message-Passing Systems
▶ Optimal Probabilistic Synchronous Byzantine Agreement
▶ Registers
▶ Set Agreement
▶ Snapshots in Shared Memory
▶ Wait-Free Synchronization

Recommended Reading

1. Aspnes J (2003) Randomized protocols for asynchronous consensus. Distrib Comput 16(2–3):165–175
2. Aspnes J, Waarts O (1992) Randomized consensus in expected $o(n \log^2 n)$ operations per processor. In: Proceedings of the 33rd symposium on foundations of computer science. IEEE Computer Society, Pittsburgh, 24–26 Oct 1992. pp 137–146
3. Attiya H, Censor K (2007) Tight bounds for asynchronous randomized consensus. In: Proceedings of the symposium on the theory of computation. ACM special interest group on algorithms and computation theory (SIGACT), San Diego, 11–13 June 2007
4. Aumann Y (1997) Efficient asynchronous consensus with the weak adversary scheduler. In: Symposium on principles of distributed computing (PODC). ACM special interest group on algorithms and computation theory (SIGACT), Santa Barbara, 21–24 Aug 1997, pp 209–218
5. Aumann Y, Kapach-Levy A (1999) Cooperative sharing and asynchronous consensus using single-reader/single-writer registers. In: Proceedings of 10th annual ACM-SIAM symposium of discrete algorithms (SODA). Society for Industrial and Applied Mathematics (SIAM), Baltimore, 17–19 Jan 1999, pp 61–70
6. Dolev D, Dwork C, Stockmeyer L (1987) On the minimal synchronism needed for distributed consensus. J ACM (JACM) 34(1):77–97
7. Fischer MJ, Lynch NA, Paterson M (1983) Impossibility of distributed consensus with one faulty process. In: Proceedings of the 2nd ACM SIGACT-SIGMOD symposium on principles of database system (PODS). Association for Computational Machinery (ACM), Atlante, 21–23 Mar 1983, pp 1–7
8. Herlihy M (1991) Wait-free synchronization. ACM Trans Program Lang Syst 13(1):124–149
9. Lynch N (1996) Distributed algorithms. Morgan Kaufmann, San Mateo

Randomized Broadcasting in Radio Networks

Alon Itai
Technion, Haifa, Israel

Keywords

Ad hoc networks; Multi-hop radio networks

Years and Authors of Summarized Original Work

1992; Reuven Bar-Yehuda, Goldreich, Itai

Problem Definition

This entry investigates deterministic and randomized protocols for achieving broadcast (distributing a message from a source to all other nodes) in arbitrary multi-hop synchronous radio networks.

The model consists of an arbitrary (undirected) network, with processors communicating in synchronous time-slots subject to the following rules. In each time-slot, each processor acts either as a *transmitter* or as a *receiver*. A processor acting as a receiver is said to receive a message in time-slot t if exactly one of its neighbors transmits in that time-slot. The message received is the one transmitted. If more than one neighbor transmits in that time-slot, a *conflict* occurs. In this case, the receiver may either get a message from one of the transmitting neighbors or get no message. It is assumed that conflicts (or "collisions") are not detected, hence a processor cannot distinguish the case in which no neighbor transmits from the case in which two or more of its neighbors transmits during that time-slot. The processors are not required to have IDs nor do they know their neighbors; in particular, the processors do not know the topology of the network.

The only inputs required by the protocol are the number of processors in the network – n, Δ – an a priori known upper bound on the maximum degree in the network, and the error bound – ϵ. (All bounds are a priori known to the algorithm.)

Broadcast is a task initiated by a single processor, called the *source*, transmitting a single *message*. The goal is to have the message reach all processors in the network.

Key Results

The main result is a randomized protocol that achieves broadcast in time which is optimal up to a logarithmic factor. In particular, with probability $1 - \epsilon$, the protocol achieves broadcast within $O((D + \log n/\epsilon) \cdot \log n)$ time-slots.

On the other hand, a linear lower bound on the deterministic time-complexity of broadcast is proved. Namely, any deterministic broadcast protocol requires $\Omega(n)$ time-slots, even if the network has diameter 3, and n is known to all

processors. These two results demonstrate an exponential gap in complexity between randomization and determinism.

Randomized Protocols

The Procedure *Decay*

The basic idea used in the protocol is to resolve potential conflicts by randomly eliminating half of the transmitters. This process of "cutting by half" is repeated each time-slot with the hope that there will exist a time-slot with a single active transmitter. The "cutting by half" process is easily implemented distributively by letting each processor decide randomly whether to eliminate itself. It will be shown that if all neighbors of a receiver follow the elimination procedure, then, with positive probability, there exists a time slot in which exactly one neighbor transmits.

What follows is a description of the procedure for sending a message m, that is executed by each processor after receiving m:

    procedure *Decay*(k, m);
    repeat at most k times (but at least once!)
    send m to all neighbors;
    set *coin*$\leftarrow$0 or 1 with equal probability.
    until *coin* = 0.

By using elementary probabilistic arguments, one can prove:

Theorem 1 *Let y be a vertex of G. Also let $d \geq 2$ neighbors of y execute Decay during the time interval $[0,k)$ and assume that they all start the execution at Time $= 0$. Then $P(k,d)$, the probability that y receives a message by Time $= k$, satisfies:*

1. $\displaystyle\lim_{k \to \infty} P(k, d) \geq \tfrac{2}{3}$;
2. For $k \geq 2\lceil \log d \rceil$, $P(k, d) > \tfrac{1}{2}$.

(All logarithms are to base 2.)

The expected termination time of the algorithm depends on the probability that coin = 0. Here, this probability is set to be one half. An analysis of the merits of using other probabilities was carried out by Hofri [4].

The Broadcast Protocol

The broadcast protocol makes several calls to *Decay* (k, m). By Theorem 1 (2), to ensure that the probability of a processor y receiving the message be at least $1/2$, the parameter k should be at least $2\log d$ (where d is the number of neighbors sending a message to y). Since d is not known, the parameter was chosen as $k = 2\lceil \log \Delta \rceil$ (recall that Δ was defined to be an upper bound on the in-degree). Theorem 1 also requires that all participants start executing *Decay* at the same time-slot. Therefore, *Decay* is initiated only at integer multiples of $2\lceil \log \Delta \rceil$.

procedure *Broadcast*;
$k = 2\lceil \log \Delta \rceil$;
   $t = 2 \lceil \log(N/\epsilon) \rceil$;
Wait until receiving a message, say m;
do t times {
Wait until $(Time \bmod k) = 0$;
Decay(k, m) ;
}

A network is said to execute the *Broadcast_scheme* if some processor, denoted s, transmits an initial message and each processor executes the abovementioned *Broadcast* procedure.

Theorem 2 *Let* $T = 2D + 5\max\{\sqrt{D},$ $\sqrt{\log(n/\epsilon)} \cdot \sqrt{\log(n/\epsilon)}$. *Assume that Broadcast_scheme starts at Time* $= 0$. *Then, with probability* $\geq 1 - 2\epsilon$, *by time* $2\lceil \log \Delta \rceil \cdot T$ *all nodes will receive the message. Furthermore, with probability* $\geq 1 - 2\epsilon$, *all the nodes will terminate by time* $2 \lceil \log \Delta \rceil \cdot (T + \lceil \log(N/\epsilon) \rceil)$.

The bound provided by Theorem 2 contains two additive terms: the first represents the diameter of the network, and the second represents delays caused by conflicts (which are rare, yet they exist).

Additional Properties of the Broadcast Protocol

- **Processor IDs** – The protocol does not use processor IDs, and thus does not require that the processors have distinct IDs (or that they know the identity of their neighbors). Furthermore, a processor is not even required to know the number of its neighbors. This property makes the protocol adaptive to changes in topology which occur throughout the execution and resilient to non-malicious faults.

- **Knowing the size of the network** – The protocol performs almost as well when given instead of the actual number of processors (i.e., n), a "good" upper bound on this number (denoted N). An upper bound polynomial in n yields the same time-complexity, up to a constant factor (since complexity is logarithmic in N).

- **Conflict detection** – The algorithm and its complexity remain valid even if no messages can be received when a conflict occurs.

- **Simplicity and fast local computation** – In each time slot, each processor performs a constant amount of local computation.

- **Message complexity** – Each processor is active for $\lceil \log(N/\epsilon) \rceil$ consecutive phases, and the average number of transmissions per phase is at most 2. Thus, the expected number of transmissions of the entire network is bounded by $2n \cdot \lceil \log(N/\epsilon) \rceil$.

- **Adaptiveness to changing topology and fault resilience** – The protocol is resilient to some changes in the topology of the network. For example, edges may be added or deleted at any time, provided that the network of unchanged edges remains connected. This corresponds to fail/stop failure of edges, thus demonstrating the resilience to some non-malicious failures.

- **Directed networks** – The protocol does not use acknowledgments. Thus it may be applied even when the communication links are not symmetric, i.e., the fact that processor v can transmit to u does not imply that u can transmit to v. (The appropriate network model is, therefore, a directed graph.) In real life this situation occurs, for instance, when v has a stronger transmitter than u.

A Lower Bound on Deterministic Algorithms

For deterministic algorithms, one can show a lower bound: for every n, there exist a family of n-node networks such that every

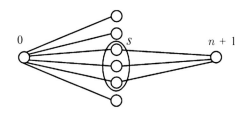

Randomized Broadcasting in Radio Networks, Fig. 1
The network used for the lower bound

deterministic broadcast scheme requires $\Omega(n)$ time. For every **non-empty** subset $S \subseteq \{1, 2, \ldots, n\}$, consider the following network G_S (Fig. 1).

Node 0 is the *source* and node $n + 1$ the *sink*. The source initiates the message and the problem of broadcast in G_S is to reach the sink. The difficulty stems from the fact that the partition of the middle layer (i.e., S) is not known a priori. The following theorem can be proved by a series of reductions to a certain "hitting game":

Theorem 3 *Every deterministic broadcast protocol that is correct for all n-node networks requires time $\Omega(n)$.*

The result of [2] depends crucially on the assumption that the nodes do not know the number and IDs of their neighbors. If this restriction is lifted, Kowalski and Pelc [5] showed how to broadcast in logarithmic time on all networks of type G_S. Moreover, they show how to broadcast in sublinear time on all n-node graphs of diameter o $(\log \log n)$.

Kowalski and Pelc also constructed a class of graphs of diameter 4, such that every broadcasting algorithm requires time $\Omega\left(\sqrt[4]{n}\right)$ on one of these graphs. Thus they showed an exponential gap for their model too.

Applications

The procedure *Decay* has been used to resolve contention in radio and cellular phone networks.

Cross-References

▶ Broadcasting in Geometric Radio Networks
▶ Deterministic Broadcasting in Radio Networks
▶ Randomized Gossiping in Radio Networks

Recommended Reading

Subsequent papers showed the optimality of the randomized algorithm:

- Alon et al. [1] showed the existence of a family of radius-2 networks on n vertices for which any broadcast schedule requires at least $\Omega(\log^2 n)$ time slots.
- Kushilevitz and Mansour [7] showed that for any randomized broadcast protocol, there exists a network in which the expected time to broadcast a message is $\Omega(D \log(N/D))$.
- Bruschi and Del Pinto [3] showed that for any deterministic distributed broadcast algorithm, any n and $D \leq n/2$ there exists a network with n nodes and diameter D such that the time needed for broadcast is $\Omega(D \log n)$.
- Kowalski and Pelc [6] discussed networks in which collisions are indistinguishable from the absence of transmission. They showed an $\Omega(n \log n / \log(n/D))$ lower bound and an $O(n \log n)$ upper bound. For this model, they also showed an $O(D \log n + \log^2 n)$ randomized algorithm, thus matching the lower bound of [1] and improving the bound of [2] for graphs for which $D = \theta(n/\log n)$.

1. Alon N, Bar-Noy A, Linial N, Peleg D (1991) A lower bound for radio broadcast. J Comput Syst Sci 43(2):290–298
2. Bar-Yehuda R, Goldreich O, Itai A (1992) On the time-complexity of broadcast in multi-hop radio networks: an exponential gap between determinism and randomization. J Comput Syst Sci 45(1): 104–126
3. Bruschi D, Del Pinto M (1997) Lower bounds for the broadcast problem in mobile radio networks. Distrib Comput 10(3):129–135
4. Hofri M (1987) A feedback-less distributed broadcast algorithm for multihop radio networks with time-varying structure. In: Computer performance and reliability: 2nd international models of communication sys-

tem, Rome, 25-29 May 1987. North Holland (1988), pp 353–368
5. Kowalski DR, Pelc A (2002) Deterministic broadcasting time in radio networks of unknown topology. In: FOCS'02: proceedings of the 43rd symposium on foundations of computer science, Washington, DC. IEEE Computer Society, pp 63–72
6. Kowalski DR, Pelc A (2005) Broadcasting in undirected ad hoc radio networks. Distrib Comput 18(1):43–57
7. Kushilevitz E, Mansour Y (1993) An $\Omega(d \log(n/d))$ lower bound for broadcast in radio networks. In: PODC'93, Los Angeles, California, pp 65–74

Randomized Contraction

Marek Cygan
Institute of Informatics, University of Warsaw, Warsaw, Poland

Keywords

Cuts; Graphs; Parameterized complexity; Randomization

Years and Authors of Summarized Original Work

2012; Chitnis, Cygan, Hajiaghayi, Pilipczuk, Pilipczuk

Problem Definition

The randomized contractions framework is often useful when designing fixed-parameter-tractable (FPT) algorithms for graph cut problems. Let us assume that we are given an undirected graph G with n vertices and m edges together with an integer k. The goal is to remove at most k edges or at most k vertices, in the edge- and vertex-deletion variants of a problem, respectively, to satisfy some problem-specific constraints. In this entry, for the sake of simplicity, we restrict our attention to edge-deletion variants only.

Examples of problems that fit in the above graph cut problem class include:

Multiway Cut
   Input: an undirected graph G, a set of terminals $T \subseteq V(G)$, and an integer k.
   Question: is there a set $X \subseteq E(G)$ of at most k edges of G, so that in $G \setminus X$, no connected component contains more than one terminal from T?
Steiner Cut
   Input: an undirected graph G, a set of terminals $T \subseteq V(G)$, and integers k, s.
   Question: is there a set $X \subseteq E(G)$ of at most k edges of G, so that in $G \setminus X$, at least s connected components contain at least one terminal from T?
Multiway Cut-Uncut
   Input: an undirected graph G, a set of terminals $T \subseteq V(G)$, an equivalence relation $\mathcal{R}$ on the set T, and an integer k.
   Question: is there a set $X \subseteq E(G)$ of at most k edges of G, so that for any $u, v \in T$, vertices u, v are in the same connected component of $G \setminus X$ iff $\mathcal{R}(u, v)$?
Unique Label Cover
   Input: an undirected graph G, a finite alphabet Σ of size s, an integer k, for each vertex $v \in V(G)$ a set $\phi_v \subseteq \Sigma$, and for each edge $e \in E(G)$ and each its endpoint v, a partial permutation $\psi_{e,v}$ of Σ, such that if $e = uv$ then $\psi_{e,u} = \psi_{e,v}^{-1}$.
   Question: is there a set $X \subseteq E(G)$ of at most k edges of G and a function $\Psi : V(G) \rightarrow \Sigma$ such that for any $v \in V(G)$ we have $\Psi(v) \in \phi_v$ and for any $uv \in E(G) \setminus X$, we have $(\Psi(u), \Psi(v)) \in \psi_{uv,u}$?

Key Results

The randomized contractions framework was obtained by Chitnis et al. [2]; however, it was inspired by an earlier work of Kawarabayashi and Thorup [4], who have shown that the k-way cut problem is fixed parameter tractable. Randomized contractions were used to obtain the first FPT algorithm for unique label cover parameterized by both the cut size and the alphabet size, as well as to improve the dependency on k in the FPT algorithms for Steiner cut and multiway cut-uncut.

To exemplify usage of randomized contractions, we use the multiway cut problem. Multiway cut is known to be FPT for a long time [5] and it admits efficient FPT algorithms with $f(k) = 4^k$ dependency on k by using important separators [1] as well as $f(k) = 2^k$ by LP-branching [3]. We use multiway cut as an illustration of usage of randomized contractions to simplify the description and magnify the most important parts of the technique.

High-Level Intuition

From now on, we assume that the given undirected graph G is connected, as otherwise one can solve the problem independently for each connected component of G. Observe that this guarantees that after removing k edges, the graph contains at most $k + 1$ connected components.

On a high level, the technique works in two phases. In the first phase, as long as the graph admits a certain type of a good edge separation, we proceed recursively and simplify the instance.

On the other hand, if the graph is well connected and does not contain a cut we are looking for, then in the second phase, we solve the problem directly, by exploiting the high connectivity of G.

Recursive Understanding

Assume that we have a set of vertices $V_1 \subseteq V(G)$, such that $G[V_1]$ is connected, V_1 contains at least $k \cdot k! + 2$ vertices, and there are at most k edges between V_1 and $V_2 = V(G) \setminus V_1$ in G. Let $B \subseteq V_1$ be the set of vertices in V_1 having at least one neighbor in V_2. In such a setting, one can show that by looking at $G[V_1]$ only (in particular without looking at $G[V_2]$), one can find an edge of $G[V_1]$ which can be safely contracted, i.e., which is not part of some solution for the whole graph G. The reason is that any solution $X \subseteq E(G)$ gives some partition of B by looking at the set of connected components o $G[V_2 \cup B] \setminus X$. There are at most $k!$ partitions of B as $|B| \leq k$. Imagine that for any such partition, we mark a set of at most k edges, which would extend the partial solution under consideration, i.e., extend $X \cap E(G[B \cup V_2])$. In total, this marking procedure would select $k \cdot k!$ edges, leaving at least

one edge unmarked, as $E(G[V_1]) \geq |V_1| - 1 \geq k \cdot k! + 1$. Such an unmarked edge can be safely contracted. The intuition behind this reasoning leads to the following definition:

Definition 1 Let G be a connected graph. A partition (V_1, V_2) of $V(G)$ is called a (q, k)-good edge separation, if

- $|V_1|, |V_2| > q$;
- $|E(V_1, V_2)| \leq k$;
- $G[V_1]$ and $G[V_2]$ are connected.

For the multiway cut problem, we would set $q = k \cdot k! + 1$. The following lemma states that we can find a (q, k)-good edge separation, if it exists:

Lemma 1 *There exists a deterministic algorithm that, given an undirected, connected graph G on n vertices along with integers q and k, in time $\mathcal{O}(2^{\mathcal{O}(\min(q,k) \log(q+k))} n^3 \log n)$ either finds a (q, k)-good edge separation or correctly concludes that no such separation exists.*

A rough sketch of the proof follows. Assume that a (q, k)-good edge separation (V_1, V_2) exists. Let E_1 be the set of edges of some subtree of $G[V_1]$ with exactly q edges; similarly let E_2 be the set of edges of some subtree of $G[V_2]$ with exactly q edges. By the definition of a (q, k)-good separation, such sets E_1, E_2 exist. Contract each edge of the graph with probability $1/2$ independently from other edges. With probability at least $1/f(k, q) = 2^{-(2q+k)}$, the following event happens (see Fig. 1):

(i) No edge between V_1 and V_2 is contracted,
(ii) All edges of $E_1 \cup E_2$ are contracted.

If we are lucky and such an event occurs, then by looking for a minimum cut between each two vertices onto which at least $q + 1$ vertices of G were contracted, we can find a (q, k)-good separation. By a better choice of contraction probability, we can improve the probability of success, whereas by using splitters [6], we can derandomize the procedure.

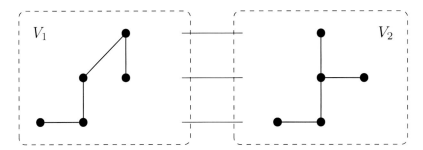

Randomized Contraction, Fig. 1 A (q, k)-good separation. In the randomized routine, we hope the thick edges to be contracted and the thin edges not to be contracted

Randomized Contraction, Fig. 2 Structure of a connected graph G that does not admit a (q, k)-good separation. After removing at most k edges, only one big connected component, C_0, remains

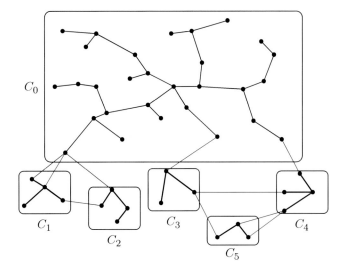

Summarizing this phase of the algorithm, we look for a (q, k)-good edge separation. If it does not exist, then we proceed to the second – high connectivity – phase of the algorithm. However, if a (q, k)-good edge separation exists, then we proceed recursively. Clearly, we are omitting some important details in this description. The most important of them is that when recursing, some vertices play a special role, as they are *border terminals* – vertices which have neighbors outside of the part of the graph under consideration. For this reason, to make the induction work, we need a stronger definition of a problem, called its border version, which for multiway cut is as follows:

Border Multiway Cut

   Input: a connected, undirected graph G, a set of terminals $T \subseteq V(G)$, an integer k, and a set $T_b \subseteq V(G)$ of at most $2k$ terminals.

Output: for each partition $\mathcal{P}$ of T_b output, a set $X_\mathcal{P}$ of size at most k (if it exists), such that in the graph $G_\mathcal{P} \setminus X$, no two terminals from T are in the same connected component, where $G_\mathcal{P} = (V(G), E(G) \cup E_\mathcal{P})$ and $E_\mathcal{P}$ contains pairs of vertices which are in the same block of $\mathcal{P}$.

High-Connectivity Phase

The second phase of the approach is usually problem specific; however, its main idea is the following. Since we know that G does not admit a (q, k)-good edge separation, if we remove any set X of at most k edges, there is at most one connected component of $G \setminus X$ containing more than q vertices (see Fig. 2). Therefore, if we independently contract each edge at random, then with good enough probability, no solution edge will be contracted and all connected components of $G \setminus X$ except possibly one will be contracted onto single vertices (again, see Fig. 2). In such a case, one can show that we can solve a cut problem under consideration either greedily or by dynamic programming.

Related Work

The currently best-known parameterized algorithm for unique label cover is due to Wahlström [7] and works in time $s^{2k} n^{\mathcal{O}(1)}$.

Recommended Reading

1. Chen J, Liu Y, Lu S (2009) An improved parameterized algorithm for the minimum node multiway cut problem. Algorithmica 55(1):1–13. doi:10.1007/s00453-007-9130-6
2. Chitnis RH, Cygan M, Hajiaghayi M, Pilipczuk M, Pilipczuk M (2012) Designing FPT algorithms for cut problems using randomized contractions. In: 53rd annual IEEE symposium on foundations of computer science (FOCS 2012), New Brunswick, 20–23 Oct 2012. IEEE Computer Society, pp 460–469. doi:10.1109/FOCS.2012.29
3. Cygan M, Pilipczuk M, Pilipczuk M, Wojtaszczyk JO (2013) On multiway cut parameterized above lower bounds. TOCT 5(1):3. doi:10.1145/2462896.2462899
4. Kawarabayashi K, Thorup M (2011) The minimum k-way cut of bounded size is fixed-parameter tractable. In: Ostrovsky R (ed) IEEE 52nd annual symposium on foundations of computer science (FOCS 2011), Palm Springs, 22–25 Oct 2011. IEEE Computer Society, pp 160–169. doi:10.1109/FOCS.2011.53
5. Marx D (2006) Parameterized graph separation problems. Theor Comput Sci 351(3):394–406. doi:10.1016/j.tcs.2005.10.007
6. Naor M, Schulman LJ, Srinivasan A (1995) Splitters and near-optimal derandomization. In: 36th annual symposium on foundations of computer science, Milwaukee, 23–25 Oct 1995. IEEE Computer Society, pp 182–191. doi:10.1109/SFCS.1995.492475
7. Wahlström M (2014) Half-integrality, LP-branching and FPT algorithms. In: Chekuri C (ed) Proceedings of the twenty-fifth annual ACM-SIAM symposium on discrete algorithms (SODA 2014), Portland, 5–7 Jan 2014. SIAM, pp 1762–1781. doi:10.1137/1.9781611973402.128

Randomized Energy Balance Algorithms in Sensor Networks

Pierre Leone[1], Sotiris Nikoletseas[2,3], and José Rolim[1]
[1]Informatics Department, University of Geneva, Geneva, Switzerland
[2]Computer Engineering and Informatics Department, University of Patras, Patras, Greece
[3]Computer Technology Institute and Press "Diophantus", Patras, Greece

Keywords

Power conservation

Years and Authors of Summarized Original Work

2005; Leone, Nikoletseas, Rolim

Problem Definition

Recent developments in wireless communications and digital electronics have led to the development of extremely small in size, low-power, low-cost sensor devices (often called smart dust). Such tiny devices integrate sensing, data processing and wireless communication capabilities. Examining each such resource constraint device individually might appear to have small utility; however, the distributed self-collaboration of large numbers of such devices into an ad hoc network may lead to the efficient accomplishment of large sensing tasks i.e., reporting data about the realization of a local event happening in the network area to a faraway control center.

The problem considered is the development of a randomized algorithm to balance energy among sensors whose aim is to detect events in the network area and report them to a sink. The network is sliced by the algorithm into layers composed of sensors at approximately equal distances from the

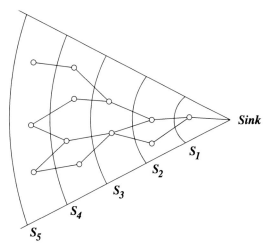

Randomized Energy Balance Algorithms in Sensor Networks, Fig. 1 The sink and five slices $S_1, \ldots, S_5$

sink [1, 2, 8] (Fig. 1). The slicing of the network depends on the communication distance. The sink initiates the process by sending a control message containing a counter, the value of which is initially 1. Sensors receiving the message assign themselves to a slice number corresponding to the counter, increment the counter and propagate the message in the network. A sensor already assigned to a slice ignores subsequent received control messages.

The strategy suggested to balance the energy among sensors consists in allowing a sensor to probabilistically choose between either sending data to a sensor in the next layer towards the sink or sending the data directly to the sink. The difference between the two choices is the energy consumption, which is much higher if the sensor decides to report to the sink directly. The energy consumption is modeled as a function of the transmission distance by assuming that the energy necessary to send data up to a distance d is proportional to d^2. Actually, more accurate models can be considered, in which the dependence is of the form d^α, with $2 \leq \alpha \leq 5$ depending on the particular environmental conditions. Although the model chosen determines the parameters of the algorithm, the particular shape of the function describing the relationship between the distance of transmission and energy consumption

is not relevant except that it might increase with distance. The distance between two successive slices is normalized to be 1. Hence, a sensor sending data to one of its neighbors consumes one unit of energy and a sensor located in slice i consumes i^2 units of energy to report to the sink directly. Small hop transmissions are cheap (with respect to energy consumption) but pass through the critical region around the sink and might strain sensors in that region, while expensive direct transmissions bypass that critical area.

Energy balance is defined as follows:

Definition 1 The network is energy-balanced if the average per sensor energy dissipation is the same for all sectors, i.e., when

$$\frac{E[\mathcal{E}_i]}{S_i} = \frac{E[\mathcal{E}_j]}{S_j}, \quad i, j = 1, \ldots, n \quad (1)$$

where $\mathcal{E}_i$ is the total energy available and S_i is the number of nodes in slice number i.

The dynamics of the network is modeled by assigning probabilities $\lambda_i, i = 1, \ldots, N$, $\sum \lambda_i = 1$, of the occurrence of an event in slice i. The protocol consists in transmitting the data to a neighbor slice with probability p_i and with probability $1 - p_i$ to the sink, for a sensor belonging to slice i. Hence, the mean energy consumption per data unit is $p_i + (1 - p_i)i^2$. A central assumption in the following is that the events are evenly generated in a given slice. Then, denoting by e_i the energy available per node in slice i (i.e., $e_i = \mathcal{E}_i / S_i$), the problem of energy-balanced data propagation can be formally stated as follows:

Given $\lambda_i, e_i, S_i, i = 1, \ldots, N$, find p_i, λ such that

$$\underbrace{\left(\lambda_i + \lambda_{i+1} p_{i+1} + \ldots + \lambda_n p_n p_{n-1} \cdots p_{i+1}\right)}_{=:x_i}$$

$$\cdot \left(p_i \frac{1}{S_i} + (1 - p_i)\frac{i^2}{S_i}\right)$$

$$= \lambda e_i, \quad i = 1, \ldots, N. \quad (2)$$

```
Initialize $\tilde{x}_0 = \lambda, \ldots, \tilde{x}_n$
Initialize NbrLoop=1
repeat forever
    Send $\tilde{x}_i$ and $\lambda$ values to the stations which compute
                                   their $p_i$ probability
    wait for a data
    for i=0 to n
        if the data passed through slice i then
            X ← 1
        else
            X ← 0
        end if
        Generate R a $\tilde{x}_i$-Bernoulli random variable
        $\tilde{x}_i$    $\tilde{x}_i + \frac{1}{NbrLoop}(X - R)$
        Increment NbrLoop by one.
    end for
end repeat
```

Randomized Energy Balance Algorithms in Sensor Networks, Fig. 2 Pseudo-code for estimation of the x_i value by the sink

Equation (2) amounts to ensuring that the mean energy dissipation for all sensors is proportional to the available energy. In turn, this ensures that sensors might, on average, run out of energy all at the same time. Notice that (2) contains the definitions of the x_i. They are the ones estimated in the pseudo-code in Fig. 2, the successive estimations being denoted as $\tilde{x}_i$. These variables are proportional to the number of messages handled by slice i.

Key Results

In [1, 2] recursive equations similar to (2) were suggested and solved in closed form under adequate hypotheses. The need for a priori knowledge of the probability of occurrence of the events, the λ_i parameters, was considered in [7], in which these parameters were estimated by the sink on the basis of the observations of the various paths the data follow. The algorithm suggested is based on recursive estimation, is computationally not expensive and converges with rate $\mathcal{O}(1/\sqrt{n})$. One might argue that the rate of convergence is slow; however, it is numerically observed that relatively quickly compared with the convergence time, the algorithm finds an estimation close enough to the final value. The estimation algorithm run by the sink (which has no energy constraints) is given in Fig. 2.

Results taken from [1, 2, 7] all assume the existence of an energy-balance solution. However, particular distributions of the events might prevent the existence of such a solution and the relevant question is no longer the computation of an energy-balance algorithm. For instance, assuming that $\lambda_N = 0$, sensors in slice N have no way of balancing energy. In [9] the problem was reformulated as finding the probability distribution $\{p_i\}_{i=1,\ldots,N}$ which leads to the maximal functional lifetime of the networks. It was proved that if an energy-balance strategy exists, then it maximizes the lifetime of the network establishing formally the intuitive reasoning which was the motivation to consider energy-balance strategies. A centralized algorithm was presented to compute the optimal parameters. Moreover, it was observed numerically that the interslice energy consumption is prone to be uneven and a spreading technique was suggested and numerically validated as being efficient to overcome this limitation of the probabilistic algorithm.

The communication graph considered is a restrictive subset of the complete communication graph and it is legitimate to wonder whether one can improve the situation by extending it. For instance, by allowing data to be sent two hops or more away. In [3, 6] it was proved that the topology in which sensors communicate only to neighbor slices and the sink is the one which maximizes the flow of data in the network. Moreover, the communication graph in which sensors send data only to their neighbors and the sink leads to a completely distributed algorithm balancing energy [6]. Indeed, as a sensor sends data to a neighbor slice, the neighbor must in turn send the data and can attach information concerning its own energy level. This information might be captured by the initial sensor since it belongs to the communication range of its neighbor (this does not hold any longer if multiple hops are allowed). Hence, a distributed strategy consists in

R

sending data to a particular neighbor only if its energy level consumption is lower, otherwise the data are sent directly to the sink.

Applications

Among the several constraints sensor networks designers have to face, energy management is central since sensors are usually battery powered, making the lifetime of the networks highly sensitive to the energy management. Besides the traditional strategy consisting in minimizing the energy consumption at sensor nodes, energy-balance schemes aim at balancing the energy consumption among sensors. The intuitive function of such schemes is to avoid energy depletion holes appearing as some sensors that run out of their available energy resources and are no longer able to participate in the global function of the networks. For instance, routing might be no longer possible if a small number of sensors run out of energy, leading to a disconnected network. This was pointed out in [5] as well as the need to develop application-specific protocols. Energy balancing is suggested as a solution in order to make the global functional lifetime of the network longer. The earliest development of dedicated protocols ensuring energy balance can be found in [4, 10, 11].

A key application is to maximize the lifetime of the network while gathering data to a sink. Besides increasing the lifetime of the networks, other criteria have to be taken into account. Indeed, the distributed algorithm might be as simple as possible owing to limited computational resources, might avoid collisions or limit the total number of transmissions, and might ensure a large enough flow of data from the sensors toward the sink. Actually, maximizing the flow of data is equivalent to maximizing the lifetime of sensor networks if some particular realizable conditions are fulfilled. Besides the simplicity of the distributed algorithm, the network deployment and the self-realization of the network structure might be possible in realistic conditions.

Cross-References

▶ Obstacle Avoidance Algorithms in Wireless Sensor Networks
▶ Probabilistic Data Forwarding in Wireless Sensor Networks

Recommended Reading

1. Efthymiou C, Nikoletseas S, Rolim J (2006) Energy balanced data propagation in wireless sensor networks. In: 4th international workshop on algorithms for wireless, mobile, ad-hoc and sensor networks (WMAN'04) IPDPS 2004. Wirel Netw J (WINET) 12(6):691–707
2. Efthymiou C, Nikoletseas S, Rolim J (2006) Energy balanced data propagation in wireless sensor networks. In: Wireless networks (WINET) Journal, Special Issue on Algorithms for Wireless, Mobile, Ad Hoc and Sensor Networks. Springer
3. Giridhar A, Kumar PR (2005) Maximizing the functional lifetime of sensor networks. In: Proceedings of the fourth international conference on information processing in sensor networks (IPSN'05). UCLA, Los Angeles, 25–27 Apr 2005
4. Guo W, Liu Z, Wu G (2003) An energy-balanced transmission scheme for sensor networks. In: 1st ACM international conference on embedded networked sensor systems (ACM SenSys 2003), Poster Session, Los Angeles, Nov 2003
5. Heinzelman W, Chandrakasan A, Balakrishnan H (2000) Energy efficient communication protocol for wireless microsensor networks. In: Proceedings of the 33rd IEEE Hawaii international conference on system sciences (HICSS 2000)
6. Jarry A, Leone P, Powell O, Rolim J (2006) An optimal data propagation algorithm for maximizing the lifespan of sensor networks. In: Second international conference, DCOSS 2006, San Francisco, June 2006. Lecture notes in computer science, vol 4026. Springer, Berlin, pp 405–421
7. Leone P, Nikoletseas S, Rolim J (2005) An adaptive blind algorithm for energy balanced data propagation in wireless sensor networks. In: First international conference on distributed computing in sensor systems (DCOSS), Marina del Rey, June/July 2005. Lecture notes in computer science, vol 3560. Springer, Berlin, pp 35–48
8. Olariu S, Stojmenovic I (2006) Design guidelines for maximizing lifetime and avoiding energy holes in sensor networks with uniform distribution and uniform reporting. In: IEEE INFOCOM, Barcelona, 24–25 Apr 2006
9. Powell O, Leone P, Rolim J (2007) Energy optimal data propagation in sensor networks. J Parallel

Distrib Comput 67(3):302–317. http://arxiv.org/abs/cs/0508052
10. Singh M, Prasanna V (2003) Energy-optimal and energy-balanced sorting in a single-hop wireless sensor network. In: Proceedings of the first IEEE international conference on pervasive computing and communications (PerCom'03), Fort Worth, 23–26 Mar 2003, pp 302–317
11. Yu Y, Prasanna VK (2003) Energy-balanced task allocation for collaborative processing in networked embedded system. In: Proceedings of the 2003 conference on language, compilers, and tools for embedded systems (LCTES'03), San Diego, 11–13 June 2003, pp 265–274

Randomized Gossiping in Radio Networks

Leszek Gąsieniec
University of Liverpool, Liverpool, UK

Keywords

All-to-all communication; Broadcast; Gossip; Total exchange of information; Wireless networks

Years and Authors of Summarized Original Work

2001; Chrobak, Gąsieniec, Rytter

Problem Definition

The two classical problems of disseminating information in computer networks are *broadcasting* and *gossiping*. In broadcasting, the goal is to distribute a message from a distinguished *source* node to all other nodes in the networks. In gossiping, each node v in the network initially contains a message m_v, and the task is to distribute each message m_v to all nodes in the network.

The radio network abstraction captures the features of distributed communication networks with multi-access channels, with minimal assumptions on the channel model and processors' knowledge. Directed edges model unidirectional links, including situations in which one of two adjacent transmitters is more powerful than the other. In particular, there is no feedback mechanism (see, for example, [6]). In some applications, collisions may be difficult to distinguish from the noise that is normally present in the channel, justifying the need for protocols that do not depend on the reliability of the collision detection mechanism (see [3, 4]). Some network configurations are subject to frequent changes. In other networks, a network topology could be unstable or dynamic, for example, when mobile users are present. In such situations, algorithms that do not assume any specific topology are more desirable.

More formally a radio network is a directed graph $G = (V, E)$, where by $|V| = n$, we denote the number of nodes in this graph. Individual nodes in V are denoted by letters $u, v, \ldots$. If there is an edge from u to v, i.e., $(u, v) \in E$, then we say that v is an *out-neighbor* of u and u is an *in-neighbor* of v. Messages are denoted by letter m, possibly with indices. In particular, the message originating from node v is denoted by m_v. The whole set of initial messages is $M = \{m_v : v \in V\}$. During the computation, each node v holds a set of messages M_v that have been received by v so far. Initially, each node v does not possess any information apart from $M_v = \{m_v\}$. Without loss of generality, whenever a node is in the transmitting mode, one can assume that it transmits the whole content of M_v.

The time is divided into discrete time steps. All nodes start simultaneously, have access to a common clock, and work synchronously. A gossiping algorithm is a protocol that for each node u, given all past messages received by u, specifies, for each time step t, whether u will transmit a message at time t, and if so, it also specifies the message. A message M transmitted at time t from a node u is sent instantly to all its out-neighbors. An out-neighbor v of u receives M at time step t only if no collision occurred, that is, if the other in-neighbors of v do not transmit at time t at all. Further, collisions cannot be distinguished from background noise. If v does not receive any message at time t, it knows that either none of its in-neighbors transmitted at time t or that at least two did, but it does not know which of these two events occurred. The *running time* of a gossiping algorithm is the smallest t such that for any network topology, and any assignment of identifiers to the nodes, all nodes receive messages originating in every other node no later than at step t.

Limited Broadcast $_v$ (k) Given an integer k and a node v, the goal of *limited broadcasting* is to deliver the message m_v (originating in v) to at least k other nodes in the network.

Distributed Coupon Collection The set of network nodes V can be interpreted as a set of n bins and the set of messages M as a set of n coupons. Each coupon has at least k copies, each copy belonging to a different bin. M_v is the set of coupons in bin v. Consider the following process. At each step, one opens every bin at random, independently, with probability $1/n$. If no bin is opened, or if two or more bins are opened, a failure occurs and no coupons are collected. If exactly one bin, say v, is opened, all coupons from M_v are collected. The task is to establish how many steps are needed to collect (a copy of) each coupon.

Key Results

Theorem 1 ([1]) *There exists a deterministic $O(k \log^2 n)$-time algorithm for limited broadcasting from any node in radio networks with an arbitrary topology.*

Theorem 2 ([1]) *Let δ be a given constant, $0 < \delta < 1$, and $s = (4n / k) \ln(n / \delta)$. After s steps of the distributed coupon collection process, with probability at least $1 - \delta$, all coupons will be collected.*

Theorem 3 ([1]) *Let ϵ be a given constant, where $0 < \epsilon < 1$. There exists a randomized $O(n \log^3 n \log(n/\epsilon))$-time Monte Carlo-type algorithm that completes radio gossiping with probability at least $1 - \epsilon$.*

Theorem 4 ([1]) *There exists a randomized Las Vegas-type algorithm that completes radio gossiping with expected running time $O(n \log^4 n)$.*

Applications

Further work on efficient randomized radio gossiping include the $O(n \log^3 n)$-time algorithm by Liu and Prabhakaran; see [5], where the deterministic procedure for limited broadcasting is replaced by its $O(k \log n)$-time randomized counterpart. This bound was later reduced to $O(n \log^2 n)$ by Czumaj and Rytter in [2], where a new randomized limited broadcasting procedure with an expected running time $O(k)$ is proposed.

Open Problems

The exact complexity of randomized radio gossiping remains an open problem. All three gossiping algorithms [1, 2, 5] are based on the concepts of limited broadcast and distributed coupon collection. The two improvements [2, 5] refer solely to limited broadcasting. Thus, very likely further reduction of the time complexity must coincide with more accurate analysis of the distributed coupon collection process or with development of a new gossiping procedure.

Recommended Reading

1. Chrobak M, Gąsieniec L, Rytter W (2004) A randomized algorithm for gossiping in radio networks. In: Proceedings of 8th annual international computing combinatorics conference, Guilin, 2001, pp 483–492; Full version in Networks 43(2):119–124 (2004)
2. Czumaj A, Rytter W (2006) Broadcasting algorithms in radio networks with unknown topology. J Algorithms 60(2):115–143
3. Ephremides A, Hajek B (1998) Information theory and communication networks: an unconsummated union. IEEE Trans Inf Theory 44:2416–2434
4. Gallager R (1985) A perspective on multiaccess communications. IEEE Trans Inf Theory 31:124–142

5. Liu D, Prabhakaran M (2002) On randomized broadcasting and gossiping in radio networks. In: Proceedings of 8th annual international computing combinatorics conference, Singapore, pp 340–349
6. Massey JL, Mathys P (1985) The collision channel without feedback. IEEE Trans Inf Theory 31: 192–204

Randomized Minimum Spanning Tree

Vijaya Ramachandran
Computer Science, University of Texas, Austin, TX, USA

Keywords

Linear time; Minimum spanning tree; Minimum spanning tree verification; Randomized algorithm

Years and Authors of Summarized Original Work

1995; Karger, Klein, Tarjan

Problem Definition

The input to the problem is a connected undirected graph $G = (V, E)$ with a weight $w(e)$ on each edge $e \in E$. The goal is to find a spanning tree of minimum weight, where for any subset of edges $E' \subseteq E$, the *weight of E'* is defined to be $w(E') = \sum_{e \in E'} w(e)$.

If the graph G is not connected, the goal of the problem is to find a *minimum spanning forest*, which is defined to be a minimum spanning tree in each connected component of G. Both problems will be referred to as the *MST* problem.

The randomized MST algorithm by Karger, Klein, and Tarjan [9] which is considered here will be called the *KKT algorithm*. Also it will be

assumed that the input graph $G = (V, E)$ has n vertices and m edges and that the edge weights are distinct.

The MST problem has been studied extensively prior to the KKT result, and several very efficient, deterministic algorithms are available from these studies. All of these are deterministic and are based on a method that greedily adds an edge to a forest that is a subgraph of the minimum spanning tree at all times. The early algorithms in this class are already efficient with a running time of $O(m \log n)$. These include the algorithms of Borůvka [1], Jarník [8] (later rediscovered by Dijkstra and Prim [5]), and Kruskal [5].

The fastest algorithm known for MST prior to the KKT algorithm runs in time $O(m \log \beta(m,n))$ [7], where $\beta(m,n) = \min\{i \mid \log^{(i)} n \le m/n\}$ [7]; here $\log^{(i)} n$ is defined as $\log n$ if $i = 1$ and as $\log \log^{(i-1)} n$ if $i > 1$. Although this running time is close to linear, it is not linear time if the graph is very sparse.

The problem of finding the minimum spanning tree efficiently is an important and fundamental problem in graph algorithms and combinatorial optimization.

Background

Some relevant background is summarized here.

- The basic step in Borůvka's algorithm [1] is the *Borůvka step*, which picks the minimum edge-weight incident on each vertex, adds it to the minimum spanning tree, and then contracts these edges. This step runs in linear time and also very efficiently in parallel. It is the backbone of the most efficient parallel algorithms for minimum spanning tree and is also used in the KKT algorithm.

- A related and simpler problem is that of *minimum spanning tree verification*. Here, given a spanning tree T of the input edge-weighted graph, one needs to determine if T is its minimum spanning tree. An algorithm that solves this problem with a linear number of edge-weight comparisons was shown by Komlós [13], and later a deterministic linear-time algorithm was given in [6] (see also [12] for a simpler algorithm).

Key Results

The main result in [9] is a randomized algorithm for the minimum spanning tree problem that runs in expected linear time. The only operations performed on the edge weights are pairwise comparisons. The algorithm does not assume any particular representation of the edge weights (i.e., integer or real values) and only assumes that any comparison between a pair of edge weights can be performed in unit time. The entry also shows that the algorithm runs in $O(m + n)$ time with the exponentially high probability $1 - \exp(-\Omega(m))$ and that its worst-case running time is $O(n + m \log n)$.

The simple and elegant *MST sampling lemma* given in Lemma 1 below is the key tool used to derive and analyze the KKT algorithm. This lemma needs a couple of definitions and facts:

1. The well-known *cycle property* for the minimum spanning tree states that the heaviest edge in any cycle in the input graph G *cannot* be in the minimum spanning tree.
2. Let F be a forest of G (i.e., an acyclic subgraph of G). An edge $e \in E$ is *F-light* if $F \cup \{e\}$ either continues to be a forest of G, or the heaviest edge in the cycle containing e is *not* e. An edge in G that is not F-light is *F-heavy*. Note that by the cycle property, an F-heavy edge cannot be in the minimum spanning tree of G, *no matter what forest F is used*. Given a forest F of G, the set of F-heavy edges can be determined in linear time by a simple modification to existing linear-time minimum spanning tree verification algorithms [6,12].

Lemma 1 (MST Sampling Lemma) *Let $H = (V, E_H)$ be formed from the input edge-weighted graph $G = (V, E)$ by including each edge with probability p independent of the other edges. Let*

F be the minimum spanning forest of H. Then, the expected number of F-light edges in G is $\leq n/p$.

The KKT algorithm identifies edges in the minimum spanning tree of G only using Borůvka steps. However, after every two Borůvka steps, it removes F-heavy edges using the minimum spanning forest F of a subgraph obtained through sampling edges with probability $p = 1/2$. As mentioned earlier, these F-heavy edges can be identified in linear time. The minimum spanning forest of the sampled graph is computed recursively.

The correctness of the KKT algorithm is immediate since every F-heavy edge it removes cannot be in the MST of G since F is a forest of G, and every edge it adds to the minimum spanning tree is in the MST since it is added through a Borůvka step.

The expected running time analysis as well as the exponentially high probability bound for the running time are surprisingly simple to derive using the MST Sampling Lemma (Lemma 1).

In summary, the entry [9] proves the following results.

Theorem 1 *The KKT algorithm is a randomized algorithm that finds a minimum spanning tree of an edge-weighted undirected graph on n nodes and m edges in $O(n + m)$ time with probability at least $1 - \exp(-\Omega(m))$. The expected running time is $O(n+m)$ and the worst-case running time is $O(n + m \log n)$.*

The model of computation used in [9] is the unit-cost RAM model since the known MST verification algorithms were for this model and not the more restrictive *pointer machine* model. More recently the MST verification result and hence the KKT algorithm have been shown to work on the pointer machine as well [2].

Lemma 1 is proved in [9] through a simulation of Kruskal's algorithm along with an analysis of the probability with which an F-light edge is not sampled. Another proof that uses a backward analysis is given in [3].

Further Comments

- Recently (and since the appearance of the KKT algorithm in 1995), two new deterministic algorithms for MST have appeared, due to Chazelle [4] and Pettie and Ramachandran [14]. The former [4] runs in $O(n + m\alpha(m, n))$ time, where α is an inverse of the Ackermann's function, whose growth rate is even smaller than the β function mentioned earlier for the best result that was known prior to the KKT algorithm [7]. The latter algorithm [14] provably runs in time that is within a constant factor of the decision-tree complexity of the MST problem and hence is optimal; its time bound is $O(n + m\alpha(m, n))$ and $\Omega(n + m)$, and the exact bound remains to be determined.
- Although the KKT algorithm runs in expected linear time (and with exponentially high probability), it is not the last word on randomized MST algorithms. A randomized MST algorithm that runs in expected linear time and uses only $O(\log {}^*n)$ random bits is given in [16, 17]. In contrast, the KKT algorithm uses a linear number of random bits.

Applications

The minimum spanning tree problems has a large number of applications, which are discussed in minimum spanning trees.

Open Problems

Some open problems that remain are the following:

1. Can randomness be removed in the KKT algorithm? A hybrid algorithm that uses the KKT algorithm within a modified version of the Pettie-Ramachandran algorithm [14] is given in [16, 17] that achieves expected linear time while reducing the number of random bits used to only $O(\log {}^*n)$. Can this tiny amount of randomness be removed as well? If all randomness can be removed from the KKT algorithm, that will establish a linear time bound for the Pettie-Ramachandran algorithm [14] and also provide another optimal deterministic MST algorithm, this one based on the KKT approach.

2. Can randomness be removed from the work-optimal *parallel algorithms* [10] for MST? A linear-work, expected logarithmic-time parallel MST algorithm for the EREW PRAM is given in [15]. This parallel algorithm is both work and time optimal. However, it uses a linear number of random bits. Another work-optimal parallel algorithm is given in [16, 17] that runs in expected polylog time using only polylog random bits. This leads to the following open questions regarding parallel algorithms for the MST problem:

   - To what extent can dependence on random bits be reduced (from the current linear bound) in a time- and work-optimal parallel algorithm for MST?
   - To what extent can the dependence on random bits be reduced (from the current polylog bound) in a work-optimal parallel algorithm with reasonable parallelism (say polylog parallel time)?

Experimental Results

Katriel, Sanders, and Träff [11] performed an experimental evaluation of the KKT algorithm and showed that it has good performance on moderately dense graphs.

Cross-References

▶ Minimum Spanning Trees

Acknowledgments This work was supported in part by NSF grant CFF-0514876.

Recommended Reading

1. Borůvka O (1926) O jistém problému minimálním. Práce Moravské Přírodovědecké Společnosti 3:37–58. (In Czech)
2. Buchsbaum A, Kaplan H, Rogers A, Westbrook JR (1998) Linear-time pointer-machine algorithms for least common ancestors, MST verification and dominators. In: Proceedings of the ACM symposium on theory of computing (STOC), Dallas, pp 279–288
3. Chan TM (1998) Backward analysis of the Karger-Klein-Tarjan algorithm for minimum spanning trees. Inf Process Lett 67:303–304
4. Chazelle B (2000) A minimum spanning tree algorithm with inverse-Ackermann type complexity. J ACM 47(6):1028–1047
5. Cormen TH, Leiserson CE, Rivest RL, Stein C (2001) Introduction to algorithms. MIT, Cambridge
6. Dixon B, Rauch M, Tarjan RE (1992) Verification and sensitivity analysis of minimum spanning trees in linear time. SIAM J Comput 21(6):1184–1192
7. Gabow HN, Galil Z, Spencer TH, Tarjan RE (1986) Efficient algorithms for finding minimum spanning trees in undirected and directed graphs. Combinatorica 6:109–122
8. Graham RL, Hell P (1985) On the history of the minimum spanning tree problem. Ann Hist Comput 7(1):43–57
9. Karger DR, Klein PN, Tarjan RE (1995) A randomized linear-time algorithm for finding minimum spanning trees. J ACM 42(2):321–329
10. Karp RM, Ramachandran V (1990) Parallel algorithms for shared-memory machines. In: van Leeuwen J (ed) Handbook of theoretical computer science. Elsevier Science, Amsterdam, pp 869–941
11. Katriel I, Sanders P, Träff JL (2003) A practical minimum spanning tree algorithm using the cycle property. In: Proceedings of the 11th annual European symposium on algorithms, Budapest. LNCS, vol 2832. Springer, Berlin, pp 679–690
12. King V (1997) A simpler minimum spanning tree verification algorithm. Algorithmica 18(2):263–270
13. Komlós J (1985) Linear verification for spanning trees. Combinatorica 5(1):57–65
14. Pettie S, Ramachandran V (2002) An optimal minimum spanning tree algorithm. J ACM 49(1):16–34
15. Pettie S, Ramachandran V (2002) A randomized time-work optimal parallel algorithm for finding a minimum spanning forest. SIAM J Comput 31(6):1879–1895
16. Pettie S, Ramachandran V (2002) Minimizing randomness in minimum spanning tree, parallel connectivity, and set maxima algorithms. In: Proceedings of the ACM-SIAM symposium on discrete algorithms (SODA), San Francisco, pp 713–722
17. Pettie S, Ramachandran V (2008) New randomized minimum spanning tree algorithms using exponentially fewer random bits. ACM Trans Algorithms 4(1):article 5

Randomized Parallel Approximations to Max Flow

Maria Serna
Department of Language and System Information, Technical University of Catalonia, Barcelona, Spain

Keywords

Approximate maximum flow construction

Years and Authors of Summarized Original Work

1991; Serna, Spirakis

Problem Definition

The work of Serna and Spirakis provides a parallel approximation schema for the Maximum Flow problem. An approximate algorithm provides a solution whose cost is within a factor of the optimal solution. The notation and definitions are the standard ones for networks and flows (see for example [2, 7]).

A *network* $N = (G, s, t, c)$ is a structure consisting of a directed graph $G = (V, E)$, two distinguished vertices, $s, t \in V$ (called the *source* and the *sink*), and $c : E \rightarrow \mathbb{Z}^+$, an assignment of an integer capacity to each edge in E. A *flow function* f is an assignment of a non-negative number to each edge of G (called the flow into the edge) such that first at no edge does the flow exceed the capacity, and second for every vertex except s and t, the sum of the flows on its incoming edges equals the sum of the flows on its outgoing edges. The *total flow* of a given flow function f is defined as the net sum of flow into the sink t. The Maximum Flow problem can be stated as

Name	Maximum Flow
Input	A network $N = (G, s, t, c)$
Output	Find a flow f for N for which the total flow is maximum.

Maximum Flows and Matchings

The Maximum Flow problem is closely related to the Maximum Matching problem on bipartite\break graphs.

Given a graph $G = (V, E)$ and a set of edges $M \subseteq E$ is a *matching* if in the subgraph (V, M) all vertices have degree at most one. A *maximum matching* for G is a matching with a maximum number of edges. For a graph $G = (V, E)$ with weight $w(e)$, the *weight* of a matching M is the sum of the weights of the edges in M. The problem can be stated as follows:

Name Maximum Weight Matching
Input A graph $G = (V, E)$ and a weight $w(e)$ for each edge $e \in E$
Output Find a matching of G with the maximum possible weight.

There is a standard reduction from the Maximum Matching problem for bipartite graphs to the Maximum Flow problem [7, 8]. In the general weighted case one has just to look at each edge with capacity $c > 1$ as c edges joining the same points each with capacity one, and transform the multigraph obtained as shown before. Notice that to perform this transformation a c value is required which is polynomially bounded. The whole procedure was introduced by Karp, Upfal, and Wigderson [5] providing the following results

Theorem 1 *The Maximum Matching problem for bipartite graphs is NC equivalent to the Maximum Flow problem on networks with polynomial capacities. Therefore, the Maximum Flow with polynomial capacities problem belongs to the class RNC.*

Key Results

The first contribution is an extension of Theorem 1 to a generalization of the problem, namely the Maximum Flow on networks with polynomially bounded maximum flow. The proof is based on the construction (in NC) of a second network which has the same maximum flow but for which

the maximum flow and the maximum capacity in the network are polynomially related.

Lemma 2 *Let $N = (G, s, t, c)$. Given any integer k, there is an NC algorithm that decides whether $f(N) \geq k$ or $f(N) < km$.*

Since Lemma 2 applies even to numbers that are exponential in size, they get

Lemma 3 *Let $N = (G, s, t, c)$ be a network, there is an NC algorithm that computes an integer value k such that $2^k \leq f(N) < m\, 2^{k+1}$.*

The following lemma establishes the NC-reduction from the Maximum Flow problem with polynomial maximum flow to the Maximum Flow problem with polynomial capacities.

Lemma 4 *Let $N = (G, s, t, c)$ be a network, there is an NC algorithm that constructs a second network $N_1 = (G, s, t, c_1)$ such that*

$$\log(\mathrm{Max}(N_1)) \leq \log(f(N_1)) + O(\log n)$$

and $f(N) = f(N_1)$.

Lemma 4 shows that the Maximum Flow problem restricted to networks with polynomially bounded maximum flow is NC-reducible to the Maximum Flow problem restricted to polynomially bounded capacities, the latter problem is a simplification of the former one, so the following results follow.

Theorem 5 *For each polynomial p, the problem of constructing a maximum flow in a network N such that $f(N) \leq p(n)$ is NC-equivalent to the problem of constructing a maximum matching in a bipartite graph, and thus it is in RNC.*

Recall that [5] gave us an $O(\log^2 n)$ randomized parallel time algorithm to compute a maximum matching. The combination of this with the reduction from the Maximum Flow problem to the Maximum Matching leads to the following result.

Theorem 6 *There is a randomized parallel algorithm to construct a maximum flow in a directed network, such that the number of processors is* *bounded by a polynomial in the number of vertices and the time used is $O((\log n)^\alpha \log f(N))$ for some constant $\alpha > 0$.*

The previous theorem is the first step towards finding an approximate maximum flow in a network N by an RNC algorithm. The algorithm, given N and an $\varepsilon > 0$, outputs a solution f' such that $f(N)/f' \leq 1 + 1/\varepsilon$. The algorithm uses a polynomial number of processors (independent of ε) and parallel time $O(\log^\alpha n(\log n + \log \varepsilon))$, where α is independent of ε. Thus, the algorithm is an RNC one as long as ε is at most polynomial in n. (Actually ε can be $O(n^{\log^\beta n})$ for some β.) Thus, being a Fully RNC approximation scheme (FRNCAS).

The second ingredient is a rough NC approximation to the Maximum Flow problem.

Lemma 7 *Let $N = (G, s, t, c)$ be a network. Let $k \geq 1$ be an integer, then there is an NC algorithm to construct a network $M = (G, s, t, c_1)$ such that $k \, f(M) \leq f(N) \leq k \, f(M) + km$.*

Putting all together and allowing randomization the algorithm can be sketched as follows:
FAST-FLOW($N = (G, s, t, c), \varepsilon$)

1. Compute k such that $2^k \leq F(N) \leq 2^{k+1}m$.
2. Construct a network N_1 such that

$$\log(\text{Max}(N_1)) \leq \log(F(N_1)) + O(\log n).$$

3. If $2^k \leq (1 + \varepsilon)m$ then $F(N) \leq (1 + \varepsilon)m^2$ so use the algorithm given in Theorem 6 to solve the Maximum Flow problem in N as a Maximum Matching and **return**
4. Let $\beta = \lfloor (2^k)/((1 + \varepsilon)m) \rfloor$. Construct N_2 from N_1 and β using the construction in Lemma 7.
5. Solve the Maximum Flow problem in N_2 as a Maximum Matching.
6. Output $F' = \beta F(M_2)$ and for all $e \in E$, $f'(e) = \beta f(e)$.

Theorem 8 *Let $N = (G, s, t, c)$ be a network. Then, algorithm FAST-FLOW is an RNC algorithm such that for all $\varepsilon > 0$ at most polynomial in the number of network vertices, the algorithm computes a legal flow of value f' such that*

$$\frac{f(N)}{f'} \leq 1 + \frac{1}{\varepsilon}.$$

Furthermore, the algorithm uses a polynomial number of processors and runs in expected parallel time $O(\log^\alpha n(\log n + \log \varepsilon))$, for some constant α, independent of ε.

Applications

The *rounding/scaling* technique is used in general to deal with problems that are hard due to the presence of large weights in the problem instance. The technique modifies the problem instance in order to produce a second instance that has no large weights, and thus can be solved efficiently. The way in which a new instance is obtained consists of computing first an estimate of the optimal value (when needed) in order to discard unnecessary high weights. Then the weights are modified, scaling them down by an appropriate factor that depends on the estimation and the allowed error. The rounding factor is determined in such a way that the so-obtained instance can be solved efficiently. Finally, a last step consisting of scaling up the value of the "easy" instance solution is performed in order to meet the corresponding accuracy requirements.

It is known that in the sequential case, the only way to construct FPTAS uses rounding/scaling and interval partition [6]. In general, both techniques can be paralyzed, although sometimes the details of the parallelization are non-trivial [1].

The Maximum Flow problem has a long history in Computer Science. Here are recorded some results about its parallel complexity. Goldschlager, Shaw, and Staples showed that the Maximum Flow problem is P-complete [3]. The P-completeness proof for Maximum Flow uses large capacities on the edges; in fact the values of some capacities are exponential in the number of network vertices. If the capacities are constrained to be no greater than some polynomial in the number of network vertices the problem is in ZNC. In the case of planar networks it is known that the Maximum Flow problem is in NC, even if arbitrary capacities are allowed [4].

Open Problems

The parallel complexity of the Maximum Weight Matching problem when the weight of the edges are given in binary is still an open problem. However, as mentioned earlier, there is a randomized NC algorithm to solve the problem in $O(\log^2 n)$ parallel steps, when the weights of the edges are given in unary. The scaling technique has been used to obtain fully randomized NC approximation schemes, for the Maximum Flow and Maximum Weight Matching problems (see [10]). The result appears to be the best possible in regard of full approximation, in the sense that the existence of an FNCAS for any of the problems considered is equivalent to the existence of an NC algorithm for perfect matching which is also still an open problem.

Cross-References

▶ Maximum Matching
▶ Online Paging and Caching

Recommended Reading

1. Díaz J, Serna M, Spirakis PG, Torán J (1997) Paradigms for fast parallel approximation. In: Cambridge international series on parallel computation, vol 8. Cambridge University Press, Cambridge
2. Even S (1979) Graph algorithms. Computer Science Press, Potomac
3. Goldschlager LM, Shaw RA, Staples J (1982) The maximum flow problem is log-space complete for P. Theor Comput Sci 21:105–111
4. Johnson DB, Venkatesan SM (1987) Parallel algorithms for minimum cuts and maximum flows in planar networks. J ACM 34:950–967
5. Karp RM, Upfal E, Wigderson A (1986) Constructing a perfect matching is in random NC. Combinatorica 6:35–48
6. Korte B, Schrader R (1980) On the existence of fast approximation schemes. Nonlinear Prog 4:415–437
7. Lawler EL (1976) Combinatorial optimization: networks and matroids. Holt, Rinehart and Winston, New York
8. Papadimitriou C (1994) Computational complexity. Addison-Wesley, Reading
9. Peters JG, Rudolph L (1987) Parallel approximation schemes for subset sum and knapsack problems. Acta Informatica 24:417–432
10. Spirakis P (1993) PRAM models and fundamental parallel algorithm techniques: part II. In: Gibbons A, Spirakis P (eds) Lectures on parallel computation. Cambridge University Press, New York, pp 41–66

Randomized Rounding

Rajmohan Rajaraman
Department of Computer Science, Northeastern University, Boston, MA, USA

Years and Authors of Summarized Original Work

1987; Raghavan, Thompson

Problem Definition

Randomized rounding is a technique for designing approximation algorithms for NP-hard optimization problems. Many combinatorial optimization problems can be represented as 0-1 integer linear programs; that is, integer linear programs in which variables take values in $\{0, 1\}$. While 0-1 integer linear programming is NP-hard, the rational relaxations (also referred to as fractional relaxations) of these linear programs are solvable in polynomial time [12, 13]. Randomized rounding is a technique to construct a provably good solution to a 0-1 integer linear program from an optimum solution to its rational relaxation by means of a randomized algorithm.

Let Π be a 0-1 integer linear program with variables $x_i \in \{0, 1\}$, $1 \leq i \leq n$. Let Π_R be the rational relaxation of Π obtained by replacing the $x_i \in \{0, 1\}$ constraints by $x_i \in [0, 1]$, $1 \leq i \leq n$. The randomized rounding approach consists of two phases:

1. Solve Π_R using an efficient linear program solver. Let the variable x_i take on value $x_i^* \in [0, 1]$, $1 \leq i \leq n$.
2. Compute a solution to Π by setting the variables x_i randomly to one or zero according to the following rule:

$$\Pr[x_i = 1] = x_i^* .$$

For several fundamental combinatorial optimization problems, the randomized rounding technique yields simple randomized approximation algorithms that yield solutions provably close to optimal. Variants of the basic approach outlined above, in which the rounding of variable x_i in the second phase is done with a probability that is some appropriate function of x_i^*, have also been studied. The analyses of algorithms based on randomized rounding often rely on Chernoff–Hoeffding bounds from probability theory [5, 11].

The work of Raghavan and Thompson [14] introduced the technique of randomized rounding for designing approximation algorithms for NP-hard optimization problems. The randomized rounding approach also implicitly proves the existence of a solution with certain desirable properties. In this sense, randomized rounding can be viewed as a variant of the probabilistic method, due to Erdös [1], which is widely used for various existence proofs in combinatorics.

Raghavan and Thompson illustrate the randomized rounding approach using three optimization problems: VLSI routing, multicommodity flow, and k-matching in hypergraphs.

Definition 1 In the **VLSI Routing** problem, we are given a two-dimensional rectilinear lattice L_n over n nodes and a collection of m nets $\{a_i : 1 \leq i \leq m\}$, where net a_i, is a set of nodes to be connected by means of a Steiner tree in

L_n. For each net a_i, we are also given a set $\mathcal{A}_i$ of *allowed* trees that can be used for connecting the nodes in that set. A solution to the problem is a set $\mathcal{T}$ of trees $\{T_i \in \mathcal{A}_i : 1 \leq i \leq m\}$. The *width* of solution $\mathcal{T}$ is the maximum, over all edges e, of the number of trees in $\mathcal{T}$ that contain the edge. The goal of the VLSI routing problem is to determine a solution with minimum width.

Definition 2 In the **Multicommodity Flow Congestiom Minimization** problem (or simply, the Congestion Minimization problem), we are given a graph $G = (V, E)$, and a set of source-destination pairs $\{(s_i, t_i) : 1 \leq i \leq k\}$. For each pair (s_i, t_i), we would like to route one unit of demand from s_i to t_i. A solution to the problem is a set $\mathcal{P} = \{P_i : 1 \leq i \leq k\}$ such that P_i is a path from s_i to t_i in G. We define the *congestion* of $\mathcal{P}$ to be the maximum, over all edges e, of the number of paths containing e. The goal of the undirected multicommodity flow problem is to determine a path set $\mathcal{P}$ with minimum congestion.

In their original work [14], Raghavan and Thompson studied the above problem for the case of undirected graphs and referred to it as the Undirected Multicommodity Flow problem. Here, we adopt the more commonly-used term of Congestion Minimization and consider both undirected and directed graphs since the results of [14] apply to both classes of graphs. Researchers have studied a number of variants of the multicommodity flow problem, which differ in various aspects of the problem such as the nature of demands (e.g., uniform vs. non-uniform), the objective function (e.g., the total flow vs. the maximum fraction of each demand), and edge capacities (e.g., uniform vs. non-uniform).

Definition 3 In the **Hypergraph Simple k-Matching** problem, we are given a hypergraph H over an n-element vertex set V. A k-matching of H is a set M of edges such that each vertex in V belongs to at most k of the edges in M. A k-matching M is simple if no edge in H occurs more than once in M. The goal of the problem is to determine a maximum-size simple k-matching of a given hypergraph H.

Key Results

Raghavan and Thompson present approximation algorithms for the above three problems using randomized rounding. In each case, the algorithm is easy to present: write a 0-1 integer linear program for the problem, solve the rational relaxation of this program, and then apply randomized rounding. They establish bounds on the quality of the solutions (i.e., the approximation ratios of the algorithm) using Chernoff–Hoeffding bounds on the tail of the sums of bounded and independent random variables [5, 11].

The VLSI Routing problem can be easily expressed as a 0-1 integer linear program, say Π_1. Let W^* denote the width of the optimum solution to the rational relaxation of Π_1.

Theorem 1 *For any ε such that $0 < \varepsilon < 1$, the width of the solution produced by randomized rounding does not exceed*

$$W^* + \left[3W^* \ln \frac{2n(n-1)}{\varepsilon} \right]^{1/2}$$

with probability at least $1 - \varepsilon$, provided $W^* \geq 3\ln(2n(n-1)/\varepsilon)$.

Since W^* is a lower bound on the width of an optimum solution to Π_1, it follows that the randomized rounding algorithm has an approximation ratio of $1 + o(1)$ with high probability as long as W^* is sufficiently large.

The Congestion Minimization problem can be easily expressed as a 0-1 integer linear program, say Π_2. Let C^* denote the congestion of the optimum solution to the linear relaxation of Π_2. This optimum solution yields a set of flows, one for each commodity i. The flow for commodity i can be decomposed into a set Γ_i of at most $|E|$ paths from s_i to t_i. The randomized rounding algorithm selects, for each commodity i, one path P_i at random from Γ_i according to the flow values determined by the flow decomposition.

Theorem 2 *For any ε such that $0 < \varepsilon < 1$, the capacity of the solution produced by randomized rounding does not exceed*

$$C^* + \left[3C^* \ln \frac{|E|}{\varepsilon} \right]^{1/2}$$

with probability at least $1 - \varepsilon$, provided $C^* \geq 2\ln|E|$.

Since C^* is a lower bound on the width of an optimum solution to Π_1, it follows that the randomized rounding algorithm achieves a constant approximation ratio with probability $1 - 1/n$ when C^* is $\Omega(\log n)$.

For both the VLSI Routing and the Congestion Minimization problems, slightly worse approximation ratios are achieved if the lower bound condition on W^* and C^*, respectively, is removed. In particular, the approximation ratio achieved is $O(\log n / \log \log n)$ with probability at least $1 - n^{-c}$ for a constant $c > 0$ whose value depends on the constant hidden in the big-Oh notation.

The hypergraph k-matching problem is different than the above two problems in that it is a packing problem with a maximization objective while the latter are covering problems with a minimization objective. Raghavan and Thompson show that randomization rounding, in conjunction with a scaling technique, yields good approximation algorithms for the hypergraph k-matching problem. They first express the matching problem as a 0-1 integer linear program, solve its rational relaxation Π_3, and then round the optimum rational solution by using appropriately scaled values of the variables as probabilities. Let S^* denote the value of the optimum solution to Π_3.

Theorem 3 *Let δ_1 and δ_2 be positive constants such that $\delta_2 > n \cdot e^{-k/6}$ and $\delta_1 + \delta_2 < 1$. Let $\alpha = 3\ln(n/\delta_2)/k$ and*

$$S' = S^* \left(1 - \frac{(\alpha^2 + 4\alpha)^{1/2} - \alpha}{2} \right).$$

Then, there exists a simple k-matching for the given hypergraph with size at least

$$S' - \left(2S' \ln \frac{1}{\delta_1} \right)^{1/2}.$$

Note that the above result is stated as an existence result. It can be modified to yield a randomized algorithm that achieves essentially the same bound with probability $1 - \varepsilon$ for a given failure probability ε.

Applications

Randomized rounding has found applications for a wide range of combinatorial optimization problems. Following the work of Raghavan and Thompson [14], Goemans and Williamson showed that randomized rounding yields an $e/(e-1)$-approximation algorithm for MAXSAT, the problem of finding an assignment that satisfies the maximum number of clauses of a given Boolean formula [7]. For the set cover problem, randomized rounding yields an algorithm with an asymptotically optimal approximation ratio of $O(\log n)$, where n is the number of elements in the given set cover instance [10]. Srinivasan has developed more sophisticated randomized rounding approaches for set cover and more general covering and packing problems [15]. Randomized rounding also yields good approximation algorithms for several flow and cut problems, including variants of undirected multicommodity flow [9] and the multiway cut problem [4].

While randomized rounding provides a unifying approach to obtain approximation algorithms for hard optimization problems, better approximation algorithms have been designed for specific problems. In some cases, randomized rounding has been combined with other algorithms to yield better approximation ratios than previously known. For instance, Goemans and Williamson showed that the better of two solutions, one obtained by randomized rounding and the other obtained by an earlier algorithm due to Johnson, yields a 4/3 approximation for MAXSAT [7].

The work of Raghavan and Thompson applied randomized rounding to a solution obtained for the relaxation of a 0-1 integer program for a given problem. In recent years, more sophisticated approximation algorithms have been obtained by applying randomized rounding to semidefinite program relaxations of the given problem. Examples include the 0.87856-approximation algorithm for MAXCUT due to Goemans and Williamson [8] and an $O(\sqrt{\log n})$-approximation algorithm for the sparsest cut problem, due to Arora, Rao, and Vazirani [3].

An excellent reference for the above and other applications of randomized rounding in approximation algorithms is the text by Vazirani [16].

Open Problems

While randomized rounding has yielded improved approximation algorithms for a number of NP-hard optimization problems, the best approximation achievable by a polynomial-time algorithm is still open for most of the problems discussed in this article, including MAXSAT, MAXCUT, the sparsest cut, the multiway cut, and several variants of the congestion minimization problem. For directed graphs, it has been shown that best approximation ratio achievable for congestion minimization in polynomial time is $\Omega(\log n/\log\log n)$, unless NP $\subset$ ZPTIME$(n^{O(\log\log n)})$, matching the upper bound mentioned in section "Key Results" up to constant factors [6]. For undirected graphs, the best known inapproximability lower bound is $\Omega(\log\log n/\log\log\log n)$ [2].

Cross-References

▶ Oblivious Routing

Recommended Reading

1. Alon N, Spencer JH (1991) The probabilistic method. Wiley, New York
2. Andrews M, Zhang L (2005) Hardness of the undirected congestion minimization problem. In: STOC'05: proceedings of the thirty-seventh annual ACM symposium on theory of computing. ACM, New York, pp 284–293
3. Arora S, Rao S, Vazirani UV (2004) Expander flows, geometric embeddings and graph partitioning. In: STOC, pp 222–231

4. Calinescu G, Karloff HJ, Rabani Y (2000) An improved approximation algorithm for multiway cut. J Comput Syst Sci 60(3):564–574
5. Chernoff H (1952) A measure of the asymptotic efficiency for tests of a hypothesis based on the sum of observations. Ann Math Stat 23:493–509
6. Chuzhoy J, Guruswami V, Khanna S, Talwar K (2007) Hardness of routing with congestion in directed graphs. In: STOC'07: proceedings of the thirty-ninth annual ACM symposium on theory of computing. ACM, New York, pp 165–178
7. Goemans MX, Williamson DP (1994) New 3/4-approximation algorithms for the maximum satisfiability problem. SIAM J Discret Math 7:656–666
8. Goemans MX, Williamson DP (1995) Improved approximation algorithms for maximum cut and satisfiability problems using semidefinite programming. J ACM 42(6):1115–1145
9. Guruswami V, Khanna S, Rajaraman R, Shepherd B, Yannakakis M (2003) Near-optimal hardness results and approximation algorithms for edge-disjoint paths and related problems. J Comput Syst Sci 67:473–496
10. Hochbaum DS (1982) Approximation algorithms for the set covering and vertex cover problems. SIAM J Comput 11(3):555–556
11. Hoeffding W (1956) On the distribution of the number of successes in independent trials. Ann Math Stat 27:713–721
12. Karmarkar N (1984) A new polynomial-time algorithm for linear programming. Combinatorica 4:373–395
13. Khachiyan LG (1979) A polynomial algorithm for linear programming. Sov Math Dokl 20:191–194
14. Raghavan P, Thompson C (1987) Randomized rounding: a technique for provably good algorithms and algorithmic proofs. Combinatorica 7
15. Srinivasan A (1995) Improved approximations of packing and covering problems. In: Proceedings of the 27th annual ACM symposium on theory of computing, pp 268–276
16. Vazirani V (2003) Approximation algorithms. Springer

Randomized Searching on Rays or the Line

Stephen R. Tate
Department of Computer Science, University of North Carolina, Greensboro, NC, USA

Keywords

Cow-path problem; Online navigation

Years and Authors of Summarized Original Work

1993; Kao, Reif, Tate

Problem Definition

This problem deals with finding a point at an unknown position on one of a set of w rays which extend from a common point (the origin). In this problem there is a *searcher*, who starts at the origin, and follows a sequence of commands such as "explore to distance d on ray i." The searcher detects immediately when the target point is crossed, but there is no other information provided from the search environment. The goal of the searcher is to minimize the distance traveled.

There are several different ways this problem has been formulated in the literature, including one called the "cow-path problem" that involves a cow searching for a pasture down a set of paths. When $w = 2$, this problem is to search for a point on the line, which has also been described as a robot searching for a door in an infinite wall or a shipwreck survivor searching for a stream after washing ashore on a beach.

Notation

The problem is as described above, with w rays. The position of the target point (or goal) is denoted (g, i) if it is at distance g on ray $i \in \{0, 1, \ldots, w - 1\}$. The standard notion of *competitive ratio* is used when analyzing algorithms for this problem: An algorithm that knows which ray the goal is on will simply travel distance g down that ray before stopping, so search algorithms are compared to this optimal, omniscient strategy.

In particular, if $\mathcal{R}$ is a randomized algorithm, then the distance traveled to find a particular goal position is a random variable denoted $distance\ (\mathcal{R}, (g, i))$, with expected value $E\,[\text{distance}\,(\mathcal{R}, (g, i))]$. Algorithm $\mathcal{R}$ has competitive ratio c if there is a constant a such that, for all goal positions (g, i),

$$E\,[\text{distance}\,(\mathcal{R}, (g, i))] \leq c.g + a. \tag{1}$$

Key Results

This problem is solved optimally using a randomized geometric sweep strategy: Search through the rays in a random (but fixed) order, with each search distance a constant factor longer than the preceding one. The initial search distance is picked from a carefully selected probability distribution, giving the following algorithm:

RAYSEARCH $_{r,w}$
$\sigma \leftarrow$ A random permutation of $\{0,1,2,\dots,w-1\}$;
$\epsilon \leftarrow$ A random real uniformly chosen from $[0,1)$;
$d \leftarrow r^{\epsilon}$;
$p \leftarrow 0$;
repeat
Explore path $\sigma(p)$ up to distance d;
if goal not found then return to origin;
$d \leftarrow d \cdot r$;
$p \leftarrow (p+1) \bmod w$;
until goal found;

The following theorems give the competitive ratio of this algorithm, show how to pick the best r, and establish the optimality of the algorithm.

Theorem 1 ([9]) *For any fixed $r > 1$, Algorithm* RAYSEARCH $_{r,w}$ *has competitive ratio*

$$R(r,w) = 1 + \frac{2}{w} \cdot \frac{1 + r + r^2 + \cdots + r^{w-1}}{\ln r},$$

Theorem 2 ([9]) *The unique solution of the equation*

$$\ln r = \frac{1 + r + r^2 + \cdots + r^{w-1}}{r + 2r^2 + 3r^3 + \cdots + (w-1)r^{w-1}} \quad (2)$$

for $r > 1$, denoted by r_w^*, gives the minimum value for $R(r,w)$.

Theorem 3 ([8, 9, 12]) *The optimal competitive ratio for any randomized algorithm for searching on w rays is*

$$\min_{r>1}\left\{1 + \frac{2}{w} \cdot \frac{1 + r + r^2 + \cdots + r^{w-1}}{\ln r}\right\}.$$

Corollary 1 *Algorithm* RAYSEARCH $_{r,w}$ *is optimally competitive.*

Using Theorem 2 and standard numerical techniques, r_w^* can be computed to any required degree of precision. The following table shows, for small values of w, approximate values for r_w^* and the corresponding optimal competitive ratio (achieved by RAYSEARCH$_{r,w}$) – the optimal deterministic competitive ratio (see [1]) is also shown for comparison (Table 1):

Theorem 4 ([9]) *The competitive ratio for algorithm* RAYSEARCH $_{r,w}$ *(with $r = r_w^*$) is $\kappa w + o(w)$, where*

$$k = \min_{s>0}\left[2\frac{e^s - 1}{s^2}\right] \approx 3.088.$$

Applications

The most direct applications of this problem are in geometric searching, such as robot navigation problems. For example, when a robot is traveling in an unknown area and encounters an obstacle, a typical first step is to find the nearest corner to go around [2, 3], which is just an instance of the ray searching problem (with $w = 2$).

In addition, any abstract search problem with a cost function that is linear in the distance to the goal reduces to ray searching. This includes applications in artificial intelligence that search for a goal in a largely unknown search space

Randomized Searching on Rays or the Line, Table 1 The asymptotic growth of the competitive ratio with w is established in the following theorem

w	r_w^*	Optimal randomized ratio	Optimal deterministic ratio
2	3.59112	4.59112	9
3	2.01092	7.73232	14.5
4	1.62193	10.84181	19.96296
5	1.44827	13.94159	25.41406
6	1.35020	17.03709	30.85984
7	1.28726	20.13033	36.30277

[11] and the construction of hybrid algorithms [8]. In hybrid algorithms, a set of algorithms $A_1, A_2, \ldots, A_w$ for solving a problem is considered – algorithm A_1 is run for a certain amount of time, and if the algorithm is not successful algorithm A_1 is stopped and algorithm A_2 is started, repeating through all algorithms as many times as is necessary to find a solution. This notion of hybrid algorithms has been used successfully for several problems (such as the first competitive algorithm for the online k-server problem [4]), and the ray search algorithm gives the optimal strategy for selecting the trial running times of each algorithm.

Open Problems

Several natural extensions of this problem have been studied in both deterministic and randomized settings, including ray searching when an upper bound on the distance to the goal is known (i.e., the rays are not infinite but are line segments) [5, 10, 12], or when a probability distribution of goal positions is known [7]. Other variations of this basic searching problem have been studied for deterministic algorithms only, such as when the searcher's control is imperfect (so distances cannot be specified precisely) [6] and for more general search spaces like points in the plane [1]. A thorough study of these variants with randomized algorithms remains an open problem.

Cross-References

▶ Alternative Performance Measures in Online Algorithms
▶ Deterministic Searching on the Line
▶ Robotics

Recommended Reading

1. Baeza-Yates RA, Culberson JC, Rawlins GJE (1993) Searching in the plane. Inf Comput 16:234–252
2. Berman P, Blum A, Fiat A, Karloff H, Rosén A, Saks M (1996) Randomized robot navigation algorithms. In: Proceedings of the seventh annual ACM-SIAM symposium on discrete algorithms (SODA), pp 75–84
3. Blum A, Raghavan P, Schieber B (1991) Navigating in unfamiliar geometric terrain. In: Proceedings 23rd ACM symposium on theory of computing (STOC), pp 494–504
4. Fiat A, Rabani Y, Ravid Y (1990) Competitive k-server algorithms. In: Proceedings 31st IEEE symposium on foundations of computer science (FOCS), pp 454–463
5. Hipke C, Icking C, Klein R, Langetepe E (1999) How to find a point on a line within a fixed distance. Discret Appl Math 93:67–73
6. Kamphans T, Langetepe E (2005) Optimal competitive online ray search with an error-prone robot. In: 4th international workshop on experimental and efficient algorithms, pp 593–596
7. Kao M-Y, Littman ML (1997) Algorithms for informed cows. In: AAAI-97 workshop on on-line search, pp 55–61
8. Kao M-Y, Ma Y, Sipser M, Yin Y (1994) Optimal constructions of hybrid algorithms. In: Proceedings 5th ACM-SIAM symposium on discrete algorithms (SODA), pp 372–381
9. Kao M-Y, Reif JH, Tate SR (1996) Searching in an unknown environment: An optimal randomized algorithm for the cow-path problem. Inf Comput 133:63–80
10. López-Ortiz A, Schuierer S (2001) The ultimate strategy to search on m rays? Theor Comput Sci 261:267–295
11. Pearl J (1984) Heuristics: intelligent search strategies for computer problem solving. Addison-Wesley, Reading
12. Schuierer S (2003) A lower bound for randomized searching on m rays. In: Computer science in perspective, pp 264–277

Randomized Self-Assembly

David Doty
Computing and Mathematical Sciences,
California Institute of Technology, Pasadena,
CA, USA

Keywords

Linear assembly; Randomized; Tile complexity

Supported by NSF grants CCF-1219274, CCF-1162589, and 1317694.

Years and Authors of Summarized Original Work

2006; Becker, Rapaport, Rémila
2008; Kao, Schweller
2009; Chandran, Gopalkrishnan, Reif
2010; Doty

Problem Definition

We use the abstract tile assembly model of Winfree [6], which models the aggregation of monomers called *tiles* that attach one at a time to a growing structure, starting from a single *seed* tile, in which bonds ("glues") on the tile are specific (glues only stick to glues of the same type on other tiles) and cooperative (so that multiple weak glues are necessary to attach a tile). The general idea of *randomized* self-assembly is to use the inherent randomness of self-assembly to help the assembly process. If multiple types of tiles are able to bind to a single binding site, then we assume that their relative concentrations determine the probability that each succeeds. With careful design, we can use the same tile set to create different structures, by changing the concentrations to affect what is likely to assemble. Another use of randomness is in reducing the number of different tile types required to assemble a shape.

Definitions

A *shape* is a finite, connected subset of $\mathbb{Z}^2$. A *tile type* is a unit square with four sides, each side consisting of a *glue label* (finite string) and a nonnegative integer *strength*. We assume a finite set T of tile types, but an infinite number of copies of each tile type, each copy referred to as a *tile*. An *assembly* is a positioning of tiles on the integer lattice $\mathbb{Z}^2$; i.e., a partial function $\alpha : \mathbb{Z}^2 \dashrightarrow T$. Write $\alpha \sqsubseteq \beta$ to denote that α is a *subassembly* of β, which means that dom $\alpha \subseteq$ dom β and $\alpha(p) = \beta(p)$ for all points $p \in$ dom α. In this case, say that β is a *superassembly* of α. Two adjacent tiles in an assembly *interact* if the glue labels on their abutting sides are equal and have positive strength. Each assembly induces a *binding graph*, a grid graph whose vertices are tiles, with an edge between two tiles if they interact. The assembly is τ-*stable* if every cut of its binding graph has strength at least τ, where the weight of an edge is the strength of the glue it represents (energy τ is required to separate the assembly). The τ-*frontier* $\partial^\tau \alpha \subset \mathbb{Z}^2 \setminus$ dom α of α (or *frontier* $\partial \alpha$ when τ is clear from context) is the set of empty locations adjacent to α at which a single tile could bind stably.

A *tile system* is a triple $\mathcal{T} = (T, \sigma, \tau)$, where T is a finite set of tile types, $\sigma : \mathbb{Z}^2 \dashrightarrow T$ is a *seed assembly* consisting of a single tile (i.e., $|\text{dom } \sigma| = 1$), and $\tau \in \mathbb{N}$ is the *temperature*. An assembly α is *producible* if either $\alpha = \sigma$ or if β is a producible assembly and α can be obtained from β by the stable binding of a single tile. In this case, write $\beta \rightarrow_1 \alpha$ (α is producible from β by the attachment of one tile), and write $\beta \rightarrow \alpha$ if $\beta \rightarrow_1^* \alpha$ (α is producible from β by the attachment of zero or more tiles). If α is producible, then there is an *assembly sequence* $\vec{\alpha} = (\alpha_i \mid 1 \le i \le k)$ such that $\alpha_1 = \sigma$, $\alpha_k = \alpha$, and, for each $i \in \{1, \ldots, k-1\}$, $\alpha_i \rightarrow_1 \alpha_{i+1}$. An assembly is *terminal* if no tile can be τ-stably attached to it. Write $\mathcal{A}[\mathcal{T}]$ to denote the set of all producible assemblies of $\mathcal{T}$, and write $\mathcal{A}_\square[\mathcal{T}]$ to denote the set of all producible, terminal assemblies of $\mathcal{T}$. We also speak of *shapes* assembled by tile assembly systems, by which we mean dom α if $\alpha \in \mathcal{A}_\square[\mathcal{T}]$, and we consider shapes to be equivalent up to translation.

We now define the semantics of incorporating randomization into self-assembly. Intuitively, there are two sources of nondeterminism in the model as defined: (1) if $|\partial \alpha| > 1$, then there are multiple binding sites, one of which is nondeterministically selected as the next site to receive a tile, and (2) if multiple tile types could bind to a *single* binding site, then one of them is nondeterministically selected. Both concepts are handled by assigning positive real-valued concentrations to each tile type; Ref. [3] gives a full definition that accounts for both of these. However, in the results we discuss, only the latter source of nondeterminism will actually affect the probabilities of various terminal assemblies being produced; the binding sites themselves can be picked in an arbitrary order without affecting these probabilities. Thus we state here a simpler definition based on this assumption.

A *tile concentration assignment* on $\mathcal{T}$ is a function $\rho : T \rightarrow [0, \infty)$. If $\rho(t)$ is not specified explicitly for some $t \in T$, then $\rho(t) = 1$. If α is a τ-stable assembly such that $t_1, \ldots, t_j \in T$ are the tiles capable of binding to the same position $\mathbf{m} \in \partial \alpha$, then for $1 \leq i \leq j$, t_i binds at position $\mathbf{m}$ with probability $\frac{\rho(t_i)}{\rho(t_1) + \cdots + \rho(t_j)}$. ρ induces a probability measure on $\mathcal{A}_\square[T]$ in a straightforward way. Formally, let $\alpha \in \mathcal{A}_\square[T]$ be a producible terminal assembly. Let $A(\alpha)$ be the set of all assembly sequences $\boldsymbol{\alpha} = (\alpha_i \mid 1 \leq i \leq k)$ such that $\alpha_k = \alpha$, with $p_{\boldsymbol{\alpha}, i}$ denoting the probability of attachment of the tile added to α_{i-1} to produce α_i (noting that $p_{\boldsymbol{\alpha}, i} = 1$ if the ith tile attached without contention). Then $\Pr[\alpha] = \sum_{\boldsymbol{\alpha} \in A(\alpha)} \prod_{i=2}^{k} \frac{1}{|\partial \alpha_i|} p_{\boldsymbol{\alpha}, i}$. Write $\mathcal{T}(\rho)$ to denote the random variable representing the producible, terminal assembly produced by $\mathcal{T}$ when using tile concentration assignment ρ.

Problems

The general problem is this: given a shape $X \subset \mathbb{Z}^2$ (a connected, finite set), set the concentrations of tile types in some tile system $\mathcal{T}$ so that $\mathcal{T}$ is likely to create a terminal assembly with shape X or "close to it." We now state formal problems that are variations on this theme. The first four problems use "concentration programming": varying the concentrations of tile types in a single tile system $\mathcal{T}$ to get it to assemble different shapes. The last two problems concern a tile system that only does one thing – assemble a line of a desired expected length – because in this setting we will require all concentrations to be equal. However, the tile system uses randomized self-assembly to do this with far fewer tile types than are needed to accomplish the same task in a deterministic tile system.

The first three problems concern the self-assembly of squares, and the problems are listed

in order of increasing difficulty. The first asks for a square with a desired expected width, the second for a guarantee that the actual width is likely to be *close* to the expected width, and finally, for a guarantee that the actual width is likely to be *exactly* the expected width.

Formally, design a tile system $\mathcal{T} = (T, \sigma, \tau)$ such that, for any $n \in \mathbb{Z}^+$, there exists a tile concentration assignment $\rho : T \rightarrow [0, \infty)$ such that...

Problem 1 ...dom $\mathcal{T}(\rho)$ is a square with expected width n.

Problem 2 ...with probability at least $1 - \delta$, dom $\mathcal{T}(\rho)$ is a square whose width is between $(1 - \epsilon)n$ and $(1 + \epsilon)n$.

Problem 3 ...with probability at least $1 - \delta$, dom $\mathcal{T}(\rho)$ is a square of width n.

The next problem generalizes the previous problems to arbitrary shapes, while making one relaxation: allowing a scaled-up version of a shape to be assembled instead of the exact shape. Formally, for $c \in \mathbb{Z}^+$ and shape $S \subset \mathbb{Z}^2$ (finite and connected), define $S^c = \{ (x, y) \in \mathbb{Z}^2 \mid (\lfloor x/c \rfloor, \lfloor y/c \rfloor) \in S \}$ to be S *scaled by factor* c.

Problem 4 Let $\delta > 0$. Design a tile system $\mathcal{T} = (T, \sigma, \tau)$ such that, for any shape $S \subset \mathbb{Z}^2$, there exists a tile concentration assignment $\rho : T \rightarrow [0, \infty)$ and $c \in \mathbb{Z}^+$ so that, with probability at least $1 - \delta$, dom $\mathcal{T}(\rho)$ is S^c.

It is easy to see that for a deterministic tile system to assemble a length n, height 1 line requires n tile types. The next problem concerns using randomization to reduce the number of tile types required, subject to the constraint that all tile type concentrations are equal. (Without this constraint, a solution to Problem 1 would trivially be a solution to the next problem, with optimal $O(1)$ tile types, but since the solution to Problem 1 uses different tile type concentrations to achieve its goal, it cannot be used directly for this purpose.)

Problem 5 Let $n \in \mathbb{Z}^+$. Design a tile system $\mathcal{T} = (T, \sigma, \tau)$ such that, with tile concentration assignment $\rho : T \to [0, \infty)$ defined by $\rho(t) = 1$ for all $t \in T$, dom $\mathcal{T}(\rho)$ is a height 1 line of expected length n.

As with the case of concentration programming, it is desirable for the line to have length likely to be close to its expected length.

Problem 6 Let $n \in \mathbb{Z}^+$ and $\delta, \epsilon > 0$. Design a tile system $\mathcal{T} = (T, \sigma, \tau)$ such that, with tile concentration assignment $\rho : T \to [0, \infty)$ defined by $\rho(t) = 1$ for all $t \in T$, dom $\mathcal{T}(\rho)$ is a height 1 line whose length is between $(1 - \epsilon)n$ and $(1 + \epsilon)n$ with probability at least $1 - \delta$.

Key Results

The solutions to Problems 1–4 use temperature 2 tile systems. The solutions to Problems 5 and 6 use a temperature 1 tile system (there is no need for cooperative binding in one dimension).

Figure 1 shows a simple tile system with three tile types that can grow a line of any desired expected length to the right of the seed tile; this is the basis for the solutions to Problems 1–4. The length of the line has a geometric distribution, with expected value controlled by the ratio of the concentrations of G and S. Figure 2 shows the solution to Problem 1, due to Becker, Remilá, and Rapaport [1]. It is essentially the tile system from

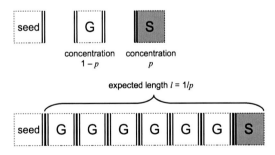

Randomized Self-Assembly, Fig. 1 A randomized temperature $\tau = 2$ tile system that can grow a line of any desired expected length l by setting $p = \frac{1}{l}$. Two tiles compete nondeterministically to bind to the right of the line (using strength 2 glues, indicated by *double black lines*), one of which stops the growth, while the other continues, giving the length of the line (not counting the seed) a geometric distribution with expected value l

Fig. 1 (tile types A and B are analogous to G and S in Fig. 1) augmented with a constant number of extra tiles that can assemble the square to be as high as the line is long.

Kao and Schweller [4] showed a solution to Problem 2, and Doty [3] improved their construction to show a solution to Problem 3. Here, we describe only the latter construction, since the two share similar ideas, and the latter construction solves both problems.

Figure 3 shows an improvement to the tile system of Fig. 1, which will be the starting point for the solution. It also can grow a line of any desired expected length. However, by using multiple independent "stages" of growth, each stage having a geometric distribution, the resulting assembly is more likely to have a length that is close to its expected length. More tile types are needed for more stages, but only a constant number of stages are required.

In particular, if the expected length is chosen to be midway between any two consecutive powers of two, i.e., midway in the interval $[2^{a-1}, 2^a)$ for arbitrary $a \in \mathbb{N}$, with $r = 113$ stages, the probability is at most 0.0025 that the actual length is outside the interval $[2^{a-1}, 2^a)$. So although the length is not controlled with exact precision, the number of bits needed to represent the length is controlled with exact precision (with high probability), using a constant number of tile types.

Figure 4 shows a tile system $\mathcal{T}$ with the following property: for any bit string s (equivalently, any natural number m if we assume the most significant bit of s is 1), there is a tile concentration assignment that causes $\mathcal{T}$ to grow an assembly of height $O(\log m)$, width $O(m^2)$, such that the tile types in the upper-right corner of the assembly encode s. The bottom row is the tile system from Fig. 3, with identical strength 2 glues on the north of the tiles (other than the final stop tile on the right).

Figure 5 shows a high-level overview of the entire tile system that assembles an $n \times n$ square, solving Problem 3. Using similar ideas to Fig. 4, one can encode three different numbers $m_1, m_2, m_3 \in \mathbb{N}$ into the tile concentrations. We choose these numbers to be such that each $m_i = O(n^{1/3})$, and each of their binary

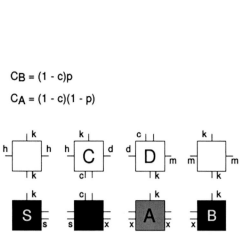

$$CB = (1 - c)p$$

$$CA = (1 - c)(1 - p)$$

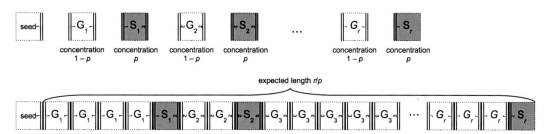

Randomized Self-Assembly, Fig. 2 A tile system that grows a square of any desired expected width (Figure taken from [4]); strength 2 glues are indicated by *two lines* between the tiles. The seed is labeled S, and C_A and C_B respectively represent the concentrations of A and B. p is used the same way as in Fig. 1, and c represents total concentration of all other tile types, since [4] assumed that concentrations of all tile types must sum to 1

Randomized Self-Assembly, Fig. 3 A tile system that grows a line of a given length with greater precision than in Fig. 1. r stages each have expected length $1/p$, making the expected total length r/p, but more tightly concentrated about that expected length than in the case of one stage

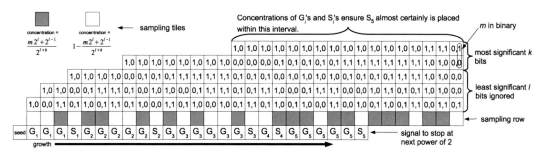

Randomized Self-Assembly, Fig. 4 Computing the binary string 10 (equivalently, the natural number $m = 2$) from tile concentrations. For brevity, glue strengths and labels are not shown. Each column increments the primary counter, represented by the bits on the left of each tile, and each *gray tile* increments the sampling counter, represented by the bits on the right of each tile. The number of bits at the end is $l + k$, where c is a constant coded into the tile set and k depends on m, and $l = k + c$. The most significant k bits of the sampling counter encode m. In this example, $k = 2$ and $c = 1$

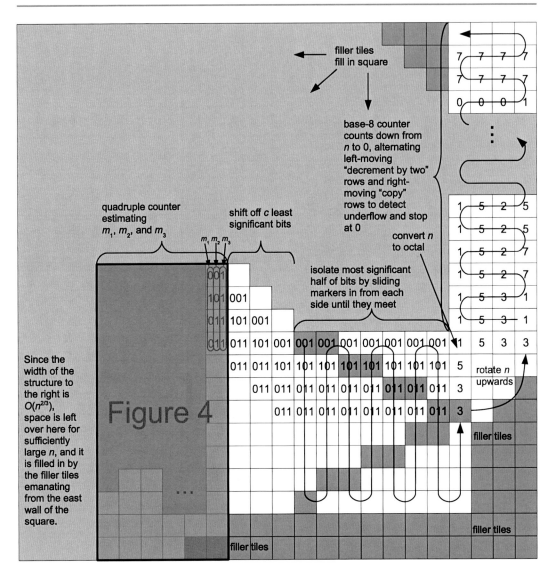

Randomized Self-Assembly, Fig. 5 High-level overview of the entire construction solving Problem 3, not at all to scale. For brevity, glue strengths and labels are not shown. The double counter number estimator of Fig. 4 is embedded with two additional counters to create a quadruple counter estimating m_1, m_2, and m_3, shown as a box labeled as "Fig. 4" in the above figure. In this example, $m_1 = 4$, $m_2 = 3$, and $m_3 = 15$, represented vertically in binary in the most significant 4 tiles at the end of the quadruple counter. Concatenating the bits of the tiles results in the string 001101011011, the binary representation of 859, which equals $n - 2k - 4$ for $n = 871$, so this example builds an 871×871 square. Once the counter ends, c tiles ($c = 3$ in this example) are shifted off the bottom, and the top half of the tiles are isolated ($k = 4$ in this example). Each remaining tile represents 3 bits of n, which are converted into octal digits, rotated to face upwards, and then used to initialize a base-8 counter that builds the east wall of the square. Filler tiles cover the remaining area of the square

expansions, interwoven into a single bit string, is the binary expansion of n. Then each tile at the upper right of Fig. 4 encodes not one but 3 bits of n, or equivalently each encodes an octal digit of n. These bits are then used to assemble a counter that counts from n down to 0 as it grows north, and a constant set of tiles (similar to Fig. 2) expand this counter to grow about as far east as the counter grows north, creating an $n \times n$ square that surrounds the assembly of Fig. 4.

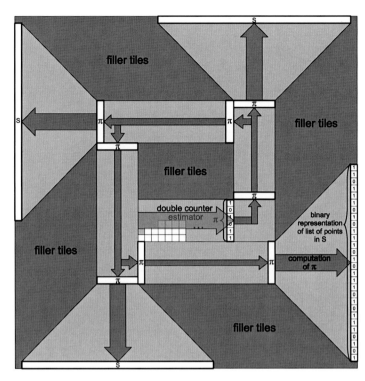

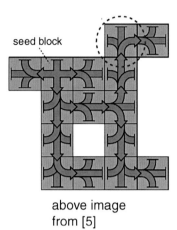

seed block

above image
from [5]

Randomized Self-Assembly, Fig. 6 On the *left* is the seed block used to replace the seed block of [5], from which the construction of [5] can assemble a scaled version of the shape S (encoded by a binary string representing the list of coordinates, also labeled "S" in the figure). S is output by the single-tape Turing machine program π. π is estimated from tile concentrations as in Fig. 4, then four copies of it are propagated to each side of the block, where it is executed in four rotated, but otherwise identical, computation regions. When completed, four copies of the binary representation of S border the seed block, which is sufficient for the construction of [5] to assemble a scaled version of S using a spanning tree of S as shown on the *right*

Since $m_i = O(n^{1/3})$, and the tiles of Fig. 4 create a structure of height $O(\log m_i)$ and width $O(m_i^2) = O(n^{2/3})$, the square is sufficiently large to contain the tiles of Fig. 4.

Finally, the tiles of Fig. 4 are used in a different way to solve Problem 4, shown in Fig. 6. Given a finite shape S, Soloveichik and Winfree [5] use an intricate construction of a "seed block" that "unpacks," from a set of tile types that depend on S, a single-tape Turing machine program $\pi \in \{0, 1\}^*$ that outputs a binary string $\mathrm{bin}(S)$ representing a list of the coordinates of S.

The width of the seed block is then c, chosen to be large enough to do the unpacking and also large enough to accommodate the simulation of π by a tile set that simulates single-tape Turing machines. Once this seed block is in place, a tile set then assembles the scaled shape by carrying bin(S) through each block. The order in which blocks are assembled is determined by a spanning tree of S, so that any blocks with an ancestor relationship have a dependency, in that the ancestor must be (mostly) assembled before the descendant, whereas blocks without an ancestor relationship can potentially assemble in parallel.

We replace the seed block tiles of [5], which depend on S, with a single tile system that produces the program π from tile concentrations, and use the remainder of the tile set of [5] unchanged. This is illustrated in Fig. 6. Choose c to be sufficiently large that π can be simulated within the trapezoidal region of the $c \times c$ block of Fig. 6 and also sufficiently large that the construction of Fig. 4 has sufficient room to estimate the binary string π from tile concentrations in the center region (the

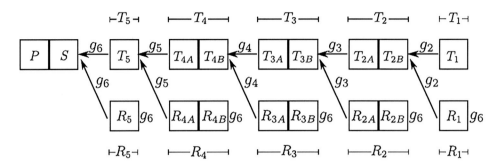

Randomized Self-Assembly, Fig. 7 Example of solution to Problem 5 for the case of expected length 92

"double counter estimator") of Fig. 6. Once this is done, the construction of [5] can take over and assemble the entire scaled shape S^c. The portion of the construction of [5] that achieves this is a constant-size tile set, so combined with the presented construction remains constant. This solves Problem 4.

Finally, Problems 5 and 6 have solutions due to Chandran, Gopalkrishnan, and Reif [2], which we now explain intuitively (the actual analysis is a bit trickier but is close to the following intuitive argument). Figure 7 shows an example of a solution to Problem 5 for the case of expected length $n = 92$. Each T_{iB} tile type has an east glue, g_i, that matches two tile types $T_{(i-1)A}$ and $R_{(i-1)A}$. There are $O(\log n)$ "stages" (five stages in this case). Each stage has probability $\frac{1}{2}$ to either decrement the stage or reset back to the highest stage. The number n is programmed into the system by choosing each stage to have either 1 or 2 tiles. Given that we are in stage i, to make it from stage i to stage 1 without resetting means that i consecutive unbiased coin flips must come up "heads," which we expect to take 2^i flips before happening. Thus we expect stage i to appear 2^i times; this means that stage i's expected contribution to the total length is either 2^i or $2 \cdot 2^i$, depending on whether it has 1 or 2 tiles. The reason this works to encode arbitrary natural numbers n is that every natural number can be expressed as $n = \sum_{i=0}^{\approx \log n} b_i 2^i$, where $b_i \in \{1, 2\}$. Since there are a constant number of tile types per stage, this implies that the number of tile types required is $O(\log n)$.

This solves Problem 5. To solve Problem 6, it suffices to concatenate k independent assemblies of the kind shown in Fig. 7, where k is a constant that, if chosen sufficiently large based on δ (the desired error probability), solves Problem 6 since it increases the number of tile types required. In addition to proving that this works, Chandran, Gopalkrishnan, and Reif [2] also show a more complex construction with even sharper bounds on the probability that the length differs very much from its expected value.

Open Problems

The construction resolving Problem 3 shows that for every $\delta, \epsilon > 0$, a tile set exists such that, for every $n \in \mathbb{N}$, appropriately programming the tile concentrations results in the self-assembly of a structure of size $O(n^\epsilon) \times O(\log n)$ whose rightmost tiles represent the value n with probability at least $1 - \delta$. (In the tile system described, $\epsilon = 2/3$, and it could be made arbitrarily close to 0 by estimating more than 3 numbers at once.) Is this optimal?

Formally, say that a tile assembly system $\mathcal{T} = (T, \sigma, 2)$ is δ-*concentration programmable* (for $\delta > 0$) if there is a (total) computable function $r : \mathcal{A}_\Box[\mathcal{T}] \to \mathbb{N}$ (the *representation function*) such that, for each $n \in \mathbb{N}$, there is a tile concentration assignment $\rho : T \to [0, \infty)$ such that $\Pr[r(\mathcal{T}(\rho)) = n] \geq 1 - \delta$. In other words, $\mathcal{T}$, programmed with concentrations ρ, almost certainly self-assembles a structure that "represents" n, according to the representation function r, and such a ρ can be found to create a high-probability representation of *any* natural number.

Question 1 Is the following statement true? For each $\delta > 0$, there is a tile assembly system $\mathcal{T}$ and a representation function $r : \mathcal{A}_{\square}[\mathcal{T}] \to \mathbb{N}$ such that $\mathcal{T}$ is δ-concentration programmable and, for each $\epsilon > 0$ and all but finitely many $n \in \mathbb{N}$, $\Pr[|\text{dom } \mathcal{T}(\rho)| < n^{\epsilon}] \geq 1 - \delta$. If so, what is the smallest bound that can be written in place of n^{ϵ}?

Cross-References

▶ Experimental Implementation of Tile Assembly
▶ Patterned Self-Assembly Tile Set Synthesis
▶ Robustness in Self-Assembly
▶ Self-Assembly at Temperature 1
▶ Self-Assembly of Fractals
▶ Self-Assembly with General Shaped Tiles
▶ Staged Assembly
▶ Temperature Programming in Self-Assembly

Recommended Reading

1. Becker F, Rapaport I, Rémila E (2006) Self-assembling classes of shapes with a minimum number of tiles, and in optimal time. In: FSTTCS 2006: foundations of software technology and theoretical computer science, Kolkata, pp 45–56
2. Chandran H, Gopalkrishnan N, Reif JH (2012) Tile complexity of linear assemblies. SIAM J Comput 41(4):1051–1073. Preliminary version appeared in ICALP 2009
3. Doty D (2010) Randomized self-assembly for exact shapes. SIAM J Comput 39(8):3521–3552. Preliminary version appeared in FOCS 2009
4. Kao M-Y, Schweller RT (2008) Randomized self-assembly for approximate shapes. In: ICALP 2008: international colloqium on automata, languages, and programming, Reykjavik. Volume 5125 of Lecture notes in computer science. Springer, pp 370–384
5. Soloveichik D, Winfree E (2007) Complexity of self-assembled shapes. SIAM J Comput 36(6):1544–1569. Preliminary version appeared in DNA 2004
6. Winfree E (1998) Algorithmic self-assembly of DNA. PhD thesis, California Institute of Technology, June 1998

Range Searching

Marc van Kreveld and Maarten Löffler
Department of Information and Computing Sciences, Utrecht University, Utrecht, The Netherlands

Keywords

Approximate range searching; Data structures; Intersection searching; Preprocessing; Query time; Ray shooting; Storage requirements

Years and Authors of Summarized Original Work

1975; Bentley
1978; Lueker
1993; Chazelle
2000; Arya, Mount

Problem Definition

Generally speaking, data structures come in two types: those that represent data and those that allow efficient searching. The collection of results in the area of *range searching* belongs to the latter type. We distinguish two computation phases: during the *preprocessing phase*, data is stored in some suitable structure, so that during the *query phase*, all data that lies inside a query range can be found and reported efficiently.

In the most basic form of range searching, the data consists of points in a one-, two-, or higher-dimensional space, and the query range is a simple shape like a rectangle, triangle, or circle. Even for this basic form, there are many different data structures and corresponding query algorithms.

Problem 1 (Range Searching)

INPUT: Set P of n points in $\mathbb{R}^d$.

OUTPUT: Description of a data structure storing P and query algorithm that will report, for any given query d-rectangle (d-

simplex, d-sphere) q, all points of P that lie inside q.

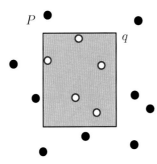

When the data is not a set of points but a set of more complex objects, such as line segments, triangles, circles, or other geometric shapes, we may not be interested in only the objects that lie completely within a query range but also all objects that *intersect* the range.

Problem 1 (Intersection Searching)

INPUT: *Set S of n non-crossing line segments in $\mathbb{R}^2$ (triangles in $\mathbb{R}^3$).*

OUTPUT: *Description of a data structure storing S and query algorithm that will report, for any given query line segment q, all line segments (triangles) of S that intersect q.*

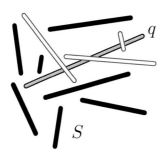

For both range searching and intersection searching, we may be interested in different types of queries. In a *counting query*, we report the number of objects in P or S that lie in the range or intersect the query object. A reporting query must spend time at least linear in the number of objects reported, whereas a counting query returns a single value. Usually, small variations

of a data structure for reporting can be used for the counting version.

Ray shooting is closely related to intersection searching. We are not interested in all objects intersected by a line segment but only the first along a directed ray.

Problem 2 (Ray Shooting)

INPUT: *Set S of n non-crossing line segments in $\mathbb{R}^2$ (triangles in $\mathbb{R}^3$).*

OUTPUT: *Description of a data structure storing S and query algorithm that will report, for any given query point q and direction in $\mathbb{R}^2$ ($\mathbb{R}^3$), the first line segment (triangle) of S that is reached when q moves in the query direction.*

A combination of a data structure and a query algorithm forms a solution to a range-searching problem. The most important aspects of efficiency are the storage requirements of the data structure and the query time. Sometimes, preprocessing time and update time are also important. If the data structure is so large that it must be stored on background storage, I/O complexity becomes relevant.

We can distinguish solutions with guaranteed efficiency and heuristics. The heuristic solutions used in practice nearly always have linear size but often have no guaranteed worst-case query time bounds. For example, R-trees [14] are among the most used data structures for range searching in practice.

One of the most interesting practical approaches for range searching with provable bounds is approximate range searching.

Problem 3 (Approximate Range Searching)

INPUT: *Set P of n points in $\mathbb{R}^d$.*

OUTPUT: *Description of a data structure storing P and query algorithm that will report, for any given query d-polyhedron q of constant complexity, all points of P that lie inside q, possibly but not necessarily some points that lie within distance ε from q, and no points that lie farther than ε from q.*

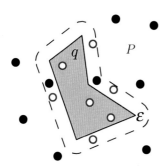

In this entry we concentrate on algorithmic results that have provable worst-case bounds for both the storage requirements and the query time.

Key Results

Orthogonal Range Searching

Range searching in one dimension is just searching in a sorted sequence of values. Standard binary search trees for one-dimensional searching can be extended in several ways to allow rectangular range-searching queries. For example, a *kd-tree* [4] is a balanced binary tree on a set of points in $\mathbb{R}^d$ that splits the point set on different coordinates in different nodes: the root splits on x_1-coordinate, its two children on x_2-coordinate, their four children on x_3-coordinate, and so on; after the splitting on the x_d-coordinate, the tree starts over by splitting on x_1-coordinate again. As soon as there is a single point left, it is stored in a leaf.

An *(orthogonal) range tree* [5, 10, 15] uses *associated structures*, a technique that has proved to be very powerful for solving various kinds of query problems. It refers to the fact that the structure has a main tree, and each internal node

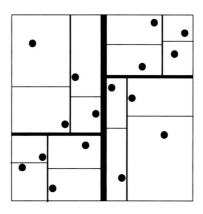

v of the tree stores – besides two pointers to children – an extra pointer to a different data structure. Suppose that in the main tree, node v is root of a subtree storing a subset S_v of the whole set S. Then the associated structure of v also stores S_v, but in a different manner.

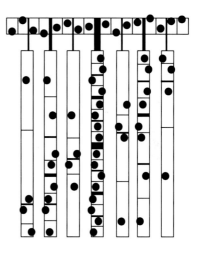

A range tree for a set S of points in d-space consists of a main tree that is a balanced binary search tree on x_d-coordinate. The leaves of the main tree store the points of S sorted on x_d-coordinate in the leaves. If $d > 1$, then each internal node v stores a pointer to a $(d - 1)$-dimensional range tree that stores S_v restricted to their first $d - 1$ coordinates.

The performance of kd-trees and range trees is given in Table 1. To achieve the stated query time for range trees, an additional technique called *fractional cascading* is needed [7]. The table also shows that in special cases, like 2-dimensional range queries in which one side of the query rectangle is unbounded (a 3-sided range), better results can be obtained using priority search trees. Other small improvements can be obtained, also depending on the machine model.

Simplex Range Searching

The range-searching problem with d-simplices is considerably harder than when the query shape is an axis-aligned d-box. There are two types of solutions: solutions with near-linear-size data structures and solutions with near-logarithmic query time. The results for d-simplex and d-half-space searching are given in Table 2.

Between the extremes of space-efficient data structures and query-efficient data structures, many other results "in between" can be obtained. For example, if for a problem in the plane one knows that a linear number of triangle range queries are needed, then one can use a data structure of size $O(n^{4/3})$ and query time close to $O(n^{1/3} + k)$, because this balances the preprocessing time (roughly the same as the size) and total query time (without the time for reporting) to something close to $O(n^{4/3})$.

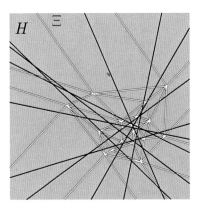

A $(1/r)$-*cutting* of a set H of hyperplanes is a set Ξ of (relatively open) disjoint simplices covering $\mathbb{R}^d$ so that each simplex intersects at most n/r hyperplanes of H. Cutting trees are based on this concept. We state the main result on cuttings as a theorem, because it has implications to multidimensional divide-and-conquer schemes as well.

Theorem 1 ([6]) *Let H be a set of n hyperplanes and $r \leq n$ a parameter. Set $k = \lceil \log_2 r \rceil$. There exist k cuttings $\Xi_1, \ldots, \Xi_k$ so that Ξ_i is a $(1/2^i)$-cutting of size $O(2^{id})$, each simplex of Ξ_i is contained in a simplex of Ξ_{i-1}, and each simplex of Ξ_{i-1} contains a constant number of simplices of Ξ_i. Moreover, $\Xi_1, \ldots, \Xi_k$ can be computed in time $O(nr^{d-1})$.*

Data structures for range searching with curved boundaries can be obtained by linearization techniques. For example, range searching with a d-ball can be done by mapping each point $(x_1, \ldots, x_d)$ from the set to a point $(x_1, \ldots, x_d, x_1^2 + \cdots + x_d^2)$ in $\mathbb{R}^{d+1}$ and storing these points in a $(d + 1)$-dimensional half-space range query structure. A d-ball with center $(b_1, \ldots, b_d)$ and radius r is mapped to the half-space $x_{d+1} \leq b_1(2x_1 - b_1) + \cdots + b_d(2x_d - b_d) + r^2$, and now the mapped points inside the mapped half-space correspond exactly to the original points inside the d-ball.

Intersection searching and ray shooting data structures are often based on the technique of associated structures mentioned before. Depending on the type of stored objects and the type of query objects (or query rays), different main trees and associated structures are combined into efficient solutions.

Approximate Range Searching

Many of the given data structures are not very useful in practice, especially in higher dimensions. One of the more interesting approaches toward a practical data structure for range searching that has performance guarantees is the approximate approach. The idea is that the query range is considered a shape with an inner boundary and a buffer zone around it. All points inside the inner boundary must be reported, all points outside the inner boundary but inside the buffer zone may

Range Searching, Table 1 Results on orthogonal range searching. n is the number of points stored and k is the number of points reported

Query range	Storage	Query time	Reference
d-box	$O(n)$	$O(n^{1-\frac{1}{d}} + k)$	kd-tree [4]
d-box	$O(n \log^{d-1} n)$	$O(\log^{d-1} n + k)$	Range tree [10]
3-sided rectangle	$O(n)$	$O(\log n + k)$	Priority search tree [13]

Range Searching, Table 2 Results on simplex and half-space range searching. n is the number of points stored, k is the number of points reported, c is some constant, and $\varepsilon > 0$ is an arbitrarily small constant

Query range	Storage	Query time	Reference
d-simplex	$O(n)$	$O(n^{1-\frac{1}{d}} + k)$	Partition trees [11]
d-simplex	$O(n^{d+\varepsilon})$	$O(\log n + k)$	Cutting trees [8]
d-half-space	$O(n \log \log n)$	$O(n^{1-\frac{1}{\lfloor d/2 \rfloor}} \log^c n + k)$	[12]
d-half-space	$O(n^{\lfloor d/2 \rfloor} \log^c n)$	$O(\log n + k)$	[2, 12]

but need not be reported, and all points outside the buffer zone may not be reported. A query will specify the inner boundary and a distance to the inner boundary that is the width of the buffer zone.

Assuming that the inner boundary has constant complexity and the buffer zone has width $\varepsilon \cdot D$, where D is the diameter of the inner boundary (ε is any positive constant), an approximate range query can be answered in $O(\log n + 1/\varepsilon^d)$ time, where d is the dimension of the space [3]. When the inner boundary is convex, the query time can be improved slightly.

Cross-References

▶ I/O-Model
▶ Point Location
▶ R-Trees

Recommended Reading

1. Agarwal PK (2004) Range searching. In: Goodman JE, O'Rourke J (eds) Handbook of discrete and computational geometry, chapter 36, 2nd edn. Chapman & Hall/CRC, Boca Raton
2. Agarwal PK, Erickson J (1998) Geometric range searching and its relatives. In: Chazelle B, Goodman J, Pollack R (eds) Advances in discrete and computational geometry. American Mathematical Society, Providence, pp 1–56
3. Arya S, Mount DM (2000) Approximate range searching. Comput Geom 17(3–4):135–152
4. Bentley JL (1975) Multidimensional binary search trees used for associative searching. Commun ACM 18(9):509–517
5. Bentley JL (1980) Multidimensional divide-and-conquer. Commun ACM 23(4):214–229
6. Chazelle B (1993) Cutting hyperplanes for divide-and-conquer. Discret Comput Geom 9:145–158
7. Chazelle B, Guibas LJ (1986) Fractional cascading: I and II. Algorithmica 1(2):133–191
8. Chazelle B, Sharir M, Welzl E (1992) Quasi-optimal upper bounds for simplex range searching and new zone theorems. Algorithmica 8(5&6):407–429
9. de Berg M, Cheong O, van Kreveld M, Overmars M (2008) Computational geometry – algorithms and applications, 3rd edn. Springer, Berlin
10. Lueker GS (1978) A data structure for orthogonal range queries. In: The annual symposium of the foundations of computer science (FOCS), Ann Arbor. IEEE Computer Society, pp 28–34
11. Matousek J (1992) Efficient partition trees. Discret Comput Geom 8:315–334
12. Matousek J (1992) Reporting points in halfspaces. Comput Geom 2:169–186
13. McCreight EM (1985) Priority search trees. SIAM J Comput 14(2):257–276
14. Samet H (2006) Foundations of multidimensional and metric data structures. Morgan Kaufmann, San Francisco
15. Willard DE (1979) The super-b-tree algorithm. Report TR-03-79, Aiken Computer Laboratory, Harvard University, Cambridge

R

Rank and Select Operations on Bit Strings

Rajeev Raman
Department of Computer Science, University of
Leicester, Leicester, UK

Keywords

Bit vectors; Predecessor search; Sets; Succinct
data structures

Years and Authors of Summarized Original Work

1974; Elias
1989; Jacobson
1998; Clark
2007; Raman, Raman, Rao
2008; Pătraşcu
2014; Golynski, Orlandi, Raman, Rao

Problem Definition

Given a static bit string $b = b_1 \ldots b_m$, the
objective is to preprocess b and to create a space-
efficient data structure that supports the following
operations rapidly:

$\text{rank}_1(i)$ takes an index i as input, $1 \leq i \leq m$,
   and returns the number of **1**s among $b_1 \ldots b_i$.
$\text{select}_1(i)$ takes an index $i \geq 1$ as input and
   returns the position of the i-th **1** in b, and -1
   if i is greater than the number of **1**s in b.

A data structure that supports the operations
above will be called a *bit vector*. The operations
rank_0 and select_0 are defined analogously for
the **0**s in b. As $\text{rank}_0(i) = i - \text{rank}_1(i)$, one
considers just rank_1 (abbreviated to rank) and
refers to select_0 and select_1 collectively as
select. In what follows, $|x|$ denotes the length
of a bit string x and $w(x)$ denotes the number of
1s in it. b is always used to denote the input bit
string, m to denote $|b|$ and n to denote $w(b)$.

Memory Usage Models

In terms of space usage, we aim not only to
store b in the minimum amount of space but
also to minimize any additional space (called the
redundancy) needed to support rank and select.
The notion of redundancy can be formalized in
two different ways.

In the *succinct index* model (also known as the
systematic model), the bit vector does not have
direct access to b, but can obtain $O(\log m)$ consecutive bits of b in $O(1)$ time. During preprocessing, one can create additional data structures
(called *succinct indices*) to allow rapid rank and
select queries. Indices allow the representation
of b to be decoupled from the auxiliary data
structure, e.g., b can be stored (in a potentially
highly compressed form) in a data structure such
as that of [6]. The redundancy in the succinct
index model is the space usage of the index.

In the *unrestricted* model, we give a "space
budget" for storing b, based upon some compressibility measure (the data structure is usually
designed to target a particular measure). We now
give some examples of space budgets:

- The obvious space budget for b is m bits, and
  this is used if b is believed to be incompress-
  ible.
- Recalling that $n = w(b)$, we define the
  space budget $B(m,n) = \lceil \log_2 \binom{m}{n} \rceil$, which is
  the information-theoretic minimum number of
  bits to store a bit string of length m with n **1**s.
  Using standard approximations of the factorial
  function, one can show [17] that $B(m,n) =
  n \log_2(m/n) + n \log_2 e + O(n^2/m)$. In par-
  ticular, if $n = o(m)$, then $B(m,n) = o(m)$.
- Yet another space budget is obtained from
  the *k-th-order empirical entropy*, denoted by
  $H_k(b)$. For any bit string s, define $\#(s)$ as the
  number of (possibly overlapping) contiguous
  occurrences of the bit string s in b. Then, for
  any $k \geq 0$,

$$H_k(b) = -\frac{1}{m} \sum_{s \in \{0,1\}^k} \left(\#(s\mathbf{0}) \log_2 \frac{\#(s\mathbf{0})}{\#(s)} \right.$$
$$\left. + \#(s\mathbf{1}) \log_2 \frac{\#(s\mathbf{1})}{\#(s)} \right) \qquad (1)$$

(take $\log_2(0/0) = 0\log_2 0 = 0$ and $\#(s) = m$ when s is the empty string). $H_k(\boldsymbol{b})$ gives the information content per bit in $\boldsymbol{b}$, when conditioned upon the previous k bits as context. The space budget is therefore $mH_k(\boldsymbol{b})$. Note that $mH_0(\boldsymbol{b}) \sim B(m,n)$, but even $H_1(\boldsymbol{b})$ can be much smaller than $H_0(\boldsymbol{b})$, and in general $H_{k+1} \leq H_k$. For example, if $\boldsymbol{b} = (01)^{m/2}$, then $H_0(\boldsymbol{b}) \sim m$ but $mH_1(\boldsymbol{b})$ vanishes.

The redundancy in the unrestricted model is the difference between the space usage of the data structure and the space budget.

Models of Computation
Three models of computation are commonly considered. One is the *word RAM* model with word size $O(\log m)$ bits [13]. The other models, which are particularly useful for proving lower bounds, are the *cell probe* and *bit probe* models. In the cell probe model, the time complexity of answering a query is the worst-case number of words of $O(\log m)$ consecutive bits of the data structure that are read by the algorithm to answer that query. All other computation is "free." The bit probe model is similar, except that we only count the number of bits of the data structure that are read when answering a query. Clearly, $O(\log m)$ bit probes can be more useful than reading $O(1)$ consecutive words, so $O(\log m)$ bit probes are more powerful than $O(1)$ cell probes. Also, $O(1)$ cell probes are more powerful than $O(1)$ time on the word RAM, since computation on values read into registers is for free in the cell probe model. Thus, an $O(t)$ upper bound in the word RAM is stronger than $O(t)$ upper bound in the cell probe model, which is stronger than an $O(t\log m)$ upper bound in the bit probe model. For lower bounds, the situation is of course reversed, with cell probe lower bounds being stronger than equivalent word RAM lower bounds.

Key Results

Relation to Predecessor Search
Given a static set $S \subseteq \{0, \ldots, m-1\}$, $|S| = n$, the *predecessor search* problem is to preprocess S to answer the query $\mathsf{pred}(x, S) = \max\{y \in S \mid y \leq x\}$. The predecessor search can easily be solved using a bit vector: we simply create a bit string $\boldsymbol{b}$ that is the characteristic vector of S, and note that (i) $|\boldsymbol{b}| = m$, (ii) $w(\boldsymbol{b}) = n = |S|$, and (iii) $\mathsf{pred}(x, S) = \mathsf{select}_1(\mathsf{rank}_1(x))$.

Clearly, if we are interested in highly space-efficient solutions, space usages of significantly more than $O(n\log m)$ bits are not of interest, since any bit string $\boldsymbol{b}$ can be represented as a set using $O(n\log m)$ bits by enumerating the positions of its $\mathbf{1}$s. However, this close connection of the bit vector problem to the predecessor search problem means that lower bounds for the predecessor search problem also apply to the bit vector problem. In particular, if rank should take $O(1)$ time and the space should be at most $O(n\log m)$ bits, then this is only possible if $n = m/(\log m)^{O(1)}$ [19]. Since constant-time rank (and select) is taken by the succinct data structure community to be a "standard" expectation, this lower bound means that we only consider moderately sparse bit strings $\boldsymbol{b}$ in this entry.

Reductions
It has been already noted that rank_0 and rank_1 reduce to each other and that operations on sets reduce to select operations on a bit string. Some other reductions, whereby one can support operations on $\boldsymbol{b}$ by performing operations on bit strings derived from $\boldsymbol{b}$, are:

Theorem 1 *(a) rank reduces to select_0 on a bit string $\boldsymbol{c}$ such that $|\boldsymbol{c}| = m + n$ and $w(\boldsymbol{c}) = n$.*

(b) If $\boldsymbol{b}$ has no consecutive $\mathbf{1}$s, then select_0 on $\boldsymbol{b}$ can be reduced to rank on a bit string $\boldsymbol{c}$ such that $|\boldsymbol{c}| = m - n$ and $w(\boldsymbol{c})$ is either $n - 1$ or n.

(c) From $\boldsymbol{b}$, one can derive two-bit string $\boldsymbol{b_0}$ and $\boldsymbol{b_1}$ such that $|\boldsymbol{b_0}| = m - n$, $|\boldsymbol{b_1}| = n$, $w(\boldsymbol{b_0}), w(\boldsymbol{b_1}) \leq \min\{m - n, n\}$, and select_0 and select_1 on $\boldsymbol{b}$ can be supported by supporting select_1 and rank on b_0 and b_1.

Parts (a) and (b) follow from Elias's observations on multiset representations, specialized to sets. For part (a), create $\boldsymbol{c}$ from $\boldsymbol{b}$ by adding a $\mathbf{0}$ after every $\mathbf{1}$. For example, if $\boldsymbol{b} = \mathbf{01100100}$, then $\boldsymbol{c} = \mathbf{01010001000}$. Then, $\mathsf{rank}_1(i)$ on $\boldsymbol{b}$ equals

$\text{select}_0(i) - i$ on c. For part (b), essentially invert the mapping of part (a). Part (c) is shown in [3].

Succinct Indices for Bit Vectors

The following is known about the sizes of succinct indices for bit vectors:

$$\begin{cases} O(m' \log(n/m')) & \text{if } n = \omega(m') \\ O(n(1 + \max\{0, \log(m'/n)\})) & \text{if } n = O(m') \end{cases} \; bits,$$

where $m' = m/\log m$, that supports rank, select_0, and select_1 in $O(1)$ time. This index size is optimal for any data structure that makes $O(\log m)$ bit probes to b.

This result generalizes an earlier result by Golynski, who showed that the index size must be $\Theta(m \log \log m / \log m)$ bits for $O(1)$ time operations [9]. The bound of Theorem 2 is asymptotically the same when n is relatively close to m, e.g., when $n = \Omega(m/(\log m)^{1/2})$, but is smaller thereafter, e.g., for $n = \Theta(m/(\log m)^2)$ the index size implied by Theorem 2 $O(m \log \log m/(\log m)^2)$ bits, which is a $\Theta(\log m)$ factor better than that given by [9].

Elias [5] previously gave an $o(m)$-bit index that supported select in $O(\log m)$ bit probes on average (where the average was computed across all select queries). Jacobson [14] gave $o(m)$-bit indices that supported rank and select in $O(\log m)$ bit probes in the worst case. Clark and Munro [2] gave the first $o(m)$-bit indices that support both rank and select in $O(1)$ time on the RAM.

Bit Vectors in the Unrestricted Model

In the unrestricted model, the best redundancy, if one is targeting the $B(m, n)$ space budget, is given by the following result due to Pătraşcu:

Theorem 3 ([18]) *A bit string b with $|b| = m$ and $w(b) = n$ can be represented using $B(m, n) + m/((\log m)/t)^t + m^{3/4}(\log m)^{O(1)}$ bits of memory, supporting rank and select_1 queries in $O(t)$ time.*

Earlier results, with a significantly higher redundancy, were given by [17, 21]. Thus, for $t =$

Theorem 2 ([11]) *Given a bit string b with $|b| = m$, $w(b) = n$, and $m/n = (\log m)^{O(1)}$, there is an index of size*

$O(1)$, the redundancy is $m/(\log m)^{O(1)}$. There is an almost matching lower bound:

Theorem 4 ([20]) *Any representation of a bit string b with $|b| = m$ and $w(b) = n$ that answers rank or select_1 queries in $O(t)$ time on the cell probe model must use $B(m, n) + m/(\log m)^t$ bits of memory.*

The case where we aim for higher-order entropy appears to be less well studied. The best-known result is as follows:

Theorem 5 ([10]) *A bit string b with $|b| = m$ can be represented using $m H_k(b) + O(mk/\log m)$ bits of memory, supporting rank and select queries in $O(1)$ time, for any $k \geq 1$.*

Applications

Bit vectors are fundamental building blocks in a huge number of space-efficient data structures, in real-world and theoretical applications such as XML document representation [1, 4, 7], text retrieval [16], bioinformatics [15], and data mining [22], to name but a few. In the Cross-References, we list the various succinct data structures that build on or are related to bit vectors.

Experimental Results

Bit vectors have been extensively experimentally evaluated. Mature implementations are available in the libraries SDSL [8] and Succinct [12]. Other libraries of note are Vigna's Sux4J (http://

sux.di.unimi.it) and Claude's `libcds` (https://github.com/fclaude/libcds).

Cross-References

▶ Compressed Document Retrieval on String Collections
▶ Compressed Range Minimum Queries
▶ Compressed Representations of Graphs
▶ Compressed Suffix Array
▶ Compressed Suffix Trees
▶ Compressed Tree Representations
▶ Compressing and Indexing Structured Text
▶ Minimal Perfect Hash Functions
▶ Monotone Minimal Perfect Hash Functions
▶ Predecessor Search
▶ Rank and Select Operations on Sequences
▶ Succinct and Compressed Data Structures for Permutations and Integer Functions
▶ Wavelet Trees

Recommended Reading

1. Arroyuelo D, Claude F, Maneth S, Mäkinen V, Navarro G, Nguyen K, Sirén J, Välimäki N (2015) Fast in-memory XPath search using compressed indexes. Softw Pract Exp 45(3):399–434
2. Clark DR (1998) Compact PAT trees. PhD thesis, University of Waterloo, Waterloo
3. Delpratt O, Rahman N, Raman R (2006) Engineering the LOUDS succinct tree representation. In: Àlvarez C, Serna MJ (eds) WEA. Lecture notes in computer science, vol 4007. Springer, Berlin/Heidelberg, pp 134–145
4. Delpratt O, Raman R, Rahman N (2008) Engineering succinct DOM. In: Kemper A, Valduriez P, Mouaddib N, Teubner J, Bouzeghoub M, Markl V, Amsaleg L, Manolescu I (eds) EDBT. ACM international conference proceeding series, vol 261. ACM, New York, pp 49–60
5. Elias P (1974) Efficient storage and retrieval by content and address of static files. J ACM 21:246–260
6. Ferragina P, Venturini R (2007) A simple storage scheme for strings achieving entropy bounds. Theor Comput Sci 372(1):115–121
7. Ferragina P, Luccio F, Manzini G, Muthukrishnan S (2009) Compressing and indexing labeled trees, with applications. J ACM 57(1)
8. Gog S, Beller T, Moffat A, Petri M (2014) From theory to practice: plug and play with succinct data structures. In: Gudmundsson J, Katajainen J (eds) Experimental algorithms – 13th international sympo-

sium, SEA 2014, Copenhagen, 29 June–1 July 2014. Proceedings. Lecture notes in computer science, vol 8504. Springer, Heidelberg, pp 326–337
9. Golynski A (2007) Optimal lower bounds for rank and select indexes. Theor Comput Sci 387(3):348–359
10. Golynski A, Raman R, Rao SS (2008) On the redundancy of succinct data structures. In: Gudmundsson J (ed) SWAT. Lecture notes in computer science, vol 5124. Springer, Heidelberg, pp 148–159
11. Golynski A, Orlandi A, Raman R, Rao SS (2014) Optimal indexes for sparse bit vectors. Algorithmica 69(4):906–924
12. Grossi R, Ottaviano G (2013) Design of practical succinct data structures for large data collections. In: Bonifaci V, Demetrescu C, Marchetti-Spaccamela A (eds) Experimental algorithms, 12th international symposium, SEA 2013, Rome, 5–7 June 2013. Proceedings. Lecture notes in computer science, vol 7933. Springer, Heidelberg/New York, pp 5–17
13. Hagerup T (1998) Sorting and searching on the word RAM. In: Morvan M, Meinel C, Krob D (eds) STACS 98, 15th annual symposium on theoretical aspects of computer science, Paris, 25–27 Feb 1998, Proceedings. Lecture notes in computer science, vol 1373. Springer, Berlin/New York, pp 366–398
14. Jacobson G (1989) Succinct static data structures. PhD thesis, Carnegie Mellon University, Pittsburgh
15. Li H, Durbin R (2009) Fast and accurate short read alignment with Burrows-Wheeler transform. Bioinformatics 25(14):1754–1760
16. Navarro G, Mäkinen V (2007) Compressed full-text indexes. ACM Comput Surv 39(1)
17. Pagh R (2001) Low redundancy in static dictionaries with constant query time. SIAM J Comput 31(2):353–363
18. Patrascu M (2008) Succincter. In: 49th annual IEEE symposium on foundations of computer science, FOCS 2008, Philadelphia, 25–28 Oct 2008. IEEE Computer Society, Los Alamitos, pp 305–313
19. Patrascu M, Thorup M (2006) Time-space trade-offs for predecessor search. In: Kleinberg JM (ed) Proceedings of the 38th annual ACM symposium on theory of computing, Seattle, 21–23 May 2006, ACM, New York, pp 232–240
20. Patrascu M, Viola E (2010) Cell-probe lower bounds for succinct partial sums. In: Charikar M (ed) Proceedings of the twenty-first annual ACM-SIAM symposium on discrete algorithms, SODA 2010, Austin, 17–19 Jan 2010. SIAM, Philadelphia, pp 117–122
21. Raman R, Raman V, Satti SR (2007) Succinct indexable dictionaries with applications to encoding k-ary trees, prefix sums and multisets. ACM Trans Algorithms 3(4):Article 43, 25pp
22. Tabei Y, Tsuda K (2011) Kernel-based similarity search in massive graph databases with wavelet trees. In: Proceedings of the eleventh SIAM international conference on data mining, SDM 2011, 28–30 Apr 2011, Mesa. SIAM/Omnipress, Philadelphia, pp 154–163

R

Rank and Select Operations on Sequences

Travis Gagie
Department of Computer Science, University of
Eastern Piedmont, Alessandria, Italy
Department of Computer Science, University of
Helsinki, Helsinki, Finland

Keywords

String data structures; Succinct and compressed
data structures

Years and Authors of Summarized Original Work

2003; Grossi, Gupta, Vitter
2006; Golynski, Munro, Rao
2007; Ferragina, Manzini, Mäkinen, Navarro
2011; Barbay, He, Munro, Rao
2012; Belazzougui, Navarro
2013; Navarro, Nekrich
2014; Barbay, Claude, Gagie, Navarro, Nekrich

Problem Definition

The query $S.\mathrm{rank}_a(i)$ on a sequence S is defined
to return the number of occurrences of the distinct
character a among the first i characters of S,
and the query $S.\mathrm{select}_a(j)$ is defined to return
the position of the jth occurrence of a in S (if
it exists). Since rank and select queries are fundamental to the field of succinct and compressed
data structures, researchers have proposed several
data structures that answer them quickly while
using little space. Most of these data structures
also support fast random access to S, and a
few of them support fast insertions and deletions of characters in S. Some of them return
$S.\mathrm{rank}_a(i)$ more quickly when the ith character
of S is itself an a; the query is then called partial
rank.

Key Results

While considering how to store trees and graphs
in small space while supporting fast navigation,
Jacobson [16] considered the problem of supporting rank and select on binary sequences. He
showed how to store an n-bit binary sequence
using $o(n)$ bits in addition to the sequence itself, such that we can answer rank and select
using $O(\log n)$ bit probes. Later authors have
considered the problem in the word-RAM model
with $\Omega(\log n)$-bit words, in which Jacobson's
implementation of rank takes $O(1)$ time; they
showed how to answer also select in this model
in $O(1)$ time while still using $o(n)$ extra bits.
Pătraşcu [20] showed how we can store an n-
bit binary sequence containing m 1s in a total
of $\lg \binom{n}{m} + O(n/\log^c n)$ bits, where c is any
constant, and still answer rank and select in $O(1)$
time.

Grossi, Gupta, and Vitter [12] described a
data structure, called a wavelet tree, that uses
rank and select on several binary sequences to
answer access, rank, and select on sequences over
larger alphabets. If S is a sequence of length
n over an alphabet of size σ and a wavelet
tree for S is implemented with uncompressed
data structures for rank and select on the binary
sequences, then it takes $n \log \sigma + o(n \log \sigma)$ bits
and answers access, rank, and select in $O(\log \sigma)$
time. With instances of Pătraşcu's data structure,
the space becomes $n H_0(S) + o(n)$ bits, where
$H_0(S)$ is the 0th-order empirical entropy of S.
To simplify, we assume throughout that $\sigma =
o(n/\log n)$.

Ferragina, Manzini, Mäkinen, and Navarro [7]
described a multiary version of the wavelet tree
that uses only $O\left(\frac{\log \sigma}{\log \log n} + 1\right)$ time for access,
rank, and select, which is $O(1)$ when $\sigma =
\lg^{O(1)} n$. Their implementation takes $n H_0(S) +
o(n)$ bits when $\sigma = \lg^{O(1)} n$ and $n H_0(S) +
o(n \log \sigma)$ bits otherwise. Golynski, Raman, and
Rao [11] reduced the space to $n H_0(S) + o(n)$ bits
in the general case.

Golynski, Munro, and Rao [10] described a
data structure that takes $n \lg \sigma + o(n \lg \sigma)$ bits

and either answers select in $O(1)$ time and access and rank in $O(\log \log \sigma)$ time, or answers access in $O(1)$ time, rank in $O(\log \log(\sigma) \log \log \log \sigma)$ time, and select in $O(\log \log \sigma)$ time. If the space is increased to $(1 + \epsilon)n \lg \sigma$ bits, where ϵ is any positive constant, then both access and select take $O(1)$ time and rank takes $O(\log \log \sigma)$ time. Golynski [9] showed that the product of the query times for access and select and the per-character redundancy in bits must be $\Omega\left(\frac{\log^2 \sigma}{w}\right)$ in general, where w is the length of a machine word.

Barbay, He, Munro, and Rao [1] described a data structure that takes $nH_k(S) + o(n \log \sigma)$ bits, where $H_k(S)$ is the kth-order empirical entropy of S, and answers access in $O(1)$ time, rank in $O(\log \log \sigma (\log \log \log \sigma)^2)$ time, and select in $O(\log \log(\sigma) \log \log \log \sigma)$ time. We assume throughout that $k = o(\log_\sigma n)$. They also reduced to $nH_0(S) + o(n \log \sigma)$ bits, the space for the version of Golynski, Munro, and Rao's data structure with $O(1)$-time select and $O(\log \log \sigma)$-time access and rank. Grossi, Orlandi, and Raman [13] reduced the space of the version with $O(1)$-time access and $O(\log \log \sigma)$-time select to $nH_k(S) + o(n \log \sigma)$ bits and reduced the time for rank to $O(\log \log \sigma)$.

Barbay, Claude, Gagie, Navarro, and Nekrich [2] combined multiary wavelet trees with the versions of Golynski, Munro and, Rao's data structure, to obtain a data structure that takes $nH_0(S) + o(n)(H_0(S) + 1)$ bits and answers one of access and select in $O(1)$ time and the other in $O(\log \log \sigma)$ time, and rank also in $O(\log \log \sigma)$ time. If the space is increased to $(1 + \epsilon)nH_0(S) + o(n)$ bits, then both access and select take $O(1)$ time. They partition the alphabet into sub-alphabets such that all the characters in each sub-alphabet have roughly the same frequency, and then store a data structure that answers access, rank, and select queries on the subsequence of characters in S from that sub-alphabet.

Belazzougui and Navarro [3, 4] showed that any data structure that takes $n \cdot w^{O(1)}$ space must use $\Omega\left(\log \frac{\log \sigma}{\log w}\right)$ time for rank. They also gave the following upper bounds:

- We can store S in $nH_0(S) + o(n)$ bits and answer access, rank, and select in $O\left(\frac{\log \sigma}{\log w} + 1\right)$ time, which is $O(1)$ when $\sigma = \lg^{O(1)} n$.
- We can store S in $nH_0(S) + o(n)(H_0(S) + 1)$ bits and answer access in $O(1)$ time and select in $O(f(n, \sigma))$ time or vice versa, where $f(n, \sigma)$ is any function in $\omega(1)$, and answer rank in $O\left(\log \frac{\log \sigma}{\log w}\right)$ time.
- We can store S in $nH_k(S) + o(n \log \sigma)$ bits and answer access in $O(1)$ time, select in $O(f(n, \sigma))$ time, and rank in $O\left(\log \frac{\log \sigma}{\log w}\right)$ time in general and in $O(f(n, \sigma))$ time when $\sigma = w^{O(1)}$.

These and the other bounds described above are summarized in Table 1. In another paper [5], Belazzougui and Navarro showed how we can add $o(n)(H_0(S) + 1)$ bits to any of these representations and answer partial rank queries in the same time as access.

Dynamic Sequences

Several authors have described data structures that store binary sequences in succinct or compressed space and support fast rank, select, and update operations, typically insertions and deletions of bits. In particular, Navarro and Sadakane [19] described data structures that store a binary sequence B in $|B|H_0(B) + o(|B|)$ bits and support rank, select, insert, and delete in $O\left(\frac{\log |B|}{\log \log |B|}\right)$ time, which is optimal [8]. These can be used in wavelet trees to obtain data structures that support rank and select on dynamic sequences over larger alphabets. Navarro and Sadakane [19] and He and Munro [15] described data structures that store a sequence S in $nH_0(S) + o(n \log \sigma)$ bits, where n is the current length of S, and support access, rank, and select queries and insertions and deletions of characters in $O\left(\frac{\log n}{\log \log n}\left(\frac{\log \sigma}{\log \log n} + 1\right)\right)$ time, which is $O\left(\frac{\log n}{\log \log n}\right)$ when $\sigma = \lg^{O(1)} n$.

Navarro and Nekrich [17, 18] recently described a data structure that stores S in $nH_0(S) + o(n \log \sigma)$ bits and supports access,

Rank and Select Operations on Sequences, Table 1
A summary of previous and current upper bounds for rank and select on a sequence S of length n over an alphabet of size $\sigma = o(n/\log n)$, with ϵ a positive constant, $k = o(\log_\sigma n)$, and $f(n, \sigma) = \omega(1)$. The bounds in the second row hold when $\sigma = \lg^{O(1)} n$ and those in the last row hold when $\sigma = w^{O(1)}$, with w the word length

Source	Space (bits)	Access	Rank	Select
[12]	$nH_0(S) + o(n)$	$O(\log \sigma)$	$O(\log \sigma)$	$O(\log \sigma)$
[7]	$nH_0(S) + o(n)$	1	1	1
[7]	$nH_0(S) + o(n \log \sigma)$	$O\left(\frac{\log \sigma}{\log \log n}\right)$	$O\left(\frac{\log \sigma}{\log \log n}\right)$	$O\left(\frac{\log \sigma}{\log \log n}\right)$
[11]	$nH_0(S) + o(n)$	$O\left(\frac{\log \sigma}{\log \log n}\right)$	$O\left(\frac{\log \sigma}{\log \log n}\right)$	$O\left(\frac{\log \sigma}{\log \log n}\right)$
[10]	$n \lg \sigma + o(n \log \sigma)$	$O(\log \log \sigma)$	$O(\log \log \sigma)$	1
[10]	$n \lg \sigma + o(n \log \sigma)$	1	$O((\log \log \sigma)^{1+\epsilon})$	$O(\log \log \sigma)$
[10]	$(1 + \epsilon)n \lg \sigma$	1	$O(\log \log \sigma)$	1
[1]	$nH_k(S) + o(n \log \sigma)$	1	$O((\log \log \sigma)^{1+\epsilon})$	$O((\log \log \sigma)^{1+\epsilon})$
[1]	$nH_0(S) + o(n \log \sigma)$	$O(\log \log \sigma)$	$O(\log \log \sigma)$	1
[13]	$nH_k(S) + o(n \log \sigma)$	1	$O(\log \log \sigma)$	$O(\log \log \sigma)$
[2]	$nH_0(S) + o(n)(H_0(S) + 1)$	1	$O(\log \log \sigma)$	$O(\log \log \sigma)$
[2]	$nH_0(S) + o(n)(H_0(S) + 1)$	$O(\log \log \sigma)$	$O(\log \log \sigma)$	1
[2]	$(1 + \epsilon)nH_0(S) + o(n)$	1	$O(\log \log \sigma)$	1
[3,4]	$nH_0(S) + o(n)$	$O\left(\frac{\log \sigma}{\log w} + 1\right)$	$O\left(\frac{\log \sigma}{\log w} + 1\right)$	$O\left(\frac{\log \sigma}{\log w} + 1\right)$
[3,4]	$nH_0(S) + o(n)(H_0(S) + 1)$	1	$O\left(\log \frac{\log \sigma}{\log w}\right)$	$O(f(n, \sigma))$
[3,4]	$nH_0(S) + o(n)(H_0(S) + 1)$	$O(f(n, \sigma))$	$O\left(\log \frac{\log \sigma}{\log w}\right)$	1
[3,4]	$nH_k(S) + o(n \log \sigma)$	1	$O\left(\log \frac{\log \sigma}{\log w}\right)$	$O(f(n, \sigma))$
[3,4]	$nH_k(S) + o(n \log \sigma)$	1	$O(f(n, \sigma))$	$O(f(n, \sigma))$

rank, select, insert, and delete in $O\left(\frac{\log n}{\log \log n}\right)$ time. This time bound is worst-case for the queries and amortized for the updates; the update times can be made worst-case as well at the cost of increasing the times for rank, insert, and delete from $O\left(\frac{\log n}{\log \log n}\right)$ to $O(\log n)$. Their structure is essentially a multiary wavelet tree built using rank and select data structures for dynamic sequences over sublogarithmic alphabets, much like He and Munro's or Navarro and Sadakane's, but they divide those component sequences into polylogarithmic-sized blocks and augment them with pointers such that they can ascend and descend the tree using only the pointers and rank and select on individual blocks.

Grossi, Raman, Rao, and Venturini [14] later reduced the time for access to $O(1)$ while using $nH_k(S) + o(n \log \sigma)$ bits but at the cost of being able only to replace characters instead of inserting and deleting them. The time for rank and select is the same.

Applications

Jacobson [16] first studied rank and select for representing unlabeled trees succinctly and planar graphs almost succinctly, while supporting fast navigation queries. Since then, rank and select on binary sequences have been used in succinct and compressed representations of several other combinatorial objects, such as binary relations and general graphs. Rank and select on sequences over larger alphabets have been used in succinct and compressed representations of labeled trees and permutations and in compressed full-text indexes such as compressed suffix arrays. Notice that with a data structure for rank and select that achieves compression in terms of 0th-order empirical entropy, we can build a full-text index that achieves compression in terms of kth-order empirical entropy.

Open Problems

The current main open problems regarding rank and select on static sequences are to answer access and select in $O(1)$ time while storing S in $nH_0(S) + o(n \log \sigma)$ bits when $\lg \sigma = o(w)$, to answer select in constant time and access in almost constant time while storing S in $nH_k(S) + o(n \log \sigma)$ bits when $k > 0$, and to answer access, rank, and select queries in $O(1)$ time while storing S in $nH_k(S) + o(n)$ bits when $\sigma = \lg^{O(1)} w$.

The current main open problems regarding rank and select on dynamic sequences are to achieve $O\left(\frac{\log n}{\log \log n}\right)$ worst-case time for all operations while still using compressed space, to achieve a similar space bound in terms of $H_k(S)$ instead of $H_0(S)$ while supporting the same operations, and to support a wider range of updates.

Experimental Results

The most recent experimental results for rank and select on static sequences are by Barbay et al. [2] and Claude, Navarro, and Ordóñez [6]. These results show that rank and select data structures can be implemented in a time- and space-efficient way in practice, even when the alphabet size is large. There are no current experimental results for rank and select on succinct or compressed dynamic sequences.

Cross-References

▶ Compressed Suffix Array
▶ Compressing and Indexing Structured Text
▶ Rank and Select Operations on Bit Strings
▶ Wavelet Trees

Recommended Reading

1. Barbay J, He M, Munro JI, Rao SS (2011) Succinct indexes for strings, binary relations and multilabeled trees. ACM Trans Algorithms 7(4):1–27
2. Barbay J, Claude F, Gagie T, Navarro G, Nekrich Y (2014) Efficient fully-compressed sequence representations. Algorithmica 69(1):232–268 [20] was presented in Philadelphia, USA
3. Belazzougui D, Navarro G (2012) New lower and upper bounds for representing sequences. In: Proceedings of the 20th European symposium on algorithms, Ljubljana, Slovenia, pp 181–192
4. Belazzougui D, Navarro G (2013) New lower and upper bounds for representing sequences. CoRR abs/1111.2621v2. To appear in ACM Transactions on Algorithms
5. Belazzougui D, Navarro G (2014) Alphabet-independent compressed text indexing. ACM Trans Algorithms 10(4):1–19
6. Claude F, Navarro G, Ordóñez A (2015) The wavelet matrix: an efficient wavelet tree for large alphabets. Inf Syst 47:15–32
7. Ferragina P, Manzini G, Mäkinen V, Navarro G (2007) Compressed representations of sequences and full-text indexes. ACM Trans Algorithms 3(2)
8. Fredman ML, Saks ME (1989) The cell probe complexity of dynamic data structures. In: Proceedings of the 21st symposium on theory of computing, Seattle, USA pp 345–354
9. Golynski A (2009) Cell probe lower bounds for succinct data structures. In: Proceedings of the 20th symposium on discrete algorithms, New York, USA, pp 625–634
10. Golynski A, Munro JI, Rao SS (2006) Rank/select operations on large alphabets: a tool for text indexing. In: Proceedings of the 17th symposium on discrete algorithms, Miami, USA, pp 368–373
11. Golynski A, Raman R, Rao SS (2008) On the redundancy of succinct data structures. In: Proceedings of the 11th scandinavian workshop on algorithm theory, Gothenburg, Sweden, pp 148–159
12. Grossi R, Gupta A, Vitter JS (2003) High-order entropy-compressed text indexes. In: Proceedings of the 14th symposium on discrete algorithms, Baltimore, USA, pp 841–850
13. Grossi R, Orlandi A, Raman R (2010) Optimal tradeoffs for succinct string indexes. In: Proceedings of the 37th international colloquium on automata, languages and programming, Bordeaux, France, pp 678–689
14. Grossi R, Raman R, Satti SR, Venturini R (2013) Dynamic compressed strings with random access. In: Proceedings of the 40th international colloquium on languages, automata and programming, Riga, Latvia, pp 504–515
15. He M, Munro JI (2010) Succinct representations of dynamic strings. In: Proceedings of the 17th symposium on string processing and information retrieval, Los Cabos, Mexico, pp 334–346
16. Jacobson G (1989) Space-efficient static trees and graphs. In: Proceedings of the 30th symposium on foundations of computer science, Research Triangle Park, North Carolina, USA, pp 549–554
17. Navarro G, Nekrich Y (2013) Optimal dynamic sequence representations. In: Proceedings of the 24th

R

symposium on discrete algorithms, New Orleans, USA, pp 865–876

18. Navarro G, Nekrich Y (2013) Optimal dynamic sequence representations. CoRR abs/1206.6982v2. To appear in SIAM Journal on Computing
19. Navarro G, Sadakane K (2014) Fully functional static and dynamic succinct trees. ACM Trans Algorithms 10(3):1–39
20. Pătraşcu M (2008) Succincter. In: Proceedings of the 49th symposium on foundations of computer science, Philadelphia, USA, pp 305–313

Ranked Matching

Kavitha Telikepalli
CSA Department, Indian Institute of Science, Bangalore, India

Keywords

Popular matching

Years and Authors of Summarized Original Work

2005; Abraham, Irving, Kavitha, Mehlhorn

Problem Definition

This problem is concerned with matching a set of *applicants* to a set of *posts*, where each applicant has a *preference list*, ranking a non-empty subset of posts in order of preference, possibly involving ties. Say that a matching M is *popular* if there is no matching M' such that the number of applicants preferring M' to M exceeds the number of applicants preferring M to M'. The ranked matching problem is to determine if the given instance admits a popular matching and if so, to compute one. There are many practical situations that give rise to such large-scale matching problems involving two sets of participants – for example, pupils and schools, doctors and hospitals – where participants of one set express preferences over the participants of the other set;

an allocation determined by a popular matching can be regarded as an optimal allocation in these applications.

Notations and Definitions

An instance of the *ranked matching problem* is a bipartite graph $G = (\mathcal{A} \cup \mathcal{P}, E)$ and a partition $E = E_1 \mathbin{\dot{\cup}} E_2 \ldots \mathbin{\dot{\cup}} E_r$ of the edge set. Call the nodes in $\mathcal{A}$ *applicants*, the nodes in $\mathcal{P}$ *posts*, and the edges in E_i the edges of rank i. If $(a, p) \in E_i$ and $(a, p') \in E_j$ with $i < j$, say that a prefers p to p'. If $i = j$, say that a is indifferent between p and p'. An instance is *strict* if the degree of every applicant in every E_i is at most one.

A matching M is a set of edges, no two of which share an endpoint. In a matching M, a node $u \in \mathcal{A} \cup \mathcal{P}$ is either *unmatched*, or *matched* to some node, denoted by $M(u)$. Say that an applicant a *prefers* matching M' to M if (i) a is matched in M' and unmatched in M, or (ii) a is matched in both M' and M, and a prefers $M'(a)$ to $M(a)$.

Definition 1 M' is *more popular than* M, denoted by $M' \succ M$, if the number of applicants preferring M' to M exceeds the number of applicants preferring M to M'. A matching M is popular if and only if there is no matching M' that is more popular than M.

Figure 1 shows an instance with $A = \{a_1, a_2, a_3\}$, $P = \{p_1, p_2, p_3\}$, and each applicant prefers p_1 to p_2, and p_2 to p_3 (assume throughout that preferences are transitive). Consider the three symmetrical matchings $M_1 = \{(a_1, p_1), (a_2, p_2), (a_3, p_3)\}$, $M_2 = \{(a_1, p_3), (a_2, p_1), (a_3, p_2)\}$ and $M_3 = \{(a_1, p_2), (a_2, p_3), (a_3, p_1)\}$. It is easy to verify that none of these matchings is popular, since $M_1 \prec M_2$, $M_2 \prec M_3$, and $M_3 \prec M_1$. In fact, this instance admits no popular matching – the problem being, of course, that the *more popular than* relation is not acyclic, and so there need not be a maximal element.

The *ranked matching problem* is to determine if a given instance admits a popular matching, and to find such a matching, if one exists. Popular matchings may have different sizes, and a largest such matching may be smaller than a maximum-

$$
\begin{array}{llll}
a_1 : & p_1 & p_2 & p_3 \\
a_2 : & p_1 & p_2 & p_3 \\
a_3 : & p_1 & p_2 & p_3
\end{array}
$$

Ranked Matching, Fig. 1 An instance for which there is no popular matching

cardinality matching. The *maximum-cardinality popular matching problem* then is to determine if a given instance admits a popular matching, and to find a *largest* such matching, if one exists.

Key Results

First consider *strict instances*, that is, instances $(\mathcal{A} \cup \mathcal{P}, E)$ where there are no ties in the preference lists of the applicants. Let n be the number of vertices and m be the number of edges in G.

Theorem 1 *For a strict instance $G = (\mathcal{A} \cup \mathcal{P}, E)$, it is possible to determine in $O(m + n)$ time if G admits a popular matching and compute one, if it exists.*

Theorem 2 *Find a maximum-cardinality popular matching of a strict instance $G = (\mathcal{A} \cup \mathcal{P}, E)$, or determine that no such matching exists, in $O(m + n)$ time.*

Next consider the general problem, where preference lists may have ties.

Theorem 3 *Find a popular matching of $G = (\mathcal{A} \cup \mathcal{P}, E)$, or determine that no such matching exists, in $O(\sqrt{n}m)$ time.*

Theorem 4 *Find a maximum-cardinality popular matching of $G = (\mathcal{A} \cup \mathcal{P}, E)$, or determine that no such matching exists, in $O(\sqrt{n}m)$ time.*

Techniques
Our results are based on a novel characterization of popular matchings. For exposition purposes, create a unique *last resort* post $l(a)$ for each applicant a and assign the edge $(a, l(a))$ a rank higher than any edge incident on a. In this way,

assume that every applicant is matched, since any unmatched applicant can be allocated to his/her last resort. From now on then, matchings are *applicant-complete*, and the size of a matching is just the number of applicants not matched to their last resort. Also assume that instances have no gaps, i.e., if an applicant has a rank i edge incident to it then it has edges of all smaller ranks incident to it. First outline the characterization in strict instances and then extend it to general instances.

Strict Instances
For each applicant a, let $f(a)$ denote the most preferred post on a's preference list. That is, $(a, f(a)) \in E_1$. Call any such post p an *f-post*, and denote by $f(p)$ the set of applicants a for which $f(a) = p$.

For each applicant a, let $s(a)$ denote the most preferred non-*f*-post on a's preference list; note that $s(a)$ must exist, due to the introduction of $l(a)$. Call any such post p an *s-post*, and remark that *f*-posts are disjoint from *s*-posts.

Using the definitions of *f*-posts and *s*-posts, show three conditions that a popular matching must satisfy.

Lemma 1 *Let M be a popular matching.*

1. *For every f-post p, (i) p is matched in M, and (ii) $M(p) \in f(p)$.*
2. *For every applicant a, $M(a)$ can never be strictly between $f(a)$ and $s(a)$ on a's preference list.*
3. *For every applicant a, $M(a)$ is never worse than $s(a)$ on a's preference list.*

It is then shown that these three necessary conditions are also sufficient. This forms the basis of the following preliminary characterization of popular matchings.

Lemma 2 *A matching M is popular if and only if (i) every f-post is matched in M, and (ii) for each applicant a, $M(a) \in \{f(a), s(a)\}$.*

Given an instance graph $G = (\mathcal{A} \cup \mathcal{P}, E)$, define the *reduced graph* $G' = (\mathcal{A} \cup \mathcal{P}, E')$

as the subgraph of G containing two edges for each applicant a: one to $f(a)$, the other to $s(a)$. The authors remark that G' need not admit an applicant-complete matching, since $l(a)$ is now isolated whenever $s(a) \neq l(a)$. Lemma 2 shows that a matching is popular if and only if it belongs to the graph G' and it matches every f-post. Recall that all popular matchings are applicant-complete through the introduction of last resorts. Hence, the following characterization is immediate.

Theorem 5 *M is a popular matching of G if and only if (i) every f-post is matched in M, and (ii) M is an applicant-complete matching of the reduced graph G'.*

The characterization in Theorem 5 immediately suggests the following algorithm for solving the popular matching problem. Construct the reduced graph G'. If G' does not admit an applicant-complete matching, then G admits no popular matching. If G' admits an applicant-complete matching M, then modify M so that every f-post is matched. So for each f-post p that is unmatched in M, let a be any applicant in $f(p)$; remove the edge $(a, M(a))$ from M and instead match a to p. This algorithm can be implemented in $O(m + n)$ time. This shows Theorem 1.

Now, consider the maximum-cardinality popular matching problem. Let $\mathcal{A}_1$ be the set of all applicants a with $s(a) = l(a)$. Let $\mathcal{A}_1$ be the set of all applicants with $s(a) = l(a)$. Our target matching must satisfy conditions (i) and (ii) of Theorem 5, and among all such matchings, allocate the fewest $\mathcal{A}_1$-applicants to their last resort. This scheme can be implemented in $O(m + n)$ time. This proves Theorem 2.

General Instances

For each applicant a, let $f(a)$ denote the *set* of first-ranked posts on a's preference list. Again, refer to all such posts p as *f-posts*, and denote by $f(p)$ the set of applicants a for which $p \in f(a)$. It may no longer be possible to match *every* f-post p with an applicant in $f(p)$ (as in Lemma 1), since, for example, there may now be more f-posts than applicants. Let M be a popular matching of some instance graph $G = (\mathcal{A} \cup \mathcal{P}, E)$. Define the *first-*

choice graph of G as $G_1 = (\mathcal{A} \cup \mathcal{P}, E_1)$, where E_1 is the set of all rank one edges. Next the authors show the following lemma.

Lemma 3 *Let M be a popular matching. Then $M \cap E_1$ is a maximum matching of G_1.*

Next, work towards a generalized definition of $s(a)$. Restrict attention to rank-one edges, that is, to the graph G_1 and using M_1, partition $\mathcal{A} \cup \mathcal{P}$ into three disjoint sets. A node v is *even* (respectively *odd*) if there is an even (respectively odd) length alternating path (with respect to M_1) from an unmatched node to v. Similarly, a node v is *unreachable* if there is no alternating path (w.r.t. M_1) from an unmatched node to v. Denote by $\mathcal{E}$, $\mathcal{O}$, and $\mathcal{U}$ the sets of even, odd, and unreachable nodes, respectively. Conclude the following facts about $\mathcal{E}$, $\mathcal{O}$, and $\mathcal{U}$ by using the well-known Gallai–Edmonds decomposition theorem.

(a) $\mathcal{E}$, $\mathcal{O}$, and $\mathcal{U}$ are pairwise disjoint. Every maximum matching in G_1 partitions the vertex set into the same partition of even, odd, and unreachable nodes.

(b)   In any maximum-cardinality matching of G_1, every node in $\mathcal{O}$ is matched with some node in $\mathcal{E}$, and every node in $\mathcal{U}$ is matched with another node in $\mathcal{U}$. The size of a maximum-cardinality matching is $|\mathcal{O}| + |\mathcal{U}|/2$.

(c)   No maximum-cardinality matching of G_1 contains an edge between two nodes in $\mathcal{O}$, or a node in $\mathcal{O}$ and a node in $\mathcal{U}$. And there is no edge in G_1 connecting a node in $\mathcal{E}$ with a node in $\mathcal{U}$.

The above facts motivate the following definition of $s(a)$: let $s(a)$ be the set of most preferred posts in a's preference list that are *even* in G_1 (note that $s(a) \neq \emptyset$, since $l(a)$ is always even in G_1). Recall that our original definition of $s(a)$ led to parts (2) and (3) of Lemma 1 which restrict the set of posts to which an applicant can be matched in a popular matching. This shows that the generalized definition leads to analogous results here.

Lemma 4 *Let M be a popular matching. Then for every applicant a, M(a) can never be strictly* *between f(a) and s(a) on a's preference list and M(a) can never be worse than s(a) in a's preference list.*

The following characterization of popular matchings is formed.

Lemma 5 *A matching M is popular in G if and only if (i) $M \cap E_1$ is a maximum matching of G_1, and (ii) for each applicant a, $M(a) \in f(a) \cup s(a)$.*

Given an instance graph $G = (\mathcal{A} \cup \mathcal{P}, E)$, we define the *reduced graph* $G' = (\mathcal{A} \cup \mathcal{P}, E')$ as the subgraph of G containing edges from each applicant a to posts in $f(a) \cup s(a)$. The authors remark that G' need not admit an applicant-complete matching, since $l(a)$ is now isolated whenever $s(a) \neq \{l(a)\}$. Lemma 11 tells us that a matching is popular if and only if it belongs to the graph G' and it is a maximum matching on rank one edges. Recall that all popular matchings are applicant-complete through the introduction of last resorts. Hence, the following characterization is immediate.

Theorem 6 *M is a popular matching of G if and only if (i) $M \cap E_1$ is a maximum matching of G_1, and (ii) M is an applicant-complete matching of G'.*

Using the characterization in Theorem 6, the authors now present an efficient algorithm for solving the ranked matching problem.

Popular-Matching $(G = (\mathcal{A} \cup \mathcal{P}, E))$

1. Construct the graph $G' = (\mathcal{A} \cup \mathcal{P}, E')$, where $E' = \{(a, p) \mid p \in f(a) \cup s(a), a \in \mathcal{A}\}$.
2. Compute a maximum matching M_1 on rank one edges i.e., M_1 is a maximum matching in $G_1 = (\mathcal{A} \cup \mathcal{P}, E_1)$.
   (M_1 is also a matching in G' because $E' \supseteq E_1$)
3. Delete all edges in G' connecting two nodes in the set $\mathcal{O}$ or a node in $\mathcal{O}$ with a node in $\mathcal{U}$, where $\mathcal{O}$ and $\mathcal{U}$ are the sets of odd and unreachable nodes of $G_1 = (\mathcal{A} \cup \mathcal{P}, E_1)$.
   Determine a maximum matching M in the modified graph G' by augmenting M_1.

4. If M is not applicant-complete, then declare that there is no popular matching in G. Else return M.

The matching returned by the algorithm Popular-Matching is an applicant-complete matching in G' and it is a maximum matching on rank one edges. So the correctness of the algorithm follows from Theorem 6. It is easy to see that the running time of this algorithm is $O(\sqrt{n}m)$. The algorithm of Hopcroft and Karp [7] is uesd to compute a maximum matching in G_1 and identify the set of edges E' and construct G' in $O(\sqrt{n}m)$ time. Repeatedly augment M_1 (by the Hopcroft–Karp algorithm) to obtain M. This proves Theorem 3.

It is now a simple matter to solve the maximum-cardinality popular matching problem. Assume that the instance $G = (\mathcal{A} \cup \mathcal{P}, E)$ admits a popular matching. (Otherwise, the process is done.) In order to compute an applicant-complete matching in G' that is a maximum matching on rank one edges and which maximizes the number of applicants not matched to their last resort, first compute an arbitrary popular matching M' and remove all edges of the form $(a, l(a))$ from M' and from the graph G'. Call the resulting subgraph of G' as H. Determine a maximum matching N in H by augmenting M'. N need not be a popular matching, since it need not be a maximum matching in the graph G'. However, this is easy to mend. Determine a maximum matching M in G' by augmenting N. It is easy to show that M is a popular matching which maximizes the number of applicants not matched to their last resort. Since the algorithm takes $O(\sqrt{n}m)$ time, Theorem 4 is shown.

Applications

The bipartite matching problem with a graded edge set is well-studied in the economics literature, see for example [1, 10, 12]. It models some important real-world problems, including the allocation of graduates to training positions [8], and families to government-owned housing [11]. The concept of a popular matching was first

R

introduced by Gardenfors [5] under the name *majority assignment* in the context of the stable marriage problem [4, 6].

Various other definitions of optimality have been considered. For example, a matching is *Pareto-optimal* [1, 2, 10] if no applicant can improve his/her allocation (say by exchanging posts with another applicant) without requiring some other applicant to be worse off. Stronger definitions exist: a matching is *rank-maximal* [9] if it allocates the maximum number of applicants to their first choice, and then subject to this, the maximum number to their second choice, and so on. A matching is *maximum utility* if it maximizes $\sum_{(a,p) \in M} u_{a,p}$, where $u_{a,p}$ is the utility of allocating post p to applicant a. Neither rank-maximal nor maximum-utility matchings are necessarily popular.

Cross-References

▶ Hospitals/Residents Problem
▶ Maximum Matching
▶ Weighted Popular Matchings

Recommended Reading

1. Abdulkadiroğlu A, Sönmez T (1998) Random serial dictatorship and the core from random endowments in house allocation problems. Econometrica 66(3):689–701
2. Abraham DJ, Cechlárová K, Manlove DF, Mehlhorn K (2004) Pareto-optimality in house al- location problems. In: Proceedings of the 15th international symposium on algorithms and computation. LNCS, vol 3341. Springer, Sanya, pp 3–15
3. Abraham DJ, Irving RW, Kavitha T, Mehlhorn K (2005) Popular matchings. In: Proceedings of the 16th ACM-SIAM symposium on discrete algorithms. SIAM, Vancouver, pp 424–432
4. Gale D, Shapley LS (1962) College admissions and the stability of marriage. Am Math Mon 69:9–15
5. Gardenfors P (1975) Match making: assignments based on bilateral preferences. Behav Sci 20:166–173
6. Guseld D, Irving RW (1989) The stable marriage problem: structure and algorithms. MIT, Cambridge
7. Hopcroft JE, Karp RM (1973) A $n^{5/2}$ algorithm for maximum matchings in Bipartite graphs. SIAM J Comput 2:225–231
8. Hylland A, Zeckhauser R (1979) The ecient allocation of individuals to positions. J Polit Econ 87(2):293–314
9. Irving RW, Kavitha T, Mehlhorn K, Michail D, Paluch K (2004) Rank-maximal matchings. In: Proceedings of the 15th ACM SIAM symposium on discrete algorithms. SIAM, New Orleans, pp 68–75
10. Roth AE, Postlewaite A (1977) Weak versus strong domination in a market with indivisible goods. J Math Econ 4:131–137
11. Yuan Y (1996) Residence exchange wanted: a stable residence exchange problem. Eur J Oper Res 90:536–546
12. Zhou L (1990) On a conjecture by Gale about one-sided matching problems. J Econ Theory 52(1):123–135

Rate-Monotonic Scheduling

Nathan Fisher[1] and Sanjoy K. Baruah[2]
[1]Department of Computer Science, Wayne State University, Detroit, MI, USA
[2]Department of Computer Science, The University of North Carolina, Chapel Hill, NC, USA

Keywords

Fixed-priority scheduling; Rate-monotonic analysis; Real-time systems; Static-priority scheduling

Years and Authors of Summarized Original Work

1973; Liu, Layland

Problem Definition

Liu and Layland [11] introduced rate-monotonic scheduling in the context of the scheduling of recurrent real-time processes upon a computing platform comprising a single preemptive processor.

The Periodic Task Model

The *periodic task* abstraction models real-time processes that make repeated requests for computation. As defined by Liu and Layland [11], each periodic task τ_i is characterized by an ordered pair of positive real-valued parameters (C_i, T_i), where C_i is the *worst-case execution requirement* and T_i the *period* of the task. The requests for computation that are made by task τ_i (subsequently referred to as *jobs* that are *generated* by τ_i) satisfy the following assumptions:

A1: τ_i's first job arrives at system start time (assumed to equal time zero), and subsequent jobs arrive every T_i time units, i.e., one job arrives at time instant $k \times T_i$ for all integer $k \geq 0$.

A2: Each job needs to execute for at most C_i time units, i.e., C_i is the maximum amount of time that a processor would require to execute each job of τ_i, without interruption.

A3: Each job of τ_i must complete before the next job arrives. That is, each job of task τ_i must complete execution by a *deadline* that is T_i time units after its arrival time.

A4: Each task is *independent* of all other tasks – the execution of any job of task τ_i is not contingent on the arrival or completion of jobs of any other task τ_j.

A5: A job of τ_i may be *preempted* on the processor without additional execution cost. In other words, if a job of τ_i is currently executing, then it is permitted that this execution be halted and a job of a different task τ_j begins execution immediately.

A periodic task system $\tau \overset{\text{def}}{=} \{\tau_1, \tau_2, \ldots, \tau_n\}$ is a collection of n periodic tasks. The *utilization* $U(\tau)$ is defined as follows:

$$U(\tau) \overset{\text{def}}{=} \sum_{i=1}^{n} C_i / T_i . \qquad (1)$$

Intuitively, this denotes the fraction of time that may be spent by the processor executing jobs of tasks in τ, in the worst case.

The Rate-monotonic Scheduling Algorithm

A (uniprocessor) scheduling algorithm determines which task executes on the shared processor at each time instant. If a scheduling algorithm is guaranteed to always meet all deadlines when scheduling a task system τ, then τ is said to be *schedulable* with respect to that scheduling algorithm.

Many scheduling algorithms work as follows: at each time instant, they assign a priority to each job and select for execution the greatest-priority job with remaining execution. A *static -priority* (often called *fixed-priority*) scheduling algorithm for scheduling periodic tasks is one in which it is required that all the jobs of each periodic task be assigned the same priority.

Liu and Layland [11] proposed the *rate-monotonic* (RM) static-priority scheduling algorithm, which assigns priority to jobs according to the period parameter of the task that generates them: *the smaller the period, the higher the priority*. Hence, if $T_i < T_j$ for two tasks τ_i and τ_j, then each job of τ_i has higher priority than all jobs of τ_j and hence any executing job of τ_j will be preempted by the arrival of one of τ_i's jobs. Ties may be broken arbitrarily, but consistently – if $T_i = T_j$, then either all jobs of τ_i are assigned higher priority than all jobs of τ_j or all jobs of τ_j are assigned higher priority than all jobs of τ_i.

Key Results

First, key results from the original paper by Liu and Layland [11] are presented. Following this, results extending the work of Liu and Layland [11] are summarized.

Results from [11]

Optimality. Liu and Layland were concerned with designing "good" static- priority scheduling algorithms. They defined a notion of optimality for such algorithms: a static-priority algorithm $\mathcal{A}$ is *optimal* if any periodic task system that is schedulable with respect to some static-priority algorithm is also schedulable with respect to $\mathcal{A}$.

Liu and Layland obtained the following result for the rate-monotonic scheduling algorithm (RM):

Theorem 1 *For periodic task systems, RM is an optimal static-priority scheduling algorithm.*

Schedulability testing. A *schedulability test* for a particular scheduling algorithm determines, for any periodic task system τ, whether τ is schedulable with respect to that scheduling algorithm. A schedulability test is said to be *exact* if it is the case that it correctly identifies all schedulable task systems and *sufficient* if it identifies some, but not necessarily all, schedulable task systems.

In order to derive good schedulability tests for the rate-monotonic scheduling algorithm, Liu and Layland considered the concept of *response time*. The response time of a job is defined as the elapsed time between the arrival of a job and its completion time in a schedule; the response time of a task is defined to be the largest response time that may be experienced by one of its jobs. For static- priority scheduling, Liu and Layland obtained the following result on the response time:

Theorem 2 *The maximum response time for a periodic task τ_i occurs when a job of τ_i arrives simultaneously with jobs of all higher-priority tasks. Such a time instant is known as the* critical instant *for task τ_i.*

Observe that the critical instant of the lowest-priority task in a periodic task system is also a critical instant for all tasks of higher priority. An immediate consequence of the previous theorem is that the response time of each task in the periodic task system can be obtained by simulating the scheduling of the periodic task system starting at the critical instant of the lowest-priority task. If the response time for each task τ_i obtained from such simulation does not exceed T_i, then the task system will always meet all deadlines when scheduled according to the given priority assignment. This argument immediately gives rise to a schedulability analysis test [9] for

any static-priority scheduling algorithm. Since the simulation may need to be carried out until $\max_{i=1}^{n}\{T_i\}$, this schedulability test has run-time pseudo-polynomial in the representation of the task system:

Theorem 3 (Lehoczky, Sha, and Ding [9]) *Exact rate-monotonic schedulability testing of a periodic task system may be done in time pseudo-polynomial in the representation in the task system.*

Liu and Layland also derived a polynomial-time sufficient (albeit not exact) schedulability test for RM, based upon the utilization of the task system:

Theorem 4 *Let n denote the number of tasks in periodic task system τ. If $U(\tau) \leq n(2^{1/n} - 1)$, then τ is schedulable with respect to the RM scheduling algorithm.*

Results Since [11]

The utilization-bound sufficient schedulability test (Theorem 4) was shown to be tight in the sense that for all n, there are unschedulable task systems comprising n tasks with utilization exceeding $n(2^{1/n} - 1)$ by an arbitrarily small amount. However, tests have been devised that exploit more knowledge about tasks' period parameters. For instance, Kuo and Mok [8] provide a potentially superior utilization bound for task systems in which the task period parameters tend to be harmonically related – exact multiples of one another. Suppose that a collection of numbers is said to comprise a *harmonic chain* if for every two numbers in the set, it is the case that one is an exact multiple of the other. Let $\tilde{n}$ denote the minimum number of harmonic chains into which the period parameters $\{T_i\}_{i=1}^{n}$ of tasks in τ may be partitioned; a sufficient condition for task system τ to be RM schedulable is that

$$U(\tau) \leq \tilde{n}(2^{1/\tilde{n}} - 1).$$

Since $\tilde{n} \leq n$ for all task systems τ, this utilization bound above is never inferior to the one in Theorem 4 and is superior for all τ for which $\tilde{n} < n$.

A different polynomial-time schedulability test was proposed by Bini, Buttazzo, and Buttazzo [4]: they showed that

$$\Pi_{i=1}^{n}((C_i / T_i) + 1) \ \leq \ 2$$

is sufficient to guarantee that the periodic task system $\{\tau_1, \tau_2, \ldots, \tau_n\}$ is rate-monotonic schedulable. This test is commonly referred to as the *hyperbolic* schedulability test for rate-monotonic schedulability. The hyperbolic test is in general known to be superior to the utilization-based test of Theorem 4 – see [4] for details.

Other work done since the seminal paper of Liu and Layland has focused on relaxing the assumptions of the periodic task model. The (implicit-deadline) *sporadic* task model relaxed assumption A17 by allowing T_i to be the *minimum* (rather than exact) separation between arrivals of successive jobs of task τ_i. It turns out that the Theorems 1–4 continue to hold for systems of such tasks as well.

A more general sporadic task model has also been studied that relaxes assumption A17 in addition to assumption A17, by allowing for the explicit specification of a deadline parameter for each task (which may differ from the task's period). The *deadline-monotonic* scheduling algorithm [10] generalizes rate-monotonic scheduling to such task systems.

Work has also been done [2, 12] in removing the independence assumption of A4, by allowing for different tasks to use critical sections to access non-preemptable serially reusable resources.

Applications

The periodic task model has been invaluable for modeling several different types of systems. For control systems, the periodic task model is well suited for modeling the periodic requests and computations of sensors and actuators. Multimedia and network applications also typically involve computation of periodically arriving packets and data.

Many of the results described in section "Key Results" above have been integrated into powerful tools, techniques, and methodologies for the design and analysis of real-time application systems [1, 7]. The general methodology framework is commonly referred to as the *rate-monotonic analysis* (RMA) methodology. Furthermore, most operating systems provide standard primitives for supporting rate-monotonic scheduling.

Open Problems

There are plenty of interesting and challenging open problems in real-time scheduling theory; however, most of these are concerned with extensions to the basic task and scheduling model considered in the original Liu and Layland paper [11]. Perhaps the most interesting open problem with respect to the task model in [11] is regarding the computational complexity of schedulability analysis of static-priority scheduling. Recent research by Eisenbrand and Rothvoß [5] has shown that determining the maximum response time of any periodic task is NP-hard. This result shows that any exact schedulability test that utilizes response time cannot run in polynomial time (unless P = NP); however, it does not settle the open question of whether there are polynomial-time schedulability tests for static-priority periodic task systems that do not (implicitly or explicitly) calculate task response time.

URLs to Code and Data Sets

Research efforts have been made to develop a standardized methodology for evaluating the efficacy and efficiency of algorithms and analysis proposed for rate-monotonic scheduling problems. Bini and Buttazzo [3] derived an unbiased method for synthetically generating random periodic task systems (http://retis. sssup.it/~bini/publications/2005BinBut.html).

Additionally, researchers have proposed suites of benchmarks as representative of embedded and real-time applications in practice. Notably, the Mälardalen WCET benchmarks [6] (http://

www.mrtc.mdh.se/projects/wcet/benchmarks. html) maintain a collection of programs that are typical for real-time applications.

Cross-References

▶ List Scheduling
▶ Online Load Balancing of Temporary Tasks
▶ Shortest Elapsed Time First Scheduling

Recommended Reading

1. Audsley N, Burns A, Wellings A (1993) Deadline monotonic scheduling theory and application. Control Eng Pract 1:71–78
2. Baker TP (1991) Stack-based scheduling of real-time processes. Real-Time Syst Int J Time-Crit Comput 3:67–100
3. Bini E, Buttazzo G (2005) Measuring the performance of schedulability tests. Real-Time Syst 30:129–154
4. Bini E, Buttazzo GC, Buttazzo GM (2003) Rate monotonic scheduling: the hyperbolic bound. IEEE Trans Comput 52:933–942
5. Eisenbrand F, Rothvoß T (2008) Static-priority real-time scheduling: response time computation is NP-hard. In: Proceedings of the IEEE real-time systems symposium, Barcelona, Nov 2008. IEEE Computer Society Press, pp 397–406
6. Gustafsson J, Betts A, Ermedahl A, Lisper B (2010) The Mälardalen WCET benchmarks – past, present and future. In: Proceedings of 10th international workshop on worst-case execution time analysis (WCET'2010), Brussels, July 2010, pp 137–147
7. Klein M, Ralya T, Pollak B, Obenza R, Harbour MG (1993) A Practitioner's handbook for real-time analysis: guide to rate monotonic analysis for real-time systems. Kluwer Academic, Boston
8. Kuo T-W, Mok AK (1991) Load adjustment in adaptive real-time systems. In: Proceedings of the IEEE real-time systems symposium, San Antonio, Dec 1991. IEEE Computer Society Press, pp 160–171
9. Lehoczky J, Sha L, Ding Y (1989) The rate monotonic scheduling algorithm: exact characterization and average case behavior. In: Proceedings of the real-time systems symposium, Santa Monica, Dec 1989. IEEE Computer Society Press, pp 166–171
10. Leung J, Whitehead J (1982) On the complexity of fixed-priority scheduling of periodic, real-time tasks. Perform Eval 2:237–250
11. Liu C, Layland J (1973) Scheduling algorithms for multiprogramming in a hard real-time environment. J ACM 20:46–61
12. Rajkumar R (1991) Synchronization in real-time systems – a priority inheritance approach. Kluwer Academic, Boston

Rectilinear Spanning Tree

Hai Zhou
Electrical Engineering and Computer Science (EECS) Department, Northwestern University, Evanston, IL, USA

Keywords

Metric minimum spanning tree; Rectilinear spanning graph

Years and Authors of Summarized Original Work

2001; Zhou, Shenoy, Nicholls

Problem Definition

Given a set of n points in a plane, a spanning tree is a set of edges that connects all the points and contains no cycles. When each edge is weighted using some distance metric of the incident points, the *metric minimum spanning tree* is a tree whose sum of edge weights is minimum. If the Euclidean distance (L_2) is used, it is called the *Euclidean minimum spanning tree*; if the rectilinear distance (L_1) is used, it is called the *rectilinear minimum spanning tree*.

Since the minimum spanning tree problem on a weighted graph is well studied, the usual approach for metric minimum spanning tree is to first define a weighted graph on the set of points and then to construct a spanning tree on it.

Much like a connection graph is defined for the maze search [4], a spanning graph can be defined for the minimum spanning tree construction.

Definition 1 Given a set of points V in a plane, an undirected graph $G = (V, E)$ is called a *spanning graph* if it contains a minimum spanning tree of V in the plane.

Since spanning graphs with fewer edges give more efficient minimum spanning tree construction, the *cardinality* of a spanning graph is defined as its number of edges. It is easy to see that a complete graph on a set of points contains all spanning trees, thus is a spanning graph. However, such a graph has a cardinality of $O(n^2)$. A rectilinear spanning graph of cardinality $O(n)$ can be constructed within $O(n \log n)$ time [6] and will be described here.

Minimum spanning tree algorithms usually use two properties to infer the inclusion and exclusion of edges in a minimum spanning tree. The first property is known as the *cut property*. It states that an edge of smallest weight crossing any partition of the vertex set into two parts belongs to a minimum spanning tree. The second property is known as the *cycle property*. It says that an edge with largest weight in any cycle in the graph can be safely deleted. Since the two properties are stated in connection with the construction of a minimum spanning tree, they are useful for a spanning graph.

Key Results

Using the terminology given in [3], the *uniqueness property* is defined as follows.

Definition 2 Given a point s, a region R has the *uniqueness property* with respect to s if for every pair of points $p, q \in R$, $\|pq\| < \max(\|sp\|, \|sq\|)$. A partition of space into a finite set of disjoint regions is said to have the uniqueness property with respect to s if each of its regions has the uniqueness property with respect to s.

The notation $\|sp\|$ is used to represent the distance between s and p under the L_1 metric. Define the *octal partition* of the plane with respect to s as the partition induced by the two rectilinear lines and the two 45° lines through s, as shown in Fig. 1a. Here, each of the regions R_1 through R_8 includes only one of its two bounding half lines as shown in Fig. 1b. It can be shown that the octal partition has the uniqueness property.

Lemma 1 *Given a point s in the plane, the octal partition with respect to s has the uniqueness property.*

Proof To show a partition has the uniqueness property, it needs to prove that each region of the partition has the uniqueness property. Since the regions R_1 through R_8 are similar to each other, a proof for R_1 will be sufficient.

The points in R_1 can be characterized by the following inequalities:

$$x \geq x_s,$$

$$x - y < x_s - y_s.$$

Suppose there are two points p and q in R_1. Without loss of generality, it can be assumed $x_p \leq x_q$. If $y_p \leq y_q$, then $\|sq\| = \|sp\| +$

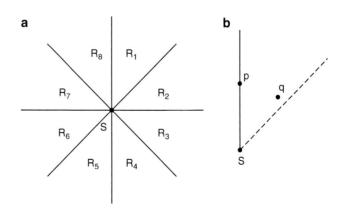

Rectilinear Spanning Tree, Fig. 1 Octal partition and the uniqueness property

$||pq|| > ||pq||$. Therefore it only needs to consider the case when $y_p > y_q$. In this case,

$$\|pg\| = |x_p - x_q| + |y_p - y_p|$$
$$= x_q - x_p + y_p - y_q$$
$$= (x_q - y_q) + y_p - x_p$$
$$< (x_s - y_s) + y_p - x_s$$
$$= y_p - y_s$$
$$\leq x_p - x_s + y_p - y_s$$
$$= \|sp\|.$$

Given two points p, q in the same octal region of point s, the uniqueness property says that $||pq|| < \max(||sp||, ||sq||)$. Consider the cycle on points s, p, and q. Based on the cycle property, only one point with the minimum distance from s needs to be connected to s. An interesting property of the octal partition is that the contour of equidistant points from s forms a line segment in each region. In regions R_1, R_2, R_5, and R_6, these segments are captured by an equation of the form $x + y = c$; in regions R_3, R_4, R_7, and R_8, they are described by the form $x - y = c$.

From each point s, the closest neighbor in each octant needs to be found. It will be described how to efficiently compute the neighbors in R_1 for all points. The case for other octant is symmetric. For the R_1 octant, a sweep line algorithm will run on all points according to nondecreasing $x + y$. During the sweep, maintained will be an *active set* consisting of points whose nearest neighbors in R_1 are yet to be discovered. When a point p is processed, all points in the active set that have p in their R_1 regions will be found. If s is such a point in the active set, since points are scanned in nondecreasing $x + y$, then p must be the nearest point in R_1 for s. Therefore, the edge sp will be added and s will be deleted from the active set. After processing those active points, the point p will be added into the active set. Each point will be added and deleted at most once from the active set.

A fundamental operation in the sweep line algorithm is to find a subset of active points

such that a given point p is in their R_1 regions. Based on the observation that point p is in the R_1 region of point s if and only if s is in the R_5 region of p, it needs to find the subset of active points in the R_5 region of p. Since R_5 can be represented as a two-dimensional range $(-\infty, x_p] \times (x_p - y_p, +\infty)$ on $(x, x - y)$, a priority search tree [1] can be used to maintain the active point set. Since each of the insertion and deletion operations takes $O(\log n)$ time, and the query operation takes $O(\log n + k)$ time where k is the number of objects within the range, the total time for the sweep is $O(n \log n)$. Since other regions can be processed in the similar way as in R_1, the algorithm is running in $O(n \log n)$ time. Priority search tree is a data structure that relies on maintaining a balanced structure for the fast query time. This works well for static input sets. When the input set is dynamic, rebalancing the tree can be quite challenging. Fortunately, the active set has a structure that can be explored for an alternate representation. Since a point is deleted from the active set if a point in its R_1 region is found, no point in the active set can be in the R_1 region of another point in the set.

Lemma 2 *For any two points p, q in the active set, it must be $x_p \neq x_q$, and if $x_p < x_q$, then $x_p - y_p \leq x_q - y_q$.*

Based on this property, the active set can be ordered in increasing order of x. This implies a nondecreasing order on $x - y$. Given a point s, the points which have s in their R_1 region must obey the following inequalities:

$$x \leq x_s,$$
$$x - y > x_s - y_s.$$

To find the subset of active points which have s in their R_1 regions, it can first find the largest x such that $x \leq x_s$ and then proceed in decreasing order of x until $x - y \geq x_s - y_s$. Since the ordering is kept on only one dimension, using any binary search tree with $O(\log n)$ insertion, deletion, and query time will also give us an $O(n \log n)$ time algorithm. Binary search trees also need to be balanced. An alternative is to use skip lists [2] which use randomization to avoid the problem of explicit balancing but provide $O(\log n)$ expected behavior.

A careful study also shows that after the sweep process for R_1, there is no need to do the sweep for R_5, since all edges needed in that phase are either connected or implied. Moreover, based on the information in R_5, the number of edge connections can be further reduced. When the sweep step processes point s, it finds a subset of active points which have s in their R_1 regions. Without lost of generality, suppose p and q are two of them. Then p and q are in the R_5 region of s, which means $||pq|| < \max(||sp||, ||sq||)$. Therefore, it needs only to connect s with the nearest active point.

Since R_1 and R_2 have the same sweep sequence, they can be processed together in one pass. Similarly, R_3 and R_4 can be processed together in another pass. Based on the above discussion, the pseudo-code of the algorithm is presented in Fig. 2.

The correctness of the algorithm is stated in the following theorem.

Theorem 1 *Given n points in the plane, the rectilinear spanning graph algorithm constructs a spanning graph in $O(n \log n)$ time, and the number of edges in the graph is $O(n)$.*

Proof The algorithm can be considered as deleting edges from the complete graph. As described, all deleted edges are redundant based on the cycle property. Thus, the output graph of the algorithm will contain at least one rectilinear minimum spanning tree.

In the algorithm, each given point will be inserted and deleted at most once from the active set for each of the four regions R_1 through R_4. For each insertion or deletion, the algorithm requires $O(\log n)$ time. Thus, the total time is upper bounded by $O(n \log n)$. The storage is needed only for active sets, which is at most $O(n)$.

Applications

Rectilinear minimum spanning tree problem has wide applications in VLSI CAD. It is frequently used as a metric of wire length estimation during placement. It is often constructed to approximate a minimum Steiner tree and is also a key step in many Steiner tree heuristics. It is also used in an approximation to the traveling salesperson problem which can be used to generate scan chains in testing. It is important to emphasize that for real-world applications, the input sizes are usually very large. Since it is a problem that will be computed hundreds of thousands times and many of them will have very large input sizes, the rectilinear minimum spanning tree problem needs a very efficient algorithm.

Experimental Results

The experimental results using the rectilinear spanning graph (RSG) followed by Kruskal's algorithm for a rectilinear minimum spanning tree were reported in Zhou et al. [5]. Two other approaches were compared. The first approach used

Rectilinear Spanning Tree, Fig. 2 The rectilinear spanning graph algorithm

Rectilinear Spanning Graph Algorithm

```
for (i = 0; i < 2; i + +) {
    if (i == 0) sort points according to x + y;
    else sort points according to x − y;
    A[1] = A[2] = ∅;
    for each point p in the order {
        find points in A[1], A[2] such that p is in their
            R_{2i+1} and R_{2i+2} regions, respectively;
        connect p with the nearest point in each subset;
        delete the subsets from A[1], A[2], respectively;
        add p to A[1], A[2];
    }
}
```

Rectilinear Spanning Tree, Table 1 Experimental results

Input		Complete		Bound degree		RSG	
Orig	Distinct	#edge	Time	#edge	Time	#edge	Time
1,000	999	498,501	5.095 s	3,878	0.299 s	2,571	0.112 s
2,000	1,996	1,991,010	24.096 s	7,825	0.996 s	5,158	0.218 s
4,000	3,995	7,978,015	2 min 7.233 s	15,761	3.452 s	10,416	0.337 s
6,000	5,991	17,943,045	5 min 54.697 s	23,704	7.515 s	15,730	0.503 s
8,000	7,981	31,844,190	13 min 7.682 s	31,624	13.141 s	21,149	0.672 s
10,000	9,962	49,615,741	–	39,510	20.135 s	26,332	0.934 s
12,000	11,948	–	–	47,424	32.300 s	31,586	1.052 s
14,000	13,914	–	–	55,251	46.842 s	36,853	1.322 s
16,000	15,883	–	–	63,089	1 min 3.759 s	42,251	1.486 s
18,000	17,837	–	–	70,876	1 min 19.812 s	47,511	1.701 s
20,000	19,805	–	–	78,723	1 min 45.792 s	52,732	1.907 s

the complete graph on the point set as the input to Kruskal's algorithm. The second approach is an implementation of concepts described in [3]; namely, for each point, scan all other points but only connect the nearest one in each quadrant region. With sizes ranging from 1,000 to 20,000, randomly generated point sets were used in the experiments. The results are reproduced here in Table 1. The first column gives the number of generated points; the second column gives the number of distinct points. For each approach, the number of edges in the given graph and the total running time are reported. For input size larger than 10,000, the complete graph approach simply runs out of memory.

Cross-References

▶ Rectilinear Steiner Tree

Recommended Reading

1. McCreight EM (1985) Priority search trees. SIAM J Comput 14:257–276
2. Pugh W (1990) Skip lists: a probabilistic alternative to balanced trees. Commun ACM 33:668–676
3. Robins G, Salowe JS (1995) Low-degree minimum spanning tree. Discret Comput Geom 14:151–165
4. Zheng SQ, Lim JS, Iyengar SS (1996) Finding obstacle-avoiding shortest paths using implicit connection graphs. IEEE Trans Comput Aided Des 15:103–110
5. Zhou H, Shenoy N, Nicholls W (2001) Efficient minimum spanning tree construction without delaunay triangulation. In: Proceedings of Asian and South Pacific design automation conference, Yokohama
6. Zhou H, Shenoy N, Nicholls W (2002) Efficient spanning tree construction without delaunay triangulation. Inf Proc Lett 81:271–276

Rectilinear Steiner Tree

Hai Zhou
Electrical Engineering and Computer Science (EECS) Department, Northwestern University, Evanston, IL, USA

Keywords

Metric minimum Steiner tree; Shortest routing tree

Years and Authors of Summarized Original Work

2003; Zhou

Problem Definition

Given n points on a plane, a Steiner minimal tree connects these points through some extra points (called Steiner points) to achieve a minimal total length. When the length between two points is measured by the rectilinear distance, the tree is called a rectilinear Steiner minimal tree.

Because of its importance, there is much previous work to solve the SMT problem. These algorithms can be grouped into two classes: exact algorithms and heuristic algorithms. Since SMT is NP-hard, any exact algorithm is expected to have an exponential worst-case running time. However, two prominent achievements must be noted in this direction. One is the *GeoSteiner* algorithm and implementation by Warme, Winter, and Zacharisen [14, 15], which is the current fastest exact solution to the problem. The other is a Polynomial Time Approximation Scheme (PTAS) by Arora [1], which is mainly of theoretical importance. Since exact algorithms have long running time, especially on large input sizes, much more previous efforts were put on heuristic algorithms. Many of them generate a Steiner tree by improving on a minimal spanning tree topology [7], since it was proved that a minimal spanning tree is a 3/2 approximation of a SMT [8]. However, since the backbones are restricted to the minimal spanning tree topology in these approaches, there is a reported limit on the improvement ratios over the minimal spanning trees. The iterated 1-Steiner algorithm by Kahng and Robins [10] is an early approach to deviate from that restriction, and an improved implementation [6] is a champion among such programs in public domain. However, the implementation in [10] has a running time of $O(n^4 \log n)$, and the implementation in [6] has a running time of $O(n^3)$. A much more efficient approach was later proposed by Borah et al. [2]. In their approach, a spanning tree is iteratively improved by connecting a point to an edge and deleting the longest edge on the created circuit. Their algorithm and implementation had a worst-case running time of $\Theta(n^2)$, even though an alternative $O(n \log n)$ implementation was also proposed. Since the backbone is no longer restricted to the minimal spanning tree topology, its performance was reported to be similar to the iterated 1-Steiner algorithm [2]. A recent effort in this direction is a new heuristic by Mandoiu et al. [11] which is based on a 3/2 approximation algorithm of the metric Steiner tree problem on quasi-bipartite graphs [12]. It performs slightly better than the iterated 1-Steiner algorithm, but its running time is also slightly longer than the iterated 1-Steiner algorithm (with the empty rectangle test [11] used). More recently, Chu [3] and Chu and Wong [4] proposed an efficient lookup table- based approach for rectilinear Steiner tree construction.

Key Results

The presented algorithm is based on the edge substitution heuristic of Borah et al. [2]. The heuristic works as follows. It starts with a minimal spanning tree and then iteratively considers connecting a point (e.g., p in Fig. 1) to a nearby edge (e.g., (a, b)) and deleting the longest edge $((b, c))$ on the circuit thus formed. The algorithm employs the spanning graph [17] as a backbone of the computation: it is first used to generate the initial minimal spanning tree and then to generate point-edge pairs for tree improvements. This kind of unification happens also in the spanning tree computation and the longest edge computation for each point-edge pair: using Kruskal's algorithm with disjoint set operations (instead of Prim's algorithm) [5] will unify these two computations.

In order to reduce the number of point-edge pair candidates from $O(n^2)$ to $O(n)$, Borah et al. suggested to use the visibility of a point from an edge, that is, only a point visible from an edge can be considered to connect to that edge. This requires a sweep line algorithm to find visibility relations between points and edges. In order to skip this complex step, the geometrical proximity information embedded within the spanning graph is leveraged. Since a point has eight nearest points connected around it, it is observed that if a point is visible to an edge, then the point is *usually* connected in the graph to at least one end point. In the algorithm, the spanning graph is used to generate point-edge pair candidates. For each edge in the current tree, all points that are neighbors of either of the end points will be considered to form point-edge pairs with the edge. Since the cardinality

Rectilinear Steiner Tree,
Fig. 1 Edge substitution
by Borah et al.

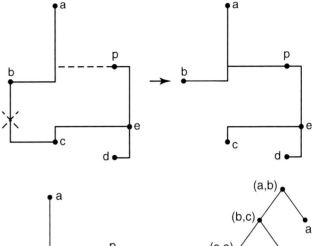

Rectilinear Steiner Tree,
Fig. 2 A minimal
spanning tree and its
merging binary tree

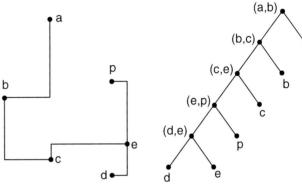

of the spanning graph is $O(n)$, the number of possible point-edge pairs generated in this way is also $O(n)$.

When connecting a point to an edge, the longest edge on the formed circuit needs to be deleted. In order to find the corresponding longest edge for each point-edge pair efficiently, it explores how the spanning tree is formed through Kruskal's algorithm. This algorithm first sorts the edges into nondecreasing lengths, and each edge is considered in turn. If the end points of the edge have been connected, then the edge will be excluded from the spanning tree; otherwise, it will be included. The structure of these connecting operations can be represented by a binary tree, where the leaves represent the points and the internal nodes represent the edges. When an edge is included in the spanning tree, a node is created representing the edge and has as its two children the trees representing the two components connected by this edge. To illustrate this, a spanning tree with its representing binary tree is shown in Fig. 2. As can be seen, the longest edge between two points is the least common

ancestor of the two points in the binary tree. For example, the longest edge between p and b in Fig. 2 is (b, c), which is the least common ancestor of p and b in the binary tree. To find the longest edge on the circuit formed by connecting a point to an edge, it needs to find the longest edge between the point and one end point of the edge that are in the same component before connecting the edge. For example, consider the pair p and (a, b); since p and b are in the same component before connecting (a, b), the edge that needs to be deleted is the longest between p and b.

Based on the above discussion, the pseudo-code of the algorithm can be described in Fig. 3. At the beginning of the algorithm, Zhou et al.'s rectilinear spanning graph algorithm [17] is used to generate the spanning graph G for the given set of points. Then, Kruskal's algorithm is used on the graph to generate a minimal spanning tree. The data structure of disjoint sets [5] is used to merge components and check whether two points are in the same component (the first **for** loop). During this process, the merging binary tree and

Rectilinear Steiner Tree (RST) Algorithm

```
T = ∅;

Generate the spanning graph G by RSG algorithm;

for (each edge (u, v) ∈ G in non-decreasing length) {
    s1 = find_set(u); s2 = find_set(v);
    if (s1 ! = s2) {
        add (u, v) in tree T;
        for (each neighbor w of u, v in G)
            if (s1 == find_set(w))
                lca_add_query(w, v, (u, v));
            else lca_add_query(w, v, (u, v));
        lca_tree_edge((u, v), s1.edge);
        lca_tree_edge((u, v), s2.edge);
        s = union_set(s1, s2); s.edge = (u, v) ;
    }
}

generate point-edge pairs by lca_answer_queries;
for (each pair (p, (a, b), (c, d)) in non-increasing positive gains)
    if ((a, b), (c, d) has not been deleted from T) {
        connect p to (a, b) by adding three edges to T;
        delete (a, b), (c, d) from T;
    }
```

the queries for least common ancestors of all point-edge pairs are also generated. Here, s, $s1$, and $s2$ represent disjoint sets, and each records the root of the component in the merging binary tree. For each edge (u, v) adding to T, each neighbor w of either u or v will be considered to connect to (u, v). The longest edge for this pair is the least common ancestor of w, u or w, v depending on which point is in the same component as w. The procedure *lca_add_query* is used to add this query. Connecting the two components by (u, v) will also be recorded in the merging binary tree by the procedure *lca_tree_edge*. After generating the minimal spanning tree, it also has the corresponding merging binary tree and the least common ancestor queries ready. Using Tarjan's off-line least common ancestor algorithm [5] (represented by *lca_answer_queries*), it can generate all longest edges for the pairs. With the longest edge for each point-edge pair, the gain of connecting the point to the edge can

be calculated. Then, each of the point to edge connections will be realized in a nonincreasing order of their gains. A connection can only be realized if both the connection edge and deletion edge have not been deleted yet.

The running time of the algorithm is dominated by the spanning graph generation and edge sorting, which take $O(n \log n)$ time. Since the number of edges in the spanning graph is $O(n)$, both Kruskal's algorithm and Tarjan's off-line least common ancestor algorithm take $O(n\alpha(n))$ time, where $\alpha(n)$ is the inverse of Ackermann's function, which grows extremely slow.

Applications

The Steiner minimal tree (SMT) problem has wide applications in VLSI CAD. A SMT is generally used in initial topology creation for noncritical nets in physical synthesis. For timing

R

critical nets, minimization of wire length is generally not enough. However, since most nets are noncritical in a design and a SMT gives the most desirable route of such a net, it is often used as an accurate estimation of congestion and wire length during floor planning and placement. This implies that a Steiner tree algorithm will be invoked millions of times. On the other hand, there exist many large pre-routes in modern VLSI design. The pre-routes are generally modeled as large sets of points, thus increasing the input sizes of the Steiner tree problem. Since the SMT is a problem that will be computed millions of times and many of them will have very large input sizes, highly efficient solutions with good performance are desired.

Experimental Results

As reported in [16], the first set of experiments were conducted on a Linux system with a 928 MHz Intel Pentium III processor and 512 M memory. The *RST* algorithm was compared with other publicly available programs: the exact algorithm *GeoSteiner* (version 3.1) by Warme, Winter, and Zacharisen [14]; the Batched Iterated 1-Steiner (*BI1S*) by Robins; and the Borah et al.'s algorithm implemented by Madden (*BOI*).

Table 1 gives the results of the first set of experiments. For each input size ranging from 100 to 5,000, 30 different test cases are randomly generated through the *rand_points* program in *GeoSteiner*. The improvement ratios of a Steiner tree *St* over its corresponding minimal spanning tree *MST* are defined as $100 \times$ (MST - St)/MST. For each input size, the average of the improvement ratios and the average running time (in seconds) on each of the programs are reported. As can be seen, *RST* always gives better improvements than *BOI* with less running times.

The second set of experiments compared *RST* with Borah's implementation of Borah et al.'s algorithm (*Borah*), Rohe's Prim-based algorithm (*Rohe*) [13], and Kahng et al.'s Batched Greedy Algorithm (*BGA*) [9]. They were run on a different Linux system with a 2.4 GHz Intel Xeon processor and 2 G memory. Besides the randomly generated test cases, the VLSI industry test cases used in [9] were also used. The results are reported in Table 2.

Cross-References

▶ Rectilinear Spanning Tree

Recommended Reading

1. Arora S (1998) Polynomial-time approximation schemes for Euclidean TSP and other geometric problem. J ACM 45:753–782
2. Borah M, Owens RM, Irwin MJ (1994) An edge-based heuristic for steiner routing. IEEE Trans Comput Aided Des 13:1563–1568

Rectilinear Steiner Tree, Table 1 Comparison with other algorithms I

Input size	GeoSteiner		BI1S		BOI		RST	
	Improve	Time	Improve	Time	Improve	Time	Improve	Time
100	11.440	0.487	10.907	0.633	9.300	0.0267	10.218	0.004
200	11.492	3.557	10.897	4.810	9.192	0.1287	10.869	0.020
300	11.492	12.685	10.931	18.770	9.253	0.2993	10.255	0.041
500	11.525	72.192	–	–	9.274	0.877	10.381	0.084
800	11.343	536.173	–	–	9.284	2.399	10.719	0.156
1,000	–	–	–	–	9.367	4.084	10.433	0.186
2,000	–	–	–	–	9.326	31.098	10.523	0.381
3,000	–	–	–	–	9.390	104.919	10.449	0.771
5,000	–	–	–	–	9.356	307.977	10.499	1.330

Rectilinear Steiner Tree, Table 2 Comparison with other algorithms II

Input size	BGA		Borah		Rohe		RST	
	Improve	Time	Improve	Time	Improve	Time	Improve	Time
Randomly generated testcases								
100	10.272	0.006	10.341	0.004	9.617	0.000	10.218	0.002
500	10.976	0.068	10.778	0.178	10.028	0.010	10.381	0.041
1,000	10.979	0.162	10.829	0.689	9.768	0.020	10.433	0.121
5,000	11.012	1.695	11.015	25.518	10.139	0.130	10.499	0.980
10,000	11.108	4.135	11.101	249.924	10.111	0.310	10.559	2.098
50,000	11.120	59.147	–	–	10.109	1.890	10.561	13.029
100,000	11.098	161.896	–	–	10.079	4.410	10.514	28.527
500,000	–	–	–	–	10.059	27.210	10.527	175.725
VLSI testcases								
337	6.434	0.035	6.503	0.037	5.958	0.010	5.870	0.016
830	3.202	0.070	3.185	0.213	3.102	0.020	2.966	0.033
1,944	7.850	0.342	7.772	2.424	6.857	0.040	7.533	0.238
2,437	7.965	0.549	7.956	4.502	7.094	0.050	7.595	0.408
2,676	8.928	0.623	8.994	3.686	8.067	0.060	8.507	0.463
12,052	8.450	4.289	8.465	232.779	7.649	0.300	8.076	2.281
22,373	9.848	11.330	9.832	1,128.365	8.987	0.570	9.462	4.605
34,728	9.046	18.416	9.010	2,367.629	8.158	0.900	8.645	5.334

3. Chu C (2004) FLUTE: Fast lookup table based wire-length estimation technique. In: Proceedings of the international conference on computer-aided design, San Jose, pp 696–701

4. Chu C, Wong YC (2005) Fast and accurate rectilinear steiner minimal tree algorithm for VLSI design. In: Proceedings of the international symposium on physical design, San Francisco, pp 28–35

5. Cormen TH, Leiserson CE, Rivest RL (1989) Introduction to algorithms. MIT, Cambridge

6. Griffith J, Robins G, Salowe JS, Zhang T (1994) Closing the gap: near-optimal steiner trees in polynomial time. IEEE Trans Comput Aided Des 13:1351–1365

7. Ho JM, Vijayan G, Wong CK (1990) New algorithms for the rectilinear steiner tree problem. IEEE Trans Comput Aided Des 9:185–193

8. Hwang FK (1976) On steiner minimal trees with rectilinear distance. SIAM J Appl Math 30:104–114

9. Kahng AB, Mandoiu II, Zelikovsky A (2003) Highly scalable algorithms for rectilinear and octilinear steiner trees. In: Proceedings of the Asia and South Pacific design automation conference, Kitakyushu, pp 827–833

10. Kahng AB, Robins G (1992) A new class of iterative steiner tree heuristics with good performance. IEEE Trans Comput Aided Des 11:893–902

11. Mandoiu II, Vazirani VV, Ganley JL (1999) A new heuristic for rectilinear steiner trees. In: Proceedings of the international conference on computer-aided design, San Jose

12. Rajagopalan S, Vazirani VV (1999) On the bidirected cut relaxation for the metric steiner tree problem. In: Proceedings of the 10th ACM-SIAM symposium on discrete algorithms, Baltimore, pp 742–751

13. Rohe A (2001) Sequential and parallel algorithms for local routing. Ph.D. thesis, Bonn University, Bonn

14. Warme DM, Winter P, Zacharisen M (2003) GeoSteiner 3.1 package. ftp://ftp.diku.dk/diku/users/martinz/geosteiner-3.1.tar.gz. Accessed Oct 2003

15. Warme DM, Winter P, Zacharisen M (1998) Exact algorithms for plane steiner tree problems: a computational study. Tech. Rep. DIKU-TR-98/11, Dept. of Computer Science, University of Copenhagen

16. Zhou H (2003) Efficient Steiner tree construction based on spanning graphs. In: ACM international symposium on physical design, Monterey

17. Zhou H, Shenoy N, Nicholls W (2002) Efficient spanning tree construction without delaunay triangulation. Inf Process Lett 81:271–276

R

Recursive Separator Decompositions for Planar Graphs

Shay Mozes
Efi Arazi School of Computer Science, The Interdisciplinary Center (IDC), Herzliya, Israel

Keywords

Divide-and-conquer; Planar separators; r-division

Years and Authors of Summarized Original Work

1987; Frederickson
1995; Goodrich
2013; Klein, Mozes, Sommer

Problem Definition

Graph decompositions are the basis for many divide-and-conquer algorithms. Two main properties make a decomposition useful. The first is *balance*, namely, that the parts of the decomposition have roughly the same size. Balanced decompositions lead to logarithmic depth recursion. The second is small *overlap* between the parts of the decomposition. The overlap affects the time it takes to combine solutions of different parts into a solution for the union of the parts.

A decomposition of a graph G is a collection of subgraphs of G, called *regions*, whose union is G. A *decomposition tree* of G is a tree $\mathcal{T}$ whose nodes correspond to subgraphs of G. The root of $\mathcal{T}$ consists of the entire graph G. For a node v of $\mathcal{T}$ that corresponds to a subgraph R, the children $v_1, v_2, \ldots, v_k$ of v correspond to subgraphs of R whose union is R. Every maximal set D of nodes of $\mathcal{T}$, such that no node in D is an ancestor of another (i.e., every maximal antichain in $\mathcal{T}$ with respect to the ancestry partial order), corresponds to a decomposition of G.

A vertex v of G that belongs to a unique region R in a decomposition is called an *interior* vertex (of R). A vertex v that belongs to more than one region is called a *boundary* vertex.

Let G be a graph with n vertices. Given a parameter $r < n$, an *r-division* of G is a decomposition of G into $\Theta(n/r)$ regions, each with at most r vertices and $O(\sqrt{r})$ boundary vertices. The bounds on the number of regions and on the number of vertices in each region imply that an r-division is a balanced decomposition of G. The $O(\sqrt{r})$ bound on the number of boundary vertices immediately implies the same bound for the overlap between different regions in the decomposition. For an increasing

sequence $\mathbf{r} = r_1, r_2, \ldots$, a *recursive* **r**-*division* is a decomposition tree $\mathcal{T}$ in which the nodes at height i form an r_i-division.

For various applications it is useful to impose additional requirements, such as requiring regions to be connected, requiring that each region share vertices with a constant number of other regions, etc. One particularly useful requirement that is relevant to planar graphs is that the boundary vertices of each region lie on a small number of faces. Formally, every region R inherits its embedding from that of the planar graph G. A hole of R is a face of R that is not a face of G. An *r-division with few holes* is an r-division in which each region has a constant number of holes.

Key Results

Balanced graph decompositions with small overlap are based on small balanced separators. An n-vertex graph G has a $f(n)$-separator if there exists a partition A, B, S of the vertices of G, such that the size of S is at most $f(n)$, the sizes of A and B are at most $2n/3$, and no edge exists between A and B. The set S is called a *separator*. The subgraphs induced on $A \cup S$ and $B \cup S$ form a balanced decomposition of G into 2 regions with $f(n)$ boundary vertices.

The best-known separator result for planar graphs is the $O(\sqrt{n})$ vertex separator of Lipton and Tarjan [16]. Consider a breadth-first-search tree T of a planar graph G. Each BFS level (i.e., the set of vertices at a specific distance from the root) is a separator of G, albeit not necessarily a small or a balanced one. Lipton and Tarjan's separator is based on the observation that it is possible to construct an $O(\sqrt{n})$ balanced separator by combining two appropriately chosen BFS levels with a fundamental cycle with respect to the BFS tree T.

Theorem 1 (Lipton-Tarjan separator) *Let G be an n-vertex planar graph, equipped with non-negative vertex weights summing to one. There exists a linear-time algorithm that returns a separation A, B, S of G such that S consists of at* most $2\sqrt{2n}$ vertices, and neither A nor B has total weight exceeding $2/3$.

Note that the formulation of the theorem allows for a balanced separation with respect to a general weight function, rather than just with respect to the number of vertices.

Miller gave an $O(\sqrt{n})$ simple cycle separator [19] for planar graphs. Miller's result can be viewed as a version of Lipton and Tarjan's separator applied to the planar dual of G.

Theorem 2 (Miller's cycle separator) *Let G be an n-vertex 2-connected planar graph, equipped with nonnegative face weights summing to 1, such that no face weighs more than 2/3. Let d denote the maximum over all face sizes in G. There exists a linear-time algorithm that returns a simple cycle C with at most $2\sqrt{2\lfloor d/2 \rfloor n}$ vertices, such that neither the interior of C nor the exterior of C has total weight exceeding 2/3.*

Similar formulations exist for vertex and edge weights.

$O(\sqrt{n})$ separators are known for other families of sparse graphs, such as graphs excluding a fixed minor [1, 11]. However, some sparse graphs (e.g., expanders) do not have small separators.

By applying a separator recursively, Frederickson [6] showed that r-divisions exist for graphs with $O(\sqrt{n})$ separators. Frederickson's construction generates a decomposition tree whose leaves correspond to the regions of an r-division. It consists of two phases. In the first phase, the separator theorem is applied to each region consisting of more than r vertices. This results in $\Theta(n/r)$ regions, each with at most r vertices and $\sqrt{r}$ boundary vertices *on average*. In a second phase, the separator theorem is applied to each region with more than $\sqrt{r}$ boundary vertices, assigning weight only to boundary vertices. Frederickson proves that this two-phase process results in an r-division. A naïve implementation of Frederickson's approach to construct an r-division takes $O(n \log n)$ time. By applying this approach to a contracted graph, and then further subdividing some of the resulting regions, an r-division can be constructed in $O(n \log r + (n/\sqrt{r}) \log n)$ time [6].

Goodrich [7] showed that for planar graphs an entire binary decomposition tree whose leaves correspond to regions with a constant number of vertices can be computed in $O(n)$ time. This is achieved by showing that, after linear time preprocessing, each invocation of Lipton and Tarjan's separator theorem can be implemented in sublinear time in the number of vertices of a region. The key components of Goodrich's algorithm are the use of a tree-cotree pair of spanning trees of G and of its planar dual to facilitate the search for balanced fundamental cycles, representing these trees using dynamic trees [22], and the use of balanced binary search trees to maintain BFS levels. At each recursive iteration a separator is found in a region with n' vertices in $O(\sqrt{n'} \log n')$ time. This leads to a total linear running time for computing a complete decomposition tree.

Subramanian and Klein [12] were the first to suggest r-divisions with few holes in planar graphs. The idea for achieving a constant number of holes is to use Miller's simple cycle separator instead of Lipton and Tarjan's. To keep the number of holes constant, one needs to alternate the separation criterion between balance with respect to the number of vertices and balance with respect to the number of holes [5]. Since a cycle separator introduces at most one new hole into each of the resulting two regions, reducing the number of holes by a constant factor every constant number of iterations ensures that the number of holes in each region is bounded by a (small) constant. Using Frederickson's approach, an r-division with few holes can be constructed in $O(n \log r + (n/\sqrt{r}) \log n)$ [9].

Klein, Mozes, and Sommer [14] presented a modified version of Miller's cycle separator and used it to obtain a linear-time algorithm for computing r-divisions with few holes in planar graphs (see also [2] for a similar result). Following Miller, their cycle separator algorithm uses BFS levels in the planar dual of G. They show a choice of a spanning tree T of G that makes their cycle separator algorithm very similar to Lipton and Tarjan's vertex separator. Using a technique similar to that of Goodrich, they use this cycle separator algorithm to generate an entire decomposition tree of G in linear time. They show that alternating between three balance criteria (number of vertices, number of boundary vertices, and number of holes) results in a decomposition tree that contains an r-division for any value of r. This is in contrast with Frederickson's two-phase construction which targets a specific value of r. As a consequence, the resulting decomposition tree also contains a recursive r-division, for practically any choice of **r**.

Theorem 3 *There exists a linear-time algorithm that, given a planar graph G and an increasing sequence* **r***, computes a recursive* **r***-division with few holes of G.*

Applications

Separator-based decompositions are wildly used in divide-and-conquer algorithms. Lipton and Tarjan [17] used their separator theorem to show a variety of approximation algorithms and subexponential-time algorithms for NP-hard problems such as the maximum independent set, as well as $O(n^{3/2})$-time algorithms for problems such as maximum matching and Gaussian elimination [18]. These recursive algorithms implicitly generate a complete binary decomposition tree of the input graph. Typically, such algorithms only use the existence of small balanced separators and do not rely on planarity. Hence, they are applicable to families of graphs other than planar graphs.

Frederickson introduced r-divisions for computing shortest paths in a planar graph with nonnegative arc lengths in $O(n\sqrt{\log n})$ time and for finding a minimum st-cut or a maximum st-flow in undirected planar graphs in $O(n \log n)$ time. Since then r-divisions were used, along with Goodrich's linear-time construction, in many algorithms and in different settings (sequential, parallel, dynamic graph problems). A very partial list includes dynamic planar graph algorithms [4], Laplacian solvers and electrical flow algorithms [15, 20], and parallel algorithms in computational geometry [7]. Henzinger et al. [8] used a recursive **r**-division with roughly

$\log^* n$ levels to compute shortest paths with nonnegative arc lengths in linear time.

Decompositions based on simple cycle separators are also wildly used in efficient algorithms for planar graphs. Examples include maximum flow [3, 10], shortest paths [13], and many others. These algorithms typically rely on additional structural properties specific to planar graphs, such as non-crossing of shortest paths (also known as the Monge property). Decompositions with few holes were introduced by Klein and Subramanian [12] to construct approximate dynamic distance oracles for planar graphs. Fakcharoenphol and Rao [5] used a complete decomposition with few holes for computing shortest paths with negative lengths in planar graphs in $O(n \log^3 n)$-time. The currently fastest algorithm for this problem uses r-divisions with few holes and runs in $O(n \log^2 n / \log \log n)$ time [21]. Italiano et al. [9] used an r-division with few holes to find a min st-cut and max st-flow in $O(n \log \log n)$ time.

Cross-References

▶ Fully Dynamic Higher Connectivity for Planar Graphs
▶ Separators in Graphs
▶ Shortest Paths in Planar Graphs with Negative Weight Edges

Recommended Reading

1. Alon N, Seymour PD, Thomas R (1990) A separator theorem for graphs with an excluded minor and its applications. In: Proceedings of the 22nd annual ACM symposium on theory of computing (STOC), Baltimore, pp 293–299
2. Arge L, van Walderveen F, Zeh N (2013) Multiway simple cycle separators and I/O-efficient algorithms for planar graphs. In: Proceedings of the 24th annual ACM-SIAM symposium on discrete algorithms (SODA), New Orleans, pp 901–918
3. Borradaile G, Klein PN, Mozes S, Nussbaum Y, Wulff-Nilsen C (2011) Multiple-source multiple-sink maximum flow in directed planar graphs in near-linear time. In: Proceedings of the 52nd annual symposium on foundations of computer science (FOCS), Palm Springs, pp 170–179

4. Eppstein D, Galil Z, Italiano GF, Spencer TH (1993) Separator based sparsification for dynamic planar graph algorithms. In: Proceedings of the 25th symposium theory of computing, San Diego. ACM, pp 208–217. http://www.acm.org/pubs/citations/proceedings/stoc/167088/p208-eppstein/
5. Fakcharoenphol J, Rao S (2006) Planar graphs, negative weight edges, shortest paths, and near linear time. J Comput Syst Sci 72(5):868–889. http://dx.doi.org/10.1016/j.jcss.2005.05.007, preliminary version in FOCS 2001
6. Frederickson GN (1987) Fast algorithms for shortest paths in planar graphs with applications. SIAM J Comput 16:1004–1022
7. Goodrich MT (1995) Planar separators and parallel polygon triangulation. J Comput Syst Sci 51(3):374–389
8. Henzinger MR, Klein PN, Rao S, Subramanian S (1997) Faster shortest-path algorithms for planar graphs. J Comput Syst Sci 55(1):3–23. doi:10.1145/195058.195092
9. Italiano GF, Nussbaum Y, Sankowski P, Wulff-Nilsen C (2011) Improved algorithms for min cut and max flow in undirected planar graphs. In: Proceedings of the 43rd annual ACM symposium on theory of computing (STOC). ACM, New York, pp 313–322. http://doi.acm.org/10.1145/1993636.1993679,http://doi.acm.org/10.1145/1993636.1993679
10. Johnson DB, Venkatesan S (1982) Using divide and conquer to find flows in directed planar networks in $O(n^{3/2} \log n)$ time. In: Proceedings of the 20th annual allerton conference on communication, control, and computing, Monticello, pp 898–905
11. Kawarabayashi K, Reed BA (2010) A separator theorem in minor-closed classes. In: 51th annual IEEE symposium on foundations of computer science (FOCS), Las Vegas, pp 153–162
12. Klein PN, Subramanian S (1998) A fully dynamic approximation scheme for shortest paths in planar graphs. Algorithmica 22(3):235–249
13. Klein PN, Mozes S, Weimann O (2010) Shortest paths in directed planar graphs with negative lengths: a linear-space $O(n \log^2 n)$-time algorithm. ACM Trans Algorithms 6(2):1–18. http://doi.acm.org/10.1145/1721837.1721846, preliminary version in SODA 2009
14. Klein PN, Mozes S, Sommer C (2013) Structured recursive separator decompositions for planar graphs in linear time. In: Symposium on theory of computing conference (STOC), Palo Alto, pp 505–514
15. Koutis I, Miller GL (2007) A linear work, o(n1/6) time, parallel algorithm for solving planar laplacians. In: Proceedings of the eighteenth annual ACM-SIAM symposium on discrete algorithms, society for industrial and applied mathematics (SODA '07), Philadelphia,pp 1002–1011. http://dl.acm.org/citation.cfm?id=1283383.1283491
16. Lipton RJ, Tarjan RE (1979) A separator theorem for planar graphs. SIAM J Appl Math 36(2):177–189
17. Lipton RJ, Tarjan RE (1980) Applications of a planar separator theorem. SIAM J Comput 9(3):615–627
18. Lipton RJ, Rose DJ, Tarjan RE (1979) Generalized nested dissection. SIAM J Numer Anal 16:346–358
19. Miller GL (1986) Finding small simple cycle separators for 2-connected planar graphs.J Comput Syst Sci 32(3):265–279. doi:10.1016/0022-0000(86)90030-9
20. Miller GL, Peng R (2013) Approximate maximum flow on separable undirected graphs. In: Proceedings of the twenty-fourth annual ACM-SIAM symposium on discrete algorithms (SODA), New Orleans, pp 1151–1170
21. Mozes S, Wulff-Nilsen C (2010) Shortest paths in planar graphs with real lengths in $O(n \log^2 n/ \log \log n)$ time. In: Proceedings of the 18th European symposium on algorithms (ESA), Liverpool, pp 206–217
22. Sleator D, Tarjan R (1983) A data structure for dynamic trees. J Comput Syst Sci 26(3):362–391. doi:10.1016/0022-0000(83)90006-5

Reducing Bayesian Mechanism Design to Algorithm Design

Yang Cai[1], Constantinos Daskalakis[2], and Matthew Weinberg[3]
[1]Computer Science, McGill University, Montreal, QC, Canada
[2]EECS, Massachusetts Institute of Technology, Cambridge, MA, USA
[3]Computer Science, Princeton University, Princeton, NJ, USA

Keywords

Equivalence of separation and optimization; Fair allocation; Job scheduling; Mechanism design; Revenue maximization

Years and Authors of Summarized Original Work

STOC2012; Cai, Daskalakis, Weinberg
FOCS2012; Cai, Daskalakis, Weinberg
SODA2013; Cai, Daskalakis, Weinberg
FOCS2013; Cai, Daskalakis, Weinberg
SODA2015; Daskalakis, Weinberg

Problem Definition

The goal is to design algorithms that succeed in models where input is reported by strategic agents (henceforth referred to as *strategic input*), as opposed to standard models where the input is directly given (henceforth referred to as *honest input*). For example, consider a resource allocation problem where a single user has m jobs to process on n self-interested machines. Each machine i can process job j in time t_{ij}, and this is privately known only to the machine. Each machine reports some processing times $\hat{t}_{ij}$ to the user, who then runs some algorithm to determine where to process the jobs. Good approximation algorithms are known when machines are honest (i.e., $\hat{t}_{ij} = t_{ij}$ for all i, j) if the user's goal is to minimize the *makespan*, the time elapsed until all jobs are completed, going back to seminal work of Lenstra, Shmoys, and Tardos [13]. However, such algorithms do not account for the strategic nature of the machines, which may want to minimize their own work: why would they report honestly their processing time for each job if they can elicit a more favorable schedule by lying? To accommodate such challenges, new algorithmic tools must be developed that draw inspiration from Game Theory.

Requiring solutions that are robust against potential strategic manipulation potentially increases the computational difficulty of whatever problem is at hand. The discussed works provide a framework with which to design such solutions (henceforth called *mechanisms*) and address the following important question.

Question 1 How much (computationally) more difficult is mechanism design than algorithm design?

Using this framework, we resolve this question with an answer of "not at all" for several important problems including job scheduling and fair allocation. Another application of our framework provides efficient algorithms and structural characterization results for multi-item revenue-optimal auction design, a central open problem in mathematical economics.

Model

Environment

1. Set $\mathcal{F}$ of feasible outcomes. Interpret $\mathcal{F}$ as the set of all (feasible) allocations of jobs to machines, allocations of items to bidders, etc.
2. n agents who all care about which outcome is chosen.

Strategic Agents

1. Each agent i has a value $t_i(x)$ for each outcome $x \in \mathcal{F}$. t_i induces a function from $\mathcal{F} \to \mathbb{R}$ and is called the agent's *type*.
2. Each t_i is drawn *independently* from some distribution $\mathcal{D}_i$ of finite support.
3. Agent i knows t_i; all other agents and the designer know only $\mathcal{D}_i$.
4. Agents are *quasi-linear* and *risk neutral*. That is, the utility of an agent of type t for a randomized outcome (distribution over outcomes) $X \in \Delta(\mathcal{F})$, when he is charged price p, is $\mathbb{E}_{x \leftarrow X}[t(x)] - p$.
5. Agents behave in a way that maximizes utility, taking into consideration beliefs about the behavior of other agents.

Designer

1. Designs an *allocation rule* A and *price rule* P. A takes as input a type profile $(t_1, \ldots, t_n)$ and outputs (possibly randomly) an outcome $A(\mathbf{t}) \in \mathcal{F}$. P takes as input a type profile and outputs (possibly randomly) a price vector $P(\mathbf{t})$. The pair (A, P) is called a (direct) *mechanism*. Note that it is without loss of generality to consider only the design of direct mechanisms by the revelation principle [14].
2. Announces A and P to agents. Invites agents to report a type. When $\mathbf{t}$ is reported, selects the outcome $A(\mathbf{t})$ and charges agent i price $P_i(\mathbf{t})$.
3. Has some objective function $\mathcal{O}$ to optimize. $\mathcal{O}$ may depend on the agents' types, the outcome selected, and the prices charged, so we write $\mathcal{O}(\mathbf{t}, x, \mathbf{P})$. Examples include:
   - Social welfare: $\mathcal{O}(\mathbf{t}, x, \mathbf{P}) = \sum_i t_i(x)$.
   - Revenue: $\mathcal{O}(\mathbf{t}, x, \mathbf{P}) = \sum_i P_i(\mathbf{t})$.
   - Makespan: $\mathcal{O}(\mathbf{t}, x, \mathbf{P}) = \max_i \{-t_i(x)\}$ (In job scheduling, agents' values from alloca-

tions are nonpositive, since they have cost for processing jobs. An agent's cost for allocation x is then $-t_i(x)$.).
- Fairness: $\mathcal{O}(\mathbf{t}, x, \mathbf{P}) = \min_i\{t_i(x)\}$.

Game Theoretic Definitions

1. The *interim rule* of a mechanism is a function that takes as input an agent i and type t_i and outputs the distribution of allocations and prices that agent i sees when reporting type t_i over the randomness of the mechanism and the other agents' types, assuming they tell the truth. So the interim allocation rule $(\pi, \mathbf{p})$ of the mechanism (A, P) satisfies:

$$\Pr[x \leftarrow \pi_i(t_i)] = \mathbb{E}_{\mathbf{t}_{-i} \leftarrow \mathcal{D}_{-i}}\left[\Pr[A(t_i; \mathbf{t}_{-i}) = x]\right].$$

$$\Pr[p \leftarrow p_i(t_i)] = \mathbb{E}_{\mathbf{t}_{-i} \leftarrow \mathcal{D}_{-i}}\left[\Pr[P_i(t_i; \mathbf{t}_{-i}) = p]\right].$$

2. A mechanism is *Bayesian Incentive Compatible (BIC)* if every agent receives at least as much utility by reporting their true type as any other type (assuming other agents report truthfully). Formally, $t_i(\pi_i(t_i)) - p_i(t_i) \geq t_i(\pi_i(t_i')) - p_i(t_i')$ for all i, t_i, t_i' (We use the shorthand $t_i(\pi_i(t_i'))$ to denote the expected value of t_i for the random allocation drawn from $\pi_i(t_i')$. Formally, $t_i(\pi_i(t_i')) = \mathbb{E}_{x \leftarrow \pi_i(t_i')}[t_i(x)]$.). A commonly used relaxation of BIC is called ϵ-*Bayesian Incentive Compatible (ϵ-BIC)*. A mechanism is ϵ-BIC if every agent derives at most ϵ less utility by reporting their true type comparing to any other type (assuming other agents report truthfully). Formally, $t_i(\pi_i(t_i)) - p_i(t_i) \geq t_i(\pi_i(t_i')) - p_i(t_i') - \epsilon$ for all i, t_i, t_i'.

3. A mechanism is *individually rational (IR)* if every agent has nonnegative expected utility by participating in the mechanism (assuming other agents report truthfully). Formally, $t_i(\pi_i(t_i)) - p_i(t_i) \geq 0$ for all i, t_i.

Bayesian Mechanism Design (BMeD)

Here we describe formally the mechanism design problem we study. BMeD is parameterized by a set of feasible outcomes $\mathcal{F}$, objective function $\mathcal{O}$, and set of possible types $\mathcal{V}$. Both $\mathcal{V}$ and $\mathcal{F}$ can be discrete or continuous. We assume that every element $v \in \mathcal{V}$ and $x \in \mathcal{F}$ can be represented by a finite bit string $\langle v \rangle$ and $\langle x \rangle$. $\mathcal{V}$ and $\mathcal{F}$ also specify how those bit strings are interpreted. For instance, $\mathcal{V}$ might be the class of all submodular functions, and the bit strings used to represent them may be interpreted as indexing a black-box value oracle. Or $\mathcal{V}$ might be the class of all subadditive functions, and the bit strings used to represent them may be interpreted as an explicit circuit. Or $\mathcal{V}$ could be the class of all additive functions, and the bit strings used to represent them may be interpreted as a vector containing values for each item. So we are parameterizing our problems both by the actual classes $\mathcal{V}$ and $\mathcal{F}$ but also by how elements of these classes are represented. Now, we are ready to formally discuss the problem BMeD$(\mathcal{F}, \mathcal{V}, \mathcal{O})$.

BMeD$(\mathcal{F}, \mathcal{V}, \mathcal{O})$:

INPUT: *For each agent $i \in [n]$, a discrete distribution $\mathcal{D}_i$ over types in $\mathcal{V}$, described explicitly by listing the support of $\mathcal{D}_i$ and the corresponding probabilities.*

OUTPUT: *A BIC, IR mechanism.*

GOAL: *Find the mechanism that optimizes $\mathcal{O}$ in expectation, with respect to all BIC, IR mechanisms (when n bidders with types drawn from $\times_i \mathcal{D}_i$ report truthfully).*

APPROXIMATION: *A mechanism is said to be an (ϵ, α)-approximation to BMeD if it outputs an ϵ-BIC mechanism whose expected value of $\mathcal{O}$ (when n bidders with types drawn from $\times_i \mathcal{D}_i$ report truthfully) is at least αOPT $- \epsilon$ (or at most αOPT $+ \epsilon$ for minimization problems).*

Generalized Objective Optimization Problem (GOOP)

Here we describe formally the algorithmic problem we show has strong connections to BMeD. GOOP is parameterized by a set of feasible outcomes $\mathcal{F}$, objective function $\mathcal{O}$, and set of possible types $\mathcal{V}$. We therefore formally discuss the problem GOOP$(\mathcal{F}, \mathcal{V}, \mathcal{O})$. Below, $\mathcal{V}^\times$ denotes the closure of $\mathcal{V}$ under linear combinations. Functions in $\mathcal{V}^\times$ are represented by a finite list of elements of $\mathcal{V}$, along with (possibly negative) scalar multipliers.

GOOP($\mathcal{F}, \mathcal{V}, \mathcal{O}$):

INPUT: *For each agent $i \in [n]$, a type $g_i \in \mathcal{V}$, multiplier $m_i \in \mathbb{R}$, and cost function $f_i \in \mathcal{V}^\times$. Additionally, an indicator bit b (The indicator bit b is included so that the optimization of just $\sum_i f_i(x)$ (without price multipliers or $\mathcal{O}$) is formally a special case of GOOP($\mathcal{F}, \mathcal{V}, \mathcal{O}$).).*

OUTPUT: *An allocation $x \in \mathcal{F}$, and price vector $\mathbf{p} \in \mathbb{R}^n$.*

GOAL: *Find $\arg\max_{x \in \mathcal{F}, \mathbf{p}}\{b \cdot \mathcal{O}(\mathbf{g}, x, \mathbf{p}) + \sum_i m_i p_i + \sum_i f_i(x)\}$ (or $\arg\min$, if $\mathcal{O}$ is a minimization objective like makespan).*

APPROXIMATION: *$(x, \mathbf{p})$ is said to be an (α, β)-approximation to GOOP if $\beta \cdot b \cdot \mathcal{O}(\mathbf{g}, x, \mathbf{p}) + \sum_i m_i p_i + \sum_i f_i(x)$ is at least/most $\alpha \cdot OPT$. Note that a $(\alpha, 1)$-approximation is the standard notion of an α-approximation. Allowing $\beta \neq 1$ boosts/discounts the value of $\mathcal{O}$ (the objective) before comparing to $\alpha \cdot OPT$. Note also that allowing $\beta \neq 1$ provides no benefit if $b = 0$.*

Key Results

We provide a poly-time black-box reduction from BMeD($\mathcal{F}, \mathcal{V}, \mathcal{O}$) to GOOP($\mathcal{F}, \mathcal{V}, \mathcal{O}$). That is, we provide a reduction from Bayesian mechanism design to traditional algorithm design.

Theorem 1 *Let G be an (α, β)-approximation algorithm for GOOP($\mathcal{F}, \mathcal{V}, \mathcal{O}$). Then for all $\epsilon > 0$, there is an $(\epsilon, \alpha/\beta)$-approximation algorithm for BMeD($\mathcal{F}, \mathcal{V}, \mathcal{O}$). If ℓ is the length of the input to a BMeD($\mathcal{F}, \mathcal{V}, \mathcal{O}$) instance, the algorithm succeeds with probability $1 - \exp(-\text{poly}(\ell, 1/\epsilon))$, makes $\text{poly}(\ell, 1/\epsilon)$ black-box calls to G on inputs of size $\text{poly}(\ell, 1/\epsilon)$, and terminates in time $\text{poly}(\ell, 1/\epsilon)$ (times the running time of each oracle call to G).*

This reduction is developed in a recent series of papers by the authors [4–7, 9]. The possibility of failure and additive error is due to a sampling procedure in the reduction. In addition to the computational aspect provided in Theorem 1, our reduction also has a structural aspect. Namely, we provide a characterization of the optimal mechanism in Bayesian settings.

Theorem 2 *For all objectives $\mathcal{O}$, feasibility constraints $\mathcal{F}$, set of possible types $\mathcal{V}$, and inputs $\mathcal{D}$ to BMeD($\mathcal{F}, \mathcal{V}, \mathcal{O}$), the optimal mechanism is a distribution over generalized objective maximizers. Formally, there exists a joint distribution Δ over an indicator bit b^δ and mappings $(f_1^\delta, \ldots, f_n^\delta)$, where each f_i^δ maps types t_i to multipliers $m_i^\delta(t_i) \in \mathbb{R}$ and cost functions $\phi_i^\delta(t_i) \in \mathcal{V}^\times$, such that the optimal mechanism first samples $(b^\delta, \mathbf{f}^\delta)$ from Δ then maps the type profile $\mathbf{t}$ to the allocation and price vector $(x(\mathbf{t}), \mathbf{p}(\mathbf{t})) = \arg\max_{x \in \mathcal{F}, \mathbf{p}}\{b^\delta \cdot \mathcal{O}(\mathbf{t}, x, \mathbf{p}) + \sum_i m_i^\delta(t_i) p_i + \sum_i \phi_i^\delta(t_i)(x)\}$.*

Perhaps the most interesting case of Theorem 2 is when the objective is revenue. In this case, we may interpret the cost functions $\phi_i^\delta \in \mathcal{V}^\times$ as the *virtual valuation function* of bidder i. By virtual valuations, we do *not* mean Myerson's specific virtual valuation functions [14], which aren't even defined for multi-item instances. Instead we simply mean *some* virtual valuation functions that may or may not be the same as the types/valuations reported by the agents. We include this and other applications of Theorems 1 and 2 below.

Applications

In this section, we apply Theorem 1 to the objectives of revenue, makespan, and fairness.

Revenue Maximization

We apply Theorem 1 to reduce the BMeD problem of optimizing revenue in multi-item settings to GOOP. In [7], it is shown that for this case, one need only consider instances of GOOP with $b = m_1 = \ldots = m_n = 0$, so the GOOP instances that must be solved require just optimization of the cost function (which we call *virtual welfare* for this application). We obtain the following computational and structural results on optimal auction design in general multi-item settings, addressing a long-standing open question following Myerson's seminal work on single-item auctions [14].

Theorem 3 (Revenue Maximization, Computational) *Let G be an α-approximation algorithm for maximizing virtual welfare over $\mathcal{F}$ when all virtual types are from $V^\times$. Then for all $\epsilon > 0$, there is an (ϵ, α)-approximation algorithm for the problem $\mathrm{BMeD}(\mathcal{F}, V, \mathrm{REVENUE})$ that makes polynomially many black-box calls to G. If ℓ is the length of the input to a $\mathrm{BMeD}(\mathcal{F},V,\mathrm{REVENUE})$ instance, the algorithm succeeds with probability $1 - \exp(-\mathrm{poly}(\ell, 1/\epsilon))$, makes $\mathrm{poly}(\ell, 1/\epsilon)$ black-box calls to G on inputs of size $\mathrm{poly}(\ell, 1/\epsilon)$, and terminates in time $\mathrm{poly}(\ell, 1/\epsilon)$ (times the running time of each oracle call to G).*

Theorem 4 (Revenue Maximization, Structural) *In any multi-item setting with arbitrary feasibility constraints and possible agent types, the allocation rule of the revenue-optimal auction is a distribution over virtual welfare maximizers. Formally, there exists a distribution Δ over mappings $(\phi_1, \ldots, \phi_n)$, where each ϕ_i maps types t_i to cost functions $f_i \in V^\times$, such that the allocation rule for the optimal mechanism first samples ϕ from Δ then maps type profile $\mathbf{t}$ to the allocation $\arg\max_{x \in \mathcal{F}}\{\sum_i \phi_i(t_i)(x)\}$.*

We further consider the following important special case: There are m items for sale to n buyers. Any allocation of items to buyers is feasible (that is, each item can be awarded to at most one buyer), so we can denote the set of feasible allocations as $\mathcal{F} = [n + 1]^m$. Furthermore, each buyer i has a value v_{ij} for item j and is *additive* across items, meaning that their value for a set S of items is $\sum_{j \in S} v_{ij}$. So we can denote the set of possible types as $\mathbb{R}_+^m$ (and have types represented as such).

Theorem 5 (Revenue Maximization for Additive Buyers, Computational) *There is a poly-time algorithm for $\mathrm{GOOP}([n + 1]^m, \mathbb{R}_+^m, \mathrm{REVENUE})$. Therefore, there is a poly-time algorithm for $\mathrm{BMeD}([n+1]^m, \mathbb{R}_+^m, \mathrm{REVENUE})$ (In this special case, no sampling is required in the reduction, so the theorem holds even for $\epsilon = 0$. Formally, this is a $(0, 1)$-approximation (an exact algorithm). See [4] for details.).*

Theorem 6 (Revenue Maximization for Additive Buyers, Structural) *In any multi-item setting with n additive buyers and m items for sale, the allocation rule of the revenue-optimal auction is a distribution over virtual welfare maximizers. Formally, there exists a distribution Δ over mappings $(\phi_1, \ldots, \phi_n)$, where each ϕ_i maps types t_i to cost functions $f_i \in \mathbb{R}^m$, such that the allocation rule for the optimal mechanism first samples ϕ from Δ then awards every item j to a buyer in $\arg\max_i\{\phi_{ij}(\mathbf{v}_i)\}$ if their virtual value for item j is nonnegative and does not allocate the item otherwise.*

Job Scheduling on Unrelated Machines

The problem of job scheduling on unrelated machines consists of m jobs and n machines, with machine i able to process job j in time t_{ij}. The goal is to find a schedule (that assigns each job to exactly one machine) minimizing the makespan. Specifically, if S_i are the jobs assigned to machine i, the makespan is $\max_i\{\sum_{j \in S_i} t_{ij}\}$. As a mechanism design problem, one considers the machines to be strategic agents who know their processing time for each job (but the designer and other machines do not). In the language of BMeD, we can denote the feasibility constraints as $[n]^m$, the set of possible types as $\mathbb{R}_+^m$, and the objective as MAKESPAN. Theorem 1 reduces $\mathrm{BMeD}([n]^m, \mathbb{R}_+^m, \mathrm{MAKESPAN})$ to $\mathrm{GOOP}([n]^m, \mathbb{R}_+^m, \mathrm{MAKESPAN})$. It is shown in [7] that for objectives that don't depend on the prices charged at all (called "allocation-only"), only instances of GOOP with $m_i = 0 \ \forall i$ need be considered. It is further shown in [9] that $\mathrm{GOOP}([n]^m, \mathbb{R}_+^m, \mathrm{MAKESPAN})$ can be interpreted as a job scheduling problem with costs. Specifically, $\mathrm{GOOP}([n]^m, \mathbb{R}_+^m, \mathrm{MAKESPAN})$ takes as input a processing time $t_{ij} \geq 0$, and monetary cost $c_{ij} \in \mathbb{R}$ for all machines i and jobs j. The goal is to find a schedule that minimizes the makespan plus cost. Formally, partition the jobs into disjoint sets S_i to minimize $\max_i\{\sum_{j \in S_i} t_{ij}\} + \sum_i \sum_j c_{ij}$. While it is NP-hard to approximate $\mathrm{GOOP}([n]^m, \mathbb{R}_+^m, \mathrm{MAKESPAN})$ within any finite factor, a result of Shmoys and Tardos from the early 1990s obtains a polynomial time $(1, 1/2)$-

approximation algorithm [15]. In combination with Theorem 1, this yields the following theorem:

Theorem 7 (Job Scheduling on Unrelated Machines) *For all $\epsilon > 0$, there is a poly-time $(\epsilon, 2)$-approximation algorithm for* BMeD($[n]^m, \mathbb{R}_+^m$, MAKESPAN). *If ℓ is the length of the input to a* BMeD($[n]^m, \mathbb{R}_+^m$, MAKESPAN) *instance, the algorithm succeeds with probability $1 - \exp(-\text{poly}(\ell, 1/\epsilon))$ and terminates in time poly$(\ell, 1/\epsilon)$.*

Fair Allocation of Indivisible Goods

The problem of fairly allocating indivisible goods consists of m indivisible goods and n children, with child i receiving value v_{ij} for good j. The goal is to find an allocation of goods (that assigns each good to at most one child) maximizing the *fairness*. Specifically, if S_i are the goods allocated to child i, the fairness is $\min_i\{\sum_{j \in S_i} v_{ij}\}$. As a mechanism design problem, one considers the children to be strategic agents who know their own value for each good (but the designer and other children do not). In the language of BMeD, we can denote the feasibility constraints as $[n+1]^m$, the set of possible types as $\mathbb{R}_+^m$, and the objective as FAIRNESS. Theorem 1 reduces BMeD($[n+1]^m, \mathbb{R}_+^m$, FAIRNESS) to GOOP($[n+1]^m, \mathbb{R}_+^m$, FAIRNESS), which can be interpreted as a fair allocation problem with costs (again, because FAIRNESS is allocation only) [7,9]. Specifically, GOOP($[n+1]^m, \mathbb{R}_+^m$, FAIRNESS) takes as input a value $v_{ij} \geq 0$ and monetary cost $c_{ij} \in \mathbb{R}$ for all children i and goods j. The goal is to find an allocation that maximizes the fairness minus cost. Formally, allocate the goods into disjoint sets S_i to maximize $\min_i\{\sum_{j \in S_i} v_{ij}\} - \sum_i \sum_j c_{ij}$. While it is NP-hard to approximate GOOP($[n+1]^m, \mathbb{R}_+^m$, FAIRNESS) within any finite factor, we develop poly-time $(1, m - n + 1)$- and $(1/2, \tilde{O}(\sqrt{n}))$-approximation algorithms for fair allocation with costs, based on algorithms of Bezáková and Dani [2] and Asadpour and Saberi [1] for fair allocation (without costs).

Theorem 8 (Fair Allocation of Indivisible Goods) *There are poly-time $(1, m - n + 1)$- and $(1/2, \tilde{O}(\sqrt{n}))$-approximation algorithms for*

GOOP($[n+1]^m, \mathbb{R}_+^m$, FAIRNESS). *Therefore, for all $\epsilon > 0$, there is a $(\epsilon, \min\{\tilde{O}(\sqrt{n}), m-n+1\})$-approximation algorithm for* BMeD($[n+1]^m, \mathbb{R}_+^m$, FAIRNESS). *If ℓ is the length of the input to a* BMeD($[n+1]^m, \mathbb{R}_+^m$, FAIRNESS) *instance, the algorithm succeeds with probability $1 - \exp(-\text{poly}(\ell, 1/\epsilon))$ and terminates in time poly$(\ell, 1/\epsilon)$.*

Tools for Convex Optimization

We prove Theorems 1 and 2 by solving a linear program over the space of possible interim allocation rules and generalizations of interim allocation rules that we do not discuss here. In doing so, we also develop new tools applicable for general convex optimization that we discuss here. We omit full details of the approach and refer the reader to a series of papers by the authors [5–7, 9] for specifics of the linear program solved and why it addresses BMeD. Seminal works of Khachiyan [12], Grötschel, Lovász, and Schrijver [10], and Karp and Papadimitriou [11] study the problems of optimization and separation over a close, convex region $P \subseteq \mathbb{R}^d$ (Below, we denote by $\alpha P = \{\alpha \mathbf{x} | \mathbf{x} \in P\}$. Also, for simplicity of exposition, we only consider P that contain the origin, so that $\alpha P \subseteq P$ for all $\alpha \leq 1$, but our results extend to all closed, convex P. See [9] for our most general results.). Formally, these problems are:

Optimize(P):

INPUT: *A direction $\mathbf{c} \in \mathbb{R}^d$.*
OUTPUT: *A point $\mathbf{x} \in P$.*
GOAL: *Find $\mathbf{x}^* \in \arg\max_{\mathbf{x} \in P}\{\mathbf{c} \cdot \mathbf{x}\}$.*

Separate(P):

INPUT: *A point $\mathbf{x} \in \mathbb{R}^d$.*
OUTPUT: *"Yes," or a direction $\mathbf{c} \in \mathbb{R}^d$.*
GOAL: *If $\mathbf{x} \in P$, output "yes." Otherwise, output any $\mathbf{c}$ such that $\mathbf{c} \cdot \mathbf{x} > \max_{\mathbf{y} \in P}\{\mathbf{c} \cdot \mathbf{y}\}$.*

Khachiyan's Ellipsoid algorithm shows that if one can solve the problem Separate(P) in time poly(d), then one can also solve Optimize(P) in time poly(d). Grötschel, Lovász, and Schrijver and independently Karp and Papadimitriou show that the other direction holds as well: if one can solve Optimize(P) in time poly(d), then one can also solve Separate(P) in time poly(d). This is colloquially called "the equivalence of separation and optimization." While separation as a means for optimization has obvious uses, optimization as a means for separation is more subtle. Still, numerous applications exist (including our results) and we refer the reader to [10, 11] for several others, including the first poly-time algorithm for submodular minimization.

In order to provide our guarantees with respect to approximation, we develop further the equivalence of separation and optimization to accommodate approximation. Specifically, consider the following problems, further parameterized by some $\alpha < 1$:

α-**Optimize**(P):

> INPUT: *A direction* $\mathbf{c} \in \mathbb{R}^d$.
> OUTPUT: *A point* $\mathbf{x} \in P$.
> GOAL: *Find* $\mathbf{x}$ *satisfying* $\mathbf{c} \cdot \mathbf{x} \geq \alpha \max_{\mathbf{y} \in P}\{\mathbf{c} \cdot \mathbf{y}\}$.

α-**Separate**(P):

> INPUT: *A point* $\mathbf{x} \in \mathbb{R}^d$.
> OUTPUT: *"Yes" and a proof that* $\mathbf{x} \in P$, *or a direction* $\mathbf{c} \in \mathbb{R}^d$ *(For formal details on exactly what constitutes a proof, we refer the reader to [6, 7, 9]. Roughly speaking,* $\mathbf{x}$ *is written as a convex combination of points known to be in* P.*)*
> GOAL: *If* $\mathbf{x} \in \alpha P$, *output "yes" and a proof that* $\mathbf{x} \in P$. *If* $\mathbf{x} \notin P$, *output a direction* $\mathbf{c}$ *such that* $\mathbf{c} \cdot \mathbf{x} > \alpha \max_{\mathbf{y} \in P}\{\mathbf{c} \cdot \mathbf{y}\}$. *If* $\mathbf{x} \in P \setminus \alpha P$, *either is acceptable.*

Theorem 9 (Approximate Equivalence of Separation and Optimization) *For all* $\alpha \leq 1$, *the problems* α-Optimize(P) *and* α-Separate(P) *are computationally equivalent. That is, if one can solve one in time* poly(d), *one can solve the other in time* poly(d) *as well.*

We also extend these results to accommodate bi-criterion approximation, via the problems below, further parameterized by some $\beta > 1$ and subset $S \subseteq [d]$ of coordinates (Below, when we write

($\beta \mathbf{x}_S, \mathbf{x}_{-S}$), we mean to take $\mathbf{x}$ and multiply each $x_i, i \in S$ by β.).

(α, β, S)-**Optimize**(P):

> INPUT: *A direction* $\mathbf{c} \in \mathbb{R}^d$.
> OUTPUT: *A point* $\mathbf{x} \in P$.
> GOAL: *Find* $\mathbf{x}$ *satisfying* $\mathbf{c} \cdot (\beta \mathbf{x}_S, \mathbf{x}_{-S}) \geq \alpha \max_{\mathbf{y} \in P}\{\mathbf{c} \cdot \mathbf{y}\}$.

(α, β, S)-**Separate**(P):

> INPUT: *A point* $\mathbf{x} \in \mathbb{R}^d$.
> OUTPUT: *"Yes" and a proof that* $\mathbf{x} \in P$, *or a direction* $\mathbf{c} \in \mathbb{R}^d$.
> GOAL: *If* $(\beta \mathbf{x}_S, \mathbf{x}_{-S}) \in \alpha P$, *output "yes" and a proof that* $\mathbf{x} \in P$. *If* $\mathbf{x} \notin P$, *output a direction* $\mathbf{c}$ *such that* $\mathbf{c} \cdot (\beta \mathbf{x}_S, \mathbf{x}_{-S}) > \alpha \max_{\mathbf{y} \in P}\{\mathbf{c} \cdot \mathbf{y}\}$. *If* $(\beta \mathbf{x}_S, \mathbf{x}_{-S}) \notin \alpha P$ *and* $\mathbf{x} \in P$, *either is acceptable (An astute reader might worry that for some* α, β, S, P, *the problem* (α, β, S)-*Separate(P) is impossible, due to the existence of an* $\mathbf{x} \notin P$ *such that* $(\beta \mathbf{x}_S, \mathbf{x}_{-S}) \in \alpha P$. *For some* α, β, S, P, *this is indeed the case, but we show that* (α, β, S)-*Optimize(P) is impossible in these cases as well.).*

Theorem 10 (Bi-Criterion Approximate Equivalence of Separation and Optimization) *For all* $\alpha \leq 1, \beta \geq 1, S \subseteq [d]$, *the problems* ($\alpha, \beta, S$)-Optimize($P$) *and* ($\alpha, \beta, S$)-Separate($P$) *are computationally equivalent. That is, if one can solve one in time* poly(d), *one can solve the other in time* poly(d) *as well.*

More formal statements and how we apply these theorems to yield our main result can be found in [9]. Finally, the theorems hold for minimization as well as maximization and without the restriction that P contains the origin (but the theorem statements are more technical).

Open Problems

Our work provides a novel computational framework for solving Bayesian mechanism design problems. We have applied our framework to solve several specific important problems, such as computing revenue-optimal auctions in multi-item settings and approximately optimal BIC mechanisms for job scheduling, but numerous important settings and objectives remain unresolved. Theorem 1 provides a concrete approach for tackling such problems, via the design of (α, β)-approximations for the purely algorithmic Generalized Objective Optimization Problem. Therefore, one important direction following our work is to apply our framework to novel settings and design algorithms for the resulting GOOP instances.

Recommended Reading

1. Asadpour A, Saberi A (2007) An approximation algorithm for max-min fair allocation of indivisible goods. In: The 39th annual ACM symposium on theory of computing (STOC), San Diego
2. Bezáková I, Dani V (2005) Allocating indivisible goods. SIGecom Exch 5(3):11–18
3. Cai Y, Daskalakis C (2011) Extreme-value theorems for optimal multidimensional pricing. In: The 52nd annual IEEE symposium on foundations of computer science (FOCS), Palm Springs
4. Cai Y, Daskalakis C, Matthew Weinberg S (2012) An algorithmic characterization of multi-dimensional mechanisms. In: The 44th annual ACM symposium on theory of computing (STOC), New York
5. Cai Y, Daskalakis C, Matthew Weinberg S (2012) Optimal multi-dimensional mechanism design: reducing revenue to welfare maximization. In: The 53rd annual IEEE symposium on foundations of computer science (FOCS), New Brunswick
6. Cai Y, Daskalakis C, Matthew Weinberg S (2013) Reducing revenue to welfare maximization: approximation algorithms and other generalizations. In: The 24th annual ACM-SIAM symposium on discrete algorithms (SODA), New Orleans
7. Cai Y, Daskalakis C, Matthew Weinberg S (2013) Understanding incentives: mechanism design becomes algorithm design. In: The 54th annual IEEE symposium on foundations of computer science (FOCS), Berkeley
8. Daskalakis C, Matthew Weinberg S (2012) Symmetries and optimal multi-dimensional mechanism design. In: The 13th ACM conference on electronic commerce (EC), Valencia
9. Daskalakis C, Matthew Weinberg S (2015) Bayesian truthful mechanisms for job scheduling from bi-criterion approximation algorithms. In: The 26th annual ACM-SIAM symposium on discrete algorithms (SODA), San Diego
10. Grötschel M, Lovász L, Schrijver A (1981) The ellipsoid method and its consequences in combinatorial optimization. Combinatorica 1(2):169–197
11. Karp RM, Papadimitriou CH (1980) On linear characterizations of combinatorial optimization problems. In: The 21st annual symposium on foundations of computer science (FOCS), Syracuse
12. Khachiyan LG (1979) A polynomial algorithm in linear programming. Sov Math Dokl 20(1):191–194
13. Lenstra JK, Shmoys DB, Tardos É (1990) Approximation algorithms for scheduling unrelated parallel machines. Math Program 46(1–3):259–271
14. Myerson RB (1981) Optimal auction design. Math Oper Res 6(1):58–73
15. Shmoys DB, Tardos É (1993) Scheduling unrelated machines with costs. In: The 4th symposium on discrete algorithms (SODA), Austin

Registers

Paul Vitányi
Centrum Wiskunde & Informatica (CWI), Amsterdam, The Netherlands

Keywords

Asynchronous communication hardware; Shared-memory (wait-free); Wait-free registers; Wait-free shared variables

Years and Authors of Summarized Original Work

1986; Lamport, Vitanyi, Awerbuch

Problem Definition

Consider a system of asynchronous processes that communicate among themselves by only executing read and write operations on a set of shared variables (also known as shared *registers*). The system has no global clock or other synchronization primitives. Every shared variable is associated with a process (called *owner*) which writes it and the other processes may read it. An execution of a write (read) operation on a shared variable will be referred to as a *Write* (*Read*) on that variable. A Write on a shared variable puts a value from a pre-determined finite domain into the variable, and a Read reports a value from the domain. A process that writes (reads) a variable is called a *writer* (*reader*) of the variable.

The goal is to construct shared variables in which the following two properties hold. (1) Operation executions are not necessarily atomic, that is, they are not indivisible but rather consist of atomic sub-operations, and (2) every operation finishes its execution within a bounded number of its own steps, irrespective of the presence of other operation executions and their relative speeds. That is, operation executions are *wait-free*. These two properties give rise to a classification of shared variables, depending on their output characteristics. Lamport [8] distinguishes three categories for 1-writer shared variables, using a precedence relation on operation executions defined as follows: for operation executions A and B, A *precedes* B, denoted $A \longrightarrow B$, if A finishes before B starts; A and B *overlap* if neither A precedes B nor B precedes A. In 1-writer variables, all the Writes are totally ordered by "$\longrightarrow$". The three categories of 1-writer shared variables defined by Lamport are the following.

1. A *safe* variable is one in which a Read not overlapping any Write returns the most recently written value. A Read that overlaps a Write may return any value from the domain of the variable.
2. A *regular* variable is a safe variable in which a Read that overlaps one or more Writes returns either the value of the most recent Write preceding the Read or of one of the overlapping Writes.
3. An *atomic* variable is a regular variable in which the Reads and Writes behave as if they occur in some total order which is an extension of the precedence relation.

A shared variable is *boolean* (Boolean variables are referred to as *bits*.) or *multivalued* depending upon whether it can hold only one out of two or one out of more than two values. A *multiwriter* shared variable is one that can be written and read (concurrently) by many processes. If there is only one writer and more than one reader it is called a *multireader* variable.

Key Results

In a series of papers starting in 1974, for details see [4], Lamport explored various notions of concurrent reading and writing of shared variables culminating in the seminal 1986 paper [8]. It formulates the notion of wait-free implementation of an atomic multivalued shared variable – written by a single writer and read by (another) single reader – from safe 1-writer 1-reader 2-valued shared variables, being mathematical versions of physical *flip-flops*, later optimized in [13]. Lamport did not consider constructions of shared variables with more than one writer or reader.

Predating the Lamport paper, in 1983 Peterson [10] published an ingenious wait-free construction of an atomic 1-writer, n-reader m-valued atomic shared variable from $n + 2$ safe 1-writer n-reader m-valued registers, $2n$ 1-writer 1-reader 2-valued atomic shared variables, and 2 1-writer n-reader 2-valued atomic shared variables. He presented also a proper notion of the wait-freedom property. In his paper, Peterson didn't tell how to construct the n-reader boolean atomic variables from flip-flops, while Lamport mentioned the open problem of doing so, and, incidentally, uses a version of Peterson's construction to bridge the algorithmically demanding step from atomic shared bits to atomic shared multivalues. On the basis of this work, N. Lynch, motivated by concurrency control of multi-user data-bases, posed around 1985 the question of how to construct wait-free multiwriter atomic variables from 1-writer multireader atomic variables. Her student Bloom [1] found in 1985 an elegant 2-writer construction, which, however, has resisted generalization to multiwriter. Vitányi and Awerbuch [14] were the first to define and explore the complicated notion of wait-free constructions of general multiwriter atomic variables, in 1986. They presented a proof method, an unbounded solution from 1-writer 1-reader atomic variables, and a bounded solution from 1-writer n-reader atomic variables. The bounded solution turned out not to be atomic, but only achieved regularity ("Errata" in [14]). The paper introduced important notions and techniques in the area, like (bounded) vector clocks, and identified open problems like the construction of atomic wait-free bounded multireader shared variables from flip-flops, and atomic wait-free bounded multiwriter shared variables from the multireader ones. Peterson who had been working on the multiwriter problem for a decade, together with Burns, tried in 1987 to eliminate the error in the unbounded construction of [14] retaining the idea of vector clocks, but replacing the obsolete-information tracking technique by repeated scanning as in [10]. The result [11] was found to be erroneous in the technical report (R. Schaffer, On the correctness of atomic multiwriter registers, Report MIT/LCS/TM-364, 1988). Neither the re-correction in Schaffer's Technical Report, nor the claimed re-correction by the authors of [11] has appeared in print. Also in 1987 there appeared at least five purported solutions for the implementation of 1-writer n-reader atomic shared variable from 1-writer 1-reader ones: [2, 7, 12] (for the others see [4]) of which [2] was shown to be incorrect (S. Haldar, K. Vidyasankar, *ACM Oper. Syst. Rev*, 26:1(1992), 87–88) and only [12] appeared in journal version. The paper [9], initially a 1987 Harvard Tech Report, resolved all multiuser constructions in one stroke: it constructs a bounded n-writer n-reader (multiwriter) atomic variable from $O(n^2)$ 1-writer 1-reader safe bits, which is optimal, and $O(n^2)$ bit-accesses per Read/Write operation which is optimal as well. It works by making the unbounded solution of [14] bounded, using a new technique, achieving a robust proof of correctness. "Projections" of the construction give specialized constructions for the implementation of 1-writer n-reader (multireader) atomic variables from $O(n^2)$ 1-writer 1-reader ones using $O(n)$ bit accesses per Read/Write operation, and for the implementation of n-writer n-reader (multiwriter) atomic variables from n 1-writer n-reader (multireader) ones. The first "projection" is optimal, while the last "projection" may not be optimal since it uses $O(n)$ control bits per writer while only a lower bound of $\Omega(\log n)$ was established. Taking up this challenge, the construction in [6] claims to achieve this lower bound.

Timestamp System

In a multiwriter shared variable it is only required that every process keeps track of which process wrote last. There arises the general question whether every process can keep track of the order of the last Writes by all processes. A. Israeli and M. Li were attracted to the area by the work in [14], and, in an important paper [5], they raised and solved the question of the more general and universally useful notion of a bounded timestamp system to track the order of events in a concurrent system. In a timestamp system every process owns an *object*, an abstraction of a set of shared variables. One of the requirements of the system is to determine the temporal order in which the objects are written. For this purpose, each object is given a *label* (also referred to as a *timestamp*) which indicates the latest (relative) time when it has been written by its owner process. The processes assign labels to their respective objects in such a way that the labels reflect the real-time order in which they are written to. These systems must support two operations, namely *labeling* and *scan*. A labeling operation execution (Labeling, in short) assigns a new label to an object, and a scan operation execution (Scan, in short) enables a process to determine the ordering in which all the objects are written, that is, it returns a set of labeled-objects ordered temporally. The concern is with those systems where operations can be executed *concurrently*, in an overlapped fashion. Moreover, operation executions must be *wait-free*, that is, each operation execution will take a bounded number of its own steps (the number of accesses to the shared space), irrespective of the presence of other operation executions and their relative speeds. Israeli and Li [5] constructed a bit-optimal bounded timestamp system for *sequential* operation executions. Their sequential timestamp system was published in the above journal reference, but the preliminary concurrent timestamp system in the conference proceedings, of which a more detailed version has been circulated in manuscript form, has not been published in final form. The first generally accepted solution of the *concurrent* case of the bounded timestamp system was from Dolev and Shavit [3]. Their construction is of the type presented in [5] and uses shared variables of size $O(n)$, where n is the number of processes in the system. Each Labeling requires $O(n)$ steps, and each Scan $O(n^2 \log n)$ steps. (A 'step' accesses an $O(n)$ bit variable.) In [4] the unbounded construction of [14] is corrected and extended to obtain an efficient version of the more general notion of a bounded concurrent timestamp system.

Applications

Wait-free registers are, together with message-passing systems, the primary interprocess communication method in distributed computing theory. They form the basis of all constructions and protocols, as can be seen in the textbooks. Wait-free constructions of concurrent timestamp systems (CTSs, in short) have been shown to be a powerful tool for solving concurrency control problems such as various types of mutual exclusion, multiwriter multireader shared variables [14], and probabilistic consensus, by synthesizing a "wait-free clock" to sequence the actions in a concurrent system. For more details see [4].

Open Problems

There is a great deal of work in the direction of register constructions that use less constituent parts, or simpler parts, or parts that can tolerate more complex failures, than previous constructions referred to above. Only, of course, if the latter constructions were not yet optimal in the parameter concerned. Further directions are work on wait-free higher-typed objects, as mentioned above, hierarchies of such objects, and probabilistic constructions. This literature is too vast and diverse to be surveyed here.

Experimental Results

Register constructions, or related constructions for asynchronous interprocess communication, are used in current hardware and software.

Cross-References

► Asynchronous Consensus Impossibility
► Atomic Broadcast
► Causal Order, Logical Clocks, State Machine Replication
► Concurrent Programming, Mutual Exclusion
► Linearizability
► Renaming
► Self-Stabilization
► Snapshots in Shared Memory
► Synchronizers, Spanners
► Topology Approach in Distributed Computing

Recommended Reading

1. Bloom B (1988) Constructing two-writer atomic registers. IEEE Trans Comput 37(12):1506–1514
2. Burns JE, Peterson GL (1987) Constructing multi-reader atomic values from non-atomic values. In: Proceedings of the 6th ACM symposium principles of distributed computing, Vancouver, 10–12 Aug 1987, pp 222–231
3. Dolev D, Shavit N (1997) Bounded concurrent timestamp systems are constructible. SIAM J Comput 26(2):418–455
4. Haldar S, Vitanyi P (2002) Bounded concurrent timestamp systems using vector clocks. J Assoc Comput Mach 49(1):101–126
5. Israeli A, Li M (1993) Bounded time-stamps. Distr Comput 6:205–209 (Preliminary, more extended, version in: proceedings of the 28th IEEE symposium on foundations of computer science, pp 371–382, 1987)
6. Israeli A, Shaham A (1992) Optimal multi-writer multireader atomic register. In: Proceedings of the 11th ACM symposium on principles distributed computing, Vancouver, 10–12 Aug 1992, pp 71–82
7. Kirousis LM, Kranakis E, Vitányi PMB (1987) Atomic multireader register. In: Proceedings of the workshop distributed algorithms. Lecture notes computer science, vol 312. Springer, Berlin, pp 278–296
8. Lamport L (1986) On interprocess communication – part I: basic formalism, part II: algorithms. Distrib Comput 1(2):77–101
9. Li M, Tromp J, Vitányi PMB (1996) How to share concurrent wait-free variables. J ACM 43(4):723–746 (Preliminary version: Li M, Vitányi PMB (1987) A very simple construction for atomic multiwriter register. Technical report TR-01-87, Computer Science Department, Harvard University, Nov 1987)
10. Peterson GL (1983) Concurrent reading while writing. ACM Trans Program Lang Syst 5(1):56–65
11. Peterson GL, Burns JE (1987) Concurrent reading while writing II: the multiwriter case. In: Proceed-

R

ings of the 28th IEEE symposium on foundations of computer science, Los Angeles, 27–29 Oct 1987, pp 383–392

12. Singh AK, Anderson JH, Gouda MG (1994) The elusive atomic register. J ACM 41(2):311–339 (Preliminary version in: proceedings of the 6th ACM symposium on principles distributed computing, 1987)

13. Tromp J (1989) How to construct an atomic variable. In: Proceedings of the workshop distributed algorithms. Lecture notes in computer science, vol 392. Springer, Berlin, pp 292–302

14. Vitányi PMB, Awerbuch B (1987) Atomic shared register access by asynchronous hardware. In: Proceedings of the 27th IEEE symposium on foundations of computer science, Los Angeles, 27–29 Oct 1987, pp 233–243 (Errata, Proceedings of the 28th IEEE symposium on foundations of computer science, Los Angeles, 27–29 Oct 1987, pp 487–487)

Regular Expression Matching

Lucian Ilie
Department of Computer Science, University of Western Ontario, London, ON, Canada

Keywords

Bit parallelism; Glushkov-McNaughton-Yamada automaton; Regular expression matching and searching

Years and Authors of Summarized Original Work

1956; Kleene
1968; Thompson
1992; Wu, Manber
2005; Navarro, Raffinot
2009; Bille, Thorup

Problem Definition

Given a *text string* T of length n and a *regular expression* R, the **regular expression matching** problem (**REM**) is to find all text positions at which an occurrence of a string in $L(R)$ ends (see below for definitions).

For an alphabet Σ, a *regular expression R* over Σ consists of elements of $\Sigma \cup \{\varepsilon\}$ (ε denotes the empty string) and operators $\cdot$ (concatenation), $|$ (union), and $*$ (iteration, i.e., repeated concatenation); the set of strings $L(R)$ represented by R is defined accordingly; see [7]. It is important to distinguish two measures for the size of a regular expression: the *size, m*, which is the total number of characters from $\Sigma \cup \{\cdot, |, *\}$, and Σ-*size, m_Σ*, which counts only the characters in Σ. As an example, for $R = (\mathtt{A} | \mathtt{T}) ((\mathtt{C} | \mathtt{CG}) *)$, the set $L(R)$ contains all strings that start with an $\mathtt{A}$ or a $\mathtt{T}$ followed by zero or more strings in the set $\{\mathtt{C}, \mathtt{CG}\}$; the size of R is $m = 8$ and the Σ-size is $m_\Sigma = 5$. Any regular expression can be processed in linear time so that $m = \mathcal{O}(m_\Sigma)$ (with a small constant); the difference becomes important when the two sizes appear as exponents.

Key Results

Finite Automata

The classical solutions for the REM problem involve finite automata which are directed graphs with the edges labelled by symbols from $\Sigma \cup \{\varepsilon\}$; their nodes are called states; see [7] for details. Unrestricted automata are called *nondeterministic finite automata (NFA)*. *Deterministic finite automata (DFA)* have no ε-labels and require that no two outgoing edges of the same state have the same label. Regular expressions and DFAs are equivalent, that is, the sets of strings represented are the same, as shown by Kleene [11]. There are two classical ways of computing an NFA from a regular expression. Thompson's construction [17] builds an NFA with up to $2m$ states and up to $4m$ edges whereas Glushkov-McNaughton-Yamada's automaton [5, 12] has the minimum number of states, $m_\Sigma + 1$, and $\mathcal{O}(m_\Sigma^2)$ edges; see Fig. 1. Any NFA can be converted into an equivalent DFA by the *subset construction*: each subset of the set of states of the NFA becomes a state of the DFA. The problem is that the DFA can have exponentially more states than the NFA.

Classical Solutions

A regular expression is first converted into an NFA or DFA which is then simulated on the text. In order to be able to search for a match starting anywhere in the text, a loop labelled by all elements of Σ is added to the initial state; see Fig. 1.

Searching with an NFA requires linear space, but many states can be active at the same time, and to update them all we need, for Thompson's NFA, $\mathcal{O}(m)$ time for each letter of the text; this gives Theorem 1. On the other hand, DFAs allow searching time that is linear in n but require more space for the automaton. Theorem 2 uses the DFA obtained from Glushkov-McNaughton-Yamada's NFA.

Theorem 1 (Thompson [17]) *The REM problem can be solved with an NFA in $\mathcal{O}(mn)$ time and $\mathcal{O}(m)$ space.*

Theorem 2 (Kleene [11]) *The REM problem can be solved with a DFA in $\mathcal{O}(n + 2^{m\Sigma})$ time and $\mathcal{O}(2^{m\Sigma})$ space.*

Lazy Construction and Modules

One heuristic to alleviate the exponential increase in the size of DFA is to build only the states reached while scanning the text, as implemented in *Gnu Grep*. Still, the space needed for the DFA remains a problem. A four-Russians approach was presented by Myers [13] where a trade-off between the NFA and DFA approaches is proposed. The syntax tree of the regular expression is divided into modules which are implemented as DFAs and are thereafter treated as leaf nodes in the syntax tree. The process continues until a single module is obtained. An $\mathcal{O}(mn/\log n)$ time and space algorithm is obtained. This bound was recently improved by Bille and Thorup [2].

Theorem 3 (Bille and Thorup [2]) *The REM problem can be solved in linear space and $\mathcal{O}\left(mn/(\log n)^{3/2}\right)$ time.*

The same authors showed in [3] that the length m of the regular expression can be essentially replaced in the complexity bounds by the number of strings (concatenations of characters) that appear in the regular expression.

Bit Parallelism

The simulation of the abovementioned modules is done by encoding all states as bits of a single computer word (called *bit mask*) so that all can be updated in a single operation. The method can be used without modules to simulate directly an NFA as done in [20] and implemented in the

Regular Expression Matching, Fig. 1
Thompson's NFA (*left*) and Glushkov-McNaughton-Yamada's NFA (*right*) for the regular expression (A|T)((C|CG)*); the initial loops labelled A,T,C,G are not part of the construction; they are needed for REM

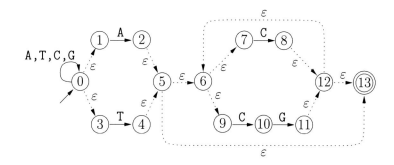

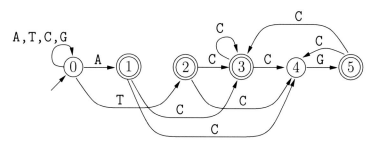

Agrep software [19]. Note that, in fact, the DFA is also simulated: a whole bit mask corresponds to a subset of states of the NFA, that is, one state of the DFA.

The bit-implementation of Wu and Manber [20] uses the property of Thompson's automaton that all Σ-labelled edges connect consecutive states, that is, they carry a bit 1 from position i to position $i + 1$. This makes it easy to deal with the Σ-labelled edges, but the ε-labelled ones are more difficult. A table of size linear in the number of states of the DFA needs to be precomputed to account for the ε-closures (set of states reachable from a given state by ε-paths).

Note that in Theorems 1, 2, and 3, the space complexity is given in words. In Theorems 4 and 5 below, for a more practical analysis, the space is given in bits and the alphabet size is also taken into consideration. For comparison, the space in Theorem 2, given in bits, is $\mathcal{O}(|\Sigma|m_\Sigma 2^{m_\Sigma})$.

Theorem 4 (Wu and Manber [20]) *Thompson's automaton can be implemented using* $2m(2^{2m+1} + |\Sigma|)$ *bits.*

Glushkov-McNaughton-Yamada's automaton has different structural properties. First, it is ε-free, that is, there are no ε-labels on edges. Second, all edges incoming to a given state are labelled the same. These properties are exploited by Navarro and Raffinot [16] to construct a bit-parallel implementation that requires less space. The result is a simple algorithm for regular expression searching which uses less space and usually performs faster than any existing algorithm.

Theorem 5 (Navarro and Raffinot [16]) *Glushkov-McNaughton-Yamada's automaton can be implemented using* $(m_\Sigma + 1)(2^{m_\Sigma+1} + |\Sigma|)$ *bits.*

All algorithms in this category run in $\mathcal{O}(n)$ time, but smaller DFA representation implies more locality of reference and thus faster algorithms in practice. An improvement of any algorithm using Glushkov-McNaughton-Yamada's automaton can be done by reducing first the automaton by merging some of its states, as done by Ilie et al. [8, 9]. The reduction can be performed in such a way that all useful properties of the automaton are preserved. The search becomes faster due to the reduction in size.

Filtration

The above approaches examine every character in the text. In [18] a multipattern search algorithm is used to search for strings that must appear inside any occurrence of the regular expression. Another technique is used in *Gnu Grep*; it extracts the longest string that must appear in any match (it can be used only when such a string exists). In [16], bit-parallel techniques are combined with a reverse factor search approach to obtain a very fast character-skipping algorithm for regular expression searching.

Related Problems

Regular expressions with *backreference* have a feature that helps remembering what was matched to be used later; the matching problem becomes NP-complete; see [1]. *Extended* regular expressions involve adding two extra operators, intersection and complement, which do not change the expressive power. The corresponding matching problem can be solved in $\mathcal{O}((n + m)^4)$ time using dynamic programming; see [7, Exercise 3.23].

Concerning finite automata construction, recall that Thompson's NFA has $\mathcal{O}(m)$ edges, whereas the ε-free Glushkov-McNaughton-Yamada's NFA can have a quadratic number of edges. It has been shown in [4] that one can always build an ε-free NFA with $\mathcal{O}(m \log m)$ edges (for fixed alphabets). However, it is the number of states which is more important in the searching algorithms.

Applications

Regular expression matching is a powerful tool in text-based applications, such as text retrieval and text editing, and in computational biology to find various motifs in DNA and protein sequences. See [6] for more details.

Open Problems

The most important theoretical problem is whether linear time and linear space can be achieved simultaneously. Characterizing the regular expressions that can be searched for using a linear-size equivalent DFA is also of interest. The expressions consisting of a single string are included here – the algorithm of Knuth, Morris, and Pratt is based on this. Also, it is not clear how much we can reduce an NFA efficiently (as done by [8,9]); the problem of finding a minimal NFA is PSPACE-complete; see [10]. Finally, for testing, it is not clear how to define random regular expressions.

Experimental Results

A disadvantage of the bit-parallel technique compared with the classical implementation of a DFA is that the former builds all possible subsets of states whereas the latter builds only the states that can be reached from the initial one (the other ones are useless). On the other hand, bit-parallel algorithms are simpler to code and more flexible (they allow also approximate matching), and there are techniques for reducing the space required. Among the bit-parallel versions, Glushkov-McNaughton-Yamada-based algorithms are better than Thompson-based ones. Modules obtain essentially the same complexity as bit-parallel ones but are more complicated to implement and slower in practice. As the number of computer words increases, bit-parallel algorithms slow down and modules may become attractive. Note also that technological progress has more impact on the bit-parallel algorithms, as opposed to classical ones, since the former depend very much on the machine word size. For details on comparison among various algorithms (including filtration based), see [15]; more recent comparisons are in [16], including the fastest algorithms to date.

URLs to Code and Data Sets

Many text editors and programming languages include regular expression search features. They are, as well, among the tools used in protein databases, such as PROSITE and SWISS-PROT, which can be found at www.expasy.org. The package *agrep* [20] can be downloaded from webglimpse.net and *nrgrep* [14] from www.dcc.uchile.cl/gnavarro/software.

Cross-References

▶ Approximate Regular Expression Matching is a more general problem where errors are allowed.

Recommended Reading

1. Aho A (1990) Algorithms for finding patterns in strings. In: van Leeuwen J (ed) Handbook of theoretical computer science. Algorithms and Complexity, vol A. MIT Press Cambridge, MA, pp 255–300
2. Bille P, Thorup M (2009) Faster regular expression matching. In: Albers S et al (eds) Automata, languages and programming. Springer, Berlin/Heidelberg, pp 171–182
3. Bille P, Thorup M (2010) Regular expression matching with multi-strings and intervals. In: Proceedings of the twenty-first annual ACM-SIAM symposium on discrete algorithms, SIAM, Philadelphia, PA, pp 1297–1308
4. Geffert V (2003) Translation of binary regular expressions into nondeterministic ε-free automata with $\mathcal{O}(n \log n)$ transitions. J Comput Syst Sci 66(3):451–472
5. Glushkov V-M (1961) The abstract theory of automata. Russ Math Surv 16:1–53
6. Gusfield D (1997) Algorithms on strings, trees and sequences. Cambridge University Press, New York, N.Y.
7. Hopcroft J, Ullman J (1979) Introduction to automata, languages, and computation. Addison-Wesley, Reading, Mass
8. Ilie L, Navarro G, Yu S (2004) On NFA reductions. In: Karhumäki J et al (eds) Theory is forever. Lecture notes in computer science vol 3113. Springer-Verlag Berlin Heidelberg, pp 112–124
9. Ilie L, Solis-Oba R, Yu S (2005) Reducing the size of NFAs by using equivalences and preorders. In: Apostolico A, Crochemore M, Park K (eds) Combinatorial pattern matching. Springer, Berlin/Heidelberg, pp 310–321
10. Jiang T, Ravikumar B (1993) Minimal NFA problems are hard. SIAM J Comput 22(6):1117–1141
11. Kleene SC (1956) Representation of events in nerve sets. In: Shannon CE, McCarthy J (eds) Automata

R

studies. Princeton University Press, Princeton, N. J., pp 3–40
12. McNaughton R, Yamada H (1960) Regular expressions and state graphs for automata. IRE Trans Electron Comput 9(1):39–47
13. Myers E (1992) A four Russians algorithm for regular expression pattern matching. J ACM 39(2):430–448
14. Navarro G (2001) Nr-grep: a fast and flexible pattern matching tool. Softw Pract Experience 31:1265–1312
15. Navarro G, Raffinot M (2002) Flexible pattern matching in strings – practical on-line search algorithms for texts and biological sequences. Cambridge University Press, Cambridge, U.K.
16. Navarro G, Raffinot M (2005) New techniques for regular expression searching. Algorithmica 41(2):89–116
17. Thompson K (1968) Regular expression search algorithm. Commun ACM 11(6):419–422
18. Watson B (1995) Taxonomies and toolkits of regular language algorithms. PhD Dissertation, Eindhoven University of Technology, The Netherlands
19. Wu S, Manber U (1992) Agrep – a fast approximate patter-matching tool. In: Proceedings of the USENIX technical conference, San Francisco, California, pp 153–162
20. Wu S, Manber U (1992) Fast text searching allowing errors. Commun ACM 35(10):83–91

Reinforcement Learning

Eyal Even-Dar
Google, New York, NY, USA

Keywords

Neuro-dynamic programming

Years and Authors of Summarized Original Work

1992; Watkins

Problem Definition

Many sequential decision problems ranging from dynamic resource allocation to robotics can be formulated in terms of stochastic control and solved by methods of reinforcement learning. Therefore, reinforcement learning (a.k.a neuro-dynamic programming) has become one of the major approaches to tackling real-life problems.

In reinforcement learning, an agent wanders in an unknown environment and tries to maximize its long-term return by performing actions and receiving rewards. The most popular mathematical models to describe reinforcement learning problems are the Markov Decision Process (MDP) and its generalization, the partially observable MDP. In contrast to supervised learning, in reinforcement learning, the agent is learning through interaction with the environment and thus influences the "future." One of the challenges that arises in such cases is the exploration-exploitation dilemma. The agent can choose either to exploit its current knowledge and perhaps not learn anything new or to explore and risk missing considerable gains.

While reinforcement learning contains many problems, due to lack of space, this entry focuses on the basic ones. For a detailed history of the development of reinforcement learning, see [1, Chapter 1]. The focus of this entry is on Q-learning and Rmax.

Notation

Markov Decision Process

A Markov decision process (MDP) formalizes the following problem. An agent is in an environment, which is composed of different states. In each time step, the agent performs an action and as a result observes a signal. The signal is composed from the reward to the agent and the state it reaches in the next time step. More formally the MDP is defined as follows:

Definition 1 A Markov decision process (MDP) M is a 4-tuple (S, A, P, R), where S is a set of states, A is a set of actions, $P_{s,s'}{}^a$ is the transition probability from state s to state s' when performing action $a \in A$ in state s, and $R(s, a)$ is the reward distribution when performing action a in state s.

A strategy for an MDP assigns, at each time t, for each state s a probability for performing action $a \in A$, given a history $F_{t-1} =$

$\{s_1, a_1, r_1, \ldots, s_{t-1}, a_{t-1}, r_{t-1}\}$ which includes the states, actions, and rewards observed until time $t - 1$. While executing a strategy π, an agent performs at time t action a_t in state s_t and observes a reward r_t (distributed according to $R(s_t, a_t)$), and a next state s_{t+1} (distributed according to $P_{s_t, \cdot}{}^a$). The sequence of rewards is combined into a single value called the *return*. The agent's goal is to maximize the return. There are several natural ways to define the return.

- *Finite horizon:* The return of policy π for a given horizon H is $\sum_{t=0}^{H} r_t$.
- *Discounted return:* For a discount parameter $\gamma \in (0, 1)$, the discounted return of policy π is $\sum_{t=0}^{\infty} \gamma^t r_t$.
- *Undiscounted return:* The return of policy π is $\lim_{t \to \infty} \frac{1}{t+1} \sum_{i=0}^{t} r_i$.

Due to lack of space, only discounted return, which is the most popular approach mainly due to its mathematical simplicity, is considered. The value function for each state s, under policy π, is defined as $V^\pi(s) = E^\pi[\sum_{i=0}^{\infty} r_i \gamma^i]$, where the expectation is over a run of policy π starting at state s. The state-action value function for using action a in state s and then following π is defined as $Q^\pi(s, a) = R(s, a) + \gamma \sum_{s'} P_{s,s'}^a V^\pi(s')$.

There exists a stationary deterministic optimal policy, π^*, which maximizes the return from any start state [11]. This implies that for any policy π and any state s, $V^{\pi*}(s) \geq V^\pi(s)$, and $\pi^*(s) = \mathrm{argmax}_a(Q^{\pi*}(s, a))$. A policy π is ε-optimal if $\|V^{\pi*} - V^\pi\|_\infty \leq \epsilon$.

Problems Formulation
The reinforcement learning problems are divided into two categories, planning and learning.

Planning
Given an MDP in its tabular form, compute the optimal policy. An MDP is given in its tabular form if the 4-tuple, (A, S, P, R) is given explicitly.

The standard methods for the planning problem in MDP are given below.

Value Iteration
Value iteration is defined as follows. Start with some initial value function, C_s, and then iterate using the Bellman operator, $TV(s) = \max_a R(s, a) + \gamma \sum_{s' \in S} P_{s,s'}^a V(s')$.

$$V_0(s) = C_s$$
$$V_{t+1}(s) = TV_t(s),$$

This method relies on the fact that the Bellman operator is contracting. Therefore, the distance between the optimal value function and current value function contracts by a factor of γ with respect to max norm (L_∞) in each iteration.

Policy Iteration
This algorithm starts with initial policy π_0 and iterates over polices. The algorithm has two phases for each iteration. In the first phase, the *value evaluation step*, a value function for π_t is calculated, by finding the fixed point of $T_{\pi t} V_{\pi t} = V_{\pi t}$, where $T_{\pi t} V = R(S, \pi_t(s)) + \gamma \sum_{s' \in S} P_{s,s'}^{\pi_t(s)} V(s')$. The second phase, *policy improvement step*, is taking the next policy π_{t+1} as a greedy policy with respect to $V_{\pi t}$. It is known that policy iteration converges with fewer iterations than value iteration. In practice the convergence of policy iteration is very fast.

Linear Programming
This approach formulates and solves an MDP as a linear program (LP). The LP variables are $V_1, \ldots, V_n$, where $V_i = V(s_i)$. The definition is:

Variables: $V_1, \ldots, V_n$
Minimize: $\sum_i V_i$
Subject to: $V_i \geq [R(s_i, a) + \gamma \sum_j P_{s_i, s_j}(a) V_j]$

$$\forall a \in A, s_i \in S.$$

Learning
Given the states and action identities, learn an (almost) optimal policy through interaction with

the environment. The methods are divided into two categories: model-free learning and model-based learning.

The widely used Q-learning [16] is a model-free algorithm. This algorithm belongs to the class of temporal difference algorithms [12]. Q-learning is an off-policy method, i.e., it does not depend on the underlying policy but, as can immediately be seen, depends on the trajectory and not on the policy generating the trajectory.

Q-Learning

The algorithm estimates the state-action value function (for discounted return) as follows:

$$Q_0(s,a) = 0$$

$$Q_{t+1}(s,a) = (1 - \alpha_t(s,a))Q_t(s,a)$$
$$+ \alpha_t(s,a)(r_t(s,a) + \gamma V_t(s'))$$

where s' is the state reached from state s when performing action a at time t, and $V_t(s) = \max_a Q_t(s,a)$. Assume that $\alpha_t(s',a') = 0$ if at time t action a' was not performed at state s'. A learning rate α_t is *well behaved* if for every state action pair (s,a): (1) $\sum_{t=1}^{\infty} \alpha_t(s,a) = \infty$ and (2) $\sum_{t=1}^{\infty} \alpha_t^2(s,a) = \infty$. As will be seen, this is necessary for the convergence of the algorithm.

The model-based algorithms are very simple to describe; they simply build an empirical model and use any of the standard methods to find the optimal policy in the empirical (approximate) model. The main challenge in these methods is in balancing exploration and exploitation and having an appropriate stopping condition. Several algorithms give a nice solution for this [3, 7]. A version of these algorithms appearing in [6] is described below.

On an intuitive level, a state will become known when it was visited "enough" times and one can estimate with high probability its parameters with good accuracy. The modified empirical model is defined as follows. All states that are not in K are represented by a single absorbing state in which the reward is maximal (which

---

Algorithm 1: A model-based algorithm

Rmax
Set $K = \emptyset$;
if $s \in K$? **then**
    Execute $\hat{\pi}(s)$
else
    Execute a random action;
    if s *becomes known* **then**
        $K = K \cup \{s\}$;
        Compute optimal policy, $\hat{\pi}$ for
        the modified empirical model
    end
end

---

causes exploration). The probability to move to the absorbing state from a state $s \in K$ is the empirical probability to move out of K from s and the probability to move between states in K is the empirical probability.

Sample complexity [6] measures how many samples an algorithm needs in order to learn. Note that the sample complexity translates into the time needed for the agent to wander in the MDP.

Key Results

The first Theorem shows that the planning problem is easy as long as the MDP is given in its tabular form, and one can use the algorithms presented in the previous section.

Theorem 1 ([10]) *Given an MDP, the planning problem is P-complete.*

The learning problem can be done also efficiently using the $R_{\max}$ algorithm as is shown below.

Theorem 2 ([3,7]) *$R_{\max}$ computes an ε-optimal policy from state s with probability at least $1 - \delta$ with sample complexity polynomial in $|A|, |S|, \frac{1}{\varepsilon}$ and $\log \frac{1}{\delta}$, where s is the state in which the algorithm halts. Also the algorithm's computational complexity is polynomial in $|A|$ and $|S|$.*

The fact that Q-learning converges in the limit to the optimal Q function (which guarantees that the greedy policy with respect to the Q function will be optimal) is now shown.

Theorem 3 ([17]) *If every state-action is visited infinitely often and the learning rate is well behaved, then Q_t converges to Q^* with probability one.*

The last statement is regarding the convergence rate of Q-learning. This statement must take into consideration some properties of the underlying policy, and assume that this policy covers the entire state space in reasonable time. The next theorem shows that the convergence rate of Q-learning can vary according to the tuning of the algorithm parameters.

Theorem 4 ([4]) *Let L be the time needed for the underlying policy to visit every state action with probability 1/2. Let T be the time until $\|Q^* - Q_T\| \leq \epsilon$ with probability at least $1 - \delta$ and $\#(s, a, t)$ be the number of times action a was performed at state s until time t. Then if $\alpha_t(s, a) = 1 / \#(s, a, t)$, then T is polynomial in $L, \frac{1}{\epsilon}, \log \frac{1}{\delta}$ and exponential in $\frac{1}{1-\gamma}$. If $\alpha_t(s, a) = 1 / \#(s, a, t)^\omega$ for $\omega \in (1/2, 1)$, then T is polynomial $L, \frac{1}{\epsilon}, \log \frac{1}{\delta}$ and $\frac{1}{1-\gamma}$.*

Applications

The biggest successes of reinforcement learning so far are mentioned here. For a list of successful applications of reinforcement learning, see http://neuromancer.eecs.umich.edu/cgi-bin/twiki/view/Main/SuccessesOfRL.

Backgammon Tesauro [14] used temporal difference learning combined with neural networks to design a player that learned to play backgammon by playing itself and resulted in a player at the level of the world's top players.

Helicopter control Ng et al. [9] used inverse reinforcement learning for autonomous helicopter flight.

Open Problems

While in this entry only MDPs given in their tabular form were discussed, much current research is dedicated to two major directions: large state space and partially observable environments.

In many real-world applications, such as robotics, the agent cannot observe the state she is in and can only observe a signal which is correlated with it. In such scenarios, the MDP framework is no longer suitable, and another model is in order. The most popular reinforcement learning for such environments is the partially observable MDP. Unfortunately, for POMDP even the planning problems are intractable (and not only for the optimal policy which is not stationary but even for the optimal stationary policy); the learning contains even more obstacles as the agent cannot repeat the same state twice with certainty, and thus, it is not obvious how she can learn. An interesting open problem is trying to characterize when a POMDP is "solvable" and when it is hard to solve according to some structure.

In most applications, the assumption that the MDP can be represented in its tabular form is not realistic and approximate methods are in order. Unfortunately not much theoretically is known under such conditions. Here are a few of the prominent directions to tackle large state space.

Function Approximation

The term "function approximation" is due to the fact that this approach takes examples from a desired function (e.g., a value function) and constructs an approximation of the entire function. Function approximation is an instance of supervised learning, which is studied in machine learning and other fields. In contrast to the tabular representation, this time a parameter vector Θ represents the value function. The challenge will be to learn the optimal vector parameter in the sense of minimum square error, i.e.,

$$\min_\Theta \sum_{s \in S} (V^\pi(s) - V(s, \Theta))^2,$$

where $V(s, \Theta)$ is the approximation function. One of the most important function approximations is the linear function approximation,

$$V_t(s, \Theta) = \sum_{i=1}^{T} \phi_s(i) \Theta_t(i),$$

where each state has a set of vector features, ϕ_s. A feature-based function approximation was analyzed and demonstrated in [2, 15]. The main goal here is designing algorithms which converge to almost optimal polices under realistic assumptions.

Factored Markov Decision Process

In an FMDP, the set of states is described via a set of random variables $X = \{X_1, \ldots, X_n\}$, where each X_i takes values in some finite domain $Dom(X_i)$. A state s defines a value $x_i \in Dom(X_i)$ for each variable X_i. The transition model is encoded using a dynamic Bayesian network. Although the representation is efficient, not only is finding an ε-optimal policy intractable [8], but it cannot be represented succinctly [1]. However, under assumptions on the FMDP structure, there exist algorithms such as [5] that have both theoretical guarantees and nice empirical results.

Cross-References

▶ Attribute-Efficient Learning
▶ Learning Automata
▶ Learning Constant-Depth Circuits
▶ PAC Learning

Recommended Reading

1. Allender E, Arora S, Kearns M, Moore C, Russell A (2002) Note on the representational incompatibility of function approximation and factored dynamics. In: Becker S, Thrun S, Obermayer K (eds) Advances in neural information processing systems 15. MIT, Cambridge
2. Bertsekas DP, Tsitsiklis JN (1996) Neuro-dynamic programming. Athena Scientific, Belmont
3. Brafman R, Tennenholtz M (2002) R-max – a general polynomial time algorithm for near optimal reinforcement learning. J Mach Learn Res 3:213–231
4. Even-Dar E, Mansour Y (2003) Learning rates for q-learning. J Mach Learn Res 5:1–25
5. Guestrin C, Koller D, Parr R, Venkataraman S (2003) Efficient solution algorithms for factored MDPs. J Artif Intell Res 19:399–468
6. Kakade S (2003) On the sample complexity of reinforcement learning. Ph.D. thesis, University College London
7. Kearns M, Singh S (2002) Near-optimal reinforcement learning in polynomial time. Mach Learn 49(2–3):209–232
8. Lusena C, Goldsmith J, Mundhenk M (2001) Nonapproximability results for partially observable Markov decision processes. J Artif Intell Res 14:83–103
9. Ng AY, Coates A, Diel M, Ganapathi V, Schulte J, Tse B, Berger E, Liang E (2006) Inverted autonomous helicopter flight via reinforcement learning. In: Ang MH Jr, Khatib O (eds) International symposium on experimental robotics. Springer tracts in advanced robotics 21. Springer, Berlin/New York
10. Papadimitriou CH, Tsitsiklis JN (1987) The complexity of Markov decision processes. Math Oper Res 12(3):441–450
11. Puterman M (1994) Markov decision processes. Wiley-Interscience, New York
12. Sutton R (1988) Learning to predict by the methods of temporal differences. Mach Learn 3:9–44
13. Sutton R, Barto A (1998) Reinforcement learning. An introduction. MIT, Cambridge
14. Tesauro GJ (1996) TD-gammon, a self-teaching backgammon program, achieves a master-level play. Neural Comput 6:215–219
15. Tsitsiklis JN, Van Roy B (1996) Feature-based methods for large scale dynamic programming. Mach Learn 22:59–94
16. Watkins C (1989) Learning from delayed rewards. Ph.D. thesis, Cambridge University
17. Watkins C, Dyan P (1992) Qlearning. Mach Learn 8(3/4):279–292

Renaming

Maurice Herlihy
Department of Computer Science, Brown University, Providence, RI, USA

Keywords

Wait-free renaming

Years and Authors of Summarized Original Work

1990; Attiya, Bar-Noy, Dolev, Peleg, Reischuk

Problem Definition

Consider a system in which $n + 1$ processes $P_0, \ldots, P_n$ communicate either by message-passing or by reading and writing a shared memory. Processes are *asynchronous*: there is no upper or lower bounds on their speeds, and up to t of them may fail undetectably by halting. In the *renaming task* proposed by Attiya, Bar-Noy, Dolev, Peleg, and Reischuk [1], each process is given a unique *input name* taken from a range $0, \ldots, N$, and chooses a unique *output name* taken from a strictly smaller range $0, \ldots, K$. To rule out trivial solutions, a process's decision function must depend only on input names, not its preassigned identifier (so that P_i cannot simply choose output name i). Attiya et al. showed that the task has no solution when $K = n$, but does have a solution when $K = N + t$. In 1993, Herlihy and Shavit [2] showed that the task has no solution when $K < N + t$.

Vertexes, simplexes, and complexes model decision tasks.(See the companion article entitled ▶ Topology Approach in Distributed Computing). A process's state at the start or end of a task is represented as a vertex $\mathbf{v}$ labeled with that process's identifier, and a value, either input or output: $\mathbf{v} = \langle P, v_i \rangle$. Two such vertexes are *compatible* if (1) they have distinct process identifiers, and (2) those process can be assigned those values together. For example, in the renaming task, input values are required to be distinct, so two input vertexes are compatible only if they are labeled with distinct process identifiers and distinct input values.

Figure 1 shows the output complex for the three-process renaming task using four names. Notice that the two edges marked A are identical, as are the two edges marked B. By identifying these edges, this task defines a simplicial complex that is topologically equivalent to a torus. Of course, after changing the number of processes or the number of names, this complex is no longer a torus.

Key Results

Theorem 1 *Let S^n be an n-simplex, and S^m a face of S^n. Let S be the complex consisting of all faces of S^m, and $\dot{S}$ the complex consisting of all*

proper faces of S^m (the boundary complex of S). If $\sigma(\dot{S})$ is a subdivision of $\dot{S}$, and $\phi: \sigma(\dot{S}) \to \mathcal{F}(S)$ a simplicial map, then there exists a subdivision $\tau(S)$ and a simplicial map $\psi: \tau(S) \to \mathcal{F}(S)$ such that $\tau(\dot{S}) = \sigma(\dot{S})$, and ϕ and ψ agree on $\sigma(\dot{S})$.

Informally, any simplicial map of an m-sphere to $\mathcal{F}$ can be "filled in" to a simplicial map of the $(m + 1)$-disk. A *span* for $\mathcal{F}(S^n)$ is a subdivision σ of the input simplex S^n together with a simplicial map $\phi: \sigma(S^n) \to \mathcal{F}(S^n)$ such that for every face S^m of S^n, $\phi: \sigma(S^m) \to \mathcal{F}(S^m)$. Spans are constructed one dimension at a time. For each $\mathbf{s} = \langle P_i, v_i \rangle \in S^n, \phi$ carries $\mathbf{s}$ to the solo execution by P_i with input $\mathbf{v}_i$. For each $S^1 = (\mathbf{s}_0, \mathbf{s}_1)$, Theorem 1 implies that $\phi(\mathbf{s}_0)$ and $\phi(\mathbf{s}_1)$ can be joined by a path in $\mathcal{F}(S^1)$. For each $S^2 = (\mathbf{s}_0, \mathbf{s}_1, \mathbf{s}_2)$, the inductively constructed spans define each face of the boundary complex $\phi: \sigma(S_{ij}^1) \to \mathcal{F}(S^1)_{ij}$, for $i, j \in \{0, 1, 2\}$. Theorem 1 implies that one can "fill in" this map, extending the subdivision from the boundary complex to the entire complex.

Theorem 2 *If a decision task has a protocol in asynchronous read/write memory, then each input simplex has a span.*

One can restrict attention to protocols that have the property that any process chooses the same name in a solo execution.

Definition 1 A protocol is *comparison-based* if the only operations a process can perform on processor identifiers is to test for equality and order; that is, given two P and Q, a process can test for $P = Q, P \le Q$, and $P \ge Q$, but cannot examine the structure of the identifiers in any more detail.

Lemma 1 *If a wait-free renaming protocol for K names exists, then a comparison-based protocol exists.*

Proof Attiya et al. [1] give a simple comparison-based wait-free renaming protocol that uses $2n + 1$ output names. Use this algorithm to assign each

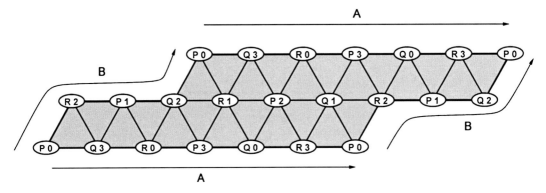

Renaming, Fig. 1 Output complex for 3-process renaming with 4 names

process an *intermediate* name, and use that intermediate name as input to the K-name protocol.□

Comparison-based algorithms are *symmetric* on the boundary of the span. Let S^n be an input simplex, $\phi: \sigma(S^n) \to \mathcal{F}(S^n)$ a span, and $\mathcal{R}$ the output complex for $2n$ names. Composing the span map ϕ and the decision map δ yields a map $\sigma(S^n) \to \mathcal{R}$. This map can be simplified by replacing each output name by its parity, replacing the complex $\mathcal{R}$ with the binary n-sphere $\mathcal{B}^n$.

$$\mu: \sigma(S^n) \to \mathcal{B}^n . \qquad (1)$$

Denote the simplex of $\mathcal{B}^n$ whose values are all zero by 0^n, and all one by 1^n.

Lemma 2 $\mu^{-1}(0^n) = \mu^{-1}(1^n) = \emptyset$.

Proof The range $0, \dots, 2n - 1$ does not contain $n + 1$ distinct even names or $n + 1$ distinct odd names. □

The *n-cylinder* C^n is the binary n-sphere without 0^n and 1^n. Informally, the rest of the argument proceeds by showing that the boundary of the span is "wrapped around" the hole in C^n a nonzero number of times.

The span $\sigma(S^n)$ (indeed any any subdivided n-simplex) is a (combinatorial) *manifold with boundary*: each $(n-1)$-simplex is a face of either one or two n-simplexes. If it is a face of two, then the simplex is an *internal simplex*, and otherwise it is a *boundary* simplex. An orientation

of S^n induces an orientation on each n-simplex of $\sigma(S^n)$ so that each internal $(n-1)$-simplex inherits opposite orientations. Summing these oriented simplexes yields a chain, denoted $\sigma_*(S^n)$, such that

$$\partial \sigma_*(S^n) = \sum_{i=0}^{n} (-1)^i \sigma_*(face_i(S^n)) .$$

The following is a standard result about the homology of spheres.

Theorem 3 *Let the chain 0^n be the simplex 0^n oriented like S^n. (1) For $0 < m < n$, any two m-cycles are homologous, and (2) every n-cycle C^n is homologous to $k \cdot \partial 0^n$, for some integer k. C^n is a boundary if and only if $k = 0$.*

Let S^m be the face of S^n spanned by solo executions of $P_0, \dots, P_m$. Let 0^m denote some m-simplex of C^n whose values are all zero. Which one will be clear from context.

Lemma 3 *For every proper face S^{m-1} of S^n, there is an m-chain $\alpha(S^{m-1})$ such that*

$$\mu_*(\sigma_*(S^m)) - 0^m - \sum_{i=0}^{m} (-1)^i \alpha(face_i(S^m))$$

is a cycle.

Proof By induction on m. When $m = 1$, $ids(S^1) = \{i, j\}$. 0^1 and $\mu_*(\sigma_*(S^1))$ are

1-chains with a common boundary $\langle P_i, 0 \rangle - \langle P_j, 0 \rangle$, so $\mu_*(\sigma_*(S^1)) - 0^1$ is a cycle, and $\alpha(\langle P_i, 0 \rangle) = \emptyset$.

Assume the claim for $m, 1 \geq m < n - 1$. By Theorem 3, every m-cycle is a boundary (for $m < n - 1$), so there exists an $(m + 1)$-chain $\alpha(S^m)$ such that

$$\mu_*(\sigma_*(S^m)) - 0^m - \sum_{i=0}^{m}(-1)^i \alpha(face_i(S^m))$$

$$= \partial\alpha(S^m).$$

Taking the alternating sum over the faces of S^{m+1}, the $\alpha(face_i(S^m))$ cancel out, yielding

$$\mu_*(\partial\sigma_*(S^{m+1})) - \partial 0^{m+1}$$

$$= \sum_{i=0}^{m+1}(-1)^i \partial\alpha(face_i(S^{m+1})).$$

Rearranging terms yields

$$\partial\left(\mu_*(\sigma_*(S^{m+1})) - 0^{m+1}\right.$$

$$\left. - \sum_{i=0}^{m+1}(-1)^i \alpha(face_i(S^{m+1}))\right) = 0,$$

implying that

$$\mu_*(\sigma_*(S^{m+1})) - 0^{m+1}$$

$$- \sum_{i=0}^{m+1}(-1)^i \alpha(face_i(S^{m+1}))$$

is an $(m + 1)$-cycle.   □

Theorem 4 *There is no wait-free renaming protocol for $(n + 1)$ processes using $2n$ output names.*

Proof Because

$$\mu_*(\sigma_*(S^{n-1})) - 0^{n-1} - \sum_{i=0}^{n}(-1)^i \alpha(face_i(S^{n-1}))$$

is a cycle, Theorem 3 implies that it is homologous to $k \cdot \partial 0^n$, for some integer k. Because μ is symmetric on the boundary of $\sigma(S^n)$, the alternating sum over the $(n - 1)$-dimensional faces of S^n yields:

$$\mu_*(\partial\sigma_*(S^n)) - \partial 0^n \sim (n + 1)k \cdot \partial 0^n$$

or

$$\mu_*(\partial\sigma_*(S^n)) \sim (1 + (n + 1)k) \cdot \partial 0^n .$$

Since there is no value of k for which $(1 + (n + 1)k)$ is zero, the cycle $\mu_*(\partial\sigma_*(S^n))$ is not a boundary, a contradiction.   □

Applications

The renaming problem is a key tool for understanding the power of various asynchronous models of computation.

Open Problems

Characterizing the full power of the topological approach to proving lower bounds remains an open problem.

Cross-References

▶ Asynchronous Consensus Impossibility
▶ Set Agreement
▶ Topology Approach in Distributed Computing

Recommended Reading

1. Attiya H, Bar-Noy A, Dolev D, Peleg D, Reischuk R (1990) Renaming in an asynchronous environment. J ACM 37(3):524–548
2. Herlihy MP, Shavit N (1993) The asynchronous computability theorem for t-resilient tasks. In: Proceedings 25th annual ACM symposium on theory of computing, pp 111–120

Revenue Monotone Auctions

Gagan Goel[1] and Mohammad Reza Khani[2]
[1]Google Inc., New York, NY, USA
[2]University of Maryland, College Park, MD, USA

Keywords

Algorithmic Game Theory; Approximation; Electronic Commerce; Incentive compatibility; Mechanism Design; Price of revenue monotonicity; Revenue monotonicity; Social welfare

Introduction

Fueled by the growth of Internet and advancements in online advertising techniques, today more and more online firms rely on advertising revenue for their business. Some of these firms include news agencies, media outlets, search engines, social and professional networks, etc. Much of this online advertising business is moving to what's called *programmatic* buying where an advertiser bids for each single impression, sometimes in real time, depending on how he values the ad opportunity. This work is motivated by the need of a desired property in the auction mechanisms that are used in these bid-based advertising systems.

A standard mechanism for most auction scenarios is the famous Vickrey-Clarke-Groves (VCG) mechanism. VCG is *incentive compatible* (IC) and maximizes *social welfare*. Incentive compatibility guarantees that the best response for each advertiser is to report its true valuation. This makes the mechanism transparent and removes the load from the advertisers to calculate the best response. *Social welfare* is the sum of the valuations of the winners. This value is treated as a proxy for how much all the participants gain from the transaction. What makes VCG mechanism versatile is that it reduces the mechanism design problem into an optimization problem for any scenario.

Even though this versatility of VCG mechanism makes it a popular choice mechanism, however, it doesn't satisfy an important property, namely, that of *revenue monotonicity*. Revenue monotonicity says that if one increases the bid values or add new bidders, the total revenue should not go down. To see that VCG is not revenue monotone, consider a simple example of two items and three bidders (A, B, and C). Say, bidder A wants only the first item and has a bid of 2. Similarly bidder B wants only the second item and has a bid of 2. Bidder C wants both the items or nothing and has a bid of 2. Now if only bidders A and B participate in the auction, then VCG gives a revenue of 2; however, if all the three bidders participate, then the revenue goes down to 0.

This lack of revenue monotonicity (which has been noted several times in the literature) is one of the serious practical drawbacks of the celebrated VCG mechanism. To think of it, an online firm that depends on advertising revenue puts significant resources in its sales efforts to attract more bidders as the general belief is that more bidders imply more competition which should lead to higher prices. Now to tell this firm that their revenue can go down if they get more bidders can be strategically very confusing for them. To see this from another perspective, say, in a search engine firm, there is a team which makes a UI change that increases the click-through probability (CTR) of the search ads. These changes are thought of as good changes in the firm as they increase the effective bid of the bidders (the effective bid of a bidder in search advertising is a function of its cost-per-click bid and the CTR of its ad). Now if after making the change, the revenue goes down, what was supposed to be a good change may seem like a bad change. The point we are trying to make is that there are many teams in a firm, and for these teams to function properly, it is important that the auction mechanisms satisfy *revenue monotonicity*.

In this entry, with a focus on auctions arising in advertising scenarios, we seek to understand mechanisms that satisfy this additional property of revenue monotonicity (RM). It is well known that for various settings (including ours), no mechanism can satisfy both IC and RM properties while attaining optimal social welfare. In fact it is known that one cannot even hope to get Pareto-optimality in social welfare while attaining both IC and RM [10]. Thus to overcome this bottleneck and develop an understanding of RM mechanisms, we relax the requirement of attaining full social welfare and define the notion of *price of revenue monotonicity* (PoRM). Price of revenue monotonicity of an IC and RM mechanism M is the ratio of optimal social welfare to the social welfare attained by the mechanism M. The goal is to design mechanisms that satisfy IC and RM properties and at the same time achieve low price of revenue monotonicity. To the best of our knowledge, this is the first work that defines and studies this notion of price of revenue monotonicity.

We study two different advertising settings in this entry. The first setting we study is the *image-text* auction. In image-text auction there is a special box designated for advertising in a publisher's website which can be filled by either k text ads or a single image ad. The second setting is the *video-pod* auction where an advertising break of a certain duration in a video content can be filled with multiple video ads of possibly different durations.

We note that revenue monotonicity is an *across-instance* constraint as it requires total revenue to behave in a certain manner across different instances, where a single instance is defined by fixing the *type* of the buyers. Note that incentive compatibility is also an across-instance constraint. A lot of research effort has gone into understanding incentive compatibility, which has resulted in useful tools for designing incentive-compatible mechanisms. Surprisingly, hardly any work has gone into understanding and building tools for designing mechanisms which satisfy the desired property of revenue monotonicity. We believe that understanding revenue monotonicity will shed new fundamental insights into the design of mechanisms for many practical scenarios.

Related Work

Ausubel and Milgrom [1] show that VCG satisfies RM if bidders' valuations satisfy *bidder submodularity*. Bidders' valuations satisfy bidders submodularity if and only if for any bidder i and any two sets of bidders S, S' with $S \subseteq S'$ we have $\text{WF}(S \cup \{i\}) - \text{WF}(S) \geq \text{WF}(S' \cup \{i\}) - \text{WF}(S')$, where $\text{WF}(S)$ is the maximum social welfare achievable using only S. Note that this is a general tool one can use to design revenue-monotone mechanisms – restrict the range of the possible allocations such that we get bidder submodularity when we run VCG on this range. However, we can show that this general tool is not so powerful by showing that for our auction scenarios, it is not possible to get a mechanism with PoRM better than $\Omega(k)$ by using the above tool.

Ausubel and Milgrom [1] also show that bidder submodularity is guaranteed when the goods are substitutes, i.e., the valuation function of each bidder is submodular over the goods. However, for many practical scenarios, including ours, the valuation function of the bidders is not submodular. Ausubel and Milgrom [1] design mechanisms which select allocations that are in the core of the exchange economy for combinatorial auctions. Here an allocation is in the core if there is no coalition of bidders and the seller to trade with each other in a way which is preferred by all the members of the coalition to the allocation. Day and Milgrom [3] show that core-selecting mechanisms that choose a core allocation which minimizes the seller's revenue satisfy RM given bidders follow so-called *best-response truncation strategy*. Therefore the core-selecting mechanism designed by [1] satisfies RM if the participants play such best-response strategy, although this mechanism is not incentive compatible.

Rastegari et al. [10] prove that no mechanism for general combinatorial auctions which satisfies IC and RM can achieve weakly maximal social welfare. An allocation is weakly maximal if it cannot be modified to make at least one participant better off without hurting anyone else. In another work [9] they design a randomized mechanism for combinatorial auctions which achieves weak maximality and expected revenue monotonicity.

Another related work is around the characterization of mechanisms that achieve the IC property. The classic result of Roberts [11] states that affine maximizers are the only social choice functions that can be implemented using IC mechanisms when bidders have unrestricted quasi-linear valuations. Subsequent works study the restricted cases [2, 6, 12, 13].

There is also an extensive body of research around designing mechanisms with good bounds on the revenue. Myerson [7] designs a mechanism which achieves the optimal expected revenue in the single parameter Bayesian setting. Goldbert et al. consider optimizing revenue in prior-free settings (see [8] for a survey on this).

Our Results

As mentioned earlier, we study two settings: (1) image-text auction and (2) video-pod auction. Both these settings can be described using the following abstract model. Say, there is a seller selling k identical items to n participants/buyers. Participant i wants either d_i items or nothing and has a valuation of v_i if it gets d_i items or 0 otherwise. Demand d_i is assumed to be public knowledge, and valuation v_i is assumed to be the private information of the participant i. We want to design a mechanism that is incentive compatible, individually rational (IR), and revenue monotone and maximizes social welfare.

For the image-text auction, the demand $d_i \in \{1, k\}$, i.e., each participant wants either 1 item (text ads) or k items (image ad). For the video-pod auction, an item corresponds to a unit time interval (say, one second), and the demand d_i could be any number between 1 and k, i.e., $d_i \in [k]$.

The first result of this entry is the following theorem.

Theorem 1 *We design a deterministic mechanism for image-text auction (MITA) which satisfies individual rationality (IR), IC, and RM with PORM of at most $\sum_{i=1}^{k} \frac{1}{i} \simeq \ln(k)$, i.e., the ratio of MITA's welfare over the optimal welfare is at most $\ln(k)$.*

The proof of Theorem 1 appears in section "Image-Text Auctions." We outline our mechanism over here: Let $v_1 \geq \ldots \geq v_{n_1}$ be the valuations of text participants and V_1 be the maximum valuation of the image participants. If $\max_{j \in [k]} j \cdot v_j$ is less than V_1, MITA gives all the items to the image participant who has valuation V_1; otherwise MITA picks the highest j^* text participants as the winners where j^* is the maximum number in $[k]$ such that $j^* \cdot v_{j^*} \geq V_1$. Note that the j that maximizes $j \cdot v_j$ might be less than the $j*$ which is the largest j such that $j \cdot v_j \geq V_1$. Also note that MITA sometimes picks less than k text ads as the winner (even if there are k or more text ads). VCG always picks the maximum number of text ads (if it decides to allocate the slot to text ads); this is one of the reasons why VCG fails to satisfy RM. When we allow lesser number of text ads to be declared as winners, intuitively, this increases the competition which boosts the revenue and thus helps in achieving RM, although this comes with a loss in social welfare.

Surprisingly, we can also show that the above mechanism achieves the optimal PORM for the image-text auction by proving a matching lower bound. We show that a mechanism that satisfies IR, IC, RM, and two additional mild assumptions of anonymity (AM) and independence of irrelevant alternatives (IIA) cannot achieve a PORM better than $\sum_{i=1}^{k} \frac{1}{i}$. Anonymity means that the auction mechanism doesn't depend on the identities of the participants (a formal definition appears in section "Image-Text Auctions"). IIA means that decreasing the bid of a losing participant shouldn't hurt any winner. Note that our mechanism satisfies both AM and IIA as well. Formally, we prove the following theorem whose proof appears in section "Image-Text Auctions."

Theorem 2 *There is no deterministic mechanism which satisfies IR, IC, RM, AM, and IIA and has PORM less than $\sum_{i=1}^{k} \frac{1}{i}$.*

Finally we prove the following theorem for video-pod auctions.

Theorem 3 *We design a mechanism for video-pod auction (MVPA) which satisfies IR, IC, and RM with PORM of at most $(\lfloor \log k \rfloor + 1) \cdot (2 + \ln k)$.*

We give the formal proof of Theorem 3 in section "Video-Pod Auctions" and outline the mechanism here. MVPA partitions the participants into $(\lfloor \log k \rfloor + 1)$ groups where each group $g \in [\log k]$ contains only the participants whose demands are in the range $[2^{g-1}, 2^g)$. MVPA selects winners only from one group. We round up the size of each participant in group g to 2^g; thus we can have at most $\frac{k}{2^g}$ number of winners from the group g. Let $v_1^{(g)} \geq \ldots \geq v_p^{(g)}$ be the sorted valuations of all the participants in group g. We define the max possible revenue of group g (MPRG(g)) to be

$$\text{MPRG}(g) = \max_{j \in [k/2^g]} j \cdot v_j^{(g)}.$$

As the name of MPRG(g) suggests, its value captures the maximum revenue we can truthfully obtain from group g without violating revenue monotonicity. Let g^* be the group with the highest MPRG value and group g' be the group whose MPRG is the second highest. The set of winners are the first j participants from group g^* where j is the largest number in $[k/2^g]$ such that $j \cdot v_j^{(g^*)}$ is greater than or equal to MPRG(g'). We show that PORM of MVPA is $(\lfloor \log k \rfloor + 1) \cdot (2 + \ln k)$.

Preliminaries

Let $N = \{1, \ldots, n\}$ be the set of all participants and k be the number of identical items. We denote the type of participant i by $\theta_i = (d_i, v_i) \in [k] \times \mathbb{R}^+$, where d_i is the number of items participant i demands and v_i is her valuation for getting d_i items. Note that the valuation of player i for getting less than d_i items is 0. Now in the image-text auction, participants have demand of either 1 or k. In the video-pod auction, participants can have arbitrary demands in $\{1, \ldots, k\}$. Let's denote the set of all possible types $[k] \times \mathbb{R}^+$ by Θ and the set of all type profiles of n participants by $\Theta^n = \underbrace{\Theta \times \ldots \times \Theta}_{n}$.

A deterministic mechanism $\mathcal{M}$ consists of an allocation rule $x : \Theta^n \to 2^n$ which maps each type profile to a subset of participants as the winners and payment rule $p : \Theta^n \to (\mathbb{R}^+)^n$ which maps each type profile to the payments of each participant.

Let $\theta = (\theta_1, \theta_2, \ldots, \theta_n) \in \Theta^n$ be a specific type profile. Also let $\mathcal{A}_\theta$ be the set of all feasible solutions, i.e.,

$$\mathcal{A}_\theta = \left\{ S \subseteq N \mid \sum_{i \in S} d_i \leq k \right\}.$$

For each feasible solution $A \in \mathcal{A}_\theta$, the social welfare of A (denoted by WF(A)) is equal to $\sum_{\theta_i \in A} v_i$. To evaluate the social welfare of a mechanism $\mathcal{M}$ on a type profile θ, we compare the welfare of its solution to the optimal solution.

Definition 1 The welfare ratio of mechanism $\mathcal{M} = (x, p)$ on type profile $\theta \in \Theta^n$ (denoted by WFR($\mathcal{M}, \theta$)) is the following:

$$\text{WFR}(\mathcal{M}, \theta) = \frac{\max_{A \in \mathcal{A}_\theta} \text{WF}(A)}{\text{WF}(x(\theta))}$$

To capture the worst-case loss in social welfare across all type profiles, we define the notion of *price of revenue monotonicity*.

Definition 2 The Price of Revenue Monotonicity of a mechanism $\mathcal{M}$ (denoted by PORM($\mathcal{M}$)) is defined as follows:

$$\text{PORM}(\mathcal{M}) = \max_{\theta \in \Theta^n} \text{WFR}(\mathcal{M}, \theta)$$

The desired goal is to design mechanisms which have low PORM value, where the best possible value is 1.

Note that since we are interested in mechanisms with bounded PORM, we restrict ourselves to mechanisms that satisfy consumer sovereignty. Consumer sovereignty says that any participant can be a winner as long as he bids high enough.

Now we will define a weakly monotone allocation rule which is used in the characterization of deterministic IC mechanisms. Let function $x_i : \Theta^n \to \{0, 1\}$ be the restriction of function x to participant i. Here $x_i(.)$ is one if participant i is a winner and zero otherwise.

Definition 3 We call allocation function x is weakly monotone if for any type profile $\theta \in \Theta^n$ and any participant $i \in [n]$ with demand

d_i, function $x_i((d_i, v_i), \theta_{-i})$ is a non-decreasing function in v_i.

Note that if a deterministic mechanism $\mathcal{M}$ satisfies consumer sovereignty and has a weakly monotone allocation function, then function $x_i((d_i, v_i), \theta_{-i})$ is a single-step function. The value at which the function $x_i((d_i, v_i), \theta_{-i})$ jumps from zero to one, i.e., the smallest value at which the participant i becomes a winner, is called *critical value*.

Definition 4 Let $\mathcal{M} = (x, p)$ be a deterministic mechanism that satisfies consumer sovereignty and has a weakly monotone allocation function; the critical value of participant i in type profile θ is $v_i^* = sup\{v_i | x_i((d_i, v_i), \theta_{-i}) = 0\}$.

The following lemma characterizes deterministic IC mechanisms (first given by [7]). We provide a proof sketch for the sake of completeness (for a complete proof, see, e.g., [8]).

Lemma 1 *Let $\mathcal{M} = (x, p)$ be a mechanism which satisfies IR. Mechanism $\mathcal{M}$ is truthful (IC) if and only if the following hold:*

1. *x is weakly monotone.*
2. *If participant i is a winner, then its payment is its critical value (v_i^*).*

Proof First we prove that if $\mathcal{M}$ is truthful, then it satisfies both conditions 1 and 2. We prove the first condition by contradiction. If x is not monotone, then there exist participant i, type profile θ, and two values $v_i^{(1)} > v_i^{(2)}$ such that i wins in type profile $\left((d_i, v_i^{(2)}), \theta_{-i}\right)$ but loses in type profile $\left((d_i, v_i^{(1)}), \theta_{-i}\right)$. This makes incentive for participant i to lie for type profile $\left((d_i, v_i^{(1)}), \theta_{-i}\right)$ and announce its valuation as $v_i^{(2)}$.

Consider an arbitrary participant i who is a winner; now we prove that the payment of participant i is its critical value. Assume for contradiction that mechanism $\mathcal{M}$ charges participant i amount c_i where $c_i < v_i^*$ in a type profile $((d_i, v_i), \theta_{-i})$. In this case, if participant i had type $(d_i, \hat{v}_i)$ where $c_i < \hat{v}_i < v_i^*$, then i is not a winner in $((d_i, \hat{v}_i), \theta_{-i})$ as v_i^* is the critical

value. Therefore, if the real type of participant i is $(d_i, \hat{v}_i)$, she has incentive to lie her type as (d_i, v_i), become a winner, and pay c_i. Hence, the payment cannot be less than v_i^*. Now suppose that there exists value v_i for which mechanism $\mathcal{M}$ charges i amount c_i which is more than v_i^*. In this case, if participant i had type $(d_i, \hat{v}_i)$ where $v_i^* < \hat{v}_i < c_i$, then i is still a winner (as v_i^* is the critical value) and pays at most $\hat{v}_i$ (as $\mathcal{M}$ satisfies IR). Therefore, she has an incentive to lie her type as $(d_i, \hat{v}_i)$, become a winner, and pay at most $\hat{v}_i$. Hence, the payment cannot be more than v_i^* for any winning valuation v_i.

For the other direction, it is easy to check that any IR mechanism that satisfies conditions 1 and 2 is truthful. $\square$

Image-Text Auctions

In this section we give our *mechanism for image-text auction* (MITA) which satisfies IR, IC, RM, and PoRM(MITA) $\leq$ ln k. Recall that in the image-text auction we have k identical items to sell and there are two groups of participants: the ones who want all the k items which we call *image participants* and the ones who want only one item which we refer to as *text participants*. As a result there are also two possible types of outcome: MITA gives all the items to an image participant; or it gives an item to each member of a subset of the text participants.

We start with explaining why VCG fails to satisfy RM and how we address this issue in MITA. Consider the type profile where we have one image participant with type $(k, 1)$ and one text participant with type $(1, 1)$. In this case either of the participants can be the winner. The payment of the winner in VCG is her critical value which is one. However if we add one more text participant with the same type $(1, 1)$, the two text participants win and each of them pays zero. The reason for the payment drop is that VCG always selects k winners from the text participants. This decreases the critical value of each text participant as the valuation of the other text participants helps her to win against image participants. In our mechanism we overcome this issue by not guaranteeing that the maximal number of text participants can win an item. In other words, in our mechanism it is possible that less than k text participants win an item even if there are more than k text participants. This way, intuitively, even if the number of text participants increases, it potentially creates more competition and hence increases the payments.

Let θ be an arbitrary-type profile where there are n_1 text participants with types $(1, v_1), \ldots,$ $(1, v_{n_1})$ and n_2 image participants with types $(k, V_1), \ldots, (k, V_{n_2})$. We define mechanism MITA $= (x^{\text{MITA}}, p^{\text{MITA}})$ by giving allocation function x^{MITA} which is weakly monotone. Given the allocation function, we obtain payment function p^{MITA} using the critical values defined in Lemma 1 which makes the mechanism truthful.

Allocation rule of MITA. Without loss of generality, we assume that $v_1 \geq v_2 \geq \ldots \geq v_{n_1}$ and $V_1 \geq V_2 \geq \ldots \geq V_{n_2}$. Also, we assume that $n_1 \geq k$; if not, we add fake text participants with value 0. For each $j \in [k]$, we consider value $j \cdot v_j$. Let candidate set C_θ contain all the values $j \in [k]$ such that $j \cdot v_j$ is greater than or equal to V_1, i.e., $C_\theta = \{j \in [k] | j \cdot v_j \geq V_1\}$. If C_θ is empty, the image participant with type (k, V_1) wins. If C_θ is nonempty, then let j^* be the maximum member of C_θ, i.e., $j^* = \max_{j \in C_\theta} j$. In this case the first j^* text participants win.

Observation 1 *Allocation function* x^{MITA} *is weakly monotone.*

Proof Recall from Definition 3, in order to prove that x^{MITA} is weakly monotone, we have to show that for any participant $i \in [n]$ with demand d_i, function $x_i((d_i, v_i), \theta_{-i})$ is a non-decreasing function in v_i.

If i is an image participant, then i wins if its valuation is larger than $\max(W, \max_{j \in [k]} j \cdot v_j)$ where W is the largest valuation of the image participants in θ_{-i}. Moreover, bidder i loses for any value smaller than or equal to $\max(W, \max_{j \in [k]} j \cdot v_j)$. Therefore x_i is weakly monotone.

If i is a text participant, then let $v'_1 \geq v'_2 \geq \ldots$ be the sorted valuations of the text participants and V_1 be the largest valuation of image participants in θ_{-i}. Let t be the smallest value such that there exist $j \in [k-1]$ where $v'_{j+1} \leq t \leq v'_j$ and $(j+1) \cdot t$ is greater than or equal to V_1. If the valuation of bidder i is larger than or equal to t, then she wins since $(j+1) \cdot t \geq V_1$; otherwise she does not win since t is the smallest value for which there exist $j \in [k-1]$ such that $(j+1) \cdot t \geq V_1$. Therefore x_i is weakly monotone. $\square$

In the following lemma we obtain the critical value (or truthful payments) of the winners in x^{MITA} using Lemma 1. The lemma also gives an intuition to why we select j^* text participants to win, which is the maximum j such that $j \cdot v_j \geq V_1$.

Lemma 2 *If* C_θ, *where* $C_\theta = \{j \in [k] | j \cdot v_j \geq V_1\}$, *is empty, then the first image participant wins all the items with critical value* $\max(V_2, \max_{j \in [k]} j \cdot v_j)$. *If* C_θ *is not empty, the first* j^* *text participants win the items where* $j^* = \max_{j \in C_\theta} j$ *and all of them have critical value* $\max(v_{k+1}, \frac{V_1}{j^*})$.

Proof We find the critical value (Definition 4) of a winner by showing that if she has any valuation larger than the critical value she wins and for any valuation less than the critical value she doesn't.

If C_θ is empty, then the first image participant (with type (k, V_1)) wins all the items. As long as V_1 is larger than $\max(V_2, \max_{j \in [k]} j \cdot v_j)$, participant (k, V_1) wins. If V_1 is less than $\max(V_2, \max_{j \in [k]} j \cdot v_j)$, then she loses to the image participant (k, V_2) if $\max(V_2, \max_{j \in [k]} j \cdot v_j) = V_2$ or loses to the text participants if $\max(V_2, \max_{j \in [k]} j \cdot v_j) = \max_{j \in [k]} j \cdot v_j$. This means that the critical value of the first image participant is $\max(V_2, \max_{j \in [k]} j \cdot v_j)$ if she is the winner.

If C_θ is nonempty, then the first j^* text participants win. Let $i \in [j^*]$ be an arbitrary winner. First we observe that for any valuation v'_i greater than or equal to $\max\left(v_{k+1}, \frac{V_1}{j^*}\right)$, participant i remains as a winner in type profile $\theta' = ((1, v'_i), \theta_{-i})$. This is because for any such change in valuation of participant i number j^* remains in set $C_{\theta'}$. Moreover, this change does not add any new number j' to $C_{\theta'}$ such that $j' > j^*$ because the valuations of the text participants with index greater than j^* are not changed in θ'.

In order to prove that for any valuation v_i' less than critical value $\max\left(v_{k+1}, \frac{V_1}{j^*}\right)$, participant i is not a winner we consider two cases: (A) when the critical value is equal to $\frac{V_1}{j^*}$ and (B) when the critical value is equal to v_{k+1}.

Case (A): We prove this case by contradiction. Let v_i' be a valuation less than $\frac{V_1}{j^*}$ for which participant i is in the set of winners in type profile $\theta' = ((1, v_i'), \theta_{-i})$. Because v_i' is less than $\frac{V_1}{j^*}$, the number of winners which contains participant i cannot be less than or equal to j^* in type profile θ'. Let $j' \in [k]$ which is greater than j^* be the number of winners in θ'. This means that there are at least j' participants whose valuation is larger than $\frac{V_1}{j'}$ in θ'. Note that all the valuations in θ are the same as θ' except v_i which is decreased to v_i'; therefore, there are also at least j' participants whose valuation is larger than $\frac{V_1}{j'}$ in θ and hence j' is in set C_θ. This contradicts with the fact that j^* is the largest member of C_θ.

Case (B): In case (B) we have $\max\left(v_{k+1}, \frac{V_1}{j^*}\right) = v_{k+1}$ which implies that $k \cdot v_{k+1}$ is larger than V_1 as $j^* \in [k]$. Therefore Case (B) can only happen when $j^* = k$. Now consider participant i decreases its valuation to value v_i' that is less than v_{k+1}; then it cannot be a winner as there are k other participants whose valuations are more than v_i' while we have only k items. $\qquad\square$

The payment function of MITA is set to the critical values of the winners as specified in Lemma 2 which by using Observation 1 and Lemma 1 implies MITA satisfies IC. Moreover, as the payments are always less than the participants' bid, IR property of MITA follows. Finally

in the following lemma, we show that MITA is revenue monotone.

Lemma 3 *Let θ' be the type profile obtained by either increasing the valuation of a participant or adding a new participant to the type profile θ; then we have* REVENUE(MITA, θ') $\geq$ REVENUE(MITA, θ).

Proof Let $v_1 \geq v_2 \geq \ldots$ be the valuations of text participants and $V_1 \geq V_2 \geq \ldots$ be the valuations of image participants in θ. Similarly let $v_1' \geq v_2' \geq \ldots$ be the valuations of text participants and $V_1' \geq V_2' \geq \ldots$ be the valuations of image participants in θ'. Note that for any i we have $v_i \leq v_i'$ and $V_i \leq V_i'$ as we have one more participant or a higher valuation in θ'. Let x be the new added participant or the participant which has higher valuation in θ'.

We prove this lemma by considering the value of REVENUE(MITA, θ) for the case when text participants win and the case when an image participant wins. If an image participant wins, then it means that $V_1 > \max_{j \in [k]} j \cdot v_j$ and she pays $\max(V_2, \max_{j \in [k]} j \cdot v_j)$ which is the total revenue.

If text participants win, then it means $V_1 \leq \max_{j \in [k]} j \cdot v_j$ and there are j^* winners where each of them pays $\max(v_{k+1}, \frac{V_1}{j^*})$. If $\max(v_{k+1}, \frac{V_1}{j^*}) = \frac{V_1}{j^*}$, then the total revenue is V_1. If $\max(v_{k+1}, \frac{V_1}{j^*}) = v_{k+1}$, it implies that $k \cdot v_{k+1}$ is larger than V_1. Remember that $C_\theta = \{j \in [k] | j \cdot v_j \geq V_1\}$ and $j^* = \max_{j \in C_\theta} j$; therefore $j^* = k$ and hence the total payment of the winners is $k \cdot v_{k+1}$.

In summary the total revenue for type profile θ is the following:

$$\text{REVENUE}(\text{MITA}, \theta) =$$
$$\begin{cases} \max(V_2, \max_{j \in [k]} j \cdot v_j) & V_1 > \max_{j \in [k]} j \cdot v_j \ (A) \\ \max(V_1, k \cdot v_{k+1}) & V_1 \leq \max_{j \in [k]} j \cdot v_j \ (B) \end{cases}$$

Similarly the total revenue for type profile θ' is the following:

$$\text{REVENUE}(\text{MITA}, \theta') =$$
$$\begin{cases} \max(V_2', \max_{j \in [k]} j \cdot v_j') & V_1' > \max_{j \in [k]} j \cdot v_j' \ (A) \\ \max(V_1', k \cdot v_{k+1}') & V_1' \leq \max_{j \in [k]} j \cdot v_j' \ (B) \end{cases}$$

Note that because for any i we have $v_i \leq v_i'$ and $V_i \leq V_i'$ the following inequalities are straightforward:

$$V_1 \leq V_1' \tag{1}$$

$$V_2 \leq V_2' \tag{2}$$

$$\max_{j \in [k]} j \cdot v_j \leq \max_{j \in [k]} j \cdot v_j' \tag{3}$$

$$k \cdot v_{k+1} \leq k \cdot v_{k+1}' \tag{4}$$

If both REVENUE(MITA, θ) and REVENUE (MITA, θ') take their value from Case (A), then

the proof of the lemma follows from Eqs. (2) and (3). Similarly if both REVENUE(MITA, θ) and REVENUE(MITA, θ') take their value from Case (B), then the proof of the lemma follows from Eqs. (1) and (4).

If REVENUE(MITA, θ) takes its value from Case (A) and REVENUE(MITA, θ') takes from Case (B), then it means that participant x is a text participant which causes $\max_{j \in [k]} j \cdot v_j'$ to be larger than V_1'. The following proves the theorem for this case:

$$\text{REVENUE}(\text{MITA}, \theta)$$

$$= \max(V_2, \max_{j \in [k]} j \cdot v_j)$$

$$< V_1 \qquad\qquad \text{REVENUE}(\text{MITA}, \theta) \text{ takes}$$

$$\text{its value from Case } (A)$$

$$= V_1' \qquad\qquad \text{participant } x \text{ is a}$$

$$\text{text-participant}$$

$$\leq \max(V_1', k \cdot v_{k+1}')$$

$$= \text{REVENUE}(\text{MITA}, \theta')$$

If REVENUE(MITA, θ) takes its value from Case (B) and REVENUE(MITA, θ') takes from Case (A), then it means that participant x is an image participant. The following proves the theorem for this case:

$$\text{REVENUE}(\text{MITA}, \theta)$$

$$= \max(V_1, k \cdot v_{k+1})$$

$$< \max_{j \in [k]} j \cdot v_j \qquad\qquad \text{REVENUE}(\text{MITA}, \theta) \text{ takes}$$

$$\text{its value from Case } (B) \text{ and}$$

$$\text{the fact that } v_k \geq v_{k+1}$$

$$= \max_{j \in [k]} j \cdot v_j' \qquad\qquad x \text{ is an image participant}$$

$$\leq \max(V_2', \max_{j \in [k]} j \cdot v_j')$$

$$= \text{REVENUE}(\text{MITA}, \theta') \qquad\qquad\qquad\qquad \square$$

In the above we proved that MITA satisfies IR, IC, and RM. In the following theorem we bound the PORM of MITA and finish this section.

Theorem 4 $\text{PORM}(\text{MITA}) \le \ln k$.

Proof Let A be the set of winner(s) which realizes the maximum social welfare in type profile θ. If A contains only one image participant with valuation V_1, then we also have $V_1 \ge \max_{j\in[k]} j \cdot v_j$. Mechanism MITA also selects an image participant with the same valuation if $V_1 > \max_{j\in[k]} j \cdot v_j$ and hence PORM(MITA) is 1. Otherwise we have $V_1 = \max_{j\in[k]} j \cdot v_j$ where MITA selects a set of text participants which overall gives social welfare V_1 and hence again the PORM(MITA) is 1.

Now we consider the case when A contains text participants. By adding enough dummy participants with value zero, and without loss of generality, we assume that set A contains the first k text participants with highest valuations $v_1 \ge v_2 \ge \ldots \ge v_k$. Mechanism MITA selects either the first j^* text participants with highest

valuations ($v_1 \ge v_2 \ge \ldots \ge v_{j^*}$) or selects an image participant with valuation V_1. Remember that j^* is the greatest number in set $C_\theta = \{j \mid j \in [k] \wedge j \cdot v_j \ge V_1\}$ which implies the following:

$$\forall j' \in \{j^* + 1, \ldots, k\} \quad v_{j'} < \frac{V_1}{j'} \qquad (5)$$

Note that if MITA selects an image participant, then Eq. (5) holds for $j^* = 0$.

Now we consider the following two cases to prove the theorem:

If MITA selects an image participant, then we have the following:

$$
\begin{aligned}
\text{PORM}(\text{MITA}) &- \frac{\sum_{j\in[k]} v_j}{V_1} \\
&\le \frac{\sum_{j\in[k]} V_1/j}{V_1} \qquad \text{Eq. (5)} \\
&\le \ln k
\end{aligned}
$$

If MITA selects the first j^* text participants, then we have the following:

$$\text{PORM}(\text{MITA}) = \frac{\sum_{j\in[k]} v_j}{\sum_{j\in[j^*]} v_j}$$

$$\le \frac{\sum_{j\in[j^*]} v_j + \sum_{j=j^*+1}^{k} v_j}{\sum_{j\in[j^*]} v_j}$$

$$\le \frac{\sum_{j\in[j^*]} v_j + \sum_{j=j^*+1}^{k} V_1/j}{\sum_{j\in[j^*]} v_j}$$

Eq. (5)

$$\le \frac{\sum_{j\in[j^*]} v_j + \sum_{j=j^*+1}^{k} \left(\sum_{j\in[j^*]} v_j\right)/j}{\sum_{j\in[j^*]} v_j}$$

$$\text{because } V_1 \le \sum_{j\in[j^*]} v_j$$

$$\le \ln k \qquad \qquad \square$$

Video-Pod Auctions

In this section we design a mechanism for video-pod auction (MVPA) which satisfies IR, IC, and RM whose PoRM is at most $(\lfloor \log k \rfloor + 1) \cdot (2 + \ln k)$. Note that all the log functions are in base 2. Let $\theta = ((d_1, v_1), \ldots, (d_n, v_n)) \in \Theta^n$ be an arbitrary-type profile of n participants. We define the allocation and payment function of MVPA for this type profile.

Mechanism MVPA partitions the participants into $\lfloor \log k \rfloor + 1$ groups $G^{(1)}, \ldots, G^{(\lfloor \log k \rfloor + 1)}$ where group $G^{(g)}$ contains all the participants whose demand is in the range $[2^{g-1}, 2^g)$. Mechanism MVPA selects winners only from one group $G^{(g)}$.

Definition 5 Let $M^{(g)}$ be equal to $\max \left(\lfloor \frac{k}{2^g} \rfloor, 1 \right)$ which is the maximum number of winners MVPA selects from group $G^{(g)}$.

Note that we can select at least $\lfloor \frac{k}{2^g} \rfloor$ winners from $G^{(g)}$ since there are k items and the demand of each participant is at most 2^g. Moreover, from the last group $G^{(\lfloor \log k \rfloor + 1)}$ we can select at least one winner although $\lfloor \frac{k}{2^{(\lfloor \log k \rfloor + 1)}} \rfloor = 0$, since we assume the demand of all the participants is from set $[k]$.

Let $\left(d_1^{(g)}, v_1^{(g)} \right), \ldots, \left(d_p^{(g)}, v_p^{(g)} \right)$ be the types of all the participants in group g where $p = |G^{(g)}|$. Here by adding enough dummy participants, we assume p is always larger than $M^{(g)}$. Also, without loss of generality we assume $v_1^{(g)} \geq v_2^{(g)} \geq \ldots \geq v_p^{(g)}$. We define the max possible revenue of group g (MPRG(g)) to be the following:

$$\text{MPRG}(g) = \max_{j \in [M^{(g)}]} j \cdot v_j^{(g)}$$

As the name MPRG suggests, we will see that its value captures the maximum revenue can be truthfully obtained from group g. Let $G^{(g^*)}$ be a group with the maximum MPRG and $G^{(g')}$ be a group with the second maximum MPRG breaking the ties arbitrarily.

The set of winners selected by MVPA is

$$\left\{ \left(d_1^{(g^*)}, v_1^{(g^*)} \right), \ldots, \left(d_j^{(g^*)}, v_j^{(g^*)} \right) \right\}$$

where j is the largest number in $[M^{(g^*)}]$ for which $j \cdot v_j^{(g^*)}$ is larger than or equal to MPRG(g'). In other words, the number of winners (j) is the largest number in $[M^{(g^*)}]$ for which $j \cdot v_j^{(g^*)} \geq \text{MPRG}(g')$.

Now we use Lemma 1 to show that MVPA is truthful and obtain the payments of winners.

Observation 2 *Allocation function* x^{MVPA} *is weakly monotone.*

Proof Note that MVPA sorts the participants according to their valuation and selects the first j participants. Therefore if any participant i increases its valuation, it only helps her to enter the winning set. Hence, the observation follows: $\quad\square$

In the rest of this section, we drop the group identifier of $M^{(g^*)}$ and simply use M unless it is about another group.

In the following lemma we find the critical value of each winner i which is actually equal to its payment $\left(p_i^{\text{MVPA}} \right)$.

Lemma 4 *Let set of winners* $x^{\text{MVPA}}(\theta)$ *contain the first j participants with highest valuations from $G^{(g^*)}$ and $v_{M+1}^{(g^*)}$ be the $(M+1)$th highest valuation in group $G^{(g^*)}$ which is zero if it does not exist. Then, the payment of participant i is the following:*

$$p_i^{\text{MVPA}}(\theta) = \begin{cases} \max \left(\dfrac{\text{MPRG}(g')}{j}, v_{M+1}^{(g^*)} \right) & i \in x^{\text{MVPA}}(\theta) \\ 0 & i \notin x^{\text{MVPA}}(\theta) \end{cases}$$

Proof If participant i is not a winner, then its payment is zero. When participant i is a winner, then we prove that its payment is equal to its critical value (Definition 4). In order to prove that value $\max\left(\frac{\text{MPRG}(g')}{j}, v_{M+1}^{(g^*)}\right)$ is the critical value of participant i, we show that for any value larger than $\max\left(\frac{\text{MPRG}(g')}{j}, v_{M+1}^{(g^*)}\right)$ participant i still wins and for any value less than it she loses.

Remember that $v_1^{(g^*)} \geq v_2^{(g^*)} \geq \ldots \geq v_p^{(g^*)}$ are the valuations of participants in group $G^{(g^*)}$ and $v_1^{(g^*)}, v_2^{(g^*)}, \ldots, v_j^{(g^*)}$ are the valuations of the winners. Because group $G^{(g^*)}$ is the group with the maximum MPRG, we have $v_j^{(g^*)} \geq \frac{\text{MPRG}(g')}{j}$. As there can be at most M winners from group $G^{(g^*)}$, we have $v_j^{(g^*)} \geq v_{M+1}^{(g^*)}$. Therefore we have

$$v_j^{(g^*)} \geq \max\left(\frac{\text{MPRG}(g')}{j}, v_{M+1}^{(g^*)}\right). \quad (6)$$

Let participant i with type profile $\left(d_i^{(g^*)}, v_i^{(g^*)}\right)$ be the ith winner in group g^* where $i \in [j]$. We show that for any valuation greater than or equal to $\max\left(\frac{\text{MPRG}(g')}{j}, v_{M+1}^{(g^*)}\right)$ participant i remains in the winning set. Equation (6) implies that there are j participants in group $G^{(g^*)}$ whose valuations are larger than $\max\left(\frac{\text{MPRG}(g')}{j}, v_{M+1}^{(g^*)}\right)$. If we decrease the valuation of participant i to $\max\left(\frac{\text{MPRG}(g')}{j}, v_{M+1}^{(g^*)}\right)$, we still have j participants in group $G^{(g^*)}$ with valuations at least $\max\left(\frac{\text{MPRG}(g')}{j}, v_{M+1}^{(g^*)}\right)$. Therefore, the value $\text{MPRG}(g^*)$ will be at least $\text{MPRG}(g')$ and group $G^{(g^*)}$ remains the winning group: hence participant i remains in the winning set.

Now we prove that if the valuation of participant i is less than the $\max\left(\frac{\text{MPRG}(g')}{j}, v_{M+1}^{(g^*)}\right)$, she cannot be in the winning set. In order to prove this, we consider two cases: (A) when $\max(\frac{\text{MPRG}(g')}{j}, v_{M+1}^{(g^*)})$ is equal to $v_{M+1}^{(g^*)}$ and

(B) when $\max\left(\frac{\text{MPRG}(g')}{j}, v_{M+1}^{(g^*)}\right)$ is equal to $\frac{\text{MPRG}(g')}{j}$.

Case (A): If $\max\left(\frac{\text{MPRG}(g')}{j}, v_{M+1}^{(g^*)}\right) = v_{M+1}^{(g^*)}$ and the valuation of participant i is less than $v_{M+1}^{(g^*)}$, then it means that there are M participants who have valuations greater than the valuation of participant i. As there can be at most M winners from group $G^{(g^*)}$, participant i cannot be a winner.

Case (B): We prove this case by contradiction. Suppose $\max\left(\frac{\text{MPRG}(g')}{j}, v_{M+1}^{(g^*)}\right) = \frac{\text{MPRG}(g')}{j}$ and $\theta' = \left((d_i^{(g^*)}, v_i^{(g^*)'}), \theta_{-i}\right)$ be a type profile in which the valuation of participant i is less than $\frac{\text{MPRG}(g')}{j}$ while she is still winner. Because the valuation of participant i $\left(v_i^{(g^*)'}\right)$ is less than $\frac{\text{MPRG}(g')}{j}$ and i is in the winning set, in order for $\text{MPRG}(g^*)$ to be larger than $\text{MPRG}(g')$, there has to be more than j winners. Let $j' > j$ be the number of winners in θ'. Having j' winners in θ' and in order for $G^{(g^*)}$ to be the group with the highest MPRG, we conclude that there are j' participants with valuation greater than $\frac{\text{MPRG}(g')}{j'}$. Note that the only difference between θ and θ' is that the valuation of participant i is higher in θ. Therefore, there are also at least j' participants with valuation greater than $\frac{\text{MPRG}(g')}{j'}$ in θ. This contradicts with the way we select the number of winners (j) in θ which is the maximum number for which $j \cdot v_j^{(g^*)}$ is larger than $\text{MPRG}(g')$. □

The allocation function x^{MVPA} is weakly monotone (Observation 2) and the payments of the winners are their critical values (Lemma 4); therefore by Lemma 1 we conclude that MVPA satisfies IC.

In the rest of this section, first we prove that MVPA satisfies RM and then bounds its PORM.

Proposition 1 *The total revenue of mechanism MVPA for type profile θ (REVENUE(MVPA, θ)) is the following:*

$$\text{REVENUE(MVPA}, \theta) = \max\left(\text{MPRG}(g'), M \cdot v_{M+1}^{(g^*)}\right)$$

where g' is a group with the second highest MPRG.

Proof From Lemma 4 we know that there are j winners and each of them pays $\max \left(\frac{\text{MPRG}(g')}{j}, v_{M+1}^{(g^*)} \right)$. Therefore the sum of payments or the revenue of MVPA is $j \cdot \max \left(\frac{\text{MPRG}(g')}{j}, v_{M+1}^{(g^*)} \right)$. The proof of the proposition follows if we show that when $\max \left(\frac{\text{MPRG}(g')}{j}, v_{M+1}^{(g^*)} \right)$ is equal to $v_{M+1}^{(g^*)}$, then the number of winners (j) is equal to M.

If $\max \left(\frac{\text{MPRG}(g')}{j}, v_{M+1}^{(g^*)} \right)$ is equal to $v_{M+1}^{(g^*)}$, then as $v_{M+1}^{(g^*)} \leq v_M^{(g^*)}$, we have $M \cdot v_M^{(g^*)} \geq \frac{\text{MPRG}(g')}{j}$. Remember that j is the maximum number in the set $[M]$ for which $j \cdot v_j^{(g^*)}$ is larger than $\text{MPRG}(g')$. Therefore j is equal to M. $\square$

Lemma 5 *Let θ' be the type profile obtained by either adding a new participant or increasing the valuation of a participant in θ. Then,*

$$\text{REVENUE}(\text{MVPA}, \theta') \geq \text{REVENUE}(\text{MVPA}, \theta).$$

Proof Let x be the new added participant or the participant which has the increased valuation in θ'. Throughout the proof we show MPRG of each group g in type profile θ by $\text{MPRG}_\theta(g)$ and in type profile θ' by $\text{MPRG}_{\theta'}(g)$. Similarly, we show the jth highest valuation of the participants of group g by $v_j^{(g^*,\theta)}$ in type profile θ and by $v_j^{(g^*,\theta')}$ in type profile θ'.

As the jth highest valuation of the participants of each group can only increase by adding participant x, we conclude

$$\forall g, \forall j \quad v_j^{(g,\theta')} \geq v_j^{(g,\theta)}. \tag{7}$$

Remember that MPRG_θ of each group g is $\max_{j \in [M^{(g)}]} j \cdot v_j^{(g,\theta)}$ and using Eq. (7) we get

$$\forall g \quad \text{MPRG}_{\theta'}(g) \geq \text{MPRG}_\theta(g). \tag{8}$$

In order to prove this lemma we consider two cases: (A) adding participant x does not change the winning group $G^{(g^*)}$ and (B) adding x changes the winning group.

Case (A): Let g'' be a group with the second highest MPRG in θ'; it is possible that g' is equal to g''.

$$\text{REVENUE}(\text{MVPA}, \theta') = \max \left(\text{MPRG}_{\theta'}(g''), M \cdot v_{M+1}^{(g^*,\theta')} \right)$$

Proposition 1

$$\geq \max \left(\text{MPRG}_{\theta'}(g'), M \cdot v_{M+1}^{(g^*,\theta')} \right)$$

definition of g''

$$\geq \max \left(\text{MPRG}_\theta(g'), M \cdot v_{M+1}^{(g^*,\theta)} \right)$$

Eqs. (7) and (8)

$$= \text{REVENUE}(\text{MVPA}, \theta)$$

Case (B): Let $G^{(g'')}$ be a group with the highest MPRG in θ'. We have

$$\text{MPRG}_\theta(g^*) \geq \text{MPRG}_\theta(g') \tag{9}$$

as g^* has the highest and g' has the second highest MPRG in θ.

$$\text{MPRG}_\theta(g^*) \geq M \cdot v_M^{(g^*,\theta)} \qquad\qquad \text{As } \text{MPRG}_\theta(g^*) = \max_{j \in [M]} j \cdot v_j^{(g^*,\theta)}$$

$$\geq M \cdot v_{M+1}^{(g^*,\theta)} \qquad\qquad \text{As } v_M^{(g^*,\theta)} \geq v_{M+1}^{(g^*,\theta)} \qquad (10)$$

Let $\hat{g}$ be the group with second highest MPRG in θ'. Because g^* is no longer the winning group in θ', it can be a candidate for the group with the second highest MPRG in θ' and hence we have the following:

$$\text{MPRG}_\theta(g^*) \leq \text{MPRG}_{\theta'}(g^*) \leq \text{MPRG}_{\theta'}(\hat{g}) \qquad (11)$$

The following equations conclude the proof of this case:

$$\text{REVENUE}(\text{MVPA}, \theta) = \max\left(\text{MPRG}_\theta(g'), M \cdot v_{M+1}^{(g^*,\theta)}\right)$$

$$\leq \text{MPRG}_\theta(g^*)$$

by Eqs. (9) and (10)

$$\leq \text{MPRG}_{\theta'}(\hat{g})$$

by Eq. (11)

$$\leq \max\left(\text{MPRG}_{\theta'}(\hat{g}), M^{(g'')} \cdot v_{M^{(g'')}+1}^{(g'',\theta)}\right)$$

$$= \text{REVENUE}(\text{MVPA}, \theta') \qquad\qquad \square$$

The following lemma which bounds PORM of MVPA finishes this section:

Theorem 5 $\text{PORM}(\text{MITA}) \leq (\lfloor \log k \rfloor + 1) \cdot (2 + \ln k)$

Proof Let $\text{WF}(g)$ to be the maximum social welfare achievable if we select the winners only from group $G^{(g)}$. Let A be a set of winner(s) which realizes the maximum welfare in type profile θ. Note that as there are $\lfloor \log k \rfloor + 1$ groups, one group $(\hat{g})$ has a subset of participants from A whose social welfare is at least $\frac{\text{WF}(A)}{\lfloor \log k \rfloor + 1}$ and hence the following:

$$\text{WF}(\hat{g}) \geq \frac{\text{WF}(A)}{\lfloor \log k \rfloor + 1} \qquad (12)$$

Now we prove the following claim about $\text{MPRG}(\hat{g})$:

Claim 1 $\text{MPRG}(\hat{g}) \geq \frac{\text{WF}(\hat{g})}{2 + \ln k}$

Proof Let B be the set of participants from group $G^{(\hat{g})}$ which give the maximum social welfare. Because the demands of all the participants of $G^{(\hat{g})}$ are in range $[2^{\hat{g}-1}, 2^{\hat{g}})$, size of B is at most $\lfloor k/2^{\hat{g}-1} \rfloor$. Remember from Definition 5 that $M^{(\hat{g})} = \max\left(\lfloor k/2^{(\hat{g})} \rfloor, 1\right)$ is the maximum number of winners that MVPA potentially selects from group $G^{(\hat{g})}$. Therefore, we have $|B| \leq 2 \cdot M^{(\hat{g})} + 1$.

Throughout the proof, we drop the superscript from $M^{(\hat{g})}$ and simply refer to it as M.

Let $v_1 \geq v_2 \geq \ldots \geq v_{2 \cdot M+1}$ be the valuations of the participants in B; if B has less than $2 \cdot M + 1$ participants, we add enough dummy participants with valuations zero. Remember that $\text{MPRG}(\hat{g}) = \max_{j \in [M]} j \cdot v_j^{(\hat{g})}$ where M is at least 1 (see Definition 5) which implies

$$v_i \leq \frac{\text{MPRG}(\hat{g})}{i} \quad \forall i \in [M] \qquad (13)$$

The following equations conclude the proof of the claim:

$$\text{WF}(\hat{g}) = \sum_{i=1}^{2 \cdot M + 1} v_i$$

$$= \sum_{i=1}^{M} v_i + \sum_{i=M+1}^{2 \cdot M + 1} v_i$$

$$\leq \sum_{i=1}^{M} v_i + \sum_{i=M+1}^{2 \cdot M + 1} v_M$$

replacing v_i with v_M for $i > M$

$$\leq \sum_{i=1}^{M} \frac{\text{MPRG}(\hat{g})}{i} + \sum_{i=M+1}^{2 \cdot M + 1} \frac{\text{MPRG}(\hat{g})}{M}$$

by Eq. (13)

$$\leq (2 + \ln k)\text{MPRG}(\hat{g})$$

□

Remember $G^{(g^*)}$ is the group with maximum MPRG value. Let j be the number for which MPRG(g^*) is equal to $j \cdot v_j^{(g^*)}$. Allocation function x^{MVPA} selects the first j^* participants from group $G^{(g^*)}$ where j^* is the maximum number for which $j^* \cdot v_{j^*}^{(g^*)}$ is larger than MPRG(g'). Therefore we can conclude that $j \leq j^*$ and hence

$$\text{WF}\left(x^{\text{MVPA}}(\theta)\right) \geq \text{MPRG}(g^*). \qquad (14)$$

The following equations conclude the proof of the theorem:

$$\text{WF}\left(x^{\text{MVPA}}(\theta)\right) \geq \text{MPRG}(g^*)$$

by Eq. (14)

$$\geq \text{MPRG}(\hat{g})$$

$G^{(g^*)}$ has the highest MPRG

$$\geq \frac{\text{WF}(\hat{g})}{2 + \ln k}$$

Claim 1

$$\geq \frac{\text{WF}(A)}{(\lfloor \log k \rfloor + 1) \cdot (2 + \ln k)}$$

by Eq. (12)

□

Lower Bound

In this section we prove Theorem 2. As mentioned earlier we need two additional mild assumptions of *anonymity* and *independence of irrelevant alternatives* (which we define below) on the class of mechanisms for which we prove our lower bound.

Definition 6 A mechanism ($\mathcal{M} = (x, p)$) is anonymous (AM) if the following holds: Suppose $\theta_1, \theta_2 \in \Theta^n$ are two type profiles which are permutations of each other (i.e., the set of type profiles are same just that the identities of participants to whom those types belongs are different). Say, $\theta_2 = \pi(\theta_1)$. Also say $x(\theta_1) = S_1$ and $x(\theta_2) = S_2$. Then $S_2 = \pi(S_1)$.

Definition 7 Let $\theta \in \Theta^n$ be an arbitrary-type profile and $i \in N$ be an arbitrary participant with type $\theta_i = (d_i, v_i)$. A mechanism ($\mathcal{M} = (x, p)$) satisfies independence of irrelevant alternatives (IIA) that if we decrease the bid of a losing participant, say, participant i, to $\hat{v}_i < v_i$, then the new set of winners is a super set of the previous one, i.e., $x(\theta) \subseteq x((d_i, \hat{v}_i), \theta_{-i})$. In other words, decreasing the bid of a losing participant does not hurt any winner.

The proof outline of Theorem 2 is the following. Let $\mathcal{M}^* = (x^*, p^*)$ be a mechanism which satisfies all the five properties and has the optimal PoRM OPT (i.e., OPT = PoRM($\mathcal{M}^*$)). We study the behavior of $\mathcal{M}^*$ in a few type profiles. Let ϵ be an arbitrary small positive real value. First we show that when there are only two participants with types $(k, 1)$ and $(k, 1 + \epsilon)$, $\mathcal{M}^*$ gives all the k items to the participant with type $(k, 1 + \epsilon)$. The revenue of $\mathcal{M}^*$ from these two participants is 1. Then, we add k

more participants to create type profile $\theta = ((1, 1 - \epsilon), (1, \frac{1}{2} - \epsilon), \ldots, (1, \frac{1}{k} - \epsilon), (k, 1), (k, 1 + \epsilon))$. The RM property requires $\mathcal{M}^*$ to make at least the same revenue for θ. From this constraint we are able to show that $\mathcal{M}^*$ assigns all the items to participant $k + 2$ with type $(k, 1+\epsilon)$ and hence gets social welfare $1+\epsilon$. Note that the maximum social welfare happens when the set of winners is $\{1, \ldots, k\}$ which implies $\text{WFR}(\mathcal{M}^*, \theta) \geq \sum_{i=1}^{k} \frac{1}{i} - k \cdot \epsilon$ (see Definition 1). Because $\text{PORM}(\mathcal{M}^*) \geq \text{WFR}(\mathcal{M}^*, \theta)$ for any $\theta \in \Theta^n$, we conclude that $\text{OPT} \geq \sum_{i=1}^{k} \frac{1}{i}$.

First we study the behavior of $\mathcal{M}^*$ when we have only two participants with types $(k, 1)$ and $(k, 1 + \epsilon)$.

Lemma 6 *Mechanism* $\mathcal{M}^*$ *in type profile* $((k, 1), (k, 1 + \epsilon))$ *gives all* k *items to the second participant and make one unit of revenue, i.e.,* $x^* ((k, 1), (k, 1 + \epsilon)) = \{2\}$ *and* $p^* ((k, 1), (k, 1 + \epsilon)) = (0, 1)$.

Proof First we study type profile $((k, v_1), (k, v_2))$ for general values $v_1, v_2 \in \mathbb{R}^+$ where $v_1 < v_2$. We prove that $\mathcal{M}^*$ gives all the items to the second participant.

Claim 2 $x^* ((k, v_1), (k, v_2)) = \{2\}$ for any $v_1, v_2 \in \mathbb{R}^+$ where $v_1 < v_2$.

Proof First note that M^* has to have a winner for this type profile because otherwise its social welfare will be zero while the maximum social welfare is v_2. This makes the social welfare ratio of $\mathcal{M}^*$ to be undefined.

Now we prove that if $x^* ((k, v_1), (k, v_2)) = \{1\}$, then $\mathcal{M}^*$ either violates IC or AM. Let call type profile $((k, v_1), (k, v_2))$ by $\theta^{(1)}$ and suppose for the sake of contradiction $x^*(\theta^{(1)}) = \{1\}$. From Lemma 1 we know that if participant 1 increases his bid to v_2, she still wins; hence $x^*(\theta^{(2)}) = \{1\}$ where $\theta^{(2)} = ((k, v_2), (k, v_2))$. Now if in type profile $\theta^{(2)}$ participant 2 decrease his bid to v_1, again from Lemma 1 we conclude that she cannot win, i.e., $x^*(\theta^{(3)}) = \{1\}$ where $\theta^{(3)} = ((k, v_2), (k, v_1))$. Type profile $\theta^{(1)}$ is $\theta^{(3)}$ with participant 1 swapped with participant 2 but in both of them the first participant wins which contradicts with AM. □

Claim 2 directly proves that the winner in type profile $((k, 1), (k, 1 + \epsilon))$ is the second participant. The only thing remains is to show that her payment (p_2) is 1. Note that payment p_2 cannot be less than one because otherwise by Lemma 1 participant 2 wins all the items in type profile $((k, 1), (k, p_2))$ which contradicts with Claim 2. Payment p_2 cannot be larger than one because otherwise for any value $1 < v_2 < p_2$ participant 2 wins all the items in type profile $((k, 1), (k, v_2))$. This contradicts with Lemma 1 which states that the payment p_2 is the smallest value for which participant 2 wins the items. □

Now we add k more participants, each of which wants only one item. In the following lemma we prove that RM forces $\mathcal{M}^*$ to assign all of the items to one of the participants who want all the items.

Lemma 7 *For the set of* $k + 2$ *participants with type profile* $\theta^{(0)} = ((1, 1 - \epsilon), (1, \frac{1}{2} - \epsilon), \ldots, (1, \frac{1}{k} - \epsilon), (k, 1), (k, 1 + \epsilon))$, *mechanism* $\mathcal{M}^*$ *assigns all the* k *items to either participant* $k + 1$ *or participant* $k + 2$, *i.e.,* $x^* (\theta^{(0)}) = \{k + 1\}$ *or* $x^* (\theta^{(0)}) = \{k + 2\}$.

Proof We prove the lemma by contradiction that if $\mathcal{M}^*$ assigns the items to a subset of the first k participants, it satisfies RM. We consider a class of k type profiles $(\theta^{(1)}, \ldots, \theta^{(k)})$ where $\theta^{(i)}$ is built from $\theta^{(i-1)}$. The only possible difference between $\theta^{(i)}$ and $\theta^{(i-1)}$ is in the valuation of participant i. If participant i is a winner in $\theta^{(i-1)}$, then we obtain $\theta^{(i)}$ by increasing the valuation of the ith participant from $\frac{1}{i} - \epsilon$ to $1 - \epsilon$. Note that the payment of participant i in $\theta^{(i-1)}$ is at most her valuation which is $\frac{1}{i} - \epsilon$ and in $\theta^{(i)}$ it remains the same by Lemma 1. If participant i is not a winner in $\theta^{(i-1)}$, then we obtain $\theta^{(i)}$ by decreasing his valuation to zero. Note that by IIA, no winner turns to a loser in $\theta^{(i)}$.

Let $j \in \{1, \ldots, k\}$ be the largest number for which participant j is a winner in $\theta^{(j-1)}$ and we increase his valuation to $1 - \epsilon$ in $\theta^{(j)}$. Note that at the start in type profile $\theta^{(0)}$, the set of winners is a nonempty subset of $\{1, \ldots, k\}$. Therefore there is at least one such j for which participant j is a winner in $\theta^{(j)}$ since decreasing the non-winners valuation does not reduce the size of the winners.

Now we prove that there is no winner in the set of participants $\{j+1, \ldots, k\}$ in type profile $\theta^{(j)}$. Assume otherwise and let $p \in \{j+1, \ldots, k\}$ be the smallest number for which participant p is a winner in $\theta^{(j)}$. Note that when we decrease the valuation of each participant $j < p' < p$ to zero to obtain $\theta^{(p')}$, participant p remains as a winner in all of them by IIA. Therefore, participant p is a winner in type profile $\theta^{(p-1)}$ and we increase his valuation in $\theta^{(p)}$ which contradicts with the fact that j is the largest number for which participant j is a winner in $\theta^{(j-1)}$.

The payment of participant j in $\theta^{(j-1)}$ is at most its valuation which is $\frac{1}{j} - \epsilon$. When we increase his bid to $1 - \epsilon$ in type profile $\theta^{(j)}$, its payment remains the same by Lemma 1. Note that by construction of $\theta^{(j)}$, the valuation of all participants in $\{1, \ldots, j\}$ is either zero or $1-\epsilon$. If the valuation of them is $1-\epsilon$ and they are winner, by AM their payment is $\frac{1}{j} - \epsilon$. Therefore the total payments or revenue of $\mathcal{M}^*$ in $\theta^{(j)}$ is at most $j \cdot (\frac{1}{j} - \epsilon) = 1 - j \cdot \epsilon$ since there is no other winner in set of participants $\{j+1, \ldots, k\}$ in type profile $\theta^{(j)}$.

Note that type profile $\theta^{(j)}$ is obtained from type profile $((k,1), (k,1+\epsilon))$ by adding k more participants. However the revenue of $\theta^{(j)}$ is $1-j \cdot \epsilon$ that is strictly less than 1 which is the revenue of $((k,1), (k,1+\epsilon))$ by Lemma 6. This contradicts with the RM property of $\mathcal{M}^*$; hence $\mathcal{M}^*$ has to assign the items to either participant $k+1$ or $k+2$. $\qquad\square$

Now we show how from Lemma 7 we can derive Theorem 2. Note that the maximum welfare for type profile $\theta^{(0)} = ((1, 1-\epsilon), (1, \frac{1}{2} - \epsilon), \ldots, (1, \frac{1}{k} - \epsilon), (k,1), (k,1+\epsilon))$ realized when we give one item to each of the first k participants for which we get the total social welfare $\sum_{i=1}^{k} \frac{1}{i} - k \cdot \epsilon$, i.e., the nominator of Definition 1 for this type profile is $\sum_{i=1}^{k} \frac{1}{i} - k \cdot \epsilon$. The denominator of Definition 1 is at most $1 + \epsilon$ by Lemma 7. Therefore the ratio of the welfare for this type profile is at least $\frac{\sum_{i=1}^{k} 1/i - k \cdot \epsilon}{1+\epsilon}$. Because OPT is the maximum ratio over all type profiles

(see Definition 2), we have OPT $\geq \frac{\sum_{i=1}^{k} 1/i - k \cdot \epsilon}{1+\epsilon}$ which results in OPT $\geq \sum_{i=1}^{k} \frac{1}{i} - \epsilon'$ where $\epsilon' = \frac{\epsilon(k - \sum_{i=1}^{k} 1/i)}{1+\epsilon}$.

Note that the value ϵ' can be made arbitrarily small by selecting a sufficiently small value for ϵ. Therefore we prove that for any positive small real value ϵ', we have OPT $\geq \sum_{i=1}^{k} \frac{1}{i} - \epsilon'$ which implies Theorem 2.

Acknowledgments The first author would like to thank Vasilis Gkatzelis for some initial fruitful discussions on this topic.

Recommended Reading

1. Ausubel LM, Milgrom P (2002) Ascending auctions with package bidding. Front Theor Econ 1(1):1–42
2. Bikhchandani S, Chatterji S, Lavi R, Mu'alem A, Nisan N, Sen A (2006) Weak monotonicity characterizes deterministic dominant-strategy implementation. Econometrica 74(4):1109–1132
3. Day R, Milgrom P (2008) Core-selecting package auctions. Int J Game Theory 36(3–4):393–407
4. Goldberg AV, Hartline JD, Karlin AR, Saks M, Wright A (2006) Competitive auctions. Games Econ Behav 55(2):242–269
5. Goldberg AV, Hartline JD, Wright A (2001) Competitive auctions and digital goods. In: Symposium on discrete algorithms, Washington, DC, pp 735–744
6. Lavi R, Mu'Alem A, Nisan N (2003) Towards a characterization of truthful combinatorial auctions. In: Foundations of computer science, Cambridge, pp 574–583
7. Myerson RB (1981) Optimal auction design. Math Oper Res 6(1):58–73
8. Nisan N, Roughgarden T, Tardos E, Vazirani VV (2007) Algorithmic game theory. Cambridge University Press, Cambridge
9. Rastegari B, Condon A, Leyton-Brown K (2009) Stepwise randomized combinatorial auctions achieve revenue monotonicity. In: Symposium on discrete algorithms, New York, pp 738–747
10. Rastegari B, Condon A, Leyton-Brown K (2011) Revenue monotonicity in deterministic, dominant-strategy combinatorial auctions. Artif Intell 175(2):441–456
11. Roberts K (1979) The characterization of implementable choice rules. Aggreg Revel Prefer 12(2):321–348
12. Rochet J-C (1987) A necessary and sufficient condition for rationalizability in a quasi-linear context. J Math Econ 16(2):191–200
13. Saks M, Yu L (2005) Weak monotonicity suffices for truthfulness on convex domains. In: Electronic commerce, Vancouver, pp 286–293

R

Reverse Search; Enumeration Algorithms

Masashi Kiyomi
International College of Arts and Sciences,
Yokohama City University, Yokohama,
Kanagawa, Japan

Keywords

Enumeration; Reverse search; Vertices of a polytope

Years and Authors of Summarized Original Work

1996; Avis, Fukuda
2003; Uno
2004; Nakano, Uno

Problem Definition

We will consider enumeration problems, i.e., we want to list all the objects that satisfy given conditions (e.g., vertices of a polytope $\{x \mid Ax \geq b\}$ or maximal cliques in a given graph). One object should not be listed twice or more.

Introduction

In this entry, we consider an enumeration scheme called *reverse search* developed by Avis and Fukuda [1]. The scheme was originally developed to enumerate all the vertices of a given polytope represented by the intersection of half spaces [1]. The scheme is very powerful, and quite many kinds of objects such as arrangements in a hyperplane, triangulations of a polygon, bases of a matroid, spanning trees, trees, or maximal cliques in a graph, plane graphs of given number of vertices, etc., can be enumerated with it [1–4, 6].

Think of a problem to enumerate (or visit) all the vertices of a given connected graph G. Most of the readers may use depth-first search

or breadth-first search algorithms. The two algorithms dynamically find tree structures in G and traverse them. Given an enumeration problem, reverse search scheme also finds some kind of tree structure on objects to enumerate dynamically and traverse it. To execute depth-first search or breadth-first search, a graph G should be given explicitly, and we have to remember which vertices have been visited. However, in most of enumeration problems, the objects that we want to enumerate are, of course, not explicitly given, nor we cannot remember all the objects that we have already output in the execution of an algorithm. For example, if we want to enumerate all the vertices of a polytope represented by the intersection of half spaces, the vertices are not given explicitly. If we want to enumerate plane graphs of 100 vertices, the number is quite large, and we do not want to remember every obtained graph. So, the scheme is designed to treat implicitly given objects and run with small amount of memory.

Key Results

When we develop an algorithm for enumerating some objects with reverse search scheme, we first think of an implicit connected graph G_{rs} of the objects. For example, when enumerating all the vertices of a given polytope, we think of a graph whose vertices correspond to the vertices of the polytope and whose edges correspond to the edges of the polytope. When enumerating spanning trees in a graph G, we think of a graph G_{rs} whose vertices correspond to the spanning trees of G, and $\{i, j\} \in E(G_{rs})$ if and only if spanning tree T_j of G corresponding to vertex j of G_{rs} can be obtained from spanning tree T_i of G corresponding to vertex i of G_{rs} by removing an edge and adding an edge. Of course, we cannot make such a graph G_{rs} explicitly without enumerating the vertices of the polytope or the spanning trees of G. However, given an object x, we can easily generate every object whose corresponding vertex is adjacent to x's corresponding vertex in G_{rs}. To put it the other way around, we require G_{rs} this property. In the above examples, G_{rs} are undirected. However, G_{rs} is sometimes directed. When enumerating (not necessarily maximal) cliques in a graph, $(i, j) \in E(G_{\mathrm{rs}})$ if and only if the clique corresponding to j is a proper subset of the clique corresponding to i.

Now we have a connected graph G_{rs} of objects to enumerate. Then for every vertex $v \in V(G_{\mathrm{rs}})$ except for a special vertex r called *root*, we define a *parent* vertex u of v such that u is adjacent to v, and no vertex of G_{rs} is a proper ancestor of itself, i.e., by iteratively moving from a vertex v to the parent of v, to the parent of the parent of v, and so on, we never come to the start vertex v again. Then we easily come to the following lemma.

Lemma 1 *For every vertex v of G_{rs}, v is a descendant of r.*

Proof Since every vertex of G_{rs} cannot be an ancestor of itself and the number of vertices in G_{rs} is bounded, every vertex of G_{rs} has its oldest ancestor. Since a vertex of G_{rs} except for r has its parent, it cannot be the oldest ancestor. Therefore, r is the oldest ancestor of every vertex of G_{rs}.   □

By the lemma above, edges in G_{rs} corresponding to the "parent-child" relations clearly induce a spanning tree (or arborescence) T_{rs} of G_{rs}. Therefore, we can enumerate every object by traversing T_{rs} from r in the depth-first manner. The whole scheme is shown below.

Procedure ENUMERATE_SUBTREE(v)
  output v
  for all w satisfying $(w, v) \in E(G_{\mathrm{rs}})$ **do**
    if v is the parent of w **then**
       ENUMERATE_SUBTREE(w)
    end if
  end for
end procedure

Procedure ENUMERATE
  find r
  ENUMERATE_SUBTREE(r)
end procedure

If the depth of T_{rs} is very deep, using a recursion needs a big amount of memory. However, since we can find the parent of each vertex of G_{rs}, we actually do not need to use a recursion. Even if we do not remember the previously visited

vertices in G_{rs}, we can go back in the tree search by finding the parents. If the time complexity for finding the parent is relatively high, the total time complexity gets high. Therefore, there is a time-space trade-off.

Examples

For enumerating all the vertices of a given polytope $P = \{Ax \geq b\}$, we use G_{rs} described in the previous section. For the sake of simplicity, we assume that P is not degenerated. First, we find a vertex x^* of P by the simplex method or the interior point method. Then, find an objective vector c such that the unique optimal solution of the linear programming problem min $c^\top x$, s. t. $Ax \geq b$ is x^*. We define a parent vertex $\overline{x}$ of a vertex x as the vertex corresponding to the basis of P obtained from the basis corresponding to x by a single pivot in the simplex method minimizing $c^\top x$ with Bland's pivot rule. The root vertex is x^*. We can easily find every vertex x' satisfying $\{x', x\} \in E(G_{\mathrm{rs}})$ by swapping a basic variable and a nonbasic variable from the basis corresponding to x'. Of course, we can easily check if vertex x is the parent of vertex x' by running the simplex method by one step from x'.

For enumerating (not necessarily maximal) cliques in a graph G, we also use (the directed graph) G_{rs} in the previous section. We define the parent of vertex v corresponding to clique C_v in G as the vertex u corresponding to clique C_u such that C_u is obtained from C_v by removing the vertex of the smallest index. The root is the empty set. Since a vertex w satisfying $(w, v) \in E(G_{\mathrm{rs}})$ corresponds to a clique obtained by adding a vertex to C_v, we can find it easily. Clearly we can check if v is the parent of u easily, too.

Note that the algorithms introduced in this section are for an easy explanation. One can develop faster algorithms for the problems.

Avoiding Long Delays

A naive implementation of the reverse search scheme sometimes causes a long delay between successive outputs of two objects. Consider the case that the depth of T_{rs} is very deep and one has to return from a leaf to the root. In order to avoid this kind of long delays, a smart method is known [5, 7]. At the odd level of recursion, we output the objects before making the recursive calls, and at the even level of the recursion, we output after the termination of the recursive calls. In this way, at least one of three iterations outputs an object when the algorithm ascends or descends the search tree T_{rs}. The algorithm is shown below.

Procedure ENUMERATE_SUBTREE(v, *parity*)
  if *parity* = 0 **then**
    output v
  end if
  for all w satisfying $(w, v) \in E(G_{rs})$ **do**
    if v is the parent of w **then**
      ENUMERATE_SUBTREE(w,*parity* $\oplus$ 1)
    end if
  end for
  if *parity* = 1 **then**
    output v
  end if
end procedure

Procedure ENUMERATE
  find r
  ENUMERATE_SUBTREE(r, 0)
end procedure

Note

For the sake of easy understanding, we introduced G_{rs}. However, most of results using the reverse search type algorithms do not treat G_{rs}. It is easy to understand that we can develop reverse search algorithms only by good definitions of parent-child relations and fast algorithms to enumerate children of given objects. If one can develop a fast children enumeration algorithm which enumerates all the children of an object in time T and if the degrees of some vertices in G_{rs} are quite large compared with T, the resulting enumeration algorithm is faster than the naive implementation described in this entry. Such examples will appear in other entries in this book.

Recommended Reading

1. Avis D, Fukuda K (1996) Reverse search for enumeration. Discret Appl Math 65:21–46
2. Makino K, Uno T (2004) New algorithms for enumerating all maximal cliques. Lect Notes Comput Sci 3111:260–272
3. Nakano S (2001) Enumerating floorplans with n rooms. Lect Notes Comput Sci 2223:107–115
4. Nakano S (2004) Efficient generation of triconnected plane triangulations. Comput Geom Theory Appl 27(2):109–122
5. Nakano S, Uno T (2004) Constant time generation of trees with specified diameter. Lect Notes Comput Sci 3353:33–45
6. Shioura A, Tamura A, Uno T (1997) An optimal algorithm for scanning all spanning trees of undirected graphs. SIAM J Comput 26(3):678–692
7. Uno T (2003) Two general methods to reduce delay and change of enumeration algorithms. National Institute of Informatics (in Japan), Technical repor

RNA Secondary Structure Boltzmann Distribution

Rune B. Lyngsø
Department of Statistics, Oxford University, Oxford, UK
Winton Capital Management, Oxford, UK

Keywords

Full partition function

Years and Authors of Summarized Original Work

2005; Miklós, Meyer, Nagy

Problem Definition

This problem is concerned with computing features of the Boltzmann distribution over RNA secondary structures in the context of the standard Gibbs free energy model used for RNA Secondary Structure Prediction by Minimum Free Energy (cf. corresponding entry). Thermodynamics state that for a system with configuration space Ω and free energy given by $E: \Omega \mapsto \mathbf{R}$, the probability of the system being in state $\omega \in \Omega$ is proportional to $e^{-E(\omega)/RT}$ where R is the universal gas constant and T the absolute temperature of the system. The normalizing factor

$$Z = \sum_{\omega \in \Omega} e^{-E(\omega)/RT} \qquad (1)$$

is called the *full partition function* of the system.

Over the past several decades, a model approximating the free energy of a structured RNA molecule by independent contributions of its secondary structure components has been developed and refined. The main purpose of this work has been to assess the stability of individual secondarystructures. However, it immediately translates into a distribution over all secondary structures. Early work focused on computing the pairing probability for all pairs of bases, i.e., the sum of the probabilities of all secondary structures containing that base pair. Recent work has extended methods to compute probabilities of base pairing probabilities for RNA heterodimers [2], i.e., interacting RNA molecules, and expectation, variance and higher moments of the Boltzmann distribution.

Notation

Let $s \in \{A, C, G, U\}^*$ denote the sequence of bases of an RNA molecule. Use $X \cdot Y$ where $X, Y \in \{A, C, G, U\}$ to denote a base pair between bases of type X and Y, and $i \cdot j$ where $1 \leq i < j \leq |s|$ to denote a base pair between bases $s[i]$ and $s[j]$.

Definition 1 (RNA Secondary Structure)

A secondary structure for an RNA sequence s is a set of base pairs $S = \{i \cdot j \mid 1 \leq i < j \leq |s| \wedge i < j - 3\}$. For $i \cdot j, i' \cdot j' \in S$ with $i \cdot j \neq i' \cdot j'$

- $\{i, j\} \cap \{i', j'\} = \emptyset$ (each base pairs with at most one other base)
- $\{s[i], s[j]\} \in \{\{A, U\}, \{C, G\}, \{G, U\}\}$ (only Watson-Crick and G, U wobble base pairs)

- $i < i' < j \Rightarrow j' < j$ (base pairs are either nested or juxtaposed but not overlapping)

The second requirement, that only canonical base pairs are allowed, is standard but not consequential in solutions to the problem. The third requirement states that the structure does not contain pseudoknots. This restriction is crucial for the results listed in this entry.

Energy Model

The model of Gibbs free energy applied, usually referred to as the nearest-neighbor model, was originally proposed by Tinoco et al. [10, 11]. It approximates the free energy by postulating that the energy of the full three dimensional structure only depends on the secondary structure, and that this in turn can be broken into a sum of independent contributions from each loop in the secondary structure.

Definition 2 (Loops) For $i \cdot j \in S$, base k is *accessible* from $i \cdot j$ iff $i < k < j$ and $\neg \exists i' \cdot j' \in S: i < i' < k < j' < j$. The *loop closed* by $i \cdot j$, $\ell_{i \cdot j}$, consists of $i \cdot j$ and all the bases accessible from $i \cdot j$. If $i' \cdot j' \in S$ and i' and j' are accessible from $i \cdot j$, then $i' \cdot j'$ is an interior base pair in the loop closed by $i \cdot j$.

Loops are classified by the number of interior base pairs they contain:

- hairpin loops have no interior base pairs
- stacked pairs, bulges, and internal loops have one interior base pair that is separated from the closing base pair on neither side, on one side, or on both sides, respectively
- multibranched loops have two or more interior base pairs

Bases not accessible from any base pair are called external. This is illustrated in Fig. 1. The free energy of structure S is

$$\Delta G(S) = \sum_{i \cdot j \in S} \Delta G(\ell_{i \cdot j}) \qquad (2)$$

RNA Secondary Structure Boltzmann Distribution, Fig. 1
A hypothetical RNA structure illustrating the different loop types. Bases are represented by *circles*, the RNA backbone by *straight lines*, and base pairs by *zigzagged lines*

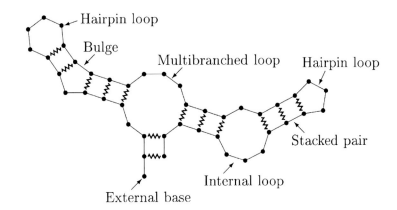

where $\Delta G(\ell_{i\cdot j})$ is the free energy contribution from the loop closed by $i \cdot j$. The contribution of S to the full partition function is

$$
e^{-\Delta G(S)/RT} = e^{-\sum_{i\cdot j \in S} \Delta G(\ell_{i\cdot j})/RT}
$$
$$
= \prod_{\ell_{i\cdot j} \in S} e^{-\Delta G(\ell_{i\cdot j})/RT}. \quad (3)
$$

Problem 1 (RNA Secondary Structure Distribution)
INPUT: RNA sequence s, absolute temperature T and specification of ΔG at T for all loops.
OUTPUT: $\sum_S e^{-\Delta G(S)/RT}$, where the sum is over all secondary structures for s.

Key Results

Solutions are based on recursions similar to those for RNA Secondary Structure Prediction by Minimum Free Energy, replacing sum and minimization with multiplication and sum (or more generally with a *merge function* and a *choice function* [8]). The key difference is that recursions are required to be non-redundant, i.e., any particular secondary structure only contributes through one path through the recursions.

Theorem 1 *Using the standard thermodynamic model for RNA secondary structures, the partition function can be computed in time $O(|s|^3)$ and space $O(|s|^2)$. Moreover, the computation can build data structures that allow $O(1)$ queries*

of the pairing probability of $i \cdot j$ for any $1 \le i < j \le |s|$ [5, 6, 7].

Theorem 2 *Using the standard thermodynamic model for RNA secondary structures, the expectation and variance of free energy over the Boltzmann distribution can be computed in time $O(|s|^3)$ and space $O(|s|^2)$. More generally, the kth moment*

$$
E_{\text{Boltzmann}}[\Delta G] = 1/Z \sum_S e^{-\Delta G(S)/RT} \Delta G^k(S),
$$
$$(4)$$

where $Z = \sum_S e^{-\Delta G(S)/RT}$ is the full partition function and the sums are over all secondary structures for s, can be computed in time $O(k^2|s|^3)$ and space $O(ks^2)$ [8].

In Theorem 2 the free energy does not hold a special place. The theorem holds for any function Φ defined by an independent contribution from each loop,

$$
\Phi(S) = \sum_{i\cdot j \in S} \phi\left(\ell_{i\cdot j}\right), \quad (5)
$$

provided each loop contribution can be handled with the same efficiency as the free energy contributions. Hence, moments over the Boltzmann distribution of e.g., number of base pairs, unpaired bases, or loops can also be efficiently computed by applying appropriately chosen indicator functions.

Applications

The original use of partition function computations was for discriminating between well defined and less well defined regions of a secondary structure. Minimum free energy predictions will always return a structure. Base pairing probabilities help identify regions where the prediction is uncertain, either due to the approximations of the model or that the real structure indeed does fluctuate between several low energy alternatives. Moments of Boltzmann distributions are used in identifying how biological RNA molecules deviates from random RNA sequences.

The data structures computed in Theorem 1 can also be used to efficiently sample secondary structures from the Boltzmann distribution. This has been used for probabilistic methods for secondary structure prediction, where the centroid of the most likely cluster of sampled structures is returned rather than the most likely, i.e., minimum free energy, structure [3]. This approach better accounts for the entropic effects of large neighborhoods of structurally and energetically very similar structures. As a simple illustration of this effect, consider twice flipping a coin with probability $p > 0.5$ for heads. The probability p^2 of heads in both flips is larger than the probability $p(1 - p)$ of heads followed by tails or tails followed by heads (which again is larger than the probability $(1 - p)^2$ of tails in both flips). However, if the order of the flips is ignored the probability of one heads and one tails is $2p(1 - p)$. The probability of two heads remains p^2 which is smaller than $2p(1 - p)$ when $p < \frac{2}{3}$. Similarly a large set of structures with fairly low free energy may be more likely, when viewed as a set, than a small set of structures with very low free energy.

Open Problems

As for RNA Secondary Structure Prediction by Minimum Free Energy, improvements in time and space complexity are always relevant. This may be more difficult for computing distributions, as the more efficient dynamic programming techniques of [9] cannot be applied. In the context of genome scans, the fact that the start and end positions of encoded RNA molecule is unknown has recently been considered [1].

Also the problem of including structures with pseudoknots, i.e., structures violating the last requirement in Definition 1, in the configuration space is an active area of research. It can be expected that all the methods of Theorems 3 through 6 in the entry on RNA Secondary Structure Prediction Including Pseudoknots can be modified to computation of distributions without affecting complexities. This may require some further bookkeeping to ensure non-redundancy of recursions, and only in [4] has this actively been considered.

Though the moments of functions that are defined as sums over independent loop contributions can be computed efficiently, it is unknown whether the same holds for functions with more complex definitions. One such function that has traditionally been used for statistics on RNA secondary structure [12] is the *order* of a secondary structure which refers to the nesting depth of multibranched loops.

URL to Code

Software for partition function computation and a range of related problems is available from www.bioinfo.rpi.edu/applications/hybrid/download.php and www.tbi.univie.ac.at/~ivo/RNA/. Software including a restricted class of structures with pseudoknots [4] is available at www.nupack.org.

Cross-References

Recommended Reading

1. Bernhart S, Hofacker IL, Stadler P (2006) Local RNA base pairing probabilities in large sequences. Bioinformatics 22:614–615
2. Bernhart SH, Tafer H, Mückstein U, Flamm C, Stadler PF, Hofacker IL (2006) Partition function and base pairing probabilities of RNA heterodimers. Algorithms Mol Biol 1:3
3. Ding Y, Chan CY, Lawrence CE (2005) RNA secondary structure prediction by centroids in a Boltzmann weighted ensemble. RNA 11:1157–1166
4. Dirks RM, Pierce NA (2003) A partition function algorithm for nucleic acid secondary structure including pseudoknots. J Comput Chem 24:1664–1677
5. Hofacker IL, Stadler PF (2006) Memory efficient folding algorithms for circular RNA secondary structures. Bioinformatics 22:1172–1176
6. Lyngsø, RB, Zuker M, Pedersen CNS (1999) Fast evaluation of internal loops in RNA secondary structure prediction. Bioinformatics 15:440–445
7. McCaskill JS (1990) The equilibrium partition function and base pair binding probabilities for RNA secondary structure. Biopolymers 29:1105–1119
8. Miklós I, Meyer IM, Nagy B (2005) Moments of the Boltzmann distribution for RNA secondary structures. Bull Math Biol 67:1031–1047
9. Ogurtsov AY, Shabalina SA, Kondrashov AS, Roytberg MA (2006) Analysis of internal loops within the RNA secondary structure in almost quadratic time. Bioinformatics 22:1317–1324
10. Tinoco I, Borer PN, Dengler B, Levine MD, Uhlenbeck OC, Crothers DM, Gralla J (1973) Improved estimation of secondary structure in ribonucleic acids. Nat New Biol 246:40–41
11. Tinoco I, Uhlenbeck OC, Levine MD (1971) Estimation of secondary structure in ribonucleic acids. Nature 230:362–367
12. Waterman MS (1978) Secondary structure of single-stranded nucleic acids. Adv Math Suppl Stud 1:167–212

RNA Secondary Structure Prediction by Minimum Free Energy

Rune B. Lyngsø
Department of Statistics, Oxford University, Oxford, UK
Winton Capital Management, Oxford, UK

Keywords

RNA folding

Years and Authors of Summarized Original Work

2006; Ogurtsov, Shabalina, Kondrashov, Roytberg

Problem Definition

This problem is concerned with predicting the set of base pairs formed in the native structure of an RNA molecule. The main motivation stems from structure being crucial for function and the growing appreciation of the importance of RNA molecules in biological processes. Base pairing is the single most important factor determining structure formation. Knowledge of the secondary structure alone also provides information about stretches of unpaired bases that are likely candidates for active sites. Early work [7] focused on finding structures maximizing the number of base pairs. With the work of Zuker and Stiegler [17], focus shifted to energy minimization in a model approximating the Gibbs free energy of structures.

Notation

Let $s \in \{A, C, G, U\}^*$ denote the sequence of bases of an RNA molecule. Use $X \cdot Y$ where $X, Y \in \{A, C, G, U\}$ to denote a base pair between bases of type X and Y and $i \cdot j$ where $1 \leq i < j \leq |s|$ to denote a base pair between bases $s[i]$ and $s[j]$.

Definition 1 (RNA Secondary Structure) A secondary structure for an RNA sequence s is a set of base pairs $\mathcal{S} = \{i \cdot j \mid 1 \leq i < j \leq |s| \land i < j - 3\}$. For $i \cdot j, i' \cdot j' \in \mathcal{S}$ with $i \cdot j \neq i' \cdot j'$:

- $\{i, j\} \cap \{i', j'\} = \emptyset$ (each base pair with at most one other base)
- $\{s[i], s[j]\} \in \{\{A, U\}, \{C, G\}, \{G, U\}\}$ (only Watson-Crick and G, U wobble base pairs)
- $i < i' < j \Rightarrow j' < j$ (base pairs are either nested or juxtaposed but not overlapping)

The second requirement that only canonical base pairs are allowed is standard but not consequen-

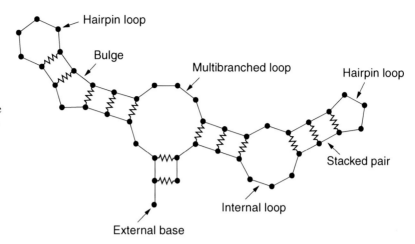

RNA Secondary Structure Prediction by Minimum Free Energy, Fig. 1 A hypothetical RNA structure illustrating the different loop types. Bases are represented by *circles*, the RNA backbone by *straight lines*, and base pairs by *zigzagged lines*

tial in solutions to the problem. The third requirement states that the structure does not contain pseudoknots. This restriction is crucial for the results listed in this entry.

Energy Model

The model of Gibbs free energy applied, usually referred to as the nearest-neighbor model, was originally proposed by Tinoco et al. [10, 11]. It approximates the free energy by postulating that the energy of the full three-dimensional structure only depends on the secondary structure and that this in turn can be broken into a sum of independent contributions from each loop in the secondary structure.

Definition 2 (Loops) For $i \cdot j \in S$, base k is *accessible* from $i \cdot j$ iff $i < k < j$ and $\neg \exists i' \cdot j' \in S : i < i' < k < j' < j$. The *loop closed by* $i \cdot j$, $\ell_{i \cdot j}$, consists of $i \cdot j$ and all the bases accessible from $i \cdot j$. If $i' \cdot j' \in S$ and i' and j' are accessible from $i \cdot j$, then $i' \cdot j'$ is an interior base pair in the loop closed by $i \cdot j$.

Loops are classified by the number of interior base pairs they contain:

- Hairpin loops have no interior base pairs.
- Stacked pairs, bulges, and internal loops have one interior base pair that is separated from the closing base pair on neither side, on one side, or on both sides, respectively.

- Multibranched loops have two or more interior base pairs.

Bases not accessible from any base pair are called external. This is illustrated in Fig. 1. The free energy of structure S is

$$\Delta G(S) = \sum_{i \cdot j \in S} \Delta G(\ell_{i \cdot j}), \qquad (1)$$

where $\Delta G(\ell_{i \cdot j})$ is the free energy contribution from the loop closed by $i \cdot j$.

Problem 1 (Minimum Free Energy Structure)

INPUT: RNA sequence s and specification of ΔG for all loops

$$\arg \min_{S} \{\Delta G(S) | S \text{ secondary structure for } s\}.$$

OUTPUT: A secondary structure achieving the minimum of free energies, taken over all possible secondary structures

Key Results

Solutions are based on using dynamic programming to solve the general recursion

$$V[i, j] = \min_{k \geq 0; i < i_1 < j_1 < \cdots < i_k < j_k < j} \left\{ \Delta G(\ell_{i \cdot j; i_1 \cdot j_1, \ldots, i_k \cdot j_k}) + \sum_{l=1}^{k} V[i_l, j_l] \right\}$$

$$W[i] = \min \left\{ W[i - 1], \min_{0 < k < i} \{W[k - 1] + V[k, i]\} \right\},$$

where $\Delta G(\ell_{i \cdot j; i_1 \cdot j_1, \ldots, i_k \cdot j_k})$ is the free energy of the loop closed by $i \cdot j$ and interior base pairs $i_1 \cdot j_1, \ldots, i_k \cdot j_k$ and with initial condition $W[0] = 0$. In the following, it is assumed that all loop energies can be computed in time $O(1)$.

Theorem 1 *If the free energy of multibranched loops is a sum of:*

- *An affine function of the number of interior base pairs and unpaired bases*
- *Contributions for each base pair from stacking with either neighboring unpaired bases in the loop or with a neighboring base pair in the loop, whichever is more favorable*

a minimum free energy structure can be computed in time $O(|s|^4)$ and space $O(|s|^2)$ [17].

With these assumptions, the time required to handle the multibranched loop parts of the recursion reduces to $O(|s|^3)$. Hence, handling the $O(|s|^4)$ possible internal loops becomes the bottleneck.

Theorem 2 *If furthermore the free energy of internal loops is a sum of:*

- *A function of the total size of the loop, i.e., the number of unpaired bases in the loop*
- *A function of the asymmetry of the loop, i.e., the difference in number of unpaired bases on the two sides of the loop*
- *Contributions from the closing and interior base pairs stacking with the neighboring unpaired bases in the loop*

a minimum free energy structure can be computed in time $O(|s|^3)$ and space $O(|s|^2)$ [5].

Under these assumptions, the time required to handle internal loops reduces to $O(|s|^3)$.

With further assumptions on the free energy contributions of internal loops, this can be reduced even further, again making the handling of multibranched loops the bottleneck of the computation.

Theorem 3 *If furthermore the size dependency is concave and the asymmetry dependency is constant for all but $O(1)$ values, a multibranched loop free minimum free energy structure can be computed in time $O(|s|^2 \log^2 |s|)$ and space $O(|s|^2)$ [8].*

The above assumptions are all based on the nature of current loop energies [6]. These energies have to a large part been developed without consideration of computational expediency and parameters determined experimentally, although understanding of the precise behavior of larger loops is limited. For multibranched loops, some theoretical considerations [4] would suggest that a logarithmic dependency would be more appropriate.

Theorem 4 *If the restriction on the dependency on number of interior base pairs and unpaired bases in Theorem 1 is weakened to any function that depends only on the number of interior base pairs, the number of unpaired bases, or the total number of bases in the loop, a minimum free energy structure can be computed in time $O(n^4)$ and space $O(n^3)$ [13].*

Theorem 5 *All the above theorems can be modified to compute a data structure that for any $1 \leq i < j \leq |s|$ allows us to compute the minimum free energy of any structure containing $i \cdot j$ in time $O(1)$ [15].*

Applications

Naturally, the key application of these algorithms is for predicting the secondary structure of

RNA molecules. This holds in particular for sequences with no homologues with common structure, e.g., functional analysis based on mutational effects and to some extent analysis of RNA aptamers. With access to structurally conserved homologues, prediction accuracy is significantly improved by incorporating comparative information [2].

Incorporating comparative information seems to be crucial when using secondary structure prediction as the basis of RNA gene finding. As it turns out, the minimum free energy of known RNA genes is not sufficiently different from the minimum free energy of comparable random sequences to reliably separate the two [9,14]. However, minimum free energy calculations are at the core of one successful comparative RNA gene finder [12].

Open Problems

Most current research is focused on refinement of the energy parametrization. The limiting factor of sequence lengths for which secondary structure prediction by the methods described here is still feasible is adequacy of the nearest-neighbor approximation rather than computation time and space. Still, improvements on time and space complexities are useful as biosequence analyses are invariably used in genome scans. In particular, improvements on Theorem 4, possibly for dependencies restricted to be logarithmic or concave, would allow for more advanced scoring of multibranched loops. A more esoteric open problem is to establish the complexity of computing the minimum free energy under the general formulation of (1), with no restrictions on loop energies except that they are computable in time polynomial in $|s|$.

Experimental Results

With the release of the most recent energy parameters [6], secondary structure prediction by finding a minimum free energy structure was found to recover approximately 73 % of the base pairs in a benchmark data set of RNA sequences with known secondary structure. Another independent assessment [1] put the recovery percentage somewhat lower at around 56 %. This discrepancy is discussed and explained in [1].

Data Sets

Families of homologous RNA sequences aligned and annotated with secondary structure are available from the Rfam database at www.sanger.ac.uk/Software/Rfam/. Three-dimensional structures are available from the Nucleic Acid Database at ndbserver.rutgers.edu/. An extensive list of this and other databases is available at www.imb-jena.de/RNA.html.

URL to Code

Software for RNA folding and a range of related problems is available at www.bioinfo.rpi.edu/applications/hybrid/download.php and www.tbi.univie.ac.at/~ivo/RNA/. Software implementing the efficient handling of internal loops of [8] is available at ftp.ncbi.nlm.nih.gov/pub/ogurtsov/Afold.

Cross-References

▶ RNA Secondary Structure Boltzmann Distribution
▶ RNA Secondary Structure Prediction Including Pseudoknots

Recommended Reading

1. Dowell R, Eddy SR (2004) Evaluation of several lightweight stochastic context-free grammars for RNA secondary structure prediction. BMC Bioinform 5:71
2. Gardner PP, Giegerich R (2004) A comprehensive comparison of comparative RNA structure prediction approaches. BMC Bioinform 30:140
3. Hofacker IL, Stadler PF (2006) Memory efficient folding algorithms for circular RNA secondary structures. Bioinformatics 22:1172–1176

4. Jacobson H, Stockmayer WH (1950) Intramolecular reaction in polycondensations. I. The theory of linear systems. J Chem Phys 18:1600–1606

5. Lyngsø RB, Zuker M, Pedersen CNS (1999) Fast evaluation of internal loops in RNA secondary structure prediction. Bioinformatics 15:440–445

6. Mathews DH, Sabina J, Zuker M, Turner DH (1999) Expanded sequence dependence of thermodynamic parameters improves prediction of RNA secondary structure. J Mol Biol 288:911–940

7. Nussinov R, Jacobson AB (1980) Fast algorithm for predicting the secondary structure of single-stranded RNA. Proc Natl Acad Sci USA 77:6309–6313

8. Ogurtsov AY, Shabalina SA, Kondrashov AS, Roytberg MA (2006) Analysis of internal loops within the RNA secondary structure in almost quadratic time. Bioinformatics 22:1317–1324

9. Rivas E, Eddy SR (2000) Secondary structure alone is generally not statistically significant for the detection of noncoding RNAs. Bioinformatics 16:583–605

10. Tinoco I, Borer PN, Dengler B, Levine MD, Uhlenbeck OC, Crothers DM, Gralla J (1973) Improved estimation of secondary structure in ribonucleic acids. Nat New Biol 246:40–41

11. Tinoco I, Uhlenbeck OC, Levine MD (1971) Estimation of secondary structure in ribonucleic acids. Nature 230:362–367

12. Washietl S, Hofacker IL, Stadler PF (2005) Fast and reliable prediction of noncoding RNA. Proc Natl Acad Sci USA 102:2454–2459

13. Waterman MS, Smith TF (1986) Rapid dynamic programming methods for RNA secondary structure. Adv Appl Math 7:455–464

14. Workman C, Krogh A (1999) No evidence that mRNAs have lower folding free energies than random sequences with the same dinucleotide distribution. Nucleic Acids Res 27:4816–4822

15. Zuker M (1989) On finding all suboptimal foldings of an RNA molecule. Science 244:48–52

16. Zuker M (2000) Calculating nucleic acid secondary structure. Curr Opin Struct Biol 10:303–310

17. Zuker M, Stiegler P (1981) Optimal computer folding of large RNA sequences using thermodynamics and auxiliary information. Nucleic Acids Res 9:133–148

RNA Secondary Structure Prediction Including Pseudoknots

Rune B. Lyngsø
Department of Statistics, Oxford University,
Oxford, UK
Winton Capital Management, Oxford, UK

Keywords

Abbreviated as *pseudoknot prediction*

Years and Authors of Summarized Original Work

2004; Lyngsø

Problem Definition

This problem is concerned with predicting the set of base pairs formed in the native structure of an RNA molecule, including overlapping base pairs also known as pseudoknots. Standard approaches to RNA secondary structure prediction only allow sets of base pairs that are hierarchically nested. Though few known real structures require the removal of more than a small percentage of their base pairs to meet these criteria, a significant percentage of known real structures contain at least a few base pairs overlapping other base pairs. Pseudoknot substructures are known to be crucial for biological function in several contexts. One of the more complex known pseudoknot structures is illustrated in Fig. 1.

Notation

Let $s \in \{A, C, G, U\}^*$ denote the sequence of bases of an RNA molecule. Use $X \cdot Y$ where $X, Y \in \{A, C, G, U\}$ to denote a base pair between bases of type X and Y and $i \cdot j$ where $1 \leq i < j \leq |s|$ to denote a base pair between bases $s[i]$ and $s[j]$.

Definition 1 (RNA Secondary Structure) A secondary structure for an RNA sequence s is a set of base pairs $\mathcal{S} = \{i \cdot j \mid 1 \leq i \leq j \leq |s| \wedge i < j - 3\}$. For $i \cdot j, i' \cdot j' \in \mathcal{S}$ with $i \cdot j \neq i' \cdot j'$:

- $\{i, j\} \cap \{i', j'\} = \emptyset$ (each base pair with at most one other base)
- $\{s[i], s[j]\} \in \{\{A, U\}, \{C, G\}, \{G, U\}\}$ (only Watson-Crick and G, U wobble base pairs)

The second requirement that only canonical base pairs are allowed is standard but not consequential in solutions to the problem.

Scoring Schemes

Structures are usually assessed by extending the model of Gibbs free energy used for ▶ RNA Secondary Structure Prediction by Minimum Free Energy (cf. corresponding entry) with ad hoc

RNA Secondary Structure Prediction Including Pseudoknots, Fig. 1 Secondary structure of the *Escherichia coli* α operon mRNA from position 16 to position 127, cf. [5], Figure 1. The backbone of the RNA molecule is drawn as *straight lines*, while base pairings are shown with *zigzagged lines*

extrapolation of multibranched loop energies to pseudoknot substructures [11] or by summing independent contributions, e.g., obtained from base pair restricted minimum free energy structures from each base pair [13]. To investigate the complexity of pseudoknot prediction, the following three simple scoring schemes will also be considered:

Number of base pairs	$\#BP(\mathcal{S}) = |\mathcal{S}|$
Number of stacking base pairs	$\#SBP(\mathcal{S}) = |\{i \cdot j \in \mathcal{S} | i+1 \cdot j-1 \in \mathcal{S} \vee i-1 \cdot j+1 \in \mathcal{S}\}|$
Number of base pair stackings	$\#BPS(\mathcal{S}) = |\{i \cdot j \in \mathcal{S} | i+1 \cdot j-1 \in \mathcal{S}\}|$

These scoring schemes are inspired by the fact that stacked pairs are essentially the only loops having a stabilizing contribution in the Gibbs free energy model.

Problem 2 (Pseudoknot Prediction)

INPUT: RNA sequence s and an appropriately specified scoring scheme

OUTPUT: A secondary structure $\mathcal{S}$ for s that is optimal under the scoring scheme specified

Key Results

Theorem 1 *The complexities of pseudoknot prediction under the three simplified scoring schemes can be classified as follows, where Σ denotes the alphabet.*

Theorem 2 *If structures are restricted to be planar, i.e., the graph with the bases of the sequence as nodes and base pairs and backbone links of*

consecutive bases as edges is required to be planar, pseudoknot prediction under the #BPS scoring scheme is NP-hard for an alphabet of size 4. Conversely, a 1/2-approximation can be found in time $O(|s|^3)$ and space $O(|s|^2)$ by observing that an optimal pseudoknot free structure is a 1/2-approximation [6].

There are no steric reasons that RNA secondary structures should be planar, and the structure in Fig. 1 is actually nonplanar. Nevertheless, known real structures have relatively simple overlapping base pair patterns with very few nonplanar structures known. Hence, planarity has been used as a defining restriction on pseudoknotted structures [2, 15]. Similar reasoning has led to the development of several algorithms for finding an optimal structure from restricted classes of structures. These algorithms tend to use more realistic scoring schemes, e.g., extensions of the Gibbs free energy model, than the three simple scoring schemes considered above.

Theorem 3 *Pseudoknot prediction for a restricted class of structures including Fig. 2a–e, but not Fig. 2f, can be done in time $O(|s|^6)$ and space $O(|s|^4)$ [11].*

Theorem 4 *Pseudoknot prediction for a restricted class of planar structures including Fig. 2a–c, but not Fig. 2d–f, can be done in time $O(|s|^5)$ and space $O(|s|^4)$ [14].*

Theorem 5 *Pseudoknot prediction for a restricted class of planar structures including Fig. 2a, b, but not Fig. 2c–f, can be done in time $O(|s|^5)$ and space $O(|s|^4)$ or $O(|s|^3)$ [1, 4] (methods differ in generality of scoring schemes that can be used).*

R

	Fixed alphabet	Unbounded alphabet
#BP [13]	Time $O(|s|^3)$, space $O(|s|^2)$	Time $O(|s|^3)$, space $O(|s|^2)$
#SBP [7]	Time $O(|s|^{|\Sigma|^2+|\Sigma|^3})$, space $O(|s|^{|\Sigma|^2+|\Sigma|^3})$	NP hard
#BPS	NP hard for $|\Sigma| = 2|$, PTAS [7] 1/3-approximation in time $O(|s|)$ [6]	NP hard [7], 1/3-approximation in time and space $O(|s|^2)$ [6]

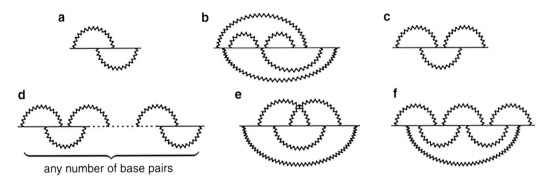

RNA Secondary Structure Prediction Including Pseudoknots, Fig. 2 RNA secondary structures illustrating restrictions of pseudoknot prediction algorithms. Backbone is drawn as a *straight line*, while base pairings are shown with *zigzagged arcs*

Theorem 6 *Pseudoknot prediction for a restricted class of planar structures including Fig. 2a, but not Fig. 2b–f, can be done in time $O(|s|^4)$ and space $O(|s|^2)$ [1, 8].*

Theorem 7 *Recognition of structures belonging to the restricted classes of Theorems 3, 5, and 6 and enumeration of all irreducible cycles (i.e., loops) in such structures can be done in time $O(|s|)$ [3, 9].*

Applications

As for the prediction of RNA secondary structures without pseudoknots, the key application of these algorithms is for predicting the secondary structure of individual RNA molecules. Due to the steep complexities of the algorithms of Theorems 3–6, these are less well suited for genome scans than prediction without pseudoknots.

Enumerating all loops of a structure in linear time also allows scoring a structure in linear time, as long as the scoring scheme allows the score of a loop to be computed in time proportional to its size. This has practical applications in heuristic searches for good structures containing pseudoknots.

Open Problems

Efficient algorithms for prediction based on restricted classes of structures with pseudoknots that still contain a significant fraction of all known structures are an active area of research. Even using the more theoretical simple #SBP scoring scheme, developing, e.g., an $O(|s|^{|\Sigma|})$ algorithm for this problem would be of practical significance. From a theoretical point of view, the complexity of planar structures is the least well understood, with results for only the #BPS scoring scheme.

Classification of realistic energy models for RNA secondary structures with pseudoknots is much less developed than for RNA secondary structures without pseudoknots. Several recent papers have been addressing this gap [3, 9, 12].

Data Sets

PseudoBase at http://biology.leidenuniv.nl/~batenburg/PKB.html is a repository of representatives of most known RNA structures with pseudoknots.

URL to Code

The method of Theorem 3 is available at http://selab.janelia.org/software.html#pknots and of one of the methods of Theorem 5 at http://www.nupack.org, and an implementation applying a slight heuristic reduction of the class of structures considered by the method of Theorem 6 is available at http://bibiserv.techfak.uni-bielefeld.de/pknotsrg/ [10].

Cross-References

▶ RNA Secondary Structure Prediction by Minimum Free Energy

Recommended Reading

1. Akutsu T (2000) Dynamic programming algorithms for RNA secondary structure prediction with pseudoknots. Discret Appl Math 104:45–62
2. Brown M, Wilson C (1996) RNA pseudoknot modeling using intersections of stochastic context free grammars with applications to database search. In: Hunter L, Klein T (eds) Proceedings of the 1st Pacific symposium on biocomputing, Big Island of Hawaii, pp 109–125
3. Condon A, Davy B, Rastegari B, Tarrant F, Zhao S (2004) Classifying RNA pseudoknotted structures. Theor Comput Sci 320:35–50
4. Dirks RM, Pierce NA (2003) A partition function algorithm for nucleic acid secondary structure including pseudoknots. J Comput Chem 24:1664–1677
5. Gluick TC, Draper DE (1994) Thermodynamics of folding a pseudoknotted mRNA fragment. J Mol Biol 241:246–262
6. Ieong S, Kao M-Y, Lam T-W, Sung W-K, Yiu S-M (2001) Predicting RNA secondary structures with arbitrary pseudoknots by maximizing the number of stacking pairs. In: Proceedings of the 2nd symposium on bioinformatics and bioengineering, Bethesda, pp 183–190
7. Lyngsø RB (2004) Complexity of pseudoknot prediction in simple models. In: Proceedings of the 31th international colloquium on automata, languages and programming (ICALP), Turku, pp 919–931
8. Lyngsø RB, Pedersen CNS (2000) RNA pseudoknot prediction in energy based models. J Comput Biol 7:409–428
9. Rastegari B, Condon A (2007) Parsing nucleic acid pseudoknotted secondary structure: algorithm and applications. J Comput Biol 14(1):16–32
10. Reeder J, Giegerich R (2004) Design, implementation and evaluation of a practical pseudoknot folding algorithm based on thermodynamics. BMC Bioinform 5:104
11. Rivas E, Eddy S (1999) A dynamic programming algorithm for RNA structure prediction including pseudoknots. J Mol Biol 285:2053–2068
12. Rødland EA (2006) Pseudoknots in RNA secondary structure: representation, enumeration, and prevalence. J Comput Biol 13:1197–1213
13. Tabaska JE, Cary RB, Gabow HN, Stormo GD (1998) An RNA folding method capable of identifying pseudoknots and base triples. Bioinformatics 14:691–699
14. Uemura Y, Hasegawa A, Kobayashi S, Yokomori T (1999) Tree adjoining grammars for RNA structure prediction. Theor Comput Sci 210:277–303
15. Witwer C, Hofacker IL, Stadler PF (2004) Prediction of consensus RNA secondary structures including pseudoknots. IEEE Trans Comput Biol Bioinform 1:66–77

Robotics

Elmar Langetepe
Department of Computer Science, University of Bonn, Bonn, Germany

Keywords

Competitive analysis; Exploration; Motion planning; Navigation; Online algorithms; Searching

Years and Authors of Summarized Original Work

1997; (Navigation) Blum, Raghavan, Schieber
1998; (Exploration) Deng, Kameda, Papadimitriou
2008; (Searching and Exploration) Fleischer, Kamphans, Klein, Langetepe, Trippen

Problem Definition

Since ancient history mankind has been fascinated by the problem of orienting itself in unknown environments. Problems like *escaping*

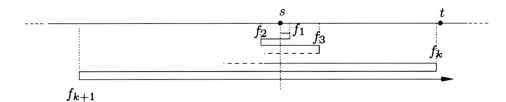

Robotics, Fig. 1 Any reasonable strategy for searching for a point on a line can be expressed as a sequence $X = f_1, f_2, f_3, \ldots$ of the search depths of the strategy

from a labyrinth or *searching for target objects* have been considered and discussed intensively. Such problems can be easily modeled in a geometric setting.

For example, let us assume that an agent is searching for an unknown door along a wall. We assume that the door is detected, if the agent exactly hits the door. Geometrically one is searching for an unknown point along a line as depicted in Fig. 1. Although the location of the point is not known and is only detected by a visit, it should be found without too much detour. This classical problem in online navigation was discussed by [2] in the early 1990s.

Since the distance and the location of the goal is not known, obviously any reasonable (deterministic) strategy should move in the two directions alternatingly and with increasing depths, f_i, until the goal is detected.

A classical result of [13] shows that it is optimal to use a search strategy that doubles the search distance in every step, i.e., $f_i = 2^i$. It can be shown that the resulting path to an arbitrary target point t is never greater than 9 times the shortest path to the target t, regardless of the position of t. There is no deterministic strategy that attains a smaller factor; see also [2].

Navigation and Exploration

We categorize three fundamental tasks: navigation, exploration, and localization. Navigation (or search) means to find a way to a predescribed location in an unknown environment as shown above. Exploration means to draw a complete map of an unknown environment or to detect or visit *all* possible targets. Localization means to determine the currently unknown position on a known and given map. In many settings, the environment is modeled geometrically as a simple polygon with or without holes. To distinguish the underlying combinatorial problems from the geometric problems, an environment may also be modeled as a graph. For an overview of online searching and exploration problems, see [25] or the more recent survey of Gal [14].

Performance Measure, Competitive Analysis

A general concept for evaluating the efficiency of an online strategy is the so-called *competitive analysis*. Formally, for a class of problems Π and any instance $P \in \Pi$, the cost, OnlAlg(P), of the online algorithm is compared to the cost, OfflOpt(P), of the optimal offline algorithm. If there are constants C and A, so that

$$\text{OnlAlg}(P) \leq C \times \text{OfflOpt}(P) + A$$

holds for any $P \in \Pi$, the online algorithm is called C *competitive*. In the case of exploration and navigation, the robot should minimize its travel distance. Therefore, the *competitive ratio* C measures the length of the detour compared to the optimal shortest tour computed under full information. An overview of efficient computations of optimal offline solutions for shortest paths problems can be found in the survey of Mitchell [23]. Many online motion planning problems were classified by the competitive analysis; see the surveys [4, 10, 25].

A randomized online algorithm against an oblivious adversary uses randomization on a fixed predetermined input (which is unknown to the online algorithm). In this case, the competitive ratio is a random variable, and it is maximized over all possible inputs. For example, an optimal randomized strategy for the introductory *point-on-a-line* problem given by Kao et al. [17] achieves an optimal competitive ratio of $4.5911\ldots$.

Different Models

The robot can be equipped with a vision system or with a local touch sensor, only. The impact of a compass is of some interest. One can consider continuous geometric settings such as (a collection of) simple polygons or a concatenation of corridors. On the other hand, the geometric environment might be given by a discrete concatenation of single cells (i.e., a grid graph environment) or is modeled by a general graph. Furthermore, we can consider a single robot or a set of k agents which are working together and exchange information to some extent. Additionally the size of the memory of the agents can be limited. Tasks for a huge set of agents with very limited abilities are related to swarm behavior which is not the topic of this overview.

Key Results

Navigation

Blum et al. [6] studied the problem of a blind robot trying to reach a goal t from a start position s (*point-to-point navigation*) in a two-dimensional scene of n non-overlapping axis-parallel rectangles of width at least one. In the *wall problem*, t is an infinite vertical line. In the *room problem*, the obstacles are within a square room with entry door s and the target t lies on the outer boundary. $O(\sqrt{n})$ competitive online algorithms have been developed. A lower bound on the competitive ratio of $\Omega(\sqrt{n})$ for the wall problem was given by [24], and for the room problem optimal $\Theta(\log n)$ competitive algorithms have been presented by [3]. For randomized strategies and point-to-point navigation, there is an $\Omega(\log\log n)$ lower bound for the model of an oblivious adversary from [18], and [5] presented randomized $O(\log n)$ competitive algorithms

for the same problem and also for the wall problem.

The introductory search problem for the line (or 2-rays) was extended to m concurrent rays where an optimal competitive ratio of $1 + 2m^m/(m-1)^{m-1}$ was shown; see [2, 13]. An optimal strategy visits the m rays alternatingly with search depth $f_i = (m/(m-1))^i$. For p agents on m rays working in parallel and exchanging information, an optimal ratio of $1 + 2(m/p - 1)(m/(m-p))^{m/p}$ can be achieved; see [21]. In a natural extension in dimension 2, the robot scans the area with a radar connected to the starting point. Gal [13] introduced this two-dimensional search problem and conjectures that a logarithmic spiral (i.e., the natural continuous extension of the doubling heuristic) gives an optimal strategy. The best logarithmic spiral attains a competitive ratio of $17.289\ldots$; finally a proof for the optimality of spiral search is given in [20].

Exploration

Deng et al. [7] introduced the online gallery route problem. We consider a simple room modeled by a simple polygon and an agent equipped with a visibility system. The task of computing the shortest roundtrip so that the agent *sees* all points in the polygon is denoted as the shortest watchman route (SWR) problem. In the case of a rectilinear simple polygon and with L_1-metric, there is an optimal (i.e., 1-competitive) online algorithm which gives a $\sqrt{2}$-approximation of the SWR for the L_2-metric in the rectilinear problem.

For general simple polygons, the problem was first solved by Icking et al. [16] with a proven competitive ratio of 26.5, whereas the greatest known lower bound is given by 1.28; see [15].

For the exploration of a geometric environment with k rectilinear obstacles, there is an $\Omega(\sqrt{k})$ lower bound on the competitive ratio for deterministic and randomized strategies; see [1].

Online graph exploration by a set of k agents means that every vertex of an unknown graph has to be visited. In some configuration, additionally all edges have to be traversed. Assume that full communication among the agents is given. Finding the optimal makespan (finishing time)

R

algorithm for k agents is an NP-hard problem even for a given tree, and there is an $O(k/\log k)$ competitive algorithm for the online exploration (edges and vertices) version; see [12]. On the other hand, a special tree construction in [9] gives an $\Omega(\log k/\log\log k)$ lower bound on the competitive ratio. For cell environments (grid graph with or without holes), optimal competitive strategies exist.

Dependency Between Searching and Exploration

From the m-concurrent rays result above, one can easily deduce that there is no constant competitive online strategy for searching a point in simple polygons with n edges where n can grow. Nevertheless in a fixed polygon, any search strategy that sees all points defines a ratio for any point and has a worst case ratio; see Fig. 2. So there have to be some optimal *search path* for any fixed polygon. We want to find a general strategy that approximates the best search path for any polygon within a constant factor, i.e., within a constant *search ratio*.

The search ratio definition was first given by Koutsoupias et al. [19]; they studied graphs with unit edge length. The result of Koutsoupias et al. is restricted to the *offline* case where the graph is completely known a priori. Only the goal remains hidden. The above concept goes beyond competitive analysis, although the definitions of the search ratio and the competitive factor are quite similar. In the competitive framework, we compare the online path from the start to the goal to the shortest s–to–t path for any possible goal. For an approximation of the optimal search ratio, we compare the online path to the best possible offline path for any goal, which – in turn – may already have a very bad competitive ratio.

The key idea for solving this problem also indicates the general dependency between *searching* and *exploration*. We make use of efficient (probably constant competitive) exploration strategies for the given environments. If they can be restricted to a bounded distance in a somewhat greater environment, we successively increase the exploration depth by a *doubling* factor.

Fleischer et al. [11] showed that it is possible to approximate an optimal search path for searching a point in a simple polygon by a factor of roughly 4 if the goal has to be visited directly. If a vision system is used, a factor of roughly 8 can be guaranteed. The result even holds when the environment is not given in advance. And the result also holds even though the optimal search path in a given simple polygon is not known.

Applications

In practice a robot can efficiently arrive at a target point (given by coordinates) in an unknown environment with obstacles by Lumelsky's BUG strategy [22]. Many such BUG-variants were developed and successfully applied, for example, in some of the mars rover expeditions. If Lumelsky's BUG algorithm is assumed to navigate between convex obstacles, in the worst case it moves at most once around every significant obstacle, which is optimal in this case. Additionally a robot with a compass can sometimes find the goal exponentially faster than a robot without a compass.

Theoretical paradigms have practical relevance in robotics; see Dudek und Jenkin [8]. The doubling heuristic is widely accepted as an approximation scheme in practice. The general concept for the approximation of the optimal search path has some influence. If somebody is searching for a target in an unknown

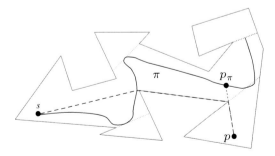

Robotics, Fig. 2 In a fixed polygon, a search path π *sees* all points and attains a ratio for any single point. At p_π the target point p is detected for the first time and defines a ratio

environment, it seems to be unavoidably that the environment has to be explored efficiently with increasing depth.

The concept also shows that there is sometimes no significant difference between a known environment with unknown targets and a fully unknown environment. Roughly speaking, if somebody is searching for a goal in unknown position, it is not important whether the corresponding environment is fully known in advance or not known at all.

Open Problems

For many settings, the precise competitive complexity of an online motion planning problems is not known. Tight lower bounds are much harder to achieve. Some examples are given below. The lower bound construction for the navigation among obstacles usually make use of arbitrary *thin* obstacles. Is it possible to get rid of such a restriction?

Exploration of a simple room with visibility: Upper bound 26.5 vs. lower bound 1.28

Exploration of graphs by k agents: Upper bound $O(k/\log k)$ vs. lower bound $\Omega(\log k/\log\log k)$

Optimality of spiral search? Upper bounds given by spiral search.

Searching for a line in the plane: Upper bound $13.81113\ldots$ vs. lower bound $\sqrt{2}\cdot 9$

Searching for a ray in the plane: Upper bound $22.513\ldots$ vs. lower bound $17.289\ldots$

Navigation among k obstacles: Does the lower bound of $\Omega(\sqrt{k})$ hold for a fixed aspect ratio of the obstacles?

Optimal search path: How to compute the optimal search path for a given polygon or graph?

Cross-References

▶ Alternative Performance Measures in Online Algorithms

▶ Deterministic Searching on the Line

▶ Metrical Task Systems

▶ Mobile Agents and Exploration

▶ Randomized Searching on Rays or the Line

Recommended Reading

1. Albers S, Kursawe K, Schuierer S (2002) Exploring unknown environments with obstacles. Algorithmica 32:123–143
2. Baeza-Yates R, Culberson J, Rawlins G (1993) Searching in the plane. Inf Comput 106:234–252
3. Bar-Eli E, Berman P, Fiat A, Yan P (1994) Online navigation in a room. J Algorithms 17:319–341
4. Berman P (1998) On-line searching and navigation. In: Fiat A, Woeginger G (eds) Competitive analysis of algorithms. Springer, London UK
5. Berman P, Blum A, Fiat A, Karloff H, Rosen A, Saks M (1996) Randomized robot navigation algorithms. In: Proceedings of the 7th ACM-SIAM symposium on discrete algorithms, Atlanta, pp 75–84
6. Blum A, Raghavan P, Schieber B (1997) Navigating in unfamiliar geometric terrain. SIAM J Comput 26(1):110–137
7. Deng X, Kameda T, Papadimitriou C (1998) How to learn an unknown environment I: the rectilinear case. J ACM 45(2):215–245
8. Dudek G, Jenkin M (2010) Computational principles of mobile robotics, 2nd edn. Cambridge University Press, New York
9. Dynia M, Łopuszański J, Schindelhauer C (2007) Why robots need maps. In: SIROCCO '07: proceedings of the 14th colloquium on structural information an communication complexity, Şirince. Lecture notes in computer science. Springer, pp 37–46
10. Fiat A, Woeginger G (eds) (1998) On-line algorithms: the state of the art. Lecture notes in computer science, vol 1442. Springer, Berlin/New York
11. Fleischer R, Kamphans T, Klein R, Langetepe E, Trippen G (2008) Competitive online approximation of the optimal search ratio. SIAM J Comput 38(3):881–898
12. Fraigniaud P, Gasieniec L, Kowalski DR, Pelc A (2006) Collective tree exploration. Networks 43(3):166–177
13. Gal S (1980) Search games. Mathematics in science and engeneering, vol 149. Academic, New York
14. Gal S, (2011) Search games. In: Wiley encyclopedia of operations research and management science, John Wiley & Sons, Inc.
15. Hagius R, Icking C, Langetepe E (2004) Lower bounds for the polygon exploration problem. In: Abstracts 20th European workshop computational geometry, Universidad de Sevilla, Sevilla, pp 135–138
16. Hoffmann F, Icking C, Klein R, Kriegel K (2001) The polygon exploration problem. SIAM J Comput 31:577–600

R

17. Kao MY, Reif JH, Tate SR (1996) Searching in an unknown environment: an optimal randomized algorithm for the cow-path problem. Inf Comput 133(1):63–79
18. Karloff H, Rabani Y, Ravid Y (1994) Lower bounds for randomized k-server and motion-planning algorithms. SIAM J Comput 23:293–312
19. Koutsoupias E, Papadimitriou CH, Yannakakis M (1996) Searching a fixed graph. In: Proceedings of the 23th international colloquium on automata language, and programming, Paderborn. Lecture notes in computer science, vol 1099, Springer, pp 280–289
20. Langetepe E (2010) On the optimality of spiral search. In: SODA 2010: proceedings of the 21st annual ACM-SIAM symposium on discrete algorithms, Austin, pp 1–12
21. López-Ortiz A, Schuierer S (2002) Online parallel heuristics and robot searching under the competitive framework. In: Proceedings of the 8th Scandinavian workshop on algorithm theory, Turku. Lecture notes in computer science, vol 2368. Springer, pp 260–269
22. Lumelsky VJ, Stepanov AA (1987) Path-planning strategies for a point mobile automaton moving amidst unknown obstacles of arbitrary shape. Algorithmica 2:403–430
23. Mitchell JSB (2000) Geometric shortest paths and network optimization. In: Sack JR, Urrutia J (eds) Handbook of computational geometry. Elsevier Science/North-Holland, Amsterdam, pp 633–701
24. Papadimitriou CH, Yannakakis M (1991) Shortest paths without a map. Theor Comput Sci 84(1):127–150
25. Rao NSV, Kareti S, Shi W, Iyengar SS (1993) Robot navigation in unknown terrains: introductory survey of non-heuristic algorithms. Technical report, ORNL/TM-12410, Oak Ridge National Laboratory

Robust Bin Packing

Kim-Manuel Klein
University Kiel, Kiel, Germany

Keywords

Approximation; Bin packing; Competitive ratio; Migration factor; Online; Robust

Years and Authors of Summarized Original Work

2009; Epstein, Levin
2013; Jansen, Klein

Problem Definition

Consider the classical online bin packing problem, where items of sizes in $(0, 1]$ arrive over time. At the arrival of each item, it has to be assigned to a bin of capacity 1 such that the total size of all items in the bin does not exceed its capacity. The objective is to minimize the number of used bins.

Online bin packing was introduced by Ullman [10] and has seen enormous research since then (see the survey of Seiden [9] for an overview). The quality of an online algorithm is typically measured by the asymptotic performance guarantee of the algorithm divided by the optimal offline solution and is called the (asymptotic) *competitive ratio*. In the case of online bin packing, the best known algorithm has an asymptotic competitive ratio of 1.58889 (see [9]). On the other hand, it was shown that no algorithm can achieve a ratio better than 1.54037 (see [1]).

To obtain algorithms with improved competitive ratio for online bin packing, one can allow to rearrange already packed items as soon as a new item arrives. The notion of robustness allows to repack a set of already packed items with limited total size whenever a new item arrives. On the one hand, we want to guarantee that we use as few bins as possible, and on the other hand, when a new item arrives, we want to minimize the total size of repacked items.

A modern way to measure the repacking costs is the notion of the *migration factor*, developed by Sanders, Sivadasan, and Skutella [8]. It is defined by the total size of all moved items divided by the size of the arriving item. Following the notation of Sanders et al., an online algorithm with (asymptotic) approximation ratio $1 + \epsilon$ is called *robust* if its migration factor is of the size $f(\frac{1}{\epsilon})$, where f is an arbitrary function that only depends on $\frac{1}{\epsilon}$.

Key Results

In the case of robust bin packing, Epstein and Levin [3] proved that the asymptotic competitive ratio of the robust bin packing problem can be arbitrarily close to the optimum. They developed an asymptotic PTAS for the problem using a migration factor of $2^{\mathcal{O}(\frac{1}{\epsilon^2}\log\frac{1}{\epsilon})}$. They also proved that there is no online algorithm for this problem which has a constant migration factor and that maintains an optimal solution. The asymptotic PTAS by Epstein and Levin was later improved by Jansen and Klein [7], who developed an asymptotic FPTAS for the problem with a migration factor of $\mathcal{O}(\frac{1}{\epsilon^4})$.

Techniques

Most robust algorithms rely on a sensitivity result for integer linear programs (ILPs) by Cook et al. [2]. It was first used by Sanders et al. [8] to develop a robust PTAS for the scheduling problem on identical machines with the objective value of minimizing the makespan. The theorem of Cook et al. roughly states that for every optimal integral solution y' of the ILP $\min\{cy \mid Ax \le b'\}$, there exists an optimal integral solution y'' of the ILP $\min\{cy \mid Ax \le b''\}$ with changed right-hand side b'' such that the distance between y' and y'' can be bounded by $\|y'' - y'\|_\infty \le n\Delta\,(\|b'' - b'\|_\infty + 2)$, where n is the number of variables and Δ is the absolute value of the largest subdeterminant of A.

The major contribution by Epstein and Levin for the robust bin packing problem was to develop a dynamic rounding technique. Based on a classical rounding by Fernandez de La Vega and Lueker [5], the dynamic rounding techniques present a way on how item sizes can be rounded in a setting where new items arrive over time. This allows to formulate an ILP of fixed dimension. As a new item arrives online, the formulated ILP changes accordingly. The changed ILP has additional columns and the right-hand side of the ILP is increased. Using the theorem of Cooks et al. [2] allows then to find a solution for the changed ILP that is close to the existing solution. This way a new packing is constructed for the bin packing instance containing the newly arrived item.

Since the number of variables n and the largest subdeterminant Δ in the ILP formulation can only be bounded by an exponential term in $\frac{1}{\epsilon}$,

the use of Cooks et al. theorem leads to an exponential migration factor. Jansen and Klein developed new LP and ILP techniques which are based on approximate solutions of the corresponding LP. Their central idea is to show that for any approximate solution x' with objective value $\|x'\|_1 \le (1 + \delta)\text{LIN}$, there is an approximate solution x'' with improved objective value $\|x'\|_1 \le (1 + \delta)\text{LIN} - 1$ such that $\|y'' - y'\|_1 = \mathcal{O}(\frac{1}{\delta})$. Based on this observation, they can avoid the use of Cooks et al. theorem to obtain an asymptotic PTAS for the bin packing problem with polynomial migration.

Open Problems

- There is the obvious open question on how much the migration factor of $\mathcal{O}(\frac{1}{\epsilon^4})$ from [7] can be improved and whether there are lower bounds for the migration factor. The existence of a robust algorithm with constant or sublinear migration is still open.
- Is there a robust approximation scheme for the case when items not only arrive but also depart? In the literature this problem is called fully dynamic bin packing and was considered by Ivković and Lloyd [6]. They developed an algorithm which achieves an asymptotic competitive ratio of $\frac{5}{4}$ using amortized $\mathcal{O}(\log n)$ shifting moves, where n is the number of packed items. A shifting move repacks one large item or a bundle of small items of bounded size.
- Epstein and Levin developed a robust asymptotic PTAS for the generalized bin packing problem, where d-dimensional cubes have to be packed into unit-sized cubes [4]. It would be interesting to find other robust approximation schemes for other packing problems like online strip packing or online bin packing with bins of different capacities.

Cross-References

▶ Approximation Schemes for Bin Packing
▶ Robust Scheduling Algorithms

Recommended Reading

1. Balogh J, Békési J, Galambos G (2010) New lower bounds for certain classes of bin packing algorithms. In: Workshop on approximation and online algorithms (WAOA), Liverpool. LNCS, vol 6534, pp 25–36
2. Cook W, Gerards A, Schrijver A, Tardos E (1986) Sensitivity theorems in integer linear programming. Math Program 34(3):251–264
3. Epstein L, Levin A (2009) A robust APTAS for the classical bin packing problem. Math Program 119(1):33–49
4. Epstein L, Levin A (2013) Robust approximation schemes for cube packing. SIAM J Optim 23(2):1310–1343
5. Fernandez de la Vega W, Lueker G (1981) Bin packing can be solved within $1 + \epsilon$ in linear time. Combinatorica 1(4):349–355
6. Ivković Z, Lloyd E (1998) Fully dynamic algorithms for bin packing: being (mostly) myopic helps. SIAM J Comput 28(2):574–611
7. Jansen K, Klein K (2013) A robust AFPTAS for online bin packing with polynomial migration. In: International colloquium on automata, languages, and programming (ICALP), Riga, pp 589–600
8. Sanders P, Sivadasan N, Skutella M (2009) Online scheduling with bounded migration. Math Oper Res 34(2):481–498
9. Seiden S (2002) On the online bin packing problem. J ACM 49(5):640–671
10. Ullman J (1971) The performance of a memory allocation algorithm. Technical report, Princeton University

Robust Geometric Computation

Chee K. Yap and Vikram Sharma
Department of Computer Science, New York
University, New York, NY, USA

Keywords

Exact geometric computation Floating-point filter; Dynamic and static filters; Topological consistency

Years and Authors of Summarized Original Work

2004; Li, Yap

Problem Definition

Algorithms in computational geometry are usually designed under the Real RAM model. In implementing these algorithms, however, fixed-precision arithmetic is used in place of exact arithmetic. This substitution introduces numerical errors in the computations that may lead to nonrobust behavior in the implementation, such as infinite loops or segmentation faults.

There are various approaches in the the literature addressing the problem of nonrobustness in geometric computations; see [9] for a survey. These approaches can be classified along two lines: the **arithmetic approach** and the **geometric approach**.

The arithmetic approach tries to address nonrobustness in geometric algorithms by handling the numerical errors arising because of fixed-precision arithmetic; this can be done, for instance, by using multi-precision arithmetic [6], or by using rational arithmetic whenever possible. In general, all the arithmetic operations, including exact comparison, can be performed on algebraic quantities. The drawback of such a general approach is its inefficiency.

The geometric approaches guarantee that certain geometric properties are maintained by the algorithm. For example, if the Voronoi diagram of a planar point set is being computed then it is desirable to ensure that the output is a planar graph as well. Other geometric approaches are finite resolution geometry [7], approximate predicates and fat geometry [8], consistency and topological approaches [4], and topology oriented approach [13]. The common drawback of these approaches is that they are problem or algorithm specific.

In the past decade, a general approach called the **Exact Geometric Computation** (EGC) [15] has become very successful in handling the issue of nonrobustness in geometric computations; strictly speaking, this approach is subsumed in the arithmetic approaches. To understand the EGC approach, it helps to understand the two parts common to all geometric computations: a *combinatorial structure* characterizing the discrete relations between geometric objects, e.g., whether a point is on a hyperplane or not; and a *numerical part* that consists of the numerical representation of the geometric objects, e.g., the coordinates of a point expressed as rational or floating-point numbers. Geometric algorithms characterize the combinatorial structure by numerically computing the discrete relations (that are embodied in geometric predicates) between geometric objects. Nonrobustness arises when numerical errors in the computations yield an incorrect characterization. The EGC approach ensures that all the geometric predicates are evaluated correctly thereby ensuring the correctness of the computed combinatorial structure and hence the robustness of the algorithm.

Notation

An **expression** E refers to a syntactic object constructed from a given set of operators over the reals $\mathbb{R}$. For example, the set of expressions on the set of operators $\{\mathbb{Z}, +, -, \times, \sqrt{}\}$ is the set of division-free radical expressions on the integers; more concretely, expressions can be viewed as directed acyclic graphs (DAG) where the internal nodes are operators with arity at least one, and the leaves are constants, i.e., operators with arity zero. The value of an expression is naturally defined using induction; note that the value may be undefined. Let E represent both the value of the expression and the expression itself.

Key Results

Following are the key results that have led to the feasibility and success of the EGC approach.

Constructive Zero Bounds

The possibility of EGC approach hinges on the computability of the sign of an expression. For determining the sign of algebraic expressions EGC libraries currently use a numerical approach based upon zero bounds. A **zero bound** $b > 0$ for an expression E is such that absolute value $|E|$ of E is greater than b if the value of E is valid and nonzero. To determine the sign of the expression E, compute an approximation $\tilde{E}$ to E such that

$|\tilde{E} - E| < \frac{b}{2}$ if E is valid, otherwise $\tilde{E}$ is also invalid. Then sign of E is the same as the sign of $\tilde{E}$ if $|\tilde{E}| \geq \frac{b}{2}$, otherwise it is zero. A **constructive zero bound** is an effectively computable function B from the set of expressions to real numbers $\mathbb{R}$ such that $B(E)$ is a zero bound for any expression E. For examples of constructive zero bounds, see [2, 11].

Approximate Expression Evaluation

Another crucial feature in developing the EGC approach is developing algorithms for approximate expression evaluation, i.e., given an expression E and a relative or absolute precision p, compute an approximation to the value of the expression within precision p. The main computational paradigm for such algorithms is the **precision-driven approach** [15]. Intuitively, this is a downward-upward process on the input expression DAG; propagate precision values down to the leaves in the downward direction; at the leaves of the DAG, assume the ability to approximate the value associated with the leaf to any desired precision; finally, propagate the approximations in the upward direction towards the root. Ouchi [10] has given detailed algorithms for the propagation of "composite precision", a generalization of relative and absolute precision.

Numerical Filters

Implementing approximate expression evaluation requires multi-precision arithmetic. But efficiency can be gained by exploiting machine floating-point arithmetic, which is fast and optimized on current hardware. The basic idea is to to check the output of machine evaluation of predicates, and fallback on multi-precision methods if the check fails. These checks are called numerical filters; they certify certain properties of computed numerical values, such as their sign. There are two main classifications of numerical filters: *static filters* are those that can be mostly computed at compile time, but they yield overly pessimistic error bounds and thus are less effective; *dynamic filters* are implemented during run time and even though they have higher costs they are much more effective than static filters, i.e., have better estimate on error bounds. See Fortune and van Wyk [5].

Applications

The EGC approach has led to the development of libraries, such as LEDA Real and CORE, that provide EGC number types, i.e., a class of expressions whose signs are guaranteed. CGAL, another major EGC Library that provides robust implementation of algorithms in computational geometry, offers various specialized EGC number types, but for general algebraic numbers it can also use LEDA Real or CORE.

Open Problems

1. An important challenge from the perspective of efficiency for EGC approach is high degree algebraic computation, such as those found in Computer Aided Design. These issues are beginning to be addressed, for instance [1].
2. The *fundamental problem of EGC* is the **zero problem**: given any set of real algebraic operators, decide whether any expression over this set is zero or not. The main focus here is on the decidability of the zero problem for non-algebraic expressions. The importance of this problem has been highlighted by Richardson [12]; recently some progress has been made for special non-algebraic problems [3].
3. When algorithms in EGC approach are embedded in larger application systems (such as mesh generation systems), the output of one algorithm needs to be cascaded as input to another; the output of such algorithms may be in high precision, so it is desirable to reduce the precision in the cascade. The geometric version of this problem is called the **geometric rounding problem**: given a consistent geometric object in high precision, "round" it to a consistent geometric object at a lower precision.
4. Recently a computational model for the EGC approach has been proposed [14]. The corresponding complexity model needs to be developed. Standard complexity analysis

based on input size is inadequate for evaluating the complexity of real computation; the complexity should be expressed in terms of the output precision.

URL to Code

1 `Core Library`: http://www.cs.nyu.edu/ exact
2 `LEDA`: http://www.mpi-sb.mpg.de/LEDA
3 `CGAL`: http://www.cgal.org

Recommended Reading

1. Berberich E, Eigenwillig A, Hemmer M, Hert S, Schmer KM, Schmer E (2002) A computational basis for conic arcs and boolean operations on conic polygons. In: 10th European symposium on algorithms (ESA'02). Lecture notes in computer science, vol 2461, pp 174–186
2. Burnikel C, Funke S, Mehlhorn K, Schirra S, Schmitt S (2001) A separation bound for real algebraic expressions. In: Lecture notes in computer science, vol 2161. Springer, pp 254–265
3. Chang EC, Choi SW, Kwon D, Park H, Yap C (2006) Shortest paths for disc obstacles is computable. Int J Comput Geom Appl 16(5–6):567–590. In: Gao XS, Michelucci D (eds) Special issue on geometric constraints. Also appeared in Proceedings of the 21st ACM symposium on computational geometry, pp 116–125 (2005)
4. Fortune SJ (1989) Stable maintenance of point-set triangulations in two dimensions. IEEE Found Comput Sci 30:494–499
5. Fortune SJ, van Wyk CJ (1993) Efficient exact arithmetic for computational geometry. In: Proceedings of the 9th ACM symposium on computational geometry, pp 163–172
6. Gowland P, Lester D (2000) A survey of exact arithmetic implementations. In: Blank J, Brattka V, Hertling P (eds) Computability and complexity in analysis. 4th International workshop, CCA 2000, selected papers. Lecture notes in computer science, No. 2064. Springer, Swansea, 17–19 Sept 2000, pp 30–47
7. Greene DH, Yao FF (1986) Finite-resolution computational geometry. IEEE Found Comput Sci 27: 143–152
8. Guibas L, Salesin D, Stolfi J (1989) Epsilon geometry: building robust algorithms from imprecise computations. ACM Symp Comput Geom 5:208–217
9. Li C, Pion S, Yap CK (2004) Recent progress in exact geometric computation. J Log Algebra Program 64(1):85–111

10. Ouchi K (1997) Real/expr: implementation of an exact computation package. Master's thesis, Department of Computer Science, Courant Institute, New York University. Jan 1997. http://cs.nyu.edu/exact/doc/
11. Pion S, Yap C (2002) Constructive root bound method for k-ary rational input numbers, Sept 2002. Extended abstract. Submitted (2003) ACM Symposium on Computational Geometry
12. Richardson D (1997) How to recognize zero. J Symb Comput 24:627–645
13. Sugihara K, Iri M, Inagaki H, Imai T (2000) Topology-oriented implementation – an approach to robust geometric algorithms. Algorithmica 27:5–20
14. Yap CK (2006) Theory of real computation according to EGC. To appear in LNCS Volume based on talks at a Dagstuhl Seminar "Reliable implementation of real number algorithms: theory and practice", Jan 8–13 2006
15. Yap CK, Dubé T (1995) The exact computation paradigm. In: Du DZ, Hwang FK (eds) Computing in euclidean geometry, 2nd edn. World Scientific Press, Singapore, pp 452–492

Robust Scheduling Algorithms

José Verschae
Departamento de Matemáticas and
Departamento de Ingeniería Industrial y de
Sistemas, Pontificia Universidad Católica de
Chile, Santiago, Chile

Keywords

Load balancing; Makespan minimization; On-line scheduling; Recourse; Robustness

Years and Authors of Summarized Original Work

2004, 2009; Sanders, Sivadasan, Skutella

Problem Definition

In the classic online scheduling model, jobs arrive one after another. At the arrival of a new job, the scheduler must immediately and irrevocably assign it to a machine. In the *parallel machine* case, we have m identical machines to process the jobs. Each job j has a processing time p_j that is revealed at the moment of its appearance. The load of a machine is the sum of processing times of jobs assigned to it. The objective is to minimize the makespan, that is, the maximum machine load.

The fact that decisions are irrevocable imposes a hard constraint on the scheduler. However, many applications allow some amount of flexibility. *Robust scheduling* algorithms take this flexibility into account: whenever a job arrives, some reassignment of jobs can be performed. More precisely, given a parameter $\beta > 0$, the arrival of job j allows to migrate a set of jobs with a total processing time of at most $\beta \cdot p_j$. The factor β is called the *migration factor* of the algorithm and it is a measure of its robustness. In this context, the quality of solutions is assessed by competitive analysis: an algorithm is α-competitive if for any sequence of job arrivals the makespan of the algorithm is at most α times the (offline) optimum cost for the set of available jobs. An important goal in this area is to understand the trade-off between the migration and competitive factors.

Key Results

Greedy Approaches
In a setting where no migration is allowed, i.e., if $\beta = 0$, a competitive ratio of $2 - \frac{1}{m}$ is achievable by a greedy *list-scheduling* algorithm [7]. Although more sophisticated algorithms have smaller competitive ratios (see, e.g., [6]), no algorithm can achieve a performance guarantee smaller than $e/(e-1) \approx 1.58$ [1, 11], even if randomization is allowed.

Sanders et al. [10] give algorithms with improved competitive ratios for small values of β. A simplified version of their most basic algorithm is as follows. Let j be an arriving job and denote by OPT the minimum makespan of the instance including j. The algorithm works as follows.

1. If $p_j \leq \text{OPT}/2$, assign job j to the machine with the smallest load.
2. Otherwise, consider a machine i in which all jobs are of size at most OPT/2. Greedily remove jobs from i until their total processing

time is at least p_j. Add job j to i and greedily reassign each removed job to the least loaded machine.

The existence of the claimed machine in Step (2) follows since there can be at most m jobs of size larger than OPT/2 in the instance. The proof that the algorithm is 3/2-competitive is a simple exercise that follows since each greedily assigned job has size at most OPT/2. By construction the algorithm has migration factor 2. The fact that the algorithm needs the value OPT as input can be avoided by trying out a handful of different solutions.

With this simple approach the competitive guarantee is already below the lower bound of 1.58 for $\beta = 0$. Sanders et al. [10] shows that a refinement of this algorithm gives the same competitive guarantee and reduces β to 4/3. They also provide more sophisticated algorithms with smaller competitive factors, for example, a 4/3-competitive algorithm with migration factor 5/2.

Robust Approximation Schemes

The algorithms above show that already small migration factors can help to significantly improve the quality of solutions. However, they tell little about the trade-off between the competitive and migration ratios. Sanders et al. [10] study this trade-off by giving a *robust polynomial time approximation scheme* (robust PTAS), that is, a family of algorithms $\{A_\varepsilon\}_{\varepsilon>0}$ such that for any constant $\varepsilon > 0$ the algorithm A_ε is $(1 + \varepsilon)$-competitive and uses a migration factor of $\beta(\varepsilon)$. We remark that β is a constant that depends only on ε and not on the specific input data.

The robust PTAS borrows ideas from the known PTAS for the offline problem [8]. At the arrival of a job j, the algorithm takes the given $(1 + \varepsilon)$-approximate solution and updates it to a schedule with the same approximation guarantee. The algorithm behaves differently depending on the size of j. If p_j is in $O(\varepsilon\text{OPT})$, where OPT denotes the optimal makespan for the current instance including j, then we can safely assign this job to the least loaded machine and maintain the approximation guarantee.

Otherwise, the processing times are rounded to simplify the instance and add symmetry to the solution. The corresponding offline minimum makespan problem for this instance can be posed as an integer program (IP) of the form $\min\{c^t x : A \cdot x = b, x \in \mathbb{N}^d\}$. A component x_C of x corresponds to the number of machines with a given *configuration* C, where each configuration is a compact description of a one machine schedule. Crucially, the number of different configurations, that is, the number of variables d in the IP, is a constant $2^{\text{poly}(1/\varepsilon)}$. Moreover, the complete instance is encoded in the right-hand side b. After a new job arrives, the corresponding IP can be updated by increasing one coordinate of b by one, obtaining a new vector b'. A sensitivity analysis result by Cook et al. [2] implies that for any optimal solution x of the IP, there exists an optimal solution x' with the right-hand side changed to b' such that $||x - x'||_1 \le d^2 \cdot \Delta(A) \cdot (||b - b'||_\infty + 2)$. Here $\Delta(A)$ is the maximum $|\det(B)|$ over all square submatrices B of A, which in this case can be bounded by $2^{\text{poly}(1/\varepsilon)}$. Therefore, the number of machines that need to be modified in order to go from schedule x to x' is $||x - x'||_1 \le 2^{\text{poly}(1/\varepsilon)}$. Since we are assuming that the new job has processing time in $\Omega(\varepsilon\text{OPT})$, and each machine has a load of $O(\text{OPT})$, we obtain an algorithm with migration factor $2^{\text{poly}(1/\varepsilon)}$.

Theorem 1 (Sanders et al. [10]) *The problem of minimum makespan on identical machines admits a robust PTAS with migration factor $\beta = 2^{\text{poly}(1/\varepsilon)}$.*

Applications

The basic technique for constructing a robust PTAS has been adapted to different related problem. Most results are based on the sensitivity analysis result mentioned above, but differ in other parts of the algorithm and analysis. In particular robust PTASs have been developed for bin packing [3] and cube packing [4]. Other objective functions for identical machine scheduling have also been considered, for example, minimizing the ℓ_p-norm of the vector of loads or maximizing the minimum machine load. These problems do not admit robust PTASs; however, it is possible to design such algorithms for an amortized analogue of the migration factor [12]. Epstein and Levin [5] consider preemptive scheduling problems on parallel machines, obtaining a 1-competitive algorithm with migration factor $1 - 1/m$ for the minimum makespan and minimum ℓ_p-norm objectives. As opposed to the previous results, this algorithm does not rely on sensitivity analysis results for IPs.

The bin packing problem was considered by Jansen and Klein [9]. They improve the result in [3] to a robust PTAS with $\beta = \text{poly}(1/\varepsilon)$. To obtain this migration factor, they develop new sensitivity analysis results aimed specifically at approximation algorithms.

Open Problems

An interesting question is to determine the precise trade-off between the competitive and migration factors for the minimum makespan problem on identical machines. In particular, determine if the migration can be made to depend polynomially on $1/\varepsilon$ for a $(1 + \varepsilon)$-competitive algorithm. Another natural question is to extend these results to *related* machines. In this setting, each machine i runs at a speed s_i, and thus the time it takes to process job j on machine i is p_j/s_i.

Other natural objective functions on parallel machine scheduling are not fully understood. The machine covering version, where we seek to maximize the minimum machine load, does not admit a robust PTAS. More specifically, a competitive ratio smaller than $20/19$ is not possible with constant migration factor [12]. It is open if this competitive ratio is indeed achievable or if the lower bound can be improved. A similar situation holds for minimizing the ℓ_p-norm for any $p > 1$ [4].

Cross-References

▶ Approximation Schemes for Bin Packing
▶ Bin Packing
▶ List Scheduling
▶ Online Load Balancing of Temporary Tasks

Recommended Reading

1. Chen B, van Vliet A, Woeginger GJ (1994) A lower bound for randomized on-line scheduling algorithms. Inf Process Lett 51:219–222
2. Cook W, Gerards AMH, Schrijver A, Tardos É (1986) Sensitivity theorems in integer linear programming. Math Program 34:251–264
3. Epstein L, Levin A (2009) A robust APTAS for the classical bin packing problem. Math Program 119:33–49
4. Epstein L, Levin A (2013) Robust approximation schemes for cube packing. SIAM J Optim 23:1310–1343
5. Epstein L, Levin A (2014) Robust algorithms for preemptive scheduling. Algorithmica 69:26–57
6. Fleischer R, Wahl M (2000) On-line scheduling revisited. J Sched 3:343–353
7. Graham RL (1966) Bounds for certain multiprocessing anomalies. Bell Syst Tech J 45:1563–1581
8. Hochbaum DS, Shmoys DB (1987) Using dual approximation algorithms for scheduling problems: theoretical and practical results. J ACM 34:144–162
9. Jansen K, Klein KM (2013) A robust AFPTAS for online bin packing with polynomial migration. In: Fomin FV, Freivalds R, Kwiatkowska M, Peleg D (eds) Automata, languages, and programming (ICALP 2013). Lecture notes in computer science, vol 7965. Springer, Berlin/Heidelberg, Riga, Latvia pp 589–600
10. Sanders P, Sivadasan N, Skutella M (2009) Online scheduling with bounded migration. Math Oper Res 34:481–498
11. Sgall J (1997) A lower bound for randomized on-line multiprocessor scheduling. Inf Process Lett 63:51–55
12. Skutella M, Verschae J (2010) A robust PTAS for machine covering and packing. In: Berg M, Meyer U (eds) Algorithms – ESA 2010. Lecture notes in computer science, vol 6346. Springer, Berlin/Heidelberg, Liverpool, UK pp 36–47

R

Robustness in Self-Assembly

Ho-Lin Chen
Department of Electrical Engineering, National Taiwan University, Taipei, Taiwan

Keywords

Error correction; Proofreading

Years and Authors of Summarized Original Work

2004; Winfree, Bekbolatov
2004; Chen, Goel

Robustness in Self-Assembly

The abstract tile assembly model (aTAM), originally proposed by Winfree [1], provides a useful framework to study algorithmic tile self-assembly. As described in other sections, many theoretical studies have shown the efficiency and computational power of aTAM.

The aTAM, although widely accepted and experimentally verified, is an overly simplified combinatorial model in describing the self-assembly of DNA tiles. In reality, several effects are observed which lead to a loss of robustness compared to the aTAM. The assembly tends to be reversible, i.e., tiles can fall off from an existing assembly, even when the total binding strength exceeds the temperature threshold τ. Also, tiles sometimes attach with a weak strength but then quickly get incorporated and locked into a growing assembly, much like defects in a crystal. However, for sophisticated combinatorial assemblies like counters, which form the basis for controlling the size of a structure, a single error can lead to assemblies drastically larger or smaller (or different in other ways) than the intended structure. An error rate of 0.5–10 % is observed in previous experimental studies.

KTAM

A more sophisticated and accurate stochastic model called the kinetic tile assembly model (kTAM) was introduced by Winfree [1]. The kTAM calculates rates for various types of attachments and removals based on thermodynamic constants. It has the following assumptions:

1. Tile concentrations are held constant throughout the self-assembly process.
2. Supertiles do not interact with each other. The only two reactions allowed are addition of a tile to a supertile and the dissociation of a tile from a supertile.
3. The forward rate constants for all tiles only depend on concentrations.
4. The reverse rate depends exponentially on the number of base-pair bonds which must be broken, and the mismatched sticky ends make no base-pair bonds.

There are two free parameters in this model, both of which are dimensionless free energies: $G_{mc} > 0$ measures the entropic cost of putting a tile at a binding site and depends on the tile concentration, and $G_{se} > 0$ measures the free energy cost of breaking a single strength-1 bond. Under this model, we can approximate the forward and reverse rates for each of the tile-supertile reactions in the process of self-assembly of DNA tiles as follows:

The rate of addition of a tile to a supertile, f, is $pe^{-G_{mc}}$. The rate of dissociation of a tile from a supertile, r_b, is $pe^{-bG_{se}}$, where b is the strength with which the tile is attached to the supertiles. The parameter p simply gives us the time scale for the self-assembly. Winfree showed that by setting appropriate tile concentrations and binding strengths such that $G_{mc} = 2G_{sc} - \epsilon$, the behavior predicted by kTAM approaches the behavior described by aTAM as $\epsilon \to 0$. However, the growth speed also goes to 0 (attachment and dissociation form an unbiased random walk) as $\epsilon \to 0$.

Problem Definition

Self-assembly processes in nature are often equipped with explicit mechanisms for both error prevention and error correction. For artificial self-assembly, these problems are even more important since we are interested in assembling large systems with great precision. Previously, a phenomenon called *insufficient attachments* has been identified to be a main source of error [3]. An insufficient attachment is the process in which a tile t first attach with total strength less than the temperature. However, before t falls off, adjacent tiles attach and secure t in place. An insufficient

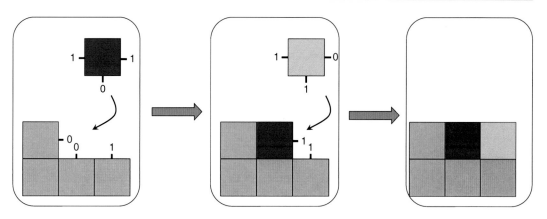

Robustness in Self-Assembly, Fig. 1 An example of growth error caused by an insufficient attachment. The *red tile* first attaches with a weaker strength. Then the *yellow tile* attaches and secures the *red tile* in place

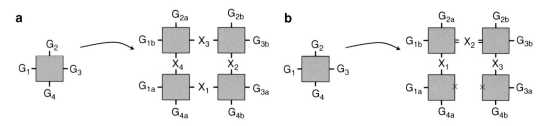

Robustness in Self-Assembly, Fig. 2 (**a**) An example of a 2 × 2 proofreading block. (**b**) An example of a 2 × 2 snaked proofreading block

attachment happening at a location at which another correct tile can attach via cooperative bindings is called a *growth error*. One example of growth error caused by insufficient attachment is illustrated in Fig. 1. An insufficient attachment happening at a location at which no tiles can attach according to aTAM is called a *facet error*. The rate at which any specific insufficient attachment happens is $e^{-2G_{mc}}/(e^{-G_{mc}} + e^{-G_{se}})$, which is roughly $e^{-G_{se}}$ times the rate of tile attachments when $G_{mc} \approx 2G_{sc}$. The main goal of this section is to introduce error-correction systems that deal with insufficient attachments.

Key Results

Proofreading Tilesets
The first error-correction scheme for the tile assembly model was the proofreading scheme proposed by Winfree and Bekbolatov [2]. The proofreading scheme turns any tile system with unidirectional growth into a new system that produces

the same pattern (with scaling). The scheme replaces each tile type t by k^2 distinct tile types. These k^2 tile types are designed to form a $k \times k$ block. All internal glues have strength 1 and are unique to this block. The glues on the boundary of the block are duplicates of glues on the original tile t. A 2 × 2 block is illustrated in Fig. 2a. This construction enforces that whenever a growth error happens, at least k growth errors must happen locally in order for the assembly process to proceed. Thus when growth error happens, the erroneous tiles are more likely to detach than to stay and wait for another $k - 1$ growth errors to happen. Since the probability that a growth error happens at any given location is roughly $e^{-G_{se}}$, the probability that a growth error happens at any proofreading block and proceeds to produce the final incorrect assembly is $O(e^{-kG_{se}})$.

Snaked Proofreading Tilesets
The abovementioned proofreading system only handles growth errors but not facet errors, which

are a major source of error both in theoretical analysis and computer simulations [3]. Chen and Goel [3] proposed the snaked proofreading scheme which handles all errors caused by insufficient attachments. Similar to the proofreading scheme, the snaked proofreading scheme replaces each tile type by a block of $2k \times 2k$ tile types. The only difference is that some internal glues have strength 0 or 2 instead of 1. A 2×2 block is illustrated in Fig. 2b. Under correct growth, the snaked proofreading block checks its two input sides alternatively. Therefore, k insufficient attachments must happen before an erroneous block "thinks" it attaches correctly and propagate toward any single direction. As a result, the snaked proofreading scheme with block size $2k \times 2k$ ensures that without k insufficient attachments happening locally, all erroneous structures have $O(k^2)$ tiles and are expected to fall off in time polynomial in k. Assuming that the thermodynamic parameters G_{mc} and G_{se} can be set arbitrarily, the following theorem characterizes the performance of the snaked proofreading system.

Theorem 1 *With a $2k \times 2k$ snaked tile system, $k = O(\log n)$, an $n \times n$ square of blocks can be assembled in expected time $\tilde{O}(n)$ and with high probability, while remaining stable for $\tilde{\Omega}(n)$ time after being assembled.*

One main drawback for the proofreading and the snaked proofreading scheme is the resolution loss. Since each tile in the original system is replaced by a $k \times k$ block, the size of the original pattern is increased by a factor of k. Chen, Goel, and Luhrs [4] showed that if the third dimension can be used, two-dimensional tile systems can be proofread with no resolution loss. Their system replaces each tile by a column in the third dimension and thus maintains the original scale on the two-dimensional plane.

Recommended Reading

1. Winfree E (1998) Algorithmic self-assembly of DNA. PhD thesis, California Institute of Technology
2. Winfree E, Bekbolatov R (2004) Proofreading tile sets: error correction for algorithmic self-assembly. In: DNA computers 9. LNCS, vol 2943. Springer, Berlin/Heidelberg, pp 126–144
3. Chen HL, Goel A (2004) Error free self-assembly using error prone tiles. In: Tenth international meeting on DNA computing, Milano
4. Chen HL, Goel A, Luhrs C (2008) Dimension augmentation and combinatorial criteria for efficient error-resistant DNA self-assembly. In: ACM-SIAM symposium on discrete algorithms, San Francisco

Routing

József Békési and Gábor Galambos
Department of Computer Science, Juhász Gyula
Teachers Training College, Szeged, Hungary

Keywords

Network flows; Oblivious routing; Routing algorithms

Years and Authors of Summarized Original Work

2003; Azar, Cohen, Fiat, Kaplan, Räcke

Problem Definition

One of the most often used techniques in modern computer networks is routing. *Routing* means selecting paths in a network along which to send data. Demands usually randomly appear on the nodes of a network, and routing algorithms should be able to send data to their destination. The transfer is done through intermediate nodes, using the connecting links, based on the topology of the network. The user waits for the network to guarantee that it has the required capacity during data transfer, meaning that the network behaves like its nodes would be connected directly by a physical line. Such service is usually called the *permanent virtual circuit (PVC)* service. To model real life situations, assume that demands arrive *on line*, given by source and destination points, and capacity (bandwidth) requirements.

Similar routing problems may occur in other environments, for example in parallel computation. In this case there are several processors connected together by wires. During an operation some data appear at given processors which should be sent to specific destinations. Thus, this also defines a routing problem. However, this paper mainly considers the network model, not the parallel computer one.

For any given situation there are several routing possibilities. A natural question is to ask which is the best possible algorithm. To find the best algorithm one must define an objective function, which expresses the effectiveness of the algorithm. For example, the aim may be to reduce the load of the network. Load can be measured in different ways, but to measure the utilization percent of the nodes or the links of the network is the most natural. In the online setting, it is interesting to compare the behavior of a routing algorithm designed for a specific instance to the best possible routing.

There are two fundamental approaches towards routing algorithms. The first approach is to route *adaptively*, i.e., depending on the actual loads of the nodes or the links. The second approach is to route *obviously*, without using any information about the current state of the network. Here the authors survey only results on oblivious routing algorithms.

Notations and Definitions

A mathematical model of the network routing problem is now presented.

Let $G(V, E, c)$ be a capacitated network, where V is the set of nodes and E is the set of edges with a capacity function $c : E \to R^+$. Let $|V| = n, |E| = m$. It can be assumed that G is directed, because if G is undirected then for each undirected edge $e = (u, v)$ two new nodes x, y and four new directed edges $e_1 = (u, x), e_2 = (v, x), e_3 = (y, u), e_4 = (y, u)$ with infinite capacity may be added to the graph. If e is considered as an undirected edge with the same capacity then a directed network equivalent to the original one is received.

Definition 1 A set of functions $f := \{f_{ij} | i, j \in V, f_{ij} : E(G) \to R^+\}$ is called a **multi-commodity flow** if

$$\sum_{e \in E_k^+} f_{ij}(e) = \sum_{e \in E_k^-} f_{ij}(e)$$

holds for all $k \neq i, k \neq j$, where $k \in V$ and E_k^+, E_k^- are the set of edges coming out from k and coming into k resp. Each function f_{ij} defines a **single-commodity flow** from i to j.

Definition 2 The **value** of a multi-commodity *flow* is an $n \times n$ matrix $T_f = (t_{ij}^f)$, where

$$t_{ij}^f = \sum_{e \in E_i^+} f_{ij}(e) - \sum_{e \in E_i^-} f_{ij}(e),$$

if $i \neq j$ and $v_{ii}^f = 0$, forall $i, j \in V$.

Definition 3 Let D be a nonnegative $n \times n$ matrix where the diagonal entries are 0. D is called as **demand matrix**. The **flow** on an edge $e \in E$ routing the demand matrix D by routing r is defined by

$$\text{flow}(e, r, D) = \sum_{i,j \in V} d_{ij} r_{ij}(e),$$

while the **edge congestion** is

$$\text{con}(e, r, D) = \frac{\text{flow}(e, r, D)}{c(e)}.$$

The **congestion** of demand D using routing r is

$$\text{con}(r, D) = \max_{e \in E} \text{con}(e, r, D).$$

Definition 4 A multi-commodity flow r is called **routing** if $t_{ij}^r = 1$, and if $i \neq j$ for all $i, j \in V$.

Routing represents a way of sending information over a network. The real load of the edges can be represented by scaling the edge congestions with the demands.

Definition 5 The **oblivious performance ratio** P_r of routing r is

$$P_r = \sup_D \left\{ \frac{\mathrm{con}(r, D)}{\mathrm{opt}(D)} \right\}$$

where $opt(D)$ is the optimal congestion which can be achieved on D. The **optimal oblivious routing ratio** for a network G is denoted by $opt(G)$, where

$$\mathrm{opt}(G) = \min_r P_r$$

Problem

Input: A capacitated network $G(V, E, c)$.
Output: An oblivious routing r, where P_r is minimal.

Key Results

Theorem 1 *There is a polynomial time algorithm that for any input network G (directed or undirected) finds the optimal oblivious routing ratio and the corresponding routing r.*

Theorem 2 *There is a directed graph G of n vertices such that $opt(G)$ is at least $\Omega(\sqrt{n})$.*

Applications

Most importantly, with these results one can efficiently calculate the best routing strategy for a network topology with capacity constraints. This is a good tool for network planning. The effectiveness of a given topology can be tested without any knowledge of the the network traffic using this analysis.

Many researchers have investigated the variants of routing problems. For surveys on the most important models and results, see [10] and [11]. Oblivious routing algorithms were first analyzed by Valiant and Brebner [15]. Here, they considered the parallel computer model and investigated specific architectures, like hypercube, square grids, etc. Borodin and Hopcroft investigated general networks [6]. They showed that such simple deterministic strategies like oblivious routing can not be very

efficient for online routing and proved a lower bound on the competitive ration of oblivious algorithms. This lower bound was later improved by Kaklamanis et al. [9], and they also gave an optimal oblivious deterministic algorithm for the hypercube.

In 2002, Räcke constructed a polylog competitive randomized algorithm for general undirected networks. More precisely, he proved that for any demand there is a routing such that the maximum edge congestion is at most polylog(n) times the optimal congestion for this demand [12]. The work of Azar et al. extends this result by giving a polynomial method for calculating the optimal oblivious routing for a network. They also prove that for directed networks no logarithmic oblivious performance ratio exists. Recently, Hajiaghayi et al. present an oblivious routing algorithm which is $O\left(\log^2 n\right)$-competitive with high probability in directed networks [8].

A special online model has been investigated in [5], where the authors define the so called "repeated game" setting, where the algorithm is allowed to chose a new routing in each day. This means that it is oblivious to the demands, that will occur the next day. They present an $1 + \varepsilon$-competitive algorithm for this model.

There are better algorithms for the adaptive case, for example in [2]. For the offline case Raghavan and Thomson gave an efficient algorithm in [13].

Open Problems

The authors investigated edge congestion in this paper, but in practice, node congestion may be interesting as well. Node congestion means the ratio of the total traffic traversing a node to its capacity. Some results can be found for this problem in [7] and in [3]. It is an open problem whether this method used for edge congestion analysis can be applied for such a model. Another interesting open question may be whether there is a more efficient algorithm to compute the optimal oblivious performance ratio of a network [1, 14].

Experimental Results

The authors applied their method on ISP network topologies and found that the calculated optimal oblivious ratios are surprisingly low, between 1.4 and 2. Other research dealing with this question found similar results [1, 14].

Cross-References

▶ Direct Routing Algorithms
▶ Mobile Agents and Exploration
▶ Oblivious Routing
▶ Online Load Balancing of Temporary Tasks
▶ Probabilistic Data Forwarding in Wireless Sensor Networks

Recommended Reading

1. Applegate D, Cohen E (2006) Making routing robust to changing traffic demands: algorithms and evaluation. IEEE/ACM Trans Networking 14(6):1193–1206. doi:10.1109/TNET.2006.886296
2. Aspnes J, Azar Y, Fiat A, Plotkin S, Waarts O (1997) On-line routing of virtual circuits with applications to load balancing and machine scheduling. J ACM 44(3):486–504
3. Azar Y, Chaiutin Y (2006) Optimal node routing. In: Proceedings of the 23rd international symposium on theoretical aspects of computer science, pp 596–607
4. Azar Y, Cohen E, Fiat A, Kaplan H, Räcke H (2003) Optimal oblivious routing in polynomial time. In: Proceedings of the thirty-fifth annual ACM symposium on theory of computing, pp 383–388
5. Bansal N, Blum A, Chawla S, Meyerson A (2003) Online oblivious routing. In: Proceedings of the 15th annual ACM symposium on parallel algorithms, pp 44–49
6. Borodin A, Hopcroft JE (1985) Routing, merging and sorting on parallel models of computation. J Comput Syst Sci 30(1):130–145
7. Hajiaghayi MT, Kleinberg RD, Leighton T, Räcke H (2005) Oblivious routing on node-capacitated and directed graphs. In: Proceedings of the 16th annual ACM-SIAM symposium on discrete algorithms, pp 782–790
8. Hajiaghayi MT, Kim JH, Leighton T, Räcke H (2005) Oblivious routing in directed graphs with random demands. In: Proceedings of the 37th annual ACM symposium on theory of computing, pp 193–201
9. Kaklamanis C, Krizanc D, Tsantilas A (1990) Tight bounds for oblivious routing in the hypercube. In: Proceedings of the 2nd annual ACM symposium on parallel algorithms and architectures, pp 31–36
10. Leighton FT (1992) Introduction to parallel algorithms and architectures arrays, trees, hypercubes. Morgan Kaufmann Publishers, San Fransisco
11. Leonardi S (1998) On-line network routing, Chapter 11. In: Fiat A, Woeginger G (eds) Online algorithms – the state of the art. Springer, Heidelberg, pp 242–267
12. Räcke H (2002) Minimizing congestions in general networks. In: Proceedings of the 43rd symposium on foundations of computer science, pp 43–52
13. Raghavan P, Thompson CD (1987) Randomized rounding: a technique for provably good algorithms and algorithmic proofs. Combinatorica 7:365–374
14. Spring N, Mahajan R, Wetherall D (2002) Measuring ISP topologies with Rocket fuel. In: Proceedings of the ACMSIGCOMM'02 conference. ACM, New York
15. Valiant LG, Brebner G (1981) Universal schemes for parallel communication. In: Proceedings of the 13th ACM symposium on theory of computing, pp 263–277

Routing in Geometric Networks

Stephane Durocher[1], Leszek Gąsieniec[2], and Prudence W.H. Wong[2]
[1]University of Manitoba, Winnipeg, MB, Canada
[2]University of Liverpool, Liverpool, UK

Synonyms

Geographic routing; Location-based routing

Keywords

Face routing; Geometric routing; Unit disk graph; Wireless communication

Years and Authors of Summarized Original Work

1999; Kranakis, Singh, Urrutia
1999; Bose, Morin, Stojmenovic, Urrutia
2003; Kuhn, Wattenhofer, Zhang, Zollinger

Problem Definition

Wireless networks are often modelled using geometric graphs. Using only local geometric information to compute a sequence of distributed forwarding decisions that send a message to its destination, routing algorithms can succeed on several common classes of geometric graphs. These graphs' geometric properties provide navigational cues that allow routing to succeed using only limited local information at each node.

Network Model

A common geometric graph model for wireless networks is to represent each node by a point in the Euclidean plane, $\mathcal{R}^2$, and to add an edge (u, v) for each pair of nodes that can communicate by direct wireless transmission. The absence of the edge (u, v) signifies that u cannot transmit directly to v, requiring a multi-hop transmission via a sequence of intermediate nodes that forms a route from u to v. The cost $c(e)$ of sending a message over an edge $e = (u, v)$ has been modeled in different ways; the most common measures include the *hop (link) metric* ($c(e) = 1$), the *Euclidean metric* ($c(e) = |e|$, where $|e| = \text{dist}(u, v)$ is the Euclidean length of the edge e), and the *energy metric* ($c(e) = |e|^\alpha$ for $\alpha \geq 2$).

In some models, transmission is assumed to be uniform in all directions and of equal range, say r, for all nodes. Under this assumption, the undirected edge (u, v) exists if and only if $\text{dist}(u, v) \leq r$. Thus, for each node v there is an edge from v to every node u that lies within a disk of radius r centered at v. This is the *unit disk graph* model for wireless networks. Common classes of geometric graphs that are used to model wireless networks include:

Unit Disk Graph. Vertices are points in $\mathcal{R}^2$ and each edge (u, v) exists if and only if $\text{dist}(u, v) \leq r$, for a given fixed $r > 0$.

Plane Graph. Vertices are points in $\mathcal{R}^2$ and no two edges cross.

Triangulation. Vertices are points in $\mathcal{R}^2$ and every interior face is a triangle.

Quasi-unit Disk Graph. Vertices are points in $\mathcal{R}^2$ and each edge (u, v) exists if $\text{dist}(u, v) \leq$ r_1, may exist if $r_1 < \text{dist}(u, v) \leq r_2$, but does not exist if $\text{dist}(u, v) > r_2$, for given fixed $r_2 > r_1 > 0$.

Unit Ball Graph. Vertices are points in $\mathcal{R}^3$ and each edge (u, v) exists if and only if $\text{dist}(u, v) \leq r$, for a given fixed $r > 0$.

Gabriel Graph. Vertices are points in $\mathcal{R}^2$ and each edge (u, v) exists if and only if the disk with diameter (u, v) does not contain any other vertices.

Other classes of geometric graphs used to model wireless networks include relative neighborhood graphs, Delaunay triangulations, Yao graphs, convex subdivisions, monotone subdivisions, edge-augmented plane graphs, and physically based models such as SINR.

A geometric graph G is *civilized* with λ-*precision* if for every pair of nodes u and v in G, $\text{dist}(u, v) \geq \lambda$ for a given fixed $\lambda > 0$, where λ is independent of n, the number of nodes in G.

Communication Protocol

In several wireless network protocols, e.g., ad hoc or wireless sensor networks, there is no fixed infrastructure for routing nor any central servers. All nodes act as hosts as well as routers. Apart from a node's immediate neighborhood, the topology of the network is unknown, i.e., each node is aware of its own location (its (x, y) coordinates) as well as the coordinates of its neighbors. Nodes must discover and maintain routes in a distributed manner without knowledge of precomputed routing tables, any particular vertex labeling (other than spatial coordinates), nor the support of a central server. Additionally, some models incorporate constraints for limited memory and power. Depending on the particular model, a limited amount of information can be stored in message headers to assist with routing. When a node receives a message, it reads the header (possibly modifying the header information) before selecting one of its neighbors to which to forward the message. A *stateless* algorithm does not modify the header. Network nodes have no memory themselves; any dynamic state information is stored in the message header.

Furthermore, no precomputed information about the network is known to the nodes.

Geometric Routing

Given the coordinates of a target node t in a (wireless) geometric network G, a source node s in G is tasked with sending a message via a multi-hop route through G from s to t. Routing proceeds by computing a sequence of distributed forwarding decisions, where each node along the route selects one of its neighbors to which to forward the message. Geometric routing is uniform in that all nodes execute the same protocol. Each node makes a forwarding decision as a function of its coordinates, the coordinates of its neighbors, the coordinates of t, and any available state bits stored in the message header. The number of state bits available is critical to guaranteeing delivery in some classes of geometric graphs by enabling the route to avoid looping and reach t. A node may modify the state bits before forwarding the message. In some models, this state information corresponds to storing data about $O(1)$ nodes, e.g., storing the coordinates of $O(1)$ nodes.

The primary objective is to guarantee message delivery to the target node t. Secondary objectives include minimizing the total cost of communication (the sum of $c(e)$ for all edges e on the route) and minimizing the worst-case or average *dilation* (the ratio of the cost of the route followed relative to that of the route of lowest cost). These secondary objectives are motivated by the need for nodes to conserve power in many wireless networking settings.

Key Results

Local geometric routing assumes only limited control information stored in message headers and local information available at each node along the route. This locality provides network independence that results in natural scalability to larger networks and continued functionality after arbitrary changes to the network. A routing algorithm is said to *succeed* on a particular class of geometric graphs if it guarantees delivery from any source node s to any target node t on any graph in the class; otherwise, the algorithm *fails* on that class of graphs.

Below we summarize key local geometric routing algorithms and their properties.

Greedy Routing. Upon receiving a message, a node forwards it to its neighbor closest to the target node t. Greedy routing is stateless. This strategy succeeds on Delaunay triangulations, but fails on more general classes of geometric graphs such as non-Delaunay triangulations, convex subdivisions, plane graphs, and unit disk graphs.

Compass Routing [7]. Upon receiving a message, a node u forwards it to its neighbor v that minimizes the angle $\angle vut$ with the target node t. Compass routing is stateless. This strategy succeeds on regular triangulations but fails on more general classes of geometric graphs such as non-regular triangulations, convex subdivisions, plane graphs, and unit disk graphs.

Greedy-Compass Routing [2]. Upon receiving a message, a node u considers its two neighbors on either side of the line segment $\overline{ut}$ (node u's *compass neighbors*) and forwards the message to the one closest to t. Greedy-compass routing is stateless. This strategy succeeds on all triangulations but fails on more general classes of geometric graphs such as convex subdivisions, plane graphs, and unit disk graphs.

Bose et al. [2] show that no stateless algorithm can succeed on convex subdivisions (including plane graphs and unit disk graphs). Therefore, to succeed on classes of geometric graphs beyond triangulations, local routing algorithms require storing one or more state bits in the message header or predecessor information, i.e., the coordinates of the node that last forwarded the message.

One State Bit [4]. Upon receiving a message, a node u chooses between forwarding the message to its clockwise or counter-clockwise compass neighbor, depending on the value of a state bit. If the compass neighbor lies opposite the vertical line through t, the state bit is flipped. This algorithm uses a single state bit. This strategy succeeds on all triangulations and convex subdivisions, but fails on more general classes of geometric graphs such as plane graphs and unit disk graphs.

Predecessor Awareness and Monotonicity [4]. Each node locally identifies its topmost left neighbor as its parent and its right neighbors as its children. With knowledge of the predecessor, the node forwards the message to its $(i + 1)$st child after receiving it from its ith child and eventually back to its parent after receiving it from its last child. The resulting route contains a depth-first traversal of a spanning tree of the network. This algorithm is stateless, but each node requires knowledge of its predecessor, i.e., the coordinates of the node that last forwarded the message. This strategy succeeds on triangulations, convex subdivisions, monotone subdivisions, and edge-augmented graphs from these classes but fails on more general classes of geometric graphs such as non-monotone plane graphs and unit disk graphs.

Face Routing [1, 7]. The message is forwarded along the perimeters of faces in the sequence of faces that intersect the line segment from the source node s to the target node t. This strategy applies the *right-hand principle*, in which each face in the sequence is traversed in a counter-clockwise direction, as if one were walking while sliding the right hand along the wall. To avoid cycling indefinitely, the algorithm must store the coordinates of $O(1)$ nodes that act as progress markers. Furthermore, each node requires knowledge of its predecessor. This strategy succeeds on plane graphs, including triangulations, convex subdivisions, and Gabriel graphs. The intersection of a unit disk graph with the Gabriel graph of a set of points is planar and remains connected if the original unit disk graph is connected. Furthermore, this subgraph can be computed locally; this property allows face routing to succeed on unit disk graphs [1], as well as quasi-unit disk graphs with bounded ratio $r_2/r_1 < \sqrt{2}$ and unit ball graphs contained within slabs of thickness less than $1/\sqrt{2}$ [6]. Although unit disk graphs are nonplanar in general, the nonplanarity is lo-

calized; face routing fails on more general classes of nonplanar geometric graphs such as quasi-unit disk graphs and unit ball graphs [6] and edge-augmented plane graphs. Face routing can have dilation $\Theta(n)$, where n is the number of network nodes.

Adaptive Face Routing (AFR) [8]. Adaptive face routing is a variant of face routing that achieves optimality on civilized unit disk graphs and civilized planar graphs with the Gabriel property. Like face routing, $O(1)$ state data are stored in the message header and each node requires knowledge of its predecessor. The algorithm attempts to estimate the length c of the shortest path from s and t by $\hat{c}$ (starting with $\hat{c} = 2|\overline{st}|$ and doubling it in every consecutive round). In each round, the face traversal is restricted to the region formed by the ellipse with the major axis $\hat{c}$ centered on $\overline{st}$. Each edge is traversed at most four times, and the dilation achieved is $\Theta(c)$.

Geometric Ad-hoc Routing (GOAFR$^+$) [9]. Combining methods from greedy routing, face routing, and adaptive face routing allows this hybrid algorithm to meet the bounds of adaptive routing on any unit disk graphs and planar graphs with the Gabriel property (not necessarily civilized). The algorithm first applies greedy routing and switches to face routing when the routed message enters a local minimum (a dead end), before again resuming greedy routing as early as possible by applying an *early fallback* technique.

General (Non-geometric) Networks

Is geometry necessary for local routing to succeed? Even with knowledge of the predecessor, stateless routing algorithms require knowledge of the induced subgraph of nodes up to distance $n/3$ away in the worst case [3]. That is, stateless routing using only *local* information is impossible. With $\Theta(\log n)$ state bits, local routing on arbitrary (not necessarily geometric) graphs is possible by deterministically recomputing a polynomial-length universal traversal sequence at each node along the route, where $\Theta(\log n)$ bits store an index into the sequence [5].

Open Problems

If a node's coordinates can be stored using $O(\log n)$ bits (e.g., if network nodes are positioned on a $n^c \times n^c$ grid), then face routing can be applied using $O(\log n)$ state bits. It remains open whether any local geometric routing algorithm can succeed on plane graphs using $o(\log n)$ state bits. Similarly, it would be interesting to characterize broad classes of geometric graphs on which local geometric routing is possible using $O(1)$ state bits. In addition to guaranteeing delivery, bounding dilation is of interest. For example, can $O(1)$ dilation be guaranteed on convex subdivisions using $O(1)$ state bits? Finally, the problem of traversing a graph (visiting all nodes) by a sequence of local forwarding decisions is interesting. Stateless algorithms are impossible for any non-Hamiltonian network. How many state bits are necessary for a local algorithm to traverse a triangulation?

Cross-References

▶ Communication in Ad Hoc Mobile Networks Using Random Walks
▶ Geographic Routing
▶ Local Computation in Unstructured Radio Networks
▶ Minimum k-Connected Geometric Networks

Recommended Reading

1. Bose P, Morin P, Stojmenovic I, Urrutia J (1999) Routing with guaranteed delivery in ad hoc wireless networks. In: Proceedings of the third international workshop on discrete algorithm and methods for mobility, Seattle, Aug 1999, pp 48–55
2. Bose P, Brodnik A, Carlsson S, Demaine ED, Fleischer R, López-Ortiz A, Morin P, Munro I (2002) Online routing in convex subdivisions. Int J Comput Geom Appl 12(4):283–295
3. Bose P, Carmi P, Durocher S (2013) Bounding the locality of distributed routing algorithms. Distrib Comput 26(1):39–58
4. Bose P, Durocher S, Mondal D, Peabody M, Skala M, Wahid MA (2015) Local routing in convex subdivisions. In: Proceedings of the forty-first international conference on current trends in theory and practice of computer science, Pec pod Sněžkou, Jan 2015, vol 8939, pp 140–151
5. Braverman M (2008) On ad hoc routing with guaranteed delivery. In: Proceedings of the twenty-seventh ACM symposium on principles of distributed computing, Toronto, vol 27, p 418
6. Durocher S, Kirkpatrick DG, Narayanan L (2010) On routing with guaranteed delivery in three-dimensional ad hoc wireless networks. Wirel Netw 16(1):227–235
7. Kranakis E, Singh H, Urrutia J (1999) Compass routing on geometric networks. In: Proceedings of the eleventh Canadian conference on computational geometry, Vancouver, Aug 1999, pp 51–54
8. Kuhn F, Wattenhofer R, Zollinger A (2002) Asymptotically optimal geometric mobile ad-hoc routing. In: Proceedings of the sixth international workshop on discrete algorithm and methods for mobility, Atlanta, Sept 2002, pp 24–33
9. Kuhn F, Wattenhofer R, Zhang Y, Zollinger A (2003) Geometric ad-hoc routing: of theory and practice. In: Proceedings of the twenty-second ACM symposium on the principles of distributed computing, Boston, July 2003, pp 63–72

Routing in Road Networks with Transit Nodes

Dominik Schultes
Institute for Computer Science, University of Karlsruhe, Karlsruhe, Germany

Keywords

Shortest paths

Years and Authors of Summarized Original Work

2007; Bast, Funke, Sanders, Schultes

Problem Definition

For a given directed graph $G = (V, E)$ with non-negative edge weights, the problem is to compute a shortest path in G from a source node s to a target node t for given s and t. Under the assumption that G does not change and that a lot of source-target queries have to be answered, it pays to invest some time for a preprocessing step that allows for very fast queries. As output, either a full description of the shortest path or only its length $d(s, t)$ is expected – depending on the application.

Dijkstra's classical algorithm for this problem [4] iteratively visits all nodes in the order of their distance from the source until the target is reached. When dealing with very large graphs, this general algorithm gets too slow for many applications so that more specific techniques are needed that exploit special properties of the particular graph. One practically very relevant case is routing in road networks where junctions are represented by nodes and road segments by edges whose weight is determined by some weighting of, for example, expected travel time, distance, and fuel consumption. Road networks are typically sparse (i.e., $|E| = O(|V|)$), almost planar (i.e., there are only a few overpasses), and hierarchical (i.e., more or less 'important' roads can be distinguished). An overview on various speedup techniques for this specific problem is given in [7].

Key Results

Transit-node routing [2, 3] is based on a simple observation intuitively used by humans: When you start from a source node s and drive to somewhere 'far away', you will leave your current location via one of only a few 'important' traffic junctions, called (forward) *access nodes* $\overrightarrow{A}(s)$. An analogous argument applies to the target t, i.e., the target is reached from one of only a few backward access nodes $\overleftarrow{A}(t)$. Moreover, the union of all forward and backward access nodes of all nodes, called *transit-node set* $\mathcal{T}$, is rather small. The two observations imply that for each node the distances to/from its forward/backward access nodes and for each transit-node pair (u, v), the distance between u and v can be stored. For given source and target nodes s and t, the length of the shortest path that passes at least one transit node is given by

$$d_\mathcal{T}(s,t) = \min\{d(s,u) + d(u,v) + d(v,t) \mid$$
$$u \in \overrightarrow{A}(s), v \in \overleftarrow{A}(t)\}.$$

Note that all involved distances $d(s, u)$, $d(u, v)$, and $d(v, t)$ can be directly looked up in the precomputed data structures. As a final ingredient, a *locality filter* $L: V \times V \to \{\text{true, false}\}$ is needed that decides whether given nodes s and t are too close to travel via a transit node. L has to fulfill the property that $\neg L(s, t)$ implies that $d(s, t) = d_\mathcal{T}(s, t)$. Note that in general the converse need not hold since this might hinder an efficient realization of the locality filter. Thus, *false positives*, i.e., "$L(s, t) \wedge d(s, t) = d_\mathcal{T}(s, t)$", may occur.

The following algorithm can be used to compute $d(s, t)$:

If $\neg L(s, t)$, then compute and return $d_\mathcal{T}(s, t)$; else, use any other routing algorithm.

Figure 1 gives an example. Knowing the length of the shortest path, a complete description of it can be efficiently derived using iterative table lookups and precomputed representations of paths between transit nodes. Provided that the above observations hold and that the percentage of false positives is low, the above algorithm is very efficient since a large fraction of all queries can be handled in line 1, $d_\mathcal{T}(s, t)$ can be computed using only a few table lookups, and source and target of the remaining queries in line 2 are quite close. Indeed, the remaining queries can be further accelerated by introducing a secondary layer of transit-node routing, based on a set of *secondary transit nodes* $\mathcal{T}_2 \supset \mathcal{T}$. Here, it is not necessary to compute and store a complete $\mathcal{T}_2 \times \mathcal{T}_2$ distance table, but it is sufficient to store only distances $\{d(u, v) \mid u, v \in \mathcal{T}_2 \wedge d(u, v) \neq d_\mathcal{T}(s, t)\}$, i.e., distances that cannot be obtained using the primary layer. Analogously, further layers can be added.

There are two different implementations: one is based on a simple geometric grid and one on *highway hierarchies*, the fastest previous approach [5, 6]. A highway hierarchy consists of a sequence of levels (Fig. 1), where level $i + 1$ is

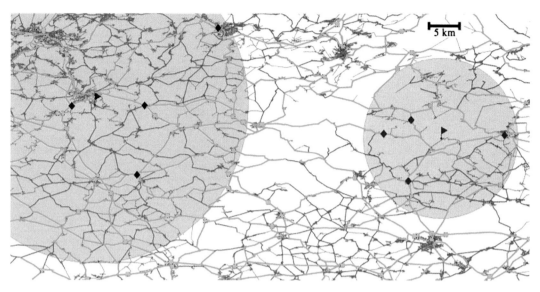

Routing in Road Networks with Transit Nodes, Fig. 1
Finding the optimal travel time between two points (*flags*) somewhere between Saarbrücken and Karlsruhe amounts to retrieving the 2 × 4 *access nodes* (*diamonds*), performing 16 table lookups between all pairs of access nodes, and checking that the two disks defining the *locality filter* do not overlap. *Transit nodes* that do not belong to the access node sets of the selected source and target nodes are drawn as small squares. The figure draws the levels of the highway hierarchy using colors *gray*, *red*, *blue*, and *green* for levels 0–1, 2, 3, and 4, respectively

constructed from level i by bypassing low-degree nodes and removing edges that never appear far away from the source or target of a shortest path. Interestingly, these levels are geometrically decreasing in size and otherwise similar to each other. The highest level contains the most 'important' nodes and becomes the primary transit-node set. The nodes of lower levels are used to form the transit-node sets of subordinated layers.

Applications

Apart from the most obvious applications in car navigation systems and server-based route planning systems, transit-node routing can be applied to several other fields, for instance to massive traffic simulations and to various optimization problems in logistics.

Open Problems

It is an open question whether one can find better transit-node sets or a better locality filter so that

the performance can be further improved. It is also not clear if transit-node routing can be successfully applied to other graph types than road networks. In this context, it would be desirable to derive some theoretical guarantees that apply to any graph that fulfills certain properties. For some practical applications, a *dynamic* version of transit-node routing would be required in order to deal with time-dependent networks or unexpected edge weight changes caused, for example, by traffic jams. The latter scenario can be handled by a related approach [8], which is, however, considerably slower than transit-node routing.

Experimental Results

Experiments were performed on road networks of Western Europe and the USA using a cost function that solely takes expected travel time into account. The results exhibit various trade-offs between average query time (5–63 μs for the USA), preprocessing time (59 min to 1,200 min), and storage overhead (21 bytes/node to 244 bytes/node). For the variant that uses three

Routing in Road Networks with Transit Nodes, Table 1 Statistics on preprocessing. The size of transit-node sets, the number of entries in distance tables,

and the average number of access nodes to the respective layer are given; furthermore, the space overhead and the preprocessing time

	Layer 1		Layer 2			Layer 3				
	$|\mathcal{T}|$	$|A|$ Avg.	$|\mathcal{T}_2|$	$|\text{Table}_2|$ [$\times 10^6$]	$|A_2|$ Avg.	$|\mathcal{T}_3|$	$|\text{Table}_3|$ [$\times 10^6$]	Space [B/node]	Time [h]
Europe	11,293	9.9	323,356	130	4.1	2,954,721	119	251	2:44
USA	10,674	5.7	485,410	204	4.2	3,855,407	173	244	3:25

Routing in Road Networks with Transit Nodes, Table 2 Performance of transit-node routing with respect to 10,000,000 random queries. The column for layer i specifies which fraction of the queries is correctly answered using only information available at

layers $\leq i$. Each box spreads from the lower to the upper quartile and contains the median, the whiskers extend to the minimum and maximum value omitting outliers, which are plotted individually

	#nodes	#edges	Layer 1	Layer 2	Layer 3	Query
Europe	18 029 721	42 199 587	99.74 %	99.9984 %	99.99981 %	5.6 μs
USA	24 278 285	58 213 192	99.89 %	99.9986 %	99.99986 %	4.9 μs

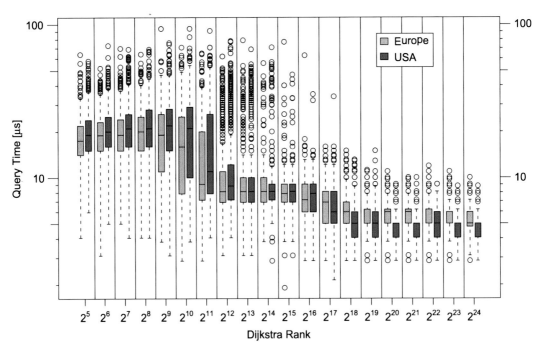

Routing in Road Networks with Transit Nodes, Fig. 2 Query time distribution as a function of Dijkstra rank–the number of iterations Dijkstra's algorithm would need to solve this instance. The distributions are represented as

box-and-whisker plots: each box spreads from the lower to the upper quartile and contains the median, the whiskers extended to the minimum and maximum value omitting, which are plotted individually

layers and is tuned for best query times, Tables 1 and 2 show statistics on the preprocessing and the query performance, respectively. The average query times of about 5 μs are six orders of magnitude faster than Dijkstra's algorithm. In addition, Fig. 2 gives for each rank r on the x-axis a dis-

tribution for 1,000 queries with random starting point s and the target node t for which Dijkstra's algorithm would need r iterations to find it. The three layers of transit-node routing with small transition zones in between can be recognized: for large ranks, it is sufficient to access only the primary layer yielding query times of about 5 μs, for smaller ranks, additional layers have to be accessed resulting in median query times of up to 20 μs.

Data Sets

The European road network has been provided by the company PTV AG, the US network has been obtained from the TIGER/Line Files [9]. Both graphs have also been used in the 9th DIMACS Implementation Challenge on Shortest Paths [1].

URL to Code

The source code might be published at some point in the future at http://algo2.iti.uka.de/schultes/hwy/.

Cross-References

▸ All Pairs Shortest Paths in Sparse Graphs
▸ All Pairs Shortest Paths via Matrix Multiplication
▸ Decremental All-Pairs Shortest Paths
▸ Engineering Algorithms for Large Network Applications
▸ Fully Dynamic All Pairs Shortest Paths
▸ Geographic Routing
▸ Implementation Challenge for Shortest Paths
▸ Shortest Paths Approaches for Timetable Information
▸ Shortest Paths in Planar Graphs with Negative Weight Edges
▸ Single-Source Shortest Paths

Recommended Reading

1. 9th DIMACS implementation challenge: shortest paths (2006). http://www.dis.uniroma1.it/~challenge9/
2. Bast H, Funke S, Matijevic D, Sanders P, Schultes D (2007) In transit to constant time shortest-path queries in road networks. In: Workshop on algorithm engineering and experiments, pp 46–59
3. Bast H, Funke S, Sanders P, Schultes D (2007) Fast routing in road networks with transit nodes. Science 316(5824):566
4. Dijkstra EW (1959) A note on two problems in connexion with graphs. Numer Math 1:269–271
5. Sanders P, Schultes D (2005) Highway hierarchies hasten exact shortest path queries. In: 13th European symposium on algorithms. LNCS, vol 3669. Springer, Berlin, pp 568–579
6. Sanders P, Schultes D (2006) Engineering highway hierarchies. In: 14th European symposium on algorithms. LNCS, vol 4168. Springer, Berlin, pp 804–816
7. Sanders P, Schultes D (2007) Engineering fast route planning algorithms. In: 6th workshop on experimental algorithms. LNCS, vol 4525. Springer, Berlin, pp 23–36
8. Schultes D, Sanders P (2007) Dynamic highway-node routing. In: 6th workshop on experimental algorithms. LNCS, vol 4525. Springer, Berlin, pp 66–79
9. U.S. Census Bureau, Washington, DC (2002) UA census 2000 TIGER/line files. http://www.census.gov/geo/www/tiger/tigerua/ua_tgr2k.html

Routing-Cost Constrained Connected Dominating Set

Hongwei Du and Haiming Luo
Department of Computer Science and Technology, Shenzhen Graduate School, Harbin Institute of Technology, Shenzhen, China

Keywords

Connected dominating set; Routing constraint; Wireless multihop networks; Wireless sensor networks

Years and Authors of Summarized Original Work

2011; Du, Ye, Wu, Lee, Li, Du, Howard
2012; Du, Ye, Zhong, Wang, Lee, Park
2013; Du, Wu, Ye, Li, Lee, Xu

Problem Definition

Connected dominating set *CDS* is typically adapted in wireless multihop networks such as wireless sensor, ad hoc networks. In order to achieve routing efficiency, a virtual backbone which is inspired by the backbone in wired networks is often used to improve routing because it can reduce the path search space and the routing table size [3]. According to [8, 10], there are many methods to construct a virtual backbone, and the competitive approach is connected dominating set(*CDS*). The detailed definition of *CDS* is as follows.

Given a connected graph $G(V, E)$ represents as a wireless sensor network, where V is the set of sensor nodes and E is the set of edges connecting sensor nodes in V. If there is a subset D($D \subseteq V$), each sensor node in V either belonging to D or adjacent to a sensor node in D, then we call D is a dominating set (*DS*). If the subgraph induced by *DS* is a connected graph, then we call *DS* is a connected dominating set(*CDS*).

Intuitively, if the size of *CDS* is smaller, the virtual backbone can play a greater role in routing. Many studies such as [1, 13, 14] and [9, 12] also aimed to construct a virtual backbone based on a *CDS* with minimum size which is called minimum connected dominating set (*MCDS*). A minimum CDS (MCDS) is a CDS that has the minimum number of nodes. For example, the gray nodes in Fig. 1a form an MCDS of the sample graph, while the black nodes in Fig. 1b make a CDS. However, these studies didn't ignore that if the size of the *CDS* is too small, some sensor nodes couldn't find a shortest routing path to their destination. Du et al. [10] analyzes how a virtual backbone based on *MCDS* makes some routing paths much longer than the shortest paths. Thus, there exists a disadvantage in *MCDS* which is

the unavailability of shortest routing paths. If we only aim to construct a virtual backbone based on *MCDS*, we couldn't achieve a guaranteed routing constraint in data delivery. Routing constraint in *CDS* becomes much more important when constructing a virtual backbone.

In this section, the problem of the routing constraint in connected dominating set (R-CDS) is given in formal by considering the wireless multihop network environment. The problem of R-CDS is defined as follows:

Given a graph $G = (V, E)$ where V represents a node set and E denotes an edge set, we would like to find a CDS D in polynomial time so that, for every pair of nodes u and v, there exists a path between u and v with intermediate nodes in D and path length at most $\alpha \cdot d(u, v)$, where α is a constant and $d(u, v)$ is the length of the shortest path between u and v. In addition, the size of the resulting CDS $|D|$ is bounded by $\beta \cdot opt_{MCDS}$, where β is a constant and opt_{MCDS} is the size of the MCDS.

The problem specified in wireless multihop networks has been considered under both general graph and unit disk graph(UDG) model and will be further discussed in the following section. With the UDG model, all nodes in the network have the same transmission range, and there does not exist any obstacle. As a result, as long as a receiving node is within the transmission range of a sending node, the receiving node will be able to receive the data successfully. With general graph model, the nodes in the network could have different transmission ranges, and obstacles might interfere with normal data communication. As a result, being in the transmission range of a sending node does not guarantee successful transmission.

Key Results

Many literatures also focus on the study of routing constraint in *CDS* which can be classified into two categories: the general graph and UDG.

In the general graph category, Ding et al. [2] studied a special connected dominating set (*CDS*) problem named minimum routing cost CDS

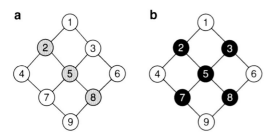

Routing-Cost Constrained Connected Dominating Set, Fig. 1 (**a**) An example MCDS. (**b**) An example CDS

(*MOC-CDS*). They proved that constructing a minimum *MOC-CDS* in general graph is *NP*-hard and proposed a distributed heuristic algorithm (called as FlagContest) for constructing *MOC-CDS* with performance ratio $(1 - ln2) + 2ln\delta$. Du et al. [8] presented a constant-approximation scheme which produces a connected dominating set D, whose size $|D|$ is within a factor α from that of the minimum connected dominating set, and each node pair exists in a routing path with all intermediate nodes in D and with length at most $5d(u, v)$, where $d(u, v)$ is the length of shortest path of this node pair. Ding et al. [3] developed an exact algorithm for minimum *CDS* with shortest path constraint called *SPCDS* and proved that finding such a minimum *SPCDS* can be achieved in polynomial time. Ding et al. [4] showed that under general graph model, α-*MOC-CDS* is NP-hard for any $\alpha \geq 1$. Ding et al. [5, 6] studied virtual backbone with guaranteed routing costs, named α minimum routing cost directional virtual backbone(α-*MOC-DVB*). They proved that the construction of a minimum α-*MOC-DVB* is an NP-hard problem in a general directed graph. Du et al. [10] proved that there is no polynomial-time constant approximation for α-*MOC-CDS* unless $P = NP$ when $\alpha \geq 2$.

In the UDG category, Wu et al. [15] studied the relationship between minimum connected dominating sets and maximal independent sets in unit disk graphs. Kim et al. [11] proposed a distributed algorithm under UDG model, *CDS-BD-D*, which constructs a *CDS* whose size and maximum path length are bounded. Du et al. [8] proposed two algorithms which are centralized algorithm and distributed algorithm to achieve constant-approximation performance ratio on MCDS and routing cost. Du et al. [7] studied a problem of minimizing the size of connected dominating set $| D |$ under constraint that for any two nodes u and v, $m_D(u, v) \leq \alpha m(u, v)$ where α is a constant, $m_D(u, v)$ is the number of intermediate nodes on a shortest path connecting u and v through D, and $m(u, v)$ is the number of intermediate nodes in a shortest path between u and v in a given unit disk graph.

In this chapter, we introduce that, under general graph model, there is no polynomial-time

---

Algorithm 1: Centralized algorithm GOC-MCDS-C

1: **Initially** Set $D \leftarrow \emptyset$.
2: **Step 1.** Construct a maximal independent set I.
3: **Step 2.** For every pair of nodes u, v in I with $d(u, v) \leq 3$, compute a shortest path $p(u, v)$ and put all intermediate nodes of $p(u, v)$ into C.
4: **Output** $D = C \cup I$.

Algorithm 2: Construct an MIS I (Stage 1)

1: **Initially** Every node is colored in white and is assigned with a positive integer ID; different nodes have different IDs.
2: **Step 1** Every white node sends its ID to its neighbors and then compares its ID with received IDs from neighbors. If its ID is smaller than every received ID from neighbors, then it turns the color from white to black.
3: **Step 2** Every black node sends message "black" to its neighbors. If a white node receives a message "black," then it turns its color from white to gray.
4: **Step 3** Go back to Step 1 until no white node exists.
5: **Output** All black nodes form a maximal independent set I

---

constant-approximation solution in terms of CDS size unless $P = NP$ shown in [10]. Under UDG model, we will present a polynomial-time constant-approximation algorithm GOC-MCDS-C which produces a CDS D with size $|D| \leq 176 \cdot opt_{MCDS} + 64$ and with a property that for any pair of nodes u and v, $d_D(u, v) \leq 7 \cdot d(u, v)$ in [10]. The distributed version of the algorithm, GOC-MCDS-D, is thoroughly analyzed.

GOC-MCDS-C: The Centralized Algorithm

Under general graph model, the existing proof is that there is no polynomial-time constant approximation for the problem under investigation unless $NP=P$ shown in [10]. However, under UDG model, polynomial-time constant-approximation algorithms do exist. Kim et al. [11] proposed a distributed algorithm, CDS-BD-D, that constructs a CDS whose size and maximum path length are bounded. In this section, we advance Kim et al.'s results by presenting the details of an innovative polynomial-time constant-approximation algorithm, GOC-MCDS-C. The proposed algorithm produces a

---

Algorithm 3: Connect the MIS I (Stage 2)

1: **Step 1** Every black node sends its ID to its neigh -bors.

2: **Step 2** Every node adds its own ID id_2 to each received ID id_1 and then sends those pairs of IDs (id_1, id_2) to all its neighbors.

3: **Step 3** Each node does the following: Suppose its ID is id^*:

    1. For each pair of IDs id_1 and id_{1^*} received in Step 1, if $id_1 < id_{1^*}$, then send a message (id_{1^*}, id^*, id_1) to the neighbor with ID id_1.

    2. For each pair of messages (id_1, id_2) and (id_{1^*}, id_{2^*}) received at Step 2, if $id_1 < id_{1^*}$, then send a message $(id_{1^*}, id_{2^*}, id^*, id_2, id_1)$ to the neighbor with ID id_2.

    3. For each message (id_1, id_2) received at Step 2 and ID id_{1^*} received at Step 1, if $id_1 < id_{1^*}$, then send a message $(id_{1^*}, id^*, id_2, id_1)$ to the neighbor with ID id_2; otherwise, send a message $(id_1, id_2, id^*, id_{1^*})$ to the neighbor with ID id_{1^*}.

4: **Step 4** When a node with ID id_2 received a message $(id_{1^*}, id_{2^*}, id^*, id_2, id_1)$ or $(id_{1^*}, id^*, id_2, id_1)$, it sends this message to its neighbor with ID id_1.

5: **Step 5** Each black node with ID id_1 collects all messages in form (id_3, id_2, id_1) or (id_4, id_3, id_2, id_1) or $(id_5, id_4, id_3, id_2, id_1)$ received in Step 3 and Step 4. Suppose those messages form a set M. Then it performs the following computation:
while $M \neq \emptyset$ **do begin**
    choose $(id_h, \ldots, id_2, id_1) \in M$;
    send message $(id_h, \ldots, id_2, id_1)$ to node with ID id_2;
    delete all messages starting with id_h from M;
end-while

6: **Step 6** When a node with ID id_i received a message $(\ldots, id_{i-1}, id_i, \ldots)$, it turns black. In addition, if id_i is not the leftmost id in the message, then it passes this message to node with ID id_{i-1}; if id_i is the leftmost id in the message, do nothing.

7: **Step 7** If no message is passed in Step 6, then stop. Otherwise, go back to Step 6.

CDS D with size $|D| \leq 176 \cdot opt_{MCDS} + 64$ and with a property that for any pair of nodes u and v, $d_D(u, v) \leq 7d(u, v)$[10]. Note that GOC-MCDS-C is a centralized algorithm. GOC-MCDS-C follows the steps of regular MCDS

construction algorithms. Namely, there are two steps in total. During the first step, an MIS is constructed. In the second step, the nodes in the MIS are connected in order to form a CDS.

GOC-MCDS-D: The Distributed Algorithm

In this section, the distributed algorithm GOC-MCDS-D is described in details. The performance of GOC-MCDS-D is the same as that of GOC-MCDS-C shown in [10]. Similar to the centralized algorithm GOC-MCDS-C, GOC-MCDS-D consists of two stages. In the first stage, an MIS is constructed using Algorithm 2. In the second stage, the MIS is connected using Algorithm 3.

Open Problems

The coverage problems in wireless sensor networks which related to the routing-cost constrained in CDS are still an open problem.

Cross-References

▶ Strongly Connected Dominating Set

Recommended Reading

1. Cheng X, Ding M, Du DH, Jia X (2006) Virtual backbone construction in multihop ad hoc wireless networks. Wirel Commun Mobile Comput 6:183–190
2. Ding L, Gao X, Wu W, Lee W, Zhu X, Du D-Z (2010) Distributed construction of connected dominating sets with minimum routing cost in wireless network. In: Proceedings of the 30th international conference on distributed computing systems (ICDCS), Genova, pp 448–457
3. Ding L, Gao X, Wu W, Lee W, Zhu X, Du D-Z (2011) An exact algorithm for minimum CDS with shortest path constraint in wireless networks. J Optim Lett 5(2):297–306
4. Ding L, Wu W, Willson J, Du H, Lee W, Du D-Z (2011) Efficient algorithms for topology control problem with routing cost constraints in wireless networks. IEEE Trans Parallel Distrib Syst 22(10):1601–1609
5. Ding L, Wu W, Willson JK, Du H (2011) Construction of directional virtual backbones with minimum routing cost in wireless networks. In: Proceedings of 30th IEEE international conference on computer communications (INFOCOM), Shanghai

6. Ding L, Wu W, Willson J, Du H, Lee W (2012) Efficient virtual backbone construction with routing cost constraint in wireless networks using directional antennas. IEEE Trans Mobile Comput 11(7): 1102–1112
7. Du H, Ye Q, Zhong J, Wang Y, Lee W, Park H (2010) PTAS for minimum connected dominating set with routing cost constraint in wireless sensor networks. J Comb Optim Appl 6508:252–259
8. Du H, Ye Q, Wu W, Lee W, Li D, Du D, Howard S (2011) Constant approximation for virtual backbone construction with guaranteed routing cost in wireless sensor networks. In: Proceedings of the 30th IEEE international conference on computer communications (INFOCOM), Shanghai
9. Funke S, Kesselman A, Meyer U (2006) A simple improved distributed algorithm for minimum CDS in unit disk graphs. ACM Trans Sens Netw 2(3):444–453
10. Hongwei Du, Weili Wu, Qiang Ye, Deying Li, Wonjun Lee, Xuepeng Xu(2013) CDS-based virtual backbone construction with guaranteed routing cost in wireless sensor networks. IEEE Trans Parallel Distrib Syst 24(4):652–661
11. Kim D, Wu Y, Li Y, Zou F, Du D-Z (2009) Constructing minimum connected dominating sets with bounded diameters in wireless networks. IEEE Trans Parallel Distrib Syste 20(2):147–157
12. Li D, Du H, Wan P-J, Gao X, Zhang Z, Wu W (2009) Construction of strongly connected dominating sets in asymmetric multihop wireless networks. Theor Comput Sci 410(8–10): 661–669
13. Min M, Du H, Jia X, Huang CX, Huang SC-H, Wu W (2006) Improving construction for connected dominating set with Steiner tree in wireless sensor networks. J Glob Optim 35(1):111–119
14. Ruan L, Du H, Jia X, Wu W, Li Y, Ko K-I (2004) A greedy approximation for minimum connected dominating sets. Theor Comput Sci 329(1–3):325–330
15. Wu W, Du H, Jia X, Li Y, Huang SC-H (2006) Minimum connected dominating sets and maximal independent sets in unit disk graphs. Theor Comput Sci 352(1–3):1–7

R-Trees

Ke Yi
Hong Kong University of Science and Technology, Hong Kong, China

Keywords

External memory data structures; R-trees; Spatial databases

Years and Authors of Summarized Original Work

2004; Arge, de Berg, Haverkort, Yi

Problem Definition

Problem statement and the I/O model. Let S be a set of N axis-parallel hypercubes in $\mathbb{R}^d$. A very basic operation in a spatial database is to answer *window queries* on the set S. A window query Q is also an axis-parallel hypercube in $\mathbb{R}^d$ that asks us to return all hypercubes in S that intersect Q. Since the set S is typically huge in a large spatial database, the goal is to design a *disk-based* or *external memory* data structure (often called an *index* in the database literature) such that these window queries can be answered efficiently. In addition, given S, the data structure should be constructed efficiently and should be able to support insertions and deletions of objects.

When external memory data structures are concerned, the standard *external memory model* [2], a.k.a. the *I/O model*, is often used as the model of computation. In this model, the machine consists of an infinite-size external memory (disk) and a main memory of size M. A block of B consecutive elements can be transferred between main memory and disk in one *I/O operation* (or simply *I/O*). An external memory data structure is a structure that is stored on disk in blocks, but computation can only occur on elements in main memory, so any operation (e.g., query, update, and construction) on the data structure must be performed using a number I/Os, which is the measure for the complexity of the operation.

R-trees. The *R-tree*, first proposed by Guttman [9], is a multi-way tree $\mathcal{T}$, very similar to a *B-tree*, that is used to store the set S such that a window query can be answered efficiently. Each node of $\mathcal{T}$ fits in one disk block. The hypercubes of S are stored only in the leaves of $\mathcal{T}$. All leaves of $\mathcal{T}$ are on the same level, and each stores $\Theta(B)$ hypercubes from S; while each internal node, except the root, has a fan-out of $\Theta(B)$. The root of $\mathcal{T}$ may have a fan-out as small as 2. For any node $u \in \mathcal{T}$, let $R(u)$ be the smallest axis-parallel hypercube, called the *minimal bounding box*, that encloses all the hypercubes stored below u. At each internal node $v \in \mathcal{T}$, whose children are denoted $v_1, \ldots, v_k$, the bounding box $R(v_i)$ is stored along with the pointer to v_i for $i = 1, \ldots, k$. Note that these bounding boxes may overlap. Please see Fig. 1 for an example of an R-tree in two dimensions.

For a window query Q, the query answering process starts from the root of $\mathcal{T}$ and visits all nodes u for which $R(u)$ intersects Q. When reaching a leaf v, it checks each hypercube stored at v to decide if it should be reported. The correctness of the algorithm is obvious, and the efficiency (the number of I/Os) is determined by the number of nodes visited.

Any R-tree occupies a linear number $O(N/B)$ disk blocks, but different R-trees might have different query, update, and construction costs. When analyzing the query complexity of window queries, the output size T is also used, in addition to N, M, and B.

Key Results

Although the structure of an R-tree is restricted, there is much freedom in grouping the hypercubes into leaves and grouping subtrees into bigger subtrees. Different grouping strategies result in different variants of R-trees. Most of the existing R-trees use various heuristics to group together hypercubes that are "close" spatially, so that a window query will not visit too many unnecessary nodes. Generally speaking, there are two ways to build an R-tree: repeated insertion and bulk loading. The former type of algorithms include the original R-tree [9], the R$^+$-tree [15], the R*-tree [6], etc. These algorithms use $O(\log_B N)$ I/Os to insert an object and hence $O(N \log_B N)$ I/Os to build the R-tree on S, which is not scalable for large N. When the set S is known in advance, it is much more efficient to bulk load the entire R-tree at once. Many

bulk-loading algorithms have been proposed, e.g., [7, 8, 11, 13]. Most of these algorithms build the R-tree with $O\left(\frac{N}{B} \log_{M/B} \frac{N}{B}\right)$ I/Os (the number of I/Os needed to sort N elements), and they typically result in better R-trees than those obtained by repeated insertion. During the past decades, there have been a large number of works devoted to R-trees from the database community, and the list here is by no means complete. The reader is referred to the book by Manolopoulos et al. [14] for an excellent survey on this subject in the database literature. However, no R-tree variant mentioned above has a guarantee on the query complexity; in fact, Arge et al. [3] constructed an example showing that some of the most popular R-trees may have to visit all the nodes without reporting a single result.

From the theoretical perspective, the following are the two main results concerning the worst-case query complexity of R-trees.

Theorem 1 ([1, 12]) *There is a set of N points in $\mathbb{R}^d$, such that for any R-tree $\mathcal{T}$ built on these points, there exists an empty window query for which the query algorithm has to visit $\Omega((N/B)^{1-1/d})$ nodes of $\mathcal{T}$.*

The *priority R-tree*, proposed by Arge et al. [3], matches the above lower bound.

Theorem 2 ([3]) *For any set S of N axis-parallel hypercubes in $\mathbb{R}^d$, the priority R-tree answers a window query with $O((N/B)^{1-1/d} + T/B)$ I/Os. It can be constructed with $O\left(\frac{N}{B} \log_{M/B} \frac{N}{B}\right)$ I/Os.*

It is also reported that the priority R-tree performs well in practice, too [3]. However, it is not known how to update it efficiently while preserving the worst-case bound. The logarithmic method was used to support insertions and deletions [3], but the resulted structure is no longer an R-tree.

Note that the lower bound in Theorem 1 only holds for R-trees. If the data structure is not restricted to R-trees, better query bounds can be obtained for the window-query problem; see e.g., [4].

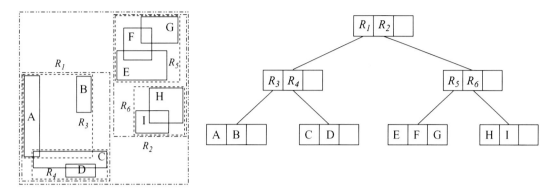

R-Trees, Fig. 1 An R-tree example in two dimensions

Applications

R-trees have been used widely in practice due to its simplicity and ability to store spatial objects of various shapes and to answer various queries. The areas of applications span from geographical information systems (GIS), computer-aided design, computer vision, and robotics. When the objects are not axis-parallel hypercubes, they are often approximated by their minimal bounding boxes, and the R-tree is then built on these bounding boxes. To answer a window query, first the R-tree is used to locate all the intersecting bounding boxes, followed by a filtering step that checks the objects exactly. The R-tree can also be used to support other kinds of queries, for example, aggregation queries, nearest neighbors, etc. In aggregation queries, each object o in S is associated with a weight $w(o) \in \mathbb{R}$, and the goal is to compute $\sum w(o)$ where the sum is taken over all objects that intersect the query range Q. The query algorithm is same as before, except that in addition it keeps running sum while traversing the R-tree and may skip an entire subtree rooted at some u if $R(u)$ is completely contained in Q. To find the nearest neighbor of a query point q, a priority queue is maintained, which stores all the nodes u that might contain an object that is closer to the current nearest neighbor found so far. The priority of u in the queue is the distance between q and $R(u)$. The search terminates when the current nearest neighbor is closer than the top

element in the priority queue. However, no worst-case guarantees are known for R-trees answering these other types of queries, although they tend to perform well in practice.

Open Problems

Several interesting problems remain open with respect to R-trees. Some of them are listed here:

- Is it possible to design an R-tree with the optimal query bound $O((N/B)^{1-1/d} + T/B)$ that can also be efficiently updated? Or prove a lower bound on the update cost for such an R-tree.
- Is there an R-tree that supports aggregation queries for axis-parallel hypercubes in $O((N/B)^{1-1/d})$ I/Os? This would be optimal because the lower bound of Theorem 1 also holds for aggregation queries on R-trees. Note that, however, no sublinear worst-case bound exists for nearest-neighbor queries, since it is not difficult to design a worst-case example for which the distance between the query point q and any bounding box is smaller than the distance between q and its true nearest neighbor.
- When the window query Q shrinks to a point, that is, the query asks for all hypercubes in

S that contain the query point, the problem is often referred to as *stabbing queries* or *point enclosure queries*. The lower bound of Theorem 1 does not hold for this special case, while a lower bound of $\Omega(\log_2 N + T/B)$ was proven in [5], which holds in the strong *indexability* model. It is intriguing to find out the true complexity for stabbing queries using R-trees, which is between $\Omega(\log_2 N + T/B)$ and $O((N/B)^{1-1/d} + T/B)$.

Experimental Results

Nearly all studies on R-trees include experimental evaluations, mostly in two dimensions. Reportedly the Hilbert R-tree [10, 11] has been shown to have good query performance while being easy to construct. The R*-tree's insertion algorithm [6] has often been used for updating the R-tree. Please refer to the book by Manolopoulos et al. [14] for more discussions on the practical performance of R-trees.

Data Sets

Besides some synthetic data sets, the TIGER/Line data (http://www.census.gov/geo/www/tiger/) from the US Census Bureau has been frequently used as real-world data to test R-trees. The R-tree portal (http://www.rtreeportal.org/) also contains many interesting data sets.

URL to Code

Code for many R-tree variants is available at the R-tree portal (http://www.rtreeportal.org/). The code for the priority R-tree is available at (http://www.cs.duke.edu/~yike/prtree/).

Cross-References

▶ B-Trees
▶ External Sorting and Permuting
▶ I/O-Model

Recommended Reading

1. Agarwal PK, de Berg M, Gudmundsson J, Hammar M, Haverkort HJ (2002) Box-trees and R-trees with near-optimal query time. Discret Comput Geom 28:291–312
2. Aggarwal A, Vitter JS (1988) The Input/Output complexity of sorting and related problems. Commun ACM 31:1116–1127
3. Arge L, de Berg M, Haverkort HJ, Yi K (2004) The priority R-tree: a practically efficient and worst-case optimal R-tree. In: Proceedings of the SIGMOD international conference on management of data, pp 347–358
4. Arge L, Samoladas V, Vitter JS (1999) On two-dimensional indexability and optimal range search indexing. In: Proceedings of the ACM symposium on principles of database systems, pp 346–357
5. Arge L, Samoladas V, Yi K (2004) Optimal external memory planar point enclosure. In: Proceedings of the European symposium on algorithms
6. Beckmann N, Seeger B (2009) A revised R*-tree in comparison with related index structures. In: Proceedings of the SIGMOD international conference on management of data, pp 799–812
7. DeWitt DJ, Kabra N, Luo J, Patel JM, Yu J-B (1994) Client-server paradise. In: Proceedings of the international conference on very large databases, pp 558–569
8. García YJ, López MA, Leutenegger ST (1998) A greedy algorithm for bulk loading R-trees. In: Proceedings of the 6th ACM symposium on advances in GIS, pp 163–164
9. Guttman A (1984) R-trees: a dynamic index structure for spatial searching. In: Proceedings of the SIGMOD international conference on management of data, pp 47–57
10. Haverkort H, van Walderveen F (2011) Four-dimensional Hilbert curves for R-trees. J Exp Algorithmics vol 16:Article no 3.4
11. Kamel I, Faloutsos C (1994) Hilbert R-tree: an improved R-tree using fractals. In: Proceedings of the international conference on very large databases, pp 500–509
12. Kanth KVR, Singh AK (1999) Optimal dynamic range searching in non-replicating index structures. In: Proceedings of the international conference on database theory. LNCS, vol 1540, pp 257–276
13. Leutenegger ST, Lopez MA, Edington J (1997) STR: a simple and efficient algorithm for R-tree packing. In: Proceedings of the 13th IEEE international conference on data engineering, pp 497–506
14. Manolopoulos Y, Nanopoulos A, Papadopoulos AN, Theodoridis Y (2005) R-trees: theory and applications. Springer, London
15. Sellis T, Roussopoulos N, Faloutsos C (1987) The R^+-tree: a dynamic index for multi-dimensional objects. In: Proceedings of the international conference on very large databases, pp 507–518

Rumor Blocking

Lidan Fan[1] and Weili Wu[2,3,4]
[1]Department of Computer Science, The
University of Texas, Tyler, TX, USA
[2]College of Computer Science and Technology,
Taiyuan University of Technology, Taiyuan,
Shanxi Province, China
[3]Department of Computer Science, California
State University, Los Angeles, CA, USA
[4]Department of Computer Science, The
University of Texas at Dallas, Richardson, TX,
USA

Keywords

Influence diffusion models; Information cascades;
Link analysis; Protector; Rumor; Submodular
function; Social networks

Years and Authors of Summarized Original Work

2008; Kimura, Saito, Motoda
2011; Budak, Agrawal, Abbadi
2013; Fan, Lu, Wu, Thuraisingham, Ma, Bi

Problem Definition

The aim is to design effective algorithms for
controlling rumor propagation in social networks.
Here, a rumor is viewed as an undesirable thing.
Social networks are represented by undirected or
directed graphs, depending on different contexts.
In these graphs, nodes denote individuals and
edges denote the influence between individuals.

A list of strategies has been proposed to limit
the spread of a rumor in a network. We group
some of the existing research works into two
categories. The first one includes the works that
launch the opposite cascade, protector, to spread
in a network, such that the number of nodes that
adopt the rumor at the end of both cascades diffusion is limited. The second contains the works

that are concerned with the network structure,
that is, they control the rumor propagation by
blocking some edges or nodes or both of them
together in a network.

In this work, we mainly introduce two specific
works belonging to the two different categories.
For each of them, we briefly introduce some
related works.

Problem 1 [2]

Two cascades, rumor (bad campaign) and protector (limiting campaign), diffuse simultaneously in
a network. Influence diffusion models are used
to capture their propagation processes. The objective is to limit the rumor propagation through
protector diffusion.

Given a directed graph $G = (V, E)$, original
rumor sources $R \subseteq V$, an integer $k > 0$, and the
time delay d (a nonnegative integer) for detecting
rumor sources, the objective is to identify k
nodes as initial protectors, such that the expected
number of nodes adopting the rumor at the end of
both rumor and protector propagation processes
is minimized, or equivalently, the reduction in the
expected number of nodes adopting the rumor is
maximized.

Two Influence Diffusion Models

Two influence diffusion models are adopted in
[2].

**Multi-campaign Independent Cascade Model
(MCICM)** In this model, a network is viewed as
a directed graph $G = (V, E)$. The initial set of
rumor sources is denoted by R, and the initial set
of protectors is denoted by P. Each node must
be in one of the three statuses: infected (by the
rumor), protected (by the protector), and inactive
(neither infected nor protected). Each edge e_{uv} is
associated with two values $0 \leq p_r(u, v) \leq 1$ and
$0 \leq p_p(u, v) \leq 1$. Once a node becomes infected
or protected, it remains so forever.

The diffusion process unfolds in discrete time
steps. In any step $t \geq 1$, when a node u first becomes infected (protected), it has a single chance
to activate each currently inactive neighbor v, and
it succeeds with probability $p_r(u, v)$ ($p_p(u, v)$)
provided no neighbor of v tries activating v at the

same step. In other words, at step $t + 1$, node v will become infected (protected) with probability $p_r(u, v)$ $(p_p(u, v))$ provided no neighbor of v tries activating v at the same step. If there are two or more nodes trying to activate v at the same step, at most one of them can succeed. If infected node(s) and protected node(s) try to activate a node at the same step, protected nodes have priority over infected nodes. The process continues until no newly infected or protected node appears.

Campaign-Oblivious Independent Cascade (COICM) This model is similar to the MCICM model; the only difference is that instead of two probabilities are associated with each edge, only one probability $0 \le p(u, v) \le 1$ is associated with each edge e_{uv}. That is, each node has the same probability to forward the two kinds of information, indicating both rumor and protector cascades pass through the same edge with the same probability.

Problem 2 [8]

A single cascade, rumor, diffuses through networks. Influence diffusion models are used to capture rumor propagation process. The objective is to limit the propagation of rumors through blocking links in networks. The aim of [8] is to minimize contamination degree by appropriately removing a fixed number of links. Here, the contamination degree of a network is used to measure how badly the rumor will contaminate the network; see its definition later.

Given a directed graph $G = (V, E)$, a positive integer k where $k < |E|$, find a subset $B^* \subset E$ with $|B^*| = k$ such that $c(G(B^*)) \le c(G(B))$ for any $B \subset E$ with $|B| = k$. Here $c(G)$ denotes the contamination degree. For any link $e \in E$, let $G(e)$ denote the graph $G(V, E \setminus e)$. And $G(e)$ is used as the graph constructed by blocking e in G. Similarly, for any $B \subset E$, let $G(B)$ denote graph $G(V, E \setminus B)$. Then $G(B)$ represents the graph constructed by blocking B in G.

Independent Cascade Model

In this model, a network is considered as a directed graph $G = (V, E)$. Each edge $e_{uv} \in$ E is assigned an influence probability $p(u, v)$, representing the possibility that node u influences node v successfully. For $e_{uv} \notin E$, let $p(u, v) = 0$. Each node can only be in one of the following two statuses: inactive or infected. Once a node becomes infected, it stays infected forever.

The diffusion process unfolds in discrete time steps. Starting with an initial set of infected nodes A_0, at any step $t \ge 1$, when node u first becomes infected in step t, it has a single chance to activate any of its currently inactive neighbors. For neighbor node v, it succeeds with probability $p(u, v)$. If u succeeds in activating v, then v will become infected in step $t + 1$, and if u fails in activating v, then v will stay inactive. If node u does not succeed in activating v, it will not have a second chance to do in all subsequent steps. The process continues until no more activations are possible. If multiple newly activated nodes are in-neighbors of the same inactive node, then their activation attempts are sequenced in an arbitrary order.

Key Results

For Problem 1

Given the set of rumor sources R, a set of initial protectors P, and rumor detection delay d, a set function $f_{Rd}(P)$ represents the number of nodes that are prevented by P with diffusion delay d from adopting R. In other words, function $f_{Rd}(P)$ denotes the nodes that will be infected by R if, instead of P, the empty set is selected as the set of protectors. Therefore, the problem is to select P such that the expectation of $f_{Rd}(P)$ is maximized.

The NP-hardness of this problem is proved. Then for the MCICM model, the high-effectiveness property where $p_p(u, v) = 1$ for edge $e_{uv} \in E$ is adopted. Then the objective functions are proved to be submodular and monotone under both the MCICM model with the high-effectiveness property and the COICM model. Therefore, Algorithm 1 is applied to provide $(1 - 1/e)$-approximation solutions for the problem. However, the objective function is not

---

Algorithm 1: The greedy algorithm

Input: Graph $G = (V, E)$, the set of initial rumor sources R, rumor detection delay d, a positive integer k, a positive number n, representing the simulation times
Output: Set P
$P = \emptyset$
for $i = 1$ *to* k **do**
   for *each* $u \in V \setminus (R \cup P)$ **do**
      $N_u = 0$
      for $j = 1$ *to* n **do**
         $N_u + = f_{Rd}(P \cup \{u\}) - f_{Rd}(P)$
      end
      $N_u = N_u/n$
   end
   $Loc = \arg\max_{u \in (V \setminus (R \cup P))}\{N_u\}$
   $P = P \cup Loc$
end
Output P

---

submodular under the MCICM model without the high-effectiveness property.

Variants of Problem 1

Several works in different contexts also consider using the diffusion of protectors to contain the spread of rumors. In comparison to Problem 1, they have adopted different influence diffusion models, as well as formulated different optimization problems.

Selection of Fixed Number of Protectors

The work of He et al. [5] studies rumors blocking maximization under an extension of the classical Linear Threshold (LT) model [6], in which they incorporate two cascades, rumor and protector. Each node in this model can be in one of the three states: infected, protected, and inactive. For each node, its currently infected neighbors and protected neighbors determine whether it will become infected, protected, or stay inactive, respectively. When a node is activated by its infected neighbors and protected neighbors at the same time, then infected neighbors have priority over protected neighbors. Each edge $e = (u, v)$ has two weights, w^r_{uv} (rumor propagation) and w^p_{uv} (protector propagation). Each node u picks two independent thresholds from $[0, 1]$; one is for rumor diffusion and the other is for protector diffusion.

Then they develop the objective function $S_R(X)$ for this problem, which represents the expected number of nodes that is saved (from being infected by rumors R) by X. This problem is shown to be NP-hard and the objective function is proved to be submodular and monotone, then the greedy algorithm with performance ratio $1 - \frac{1}{e}$ is applied. To efficiently compute the values of $S_R(X)$, the authors propose the CLDAG algorithm.

Instead of choosing initial protectors from nodes not in rumor sources, the authors in [10] select a fixed number of nodes from initial infected nodes (rumor sources) and the rest of the nodes in a network as initial protectors, such that the number of nodes protected during T time steps is maximized. They study this problem under the LT model and the IC model. Two approximation algorithms are proposed.

Protection of a Subset of Nodes

Instead of limiting rumor diffusion through launching a fixed number of protectors, the work of [12] exploits the problem which aims to select the smallest set of influential people as protectors, such that the diffusion process starting from these protectors limits the propagation of rumors R in a fraction $0 \le \alpha \le 1$ of the whole network in T time steps. They study four variants of this problem, which are the combinations of the two parameters: R (can be unknown or known) and T (can be constrained or unconstrained). These problems are studied under the extensions of the IC model and the LT model, in which two cascades, rumors and protectors, are considered. For each edge, both of them have the same influence probability (IC) or influence weight (LT). For each node, they have the same threshold (LT). The key point is that when the two cascades try to activate a node at the same time, protectors have priority over rumors.

The authors prove the NP-hardness of the four problems under the proposed models. For the variant that R is unknown and T is unconstrained, the Greedy Viral Stopper (GVS) algorithm is adopted to select the protectors, and the solution obtained is within a constant factor (in terms of the number of nodes in the network) extra from the optimal solution. The GVS algorithm can be used in the variants that R is known and T is either constrained or unconstrained. These variants are shown to be hard to approximate to a logarithmic factor in terms of the number of nodes in the network. To get a good solution within short time, the Community-Based Heuristic algorithm is proposed.

Noticing the community structure of social networks, the work of Fan et al. [3] contains rumor propagation by selecting a minimal set of initial protectors to protect a special kind of vertex set, which play the role as the "gates" of rumors' neighborhood communities. Two variants of the problem are studied under two different models, both variants are shown to be NP-hard, and approximation algorithms are developed to obtain good solutions.

Game Theory Aspect

The rumor blocking is also studied from the game theory aspect [13], where it uses graphs with nodes representing the tribal leaders and edges representing possible transmission of influence. Under this context, rumor blocking is viewed as a two-player game, in which one player, the rumor, will attempt to maximize the number of nodes accepting it while the second player, the protector, will attempt to minimize the rumor's influence. Both the rumor and the protector will choose their action sources (initial rumor sources and initial protector sources). In the zero-sum game context, the rumor's payoff is equal to the expected number of nodes infected, and the protector's payoff is the opposite of the rumor's payoff. The authors propose a double oracle algorithm for this game.

For Problem 2

Under the IC model, given an initial active set X, define the number of active nodes at the end of the influence diffusion process on G as $f(X; G)$. Let $\sigma(X; G)$ denote the expected value of $f(X; G)$. $\sigma(X; G)$ is called as the influence degree of node set X on graph G. Two notions of containment degrees are defined. One is called *Average Contamination Degree*, representing the average of influence degree of all the

---

Algorithm 2: The greedy algorithm - IC

Input: Graph $G_0 = (V_0, E_0)$, a positive integer $k < |E_0|$
Output: The set of links blocked
Initialize a subset $L \subseteq E_0$ as $L = $ Initialize a graph $G = (V, E)$ as $V = V_0, E = E_0$
while $|L| < k$ **do**
     select a link e^*, such that
     $e^* = \arg\min_{e \in E} c(G(e))$
     $L = L \cup \{e^*\}$
     $E = E \setminus \{e^*\}$
end
Output L

---

nodes in G, denoted as $c_0(G)$. Its definition is $c_0(G) = \frac{1}{|V|} \sum_{v \in V} \sigma(v; G)$. The other is called *Worst Contamination Degree*, representing the maximum of influence degree of all the nodes in G, denoted as $c_+(G)$. Its definition is: $c_+(G) = \max_{v \in V} \sigma(v; G)$. Approximation algorithms are proposed to find good solution for the problem.

For a given graph $G = (V, E)$, exactly computing influence degree $c(G(e)); e \in E$ in Algorithm 2 is an open problem. Therefore, heuristic strategies are proposed to estimate $c(G(e)); e \in E$. These estimations are based on the *Bond Percolation Method* proposed by Kimura in [7], which we describe below.

Bond Percolation Method [7]

Assume there are propagation probabilities $\{p_e; e \in E\}$ on a graph $G = (V, E)$. In terms of information diffusion on a network, the *occupied* links represent the links that the information propagates, and the *unoccupied* links represent the links that the information does not propagates. The bond percolation process with occupation probabilities $\{p_e; e \in E\}$ on a graph $G = (V, E)$ is a stochastic process in which the probability of each link $e \in E$ becomes occupied is p_e.

Construct N graphs through the bond percolation process, that is, $\{G_n = (V, E_n); n = 1, \ldots, N\}$. For any $u \in V'$ on graph $G' = (V', E')$, let $F(u; G')$ represent the set of all the nodes that are *reachable* from u on G'. A node v is said to be reachable from u if there is a path from u to v through the links on G'. Define function $g(u; G, N) =$

$\frac{1}{N}\sum_{n=1}^{N}|F(u;G^n)|$. Then, $g(u;G,N)$ can be used to estimate $\sigma(u;G)$, where $u \in V$ and if N is sufficiently large. Decompose each G_n into the *strongly connected components (SCC)* as $V = \bigcup_{i=1}^{I_n} SCC(u_i^n;G_n)$, where $u_i^n \in V$ and $SCC(u_i^n;G_n)$ denotes the SCC of graph G_n that contains u_i^n. I_n is the number of the SCC of graph G_n, using the fact that $|F(u;G^n)| = |F(u_i^n;G_n)|$ for all $u \in V$ to calculate $\{|F(u;G^n)|;u \in V, n = 1,\ldots,N\}$, then compute $g(u;G,N)$, and finally $\sigma(u;G)$ can be calculated, where $u \in V$.

Estimation Method

We now describe how to estimate $c(G(e));e \in E$ in [8]. For a graph $G = (V,E)$, first, construct N sample graphs through the bond percolation process as $\{G_n = (V,E_n);n = 1,\ldots,N\}$. Next, for each $e \in E$, identify the subset of N, which is denoted as $S_N(e)$ and satisfies $S_N(e) = \{n \in \{1,\ldots,N\};e \notin E_n\}$. Now apply the bond percolation process on the graph $G(e) = (V, E \setminus \{e\})$ for $|S_N(e)|$ times, then $|S_N(e)|$ graphs are obtained by the occupied links, denote them as $\{G(e)^n; n = 1,\ldots,|S_N(e)|\}$. Given that N is large enough to ensure $|S_N(e)|$ sufficiently large, then the function $g(u;G(e),|S_N(e)|)$ equals to $\frac{1}{|S_N(e)|}\sum_{n=1}^{|S_N(e)|}|F(u;G(e)^n)|$, where $u \in V$, can be used to estimate $\sigma(u;G(e))$. Since each link of the graph G is independently declared occupied in the bond percolation process, then an alternative, $g'(u,e) = \frac{1}{|S_N(e)|}\sum_{n\in S_N(e)}|F(u;G^n)|$, is used to estimate $\sigma(u;G(e))$.

Variants of Problem 2

The authors in [9] adapt the method they used for the IC model to study Problem 2 under the LT model [6]. In [1], the authors incorporate the trust among users in the information propagation process, and they propose a measure to compute trust between a pair of users. Then a Weighted Trust Network (WTN) is built, and the objective of the problem is to find the Maximum Spanning Tree (MST) in the WTN and, finally, immunize all the edges in the MST of the WTN. Another method that controls rumor spread through block-ing nodes and links simultaneously can be found in [4].

Nguyen et al. [11] study the rumor block-ing problem under a dynamic social network structure. They propose to distribute patches to the most influential nodes in the social network, such that the number of nodes influenced by rumors is limited. In their work, they first take into account the network community structure and adaptively keeps it updated as the social network evolves, and then select most influen-tial individuals from each communities to be patched.

The work of Zhu [14] focuses on the rumor blocking problem in cellular networks. First, a social relationship graph between mobile phones is obtained based on network traffic; the au-thors develop two graph-partitioning algorithms to partition the graph into many separate parts as possible and contain the rumor diffusion within each part. Then a minimum set of *key nodes*, which separate these different parts, is selected to be patched. The intuition is that the infected nodes in a part need to go through some of these key nodes to influence nodes in another part. Once these nodes are patched, it is impossi-ble for the influence propagates among different parts.

Applications

Practical applications can be seen in control-ling: propagation of computer viruses and worms propagates over computer networks, spread of malicious rumors through social networks, diffu-sion of infections or epidemics (such as swine flu) among groups of people, propagation of mobile worm in cellular networks, and so on.

Open Problems

There are many interesting directions that deserve further explorations. One direction is to improve existing influence diffusion models by consider-ing continuous time influence diffusion, users' preferences to different kinds of information, factors influencing users' threshold in adopting a kind of information, etc. Another direction is to design efficient strategies to control the spread of rumors when only partial of a network structure is observable. Another research issue is incorporating the detection of rumor sources into rumor blocking and continuous time delay of protectors.

Cross-References

▶ Influence and Profit
▶ Influence Maximization

Recommended Reading

1. Bao Y, Niu Y, Yi C, Xue Y (2014) Effective immunization strategy for rumor propagation based on maximum spanning tree. In: Computing, Networking and Communications (ICNC), 2014 International Conference on, pp 11–15, DOI:10.1109/ICCNC.2014.6785296
2. Budak C, Agrawal D, El Abbadi A (2011) Limiting the spread of misinformation in social networks. In: Srinivasan S, Ramamritham K, Kumar A, Ravindra MP, Bertino E, Kumar R (eds) WWW, ACM, pp 665–674
3. Fan L, Lu Z, Wu W, Thuraisingham BM, Ma H, Bi Y (2013) Least cost rumor blocking in social networks. In: ICDCS, IEEE, pp 540–549
4. He J, Liang H, Yuan H (2011) Controlling infection by blocking nodes and links simultaneously. In: Chen N, Elkind E, Koutsoupias E (eds) WINE, Springer, Lecture Notes in Computer Science, vol 7090, pp 206–217
5. He X, Song G, Chen W, Jiang Q (2012) Inuence blocking maximization in social networks under the competitive linear threshold model. In: SDM, SIAM / Omnipress, pp 463–474
6. Kempe D, Kleinberg JM, Tardos E (2003) Maximizing the spread of inuence through a social network. In: Getoor L, Senator TE, Domingos P, Faloutsos C (eds) KDD, ACM, pp 137–146
7. Kimura M, Saito K, Nakano R (2007) Extracting inuential nodes for information diusion on a social network. In: AAAI, AAAI Press, pp 1371–1376
8. Kimura M, Saito K, Motoda H (2008) Minimizing the spread of contamination by blocking links in a network. In: Fox D, Gomes CP (eds) AAAI, AAAI Press, pp 1175–1180
9. Kimura M, Saito K, Motoda H (2008) Solving the contamination minimization problem on networks for the linear threshold model. In: Ho TB, Zhou ZH (eds) PRICAI, Springer, Lecture Notes in Computer Science, vol 5351, pp 977–984
10. Li S, Zhu Y, Li D, Kim D, Huang H (2013) Rumor restriction in online social networks. In: IPCCC, IEEE, pp 1–10
11. Nguyen N, Xuan Y, Thai M (2010) A novel method for worm containment on dynamic social networks. In: MILITARY COMMUNICATIONS CONFERENCE, 2010 - MILCOM 2010, pp 2180–2185, DOI:10.1109/MILCOM.2010.5680488
12. Nguyen NP, Yan G, Thai MT, Eidenbenz S (2012) Containment of misinformation spread in online social networks. In: Contractor NS, Uzzi B, Macy MW, Nejdl W (eds) WebSci, ACM, pp 213–222
13. Tsai J, Nguyen TH, Tambe M (2012) Security games for controlling contagion. In: Homann J, Selman B (eds) AAAI, AAAI Press
14. Zhu Z, Cao G, Zhu S, Ranjan S, Nucci A (2009) A social network based patching scheme for worm containment in cellular networks. In: INFOCOM, IEEE, pp 1476-1484

S

Schedulers for Optimistic Rate Based Flow Control

Panagiota Fatourou
Department of Computer Science, University of
Ioannina, Ioannina, Greece

Keywords

Bandwidth allocation; Rate adjustment; Rate allocation

Years and Authors of Summarized Original Work

2005; Fatourou, Mavronicolas, Spirakis

Problem Definition

The problem concerns the design of efficient rate-based flow control algorithms for virtual-circuit communication networks where a connection is established by allocating a fixed path, called *session*, between the source and the destination. Rate-based flow-control algorithms repeatedly adjust the transmission rates of different sessions in an end-to-end manner with primary objectives to optimize the network utilization and achieve some kind of fairness in sharing bandwidth between different sessions.

A widely-accepted fairness criterion for flow-control is *max-min fairness* which requires that the rate of a session can be increased only if this increase does not cause a decrease to any other session with smaller or equal rate. Once max-min fairness has been achieved, no session rate can be increased any further without violating the above condition or exceeding the bandwidth *capacity* of some link. Call *max-min* rates the session rates when max-min fairness has been reached.

Rate-based flow control algorithms perform rate adjustments through a sequence of *operations* in a way that the capacities of network links are never exceeded. Some of these algorithms, called *conservative* [3, 6, 10, 11, 12], employ operations that gradually increase session rates until they converge to the max-min rates without ever performing any rate decreases. On the other hand, *optimistic* algorithms, introduced more recently by Afek, Mansour, and Ostfeld [1], allow for decreases, so that a session's rate may be intermediately be larger than its final max-min rate.

Optimistic algorithms [1, 7] employ a specific rate adjustment operation, called *update operation* (introduced in [1]). The goal of an update operation is to achieve fairness among a set of neighboring sessions and optimize the network utilization in a local basis. More specifically, an update operation calculates an increase for the rate of a particular session (the *updated* session) for each link the session traverses. The calculated increase on a particular link is the maximum possible that respects the max-min fairness condition between the sessions traversing the link; that

© Springer Science+Business Media New York 2016
M.-Y. Kao (ed.), *Encyclopedia of Algorithms*,
DOI 10.1007/978-1-4939-2864-4

is, this increase should not cause a decrease to the rate of any other session traversing the link with smaller rate than the rate of the updated session after the increase. Once the maximum increase on each link has been calculated the minimum among them is applied to the session's rate (let e be the link for which the minimum increase has been calculated). This causes the decrease of the rates of those sessions traversing e which had larger rates than the increased rate of the updated session to the new rate. Moreover, the update operation guarantees that all the capacity of link e is allocated to the sessions traversing it (so the bandwidth of this link is fully utilized).

One important performance parameter of a rate-based flow control algorithm is its *locality* which is characterized by the amount of knowledge the algorithm requires to decide which session's rate to update next. *Oblivious* algorithms do not assume any knowledge of the network topology or the current session rates. *Partially oblivious* algorithms have access to session rates but they are unaware of the network topology, while *non-oblivious* algorithms require full knowledge of both the network topology and the session rates. Another crucial performance parameter of rate-based flow control algorithms is the *convergence complexity* measured as the maximum number of rate-adjustment operations performed in any execution until max-min fairness is achieved.

Key Results

Fatourou, Mavronicolas and Spirakis [7] have studied the convergence complexity of optimistic rate-based flow control algorithms under varying degrees of locality. More specifically, they have proved lower and upper bounds on the convergence complexity of oblivious, partially-oblivious and non-oblivious algorithms. These bounds are expressed in terms of n the number of sessions laid out on the network.

Theorem 1 (Lower Bound for Oblivious Algorithms, Fatourou, Mavronicolas and Spirakis [7]) *Any optimistic, oblivious, rate-based flow control algorithm requires $\Omega(n^2)$ update operations to compute the max-min rates.*

Fatourou, Mavronicolas and Spirakis [7] have presented algorithm RoundRobin, which applies update operations to sessions in a round robin order. Obviously, RoundRobin is oblivious. It has been proved [7] that the convergence complexity of RoundRobin is $O(n^2)$. This shows that the lower bound for oblivious algorithms is tight.

Theorem 2 (Upper Bound for Oblivious Algorithms, Fatourou, Mavronicolas and Spirakis [7]) RoundRobin *computes the max-min rates after performing $O(n^2)$ update operations.*

RoundRobin belongs to a class of oblivious algorithms, called Epoch [7]. Each algorithm of this class repeatedly chooses some permutation of all session indices and applies update operations on the sessions in the order determined by this permutation. This is performed n times. Clearly, Epoch is a class of oblivious algorithms. It has been proved [7] that each of the algorithms in this class has convergence complexity $O(n^2)$.

Another oblivious algorithm, called Arbitrary, has been presented in [1]. The algorithm works in a very simple way by choosing the next session to be updated in an arbitrary way, but it requires an exponential number of update operations to compute the max-min rates.

Fatourou, Mavronicolas and Spirakis [7] have proved that partially-oblivious algorithms do not achieve better convergence complexity than oblivious algorithms despite the knowledge they employ.

Theorem 3 (Lower Bound for Partially Oblivious Algorithms, Fatourou, Mavronicolas and Spirakis [7]) *Any optimistic, partially oblivious, rate-based flow control algorithm* requires $\Omega(n^2)$ *update operations to compute the max-min rates.*

Afek, Mansour and Ostfeld [1] have presented a partially oblivious algorithm, called GlobalMin. The algorithm chooses as the session to update next the one with the minimum rate among all sessions. The convergence complexity of GlobalMin is $O(n^2)$ [1]. This shows that the lower bound for partially-oblivious algorithms is tight.

Theorem 4 (Upper Bound for Partially Oblivious algorithms, Afek, Mansour and Ostfeld [1]) GlobalMin *computes the max-min rates after performing $O(n^2)$ update operations.*

Another partially-oblivious algorithm, called LocalMin, is also presented in [1]. The algorithm chooses to schedule next a session which has a minimum rate among all the sessions that share a link with it. LocalMin has time complexity $O(n^2)$.

Fatourou, Mavronicolas and Spirakis [7] have presented a non-oblivious algorithm, called Linear, that exhibits linear convergence complexity. Linear follows the classical idea [3, 12] of selecting as the next updated session one of the sessions that traverse the most congested link in the network. To discover such a session, Linear requires knowledge of the network topology and the session rates.

Theorem 5 (Upper Bound for Non-Oblivious Algorithms, Fatourou, Mavronicolas and Spirakis [7]) Linear *computes the max-min rates after performing $O(n)$ update operations.*

The convergence complexity of Linear is optimal, since n rate adjustments must be performed in any execution of an optimistic rate-based flow control algorithm (assuming that the initial session rates are zero). However, this comes at a remarkable cost in locality which makes Linear impractical.

Applications

Flow control is the dominant technique used in most communication networks for preventing data traffic congestion when the externally injected transmission load is larger than what can be handled even with optimal routing. Flow control is also used to ensure high network utilization and fairness among the different connections. Examples of networking technologies where flow control techniques have been extensively employed to achieve these goals are TCP streams [5] and ATM networks [4]. An overview of flow control in practice is provided in [3].

The idea of controlling the rate of a traffic source originates back to the data networking protocols of the ANSI Frame Relay Standard. Rate-based flow control is considered attractive due to its simplicity and its low hardware requirements. It has been chosen by the ATM Forum on Traffic Management as the best suited technique for the goals of ABR service [4].

A substantial amount of research work has been devoted in past to conservative flow control algorithms [3, 6, 10, 11, 12]. The optimistic framework has been introduced much later by Afek et al. [1] as a more suitable approach for real dynamic networks where decreases of session rates may be necessary (e.g., for accommodating the arrival of new sessions). The algorithms presented in [7] improve upon the original algorithms proposed in [1] in terms of either convergence complexity, or locality, or both. Moreover, they identify that certain classical scheduling techniques, such as round-robin [11], or adjusting the rates of sessions traversing one of the most congested links [3, 12] can be efficient under the optimistic framework. The first general lower bounds on the convergence complexity of rate-based flow control algorithms are also presented in [7].

The performance of optimistic algorithms has been theoretically analyzed in terms of an abstraction, namely the update operation, which has been designed to address most of the intricacies encountered by rate-based flow control algorithms. However, the update operation masks low-level implementation details, while it may incur non-trivial, local computations on the switches of the network. Fatourou, Mavronicolas and Spirakis [9] have studied the impact on the efficiency of optimistic algorithms of local computations required at network switches in order to implement the update operation, and proposed a distributed scheme that implements a broad class of such algorithms. On a different avenue, Afek, Mansour and Ostfeld [2] have proposed a simple flow control scheme, called Phantom, which employs the idea of considering an imaginary session on each link [10, 12], and they have discussed how Phantom can be applied to ATM networks and networks of TCP routers.

A broad class of modern distributed applications (e.g., remote video, multimedia conferencing, data visualization, virtual reality, etc.) exhibit highly differing bandwidth requirements and need some kind of quality of service guarantees. To efficiently support a wide diversity of applications sharing available bandwidth, a lot of research work has been devoted on incorporating priority schemes on current networking technologies. Priorities offer a basis for modeling the diverse resource requirements of modern distributed applications, and they have been used to accommodate the needs of network management policies, traffic levels, or pricing. The first efforts for embedding priority issues into max-min fair, rate-based flow control were performed in [10, 12]. An extension of the classical theory of max-min fair, rate-based flow control to accommodate priorities of different sessions has been presented in [8]. (A number of other papers addressing similar generalizations of max-min fairness to account for priorities and utility have been presented after the original publication of [8].)

Many modern applications are not based solely on point-to-point communication but they rather require multipoint-to-multipoint transmissions. A max-min fair rate-based flow control algorithm for multicast networks is presented in [14]. Max-min fair allocation of bandwidth in wireless adhoc networks is studied in [15].

Open Problems

The research work on optimistic, rate-based flow control algorithms leaves open several interesting questions. The convergence complexity of the proposed optimistic algorithms has been analyzed only for a static set of sessions laid out on the network. It would be interesting to evaluate these algorithms under a dynamic network setting, and possibly extend the techniques they employ to efficiently accommodate arriving and departing sessions.

Although max-min fairness has emerged as the most frequently praised fairness criterion for flow control algorithms, achieving it might be expensive in highly dynamic situations. Afek et al. [1] have proposed a modified version of the update operation, called *approximate* update, which applies an increase to some session only if it is larger than some quantity $\delta > 0$. An *approximate* optimistic algorithm uses the approximate update operation and terminates if no session rate can be increased by more than δ. Obviously such an algorithm does not necessarily reach max-min fairness. It has been proved [1] that for some network topologies every approximate optimistic algorithm may converge to session rates that are away from their max-min counterparts by an exponential factor. The consideration of other versions of update operation or different termination conditions might lead to better max-min fairness approximations and deserves more study; different choices may also significantly impact the convergence complexity of approximate optimistic algorithms. It would be also interesting to derive trade-off results between the convergence complexity of such algorithms and the distance of the terminating rates they achieve to the max-min rates.

Fairness formulations that naturally approximate the max-min condition have been proposed by Kleinberg et al. [13] as suitable fairness criteria for certain routing and load balancing applications. Studying these formulations under the rate-based flow control setting is an interesting open problem.

Cross-References

▶ Multicommodity Flow, Well-Linked Terminals and Routing Problems

Recommended Reading

1. Afek Y, Mansour Y, Ostfeld Z (1999) Convergence complexity of optimistic rate based flow control algorithms. J Algorithms 30(1):106–143
2. Afek Y, Mansour Y, Ostfeld Z (2000) Phantom: a simple and effective flow control scheme. Comput Netw 32(3):277–305
3. Bertsekas DP, Gallager RG (1992) Data networks, 2nd edn. Prentice Hall, Englewood Cliffs
4. Bonomi F, Fendick K (1995) The rate-based flow control for available bit rate ATM service. IEEE/ACM Trans Netw 9(2):25–39
5. Brakmo LS, Peterson L (1995) TCP vegas: end-to-end congestion avoidance on a global internet. IEEE J Sel Areas Commun 13(8):1465–1480
6. Charny A (1994) An algorithm for rate-allocation in a packet switching network with feedback. Technical report MIT/LCS/TR-601, Massachusetts Institute of Technology, Apr 1994
7. Fatourou P, Mavronicolas M, Spirakis P (2005) Efficiency of oblivious versus non-oblivious schedulers for optimistic, rate-based flow control. SIAM J Comput 34(5):1216–1252
8. Fatourou P, Mavronicolas M, Spirakis P (2005) Max-min fair flow control sensitive to priorities. J Interconnect Netw 6(2):85–114, Also in Proceedings of the 2nd international conference on principles of distributed computing, pp 45–59 (1998)
9. Fatourou P, Mavronicolas M, Spirakis P (1998) The global efficiency of distributed, rate-based flow control algorithms. In: Proceedings of the 5th colloquium on structural information and communication complexity, June 1998, pp 244–258
10. Gafni E, Bertsekas D (1984) Dynamic control of session input rates in communication networks. IEEE Trans Autom Control 29(11):1009–1016
11. Hahne E (1991) Round Robin scheduling for max-min fairness in data networks. IEEE J Sel Areas Commun 9(7):1024–1039
12. Jaffe J (1981) Bottleneck flow control. IEEE Trans Commun 29(7):954–962
13. Kleinberg J, Rabani Y, Tardos É (1999) Fairness in routing and load balancing. In: Proceedings of the 40th annual IEEE symposium on foundations of computer science, Oct 1999, pp 568–578
14. Sarkar S, Tassiulas L (2005) Fair distributed congestion control in multirate multicast networks. IEEE/ACM Trans Netw 13(1):121–133
15. Tassiulas L, Sarkar S (2005) Maxmin fair scheduling in wireless adhoc networks. IEEE J Sel Areas Commun 23(1):163–173

Scheduling in Data Broadcasting

Xiaofeng Gao
Department of Computer Science, Shanghai Jiao Tong University, Shanghai, China

Keywords

Algebraic algorithm; Data retrieval; Scheduling; Wireless data broadcast

Years and Authors of Summarized Original Work

2011; Gao, Lu, Wu, Fu
2013; Gao, Lu, Wu, Fu

Problem Definition

Wireless data broadcasting means a set of data are repeatedly broadcast from a base station to a mass number of wireless and mobile clients. If a client wants a specific datum, it will access onto the broadcasting channel, get the location (appearance time) of the datum with the help indices, and wait until the datum has been broadcast. The scheduling problem in data broadcasting deals with the design of an efficient permutation strategy for a client to download a required subset of data from an multichannel broadcasting system, with both time and energy constraints. Here time constraint means the client wants the minimum downloading time from when it starts the query until the moment it has successfully download each piece of datum, while the energy constraint means the client wants the minimum switching numbers among channels to reduce extra battery consumption. Correspondingly, we can define the scheduling problem formally as follows:

A client wants to download a group of k data items $D = \{d_1, d_2, \ldots, d_k\}$, each with different sizes. Those data items are broadcasted on n different channels $C = \{c_1, c_2, \ldots, c_n\}$ repeatedly together with many other data items. Each channel may have different bandwidth and

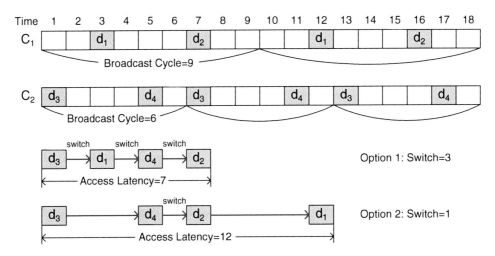

Scheduling in Data Broadcasting, Fig. 1 Example of possible objective contradiction

broadcast cycle length. Let the time to download the smallest transmission packet be a unit time, and the length of d_i can be represented as l_i (also referring as downloading time).

Assume the client knows the locations (channel id and time offsets) of the required data set beforehand at the starting time $t = 0$ (with the help of indices, which is beyond the scope of our problem), and the target is how to download k known data from n channels efficiently with minimum downloading time (we also refer it as *access latency*) and minimum switching numbers.

Unfortunately, the two objectives in this problem are conflicting to each other. Figure 1 is an example to illustrate this phenomenon. In Fig. 1, there are two channels broadcasting 15 data items repeatedly. Suppose the gray data items {1,2,3,4} are of the request. The starting point of the client retrieving process is at $t = 0$. If we want to minimize the access latency, the request should be retrieved in the order of "3 → 1 → 4 → 2" which takes only 7 time units but needs 3 switches (shown as Option 1 in Fig. 1). However, if we want to minimize the switches, the best retrieving order should be "3 → 4 → 2 → 1" which needs only 1 switch but takes 12 time units (shown as Option 2 in Fig. 1). This example exhibits that access latency and number of switches cannot be minimized at the same time. They are contradictory factors.

As a consequence, we want to fix one factor and minimize another objective, and thus have the following objective:

Objective

We hope to design a data downloading order for a client to download k data items from n broadcasting channels, such that the access latency t is minimized if we restrict the switch number among channels (denoted as h); otherwise, we will minimize the number of switches h once the access latency t is bounded.

Constraints

1. **Switch Constraint**: Note that if a client is downloading a data from channel c_i at time t_0, then it cannot switch to channel c_j, where $j \neq i$, to download another data at time $t_0 + 1$ due to connection protocols. Thus, we assume if a client wants to download data from another channel, it needs at least one time unit for channel switching. Figure 2 gives a typical process of data retrieval in multichannel broadcast environments. The query data set is {d_1, d_3, d_5}, and a user can download data object d_1 and d_3 from channel c_1 and then switch to channel c_3 at time $t = 6$ to download data object d_5 at time $t = 7$. However, after time $t = 5$, the user cannot switch from channel c_1 to c_2 to download data

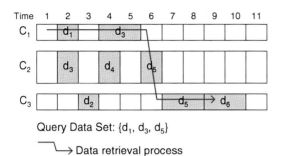

Scheduling in Data Broadcasting, Fig. 2 Switch constraint

d_5 at time $t = 6$. From Fig. 2, we also get that the bandwidths of different channels are not necessarily the same. Actually, the bandwidth of channel c_2 is twice as that of c_1 or c_3, thus d_3 or d_5, which take two time slots on c_1 or c_3, can be broadcasted in one time unit by c_2.

2. **Objective Constraint**: We have to setup a reasonable threshold for latency constraint t and switch constraint h, such that we would achieve a feasible solution for the corresponding minimized switches and shortest access latency.

Problem 1 (Scheduling in Data Broadcasting)

INPUT: *The required data subset $D = \{d_1, d_2, \ldots, d_k\}$ broadcast on n different channels with their locations and downloading time l_i, a switch constraint h or latency constraint t.*

OUTPUT: *A permutation of D such that if starting from time slot zero, a client would achieve the shortest access latency (with switch threshold h) or the minimum switch numbers (with latency threshold t) if it follows this permutation to download each data item sequentially with switch constraint.*

Key Results

Scheduling is an important part in the wireless data broadcast system. Researchers tend to divide the scheduling problems into two subproblems. The first one is the data allocation problem in

the server side, while the other one is the data retrieval problem in the client size.

With respect to server side scheduling, several works have been proposed to improve the system performance [1–5]. Acharya et al. [1] first dealt with the data allocation problem for single-channel environment. He proposed a scheduling algorithm considering data access frequencies and allowed frequent accessed data to be broadcasted more often. Most works concerned multi-channel environment. For data set with uniform length, Yee et al. [2] proposed an $O(t^2m)$ time-complexity dynamic programming algorithm to find the optimal schedule and also a near optimal greedy algorithm to reduce the time complexity. For nonuniform lengths case, Ardizzoni et al. [3] proved that this problem is strong NP-hard. Ardizzoni et al. [3], Anticaglia et al. [4], and Kenyon et al. [5] designed algorithms based on greedy and heuristic strategy.

Also most of the literature discussed the data allocation problem from server's point of view; several works [6–10] considered the data retrieval scheduling problem from the client's point of view. Shi et al. [6] defined the data retrieval problem in MIMO environment as parallel data retrieval scheduling with MIMO Antennae (PADRS-MIMO) and proposed two greedy heuristics to guarantee minimum switchings among channels or reduce the downloading time when the number of antennae in the mobile devices are limited. Lu et al. [7, 8] defined the largest number data retrieval (LNDR) and maximum cost data retrieval (MCDR) problems and considered the hopping cost. He also proved that when the hopping cost cannot be ignored, LNDR is NP-hard and designed a 1/2-approximation algorithm. Gao et al. [9, 10] designed a randomized algebraic algorithm that takes both energy cost and access time into consideration to schedule the data retrieval process in multichannel environments. The algorithm proposed can detect whether a given data retrieval problem has a solution with access time t and number of switchings h in $O\left(2^k (hnt)^{O(1)}\right)$ time, where n is the number of channels and k is the number of requested data items.

S

**Scheduling in Data
Broadcasting, Fig. 3**
Example of VC reduction

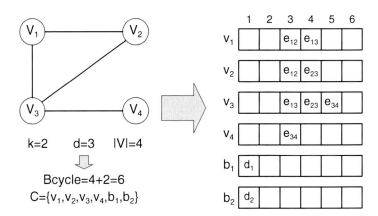

Bcycle=4+2=6
C={v_1,v_2,v_3,v_4,b_1,b_2}

Hardness Analysis

Define a tuple $s = \{i_s, j_s, t_s, t'_s\}$ to denote the datum d_{i_s}, which can be downloaded from channel c_{j_s} during the time span $[t_s, t'_s]$; then it is clear that a valid data retrieval schedule is a sequence of k intervals $s_1, s_2, \ldots, s_k$, each tuple corresponds to a distinct data item in D, and there are no conflicts between any two of the k tuples. To analyze the NP-hardness, we then define the decision problem of MCDR.

Definition 1 (Decision MCDR) Given a data set D, a channel set C, a time threshold t, and a switching threshold h, find a valid data retrieval schedule to download all the data in D from C before time t with at most h switchings.

Theorem 1 *MCDR problem is NP-hard.*

Proof We use $VC \leq_p MCDR$ to prove this theorem. Here VC is the decision problem of vertex cover, say, given a graph $G = (V, E)$, we want to find a minimum size vertex subset $VC \subseteq V$ such that for any edge $(v_i, v_j) \in E$, either $v_i \in VC$ or $v_j \in VC$. An instance of vector cover is: given a graph $G = (V, E)$ and integer k, does it have a vertex cover VC with size k? Then we construct an instance of MCDR from G and k as follows:

1. For each vertex $v_i \in V$, define a channel v_i. Define another k channels $b_1, \ldots, b_k$. Then the channel set is $C = \{v_1, \ldots, v_{|V|}, b_1, \ldots,$

$b_k\}$. Totally $|V| + k$ channels. Let δ be the maximum vertex degree in G, and then each channel has broadcast cycle length $\delta + 3$.

2. For each edge $(v_i, v_j) \in E$, define a unit length data item e_{ij} in data set D_e and append it on channel c_i and c_j (the order can be arbitrary and starting from the third time unit).

3. For each channel b_i, define a unit length data item d_i in data set D_d and allocate it on the first time unit of channel b_i.

4. The data set $D = D_e \cup D_b$.

Figure 3 is an example to show how to construct the broadcast system. In this figure, $\delta = 3$, $k = 2$, $|V| = 4$; thus, the channel set should be $\{v_1, v_2, v_3, v_4, b_1, b_2\}$, each having broadcast length $\delta + 3 = 6$. Each e_{ij} represents an edge (v_i, v_j), and it is clear that if we download all data items from channel v_i, then it means we cover the edges connecting node v_i.

Next, we prove that G has a vertex cover with size k if and only if there is a valid data retrieval schedule S such that $t = k(\delta+3)$ and $h = 2k-1$.

$\Longrightarrow$: If G has a vertex cover VC with size k, then we can select the corresponding k channels in $\{v_i | v_i \in VC\}$ to receive all the data in k cycles. At the beginning of ith cycle (iteration), the client will visit b_i at $t = 1$, and hop to some $v_i \in VC$ channel, stay on this channel till the last time unit of the broadcast cycle, and then hop to b_{i+1}. There are k b_is, so each iteration client will download one of them. VC is a vertex cover, so following all

$v_i \in VC$ we must download every e_{ij}. The length of each broadcast cycle is $\delta + 3$, so the total access latency is $k(\delta + 3)$. In each broadcast cycle, the client will switch twice (except the last cycle), so $h = 2k - 1$.

$\Longleftarrow$: Assume MCDR has a valid schedule S with $t = k(\delta + 3)$ and $h = 2k - 1$. Let us consider D_b first. There are k b_i's located at the first time unit on k different channels. It means we have to switch at least $k - 1$ hops to download D_b, and then we only have another k hops for D_e, which means we can visit at most k channels in $\{v_i\}$. At the beginning of each broadcast cycle, we always stay at some channel b_i to download d_i, and then we switch to some v_i, and at the end of this cycle, we have to switch to channel b_{i+1} for d_{i+1}. This means we cannot switch to two vertex channels within one broadcast cycle, otherwise we cannot download $D = D_e \cup D_b$ in k iterations. Since S is valid, we visit k vertex channels and download all D_e data items, it means these k vertices form a vertex cover with size k.

This reduction can be done in polynomial time, and we can conclude that MCDR is NP-hard.

Randomized Algebraic Algorithm

To solve the above decision problem, we developed a randomized algebraic algorithm. It can detect if a given problem has a schedule to download all the requested data before time t and with at most h channel switchings in $O\left(2^k (nht)^{O(1)}\right)$ time, where n is the number of channels and k is the number of required data items. We also provide a fixed parameter tractable (FPT) algorithm with computational time $O\left(2^l (nht)^{O(1)}\right)$. It can determine whether there is a scheduling to download l data items from D in at most n time slots and at most h channel switches. Service provider can adjust n and h freely to fit their own requirement. We firstly give some preliminaries and then present our algorithms in detail.

Preliminaries

Here we introduce some notions about group algebra which are not often used in algorithm design.

Definition 2 Assume that $x_1, \ldots, x_k$ are variables in group algebra. Then,

1. A *monomial* has format $x_1^{a_1} x_2^{a_2} \ldots x_k^{a_k}$.
2. A *multilinear monomial* is a monomial such that each variable has degree exactly one. For example, $x_3 x_5 x_6$ is a multilinear monomial, but $x_3 x_5^2 x_6^3$ is not.
3. For a *polynomial* $p(x_1, \ldots, x_k)$, its *sum of product expansion* is $\sum_j p_j(x_1, \ldots, x_k)$, where each p_j is a monomial, which has a format $c_j x_1^{a_{j_1}} \ldots x_k^{a_{j_k}}$ with c_j respect to its coefficient.
4. $G_2 = (\{0, 1\}, +, \cdot)$ is a field with two elements $\{0, 1\}$ and two operations $+$ and $\cdot$. The addition operation is under the modular of 2 (mod 2).
5. Z_2^k is the group of binary k-vectors. Let w_0 denote the all-zero vector, which is the identity of Z_2^k, and then for every $v \in Z_2^k$, $v^2 = w_0$, $v \cdot w_0 = v$.

The operations between elements in the group algebra are standard.

Algorithm Description

The basic idea of our algebraic algorithm is that for each item $d_i \in D$, where D is the query data set, we create a variable x_i to represent it. Therefore, given $D = \{d_1, d_2, \ldots, d_k\}$, we construct a variable set $X = \{x_1, x_2, \ldots, x_k\}$. We then design a circuit $H_{t,h,n}$ such that a schedule without conflict will be generated by a multilinear monomial in the sum of product expansion of the circuit. The existence of schedules to download all the data items in D from the multiple channel set C is converted into the existence of multilinear monomials of $H_{t,h,n}$. Replace each variable by a specified binary vector which can remove all of the non-multilinear monomials by converting them to zero. Thus, the data retrieval problem is transformed into testing if a multivariate polynomial is zero. It is well known that randomized algorithms can be used to check if a circuit is identical to zero in polynomial time. Thus, we have the following statements.

Lemma 1 *There is a polynomial time algorithm such that given a channel c_i, a time interval $[t_1, t_2]$, and an integer m, it constructs a circuit of polynomial $P_{i,t_1,t_2,m}$ such that for any subset $D' = \{d_{i_1}, \ldots, d_{i_m}\} \subseteq D$ which has a size of m and is downloadable in the time interval $[t_1, t_2]$ from channel c_i, the product expansion of $P_{i,t_1,t_2,m}$ contains a multilinear monomial $x_{i_1} x_{i_2} \ldots x_{i_m}$.*

Proof We can use a recursive way to compute the circuit $P_{i,t_1,t_2,m}$ in polynomial time.

1. $P_{i,t_1,t_2,0} = 0$.
2. $P_{i,t_1,t_2,1} = \sum_j x_j$, $x_j \subseteq X$, and the corresponding data d_j is entirely in the time interval $[t_1, t_2]$ of channel c_i.
3. $P_{i,t_1,t_2,l+1} = \sum_j x_j \cdot P_{i,t_1,t_2',l} + P_{i,t_1,t_2',l+1}$, d_j starts at time $t_2' + 1$ and ends before time t_2 on channel c_i.

When computing $P_{i,t_1,t_2,l+1}$, x_j multiplies $P_{i,t_1,t_2',l}$ is based on the case that d_j is downloadable from time $t_2' + 1$ to t_2 in the final phase, and the other l data items are downloadable before time t_2'. The term $P_{i,t_1,t_2',l+1}$ is the case that $l + 1$ items are downloaded before time t_2'. Note that the parameter m in $P_{i,t1,t2,m}$ controls the total number of data to be downloaded.

Definition 3 A subset data items $D' = \{d_{i_1}, \ldots, d_{i_m}\} \subseteq D$ is (i, t, h)-downloadable if we can download all data items in D' before time t, the total number of channel switches is at most h, and the last downloaded item is from channel c_i.

Lemma 2 *Given two integers t and h, there is a polynomial time algorithm to construct a circuit of polynomial $F_{i,t,h,m}$ such that for any (i, t, h)-downloadable subset $D' = \{d_{i_1}, \ldots, d_{i_m}\} \subseteq D$, the product expansion of $F_{i,t,h,m}$ contains a multilinear monomial $\left(x_{i_1}, \ldots, x_{i_m}\right) Y$, where Y*

is a multilinear monomial which does not include any variable in X.

Proof We still use a recursive way to construct the circuit. Some additional variables are used as needed. Without loss of generality, we assume the data retrieval process start at time 0.

1. $F_{i,t,0,0} = 0$.
2. $F_{i,t,0,1} = P_{i,1,t,1} \cdot y_{i,t,0,1}$.
3. $F_{i,t,h'+1,m'+1} = y_{i,t,h'+1,m'+1,0} \left(\sum_{t'<t} F_{i,t',h'+1,m'} \cdot P_{i,t'+1,t,1} \right) + y_{i,t,h'+1,m'+1,1} \left(\sum_{j \neq i} \sum_{t'<t} F_{i,t'-1,h',m'} \cdot P_{i,t'+1,t,1} \right)$.

Then we can get Lemma 2 immediately.

The computation of $F_{i,t,h'+1,m'+1}$ is based on two cases, and we use two variables, $y_{i,t,h'+1,m'+1,0}$ and $y_{i,t,h'+1,m'+1,1}$, to mark them respectively. We now present an algorithm that involves one layer randomization to determine if there is a schedule to download all the data items in D before time t and with at most h channel switchings.

Theorem 2 *There is an $O\left(2^k (hnt)^{O(1)}\right)$ time randomized algorithm to determine if there is a scheduling to download $|D| = k$ data items before time t and the number of channel switches is at most h, where n is the total number of channels.*

Proof By Lemma 2, we can construct a circuit $H_{t,h,n} = \sum_{i=1}^n F_{i,t,h,k}$ in polynomial time. It is easy to see there is a scheduling for downloading the k data items before time t and with h channel switches, if and only if the sum product expansion of $H_{t,h,n}$ has a multilinear monomial $(x_1, \ldots, x_k) Y$.

We can replace each s_i by a vector $w_i = w_0^T + v_i^T$, where w_0 is the all-zero vector of dimension k and v_i is a binary vector of dimension k with its ith element being 1 and all other elements being 0. Assume $k = 3$, we define the following operations:

$$v_a \cdot v_b = \begin{pmatrix} a_1 \\ a_2 \\ a_3 \end{pmatrix} \cdot \begin{pmatrix} b_1 \\ b_2 \\ b_3 \end{pmatrix} = \begin{pmatrix} (a_1 + b_1)(\bmod 2) \\ (a_1 + b_2)(\bmod 2) \\ (a_1 + b_3)(\bmod 2) \end{pmatrix} \tag{1}$$

$$(v_a + v_b) \cdot v_c = v_a \cdot v_c + v_b \cdot v_c \tag{2}$$

By Eqs. 1 and 2, for any k-dimensional binary vector $w' = w_0 + v$, we have $w'^2 = w_0^2 + 2w_0 \cdot v + v^2 = w_0 + 2(w_0 \cdot v) + w_0 = 2(w_0 \cdot v) + 2w_0 = 0$, because of the coefficients are in the field of G_2. The replacement $x_i = w_i (i = 1, \ldots, m)$ makes all the non-multilinear monomials become zero. Meanwhile, all the multilinear monomials remain nonzero. Hence, it is clear that there is a scheduling to download all the data items in D before time t and with at most h channel switchings if and only if $H_{t,h,n}|x_i=w_i (i=1,\ldots,k)$ is a nonzero polynomial. The variables in Y makes it impossible to have cancelation when adding two identical multilinear monomials, which can be generated from different paths with variables in $\{x_1, \ldots, x_k\}$. It is well known that randomized algorithms can be used to check if a circuit is identical to zero in polynomial time [11, 12].

The algorithm generates less than 2^k terms during the computing process since there are at most 2^k distinct binary vectors. Therefore, the computational time is $O\left(2^k (nht)^{O(1)}\right)$.

Example Let $H_1 = x_1 x_2 y_1 + x_2^2 y_2$ and $H_2 = x_1^2 y_1 + x_2^2 y_2$. Consider the replacement $x_1 = w_1 = \binom{0}{0} + \binom{1}{0}$ and $x_2 = w_2 = \binom{0}{0} + \binom{0}{1}$. We have the following steps of operations.

$H_1|x_1 = w_1, x_2 = w_2$

$$= \left(\binom{0}{0} + \binom{1}{0} \right) \left(\binom{0}{0} + \binom{0}{1} \right) y_1$$

$$+ \left(\binom{0}{0} + \binom{0}{1} \right)^2 y_2$$

$$= \left(\binom{0}{0} + \binom{0}{1} + \binom{1}{0} + \binom{1}{1} \right) y_1$$

$$+ \left(\binom{0}{0} + \binom{0}{1} + \binom{0}{1} + \binom{0}{0} \right) y_2$$

$$= \left(\binom{0}{0} + \binom{0}{1} + \binom{1}{0} + \binom{1}{1} \right) y_1$$

$$+ \left(2\binom{0}{0} + 2\binom{0}{1} \right) y_2$$

$$= \left(\binom{0}{0} + \binom{0}{1} + \binom{1}{0} + \binom{1}{1} \right) y_1 + (0 + 0)y_2$$

$$= \left(\binom{0}{0} + \binom{0}{1} + \binom{1}{0} + \binom{1}{1} \right) y_1 + 0$$

$$= \left(\binom{0}{0} + \binom{0}{1} + \binom{1}{0} + \binom{1}{1} \right) y_1$$

$$\neq 0$$

$H_2|x_1 = w_1, x_2 = w_2$

$$= \left(\binom{0}{0} + \binom{1}{0} \right)^2 y_1 + \left(\binom{0}{0} + \binom{0}{1} \right)^2 y_2$$

$$= \left(\binom{0}{0} + \binom{1}{0} + \binom{1}{0} + \binom{0}{0} \right) y_1$$

$$+ \left(\binom{0}{0} + \binom{0}{1} + \binom{0}{1} + \binom{0}{0} \right) y_2$$

$$= \left(2\binom{0}{0} + 2\binom{1}{0} \right) y_1 + \left(2\binom{0}{0} + 2\binom{0}{1} \right) y_2$$

$$= (0 + 0)y_1 + (0 + 0)y_2$$

$$= 0$$

H_1 is a polynomial that contains a multilinear monomial. It becomes nonzero after replacement. H_2 is a polynomial that is without multilinear monomials. It becomes zero after the replacement. If we just down a subset of l data items in set D, we have the following theorem that involves two layers of randomization.

Theorem 3 *There is an $O\left(2^l (hnt)^{O(1)}\right)$ time randomized algorithm to determine if there is a scheduling to download l data items from D in at most t time units and at most h channel switches.*

Proof By Lemma 2, we can construct a polynomial $H_{t,h,l} = \sum_{i=1}^{l} F_{i,t,h,l}$ in polynomial time. Replace each x_i with a vector $w_i = w_0^T + v_i^T$, where w_0 is the all-zero vector of dimension l and v_i is a random distinct vector of dimension l. The replacement $x_i = w_i$ $(i = 1,\ldots,k)$ makes all monomials which has non-multilinear monomial at x part become zero.

Therefore, there is a scheduling to download l data items of D before time t and with the number of switches no more than h if and only if $H_{t,h,k|x_i=w_i}(i=1,\ldots,k)$ is not a zero polynomial in the field of G_2. Assume that the product expansion of $H_{t,h,l}$ has a multilinear monomial $(x_{i_1},\ldots,x_{i_l})Y$, where Y is a multilinear monomial with variables not in $x_1,\ldots,x_k$. For a series of randomly assigned vectors with dimension l: $v_{j_1},\ldots,v_{j_l}$, the probability that v_{j_i} is a linear combination of $v_{j_1},\ldots,v_{j_{i-1}}$ is at most $\dfrac{2^{i-1}}{2^l} = \dfrac{1}{2^{l-i+1}}$. Therefore, with probability at most $\sum_{i=1}^{l} \dfrac{1}{2^{l-i+1}}$, v_{j_i} is a linear combination of $v_{j_1},\ldots,v_{j_{i-1}}$ for some $i \leq l$. When $v_{j_i},\ldots,v_{j_l}$ are linearly independent, the product of $v_{j_1},\ldots,v_{j_l}$ is nonzero. Every multilinear monomial in the product expansion has different variables to form Y since it is determined by a unique path to generate the polynomial. Therefore, for those random vectors v_i, every multilinear monomial has a chance at least $1 - \frac{3}{4} = \frac{1}{4}$ to be nonzero. Therefore, if there is a solution, $H_{t,h,k|x_i=w_i}(i,1,\ldots,l)$ with random assignment Y is not zero in the field of G_2 with probability at least $\frac{1}{4}$.

After the replacements, it generates less than $2l$ terms since there are at most $2k$ different vectors for a group of Z_2^l. The coefficient of each vector is kept as a polynomial size circuit. Therefore, the computational time of our algorithm is $O\left(2^l (hnt)^{O(1)}\right)$, and if we run it 30 times, the error rate is $\left(\frac{3}{4}\right)^{30} < 0.0002$.

Applications

Scheduling problem is one of the most fundamental problems in combinatorial optimization, which could model various real-world practical applications. Especially, scheduling problem at client hand side would be very useful for data retrieval problem in wireless data broadcasting or data streaming environment to reduce energy consumption and improve query efficiency. Such problem would also be helpful for parallel query applications in distributed storage systems.

Open Problems

How to download data items efficiently in wireless data broadcast environment can usually be formulized as NP-hard problems with different constraints, and can be categorized into two kinds: single channel process and multiple channel process. The best known result for the former problem is constant-factor approximations, while currently there is no polynomial time approximation scheme (PTAS) for both of them. The results for this problem is also helpful for parallel data retrieval problem in distributed data storage system and cloud system.

Experimental Results

Many literature proposed experimental results for scheduling problem in data broadcasting. Shi et al. [6] simulated a base station with n broadcast channels and 10,000 items, each of size 1KB, and multiple clients with various requests of data. The access probability of the database follows Zipf distribution, n varies from 5 to 30, the number of antennae varies from 1 to 10, and the size of a request varies from 10 to 1,000. For each experiment, they generated 100 requests to get the average access latency and number of switchings during data retrieval. Lv et al. [7, 8] constructed two types of broadcast programs: special data broadcast without channel switching time (SDB) and general data broadcast with channel switching time (GDB). In both types of programs, they simulated a base station with n broadcast channels; the bandwidth of each channel is 1Mbit/sec. The database to be broadcasted has N data items, each of size 512 bytes. The time duration is denoted by t. The data items of query data set D

is generated with access probabilities following the Zipf distribution.

URLs to Code and Data Sets

Shi et al. [6] provided the program for users to test parameter setting for their own data sets and available channels (http://theory.utdallas.edu/ dataengineering).

Cross-References

► Efficient Polynomial Time Approximation Scheme for Scheduling Jobs on Uniform Processors

Recommended Reading

1. Acharya S, Alonso R, Franklin M, Zdonik S (1995) Broadcast disks: data management for asymmetric communication environments. In: The ACM special interest group on management of data conference (SIGMOD), San Jose, 22–25 May 1995, pp 199–210
2. Yee W, Navathe S, Omiecinski E, Jermaine C (2002) Efficient data allocation over multiple channels at broadcast servers. IEEE Trans Comput 51(10):1231–1236
3. Ardizzoni E, Bertossi A, Pinotti M, Ramaprasad S, Rizzi R, Shashanka M (2005) Optimal skewed data allocation on multiple channels with flat broadcast per channel. IEEE Trans Comput 54(5):558–572
4. Anticaglia S, Barsi F, Bertossi A, Iamele L, Pinotti M (2008) Efficient heuristics for data broadcasting on multiple channels. Wirel Netw 14(2):219–231
5. Kenyon C, Schabanel N (1999) The data broadcast problem with non-uniform transmission times. In: Proceedings of the tenth annual ACM-SIAM symposium on discrete algorithms (SODA), Baltimore, 17–19 Jan 1999, pp 547–556
6. Shi Y, Gao X, Zhong J, Wu W (2010) Efficient parallel data retrieval protocols with MIMO antennae for data broadcast in 4G wireless communications. In: The 21st international conference on database and expert systems applications (DEXA), Bilbao, 30 Aug–3 Sept 2010, pp 80–95
7. Lu Z, Shi Y, Wu W, Fu B (2012) Efficient data retrieval scheduling for multi-channel wireless data broadcast. In: International conference on computer communications (INFOCOM), Orlando, 25–30 Mar 2012, pp 891–899
8. Lu Z, Shi Y, Wu W, Fu B (2014) Data retrieval scheduling for multi-Item requests in multi-channel wirelessBroadcast environments. IEEE Trans Mobile Comput 13(4):752–765
9. Gao X, Lu Z, Wu W, Fu B (2011) Algebraic algorithm for scheduling data retrieval in multi-channel wireless data broadcast environments. In: The 6th annual international conference on combinatorial optimization and applications (COCOA), Zhangjiajie, 4–6 Aug 2011, pp 74–81
10. Gao X, Lu Z, Wu W, Fu B (2013) Algebraic data retrieval algorithms for multi-channel wireless data broadcast. Theor Comput Sci 497:123–130
11. Williams R (2009) Finding paths of length k in $O*$ $(2^k)time$. Inf Process Lett 109(6):315–318
12. Koutis I (2008) Faster algebraic algorithms for path and packing problems. In: The 2008 international colloquium on automata, languages and programming (ICALP), Reykjavik,6– 13 July 2008, pp 575–586

Scheduling with a Reordering Buffer

Matthias Englert[1] and Matthias Westermann[2]
[1] Department of Computer Science, University of Warwick, Coventry, UK
[2] Department of Computer Science, TU Dortmund University, Dortmund, Germany

Keywords

Machine scheduling; Minimum makespan scheduling; Online algorithms; Reordering buffer; Sorting buffer

Years and Authors of Summarized Original Work

2002; Räcke, Sohler, Westermann
2009; Gamzu, Segev
2010; Englert, Räcke, Westermann
2011; Adamaszek, Czumaj, Englert, Räcke
2011; Dósa, Epstein
2013; Avigdor-Elgrabli, Rabani
2014; Englert, Özmen, Westermann

Problem Definition

The problem known as the reordering buffer problem or as the sorting buffer problem is concerned with sorting a sequence of colored items according to their color using a limited size buffer. More precisely the items are to be processed and arrive one by one. Arriving items must first be placed into a buffer which can hold up to k items. Once the buffer is completely filled, an algorithm has to free space by selecting one of the items in the buffer for processing and removing that item from the buffer. After items stop arriving, the remaining items in the buffer may be processed in any order. Whenever an item is processed that has a different color than the item processed in the step before, this generates a cost of 1. The objective is to minimize the total cost.

Metric Space Generalization

This problem can be further generalized. Items, instead of having a color, correspond to points in a metric space. A single server must process all items. In order to process an item, the server has to move to the corresponding point in the metric space. At every point, the server has to chose one of the first k as of yet unprocessed items for processing and move the server accordingly. The goal is to minimize the total distance the server travels.

The uniform metric in which any two points either have distance 0 or distance 1 from one another corresponds to the original "color sorting" setting. Other metrics studied include line metrics and "star" metrics which are the distance metrics over weighted undirected trees of diameter 2.

Block Operation Setting

Another variant is the so-called block operation setting. Once again, the input consists of a sequence of colored items. The first k items are placed in a buffer. In each step, an algorithm selects one of the colors and processes all items of that color currently stored in the buffer, incurring a cost of 1. This is called a block operation. The processed items are removed from the buffer and replaced with the next items from the input

sequence (if there are any). The goal is once again to minimize the total cost.

The difference between this block operation setting and the original setting is most pronounced for an input sequence consisting of ℓ items of a single color. While in the original setting such a sequence would not produce any cost, the cost in the block device setting would be ℓ/k since only k items can be processed per block operation.

Minor Variants Found in the Literature

In some cases, there are slight differences in which these problems are defined in the literature. Does a cost incur for the first ever processed item or, similarly, is the first position of the server in the metric space part of the input or does the algorithm get to chose that position (without incurring any cost)? Does an item first have to be placed in the buffer or can the algorithm process an arriving item directly, thereby bypassing the buffer? Do we need to remove the remaining items in the buffer once new items stopped arriving? It turns out however that these details are inconsequential for most of the results we are interested in.

Key Results

The main focus of study in the area of scheduling with a reordering buffer has been on online algorithms. In the online setting, the algorithm's decisions have to be based solely on the items that arrived in the past and must not depend on items arriving in the future. An online algorithm is called c-competitive if the cost of the algorithm is at most c times that of an optimal off-line solution.

The Online Problem

Deterministic Algorithms
Räcke, Sohler, and Westermann [29] first introduced the problem for the uniform metric

and gave a $O(\log^2 k)$-competitive online algorithm. After further improvements to $O(\log k)$ [18] and $O(\log k/\log \log k)$ [6] eventually, an $O(\sqrt{\log k})$-competitive online algorithms was designed [2]. This is almost optimal since a lower bound of $\Omega(\sqrt{\log k/\log \log k})$ is known [2]. Many of these upper bounds also generalize from the uniform metric to star metrics.

While the proof techniques for some of these results differ significantly, the basic idea behind all the algorithms is the same. As long as the buffer contains an item of the same color as the previously processed item, such an item is processed next. Otherwise, the algorithm has to pick a different color and performs a color switch which incurs a cost of 1. In order to decide which color to switch to, each color is assigned a "penalty" counter which is initially set to 0 and is reset to 0 whenever the color is selected for processing. If there is a color with penalty at least k, then an item with that color is selected next. Otherwise, an arbitrary color is selected and the penalty counters for each color are increased proportional to the number of items of that color that are stored in the buffer. For the $O(\sqrt{\log k})$-competitive algorithm, instead of picking an arbitrary color, a more sophisticated rule is used.

Randomized Algorithms
Randomized algorithms can achieve much smaller competitive ratios. The first randomized algorithm with a competitive ratio of $O(\log \log k)$ was given for the block operation model [3]. Shortly afterward, a randomized algorithm with the same competitive ratio was presented for the original model [8]. This is best possible since a matching lower bound is known [2]. These randomized algorithms are based on online primal-dual LP schemes [11].

Other Metric Spaces
Apart from the uniform metric, line metrics have received some attention. After a randomized $O(\log^2 n)$-competitive online algorithm for n equally spaced points on a line [27], an improved deterministic $O(\log n)$-competitive algorithm was given [24]. A deterministic $O(\log N \log \log N)$-competitive algorithm for a line metric with not necessarily equally spaced points was also given, but here N refers to the number of items in the input sequence [24]. An easy observation, however, shaves off the $\log \log N$ factor and improves the analysis to show $O(\log N)$ competitiveness (Cygan, Mucha, Private communication, 2011). There is still a significant gap between this upper bound and the best known lower bound of about 2.154 [24].

For general metric spaces, a randomized $O(\log^2 k \log n)$-competitive online algorithm is known, where n is the number of points in the metric space [19]. This result is based on a deterministic algorithm for trees that is turned into an algorithm for general metrics by using a metric embedding [23].

Stochastic Inputs
In a setting where the input is not adversarial constructed but where the colors of the items are drawn i.i.d. from an unknown distribution, a constant competitive ratio is achievable [22]. This result also holds when the colors of the items are fixed by an adversary but the order in which the items arrive is random. The proof is based on the fact that a constant competitive online algorithm is known for adversarial inputs, if the online algorithm can use a buffer that is four times as large as the one used by the optimal off-line algorithm. In the stochastic input setting, this difference in buffer size does not lead to significantly different cost, i.e., the cost of an optimal algorithm with buffer size k is only by a constant factor larger than the cost of an optimal algorithm with buffer size $4k$. This is not true for adversarial inputs [1].

The Off-Line Problem

The reordering buffer problem is NP-hard [5, 12] for the uniform metric, and the complexity for line metrics is unknown. Therefore, several papers focus on approximating the off-line scenario.

A constant factor approximation is known for the uniform metric [7]. For star metrics, the best known approximation factor of $O(\log \log k\gamma)$ is achieved by a randomized algorithm, where γ denotes the ratio of the maximum to the minimum weight [25]. Both results are based on the intricate rounding of the solution to an LP relaxation of the corresponding problem.

Bicriteria Approximations

For more general metric spaces, the best approximation ratios are achieved by bicriteria approximations, i.e., the approximation algorithm can make use of more buffer capacity than an optimal algorithm. For metric spaces given by the distance metric over a weighted undirected tree, a bicriteria approximation with approximation factor 9 to cost and $4 + 1/k$ to buffer size is known [10]. Using metric embeddings [23], this implies a randomized bicriteria approximation with approximation factor $O(\log n)$ to cost and $O(1)$ to buffer size, where n denotes the number of points in the metric space.

The Maximization Problem

In the maximization version of the problem, the goal is to maximize the total cost savings that result from reordering the input sequence. In terms of an optimal solution, the minimization and maximization scenario are identical. However, in terms of approximation, they behave quite differently in the sense that a c-approximate solution for the maximization problem usually has very different cost from a c-approximate solution for the minimization problem. For the uniform metric, the first result was an approximation algorithm with an approximation factor of 20 [28]. This was later improved to a factor of 9 [9].

Online Minimum Makespan Scheduling

Reordering buffers have also been studied in connection with other scheduling problems, in particular online minimum makespan scheduling. As in the classic problem without reordering, the input consists of a sequence of jobs with processing times, and a scheduling algorithm has to assign the jobs to m parallel machines, with the objective to minimize the makespan, which is the time it takes until all jobs are processed. However, it is not required that each arriving job has to be assigned immediately to one of the machines. A reordering buffer can be used to reorder the input sequence of jobs. At each point in time, the reordering buffer contains the first k jobs of the input sequence that have not been assigned so far. An online scheduling algorithm has to decide which job to assign to which machine next. Upon its decision, the corresponding job is removed from the buffer and assigned to the corresponding machine, and thereafter the next job in the input sequence takes its place.

Non-preemptive Scheduling

For non-preemptive scheduling, Englert, Özmen, and Westermann [20] give, for m identical machines, a tight bound on the competitive ratio. Depending on m, the achieved competitive ratio lies between 4/3 and 1.4659. This optimal ratio is achieved with a buffer of size of at most $\lceil 2.5 \cdot m \rceil + 2$. They show that larger buffer sizes do not result in an additional advantage and that a buffer of size $\Omega(m)$ is necessary to achieve this competitive ratio. This improves upon an optimal algorithm for two identical machines [26].

Further, they present several algorithms for different buffer sizes. In addition, for m uniformly related machines, they give a scheduling algorithm that achieves a competitive ratio of 2 with a reordering buffer of size m.

Subsequently to [20], a variety of related papers appeared (compare, e.g., [4, 14–16, 21]). For 2 uniformly related machines with speed ratio $s \geq 1$, it is shown that, for any $s > 1$, a buffer of size 3 is sufficient to achieve an optimal competitive ratio, and in the case $s \geq 2$, a buffer of size 2 already allows to achieve an optimal ratio [15].

Job Migrations

The results of [20] can be generalized to the problem of online minimum makespan scheduling with job migrations, i.e., where no reordering buffer is available, but a limited number of job reassignments may be performed. For m identical machines, the same competitive ratio as in [20] can be achieved [4]. The algorithm uses, for $m \geq 11$, at most $7m$ migration operations and, for smaller m, $8m$ to $10m$ migration operations. A number of papers consider similar models (compare, e.g., [13, 17, 30, 31]).

Preemptive Scheduling

For preemptive scheduling on m identical machines, tight bounds on the competitive ratio can be achieved for any m. This bound is $4/3$ for even values of m and slightly lower for odd values of m [16]. A buffer of size $\Theta(m)$ is sufficient to achieve this bound, but a buffer of size $o(m)$ does not reduce the best overall competitive ratio $e/(e - 1)$ that is known for the case without reordering [16].

Cross-References

▶ Approximation Schemes for Makespan Minimization
▶ Efficient Polynomial Time Approximation Scheme for Scheduling Jobs on Uniform Processors
▶ Online Preemptive Scheduling on Parallel Machines

Recommended Reading

1. Aboud A (2008) Correlation clustering with penalties and approximating the reordering buffer management problem. Master's thesis, Computer Science Department, The Technion—Israel Institute of Technology
2. Adamaszek A, Czumaj A, Englert M, Räcke H (2011) Almost tight bounds for reordering buffer management. In: Proceedings of the 43rd ACM symposium on theory of computing (STOC), San Jose, pp 607–616
3. Adamaszek A, Czumaj A, Englert M, Räcke H (2012) Optimal online buffer scheduling for block devices. In: Proceedings of the 44th ACM symposium on theory of computing (STOC), New York, pp 589–598
4. Albers S, Hellwig M (2012) On the value of job migration in online makespan minimization. In: Proceedings of the 20th European symposium on algorithms (ESA), Ljubljana, pp 84–95
5. Asahiro Y, Kawahara K, Miyano E (2012) NP-hardness of the sorting buffer problem on the uniform metric. Discret Appl Math 160(10–11):1453–1464
6. Avigdor-Elgrabli N, Rabani Y (2010) An improved competitive algorithm for reordering buffer management. In: Proceedings of the 21st ACM-SIAM symposium on discrete algorithms (SODA), Austin, pp 13–21
7. Avigdor-Elgrabli N, Rabani Y (2013) A constant factor approximation algorithm for reordering buffer management. In: Proceedings of the 24th ACM-SIAM symposium on discrete algorithms (SODA), New Orleans, pp 973–984
8. Avigdor-Elgrabli N, Rabani Y (2013) An optimal randomized online algorithm for reordering buffer management. In: Proceedings of the 54th IEEE symposium on foundations of computer science (FOCS), Berkeley, pp 1–10
9. Bar-Yehuda R, Laserson J (2007) Exploiting locality: approximating sorting buffers. J Discret Algorithms 5(4):729–738
10. Barman S, Chawla S, Umboh S (2012) A bicriteria approximation for the reordering buffer problem. In: Proceedings of the 20th European symposium on algorithms (ESA), Ljubljana, pp 157–168
11. Buchbinder N, Naor J (2009) The design of competitive online algorithms via a primal-dual approach. Found Trends Theor Comput Sci 3(2–3):93–263
12. Chan H, Megow N, Sitters R, van Stee R (2012) A note on sorting buffers offline. Theor Comput Sci 423:11–18
13. Chen X, Lan Y, Benko A, Dósa G, Han X (2011) Optimal algorithms for online scheduling with bounded rearrangement at the end. Theor Comput Sci 412(45):6269–6278
14. Ding N, Lan Y, Chen X, Dósa G, Guo H, Han X (2014) Online minimum makespan scheduling with a buffer. Int J Found Comput Sci 25(5):525–536
15. Dósa G, Epstein L (2010) Online scheduling with a buffer on related machines. J Comb Optim 20(2):161–179
16. Dósa G, Epstein L (2011) Preemptive online scheduling with reordering. SIAM J Discret Math 25(1):21–49
17. Dósa G, Wang Y, Han X, Guo H (2011) Online scheduling with rearrangement on two related machines. Theor Comput Sci 412(8–10):642–653
18. Englert M, Westermann M (2005) Reordering buffer management for non-uniform cost models. In: Proceedings of the 32nd international colloquium on automata, languages and programming (ICALP), Lisbon, pp 627–638
19. Englert M, Räcke H, Westermann M (2010) Reordering buffers for general metric spaces. Theory Comput 6(1):27–46
20. Englert M, Özmen D, Westermann M (2014) The power of reordering for online minimum makespan scheduling. SIAM J Comput 43(3):1220–1237
21. Epstein L, Levin A, van Stee R (2011) Max-min online allocations with a reordering buffer. SIAM J Discret Math 25(3):1230–1250

22. Esfandiari H, Hajiaghayi M, Khani MR, Liaghat V, Mahini H, Räcke H (2014) Online stochastic reordering buffer scheduling. In: Proceedings of the 41st international colloquium on automata, languages and programming (ICALP), Copenhagen, pp 465–476

23. Fakcharoenphol J, Rao SB, Talwar K (2004) A tight bound on approximating arbitrary metrics by tree metrics. J Comput Syst Sci 69(3):485–497

24. Gamzu I, Segev D (2009) Improved online algorithms for the sorting buffer problem on line metrics. ACM Trans Algorithms 6(1):15:1–15:14

25. Im S, Moseley B (2014) New approximations for reordering buffer management. In: Proceedings of the 25th ACM-SIAM symposium on discrete algorithms (SODA), Portland, pp 1093–1111

26. Kellerer H, Kotov V, Speranza MG, Tuza Z (1997) Semi on-line algorithms for the partition problem. Oper Res Lett 21(5):235–242

27. Khandekar R, Pandit V (2010) Online and offline algorithms for the sorting buffers problem on the line metric. J Discret Algorithms 8(1):24–35

28. Kohrt JS, Pruhs K (2004) A constant factor approximation algorithm for sorting buffers. In: Proceedings of the 6th Latin American symposium on theoretical informatics (LATIN), Buenos Aires, pp 193–202

29. Räcke H, Sohler C, Westermann M (2002) Online scheduling for sorting buffers. In: Proceedings of the 10th European symposium on algorithms (ESA), Rome, pp 820–832

30. Tan Z, Yu S (2008) Online scheduling with reassignment. Oper Res Lett 36(2):250–254

31. Wang Y, Benko A, Chen X, Dósa G, Guo H, Han X, Sik-Lányi C (2012) Online scheduling with one rearrangement at the end: revisited. Inform Process Lett 112(16):641–645

Secretary Problems and Online Auctions

MohammadHossein Bateni
Google Inc., New York, NY, USA

Keywords

Competitive analysis; Knapsack constraint; Matroid constraint; Mechanism design; Online algorithm; Secretary problem; Strategyproof; Submodularity; Truthfulness

Years and Authors of Summarized Original Work

2004; Hajiaghayi, Kleinberg, Parkes
2008; Babaioff, Immorlica, Kempe, Kleinberg
2013; Bateni, Hajiaghayi, Zadimoghaddam

Problem Definition

The classic secretary problem, a prime example of stopping theory, has been studied extensively in the computer science literature. Consider the scenario where an employer is interested in hiring one secretary out of a pool of candidates. The difficulty is that, although the employer does not know the utility of a candidate before she is interviewed, the irrevocable hiring decision for each candidate has to be made right after the interview and prior to interviewing the subsequent candidates. The goal is nonetheless to pick the best candidate or maximize the probability of achieving this.

Optimization Angle

The above scenario is hopeless from an algorithmic point of view since an adversarial input makes it impossible to hire the best candidate. We can take either of two paths to make the problem tractable: restrict the set of utilities or the arrival order of candidates. The former path yields, for instance, the stochastic variant of the problem. However, we follow the second idea here that leads to the classic secretary problem. The extra assumption, then, is that the candidates arrive in a random error; i.e., although each candidate may have an arbitrary adversarial utility, every permutation of the candidates is equally likely to be the arrival order.

A folklore solution to the problem, often attributed to [3], is to look into the first $\frac{1}{e}$ fraction of the candidates (called the "tuning set"), without giving them any offers, and then hire the first candidate with utility more than every one in the tuning set. It is not difficult to show that this approach hires the best candidate with probability at least $\frac{1}{e}$. Indeed, it is known that this is the best possible performance.

There are two questions to be answered, once we extend the problem to multiple secretaries.

1. *What subsets of secretaries can be hired together?* The simplest answer is to allow at most k secretaries to be hired. Alternately, we can place (several) knapsack and/or matroid constraints on the feasible set. The former assigns a cost to each hire – say, the requested salary – that is to be paid out of a given budget. The latter permits only those combinations that form an independent set according to a given matroid. It is easy to see that both generalize the cardinality constraint.
2. *How do we compute the utility of a set?* The utility of a set can be defined as the sum of the utilities of individual secretaries in the set. More generally, a submodular or subadditive function may be employed to describe the utility of a set.

We then attempt to hire a feasible set of secretaries of maximum expected utility.

Mechanism Design Angle

Mechanism design literature has looked at this problem from a slightly different angle. In this setting, the players that arrive in a random order declare a *bid* – i.e., how much they value the item being sold – and then the seller decides who should get the item (or items) and how much they should be charged. Such decisions are to be taken irrevocably as in the optimization problem discussed above.

The players can play strategically, though, by declaring higher or lower bids in order to increase their chances of winning the item or to reduce the price they pay. In addition, they may declare their arrival/departure time untruthfully to achieve a better result. We want to design a "truthful" auction that precludes such undesirable outcomes. Although we allow the player to declare any nonnegative bid (if it is in her favor), we do not let them state an arrival time that is earlier than their actual one. (Presence intervals

may be overlapping and/or nested.) We say that a mechanism is value-strategyproof if no player can benefit from declaring a bid different from her real value. Similarly the mechanism is called time-strategyproof if there is no benefit in stating the arrival/departure times untruthfully. We look for mechanisms that are both time- and value-strategyproof.

Key Results

Optimization

Kleinberg [6] studies the multiple-choice generalization where the goal is to hire k candidates, whose total utility (defined as the sum of the individual utilities) is maximized. He presents a tight performance guarantee of $1 + \Theta\left(\frac{1}{\sqrt{k}}\right)$ for the problem. In the case of $k = 1$, this is equivalent to the classic secretary problem. (The nontrivial direction follows from a construction where the utilities are hugely different.) Kleinberg's algorithm partitions the set of candidates into two (almost) equal pieces, recursively hires $\frac{k}{2}$ secretaries in the first, sets the threshold for the second piece by looking at the solution to the first piece, and picks as many as $\frac{k}{2}$ secretaries in the second piece who are better than threshold.

Babaioff et al. [1] look at the generalization where there is a restriction on the set of candidates that can be hired together; the restriction is in the form of a matroid. They present an $O(\log n)$ competitive ratio in this case along with improved bounds when the matroid has a special form. Their general matroid algorithm partitions the items into logarithmically many sets of almost equal utility and focuses (randomly) on one such set, which reduces the problem into that of maximizing the cardinality of the solution (solved via the greedy method).

The case of submodular utilities is discussed in Bateni et al. [2]: several matroid or knapsack constraints can be placed on the set of feasible candidates, and the total utility of a set is computed by a submodular function of the participating candidates. They provide constant competitive ratios as long as a fixed number of knapsack constraints are present. When (a constant number of) matroid constraints are involved, too, their performance guarantees grow to $O(\log^2 k)$ where k is the rank of the matroid. They divide the input into different pieces where at most one secretary should be picked from each, not losing too much utility in the process. As a result, the submodular function collapses to an additive one within each piece (by taking the marginal values of secretaries with respect to the current solution). The classic algorithm is then used inside each piece. The main idea behind the matroid algorithm is that we only need to show that, whatever choices we have already committed to, there are enough options left that can appropriately augment the current solution. The argument goes by proving the existence of a magical solution with k' secretaries any of whose $\frac{k'}{2}$-size subsets has significant contribution (say, at least a $\frac{1}{\log k}$ fraction of the optimum) in the submodular function. Had we known k', a simple greedy algorithm would have sufficed to find a solution similar to the magical set. At the cost of another factor $O(\log k)$, we can guess k'.

Furthermore, Bateni et al. show that subadditive utility functions make the problem much more difficult. In particular, they provide matching $\Theta(\sqrt{k})$ competitive ratios.

Mechanism Design

The Dynkin's algorithm for the classic secretary problem can be readily turned into an auction: set the price after observing the tuning set, and then sell to anyone with a higher bid. This mechanism is not truthful, though, since high-bid players spanning across the time threshold have an incentive to declare later arrival time (i.e., after the threshold); this way, they will win the item but do not set the price.

Nevertheless, Hajiaghayi et al. [5] show how one can modify the mechanism slightly to make it truthful: after the threshold, consider the option of selling the item to the agent with the highest bid so far – if she is still present – and charge her the second-highest bid so far. Their method achieves constant competitiveness for both efficiency and revenue. Their $1/e$ competitiveness for efficiency is best possible since it generalizes

the optimization problem; however, when comparing the revenue to that achieved by the Vickrey auction, their upper bound of $1/e^2$ for competitiveness fares against a lower bound of $1/e$. (It is possible, they show, to modify the mechanism slightly to trade efficiency loss for revenue gain; for instance, simultaneous 4 competitiveness for both objectives is possible.)

The general idea for the transformation is to define a "tuning period" where the price is set for everyone. Then, not only a simple auction-like mechanism is employed in the "hiring phase" to obtain a strategyproof mechanism, but also extra care should be given to the "transition phase" (from tuning to hiring) so as not to incentivize untruthful declaration of arrival time for those whose presence spans the transition. The same approach can be applied to the multiple-choice secretary problem to obtain constant-factor competitive mechanisms (for efficiency and revenue), but this bound is far from the one achieved in the optimization setting by Kleinberg [6].

Open Problems

Though there has been some improvements on the matroid case, we still do not know which cases are hard and admit no constant-factor competitive ratio. For submodular utilities (and simple cardinality constraints), in particular, there is a gap between $\left(1 - \frac{1}{e}\right)/(e+1)$ algorithmic result [4] and the $1 - \frac{1}{e}$ $\left(\text{or } 1 - \frac{1}{\sqrt{k}}\right)$ target known for linear utilities.

Cross-References

▶ Algorithmic Mechanism Design

Recommended Reading

1. Babaioff M, Immorlica N, Kempe D, Kleinberg R (2008) Online auctions and generalized secretary problems. SIGecom Exch 7(2):1–11. doi:http://doi.acm.org/10.1145/1399589.1399596

2. Bateni M, Hajiaghayi MT, Zadimoghaddam M (2013) Submodular secretary problem and extensions. ACM Trans Algorithms 9(4):32
3. Dynkin EB (1963) The optimum choice of the instant for stopping a markov process. Sov Math Dokl 4:627–629
4. Feldman M, Naor J, Schwartz R (2011) Improved competitive ratios for submodular secretary problems (extended abstract). In: APPROX, Princeton, pp 218–229
5. Hajiaghayi MT, Kleinberg R, Parkes DC (2004) Adaptive limited-supply online auctions. In: EC, New York, pp 71–80
6. Kleinberg R (2005) A multiple-choice secretary algorithm with applications to online auctions. In: SODA, Vancouver, pp 630–631

Self-Assembly at Temperature 1

Pierre-Étienne Meunier
Le Laboratoire d'Informatique Fondamentale de Marseille (LIF), Aix-Marseille Université, Marseille, France

Keywords

Computational geometry; Concurrency; Self-assembly

Years and Authors of Summarized Original Work

2000; Rothemund, Winfree
2009; Doty, Patitz, Summers
2011; Cook, Fu, Schweller
2014; Meunier, Patitz, Summers, Theyssier, Winslow, Woods

Problem Definition

Temperature 1 (also called *noncooperative*) self-assembly is a model of the formation of structures by growing and branching tips. Despite its ubiquity in nature (in systems such as plants and mycelium or percolation processes) and apparent dynamic simplicity, it is one of the least understood models of self-assembly.

This model was introduced in a broader framework called the *abstract Tile Assembly Model* (aTAM) [10]. In the aTAM, we consider *tile assembly systems*, which are defined by a finite set T of square or cubic *tile types*, an initial *seed assembly* σ (one or more tiles stuck together), and an integer temperature $\tau = 1, 2, 3, \ldots$. All tiles, on each of their sides, have *glues* with an integer *color* and an integer *strength*.

The dynamics of tile self-assembly starts from the seed assembly and proceeds one tile at a time, asynchronously and nondeterministically. A tile can stick to an existing assembly if it can be placed so that the sum of the strengths on its sides matching the existing assembly is at least the temperature. In the case of temperature 1, this means that tiles can be placed as soon as one of their sides matches the existing assembly. At higher temperatures, we can require that newly placed tiles match *several* of their neighbors to attach.

Ultimately, after a countable (potentially infinite) number of steps, no tile can be added to the assembly, in which case we call it *terminal*. Like in Wang tilings, tiles cannot overlap, be rotated, or be flipped. However, tiles can have *mismatches* with their neighbors (Fig. 1).

Key Results

The first comparison between temperatures 1 and 2 was shown by Rothemund and Winfree [9], with the motivation of computing and efficiently

Non-cooperative $(\tau = 1)$ Cooperative $(\tau \geq 2)$

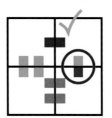 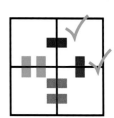

Self-Assembly at Temperature 1, Fig. 1 In the non-cooperative model, tiles can attach as soon as one side matches the neighborhood

building arbitrary shapes at the nanoscale. In this context, the generally accepted definition of "efficient" is *with significantly less tile types than the size of the output.*

Assembling Simple Shapes Efficiently

The first step toward these goals is the programming of simple shapes like squares or trees. At temperature ≥ 2, constructions with Turing machines can be used to show the following bound:

Theorem 1 (from [9]) *The smallest two-dimensional tileset T_n producing only squares of size $n \times n$ from a single-tile seed is of size $\theta \left(\frac{\log n}{\log \log n} \right).$*

The smallest number of tile types that can assemble exactly a set of shapes is called the *tile complexity* of that set. In the noncooperative model, the following upper bound is known:

Theorem 2 (from [9]) *For all integer n, there is a tileset T_n, of size $2n - 1$, that produces only squares of size $n \times n$ from a single-tile seed.*

Whether this upper bound is optimal is still one of the major open problems of the model, and little progress has been made since its identification. The real motivation behind this question is whether we (or natural systems) can perform useful computations with this model.

Finding the smallest tileset for assembling an input shape can also be treated as an optimization problem: see Adleman et al. [1] for the case of tree shapes.

The Role of Geometry

A partial answer to this question was found by Cook, Fu, and Schweller [2], who tried to "fake" cooperation by blocking the growth of some parts of the assembly. They introduced two different ways to do this: removing the planarity constraint and allowing errors.

In both cases, "faking" cooperation means producing the same assemblies as a temperature 2 tile assembly systems up to rescaling by a constant factor.

Three-Dimensional Noncooperative Self-Assembly

In three dimensions, temperature 1 self-assembly is able to simulate Turing computations:

Theorem 3 (from [2]) *There is a three-dimensional tileset T such that for all Turing machine $\mathcal{M}$ and input $x \in \mathbb{N}$, there is a computable seed assembly $\sigma_{\mathcal{M},x}$ and a tile $t \in T$, such that all terminal assemblies of $(T, \sigma_{\mathcal{M},x}, 1)$ contain t if and only if $\mathcal{M}$ accepts input x.*

The construction simulates a Turing-universal cooperative tile assembly system called a *zigzag system*, in which rows grow on top of each other, alternatively to the left and to the right, using cooperation to copy and update the previous row (Fig. 2).

The idea is pictured on Fig. 3. A "main" path grows on each row, building "bridges" and "blockers" (in blue on Fig. 3) that encode bits. These bits can be read by the next row: before reading a bit, the main path (in orange on Fig. 3) of the row forks into two branches, respectively probing for a bridge (encoding a 1) and a blocker (encoding a 0). Exactly one branch passes through and can accumulate successive bits in its state, until a full tile has been read. Then, it rewrites bits encoding the next tile for the row above.

Allowing Erroneous Blocking

Adapting the mechanism used in the 3D construction to the planar case is widely conjectured impossible [9], because allowing the "wrong" branch to grow and collide against a previous

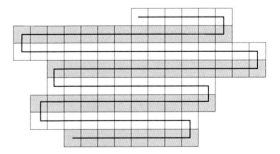

Self-Assembly at Temperature 1, Fig. 2 An example zigzag system (Figure from [2])

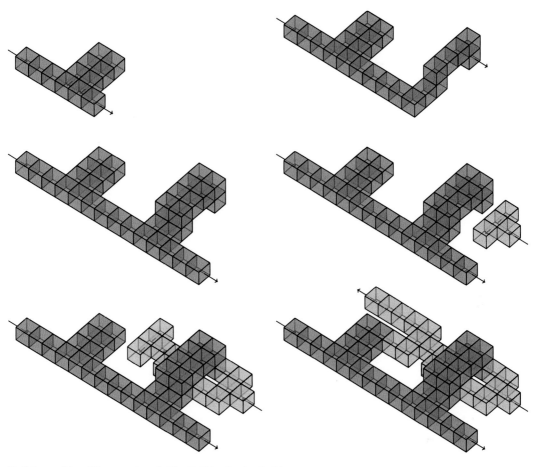

Self-Assembly at Temperature 1, Fig. 3 Bit selection in 3d

part of the assembly, in Fig. 3, *encloses* the other "correct" branch inside a finite portion of the plane.

However, it becomes possible if we consider a stochastic assembly schedule, where at each time step, exactly one tile attaches, and all tiles that can attach do so with equal probability. If we repeat the above construction k times consecutively, only one needs to succeed. We can therefore lower the probability of failure of each bit selection to 2^{-k}:

Theorem 4 (from [2]) *For all $\varepsilon > 0$ and all zigzag tile systems $\mathcal{T} = (T, s, 2)$, whose producible assemblies have size at most some constant r, there is a planar temperature 1 probabilistic tile assembly system $\mathcal{S}$ that simulates $\mathcal{T}$ without error with probability at least $1 - \varepsilon$.*

Of course, this construction means that the number of tile types and scaling factor will increase by a factor depending on ε and r.

Simulation up to Rescaling

One of the latest developments of tile assembly is the notion of *intrinsic universality* [3, 4, 11], a notion of simulation by rescaling only between tile assembly systems.

This idea is useful in particular to compare different models, because it provides qualitative properties to check, as opposed to quantitative properties such as tile complexity. The general argument is:

- At temperature 2 in two dimensions, there is a tileset known from [4] to be able to simu-

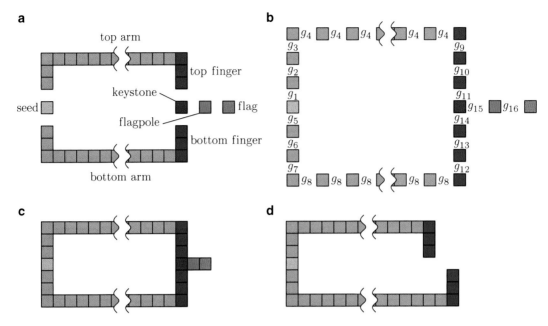

Self-Assembly at Temperature 1, Fig. 4 Tile assembly system $\mathcal{T}$ (Figure from [7])

late any other tile assembly system, modulo rescaling.

- However, there is a tile assembly system $\mathcal{T}$ that no tileset in model X (in our case, temperature 1) can simulate without errors.
- Therefore, model X is not as powerful as planar temperature 2.

This argument was used, for instance, to prove the first separation result between temperature 2 and the fully general model of temperature 1 [7]:

Theorem 5 (from [7]) *There is a planar (temperature 2) tile assembly system $\mathcal{T}$ (whose productions are pictured on Fig. 4) that no (two- or three- dimensional) tile assembly system $(A, \alpha, 1)$ can simulate up to rescaling.*

The proof uses a combinatorial argument (called the *window movie lemma*) to show that if there were a tile assembly system simulating all productions of $\mathcal{T}$, then it would also be able to produce other "illegal" assemblies (see Fig. 5) that do not represent any of $\mathcal{T}$'s producible assemblies.

Important Particular Cases

Noncooperative self-assembly, when restricted to dimension one, is similar to nondeterministic finite automata. It is therefore natural to look for a *pumping lemma*.

The first result in this direction was proven by Doty, Patitz, and Summers [5], who introduced the notion of *pumpable paths*: a path P is *pumpable* if it contains a subsegment $P_{i,i+1,\dots,j}$ that can be repeated arbitrarily many (consecutive) times along $\overrightarrow{P_i P_j}$ while remaining self-avoiding.

Theorem 6 (from [5]) *Let $\mathcal{T}$ be a tile assembly system that assembles exactly one (potentially infinite) terminal assembly α. If any path in α, longer than a constant c, is pumpable, then there are finite families of vectors $\mathbf{b_1}, \dots, \mathbf{b_n}$, $\mathbf{u_1}, \dots, \mathbf{u_n}$, and $\mathbf{v_1}, \dots, \mathbf{v_n} \in \mathbb{Z}^2$, such that:*

$$\mathrm{dom}(\alpha) = \bigcup_{1 \le i \le n} \{\mathbf{b_i} + j\,\mathbf{u_i} + k\,\mathbf{v_i} | j, k \in \mathbb{N}\}$$

In [5], examples were identified, of paths with segments that could be repeated, but only finitely many times due to collisions. The formalization of these examples was later done by Manuch,

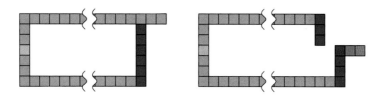

Self-Assembly at Temperature 1, Fig. 5 Illegal productions, that a temperature 1 system that can simulate all productions of $\mathcal{T}$ must also be able to simulate (Figure modified from [7])

Stacho, and Stoll [6], proving lower bounds under the hypothesis that no mismatches can occur.

This approach was then extended by Reif and Song [8], to show that tile assembly systems without mismatches have a recursive set of productions. However, the decidability of the "no mismatches" hypothesis is still an open problem.

Applications

Given the successful experimental applications of tile self-assembly, particularly in the field of DNA nanotechnologies, it seems natural to try to implement them: indeed, intuition suggests that they would make no errors in cooperation tiles. However, no successful construction of noncooperative experiments has been reported; the reason might be that ensuring uniqueness of the seed is impossible, as any two tiles in solution together might bind, without any of them being bound to the seed.

Open Problems

Aside from understanding the exact geometric requirements for Turing universality, a number of open problems have been identified in this model:

1. From [9]: What is the tile complexity of squares of size $n \times n$ in the planar, temperature 1 model?

    A related problem, which has been in the folklore for some time, is the existence of a shape of tile complexity arbitrarily smaller than its Manhattan diameter.

2. From [5]: If $\mathcal{T}$ is a tile assembly system with exactly one terminal assembly, is there a constant c such that any path longer than c is pumpable?

3. From [7]: Is there a temperature 1 tile assembly system with a non-recursive set of productions?

4. From [7]: Is there a single tileset able to simulate any temperature 1 tile assembly system up to rescaling, using only noncooperative bindings?

Cross-References

▶ Experimental Implementation of Tile Assembly

▶ Intrinsic Universality in Self-Assembly

Recommended Reading

1. Adleman LM, Cheng Q, Goel A, Huang MDA, Kempe D, de Espanés PM, Rothemund PWK (2002) Combinatorial optimization problems in self-assembly. In: Proceedings of the thirty-fourth annual ACM symposium on theory of computing (STOC), Montréal, pp 23–32
2. Cook M, Fu Y, Schweller RT (2011) Temperature 1 self-assembly: deterministic assembly in 3D and probabilistic assembly in 2D. In: Proceedings of the 22nd annual ACM-SIAM symposium on discrete algorithms (SODA), San Francisco, pp 570–589, arxiv preprint: arXiv:0912.0027
3. Doty D, Lutz JH, Patitz MJ, Summers SM, Woods D (2009) Intrinsic universality in self-assembly. In: Proceedings of the 27th international symposium on theoretical aspects of computer science (STACS), Nancy, pp 275–286. arxiv preprint: arXiv:1001.0208
4. Doty D, Lutz JH, Patitz MJ, Schweller RT, Summers SM, Woods D (2012) The tile assembly model is intrinsically universal. In: Proceedings of the 53rd

annual IEEE symposium on foundations of computer science (FOCS), New Brunswick, pp 439–446. arxiv preprint: arXiv:1111.3097

5. Doty D, Patitz MJ, Summers SM (2009) Limitations of self-assembly at temperature 1. In: Proceedings of the fifteenth international meeting on DNA computing and molecular programming, Fayetteville, 8–11 June 2009, pp 283–294. arxiv preprint: arXiv:0906.3251

6. Maňuch J, Stacho L, Stoll C (2010) Two lower bounds for self-assemblies at temperature 1. J Comput Biol 17(6):841–852

7. Meunier PE, Patitz MJ, Summers SM, Theyssier G, Winslow A, Woods D (2014) Intrinsic universality in tile self-assembly requires cooperation. In: Proceedings of the 25th annual ACM-SIAM symposium on discrete algorithms (SODA), Portland, pp 752–771. arxiv preprint: arXiv:1304.1679

8. Reif JH, Song T (2013) Complexity and computability of temperature-1 tilings. In: FNANO 2013, poster abstract

9. Rothemund PWK, Winfree E (2000) The program-size complexity of self-assembled squares (extended abstract). In: Proceedings of the thirty-second annual ACM symposium on theory of computing (STOC), Portland. ACM, pp 459–468. doi:http://doi.acm.org/10.1145/335305.335358

10. Winfree E (1998) Algorithmic self-assembly of DNA. PhD thesis, California Institute of Technology

11. Woods D (2013) Intrinsic universality and the computational power of self-assembly. In: Neary T, Cook M (eds) MCU, Zürich. EPTCS, vol 128, pp 16–22

Self-Assembly of Fractals

Matthew J. Patitz
Department of Computer Science and Computer Engineering, University of Arkansas, Fayetteville, AR, USA

Keywords

Discrete self-similar fractals; Fractal dimension; Self-assembly; Tile Assembly Model

Years and Authors of Summarized Original Work

2009; Kautz, Lathrop
2009; Lathrop, Lutz, Summers
2010; Patitz, Summers
2012; Lutz, Shutters
2013; Kautz, Shutters
2014; Barth, Furcy, Summers, Totzke

Problem Definition

This problem is concerned with the self-assembly fractal patterns and structures. More specifically, it deals with discrete self-similar fractals and different notions of them self-assembling from tiles in the abstract Tile Assembly Model (aTAM) and derivative models. The self-assembly of fractals and fractal-like structures is particularly interesting due to their pervasiveness in nature, as well their complex aperiodic structures which result in them occupying less dimensional space than the space they are embedded within.

Using the terminology from [1], we define $\mathbb{N}_g$ as the subset $\{0, 1, \ldots, g-1\}$ of $\mathbb{N}$, and if $A, B \subseteq \mathbb{N}^2$ and $k \in \mathbb{N}$, then $A + kB = \{\mathbf{m} + k\mathbf{n} | \mathbf{m} \in A$ and $\mathbf{n} \in B\}$. We then define discrete self-similar fractals as follows.

We say that $\mathbf{X} \subset \mathbb{N}^2$ is a *discrete self-similar fractal* (or *dssf* for short) if there exist $1 < g \in \mathbb{N}$ and a set $\{(0,0)\} \subset G \subset \mathbb{N}_g^2$ with at least one point in every row and column, such that $\mathbf{X} = \bigcup_{i=1}^{\infty} X_i$, where X_i, the ith stage of $\mathbf{X}$, is defined by $X_1 = G$ and $X_{i+1} = X_i + g^i G$. We say that G is the generator of $\mathbf{X}$.

Figure 1 shows, as an example, the first 5 stages of the discrete self-similar fractal known as the Sierpinski triangle. In this example, $G = \{(0,0), (1,0), (0,1)\}$.

In general, we ask whether or not a given dssf $\mathbf{X}$ can self-assemble within a given model.

Variants

The general problem of determining whether or not a discrete self-similar fractal self-assembles within a given model has several variants, which determine the way in which the fractal shape is represented within a resulting assembly.

1. **Weak self-assembly.** If a dssf $\mathbf{X}$ weakly self-assembles using a tile set T, then there exists a subset of tile types $B \subseteq T$ such that, in

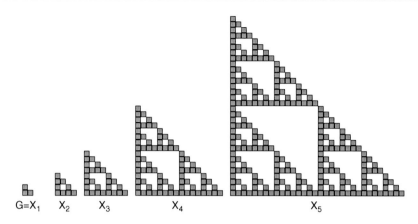

$$G = X_1 \quad X_2 \quad X_3 \quad\quad X_4 \quad\quad\quad\quad X_5$$

Self-Assembly of Fractals, Fig. 1 Example discrete self-similar fractal: the first 5 stages of the Sierpinski triangle

the terminal assembly α, for every point $\mathbf{p} \in$ dom α such that $\mathbf{p} \in \mathbf{X}$, the tile type at location $\mathbf{p}$ in α is a type within B, and for every point $\mathbf{p} \in$ dom α such that $\mathbf{p} \notin \mathbf{X}$, the tile type at location $\mathbf{p}$ in α is not within B. That is, the tile types in the subset B precisely "paint a picture" of $\mathbf{X}$, while tiles of types not in B may appear in locations outside of $\mathbf{X}$.

2. **Strict self-assembly.** If a dssf $\mathbf{X}$ strictly self-assembles, then it weakly self-assembles with $B = T$, i.e., the locations that exist in the domain of the terminal assembly are exactly those of $\mathbf{X}$.

3. **Approximate self-assembly.** A dssf $\mathbf{X}$, and thus a strictly self-assembled version of $\mathbf{X}$, has fractal dimension (i.e., zeta-dimension [3]) <2, and a weakly self-assembled version has dimension 2. Since it appears to be difficult if not impossible to strictly self-assemble many (or all) dssf's, it is interesting to consider if an approximation of a dssf $\mathbf{X}$ which retains the same fractal dimension as $\mathbf{X}$ can strictly self-assemble.

Key Results

Self-assembly of dssf's has been studied in all of the above variants and within the aTAM, 2HAM, and STAM [9]. As previously mentioned, the complexity of dssf's makes them interesting to study since they are infinite, aperiodic structures.

This requires any system in which they self-assemble to rely on algorithmic self-assembly (rather than unique tile types hard coded to each position of the shape), and for this reason early experimental results even included the weak self-assembly of the initial few stages of the Sierpinski triangle [11] as a proof of concept that DNA-based tile implementations of the aTAM are capable of algorithmic self-assembly. Nonetheless, as infinite structures, dssf's are more often the focus of theoretical studies.

Weak Self-Assembly

As seen in [11], it is possible for a very simple tile set of only 7 tile types to weakly self-assemble the Sierpinski triangle. This tile set can essentially be thought of as computing the xor function on two inputs (i.e., $00 \to 0$, $01 \to 1$, $10 \to 1$, and $11 \to 0$), with the glues with which a tile initially binds to an assembly encoding the input bits and those to which tiles later attach encoding the output bits.

In [4] it was noted that another characterization of the Sierpinski triangle is as the nonzero residues modulo 2 of Pascal's triangle. They then provided a characterization of an infinite class of dssf's, known as *generalized Sierpinski carpets*, which can be defined as the residues, modulo a prime number, of the entries in a two-dimensional matrix generated by a simple recursive equation. (A well-known example among this class of dssf's is the Sierpinski carpet.) They then proved that all generalized Sierpinski carpets weakly self-assemble in the aTAM.

Strict Self-Assembly

Although weak self-assembly of many dssf's can be achieved with very simple tile sets, it turns out that strict self-assembly is an entirely different, and much more difficult, problem. In fact, in [6] they proved that it is impossible for the Sierpinski triangle to strictly self-assemble in the aTAM. Furthermore, their proof showed that to be the case regardless of the temperature parameter. This result was extended in [10] to a proof that an infinite class of "pinch-point" dssf's, which includes the Sierpinski triangle, cannot strictly self-assemble in the aTAM at any temperature. Pinch-point fractals are those whose generators have exactly one point in their topmost row, the leftmost, and one in their eastmost column, the bottommost. Yet another extension was provided in [1], where the authors defined "tree" fractals, which again include the Sierpinski triangle, as those with generators which are trees and which have a single point in their topmost row and a single point in their rightmost column. They then proved that, regardless of the temperature or even of the scale factor, no tree fractal strictly self-assembles in the aTAM.

Additional results related to strict self-assembly of dssf's include the proof in [10] that in the aTAM at temperature 1 (i.e., systems with $\tau = 1$), it is impossible for any dssf to self-assemble within a locally deterministic system (see [12] for a definition of local determinism), and in [2] it was proven that the Sierpinski triangle also cannot self-assemble in the 2-Handed Assembly Model, at any temperature.

To date, the single positive result related to the strict self-assembly of a dssf is for an "active" model of self-assembly, where tiles are allowed to change the states of their glues during assembly, called the Signal-passing Tile Assembly Model (STAM). In [9] they gave a construction proving that the Sierpinski triangle can self-assemble within the STAM at temperature 1 and scale factor 2. By the result of [1], this is impossible

in the aTAM and demonstrates the power of the active nature of the STAM, as that construction essentially builds stages of the Sierpinski triangle in a manner analogous to weak self-assembly, but then causes the unwanted interior portions to dissociate and then break apart.

Approximate Self-Assembly

It has been shown that an infinite subset of dssf's can weakly self-assemble in the aTAM, while another infinite subset cannot strictly self-assemble. Recall also that dssf's have fractal dimension <2, and since their strictly self-assembled versions retain their original fractal dimensions, so do they. However, their weakly self-assembled versions have dimension 2. Therefore, the question arises about whether or not some transformation of a dssf (especially, a dssf which cannot strictly self-assemble), which visually approximates the original dssf while retaining its fractal dimension, can strictly self-assemble in the aTAM.

This question was first answered positively in [6], where they defined a transformation for the Sierpinski triangle which they called "fibering," and they then gave a construction proving that the so-called fibered Sierpinski triangle does strictly self-assemble in the aTAM while maintaining the Sierpinski triangle's fractal dimension of ≈ 1.585. An example can be seen in Fig. 2b, showing how the fibering consists an additional row of tiles added to the south and west borders of each copy of each subsequent stage of the fractal. In [10] they extended the technique of fibering to include an infinite subclass of dssf's (which again includes the Sierpinski triangle) which they called "nice" dssf's. Nice dssf's are those whose generators are connected and contain all points on the west and south boundaries.

While the fibering technique creates visual approximations of fractals, it results in subsequent stages being further and further separated from each other. To counter this drawback, in [8] they introduced a technique for fibering the Sierpinski triangle "in place." An example can be seen in Fig. 2c, showing how this version of fibering only uses space on the interior of each stage of the fractal, thus allowing the stages to remain in the same positions relative to each other.

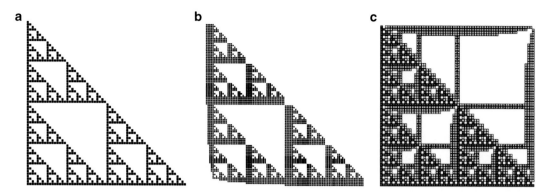

Self-Assembly of Fractals, Fig. 2 Various patterns corresponding to the Sierpinski triangle. (**a**) A portion of the discrete Sierpinski triangle. (**b**) A portion of the fibered Sierpinski triangle of [7] (Figure from [7]). (**c**) A portion of the in-place fibered Sierpinski triangle of [8] (Figure from [8])

Furthermore, this technique retains the same fractal dimension as the Sierpinski triangle, and they showed that it is impossible to use asymptotically less space than their construction while strictly self-assembling a shape which contains the Sierpinski triangle as a subset. In [5] this technique was extended to strictly self-assemble approximations for every generalized Sierpinski carpet.

Open Problems

1. Does there exist a discrete self-similar fractal which can strictly self-assemble in the aTAM, or conversely, can it be shown that none does?
2. What is the class of discrete self-similar fractals for which an approximation, such as fibering or in-place fibering, which maintains the original fractal dimension, strictly self-assembles in the aTAM?

URLs to Code and Data Sets

ISU TAS simulation software for the aTAM, kTAM, and 2HAM (http://self-assembly.net/wiki/index.php?title=ISU_TAS) and the Fibered Fractal Tiler for defining discrete self-similar fractals which can be fibered and generating the corresponding aTAM tile sets (http://

self-assembly.net/wiki/index.php?title=Fibered_Fractal_Tiler).

Cross-References

▶ Experimental Implementation of Tile Assembly
▶ Self-Assembly at Temperature 1
▶ Self-Assembly of Squares and Scaled Shapes

Recommended Reading

1. Barth K, Furcy D, Summers SM, Totzke P (2014) Scaled tree fractals do not strictly self-assemble. In: Unconventional computation & natural computation (UCNC) 2014, University of Western Ontario, London, 14–18 July 2014 (to appear)
2. Cannon S, Demaine ED, Demaine ML, Eisenstat S, Patitz MJ, Schweller R, Summers SM, Winslow A (2012) Two hands are better than one (up to constant factors). Technical report 1201.1650, Computing Research Repository. http://arxiv.org/abs/1201.1650
3. Doty D, Gu X, Lutz JH, Mayordomo E, Moser P (2005) Zeta-dimension. In: Proceedings of the thirtieth international symposium on mathematical foundations of computer science, Gdansk. Springer, pp 283–294
4. Kautz SM, Lathrop JI (2009) Self-assembly of the Sierpinski carpet and related fractals. In: Proceedings of the fifteenth international meeting on DNA computing and molecular programming, Fayetteville, 8–11 June 2009, pp 78–87
5. Kautz S, Shutters B (2013) Self-assembling rulers for approximating generalized Sierpinski carpets.

S

Algorithmica 67(2):207–233. doi:10.1007/s00453-012-9691-x, http://dx.doi.org/10.1007/s00453-012-9691-x

6. Lathrop JI, Lutz JH, Summers SM (2007) Strict self-assembly of discrete Sierpinski triangles. In: Proceedings of the third conference on computability in Europe, Siena, 18–23 June 2007

7. Lathrop JI, Lutz JH, Summers SM (2009) Strict self-assembly of discrete Sierpinski triangles. Theor Comput Sci 410:384–405

8. Lutz JH, Shutters B (2012) Approximate self-assembly of the Sierpinski triangle. Theory Comput Syst 51(3):372–400

9. Padilla JE, Patitz MJ, Pena R, Schweller RT, Seeman NC, Sheline R, Summers SM, Zhong X (2013) Asynchronous signal passing for tile self-assembly: fuel efficient computation and efficient assembly of shapes. In: UCNC, Milan, pp 174–185

10. Patitz MJ, Summers SM (2008) Self-assembly of discrete self-similar fractals (extended abstract). In: Proceedings of the fourteenth international meeting on DNA computing, Prague, 2–6 June 2008 (to appear)

11. Rothemund PWK, Papadakis N, Winfree E (2004) Algorithmic self-assembly of DNA Sierpinski triangles. PLoS Biol 2(12):e424. doi:10.1371/journal.pbio.0020424, http://dx.doi.org/10.1371

12. Soloveichik D, Winfree E (2007) Complexity of self-assembled shapes. SIAM J Comput 36(6):1544–1569

Self-Assembly of Squares and Scaled Shapes

Robert Schweller
Department of Computer Science, University of Texas Rio Grande Valley, Edinburg, TX, USA

Keywords

Algorithmic self-assembly; Kolmogorov complexity; Scaled shapes; Self-assembly; Tile assembly model; Tile complexity

Years and Authors of Summarized Original Work

2000; Rothemund, Winfree
2001; Adleman, Cheng, Goel, Huang
2005; Cheng, Aggarwal, Goldwasser, Kao, Schweller, Espanes
2007; Soloveichik, Winfree

Problem Definition

Abstract Tile Assembly Model

The abstract Tile Assembly Model (aTAM) [3] is a mathematical model of self-assembly in which system components are four-sided Wang tiles with glue types assigned to each tile edge. Any pair of glue types are assigned some nonnegative interaction strength denoting how strongly the pair of glues bind. An aTAM system is an ordered triplet (T, τ, σ) consisting of a set of tiles T, a positive integer threshold parameter τ called the system's *temperature*, and a special tile $\sigma \in T$ denoted as the *seed* tile. Assembly proceeds by attaching copies of tiles from T to a growing seed assembly whenever the placement of a tile on the 2D grid achieves a total strength of attachment from abutting edges, determined by the sum of pairwise glue interactions, that meets or exceeds the temperature parameter τ. The pairwise strength assignment between glues on tile edges is often restricted to be "linear" in that identical glue pairs may be assigned arbitrary positive values, while non-equal pairs are required to have interaction strengths of 0. We denote this restricted version of the model as the *standard* aTAM. When this restriction is not applied, i.e., any pair of glues may be assigned any positive integer strength, we call the model the *flexible glue* aTAM.

Given the aTAM's model of growth, we may consider the problem of designing an aTAM system which is guaranteed to grow into a target shape S, given by a set of 2D integer coordinates, and stop growing. Such systems are guaranteed to exist for any finite shape S, but solutions will typically vary in the number of tiles $|T|$ used. For a given shape S, an interesting problem is to design a system that assembles S while using the fewest, or close to the fewest, number of tiles $|T|$ possible. This fewest possible number of tiles required for the assembly of a given shape S is termed the *program-size* complexity of S.

Problem 1 Let $K_{SA}(n)$ and $K_{\widetilde{SA}}(n)$ denote the program-size complexity of an $n \times n$ square for the standard aTAM and the flexible glue aTAM, respectively. What are $K_{SA}(n)$ and $K_{\widetilde{SA}}(n)$?

Problem 2 Let $K_{SA}(n,k)$ and $K_{\widetilde{SA}}(n,k)$ denote the program-size complexity of a $k \times n$ rectangle for the standard aTAM and the flexible glue aTAM, respectively. What are $K_{SA}(n,k)$ and $K_{\widetilde{SA}}(n,k)$?

Problem 3 For an arbitrary given shape S, what is the program-size complexity of S? Let the *scale-free* program size of S be the smallest tile set system that uniquely builds some scaled-up version S. Let $K_{SA}(S)$ and $K_{\widetilde{SA}}(S)$ denote the *scale-free* program size of S for the standard aTAM and the flexible glue aTAM, respectively. What are $K_{SA}(S)$ and $K_{\widetilde{SA}}(S)$?

Key Results

The best known bounds for program-size complexity for squares, rectangles, and general scaled shapes are presented in this section.

$n \times n$ Squares

The efficient self-assembly of $n \times n$ squares has served as a benchmark for self-assembly algorithms within the aTAM and more general tile assembly models. Within the aTAM, the problem is well understood up to constant factors. The first result states a general upper bound for the program size of self-assembled squares for general n, which is matched by an information-theoretic lower bound that holds for almost all integers n. The precise bounds differ between the standard and flexible glue models but are tight in both cases. The lower bound of inequality (1) is proven in [3] and is based on the Kolmogorov complexity of the integer n. The lower bound of (2) is proven in [2] by the same approach. The upper bound of (1) is proven in [1] and offers an improvement over the initial upper bound of $O(\log n)$ from [3]. The $O(\log n)$ result of [3] is achieved by implementing a key primitive in tile self-assembly: a binary counter of $\log n$ tiles that grows to length n. The improvement of [1] is achieved by modifying the counter concept to work with an optimal, variable base. The upper bound of (2) is proven in [2] and is obtained by combining the aTAM counter primitives with

a scheme for efficiently seeding the counter by extracting bits from the values of the flexible glue interactions.

Theorem 1 *There exist positive constants c_1 and c_2 such that for almost all integers $n \in \mathbb{N}$, the following inequalities hold. Moreover, the upper bounds hold for all $n \in \mathbb{N}$.*

$$c_1 \frac{\log n}{\log \log n} \le K_{SA}(n) \le c_2 \frac{\log n}{\log \log n}. \quad (1)$$

$$c_1 \sqrt{\log n} \le K_{\widetilde{SA}}(n) \le c_2 \sqrt{\log n}. \quad (2)$$

While the above theorem presents a tight understanding of the program-size complexity for most self-assembled squares, the information-theoretic lower bound allows for special values of n to be assembled with a much smaller program size. The program size is in fact as small as one could reasonably hope for. In [3], a tile system is presented that simulates a Busy Beaver Turing Machine and assembles correspondingly large squares for each tile set size. This construction yields the following theorem implying that the largest self-assembled square for a given number of tiles grows faster than any computable function!

Theorem 2 *There exists a positive constant c such that for infinitely many n, $K_{SA}(n) \le cf(n)$ for $f(n)$ any nondecreasing unbounded computable function.*

Thin Rectangles

The program size of self-assembled squares and other thick rectangles is dictated by information-theoretic bounds which stem from the aTAM's ability to simulate arbitrary Turing machines given enough geometric space to work within. When this space is cut down, such as in the case of building a thin $k \times n$ rectangle, the program size is limited by geometric factors. The following upper and lower bounds are shown in [2] and represent the best known bounds for *thin* $k \times n$ rectangles in which $k = O(\log / \log \log n)$. The lower bound is achieved by a pigeon-hole pumping argument on the types of tiles placed, along with their order of placement, along a width k column of the target rectangle. The upper bound is based on the construction of a general-base, general-width counter, which generalizes the binary counter concept of [3].

Theorem 3 *There exist positive constants c_1 and c_2 such that for any $n, k \in \mathbb{N}$, the following inequalities hold.*

$$c_1 \frac{n^{1/k}}{k} \le K_{\widetilde{SA}}(n, k) \le K_{SA}(n, k) \le c_2 (n^{1/k} + k).$$

Scaled Shapes

The program size of general shapes is difficult to analyze as it is highly dependent on geometric features of the target shape. However, if we consider the assembly of an arbitrarily scaled-up version of a target shape, these geometric difficulties can be eliminated and a very general result can be achieved. The next result from [4] shows that the scale-free program size of S is closely related to the Kolmogorov complexity of S. In particular, the scale-free program-size complexity of S is a log factor less than the Kolmogorov complexity of S for the standard model, and the scale-free program size complexity of S is the square root of the Kolmogorov complexity of S for the flexible glue model. The standard model result is shown in [4] and is achieved by encoding a compressed description of S in a small tile set which is extracted by a set of tiles simulating a Turing machine that extracts the pixels of S from this compressed representation. The need for the scale factor increase of S is to allow room for the Turing machine simulation. In fact, the required scale factor is the run time of the Turing machine that decompresses the optimal encoding of S. The flexible glue result is achieved by combining portions of the flexible glue construction for squares [2] with the construction of [4]. In the following theorem, $K(S)$ denotes the Kolmogorov complexity of S with respect to some fixed universal Turing machine.

Theorem 4 *For any shape S, there exist positive constants c_1 and c_2 such that*

$$c_1 \frac{K(S)}{\log K(S)} \le K_{SA}(S) \le c_2 \frac{K(S)}{\log K(S)}. \quad (3)$$

$$c_1 \sqrt{K(S)} \le K_{\widetilde{SA}}(S) \le c_2 \sqrt{K(S)}. \quad (4)$$

Open Problems

A few important open problems in this area are as follows. In the case of squares, the program size is well understood as long as the temperature of the system is at least two. A long-standing open problem has been to determine the program-size complexity of $n \times n$ squares for temperature-1 self-assembly in which each positive glue force alone is sufficient to cause a tile attachment. To date, no known method is able to achieve $o(n)$ tile complexity at temperature-1 for an $n \times n$ square, but no proof exists that this cannot be done. With respect to thin $k \times n$ rectangles, the best upper and lower bound have a gap with respect to variable k. Does there exist a more efficient rectangle construction, or can a higher lower bound be derived? Finally, while the scale-free program-size complexity of general shapes is well understood, little is known about the (unscaled) program size of general shapes. What new tools and geometric classifications can be developed to analyze and bound this complexity for general shapes?

Cross-References

Recommended Reading

1. Adleman L, Cheng Q, Goel A, Huang, M-D (2001) Running time and program size for self-assembled squares. In Proceedings of the thirty-third annual ACM symposium on theory of computing, New York. ACM, pp 740–748
2. Cheng Q, Aggarwal G, Goldwasser MH, Kao M-Y, Schweller RT, de Espanés PM (2005) Complexities for generalized models of self-assembly. SIAM J Comput 34:1493–1515
3. Rothemund PWK, Winfree E (2000) The program-size complexity of self-assembled squares (extended abstract). In Proceedings of the 32nd ACM symposium on theory of computing, STOC'00, Portland, pp 459–468
4. Soloveichik D, Winfree E (2007) Complexity of self-assembled shapes. SIAM J Comput 36(6):1544–1569

Self-Assembly with General Shaped Tiles

Andrew Winslow
Department of Computer Science, Tufts University, Medford, MA, USA

Keywords

Cellular automata; DNA computing; Geometric computing; Natural computing; Turing universality

Years and Authors of Summarized Original Work

2012; Fu, Patitz, Schweller, Sheline
2014; Demaine, Demaine, Fekete, Patitz, Schweller, Winslow, Woods

Problem Definition

Self-assembly is an asynchronous, decentralized process in which particles aggregate to form superstructures according to localized interactions. The most well-studied models of these particle systems, e.g., the abstract Tile Assembly Model of Winfree [11], utilize square-shaped particles arranged on a lattice by attaching edgewise. Particles attach to form larger *assemblies*, and a pair of assemblies or tiles can attach if they can translate to a nonoverlapping configuration with a set of k coincident edges, where $k \geq \tau$, a parameter of the system called the *temperature*.

In *seeded* assembly, individual particles attach to a growing *seed assembly*. This assembly may begin as a single-tile or a multi-tile assembly. In *unseeded* assembly (also called *hierarchical* [3], *two-handed* [2], or *polyomino* [7] assembly), there is no such restriction. The set of assemblies to which a single tile cannot attach (in seeded assembly) or that cannot attach to any other assembly (in unseeded assembly) are the *terminal assemblies* of the system.

Objectives In general, the goal is to design a system of minimal complexity that assembles into a unique terminal assembly with a desired shape or property. In models using square tiles, this is equivalent to designing a system using the fewest tile types. When tiles are allowed to be more general shapes, then the option of trading tile types for tile complexity becomes available. The motivation for this work is to understand how more complex tile shapes can be used to reduce the number of tile types in a system, and two benchmark problems regarding the computational power and efficiency of tile systems are considered in the context of systems of non-square tiles:

Problem 1 (Square Assembly)

INPUT*: A natural number n.*
OUTPUT*: A self-assembly system with a unique terminal assembly consisting of n^2 tiles in a $n \times n$ square shape.*

Problem 2 (Computational Power) *What systems of non-square tiles are capable of simulating computation, and to what extent?*

Key Results

In general, it is the case that allowing non-square tiles permits an asymptotic reduction in the number of tile types, and systems of very few non-square tiles are capable of universal computation. At a high-level, such reductions are achieved by simulating many tiles via translations and rotations of a single tile type.

Models

Fu, Patitz, Schweller, and Sheline [6] introduce two models of general shaped types. The first, called the *geometric Tile Assembly Model (gTAM)*, is a model of seeded, translation-only assembly where tiles are polyomino-shaped – equivalent to prebuilt assemblies of square tiles. The second is an unseeded version they call the *Two-Handed Planar Geometric Tile Assembly Model (2GAM)*, which has the added restriction that assemblies can only attach if there exists a continuous motion bringing the two assemblies together during which they remain disjoint. This can be thought of as a restriction that the assemblies live in the plane and do not make use of the third dimension to maneuver into place.

Demaine et al. [4] introduce the *polygonal free-body Tile Assembly Model (pfbTAM)* in which tiles may have arbitrary simple polygonal shapes, attaching edgewise along equal-length edges. Systems without and without rotation are both permitted – we note that rotation is forbidden in the gTAM and 2GAM (as well as the aTAM).

Efficient Construction

Fu, Patitz, Schweller, and Sheline [6] prove that both the gTAM and 2GAM allow an asymptotic reduction in the number of tiles needed to assemble an $n \times n$ square of tiles. For the gTAM, they prove that such a square can be assembled using a temperature-1 system of $O(\sqrt{\log n})$ tile types, beating the optimal (temperature-2) system of $\Omega(\log n / \log\log n)$ square tiles by Adleman et al. [1]. This is a reduction in both the number of tile types (by a quadratic factor) and temperature. The temperature reduction is especially significant, as a lower bound of $\Omega(n)$ for temperature-1 aTAM systems is a widely believed conjecture [8–10].

For the 2GAM, they reduce the number of tile types even further, using a temperature-2 system $O(\log\log n)$ tile types to assemble an $n \times n$

square. However, this system comes with the caveat that system makes use of either a disconnected tile shape or a slightly three-dimensional shape.

Computational Power

Positive results on the computational power of general shaped tile systems fall into two categories: Turing universality and bounded-time computation. Fu, Patitz, Schweller, and Sheline [6] prove that any Turing machine computation can be carried out by a temperature-1 gTAM system. As with the temperature-1 construction of squares, this result is surprising due to the open conjecture regarding the computational power of square tile systems at temperature 1.

Demaine et al. [4] prove that any Turing machine computation can be carried out by a temperature-2 pfbTAM system (with rotation) consisting of a single tile. Their result actually proves that any aTAM system can be simulated by such a system, and thus Turing universality is achieved by simulating aTAM systems carrying out computation. Combined with the intrinsic universality result of Doty et al. [5], this result can be extended to prove that a *single* temperature-2 pfbTAM system (with rotation) consisting of a single tile can carry out any Turing machine computation, given an appropriate seed assembly consisting of copies of this tile.

Finally, Demaine et al. also prove that temperature-3 pfbTAM systems (*without* rotation) consisting of a single tile can carry out simulation of computationally universal cellular automata for a number of steps limited by the size of the seed assembly. Specifically, they prove that n steps can be carried out using a seed assembly of $O(n)$ tiles. A loose lower bound is also proved, namely, that more than three tiles are needed to carry out any computation.

Applications

The generic ability to reduce the number of tile types in a system by increasing the geometric complexity of these tiles extends many other constructions in theoretical tile assembly. Additionally, there may be practical barriers to systems of many tile types, e.g., additional cost of manufacturing or longer assembly time due to heterogenous combinations of many particle types, that can be reduced or eliminated by replacing these systems with systems of fewer, more complex tile.

Open Problems

Obtaining an upper bound on the number of steps of a cellular automaton simulable by single-tile translation-only systems remains open. It is conjectured that a seed assembly of size n can only carry out $O(n^2)$ steps.

Cross-References

▶ Combinatorial Optimization and Verification in Self-Assembly
▶ Experimental Implementation of Tile Assembly
▶ Self-Assembly at Temperature 1

Recommended Reading

1. Adleman L, Cheng Q, Goel A, Huang MD (2001) Running time and program size for self-assembled squares. In: Proceedings of symposium on theory of computing (STOC), Heraklion
2. Cannon S, Demaine ED, Demaine ML, Eisenstat S, Patitz MJ, Schweller RT, Summers SM, Winslow A (2013) Two hands are better than one (up to constant factors): self-assembly in the 2HAM vs. aTAM. In: STACS 2013, Kiel, LIPIcs, vol 20. Schloss Dagstuhl–Leibniz-Zentrum fuer Informatik, pp 172–184
3. Chen H, Doty D (2012) Parallelism and time in hierarchical self-assembly. In: Proceedings of the 23rd annual ACM-SIAM symposium on discrete algorithms (SODA), Kyoto, pp 1163–1182
4. Demaine ED, Demaine ML, Fekete SP, Patitz MJ, Schweller RT, Winslow A, Woods D (2014) One tile to rule them all: simulating any tile assembly system with a single universal tile. In: Esparza J, Fraigniaud P, Husfeldt T, Koutsoupias E (eds) Automata, languages and programming (ICALP), Copenhagen. LNCS, vol 8572. Springer, Berlin/Heidelberg, pp 368–379
5. Doty D, Lutz JH, Patitz MJ, Schweller RT, Summers SM, Woods D (2012) The tile assembly model is intrinsically universal. In: Proceedings of the 53rd annual symposium on foundations of computer science (FOCS), New Brunswick, pp 302–310
6. Fu B, Patitz MJ, Schweller RT, Sheline B (2012) Self-assembly with geometric tiles. In: Czumaj A, Mehlhorn K, Pitts A, Wattenhofer R (eds) Automata, languages and programming (ICALP), Warwick. LNCS, vol 7391. Springer, Berlin/New York, pp 714–725
7. Luhrs C (2010) Polyomino-safe DNA self-assembly via block replacement. Nat Comput 9(1):97–109
8. Meunier PE, Patitz MJ, Summers SM, Theyssier G, Winslow A, Woods D (2014) Intrinsic universality in tile self-assembly requires cooperation. In: Proceedings of the 25th annual ACM-SIAM symposium on discrete algorithms (SODA), Portland, pp 752–771
9. Rothemund PWK, Winfree E (2000) The program-size complexity of self-assembled squares (extended abstract). In: Proceedings of ACM symposium on theory of computing (STOC), Portland, pp 459–468
10. Summers SM (2010) Universality in algorithmic self-assembly. PhD thesis, Iowa State University
11. Winfree E (1998) Algorithmic self-assembly of DNA. PhD thesis, Caltech

Selfish Bin Packing Problems

Leah Epstein
Department of Mathematics, University of Haifa, Haifa, Israel

Keywords

Bin packing; Price of anarchy; Selfish agents

Years and Authors of Summarized Original Work

2006; Bilò
2008, 2011; Epstein, Kleiman

Problem Definition

In bin packing games with selfish items, n items are to be packed into (at most) n bins, where each item chooses a bin that it wishes to be packed into. The cost of an item i of size $0 < s_i \leq 1$ is defined based on its size and the contents of its bin. Nash equilibria (NE) are defined as solutions where there is no item that can change its choice unilaterally and gain from this change. Bin packing games were inspired by the well-known bin packing problem [2]. In this problem, a set of items, each of size in $(0, 1]$, is given. The goal is to partition (or pack) the items into a minimum number of subsets that are called *bins*. Each bin has unit capacity, and the load of a bin is defined to be the total size of items packed into it (where the load cannot exceed 1). The problem is NP-hard in the strong sense, and thus theoretical research has focused on studying and developing approximation algorithms, which allow to design nearly optimal solutions, and on online algorithms, which receive the items one by one and must assign each item to a bin immediately and irrevocably (without any information on further items).

In a bin packing game, every item is operated by a selfish player. There are n bins, and the strategy of a player is the bin that it selects. If the resulting packing is valid (i.e., the load of no bin exceeds 1), then the set of items sharing a bin share its cost proportionally, i.e., let B be a bin (a subset of items). The cost of $i \in B$ is $s_i / \left(\sum_{j \in B} s_j \right)$. If the resulting packing is invalid, any item packed into an invalid bin has infinite cost. We are interested in pure Nash equilibria, and by the term NE, we refer to such an equilibrium. The problem was presented by Bilò [1].

There are several directions which can be explored. First, one would like to find out if any bin packing game has an NE. If this is the case, other kinds of equilibria might be of interest as well. For a class of games (such that each of them has an NE), a process of convergence is defined as follows. The process starts with an arbitrary configuration, and at each time, an item that can reduce its cost is selected and moved to another bin (where the cost of this item will be smaller than its cost before it is moved). Such a process can also be seen as local search. Items are moved one at a time; a single move (for one

item) is called a step. Note that one item can participate in multiple steps. The questions which can be asked are whether the process converges for any initial packing (i.e., reaches a state that no further step can be applied) and how large can the number of steps be. As it turns out, any bin packing game has at least one NE, and the processes described here always converge [1, 8]. Since it is possible that the process converges in exponential time, it is of interest to develop a polynomial time algorithm that computes NE packings. Such an algorithm for this problem defined above was designed by Yu and Zhang [9]. Finally, once the existence of NE packing has been established, the primary goal becomes the study of the quality of worst-case equilibria. This concept is called *price of anarchy*. For a given game G (i.e., a set of items which is an input for bin packing), the price of anarchy of this game, denoted by $POA(G)$, is the ratio between the maximum number of nonempty bins in any NE packing and the minimum number of bins in any packing (the number of bins in an optimal packing, also called the social optimum, denoted by $OPT(G)$). The price of stability is similar, but best-case equilibria are studied, and as Bilò [1] proved that any game has a social optimum that is an NE, the price of stability is 1 for any game.

The price of anarchy (POA) of a class of games (here, the class of all bin packing games) is defined to be the supremum POA over all games in the class. However, as bin packing is typically studied with respect to the asymptotic approximation ratio, the POA for the bin packing class of games is defined, similarly to the asymptotic approximation ratio, as $\lim_{M \to \infty} \sup_{\{G : OPT(G) \geq M\}} POA(G)$.

Key Results

The POA was studied already in [1], where Bilò provided the first bounds on it, a lower bound of $\frac{8}{5}$ and an upper bound of $\frac{5}{3}$. The quality of NE solutions was further investigated in [4], where nearly tight bounds for the PoA were given, an upper bound of 1.6428 and a lower bound of 1.6416 (see also [9]). The parametric POA, which is the POA for subclasses of games where the size of no item exceeds a given value, was considered as well [5].

NE packings are related to outputs of the algorithm First Fit (FF) for bin packing [7]. FF is in fact an online algorithm that packs each item, in turn, into a minimum index bin where it fits (using an empty bin if there is no other option). It is not difficult to see that every NE is an output of FF; sort the bins of the NE by non-increasing loads, and create a list of items according to the ordering of bins. FF will create exactly the bins of the original packing. Interestingly, the POA is significantly smaller than the asymptotic approximation ratio of FF (which is equal to 1.7 [7]). Note that the PoA is not equal to the approximation ratio of any natural algorithm for bin packing.

Some intuition regarding the difference between the asymptotic approximation ratio of First Fit and the POA of this class of games can be shown using a small example. Consider items of the following sizes (for a sufficiently small $\varepsilon > 0$): $\frac{1}{6} - 2\varepsilon$ (small items), $\frac{1}{3} + \varepsilon$ (medium items), and $\frac{1}{2} + \varepsilon$ (large items). The worst-case examples for FF are similar to this example, though the items of the first two types have a number of different sizes; small items can be slightly smaller or slightly larger than $\frac{1}{6}$, and medium items can be slightly smaller or slightly larger than $\frac{1}{3}$. Given the item types defined above, assume that there are $6N$ items of each type (for some positive integer N), when FF receives this input (sorted by non-decreasing size), it creates N bins with six small items packed into each bin, $3N$ bins with two medium items packed into each bin, and the remaining items are packed into dedicated bins. This packing is not an NE, as a medium item reduces its cost from $\frac{1}{2}$ to approximately $\frac{2}{5}$ if it joins a large item. Indeed, roughly speaking, if an NE packing consists of a large number of bins (compared to an optimal solution), a bin of this NE packing either has an item whose size exceeds $\frac{1}{2}$ or its load cannot be as small as approximately $\frac{2}{3}$. This allows a tighter analysis. Interestingly, in

worst-case examples for the POA, medium items have sizes that are close to $\frac{1}{4}$ instead of $\frac{1}{3}$.

Related Results

Bin packing games, where the cost of an item is defined differently, were studied. One option is to assign equal costs to all players (which are packed together into a valid bin) [3, 6]. A generalized version where each item has a positive weight, and costs are based on cost sharing proportional to the weights of items that share a bin [3] was studied as well. The weights of items in the games described above (those of [1, 4]) are equal to their sizes. These are two classes of games, for which the POA turns out to be of interest. The POA for the class of games with equal weights is slightly (strictly) below 1.7, and in the case of general weights, the POA is equal to 1.7 [3]. Another topic of interest is the quality of other kinds of equilibria. Those are strong equilibria, which are solutions that are also resilient to deviations of subsets of items reducing their costs, and Pareto optimal equilibria, where the solution is required to be weakly (or strictly) Pareto optimal, that is, there is no alternative packing where all items reduce their costs (or a packing where no item increases its cost and at least one item reduces it) [3]. For these last kinds of equilibria, the POA is still above 1.6 (but at most 1.7).

Cross-References

▶ Subset Sum Algorithm for Bin Packing

Recommended Reading

1. Bilò V (2006) On the packing of selfish items. In: Proceedings of the 20th international parallel and distributed processing symposium (IPDPS2006). IEEE, Rhodes, Greece, 9pp
2. Coffman E Jr, Csirik J (2007) Performance guarantees for one-dimensional bin packing. In: Gonzalez TF (ed) Handbook of approximation algorithms and metaheuristics, chap 32. Chapman & Hall/CRC, Boca Raton, pp (32–1)–(32–18)

3. Dósa Gy, Epstein L (2012) Generalized selfish bin packing. CoRR, abs/1202.4080
4. Epstein L, Kleiman E (2011) Selfish bin packing. Algorithmica 60(2):368–394
5. Epstein L, Kleiman E, Mestre J (2009) Parametric packing of selfish items and the subset sum algorithm. In: Proceedings of the 5th international workshop on internet and network economics (WINE2009), Rome, Italy, pp 67–78
6. Han X, Dósa Gy, Ting HF, Ye D, Zhang Y (2013) A note on a selfish bin packing problem. J Glob Optim 56(4):1457–1462
7. Johnson DS, Demers AJ, Ullman JD, Garey MR, Graham RL (1974) Worst-case performance bounds for simple one-dimensional packing algorithms. SIAM J Comput 3(4):299–325
8. Miyazawa FK, Vignatti AL (2009) Convergence time to Nash equilibrium in selfish bin packing. Electron Notes Discret Math 35:151–156
9. Yu G, Zhang G (2008) Bin packing of selfish items. In: The 4th international workshop on internet and network economics (WINE2008), Shanghai, China, pp 446–453

Selfish Unsplittable Flows: Algorithms for Pure Equilibria

Paul (Pavlos) Spirakis
Computer Engineering and Informatics, Research and Academic Computer Technology Institute, Patras University, Patras, Greece
Computer Science, University of Liverpool, Liverpool, UK
Computer Technology Institute (CTI), Patras, Greece

Keywords

Atomic network congestion games; Cost of anarchy

Years and Authors of Summarized Original Work

2005; Fotakis, Kontogiannis, Spirakis

Problem Definition

Consider having a set of resources E in a system. For each $e \in E$, let $d_e(\cdot)$ be the delay per user

that requests its service, as a function of the total usage of this resource by all the users. Each such function is considered to be non–decreasing in the total usage of the corresponding resource. Each resource may be represented by a pair of points: an entry point to the resource and an exit point from it. So, each resource is represented by an arc from its entry point to its exit point and the model associates with this arc the cost (e.g., the delay as a function of the load of this resource) that each user has to pay if she is served by this resource. The entry/exit points of the resources need not be unique; they may coincide in order to express the possibility of offering joint service to users, that consists of a sequence of resources. Here, denote by V the set of all entry/exit points of the resources in the system. Any nonempty collection of resources corresponding to a directed path in $G \equiv (V, E)$ comprises an *action* in the system.

Let $N \equiv [n]$ be a set of users, each willing to adopt some action in the system. $\forall i \in N$, let w_i denote user i's *demand* (e.g., the flow rate from a source node to a destination node), while $\Pi_i \subseteq 2^E \setminus \emptyset$ is the collection of actions, any of which would satisfy user i (e.g., alternative routes from a source to a destination node, if G represents a communication network). The collection Π_i is called the *action set* of user i and each of its elements contains at least one resource. Any vector $r = (r_1, \ldots, r_n) \in \Pi \equiv \times_{i=1}^{n} \Pi_i$ is a *pure strategies profile*, or a *configuration* of the users. Any vector of real functions $p = (p_1, p_2, \ldots, p_n)$ s.t. $\forall i \in [n]$, $p_i : \Pi_i \to [0, 1]$ is a probability distribution over the set of allowable actions for user i (i.e., $\sum_{r_i \in \Pi_i} p_i(r_i) = 1$), and is called a *mixed strategies profile* for the n users.

A congestion model typically deals with users of identical demands, and thus, user cost function depending on the *number* of users adopting each action [1, 4, 6]. In this work the more general case is considered, where a *weighted congestion model* is the tuple $((w_i)_{i \in N}, (\Pi_i)_{i \in N}, (d_e)_{e \in E})$. That is, the users are allowed to have different demands for service from the whole system, and thus affect the resource delay functions in a different way, depending on their own weights.

A *weighted congestion game* associated with this model, is a game in strategic form with the set of users N and user demands $(w_i)_{i \in N}$, the action sets $(\Pi_i)_{i \in N}$ and cost functions $(\lambda_{r_i}^i)_{i \in N, r_i \in \Pi_i}$ defined as follows: For any configuration $r \in \Pi$ and $\forall e \in E$, let $\Lambda_e(r) = \{i \in N : e \in r_i\}$ be the set of users exploiting resource e according to r (called the *view* of resource e wrt configuration r). The *cost* $\lambda^i(r)$ *of user* i *for adopting strategy* $r_i \in \Pi_i$ in a given configuration r is equal to the cumulative *delay* $\lambda_{r_i}(r)$ along this path:

$$\lambda^i(r) = \lambda_{r_i}(r) = \sum_{e \in r_i} d_e(\theta_e(r)) \qquad (1)$$

where, $\forall e \in E, \theta_e(r) \equiv \sum_{i \in \Lambda_e(r)} w_i$ is the load on resource e wrt the configuration r.

On the other hand, for a mixed strategies profile p, the *expected cost of user* i *for adopting strategy* $r_i \in \Pi_i$ is

$$\lambda_{r_i}^i(p) = \sum_{r^{-i} \in \Pi^{-i}} P(p^{-i}, r^{-i}) \cdot \\ \sum_{e \in r_i} d_e\left(\theta_e(r^{-i} \oplus r_i)\right) \qquad (2)$$

where, r^{-i} is a configuration of all the users except for user i, p^{-i} is the mixed strategies profile of all users except for i, $r^{-i} \oplus r_i$ is the new configuration with user i choosing strategy r_i, and $P(p^{-i}, r^{-i}) \equiv \prod_{j \in N \setminus \{i\}} p_j(r_j)$ is the occurrence probability of r^{-i}.

Remark 1 Here notation is abused a little bit and the model considers the user costs $\lambda_{r_i}^i$ as functions whose exact definition depends on the other users' strategies: In the general case of a mixed strategies profile p, (2) is valid and expresses the expected cost of user i wrt p, conditioned on the event that i chooses path r_i. If the other users adopt a pure strategies profile r^{-i}, we get the special form of (1) that expresses the exact cost of user i choosing action r_i.

A congestion game in which all users are indistinguishable (i.e., they have the same user cost functions) and have the same action set, is

called *symmetric*. When each user's action set Π_i consists of sets of resources that comprise (simple) paths between a unique origin-destination pair of nodes (s_i, t_i) in a network $G = (V, E)$, the model refers to a *network congestion game*. If additionally all origin-destination pairs of the users coincide with a unique pair (s, t) one gets a *single commodity network congestion game* and then all users share exactly the same action set. Observe that a single-commodity network congestion game is not necessarily symmetric because the users may have different demands and thus their cost functions will also differ.

Selfish Behavior

Fix an arbitrary (mixed in general) strategies profile p for a congestion game $((w_i)_{i \in N}, (\Pi_i)_{i \in N}, (d_e)_{e \in E})$. We say that p is a *Nash Equilibrium (NE)* if and only if $\forall i \in N, \forall r_i, \pi_i \in \Pi_i, p_i(r_i) > 0 \Rightarrow \lambda_{r_i}^i(p) \leq \lambda_{\pi_i}^i(p)$. A configuration $r \in \Pi$ is a *Pure Nash Equilibrium (PNE)* if and only if $(\forall i \in N, \forall \pi_i \in \Pi_i, \lambda_{r_i}(r) \leq \lambda_{\pi_i} (r^{-i} \oplus \pi_i)$ where, $r^{-i} \oplus \pi_i$ is the same configuration with r except for user i that now chooses action π_i.

Key Results

In this section the article deals with the existence and tractability of PNE in weighted network congestion games. First, it is shown that it is not always the case that a PNE exists, even for a weighted single-commodity network congestion game with only linear and 2-wise linear (e.g., the maximum of two linear functions) resource delays. In contrast, it is well known [1, 6] that any unweighted (not necessarily single-commodity, or even network) congestion game has a PNE, for any kind of nondecreasing delays. It should be mentioned that the same result has been independently proved also by [3].

Lemma 1 *There exist instances of weighted single–commodity network congestion games with resource delays being either linear or 2–wise linear functions of the loads, for which there is no PNE.*

Theorem 2 *For any weighted multi–commodity network congestion game with linear resource delays, at least one PNE exists and can be computed in pseudo-polynomial time.*

Proof Fix an arbitrary network $G = (V, E)$ with linear resource/edge delays $d_e(x) = a_e x + b_e$, $e \in E$, $a_e, b_e \geq 0$. Let $r \in \Pi$ be an arbitrary configuration for the corresponding weighted multi–commodity congestion game on G. For the configuration r consider the potential $\Phi(r) = C(r) + W(r)$, where

$$C(r) = \sum_{e \in E} d_e(\theta_e(r))\theta_e(r)$$

$$= \sum_{e \in E} [a_e \theta_e^2(r) + b_e \theta_e(r)],$$

and

$$W(r) = \sum_{i=1}^{n} \sum_{e \in r_i} d_e(w_i)w_i$$

$$= \sum_{e \in E} \sum_{i \in \tilde{e}(r)} d_e(w_i)w_i$$

$$= \sum_{e \in E} \sum_{i \in \tilde{e}(r)} (a_e w_i^2 + b_e w_i)$$

one concludes that

$$\Phi(r') - \Phi(r) = 2w_i [\lambda^i(r') - \lambda^i(r)],$$

Note that the potential is a global system function whose changes are proportional to selfish cost improvements of any user. The global minima of the potential then correspond to configurations in which no user can improve her cost acting unilaterally. Therefore, any weighted multi–commodity network congestion game with linear resource delays admits a PNE.   □

Applications

In [5] many experiments have been conducted for several classes of pragmatic networks. The experiments show even faster convergence to pure Nash Equilibria.

Open Problems

The Potential function reported here is polynomial on the loads of the users. It is open whether one can find a purely combinatorial potential, which will allow strong polynomial time for finding Pure Nash equilibria.

Cross-References

▶ Best Response Algorithms for Selfish Routing
▶ Computing Pure Equilibria in the Game of Parallel Links
▶ General Equilibrium

Recommended Reading

1. Fabrikant A, Papadimitriou C, Talwar K (2004) The complexity of pure nash equilibria. In: Proceedings of the 36th ACM symposium on theory of computing (STOC'04). ACM, Chicago
2. Fotakis D, Kontogiannis S, Spirakis P (2005) Selfish unsplittable flows. J Theory Comput Sci 348:226–239
3. Libman L, Orda A (2001) Atomic resource sharing in noncooperative networks. Telecommun Syst 17(4):385–409
4. Monderer D, Shapley L (1996) Potential games. Game Econ Behav 14:124–143
5. Panagopoulou P, Spirakis P (2006) Algorithms for pure nash equilibrium in weighted congestion games. ACM J Exp Algorithms 11:2.7
6. Rosenthal RW (1973) A class of games possessing pure-strategy nash equilibria. Int J Game Theory 2:65–67

Self-Stabilization

Ted Herman
Department of Computer Science, University of Iowa, Iowa City, IA, USA

Keywords

Autonomic system control; Autopoesis; Homeostasis

Years and Authors of Summarized Original Work

1974; Dijkstra

Problem Definition

An algorithm is self-stabilizing if it eventually manifests correct behavior regardless of initial state. The general problem is to devise self-stabilizing solutions for a specified task. The property of self-stabilization is now known to be feasible for a variety of tasks in distributed computing. Self-stabilization is important for distributed systems and network protocols subject to transient faults. Self-stabilizing systems automatically recover from faults that corrupt state.

The operational interpretation of self-stabilization is depicted in Fig. 1. Part (a) of the figure is an informal presentation of the behavior of a self-stabilizing system, with time on the x-axis and some informal measure of correctness on the y-axis. The curve illustrates a system trajectory, through a sequence of states, during execution. At the initial state, the system state is incorrect; later, the system enters a correct state, then returns to an incorrect state, and subsequently stabilizes to an indefinite period where all states are correct. This period of stability is disrupted by a transient fault that moves the system to an incorrect state, after which the scenario above repeats. Part (b) of the figure illustrates the scenario in terms of state predicates. The box represents the predicate *true*, which characterizes all possible states. Predicate $\mathcal{C}$ characterizes the correct states of the system, and $\mathcal{L} \subset \mathcal{C}$ depicts the closed *legitimacy* predicate. Reaching a state in $\mathcal{L}$ corresponds to entering a period of stability in part (a). Given an algorithm A with this type of behavior, it is said that A self-stabilizes to $\mathcal{L}$; when $\mathcal{L}$ is implicitly understood, the statement is simplified to: A is self-stabilizing.

Problem [3]. The first setting for self-stabilization posed by Dijkstra is a ring of n processes numbered 0 through $n - 1$. Let the state of process i be denoted by $x[i]$. Communication is unidirectional in the ring using a *shared state* model. An atomic step of process i can be expressed by a guarded assignment of the form $g(x[i \ominus 1], x[i]) \rightarrow x[i] := f(x[i \ominus 1], x[i])$. Here, $\ominus$ is subtraction modulo n, so that $x[i \ominus 1]$ is the state of the previous process in the ring with respect to process i. The guard g is a boolean expression; if $g(x[i \ominus 1], x[i])$ is *true*, then process i is said to be *privileged* (or enabled). Thus in one atomic step, privileged process i reads the state of the previous process and computes a new state. Execution scheduling is controlled by a *central daemon*, which fairly chooses one among all enabled processes to take the next step. The problem is to devise g and f so that, regardless of initial states of $x[i]$, $0 \le i < n$, eventually there is one privilege and every process enjoys a privilege infinitely often.

Complexity Metrics

The complexity of self-stabilization is evaluated by measuring the resource needed for convergence from an arbitrary initial state. Most prominent in the literature of self-stabilization are metrics for worst-case time of convergence and space required by an algorithm solving the given task. Additionally, for reactive self-stabilizing algorithms, metrics are evaluated for the stable behavior of the algorithm, that is, starting from a legitimate state, and compared to non-stabilizing algorithms, to measure costs of self-stabilization.

Key Results

Composition

Many self-stabilizing protocols have a layered construction. Let $\{A_i\}_{i=0}^{m-1}$ be a set programs with the property that for every state variable x, if program A_i writes x, then no program A_j, for $j > i$, writes x. Programs in $\{A_j\}_{j=i+1}^{m-1}$ may read variables written by A_i, that is, they use the output of A_i as input. Fair composition of programs B and C, written $B [] C$, assumes fair scheduling of steps of B and C. Let X_j be the set

S

Self-Stabilization, Fig. 1
Self-stabilization
trajectories

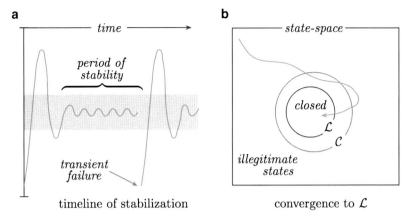

timeline of stabilization   convergence to $\mathcal{L}$

of variables read by A_j and possibly written by $\{A_i\}_{i=0}^{j-1}$.

Theorem 1 (Fair Composition [4]) *Suppose A_i is self-stabilizing to $\mathcal{L}_i$ under the assumption that all variables in X_i remain constant throughout any execution; then $A_0\,[\,]\,A_1[\,] \cdots [\,]\,A_{m-1}$ self-stabilizes to $\{\mathcal{L}_i\}_{i=0}^{m-1}$.*

Fair composition with a layered set $\{A_i\}_{i=0}^{m-1}$ corresponds to sequential composition of phases in a distributed algorithm. For instance, let B be a self-stabilizing algorithm for mutual exclusion in a network that assumes the existence of a rooted, spanning tree and let algorithm C be a self-stabilizing algorithm to construct a rooted spanning tree in a connected network; then $B\,[\,]\,C$ is a self-stabilizing mutual exclusion algorithm for a connected network.

Synchronization Tasks

One question related to the problem posed in section "Problem Definition" is whether or not there can be a *uniform* solution, where all processes have identical algorithms. Dijkstra's result for the unidirectional ring is a semi-uniform solution (all but one process have the same algorithm), using n states per process. The state of each process is a counter: process 0 increments the counter modulo k, where $k \geq n$ suffices for convergence; the other processes copy the counter of the preceding process in the ring. At a legitimate state, each time process 0 increments the counter, the resulting value is different from all other counters in

the ring. This ring algorithm turns out to be self-stabilizing for the *distributed daemon* (any subset of privileged processes may execute in parallel) when $k > n$. Subsequent results have established that mutual exclusion on a unidirection ring is $\Theta(1)$ space per process with a non-uniform solution. Deterministic uniform solutions to this task are generally impossible, with the exceptional case where n is and prime. Randomized uniform solutions are known for arbitrary n, using $O(\lg \alpha)$ space where α is the smallest number that does not divide n. Some lower bounds on space for uniform solutions are derived in [7]. Time complexity of Dijkstra's algorithm is $O(n^2)$ rounds, and some randomized solutions have been shown to have expected $O(n^2)$ convergence time.

Dijkstra also presented a solution to mutual exclusion for a linear array of processes, using $O(1)$ space per process [3]. This result was later generalized to a rooted tree of processes, but with mutual exclusion relaxed to having one privilege along any path from root to leaf. Subsequent research built on this theme, showing how tasks for distributed wave computations have self-stabilizing solutions. Tasks of phase synchronization and clock synchronization have also been solved. See reference [9] for an example of self-stabilizing mutual exclusion in a multiprocessor shared memory model.

Graph Algorithms

Communication networks are commonly represented with graph models and the need for distributed graph algorithms that tolerate transient faults motivates study of such tasks. Specific results in this area include self-stabilizing algorithms for spanning trees, center-finding, matching, planarity testing, coloring, finding independent sets, and so forth. Generally, all graph tasks can be solved by self-stabilizing algorithms: tasks that have network topology and possibly related factors, such as edge weights, for input, and define outputs to be a function of the inputs, can be solved by general methods for self-stabilization. These general methods require considerable space and time resource, and may also use stronger model assumptions than needed for specific tasks, for instance unique process identifiers and an assumed bound on network diameter. Therefore research continues on graph algorithms.

One discovery emerging from research on self-stabilizing graph algorithms is the difference between algorithms that terminate and those that continuously change state, even after outputs are stable. Consider the task of constructing a spanning tree rooted at process r. Some algorithms self-stabilize to the property that, for every $p \neq r$, the variable u_p refers to p's parent in the spanning tree and the state remains unchanged. Other algorithms are self-stabilizing protocols for token circulation with the side-effect that the circulation route of the token establishes a spanning tree. The former type of algorithm has $O(\lg n)$ space per process, whereas the latter has $O(\lg \delta)$ where δ is the degree (number of neighbors) of a process. This difference was formalized in the notion of *silent* algorithms, which eventually stop changing any communication value; it was shown in [5] for the link register model that silent algorithms for many graph tasks have $\Omega(\lg n)$ space.

Transformation

The simple presentation of [3] is enabled by the abstract computation model, which hides details of communication, program control, and atomicity. Self-stabilization becomes more complicated when considering conventional architectures that have messages, buffers, and program counters. A natural question is how to transform or refine self-stabilizing algorithms expressed in abstract models to concrete models closer to practice. As an example, consider the problem of transforming algorithms written for the central daemon to the distributed daemon model. This transformation can be reduced to finding a self-stabilizing token-passing algorithm for the distributed daemon model such that, eventually, no two neighboring processes concurrently have a token; multiple tokens can increase the efficiency of the transformation.

General Methods

The general problem of constructing a self-stabilizing algorithm for an input nonreactive task can be solved using standard tools of distributed computing: snapshot, broadcast, system reset, and synchronization tasks are building blocks so that the global state can be continuously validated (in some fortunate cases $\mathcal{L}$ can be locally checked and corrected). These building blocks have self-stabilizing solutions, enabling the general approach.

Fault Tolerance

The connection between self-stabilization and transient faults is implicit in the definition. Self-stabilization is also applicable in executions that asynchronously change inputs, silently crash and restart, and perturb communication [10]. One objection to the mechanism of self-stabilization, particularly when general methods are applied, is that a small transient fault can lead to a system-wide correction. This problem has been investigated, for example in [8], where it is shown how convergence can be optimized for a limited number of faults. Self-stabilization has also been combined with other types of failure tolerance, though this is not always possible: the task of counting the number of processes in a ring has no self-stabilizing solution in the shared state model if a process may crash [1], unless a failure detector is provided.

Applications

Many network protocols are self-stabilizing by the following simple strategy: periodically, they discard current data and regenerate it from trusted information sources. This idea does not work in purely asynchronous systems; the availability of real-time clocks enables the simple strategy. Similarly, watchdogs with hardware clocks can provide an effective basis for self-stabilization [6].

Cross-References

► Concurrent Programming, Mutual Exclusion

Recommended Reading

1. Anagnostou E, Hadzilacos V (1993) Tolerating transient and permanent failures. In: Distributed algorithms 7th international workshop. LNCS, vol 725. Springer, Heidelberg, pp 174–188
2. Cournier A, Datta AK, Petit F, Villain V (2002) Snap-stabilizing PIF algorithm in arbitrary networks. In: Proceedings of the 22nd international conference distributed computing systems, Vienna, July 2002, pp 199–206
3. Dijkstra EW (1974) Self stabilizing systems in spite of distributed control. Commun ACM 17(11):643–644. See also EWD391 (1973) In: Selected writings on computing: a personal perspective. Springer, New York, pp 41–46 (1982)
4. Dolev S (2000) Self-stabilization. MIT, Cambridge
5. Dolev S, Gouda MG, Schneider M (1996) Memory requirements for silent stabilization. In: Proceedings of the 15th annual ACM symposium on principles of distributed computing, Philadelphia, May 1996, pp 27–34
6. Dolev S, Yagel R (2004) Toward self-stabilizing operating systems. In: 2nd international workshop on self-adaptive and autonomic computing systems, Zaragoza, Aug 2004, pp 684–688
7. Israeli A, Jalfon M (1990) Token management schemes and random walks yield self-stabilizing mutual exclusion. In: proceedings of the 9th annual ACM symposium on principles of distributed computing, Quebec City, Aug 1990, pp 119–131
8. Kutten S, Patt-Shamir B (1997) Time-adaptive self stabilization. In: Proceedings of the 16th annual ACM symposium on principles of distributed computing, Santa Barbara, Aug 1997, pp 149–158
9. Lamport L (1986) The mutual exclusion problem: part II-statement and solutions. J ACM 33(2):327–348
10. Varghese G, Jayaram M (2000) The fault span of crash failures. J ACM 47(2):244–293

Semi-supervised Learning

Avrim Blum
School of Computer Science, Carnegie Mellon University, Pittsburgh, PA, USA

Keywords

Co-training; Learning from labeled and unlabeled data; Semi-supervised SVM

Years and Authors of Summarized Original Work

1998; Blum, Mitchell
1999; Joachims
2010; Balcan, Blum

Problem Definition

Semi-supervised learning [1, 4, 5, 8, 12] refers to the problem of using a large unlabeled data set U together with a given labeled data set L in order to generate prediction rules that are more accurate on new data than would have been achieved using just L alone. Semi-supervised learning is motivated by the fact that in many settings (e.g., document classification, image classification, speech recognition), unlabeled data is plentiful but labeled data is more limited or expensive, e.g., due to the need for human labelers. Therefore, one would like to make use of the unlabeled data if possible.

The general idea behind semi-supervised learning is that unlabeled data, while missing the labels, nonetheless often contains useful information. As an example, suppose one believes the correct decision boundary for some classification problem should be a linear separator that separates most of the data by a large margin. By observing enough unlabeled data to estimate the probability mass near to any given linear separator, one could in principle then discard separators in advance that slice through dense regions and instead focus attention on just those that indeed separate most of the distribution by a large margin. This is the high-level idea behind semi-supervised SVMs. Alternatively, suppose data objects can be described by two different "kinds" of features, and one believes that each kind should be sufficient to produce an accurate classifier. Then one might want to train a *pair* of classifiers and use unlabeled data for which one classifier is confident but the other is not to bootstrap, labeling such examples with the confident classifier and then feeding them as training data to the less-confident classifier. This is the high-level idea behind Co-Training. Or, if one believes "similar examples should generally have the same label," one might construct a graph with an edge between examples that are sufficiently similar and aim for a classifier that is correct on the labeled data and has a small cut value on the unlabeled data; this is the high-level idea behind graph-based methods. (These will all be discussed in more detail later.) General surveys of semi-supervised learning appear in [5, 12].

A Formal Framework

We now present a formal model for analyzing semi-supervised learning due to Balcan and Blum [1]. This model was developed to provide a unified explanation for a wide range of semi-supervised learning algorithms including the semi-supervised SVMs, Co-Training, and graph-based methods mentioned above. Before describing it, however, we first describe the classic PAC and agnostic learning models for *supervised* learning that this model builds on.

In the PAC and agnostic learning models, data is assumed to be drawn iid from some fixed but initially unknown distribution D over an instance space $\mathcal{X}$ and labeled by some unknown target function $c^* : \mathcal{X} \to \{0, 1\}$. The *error* of some hypothesis function h is defined as $\mathrm{err}(h) = \mathrm{Pr}_{x \sim D}[h(x) \neq c^*(x)]$. In the PAC model (also known as the *realizable case*), we assume that c^* is a member of some known class of functions $\mathcal{C}$, and we say that an algorithm PAC-learns $\mathcal{C}$ if for any given $\epsilon, \delta > 0$, with probability $\geq 1 - \delta$, it produces a hypothesis h such that $\mathrm{err}(h) \leq \epsilon$. In the *agnostic case*, we do not assume that

$c^* \in \mathcal{C}$ and instead aim to achieve error close to $\inf_{f \in \mathcal{C}}[\mathrm{err}(f)]$.

The PAC and agnostic learning models in essence assume that one's prior beliefs about the target be described in terms of a class of functions $\mathcal{C}$. In order to capture the reasoning used in semi-supervised learning, however, we need to also describe beliefs about the *relation* between the target function and the data distribution. This is done in the model of Balcan and Blum [1] via a *notion of compatibility* χ between a hypothesis h and a distribution D. Formally, χ maps pairs (h, D) to $[0, 1]$ with $\chi(h, D) = 1$ meaning that h is highly compatible with D and $\chi(h, D) = 0$ meaning that h is very *incompatible* with D. The quantity $1 - \chi(h, D)$ is called the *unlabeled error rate* of h and denoted $\mathrm{err}_{\mathrm{unl}}(h)$. Note that for χ to be useful, it must be estimatable from a finite sample; to this end, χ is further required to be an expectation over individual examples. That is, overloading notation for convenience, we require $\chi(h, D) = \mathbf{E}_{x \sim D}[\chi(h, x)]$, where $\chi : \mathcal{C} \times \mathcal{X} \to [0, 1]$. As with the class $\mathcal{C}$, one can either assume that the target is fully compatible ($\mathrm{err}_{\mathrm{unl}}(c^*) = 0$) or instead aim to do well as a function of how compatible the target is. The case that we assume $c^* \in \mathcal{C}$ and $\mathrm{err}_{\mathrm{unl}}(c^*) = 0$ is termed the "doubly realizable case." The concept class $\mathcal{C}$ and compatibility notion χ are both viewed as *known*.

Examples

Suppose we believe the target should separate most data by a large margin γ. We can represent this belief by defining $\chi(h, x) = 0$ if x is within distance γ of the decision boundary of h and $\chi(h, x) = 1$ otherwise. In this case, $\mathrm{err}_{\mathrm{unl}}(h)$ will denote the probability mass of D within distance γ of h's decision boundary. Alternatively, if we do not wish to commit to a specific value of γ, we could define $\chi(h, x)$ to be a smooth function of the distance of x to the separator defined by h. As a very different example, in co-training (described in more detail below), we assume each example can be described using two "views" that each are sufficient for classification, that is, there exist c_1^*, c_2^* such that for each example $x = \langle x_1, x_2 \rangle$, we have $c_1^*(x_1) = c_2^*(x_2)$. We can represent this belief by defining a hypothesis

$h = \langle h_1, h_2 \rangle$ to be compatible with an example $\langle x_1, x_2 \rangle$ if $h_1(x_1) = h_2(x_2)$ and incompatible otherwise; $\mathrm{err}_{\mathrm{unl}}(h)$ is then the probability mass of examples, under D, where the two halves of h disagree.

Intuition

In this framework, the way that unlabeled data helps in learning can be intuitively described as follows. Suppose one is given a concept class C (such as linear separators) and a compatibility notion χ (such as penalizing h for points within distance γ of the decision boundary). Suppose also that one believes $c^* \in C$ (or at least is close) and that $\mathrm{err}_{\mathrm{unl}}(c^*) = 0$ (or at least is small). Then, unlabeled data can help by allowing one to estimate the *unlabeled error rate* of all $h \in C$, thereby in principle reducing the search space from C (all linear separators) down to just the subset of C that is highly compatible with D. The key challenge is how this can be done efficiently (in theory, in practice, or both) for natural notions of compatibility, as well as identifying types of compatibility that data in important problems can be expected to satisfy.

Key Results

The following, from [1], illustrate formally how unlabeled data can help in this model. Fix some concept class C and compatibility notion χ. Given a labeled sample L, define $\widehat{\mathrm{err}}(h)$ to be the fraction of mistakes of h on L. Given an unlabeled sample U, define $\chi(h, U) = \mathbf{E}_{x \sim U}[\chi(h, x)]$ and define $\widehat{\mathrm{err}}_{\mathrm{unl}}(h) = 1 - \chi(h, U)$. That is, $\widehat{\mathrm{err}}(h)$ and $\widehat{\mathrm{err}}_{\mathrm{unl}}(h)$ are the empirical error rate and unlabeled error rate of h, respectively. Finally, given $\alpha > 0$, define $C_{D,\chi}(\alpha)$ to be the set of functions $f \in C$ such that $\mathrm{err}_{\mathrm{unl}}(f) \leq \alpha$.

Theorem 1 ([1]) *If $c^* \in C$ then with probability at least $1 - \delta$, for a random labeled set L and unlabeled set U, the $h \in C$ that optimizes $\widehat{\mathrm{err}}_{\mathrm{unl}}(h)$ subject to $\widehat{\mathrm{err}}(h) = 0$ will have $\mathrm{err}(h) \leq \epsilon$ for*

$$|U| \geq \frac{2}{\epsilon^2}\left[\ln |C| + \ln \frac{4}{\delta}\right],$$

$$|L| \geq \frac{1}{\epsilon}\left[\ln |C_{D,\chi}(\mathrm{err}_{\mathrm{unl}}(c^*) + 2\epsilon)| + \ln \frac{2}{\delta}\right].$$

Equivalently, for $|U|$ satisfying the above bound, for any $|L|$, with probability at least $1 - \delta$, the $h \in C$ that optimizes $\widehat{\mathrm{err}}_{\mathrm{unl}}(h)$ subject to $\widehat{\mathrm{err}}(h) = 0$ has

$$\mathrm{err}(h) \leq \frac{1}{|L|}\left[\ln |C_{D,\chi}(\mathrm{err}_{\mathrm{unl}}(c^*) + 2\epsilon)| + \ln \frac{2}{\delta}\right].$$

One can view Theorem 1 as bounding the number of labeled examples needed to learn well as a function of the "helpfulness" of the distribution D with respect to χ, for sufficiently large U. Namely, a helpful distribution is one in which $C_{D,\chi}(\alpha)$ is small for α slightly larger than the compatibility of the true target function, so we do not need much labeled data to identify a good function among those in $C_{D,\chi}(\alpha)$.

For infinite hypothesis classes, one needs to consider both the complexity of the class C and the complexity of the compatibility notion χ. Specifically, given $h \in C$, define $\chi_h(x) = \chi(h, x)$ and let $\mathrm{VCdim}(\chi(C))$ denote the VC-dimension of the set $\{\chi_h | h \in C\}$. A sample complexity bound from [1] based on ϵ-cover size is the following.

Theorem 2 ([1]) *Assume $c^* \in C$ and let p be the size of the smallest set of functions H such that every function in $C_{D,\chi}(\mathrm{err}_{\mathrm{unl}}(c^*) + \epsilon/3)$ is $\epsilon/6$-close to some function in H. Then $|U| = O\left(\frac{\max[\mathrm{VCdim}(C), \mathrm{VCdim}(\chi(C))]}{\epsilon^2} \ln \frac{1}{\epsilon} + \frac{1}{\epsilon^2} \ln \frac{2}{\delta}\right)$ and $|L| = O\left(\frac{1}{\epsilon} \ln \frac{p}{\delta}\right)$ is sufficient to identify a function $f \in C$ of error at most ϵ with probability at least $1 - \delta$.*

Finally, for the general (agnostic) case that $c^* \notin C$, we can define a regularizer based on empirical unlabeled error rates, and then get good bounds for optimizing a combination of the empirical labeled error and the regularization term. Specifically, for a hypothesis h, define $\hat{N}(h)$ to be the number of ways of partitioning the first $|L|$ points in U using $\{f \in C : \widehat{\mathrm{err}}_{\mathrm{unl}}(f) \leq \widehat{\mathrm{err}}_{\mathrm{unl}}(h)\}$. Then we have

Theorem 3 ([1]) *With probability at least $1 - \delta$, the hypothesis*

$$h = \arg\min_{h' \in C}[\widehat{\mathrm{err}}(h') + R(h')], \quad where$$

$$R(h') = \sqrt{\frac{24\ln(\hat{N}(h'))}{|L|}},$$

satisfies

$$\mathrm{err}(h) \leq \min_{h' \in C}\left[\mathrm{err}(h') + R(h')\right] + 5\sqrt{\frac{\ln(8/\delta)}{|L|}}.$$

Co-Training

Co-Training is a semi-supervised learning method due to [4] for settings in which examples can be thought of as having two "views," that is, two distinct types of information. For example, in classifying webpages (e.g., into student home page, faculty member home page, course home page, etc.), one could use the words on the page itself, but one could also use information from links pointing *to* that page [4]. Or, in classifying visual images, one might have two cameras or even two different filters or preprocessing steps on images from the same camera [9]. Or, in understanding video, one can use visual images and spoken dialogue [7]. In such settings, one can think of an example x as a pair $x = \langle x_1, x_2 \rangle$. The idea of Co-Training is that if each view is in principle enough to achieve a good classification by itself, but each provides somewhat different information, then one can hope to improve performance using unlabeled data. Specifically, in Co-Training, one maintains two hypotheses, one for each view (e.g., a hypothesis that classifies webpages based on the text on the page itself and one that classifies webpages based on information from links pointing to the page). A hypothesis pair $h = \langle h_1, h_2 \rangle$ is compatible with an example $\langle x_1, x_2 \rangle$ if $h_1(x_1) = h_2(x_2)$ and is incompatible otherwise. So the unlabeled error rate of a hypothesis (pair) $h = \langle h_1, h_2 \rangle$ is the probability mass of examples $\langle x_1, x_2 \rangle$ on which the two parts of h disagree.

In practice, there are two primary ways that this notion of compatibility is used to learn from a small amount of labeled data and a large amount of unlabeled data. The first is *iterative*

co-training, introduced in [4]. In iterative co-training, a small labeled sample L is used to produce predictors for each view that are confident in some part of their respective input spaces and not confident in other parts. Then, the algorithm searches through the (large) unlabeled set U to find examples $x = \langle x_1, x_2 \rangle$ for which one classifier is confident and the other is not. These examples are labeled by the confident classifier and handed to the less-confident classifier to improve its predictor. The other primary method is to optimize a global objective that combines accuracy over the labeled sample L with agreement over the unlabeled sample U. That is, one searches for the hypothesis pair h that minimizes $\widehat{\mathrm{err}}(h) + \lambda\widehat{\mathrm{err}}_{\mathrm{unl}}(h)$ for some regularization parameter λ [6, 10]. This is generally a non-convex optimization problem, and so various heuristics are typically applied to perform the optimization.

Theoretically, the guarantees for Co-Training are strongest when the data satisfies independence given the label (with some probability p, a random positive example $\langle x_1, x_2 \rangle$ is drawn from $D_1^+ \times D_2^+$, and with probability $1 - p$, a random negative example is drawn from $D_1^- \times D_2^-$) and in the realizable case (there exist targets $c_1^*, c_2^* \in C$ such that all examples $\langle x_1, x_2 \rangle$ in the support of the distribution satisfy $c_1^*(x_1) = c_2^*(x_2)$). Specifically, two key results are

Theorem 4 ([4]) *Any class C that is efficiently PAC-learnable from random classification noise is efficiently learnable from unlabeled data alone in the realizable Co-Training setting, if data satisfies independence given the label and one is given an initial weakly useful predictor $h_1(x_1)$.*

Here, h is a *weakly useful predictor* of a function f if for some $\epsilon > 1/poly(n)$ we have both (a) $\Pr_{x \sim D}[h(x) = 1] \geq \epsilon$ and (b) $\Pr_{x \sim D}[f(x) = 1 | h(x) = 1] \geq \Pr_{x \sim D}[f(x) = 1] + \epsilon$. Theorem 4 implies that if one is able to use a small labeled sample to produce an initial hypothesis that gives a slight "edge" in predicting the target beyond just the overall class probabilities, then under independence given the label one can boost that to a high-accuracy predictor from just unlabeled data.

Furthermore, ignoring computation time, under independence given the label, any class of finite VC-dimension is learnable from a single labeled example. In the case of linear separators, this can be done computationally efficiently.

Theorem 5 ([1]) *Any class $\mathcal{C}$ of finite VC-dimension is learnable from polynomially many unlabeled examples and a single labeled example if D satisfies independence given the label. Furthermore, for linear separators this can be done in polynomial time.*

Semi-supervised SVMs

Semi-supervised SVMs (also called transductive SVMs) [8, 11] aim to find a linear separator that separates both the labeled sample L and the unlabeled sample U by the largest possible margin. That is, one wants to find a separator such that for γ as large as possible, all labeled examples are on the correct side of the separator by distance at least γ and all unlabeled examples are on *some* side of the separator by distance at least γ. In practice, one combines a large-margin objective with a hinge-loss penalty for labeled examples that fail to satisfy the condition, and a "hat-loss" penalty for unlabeled examples that fail to satisfy the condition. Formally, the goal is to minimize $c_1 w^T w + c_2 \sum_{(x_i, y_i) \in L} \alpha_i + c_3 \sum_{x_j \in U} \beta_j$ subject to $(w^T x_i) y_i \geq 1 - \alpha_i$ for all $(x_i, y_i) \in L$ and $(w^T x_j) \tilde{y}_j \geq 1 - \beta_j$ for all $x_j \in U$ (and $\alpha_i, \beta_j \geq 0$), where $y_i \in \{-1, 1\}$ is the (known) label of $x_i \in L$ and $\tilde{y}_j \in \{-1, 1\}$ is a variable representing the algorithm's guess of the label of $x_j \in U$. While the optimization problem is NP-hard, a number of heuristics have been developed. For example, Joachims [8] uses an iterative labeling heuristic to approximately optimize the objective. Semi-supervised SVMs have been shown to achieve high accuracy in a number of text classification domains where unlabeled data is plentiful [8].

Graph-Based Methods

Graph-based methods [3, 13] can be viewed as a (transductive) semi-supervised version of

nearest-neighbor learning. In these methods, one creates a graph with a vertex for each example in $L \cup U$ and an edge between two examples x, x' if they are deemed to be sufficiently "similar" (or with edge weights based on *how* similar they are deemed to be). Similarity can be directly based on distance between the examples in the input space or given by some provided kernel function $k(x, x')$. Given the labels for the examples in L, one then finds a "most compatible" labeling for the examples in U, based on the belief that similar examples will typically have the same label. Specifically, in the mincut approach of [3], the labeling h produced is the cut of least total weight subject to agreeing with the known labels on examples in L or equivalently the cut that agrees with L minimizing $\sum_{e=(x,x')} w_e |h(x) - h(x')|$. In the algorithm of [13], in order to produce a smoother solution, the algorithm instead views the graph as an electrical network, finding the cut agreeing with L that minimizes $\sum_{e=(x,x')} w_e (h(x) - h(x'))^2$.

Open Problems

There are a number of open problems in developing computationally efficient semi-supervised learning algorithms. For example, can one extend the algorithm of Theorem 5 for Co-Training with linear separators to weaker conditions than independence given the label, while maintaining computational efficiency? (Note: A number of weaker conditions are known to produce good sample bounds if computational considerations are ignored [2].) More broadly, can one develop efficient algorithms for other classes or notions of compatibility that meet the cover-based sample complexity bounds of Theorem 2? Additional open problems are given in [1].

Recommended Reading

1. Balcan MF, Blum A (2010) A discriminative model for semi-supervised learning. J ACM

57(3):19:1–19:46.   doi:10.1145/1706591.1706599.
http://doi.acm.org/10.1145/1706591.1706599

2. Balcan MF, Blum A, Yang K (2004) Co-training and expansion: towards bridging theory and practice. In: Proceedings of 18th conference on neural information processing systems, Vancouver

3. Blum A, Chawla S (2001) Learning from labeled and unlabeled data using graph mincuts. In: Proceedings of 18th international conference on machine learning, Williams College

4. Blum A, Mitchell TM (1998) Combining labeled and unlabeled data with co-training. In: Proceedings of the 11th annual conference on computational learning theory, Madison, pp 92–100

5. Chapelle O, Schölkopf B, Zien A (eds) (2006) Semi-supervised learning. MIT, Cambridge. http://www.kyb.tuebingen.mpg.de/ssl-book

6. Collins M, Singer Y (1999) Unsupervised models for named entity classification. In: Proceedings of the joint SIGDAT conference on empirical methods in natural language processing and very large corpora, College Park, pp 189–196

7. Gupta S, Kim J, Grauman K, Mooney R (2008) Watch, listen & learn: co-training on captioned images and videos. In: Machine learning and knowledge discovery in databases (ECML PKDD). Lecture notes in computer science, vol 5211. Springer, Berlin/Heidelberg, pp 457–472. DOI 10.1007/978-3-540-87479-9_48. http://dx.doi.org/10.1007/978-3-540-87479-9_48

8. Joachims T (1999) Transductive inference for text classification using support vector machines. In: Proceedings of 16th international conference on machine learning, Bled, pp 200–209

9. Levin A, Viola P, Freund Y (2003) Unsupervised improvement of visual detectors using co-training. In: Proceedings of the ninth IEEE international conference on computer vision, ICCV '03, vol 2, Nice. IEEE Computer Society, Washington, DC, pp 626–633. http://dl.acm.org/citation.cfm?id=946247.946615

10. Nigam K, Ghani R (2000) Analyzing the effectiveness and applicability of co-training. In: Proceedings of ACM CIKM international conference on information and knowledge management, McLean, pp 86–93

11. Vapnik V (1998) Statistical learning theory, vol 2. Wiley, New York

12. Zhu X (2006) Semi-supervised learning literature survey Computer sciences TR 1530 University of Wisconsin, Madison

13. Zhu X, Ghahramani Z, Lafferty J (2003) Semi-supervised learning using Gaussian fields and harmonic functions. In: Proceedings of 20th international conference on machine learning, Washington, DC, pp 912–919

Separators in Graphs

Goran Konjevod
Department of Computer Science and
Engineering, Arizona State University, Tempe,
AZ, USA

Keywords

Balanced cuts

Years and Authors of Summarized Original Work

1998; Leighton, Rao
1999; Leighton, Rao

Problem Definition

The (balanced) separator problem asks for a cut of minimum (edge)-weight in a graph, such that the two shores of the cut have approximately equal (node)-weight.

Formally, given an undirected graph $G = (V, E)$, with a nonnegative edge-weight function $c : E \to \mathbb{R}_+$, a nonnegative node-weight function $\pi : V \to \mathbb{R}_+$, and a constant $b \leq 1/2$, a cut $(S : V \setminus S)$ is said to be b-balanced, or a $(b, 1-b)$-separator, if $b\pi(V) \leq \pi(S) \leq (1 - b)\pi(V)$ (where $\pi(S)$ stands for $\sum_{v \in S} \pi(v)$).

Problem 1 (b-balanced separator)

INPUT: Edge- and node-weighted graph $G = (V, E, c, \pi)$, constant $b \leq 1/2$.
OUTPUT: A b-balanced cut $(S : V \setminus S)$. Goal: minimize the edge weight $c(\delta(S))$.

Closely related is the *product sparsest cut problem*.

Problem 2 ((Product) Sparsest cut)

INPUT: Edge- and node-weighted graph $G = (V, E, c, \pi)$.
OUTPUT: A cut $(S : V \setminus S)$ minimizing the ratio-cost $\frac{1}{c}(\delta(S)))/(\pi(S)\pi(V \setminus S))$.

Problem 2 is the most general version of sparsest cut solved by Leighton and Rao. Setting all node weights are equal to 1 leads to the uniform version, Problem 3.

Problem 3 ((Uniform) Sparsest cut)

INPUT: Edge-weighted graph $G = (V, E, c)$.
OUTPUT: A cut $(S : V \setminus S)$ minimizing the ratio-cost $(c(\delta(S)))/(|S||V \setminus S|)$.

Sparsest cut arises as the (integral version of the) linear programming dual of *concurrent multicommodity flow* (Problem 4). An instance of a multicommodity flow problem is defined on an edge-weighted graph by specifying for each of k *commodities* a *source* $s_i \in V$, a *sink* $t_i \in V$, and a *demand* D_i. A feasible solution to the multicommodity flow problem defines for each commodity a flow function on E, thus routing a certain amount of flow from s_i to t_i. The edge weights represent capacities, and for each edge e, a capacity constraint is enforced: the sum of all commodities' flows through e is at most the capacity $c(e)$.

Problem 4 (Concurrent multicommodity flow)

INPUT: Edge-weighted graph $G = (V, E, c)$, commodities $(s_1, t_1, D_1), \ldots (s_k, t_k, D_k)$.
OUTPUT: A multicommodity flow that routes $f D_i$ units of commodity i from s_i to t_i for each i simultaneously, without violating the capacity of any edge. Goal: maximize f.

Problem 4 can be solved in polynomial time by linear programming, and approximated arbitrarily well by several more efficient combinatorial algorithms (section "Implementation"). The maximum value f for which there exists a multicommodity flow is called the *max-flow* of the instance. The *min-cut* is the minimum ratio $(c(\delta(S)))/(D(S, V \setminus S))$, where $D(S, V \setminus S) = \sum_{i:|\{s_i, t_i\} \cap S| = 1} D_i$. This dual interpretation motivates the most general version of the problem, the *nonuniform sparsest cut* (Problem 5).

Problem 5 ((Nonuniform) Sparsest cut)

INPUT: Edge-weighted graph $G = (V, E, c)$, commodities $(s_1, t_1, D_1), \ldots (s_k, t_k, D_k)$.

OUTPUT: A min-cut $(S : V \setminus S)$, that is, a cut of minimum ratio-cost $(c(\delta(S)))/(D(S, V \setminus S))$.

(Most literature focuses on either the uniform or the general nonuniform version, and both of these two versions are sometimes referred to as just the "sparsest cut" problem.)

Key Results

Even when all (edge- and node-) weights are equal to 1, finding a minimum-weight b-balanced cut is NP-hard (for $b = 1/2$, the problem becomes *graph bisection*). Leighton and Rao [23, 24] give a pseudo-approximation algorithm for the general problem.

Theorem 1 *There is a polynomial-time algorithm that, given a weighted graph $G = (V, E, c, \pi)$, $b \le 1/2$ and $b' < \min\{b, 1/3\}$, finds a b'-balanced cut of weight $O((\log n)/(b - b'))$ times the weight of the minimum b-balanced cut.*

The algorithm solves the sparsest cut problem on the given graph, puts aside the smaller-weight shore of the cut, and recurses on the larger-weight shore until both shores of the sparsest cut found have weight at most $(1 - b')\pi(G)$. Now the larger-weight shore of the last iteration's sparsest cut is returned as one shore of the balanced cut, and everything else as the other shore. Since the sparsest cut problem is itself NP-hard, Leighton and Rao first required an approximation algorithm for this problem.

Theorem 2 *There is a polynomial-time algorithm with approximation ratio $O(\log p)$ for product sparsest cut (Problem 2), where p denotes the number of nonzero-weight nodes in the graph.*

This algorithm follows immediately from Theorem 3.

Theorem 3 *There is a polynomial-time algorithm that finds a cut $(S : V \setminus S)$ with ratio-cost $(c(\delta(S)))/(\pi(S)\pi(V \setminus S)) \in O(f \log p)$, where* *f is the max-flow for the product multicommodity flow and p the number of nodes with nonzero weight.*

The proof of Theorem 3 is based on solving a linear programming formulation of the multicommodity flow problem and using the solution to construct a sparse cut.

Related Results

Shahrokhi and Matula [27] gave a max-flow min-cut theorem for a special case of the multicommodity flow problem and used a similar LP-based approach to prove their result. An $O(\log n)$ upper bound for arbitrary demands was proved by Aumann and Rabani [6] and Linial et al. [26]. In both cases, the solution to the dual of the multicommodity flow linear program is interpreted as a finite metric and embedded into ℓ_1 with distortion $O(\log n)$, using an embedding due to Bourgain [10]. The resulting ℓ_1 metric is a convex combination of cut metrics, from which a cut can be extracted with sparsity ratio at least as good as that of the combination.

Arora et al. [5] gave an $O(\sqrt{\log n})$ pseudo-approximation algorithm for (uniform or product-weight) balanced separators, based on a semidefinite programming relaxation. For the nonuniform version, the best bound is $O(\sqrt{\log n} \log \log n)$ due to Arora et al. [4]. Khot and Vishnoi [18] showed that, for the nonuniform version of the problem, the semidefinite relaxation of [5] has an integrality gap of at least $(\log \log n)^{1/6-\delta}$ for any $\delta > 0$, and further, assuming their Unique Games Conjecture, that it is NP-hard to (pseudo)-approximate the balanced separator problem to within any constant factor. The SDP integrality gap was strengthened to $\Omega(\log \log n)$ by Krauthgamer and Rabani [20]. Devanur et al. [11] show an $\Omega(\log \log n)$ integrality gap for the SDP formulation even in the uniform case.

Implementation

The bottleneck in the balanced separator algorithm is solving the multicommodity flow linear program. There exists a substantial amount of work on fast approximate solutions to such linear programs [19, 22, 25]. In most of the following results, the algorithm produces a $(1 + \epsilon)$-approximation, and its hidden constant depends on ϵ^{-2}. Garg and Könemann [15], Fleischer [14] and Karakostas [16] gave efficient approximation schemes for multicommodity flow and related problems, with running times $\tilde{O}((k + m)m)$ [15] and $\tilde{O}(m^2)$ [14, 16]. Benczúr and Karger [7] gave an $O(\log n)$ approximation to sparsest cut based on randomized minimum cut and running in time $\tilde{O}(n^2)$. The current fastest $O(\log n)$ sparsest cut (balanced separator) approximation is based on a primal-dual approach to semidefinite programming due to Arora and Kale [3], and runs in time $O(m + n^{3/2})(\tilde{O}(m + n^{3/2})$, respectively). The same paper gives an $O(\sqrt{\log n})$ approximation in time $O(n^2)(\tilde{O}(n^2)$, respectively), improving on a previous $\tilde{O}(n^2)$ algorithm of Arora et al. [2]. If an $O(\log^2 n)$ approximation is sufficient, then sparsest cut can be solved in time $\tilde{O}(n^{3/2})$, and balanced separator in time $\tilde{O}(m + n^{3/2})$ [17].

Applications

Many problems can be solved by using a balanced separator or sparsest cut algorithm as a subroutine. The approximation ratio of the resulting algorithm typically depends directly on the ratio of the underlying subroutine. In most cases, the graph is recursively split into pieces of balanced size. In addition to the $O(\log n)$ approximation factor required by the balanced separator algorithm, this leads to another $O(\log n)$ factor due to the recursion depth. Even et al. [12] improved many results based on balanced separators by using *spreading metrics*, reducing the approximation guarantee to $O(\log n \log \log n)$ from $O(\log^2 n)$.

Some applications are listed here; where no reference is given, and for further examples, see [24].

- Minimum cut linear arrangement and minimum feedback arc set. One single algorithm provides an $O(\log^2 n)$ approximation for both of these problems.

- Minimum chordal graph completion and elimination orderings [1]. Elimination orderings are useful for solving sparse symmetric linear systems. The $O(\log^2 n)$ approximation algorithm of [1] for chordal graph completion has been improved to $O(\log n \log \log n)$ by Even et al. [12].
- Balanced node cuts. The cost of a balanced cut may be measured in terms of the weight of nodes removed from the graph. The balanced separator algorithm can be easily extended to this node-weighted case.
- VLSI layout. Bhatt and Leighton [8] studied several optimization problems in VLSI layout. Recursive partitioning by a balanced separator algorithm leads to polylogarithmic approximation algorithms for crossing number, minimum layout area and other problems.
- Treewidth and pathwidth. Bodlaender et al. [9] showed how to approximate treewidth within $O(\log n)$ and pathwidth within $O(\log^2 n)$ by using balanced node separators.
- Bisection. Feige and Krauthgamer [13] gave an $O(\alpha \log n)$ approximation for the minimum bisection, using any α-approximation algorithm for sparsest cut.

Experimental Results

Lang and Rao [21] compared a variant of the sparsest cut algorithm from [24] to methods used in graph decomposition for VLSI design.

Cross-References

▶ Fractional Packing and Covering Problems
▶ Minimum Bisection
▶ Sparsest Cut

Recommended Reading

Further details and pointers to additional results may be found in the survey [28].

1. Agrawal A, Klein PN, Ravi R (1993) Cutting down on fill using nested dissection: provably good elimination orderings. In: Brualdi RA, Friedland S, Klee V (eds) Graph theory and sparse matrix computation. IMA volumes in mathematics and its applications. Springer, New York, pp 31–55
2. Arora S, Hazan E, Kale S (2004) $O(\sqrt{\log n})$ approximation to sparsest cut in $\tilde{O}(n^2)$ time. In: FOCS '04: proceedings of the 45th annual IEEE symposium on foundations of computer science (FOCS'04). IEEE Computer Society, Washington, pp 238–247
3. Arora S, Kale S (2007) A combinatorial, primal-dual approach to semidefinite programs. In: STOC '07: proceedings of the 39th annual ACM symposium on theory of computing. ACM, pp 227–236
4. Arora S, Lee JR, Naor A (2005) Euclidean distortion and the sparsest cut. In: STOC '05: proceedings of the thirty-seventh annual ACM symposium on theory of computing. ACM, New York, pp 553–562
5. Arora S, Rao S, Vazirani U (2004) Expander flows, geometric embeddings and graph partitioning. In: STOC '04: proceedings of the thirty-sixth annual ACM symposium on theory of computing. ACM, New York, pp 222–231
6. Aumann Y, Rabani Y (1998) An (log) approximate min-cut maxflow theorem and approximation algorithm. SIAM J Comput 27(1):291–301
7. Benczúr AA, Karger DR (1996) Approximating s-t minimum cuts in $\tilde{O}(n^2)$ time. In: STOC '96: proceedings of the twenty-eighth annual ACM symposium on theory of computing. ACM, New York, pp 47–55
8. Bhatt SN, Leighton FT (1984) A framework for solving vlsi graph layout problems. J Comput Syst Sci 28(2):300–343
9. Bodlaender HL, Gilbert JR, Hafsteinsson H, Kloks T (1995) Approximating treewidth, pathwidth, frontsize, and shortest elimination tree. J Algorithms 18(2):238–255
10. Bourgain J (1985) On Lipshitz embedding of finite metric spaces in Hilbert space. Isr J Math 52:46–52
11. Devanur NR, Khot SA, Saket R, Vishnoi NK (2006) Integrality gaps for sparsest cut and minimum linear arrangement problems. In: STOC '06: proceedings of the thirty-eighth annual ACM symposium on theory of computing. ACM, New York, pp 537–546
12. Even G, Naor JS, Rao S, Schieber B (2000) Divide-and-conquer approximation algorithms via spreading metrics. J ACM 47(4):585–616
13. Feige U, Krauthgamer R (2002) A polylogarithmic approximation of the minimum bisection. SIAM J Comput 31(4):1090–1118
14. Fleischer L (2000) Approximating fractional multicommodity flow independent of the number of commodities. SIAM J Discret Math 13(4):505–520
15. Garg N, Könemann J (1998) Faster and simpler algorithms for multicommodity flow and other fractional packing problems. In: FOCS '98: proceedings of the 39th annual symposium on foundations of computer science. IEEE Computer Society, Washington, p 300

16. Karakostas G (2002) Faster approximation schemes for fractional multicommodity flow problems. In: SODA '02: proceedings of the thirteenth annual ACM-SIAM symposium on discrete algorithms. Society for Industrial and Applied Mathematics, Philadelphia, pp 166–173

17. Khandekar R, Rao S, Vazirani U (2006) Graph partitioning using single commodity flows. In: STOC '06: proceedings of the thirty-eighth annual ACM symposium on theory of computing. ACM, New York, pp 385–390

18. Khot S, Vishnoi NK (2005) The unique games conjecture, integrality gap for cut problems and embeddability of negative type metrics into l_1. In: FOCS '07: proceedings of the 46th annual IEEE symposium on foundations and computer science. IEEE Computer Society, pp 53–62

19. Klein PN, Plotkin SA, Stein C, Tardos É (1994) Faster approximation algorithms for the unit capacity concurrent flow problem with applications to routing and finding sparse cuts. SIAM J Comput 23(3):466–487

20. Krauthgamer R, Rabani Y (2006) Improved lower bounds for embeddings into l_1. In: SODA '06: proceedings of the seventeenth annual ACM-SIAM symposium on discrete algorithm. ACM, New York, pp 1010–1017

21. Lang K, Rao S (1993) Finding near-optimal cuts: an empirical evaluation. In: SODA '93: proceedings of the fourth annual ACM SIAM symposium on discrete algorithms. Society for Industrial and Applied Mathematics, Philadelphia, pp 212–221

22. Leighton FT, Makedon F, Plotkin SA, Stein C, Stein É, Tragoudas S (1995) Fast approximation algorithms for multicommodity flow problems. J Comput Syst Sci 50(2):228–243

23. Leighton T, Rao S (1988) An approximate max-flow min-cut theorem for uniform multicommodity flow problems with applications to approximation algorithms. In: Proceedings of the 29th annual symposium on foundations of computer science. IEEE Computer Society, Washington, DC, pp 422–431

24. Leighton T, Rao S (1999) Multicommodity max-flow min-cut theorems and their use in designing approximation algorithms. J ACM 46(6):787–832

25. Leong T, Shor P, Stein C (1991) Implementation of a combinatorial multicommodity flow algorithm. In: Johnson DS, McGeoch CC (eds) Network flows and matching. DIMACS series in discrete mathematics and theoretical computer science, vol 12, AMS, Providence, pp 387–406

26. Linial N, London E, Rabinovich Y (1995) The geometry of graphs and some of its algorithmic applications. Combinatorica 15(2):215–245

27. Shahrokhi F, Matula DW (1990) The maximum concurrent flow problem. J ACM 37(2):318–334

28. Shmoys DB (1997) Cut problems and their applications to divide-and-conquer. In: Hochbaum DS (ed) Approximation algorithms for NP-hard problems. PWS Publishing Company, pp 192–235

Sequence and Spatial Motif Discovery in Short Sequence Fragments

Jie Liang[1] and Ronald Jackups[2]
[1]Department of Bioengineering, University of Illinois, Chicago, IL, USA
[2]Department of Pediatrics, Washington University, St. Louis, MO, USA

Keywords

Internally random model; Permutation model; Positional null model; Sequence motif; Spatial motif; String pairing pattern; String pattern

Years and Authors of Summarized Original Work

2005; Jackups, Jr and Liang
2006; Jackups, Jr, Cheng, Liang
2010; Jackups, Jr and Liang

Problem Definition

The problem is to detect specific patterns in string and patterns in string pairs for discovery of sequence and spatial motifs in membrane proteins. A *spatial interaction* motif of residue pair X-Y is defined as a pattern in which a character or residue of type X is found interacting with a residue of type Y on two strings or sequences (Fig. 1a). We define a *sequence pair* XYk as a pattern in which a residue of type Y is found at the k-th position from a residue of type X along a single sequence (Fig. 1b). The propensity $P(X, Y)$ of residue pairing XY is $P(X, Y) = \frac{f_{\text{obs}}(X,Y)}{\mathbb{E}[f(X,Y)]}$, where $f_{\text{obs}}(X, Y)$ is the observed count of XY patterns and $\mathbb{E}[f(X, Y)]$ is the expected count of XY patterns according to some random null model. We define a *motif* as a residue pair with propensity >1.0 (or greater than some other predefined limit) and statistically significant.

a

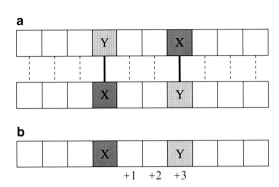

b

Sequence and Spatial Motif Discovery in Short Sequence Fragments, Fig. 1 Examples of spatial and sequence patterns. (a) Two X-Y spatial patterns on interacting sequences. (b) An $XY3$ sequence pattern

The null model for calculating $\mathbb{E}[f(X, Y)]$ is critical for motif detection. For short sequence fragments, the null model for spatial motif detection cannot be the χ^2 distribution as was used in [13], since the assumption of Gaussian distribution is not valid for short sequences. The null model for sequence motif detection cannot be the binomial distribution as was used in [4, 10], since the assumption of drawing from a universal population with replacement is unrealistic for short sequence fragments. Instead, we use a combinatorial model called the *permutation model* more effective for discoveries of motifs [5–7]. This null model is similar for both pair types: the residues within each sequence are exhaustively and independently permuted *without replacement*, and each permutation occurs with equal probability. This model has been called the *internally random* model [6]. This permutation model is further extended to positional null model to correct position-specific bias in residue distributions [6].

Objective. Our task is to determine explicit formulas to calculate $\mathbb{E}[f(X, Y)]$ for each pair type under different conditions. Explicit probability distributions for $f(X, Y)$ can also be found for many special cases, which will allow for the calculation of statistical significance p-values. These formulas can also be used to study whole datasets of short sequences.

Key Results

Spatial Motifs by Permutation Model

Expectation for Interacting Residues of the Same Type

For cases in which X is the same as Y (i.e., X-X pairs), let x_1 be the number of residues of type X in the first sequence, x_2 the number of residues of type X in the second sequence, and l the common length of the sequence pair. The probability $\mathbb{P}_{XX}(i)$ of exactly $i = f(X, X)$ number of X-X contacts follows a hypergeometric distribution: $\mathbb{P}_{XX}(i) = \binom{x_1}{i}\binom{l-x_1}{x_2-i}/\binom{l}{x_2}$. Its expectation $\mathbb{E}[f(X, X)]$ is then:

$$\mathbb{E}[f(X, X)] = \frac{x_1 x_2}{l}.$$

Expectation for Interacting Residues of Different Types

When $X \neq Y$, the number of X-Y contacts in the permutation model for one sequence pair is the sum of two dependent hypergeometric variables, one variable for type X residues in the first sequence s_1 and type Y in the second sequence s_2, and another variable for type Y residues in s_1 and type X in s_2. The expected number of X-Y contacts $\mathbb{E}[f(X, Y)]$ is the sum of the two expected values $\mathbb{E}[f(X, Y|X \in s_1, Y \in s_2)] + \mathbb{E}[f(X, Y|Y \in s_1, X \in s_2)]$:

$$\mathbb{E}[f(X, Y)] = \frac{x_1 y_2}{l} + \frac{y_1 x_2}{l},$$

where x_1 and x_2 are the numbers of residues of type X in the first and second sequence, respectively, y_1 and y_2 are the numbers of residues of type Y in the first and second sequence, respectively, and l is the length of the sequence pair.

Significance of Spatial Motifs

To calculate the statistical significance in the form of p-value of interacting residues of the same type, two-tailed p-values can be calculated using the hypergeometric distribution for a dataset of sequence pairs.

For interacting residues of different types, the formula to determine the p-value for a specific observed number of X-Y contacts is more complex because of the dependency. We define a 3-element multinomial function $M(a,b,c) \equiv \frac{a!}{b!c!(a-b-c)!}$, where $M(a,b,c) = 0$ if $a-b-c < 0$. This represents the number of distinct permutations, without replacement, in a multiset of size a containing three different types of elements, with number count b, c, and $a-b-c$ of each of the three element types.

The probability $\mathbb{P}(h,i,j,k)$ of inter-sequence matches, namely, the probability of h X-X contacts, i X-Y contacts, j Y-X contacts, and k Y-Y contacts occurring in a random permutation is (Fig. 2)

$$\mathbb{P}(h,i,j,k) = \frac{M(x_1,h,i) \cdot M(y_1,j,k) \cdot M(l-x_1-y_1, x_2-h-j, y_2-i-k)}{M(l,x_2,y_2)}.$$

The marginal probability $\mathbb{P}_{XY}(m)$ that there are a total of $i+j = m$ X-Y contacts is

$$\mathbb{P}_{XY}(m) = \sum_{h=0}^{x_1} \sum_{i=0}^{x_1-h} \sum_{k=0}^{y_1-(m-i)} \mathbb{P}(h,i,m-i,k).$$

There are x_1 possible values for h, one for each residue of type X on sequence 1; $x_1 - h$ possible values for i, once h has been determined; and $y_1 - j = y_1 - (m-i)$ possible values for k, once i has been determined. The i number of X-Y contacts plus the $m-i$ number of Y-X contacts will sum to the m number of contacts desired.

This closed-form formula allows calculation of p-values analytically. The running time is $O(l^4)$, due to the presence of 3 summations and $l!$ in the summand. For short sequences, the computing cost is not prohibitive.

Sequences of Different Lengths

The requirement for interacting sequences to be of the same length may be relaxed by introducing a 21st "dummy" amino acid type. All unpaired residues in the longer member of a sequence pair will be paired to this extra amino acid type, and our standard method can be applied to determine the propensity of unpaired amino acids (i.e., residues paired with the "dummy" amino acid type).

Sequence Motifs by Permutation Model

The propensity $P(X,Y|k)$ for the XYk pattern of two ordered intrasequence residues of type X and type Y that are k positions away on the same sequence (Fig. 1b) is $P(X,Y|k) = \frac{f_{\text{obs}}(X,Y|k)}{\mathbb{E}[f(X,Y|k)]}$, where $f_{\text{obs}}(X,Y|k)$ is the observed count of XYk patterns, and $\mathbb{E}[f(X,Y|k)]$ is the expected count of XYk patterns.

Expectation of XYk and XXk Two-Residue Motifs

We can regard $f(X,Y|k)$ as the sum of identical Bernoulli variables $f_t(X,Y|k)$, each of which equals 1 if one of the x number of residues of type X occurs at position t in the sequence and one of the y number of residues of type Y occurs

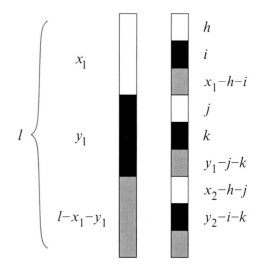

Sequence and Spatial Motif Discovery in Short Sequence Fragments, Fig. 2 Division of residues in spatial motif analysis when $X \neq Y$. *White = X, black = Y, gray = "neither" X or Y*

at position $t + k$ or equals 0 otherwise. Since an XYk pattern cannot occur if $t > l - k$, we concern ourselves only with the first $l - k$ positions. We have: $\mathbb{E}[f_t(X, Y|k)] = \mathbb{P}[f_t(X, Y|k) = 1] = \frac{x}{l} \cdot \frac{y}{(l-1)}$ if $t \leq l - k$. There are $l - k$ such identical variables, and their expectations may be summed as

$$\mathbb{E}[f(X, Y|k)] = (l - k)\frac{xy}{l(l-1)}, \qquad (1)$$

where l is the length of the sequence, x is the number of residues of type X, and y is the number of residues of type Y.

For XXk patterns, the expectation is calculated as

$$\mathbb{E}[f(X, X|k)] = (l - k)\frac{x(x-1)}{l(l-1)}, \qquad (2)$$

as there will be $x - 1$ residues available to place the second X residue at position $t + k$ after the first X residue is placed at t. Although these Bernoulli random variables are dependent (i.e., the placement of one XYk pattern will affect the probability of another XYk pattern), their expectations may be summed, because expectation is a linear operator.

Significance of XYk and XXk Two-Residue Sequence Motifs

To calculate statistical significance p-values, several formulas have been derived to determine $\mathbb{P}_{XYk}(i)$, the probability of the occurrence of $i = f(X, Y|k)$ XYk patterns for different k values.

1. **Sequence motifs when $k = 1$.** We have

$$\mathbb{P}_{XY1}(i) = \frac{\binom{l-y}{x}\binom{x}{i}\binom{l-x}{y-i}}{\frac{l!}{x!y!(l-x-y)!}} = \frac{\binom{x}{i}\binom{l-x}{y-i}}{\binom{l}{y}}, \quad \text{and}$$

$$\mathbb{P}_{XX1}(i) = \frac{\binom{l-x+1}{x-i}\binom{x-1}{i}}{\binom{l}{x}},$$

with the convention that $\binom{n}{r} = 0$ if $n < r$.
2. **Sequence motifs with residues of different types and if $x \leq 2$ or $y \leq 2$.**
   - If either $x = 1$ or $y = 1$, we have

$$\mathbb{P}_{XYk}(1) = (l - k)\frac{xy}{l(l-1)}.$$

For $i = 0$, we have simply $\mathbb{P}_{XYk}(0) = 1 - \mathbb{P}_{XYk}(1)$.
   - If $x = 2$ or $y = 2$, the probability of two XYk patterns is

$$\mathbb{P}_{XYk}(2) = \frac{\left[\binom{l-k}{2} - (l - 2k)\right]}{\frac{l(l-1)(l-2)(l-3)}{x(x-1)y(y-1)}}.$$

We also have for the probabilities of exactly one XYk pattern or zero pattern:

$$\mathbb{P}_{XYk}(1) = \mathbb{E}[f(XYk)] - 2\mathbb{P}_{XYk}(2) \quad \text{and}$$

$$\mathbb{P}_{XYk}(0) = 1 - [\mathbb{P}_{XYk}(1) + \mathbb{P}_{XYk}(2)].$$

3. **Sequence motifs with residues of the same type if $x \leq 3$.**
   - If $x = 2$, the probability of one XXk pattern is

$$\mathbb{P}_{XXk}(1) = \mathbb{E}[f(XXk)] = (l - k)\frac{x(x-1)}{l(l-1)},$$

The probability of no XXk pattern is

$$\mathbb{P}_{XXk}(0) = 1 - \mathbb{P}_{XXk}(1).$$

   - If $x = 3$, the probability of exactly two XXk patterns is

$$\mathbb{P}_{XXk}(2) = \frac{l - 2k}{\binom{l}{x}},$$

4. **Sequence motifs with $k > 1$, $x > 2$, and $y > 2$.** When $k > 1$, $x > 2$, and $y > 2$, the analytical formulas for $\mathbb{P}_{XYk}(i)$ become very complicated. However, when the sequences in the dataset used are short, it is possible to fully enumerate all permutations of a sequence and calculate $\mathbb{P}_{XYk}(i)$ and p-values exactly, as shown by Senes et al. [11]. Because x and y are usually small in short sequences, the computation time needed for motif analysis of short sequences is not prohibitive.

$$X_0 \quad X_1 \ X_2 \quad X_3 \quad X_4$$
$$k_1 \ k_2 \qquad k_3 \qquad k_4$$

Sequence and Spatial Motif Discovery in Short Sequence Fragments, Fig. 3 Example of a multi-residue sequence pattern as described in the text. This pattern contains five specified residues in a span of ten residues. Here, X_0, X_1, X_2, X_3, and X_4 are specified amino acid types, and the corresponding k values are counted as the distance from the first position of the sequence (i.e., the position occupied by X_0). Thus, $k_1 = 2$, $k_2 = 3$, $k_3 = 6$, and $k_4 = 9$. All other residues (in *white*) are unspecified and may be any amino acid type. This pattern is written as $(X_0, X_1, X_2, X_3, X_4 \mid 2, 3, 6, 9)$

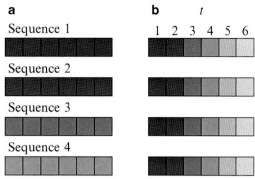

Sequence and Spatial Motif Discovery in Short Sequence Fragments, Fig. 4 Difference between (**a**) a *permutation* null model for sequence motif analysis and (**b**) a *position-dependent* null model. In both cases, only residues of the same shade are permuted with each other. In (**a**), residues are permuted only within each sequence individually, while in (**b**), residues are permuted across sequences but only within their specified position t

Propensity of Multi-residue Sequence Motifs

We now discuss the expected number $\mathbb{E}[f(X_0, X_1, X_2, \ldots, X_n \mid k_1, k_2, \ldots, k_n)]$ of a specific pattern containing $n + 1$ residues placed in a contiguous subsequence of $k_n + 1$ residues ($k_n \geq n$). Here X_i is the residue type of the i-th fixed residue in the pattern and k_i is the position of this residue from the 0-th residue ($k_0 = 0$). Positions not specified by k_i can be any residue type. For example, the pattern $(A, L, Y \mid 2, 4)$ is written as AL2Y4 and represents AxLxY. A graphic example is shown in Fig. 3. Many examples of these multi-residue sequence motifs in proteins have been discovered, including the GxGxxG NADH binding motif [1] and the RSxSxP 14-3-3 binding motif [14].

The expected value can be calculated as:

$$\mathbb{E}[f(X_0, X_1, X_2, \ldots, X_n \mid k_1, k_2, \ldots, k_n)]$$
$$= (l - k_n) \frac{\prod_{i=0}^{n} [x_i - \#(\mathbb{I}(X_i))]}{\frac{l!}{(l-n-1)!}}, \qquad (3)$$

where x_i is the number of residues of type X_i, l is the length of the sequence, and $\#(\mathbb{I}(X_i))$ is the number of times residue type X_i appears in the "subpattern" $\{X_0, X_1, X_2, \ldots, X_{i-1}\}$.

Remark

The above discussions are for determining motifs in a single short sequence or sequence pair. This can be extended so analysis can be performed on a dataset of multiple short sequences to attain sufficient statistical significance. This has the advantage of capturing within-sequence relationships on a scale large enough to obtain reliable p-values. Details can be found in [6].

Spatial Motifs by Positional Null Model

When there are significant biases in residue preferences for certain positions in a sequence known a priori, e.g., the enrichment of aromatic residues at either end of a transmembrane α-helix or β-strand [12], these single-residue biases may confound two-residue propensities. The *positional null model* should be used for motif detection in such cases [6]. Instead of permuting residues across all positions within individual sequences, we permute residues across all sequences in a dataset within specific positions (Fig. 4).

Expectation and Significance of Interacting Residue Pairs

We allocate residues into regions, which do not overlap. Regions may have different lengths along the sequences. Interacting regions within a sequence pair are assumed to have equal length. If a residue in region r interacts with a residue in region s on a spatially adjacent sequence fragment, all residues in region r in the dataset must only interact with residues in region s. For example, for interacting antiparallel β-strands, we divide each strand into three regions, the N-terminal, central core, and C-terminal regions, and all interacting strand pairs into two spatial pair types, N-terminal with C-terminal and core with core. We require that no core residue interact with an N-terminal or C-terminal residue.

The null model for position-dependent spatial motifs differs depending on whether paired residues are from the same region ($r = s$) or different regions ($r \neq s$), and whether the residue types in the pair are the same ($X = Y$) or different ($X \neq Y$).

1. **When $r = s$ and $X = Y$.** The expected value of X-X pairs in region r is

$$\mathbb{E}(X, X \,|\, rr) = \frac{\binom{x_r}{2}}{\binom{n_r}{2}} \cdot \frac{n_r}{2} = \frac{x_r(x_r - 1)}{2(n_r - 1)},$$

where n_r is the number of residues in region r. The probability $\mathbb{P}_{XX|rr}(i)$ of i X-X interacting pairs in region r in the dataset for p-values calculation calculated is

$$\mathbb{P}_{XX|rr}(i) = \frac{M(\frac{n_r}{2}, i, x_r - 2i) \cdot 2^{x_r - 2i}}{\binom{n_r}{x_r}},$$

where the 3-element multinomial function $M(a, b, c)$ is as defined before.

2. **When $r = s$ and $X \neq Y$.** The expected value when $X \neq Y$ is

$$\mathbb{E}(XY \,|\, rr) = \frac{x_r y_r}{\binom{n_r}{2}} \cdot \frac{n_r}{2} = \frac{x_r y_r}{n_r - 1}.$$

The probability $\mathbb{P}(i, j, k)$ of each combination of i, j, and k pairs of type X-Y, X-X, and Y-Y interactions, respectively, is

$$\mathbb{P}(i, j, k) = \frac{M(\frac{n_r}{2}, i, j, k, x_r - i - 2j, y_r - i - 2k) \cdot 2^{x_r + y_r - i - 2j - 2k}}{M(n_r, x_r, y_r)},$$

where the 6-variable multinomial function $M(a,b,c,d,e,f) \equiv \frac{a!}{b!c!d!e!f!(a-b-c-d-e-f)!}$. The probability $\mathbb{P}_{XY|rr}(i)$ of i X-Y pairs in the dataset is then

$$\mathbb{P}_{XY|rr}(i) = \sum_{j=0}^{\frac{x_r - i}{2}} \sum_{k=0}^{\frac{y_r - i}{2}} \mathbb{P}(i, j, k).$$

3. **When $r \neq s$.** We distinguish X_r, a residue of type X occurring in region r in one sequence, and X_s, a residue of type X occurring in region s in the other sequence. Thus, an X-Y pair, which we define as an $X_r - Y_s$ pair, is different from a Y-X pair, which is $Y_r - X_s$. Because there is a one-to-one correspondence between residues in region r and region s, $n_r = n_s$ is the total number of $r - s$ pairs.

In order for exactly i X-Y pairs to occur, i X_r residues must be drawn from a possible x_r residues of type X to match i Y_s residues

drawn from a possible y_s residues of type Y. This can be modeled with a simple hypergeometric distribution. The expected value can be calculated as

$$\mathbb{E}(XY \,|\, rs) = \frac{x_r y_s}{n_r}.$$

The $\mathbb{P}_{XY|rs}(i)$ of i X-Y pairs is

$$\mathbb{P}_{XY|rs}(i) = \frac{\binom{x_r}{i}\binom{n_r - x_r}{y_s - i}}{\binom{n_r}{y_s}}.$$

Expectation and Significance of Sequence Motifs

We define the *positional residue frequency* x_t as the number of residues of type X occupying the t-th position of all sequences in the dataset. If sequences of different lengths are represented in the dataset, it is necessary to normalize t to be within an appropriate range $[1, l]$, to approximate an average or predetermined sequence length of l:

$$t = \lceil \frac{l(t_{obs} - 0.5)}{l_{obs}} \rceil,$$

where $t_{obs} \in \{1, 2, 3, \cdots, l_{obs}\}$ is the actual position of the residue within its sequence, l_{obs} is the actual length of the sequence, $\lceil x \rceil$ represents the ceiling function, equal to the lowest integer greater than or equal to x, and the 0.5 factor is a correction for continuity to round to the next integer. This ensures that $1 \leq t \leq l$, no residues are removed from the model by truncation, and each position t will be represented by nearly the same number of residues.

For sequence motif, we use the model of permutation within each position in a sequence *with replacement* across all sequences. Although all other null models in this study rely on permutation without replacement, this model is based on datasets of multiple sequences instead of individual sequences, and the approximation of sampling without replacement will not be problematic once a sufficiently large sample of sequences is assembled.

1. XYk **motif at position** t. When $t \leq 1-k$, the probability of an XYk pattern at position t is

$$\mathbb{P}(X, Y|k, t) = \frac{x_t}{n_t} \cdot \frac{y_{t+k}}{n_{t+k}},$$

where x_t is the number of residues of type X in position t on all sequences, y_t is the number of residues of type Y in position t, and n_t is the number of all residues of all types in position t. This null model can be represented as a binomial distribution.

The expected frequency of XYk patterns at position t is

$$\mathbb{E}[f(X, Y|k, t)] = n_t \cdot \mathbb{P}(X, Y|k, t).$$

The probability of i XYk patterns at position t in the dataset is

$$\mathbb{P}_{XYk|t}(i) = \binom{n_t}{i} \mathbb{P}(X, Y|k, t)^i$$

$$[1 - \mathbb{P}(X, Y|k, t)]^{n_t - i}.$$

Note that the probability that an XYk pattern appears at position t is 0 if $t > l - k$, as an XYk pattern would span across the end of a sequence of length l.

2. XYk **motif at any arbitrary position.** To calculate the dataset-wide probability of an XYk pattern at any arbitrary position of the sequence, we average $\mathbb{P}(X, Y|k, t)$ over all $l - k$ possible positions:

$$\mathbb{P}(X, Y|k) = \frac{1}{l - k} \sum_{t=1}^{l-k} \mathbb{P}(X, Y|k, t).$$

This can similarly be represented as a binomial distribution with probability distribution function: $\mathbb{P}_{XYk}(i) = \binom{n_k}{i} \mathbb{P}(X, Y|k)^i [1 - \mathbb{P}(X, Y|k)]^{n_k - i}$, where n_k is the number of all pairs of all residue types k residues apart in the dataset. The expected value is

$$\mathbb{E}[f(X, Y|k)] = n_k \cdot \mathbb{P}(X, Y|k).$$

Unlike the situation where only one position t is concerned, this distribution represents the sum of dependent Bernoulli variables. Methods of accounting for this dependence can be found in Robin et al. [10].

Applications

Several spatial and sequence motifs have been discovered using the approach discussed here [5–7]. The estimated propensities have also been used to develop empirical potential function for prediction of oligomerization stated [8], protein-protein interaction interfaces [3, 8], engineering of thermal resistance [2], and in predicting structures of β-barrel membrane proteins [9].

Open Problems

General analytical formulas for calculating probabilities of two-residue and multi-residue motifs under the permutation model are unknown.

Recommended Reading

1. Baker PJ, Britton KL, Rice DW, Rob A, Stillman TJ (1992) Structural consequences of sequence patterns in the fingerprint region of the nucleotide binding fold. Implications for nucleotide specificity. J Mol Biol 228(2):662–671
2. Gessmann D, Mager F, Naveed H, Arnold T, Weirich S, Linke D, Liang J, Nussberger S (2011) Improving the resistance of a eukaryotic beta-barrel protein to thermal and chemical perturbations. J Mol Biol 413(1):150–161. doi:10.1016/j.jmb.2011.07.054
3. Geula S, Naveed H, Liang J, Shoshan-Barmatz V (2012) Structure-based analysis of VDAC1 protein: defining oligomer contact sites. J Biol Chem 287(3):2179–2190
4. Hart R, Royyuru A, Stolovitzky G, Califano A (2000) Systematic and fully automated identification of protein sequence patterns. J Comput Biol 7(3–4):585–600
5. Jackups R Jr, Liang J (2005) Interstrand pairing patterns in beta-barrel membrane proteins: the positive-outside rule, aromatic rescue, and strand registration prediction. J Mol Biol 354(4):979–993
6. Jackups R Jr, Liang J (2010) Combinatorial analysis for sequence and spatial motif discovery in short sequence fragments. IEEE/ACM Trans Comput Biol Bioinform 7(3):524–536
7. Jackups R Jr, Cheng S, Liang J (2006) Sequence motifs and antimotifs in beta barrel membrane proteins from a genome-wide analysis: the ala-tyr dichotomy and chaperone binding motifs. J Mol Biol 363(2):611–623
8. Naveed H, Jackups R Jr, Liang J (2009) Predicting weakly stable regions, oligomerization state, and protein-protein interfaces in transmembrane domains of outer membrane proteins. Proc Natl Acad Sci USA 106(31):12735–12740. doi:10.1073/pnas.0902169106
9. Naveed H, Xu Y, Jackups R, Liang J (2012) Predicting three-dimensional structures of transmembrane domains of β-barrel membrane proteins. J Am Chem Soc 134(3):1775–1781
10. Robin S, Rodophe F, Schbath S (2005) DNA, words and models: statistics of exceptional words. Cambridge University Press, Cambridge/New York
11. Senes A, Gerstein M, Engelman DM (2000) Statistical analysis of amino acid patterns in transmembrane helices: the GxxxG motif occurs frequently and in association with β-branched residues at neighboring positions. J Mol Biol 296:921–936
12. Wimley WC (2002) Toward genomic identification of β-barrel membrane proteins: composition and architecture of known structures. Protein Sci 11:301–312
13. Wouters MA, Curmi PM (1995) An analysis of side chain interactions and pair correlations within antiparallel β-sheets: the differences between backbone hydrogen-bonded and non-hydrogen-bonded residue pairs. Proteins 22:119–131
14. Yaffe MB, Rittinger K, Volinia S, Caron PR, Aitken A, Leffers H, Gamblin SJ, Smerdon SJ, Cantley LC (1997) The structural basis for 14-3-3:phosphopeptide binding specificity. Cell 91(7):961–971

Sequential Circuit Technology Mapping

Peichen Pan
Xilinx, Inc., San Jose, CA, USA

Keywords

Boolean network; EDA; FPGA; Retiming; Technology mapping; VLSI CAD

Years and Authors of Summarized Original Work

1996; Pan, Liu
1998; Pan, Liu
1998; Pan, Lin

Problem Definition

One of the key steps in a VLSI design flow is technology mapping that converts a Boolean network of technology-independent logic gates and D-flipflops (FFs) into an equivalent one comprised of cells from a technology library [1, 4]. Technology mapping can be formulated as a covering problem where logic gates are covered by cells from the technology library. For ease of discussion, it is assumed that the cell library contains only one cell, a K-input lookup table (K-LUT) with one unit of delay. A K-LUT can implement any Boolean function with up to K inputs as is the case in field-programmable gate arrays (FPGAs) [1, 3].

Figure 1 shows an example of technology mapping. The original network in (1) with three FFs and four gates is covered by three 3-input cones as indicated in (2). The corresponding

Sequential Circuit Technology Mapping, Fig. 1 Technology mapping: (**1**) original network, (**2**) covering, (**3**) mapping solution

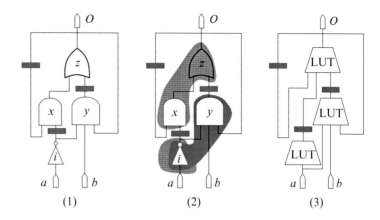

(1)   (2)   (3)

Sequential Circuit Technology Mapping, Fig. 2 Retiming and mapping: (**1**) retiming and covering, (**2**) mapping solution, (**3**) retimed solution

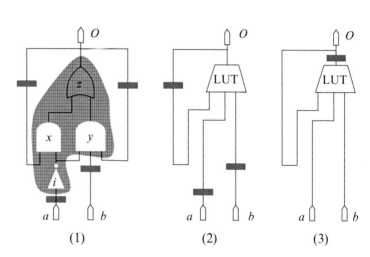

(1)   (2)   (3)

mapping solution using 3-LUTs is shown in (3). Note that gate i is covered by two cones so it will be replicated. The mapping solution has a *cycle time* (or *clock period*) of two units, which is the maximum number of LUTs on all paths without FFs.

Retiming relocates FFs in a network by moving FFs across logic nodes backward or forward [5]. Retiming does not alter the functionality of a network. Figure 2 (1) shows the network obtained from the one in Fig. 1 (1) by moving the FFs at the output of gates y and i to their inputs. It can now be covered with just one 3-input cone as indicated in (1). The corresponding mapping solution shown in (2) is better in both cycle time and area than the one in Fig. 1 (3) obtained without retiming.

A *K-bounded* network is one in which each gate has at most K inputs. The sequential

circuit technology mapping problem can be defined as follows: *Given a K-bounded Boolean network N and a target cycle time ϕ, find a mapping solution with a cycle time of ϕ, assuming FFs can be relocated using retiming.*

Key Results

The first polynomial time algorithm for the problem was proposed in [9, 10]. An improved algorithm was proposed in [2] to reduce runtime. Both algorithms are based on min-cost flow computation.

In [8], an efficient algorithm was proposed to take advantage of the fact that K is a small integer usually between 3 and 6. The algorithm is based

Sequential Circuit Technology Mapping, Fig. 3 Cut enumeration procedure

---

FindAllCuts(N, K)

   foreach node v in N **do** $C(v) \Leftarrow \{\{v^0\}\}$

   while (*new cuts discovered*) **do**

      foreach node v in N **do** $C(v) \Leftarrow merge(C(u_1),....,C(u_t))$

---

on enumerating all K-input cones for each gate and will be described next.

Cut Enumeration

A Boolean network can be represented as an edge-weighted directed graph where the nodes denote logic gates and primary inputs/outputs. There is a directed edge (u, v) with weight d if u, after going through d FFs, drives v.

A *cone* for a node can be captured by a *cut* consisting of inputs to the cone. An element in a cut for v consists of the driving node u and the total weight d on the paths from u to v, denoted by u^d. If u can reach v on several paths with different FF counts, u will appear in the cut multiple times with different ds. For the cone for z in Fig. 2 (2), the corresponding cut is $\{z^1, a^1, b^1\}$. A cut of size K is called a K-cut.

Let (u_i, v), where $i = 1, \ldots, t$, be all input edges to v in N. Further assume the weight of (u_i, v) is d_i and $C(u_i)$ is a set of K-cuts for u_i. Let $merge(C(u_1), \ldots, C(u_t))$ denote the following set operation:

$$\{\{v^0\}\} \cup \{c_1^{d_1} \cup \cdots \cup c_t^{d_t} | c_1 \in (u_1), \ldots, c_t$$
$$\in C(u_t), |c_1^{d_1} \cup \cdots \cup c_t^{d_t}| \leq K\}$$

where $c_i^{d_i} = \{u^{d+d_i} | u^d \in c_i\}$ for $i = 1, \ldots, t$. It is obvious that $merge(C(u_1), \ldots, C(u_t))$ is a set of K-cuts for v.

If the network N does not contain cycles, the K-cuts of all nodes can be determined using the merge operation in a topological order starting from the PIs. For general networks, Fig. 3 outlines the iterative cut computation procedure proposed in [8].

Figure 4 depicts the iterations in enumerating the 3-cuts for the network in Fig. 1 (1) where cuts are merged in the order i, x, y, z, and o. At the beginning, every node has a trivial cut formed by itself (Row 0). Row 1 shows the new cuts discovered in the first iteration. In second iteration, two more cuts are discovered (for x). After that, further merging does not yield any new cut and the procedure stops.

Lemma 1 *After at most Kn iterations, the cut enumeration procedure will find all the K-cuts for every node in N.*

Techniques have been proposed to speed up the procedure [8]. For practical networks, the cut enumeration procedure typically converges in just a few iterations.

Label Computation

After obtaining all K-cuts, the cuts are evaluated based on sequential arrival times (or l-values), which is an extension of traditional arrival times, to consider the effect of retiming [7, 9].

The labeling procedure tries to find a label for each node as outlined in Fig. 5, where w_v denotes the weight of the shortest paths from PIs to node v.

Figure 6 shows the iterations for label computation for the network in Fig. 1 (1), assuming that the target cycle time $\phi = 1$ and the nodes are evaluated in the order of i, x, y, z, and o. In the table, the current label as well as a corresponding cut for each node is listed. In this example, after the first iteration, none of the labels will change and the procedure stops.

It can be shown that the labeling procedure will stop after at most $n(n - 1)$ iterations [10]. The following lemma relates labels to mapping:

Iter	a	B	I	x	y	z	o
0	$\{a^0\}$	$\{b^0\}$	$\{i^0\}$	$\{x^0\}$	$\{y^0\}$	$\{z^0\}$	$\{o^0\}$
1			$\{a^0\}$	$\{i^1,z^1\}$	$\{i^0,b^0,z^0\}$	$\{x^0,y^1\}$	$\{z^0\}$
				$\{a^1,z^1\}$	$\{a^0,b^0,z^0\}$	$\{i^1,z^1,b^1\}$	
						$\{a^1,z^1,b^1\}$	
						$\{i^1,z^1,y^1\}$	
						$\{a^1,z^1,y^1\}$	
2				$\{i^1,x^1,y^2\}$			
				$\{a^1,x^1,y^2\}$			

Sequential Circuit Technology Mapping, Fig. 4 Cut enumeration example

Sequential Circuit Technology Mapping, Fig. 5 Label computation procedure

---

FindMinLabels(N)

    foreach node v in N **do** $l(v) \Leftarrow -w_v \cdot \phi$

    while (*there are label updates*) **do**

        foreach node v in N **do**

            $l(v) \Leftarrow \min_{c \in C(v)} \{\max\{l(u) - d \cdot \phi + 1 \mid u^d \in c\}\}$

            if v is a primary output and $l(v) > \phi$, **return** failure

    return success

---

iter	a	B	I	x	y	z	o
0	$\{a^0\}$:0	$\{b^0\}$:0	$\{i^0\}$:0	$\{x^0\}$:-1	$\{y^0\}$:0	$\{z^0\}$:-1	$\{o^0\}$:-1
1			$\{a^0\}$:1	$\{a^1,z^1\}$:0	$\{a^0,b^0,z^0\}$:1	$\{a^1,z^1,b^1\}$:0	$\{z^0\}$:0

Sequential Circuit Technology Mapping, Fig. 6 Label computation example

Lemma 2 *N has a mapping solution with cycle time ϕ iff the labeling procedure returns "success."*

Mapping Solution Generation

Once the labels for all nodes are computed successfully, a mapping solution can be constructed starting from primary outputs. At each node v, the procedure selects the cut that realizes the label of the node and then moves on to select a cut for u if u^d is in the cut selected for v. On the edge from u to v, d FFs are inserted. For the network in Fig. 1 (1), the mapping solution generated based on the labels found in Fig. 6 is exactly the one in Fig. 2 (2).

To obtain a mapping solution with the target cycle time ϕ, v will be retimed by a value of $\lceil l(v)/\phi \rceil - 1$. For the network in Fig. 1 (1), the final mapping solution after retiming is shown in Fig. 2 (3) which has a cycle time of 1.

Applications

The algorithm can be used to map a technology-independent Boolean network to a network consisting of cells from a target technology library. The concepts and framework are generally enough to be adapted to study other circuit optimizations such as sequential circuit clustering and sequential circuit restructuring [6].

Cross-References

▶ Circuit Retiming
▶ Circuit Retiming: An Incremental Approach
▶ FPGA Technology Mapping
▶ Technology Mapping

Recommended Reading

1. Cong J, Ding Y (1994) FlowMap: an optimal technology mapping algorithm for delay optimization in lookup-table based FPGA designs. IEEE Trans Comput Aided Des Integr Circuits Syst 13(1):1–12
2. Cong J, Wu C (1997) FPGA synthesis with retiming and pipelining for clock period minimization of sequential circuits. In: ACM/IEEE design automation conference, Anaheim
3. Cong J, Wu C, Ding Y (1999) Cut ranking and pruning: enabling a general and efficient FPGA mapping solution. In: ACM international symposium on field-programmable gate arrays, Monterey
4. Keutzer K (1987) DAGON: technology binding and local optimization by DAG matching. In: ACM/IEEE design automation conference, Miami Beach
5. Leiserson CE, Saxe JB (1991) Retiming synchronous circuitry. Algorithmica 6:5–35
6. Mishchenko A, Chatterjee S, Brayton R, Pan P (2006) Integrating logic synthesis, technology mapping, and retiming. ERL technical report, EECS Department, UC Berkeley
7. Pan P (1997) Continuous retiming algorithms and applications. In: IEEE international conference on computer design, Austin
8. Pan P, Lin CC (1998) A new retiming-based technology mapping algorithm for LUT-based FPGAs. In: ACM international symposium on field-programmable gate arrays, Monterey
9. Pan P, Liu CL (1996) Optimal clock period FPGA technology mapping for sequential circuits. In: ACM/IEEE design automation conference, Las Vegas
10. Pan P, Liu CL (1998) Optimal clock period FPGA technology mapping for sequential circuits. ACM Trans Des Autom Electron Syst 3(3):437–462

Set Agreement

Michel Raynal
Institut Universitaire de France and IRISA,
Université de Rennes, Rennes, France

Keywords

Distributed coordination

Years and Authors of Summarized Original Work

1993; Chaudhuri

Problem Definition

Short History

The k-set agreement problem is a paradigm of coordination problems. Defined in the setting of systems made up of processes prone to failures, it is a simple generalization of the consensus problem (that corresponds to the case $k = 1$). That problem was introduced in 1993 by Chaudhuri [2] to investigate how the number of choices (k) allowed for the processes is related to the maximum number of processes that can crash. (After it has crashed, a process executes no more steps: a crash is a premature halting.)

Definition

Let S be a system made up of n processes where up to t can crash and where each process has an input value (called a *proposed* value). The problem is defined by the three following properties (i.e., any algorithm that solves that problem has to satisfy these properties):

1. **Termination.** Every nonfaulty process decides a value.
2. **Validity.** A decided value is a proposed value.
3. **Agreement.** At most k different values are decided.

The Trivial Case

It is easy to see that this problem can be trivially solved if the upper bound on the number of process failures t is smaller than the allowed number of choices k, also called the *coordination degree*. (The trivial solution consists in having $t + 1$ predetermined processes that send their proposed values to all the processes, and a process deciding the first value it ever receives.) So, $k \leq t$ is implicitly assumed in the following.

Key Results

Key Results in Synchronous Systems

The Synchronous Model

In this computation model, each execution consists of a sequence of rounds. These are identified by the successive integers $1, 2$, etc. For the processes, the current round number appears as a global variable whose global progress entails their own local progress.

During a round, a process first broadcasts a message, then receives messages, and finally executes local computation. The fundamental synchrony property the a synchronous system provides the processes with is the following: a message sent during a round r is received by its destination process during the very same round r. If during a round, a process crashes while sending a message, an arbitrary subset (not known in advance) of the processes receive that message.

Main Results

The k-set agreement problem can always be solved in a synchronous system. The main result is for the minimal number of rounds (R_t) that are needed for the nonfaulty processes to decide in the worst-case scenario (this scenario is when exactly k processes crash in each round). It was shown in [3] that $R_t = \lfloor \frac{t}{k} \rfloor + 1$. A very simple algorithm that meets this lower bound is described in Fig. 1.

Although failures do occur, they are rare in practice. Let f denote the number of processes that crash in a given run, $0 \leq f \leq t$. We are interested in synchronous algorithms that terminate in at most R_t rounds when t processes crash in the current run, but that allow the nonfaulty processes to decide in far fewer rounds when there are few failures. Such algorithms are called *early-deciding* algorithms. It was shown in [4] that, in the presence of f process crashes, any early-deciding k-set agreement algorithm has runs in which no process decides before the round $R_f = \min(\lfloor \frac{f}{k} \rfloor + 2, \lfloor \frac{t}{k} \rfloor + 1)$. This lower bound shows an inherent tradeoff linking the coordination degree k, the maximum number of process failures t, the actual number of process failures f, and the best time complexity that can be achieved. Early-deciding k-set agreement algorithms for the synchronous model can be found in [4, 12].

Other Failure Models

In the send omission failure model, a process is faulty if it crashes or forgets to send messages. In the general omission failure model, a process is faulty if it crashes, forgets to send messages, or forgets to receive messages. (A send omission failure models the failure of an output buffer, while a receive omission failure models the failure of an input buffer.) These failure models were introduced in [11].

The notion of *strong* termination for set agreement problems was introduced in [13]. Intuitively, that property requires that as many processes as possible decide. Let a *good* process be a process that neither crashes nor commits receive omission failures. A set agreement algorithm is strongly terminating if it forces all the good processes to decide. (Only the processes that crash during the execution of the algorithm, or that do not receive enough messages, can be prevented from deciding.)

An early-deciding k-set agreement algorithm for the general omission failure model was described in [13]. That algorithm, which requires $t < n/2$, directs a good process to decide and stop in at most $R_f = \min(\lfloor \frac{f}{k} \rfloor + 2, \lfloor \frac{t}{k} \rfloor + 1)$ rounds. Moreover, a process that is not a good

Set Agreement, Fig. 1
A simple k-set agreement
synchronous algorithm
(code for p_i)

```
Function k-set_agreement (vᵢ)
(1)   estᵢ ← vᵢ;
(2)   when r = 1, 2, . . . , ⌊t/k⌋ + 1 do % r: round number %
(3)   begin_round
(4)         send (estᵢ) to all; % including pᵢ itself %
(5)         estᵢ ← min({estⱼ values received during
                                      the current round r});
(6)   end_round;
(7)   return (estᵢ)
```

process executes at most $R_f(not\ good)$ $\min(\lceil \frac{f}{k} \rceil + 2, \lfloor \frac{t}{k} \rfloor + 1)$ rounds.

As R_f is a lower bound for the number of rounds in the crash failure model, the previous algorithm shows that R_f is also a lower bound for the nonfaulty processes to decide in the more severe general omission failure model. Proving that $R_f(not\ good)$ is an upper bound for the number of rounds that a nongood process has to execute remains an open problem.

It was shown in [13] that, for a given coordination degree k, $t < \frac{k}{k+1}n$ is an upper bound on the number of process failures when one wants to solve the k-set agreement problem in a synchronous system prone to process general omission failures. A k-set agreement algorithm that meets this bound was described in [13]. That algorithm requires the processes execute $R = t + 2 - k$ rounds to decide. Proving (or disproving) that R is a lower bound when $t < \frac{k}{k+1}n$ is an open problem. Designing an early-deciding k-set agreement algorithm for $t < \frac{k}{k+1}n$ and $k > 1$ is another problem that remains open.

Key Results in Asynchronous Systems

Impossibility
A fundamental result of distributed computing is the impossibility to design a deterministic algorithm that solves the k-set agreement problem in asynchronous systems when $k \leq t$ [1, 7, 15]. Compared with the impossibility of solving asynchronous consensus despite one process crash, that impossibility is based on deep combinatorial arguments. This impossibility has opened

new research directions for the connection between distributed computing and topology. This topology approach has allowed the discovery of links relating asynchronous k-set agreement with other distributed computing problems such as the *renaming* problem [5].

Circumventing the Impossibility
Several approaches have been investigated to circumvent the previous impossibility. These approaches are the same as those that have been used to circumvent the impossibility of asynchronous consensus despite process crashes.

One approach consists in replacing the "deterministic algorithm" by a "randomized algorithm." In that case, the termination property becomes "the probability for a correct process to decide tends to 1 when the number of rounds tends to $+\infty$." That approach was investigated in [9].

Another approach that has been proposed is based on failure detectors. Roughly speaking, a failure detector provides each process with a list of processes suspected to have crashed. As an example, the class of failure detectors denoted $\Diamond S_x$ includes all the failure detectors such that, after some finite (but unknown) time, (1) any list contains the crashed processes and (2) there is a set Q of x processes such that Q contains one correct process and that correct process is no longer suspected by the processes of Q (let us observe that correct processes can be suspected intermittently or even forever). Tight bounds for the k-set agreement problem in asynchronous systems equipped with such failure detectors, conjectured in [9], were proved in [6]. More precisely, such a failure detector class allows the k-set agreement problem to be solved for $k \geq t - x + 2$ [9], and cannot solve it when $k < t - x + 2$ [6].

Another approach that has been investigated is the combination of failure detectors and conditions [8]. A condition is a set of input vectors, and each input vector has one entry per process. The entries of the input vector associated with a run contain the values proposed by the processes in that run. Basically, such an approach guarantees that the nonfaulty processes always decide when the actual input vector belongs to the condition the k-set algorithm has been instantiated with.

Applications

The set agreement problem was introduced to study how the number of failures and the synchronization degree are related in an asynchronous system; hence, it is mainly a theoretical problem. That problem is used as a canonical problem when one is interested in asynchronous computability in the presence of failures. Nevertheless, one can imagine practical problems the solutions of which are based on the set agreement problem (e.g., allocating a small shareable resources – such as broadcast frequencies – in a network).

Cross-References

► Asynchronous Consensus Impossibility
► Failure Detectors
► Renaming
► Topology Approach in Distributed Computing

Recommended Reading

1. Borowsky E, Gafni E (1993) Generalized FLP impossibility results for t-resilient asynchronous computations. In: Proceedings of the 25th ACM symposium on theory of computation, California, pp 91–100
2. Chaudhuri S (1993) More choices allow more faults: set consensus problems in totally asynchronous systems. Inf Comput 105:132–158
3. Chaudhuri S, Herlihy M, Lynch N, Tuttle M (2000) Tight bounds for k set agreement. J ACM 47(5):912–943
4. Gafni E, Guerraoui R, Pochon B (2005) From a static impossibility to an adaptive lower bound: the complexity of early deciding set agreement. In: Proceedings of the 37th ACM symposium on theory of computing (STOC 2005). ACM, New York, pp 714–722
5. Gafni E, Rajsbaum S, Herlihy M (2006) Subconsensus tasks: renaming is weaker than set agreement. In: Proceedings of the 20th international symposium on distributed computing (DISC'06). LNCS, vol 4167. Springer, Berlin, pp 329–338
6. Herlihy MP, Penso LD (2005) Tight bounds for k set agreement with limited scope accuracy failure detectors. Distrib Comput 18(2):157–166
7. Herlihy MP, Shavit N (1999) The topological structure of asynchronous computability. J ACM 46(6):858–923
8. Mostefaoui A, Rajsbaum S, Raynal M (2005) The combined power of conditions and failure detectors to solve asynchronous set agreement. In: Proceedings of the 24th ACM symposium on principles of distributed computing (PODC'05). ACM, New York, pp 179–188
9. Mostefaoui A, Raynal M (2000) k set agreement with limited scope accuracy failure detectors. In: Proceedings of the 19th ACM symposium on principles of distributed computing. ACM, New York, pp 143–152
10. Mostefaoui A, Raynal M (2001) Randomized set agreement. In: Proceedings of the 13th ACM symposium on parallel algorithms and architectures (SPAA'01), Hersonissos (Crete). ACM, New York, pp 291–297
11. Perry KJ, Toueg S (1986) Distributed agreement in the presence of processor and communication faults. IEEE Trans Softw Eng SE-12(3):477–482
12. Raipin Parvedy P, Raynal M, Travers C (2005) Early-stopping k-set agreement in synchronous systems prone to any number of process crashes. In: Proceedings of the 8th international conference on parallel computing technologies (PaCT'05). LNCS, vol 3606. Springer, Berlin, pp 49–58
13. Raipin Parvedy P, Raynal M, Travers C (2006) Strongly-terminating early-stopping k-set agreement in synchronous systems with general omission failures. In: Proceedings of the 13th colloquium on structural information and communication complexity (SIROCCO'06). LNCS, vol 4056. Springer, Berlin, pp 182–196
14. Raynal M, Travers C (2006) Synchronous set agreement: a concise guided tour (including a new algorithm and a list of open problems). In: Proceedings of the 12th international IEEE pacific rim dependable computing symposium (PRDC'2006). IEEE Computer Society, Los Alamitos, pp 267–274
15. Saks M, Zaharoglou F (2000) Wait-free k-set agreement is impossible: the topology of public knowledge. SIAM J Comput 29(5):1449–1483

S

Set Cover with Almost Consecutive Ones

Michael Dom
Department of Mathematics and Computer Science, University of Jena, Jena, Germany

Keywords

Hitting set

Years and Authors of Summarized Original Work

2004; Mecke, Wagner

Problem Definition

The SET COVER problem has as input a set R of m items, a set C of n subsets of R and a weight function $w: C \rightarrow \mathbb{Q}$. The task is to choose a subset $C' \subseteq C$ of minimum weight whose union contains all items of R.

The sets R and C can be represented by an $m \times n$ binary matrix A that consists of a row for every item in R and a column for every subset of R in C, where an entry $a_{i,j}$ is 1 iff the ith item in R is part of the jth subset in C. Therefore, the SET COVER problem can be formulated as follows.

> **Input**: An $m \times n$ binary matrix A and a weight function w on the columns of A.
> **Task:** Select some columns of A with minimum weight such that the submatrix A' of A that is induced by these columns has at least one 1 in every row.

While SET COVER is NP-hard in general [4], it can be solved in polynomial time on instances whose columns can be permuted in such a way that in every row the ones appear consecutively, that is, on instances that have the *consecutive ones property* (*C1P*). (The C1P can be defined symmetrically for columns; this article focuses on rows. SET COVER on instances with the C1P can be solved in polynomial time, e.g., with a linear programming approach, because the cor-

responding coefficient matrices are totally unimodular (see [9]).

Motivated by problems arising from railway optimization, Mecke and Wagner [7] consider the case of SET COVER instances that have "almost the C1P". Having almost the C1P means that the corresponding matrices are similar to matrices that have been generated by starting with a matrix that has the C1P and replacing randomly a certain percentage of the 1's by 0's [7]. For Ruf and Schöbel [8], in contrast, having almost the C1P means that the average number of blocks of consecutive 1's per row is much smaller than the number of columns of the matrix. This entry will also mention some of their results.

Notation

Given an instance (A, w) of SET COVER, let R denote the row set of A and C its column set. A column c_j *covers* a row r_i, denoted by $r_i \in c_j$, if $a_{i,j} = 1$.

A binary matrix has the *strong C1P* if (without any column permutation) the 1's appear consecutively in every row. A *block of consecutive 1's* is a maximal sequence of consecutive 1's in a row. It is possible to determine in linear time if a matrix has the C1P, and if so, to compute a column permutation that yields the strong C1P [2, 3, 6]. However, note that it is NP-hard to permute the columns of a binary matrix such that the number of blocks of consecutive 1's in the resulting matrix is minimized [1, 4, 5].

A *data reduction rule* transforms in polynomial time a given instance I of an optimization problem into an instance I' of the same problem such that $|I'| < |I|$ and the optimal solution for I' has the same value (e.g., weight) as the optimal solution for I. Given a set of data reduction rules, *to reduce* a problem instance means to repeatedly apply the rules until no rule is applicable; the resulting instance is called *reduced*.

Key Results

Data Reduction Rules

For SET COVER there exist well-known data reduction rules:

Row domination rule: If there are two rows $r_{i_1}, r_{i_2} \in R$ with $\forall c \in C : r_{i_1} \in c$ implies $r_{i_2} \in c$, then r_{i_2} is *dominated* by r_{i_1}. Remove row r_{i_2} from A.

Column domination rule: If there are two columns $c_{j_1}, c_{j_2} \in C$ with $w(c_{j_1}) \geq w(c_{j_2})$ and $\forall r \in R : r \in c_{j_1}$ implies $r \in c_{j_2}$, then c_{j_1} is *dominated* by c_{j_2}. Remove c_{j_1} from A.

In addition to these two rules, a column $c_{j_1} \in C$ can also be dominated by a subset $C' \subseteq C$ of the columns instead of a single column: If there is a subset $C' \subseteq C$ with $w(c_{j_1}) \geq \sum_{c \in C'} w(c)$ and $\forall r \in R : r \in c_{j_1}$ implies $(\exists c \in C' : r \in c)$, then remove c_{j_1} from A. Unfortunately, it is NP-hard to find a dominating subset C' for a given set c_{j_1}. Mecke and Wagner [7], therefore, present a restricted variant of this generalized column domination rule.

For every row $r \in R$, let $c_{\min}(r)$ be a column in C that covers r and has minimum weight under this property. For two columns $c_{j_1}, c_{j_2} \in C$, define $X(c_{j_1}, c_{j_2}) := \{c_{\min}(r) \mid r \in c_{j_1} \wedge r \notin c_{j_2}\}$. The new data reduction rule then reads as follows.

Advanced column domination rule: If there are two columns $c_{j_1}, c_{j_2} \in C$ and a row that is covered by both c_{j_1} and c_{j_2}, and if $w(c_{j_1}) \geq w(c_{j_2}) + \sum_{c \in X(c_{j_1}, c_{j_2})} w(c)$, then c_{j_1} is *dominated* by $\{c_{j_2}\} \cup X(c_{j_1}, c_{j_2})$. Remove c_{j_1} from A.

Theorem 1 ([7]) *A matrix A can be reduced in $O(Nn)$ time with respect to the column domination rule, in $O(Nm)$ time with respect to the row domination rule, and in $O(Nmn)$ time with respect to all three data reduction rules described above, when N is the number of 1's in A.*

In the databases used by Ruf and Schöbel [8], matrices are represented by the column indices of the first and last 1's of its blocks of consecutive 1's. For such matrix representations, a fast data reduction rule is presented [8], which eliminates "unnecessary" columns and which, in the implementations, replaces the column domination rule. The new rule is faster than the column domination rule (a matrix can be reduced in $O(mn)$ time with respect to the new rule), but not

as powerful: Reducing a matrix A with the new rule can result in a matrix that has more columns than the matrix resulting from reducing A with the column domination rule.

Algorithms

Mecke and Wagner [7] present an algorithm that solves SET COVER by enumerating all feasible solutions.

Given a row r_i of A, a *partial solution for the rows $r_1, \ldots, r_i$* is a subset $C' \subseteq C$ of the columns of A such that for each row r_j with $j \in \{1, \ldots, i\}$ there is a column in C' that covers row r_j.

The main idea of the algorithm is to find an optimal solution by iterating over the rows of A and updating in every step a data structure S that keeps *all* partial solutions for the rows considered so far. More exactly, in every iteration step the algorithm considers the first row of A and updates the data structure S accordingly. Thereafter, the first row of A is deleted. The following code shows the algorithm.

```
1 Repeat m times: {
2    for every partial solution C' in S that does not
                                cover the first row of A: {
3       for every column c of A that covers the first row
                                            of A: {
4          Add {c} ∪ C' to S; }
5    Delete C' from S; }
6 Delete the first row of A; }
```

This straightforward enumerative algorithm could create a set S of exponential size. Therefore, the data reduction rules presented above are used to delete after each iteration step partial solutions that are not needed any more. To this end, a matrix B is associated with the set S, where every row corresponds to a row of A and every column corresponds to a partial solution in S–an entry $b_{i,j}$ of B is 1 iff the jth partial solution of B contains a column of A that covers the row r_i. The algorithm uses the matrix $C := \left(\begin{array}{c|c} A & B \\ \hline 0 \ldots 0 & 1 \ldots 1 \end{array} \right)$, which is updated together with S in every iteration step. (The last row of C allows to distinguish the columns belonging to A from those belonging to B.) Line 6 of the code shown above is replaced by the following two lines:

6 Delete the first row of the matrix C;
7 Reduce the matrix C and update S accordingly;}

At the end of the algorithm, S contains exactly one solution, and this solution is optimal. Moreover, if the SET COVER instance is nicely structured, the algorithm has polynomial running time:

Theorem 2 ([7]) *If A has the strong C1P, is reduced, and its rows are sorted in lexicographic order, then the algorithm has a running time of $O(M^{3n})$ where M is the maximum number of 1's per row and per column.*

Theorem 3 ([7]) *If the distance between the first and the last 1 in every column is at most k, then at any time throughout the algorithm the number of columns in the matrix B is $O(2^{kn})$, and the running time is $O(2^{2k} kmn^2)$.*

Ruf and Schöbel [8] present a branch and bound algorithm for SET COVER instances that have a small average number of blocks of consecutive 1's per row.

The algorithm considers in each step a row r_i of the current matrix (which has been reduced with data reduction rules before) and branches into bl_i cases, where bl_i is the number of blocks of consecutive 1's in r_i. In each case, one block of consecutive 1's in row r_i is selected, and the 1's of all other blocks in this row are replaced by 0's. Thereafter, a lower and an upper bound on the weight of the solution for each resulting instance is computed. If a lower bound differs by a factor of more than $1 + \epsilon$, for a given constant ϵ, from the best upper bound achieved so far, the corresponding instance is subjected to further branchings. Finally, the best upper bound that was found is returned.

In each branching step, the bl_i instances that are newly generated are "closer" to have the (strong) C1P than the instance from which they descend. If an instance has the C1P, the lower and upper bound can easily be computed by exactly solving the problem. Otherwise, standard heuristics are used.

Applications

SET COVER instances occur e.g., in railway optimization, where the task is to determine where new railway stations should be built. Each row then corresponds to an existing settlement, and each column corresponds to a location on the existing trackage where a railway station could be build. A column c covers a row r, if the settlement corresponding to r lies within a given radius around the location corresponding to c.

If the railway network consisted of one straight line rail track only, the corresponding SET COVER instance would have the C1P; instances arising from real world data are close to have the C1P [7, 8].

Experimental Results

Mecke and Wagner [7] make experiments on real-world instances as described in the Applications section and on instances that have been generated by starting with a matrix that has the C1P and replacing randomly a certain percentage of the 1's by 0's. The real-world data consists of a railway graph with 8,200 nodes and 8,700 edges, and 30,000 settlements. The generated instances consist of 50–50,000 rows with 10–200 1's per row. Up to 20 % of the 1's are replaced by 0's.

In the real-world instances, the data reduction rules decrease the number of 1's to between 1 % and 25 % of the original number of 1's without and to between 0.2 % and 2.5 % with the advanced column reduction rule. In the case of generated instances that have the C1P, the number of 1's is decreased to about 2 % without and to 0.5 % with the advanced column reduction rule. In instances with 20 % perturbation, the number of 1's is decreased to 67 % without and to 20 % with the advanced column reduction rule.

The enumerative algorithm has a running time that is almost linear for real-world instances and most generated instances. Only in the case of generated instances with 20 % perturbation, the running time is quadratic.

Ruf and Schöbel [8] consider three instance types: real-world instances, instances arising from Steiner triple systems, and randomly generated instances. The latter have a size of 100×100 and contain either 1–5 blocks of consecutive 1's in each row, each one consisting of between one and nine 1's, or they are generated with a probability of 3 % or 5 % for any entry to be 1.

The data reduction rules used by Ruf and Schöbel turn out to be powerful for the real-world instances (reducing the matrix size from about $1,100 \times 3,100$ to 100×800 in average), whereas for all other instance types the sizes could not be reduced noticeably.

The branch and bound algorithm could solve almost all real-world instances up to optimality within a time of less than a second up to one hour. In all cases where an optimal solution has been found, the first generated subproblem had already provided a lower bound equal to the weight of the optimal solution.

Cross-References

▶ Greedy Set-Cover Algorithms

Recommended Reading

1. Atkins JE, Middendorf M (1996) On physical mapping and the consecutive ones property for sparse matrices. Discret Appl Math 71(1–3):23–40
2. Booth KS, Lueker GS (1976) Testing for the consecutive ones property, interval graphs, and graph planarity using PQ-tree algorithms. J Comput Syst Sci 13:335–379
3. Fulkerson DR, Gross OA (1965) Incidence matrices and interval graphs. Pac J Math 15(3):835–855
4. Garey MR, Johnson DS (1979) Computers and intractability: a guide to the theory of NP-completeness. Freeman, New York
5. Goldberg PW, Golumbic MC, Kaplan H, Shamir R (1995) Four strikes against physical mapping of DNA. J Comput Biol 2(1):139–152
6. Hsu WL, McConnell RM (2003) PC trees and circular-ones arrangements. Theor Comput Sci 296(1):99–116
7. Mecke S, Wagner D (2004) Solving geometric covering problems by data reduction. In: Proceedings of the 12th annual European symposium on algorithms (ESA '04). LNCS, vol 3221. Springer, Berlin, pp 760–771
8. Ruf N, Schöbel A (2004) Set covering with almost consecutive ones property. Discret Optim 1(2):215–228
9. Schrijver A (1986) Theory of linear and integer programming. Wiley, Chichester

Shadowless Solutions for Fixed-Parameter Tractability of Directed Graphs

Rajesh Chitnis and Mohammad Taghi Hajiaghayi
Department of Computer Science, University of Maryland, College Park, MD, USA

Keywords

Directed graphs; Fixed-parameter tractability; Important Separators; Shadowless solutions; Transversal problems

Years and Authors of Summarized Original Work

2012; Chitnis, Hajiaghayi, Marx
2012; Chitnis, Cygan, Hajiaghayi, Marx

Problem Definition

The study of the parameterized complexity of problems on directed graphs has been hitherto relatively unexplored. Usually the directed version of the problems require significantly different and more involved ideas than the ones for the undirected version. Furthermore, for directed graphs there are no known algorithmic meta-techniques: for example, there is no known algorithmic analogue of the Graph Minor Theory of Robertson and Seymour for directed graphs. As a result, the fixed-parameter tractability status of the directed versions of several fundamental problems such as Multiway Cut, Multicut, Feedback Vertex Set, etc., was open for a long time. The problem of Feedback Vertex Set best illustrates this gulf between undirected and directed graphs with respect to parameterized complexity. In this problem, we are given a graph and the question is whether there exists a set of size at most k whose deletion makes the graph acyclic. The undirected version was known to be FPT since 1984 [10]. However, the directed version was a long-standing open problem until it was shown to be FPT in 2008 [1].

The framework of *shadowless solutions* aims to bridge this gap by providing an important first step in designing FPT algorithms for a general class of transversal problems on directed graphs. In undirected graphs, the framework of *shadowless solutions* was introduced in [9] and has since been used in [4, 6, 7]. It was adapted and generalized to directed graphs in [2, 3] for the following general class of problems:

Finding an $\mathcal{F}$-transversal for some T-connected $\mathcal{F}$

Input: A directed graph $G = (V, E)$, a positive integer k, a set $T \subseteq V$, and a set $\mathcal{F} = \{F_1, F_2, \ldots, F_q\}$ of subgraphs such that $\mathcal{F}$ is T-connected, i.e., $\forall\, i \in [q]$ each vertex of F_i can reach some vertex of T by a walk completely contained in $G[F_i]$ and is reachable from some vertex of T by a walk completely contained in $G[F_i]$.

Parameter: k

Question: Is there an $\mathcal{F}$-transversal $W \subseteq V$ with $|W| \le k$, i.e., a set W such that $F_i \cap W \ne \emptyset$ for every $i \in [q]$?

The collection $\mathcal{F}$ is implicitly defined in a problem-specific way and need not be given explicitly in the input. In fact, it is possible that $\mathcal{F}$ is exponentially large. The *shadow* of a solution X is the set of vertices that are disconnected from T (in either direction) after the removal of X. More formally, the reverse shadow of X is given by $r_T(X) = \{v : X \text{ is a } v \rightarrow T \text{ separator}\}$. Similarly, the forward shadow of X is given by $f_T(X) = \{v : X \text{ is a } T \rightarrow v \text{ separator}\}$. The shadow of X is given by the union of its reverse and forward shadows, i.e., shadow$(X) = r(X) \cup f(X)$. A set X is said to be *shadowless* if its shadow is empty.

The aim is to ensure first that there is a solution whose shadow is empty, as finding such a shadowless solution can be a significantly easier task.

Key Results

For the $\mathcal{F}$-transversal problem defined above, [2] shows how to invoke the technique of *random sampling of important separators* and obtain a set Z which is disjoint from a minimum solution X and covers its shadow.

Theorem 1 (randomized covering of the shadow) *Let* $T \subseteq V(G)$. *There is an algorithm* $\mathrm{RandomSet}(G, T, k)$ *that runs in* $4^k \cdot n^{O(1)}$ *time and returns a set* $Z \subseteq V(G)$ *such that for any set* $\mathcal{F}$ *of* T-connected subgraphs, if there exists an $\mathcal{F}$-transversal of size $\le k$, then the following holds with probability $2^{-2^{O(k)}}$: there is an $\mathcal{F}$-transversal X of size $\le k$ such that

1. $X \cap Z = \emptyset$ and
2. Z covers the shadow of X, i.e., $r(X) \cup f(X) \subseteq Z$.

The set $\mathcal{F}$ is *not* an input of the algorithm described by Theorem 1: the set Z constructed in the above theorem works for *every* T-connected set $\mathcal{F}$ of subgraphs. Therefore, issues related to the representation of $\mathcal{F}$ do not arise. Theorem 1 can be derandomized using the theory of splitters [11]:

Theorem 2 (deterministic covering of the shadow) *Let* $T \subseteq V(G)$. *We can construct a set* $\{Z_1, Z_2, \ldots, Z_t\}$ *with* $t = 2^{2^{O(k)}} \cdot \log^2 n$ *in time* $2^{2^{O(k)}} \cdot n^{O(1)}$ *such that for any set* $\mathcal{F}$ *of* T-connected, if there exists an $\mathcal{F}$-transversal of size $\le k$, then there is an $\mathcal{F}$-transversal X of size $\le k$ such that for at least one $1 \le i \le t$ we have

1. $X \cap Z_i = \emptyset$ and
2. Z_i covers the shadow of X, i.e., $r(X) \cup f(X) \subseteq Z_i$.

Consider one such set Z_i for some $1 \le i \le 2^{2^{O(k)}} \cdot \log^2 n$. Since this set Z_i is disjoint from a minimum solution X, it can be removed from the graph. However, we need to remember the structure that the set Z_i imposed on the problem. This structure is problem specific, and the reduced (equivalent) instance is obtained on a supergraph of $G \setminus Z_i$ via the *torso operation*. It can be shown that the original instance G has a solution if and only if the reduced instance has a shadowless solution. Therefore, one can focus on the simpler task of finding a shadowless solution or more precisely, finding any solution under the guarantee that a shadowless solution exists.

Applications

The first FPT algorithms for the Directed Multiway Cut problem [3] and the Directed Subset Feedback Vertex Set problem [2] were obtained via the framework of shadowless solutions.

Directed Multiway Cut
In the Directed Multiway Cut problem, given a directed graph $G = (V, E)$, an integer k, and a set of terminals $T = \{t_1, t_2, \ldots, t_p\}$, the objective is to find whether there exists a set $X \subseteq V(G)$ of size at most k such that $G \setminus X$ has no $t_i \to t_j$ path for any $1 \le i \ne j \le p$. Let $\mathcal{F}$ be the set of all paths between pairs of (distinct) terminals. Then it is easy to show that $\mathcal{F}$ is T-connected, and the problem of finding an $\mathcal{F}$-transversal is exactly the same as the Directed Multiway Cut problem. It is shown in [3] that a shadowless solution of Directed Multiway Cut is also a solution of the underlying undirected instance of Multiway Cut, which is known to be FPT [8] parameterized by k. Combining with Theorem 2, this gives an FPT algorithm for the Directed Multiway Cut problem.

Directed Subset Feedback Vertex Set
In the Directed Subset Feedback Vertex Set problem, given a directed graph $G = (V, E)$, an integer k, and a set $S \subseteq V(G)$, the objective is to find whether there exists a set $X \subseteq V(G)$ of size at most k such that $G \setminus X$ has no S-cycles, i.e., cycles containing at least one vertex of S. The special case when $S = V(G)$ is the Directed Feedback Vertex Set problem. Let $\mathcal{F}$ be the set of all S-cycles and T be a solution of

size $k + 1$ (which can be obtained via *iterative compression*). Then it is easy to show that $\mathcal{F}$ is T-connected, and the problem of finding an $\mathcal{F}$-transversal is exactly the same as the Directed Subset Feedback Vertex Set problem. It is shown in [2] that a shadowless solution of Directed Subset Feedback Vertex Set can be found in FPT time. Combining with Theorem 2, this gives an FPT algorithm for the Directed Subset Feedback Vertex Set problem. This generalizes the FPT algorithm for Directed Feedback Vertex Set [1].

Open Problems

The two main open problems which fit within the framework of "Finding an $\mathcal{F}$-transversal for some T-connected $\mathcal{F}$" are Directed Multicut and Directed Odd Cycle Transversal. Unfortunately, the structure of shadowless solutions is not yet understood well enough to be able to find them in FPT time.

Directed Multicut
In the Directed Multicut problem, given a directed graph $G = (V, E)$, an integer k, and a set of terminal pairs $T = \{(s_1, t_1), (s_2, t_2), \ldots, (s_p, t_p)\}$, the objective is to find whether there exists a set $X \subseteq V(G)$ of size at most k such that $G \setminus X$ has no $s_i \to t_i$ path for any $1 \le i \le p$. Let $\mathcal{F}$ be the union of set of all $s_i \to t_i$ paths for $1 \le i \le p$. Then it is easy to show that $\mathcal{F}$ is T-connected, and the problem of finding an $\mathcal{F}$-transversal is exactly the same as the Directed Multicut problem. It is known [9] that Directed Multicut parameterized by k is W[1]-hard. However, for the special case of $p = 2$ terminal pairs, the problem can be reduced to Directed Multiway Cut and is hence FPT parameterized by k [3]. The complexity for $p = 3$ parameterized by k is an important open problem. With respect to the bigger parameter $p + k$, the problem is known [5] to be FPT on directed acyclic graphs. However, this algorithm heavily uses the properties of a topological ordering, and the complexity parameterized by $p + k$ on general graphs is another important open problem.

Directed Odd Cycle Transversal

In the Directed Odd Cycle Transversal problem, given a directed graph $G = (V, E)$ and an integer k, the objective is to find whether there exists a set $X \subseteq V(G)$ of size at most k such that $G \setminus X$ has no cycle of odd length. Let $\mathcal{F}$ be the set of all odd cycles in G and T be a solution of size $k + 1$ (which can be obtained via *iterative compression* [12]). Then it is easy to show that $\mathcal{F}$ is T-connected, and the problem of finding an $\mathcal{F}$-transversal is exactly the same as the Directed Odd Cycle Transversal problem. The complexity parameterized by k is open. Moreover, it is known that Directed Odd Cycle Transversal problem generalizes the Directed Feedback Vertex Set problem [1] and the Undirected Odd Cycle Transversal problem [12]. Hence, an FPT algorithm for Directed Odd Cycle Transversal would have to generalize the ideas used to obtain FPT algorithms for these two problems.

Cross-References

▶ Bidimensionality
▶ Undirected Feedback Vertex Set

Recommended Reading

1. Chen J, Liu Y, Lu S, O'Sullivan B, Razgon I (2008) A fixed-parameter algorithm for the directed feedback vertex set problem. In: STOC, Victoria, pp 177–186
2. Chitnis RH, Cygan M, Hajiaghayi MT, Marx D (2012) Directed subset feedback vertex set is fixed-parameter tractable. In: ICALP (1), Warwick, pp 230–241
3. Chitnis RH, Hajiaghayi M, Marx D (2012) Fixed-parameter tractability of directed multiway cut parameterized by the size of the cutset. In: SODA, Kyoto, pp 1713–1725
4. Chitnis RH, Egri L, Marx D (2013) List H-coloring a graph by removing few vertices. In: ESA, Sophia Antipolis, pp 313–324
5. Kratsch S, Pilipczuk M, Pilipczuk M, Wahlström M (2012) Fixed-parameter tractability of multicut in directed acyclic graphs. In: ICALP (1), Warwick, pp 581–593
6. Lokshtanov D, Marx D (2011) Clustering with local restrictions. In: ICALP (1), Zurich, pp 785–797
7. Lokshtanov D, Ramanujan MS (2012) Parameterized tractability of multiway cut with parity constraints. In: ICALP (1), Warwick, pp 750–761
8. Marx D (2006) Parameterized graph separation problems. Theor Comput Sci 351(3):394–406
9. Marx D, Razgon I (2011) Fixed-parameter tractability of multicut parameterized by the size of the cutset. In: STOC, San Jose, pp 469–478
10. Mehlhorn K (1984) Data structures and algorithms 2: graph algorithms and NP-completeness. Springer, Berlin/New York
11. Naor M, Schulman LJ, Srinivasan A (1995) Splitters and near-optimal derandomization. In: FOCS, Milwaukee, pp 182–191
12. Reed BA, Smith K, Vetta A (2004) Finding odd cycle transversals. Oper Res Lett 32(4):299–301

Shortest Elapsed Time First Scheduling

Nikhil Bansal
Eindhoven University of Technology, Eindhoven, The Netherlands

Keywords

Feedback queues; MLF algorithm; Response time; Scheduling with unknown job sizes; Sojourn time

Years and Authors of Summarized Original Work

2003; Bansal, Pruhs

Problem Definition

The problem is concerned with scheduling dynamically arriving jobs in the scenario when the processing requirements of jobs are unknown to the scheduler. The lack of knowledge of how long a job will take to execute is a particularly attractive assumption in real systems where such information might be difficult or impossible to obtain. The goal is to schedule jobs to provide good quality of service to the users. In particular the goal is to design algorithms that have good average performance and are also fair in the sense that no subset of users experiences substantially worse performance than others.

Notations

Let $\mathcal{J} = \{1, 2, \ldots, n\}$ denote the set of jobs in the input instance. Each job j is characterized by its release time r_j and its processing requirement p_j. In the online setting, job j is revealed to the scheduler only at time r_j. A further restriction is the *non-clairvoyant* setting, where only the existence of job j is revealed at r_j, in particular the scheduler does not know p_j until the job meets its processing requirement and leaves the system. Given a schedule, the completion time c_j of a job is the earliest time at which job j receives p_j amount of service. The flow time f_j of j is defined as $c_j - r_j$. The stretch of a job is defined as the ratio of its flow time divided by its size. Stretch is also referred to as normalized flow time or slowdown and is a natural measure of fairness as it measures the waiting time of a job per unit of service received. A schedule is said to be preemptive, if a job can be interrupted arbitrarily, and its execution can be resumed later from the point of interruption without any penalty. It is well known that preemption is necessary to obtain reasonable guarantees for flow time even in the offline setting [6].

Recall that the online shortest remaining processing time (SRPT) algorithm that at any time works on the job with the least remaining processing is optimum for minimizing average flow time. However, a common critique of SRPT is that it may lead to starvation of jobs, where some jobs may be delayed indefinitely. For example, consider the sequence where a job of size 3 arrives at time $t = 0$ and one job of size 1 arrives every unit of time starting $t = 1$ for a long time. Under SRPT, the size 3 job will be delayed until the size 1 jobs stop arriving. On the other hand, if the goal is to minimize the maximum flow time, then it is easily seen that first in first out (FIFO) is the optimum algorithm. However, FIFO can perform very poorly with respect to average flow time (e.g., many small jobs could be stuck behind a very large job that arrived just earlier). A natural way to balance both the average and worst case performance is to consider the ℓ_p norms of flow time and stretch, where the ℓ_p norm of the sequence $x_1, \ldots, x_n$ is defined as $\left(\sum_i x_i^p \right)^{1/p}$.

The shortest elapsed time first (SETF) is a non-clairvoyant algorithm that at any time works on the job that has received the least amount of service thus far. This is a natural way to favor short jobs given the lack of knowledge of job sizes. In fact, SETF is the continuous version of the multilevel feedback (MLF) algorithm. Unfortunately, SETF (or any other deterministic non-clairvoyant algorithm) performs poorly in the framework of competitive analysis, where an algorithm is called c-competitive if for every input instance, its performance is no worse than c times that of the optimum offline (clairvoyant) solution for that instance [7]. However, competitive analysis can be overly pessimistic in its guarantee. A way around this problem was proposed by Kalyanasundaram and Pruhs [5] who allowed the online scheduler a slightly faster processor to make up for its lack of knowledge of future arrivals and job sizes. Formally, an algorithm Alg is said to be s-speed, c-speed competitive where c is the worst case ratio over all instance I, of $\mathrm{Alg}_s(I)/\mathrm{Opt}_1(I)$, where Alg_s is the value of solution produced by Alg when given an s-speed processor, and Opt_1 is the optimum value using a speed 1 processor. Typically the most interesting results are those where c is small and $s = (1 + \epsilon)$ for any arbitrary $\epsilon > 0$.

Key Results

In their seminal paper [5], Kalyanasundaram and Pruhs showed the following.

Theorem 1 ([5]) *SETF is a* $(1 + \epsilon)$-*speed,* $(1 + 1/\epsilon)$-*competitive non-clairvoyant algorithm for minimizing the average flow time on a single machine with preemptions.*

For minimizing the average stretch, Muthukr-ishnan, Rajaraman, Shaheen, and Gehrke [6] considered the clairvoyant setting and showed that SRPT is 2-competitive for a single machine and 14-competitive for multiple machines. The non-clairvoyant setting was consider by Bansal, Dhamdhere, Konemann, and Sinha [7]. They showed that

Theorem 2 ([1]) *SETF is a* $(1 + \epsilon)$*-speed,* $0(\log^2 P)$*-competitive for minimizing average stretch, where* P *is the ratio of the maximum to minimum job size. On the other hand, even with* $O(1)$*-speed, any non-clairvoyant algorithm is at least* $\Omega(\log P)$*-competitive. Interestingly, in terms of* n*, any non-clairvoyant algorithm must be* $\Omega(n)$*-competitive even with* $O(1)$*-speedup. Moreover, SETF is* $O(n)$*-competitive (even without extra speedup). For the special case when all jobs arrive at time 0, SETF is optimum up to constant factors. It is* $O(\log P)$*-competitive (without any extra speedup). Moreover, any non-clairvoyant must be* $\Omega(\log P)$*-competitive even with factor* $O(1)$*-speedup.*

The key idea of the above result was a connection between SETF and SRPT. First, at the expense of $(1 + \epsilon)$-speedup, it can be seen that SETF is no worse than MLF where the thresholds are powers of $(1 + \epsilon)$. Second, the behavior of MLF on an instance I can be related to the behavior of shortest job first (SJF) algorithm on another instance I' that is obtained from/by dividing each job into logarithmically many jobs with geometrically increasing sizes. Finally, the performance of SJF is related to SRPT using another $(1 + \epsilon)$ factor speedup.

Bansal and Pruhs [2] considered the problem of minimizing the ℓ_p norms of flow time and stretch on a single machine. They showed the following.

Theorem 3 ([2]) *In the clairvoyant setting, SRPT and SJF are* $(1 + \epsilon)$*-speed,* $O(1/\epsilon)$*-competitive for minimizing the* ℓ_p *norms of both flow time and stretch. On the other hand, for* $1 < p < \infty$*, no online algorithm (possibly clairvoyant) can be* $O(1)$*-competitive for minimizing* ℓ_p *norms of stretch or flow time*

without speedup. In particular, any randomized online algorithm is at least $\Omega(n^{(p-1)/3p^2})$*-competitive for* ℓ_p *norms of stretch and is at least* $\Omega(n^{(p-1)/p(3p-1)})$*-competitive for* ℓ_p *norms of flow time.*

The above lower bounds are somewhat surprising, since SRPT and FIFO are optimum for the case $p = 1$ and $p = \infty$ for flow time.

Bansal and Pruhs [2] also consider the non-clairvoyant case.

Theorem 4 ([2]) *In the non-clairvoyant setting, SETF is* $(1 + \epsilon)$*-speed,* $O(1/\epsilon^{2+2/P})$*-competitive for minimizing the* ℓ_p *norms of flow time. For minimizing* ℓ_p *norms of stretch, SETF is* $(1 + \epsilon)$*-speed,* $O(1/\epsilon^{3+1/p} \cdot \log^{1+1/p} P)$*-competitive*

Finally, Bansal and Pruhs also consider round robin (RR) or processor sharing that at any time splits the processor equally among the unfinished jobs. RR is considered to be an ideal fair strategy since it treats all unfinished jobs equally. However, they show that

Theorem 5 *For any* $p \geq 1$*, there is an* $\epsilon > 0$ *such that even with a* $(1 + \epsilon)$ *times faster processor, RR is not* $n^{o(1)}$*-competitive for minimizing the* ℓ_p *norms of flow time. In particular, for* $\epsilon < 1/2p$*, RR is* $(1 + \epsilon)$*-speed,* $\Omega(n^{(1-2\epsilon p)/p})$*-competitive. For* ℓ_p *norms of stretch, RR is* $\Omega(n)$*-competitive as is in fact any randomized non-clairvoyant algorithm.*

The results above have been extended in a couple of directions. Bansal and Pruhs [3] extend these results to *weighted* ℓ_p norms of flow time and stretch. Chekuri, Khanna, Kumar, and Goel [4] have extended these results to the multiple machines case. Their algorithms are particularly elegant: Each job is assigned to some machine at random, and all jobs at a particular machine are processed using SRPT or SETF (as applicable).

Applications

SETF and its variants such as MLF are widely used in operating systems [9,10]. Note that SETF is not really practical since each job could be preempted infinitely often. However, variants of SETF with fewer preemptions are quite popular.

Open Problems

It would be interesting to explore other notions of fairness in the dynamic scheduling setting. In particular, it would be interesting to consider algorithms that are both fair and have a good average performance.

An immediate open problem is whether the gap between $O(\log^2 P)$ and $\Omega(\log P)$ can be closed for minimizing the average stretch in the non-clairvoyant setting.

Cross-References

▶ Flow Time Minimization
▶ Minimum Flow Time
▶ Multilevel Feedback Queues

Recommended Reading

1. Bansal N, Dhamdhere K, Konemann J, Sinha A (2004) Non-clairvoyant scheduling for minimizing mean slowdown. Algorithmica 40(4):305–318
2. Bansal N, Pruhs K (2003) Server scheduling in the Lp norm: a rising tide lifts all boat. In: Symposium on theory of computing (STOC), San Diego, pp 242–250
3. Bansal N, Pruhs K (2004) Server scheduling in the weighted Lp norm. In: LATIN, Buenos Aires, pp 434–443
4. Chekuri C, Goel A, Khanna S, Kumar A (2004) Multi-processor scheduling to minimize flow time with epsilon resource augmentation. In: Symposium on theory of computing (STOC), Chicago, pp 363–372
5. Kalyanasundaram B, Pruhs K (2000) Speed is as powerful as clairvoyance. J ACM 47(4):617–643
6. Kellerer H, Tautenhahn T, Woeginger GJ (1999) Approximability and nonapproximability results for minimizing total flow time on a single machine. SIAM J Comput 28(4):1155–1166
7. Motwani R, Phillips S, Torng E (1994) Non-clairvoyant scheduling. Theor Comput Sci 130(1):17–47
8. Muthukrishnan S, Rajaraman R, Shaheen A, Gehrke J (2004) Online scheduling to minimize average stretch. SIAM J Comput 34(2):433–452
9. Nutt G (1999) Operating system projects using Windows NT. Addison Wesley, Reading
10. Tanenbaum AS (1992) Modern operating systems. Prentice-Hall, Englewood Cliffs

Shortest Paths Approaches for Timetable Information

Riko Jacob
Institute of Computer Science,
Technical University of Munich, Munich,
Germany
IT University of Copenhagen, Copenhagen,
Denmark

Keywords

Journey planner; Passenger information system; Timetable lookup; Trip planner

Years and Authors of Summarized Original Work

2004; Pyrga, Schulz, Wagner, Zaroliagis

Problem Definition

Consider the route-planning task for passengers of scheduled public transportation. Here, the running example is that of a train system, but the discussion applies equally to bus, light-rail and similar systems. More precisely, the task is to construct a timetable information system that, based upon the detailed schedules of all trains, provides passengers with good itineraries, including the transfer between different trains.

Solutions to this problem consist of a model of the situation (e.g., can queries specify a limit on the number of transfers?), an algorithmic approach, its mathematical analysis (does it always return the best solution? Is it guaranteed to work fast in all settings?), and an evaluation in the real world (Can travelers actually use the produced itineraries? Is an implementation fast enough on current computers and real data?).

Key Results

The problem is discussed in detail in a recent survey article [6].

Modeling

In a simplistic model, it is assumed that a transfer between trains does not take time. A more realistic model specifies a certain minimum transfer time per station. Furthermore, the objective of the optimization problem needs to be defined. Should the itinerary be as fast as possible, or as cheap as possible, or induce the least possible transfers? There are different ways to resolve this as surveyed in [6], all originating in multi-objective optimization, like resource constraints or Pareto-optimal solutions. From a practical point of view, the preferences of a traveler are usually difficult to model mathematically, and one might want to let the user choose the best option among a set of reasonable itineraries himself. For example, one can compute all itineraries that are not inferior to some other itinerary in all considered aspects. As it turns out, in real timetables the number of such itineraries is not too big, such that this approach is computationally feasible and useful for the traveler [5]. Additionally, the fare structure of most railways is fairly complicated [4], mainly because fares usually are not additive, i.e., are not the sum of fares of the parts of a trip.

Algorithmic Models

The current literature establishes two main ideas how to transform the situation into a shortest path problem on a graph. As an example, consider the simplistic modeling where transfer takes no time, and where queries specify starting time and station to ask for an itinerary that achieves the earliest arrival time at the destination.

In the time-expanded model [11], every arrival and departure event of the timetable is a vertex of the directed graph. The arcs of the graph represent consecutive events at one station, and direct train connections. The length of an arc is given by the time difference of its end vertices. Let s be the vertex at the source station whose time is directly after the starting time. Now, a shortest path from s to any vertex of the destination station is an optimal itinerary.

In the time-dependent model [3, 7, 9, 10], the vertices model stations, and the arcs stand for the existence of a direct (non-stop) train connection. Instead of edge length, the arcs are labeled with edge-traversal functions that give the arrival time at the end of the arc in dependence on the time a passenger starts at the beginning of the arc, reflecting the times when trains actually run. To solve this time-dependent shortest path problem, a modification of Dijkstra's algorithm can be used. Further exploiting the structure of this situation, the graph can be represented in a way that allows constant time evaluation of the link traversal functions [3]. To cope with more realistic transfer models, a more complicated graph can be used.

Additionally, many of the speed-up techniques for shortest path computations can be applied to the resulting graph queries.

Applications

The main application are timetable information systems for scheduled transit (buses, trains, etc.). This extends to route planning where trips in such systems are allowed, as for example in the setting of fine-grained traffic simulation to compute fastest itineraries [2].

Open Problems

Improve computation speed, in particular for fully integrated timetables and the multi-criteria case. Extend the problem to the dynamic case, where the current real situation is reflected, i.e., delayed or canceled trains, and otherwise temporarily changed timetables are reflected.

Experimental Results

In the cited literature, experimental results usually are part of the contribution [2, 4, 5, 6, 7, 8, 9, 10, 11]. The time-dependent approach can be significantly faster than the time-expanded approach. In particular for the simplistic models speed-ups in the range 10–45 are observed [8, 10]. For more detailed models, the performance of the two approaches becomes comparable [6].

Cross-References

▶ Implementation Challenge for Shortest Paths
▶ Routing in Road Networks with Transit Nodes
▶ Single-Source Shortest Paths

Acknowledgments I want to thank Matthias Müller-Hannemann, Dorothea Wagner, and Christos Zaroliagis for helpful comments on an earlier draft of this entry.

Recommended Reading

1. Gerards B, Marchetti-Spaccamela A (eds) (2004) Proceedings of the 3rd workshop on algorithmic methods and models for optimization of railways (ATMOS'03) 2003. Electronic notes in theoretical computer science, vol 92. Elsevier
2. Barrett CL, Bisset K, Jacob R, Konjevod G, Marathe MV (2002) Classical and contemporary shortest path problems in road networks: implementation and experimental analysis of the TRANSIMS router. In: Algorithms – ESA 2002: 10th annual European symposium, Rome, 17–21 Sept 2002. Lecture notes computer science, vol 2461. Springer, Berlin, pp 126–138
3. Brodal GS, Jacob R (2003) Time-dependent networks as models to achieve fast exact time-table queries. In: Proceedings of the 3rd workshop on algorithmic methods and models for optimization of railways (ATMOS'03), [1], pp 3–15
4. Müller-Hannemann M, Schnee M (2006) Paying less for train connections with MOTIS. In: Kroon LG, Möhring RH (eds) Proceedings of the 5th workshop on algorithmic methods and models for optimization of railways (ATMOS'05), Dagstuhl, Internationales Begegnungs- und Forschungszentrum fuer Informatik (IBFI), Schloss Dagstuhl. Dagstuhl Seminar Proceedings, no. 06901
5. Müller-Hannemann M, Schnee M (2007) Finding all attractive train connections by multi-criteria pareto search. In: Geraets F, Kroon LG, Schöbel A, Wagner D, Zaroliagis CD (eds) Algorithmic methods for railway optimization, international Dagstuhl workshop, Dagstuhl Castle, 20–25 June 2004, 4th international workshop, ATMOS 2004, Bergen, 16–17 Sept 2004, revised selected papers. Lecture notes in computer science, vol 4359. Springer, Berlin, pp 246–263
6. Müller-Hannemann M, Schulz F, Wagner D, Zaroliagis CD (2007) Timetable information: models and algorithms. In: Geraets F, Kroon LG, Schöbel A, Wagner D, Zaroliagis CD (eds) Algorithmic methods for railway optimization, international Dagstuhl workshop, Dagstuhl Castle, 20–25 June 2004, 4th International Workshop, ATMOS 2004, Bergen, 16–17 Sept 2004, revised selected papers. Lecture notes in computer science, vol 4359. Springer, pp 67–90
7. Nachtigall K (1995) Time depending shortest-path problems with applications to railway networks. Eur J Oper Res 83:154–166
8. Pyrga E, Schulz F, Wagner D, Zaroliagis C (2004) Experimental comparison of shortest path approaches for timetable information. In: Proceedings 6th workshop on algorithm engineering and experiments (ALENEX). Society for Industrial and Applied Mathematics, pp 88–99
9. Pyrga E, Schulz F, Wagner D, Zaroliagis C (2003) Towards realistic modeling of time-table information through the time-dependent approach. In: Proceedings of the 3rd workshop on algorithmic methods and models for optimization of railways (ATMOS'03), [1], pp 85–103
10. Pyrga E, Schulz F, Wagner D, Zaroliagis C (2007) Efficient models for timetable information in public transportation systems. J Exp Algorithm 12:2.4
11. Schulz F, Wagner D, Weihe K (2000) Dijkstra's algorithm on-line: an empirical case study from public railroad transport. J Exp Algorithm 5:1–23

Shortest Paths in Planar Graphs with Negative Weight Edges

Jittat Fakcharoenphol[1] and Satish Rao[2]
[1]Department of Computer Engineering, Kasetsart University, Bangkok, Thailand
[2]Department of Computer Science, University of California, Berkeley, CA, USA

Keywords

Shortest paths in planar graphs with arbitrary arc weights; Shortest paths in planar graphs with general arc weights

Years and Authors of Summarized Original Work

2001; Fakcharoenphol, Rao

Problem Definition

This problem is to find shortest paths in planar graphs with general edge weights. It is known that shortest paths exist only in graphs that contain no negative weight cycles. Therefore, algorithms that work in this case must deal with the presence of negative cycles, i.e., they must be able to detect negative cycles.

In general graphs, the best known algorithm, the Bellman-Ford algorithm, runs in time $O(mn)$ on graphs with n nodes and m edges, while algorithms on graphs with no negative weight edges run much faster. For example, Dijkstra's algorithm implemented with the Fibonacchi heap runs in time $O(m + n \log n)$, and, in case of integer weights Thorup's algorithm runs in linear time. Goldberg [5] also presented an $O(m \sqrt{n} \log L)$-time algorithm where L denotes the absolute value of the most negative edge weights. Note that his algorithm is weakly polynomial.

Notations

Given a directed graph $G = (V, E)$ and a weight function $w : E \rightarrow \mathbb{R}$ on its directed edges, a *distance labeling* for a source node s is a function $d : V \rightarrow \mathbb{R}$ such that $d(v)$ is the minimum length over all s-to-v paths, where the *length of path P* is $\sum_{e \in P} w(e)$.

Problem 1 (Single-Source-Shortest-Path)
INPUT: A directed graph $G = (V, E)$, weight function $w : E \rightarrow \mathbb{R}$, source node $s \in V$.
OUTPUT: If G does not contain negative length cycles, output a distance labeling d for source node s. Otherwise, report that the graph contains some negative length cycle.

The algorithm by Fakcharoenphol and Rao [4] deals with the case when G is planar. They gave an $O(n \log^3 n)$-time algorithm, improving on an $O(n^{3/2})$-time algorithm by Lipton, Rose, and Tarjan [9] and an $O(n^{4/3} \log nL)$-time algorithm by Henzinger, Klein, Rao, and Subramanian [6].

Their algorithm, as in all previous algorithms, uses a recursive decomposition and constructs a data structure called a dense distance graph, which shall be defined next.

A *decomposition* of a graph is a set of subsets $P_1, P_2, \ldots, P_k$ (not necessarily disjoint) such that the union of all the sets is V and for all $e = (u, v) \in E$, there is a unique P_i that contains e. A node v is a *border node* of a set P_i if $v \in P_i$ and there exists an edge $e = (v, x)$ where $x \notin P_i$. The subgraph induced on a subset P_i is referred to as a *piece* of the decomposition.

The algorithm works with a *recursive decomposition* where at each level, a piece with n nodes and r border nodes is divided into two subpieces such that each subpiece has no more than $2n/3$ nodes and at most $2r/3 + c \sqrt{n}$ border nodes, for some constant c. In this recursive context, a border node of a subpiece is defined to be any border node of the original piece or any new border node introduced by the decomposition of the current piece.

With this recursive decomposition, the *level of a decomposition* can be defined in the natural way, with the entire graph being the only piece in the level 0 decomposition, the pieces of the decomposition of the entire graph being the level 1 pieces in the decomposition, and so on.

For each piece of the decomposition, the all-pair shortest path distances between all its border nodes along paths that lie entirely inside the piece are recursively computed. These all-pair distances form the edge set of a non-planar graph representing shortest paths between border nodes. The dense distance graph of the planar graph is the union of these graphs over all the levels.

Using the dense distance graph, the shortest distance queries between pairs of nodes can be answered.

Problem 2 (Shortest-Path-Distance-Data-Structure)
INPUT: A directed graph $G = (V, E)$, weight function $w : E \rightarrow \mathbb{R}$, source node $s \in V$.
OUTPUT: If G does not contain negative length cycles, output a data structure that support distance queries between pairs of nodes. Otherwise, report that the graph contains some negative length cycle.

The algorithm of Fakcharoenphol and Rao relies heavily on planarity, i.e., it exploits properties regarding how shortest paths on each piece intersect. Therefore, unlike previous algorithms that require only that the graph can be recursively decomposed with small numbers of border nodes [10], their algorithm also requires that each piece has a nice embedding.

Given an embedding of the piece, a *hole* is a bounded face where all adjacent nodes are border nodes. Ideally, one would hope that there is a planar embedding of any piece in the recursive decomposition where all the border nodes are on a single face and are circularly ordered, i.e., there is no holes in each piece. Although this is not always true, the algorithm works with any decomposition with a constant number of holes in each piece. This decomposition can be found in $O(n \log n)$ time using the simple cycle separator algorithm by Miller [12].

Key Results

Theorem 1 *Given a recursive decomposition of a planar graph such that each piece of the decomposition contains at most a constant number of holes, there is an algorithm that constructs the dense distance graph is $O(n \log^3 n)$ time.*

Given the procedure that constructs the dense distance graph, the shortest paths from a source s can be computed by first adding s as a border node in every piece of the decomposition, computing the dense distance graph, and then extending the distances into all internal nodes on every piece. This can be done in time $O(n \log^3 n)$.

Theorem 2 *The single-source shortest path problem for an n-node planar graph with negative weight edges can be solved in time $O(n \log^3 n)$.*

The dense distance graph can be used to answer distance queries between pairs of nodes.

Theorem 3 *Given the dense distance graph, the shortest distance between any pair of nodes can be found in $O(\sqrt{n} \log^2 n)$ time.*

It can also be used as a dynamic data structure that answers shortest path queries and allows edge cost updates.

Theorem 4 *For planar graphs with only non-negative weight edges, there is a dynamic data structure that supports distance queries and update operations that change edge weights in amortized $O(n^{2/3} \log^{7/3} n)$ time per operation. For planar graph with negative weight edges, there is a dynamic data structures that supports the same set of operations in amortized $O(n^{4/5} \log^{13/5} n)$ time per operation.*

Note that the dynamic data structure does not support edge insertions and deletions, since these operations might destroy the recursive decomposition.

Applications

The shortest path problem has long been studied and continues to find applications in diverse areas. There are a many problems that reduce to the shortest path problem where negative weight edges are required, for example the minimum-mean length directed circuit. For planar graphs, the problem has wide application even when the underlying graph is a grid. For example, there are recent image segmentation approaches that use negative cycle detection [2, 3]. Some of other applications for planar graphs include separator algorithms [13] and multi-source multi-sink flow algorithms [11].

Open Problems

Klein [8] gives a technique that improves the running time of the construction of the dense distance graph to $O(n \log^2 n)$ when all edge weights are non-negative; this also reduces the amortized running time for the dynamic case down to $O(n^{2/3} \log^{5/3} n)$. Also, for planar graphs with no negative weight edges, Cabello [1] gives a faster algorithm for computing the shortest distances between k pairs of nodes. However, the problem for improving the bound of $O(n \log^3 n)$ for finding shortest paths in planar graphs with general edge weights remains opened.

It is not known how to handle edge insertions and deletions in the dynamic data structure. A new data structure might be needed instead of the dense distance graph, because the dense distance graph is determined by the decomposition.

Cross-References

▶ All Pairs Shortest Paths in Sparse Graphs
▶ All Pairs Shortest Paths via Matrix Multiplication
▶ Approximation Schemes for Planar Graph Problems
▶ Decremental All-Pairs Shortest Paths
▶ Fully Dynamic All Pairs Shortest Paths
▶ Implementation Challenge for Shortest Paths
▶ Negative Cycles in Weighted Digraphs
▶ Planarity Testing
▶ Shortest Paths Approaches for Timetable Information
▶ Single-Source Shortest Paths

Recommended Reading

1. Cabello S (2006) Many distances in planar graphs. In: SODA '06: proceedings of the seventeenth annual ACM-SIAM symposium on discrete algorithm. ACM, New York, pp 1213–1220
2. Cox IJ, Rao SB, Zhong Y (1996) 'Ratio Regions': a technique for image segmentation. In: Proceedings international conference on pattern recognition, Aug 1996. IEEE, pp 557–564
3. Geiger LCD, Gupta A, Vlontzos J (1995) Dynamic programming for detecting, tracking and matching elastic contours. IEEE Trans Pattern Anal Mach Intell
4. Fakcharoenphol J, Rao S (2006) Planar graphs, negative weight edges, shortest paths, and near linear time. J Comput Syst Sci 72:868–889
5. Goldberg AV (1992) Scaling algorithms for the shortest path problem. SIAM J Comput 21:140–150
6. Henzinger MR, Klein PN, Rao S, Subramanian S (1997) Faster shortest-path algorithms for planar graphs. J Comput Syst Sci 55:3–23
7. Johnson D (1977) Efficient algorithms for shortest paths in sparse networks. J Assoc Comput Mach 24:1–13
8. Klein PN (2005) Multiple-source shortest paths in planar graphs. In: Proceedings of the 16th ACM-SIAM symposium on discrete algorithms, pp 146–155
9. Lipton R, Rose D, Tarjan RE (1979) Generalized nested dissection. SIAM J Numer Anal 16:346–358
10. Lipton RJ, Tarjan RE (1979) A separator theorem for planar graphs. SIAM J Appl Math 36:177–189
11. Miller G, Naor J (1995) Flow in planar graphs with multiple sources and sinks. SIAM J Comput 24:1002–1017
12. Miller GL (1986) Finding small simple cycle separators for 2-connected planar graphs. J Comput Syst Sci 32:265–279
13. Rao SB (1992) Faster algorithms for finding small edge cuts in planar graphs (extended abstract). In: Proceedings of the twenty-fourth annual ACM symposium on the theory of computing, May 1992, pp 229–240
14. Thorup M (2004) Compact oracles for reachability and approximate distances in planar digraphs. J ACM 51:993–1024

Shortest Vector Problem

Daniele Micciancio
Department of Computer Science, University of California, San Diego, La Jolla, CA, USA

Keywords

Closest vector problem; Lattice basis reduction; LLL algorithm; Nearest vector problem; Minimum distance problem

Years and Authors of Summarized Original Work

1982; Lenstra, Lenstra, Lovasz

Problem Definition

A *point lattice* is the set of all integer linear combinations

$$\mathcal{L}(\mathbf{b}_1, \ldots, \mathbf{b}_n) = \left\{ \sum_{i=1}^{n} x_i \mathbf{b}_i : x_1, \ldots, x_n \in \mathbb{Z} \right\}$$

of n linearly independent vectors $\mathbf{b}_1, \ldots, \mathbf{b}_n \in \mathbb{R}^m$ in m-dimensional Euclidean space. For computational purposes, the lattice vectors $\mathbf{b}_1, \ldots, \mathbf{b}_n$ are often assumed to have integer (or rational) entries, so that the lattice can be represented by an integer matrix $\mathbf{B} = [\mathbf{b}_1, \ldots, \mathbf{b}_n] \in \mathbb{Z}^{m \times n}$ (called *basis*) having the generating vectors as columns. Using matrix notation, lattice points in $\mathcal{L}(\mathbf{B})$ can be conveniently represented as $\mathbf{Bx}$ where $\mathbf{x}$ is an integer vector. The integers m and n are called the *dimension* and *rank* of the lattice respectively. Notice that any lattice admits multiple bases, but they all have the same rank and dimension.

The main computational problems on lattices are the *Shortest Vector Problem*, which asks to find the shortest nonzero vector in a given lattice, and the *Closest Vector Problem*, which asks to find the lattice point closest to a given target. Both problems can be defined with respect to any norm, but the Euclidean norm $\|\mathbf{v}\| = \sqrt{\sum_i v_i^2}$ is the most common. Other norms typically found in computer science applications are the ℓ_1 norm $\|\mathbf{v}\|_1 = \sum_i |v_i|$ and the *max* norm $\|\mathbf{v}\|_\infty = \max_i |v_i|$. This entry focuses on the Euclidean norm.

Since no efficient algorithm is known to solve SVP and CVP exactly in arbitrary high dimension, the problems are usually defined in their approximation version, where the approximation factor $\gamma \geq 1$ can be a function of the dimension or rank of the lattice.

Definition 1 (Shortest Vector Problem, SVP_γ) Given a lattice $\mathcal{L}(\mathbf{B})$, find a nonzero lattice vector $\mathbf{Bx}$ (where $\mathbf{x} \in \mathbb{Z}^n \setminus \{\mathbf{0}\}$) such that $\|\mathbf{Bx}\| \leq \gamma \cdot \|\mathbf{By}\|$ for any $\mathbf{y} \in \mathbb{Z}^n \setminus \{\mathbf{0}\}$.

Definition 2 (Closest Vector Problem, CVP_γ) Given a lattice $\mathcal{L}(\mathbf{B})$ and a target point $\mathbf{t}$, find a lattice vector $\mathbf{Bx}$ (where $\mathbf{x} \in \mathbb{Z}^n$) such that $\|\mathbf{Bx} - \mathbf{t}\| \leq \gamma \cdot \|\mathbf{By} - \mathbf{t}\|$ for any $\mathbf{y} \in \mathbb{Z}^n$.

Lattices have been investigated by mathematicians for centuries in the equivalent language of quadratic forms, and are the main object of study in the *geometry of numbers*, a field initiated by Minkowski as a bridge between geometry and number theory. For a mathematical introduction to lattices see [3]. The reader is referred to [6, 12] for an introduction to lattices with an emphasis on computational and algorithmic issues.

Key Results

The problem of finding an efficient (polynomial time) solution to SVP_γ for lattices in arbitrary dimension was first solved by the celebrated *lattice reduction* algorithm of Lenstra, Lenstra and Lovász [11], commonly known as the *LLL* algorithm.

Theorem 1 *There is a polynomial time algorithm to solve* SVP_γ *for* $\gamma = (2/\sqrt{3})^n$, *where* n *is the rank of the input lattice.*

The LLL algorithm achieves more than just finding a relatively short lattice vector: it finds a so-called *reduced basis* for the input lattice, i.e., an entire basis of relatively short lattice vectors. Shortly after the discovery of the LLL algorithm, Babai [2] showed that reduced bases can be used to efficiently solve CVP_γ as well within similar approximation factors.

Corollary 1 There is a polynomial time algorithm to solve CVP_γ for $\gamma = O(2/\sqrt{3})^n$, where n is the rank of the input lattice.

The reader is referred to the original papers [2, 11] and [12, chap. 2] for details. Introductory presentations of the LLL algorithm can also be found in many other texts, e.g., [5, chap. 16] and [15, chap. 27]. It is interesting to note that CVP is at least as hard as SVP (see [12, chap 2]) in the sense that any algorithm that solves CVP_γ can be efficiently adapted to solve SVP_γ within the same approximation factor.

Both SVP_γ and CVP_γ are known to be NP-hard in their exact ($\gamma = 1$) or even approximate versions for small values of γ, e.g., constant γ independent of the dimension. (See [13, chaps. 3 and 4] and [4, 10] for the most recent results.) So, no efficient algorithm is likely to exist to solve the problems exactly in arbitrary dimension. For any fixed dimension n, both SVP and CVP can be solved exactly in polynomial time using an algorithm of Kannan [9]. However, the dependency of the running time on the lattice dimension is $n^{O(n)}$. Using randomization, exact SVP can be solved probabilistically in $2^{O(n)}$ time and space using the *sieving* algorithm of Ajtai, Kumar and Sivakumar [1].

As for approximate solutions, the LLL lattice reduction algorithm has been improved both in terms of running time and approximation guarantee. (See [14] and references therein.) Currently, the best (randomized) polynomial time approximation algorithm achieves approximation factor $\gamma = 2^{O(n \log \log n / \log n)}$.

Applications

Despite the large (exponential in n) approximation factor, the LLL algorithm has found numerous applications and lead to the solution of many algorithmic problems in computer science. The number and variety of applications is too large to give a comprehensive list. Some of the most representative applications in different areas of computer science are mentioned below.

The first motivating applications of lattice basis reduction were the solution of integer programs with a fixed number of variables and the factorization of polynomials with rationals coefficients. (See [11, 8], and [15, chap. 16].) Other classic applications are the solution of random instances of low-density subset-sum problems, breaking (truncated) linear congruential pseudorandom generators, simultaneous Diophantine approximation, and the disproof of Mertens' conjecture. (See [8] and [5, chap. 17].)

More recently, lattice basis reduction has been extensively used to solve many problems in cryptanalysis and coding theory, including breaking several variants of the RSA cryptosystem and the DSA digital signature algorithm, finding small solutions to modular equations, and list decoding of CRT (Chinese Reminder Theorem) codes. The reader is referred to [7, 13] for a survey of recent applications, mostly in the area of cryptanalysis.

One last class of applications of lattice problems is the design of cryptographic functions (e.g., collision resistant hash functions, public key encryption schemes, etc.) based on the appar-

ent intractability of solving SVP_γ within small approximation factors. The reader is referred to [12, chap. 8] and [13] for a survey of such applications, and further pointers to relevant literature. One distinguishing feature of many such lattice based cryptographic functions is that they can be proved to be hard to break *on the average*, based on a *worst-case* intractability assumption about the underlying lattice problem.

Open Problems

The main open problems in the computational study of lattices is to determine the complexity of approximate SVP_γ and CVP_γ for approximation factors $\gamma = n^c$ polynomial in the rank of the lattice. Specifically,

- Are there polynomial time algorithm that solve SVP_γ or CVP_γ for polynomial factors $\gamma = n^c$? (Finding such algorithms even for very large exponent c would be a major breakthrough in computer science.)
- Is there an $\epsilon > 0$ such that approximating SVP_γ or CVP_γ to within $\gamma = n^\epsilon$ is NP-hard? (The strongest known inapproximability results [4] are for factors of the form $n^{O(1/\log \log n)}$ which grow faster than any poly-logarithmic function, but slower than any polynomial.)

There is theoretical evidence that for large polynomials factors $\gamma = n^c$, SVP_γ and CVP_γ are not NP-hard. Specifically, both problems belong to complexity class coAM for approximation factor $\gamma = O(\sqrt{n / \log n})$. (See [12, chap. 9].) So, the problems cannot be NP-hard within such factors unless the polynomial hierarchy PH collapses.

URL to Code

The LLL lattice reduction algorithm is implemented in most library and packages for computational algebra, e.g.,

- GAP (http://www.gap-system.org)
- LiDIA (http://www.cdc.informatik.tu-darmstadt. de/TI/LiDIA/)
- Magma   (http://magma.maths.usyd.edu.au/ magma/)
- Maple (http://www.maplesoft.com/)
- Mathematica   (http://www.wolfram.com/ products/mathematica/index.html)
- NTL (http://shoup.net/ntl/).

NTL also includes an implementation of Block Korkine-Zolotarev reduction that has been extensively used for cryptanalysis applications.

Cross-References

▶ Cryptographic Hardness of Learning

▶ Knapsack

▶ Learning Heavy Fourier Coefficients of Boolean Functions

▶ Quantum Algorithm for the Discrete Logarithm Problem

▶ Quantum Algorithm for Factoring

▶ Sphere Packing Problem

Recommended Reading

1. Ajtai M, Kumar R, Sivakumar D (2001) A sieve algorithm for the shortest lattice vector problem. In: Proceedings of the thirty-third annual ACM symposium on theory of computing – STOC 2001, Heraklion, July 2001. ACM, New York, pp 266–275
2. Babai L (1986) On Lovasz' lattice reduction and the nearest lattice point problem. Combinatorica 6(1):1–13, Preliminary version in STACS 1985
3. Cassels JWS (1971) An introduction to the geometry of numbers. Springer, New York
4. Dinur I, Kindler G, Raz R, Safra S (2003) Approximating CVP to within almost-polynomial factors is NP-hard. Combinatorica 23(2):205–243, Preliminary version in FOCS 1998
5. von zur Gathen J, Gerhard J (2003) Modern computer algebra, 2nd edn. Cambridge
6. Grotschel M, Lovász L, Schrijver A (1993) Geometric algorithms and combinatorial optimization. Algorithms and combinatorics, vol 2, 2nd edn. Springer
7. Joux A, Stern J (1998) Lattice reduction: a toolbox for the cryptanalyst. J Cryptol 11(3):161–185
8. Kannan R (1987) Algorithmic geometry of numbers. In: Annual reviews of computer science, vol 2. Annual Review, Palo Alto, pp 231–267
9. Kannan R (1987) Minkowski's convex body theorem and integer programming. Math Oper Res 12(3):415–440
10. Khot S (2005) Hardness of approximating the shortest vector problem in lattices. J ACM 52(5):789–808, Preliminary version in FOCS 2004
11. Lenstra AK, Lenstra HW Jr, Lovász L (1982) Factoring polynomials with rational coefficients. Math Ann 261:513–534
12. Micciancio D, Goldwasser S (2002) Complexity of lattice problems: a cryptographic perspective, vol 671, The Kluwer international series in engineering and computer science. Kluwer Academic, Boston
13. Nguyen P, Stern J (2001) The two faces of lattices in cryptology. In: Silverman J (ed) Cryptography and lattices conference – CaLC 2001, Providence, Mar 2001. Lecture notes in computer science, vol 2146. Springer, Berlin, pp 146–180
14. Schnorr CP (2006) Fast LLL-type lattice reduction. Inf Comput 204(1):1–25
15. Vazirani VV (2001) Approximation algorithms. Springer

Similarity Between Compressed Strings

Jin Wook Kim[1], Amihood Amir[2,5], Gad M. Landau[3], and Kunsoo Park[4]
[1] HM Research, Seoul, Korea
[2] Department of Computer Science, Bar-Ilan University, Ramat-Gan, Israel
[3] Department of Computer Science, University of Haifa, Haifa, Israel
[4] School of Computer Science and Engineering, Seoul National University, Seoul, Korea
[5] Department of Computer Science, Johns Hopkins University, Baltimore, MD, USA

Keywords

Alignment between compressed strings; Compressed approximate string matching; Similarity between compressed strings

Years and Authors of Summarized Original Work

2005; Kim, Amir, Landau, Park

Similarity Between Compressed Strings, Table 1 Various scoring metrics

Metric	Match	Mismatch	Indel	Indel of k characters
Longest common subsequence	1	0	0	0
Levenshtein distance	0	1	1	k
Weighted edit distance	0	δ	μ	$k\mu$
Affine gap penalty	1	$-\delta$	$-\gamma - \mu$	$-\gamma - k\mu$

Problem Definition

The problem of computing similarity between two strings is concerned with comparing two strings using some scoring metric. There exist various scoring metrics and a popular one is the Levenshtein distance (or edit distance) metric. The standard solution for the Levenshtein distance metric was proposed by Wagner and Fischer [13], which is based on dynamic programming. Other widely used scoring metrics are the longest common subsequence metric, the weighted edit distance metric, and the affine gap penalty metric. The affine gap penalty metric is the most general, and it is a quite complicated metric to deal with. Table 1 shows the differences between the four metrics.

The problem considered in this entry is the similarity between two compressed strings. This problem is concerned with efficiently computing similarity without decompressing two strings. The compressions used for this problem in the literature are run-length encoding and Lempel-Ziv (LZ) compression [14].

Run-Length Encoding

A string S is run-length encoded if it is described as an ordered sequence of pairs (σ, i), often denoted "σ^i", each consisting of an alphabet symbol, σ, and an integer, i. Each pair corresponds to a *run* in S, consisting of i consecutive occurrences of σ. For example, the string $aaabbbbacccccbb$ can be encoded $a^3b^4a^1c^4b^2$ or, equivalently, $(a, 3)(b, 4)(a, 1)(c, 4)(b, 2)$. Let A and B be two strings with lengths n and m, respectively. Let A' and B' be the run-length encoded strings of A and B, and n' and m' be the lengths of A' and B', respectively.

Problem 1

INPUT: Two run-length encoded strings A' and B', a scoring metric d.
OUTPUT: The similarity between A' and B' using d.

LZ Compression

Let X and Y be two strings with length $O(n)$. Let X' and Y' be the LZ compressed strings of X and Y, respectively. Then the lengths of X' and Y' are $O(hn/\log n)$, where $h \leq 1$ is the entropy of strings X and Y.

Problem 2

INPUT: Two LZ compressed strings X' and Y', a scoring metric d.
OUTPUT: The similarity between X' and Y' using d.

Block Computation

To compute similarity between compressed strings efficiently, one can use a block computation method. Dynamic programming tables are divided into submatrices, which are called "*blocks*". For run-length encoded strings, a block is a submatrix made up of two runs – one of A and one of B. For LZ compressed strings, a block is a submatrix made up of two phrases – one phrase from each string. See [5] for more details. Then, blocks are computed from left to right and from top to bottom. For each block, only the bottom row and the rightmost column are computed. Figure 1 shows an example of block computation.

Key Results

The problem of computing similarity of two run-length encoded strings, A' and B', has been

Similarity Between Compressed Strings, Fig. 1 Dynamic programming table for strings $a^r c^p b^t$ and $a^s b^q c^u$ is divided into 9 blocks. For one of the blocks, e.g., B, only the bottom row C and the rightmost column D are computed from E and F

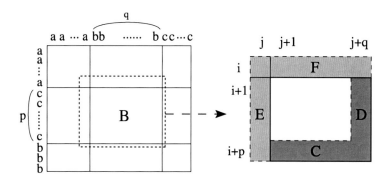

studied for various scoring metrics. Bunke and Csirik [4] presented the first solution to Problem 1 using the longest common subsequence metric. The algorithm is based on block computation of the dynamic programming table.

Theorem 1 (Bunke and Csirik [4]) *A longest common subsequence of run-length encoded strings A' and B' can be computed in $O(nm' + n'm)$ time.*

For the Levenshtein distance metric, Arbell, Landau, and Mitchell [2] and Mäkinen, Navarro, and Ukkonen [10] presented $O(nm' + n'm)$ time algorithms, independently. These algorithms are extensions of the algorithm of Bunke and Csirik.

Theorem 2 (Arbell, Landau, and Mitchell [2], Mäkinen, Navarro, and Ukkonen [10]) *The Levenshtein distance between run-length encoded strings A' and B' can be computed in $O(nm' + n'm)$ time.*

For the weighted edit distance metric, Crochemore, Landau, and Ziv-Ukelson [6] and Mäkinen, Navarro, and Ukkonen [11] gave $O(nm' + n'm)$ time algorithms using techniques completely different from each other. The algorithm of Crochemore, Landau, and Ziv-Ukelson [6] is based on the technique which is used in the LZ compressed pattern matching algorithm [6], and the algorithm of Mäkinen, Navarro, and Ukkonen [11] is an extension of the algorithm for the Levenshtein distance metric.

Theorem 3 (Crochemore, Landau, and Ziv-Ukelson [6] Mäkinen, Navarro, and Ukkonen [11]) *The weighted edit distance between* run-length encoded strings A' and B' can be computed in $O(nm' + n'm)$ time.

For the affine gap penalty metric, Kim, Amir, Landau, and Park [8] gave an $O(nm' + n'm)$ time algorithm. To compute similarity in this metric efficiently, the problem is converted into a path problem on a directed acyclic graph and some properties of maximum paths in this graph are used. It is not necessary to build the graph explicitly since they came up with new recurrences using the properties of the graph.

Theorem 4 (Kim, Amir, Landau, and Park [8]) *The similarity between run-length encoded strings A' and B' in the affine gap penalty metric can be computed in $O(nm' + n'm)$ time.*

The above results show that comparison of run-length encoded strings using the longest common subsequence metric is successfully extended to more general scoring metrics.

For the longest common subsequence metric, there exist improved algorithms. Apostolico, Landau, and Skiena [1] gave an $O(n'm' \log(n'm'))$ time algorithm. This algorithm is based on tracing specific optimal paths.

Theorem 5 (Apostolico, Landau, and Skiena [1]) *A longest common subsequence of run-length encoded strings A' and B' can be computed in $O(n'm' \log(n' + m'))$ time.*

Mitchell [12] obtained an $O((d + n' + m') \log(d + n' + m'))$ time algorithm, where d is the number of matches of compressed characters. This algorithm is based on computing geometric shortest paths using special convex distance functions.

Theorem 6 (Mitchell [12]) *A longest common subsequence of run-length encoded strings A' and B' can be computed in $O((d + n' + m') \log(d + n' + m'))$ time, where d is the number of matches of compressed characters.*

Mäkinen, Navarro, and Ukkonen [11] conjectured an $O(n'm')$ time algorithm on average under the assumption that the lengths of the runs are equally distributed in both strings.

Conjecture 1 (Mäkinen, Navarro, and Ukkonen [11]) A longest common subsequence of run-length encoded strings A' and B' can be computed in $O(n'm')$ time on average.

For Problem 2, Crochemore, Landau, and Ziv-Ukelson [6] presented a solution using the additive gap penalty metric. The additive gap penalty metric consists of 1 for match, $-\delta$ for mismatch, and $-\mu$ for indel, which is almost the same as the weighted edit distance metric.

Theorem 7 (Crochemore, Landau, and Ziv-Ukelson [6]) *The similarity between LZ compressed strings X' and Y' in the additive gap penalty metric can be computed in $O(hn^2 / \log n)$ time, where $h \leq 1$ is the entropy of strings X and Y.*

Applications

Run-length encoding serves as a popular image compression technique, since many classes of images (e.g., binary images in facsimile transmission or for use in optical character recognition) typically contain large patches of identically-valued pixels. Approximate matching on images can be a useful tool to handle distortions. Even a one-dimensional compressed approximate matching algorithm would be useful to speed up two-dimensional approximate matching allowing mismatches and even rotations [3, 7, 9].

Open Problems

The worst-case complexity of the problem is not fully understood. For the longest common subsequence metric, there exist some results whose time complexities are better than $O(nm' + n'm)$ to compute the similarity of two run-length encoded strings [1, 11, 12]. It remains open to extend these results to the Levenshtein distance metric, the weighted edit distance metric and the affine gap penalty metric.

In addition, for the longest common subsequence metric, it is an open problem to prove Conjecture 1.

Cross-References

▶ Approximate String Matching
▶ Local Alignment (with Affine Gap Weights)
▶ Multidimensional Compressed Pattern Matching

Recommended Reading

1. Apostolico A, Landau GM, Skiena S (1999) Matching for run length encoded strings. J Complex 15(1):4–16
2. Arbell O, Landau GM, Mitchell J (2002) Edit distance of run-length encoded strings. Inf Process Lett 83(6):307–314
3. Baeza-Yates R, Navaro G (2000) New models and algorithms for multidimensional approximate pattern matching. J Discret Algorithm 1(1):21–49
4. Bunke H, Csirik H (1995) An improved algorithm for computing the edit distance of run length coded strings. Inf Process Lett 54:93–96
5. Crochemore M, Landau GM, Schieber B, Ziv-Ukelson M (2005) Re-use dynamic programming for sequence alignment: an algorithmic toolkit. In: Iliopoulos CS, Lecroq T (eds) String algorithmics. King's College London Publications, London, pp 19–59
6. Crochemore M, Landau GM, Ziv-Ukelson M (2003) A subquadratic sequence alignment algorithm for unrestricted scoring matrices. SIAM J Comput 32(6):1654–1673
7. Fredriksson K, Navarro G, Ukkonen E (2005) Sequential and indexed two-dimensional combinatorial template matching allowing rotations. Theor Comput Sci 347(1–2):239–275

8. Kim JW, Amir A, Landau GM, Park K (2005) Computing similarity of run-length encoded strings with affine gap penalty. In: Proceedings of the 12th symposium on string processing and information retrieval (SPIRE'05), LNCS, vol 3772, pp 440–449

9. Krithivasan K, Sitalakshmi R (1987) Efficient two-dimensional pattern matching in the presence of errors. Inf Sci 43:169–184

10. Mäkinen V, Navarro G, Ukkonen E (2001) Approximate matching of run-length compressed strings. In: Proceedings of the 12th symposium on combinatorial pattern matching (CPM'01). LNCS, vol 2089, pp 31–49

11. Mäkinen V, Navarro G, Ukkonen E (2003) Approximate matching of run-length compressed strings. Algorithmica 35:347–369

12. Mitchell J (1997) A geometric shortest path problem, with application to computing a longest common subsequence in run-length encoded strings. Technical report, Department of Applied Mathematics, SUNY Stony Brook

13. Wagner RA, Fischer MJ (1974) The string-to-string correction problem. J ACM 21(1):168–173

14. Ziv J, Lempel A (1978) Compression of individual sequences via variable rate coding. IEEE Trans Inf Theory 24(5):530–536

Simple Algorithms for Spanners in Weighted Graphs

Surender Baswana[1] and Sandeep Sen[2]
[1]Department of Computer Science and Engineering, Indian Institute of Technology (IIT), Kanpur, Kanpur, India
[2]Indian Institute of Technology (IIT) Delhi, Hauz Khas, New Delhi, India

Keywords

Graph algorithms; Randomized algorithms; Shortest path; Spanner

Years and Authors of Summarized Original Work

2003; Baswana, Sen

Problem Definition

A spanner is a *sparse* subgraph of a given undirected graph that preserves approximate distance between each pair of vertices. More precisely, a t-spanner of a graph $G = (V, E)$ is a subgraph $(V, E_S), E_S \subseteq E$ such that, for any pair of vertices, their distance in the subgraph is at most t times their distance in the original graph, where t is called the *stretch factor*. The spanners were defined formally by Peleg and Schäffer [15] though the associated notion was used implicitly by Awerbuch [3] in the context of network synchronizers.

Computing t-spanner of smallest size for a given graph is a well-motivated combinatorial problem with many applications. However, computing t-spanner of smallest size for a graph is NP-hard. In fact, for $t > 2$, it is NP-hard [11] even to approximate the smallest size of t-spanner of a graph with ratio $O(2^{(1-\mu)\ln n})$ for any $\mu > 0$. Having realized this fact, researchers have pursued another direction which is quite interesting and useful. Let S_G^t be the size of the sparsest t-spanner of a graph G, and let S_n^t be the maximum value of S_G^t over all possible graphs on n vertices. Does there exist a polynomial time algorithm which computes, for any weighted graph and parameter t, its t-spanner of size $O(S_n^t)$? Such an algorithm would be the best one can hope for given the hardness of the original t-spanner problem. Naturally, the question arises as to how large can S_n^t be? A 43-year-old girth lower bound conjecture by Erdös [13] implies that there are graphs on n vertices whose $2k-$ as well as $(2k-1)$-spanner will require $\Omega(n^{1+1/k})$ edges. This conjecture has been proved for $k = 1, 2, 3$, and 5. Note that a $(2k-1)$-spanner is also a $2k$-spanner, and the lower bound on the size is the same for both $2k$-spanner and $(2k-1)$-spanner. So the objective is to design an algorithm that, for any weighted graph on n vertices, computes a $(2k-1)$-spanner of $O(n^{1+1/k})$ size. Needless to say, one would like to design the fastest algorithm for this problem, and the most ambitious aim would be to achieve the linear time complexity.

Key Results

The key results of this entry are two very simple algorithms which compute a $(2k-1)$-spanner of a given weighted graph $G = (V, E)$. Let n and m

denote, respectively, the number of vertices and edges of G. The first algorithm, due to Althöfer et al. [2], is based on a greedy strategy and runs in $O(mn^{1+1/k})$ time. The second algorithm [6] is based on a very local approach and runs in an expected $O(m)$ time. To start with, consider the following simple observation. Suppose there is a subset $E_S \subset E$ that ensures the following proposition for every edge $(x, y) \in E \backslash E_S$.

$\mathcal{P}_t(x, y)$: the vertices x and y are connected in the subgraph (V, E_S) by a path consisting of at most t edges, and the weight of each edge on this path is not more than that of the edge (x, y).

It follows easily that the subgraph (V, E_S) will be a t-spanner of G. The two algorithms for computing $(2k - 1)$-spanner eventually compute such set E_S based on two completely different approaches.

Algorithm I

This algorithm selects edges for spanner in a greedy fashion and is similar to Kruskal's algorithm for computing a minimum spanning tree. The edges of the graph are processed in the increasing order of their weights. To begin with, the spanner $E_S = \emptyset$; and the algorithm adds edges to it gradually. The decision as to whether an edge, say (u, v), has to be added (or not) to E_S is made as follows:

> If the distance between u and v in the subgraph induced by the current spanner edges E_S is more than $t \cdot weight(u, v)$, then add the edge (u, v) to E_S; otherwise, discard the edge.

It follows that $\mathcal{P}_t(x, y)$ would hold for each edge of E missing in E_S, and so at the end, the subgraph (V, E_S) will be a t-spanner. A well-known result in elementary graph theory states that a graph with more than $n^{1+1/k}$ edges must have a cycle of length at most $2k$. It follows from the above algorithm that the length of any cycle in the subgraph (V, E_S) has to be at least $t + 1$. Hence, for $t = 2k - 1$, the number of edges in the subgraph (V, E_S) will be less than $n^{1+1/k}$. Thus, the algorithm I described above

computes a $(2k - 1)$-spanner of size $O(n^{1+1/k})$, which is indeed optimal based on the lower bound mentioned earlier.

A simple $O(mn^{1+1/k})$ implementation of algorithm I follows based on Dijkstra's algorithm. Cohen [10] and later Thorup and Zwick [19] designed algorithms for $(2k - 1)$-spanner with an improved running time of $O(kmn^{1+1/k})$. These algorithms relied on several calls to Dijkstra's single-source Shortest path algorithm for distance computation and therefore were far from achieving linear time. On the other hand, since a spanner must approximate all-pairs distances in a graph, it appears difficult to compute a spanner by avoiding explicit distance information. Somewhat surprisingly, algorithm II, described in the following section, avoids any sort of distance computation and achieves expected linear time.

Algorithm II

This algorithm employs a novel clustering based on a very local approach and establishes the following result for the spanner problem:

> Given a weighted graph $G = (V, E)$ and an integer $k > 1$, a spanner of $(2k - 1)$ stretch and $O(kn^{1+1/k})$ size can be computed in expected $O(km)$ time.

The algorithm executes in $O(k)$ rounds, and in each round it essentially explores adjacency list of each vertex to prune dispensable edges. As a testimony of its simplicity, we will present the entire algorithm for 3-spanner and its analysis in the following section. The algorithm can be easily adapted in other computational models (parallel, external memory, distributed) with nearly optimal performance (see [6] for more details).

Computing a 3-Spanner in Linear Time

To meet the size constraint of a 3-spanner, a vertex, on an average, contributes $\sqrt{n}$ edges to the spanner. So the vertices with degree $O(\sqrt{n})$ are easy to handle since all their edges can be selected in the spanner. For vertices with higher degree, a clustering (groupings) scheme is employed to tackle this problem which has its basis in *dominating sets*.

To begin with, there is a set of edges E' initialized to E and empty spanner E_S. The algorithm processes the edges E', moves some of them to the spanner E_S, and discards the remaining ones. It does so in the following two phases:

1. *Forming the clusters*

   A sample $\mathcal{R} \subset V$ is chosen by picking each vertex independently with probability $\frac{1}{\sqrt{n}}$. The clusters will be formed around these sampled vertices. Initially, the clusters are $\{\{u\}|u \in \mathcal{R}\}$. Each $u \in \mathcal{R}$ is called the *center* of its cluster. Each unsampled vertex $v \in V - \mathcal{R}$ is processed as follows:

   (a) If v is not adjacent to any sampled vertex, then every edge incident on v is moved to E_S.

   (b) If v is adjacent to one or more sampled vertices, let $\mathcal{N}(v, \mathcal{R})$ be the sampled neighbor that is nearest (Ties can be broken arbitrarily. However, it helps conceptually to assume that all weights are distinct) to v. The edge $(v, \mathcal{N}(v, \mathcal{R}))$ along with every edge that is incident on v with weight less than this edge is moved to E_S. The vertex v is added to the cluster centered at $\mathcal{N}(v, \mathcal{R})$.

   As a last step of the first phase, all those edges (u, v) from E' where u and v are not sampled and belong to the same cluster are discarded.

   Let V' be the set of vertices corresponding to the endpoints of the edges E' left after the first phase. It follows that each vertex from V' is either a sampled vertex or adjacent to some sampled vertex, and step 1(b) has partitioned V' into disjoint clusters each centered around some sampled vertex. Also note that, as a consequence of the last step, each edge of the set E' is an intercluster edge. The graph (V', E'), and the corresponding clustering of V', is passed onto the following (second) phase.

2. *Joining vertices with their neighboring clusters*

   Each vertex v of graph (V', E') is processed as follows. Let $E'(v, c)$ be the edges from the set E' incident on v from a cluster c. For each cluster c neighboring to v, the least-weight edge from $E'(v, c)$ is moved to E_S, and the remaining edges are discarded.

The number of edges added to the spanner E_S during the algorithm described above can be bounded as follows. Note that the sample set $\mathcal{R}$ is formed by picking each vertex randomly independently with probability $\frac{1}{\sqrt{n}}$. It thus follows from elementary probability that for each vertex $v \in V$, the expected number of incident edges with weight less than that of $(v, \mathcal{N}(v, \mathcal{R}))$ is at most $\sqrt{n}$. Thus, the expected number of edges contributed to the spanner by each vertex in the first phase of the algorithm is at most $\sqrt{n}$. The number of edges added to the spanner in the second phase is $O(n|\mathcal{R}|)$. Since the expected size of the sample $\mathcal{R}$ is $\sqrt{n}$, therefore, the expected number of edges added to the spanner in the second phase is at most $n^{3/2}$. Hence, the expected size of the spanner E_S at the end of the algorithm described above is at most $2n^{3/2}$. The algorithm is repeated if the size of the spanner exceeds $3n^{3/2}$. It follows using Markov's inequality that the expected number of such repetitions will be $O(1)$.

We now establish that E_S is a 3-spanner. Note that for every edge $(u, v) \notin E_S$, the vertices u, v belong to some cluster in the first phase. There are two cases now.

Case 1 : *(u and v belong to the same cluster)*

Let u and v belong to the cluster centered at $x \in \mathcal{R}$. It follows from the first phase of the algorithm that there is a 2-edge path $u - x - v$ in the spanner with each edge not heavier than the edge (u, v). (This provides a justification for discarding all intracluster edges at the end of the first phase.)

Case 2 : *(u and v belong to different clusters)*

Clearly, the edge (u, v) was removed from E' during phase 2, and suppose it was removed while processing the vertex u. Let v belong to the cluster centered at $x \in \mathcal{R}$.

In the beginning of the second phase, let $(u, v') \in E'$ be the least-weight edge among all the edges incident on u from the vertices of the cluster centered at x. So it must be that $weight(u, v') \leq weight(u, v)$. The processing of vertex u during the second phase of our algorithm ensures that the edge (u, v') gets added to E_S. Hence, there is a path $\Pi_{uv} = u - v' - x - v$ between u and v in the spanner E_S, and its weight can be bounded as $weight(\Pi_{uv}) = weight(u, v') + weight(v', x) + weight(x, v)$. Since (v', x) and (v, x) were chosen in the first phase, it follows that $weight(v', x) \leq weight(u, v')$ and $weight(x, v) \leq weight(u, v)$. It follows that the spanner (V, E_S) has stretch 3. Moreover, both phases of the algorithm can be executed in $O(m)$ time using elementary data structures and bucket sorting.

The algorithm for computing a $(2k - 1)$-spanner executes k iterations where each iteration is similar to the first phase of the 3-spanner algorithm. For details and formal proofs, the reader may refer to [6].

Other Related Works

The notion of a spanner has been generalized in the past by many researchers.

Additive Spanners

A t-spanner as defined above approximates pairwise distances with multiplicative error and can be called a multiplicative spanner. In an analogous manner, one can define spanners that approximate pairwise distances with additive error. Such a spanner is called an additive spanner, and the corresponding error is called *surplus*. Aingworth et al. [1] presented the first additive spanner of size $O(n^{3/2} \log n)$ with surplus 2. Baswana et al. [7] presented a construction of $O(n^{4/3})$-size additive spanner with surplus 6. Recently, Chechik [9] presented a construction of $O(n^{7/5})$-size additive spanner with surplus 4. It is a major open problem if there exists any sparser additive spanner.

(α, β)-Spanner

Elkin and Peleg [12] introduced the notion of (α, β)-spanner for unweighted graphs, which can be viewed as a hybrid of multiplicative and additive spanners. An (α, β)-spanner is a subgraph such that the distance between any pair of vertices $u, v \in V$ in this subgraph is bounded by $\alpha\delta(u, v) + \beta$, where $\delta(u, v)$ is the distance between u and v in the original graph. Elkin and Peleg showed that an $(1 + \epsilon, \beta)$-spanner of size $O(\beta n^{1+\delta})$, for arbitrarily small $\epsilon, \delta > 0$, can be computed at the expense of sufficiently large surplus β. Recently, Thorup and Zwick [20] introduced a spanner where the additive error is sublinear in terms of the *distance* being approximated.

Other interesting variants of spanner include *distance preserver* proposed by Bollobás et al. [8] and *lightweight* spanner proposed by Awerbuch et al. [4]. A subgraph is said to be a d-preserver if it preserves exact distances for each pair of vertices which are separated by distance at least d. A lightweight spanner tries to minimize the number of edges as well as the total edge weight. A *lightness* parameter is defined for a subgraph as the ratio of total weight of all its edges and the weight of the minimum spanning tree of the graph. Awerbuch et al. [4] showed that for any weighted graph and integer $k > 1$, there exists a polynomially constructible $O(k)$-spanner with $O(k\rho n^{1+1/k})$ edges and $O(k\rho n^{1/k})$ lightness, where $\rho = \log(diameter)$.

In addition to the above work on the generalization of spanners, a lot of work has also been done on computing spanners for special classes of graphs, e.g., chordal graphs, unweighted graphs, and Euclidean graphs. For chordal graphs, Peleg and Schäffer [15] designed an algorithm that computes a 2-spanner of size $O(n^{3/2})$ and a 3-spanner of size $O(n \log n)$. For unweighted graphs, Halperin and Zwick [14] gave an $O(m)$ time algorithm for this problem. Salowe [18] presented an algorithm for computing a $(1 + \epsilon)$-spanner of a d-dimensional complete Euclidean graph in $O(n \log n + \frac{n}{\epsilon^d})$ time. However, none of the algorithms for these special classes of graphs seem to extend to general weighted undirected graphs.

Applications

Spanners are quite useful in various applications in the area of distributed systems and communication networks. In these applications, spanners appear as the underlying graph structure. In order to build compact routing tables [17], many existing routing schemes use the edges of a sparse spanner for routing messages. In distributed systems, spanners play an important role in designing *synchronizers*. Awerbuch [3] and Peleg and Ullman [16] showed that the quality of a spanner (in terms of stretch factor and the number of spanner edges) is very closely related to the time and communication complexity of any synchronizer for the network. The spanners have also been used implicitly in a number of algorithms for computing all-pairs approximate shortest paths [5, 10, 19, 21]. For a number of other applications, please refer to the papers [2, 3, 15, 17].

Open Problems

The running time as well as the size of the $(2k - 1)$-spanner computed by the algorithm described above are away from their respective worst-case lower bounds by a factor of k. For any constant value of k, both these parameters are optimal. However, for the extreme value of k, that is, for $k = \log n$, there is deviation by a factor of $\log n$. Is it possible to get rid of this multiplicative factor of k from the running time of the algorithm and/or the size of the $(2k-1)$-spanner computed? It seems that a more careful analysis coupled with advanced probabilistic tools might be useful in this direction.

Recommended Reading

1. Aingworth D, Chekuri C, Indyk P, Motwani R (1999) Fast estimation of diameter and shortest paths(without matrix multiplication). SIAM J Comput 28:1167–1181
2. Althöfer I, Das G, Dobkin DP, Joseph D, Soares J (1993) On sparse spanners of weighted graphs. Discret Comput Geom 9:81–100
3. Awerbuch B (1985) Complexity of network synchronization. J Assoc Comput Mach 32(4):804–823
4. Awerbuch B, Baratz A, Peleg D (1992) Efficient broadcast and light weight spanners. Tech. Report CS92-22, Weizmann Institute of Science
5. Awerbuch B, Berger B, Cowen L, Peleg D (1998) Near-linear time construction of sparse neighborhod covers. SIAM J Comput 28:263–277
6. Baswana S, Sen S (2007) A simple and linear time randomized algorithm for computing sparse spanners in weighted graphs. Random Struct Algorithms 30(4):532–563
7. Baswana S, Kavitha T, Mehlhorn K, Pettie S (2010) Additive spanners and (alpha, beta)-spanners. ACM Trans Algorithms 7(1):5
8. Bollobás B, Coppersmith D, Elkin M (2003) Sparse distance preserves and additive spanners. In Proceedings of the 14th annual ACM-SIAM symposium on discrete algorithms (SODA), Baltimore, pp 414–423
9. Chechik S (2013) New additive spanners. In Proceedings of the twenty-fourth annual ACM-SIAM symposium on discrete algorithms, SODA 2013, New Orleans, 6–8 Jan 2013, pp 498–512
10. Cohen E (1998) Fast algorithms for constructing t-spanners and paths with stretch t. SIAM J Comput 28:210–236
11. Elkin M, Peleg D (2000) The hardness of approximating spanner problems. In STACS 2000, 17th annual symposium on theoretical aspects of computer science, Lille, Feb 2000, proceedings, pp 370–381
12. Elkin M, Peleg D (2004) $(1 + \epsilon, \beta)$-spanner construction for general graphs. SIAM J Comput 33:608–631
13. Erdős P (1964) Extremal problems in graph theory. In Theory of graphs and its applications (Proc. Sympos. Smolenice,1963), pp 29–36, Publ. House Czechoslovak Acad. Sci., Prague
14. Halperin S, Zwick U (1996) Linear time deterministic algorithm for computing spanners for unweighted graphs. Unpublished manuscript
15. Peleg D, Schaffer AA (1989) Graph spanners. J Graph Theory 13:99–116
16. Peleg D, Ullman JD (1989) An optimal synchronizer for the hypercube. SIAM J Comput 18:740–747
17. Peleg D, Upfal E (1989) A trade-off between space amd efficiency for routing tables. J Assoc Comput Mach 36(3):510–530
18. Salowe JD (1991) Construction of multidimensional spanner graphs, with application to minimum spanning trees. In ACM symposium on computational geometry, North Conway, pp 256–261
19. Thorup M, Zwick U (2005) Approximate distance oracles. J Assoc Comput Mach 52:1–24
20. Thorup M, Zwick U (2006) Spanners and emulators with sublinear distance errors. In Proceedings of 17th annual ACM-SIAM symposium on discrete algorithms, Miami, 22-26 Jan 2006, pp 802–809

S

21. Wulff-Nilsen C (2012) Approximate distance oracles with improved preprocessing time. In Proceedings of the twenty-third annual ACM-SIAM symposium on discrete algorithms, SODA 2012, Kyoto, 17–19 Jan 2012, pp 202–208

Simpler Approximation for Stable Marriage

Zoltán Király
Department of Computer Science, Eötvös Loránd University, Budapest, Hungary
Egerváry Research Group (MTA-ELTE), Eötvös Loránd University, Budapest, Hungary

Keywords

Approximation; Local algorithm; SMTI; Stable marriage

Years and Authors of Summarized Original Work

2013; Király

Introduction

We have a two-sided market, one side is a set U of men, the other side is a set V of women. The first part of the input also contains the mutually acceptable man-woman pairs E. This makes up a bipartite graph $G(U \cup V, E)$. The second part of the input contains the preference lists of each person, that is a weak order (may contain ties) on his/her acceptable pairs.

A *matching* is a set of mutually disjoint acceptable man-woman pairs. Given a matching M, a man m and a woman w form a *blocking* pair, if they are an acceptable pair but are not partners in M, and they both prefer each other to their partner, or have no partner in M. That is either w is unmatched in M or w prefers m to her M-partner, *and* either m is unmatched in M or m

prefers w to his M-partner. A matching M is *stable* if there are no blocking pairs.

We consider a two-sided market under incomplete preference lists with ties (SMTI), where the goal is to find a maximum size stable matching (MAX-SMTI).

Problem Definition

Problem 1 (MAX-SMTI)

INPUT: Set U of men, and set V of women and each person's preference list.
OUTPUT: A stable matching of maximum size.

Input format A list of an agent a consists of pairs (a_1, p_1), (a_2, p_2), ..., (a_d, p_d), where a_i are the acceptable persons from the other gender and $1 \leq p_i \leq \max(|U|, |V|)$ are integers with ordering $p_1 \geq p_2 \geq \cdots \geq p_d$. Agent a strictly prefers a_i to a_j if $p_i > p_j$ and is indifferent between a_i and a_j if $p_i = p_j$. Moreover women needs a black-box procedure, which on input a_i outputs in constant time p_i (we assume that this procedure is also a part of the input). The size of the input is the number of agents plus the total length of the lists.

Definition of approximation ratios A goodness measure of an approximation algorithm A for a maximization problem is defined as follows: the *approximation ratio* of A is $\max\{\text{opt}(I)/A(I)\}$ over all instances I, where $\text{opt}(I)$ and $A(I)$ are the size of the optimal and the algorithm's solution on instance I, respectively.

Short history It was shown in [4] that finding the optimal solution is NP-hard; moreover, it is APX-hard [3]. The original Deferred Acceptance Algorithm of Gale and Shapley gives a 2-approximation; the first approximation algorithm with a strictly better ratio was presented in [5], where the approximation ratio was 15/8. This was improved in [6] to a 5/3-approximation and later in [9] to a 3/2-approximation; this latter algorithm had nonlinear running time. Recently in [10] and in [7], linear time 3/2-approximation algorithms were given.

Key Results

A simple variation of the famous Deferred Acceptance Algorithm of Gale and Shapley is presented; which also runs in linear time and gives a 3/2-approximation for the problem MAX-SMTI. This algorithm is local; no central agent or knowledge about the global input is needed.

Algorithm

Preliminary Definitions and Concepts for the Algorithm

During the algorithm, the agents may have different statuses, and some Boolean properties described below, and also varying actual preferences.

A status of a man can be either a **lad** or a **bachelor** or an **old bachelor**. A man can be **active** or inactive. A man is active, if he is not an old bachelor and he is not **engaged** (i.e., he has actually no partner). A man can also be **uncertain**, described later. Initially every man is an active lad.

A status of a woman can be either **maiden** or **engaged**. An engaged woman is **flighty**, if her fiancé is *uncertain*. Initially every woman is maiden.

The actual preferences a man m is described as follows. If women w_1 and w_2 are indifferent on m's list, and w_1 is maiden but w_2 is engaged, then m **prefers maiden** w_1 to **engaged** w_2. An engaged lad is **uncertain** if his list contains a woman he prefers to his actual fiancée (this can happen, if there were two maidens with the same highest priority on m's list, and m became engaged to one of them).

The actual preferences a woman w is described as follows. If there are two men, m_1 and m_2 with the same priority in w's list, and m_1 is a lad, but m_2 is a bachelor, then w **prefers bachelor** m_2 to **lad** m_1. If w is flighty, then she prefers a man who is not uncertain, to a man who is uncertain (regardless of her original preferences).

The Algorithm

While there exists an active man m, he proposes to his favorite woman w. If w accepts his proposal, they become engaged. If w rejects him, m deletes w from his list and remain active.

When a woman w gets a new proposal from man m, she accepts this proposal if she (actually) prefers m to her current fiancé. Otherwise she rejects m.

If w accepted m, then she rejects her previous fiancé, if there was one (breaks off her engagement), and becomes engaged to m.

If m was engaged to a woman w and later w rejects him, then m becomes active again and deletes w from his list, except if m is uncertain, in this case m keeps w on the list.

If the list of m becomes empty for the first time, he turns into a bachelor, his original list is recovered, and he reactivates himself. If the list of m becomes empty for the second time, he will turn into an old bachelor and will remain inactive forever.

After the algorithm finishes, the engaged pairs get married and form matching M.

Theorem 1 ([7]) *The algorithm always gives a stable matching M and it is 3/2-approximating, i.e., the stable matching given has size at least 2/3 of the maximum size stable matching.*

Running Time, Locality

This algorithm runs in linear time using the assumptions on the input format. Though it is clear that along every edge at most three proposals happen, the technical details must be worked out; see [7] for details.

Local algorithm Each agent (a man or woman) always makes a greedy decision based only on local information (his/her preference list, and provided by some communication with his/her acceptable partners). A local algorithm is linear if every agent communicates with each acceptable partner only a constant time during the algorithm.

The algorithm presented is a linear time local algorithm (using the appropriate data structures); see [7] for details.

Cross-References

▶ Hospitals/Residents Problem
▶ Maximum Cardinality Stable Matchings
▶ Stable Marriage
▶ Stable Marriage with One-Sided Ties
▶ Stable Marriage with Ties and Incomplete Lists

Recommended Reading

1. Gale D, Shapley LS (1962) College admissions and the stability of marriage. Am Math Mon 69:9–15
2. Gusfield D, Irving RW (1989) The stable marriage problem: structure and algorithms. MIT, Boston
3. Halldórsson MM, Irving RW, Iwama K, Manlove DF, Miyazaki S, Morita Y, Scott S (2003) Approximability results for stable marriage problems with ties. Theor Comput Sci 306:431–447
4. Iwama K, Manlove DF, Miyazaki S, Morita Y (1999) Stable marriage with incomplete lists and ties. In: Proceedings of the 26th international colloquium on automata, languages and programming (ICALP 1999), Prague. LNCS, vol 1644, pp 443–452
5. Iwama K, Miyazaki S, Yamauchi N (2007) A 1.875-approximation algorithm for the stable marriage problem. In: SODA '07: Proceedings of the eighteenth annual ACM-SIAM symposium on discrete algorithms, pp 288–297
6. Király Z (2009(online), 2011) Better and simpler approximation algorithms for the stable marriage problem. Algorithmica 60(1):3–20
7. Király Z (2013) Linear time local approximation algorithm for maximum stable marriage. Algorithms 6(3):471–484
8. Manlove D (2013) Algorithmics of matching under preferences. World Scientific Publishing, Singapore
9. McDermid EJ (2009) A $\frac{3}{2}$-approximation algorithm for general stable marriage. In: Proceedings of the 36th international colloquium automata, languages and programming (ICALP 2009), Rhodes, pp 689–700
10. Paluch K (2014) Faster and simpler approximation of stable matchings. Algorithms 7(2):189–202

Single and Multiple Buffer Processing

Sergey I. Nikolenko[1] and Kirill Kogan[2]
[1]Laboratory of Mathematical Logic, Steklov Institute of Mathematics, St. Petersburg, Russia
[2]IMDEA Networks, Madrid, Spain

Keywords

Admission control; Buffer management policies; Online algorithms

Years and Authors of Summarized Original Work

2004; Kesselman, Mansour
2012; Keslassy, Kogan, Scalosub, Segal
2012; Kesselman, Kogan, Segal
2012, 2013; Kogan, López-Ortiz, Nikolenko, Sirotkin
2014; Eugster, Kogan, Nikolenko, Sirotkin

Problem Definition

Buffer management policies are online algorithms that control a limited buffer of packets with homogeneous or heterogeneous characteristics, deciding whether to accept new packets when they arrive, which packets to process and transmit, and possibly whether to push out packets already residing in the buffer. Although settings differ, the problem is always to achieve the best possible competitive ratio, i.e., find a policy with good worst-case guarantees in comparison with an optimal offline clairvoyant algorithm. The policies themselves are often simple, simplicity being an important advantage for implementation in switches; the hard problem is to find proofs of lower and especially upper bounds for their competitive

The work of Sergey Nikolenko was partially supported by the Government of the Russian Federation (grant 14.Z50.31.0030).

ratios. Thus, this problem is more theoretical in nature, although the resulting throughput guarantees are important tools in the design of network elements. Comprehensive surveys of this field have been given in the past by Goldwasser [9] and Epstein and van Stee [7].

General Model Description

We assume discrete slotted time. A packet is *fully processed* if the processing unit has scheduled the packet for processing for at least its required number of cycles. Each packet may have the following characteristics: (i) *required processing*, i.e., how many processing cycles the packet has to go through before it can be transmitted; (ii) *value*, i.e., how much the packet contributes to the objective function when it is transmitted; (iii) *output port*, i.e., where the packet is headed (in settings with multiple output ports, it is usually assumed that processing occurs independently at each port, so it becomes advantageous to have more busy output ports at a time); and (iv) *size*, i.e., how many slots (bytes) a packet occupies in the buffer. The objective of a buffer management policy is to maximize the total value of transmitted packets. Different settings may assume that some characteristics are uniform.

Competitive Analysis

Competitive analysis provides a uniform throughput guarantee for online algorithms across all traffic patterns. An online algorithm ALG is said to be α-*competitive* with respect to some objective function f (for some $\alpha \geq 1$ which is called the *competitive ratio*) if for any arrival sequence σ the objective function value on the result of ALG is at least $1/\alpha$ times the objective function value on the solution obtained by an offline clairvoyant algorithm, denoted OPT.

Problem 1 (Competitive Ratio) For a given switch architecture, packet characteristics, and an online algorithm ALG in a given setting, prove lower and upper bounds on its competitive ratio with respect to weighted throughput (total value of packets transmitted by an algorithm).

Key Results

Policies and lower and upper bounds on their competitive ratios are outlined according to problem settings; the latter differ in which packet characteristics they assume to be uniform and which are allowed to vary, and additional restrictions may be imposed on admission, processing and/or transmission order, and admissible packet characteristics.

Uniform Processing, Uniform Value, Shared Memory Switch

Since all packets are identical, the problem for a single queue with one output port is trivial. We consider an $M \times N$ shared memory switch that can hold B packets, with a separate processor on each output port. All packets require a single processing cycle and have equal value; the goal is to maximize the number of transmitted packets. Each packet is labeled with an output port where it has to be processed and transmitted.

Non-Push-Out Policies

Kesselman and Mansour [14] show an adversarial logarithmic lower bound: no non-push-out policy can achieve competitive ratio better than $d/2$ for $d = \log_d N$. On the positive side, they present the Harmonic policy that allocates approximately $1/i$ of the buffer to the ith largest queue and, for its variant, the Parametric Harmonic policy, show an upper bound of $c \log_c N + 1$.

Push-Out Policies

The best known policy is Longest Queue Drop (LQD): accept packets greedily if the buffer is not full; if it is, accept the new packet and then drop a packet from the longest queue (destined to the output port with the most packets assigned to it). Aiello et al. [1,10] show that the competitive ratio of LQD is between $\sqrt{2}$ and 2; they also provide nonconstant lower bounds for other popular policies and a general adversarial lower bound of $\frac{4}{3}$ on the competitive ratio of any online algorithm.

Uniform Processing, Uniform Value, Multiple Separated Queues

In an $N \times 1$ switch where each of N input queues has a separate independent buffer of size B, a policy must select which input queue to take a packet from and set admission policies for input queues. For uniform values, the problem was closed by Azar and Litichevskey [3] with a deterministic policy with competitive ratio converging to $\frac{e}{e-1} \approx 1.582$ for arbitrary B; a matching lower bound was shown by Azar and Richter [4].

Uniform Processing, Variable Values, Single Queue

Here, there is only one output port (a single queue), and each packet is fully processed in one cycle; however, packets have different values, making it desirable to drop packets with smaller value and process packets of larger value. It is easy to show that the Priority Queue (PQ) policy that sorts packets with respect to values and pushes out smaller values for larger ones is optimal. Research has concentrated on models with additional constraints: non-push-out policies that are not allowed to push admitted packets out and the FIFO model where packets have to be transmitted in order of arrival. Another important special case considers two possible values: 1 and $V > 1$.

Non-Push-Out Policies

Aiello et al. [2] consider five online policies for the two-valued case, considering the specific cases of $V = 1$, $V = 2$, and $V = \infty$. Andelman, Mansour, and Zhu provide a deterministic policy (Ratio Partition) that achieves optimal $\left(2 - \frac{1}{V}\right)$-competitiveness [26]. In the case of arbitrary values between 1 and $V > 1$, they show that the optimal competitive ratio is $\ln V$, proving tightly matching bounds of $1 + \ln V$ and $2 + \ln V + O(\ln^2 V / B)$ [2, 26].

Push-Out Policies

In the FIFO model, there has been a line of adversarial lower bounds culminating in the lower bound of 1.419 shown by Kesselman, Mansour, and van Stee [18] that applies to all algorithms, with a stronger bound of 1.434 for $B = 2$

[2, 26]. As for upper bounds, in this simple model the FIFO greedy push-out policy (accept every packet to end of queue, then push out the packet with smallest value if buffer has overflown) has been shown by Kesselman et al. to be 2-competitive [17]; in the two-valued case, they provide an adversarial lower bound of 1.282, and a long line of improvements for the upper bound has led to the optimal Account Strategy policy of Englert and Westermann [6]. They show an adversarial lower bound of $r = \frac{1}{2}(\sqrt{13} - 1) \approx 1.303$ for any $B \geq 2$ and $r_\infty = \sqrt{2} - \frac{1}{2}(\sqrt{5 + 4\sqrt{2}} - 3) \approx 1.282$ for $B \to \infty$ and show that Account Strategy achieves competitive ratio r for arbitrary B and r_∞ for $B \to \infty$. Thus, in the push-out two-valued case, the gap between lower and upper bounds has been closed completely.

Uniform Processing, Variable Values, Multiple Separated Queues

Kawahara et al. [11] consider an $N \times 1$ switch with N separated queues, each of which has a distinct buffer of size B and has a value α_j associated with it, $1 = \alpha_1 \leq \ldots \leq \alpha_N = \alpha$. A policy selects one of N queues, maximizing total transmitted value; [11] provides matching lower and upper bounds for the PQ policy as $1 + \frac{\sum_{j=1}^{n'} \alpha_j}{\sum_{j=1}^{n'+1} \alpha_j}$, where $n' = \arg\max_n \frac{\sum_{j=1}^{n} \alpha_j}{\sum_{j=1}^{n+1} \alpha_j}$, and an adversarial lower bound $1 + \frac{\alpha^3 + \alpha^2 + \alpha}{\alpha^4 + 4\alpha^3 + 3\alpha^2 + 4\alpha + 1}$ for any online algorithm. Azar and Richter [4] show that any r-competitive policy for a FIFO queue with variable values yields a $2r$-competitive policy for multiple queues. Kobayashi et al. [21] show that an r-competitive policy for unit values and multiple queues yields a $\min\left\{Vr, \frac{Vr(2-r)+r^2-2r+2}{V(2-r)+r-1}\right\}$-competitive policy for the two-valued case.

Uniform Processing, Variable Values, Shared Memory Switch

Several output queues, each with a processor, share a buffer of size B, and each unit-sized packet is labeled with an output port and an intrinsic value from 1 to V. Eugster, Kogan, Nikolenko, and Sirotkin [8] show a $(\sqrt[3]{V} - o(\sqrt[3]{V}))$ lower bound for the LQD (Longest Queue Drop) policy, an $\frac{1}{2}(\min\{V, B\} - 1)$ lower bound for the MVD (Minimal Value Drop) policy, and a $\frac{4}{3}$ lower bound for the MRD (Maximal Ratio Drop) policy.

Uniform Processing, CIOQ Switches

In CIOQ (Combined Input–Output Queued) switches, one maintains at each input a separate queue for each output (also called Virtual Output Queuing, VOQ). To get delay guarantees of an input queuing (IQ) switch closer to those of an output queuing switch (OQ), one usually assumes increased *speedup* S: the switching fabric runs S times faster than each of the input or the output ports. Hence, an OQ switch has a speedup of N (where N is the number of input/output ports), whereas an IQ switch has a speedup of 1; for $1 < S < N$, packets need to be buffered at the inputs before switching as well as at the outputs after switching. This architecture is called a CIOQ switch.

Uniform Values

Consider an $N \times N$ CIOQ switch with speedup S. Packets of equal size arrive at input ports, each labeled with the output port where it has to leave the switch. Each packet is placed in the input queue corresponding to its output port; when it crosses the switch fabric, it is placed in the output queue and resides there until it is sent on the output link. For unit-valued packets, Kesselman and Rosén [15] proposed a non-push-out policy which is 3-competitive for any S and 2-competitive for $S = 1$. Kesselman, Kogan, and Segal [13] show an upper bound of 4 on the competitiveness of a simple greedy policy.

Variable Values

For up to m packet values in $[1, V]$, Kesselman and Rosén [15] show two push-out policies to be $4S$- and $8 \min\{m, 2 \log V\}$-competitive. Azar and Richter [5] propose a push-out policy β-PG with parameter β; Kesselman et al. [20] show that the competitive ratio of β-PG is at most 7.5 for $\beta = 3$ and at most 7.47 for $\beta = 2.8$. Kesselman

and Rosén [16] consider CIOQ switches with PQ buffers (transmit the highest value packet) and show that this policy is 6-competitive for any S.

Uniform Processing, Crossbar Switches

In the buffered crossbar switch architecture, a small buffer is placed on each crosspoint in addition to input and output queues, which greatly simplifies the scheduling process. For packets with unit length and value, Kesselman et al. [20] introduce a greedy switch policy with competitive ratio between $\frac{3}{2}$ and 4 and show a general lower bound of $\frac{3}{2}$ for unit-size buffers. For variable values and PQ buffers, they propose a push-out greedy switch policy with preemption factor β with competitive ratio between $(2\beta - 1)/(\beta - 1)$ (3.87 for $\beta = 1.53$) and $(\beta + 2)^2 + 2/(\beta - 1)$ (16.24 for $\beta = 1.53$). For variable values and FIFO buffers, they propose a β-push-out greedy switching policy with competitive ratio $6 + 4\beta + \beta^2 + 3/(\beta - 1)$ (19.95 for $\beta = 1.67$) [19].

Uniform Values, Variable Processing, Single Queue

In this setting, each packet contributes one unit to the objective function, but different packets have different processing requirements, i.e., they spend a different number of time slots at the processor. We denote maximal possible required processing by k.

Non-Push-Out Policies

For a single queue and packets with heterogeneous processing, non-push-out policies have not been considered in any detail. Kogan, López-Ortiz, Nikolenko, and Sirotkin [23] have shown that any greedy non-push-out policy is at least $\frac{1}{2}(k + 1)$-competitive. It remains an open problem to find non-push-out policies with sublinear competitive ratios or show that none exists.

Push-Out Policies

Keslassy et al. [12] showed that again, for a single queue, PQ (Priority Queue) that sorts packets with respect to required processing (smallest first) is optimal; research has concentrated on the FIFO case, where packets have to be transmitted in order of arrival. Kogan et al. [24] introduced *lazy* policies that process packets down to a single cycle but then delay their transmission until the entire queue consists of such packets; then all packets are transmitted out in as many time slots as there are packets in the queue. In [24], LPO (Lazy Push-Out) was proven to be at most $(\max\{1, \ln k\} + 2 + o(1))$-competitive; [24] also provides a lower bound of $\lfloor \log_B k \rfloor + 1 - O(1/B)$ for both PO (push-out FIFO) and LPO; for large k this bound matches the upper bound up to a factor of $\log B$. Proving a matching upper bound for the PO policy remains an important open problem. In the two-valued case, when packets may have required processing only 1 or k, LPO has a lower bound of $2 - \frac{1}{k}$ and a matching upper bound of $2 + \frac{1}{B}$ [24]. Kogan, López-Ortiz, Nikolenko, and Sirotkin [23] introduce *semi-FIFO* policies, separating processing order from transmission order so that transmission can conform to FIFO constraints while processing order remains arbitrary. Lazy policies thus become a special case of semi-FIFO policies. The authors show a general upper bound of $\frac{1}{B} \log_{\frac{B}{B-1}} k + 3$ on the competitive ratio of any lazy policy and a matching lower bound of $\frac{1}{B} \log_{\frac{B}{B-1}} k + 1$ for several processing orders. In the two-valued case, when processing is only 1 or k, this upper bound improves to $2 + \frac{1}{B}$, so any lazy policy has constant competitiveness. LPQ (Lazy Priority Queue) also falls in the semi-FIFO class; its competitiveness is between $\left(2 - \frac{1}{B} \lceil \frac{B}{k} \rceil\right)$ and 2 even for arbitrary processing requirements. Kogan et al. [22] consider a generalization with packets of varying size, considering several natural policies and showing an upper bound of $4L$ for one of PO policies, where L is the maximal packet size.

Copying Cost
An important generalization of the heterogeneous processing model was introduced by Keslassy et al. [12]. They attach a penalty α called copying cost to admitting a packet in the queue; thus, the objective function is now $T - \alpha A$, where T is the number of transmitted packets and A is the number of accepted ones, and it becomes less advantageous to push packets out. To deal with copying cost, the authors propose to use β-push-out policies that push a packet out only

if its required processing is at least $\beta > 1$ times less than the required processing of a packet which is being pushed out. Keslassy et al. [12] consider the PQ$_\beta$ policy (Priority Queue with β-push-out) and show that it is at most
$$\frac{1}{1 - \alpha \log_\beta k}\left(1 + \log_{\frac{\beta}{\beta-1}} \frac{k}{2} + 2 \log_\beta k\right)(1 - \alpha)\text{-}$$
competitive. Kogan, López-Ortiz, Nikolenko, and Sirotkin [23] show that for any processing order, a β-push-out lazy policy LA$_\beta$ has competitive ratio at most $\left(3 + \frac{1}{B} \log_{\frac{\beta B}{\beta B - 1}} k\right) \frac{1 - \alpha}{1 - \alpha \log_\beta k}$. They show a lower bound $\frac{1 - \alpha}{1 - \alpha \log_\beta k}$ on the competitive ratio of any β-push-out policy, which matches the additional factor in the upper bound. In the two-valued case, the upper bound becomes $\left(2 + \frac{1}{B}\right) \frac{1 - \alpha}{1 - 2\alpha}$, and the authors also show a matching lower bound of $\frac{(2B - 2)(1 - \alpha)}{(B - 1)(1 - 2\alpha) + (1 - \alpha)}$.

Uniform Values, Variable Processing, Multiple Separated Queues
Consider k separate queues of size B each; packets with required processing i fall into the ith queue, and the processor chooses which queue to process on a given time slot. Push-out is irrelevant since queues are independent and packets in a queue are identical. Kogan, López-Ortiz, Nikolenko, and Sirotkin [25] show linear lower bounds for several seemingly attractive policies: $\frac{1}{2} \min\{k, B\}$ for LQF (Longest Queue First), k for SQF (Shortest Queue First), $\frac{3k(k+2)}{4k+16}$ for PRR (Packet Round Robin), and an almost linear lower bound of $\frac{k}{H(k)}$, where $H(k) = \sum_{i=1}^{k} \frac{1}{i} \approx \ln k + \gamma$, for CRR (Cycle Round Robin). They introduce a policy called MQF (Minimal Queue First) that processes packets from a nonempty queue with minimal processing requirement. They show that MQF is at least $\left(1 + \frac{k-1}{2k}\right)$-competitive and prove a constant upper bound of 2. For the two-valued case with two queues, 1 and k, Kogan et al. [25] show exactly matching lower and upper bounds for MQF of $1 + \left(1 + \lfloor \frac{aB-1}{b} \rfloor\right) / \left(B + \lceil \frac{1}{a}\left(b \lfloor \frac{aB-1}{b} \rfloor + 1\right) \rceil\right)$.

Uniform Values, Variable Processing, Shared Memory Switch
In this setting, multiple queues with shared memory are implemented in the same way as for

uniform processing and heterogeneous values: there are N output ports, each output port manages a single output queue Q_i, and each output queue collects packets with the same processing requirement (so packets in a given queue are identical).

Non-Push-Out Policies

Eugster, Kogan, Nikolenko, and Sirotkin [8] consider non-push-out policies and show that NHST (Non-Push-Out Harmonic with Static Threshold: $|Q_i|$ is bounded by $\frac{B}{r_i Z}$) is $(kZ + o(kZ))$-competitive, NEST (Non-Push-Out with Equal Static Threshold: $|Q_i|$ is bounded by B/n) is $(N + o(N))$-competitive, NHDT (Non-Push-Out with Harmonic Dynamic Threshold: accept into Q_i if $\sum_{s=1}^{m} |Q_{j_s}| < \frac{B}{H_k}\left(1 + \frac{1}{2} + \ldots + \frac{1}{m}\right)$, where $j_1 \ldots j_m = i$ are queues for which $|Q_j| \geq |Q_i|$) is $(\frac{1}{2}\sqrt{k \ln k} - o(\sqrt{k \ln k}))$-competitive; finding better non-push-out policies is an open problem.

Push-Out Policies

The work [8] also shows lower bounds on the competitive ratio of well-known policies: $(\sqrt{k} - o(\sqrt{k}))$ for LQD (Longest Queue Drop), $(\ln k + \gamma)$ for BQD (Biggest Packet Drop), and $(\frac{4}{3} - \frac{6}{B})$ for LWD (Largest Work Drop). The main result of [8] is that LWD is at most 2-competitive.

Open Problems

1. Close the gap between competitive ratios $\frac{4}{3}$ (lower bound for any policy) and 2 (upper bound for LQD) in the uniform processing, uniform value case.

2. Do there exist policies with constant competitive ratio in the uniform processing, variable values, shared memory multiple output queues setting?

3. Do there exist non-push-out policies with sublinear competitive ratio in the case of a single queue with packets with variable processing and uniform values?

4. Prove an upper bound on the competitiveness of PO (push-out) policy in the single-queue

FIFO model with heterogeneous required processing and uniform values.

5. Do there exist non-push-out policies with logarithmic competitive ratio in the case of multiple output ports with shared memory that contain packets with variable processing and uniform values?

6. Design efficient policies for CIOQ and crossbar switches with packets with heterogeneous processing and uniform values; prove bounds on their competitive ratios.

7. Design efficient policies and prove bounds on their competitive ratios for the case of packets with both variable values and heterogeneous processing requirements in all of the above settings.

Cross-References

▶ Packet Switching in Multi-queue Switches
▶ Packet Switching in Single Buffer

Recommended Reading

1. Aiello W, Kesselman A, Mansour Y (2008) Competitive buffer management for shared-memory switches. ACM Trans Algorithms 5(1):1–16
2. Andelman N, Mansour Y, Zhu A (2003) Competitive queueing policies for QoS switches. In: Proceedings of the 4th annual ACM-SIAM symposium on discrete algorithms, Baltimore, pp 761–770
3. Azar Y, Litichevskey A (2006) Maximizing throughput in multi-queue switches. Algorithmica 45(1):69–90
4. Azar Y, Richter Y (2005) Management of multi-queue switches in QoS networks. Algorithmica 43(1-2):81–96
5. Azar Y, Richter Y (2006) An improved algorithm for CIOQ switches. ACM Trans Algorithms 2(2):282–295
6. Englert M, Westermann M (2009) Lower and upper bounds on FIFO buffer management in QoS switches. Algorithmica 53(4):523–548
7. Epstein L, van Stee R (2004) Buffer management problems. SIGACT News 35(3):58–66
8. Eugster P, Kogan K, Nikolenko SI, Sirotkin AV (2014) Shared-memory buffer management for heterogeneous packet processing. In: Proceedings of the 34th international conference on distributed computing systems, Madrid
9. Goldwasser M (2010) A survey of buffer management policies for packet switches. SIGACT News 41(1):100–128

S

10. Hahne EL, Kesselman A, Mansour Y (2001) Competitive buffer management for shared-memory switches. In: 13th ACM symposium on parallel algorithms and architectures, Crete Island, pp 53–58
11. Kawahara J, Kobayashi K, Maeda T (2012) Tight analysis of priority queuing policy for egress traffic. CoRR abs/1207.5959
12. Keslassy I, Kogan K, Scalosub G, Segal M (2012) Providing performance guarantees in multipass network processors. IEEE/ACM Trans Netw 20(6):1895–1909
13. Kesselman A, Kogan K, Segal M (2008) Best effort and priority queuing policies for buffered crossbar switches. In: Structural information and communication complexity, 15th international colloquium (SIROCCO 2008), Villars-sur-Ollon, 170–184 http://dx.doi.org/10.1007/978-3-540-69355-0_15
14. Kesselman A, Mansour Y (2004) Harmonic buffer management policy for shared memory switches. Theor Comput Sci 324(2-3):161–182
15. Kesselman A, Rosén A (2006) Scheduling policies for CIOQ switches. J Algorithms 60(1):60–83
16. Kesselman A, Rosén A (2008) Controlling CIOQ switches with priority queuing and in multistage interconnection networks. J Interconnect Netw 9(1/2):53–72
17. Kesselman A, Lotker Z, Mansour Y, Patt-Shamir B, Schieber B, Sviridenko M (2004) Buffer overflow management in QoS switches. SIAM J Comput 33(3):563–583
18. Kesselman A, Mansour Y, van Stee R (2005) Improved competitive guarantees for QoS buffering. Algorithmica 43(1-2):63–80
19. Kesselman A, Kogan K, Segal M (2010) Packet mode and QoS algorithms for buffered crossbar switches with FIFO queuing. Distrib Comput 23(3):163–175
20. Kesselman A, Kogan K, Segal M (2012) Improved competitive performance bounds for CIOQ switches. Algorithmica 63(1–2):411–424
21. Kobayashi K, Miyazaki S, Okabe Y (2009) Competitive buffer management for multi-queue switches in QoS networks using packet buffering algorithms. In: Proceedings of the 21st ACM symposium on parallelism in algorithms and architectures (SPAA), Portland, OR, USA, pp 328–336
22. Kogan K, López-Ortiz A, Nikolenko S, Scalosub G, Segal M (2014) Balancing Work and Size with Bounded Buffers. In: Proceedings of the 6th international conference on communication systems and networks (COMSNETS 2014), Bangalore, pp 1–8
23. Kogan K, López-Ortiz A, Nikolenko SI, Sirotkin AV (2012) A taxonomy of semi-FIFO policies. In: Proceedings of the 31st IEEE international performance computing and communications conference (IPCCC2012), Austin, pp 295–304
24. Kogan K, López-Ortiz A, Nikolenko SI, Sirotkin AV, Tugaryov D (2012) FIFO queueing policies for packets with heterogeneous processing. In: Proceedings of the 1st Mediterranean conference on algorithms (MedAlg 2012), Ein Gedi. Lecture notes in computer science, vol 7659, pp 248–260
25. Kogan K, López-Ortiz A, Nikolenko SI, Sirotkin A (2013) Multi-queued network processors for packets with heterogeneous processing requirements. In: Proceedings of the 5th international conference on communication systems and networks (COMSNETS 2013), Bangalore, pp 1–10
26. Zhu A (2004) Analysis of queueing policies in QoS switches. J Algorithms 53(2):137–168

Single-Source Fully Dynamic Reachability

Camil Demetrescu[1,2] and Giuseppe F. Italiano[1,2]
[1]Department of Computer and Systems Science, University of Rome, Rome, Italy
[2]Department of Information and Computer Systems, University of Rome, Rome, Italy

Keywords

Single-source fully dynamic transitive closure

Years and Authors of Summarized Original Work

2005; Demetrescu, Italiano

Problem Definition

A dynamic graph algorithm maintains a given property $\mathcal{P}$ on a graph subject to dynamic changes, such as edge insertions, edge deletions and edge weight updates. A dynamic graph algorithm should process queries on property $\mathcal{P}$ quickly, and perform update operations faster than recomputing from scratch, as carried out by the fastest static algorithm. An algorithm is *fully dynamic* if it can handle both edge insertions and edge deletions and *partially dynamic* if it can handle either edge insertions or edge deletions, but not both.

Given a graph with n vertices and m edges, the *transitive closure* (or *reachability*) problem consists of building an $n \times n$ Boolean matrix M such that $M[x, y] = 1$ if and only if there is a directed path from vertex x to vertex y in the graph. The fully dynamic version of this problem can be defifined as follows:

Definition 1 (Fully dynamic reachability problem) The *fully dynamic reachability problem* consists of maintaining a directed graph under an intermixed sequence of the following operations:

- `insert(u,v)`: insert edge (u,v) into the graph.
- `delete(u,v)`: delete edge (u,v) from the graph.
- `reachable(x,y)`: return *true* if there is a directed path from vertex x to vertex y, and *false* otherwise.

This entry addresses the *single-source* version of the fully-dynamic reachability problem, where one is only interested in queries with a fixed source vertex s. The problem is defined as follows:

Definition 2 (Single-source fully dynamic reachability problem) The *fully dynamic single-source reachability problem* consists of maintaining a directed graph under an intermixed sequence of the following operations:

- `insert(u,v)`: insert edge (u,v) into the graph.
- `delete(u,v)`: delete edge (u,v) from the graph.
- `reachable(y)`: return *true* if there is a directed path from the source vertex s to vertex y, and *false* otherwise.

Approaches

A simple-minded solution to the problem of Definition would be to keep explicit reachability information from the source to all other vertices and update it by running any graph traversal algorithm from the source s after each insert or delete. This takes $O(m + n)$ time per operation, and then reachability queries can be answered in constant time.

Another simple-minded solution would be to answer queries by running a point-to-point reachability computation, without the need to keep explicit reachability information up to date after each insertion or deletion. This can be done in $O(m + n)$ time using any graph traversal algorithm. With this approach, queries are answered in $O(m + n)$ time and updates require constant time. Notice that the time required by the slowest operation is $O(m+n)$ for both approaches, which can be as high as $O(n^2)$ in the case of dense graphs.

The first improvement upon these two basic solutions is due to Demetrescu and Italiano, who showed how to support update operations in $O(n^{1.575})$ time and reachability queries in $O(1)$ time [1] in a directed acyclic graph. The result is based on a simple reduction of the single-source problem of Definition to the all-pairs problem of Definition. Using a result by Sankowski [2], the bounds above can be extended to the case of general directed graphs.

Key Results

This Section presents a simple reduction presented in [1] that allows it to keep explicit single-source reachability information up to date in subquadratic time per operation in a directed graph subject to an intermixed sequence of edge insertions and edge deletions. The bounds reported in this entry were originally presented for the case of directed acyclic graphs, but can be extended to general directed graphs using the following theorem from [2]:

Theorem 1 *Given a general directed graph with n vertices, there is a data structure for the fully dynamic reachability problem that supports each insertion/deletion in $O(n^{1.575})$ time and each reachability query in $O(n^{0.575})$ time. The algorithm is randomized with one-sided error.*

The idea described in [1] is to maintain reachability information from the source vertex s to all other vertices explicitly by keeping a Boolean array R of size n such that $R[y] = 1$ if and only if there is a directed path from s to y. An instance D of the data structure for fully dynamic reachability of Theorem is also maintained. After each insertion or deletion, it is possible to update D in $O(n^{1.575})$ time and then rebuild R in $O(n \cdot n^{0.575}) = O(n^{1.575})$ time by letting $R[y] \leftarrow D$. reachable (s,y) for each vertex y. This yields the following bounds for the single-source fully dynamic reachability problem:

Theorem 2 *Given a general directed graph with n vertices, there is a data structure for the single-source fully dynamic reachability problem that supports each insertion/deletion in $O(n^{1.575})$ time and each reachability query in $O(1)$ time.*

Applications

The graph reachability problem is particularly relevant to the field of databases for supporting transitivity queries on dynamic graphs of relations [3]. The problem also arises in many other areas such as compilers, interactive verification systems, garbage collection, and industrial robotics.

Open Problems

An important open problem is whether one can extend the result described in this entry to maintain fully dynamic single-source shortest paths in subquadratic time per operation.

Cross-References

▶ Trade-Offs for Dynamic Graph Problems

Recommended Reading

1. Demetrescu C, Italiano G (2005) Trade-offs for fully dynamic reachability on dags: breaking through the $O(n^2)$ barrier. J Assoc Comput Mach 52: 147–156
2. Sankowski P (2004) Dynamic transitive closure via dynamic matrix inverse. In: FOCS '04: proceedings of the 45th annual IEEE symposium on foundations of computer science (FOCS'04). IEEE Computer Society, Washington, DC, pp 509–517
3. Yannakakis M (1990) Graph-theoretic methods in database theory. In: Proceedings of the 9-th ACM SIGACT-SIGMOD-SIGART symposium on principles of database systems, Nashville, pp 230–242

Single-Source Shortest Paths

Seth Pettie
Electrical Engineering and Computer Science (EECS) Department, University of Michigan, Ann Arbor, MI, USA

Keywords

Shortest route; Quickest route

Years and Authors of Summarized Original Work

1999; Thorup

Problem Definition

The *single-source* shortest path problem (SSSP) is, given a graph $G = (V, E, l)$ and a *source* vertex $s \in V$, to find the shortest path from s to every $v \in V$. The difficulty of the problem depends on whether the graph is directed or undirected and the assumptions placed on the length function ℓ. In the most general situation, $l : E \to \mathbb{R}$ assigns arbitrary (positive and negative) real lengths. The algorithms of Bellman-Ford and Edmonds [1, 4] may be applied in this situation and have running times of roughly $O(mn)$, (Edmonds's algorithm works for undirected graphs and presumes that there are no negative length simple cycles.) where $m = |E|$ and $n = |V|$ are the number of edges and vertices. If ℓ assigns only *nonnegative* real edge lengths, then the algorithms of Dijkstra and Pettie-Ramachandran [4, 13] may be applied on directed and undirected graphs, respectively. These algorithms include a *sorting bottleneck* and, in the worst case, take $\Omega(m + n \log n)$ time. (The [13] algorithm actually runs in $O(m + n \log \log n)$ time if the ratio of any two edge lengths is polynomial in n).

A common assumption is that ℓ assigns *integer* edge lengths in the range $\{0, \ldots, 2^w - 1\}$ or $\{-2^{w-1}, \ldots, 2^{w-1} - 1\}$ and that the machine is a w-bit *word RAM*; that is, each edge length fits in one register. For general integer edge lengths, the best SSSP algorithms improve on Bellman-Ford and Edmonds by a factor of roughly $\sqrt{n}$ [6]. For nonnegative integer edge lengths, the best SSSP algorithms are faster than Dijkstra and Pettie-Ramachandran by up to a logarithmic factor. They are frequently based on integer priority queues [9].

Key Results

Thorup's primary result [16] is an optimal linear time SSSP algorithm for undirected graphs with integer edge lengths. This is the first and only linear time shortest path algorithm that does not make serious assumptions on the class of input graphs.

Theorem 1 *There is a SSSP algorithm for integer-weighted undirected graphs that runs in $O(m)$ time.*

Thorup avoids the sorting bottleneck inherent in Dijkstra's algorithm by precomputing (in linear time) a *component hierarchy*. The algorithm of [16] operates in a manner similar to Dijkstra's algorithm [4] but uses the component hierarchy to identify groups of vertices that can be visited in any order. In later work, Thorup [17] extended this approach to work when the edge lengths are floating-point numbers. (There is some flexibility in the definition of *shortest path* since floating-point addition is neither commutative nor associative).

Thorup's hierarchy-based approach has since been extended to directed and/or real-weighted graphs and to solve the *all pairs* shortest path (APSP) problem [11–13]. The generalizations to related SSSP problems are summarized below. See [11, 12] for hierarchy-based APSP algorithms.

Theorem 2 (Hagerup [8], 2000) *A component hierarchy for a directed graph $G = (V, E, l)$, where $l : E \to \{0, \ldots, 2^w - 1\}$, can be constructed in $O(m \log w)$ time. Thereafter, SSSP from any source can be computed in $O(m + n \log \log n)$ time.*

Theorem 3 (Pettie and Ramachandran [13], 2005) *A component hierarchy for an undirected graph $G = (V, E, l)$, where $l : E \to \mathbb{R}^+$, can be constructed in $O(m\alpha(m, n) + \min\{n \log \log r, n \log n\})$ time, where r is the ratio of the maximum-to-minimum edge length. Thereafter, SSSP from any source can be computed in $O(m \log \alpha(m, n))$ time.*

The algorithms of Hagerup [8] and Pettie-Ramachandran [13] take the same basic approach as Thorup's algorithm: use some kind of component hierarchy to identify groups of vertices that can safely be visited in any order. However, the assumption of directed graphs [8] and real edge lengths [13] renders Thorup's hierarchy inapplicable or inefficient. Hagerup's component hierarchy is based on a directed analogue of the minimum spanning tree. The Pettie-Ramachandran algorithm enforces a certain degree of balance in its component hierarchy and, when computing SSSP, uses a specialized priority queue to take advantage of this balance.

Applications

Shortest path algorithms are frequently used as a subroutine in other optimization problems, such as flow and matching problems [1] and facility location [18]. A widely used commercial application of shortest path algorithms is finding efficient routes on road networks, e.g., as provided by Google Maps, MapQuest, or Yahoo Maps.

S

Open Problems

Thorup's SSSP algorithm [16] runs in linear time and is therefore optimal. The main open problem is to find a linear time SSSP algorithm that works on *real*-weighted *directed* graphs. For real-weighted undirected graphs, the best running time is given in Theorem 3. For integer-weighted directed graphs, the fastest algorithms are based on Dijkstra's algorithm (not Theorem 2) and run in $O(m\sqrt{\log\log n})$ time (randomized) and deterministically in $O(m + n\log\log n)$ time.

Problem 1 Is there an $O(m)$ time SSSP algorithm for integer-weighted directed graphs?

Problem 2 Is there an $O(m) + o(n\log n)$ time SSSP algorithm for real-weighted graphs, either directed or undirected?

The complexity of SSSP on graphs with positive and negative edge lengths is also open.

Experimental Results

Asano and Imai [2] and Pettie et al. [14] evaluated the performance of the hierarchy-based SSSP algorithms [13,16]. There have been a number of studies of SSSP algorithms on integer-weighted directed graphs; see [7] for the latest and references to many others. The trend in recent years is to find practical preprocessing schemes that allow for very quick point-to-point shortest path queries. See [3, 10, 15] for recent work in this area.

Data Sets

See [5] for a number of US and European road networks.

URL to Code

See [5].

Cross-References

▶ All Pairs Shortest Paths via Matrix Multiplication

Recommended Reading

1. Ahuja RK, Magnati TL, Orlin JB (1993) Network flows: theory, algorithms, and applications. Prentice Hall, Englewood Cliffs
2. Asano Y, Imai H (2000) Practical efficiency of the linear-time algorithm for the single source shortest path problem. J Oper Res Soc Jpn 43(4):431–447
3. Bast H, Funke S, Matijevic D, Sanders P, Schultes D (2007) In transit to constant shortest-path queries in road networks. In: Proceedings 9th workshop on algorithm engineering and experiments (ALENEX), New Orleans
4. Cormen TH, Leiserson CE, Rivest RL, Stein C (2001) Introduction to algorithms. MIT, Cambridge
5. Demetrescu C, Goldberg AV, Johnson D (2006) 9th DIMACS implementation challenge-shortest paths. http://www.dis.uniroma1.it/~challenge9/
6. Goldberg AV (1995) Scaling algorithms for the shortest paths problem. SIAM J Comput 24(3):494–504
7. Goldberg AV (2001) Shortest path algorithms: engineering aspects. In: Proceedings of the 12th international symposium on algorithms and computation (ISAAC), Christchurch. LNCS, vol 2223. Springer, Berlin, pp 502–513
8. Hagerup T (2000) Improved shortest paths on the word RAM. In: Proceedings of the 27th international colloquium on automata, languages, and programming (ICALP), Geneva. LNCS, vol 1853. Springer, Berlin, pp 61–72
9. Han Y, Thorup M (2002) Integer sorting in $O(n\sqrt{\log\log n})$ expected time and linear space. In: Proceedings of the 43rd symposium on foundations of computer science (FOCS), Vancouver, pp 135–144
10. Knopp S, Sanders P, Schultes D, Schulz F, Wagner D (2007) Computing many-to-many shortest paths using highway hierarchies. In: Proceedings of the 9th workshop on algorithm engineering and experiments (ALENEX), New Orleans
11. Pettie S (2002) On the comparison-addition complexity of all-pairs shortest paths. In: Proceedings of the 13th international symposium on algorithms and computation (ISAAC), Vancouver, pp 32–43
12. Pettie S (2004) A new approach to all-pairs shortest paths on real-weighted graphs. Theor Comput Sci 312(1):47–74
13. Pettie S, Ramachandran V (2005) A shortest path algorithm for real-weighted undirected graphs. SIAM J Comput 34(6):1398–1431
14. Pettie S, Ramachandran V, Sridhar S (2002) Experimental evaluation of a new shortest path algorithm. In: Proceedings of the 4th workshop on algorithm

engineering and experiments (ALENEX), San Francisco, pp 126–142
15. Sanders P, Schultes D (2006) Engineering highway hierarchies. In: Proceedings of the 14th European symposium on algorithms (ESA), Zurich, pp 804–816
16. Thorup M (1999) Undirected single-source shortest paths with positive integer weights in linear time. J ACM 46(3):362–394
17. Thorup M (2000) Floats, integers, and single source shortest paths. J Algorithms 35:189–201
18. Thorup M (2003) Quick and good facility location. In: Proceedings of the 14th annual ACM-SIAM symposium on discrete algorithms (SODA), Baltimore, pp 178–185

Ski Rental Problem

Mark S. Manasse
Microsoft Research, Mountain View, CA, USA

Keywords

Metrical task systems; Oblivious adversaries; Worst-case approximation

Years and Authors of Summarized Original Work

1990; Karlin, Manasse, McGeogh, Owicki

Problem Definition

The ski rental problem was developed as a pedagogical tool for understanding the basic concepts in some early results in online algorithms. (In the interest of full disclosure, the earliest presentations of these results described the problem as the wedding-tuxedo-rental problem. Objections were presented that this was a gender-biased name for the problem, since while groomsmen can rent their wedding apparel, bridesmaids usually cannot. A further complication, owing to the difficulty of instantaneously producing fitted garments or ski equipment outlined below, suggests that some complications could have been avoided by focusing on the dilemma of choosing between daily lift passes or season passes, although this leads to the pricing complexities of purchasing season passes well in advance of the season, as opposed to the higher cost of purchasing them at the mountain during the ski season. A similar problem could be derived from the question as to whether to purchase the daily newspaper at a newsstand or to take a subscription, after adding the challenge that one's peers will treat one contemptuously if one has not read the news on days on which they have.) The ski rental problem considers the plight of one consumer who, in order to socialize with peers, is forced to engage in a variety of athletic activities, such as skiing, bicycling, windsurfing, rollerblading, sky diving, scuba-diving, tennis, soccer, and ultimate Frisbee, each of which has a set of associated apparatus, clothing, or protective gear.

In all of these, it is possible either to purchase the accoutrements needed or to rent them. For the purpose of this problem, it is assumed that one-time rental is less expensive than purchasing. It is also assumed that purchased items are durable, and suitable for reuse for future activities of the same type without further expense, until the items wear out (which occurs at the same rate for all users), are outgrown, become unfashionable, or are disposed of to make room for other purchased items. The social consumer must make the decision to rent or buy for each event, although it is assumed that the consumer is sufficiently parsimonious as to abjure rental if already in possession of serviceable purchased equipment. Whether purchases are as easy to arrange as rentals, or whether some advance planning is required (e.g., to mount bindings on a ski) is a further detail considered in this problem. It is assumed that the social consumer has no particular independent interest in these activities, and engages in these activities only to socialize with peers who choose to engage in these activities disregarding the consumer's desires.

These putative peers are more interested in demonstrating the superiority of their financial acumen to that of the social consumer in question than they are in any particular activity. To that end, the social consumer is taunted mercilessly based on the ratio of his/her total expenses on rentals and purchases to theirs. Consequently, the peers endeavor to invite the social consumer to engage in events while they are costly to him/her, and once the activities are free to the social consumer, if continued activity would be costly to them, cease. But, to present an illusion of fairness, skis, both rented and purchased, have the same cost for the peers as they do for the social consumer in question. The ski rental problem takes a very restricted setting. It assumes that purchased ski equipment never needs replacement, and that there are no costs to a ski trip other than the skis (thus, no cost for the gasoline, for the lift and/or speeding tickets, for the hot chocolates during skiing, or for the après-ski liqueurs and meals). It is assumed that the social consumer experiences no physical disabilities preventing him/her from skiing and has no impending restrictions to his/her participation in ski trips (obviously, a near-term-fatal illness or an anticipated conviction leading to confinement for life in a penitentiary would eliminate any potential interest in purchasing alpine equipment – when the ratio of purchase to rental exceeds the maximum need for equipment, one should always rent). It is assumed that the social consumer's peers have disavowed any interest in activities other than skiing, and that the closet, basement, attic, garage, or storage locker included in the social consumer's rent or mortgage (or necessitated by other storage needs) has sufficient capacity to hold purchased ski equipment without entailing the disposal of any potentially useful items. Bringing these complexities into consideration brings one closer to the hardware-based problems which initially inspired this work.

The impact of invitations issued with sufficient time allowed for purchasing skis, as well as those without, will be considered.

Given all of that, what ratio of expenses can the social consumer hope to attain? What ratio can the social consumer not expect to beat? These are the basic questions of competitive analysis.

The impact of keeping secrets from one's peers is further considered. Rather than a fixed strategy for when to purchase skis, the social consumer may introduce an element of chance into the process. If the peers are able to observe his/her ski equipment and notice when it changes from rented skis to purchased skis, and change their schedule for alpine recreation in light of this observation, randomness provides no advantages. If, on the other hand, the social consumer announces to the peers, in advance of the first trip, how he/she will decide when the time is right for purchasing skis, including any use of probabilistic techniques, and they then decide on the schedule for ski trips for the coming winter, a deterministic decision procedure generally produces a larger competitive ratio than does a randomized procedure.

Key Results

Given an unbounded sequence of skiing trips, one should eventually purchase skis if the cost of renting skis, r, is positive. In particular, let the cost of purchasing skis be some number $p \geq r$. If one never intends to make a purchase, one's cost for the season will be rn, where n is the number of ski trips in which one participates. If n exceeds p/r, one's cost will exceed the price of purchasing skis; as n continues to increase, the ratio of one's costs to those of one's peers increases to nr/p, which grows unboundedly with n, since your peers, knowing that n exceeds p/r, will have purchased skis prior to the first trip.

On the other hand, if one rushes out to purchase skis upon being told that the ski season is approaching, one's peers will decide that this season looks inopportune, and that skiing is passé, leaving their costs at zero, and one's costs at p, leaving an infinite ratio between one's costs and theirs; if one chooses to defer the purchase until after one's first ski trip, this produces the less unfavorable ratio p/r or $1 + p/r$, depending on whether the invitation left one time to purchase skis before the first trip or not.

Suppose one chooses, instead, to defer one's purchase until after one has made k rentals, but before ski trip $k + 1$. One's costs are then bounded by $kr + p$. After k ski trips, the cost to one's peers will be the lesser of kr and p (as one's peers will have decided whether to rent or buy for the season upon knowing one's plans, which in this case amounts to knowing k), for a ratio equal to the larger of $1 + kr/p$ and $1 + p/kr$. Were they to choose to terminate the activity earlier (so $n < k$), the ratio would be only the greater of kr/p and 1, which is guaranteed to be less than the sum of the two – one's peers would be shirking their opportunity to make one's behavior look foolish were they to allow one to stop skiing prior to one's purchase of a pair of skis!

It is certain, since kr/p and p/kr are reciprocals, that one of them is at least equal to 1, ensuring that one will be compelled to spend at least twice as much as one's peers.

The analysis above applies to the case where ski trips are announced without enough warning to leave one time to buy skis. Purchases in that case are not instantaneous; in contrast, if one is able to purchase skis on demand, the cost to one's peers changes to the lesser of $(k + 1)r$ and p. The overall results are not much different; the ratio choices become the larger of $1 + kr/p$ and $1 + (p - r)/((k + 1)r)$.

When probabilistic algorithms are considered with oblivious frenemies (those who know the way in which random choices will affect one's purchasing decisions, but who do not take time to notice that one's skis are no longer marked with the name and phone number of a rental agency), one can appear more thrifty.

A randomized algorithm can be viewed as a distribution over deterministic algorithms. No good algorithm can purchase skis prior to the first invitation, lest it exhibit infinite regrettability (some positive cost compared to zero). A good algorithm must purchase skis by the time one's peers will have; otherwise, one's cost ratio continues to increase with the number of ski trips. Moreover, the ratio should be the same after every ski trip; if not, then there is an earliest ratio not equal to the largest, and probabilities can be adjusted to change this earliest ratio to be closer to the largest while decreasing all larger ratios.

Consider, for example, the case of $p = 2r$, with purchases allowed at the time of an invitation. The best deterministic ratio in this case is 1.5. It is only necessary to choose a probability q, the probability of purchasing at the time of

the first invitation. The cost after one trip is then $(1-q)r+2qr = r(1+q)$, for a ratio of $1+q$, and after two trips the cost is $q(2r) + (1 - q)(3r) = 3 - q)r$, producing a ratio of $(3 - q)/2$. Setting these to be equal yields $q = 1/3$, for a ratio of 4/3.

If insufficient time is allowed for purchases before skiing, the best deterministic ratio is 2. Purchasing after the first ski trip with probability q (and after the second with probability $1 - q$) leads to expected costs of $(1-q)r+3qr = r(1+2q)$ after the first trip, and $(1-q)(2+2)r+3qr = r(4 + q)$, leading to a ratio of $2 - q/2$. Setting $1 + 2q = 2 - q/2$ yields $q = 2/5$, for a ratio of 9/5.

More careful analysis, for which readers are referred to the references and the remainder of this volume, shows that the best achievable ratio approaches $\epsilon/(\epsilon - 1) \approx 1.58197$ as p/r increases, approaching the limit from below if sufficient warning time is offered, and from above otherwise.

Applications

The primary initial results were directed towards problems of computer architecture; in particular, design questions for capacity conflicts in caches, and shared memory design in the presence of a shared communication channel. The motivation for these analyses was to find designs which would perform reasonably well on as-yet-unknown workloads, including those to be designed by competitors who may have chosen alternative designs which favor certain cases. While it is probably unrealistic to assume that precisely the least-desirable workloads will occur in ordinary practice, it is not unreasonable to assume that extremal workloads favoring either end of a decision will occur.

History and Further Reading

This technique of finding algorithms with bounded worst-case performance ratios is common in analyzing approximation algorithms.

The initial proof techniques used for such analyses (the method of amortized analysis) were first presented by Sleator and Tarjan.

The reader is advised to consult the remainder of this volume for further extensions and applications of the principles of competitive online algorithms.

Cross-References

▶ Algorithm DC-TREE for k-Servers on Trees
▶ Metrical Task Systems
▶ Online List Update
▶ Online Paging and Caching
▶ Work-Function Algorithm for k-Servers

Recommended Reading

1. Karlin AR, Manasse MS, Rudolph L, Sleator DD (1988) Competitive snoopy caching. Algorithmica 3:77–119 (Conference version: FOCS 1986, pp 244–254)
2. Karlin AR, Manasse MS, McGeoch LA, Owicki SS (1994) Competitive randomized algorithms for nonuniform problems. Algorithmica 11(6):542–571 (Conference version: SODA 1990, pp 301–309)
3. Reingold N, Westbrook J, Sleator DD (1994) Randomized competitive algorithms for the list update problem. Algorithmica 11(1):15–32 (Conference version included author Irani S: SODA 1991, pp 251–260)

Slicing Floorplan Orientation

Evangeline F.Y. Young
Department of Computer Science and Engineering, The Chinese University of Hong Kong, Hong Kong, China

Keywords

Shape curve computation

Years and Authors of Summarized Original Work

1983; Stockmeyer

Problem Definition

This problem is about finding the optimal orientations of the cells in a slicing floorplan to minimize the total area. In a floorplan, cells represent basic pieces of the circuit which are regarded as indivisible. After performing an initial placement, for example, by repeated application of a min-cut partitioning algorithm, the relative positions between the cells on a chip are fixed. Various optimizations can then be done on this initial layout to optimize different cost measures such as chip area, interconnect length, routability, etc. One such optimization, as mentioned in Lauther [3], Otten [4], and Zibert and Saal [13], is to determine the best orientation of each cell to minimize the total chip area. This work by Stockmeyer [8] gives a polynomial time algorithm to solve the problem optimally in a special type of floorplans called *slicing floorplans* and shows that this orientation optimization problem in general non-slicing floorplans is NP-complete.

Slicing Floorplan

A floorplan consists of an enclosing rectangle subdivided by horizontal and vertical line segments into a set of non-overlapping *basic rectangles*. Two different line segments can meet but not cross. A floorplan F is characterized by a pair of planar acyclic directed graphs A_F and L_F defined as follows. Each graph has one source and one sink. The graph A_F captures the "above" relationships and has a vertex for each horizontal line segment, including the top and the bottom of the enclosing rectangle. For each basic rectangle R, there is an edge e_r directed from segment σ to segment σ' if and only if σ (or part of σ) is the top of R and σ' (or part of σ') is the bottom of R. There is a one-to-one correspondence between the basic rectangles and the edges in A_F. The graph L_F is defined similarly for the

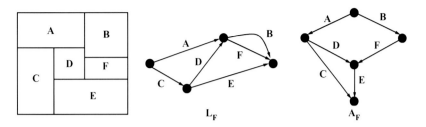

Slicing Floorplan Orientation, Fig. 1 A floorplan F and its A_F and L_F representing the above and left relationships

Slicing Floorplan Orientation, Fig. 2 A slicing floorplan F and its slicing tree representation

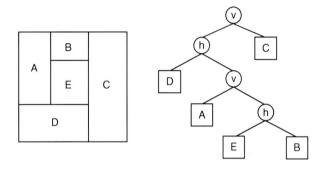

"left" relationships of the vertical segments. An example is shown in Fig. 1. Two floorplans F and G are equivalent if and only if $A_F = A_G$ and $L_F = L_G$. A floorplan F is slicing if and only if both its A_F and L_F are series parallel.

Slicing Tree

A slicing floorplan can also be described naturally by a rooted binary tree called *slicing tree*. In a slicing tree, each internal node is labeled by either an h or a v, indicating a horizontal or a vertical slice respectively. Each leaf corresponds to a basic rectangle. An example is shown in Fig. 2. There can be several slicing trees describing the same slicing floorplan, but this redundancy can be removed by requiring the label of an internal node to differ from that of its right child [12]. For the algorithm presented in this work, a tree of smallest depth should be chosen, and this depth minimization process can be done in $O(n \log n)$ time using the algorithm by Golumbic [2].

Orientation Optimization

In optimization of a floorplan layout, some freedom in moving the line segments and in choosing the dimensions of the rectangles are allowed. In the input, each basic rectangle R has two positive

integers a_R and b_R, representing the dimensions of the cell that will be fit into R. Each cell has two possible orientations resulting in either the side of length a_R or b_R being horizontal. Given a floorplan F and an orientation p, each edge e in A_F and L_F is given a label $l(e)$ representing the height or the width of the cell corresponding to e depending on its orientation. Define an (F, ρ)-*placement* to be a labeling l of the vertices in A_F and L_F such that (i) the sources are labeled by zero and (ii) if e is an edge from vertex σ to $\sigma', l(\sigma') \geq l(\sigma) + l(e)$. Intuitively, if σ is a horizontal segment, $l(\sigma)$ is the distance of σ from the top of the enclosing rectangle, and the inequality constraint ensures that the basic rectangle corresponding to e is tall enough for the cell contained in it and similarly for the vertical segments. Now, $h_F(\rho)$ (resp. $w_F(\rho)$) is defined to be the minimum label of the sink in $A_F(\rho)$ (resp. $L_F(\rho)$) over all (F, ρ)-placements, where $A_F(\rho)$ (resp. $L_F(\rho)$) is obtained from A_F (resp. L_F) by labeling the edges and vertices as described above. Intuitively, $h_F(\rho)$ and $w_F(\rho)$ give the minimum height and width of a floorplan F given an orientation ρ of all the cells such that each cell fits well into its associated basic rectangle.

The orientation optimization problem can be defined formally as follows:

Problem 1 (Orientation Optimization Problem for Slicing Floorplan)

INPUT: A slicing floorplan F of n cells described by a slicing tree T, the widths and heights of the cells a_i and b_i for $i = 1 \ldots n$, and a cost function $\psi(h, w)$.

OUTPUT: An orientation ρ of all the cells that minimizes the objective function $\psi(h_F(\rho), w_F(\rho))$ over all orientations ρ.

For this problem, Lauther [3] has suggested a greedy heuristic. Zibert and Saal [13] use integer programming methods to do rotation optimization and several other optimization simultaneously for general floorplans. In the following sections, an efficient algorithm will be given to solve the problem optimally in $O(nd)$ time where n is the number of cells and d is the depth of the given slicing tree.

Key Results

In the following algorithm, $F(u)$ denotes the floorplan described by the subtree rooted at u in the given slicing tree T, and let $L(u)$ be the set of leaves in that subtree. For each node u of T, the algorithm constructs recursively a list of pairs:

$$\{(h_1, w_1), (h_2, w_2), \ldots, (h_m, w_m)\}$$

where (1) $m \leq |L(u)| + 1$, (2) $h_i > h_{i+1}$ and $w_i < w_{i+1}$ for $i = 1 \ldots m - 1$, (3) there is an orientation ρ of the cells in $L(u)$ such that $(h_i, w_i) = (h_{F(u)}(\rho), w_{F(u)}(\rho))$ for each $i = 1 \ldots m$, and (4) for each orientation ρ of the cells in $L(u)$, there is a pair (h_i, w_i) in the list such that $h_i \leq h_{F(u)}(\rho)$ and $w_i \leq w_{F(u)}(\rho)$.

$L(u)$ is thus a non-redundant list of all possible dimensions of the floorplan described by the subtree rooted at u. Since the cost function ψ is non-decreasing, it can be minimized over all orientations by finding the minimum $\psi(h_i, w_i)$ over all the pairs (h_i, w_i) in the list constructed at the root of T. At the beginning, a list is constructed at each leaf node of T representing

the possible dimensions of the cell. If a leaf cell has dimensions a and b with $a > b$, the list is $\{(a, b), (b, a)\}$. If $a = b$, there will just be one pair (a, b) in the list. (If the cell has a fixed orientation, there will also be just one pair as defined by the fixed orientation.) Notice that the condition (1) above is satisfied in these leaf node lists. The algorithm then works its way up the tree and constructs the list at each node recursively. In general, assume that u is an internal node with children v and v' and u represents a vertical slice. Let $\{(h_1, w_1) \ldots (h_k, w_k)\}$ and $\{(h'_1, w'_1) \ldots (h'_m, w'_m)\}$ be the lists at v and v' respectively where $k \leq |L(v)| + 1$ and $m \leq |L(v')| + 1$. A pair (h_i, wi) from v can be put together by a vertical slice with a pair (h'_j, w'_j) from v' to give a pair:

$$\text{join}((h_i, w_i), (h'_j, w'_j)) = (\max(h_i, h'_j), w_i + w'_j)$$

in the list of u (see Fig. 3). The key fact is that most of the km pairs are sub-optimal and do not need to be considered. For example, if $h_i > h'_j$, there is no need to join (h_i, w_i) with (h'_z, w'_z) for any $z > j$ since

$$\max(h_i, h'_z) = \max(h_i, h'_j) = h_i,$$
$$w_i + w'_z > w_i + w'_j$$

Similarly, if node u represents a horizontal slice, the join operation will be

$$\text{join}((h_i, w_i), (h'_j, w'_j)) = (h_i + h'_j, \max(w_i, w'_j))$$

The algorithm also keeps two pointers for each element in the lists in order to construct back the optimal orientation at the end. The algorithm is summarized by the following pseudocode:

Pseudocode *Stockmeyer()*
1. Initialize the list at each leaf node.
2. Traverse the tree in postorder. At each internal node u with children v and v', construct a list at node u as follows:
3. Let $\{(h_1, w_1) \ldots (h_k, w_k)\}$ and $\{(h'_1, w'_1) \ldots (h'_m, w'_m)\}$ be the lists at v and v' respectively.
4. Initialize i and j to one.

Slicing Floorplan Orientation, Fig. 3 An illustration of the merging step

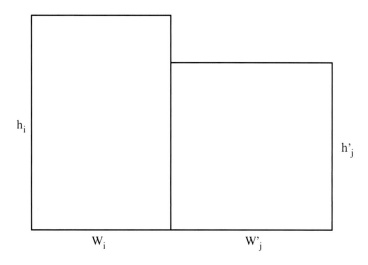

5. If $i > k$ or $j > m$, the whole list at u is constructed.
6. Add $\text{join}((h_i, w_i), (h'_j, w'_j))$ to the list with pointers pointing to (h_i, w_i) and (h'_j, w'_j) in $L(v)$ and $L(v')$ respectively.
7. If $h_i > h'_j$, increment i by 1.
8. If $h_i > h'_j$, increment j by 1.
9. If $h_i > h'_j$, increment both i and j by 1.
10. Go to step 5
11. Compute $\psi(h_i, w_i)$ for each pair (h_i, w_i) in the list L_r at the root r of T.
12. Return the minimum $\psi(h_i, w_i)$ for all (h_i, w_i) in L_r and construct back the optimal orientation by following the pointers.

Correctness

The algorithm is correct since at each node u, a list is constructed that records all the possible non-redundant dimensions of the floorplan described by the subtree rooted at u. This can be proved easily by induction starting from the leaf nodes and working up the tree recursively. Since the cost function ψ is non-decreasing, it can be minimized over all orientations of the cells by finding the minimum $\psi(h_i, w_i)$ over all the pairs (h_i, w_i) in the list L_r constructed at the root r of T.

Runtime

At each internal node u with children v and v'. If the lengths of the lists at v and v' are k and m respectively, the time spent at u to combine the

two lists is $O(k + m)$. Each possible dimension of a cell will thus invoke one unit of execution time at each node on its path up to the root in the postorder traversal. The total runtime is thus $O(d \times N)$ where N is the total number of realizations of all the n cells, which is equal to $2n$ in the orientation optimization problem. Therefore, the runtime of this algorithm is $O(nd)$.

Theorem 1 *Let* $\psi(h, w)$ *be non-decreasing in both arguments, i.e., if* $h \leq h'$ *and* $w \leq w'$, $\psi(h, w) \leq \psi(h', w')$, *and computable in constant time. For a slicing floorplan* F *described by a binary slicing tree* T, *the problem of minimizing* $\psi(h_F(\rho), w_F(\rho))$ *over all orientations* ρ *can be solved in time* $O(nd)$ *time, where* n *is the number of leaves of* T *(equivalently, the number of cells of* F) *and* d *is the depth of* T.

Applications

Floorplan design is an important step in the physical design of VLSI circuits. Stockmeyer's optimal orientation algorithm [8] has been generalized to solve the area minimization problem in slicing floorplans [7], in hierarchical non-slicing floorplans of order five [6,9], and in general floorplans [5]. The floorplan area minimization problem is similar except that each *soft cell* now has a number of possible realizations, instead of just two different orientations. The same technique can be applied immediately to solve optimally the area minimization problem for slicing floorplans in $O(nd)$ time where n is the total number of realizations of all the cells in a given floorplan F and d is the depth of the slicing tree of F. Shi [7] has further improved this result to $O(n \log n)$ time. This is done by storing the list of nonredundant pairs at each node in a balanced binary search tree structure called *realization tree* and using a new merging algorithm to combine two such trees to create a new one. It is also proved in [7] that this $O(n \log n)$ time complexity is the lower bound for this area minimization problem in slicing floorplans.

For hierarchical non-slicing floorplans, Pan et al. [6] prove that the problem is NP-complete. Branch-and-bound algorithms are developed by Wang and Wong [9], and pseudopolynomial time algorithms are developed by Wang and Wong [10] and Pan et al. [6]. For general floorplans, Stockmeyer [8] has shown that the problem is strongly NP-complete. It is therefore unlikely to have any pseudopolynomial time algorithm. Wimer et al. [11] and Chong and Sahni [1] propose branch-and-bound algorithms. Pan et al. [5] develop algorithms for general floorplans that are approximately slicing.

Recommended Reading

1. Chong K, Sahni S (1993) Optimal realizations of floorplans. IEEE Trans Comput Aided Des 12(6):793–901
2. Golumbic MC (1976) Combinatorial merging. IEEE Trans Comput C-25:1164–1167
3. Lauther U (1980) A Min-cut placement algorithm for general cell assemblies based on a graph representation. J Digit Syst 4:21–34
4. Otten RHJM (1982) Automatic floorplan design. In: Proceedings of the 19th design automation conference, Las Vegas, pp 261–267
5. Pan P, Liu CL (1995) Area minimization for floorplans. IEEE Trans Comput Aided Des 14(1):123–132
6. Pan P, Shi W, Liu CL (1996) Area minimization for hierarchical floorplans. Algorithmica 15(6):550–571
7. Shi W (1996) A fast algorithm for area minimization of slicing floorplan. IEEE Trans Comput Aided Des 15(12):1525–1532
8. Stockmeyer L (1983) Optimal orientations of cells in slicing floorplan designs. Infect Control 59:91–101
9. Wang TC, Wong DF (1992) Optimal floorplan area optimization. IEEE Trans Comput Aided Des 11(8):992–1002
10. Wang TC, Wong DF (1993) A note on the complexity of Stockmeyer's floorplan optimization technique. In: Algorithmic aspects of VLSI layout. Lecture notes series on computing, vol 2. World Scientific, Singapore, pp 309–320
11. Wimer S, Koren I, Cederbaum I (1989) Optimal aspect ratios of building blocks in VLSI. IEEE Trans Comput Aided Des 8(2):139–145
12. Wong DF, Liu CL (1986) A new algorithm for floorplan design. In: Proceedings of the 23rd ACM/IEEE design automation conference, Las Vegas, pp 101–107
13. Zibert K, Saal R (1974) On computer aided hybrid circuit layout. In: Proceedings of the IEEE international symposium on circuits and systems, San Francisco, pp 314–318

Sliding Window Algorithms

Vladimir Braverman
Department of Computer Science, Johns Hopkins University, Baltimore, MD, USA

Keywords

Data streams; Histograms; Randomized algorithms; Sampling; Sketching

Years and Authors of Summarized Original Work

2007; Braverman, Ostrovsky

Problem Definition

In the last decade, the theoretical study of the sliding window model was developed to advance applications with very large input and time-sensitive output. In some practical situations, input might be seen as an ordered sequence, and it is useful to restrict computations to recent portions of the input. Examples include the analysis of recent tweets and time series of the stock market.

To address the aforementioned practical situations, Datar et al. [20] introduced the sliding window model that assumes that the input is a stream (i.e., the ordered sequence) of data elements and divides the data elements into two categories: *active* elements and *expired* elements. Typically, a recent portion (i.e., a suffix) of the stream defines the window of active elements, and the reminder (i.e., a complimenting prefix) of the stream defines the set of expired elements. When a new data element arrives, the set of active elements expands to include the new element, but the set might also shrink by discarding some portion of oldest active elements. This process of additions and expirations reminds one of the movements of an interval (or a window) along a line and explains the name of the model. The number of active elements N is often called a *size of the sliding window*. There are two popular variants of the sliding window model. The variant of a *sequence-based* window fixes the number of active elements N, and every insertion (or arrival) of a new element corresponds to a deletion (or expiration) of the oldest active element (after the size of the stream becomes larger than N). For example, a sequence-based window on a stream of IP packets is a set of last N packets. The variant of a *timestamp-based* window associates each element with a nondecreasing timestamp, and the window contains all elements with timestamps larger than a certain value. Thus, there is no obvious dependence between the number of elements that arrive and expire. In the previous example, the timestamp-based window might be defined as a set of all packets that arrived within the last t seconds.

Formal Definition

We denote the stream D by a sequence of elements $\{p_i\}_{i=1}^m$ where $p_i \in [n]$. It is important to note that m is incremented for each new arrival. A *bucket* $B(x, y) = \{p_i, i \in [x, y]\}$ is the set of all stream elements between p_x and p_y, inclusively. A sequence-based window is defined $W = B(m - N + 1, m)$ where N is a predefined parameter. Consider a nondecreasing *timestamp* function $T : [m] \to R$ and let t be a parameter. Given T and t, a timestamp-based

window is defined as $W = B(l(t), m)$ where $l(t) = \min\{i : T(i) \geq T(m) - t\}$. Consider function f that is defined on buckets. An algorithm maintains a $(1 \pm \epsilon)$-approximation of f on W if, at any moment, the algorithm outputs X s.t. $|f(W) - X| \leq \epsilon f(W)$. Similarly, a randomized algorithm maintains a $(1 \pm \epsilon, \delta)$-approximation if $P(|f(W) - X| > \epsilon f(W)) \leq \delta$. It is often the case that f can be computed precisely if the entire window is available, but sublinear-space approximations, i.e., computation when the size of the available memory is $o(N + n)$, might be challenging. For example, Datar et al. [20] show linear space is required to maintain a $(1 \pm \epsilon, \delta)$-approximation of a sum of active elements if $p_i \in \{1, 0, -1\}$. A typical question in the sliding window model is the following: given function f, what are the upper and lower bounds on the space complexity of maintaining $(1 \pm \epsilon, \delta)$-approximation of f.

History

In their pioneering papers, Datar et al. [20, 21] and Babcock et al. [3] gave the first formal definition of the sliding window model. The model arose in the context of relational databases as a special case of time-sensitive queries in temporal databases [3]. Below we give a short survey of a subset of known results. A survey of Datar and Motwani [1] provides additional details. Datar et al. [20] gave the first algorithms for estimating the count and sum of positive integers, average, L_p for $p \in [1, 2]$, and a wide class of *weakly additive functions*. Gibbons and Tirthapura [24] provided further improvements to count and sum and gave the first methods for distributed computations. Lee and Ting [29] provided an optimal solution for a relaxed version of the counting problem, where the correct answer is provided only if it is comparable with the window's size. Braverman and Ostrovsky [6, 7] extended the results in [20] to a wider class of *smooth functions*. Chi et al. [15] considered a problem of frequent itemsets. Arasu and Manku [2], Lee and Ting [30], and Golab et al. [26] considered the problem of finding frequent elements, frequency counts, and quantiles. Babcock, Datar, Motwani, and O'Callaghan [5] provided first algorithms for variance and k-medians problems. Feigenbaum, Kannan and Zhang [22] presented an efficient solution for the diameter of a data set in multidimensional space. Later, Chan and Sadjad [23] presented optimal solutions for this and other geometric problems. Babcock, Datar and Motwani [4] presented algorithms for uniform random sampling from sliding windows.

Recently, Crouch et al. [17] presented the first approximation algorithms for important graph problems such as combinatorial sparsifiers and spanners, graph matching, and minimum spanning tree. Among other results, the methods in [17] allow non-smooth statistics using a modified smooth histogram to be computed. McGregor provided a detailed survey of these and other graph algorithms [32]. Datar and Muthukrishnan [19] solved problems of rarity and similarity. Braverman et al. [11] gave improved algorithms for rarity, similarity, and L_2-heavy hitters. Cormode and Yi developed several first algorithms for sliding windows in distributed streams [16]. Babcock et al. [4] gave the first method of sampling an element with constant expected space complexity. Braverman et al. [9, 10] gave a solution with a space complexity that is a constant in the worst case. Tatbul and Zdonik [35] considered the problem of load shedding for aggregation queries. Golab and Özsu [25] gave the first algorithm for approximating multi-joins. Recently, Braverman et al. [13] extended the zero-one law for increasing frequency-based functions [8] to sliding windows.

Key Results

Smooth Histogram

Extending the results in [20], Braverman and Ostrovsky [6, 7] introduced a notion of a *smooth function* and presented techniques for approximating smooth functions over sliding windows. Denote by $B \subseteq_r A$ the event when bucket B is a suffix of A; i.e., if $A = \{p_{n_1}, \ldots, p_{n_2}\}$ (for some $n_1 < n_2$), then $B = \{p_{n_3}, \ldots, p_{n_2}\}$, where $n_1 \leq n_3 \leq n_2$. Denote by $A \cup C$ the union of adjacent buckets A and C.

Definition 1 Function f is (α, β)-*smooth* if it preserves the following properties:

1. $f(A) \geq 0$.
2. $f(A) \geq f(B)$ for $B \subseteq_r A$.
3. $f(A) \leq poly(|A|)$.
4. For any $0 < \epsilon < 1$, there exist $\alpha = \alpha(\epsilon, f)$ and $\beta = \beta(\epsilon, f)$ such that
   - $0 < \beta \leq \alpha < 1$.
   - If $B \subseteq_r A$ and $(1 - \beta)f(A) \leq f(B)$, then $(1 - \alpha)f(A \cup C) \leq f(B \cup C)$ for any adjacent C.

In other words, a nonnegative, nondecreasing, and polynomially bounded function f is (α, β)-smooth if the following is true. If $f(B)$ is a $(1 \pm \beta)$-approximation of $f(A)$, then $f(B \cup C)$ is $(1 \pm \alpha)$-approximation of $f(A \cup C)$ for any $B \subseteq_r A$ and C. The main technical result of [7] is a new data structure called "smooth histogram" that allows algorithms for insertion-only streams to be extended to sliding windows with space complexity increased by a polylogarithmic factor. If there exists an algorithm that computes f precisely using g space and h time per element, then a smooth histogram can be used to maintain a $(1 \pm \alpha)$-approximation of f over sliding windows, using $O\left(\frac{1}{\beta} \log n(g + \log n)\right)$ bits and $O\left(\frac{1}{\beta} h \log n\right)$ time. Further, $(1 \pm \rho)$-approximation of f on D results in $(1 \pm (\alpha + \rho))$-approximation of f over sliding windows. Examples of smooth functions include sum, count, min, diameter, weakly additive functions, L_p norms, frequency moments, length of longest subsequence, and geometric mean.

Let f be (α, β)-smooth for which there exists an algorithm Λ that calculates f on D using g space and h operation per element. To maintain f on sliding windows, we construct a data structure that we call *smooth histogram*. It consists of a set of indexes $x_1 < x_2 < \cdots < x_s = N$ and instances of Λ for each bucket $B(x_i, N)$. Informally, the smooth histogram ensures the following properties of the sequence. The first two elements of the sequence always "sandwich" the window, i.e., $x_1 \leq N - n < x_2$. This requirement and the monotonicity of f give us useful bounds for the sliding window W: $f(x_2, N) \leq f(W) \leq f(x_1, N)$. Also, f should slowly but constantly decrease with i, i.e., $f(x_{i+2}, N) < (1 - \beta) f(x_i, N)$. This gradual decrease, together with the fact that f is polynomially bounded, ensures that the sequence is short, i.e., $s = O\left(\frac{1}{\beta} \log n\right)$. Finally, the values of f on successive buckets were close in the past, i.e., $f(x_{i+1}, N') \geq (1 - \beta) f(x_i, N')$ for some $N' \leq N$. This represents our key idea and exploits the properties of smoothness. Indeed, $f(x_2, N') \geq (1-\beta) f(x_1, N')$ for some $N' \leq N$; thus, by the (α, β)-smoothness of f, we have $f(x_2, N) \geq (1 - \alpha) f(x_1, N) \geq (1 - \alpha) f(W)$. We refer a reader to [7] for further technical details.

Applications

There are several applications of the theoretical methods for the sliding window model, for example, [15, 18, 31, 33, 36].

Open Problems

We list several interesting open problems. It would be important to understand the difference between the sliding window model and other streaming models such as the insertion-only model, the turnstile, and decay models. This is perhaps one of the most important unresolved open problems; see, e.g., Sohler [34]. In particular, it would be nice to understand the exact space complexity of the frequency moments that are well understood in the other streaming models [12, 27, 28]. Also, it would be interesting to extend the coreset methods [14] to sliding windows, obtain polylogarithmic solutions for clustering, and improve the first clustering algorithm in [5]. Also, it would be nice to further develop graph methods [17]. Improving the approximation ratio of the maximum matching and obtaining the $O(n^{1+1/t})$

space bound for $(2t - 1)$-spanners are important open problems.

Acknowledgments This material is based upon work supported in part by the National Science Foundation under Grant No. 1447639.

Recommended Reading

1. Aggarwal C (2007) Data streams: models and algorithms. Advances in database systems. http://www.springer.com/west/home/default?SGWID=4-40 356-22-107949228-0
2. Arasu A, Manku GS (2004) Approximate counts and quantiles over sliding windows. In: Proceedings of the twenty-third ACM SIGMOD-SIGACT-SIGART symposium on principles of database systems (PODS'04). ACM, New York, pp 286–296. doi:10.1145/1055558.1055598
3. Babcock B, Babu S, Datar M, Motwani R, Widom J (2002) Models and issues in data stream systems. In: Proceedings of the twenty-first ACM SIGMOD-SIGACT-SIGART symposium on principles of database systems (PODS'02). ACM, New York, pp 1–16. doi:10.1145/543613.543615
4. Babcock B, Datar M, Motwani R (2002) Sampling from a moving window over streaming data. In: Proceedings of the thirteenth annual ACM-SIAM symposium on discrete algorithms (SODA'02). Society for Industrial and Applied Mathematics, Philadelphia, pp 633–634. http://dl.acm.org/citation.cfm?id=545381.545465
5. Babcock B, Datar M, Motwani R, O'Callaghan L (2003) Maintaining variance and k-medians over data stream windows. In: Proceedings of the twenty-second ACM SIGMOD-SIGACT-SIGART symposium on Principles of database systems (PODS'03). ACM, New York, pp 234–243. doi:10.1145/773153.773176
6. Braverman V, Ostrovsky R (2007) Smooth histograms for sliding windows. In: Proceedings of the 48th annual IEEE symposium on foundations of computer science (FOCS'07). IEEE Computer Society, Washington, DC, pp 283–293. doi:10.1109/FOCS.2007.63
7. Braverman V, Ostrovsky R (2010) Effective computations on sliding windows. SIAM J Comput 39(6):2113–2131. doi:10.1137/090749281
8. Braverman V, Ostrovsky R (2010) Zero-one frequency laws. In: Proceedings of the 42nd ACM symposium on theory of computing (STOC'10). ACM, New York, pp 281–290. doi:10.1145/1806689.1806729
9. Braverman V, Ostrovsky R, Zaniolo C (2009) Optimal sampling from sliding windows. In: Proceedings of the twenty-eighth ACM SIGMOD-SIGACT-

S

SIGART symposium on principles of database systems (PODS'09). ACM, New York, pp 147–156. doi:10.1145/1559795.1559818

10. Braverman V, Ostrovsky R, Zaniolo C (2012) Optimal sampling from sliding windows. J Comput Syst Sci 78(1):260–272. doi:10.1016/j.jcss.2011.04.004

11. Braverman V, Gelles R, Ostrovsky R (2013) How to catch 1 2-heavy-hitters on sliding windows. In: Du DZ, Zhang G (eds) Computing and combinatorics. Lecture notes in computer science, vol 7936. Springer, Berlin/Heidelberg, pp 638–650. doi:10.1007/978-3-642-38768-5_56

12. Braverman V, Katzman J, Seidell C, Vorsanger G (2014) An optimal algorithm for large frequency moments using $O(n^{1-2/k})$ bits. In: Proceedings of the 18th international workshop on randomization and computation (RANDOM'2014)

13. Braverman V, Ostrovsky R, Roytman A (2014) Universal Streaming. ArXiv e-prints 1408.2604

14. Chen K (2009) On coresets for k-median and k-means clustering in metric and euclidean spaces and their applications. SIAM J Comput 39(3):923–947. doi:http://dx.doi.org/10.1137/070699007

15. Chi Y, Wang H, Yu PS, Muntz RR (2004) Moment: Maintaining closed frequent itemsets over a stream sliding window. In: ICDM, pp 59–66

16. Cormode G, Yi K (2012) Tracking distributed aggregates over time-based sliding windows. In: Proceedings of the 24th international conference on scientific and statistical database management (SSDBM'12). Springer, Berlin/Heidelberg, pp 416–430. doi:10.1007/978-3-642-31235-9_28

17. Crouch MS, McGregor A, Stubbs D (2013) Dynamic graphs in the sliding-window model. In: ESA, pp 337–348

18. Dang XH, Lee VC, Ng WK, Ong KL (2009) Incremental and adaptive clustering stream data over sliding window. In: Proceedings of the 20th international conference on database and expert systems applications (DEXA'09). Springer, Berlin/Heidelberg, pp 660–674. doi:10.1007/978-3-642-03573-9_55

19. Datar M, Muthukrishnan MS (2002) Estimating rarity and similarity over data stream windows. In: Proceedings of the 10th annual European symposium on algorithms (ESA'02). Springer, London, pp 323–334. http://dl.acm.org/citation.cfm?id=647912.740833

20. Datar M, Gionis A, Indyk P, Motwani R (2002) Maintaining stream statistics over sliding windows. SIAM J Comput 31(6):1794–1813

21. Datar M, Gionis A, Indyk P, Motwani R (2002) Maintaining stream statistics over sliding windows: (extended abstract). In: Proceedings of the thirteenth annual ACM-SIAM symposium on discrete algorithms (SODA'02). Society for Industrial and Applied Mathematics, Philadelphia, pp 635–644. http://dl.acm.org/citation.cfm?id=545381.545466

22. Feigenbaum J, Kannan S, Zhang J (2004) Computing diameter in the streaming and sliding-window models. Algorithmica 41(1):25–41

23. Feigenbaum J, Kannan S, Zhang J (2005) Computing diameter in the streaming and sliding-window models. Algorithmica 41:25–41

24. Gibbons PB, Tirthapura S (2002) Distributed streams algorithms for sliding windows. In: Proceedings of the fourteenth annual ACM symposium on parallel algorithms and architectures (SPAA'02). ACM, New York, pp 63–72. doi:10.1145/564870.564880

25. Golab L, Özsu MT (2003) Processing sliding window multi-joins in continuous queries over data streams. In: Proceedings of the 29th international conference on Very large data bases (VLDB'03), vol 29. VLDB Endowment, pp 500–511. http://dl.acm.org/citation.cfm?id=1315451.1315495

26. Golab L, DeHaan D, Demaine ED, Lopez-Ortiz A, Munro JI (2003) Identifying frequent items in sliding windows over on-line packet streams. In: Proceedings of the 3rd ACM SIGCOMM conference on internet measurement (IMC'03). ACM, New York, pp 173–178. doi:10.1145/948205.948227

27. Kane DM, Nelson J, Woodruff DP (2010) On the exact space complexity of sketching and streaming small norms. In: Proceedings of the 21st annual ACM-SIAM symposium on discrete algorithms (SODA'10)

28. Kane DM, Nelson J, Woodruff DP (2010) An optimal algorithm for the distinct elements problem. In: Proceedings of the twenty-ninth ACM SIGMOD-SIGACT-SIGART symposium on Principles of database systems (PODS'10). ACM, New York, pp 41–52. doi:10.1145/1807085.1807094

29. Lee LK, Ting HF (2006) Maintaining significant stream statistics over sliding windows. In: SODA '06: Proceedings of the seventeenth annual ACM-SIAM symposium on Discrete algorithm. ACM, New York, pp 724–732. doi:http://doi.acm.org/10.1145/1109557.1109636

30. Lee LK, Ting HF (2006) A simpler and more efficient deterministic scheme for finding frequent items over sliding windows. In: Proceedings of the twenty-fifth ACM SIGMOD-SIGACT-SIGART symposium on Principles of database systems (PODS'06). ACM, New York, pp 290–297. doi:http://doi.acm.org/10.1145/1142351.1142393

31. Li J, Maier D, Tufte K, Papadimos V, Tucker PA (2005) No pane, no gain: efficient evaluation of sliding-window aggregates over data streams. SIGMOD Rec 34(1):39–44. doi:10.1145/1058150.1058158

32. McGregor A (2014) Graph stream algorithms: A survey. SIGMOD Rec 43(1):9–20. doi:10.1145/2627692.2627694

33. Ren J, Ma R, Ren J (2009) Density-based data streams clustering over sliding windows. In: Proceedings of the 6th international conference on Fuzzy systems and knowledge discovery (FSKD'09), vol 5. IEEE Press, Piscataway, pp 248–252. http://dl.acm.org/citation.cfm?id=1801874.1801929

34. Sohler C (2006) List of open problems in sublinear algorithms: problem 20. http://sublinear.info/20

35. Tatbul N, Zdonik S (2006) Window-aware load shedding for aggregation queries over data streams. In: Proceedings of the 32nd international conference on very large data bases (VLDB'06). VLDB Endowment, pp 799–810. http://dl.acm.org/citation.cfm?id=1182635.1164196

36. Zhang L, Li Z, Yu M, Wang Y, Jiang Y (2005) Random sampling algorithms for sliding windows over data streams. In: Proceedings of the 11th joint international computer conference, pp 572–575

Smooth Surface and Volume Meshing

Tamal Krishna Dey
Department of Computer Science and Engineering, The Ohio State University, Columbus, OH, USA

Keywords

Delaunay mesh; Delaunay refinement; Surface mesh; Topology; Volume mesh

Years and Authors of Summarized Original Work

2001; Cheng, Dey, Edelsbrunner, Sullivan
2003; Boissonnat, Oudot
2004; Cheng, Dey, Ramos
2005; Oudot, Rienau, Yvinec
2012; Cheng, Dey, Shewchuk

Problem Definition

Given a smooth surface $S \subset \mathbb{R}^3$, we are required to compute a set of points $P \subset S$ and connect them with edges and triangles so that the resulted triangulation T is *geometrically* close and is *topologically* equivalent to S.

The output triangulation T is a simplicial 2-complex whose vertices are the points in P. Its underlying space, which is the pointwise union of the simplices (vertices, edges, triangles), is denoted with $|T|$. Geometric proximity is often characterized by Hausdorff distance between S

and the underlying space $|T|$ of T. It is also desired that the triangle normals in T closely approximate the surface normals at its vertices. Topological equivalence is characterized by the existence of a *homeomorphism* between S and $|T|$. In some cases, the topological guarantee can be given in terms of *isotopy* which is stronger than homeomorphism. It is important to notice that, unlike polyhedral surfaces, a smooth surface cannot be represented *exactly* and hence needs to be approximated with a finite triangulation. This approximation requires that the mesh generation algorithms guarantee topological fidelity in addition to the geometric proximity.

In volume mesh generation, the space bounded by a smooth surface S is required to be tessellated with tetrahedra which form a simplicial 3-complex T. Similar to the surface case, it is required that the underlying space $|T|$ is geometrically close and topologically equivalent to the space bounded by S. It turns out that if the underlying space of the boundary 2-complex of T is geometrically close and has an isotopy to S, then so is $|T|$.

In both surface and volume meshes, it is desirable that the triangles and tetrahedra have good aspect ratio. This is often achieved by bounding the circumradius to shortest edge length ratios for triangles. Unfortunately, for tetrahedra, a bounded radius-edge ratio does not necessarily imply a bounded aspect ratio though most poor quality tetrahedra except slivers [4] are eliminated by bounded radius-edge ratio. Figure 1 shows an example of a surface and a volume mesh.

Key Results

Theoretically sound algorithms for surface meshing use the technique of Delaunay refinement originally proposed by Chew [8]. For a point set $P \subset \mathbb{R}^3$, let Vor P and Del P denote the Voronoi diagram and Delaunay triangulation of P, respectively. A typical Delaunay refinement algorithm iteratively samples the space to be meshed with a *locally furthest point* strategy that inserts points where a Voronoi face of appropriate dimension intersects the space. The decision of

Smooth Surface and Volume Meshing, Fig. 1 A knotted torus, its surface mesh, and its volume mesh

which points to be inserted is guided by certain desirable properties of the output such as topological equivalence, simplex radius-edge ratios, geometric proximity, and so on.

In both surface and volume meshing, the features of the surface S play an important role because regions of small features need to be sampled relatively densely to capture the geometry and topology of S. The definition of *local feature size* and ε-sample given by Amenta, Bern, and Eppstein [2] captures this idea.

Let S be a smooth, closed surface, that is, S is compact, C^2-smooth, and has no boundary. The *medial axis* $M(S)$ of S is defined as the closure of the set of points $x \in \mathbb{R}^3$ so that the distance $d(x, S)$ is realized by two or more points in S. The *local feature size* is defined as

$$f(x) = d(x, M).$$

A set of points $P \subset S$ is called an ε-sample of S if every point $x \in S$ has a sample point in P within $\varepsilon f(x)$ distance.

It turns out that if P is an ε-sample of S for a sufficiently small value of ε, a subcomplex of the Delaunay triangulation of this sample captures the topology of S. We define this subcomplex in generality and then specialize it to S.

Let V_ξ denote the dual Voronoi face of a Delaunay simplex ξ in Del P. The restricted Voronoi face of V_ξ with respect to $\mathbb{X} \subset \mathbb{R}^3$ is the intersection $V_\xi|_\mathbb{X} = V_\xi \cap \mathbb{X}$. The *restricted Voronoi diagram* and *restricted Delaunay triangulation* of P with respect to $\mathbb{X}$ are

$$\text{Vor } P|_\mathbb{X} = \{V_\xi|_\mathbb{X} \mid V_\xi|_\mathbb{X} \neq \emptyset\} \text{ and Del } P|_\mathbb{X}$$
$$= \{\xi \mid V_\xi|_\mathbb{X} \neq \emptyset\} \text{ respectively.}$$

In words, Del $P|_\mathbb{X}$ consists of those Delaunay simplices in Del P whose dual Voronoi face intersects $\mathbb{X}$. We call these simplices *restricted*.

Now consider a sample P on the surface S. The restricted Delaunay triangulation of P with respect to S is Del $P|_S$. It is known that if P is an ε-sample of S for $\varepsilon \leq 0.09$, then Del $P|_S$ has its underlying space homeomorphic to S [1, 9]. To use this result one requires computing an ε-sample of S. A computation of local feature size or its approximation is necessary to determine if a sample is an ε-sample for a predetermined ε. Even if one is allowed to assume the availability of the local feature size at any given point, it is not immediately obvious how to place points on S so that they become ε-sample for a given $\varepsilon > 0$.

Surface Meshing

The following theorem about the fidelity of the restricted Delaunay triangulation of a dense sample on a smooth closed surface is the basis of provable surface meshing algorithms. It has been proved in various versions in [1,5,7,9].

Theorem 1 *Let P be an ε-sample of a smooth, compact, boundary-less surface $S \subset \mathbb{R}^3$. The restricted Delaunay complex $T = $ Del $P|_S$ satisfies the following properties for $\varepsilon \leq 0.09$:*

1. *The underlying space $|T|$ is homeomorphic to S (actually, there is an ambient isotopy taking $|T|$ to S).*
2. *Every point in $|T|$ has a point $x \in S$ so that $d(p, x) \leq O(\varepsilon) f(x)$. Similarly, every point x in S has a point p in $|T|$ so that $d(p, x) \leq O(\varepsilon) f(x)$.*

3. *Each triangle $t \in T$ has a normal making an angle $O(\varepsilon)$ with the normal to the surface S at any of its vertices.*

Cheng, Dey, Edelsbrunner, and Sullivan [5] applied Chew's furthest point placement strategy [8] to maintain a dynamic surface mesh of a special type of surface called *skin surface* for which they computed the local feature size explicitly. The above theorem then allowed them to argue the geometric and topological fidelity of the output. Boissonnat and Oudot [3] used similar point placement strategy assuming that the local feature sizes are available, but they suggested how to initialize the meshing procedure for general surfaces. For a restricted triangle $t \in \text{Del } P|_S$, the dual Voronoi edge intersects S possibly at multiple points. Each ball centering such an intersection point and circumscribing vertices of t is called a *surface Delaunay ball* of t. Boissonnat and Oudot observed that if every surface Delaunay ball of each restricted triangle has small radius, say at most 0.05 times the local feature size at the center, then P a 0.09-sample of S. It follows that Del $P|_S$ at this point satisfies the properties stated in Theorem 1. The deduction of this conclusion also requires that every component of S has at least one Voronoi edge intersecting it which Boissonnat and Oudot ensure with *persistent triangles*.

When local feature sizes are not known, we cannot use the method of Boissonnat and Oudot [3]. Instead, we fall back upon a different strategy to drive the Delaunay refinement. A result of Edelsbrunner and Shah [10] says that if Voronoi faces intersect S in a closed topological ball of appropriate dimension, then the underlying space of the restricted Delaunay triangulation becomes homeomorphic to S. In fact, this is the basis of the proof of Theorem 1. Therefore, a Delaunay refinement driven by the violation of the topological ball conditions provides a viable strategy for meshing with topological guarantees. This strategy is followed by Cheng, Dey, Ramos, and Ray [6].

The algorithm of Cheng et al. avoids computing local feature sizes or their approximation; however, it needs to compute critical points of

certain functions on the surface, which may not be easily computable. In a recent book on Delaunay mesh generation [7], Cheng, Dey, and Shewchuk have suggested a strategy that is more practical which leverages on both algorithms of Boissonnat and Oudot [3] and Cheng et al. [6]. It operates with an input parameter $\lambda > 0$. As long as the surface Delaunay balls of the restricted triangles are not all smaller than a ball of radius λ, the algorithm refines. It also refines if the restricted triangles around each vertex do not form a topological disk. The algorithm can be shown to terminate and has the following guarantees.

Theorem 2 ([7]) *There is a Delaunay refinement algorithm that runs with a parameter $\lambda > 0$ on an input smooth, compact, boundary-less surface S with the following guarantees:*

1. *The output mesh is a Delaunay subcomplex and is a 2-manifold for all values of λ.*
2. *If λ is sufficiently small, then the output mesh has similar guarantees with respect to the input surface S as in Theorem 1 (replace ε with λ).*

It should be noted that in any of the above algorithms, one may introduce the condition that the output triangles have radius-edge ratio of at most 1 without loosing any of the geometric or topological guarantees. Even a graded mesh can be guaranteed by supplying an appropriate grading function as input. For details see [7].

Volume Meshing

Let $\mathcal{O}$ denote the volume enclosed by a smooth surface S. Consider the surface mesh of S produced by one of the algorithms mentioned above. The volume enclosed by this surface mesh is already triangulated with Delaunay tetrahedra. We can further refine them for quality using the radius-edge ratio condition. The circumcenters of skinny tetrahedra can be added as long as they do not disturb the surface triangulation. One easy approach is to skip adding those circumcenters who encroach the surface Delaunay balls meaning that they lie inside these balls. This ensures that all surface triangles remain intact. The trade-off of this easy fix is that the tetrahedra near the boundary may not have bounded radius-edge ratios. To ensure the quality for all tetrahedra, additional effort is required to maintain the surface. Oudot, Rineau, and Yvinec [11] proposed an algorithm for guaranteed quality volume meshing.

The algorithm first runs the algorithm of [3] to obtain a surface triangulation with a vertex set P on the surface. It uses two parameters ε and ρ where ε controls the level of refinement and ρ controls the aspect ratios of the tetrahedra and triangles. It ensures that all restricted triangles on the surface have vertices from S. It refines surface triangles as in surface meshing algorithm. Then, it refines the tetrahedra. Refinement of surface triangles is given priority over the tetrahedra. Oudot et al. [11] prove that their algorithm terminates and has the following geometric and topological guarantees.

Theorem 3 ([11]) *Given a volume $\mathcal{O}$ bounded by a smooth surface S, for $\varepsilon \leq 0.05$ and $\rho > 1$, there is an algorithm that produces $T = \mathrm{Del}\, P|_{\mathcal{O}}$ where each tetrahedron in T has radius-edge ratio at most ρ and $|T|$ is homeomorphic (isotopic) to $\mathcal{O}$ and the boundary of T is $\mathrm{Del}\, P|_S$. Furthermore, the isotopy moves a point $x \in S$ by at most $O(\varepsilon^2) f(x)$ distance.*

An improved version of the algorithm and its analysis in presented in the book [7].

URLs to Code and Data Sets

CGAL(http://cgal.org), a library of geometric algorithms, contains software for surface and volume mesh generation. The DelPSC software that implements the surface and volume meshing algorithms as described in [7] is also available from http://web.cse.ohio-state.edu/~tamaldey/delpsc.html.

Cross-References

▸ Manifold Reconstruction
▸ Meshing Piecewise Smooth Complexes
▸ Surface Reconstruction

Recommended Reading

1. Amenta N, Bern M (1999) Surface reconstruction by Voronoi filtering. Discret Comput Geom 22:481–504
2. Amenta N, Bern M, Eppstein D (1998) The crust and the beta-skeleton: combinatorial curve reconstruction. Graph Models Image Process 60(2:2):125–135
3. Boissonnat J-D, Oudot S (2005) Provably good surface sampling and meshing of surfaces. Graph Models 67:405–451. Conference version 2003
4. Cheng S-W, Dey TK, Edelsbrunner H, Teng SH (2000) Sliver exudation. J ACM 47:883–904
5. Cheng H-L, Dey TK, Edelsbrunner H, Sullivan J (2001) Dynamic skin triangulation. Discret Comput Geom 25:525–568
6. Cheng S-W, Dey TK, Ramos EA, Ray T (2007) Sampling and meshing a surface with guaranteed topology and geometry. SIAM J Comput 37:1199–1227. Conference version 2004
7. Cheng S-W, Dey TK, Shewchuk JR (2012) Delaunay mesh generation. CRC Press, Boca Raton
8. Chew LP (1993) Guaranteed-quality mesh generation for curved surfaces. In: Proceedings of the 9th annual symposium on computational geometry, San Diego, pp 274–280
9. Dey TK (2006) Curve and surface reconstruction: algorithms with mathematical analysis. Cambridge University Press, New York
10. Edelsbrunner H, Shah N (1997) Triangulating topological spaces. Int J Comput Geom Appl 7:365–378
11. Oudot S, Rineau L, Yvinec M (2005) Meshing volumes bounded by smooth surfaces. In: Proceedings of the 14th international meshing roundtable, pp 203–219

Smoothed Analysis

Heiko Röglin
Department of Computer Science, University of Bonn, Bonn, Germany

Keywords

Computational complexity; Linear programming; Probabilistic analysis

Years and Authors of Summarized Original Work

2001; Spielman, Teng
2004; Beier, Vöcking

Problem Definition

Smoothed analysis has originally been introduced by Spielman and Teng [22] in 2001 to explain why the simplex method is usually fast in practice despite its exponential worst-case running time. Since then it has been applied to a wide range of algorithms and optimization problem. In smoothed analysis, inputs are generated in two steps: first, an adversary chooses an arbitrary instance, and then this instance is slightly perturbed at random. The smoothed performance of an algorithm is defined to be the worst expected performance the adversary can achieve. This model can be viewed as a less pessimistic worst-case analysis, in which the randomness rules out pathological worst-case instances that are rarely observed in practice but dominate the worst-case analysis. If the smoothed running time of an algorithm is low (i.e., the algorithm is efficient in expectation on any perturbed instance) and inputs are subject to a small amount of random noise, then it is unlikely to encounter an instance on which the algorithm performs poorly. In practice, random noise can stem, for example, from measurement errors, numerical imprecision, or rounding errors. It can also model arbitrary influences, which we cannot quantify exactly, but for which there is also no reason to believe that they are adversarial. After its invention smoothed analysis has been applied in a variety of different contexts, e.g., linear programming [8, 19, 21, 23], multi-objective optimization [5, 10, 17, 18], online and approximation algorithms [4, 7, 20], searching and sorting [3, 12, 15, 16], game theory [9, 11], and local search [1, 2, 13, 14].

Key Results

Simplex Method

Spielman and Teng [22] considered linear programs of the form

$$\text{maximize } c^T x$$
$$\text{subject to } (\overline{A} + G)x \le (\overline{b} + h),$$

where $\overline{A} \in \mathbb{R}^{n \times d}$ and $\overline{b} \in \mathbb{R}^n$ are chosen arbitrarily by an adversary and the entries of the

matrix $G \in \mathbb{R}^{n \times d}$ and the vector $h \in \mathbb{R}^n$ are independent Gaussian random variables that represent the perturbation. These Gaussian random variables have mean 0 and standard deviation $\sigma \cdot (\max_i \|(\overline{b}_i, \overline{a}_i)\|)$, where the vector $(\overline{b}_i, \overline{a}_i) \in \mathbb{R}^{d+1}$ consists of the i-th component of $\overline{b}$ and the i-th row of $\overline{A}$ and $\|\cdot\|$ denotes the Euclidean norm. Without loss of generality, we can scale the linear program specified by the adversary and assume that $\max_i \|(\overline{b}_i, \overline{a}_i)\| = 1$. Then the perturbation consists of adding an independent Gaussian random variable with standard deviation σ to each entry of $\overline{A}$ and $\overline{b}$. The smaller σ is chosen, the more concentrated are the random variables, and hence, the better worst-case instances can be approximated by the adversary. Intuitively, σ can be seen as a measure specifying how close the analysis is to a worst-case analysis.

Spielman and Teng analyzed the smoothed running time of the simplex algorithm using the *shadow vertex pivot rule*. This pivot rule has a simple and intuitive geometric description which makes probabilistic analyses feasible. Let x_0 denote the given initial vertex of the polytope $\mathcal{P}$ of feasible solutions. Since x_0 is a vertex of the polytope, there exists an objective function $u^T x$ which is maximized by x_0 subject to the constraint $x \in \mathcal{P}$. In the first step, the shadow vertex pivot rule computes an objective function $u^T x$ with this property. If x_0 is not an optimal solution of the linear program, then the vectors c and u are linearly independent and span a plane. The shadow vertex method projects the polytope $\mathcal{P}$ onto this plane. The *shadow*, that is, the projection of $\mathcal{P}$ onto this plane is a possibly open polygon. One can show that both x_0 and the optimal solution x^* are projected onto vertices of the polygon and that each path between the projections of x_0 and x^* in the polygon corresponds to a path between x_0 and x^* in the polytope. Hence, one only needs to follow the edges of the polygon starting from the projection of x_0 to (the projection of) x^*.

The number of steps performed by the simplex method with shadow vertex pivot rule is upper bounded by the number of vertices of the two-dimensional projection of the polytope. Hence, bounding the expected number of vertices on the polygon is the crucial step for bounding the expected running time of the simplex method with shadow vertex pivot rule. Spielman and Teng first consider the case that the polytope $\mathcal{P}$ is projected onto a fixed plane specified by two fixed vectors c and u. They show that the expected number of vertices of the polygon is polynomially bounded in d, n, and $1/\sigma$. Though this result is the main ingredient of the analysis, alone it does not yield a polynomial bound on the smoothed running time of the simplex method. We have, for example, not yet described how the initial solution x_0 is found. It is also problematic that the vector u is not independent of the constraints because it is determined by x_0 which in turn is determined by a subset of the constraints. Spielman and Teng showed in a very involved analysis the following theorem.

Theorem 1 *The smoothed running time of the shadow vertex simplex method is bounded polynomially in d, n, and $1/\sigma$.*

Later, this analysis was substantially improved and simplified by Vershynin [23], who proved that the smoothed running time is even polynomially bounded in d, $\log n$, and $1/\sigma$.

Binary Optimization Problems

Beier and Vöcking [6] studied the question which *linear binary optimization problems* have *polynomial smoothed complexity*. Intuitively these are the problems that can be solved efficiently on perturbed inputs. An instance I of such an optimization problem Π consists of a set of feasible solutions $\mathcal{S} \subseteq \{0, 1\}^n$ and a linear objective function $f : \{0, 1\}^n \to \mathbb{R}$ of the form maximize (or minimize) $f(x) = c^T x$ for some $c \in \mathbb{R}^n$. Many well-known optimization problems can be formulated this way, e.g., the problem of finding a *Minimum Spanning Tree*, the *Knapsack Problem*, and the *Traveling Salesman Problem*.

It is assumed that an adversary is allowed to choose the coefficients of the objective function from the interval $[-1, 1]$. In the second step, these coefficients are perturbed by adding independent Gaussian random variables with mean 0 and standard deviation σ to them. Naturally one might define that a problem Π has

polynomial smoothed complexity if there exists an algorithm A for Π whose expected running time $\mathbf{E}[T_A(I)]$ is bounded polynomially in the input size $|I|$ and $1/\sigma$. This definition, however, is not sufficiently robust as it depends on the machine model. An algorithm with expected polynomial running time on one machine model might have expected exponential running time on another machine model even if the former can be simulated by the latter in polynomial time. In contrast, the definition from [6] yields a notion of polynomial smoothed complexity that does not vary among classes of machines admitting polynomial time simulations among each other. It states that a problem Π has polynomial smoothed complexity if there exists an algorithm A for Π and some $\alpha > 0$ such that $\mathbf{E}[T_A(I)^\alpha]$ is bounded polynomially in the input size $|I|$ and $1/\sigma$.

Beier and Vöcking proved the following theorem that characterizes the class of linear binary optimization problems with polynomial smoothed complexity.

Theorem 2 *A linear binary optimization problem Π has polynomial smoothed complexity if and only if there exists a randomized algorithm for solving Π whose expected worst-case running time is pseudo-polynomial with respect to the coefficients in the objective function.*

For example, the knapsack problem, which can be solved by dynamic programming in pseudo-polynomial time, has polynomial smoothed complexity even if the weights are fixed and only the profits are randomly perturbed. Moreover, the traveling salesman problem does not have polynomial smoothed complexity when only the distances are randomly perturbed, unless P = NP, since a simple reduction from Hamiltonian cycle shows that it is strongly NP-hard.

Open Problems

An interesting open question is whether or not other pivot rules for the simplex method also have polynomial smoothed running time. It would also be interesting to see whether the insights gained from smoothed analysis can be used to improve existing algorithms.

Cross-References

▶ Knapsack

Recommended Reading

1. Arthur D, Vassilvitskii S (2009) Worst-case and smoothed analysis of the ICP algorithm, with an application to the k-means method. SIAM J Comput 39(2):766–782
2. Arthur D, Manthey B, Röglin H (2011) Smoothed analysis of the k-means method. J ACM 58(5)
3. Banderier C, Beier R, Mehlhorn K (2003) Smoothed analysis of three combinatorial problems. In: Proceedings of the 28th international symposium on mathematical foundations of computer science (MFCS), Bratislava. Lecture notes in computer science, vol 2747. Springer, pp 198–207
4. Becchetti L, Leonardi S, Marchetti-Spaccamela A, Schäfer G, Vredeveld T (2006) Average case and smoothed competitive analysis of the multilevel feedback algorithm. Math Oper Res 31(1):85–108
5. Beier R, Vöcking B (2004) Random knapsack in expected polynomial time. J Comput Syst Sci 69(3):306–329
6. Beier R, Vöcking B (2006) Typical properties of winners and losers in discrete optimization. SIAM J Comput 35(4):855–881
7. Bläser M, Manthey B, Rao BVR (2011) Smoothed analysis of partitioning algorithms for Euclidean functionals. In: Proceedings of the 12th workshop on algorithms and data structures (WADS), New York. Lecture notes in computer science. Springer, pp 110–121
8. Blum AL, Dunagan JD (2002) Smoothed analysis of the perceptron algorithm for linear programming. In: Proceedings of the 13th annual ACM-SIAM symposium on discrete algorithms (SODA), San Francisco. SIAM, pp 905–914
9. Boros E, Elbassioni K, Fouz M, Gurvich V, Makino K, Manthey B (2011) Stochastic mean payoff games: smoothed analysis and approximation schemes. In: Proceedings of the 38th international colloquium on automata, languages and programming (ICALP), Zurich, Part I. Lecture notes in computer science, vol 6755. Springer, pp 147–158
10. Brunsch T, Röglin H (2012) Improved smoothed analysis of multiobjective optimization. In: Proceedings of the 44th annual ACM symposium on theory of computing (STOC), New York, pp 407–426
11. Chen X, Deng X, Teng SH (2009) Settling the complexity of computing two-player Nash equilibria. J ACM 56(3)
12. Damerow V, Manthey B, auf der Heide FM, Räcke H, Scheideler C, Sohler C, Tantau T (2012) Smoothed analysis of left-to-right maxima with applications. ACM Trans Algorithms 8(3): Article no 30
13. Englert M, Röglin H, Vöcking B (2007) Worst case and probabilistic analysis of the 2-Opt algorithm for the TSP. In: Proceedings of the 18th annual ACM-SIAM symposium on discrete algorithms (SODA), New Orleans. SIAM, pp 1295–1304
14. Etscheid M, Röglin H (2014) Smoothed analysis of local search for the maximum-cut problem. In: Proceedings of the 25th annual ACM-SIAM symposium on discrete algorithms (SODA), Portland pp 882–889
15. Fouz M, Kufleitner M, Manthey B, Zeini Jahromi N (2012) On smoothed analysis of quicksort and Hoare's find. Algorithmica 62(3–4):879–905
16. Manthey B, Reischuk R (2007) Smoothed analysis of binary search trees. Theor Comput Sci 378(3):292–315
17. Moitra A, O'Donnell R (2012) Pareto optimal solutions for smoothed analysts. SIAM J Comput 41(5):1266–1284
18. Röglin H, Teng SH (2009) Smoothed analysis of multiobjective optimization. In: Proceedings of the 50th annual IEEE symposium on foundations of computer science (FOCS), Atlanta. IEEE, pp 681–690
19. Sankar A, Spielman DA, Teng SH (2006) Smoothed analysis of the condition numbers and growth factors of matrices. SIAM J Matrix Anal Appl 28(2):446–476
20. Schäfer G, Sivadasan N (2005) Topology matters: smoothed competitiveness of metrical task systems. Theor Comput Sci 241(1–3):216–246
21. Spielman DA, Teng SH (2003) Smoothed analysis of termination of linear programming algorithms. Math Program 97(1–2):375–404
22. Spielman DA, Teng SH (2004) Smoothed analysis of algorithms: why the simplex algorithm usually takes polynomial time. J ACM 51(3):385–463
23. Vershynin R (2009) Beyond Hirsch conjecture: walks on random polytopes and smoothed complexity of the simplex method. SIAM J Comput 39(2):646–678

Snapshots in Shared Memory

Eric Ruppert
Department of Computer Science and Engineering, York University, Toronto, ON, Canada

Keywords

Atomic scan

Years and Authors of Summarized Original Work

1993; Afek, Attlya, Dolev, Gatni, Merritt, Shavit

Problem Definition

Implementing a snapshot object is an abstraction of the problem of obtaining a consistent view of several shared variables while other processes are concurrently updating those variables.

In an asynchronous shared-memory distributed system, a collection of n processes communicate by accessing shared data structures, called *objects*. The system provides basic types of shared objects; other needed types must be built from them. One approach uses locks to guarantee exclusive access to the basic objects, but this approach is not fault-tolerant, risks deadlock or livelock, and causes delays when a process holding a lock runs slowly. Lock-free algorithms avoid these problems but introduce new challenges. For example, if a process reads two shared objects, the values it reads may not be consistent if the objects were updated between the two reads.

A *snapshot object* stores a vector of m values, each from some domain D. It provides two operations: scan and update(i, v), where $1 \leq i \leq m$ and $v \in D$. If the operations are invoked sequentially, an update(i, v) operation changes the value of the ith component of the stored vector to v, and a scan operation returns the stored vector.

Correctness when snapshot operations by different processes overlap in time is described by the *linearizability* condition, which says operations should appear to occur instantaneously. More formally, for every execution, one can choose an instant of time for each operation (called its *linearization point*) between the invocation and the completion of the operation. (An incomplete operation may either be assigned no linearization point or given a linearization point at any time after its invocation.) The responses returned by all completed operations in the execution must return the same result as they would if all operations were executed sequentially in the order of their linearization points.

An implementation must also satisfy a progress property. *Wait-freedom* requires that each process completes each scan or update in a finite number of its own steps. The weaker *non-blocking* progress condition says the system cannot run forever without some operation completing.

This article describes implementations of snapshots from more basic types, which are also linearizable, without locks. Two types of snapshots have been studied. In a *single-writer* snapshot, each component is owned by a process, and only that process may update it. (Thus, for single-writer snapshots, $m = n$.) In a *multi-writer* snapshot, any process may update any component. There also exist algorithms for *single-scanner* snapshots, where only one process may scan at a time [10, 13, 14, 16]. Snapshots were introduced by Afek et al. [1], Anderson [2] and Aspnes and Herlihy [4].

Space complexity is measured by the number of basic objects used and their size (in bits). Time complexity is measured by the maximum number of steps a process must do to finish a scan or update, where a step is an access to a basic shared object. (Local computation and local memory accesses are usually not counted.) Complexity bounds will be stated in terms of $n, m, d = \log |D|$ and k, the number of operations invoked in an execution. Ordinarily, there is no bound on k.

Most of the algorithms below use read-write registers, the most elementary shared object type. A *single-writer* register may only be written by one process. A *multi-writer* register may be written by any process. Some algorithms using stronger types of basic objects are discussed in section "Wait-Free Implementations from Small, Stronger Objects".

Key Results

A Simple Non-blocking Implementation from Small Registers

Suppose each component of a single-writer snapshot object is represented by a single-writer register. Process i does an update(i, v) by writing v and a sequence number into register i, and incrementing its sequence number. Performing a scan operation is more difficult than merely reading each of the m registers, since some registers might change while these reads are done. To scan, a process repeatedly reads all the registers. A sequence of reads of all the registers is called a *collect*. If two collects return the same vector, the scan returns that vector (with the sequence numbers stripped away). The sequence numbers ensure that, if the same value is read in a register twice, the register had that value during the entire interval between the two reads. The scan can be assigned a linearization point between the two identical collects, and updates are linearized at the write. This algorithm is non-blocking, since a scan continues running only if at least one update operation is completed during each collect. A similar algorithm, with process identifiers appended to the sequence numbers, implements a non-blocking multi-writer snapshot from m multi-writer registers.

Wait-Free Implementations from Large Registers

Afek et al. [1] described how to modify the non-blocking single-writer snapshot algorithm to make it wait-free using scans embedded within the updates. An update(i, v) first does a scan and then writes a triple containing the scan's result, v and a sequence number into register i. While a process P is repeatedly performing collects to do a scan, either two collects return the same vector (which P can return) or P will eventually have seen three different triples in the register of some other process. In the latter case, the third triple that P saw must contain a vector that is the result of a scan that started after P's scan, so P's scan outputs that vector. Updates and scans that terminate after seeing two identical collects are assigned linearization points as before. If one scan obtains its output from an embedded scan, the two scans are given the same linearization point. This is a wait-free single-writer snapshot implementation from n single-writer registers of $(n + 1)d + \log k$ bits each. Operations complete within $O(n^2)$ steps. Afek et al. [1] also describe how to replace the unbounded sequence numbers with handshaking bits. This requires $n \Theta(nd)$-bit registers and n^2 1-bit registers. Operations still complete in $O(n^2)$ steps.

The same idea can be used to build multi-writer snapshots from multi-writer registers. Using unbounded sequence numbers yields a wait-free algorithm that uses m registers storing $\Theta(nd + \log k)$ bits each, in which each operation completes within $O(mn)$ steps. (This algorithm is given explicitly in [9].) No algorithm can use fewer than m registers if $n \geq m$ [9]. If handshaking bits are used instead, the multi-writer snapshot algorithm uses n^2 1-bit registers, $m(d + \log n)$-bit registers and n (md)-bit registers, and each operation uses $O(nm + n^2)$ steps [1].

Guerraoui and Ruppert [12] gave a similar wait-free multi-writer snapshot implementation that is anonymous, i.e., it does not use process identifiers and all processes are programmed identically.

Anderson [3] gave an implementation of a multi-writer snapshot from a single-writer snapshot. Each process stores its latest update to each component of the multi-writer snapshot in the single-writer snapshot, with associated timestamp information computed by scanning the single-writer snapshot. A scan is done using just one scan of the single-writer snapshot. An update requires scanning and updating the single-writer snapshot twice. The implementation involves some blow-up in the size of the components, i.e., to implement a multi-writer snapshot with domain D requires a single-writer snapshot with a much larger domain D'. If the goal is to implement multi-writer snapshots from single-writer registers (rather than multi-writer registers), Anderson's construction gives a more efficient solution than that of Afek et al.

Attiya, Herlihy and Rachman [7] defined the *lattice agreement* object, which is very closely linked to the problem of implementing a single-writer snapshot when there is a known upper bound on k. Then, they showed how to construct a single-writer snapshot (with no bound on k) from an infinite sequence of lattice agreement objects. Each snapshot operation accesses the lattice agreement object twice and does $O(n)$ additional steps. Their implementations of lattice agreement are discussed in section "Wait-Free Implementations from Small, Stronger Objects".

Attiya and Rachman [8] used a similar approach to give a single-writer snapshot implementation from large single-writer registers using $O(n \log n)$ steps per operation. Each update has an associated sequence number. A scanner traverses a binary tree of height $\log k$ from root to leaf (here, a bound on k is required). Each node has an array of n single-writer registers. A process arriving at a node writes its current vector into a single-writer register associated with the node and then gets a new vector by combining information read from all n registers. It proceeds to the left or right child depending on the sum of the sequence numbers in this vector. Thus, all scanners can be linearized in the order of the leaves they reach. Updates are performed by doing a similar traversal of the tree. The bound on k can be removed as in [7]. Attiya and Rachman also give a more direct implementation that achieves this by recycling the snapshot object that assumes a bound on k. Their algorithm has also been adapted to solve condition-based consensus [15].

Attiya, Fouren and Gafni [6] described how to adapt the algorithm of Attiya and Rachman [8] so that the number of steps required to perform an operation depends on the number of processes that actually access the object, rather than the number of processes in the system.

Attiya and Fouren [5] solve lattice agreement in $O(n)$ steps. (Here, instead of using the terminology of lattice agreement, the algorithm is described in terms of implementing a snapshot in which each process does at most one snapshot operation.) The algorithm uses, as a data structure, a two-dimensional array of $O(n^2)$ *reflectors*. A reflector is an object that can be used by two processes to exchange information. Each reflector is built from two large single-writer registers. Each process chooses a path through the array of reflectors, so that at most two processes visit each reflector. Each reflector in column i is used by process i to exchange information with one process $j < i$. If process i reaches the reflector first, process j learns about i's update (if any). If process j reaches it first, then process i learns all the information that j has already gathered. (If both reach it at about the same time, both

processes learn the information described above.) As the processes move from column $i - 1$ to column i, a process that enters column i at some row r will have gathered all the information that has been gathered by any process that enters column i below row r (and possibly more). This invariant is maintained by ensuring that if process i passes information to any process $j < i$ in row r of column i, it also passes that information to all processes that entered column i above row r. Furthermore, process i exits column i at a row that matches the amount of information it learns while traveling through the column. When processes have reached the rightmost column of the array, the ones in higher rows know strictly more than the ones in lower rows. Thus, the linearization order of their scans is the order in which they exit the rightmost column, from bottom to top. The techniques of Attiya, Herlihy and Rachman [7, 8], mentioned above, can be used to remove the restriction that each process performs at most one operation. The number of steps per operation is still $O(n)$.

Wait-Free Implementations from Small, Stronger Objects

All of the wait-free implementations described above use registers that can store $\Omega(m)$ bits each, and are therefore not practical when m is large. Some implementations from smaller objects equipped with stronger synchronization operations, rather than just reads and writes, are described in this section. An object is considered to be small if it can store $O(d + \log n + \log k)$ bits. This means that it can store a constant number of component values, process identifiers and sequence numbers.

Attiya, Herlihy and Rachman [7] gave an elegant divide-and-conquer recursive solution to the lattice agreement problem. The division of processes into groups for the recursion can be done dynamically using test&set objects. This provides a snapshot algorithm that runs in $O(n)$ time per operation, and uses $O(k n^2 \log n)$ small single-writer registers and $O(k n \log^2 n)$ test&set objects. (This requires modifying their implementation to replace those registers that are large, which are written only once, by many small registers.) Using randomization, each test&set object can be replaced by single-writer registers to give a snapshot implementation from registers only with $O(n)$ expected steps per operation.

Jayanti [13] gave a multi-writer snapshot implementation from $O(mn^2)$ small compare & swap objects where updates take $O(1)$ steps and scans take $O(m)$ steps. He began with a very simple single-scanner, single-writer snapshot implementation from registers that uses a secondary array to store a copy of recent updates. A scan clears that array, collects the main array, and then collects the secondary array to find any overlooked updates. Several additional mechanisms are introduced for the general, multi-writer, multi-scanner snapshot. In particular, compare & swap operations are used instead of writes to coordinate writers updating the same component and multiple scanners coordinate with one another to simulate a single scanner. Jayanti's algorithm builds on an earlier paper by Riany, Shavit and Touitou [16], which gave an implementation that achieved similar complexity, but only for a single-writer snapshot.

Applications

Applications of snapshots include distributed databases, storing checkpoints or backups for error recovery, garbage collection, deadlock detection, debugging distributed programmes and obtaining a consistent view of the values reported by several sensors. Snapshots have been used as building blocks for distributed solutions to randomized consensus and approximate agreement. They are also helpful as a primitive for building other data structures. For example, consider implementing a counter that stores an integer and provides increment, decrement and read operations. Each process can store the number of increments it has performed minus the number of its decrements in its own component of a single-writer snapshot object, and the counter may be read by summing the values from a scan. See [10] for references on many of the applications mentioned here.

Open Problems

Some complexity lower bounds are known for implementations from registers [9], but there remain gaps between the best known algorithms and the best lower bounds. In particular, it is not known whether there is an efficient wait-free implementation of snapshots from small registers.

Experimental Results

Riany, Shavit and Touitou gave performance evaluation results for several implementations [16].

Cross-References

▸ Implementing Shared Registers in Asynchronous Message-Passing Systems
▸ Linearizability
▸ Registers

Recommended Reading

See also Fich's survey paper on the complexity of implementing snapshots [11].

1. Afek Y, Attiya H, Dolev D, Gafni E, Merritt M, Shavit N (1993) Atomic snapshots of shared memory. J Assoc Comput Mach 40:873–890
2. Anderson JH (1993) Composite registers. Distrib Comput 6:141–154
3. Anderson JH (1994) Multi-writer composite registers. Distrib Comput 7:175–195
4. Aspnes J, Herlihy M (1990) Wait-free data structures in the asynchronous PRAM model. In: Proceedings of the 2nd ACM symposium on parallel algorithms and architectures, Crete, July 1990. ACM, New York, pp 340–349
5. Attiya H, Fouren A (2001) Adaptive and efficient algorithms for lattice agreement and renaming. SIAM J Comput 31:642–664
6. Attiya H, Fouren A, Gafni E (2002) An adaptive collect algorithm with applications. Distrib Comput 15:87–96
7. Attiya H, Herlihy M, Rachman O (1995) Atomic snapshots using lattice agreement. Distrib Comput 8:121–132
8. Attiya H, Rachman O (1998) Atomic snapshots in O(n log n) operations. SIAM J Comput 27:319–340

S

9. Ellen F, Fatourou P, Ruppert E (2007) Time lower bounds for implementations of multi-writer snapshots. J Assoc Comput Mach 54(6), 30
10. Fatourou P, Kallimanis ND (2006) Single-scanner multi-writer snapshot implementations are fast! In: Proceedings of the 25th ACM symposium on principles of distributed computing, Colorado, July 2006. ACM, New York, pp 228–237
11. Fich FE (2005) How hard is it to take a snapshot? In: SOFSEM 2005: theory and practice of computer science, Liptovský Ján, Jan 2005. LNCS, vol 3381. Springer, pp 28–37
12. Guerraoui R, Ruppert E (2007) Anonymous and fault-tolerant shared-memory computing. Distrib Comput 20(3):165–177
13. Jayanti P (2005) An optimal multi-writer snapshot algorithm. In: Proceedings of the 37th ACM symposium on theory of computing, Baltimore, May 2005. ACM, New York, pp 723–732
14. Kirousis LM, Spirakis P, Tsigas P (1996) Simple atomic snapshots: a linear complexity solution with unbounded time-stamps. Inf Process Lett 58:47–53
15. Mostéfaoui A, Rajsbaum S, Raynal M, Roy M (2004) Conditionbased consensus solvability: a hierarchy of conditions and efficient protocols. Distrib Comput 17:1–20
16. Riany Y, Shavit N, Touitou D (2001) Towards a practical snapshot algorithm. Theor Comput Sci 269:163–201

Sorting by Transpositions and Reversals (Approximate Ratio 1.5)

Chin Lung Lu
Institute of Bioinformatics and Department of Biological Science and Technology, National Chiao Tung University, Hsinchu, Taiwan

Keywords

Genome rearrangements

Problem Definition

One of the most promising ways to determine evolutionary distance between two organisms is to compare the order of appearance of identical (e.g., orthologous) genes in their genomes. The resulting genome rearrangement problem calls for finding a shortest sequence of rearrangement operations that sorts one genome into the other.

In this work [8], Hartman and Sharan provide a 1.5-approximation algorithm for the problem of sorting by transpositions, transreversals, and revrevs, improving on a previous 1.75 ratio for this problem. Their algorithm is also faster than current approaches and requires $O(n^{3/2}\sqrt{\log n})$ time for n genes.

Notations and Definition

A *signed permutation* $\pi = [\pi_1, \pi_2, \ldots, \pi_n]$ on $n(\pi) \equiv n$ elements is a permutation in which each element is labeled by a sign of plus or minus. A *segment* of π is a sequence of consecutive elements $\pi_i, \pi_{i+1}, \ldots, \pi_k$, where $1 \leq i \leq k \leq n$. A *reversal* ρ is an operation that reverses the order of the elements in a segment and also flips their signs. Two segments $\pi_i, \pi_{i+1}, \ldots, \pi_k$ and $\pi_j, \pi_{j+1}, \ldots, \pi_l$ are said to be *contiguous* if $j = k+1$ or $i = l+1$. A *transposition* τ is an operation that exchanges two contiguous (disjoint) segments. A *transreversal* $\tau\rho_{A,B}$ (respectively, $w\tau\rho_{B,A}$) is a transposition that exchanges two segments A and B and also reverses A (respectively, B). A *revrev* operation $\rho\rho$ reverses each of the two contiguous segments (without transposing them). The problem of finding a shortest sequence of transposition, transreversal, and revrev operations that transforms a permutation into the identity permutation is called *sorting by transpositions, transreversals, and revrevs*. The *distance* of a permutation π, denoted by $d(\pi)$, is the length of the shortest sorting sequence.

Key Results

Linear vs. Circular Permutations

An operation is said to *operate* on the affected segments as well as on the elements in those segments. Two operations μ and μ' are *equivalent* if they have the same rearrangement result, i.e., $\mu \cdot \pi = \mu' \cdot \pi$ for all π. In this work [8], Hartman and Sharan showed that for an element x of a circular permutation π, if μ is an operation that operates on x, then there exists an equivalent operation μ' that does not operate on x. Based on this property, they further proved that the problem of sorting by transpositions, transreversals, and revrevs is

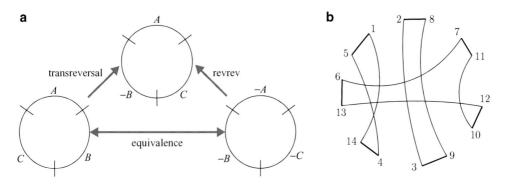

Sorting by Transpositions and Reversals (Approximate Ratio 1.5), Fig. 1 (a) The equivalence of transreversal and revrev on circular permutations. (b) The breakpoint graph $G(\pi)$ of the permutation $\pi = [1, -4, 6, -5, 2, -7, -3]$, for which $f(\pi) = [1, 2, 8, 7, 11, 12, 10, 9, 3, 4, 14, 13, 6, 5]$. It is convenient to draw $G(\pi)$ on a circle such that *black edges* (i.e., *thick lines*) are on the circumference and *gray edges* (i.e., *thin lines*) are chords

equivalent for linear and circular permutations. Moreover, they observed that revrevs and transreversals are equivalent operations for circular permutations (as illustrated in Fig. 1a), implying that the problem of sorting a linear/circular permutation by transpositions, transreversals, and revrevs can be reduced to that of sorting a circular permutation by transpositions and transreversals only.

The Breakpoint Graph

Given a signed permutation π on $\{1,2,\ldots,n\}$ of n elements, it is transformed into an unsigned permutation $f(\pi) = \pi' = [\pi'_1, \pi'_2, \ldots, \pi'_{2n}]$ on $\{1,2,\ldots,2n\}$ of $2n$ elements by replacing each positive element i with two elements $2i - 1, 2i$ (in this order) and each negative element $-i$ with $2i, 2i - 1$. The extended $f(\pi)$ is considered here as a circular permutation by identifying $2n + 1$ and 1 in both indices and elements. To ensure that every operation on $f(\pi)$ can be mimicked by an operation on π, only operations that cut before odd position are allowed for $f(\pi)$. The *breakpoint graph* $(G\pi)$ is an edge-colored graph on $2n$ vertices $\{1, 2, \ldots, 2n\}$, in which for every $1 \le i \le n$, π'_{2i} is joined to π'_{2i+1} by a black edge and $2i$ is joined to $2i + 1$ by a gray edge (e.g., see Fig. 1b). Since the degree of each vertex in $G(\pi)$ is exactly 2, $G(\pi)$ uniquely decomposes into cycles. A k-cycle (i.e., a cycle of *lengthk*) is a cycle with k black edges, and it is *odd* if k is odd.

The number of odd cycles in $G(\pi)$ is denoted by $c_{odd}(\pi)$. It is not hard to verify that $G(\pi)$ consists of n 1-cycles, and hence, $c_{odd}(\pi) = n$, if π is an identity permutation $[1, 2, \ldots, n]$. Gu et al. [5] have shown that $c_{odd}(\mu \cdot \pi) \le c_{odd}(\pi) + 2$ for all linear permutations π and operations μ. In this work [8], Hartman and Sharan further noted that the above result holds also for circular permutations and proved that the lower bound of $d(\pi)$ is $(n(\pi) - c_{odd}(\pi))/2$.

Transformation into 3-Permutations

A permutation is called *simple* if its breakpoint graph contains only k-cycle, where $k \le 3$. A simple permutation is also called a *3-permutation* if it contains no 2-cycles. A transformation from π to $\hat{\pi}$ is said to be *safe* if $n(\pi) - c_{odd}(\pi) = n(\hat{\pi}) - c_{odd}(\hat{\pi})$. It has been shown that every permutation π can be transformed into a simple one π' by safe transformations and, moreover, every sorting of π' mimics a sorting of π with the same number of operations [6, 11]. Here, Hartman and Sharan [8] further showed that every simple permutation π' can be transformed into a 3-permutation $\hat{\pi}$ by safe paddings (of transforming those 2-cycles into 1-twisted 3-cycles) and, moreover, every sorting of $\hat{\pi}$ mimics a sorting of π' with the same number of operations. Hence, based on these two properties, an arbitrary permutation π can be transformed into a 3-permutation $\hat{\pi}$ such that every sorting of $\hat{\pi}$ mimics a sorting of π with the

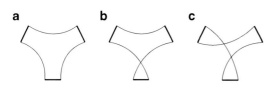 

Sorting by Transpositions and Reversals (Approximate Ratio 1.5), Fig. 2 Configurations of 3-cycles. (**a**) Unoriented, 0-twisted 3-cycle. (**b**) Unoriented, 1-twisted 3-cycle. (**c**) Oriented, 2-twisted 3-cycle. (**d**) Oriented, 3-twisted 3-cycle. (**e**) A pair of intersecting 3-cycles. (**f**) A pair of interleaving 3-cycles

same number of operations, suggesting that one can restrict attention to circular 3-permutations only.

Cycle Types

An operation that cuts some black edges is said to *act* on these edges. An operation is further called a *k-operation* if it increases the number of odd cycles by k. A *(0, 2,2)-sequence* is a sequence of three operations, of which the first is a 0-operation and the next two are 2-operations. An odd cycle is called *oriented* if there is a 2-operation that acts on three of its black edges; otherwise, it is *unoriented*. A *configuration* of cycles is a subgraph of the breakpoint graph that contains one ore more cycles. As shown in Fig. 2a–d, there are four possible configurations of single 3-cycles. A black edge is called *twisted* if its two adjacent gray edges cross each other in the circular breakpoint graph. A cycle is k-twisted if k of its black edges is twisted. For example, the 3-cycles in Fig. 2a–d are 0-, 1-, 2-, and 3-twisted, respectively. Hartman and Sharan observed that a 3-cycle is oriented if and only if it is 2- or 3-twisted.

Cycle Configurations

Two pairs of black edges are called *intersecting* if they alternate in the order of their occurrence along the circle. A pair of black edges *intersects* with cycle C, if it intersects with a pair of black edges that belong to C. Cycles C and D *intersect* if there is a pair of black edges in C that intersects with D (see Fig. 2e). Two intersecting cycles are called *interleaving* if their black edges alternate in their order of occurrence along the circle (see Fig. 2f). Clearly, the relation between two cycles is one of (1) nonintersecting, (2) intersecting but non-interleaving, and (3) interleaving. A pair of black edges is *coupled* if they are connected by a gray edge and when reading the edges along

the cycle, they are read in the same direction. For example, all pairs of black edges in Fig. 2a are coupled. Gu et al. [5] have shown that given a pair of coupled black edges (b_1, b_2), there exists a cycle C that intersects with (b_1, b_2). A *1-twisted pair* is a pair of 1-twisted cycles, whose twists are consecutive on the circle in a configuration that consists of these two cycles only. A 1-twisted cycle is called *closed* in a configuration if its two coupled edges intersect with some other cycle in the configuration. A configuration is *closed* if at least one of its 1-twisted cycles is closed; otherwise, it is called *open*.

The Algorithm

The basic ideas of the Hartman and Sharan's 1.5-approximation algorithm [8] for the problem of sorting by transpositions, transreversals, and revrevs are as follows. Hartman and Sharan reduced the problem to that of sorting a circular 3-permutation by transpositions and transreversals only and then focused on transforming the 3-cycles into 1-cycles in the breakpoint graph of this 3-permutation. By definition, an oriented (i.e., 2- or 3-twisted) 3-cycle admits a 2-operation and, therefore, they continued to consider unoriented (i.e., 0- or 1-twisted) 3-cycles only. Since configurations involving only 0-twisted 3-cycles were handled with (0,2,2)-sequences in [7], Hartman and Sharan restricted their attention to those configurations that consist of 0- and 1-twisted 3-cycles. They showed that these configurations are all closed and that it can be sorted by a (0,2,2)-sequence of operations for each of the following five possible closed configurations: (1) a closed configuration with two unoriented, interleaving 3-cycles that do not form a 1-twisted pair; (2) a closed configuration with two intersecting, 0-twisted 3-cycles; (3) a closed configuration with two intersecting, 1-twisted 3-cycles; (4) a closed configuration with a 0-twisted 3-cycles that intersects with the coupled edges of a 1-twisted 3-cycle; and (5) a closed configuration that contains $k \geq 2$ mutually interleaving 1-twisted 3-cycles such that all their twists are consecutive on the circle and k is maximal with this property. As a result, the sequence of operations used by Hartman and Sharan in their algorithm contains only 2-operations and (0,2,2)-sequences. Since every sequence of three operations increases the number of odd cycles by at least 4 out of 6 possible in 3 steps, the ratio of their approximation algorithm is 1.5. Furthermore, Hartman and Sharan showed that their algorithm can be implemented in $O(n^{3/2} \sqrt{\log n})$ time using the data structure of Kaplan and Verbin [10], where n is the number of elements in the permutation.

Theorem 1 *The problem of sorting linear permutations by transpositions, transreversals, and revrevs is linearly equivalent to the problem of sorting circular permutations by transpositions, transreversals, and revrevs.*

Theorem 2 *There is a 1.5-approximation algorithm for sorting by transpositions, transreversals, and revrevs, which runs in $O(n^{3/2} \sqrt{\log n})$ time.*

Applications

When trying to determine evolutionary distance between two organisms using genomic data, biologists may wish to reconstruct the sequence of evolutionary events that have occurred to transform one genome into the other. One of the most promising ways to do this phylogenetic study is to compare the order of appearance of identical (e.g., orthologous) genes in two different genomes [9, 12]. This comparison of computing global rearrangement events (such as reversals, transpositions, and transreversals of genome segments) may provide more accurate and robust clues to the evolutionary process than the analysis of local point mutations (i.e., substitutions, insertions, and deletions of nucleotides/amino acids). Usually, the two genomes being compared are represented by signed permutations, with each element standing for a gene and its

sign representing the (transcriptional) direction of the corresponding gene on a chromosome. Then the goal of the resulting genome rearrangement problem is to find a shortest sequence of rearrangement operations that transforms (or, equivalently, *sorts*) one permutation into the other. Previous work focused on the problem of sorting a permutation by reversals. This problem has been shown by Capara [2] to be NP-hard, if the considered permutation is unsigned. However, for signed permutations, this problem becomes tractable and Hannenhalli and Pevzer [6] gave the first polynomial-time algorithm for it. On the other hand, there has been less progress on the problem of sorting by transpositions. Thus far, the complexity of this problem is still open, although several 1.5-approximation algorithms [1, 3, 7] have been proposed for it. Recently, the approximation ratio of sorting by transpositions was further improved to 1.375 by Elias and Hartman [4]. Gu et al. [5] and Lin and Xue [11] gave quadratic-time 2-approximation algorithms for sorting signed, linear permutations by transpositions and transreversals. In [11], Lin and Xue considered the problem of sorting signed, linear permutations by transpositions, transreversals, and revrevs and proposed a quadratic-time 1.75-approximation algorithm for it. In this work [8], Hartman and Sharan further showed that this problem is equivalent for linear and circular permutations and can be reduced to that of sorting signed, circular permutations by transpositions and transreversals only. In addition, they provided a 1.5-approximation algorithm that can be implemented in $O(n^{3/2} \sqrt{\log n})$ time.

Cross-References

▶ Sorting Signed Permutations by Reversal (Reversal Distance)
▶ Sorting Signed Permutations by Reversal (Reversal Sequence)

Recommended Reading

1. Bafna V, Pevzner PA (1998) Sorting by transpositions. SIAM J Discret Math 11:224–240

2. Caprara A (1999) Sorting permutations by reversals and Eulerian cycle decompositions. SIAM J Discret Math 12:91–110
3. Christie DA (1999) Genome rearrangement problems. Ph.D. thesis, Department of Computer Science. University of Glasgow, U.K.
4. Elias I, Hartman T (2006) A 1.375-approximation algorithm for sorting by transpositions. IEEE/ACM Trans Comput Biol Bioinform 3:369–379
5. Gu QP, Peng S, Sudborough H (1999) A 2-approximation algorithm for genome rearrangements by reversals and transpositions. Theor Comput Sci 210:327–339
6. Hannenhalli S, Pevzner PA (1999) Transforming cabbage into turnip: polynomial algorithm for sorting signed permutations by reversals. J Assoc Comput Mach 46:1–27
7. Hartman T, Shamir R (2006) A simpler and faster 1.5-approximation algorithm for sorting by transpositions. Inf Comput 204:275–290
8. Hartman T, Sharan R (2004) A 1.5-approximation algorithm for sorting by transpositions and transreversals. In: Proceedings of the 4th workshop on algorithms in bioinformatics (WABI'04), Bergen, pp 50–61, 17–21 Sept (2004)
9. Hoot SB, Palmer JD (1994) Structural rearrangements, including parallel inversions, within the chloroplast genome of Anemone and related genera. J Mol Evol 38:274–281
10. Kaplan H, Verbin E (2003) Efficient data structures and a new randomized approach for sorting signed permutations by reversals. In: Proceedings of the 14th annual symposium on combinatorial pattern matching (CPM'03), Morelia, pp 170–185, 25–27 June (2003)
11. Lin GH, Xue G (2001) Signed genome rearrangements by reversals and transpositions: models and approximations. Theor Comput Sci 259:513–531
12. Palmer JD, Herbon LA (1986) Tricircular mitochondrial genomes of Brassica and Raphanus: reversal of repeat configurations by inversion. Nucleic Acids Res 14:9755–9764

Sorting Signed Permutations by Reversal (Reversal Distance)

David A. Bader
College of Computing, Georgia Institute of Technology, Atlanta, GA, USA

Keywords

Inversion distance; Reversal distance; Sorting by reversals

Years and Authors of Summarized Original Work

2001; Bader, Moret, Yan

Problem Definition

This entry describes algorithms for finding the minimum number of steps needed to sort a signed permutation (also known as inversion distance, reversal distance). This is a real-world problem and, for example, is used in computational biology.

Inversion distance is a difficult computational problem that has been studied intensively in recent years [1, 4, 6–10]. Finding the inversion distance between unsigned permutations is NP-hard [7], but with signed ones, it can be done in linear time [1].

Key Results

Bader et al. [1] present the first worst-case linear-time algorithm for computing the reversal distance that is simple and practical and runs faster than previous methods. Their key innovation is a new technique to compute connected components of the overlap graph using only a stack, which results in the simple linear-time algorithm for computing the inversion distance between two signed permutations. Bader et al. provide ample experimental evidence that their linear-time algorithm is efficient in practice as well as in theory: they coded it as well as the algorithm of Berman and Hannenhalli, using the best principles of algorithm engineering to ensure that both implementations would be as efficient as possible and compared their running times on a large range of instances generated through simulated evolution.

Bafna and Pevzner introduced the cycle graph of a permutation [3], thereby providing the basic data structure for inversion distance computations. Hannenhalli and Pevzner then developed the basic theory for expressing the inversion distance in easily computable terms (number of breakpoints minus number of cycles plus number of hurdles plus a correction factor for a fortress [3, 15]-hurdles and fortresses are easily detectable from a connected component analysis). They also gave the first polynomial-time algorithm for sorting signed permutations by reversals [9]; they also proposed a $O(n^4)$ implementation of their algorithm which runs in quadratic time when restricted to distance computation. Their algorithm requires the computation of the connected components of the overlap graph, which is the bottleneck for the distance computation. Berman and Hannenhalli later exploited some combinatorial properties of the cycle graph to give a $O(n\alpha(n))$ algorithm to compute the connected components, leading to a $O(n^2\alpha(n))$ implementation of the sorting algorithm [6], where α is the inverse Ackerman function. (The later Kaplan-Shamir-Tarjan (KST) algorithm [10] reduces the time needed to compute the shortest sequence of inversions, but uses the same algorithm for computing the length of that sequence.)

No algorithm that actually builds the overlap graph can run in linear time, since that graph can be of quadratic size. Thus, Bader's key innovation is to construct an *overlap forest* such that two vertices belong to the same tree in the forest exactly when they belong to the same connected component in the overlap graph. An overlap forest (the composition of its trees is unique, but their structure is arbitrary) has exactly one tree per connected component of the overlap graph and is thus of linear size. The linear-time step for computing the connected components scans the permutation twice. The first scan sets up a trivial forest in which each node is its own tree, labeled with the beginning of its cycle. The second scan carries out an iterative refinement of this first forest, by adding edges and so merging trees in the forest; unlike a Union-Find, however, this algorithm does not attempt to maintain the trees within certain shape parameters. This step is the key to Bader's linear-time algorithm for computing the reversal distance between signed permutations.

Applications

Some organisms have a single chromosome or contain single-chromosome organelles (such as mitochondria or chloroplasts), the evolution of which is largely independent of the evolution of the nuclear genome. Given a particular strand from a single chromosome, whether linear or circular, we can infer the ordering and directionality of the genes, thus representing each chromosome by an ordering of oriented genes. In many cases, the evolutionary process that operates on such single-chromosome organisms consists mostly of inversions of portions of the chromosome; this finding has led many biologists to reconstruct phylogenies based on gene orders, using as a measure of evolutionary distance between two genomes the inversion distance, i.e., the smallest number of inversions needed to transform one signed permutation into the other [11, 12, 14].

The linear-time algorithm is in wide use (as it has been cited nearly 200 times within the first several years of its publication). Examples include the handling multichromosomal genome rearrangements [16], genome comparison [5], parsing RNA secondary structure [13], and phylogenetic study of the HIV-1 virus [2].

Open Problems

Efficient algorithms for computing minimum distances with weighted inversions, transpositions, and inverted transpositions are open.

Experimental Results

Bader et al. give experimental results in [1].

URL to Code

An implementation of the linear-time algorithm is available as C code from www.cc.gatech.edu/~bader. Two other dominated implementations are available that are designed to compute the shortest sequence of inversions as well as its length:

one, due to Hannenhalli that implements his first algorithm [9], which runs in quadratic time when computing distances, while the other, a Java applet written by Mantin (http://www.math.tau.ac.il/~rshamir/GR/), that implements the KST algorithm [10], but uses an explicit representation of the overlap graph and thus also takes quadratic time. The implementation due to Hannenhalli is very slow and implements the original method of Hannenhalli and Pevzner and not the faster one of Berman and Hannenhalli. The KST applet is very slow as well since it explicitly constructs the overlap graph.

Cross-References

For finding the actual sorting sequence, see the entry:

▶ Sorting Signed Permutations by Reversal (Reversal Sequence)

Recommended Reading

1. Bader DA, Moret BME, Yan M (2001) A linear-time algorithm for computing inversion distance between signed permutations with an experimental study. J Comput Biol 8(5):483–491. An earlier version of this work appeared In: The proceedings of 7th Int'l workshop on algorithms and data structures (WADS 2001)
2. Badimo A, Bergheim A, Hazelhurst S, Papathanasopolous M, Morris L (2003) The stability of phylogenetic tree construction of the HIV-1 virus using genome-ordering data versus env gene data. In: Proceedings of ACM annual research conference of the South African institute of computer scientists and information technologists on enablement through technology (SAICSIT 2003), Port Elizabeth, Sept 2003, vol 47. ACM, Fourways, South Africa, pp 231–240
3. Bafna V, Pevzner PA (1993) Genome rearrangements and sorting by reversals. In: Proceedings of 34th annual IEEE symposium on foundations of computer science (FOCS93), Palo Alto, CA, pp 148–157. IEEE Press
4. Bafna V, Pevzner PA (1996) Genome rearrangements and sorting by reversals. SIAM J Comput 25:272–289
5. Bergeron A, Stoye J (2006) On the similarity of sets of permutations and its applications to genome comparison. J Comput Biol 13(7):1340–1354
6. Berman P, Hannenhalli S (1996) Fast sorting by reversal. In: Hirschberg DS, Myers EW (eds) Proceedings of 7th annual symposium combinatorial pattern matching (CPM96), Laguna Beach, June 1996. Lecture notes in computer science, vol 1075. Springer, pp 168–185
7. Caprara A (1997) Sorting by reversals is difficult. In: Proceedings of 1st conference on computational molecular biology (RECOMB97), Santa Fe. ACM, pp 75–83
8. Caprara A (1999) Sorting permutations by reversals and Eulerian cycle decompositions. SIAM J Discret Math 12(1):91–110
9. Hannenhalli S, Pevzner PA (1995) Transforming cabbage into turnip (polynomial algorithm for sorting signed permutations by reversals). In: Proceedings of 27th annual symposium on theory of computing (STOC95), Las Vegas. ACM, pp 178–189
10. Kaplan H, Shamir R, Tarjan RE (1999) A faster and simpler algorithm for sorting signed permutations by reversals. SIAM J Comput 29(3):880–892. First appeared In: Proceedings of 8th annual symposium on discrete algorithms (SODA97), New Orleans. ACM, pp 344–351
11. Olmstead RG, Palmer JD (1994) Chloroplast DNA systematics: a review of methods and data analysis. Am J Bot 81:1205–1224
12. Palmer JD (1992) Chloroplast and mitochondrial genome evolution in land plants. In: Herrmann R (ed) Cell organelles. Springer, Vienna, pp 99–133
13. Rastegari B, Condon A (2005) Linear time algorithm for parsing RNA secondary structure. In: Casadio R, Myers E (eds) Proceedings of 5th workshop algorithms in bioinformatics (WABI'05), Mallorca. Lecture notes in computer science, vol 3692. Springer, Mallorca, Spain, pp 341–352
14. Raubeson LA, Jansen RK (1992) Chloroplast DNA evidence on the ancient evolutionary split in vascular land plants. Science 255:1697–1699
15. Setubal JC, Meidanis J (1997) Introduction to computational molecular biology. PWS, Boston
16. Tesler G (2002) Efficient algorithms for multichromosomal genome rearrangements. J Comput Syst Sci 63(5):587–609

Sorting Signed Permutations by Reversal (Reversal Sequence)

Eric Tannier
LBBE Biometry and Evolutionary Biology, INRIA Grenoble Rhône-Alpes, University of Lyon, Lyon, France

Keywords

Bioinformatics; Computational biology; Genome evolution; Genome rearrangements; Inversion distance; Sorting by inversions

Years and Authors of Summarized Original Work

2004; Tannier, Sagot

Problem Definition

A *signed permutation* π of size n is a permutation over $\{-n, \ldots, -1, 1 \ldots n\}$, where $\pi_{-i} = -\pi_i$ for all i. We note $\pi = (\pi_1, \ldots, \pi_n)$.

The *reversal* $\rho = \rho_{i,j} (1 \leq i \leq j \leq n)$ is an operation that reverses the order and flips the signs of the elements $\pi_i, \ldots, \pi_j$ in a permutation π:

$$\pi \cdot \rho$$
$$= (\pi_1, \ldots, \pi_{i-1}, -\pi_j, \ldots, -\pi_i \pi_{j+1}, \ldots, \pi_n).$$

If $\rho_1, \ldots, \rho_k$ is a sequence of reversals, it is said to *sort* a permutation π if $\pi \cdots \rho_1 \cdots \rho_k = Id$, where $Id = (1, 2, \ldots, n)$ is the identity permutation. The length of a shortest sequence of reversals sorting π is called the *reversal distance* of π and is denoted by $d(\pi)$.

If the computation of $d(\pi)$ is solved in linear time [3] (see the entry ▸ Sorting Signed Permutations by Reversal (Reversal Distance)), the computation of a sequence $\rho^1, \ldots, \rho^k$ of size $k = d(\pi)$ that sorts π is more complicated, and no linear time algorithm is known so far. The best complexity is currently achieved by the subquadratic solution of Tannier and Sagot [17], which has later been improved by Tannier, Bergeron and Sagot [18], and Han [9].

Key Results

The $O(n^4)$ Self-Reduction

Recall there is a linear algorithm to compute the reversal distance thanks to the formula $d(\pi) = n + 1 - c(\pi) + t(\pi) + h(\pi) + f(\pi)$, where $c(\pi)$ is the number of cycles in the breakpoint graph and $h(\pi) + f(\pi)$ is computed from the unoriented components of the permutation (see the entry ▸ Sorting Signed Permutations by Re-

versal (Reversal Distance)). Once this is known, the self-reduction technique trivially computes a sequence of size $d(\pi)$: try every possible reversal ρ at one step, until you find one such that $d(\pi \cdot \rho) = d(\pi) - 1$. Such a reversal is called a *sorting reversal*. This necessitates $O(n)$ computations for every possible reversal. There are at most $n(n + 1)/2 = O(n^2)$ reversals to try, so iterating this to find a sequence yields an $O(n^4)$ algorithm.

The first polynomial algorithm by Hannenhalli and Pevzner [10] was not achieving a better complexity, and the algorithmic study of finding the shortest sequences of reversals began its history.

The Quadratic Roof

All the published solutions for the computations of a sorting sequence are divided into two, following the division of the distance formula into its parameters: a first part computes a sequence of reversals so that the resulting permutation has no unoriented component, and a second part sorts all oriented components.

The first part was given its best solution by Kaplan, Shamir, and Tarjan [12], whose algorithm runs in linear time when coupled with the linear distance computation [3], and it is based on Hannenhalli and Pevzner's [10] early results.

The second part is the bottleneck of the whole procedure. At this point, if there is no unoriented component, the distance is $d(\pi) = n + 1 - c(\pi)$, so a sorting reversal is one that increases $c(\pi)$ and does not create unoriented components.

A reversal that increases $c(\pi)$ is called *oriented*. Finding an oriented reversal is an easy part: any two consecutive numbers that have different signs in the permutation define one. This can easily be done in linear time or sublinear with ad hoc data structures to maintain the permutation during the scenario. The hard part is to make sure it does not create unoriented components.

The quadratic solutions (see, e.g., the one of Kaplan, Shamir, and Tarjan [12]) are based on the linear recognition of sorting reversals. No better algorithm is known so far to recognize sorting reversals, and it seemed that a lower bound had been reached, as witnessed by a survey of Ozery-Flato and Shamir [15] in which they wrote that "a central question in the study of genome rearrangements is whether one can obtain a subquadratic algorithm for sorting by reversals." This was obtained by Tannier and Sagot [17], who proved that the recognition of sorting reversal at each step is not necessary, but only the recognition of oriented reversals.

A Promising New but Still Quadratic Method

The algorithm is based on the following theorem, taken from [18]. A sequence of oriented reversals $\rho_1, \ldots, \rho_k$ is said to be *maximal* if there is no oriented reversal in $\pi \cdot \rho_1 \ldots \rho_k$. In particular a sorting sequence is maximal, but the converse is not true.

Theorem 1 *If S is a maximal but not a sorting sequence of oriented reversals for a permutation, then there exists a nonempty sequence S' of oriented reversals such that S may be split into two parts $S = S_1, S_2$, and S_1, S', S_2 is a sequence of oriented reversal.*

This allows to construct sequences of oriented reversals instead of sorting reversals, increase their size by adding reversals inside the sequence instead of at the end, and obtain a sorting sequence.

This algorithm, with a classical data structure to represent permutations (e.g., as an array), has still an $O(n^2)$ complexity, because at each step it has to test the presence of an oriented reversal and apply it to the permutation.

Composing with Data Structures
The slight modification of a data structure invented by Kaplan and Verbin [11] allows to pick and apply an oriented reversal in $O(\sqrt{n \log n})$, and using this, Tannier and Sagot's algorithm achieves $O(n^{3/2} \sqrt{\log n})$ time complexity.

Han [9] announced another data structure that allows to pick and apply an oriented reversal in $O(\sqrt{n})$ time, and integrating this to the algorithm can plausibly decrease the complexity of the overall method to $O(n^{3/2})$. Swenson et al. [16]

gave an $O(n \log n)$ solution for picking oriented reversals, but their attempts of integrating it to the overall procedure seems to fail on worst cases.

Extensions

Once sorting by reversals has reached its best solutions, there are natural extensions guided by the main motivation for the problem in computational biology: sample among optimal solutions, and handle several permutations and more operations than just the reversal.

Counting optimal solutions is conjectured to be #P-complete [14], but sampling almost uniformly from the solution space is still open, and has been given a heuristic solution [14], including suboptimal solutions in the sample.

Algorithms to enumerate all sorting reversals at one step have also been worked out [4], which provides a way for enumeration. A structure of the solution space was proposed, but with a possibly exponential number of objects to enumerate [5].

The median problem consists in handling more than one permutation and is a particular case of the so-called small parsimony problem, which consists in reconstructing ancestral states in a phylogenetic context. Additional operations can be transpositions, duplications, or many others. Many generalizations and variants have been listed in a book on Combinatorics of Genome Rearrangements [8]. Almost all are NP-hard.

Applications

The motivation as well as the main application of this problem is in computational biology. Signed permutations are an adequate object to model the relative position and orientation of homologous segments of DNA in two species.

Reversal scenarios were used to test some evolutionary properties, like the propension of rearrangement to cut around the replication origin [1] or the fragility of certain genomic regions [2]. But evolutionary hypotheses can hardly be tested from a single optimal solution; this would necessitate a better view of the solution space.

The gain of complexity for sorting by reversals inspired many other algorithmic works, and several problems in genome rearrangement found a better solution thanks to the subquadratic gain described here. But the computational difficulties of the problem (parameters $h(\pi)$ and $f(\pi)$, additional complexity for generating a scenario compared to the distance calculation, NP-completeness of every generalization with more operations, more permutations, more realistic models) lead most computational biologists to progressively abandon the reversal model for simpler ones (DCJ [19], SCJ [7]).

Sometimes heroic gains in complexity are worth for computer science but seem just like going a bit further in a dead end for applications. Research consists in breaking walls without always knowing if behind there is a space for a community to work in or another thicker wall.

Open Problems

Still there are a couple of questions that remain unsolved before closing (or reopening?) this entry:

- I conjecture that the "real" complexity of giving a reversal scenario is $O(n \log n)$. It is more or less what Swenson et al. [16] also claim, but without giving a full proof.
- Counting and sampling, even approximately, are open. I learned this interesting conjecture from Istvan Miklos: is it possible to walk in the entire space of sequences of sorting reversals by small transformations of scenarios, consisting at each step to change at most 4 reversals? This would be a first step to design an almost uniform sampler.

Experimental Results

To my knowledge the data structure that allows the subquadratic complexity described in this entry has never been implemented. The size of the data, as well as the limited possibilities of applications of handling only two genomes and a single optimal solution, makes the subquadratic version, while a good piece of algorithmics, not really worth for applications.

URL to Code

- There are a few old programs still able to give a sorting sequence of reversals: in San Diego http://grimm.ucsd.edu/GRIMM/, New Mexico www.cs.unm.edu/~moret/GRAPPA/, or Tel Aviv www.math.tau.ac.il/~rshamir/GR/ and more recent ones in Lyon http://doua.prabi.fr/software/luna or Bielefeld http://bibiserv.techfak.uni-bielefeld.de/dcj/wel come.html.
- The standard software for Bayesian sampling in the space of sorting sequences (including nonoptimal ones) is Badger http://bibiserv.techfak.uni-bielefeld.de/dcj/welcome.html, and there is also one biased to optimal solutions called DCJ2HP http://www.renyi.hu/~miklosi/DCJ2HP/ that uses a parallel tempering between DCJ solutions (easier to sample) and reversals solutions.

Cross-References

▶ Sorting Signed Permutations by Reversal (Reversal Distance)

Recommended Reading

1. Ajana Y, Lefebvre J-F, Tillier E, El-Mabrouk N (2002) Exploring the set of all minimal sequences of reversals – an application to test the replication-directed reversal hypothesis. In: Proceedings of the second workshop on algorithms in bioinformatics. Lecture notes in computer science, vol 2452. Springer, Berlin, pp 300–315
2. Attie O, Darling A, Yancopoulos Y (2011) The rise and fall of breakpoint reuse depending on genome resolution. BMC Bioinform 12(supp 9):S1
3. Bader DA, Moret BME, Yan M (2001) A linear-time algorithm for computing inversion distance between

signed permutations with an experimental study. J Comput Biol 8(5):483–491

4. Badr G, Swenson KM, Sankoff D (2011) Listing all parsimonious **reversal** sequences: new algorithms and perspectives. J Comput Biol 18:1201–1210

5. Braga MDV, Sagot MF, Scornavacca C, Tannier E (2008) Exploring the solution space of sorting by reversals with experiments and an application to evolution. IEEE-ACM Trans Comput Biol Bioinform 5:348–356

6. Darling AE, Miklós I, Ragan MA (2008) Dynamics of genome rearrangement in bacterial populations. PLoS Genet (7):e1000128

7. Feijão P, Meidanis J (2011) SCJ: a breakpoint-like distance that simplifies several rearrangement problems. IEEE/ACM Trans Comput Biol Bioinform 8(5):1318–1329

8. Fertin G, Labarre A, Rusu I, Tannier E, Vialette S (2009) Combinatorics of genome rearrangements. MIT, Cambridge

9. Han Y (2006) Improving the efficiency of sorting by reversals. In: Proceedings of the 2006 international conference on bioinformatics and computational biology, Las Vegas

10. Hannenhalli S, Pevzner PA (1999) Transforming cabbage into turnip: polynomial algorithm for sorting signed permutations by reversals. J ACM (JACM) 46:1–27

11. Kaplan H, Verbin E (2003) Efficient data structures and a new randomized approach for sorting signed permutations by reversals. In: Proceedings of CPM'03. Lecture notes in computer science, vol 2676. Springer, Berlin/Heidelberg, pp 170–185

12. Kaplan H, Shamir R, Tarjan RE (1999) Faster and simpler algorithm for sorting signed permutations by reversals. SIAM J Comput 29:880–892

13. Larget B, Simon DL, Kadane JB (2002) On a Bayesian approach to phylogenetic inference from animal mitochondrial genome arrangements (with discussion). J R Stat Soc B 64:681–693

14. Miklós I, Tannier E (2010) Bayesian sampling of genomic rearrangement scenarios via double cut and join. Bioinformatics 26:3012–3019

15. Ozery-Flato M, Shamir R (2003) Two notes on genome rearrangement. J Bioinform Comput Biol 1:71–94

16. Swenson KM, Rajan V, Lin Y, Moret BME (2010) Sorting signed permutations by inversions in $O(n \log n)$ time. J Comput Biol 17:489–501

17. Tannier E, Sagot M-F (2004) Sorting by reversals in subquadratic time. In: Proceedings of CPM'04. Lecture notes in computer science, vol 3109. Springer, Berlin/Heidelberg, pp 1–13

18. Tannier E, Bergeron A, Sagot M-F (2006) Advances on sorting by reversals. Discret Appl Math 155:881–888

19. Yancopoulos S, Attie O, Friedberg R (2005) Efficient sorting of genomic permutations by translocation, inversion and block interchange. Bioinformatics 21:3340–3346

Spanning Trees with Low Average Stretch

Ittai Abraham[1] and Ofer Neiman[2]
[1]Microsoft Research, Silicon Valley, Palo Alto, CA, USA
[2]Department of Computer Science, Ben-Gurion University of the Negev, Beer Sheva, Israel

Keywords

Embedding; Spanning tree; Stretch

Years and Authors of Summarized Original Work

2012; Abraham, Neiman

Problem Definition

Let $G = (V, E)$ be an undirected graph, with nonnegative weights on the edges $w : E \to \mathbb{R}_+$. Let d_G be the shortest-path metric on G, with respect to the weights. For a spanning (subgraph) tree T of G, define the stretch of an edge $\{u, v\} \in E$ in T as $\text{stretch}_T(u, v) = \frac{d_T(u,v)}{d_G(u,v)}$ and the average stretch as

$$\text{avg} - \text{stretch}_T(G) = \frac{1}{|E|} \sum_{e \in E} \text{stretch}_T(e) .$$

We shall consider the problem of finding a tree T whose average stretch is small. We also study the problem of finding a distribution over spanning trees, such that for all $e \in E$, $\mathbb{E}_T[\text{stretch}_T(e)]$ is small.

Key Results

Low-stretch spanning trees were first studied by [3], who showed that any graph on n vertices has a spanning tree with average stretch $2^{O(\sqrt{\log n \log \log n})}$ and showed a family of graphs that requires $\Omega(\log n)$ average stretch. Their result was substantially improved by [7], who showed an upper bound of $O(\log^2 n \log \log n)$, and later [1] improved this to a near optimal $\tilde{O}(\log n)$.

The main result discussed here is from [2]:

Theorem 1 *For any graph G with n vertices and m edges, there is a deterministic algorithm that constructs a spanning tree T, such that $\mathrm{avg} - \mathrm{stretch}_T(G) \leq O(\log n \log \log n)$. The running time of the algorithm is $O(m \log n \log \log n)$.*

We also show an efficient algorithm to sample from a distribution over spanning trees, such that the expected stretch of any edge is bounded by $O(\log n \log \log n \log \log \log n)$.

Applications

An important problem in algorithm design is obtaining fast algorithms for solving linear systems. For many applications, the matrix is sparse, and while little is known for general sparse matrices, the case of symmetric diagonally dominant (SDD) matrices has received a lot of attention recently. In a seminal sequence of results, Spielman and Teng [12] showed a near-linear time solver for this important case. This solver has proven a powerful algorithmic tool and is used to calculate eigenvalues, obtain spectral graph sparsifiers [11], and approximate maximum flow [6] and many other applications. A basic step in solving these systems $Ax = b$ is combinatorial preconditioning. If one uses the Laplacian matrix corresponding to a spanning tree (and a few extra edges) of the graph whose Laplacian matrix is A, then the condition number depends on the total stretch of the tree. This will improve the run-time of iterative methods, such as conjugate gradient or Chebyshev iterations. See [9, 10] for the latest progress on this direction. In this work we show that one can construct such a spanning tree with both run-time and total stretch bounded by $O(m \log n \log \log n)$.

Probabilistic embedding into trees, introduced by [4], has been a successful paradigm in algorithm design. Many hard optimization problems on graphs can be reduced, via embedding, to a similar problem on a tree, which is often considerably easier. This framework can be applied to approximation algorithms, online algorithms, network design, and other settings. Some of the notable examples are metrical task system, buy-at-bulk network design, the k-server problem, group Steiner tree, etc. An asymptotical optimal result of expected $O(\log n)$ distortion for probabilistic embedding into trees was given by [8]. The trees in the support of the FRT distribution are not subgraphs of the input graph and may contain Steiner nodes and new edges. While this is fine for most applications, there are some that must have trees which are subgraphs, such as minimum cost communication spanning tree: Given a weighted graph $G = (V, E)$ and a requirement matrix $R = (r_{uv})$, the objective is to find a spanning tree T that minimizes $\sum_{u,v \in V} r_{uv} \cdot d_T(u, v)$. Our result implies a $\tilde{O}(\log n)$ approximation.

Petal Decomposition

A basic tool that is often used in constructing tree metrics and spanning trees with low stretch is *sparse graph decomposition*. The idea is to partition the graph into small diameter pieces, such that few edges are cut. Each cluster of the decomposition is partitioned recursively, which yields a hierarchical decomposition. Creating a tree recursively on each cluster of the decomposition, and connecting these in a tree structure, will yield a spanning tree of the graph. The edges cut by the decomposition are potentially stretched by a factor proportional to the diameter of the created tree. The construction has to balance between these two goals: cut a small number of edges and maintain small diameter in the created tree.

One of the main difficulties in such a spanning tree construction is that the radius (The radius of a graph is the maximal distance from a designated center.) may increase by a small factor at every application of the decomposition, which translates to increased stretch. If we drop the requirement that the tree is a *spanning tree* of the graph and just require a tree metric, then this difficulty does not appear, and indeed, optimal $\Theta(\log n)$ bound is known on the average stretch [5, 8]. Our *petal decomposition* allows essentially optimal control on the radius increase of the spanning tree; it increases by at most a factor of 4 over all the recursion levels.

Highways

One of the components in the decomposition scheme is highways. Each cluster $X \subseteq V$ in our decomposition scheme has a designated center $x_0 \in X$ and a "target" $t \in X$. It is guaranteed that the shortest path from x_0 to t will be fully contained in the *final* spanning tree T. This path is called the petal's highway. Intuitively, the highway will provide short paths from the center x_0 to many of the points in the cluster.

Cones and Petals

A cone is a generalization of a ball; the notion of cones was introduced in [7] and was used also in [1] for low-stretch spanning trees. Informally, a cone $C(t, r)$ of radius r centered at t (with respect to the cluster center x_0) contains all the points $z \in X$ such that $d(z, t) + d(t, x_0) \leq d(z, x_0) + r$ (here d is the shortest-path metric on X). In other words, the cone contains all the points for which the path to x_0 through t is not much longer than the direct shortest path to x_0. The parameter r is a bound on the radius increase in the current decomposition.

One way to define a petal is as a *union of cones*. The petal $P(t, r)$ around a target t with radius r is defined as $\bigcup_{0 \leq k \leq r} C(p_k, k/2)$, where p_k is the point of distance $r - k$ from t on the shortest path from t to x_0. The center of the petal is defined as $x = p_0$, and the path from x to t is the petal's highway. The petal-decomposition algorithm iteratively picks an arbitrary target of distance at least $3\Delta/4$ (where Δ is the radius of X) away from x_0, generates a petal for it, and removes the petal from the graph. When there are no longer such points, the remaining points will form the central cluster (the stigma). The first petal

requires extra care in its target choice, as it may contain the designated target of the cluster, which implies we cannot allow the shortest path to this target to be cut by this or subsequent petals. The radii of the petals are chosen by a region-growing argument that cuts few edges, where the length of the possible range for the radius is $\approx \Delta$. This is in contrast with the previous work, where in order to give an appropriate bound on the radius increase, the range was much smaller than Δ, which immediately translates to a loss in the stretch. The precise method for choosing r is essentially given in [7], and we also give a randomized version similar in spirit to the one in [1].

Fast Petal Construction

The alternative way to define petals and cones is as balls in an appropriately defined *directed* graph created from G. This suggests that we can use a variant of Dijkstra to compute a petal in nearly linear time in the sum of degrees of its vertices. Let $\tilde{G} = (V, A, \bar{w})$ be the weighted directed graph induced by adding the two directed edges $(u, v), (v, u) \in A$ for each $\{u, v\} \in E$ and setting $\bar{w}(u, v) = d(u, v) - (d(v, x_0) - d(u, x_0))$. The cone $C(t, r)$ is simply the ball around t of radius r in $\tilde{G}$. The petal $P(t, r)$ is the ball around t of radius $r/2$ in $\tilde{G}$ with one change: the weight of each edge on the path from t to $x = p_0$ is changed to be $1/2$ of its original weight (i.e., $1/2$ of its weight in G).

Ideas in the Analysis

Informally, the crucial property of a petal and its highway is the following: Assume $z \in P(t, r)$, and $P_{x_0 z}$ is the shortest path from the original center x_0 to z. By forming the petal, we remove all edges between $P(t, r)$ and $X \setminus P(t, r)$ except for the edge from the petal center x toward x_0. Hence, any path from x_0 to z must go through the petal center x. If the new shortest path $P'_{x_0 z}$ (after forming the petal) is (additively) $k/2$ longer than the length of $P_{x_0 z}$, then $z \in C(p_k, k/2)$ and so $P'_{x_0 z}$ will contain part of the new petal's highway of length at least k. Such a property could allow the following wishful thinking: Suppose that in each iteration we increase the distance of a point to the center by at most α but also mark a new portion of the path of length 2α as edges that are guaranteed to appear in the final tree (part of a highway). In such a case, it is easy to see that the final path will have stretch at most 2: If the original distance was b, once the total increase is b, we have marked $2b$ – all of the path – as a highway that will appear in the tree. Unfortunately, the path from x to z in the final tree may not use the prescribed highway of the parent cluster so the above "wishful thinking" argument does not work.

The key algorithmic idea to alleviate this problem is to decrease the weight of an edge by half when it becomes part of a highway (we ensure that this happens at most once for every edge). This reweighting signals later iterations to use the prescribed highway, as this must remain the shortest path. We maintain the invariant that in every cluster, the highway edges are the only cluster edges which have been reweighted. Now, in every petal (except for maybe the first), we create a *new* petal highway when we form $P(t, r)$. For any $z \in P(t, r)$, *the length of the path from x_0 to z does not increase at all* (after reweighting the highway): For some $k \le r$, it increased by at most $k/2$, but a highway length of at least k was reduced by $1/2$.

We have to take care of radius increase generated by the very first petal as well, where it could be that no new highway is created (this petal's highway may be a part of the highway of the original cluster). In this case, we use the fact that the path from x_0 to x_1 (the center of the first petal) must also be on the highway of the original cluster and that its length is at least $\Delta/2$. This implies that even though we may have increased the radius, at least half of the path is guaranteed not to increase ever again. We use a subtle inductive argument to make this intuition precise, and in fact we lose a factor of 2 for each of these cases, so the maximal increase is by a factor of 4.

Recommended Reading

1. Abraham I, Bartal Y, Neiman O (2008) Nearly tight low stretch spanning trees. In: FOCS '08: Proceedings of the 2008 49th annual IEEE symposium on foundations of computer science, Philadelphia. IEEE Computer Society, Washington, DC, pp 781–790
2. Abraham I, Neiman O (2012) Using petal-decompositions to build a low stretch spanning tree. In: Proceedings of the forty-fourth annual ACM symposium on theory of computing, STOC '12, New York. ACM, pp 395–406
3. Alon N, Karp RM, Peleg D, West D (1995) A graph-theoretic game and its application to the k-server problem. SIAM J Comput 24(1):78–100
4. Bartal Y (1996) Probabilistic approximation of metric spaces and its algorithmic applications. In: Proceedings of the 37th annual symposium on foundations of computer science, Burlington. IEEE Computer Society, Washington, DC, pp 184–
5. Bartal Y (2004) Graph decomposition lemmas and their role in metric embedding methods. In: Albers S, Radzik T (eds) Proceedings of the 12th annual European symposium on algorithms – ESA 2004, Bergen, 14–17 Sept 2004. Volume 3221 of Lecture notes in computer science. Springer, pp 89–97
6. Christiano P, Kelner JA, Madry A, Spielman DA, Teng S-H (2011) Electrical flows, Laplacian systems, and faster approximation of maximum flow in undirected graphs. In: Proceedings of the 43rd annual ACM symposium on theory of computing, STOC '11, San Jose. ACM, New York, pp 273–282
7. Elkin M, Emek Y, Spielman DA, Teng S-H (2005) Lower-stretch spanning trees. In: Proceedings of the thirty-seventh annual ACM symposium on theory of computing, STOC '05, Baltimore. ACM, New York, pp 494–503
8. Fakcharoenphol J, Rao S, Talwar K (2003) A tight bound on approximating arbitrary metrics by tree metrics. In: Proceedings of the thirty-fifth annual ACM symposium on theory of computing, STOC '03, San Diego. ACM, New York, pp 448–455
9. Koutis I, Miller GL, Peng R (2010) Approaching optimality for solving SDD linear systems. In: 51th annual IEEE symposium on foundations of computer science, 23–26 Oct 2010, Las Vegas, pp 235–244
10. Koutis I, Miller GL, Peng R (2011) A nearly $O(m \log n)$ time solver for SDD linear systems. In: 52th annual IEEE symposium on foundations of computer science, Palm Springs
11. Spielman DA, Srivastava N (2008) Graph sparsification by effective resistances. In: Proceedings of the 40th annual ACM symposium on theory of computing, STOC '08, Victoria. ACM, New York, pp 563–568
12. Spielman DA, Teng S-H (2004) Nearly-linear time algorithms for graph partitioning, graph sparsification, and solving linear systems. In: Proceedings of the thirty-sixth annual ACM symposium on theory of computing, STOC '04, Chicago. ACM, New York, pp 81–90

Sparse Fourier Transform

Eric Price
Department of Computer Science, The
University of Texas, Austin, TX, USA

Keywords

Compressed sensing; Sparse recovery

Years and Authors of Summarized Original Work

2012; Hassanieh, Indyk, Katabi, Price

Problem Definition

Suppose that we have access to a vector $x \in \mathbb{C}^n$. How much time does it take to compute its Fourier transform $\hat{x}$? One can do this with the Fast Fourier Transform (FFT) in $O(n \log n)$ time. But can we do better?

We do not know the answer in general, but some classes of algorithms cannot do better [1, 20] and certainly one cannot do better than $O(n)$ time for arbitrary signals x. But the Fourier transform is ubiquitous in signal processing, appearing in compression of audio, images, and video, in manipulation of audio, and in recovery of radio or MRI signals, so we would really like to do better. If we cannot improve on the FFT in general, then perhaps we can for the signals commonly seen in these applications. To do this, we need some notion for how the signals we typically see are "easier" than arbitrary ones.

One such notion is *sparsity*. The main reason to use the Fourier transform in compression is because it concentrates the energy of the signal into a few large (or "heavy") coordinates and many small ones; signals with such concentrated coordinates are called *sparse*. One can then throw out the small coordinates and only store the heavy ones; this is the main principle behind lossy compression such as MP3 or JPEG. In fact,

in all the applications discussed in the previous paragraph, the signals typically have an approximately sparse Fourier transform. This brings us to the problem described in this entry: can we speed up the Fourier transform for signals when the result is approximately sparse?

Moreover, as with lossy compression, we often only care about the heavy coordinates and are willing to tolerate an error proportional to the energy in the small coordinates. This relaxation will allow us to compute the sparse Fourier transform in *sublinear* time.

Formal Definition

The discrete Fourier transform $\hat{x} \in \mathbb{C}^n$ of a vector $x \in \mathbb{C}^n$ is given by

$$\hat{x}_j = \sum_{i=1}^{n} \omega^{ij} \text{ for } \omega = e^{2\pi i/n}$$

We say that $\hat{x}$ is *exactly k-sparse* if it has at most k nonzero coordinates, i.e., $|\text{supp}(x)| \leq k$. We say that $\hat{x}$ is *approximately k-sparse* if most of the energy is contained in the heaviest k coordinates, in particular

$$\text{Err}_k(x) := \min_{k\text{-sparse } \hat{y}} \|\hat{x} - \hat{y}\|_2$$

is small relative to $\|\hat{x}\|_2$. A sparse Fourier transform algorithm can access $x \in \mathbb{C}^n$ in arbitrary positions and outputs a vector $\hat{x}'$ such that

$$\|\hat{x} - \hat{x}'\|_2 \leq C \, \text{Err}_k(x) + \delta \|\hat{x}\|_2 \qquad (1)$$

for some approximation factor $C > 1$ and $\delta \ll 1$. An algorithm for the exactly sparse case would do this for $C = \infty$, while robust algorithms can achieve $C = O(1)$ or even $C = 1 + \varepsilon$. The algorithms we will discuss will feature a logarithmic dependence on $1/\delta$, so one typically sets $\delta = 1/\text{poly}(n)$, and for typical signals, the right-hand side of (1) will be dominated by the $C \, \text{Err}_k(x)$ term; we will assume this for the rest of the entry.

We would like to optimize both the sample complexity – the number of positions of x that are accessed by the algorithm – and the running time. Optimizing sample complexity is important for applications such as spectrum sensing or MRIs, which do not have the input x in memory but must sample it at some expense.

We also allow the algorithm to be randomized and to fail with some small probability p. For simplicity we set p to a small constant; for any algorithm one can amplify this probability with a $O(\log \frac{1}{p})$ overhead in sample complexity and time. It is an open question whether the algorithms that achieve the best known time and sample complexities can be modified to avoid this overhead.

Related Work

The modern research on sparse Fourier transforms is closely related to work on sparse recovery from general linear measurements. In this problem, one would like to (approximately) recover an (approximately) sparse vector x from linear measurements Ax for some "measurement" matrix A with fewer rows than columns. The sparse Fourier transform is precisely this where A is a subset of rows of the inverse Fourier matrix.

Broadly speaking, there are two conceptual classes of algorithms and results for the general linear measurement setting. The first class, often called *compressed sensing* and first studied in [2, 3, 7], generally (1) involves independent random linear measurements,;(2) shows with high probability, the measurement matrix gives good recovery *for all* vectors x; (3) optimizes the sample complexity but not the running time, which is superlinear or polynomial in n; and (4) give algorithms that work for general classes of measurements and work for both random Gaussian and random Fourier matrices at the same time. These papers often refer to properties like the restricted isometry property that measurement matrices may have and use either convex optimization (e.g., L1 minimization or the LASSO) or iterative greedy methods (e.g., IHT or CoSaMP) to perform the recovery.

The second class, more often called *sparse recovery*, is largely an outgrowth of the streaming

algorithms literature [4, 5]. These results generally (1) involve more structured linear measurements that use randomness and also have dependencies among the samples; (2) show *for each* vector x that, with high probability, the measurement matrix gives good recovery; (3) optimize both the sample complexity and the running time, so both may be sublinear in n; and (4) give algorithms that are closely connected to the measurement matrix and would not work for matrices with different structure. These papers often construct the matrix to emulate hash tables and use medians to perform robust recovery.

These statements are generalizations, and not every algorithm matches the trend in all four ways, but they hold more often than not. Our algorithm falls in the second class, which for Fourier measurements can achieve both better sample complexity and better running time than algorithms in the first class.

There's a much older collection of algorithms that can do sparse Fourier transforms in the exact setting when $|\text{supp}(\hat{x})| \leq k$. These include Prony's method from 1795, the matrix pencil method, and Berlekamp-Massey syndrome decoding. These can achieve the optimal sample complexity of $2k$ and recovery time $\text{poly}(k)$ (down to $O(k^2 + k \log^c \log n)$ [8]). Additionally, they use a deterministic set of samples and work for all vectors x. However, it is not known how to make the techniques in these algorithms robust to approximately sparse signals, so they do not apply to the signals appearing in typical applications.

Noise-tolerant sparse Fourier transforms were first studied over the Boolean cube, also known as the Hadamard transform. In this setting, Goldreich and Levin [12, 18] showed how to get $O(k \log(n/k))$ samples and $O(k \log^c n)$ time, which is essentially optimal. Mansour [19] extended this to the $\mathbb{C}^n$ setting that we consider in this entry but with more than k^2 sample complexity. Over the next couple decades, a number of subsequent works, including [9, 10, 13, 14, 16], have improved our understanding of the $\mathbb{C}^n$ setting.

Key Results

At present, the two best sparse Fourier transform algorithms are [13], which is fastest at $O(k \log(n/k) \log n)$ time and sample complexity, and [14], which has nearly optimal $O(k \log n)$ sample complexity at the cost of $\tilde{O}(n)$ running time. These works build on [9, 10, 17].

We know that the optimal *nonadaptive* sample complexity – that is, among algorithms that choose the sample set Ω independently of the vector x – is $\Omega(k \log(n/k))$ [6], which matches [14] for $k < n^{0.99}$. One could imagine constructing an algorithm that uses adaptive samples, where one uses the first few samples to decide where to look in future samples. In the general sparse recovery setting, this adaptivity can lead to significant improvements [15], but we know that $\Omega(k \log(n/k)/ \log \log n)$ Fourier samples remain necessary in the adaptive setting [13].

Algorithm Overview

At a high level, sparse recovery algorithms are built in three stages: one-sparse recovery, where we solve the problem for $k = 1$; partial k-sparse recovery, where we find and estimate *most* (say, 90 %) of the heavy coordinates or of the energy; and full k-sparse recovery, where we get a good approximation to the entire signal and achieve (1). Each stage uses the previous as (nearly) a black box. This architecture generally holds for the class of "sparse recovery" algorithms; in the sparse Fourier transform setting, the

pieces change, but the architecture does not. We will go through each in turn.

One-Sparse Recovery

Let us consider the one-sparse setting for $C = O(1)$. We have access to $x_j = v\omega^{i^* j} + g_j$ for some "signal" $(v, i^*) \in \mathbb{C} \times [n]$ and "noise" $g \in \mathbb{C}^n$ with $\|g\|_2 \leq c|v|\sqrt{n}$ for a sufficiently small constant c. To satisfy (1), we would like to find i^* exactly and find v to within $O(\|g\|/\sqrt{n})$.

The tricky bit is to find i^*; once we know i^*, then $x_j \omega^{-i^* j}$ is a good estimator of v. In particular, for a random $j \in [n]$, we have $\mathbb{E}_j |x_j \omega^{-i^* j} - v|^2 = \|g\|^2/n$, so taking the median of several such estimates will have $O(\|g\|/\sqrt{n})$ error with large probability. So that just leaves us to find i^*.

As a first step, consider for a fixed $a \in [n]$ looking at the random variable

$$y_a := x_{a+j}/x_j \approx \omega^{i^* a}$$

as a distribution over random $j \in [n]$, where addition of indices is taken modulo n. This allows us to remove the influence of v and focus on i^*. We can show that $y_a - \omega^{i^* a} < O(c)$ with large (say, 3/4) probability. Suppose this were instead true with probability 1.

By knowing $\omega^{i^* a}$ to within $O(c)$, we know $i^* a \bmod n$ to with $\pm O(cn)$. For small enough c, this is within $\pm n/4$. Then we could look at y_1 to learn i^* to within $\pm n/4$, y_2 to refine the estimate to $\pm n/8$, and y_4 to refine to $\pm n/16$, until we identify i^* using $\log n$ different y_a. This is illustrated in Fig. 1.

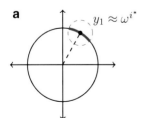 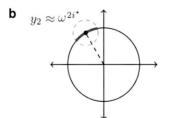 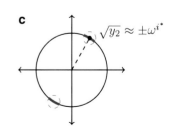

Sparse Fourier Transform, Fig. 1 The first two steps of estimating i^* using y_1 and y_2. Using y_1 we can identify i^* to an $O(cn)$ size region. With y_2 we learn $2i^* \bmod n$ to within $O(cn)$, which tells us that i^* is within one of two antipodal regions of half the size. Based on y_1, we can throw out the spurious region and narrow our estimate of i^* (**a**) Error in y_1. (**b**) Error in y_2. (**c**) Set of ω^{i^*} consistent with y_2

Filter (time): Gaussian · sinc

Filter (frequency): Gaussian * rectangle

Sparse Fourier Transform, Fig. 2 Filters used in [13]. (**a**) In time domain: $O(k \log n)$ sparse. (**b**) In frequency domain: width $O(n/k)$ rectangle

In reality y_a has a small constant chance of failure at each stage. One could fix this by taking $O(\log \log n)$ different samples of y_a at each stage and using the median, which would give an algorithm with $O(\log n \log \log n)$ sample complexity and time. An alternative, as used in [13], is to learn i^* in chunks of $O(\log \log n)$ bits at a time, which gets the optimal $O(\log n)$ sample complexity using $O(\log^{1.1} n)$ running time.

Partial k-Sparse Recovery

The goal of partial k-sparse recovery is to find *most* of the heavy coordinates of $\hat{x}$. The general idea is to "hash" the coordinates randomly into $B = O(k)$ bins in a way that lets us take measurements of the signal restricted to frequencies within each bin. By taking the measurements corresponding to the one-sparse recovery algorithm, we recover frequencies that are alone in their bin. This will happen with a large constant (say, 90 %) probability for each heavy frequency, so we recover *most* of the heavy frequencies well.

To see how this is done, we start with a deterministic way of hashing the frequencies into bins and then show how to randomize it. Hashing is based on filters that are sparse in both time and frequency domain. The filter F is designed to be as close as possible to a rectangular filter in frequency domain while still being sparse in frequency domain. Figure 2 demonstrates the filter used in [13], where F is a sinc function times a (truncated) Gaussian with support size $O(k \log n)$. In frequency domain, $\hat{F}$ approximates a rectangle of width $O(n/k)$, matching

it up to a small transition region between the passband and the stopband and with $1/n^c$ error inside the passband and stopband.

Using these filters, Fig. 3 demonstrates a method for learning information about the signal. Given the signal x, we compute the $O(k \log n)$-size vector $F \cdot x$. We then "alias" it down to $B = O(k)$ elements – adding up terms $1, B + 1, 2B + 1, \dots$ – and take the B-dimensional DFT. This lets us compute the red points in Fig. 3f in $O(k \log n + B \log B) = O(k \log n)$ time. The red points are B evenly spaced samples of $\hat{x} * \hat{F}$.

We can think of the ith red point in a different way. The ith red point is the sum of all the entries of $\hat{x} \cdot \text{shift}(\hat{F}, in/B)$, where $\text{shift}(\hat{F}, in/B)$ denotes shifting $\hat{F}$ to the right by in/B. This equals the zeroth time domain coefficient of the vector with Fourier coefficients given by $\hat{x} \cdot \text{shift}(\hat{F}, in/B)$. And if our algorithm looks not at $y_j = F_j x_j$ but $y_j^{(a)} = F_j x_{j+a}$ when computing the red points, then the ith red point will equal the ath time domain coefficient of the vector with Fourier coefficients given by $\hat{x} \cdot \text{shift}(\hat{F}, in/B)$. This lets us sample from the time domain representation of the vectors with Fourier coefficients given by $\hat{x} \cdot \text{shift}(\hat{F}, in/B)$ for $i \in [B]$. It takes $O(k \log n)$ time to get these samples, for $O(\log n)$ overhead (in time and samples) per "effective" sample.

Now, we simply choose our samples a from the distribution requested by the one-sparse recovery algorithm. In every bucket for which $\hat{x} \cdot \text{shift}(\hat{F}, in/B)$ is one-sparse, this procedure will let us recover the heavy frequency. Because the different shifts of F give B different buckets,

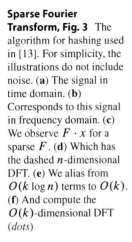

**Sparse Fourier
Transform, Fig. 3** The
algorithm for hashing used
in [13]. For simplicity, the
illustrations do not include
noise. (**a**) The signal in
time domain. (**b**)
Corresponds to this signal
in frequency domain. (**c**)
We observe $F \cdot x$ for a
sparse F. (**d**) Which has
the dashed n-dimensional
DFT. (**e**) We alias from
$O(k \log n)$ terms to $O(k)$.
(**f**) And compute the
$O(k)$-dimensional DFT
(*dots*)

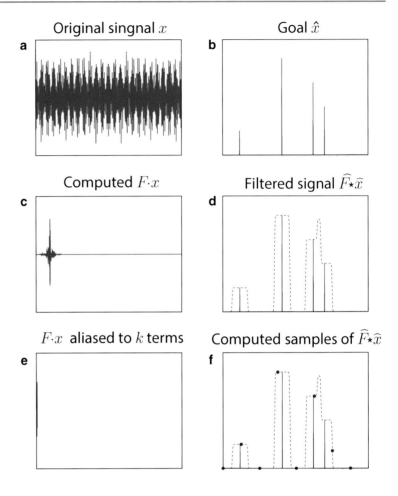

if the frequencies were randomly distributed, this technique would get us partial sparse recovery. The one-sparse recovery algorithm only takes $O(\log(n/k))$ samples because each frequency is known to lie within an $n/B = O(n/k)$ size region; hence the overall method takes $O(k \log n \log(n/k))$ time and samples.

We would like the algorithm to work for arbitrary input signals, so we need a way of randomizing the frequencies. To do this, we further refine the algorithm to choose a random $\sigma, b \in [n]$ with σ relatively prime to n. Then we have the algorithm look at $y_j^{(a)} = F_j x_{\sigma(j+a)} \omega^{-\sigma j b}$. The effect of σ, b is to apply an hash function $j \rightarrow \sigma^{-1} j + b$ in frequency domain; this is approximately pairwise independent, so the frequencies become effectively randomly distributed. Each frequency then has a good chance of landing alone in its bucket, so we can recover

most frequencies in $O(k \log(n/k) \log n)$ time and samples.

Full k-Sparse Recovery

Once we have partial k-sparse recovery, one can naively achieve full k-sparse recovery by repeating the algorithm $O(\log k)$ times. Since each heavy frequency is recovered with 90% probability in each stage, the median of all the estimations will recover all the heavy frequencies – and in fact achieve (1) – with high probability. This method is simple but loses a $\log k$ factor in running time and sample complexity, which more intricate techniques can avoid.

One such technique, used in [13] and based off [11], is to use smaller and smaller k in successive iterations. Once we have performed partial sparse recovery on $\hat{x}$ to get $\hat{x}^{(1)}$ that contains

90 % of the heavy hitters, we can then perform sparse recovery on the residual $\hat{x} - \hat{x}^{(1)}$. The residual is then roughly $k/10$-sparse, so we run a partial $k/10$-sparse recovery algorithm in the second stage that is much faster than in the first stage. Similar geometric decay happens in later stages, so the total time spent will be dominated by the first stage. This gives $O(k \log(n/k) \log n)$ time and sample complexity for the problem.

Recommended Reading

1. Ailon N (2014) An n\log n lower bound for Fourier transform computation in the well conditioned model. CoRR. abs/1403.1307, http://arxiv.org/abs/1403.1307
2. Candes E, Tao T (2006) Near optimal signal recovery from random projections: universal encoding strategies. IEEE Trans Info Theory
3. Candes E, Romberg J, Tao T (2006) Robust uncertainty principles: exact signal reconstruction from highly incomplete frequency information. IEEE Trans Info Theory 52:489–509
4. Charikar M, Chen K, Farach-Colton M (2002) Finding frequent items in data streams. In: ICALP
5. Cormode G, Muthukrishnan S (2004) Improved data stream summaries: the count-min sketch and its applications. In: LATIN
6. Do Ba K, Indyk P, Price E, Woodruff D (2010) Lower bounds for sparse recovery. In: SODA
7. Donoho D (2006) Compressed sensing. IEEE Trans Info Theory 52(4):1289–1306
8. Ghazi B, Hassanieh H, Indyk P, Katabi D, Price E, Shi L (2013) Sample-optimal average-case sparse Fourier transform in two dimensions. In: Allerton
9. Gilbert A, Guha S, Indyk P, Muthukrishnan M, Strauss M (2002) Near-optimal sparse Fourier representations via sampling. In: STOC
10. Gilbert A, Muthukrishnan M, Strauss M (2005) Improved time bounds for near-optimal space Fourier representations. In: SPIE conference on wavelets
11. Gilbert AC, Li Y, Porat E, Strauss MJ (2010) Approximate sparse recovery: optimizing time and measurements. In: STOC, pp 475–484
12. Goldreich O, Levin L (1989) A hard-corepredicate for allone-way functions. In: STOC, pp 25–32
13. Hassanieh H, Indyk P, Katabi D, Price E (2012) Nearly optimal sparse Fourier transform. In: STOC
14. Indyk P, Kapralov M (2014) Sample-optimal Fourier sampling in any constant dimension. In: 2014 IEEE 55th annual symposium on foundations of computer science (FOCS). IEEE, pp 514–523
15. Indyk P, Price E, Woodruff D (2011) On the power of adaptivity in sparse recovery. In: FOCS
16. Iwen MA (2010) Combinatorial sublinear-time Fourier algorithms. Found Comput Math 10:303–338
17. Kushilevitz E, Mansour Y (1991) Learning decision trees using the Fourier spectrum. In: STOC
18. Levin L (1993) Randomness and non-determinism. J Symb Logic 58(3):1102–1103
19. Mansour Y (1992) Randomized interpolation and approximation of sparse polynomials. In: ICALP
20. Morgenstern J (1973) Note on a lower bound on the linear complexity of the fast Fourier transform. J ACM (JACM) 20(2):305–306

Sparse Graph Spanners

Michael Elkin
Department of Computer Science, Ben-Gurion University, Beer-Sheva, Israel

Keywords

$(1 + \epsilon, \beta)$-spanners; Almost additive spanners

Years and Authors of Summarized Original Work

2004; Elkin, Peleg

Problem Definition

For a pair of numbers α, β, $\alpha \geq 1$, $\beta \geq 0$, a subgraph $G' = (V, H)$ of an unweighted undirected graph $G = (V, E)$, $H \subseteq E$, is an (α, β)-spanner of G if for every pair of vertices $u, w \in V$, $\text{dist}_{G'}(u, w) \leq \alpha \cdot \text{dist}_G(u, w) + \beta$, where $\text{dist}_G(u, w)$ stands for the distance between u and w in G. It is desirable to show that for every n-vertex graph there exists a sparse (α, β)-spanner with as small values of α and β as possible. The problem is to determine asymptotic tradeoffs between α and β on one hand, and the sparsity of the spanner on the other.

Key Results

The main result of Elkin and Peleg [8] establishes the existence and efficient constructibility of

$(1 + \epsilon, \beta)$-spanners of size $O(\beta n^{1+1/\kappa})$ for every n-vertex graph G, where $\beta = \beta(\epsilon, \kappa)$ is constant whenever κ and ϵ are. The specific dependence of β on κ and ϵ is $\beta(\kappa, \epsilon) = \kappa^{\log \log \kappa - \log \epsilon}$.

An important ingredient of the construction of [8] is a partition of the graph G into regions of small diameter in such a way that the supergraph induced by these regions is sparse. The study of such partitions was initiated by Awerbuch [3], that used them for network synchronization. Peleg and Schäffer [10] were the first to employ such partitions for constructing spanners. Specifically, they constructed $(O(\kappa), 1)$-spanners with $O(n^{1+1/\kappa})$ edges. Althofer et al. [2] provided an alternative proof of the result of Peleg and Schäffer that uses an elegant greedy argument. This argument also enabled Althofer et al. to extend the result to weighted graphs, to improve the constant hidden by the O-notation in the result of Peleg and Schäffer, and to obtain related results for planar graphs.

Applications

Efficient algorithms for computing sparse $(1 + \epsilon, \beta)$-spanners were devised in [7] and [13]. The algorithm of [7] was used in [7, 9, 12] for computing almost shortest paths in centralized, distributed, streaming, and dynamic centralized models of computations. The basic approach used in these results is to construct a sparse spanner, and then to compute exact shortest paths on the constructed spanner. The sparsity of the latter guarantees that the computation of shortest paths in the spanner is far more efficient than in the original graph.

Open Problems

The main open question is whether it is possible to achieve similar results with $\epsilon = 0$. More formally, the question is: Is it true that for any $\kappa \geq 1$ and any n-vertex graph G there exists $(1, \beta(\kappa))$-spanner of G with $O(n^{1+1/\kappa})$ edges?

This question was answered in affirmitive for κ equal to 2, 5/2, and 3 [1, 4-6, 8]. Some lower bounds were recently proved by Woodruff [14].

A less challenging problem is to improve the dependence of β on ϵ and κ. Some progress in this direction was achieved by Thorup and Zwick [13], and very recently by Pettie [11].

Cross-References

▶ Synchronizers, Spanners

Recommended Reading

1. Aingworth D, Chekuri D, Indyk P, Motwani R (1999) Fast estimation of diameter and shortest paths (without matrix multiplication). SIAM J Comput 28(4):1167–1181
2. Althofer I, Das G, Dobkin DP, Joseph D, Soares J (1993) On sparse spanners of weighted graphs. Discret Comput Geom 9:81–100
3. Awerbuch B (1985) Complexity of network synchronization. J ACM 4:804–823
4. Baswana S, Kavitha T, Mehlhorn K, Pettie S (2010) Additive spanners and (alpha, beta)-spanners. ACM Trans Algorithms 7:Article 1
5. Chechik S (2013) New additive spanners. In: Proceedings 24th annual ACM-SIAM symposium on discrete algorithms, New Orleans, Jan 2013, pp 498–512.
6. Dor D, Halperin S, Zwick U (2000) All pairs almost shortest paths. SIAM J Comput 29:1740–1759
7. Elkin M (2005) Computing almost shortest paths. Trans Algorithm 1(2):283–323
8. Elkin M, Peleg D (2004) $(1 + \epsilon, \beta)$-spanner constructions for general graphs. SIAM J Comput 33(3):608–631
9. Elkin M, Zhang J (2006) Efficient algorithms for constructing $(1 + \epsilon, \beta)$-spanners in the distributed and streaming models. Distrib Comput 18(5):375–385
10. Peleg D, Schäffer A (1989) Graph spanners. J Graph Theory 13:99–116
11. Pettie S (2009) Low-distortion spanners. ACM Trans Algorithms 6:Article 1
12. Roditty L, Zwick U (2004) Dynamic approximate all-pairs shortest paths in undirected graphs. In: Proceedings of symposium on foundations of computer science, Rome, Oct 2004, pp 499–508
13. Thorup M, Zwick U (2006) Spanners and emulators with sublinear distance errors. In: Proceedings of symposium on discrete algorithms, Miami, Jan 2006, pp 802–809

14. Woodruff D (2006) Lower bounds for additive span-
    ners, emulators, and more. In: Proceedings of sympo-
    sium on foundations of computer science, Berckeley,
    Oct 2006, pp 389–398

Sparsest Cut

Shuchi Chawla
Department of Computer Science, University of
Wisconsin–Madison, Madison, WI, USA

Keywords

Minimum ratio cut

Years and Authors of Summarized Original Work

2004; Arora, Rao, Vazirani

Problem Definition

In the Sparsest Cut problem, informally, the goal
is to partition a given graph into two or more large
pieces while removing as few edges as possible.
Graph partitioning problems such as this one occupy a central place in the theory of network flow,
geometric embeddings, and Markov chains, and
form a crucial component of divide-and-conquer
approaches in applications such as packet routing, VLSI layout, and clustering.

Formally, given a graph $G = (V, E)$, the *sparsity* or *edge expansion* of a non-empty set $S \subset V$,
$|S| \leq \frac{1}{2}|V|$, is defined as follows:

$$\alpha(S) = \frac{|E(S, V \setminus S)|}{|S|}.$$

The sparsity of the graph, $\alpha(G)$, is then defined as
follows:

$$\alpha(G) = \min_{S \subseteq V, |S| \leq \frac{1}{2}|V|} \alpha(S).$$

The goal in the Sparsest Cut problem is to find
a subset $S \subset V$ with the minimum sparsity, and
to determine the sparsity of the graph.

The first approximation algorithm for the
Sparsest Cut problem was developed by Leighton
and Rao in 1988 [13]. Employing a linear
programming relaxation of the problem, they
obtained an $O(\log n)$ approximation, where n is
the size of the input graph. Subsequently Arora,
Rao and Vazirani [4] obtained an improvement
over Leighton and Rao's algorithm using
a semi-definite programming relaxation, approximating the problem to within an $O(\sqrt{\log n})$
factor.

In addition to the Sparsest Cut problem, Arora
et al. also consider the closely related Balanced
Separator problem. A partition $(S, V \setminus S)$ of the
graph G is called a c-balanced separator for
$0 < c \leq \frac{1}{2}$, if both S and $V \setminus S$ have at least
$c|V|$ vertices. The goal in the Balanced Separator
problem is to find a c-balanced partition with
the minimum sparsity. This sparsity is denoted
$\alpha_c(G)$.

Key Results

Arora et al. provide an $O(\sqrt{\log n})$ *pseudo-
approximation* to the balanced-separator problem
using semi-definite programming. In particular,
given a constant $c \in (0, \frac{1}{2}]$, they produce
a separator with balance c' that is slightly worse
than c (that is, $c' < c$), but sparsity within an
$O(\sqrt{\log n})$ factor of the sparsity of the optimal
c-balanced separator.

Theorem 1 *Given a graph $G = (V, E)$, let
$\alpha_c(G)$ be the minimum edge expansion of
a c-balanced separator in this graph. Then
for every fixed constant $a < 1$, there exists
a polynomial-time algorithm for finding a c'-
balanced separator in G, with $c' \geq ac$, that has
edge expansion at most $O(\sqrt{\log n}\alpha_c(G))$.*

Extending this theorem to include unbalanced
partitions, Arora et al. obtain the following:

Theorem 2 *Let $G = (V, E)$ be a graph with
sparsity $\alpha(G)$. Then there exists a polynomial-time*

algorithm for finding a partition $(S, V \setminus S)$, *with* $S \subset V$, $S \neq \emptyset$, *having sparsity at most* $O(\sqrt{\log n}\alpha(G))$.

An important contribution of Arora et al. is a new geometric characterization of vectors in n-dimensional space endowed with the squared-Euclidean metric. This result is of independent significance and has lead to or inspired improved approximation factors for several other partitioning problems (see, for example, [1, 5, 6, 7, 11]).

Informally, the result says that if a set of points in n-dimensional space is randomly projected on to a line, a good separator on the line is, with high probability, a good separator (in terms of squared-Euclidean distance) in the original high-dimensional space. Separation on the line is related to separation in the original space via the following definition of stretch.

Definition 1 (Def. 4 in [4]) Let $\vec{x}_1, \vec{x}_2, \ldots, \vec{x}_n$ be a set of n points in $\mathcal{R}^n$, equipped with the squared-Euclidean metric $d(x, y) = \|x - y\|_2^2$. The set of points is said to be (t, γ, β) *-stretched at scale* ℓ, if for at least a γ fraction of all the n-dimensional unit vectors u, there is a partial matching $M_u = \{(x_i, y_i)\}_i$ among these points, with $|M_u| \geq \beta n$, such that for all $(x, y) \in M_u$, $d(x, y) \leq \ell^2$ and $\langle u, \vec{x} - \vec{y} \rangle \geq t\ell/\sqrt{n}$. Here $\langle \cdot, \cdot \rangle$ denotes the dot product of two vectors.

Theorem 3 *For any* $\gamma, \beta > 0$, *there is a constant* $C = C(\gamma, \beta)$ *such that if* $t > C \log^{1/3} n$, *then no set of n points in $\mathcal{R}^n$ can be* (t, γ, β)-*stretched for any scale* ℓ.

In addition to the SDP-rounding algorithm, Arora et al. provide an alternate algorithm for finding approximate sparsest cuts, using the notion of expander flows. This result leads to fast (quadratic time) implementations of their approximation algorithm [3].

Applications

One of the main applications of balanced separators is in improving the performance of divide and conquer algorithms for a variety of optimization problems.

One example is the Minimum Cut Linear Arrangement problem. In this problem, the goal is to order the vertices of a given n vertex graph G from 1 through n in such a way that the capacity of the largest of the cuts $(\{1, 2, \cdots, i\}, \{i + 1, \cdots, n\})$, $i \in [1, n]$, is minimized. Given a ρ-approximation to the balanced separator problem, the following divide and conquer algorithm gives an $O(\rho \log n)$-approximation to the Minimum Cut Linear Arrangement problem: find a balanced separator in the graph, then recursively order the two parts, and concatenate the orderings. The approximation follows by noting that if the graph has a balanced separator with expansion $\alpha_c(G)$, only $O(\rho n \alpha_n(G))$ edges are cut at every level, and given that a balanced separator is found at every step, the number of levels of recursion is at most $O(\log n)$.

Similar approaches can be used for problems such as VLSI layout and Gaussian elimination. (See the survey by Shmoys [14] for more details on these topics.)

The Sparsest Cut problem is also closely related to the problem of embedding squared-Euclidean metrics into the Manhattan (ℓ_1) metric with low distortion. In particular, the integrality gap of Arora et al.'s semi-definite programming relaxation for Sparsest Cut (generalized to include weights on vertices and capacities on edges) is exactly equal to the worst-case distortion for embedding a squared-Euclidean metric into the Manhattan metric. Using the technology introduced by Arora et al., improved embeddings from the squared-Euclidean metric into the Manhattan metric have been obtained [5, 7].

Open Problems

Hardness of approximation results for the Sparsest Cut problem are fairly weak. Recently Chuzhoy and Khanna [9] showed that this problem is APX-hard, that is, there exists a constant $\epsilon > 0$, such that a $(1 + \epsilon)$-approximation algorithm for Sparsest Cut would imply P $=$ NP. It is conjectured that the weighted version of the problem is NP-hard to approximate better than $O((\log \log n)^c)$ for some constant c, but this is only known to hold true assuming a version of the so-called Unique Games conjecture [8, 12]. On the other hand, the semi-definite programming relaxation of Arora et al. is known to have an integrality gap of $\Omega(\log \log n)$ even in the unweighted case [10]. Proving an unconditional super-constant hardness result for weighted or unweighted Sparsest Cut, or obtaining $o(\sqrt{\log n})$-approximations for these problems remain open.

The directed version of the Sparset Cut problem has also been studied, and is known to be hard to approximate within a $2^{\Omega(\log^{1-\epsilon} n)}$ factor [9]. On the other hand, the best approximation known for this problem only achieves a polynomial factor of approximation–a factor of $O(n^{11/23} \log^{O(1)} n)$ due to Aggarwal, Alon and Charikar [2].

Recommended Reading

1. Agarwal A, Charikar M, Makarychev K, Makarychev Y (2005) Proceedings of the 37th ACM symposium on theory of computing (STOC), Baltimore, May 2005, pp 573–581
2. Aggarwal A, Alon N, Charikar M (2007) Improved approximations for directed cut problems. In: Proceedings of the 39th ACM symposium on theory of computing (STOC), San Diego, June 2007, pp 671–680
3. Arora S, Hazan E, Kale S (2004) Proceedings of the 45th IEEE symposium on foundations of computer science (FOCS), Rome, 17–19 Oct 2004, pp 238–247
4. Arora S, Rao S, Vazirani U (2004) Expander flows, geometric embeddings, and graph partitionings. In: Proceedings of the 36th ACM symposium on theory of computing (STOC), Chicago, June 2004, pp 222–231
5. Arora S, Lee J, Naor A (2005) Euclidean distortion and the sparsest cut. In: Proceedings of the 37th ACM Symposium on Theory of Computing (STOC), Baltimore, May 2005, pp 553–562
6. Arora S, Chlamtac E, Charikar M (2006) New approximation guarantees for chromatic number. In: Proceedings of the 38th ACM symposium on theory of computing (STOC), Seattle, May 2006, pp 215–224
7. Chawla S, Gupta A, Räcke H (2005) Embeddings of negative-type metrics and an improved approximation to generalized sparsest cut. In: Proceedings of the ACM-SIAM symposium on discrete algorithms (SODA), Vancouver, Jan 2005, pp 102–111
8. Chawla S, Krauthgamer R, Kumar R, Rabani Y, Sivakumar D (2005) On the hardness of approximating sparsest cut and multicut. In: Proceedings of the 20th IEEE conference on computational complexity (CCC), San Jose, June 2005, pp 144–153
9. Chuzhoy J, Khanna S (2007) Polynomial flow-cut gaps and hardness of directed cut problems. In: Proceedings of the 39th ACM symposium on theory of computing (STOC), San Diego, June 2007, pp 179–188
10. Devanur N, Khot S, Saket R, Vishnoi N (2006) Integrality gaps for sparsest cut and minimum linear arrangement problems. In: Proceedings of the 38th ACM symposium on theory of computing (STOC), Seattle, May 2006, pp 537–546
11. Feige U, Hajiaghayi M, Lee J (2005) Improved approximation algorithms for minimum-weight vertex separators. In: Proceedings of the 37th ACM symposium on theory of computing (STOC), Baltimore, May 2005, pp 563–572
12. Khot S, Vishnoi N (2005) Proceedings of the 46th IEEE symposium on foundations of computer science (FOCS), Pittsburgh, Oct 2005, pp 53–62
13. Leighton FT, Rao SB (1988) An approximate max-flow min-cut theorem for uniform multicommodity flow problems with applications to approximation algorithms. In: Proceedings of the 29th IEEE symposium on foundations of computer science (FOCS), White Plains, Oct 1988, pp 422–431
14. Shmoys DB (1997) Cut problems and their application to divide-and-conquer. In: Hochbaum DS (ed) Approximation algorithms for NP-hard problems. PWS Publishing, Boston, pp 192–235

Speed Scaling

Kirk Pruhs
Department of Computer Science, University of Pittsburgh, Pittsburgh, PA, USA

Keywords

Frequency scaling; Speed scaling; Voltage scaling

Years and Authors of Summarized Original Work

1995; Yao, Demers, Shenker

Problem Definition

Speed scaling is a power management technique in modern processor that allows the processor to run at different speeds. There is a power function $P(s)$ that specifies the power, which is energy used per unit of time, as a function of the speed. In CMOS-based processors, the cube-root rule states that $P(s) \approx s^3$. This is usually generalized to assume that $P(s) = s^\alpha$ form some constant α. The goals of power management are to reduce temperature and/or to save energy. Energy is power integrated over time. Theoretical investigations to date have assumed that there is a fixed ambient temperature and that the processor cools according to Newton's law, that is, the rate of cooling is proportional to the temperature difference between the processor and the environment.

In the resulting scheduling problems, the scheduler must not only have a job-selection policy to determine the job to run at each time, but also a speed scaling policy to determine the speed at which to run that job. The resulting problems are generally dual objective optimization problems. One objective is some quality of service measure for the schedule, and the other objective is temperature or energy.

We will consider problems where jobs arrive at the processor over time. Each job i has a release time r_i when it arrives at the processor, and a work requirement w_i. A job i run at speed s takes w_i/s units of time to complete.

Key Results

Yao et al. [5] initiated the theoretical algorithmic investigation of speed scaling problems. Yao et al. [5] assumed that each job i had a deadline d_i, and that the quality of service measure was deadline feasibility (each job completes by its deadline). Yao et al. [5] gives a greedy algorithm

YDS to find the minimum energy feasible schedule. The job selection policy for YDS is to run the job with the earliest deadline. To understand the speed scaling policy for YDS, define the intensity of a time interval to be the work that must be completed in this time interval divided by the length of the time interval. YDS then finds the maximum intensity interval, runs the jobs that must be run in this interval at constant speed, eliminates these jobs and this time interval from the instance, and proceeds recursively. Yao et al. [5] gives two online algorithms: OA and AVR. In OA the speed scaling policy is the speed that YDS would run at, given the current state and given that no more jobs will be released in the future. In AVR, the rate at which each job is completed is constant between the time that a job is released and the deadline for that job. Yao et al. [5] showed that AVR is $2^{\alpha-1}\alpha^\alpha$-competitive with respect to energy.

The results in [5] were extended in [2]. Bansal et al. [2] showed that OA is α^α-competitive with respect to energy. Bansal et al. [2] proposed another online algorithm, BKP. BKP runs at the speed of the maximum intensity interval containing the current time, taking into account only the work that has been released by the current time. They show that the competitiveness of BKP with respect to energy is at most $2(\alpha/(\alpha-1))^\alpha e^\alpha$. They also show that BKP is e-competitive with respect to the maximum speed.

Bansal et al. [2] initiated the theoretical algorithmic investigation of speed scaling to manage temperature. Bansal et al. [2] showed that the deadline feasible schedule that minimizes maximum temperature can in principle be computed in polynomial time. Bansal et al. [2] showed that the competitiveness of BKP with respect to maximum temperature is at most $2^{\alpha+1} e^\alpha (6(\alpha/(\alpha-1))^\alpha + 1)$.

Pruhs et al. [4] initiated the theoretical algorithmic investigation into speed scaling when the quality-of-service objective is average/total flow time. The flow time of a job is the delay from when a job is released until it is completed. Pruhs et al. [4] give a rather complicated polynomial-time algorithm to find the optimal flow time schedule for unit work jobs, given a bound on the energy available. It is easy to see that no $O(1)$-competitive algorithm exists for this problem.

Albers and Fujiwara [1] introduce the objective of minimizing a linear combination of energy used and total flow time. This has a natural interpretation if one imagines the user specifying how much energy he is willing to use to increase the flow time of a job by a unit amount. Albers and Fujiwara [1] give an $O(1)$-competitive online algorithm for the case of unit work jobs. Bansal et al. [3] improves upon this result and gives a 4-competitive online algorithm. The speed scaling policies of the online algorithms in [1] and [3] essentially run as power equal to the number of unfinished jobs (in each case modified in a particular way to facilitate analysis of the algorithm). Bansal et al. [3] extend these results to apply to jobs with arbitrary work, and even arbitrary weight. The speed scaling policy is essentially to run at power equal to the weight of the unfinished work. The expression for the resulting competitive ratio is a bit complicated but is approximately 8 when the cube-root rule holds.

The analysis of the online algorithms in [2] and [3] heavily relied on amortized local competitiveness. An online algorithm is locally competitive for a particular objective if for all times the rate of increase of that objective for the online algorithm, plus the rate of change of some potential function, is at most the competitive ratio times the rate of increase of the objective in any other schedule.

Applications

None

Open Problems

The outstanding open problem is probably to determine if there is an efficient algorithm to compute the optimal flow time schedule given a fixed energy bound.

Recommended Reading

1. Albers S, Fujiwara H (2006) Energy-efficient algorithms for flow time minimization. In: STACS. Lecture notes in computer science, vol 3884. Springer, Berlin, pp 621–633
2. Bansal N, Kimbrel T, Pruhs K (2007) Speed scaling to manage energy and temperature. J ACM 54(1)
3. Bansal N, Pruhs K, Stein C (2007) Speed scaling for weighted flow. In: ACM/SIAM symposium on discrete algorithms
4. Pruhs K, Uthaisombut P, Woeginger G (2004) Getting the best response for your ERG. In: Scandanavian workshop on algorithms and theory
5. Yao F, Demers A, Shenker S (1995) A scheduling model for reduced CPU energy. In: IEEE symposium on foundations of computer science, p 374

Sphere Packing Problem

Danny Z. Chen
Department of Computer Science and Engineering, University of Notre Dame, Notre Dame, IN, USA

Keywords

Ball packing; Disk packing

Years and Authors of Summarized Original Work

2001; Chen, Hu, Huang, Li, Xu

Problem Definition

The sphere packing problem seeks to pack spheres into a given geometric domain. The problem is an instance of geometric packing. Geometric packing is a venerable topic in mathematics. Various versions of geometric packing problems have been studied, depending on the shapes of packing domains, the types of packing objects, the position restrictions on the objects, the optimization criteria, the dimensions, etc. It also arises in numerous applied areas. The sphere packing problem under consideration here finds applications in radiation cancer treatment using Gamma Knife systems. Unfortunately, even very restricted versions of geometric packing problems (e.g., regular-shaped objects and domains in lower dimensional spaces) have been proved to be NP-hard. For example, for *congruent packing* (i.e., packing copies of the same object), it is known that the 2-D cases of packing fixed-sized congruent squares or disks in a simple polygon are NP-hard [7]. Baur and Fekete [2] considered a closely related *dispersion problem* of packing k congruent disks in a polygon of n vertices such that the radius of the disks is maximized; they proved that the dispersion problem cannot be approximated arbitrarily well in polynomial time unless P = NP, and gave a $\frac{2}{3}$-approximation algorithm for the L_∞ disk case with a time bound of $O(n^{38})$.

Chen et al. [4] proposed a practically efficient heuristic scheme, called *pack-and-shake*, for the **congruent sphere packing** problem, based on computational geometry techniques. The problem is defined as follows.

The Congruent Sphere Packing Problem

Given a d-D polyhedral region $R(d = 2, 3)$ of n vertices and a value $r > 0$, find a packing SP of R using spheres of radius r, such that (i) each sphere is contained in R, (ii) no two distinct spheres intersect each other in their interior, and (iii) the ratio (called the packing density) of the covered volume in R by SP over the total volume of R is maximized.

In the above problem, one can view the spheres as "solid" objects. The region R is also called the *domain* or *container*. Without loss of generality, let $r = 1$.

Much work on congruent sphere packing studied the case of packing spheres into an unbounded domain or even the whole space [5]. There are also results on packing congruent spheres into a bounded region. Hochbaum and Maass [8] presented a unified and powerful *shifting technique* for designing pseudo-polynomial time approximation schemes for packing congruent squares

into a rectilinear polygon. But, the high time complexities associated with the resulting algorithms restrict their applicability in practice. Another approach is to formulate a packing problem as a non-linear optimization problem, and resort to an available optimization software to generate packings; however, this approach works well only for small problem sizes and regular-shaped domains.

To reduce the running time yet achieve a dense packing, a common idea is to consider objects that form a certain lattice or double-lattice. A number of results were given on lattice packing of congruent objects in the whole (especially high dimensional) space [5]. For a bounded rectangular 2-D domain, Milenkovic [10] adopted a method that first finds the densest translational lattice packing for a set of polygonal objects in the whole plane, and then uses some heuristics to extract the actual bounded packing.

Key Results

The *pack-and-shake* scheme of Chen et al. [4] for packing congruent spheres in an irregular-shaped 2-D or 3-D bounded domain R consists of three phases. In the first phase, the d-D domain R is partitioned into a set of convex subregions (called *cells*). The resulting set of cells defines a dual graph G_D, such that each vertex v of G_D corresponds to a cell $C(v)$ and an edge connects two vertices if and only if their corresponding cells share a $(d-1)$-D face. In the second phase, the algorithm repeats the following *trimming and packing* process until $G_D = \emptyset$: Remove the lowest degree vertex v from G_D and pack the cell $C(v)$. In the third phase, a *shake* procedure is applied to globally adjust the packing to obtain a denser one.

The objective of the trimming and packing procedure is that after each cell is packed, the remaining "packable" subdomain R' of R is always kept as a connected region. The rationale for maintaining the connectivity of R' is as follows. To pack spheres in a bounded domain R, two typical approaches have been used: (a) packing spheres layer by layer going from the boundary of R towards its interior [9], and (b) packing spheres starting from the "center" of R, such as its medial axis, towards its boundary [3, 13, 14]. Due to the shape irregularity of R, both approaches may fragment the remaining "packable" subdomain R' into more and more disconnected regions; however, at the end of packing each such region, a small "unpackable" area may eventually remain that allows no further packing. It could fit more spheres if the "packable" subdomain R' is lumped together instead of being divided into fragments, which is what the trimming and packing procedure aims to achieve.

Due to the packing of its adjacent cells that have been done by the trimming and packing procedure, the boundary of a cell $C(v)$ that is to be packed may consist of both line segments and arcs (from packed spheres). Hence, a key problem is to pack spheres in a cell bounded by curves of low degrees. Chen et al.'s algorithms [4] for packing each cell are based on certain lattice structures and allow the cell to both translate and rotate. Their algorithms have fairly low time bounds. In certain cases, they even run in nearly linear time.

An interesting feature of the cell packings generated by the trimming and packing procedure is that the resulted spheres cluster together in the middle of the cells of the domain R, leaving some small unpackable areas scattered along the boundary of R. The "shake" procedure in [4] thus seeks to collect these small areas together by "pushing" the spheres towards the boundary of R, in the hope of obtaining some "packable" region in the middle of R.

The approach in [4] is to first obtain a densest lattice unit sphere packing $LSP(C)$ for each cell C of R, and then use a "shake" procedure to globally adjust the resulting packing of R to generate a denser packing SP in R. Suppose the plane P is already packed by infinitely many unit spheres whose center points form

a lattice (e.g., the hexagonal lattice). To obtain a densest packing $LSP(C)$ for a cell C from the lattice packing of the plane P, a position and orientation of C on P need to be computed such that C contains the maximum number of spheres from the lattice packing of P. There are two types of algorithms in [4] for computing an optimal placement of C on P: translational algorithms that allow C to be translated only, and translational/rotational algorithms that allow C to be both translated and rotated.

Let $n = |C|$, the number of bounding curves of C, and m be the number of spheres along the boundary of C in a sought optimal packing of C.

Theorem 1 *Given a polygonal region C bounded by n algebraic curves of constant degrees, a densest lattice unit sphere packing of C based only on translational motion can be computed in $O(N \log N + K)$ time, where $N = f(n, m)$ is a function of n and m, and K is the number of intersections between N planar algebraic curves of constant degrees that are derived from the packing instance.*

Note: In the worst case, $N = f(n, m) = n \times m$. But in practice, N may be much smaller. The N planar algebraic curves in Theorem 1 form a structure called *arrangement*. Since all these curves are of a constant degree, any two such curves can intersect each other at most a constant number of times. In the worst case, the number K of intersections between the N algebraic curves, which is also the *size* of the arrangement, is $O(N^2)$. The arrangement of these curves can be computed by the algorithms [1, 6] in $O(N \log N + K)$ time.

Theorem 2 *Given a polygonal region C bounded by n algebraic curves of constant degrees, a densest lattice unit sphere packing of C based on both translational and rotational motions can be computed in $O(T(n) + (N + K') \log N)$ time, where $N = f(n, m)$ is a function of n and m, K' is the size of the arrangement of N*

pseudo-plane surfaces in 3-D that are derived from the packing instance, and T(n) is the time for solving $O(n^2)$ quadratic optimization problem instances associated with the packing instance.

In Theorem 2, $K' = O(N^3)$ in the worst case. In practice, K' can be much smaller.

The results on 2-D sphere packing in [4] can be extended to d-D for any constant integer $d \geq 3$, so long as a good d-D lattice packing of the d-D space is available.

Applications

Recent interest in the considered congruent sphere packing problem was motivated by medical applications in Gamma Knife radiosurgery [4, 11, 12]. Radiosurgery is a minimally invasive surgical procedure that uses radiation to destroy tumors inside human body while sparing the normal tissues. The Gamma Knife is a radiosurgical system that consists of 201 Cobalt-60 sources [3, 14]; the gamma-rays from these sources are all focused on a common center point, thus creating a spherical volume of radiation field. The Gamma Knife treatment normally applies high radiation dose. In this setting, overlapping spheres may result in overdose regions (called *hot spots*) in the target treatment domain, while a low packing density may cause underdose regions (called *cold spots*) and a non-uniform dose distribution. Hence, one may view the spheres used in Gamma Knife packing as "solid" spheres. Therefore, a key geometric problem in Gamma Knife treatment planning is to fit multiple spheres into a 3-D irregular-shaped tumor [3, 13, 14]. The total treatment time crucially depends on the number of spheres used. Subject to a given packing density, the minimum number of spheres used in the packing (i.e., treatment) is desired. The Gamma Knife currently produces spheres of four different radii (4, 8, 14, and 18 mm), and hence the Gamma Knife sphere packing is in general not congruent. In practice, a commonly used approach is to pack larger spheres first,

and then fit smaller spheres into the remaining subdomains, in the hope of reducing the total number of spheres involved and thus shortening the treatment time. Therefore, congruent sphere packing can be used as a key subroutine for such a common approach.

Open Problems

An open problem is to analyze the quality bounds of the resulting packing for the algorithms in [4]; such packing quality bounds are currently not yet known. Another open problem is to reduce the running time of the packing algorithms in [4], since these algorithms, especially for sphere packing problems in higher dimensions, are still very time-consuming. In general, it is highly desirable to develop efficient sphere packing algorithms in d-D ($d \geq 2$) with guaranteed good packing quality.

Experimental Results

Some experimental results of the 2-D pack-and-shake sphere packing algorithms were given in [4]. The planar hexagonal lattice was used for the lattice packing. On packings whose sizes are in the hundreds, the C++ programs of the algorithms in [4] based only on translational motion run very fast (a few minutes), while those of the algorithms based on both translation and rotation take much longer time (hours), reflecting their respective theoretical time bounds, as expected. On the other hand, the packing quality of the translation-and-rotation based algorithms is a little better than the translation based algorithms. The packing densities of all the algorithms in the experiments are well above 70 % and some are even close to or above 80 %. Comparing with the nonconvex programming methods, the packing algorithms in [4] seemed to run faster based on the experiments.

Cross-References

▶ Local Approximation of Covering and Packing Problems

Recommended Reading

1. Amato NM, Goodrich MT, Ramos EA (2000) Computing the arrangement of curve segments: divide-and-conquer algorithms via sampling. In: Proceedings of the 11th annual ACM-SIAM symposium on discrete algorithms, pp 705–706
2. Baur C, Fekete SP (2001) Approximation of geometric dispersion problems. Algorithmica 30(3):451–470
3. Bourland JD, Wu QR (1996) Use of shape for automated, optimized 3D radiosurgical treatment planning. In: Proceedings of the SPIE international symposium on medical imaging, pp 553–558
4. Chen DZ, Hu X, Huang Y, Li Y, Xu J (2001) Algorithms for congruent sphere packing and applications. In: Proceedings of the 17th annual ACM symposium on computational geometry, pp 212–221
5. Conway JH, Sloane NJA (1988) Sphere packings, lattices and groups. Springer, New York
6. Edelsbrunner H, Guibas LJ, Pach J, Pollack R, Seidel R, Sharir M (1992) Arrangements of curves in the plane: topology, combinatorics, and algorithms. Theor Comput Sci 92:319–336
7. Fowler RJ, Paterson MS, Tanimoto SL (1981) Optimal packing and covering in the plane are NP-complete. Inf Process Lett 12(3):133–137
8. Hochbaum DS, Maass W (1985) Approximation schemes for covering and packing problems in image processing and VLSI. J ACM 32(1):130–136
9. Li XY, Teng SH, Üngör A (2000) Biting: advancing front meets sphere packing. Int J Numer Methods Eng 49(1–2):61–81
10. Milenkovic VJ (2000) Densest translational lattice packing of nonconvex polygons. In: Proceedings of the 16th ACM annual symposium on computational geometry, pp 280–289
11. Shepard DM, Ferris MC, Ove R, Ma L (2000) Inverse treatment planning for Gamma Knife radiosurgery. Med Phys 27(12):2748–2756
12. Sutou A, Dai Y (2002) Global optimization approach to unequal sphere packing problems in 3D. J Optim Theory Appl 114(3):671–694
13. Wang J (1999) Medial axis and optimal locations for min-max sphere packing. J Comb Optim 3:453–463
14. Wu QR (1996) Treatment planning optimization for Gamma unit radiosurgery. Ph.D. thesis, The Mayo Graduate School

Split Decomposition via Graph-Labelled Trees

Christophe Paul
CNRS, Laboratoire d'Informatique Robotique et Microélectronique de Montpellier, Université Montpellier 2, Montpellier, France

Keywords

Circle graphs; Distance hereditary graphs; LexBFS; Parity graphs; Permutation graphs; Split decomposition

Years and Authors of Summarized Original Work

2012; Gioan, Paul
2014; Gioan, Paul, Tedder, Corneil

Problem Definition

Introduced by Cunningham and Edmonds [11], the *split decomposition*, also known as the *join (or 1-join) decomposition*, ranges among the classical graph decomposition schemes. Given a graph $G = (V, E)$, a bipartition (A, B) of the vertex set V (with $|A| \geqslant 2$ and $|B| \geqslant 2$) is a *split* if there are subsets $A' \subseteq A$ and $B' \subseteq B$, called *frontiers*, such that there is an edge between a vertex $u \in A$ and $v \in B$ if and only if $u \in A'$ and $v \in B'$ (see Fig. 1). A graph is *prime* if it does not contain any split. Observe that an induced cycle of length at least 5 is a prime graph. A graph is *degenerate* if every bipartition (A, B) with $|A| \geqslant 2$ and $|B| \geqslant 2$ is a split. It can be shown that a degenerate graphs are either *cliques* or *stars*. The split decomposition consists in recursively decompose a graph into a set of disjoint graphs $\{G_1, \ldots G_k\}$, called *split components*, each of which is either prime or degenerate. There are two cases:

1. If G is prime or degenerate, then return the set $\{G\}$;

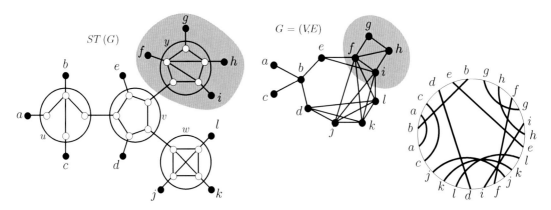

Split Decomposition via Graph-Labelled Trees, Fig. 1 A circle graph G with a chord diagram on the *right* and its split decomposition tree $ST(G)$ on the *left*. The nodes v and y are prime nodes, whereas u is a star node and w a clique node. The bipartition $(\{f, g, h, i\}, V \setminus \{f, g, h, i\})$ forms a split of G and corresponds to a tree edge of $ST(G)$. The frontiers are $\{f, i\}$ on one side and $\{e, j, k, l\}$ on the other. Observe that $(\{k, l\}, V \setminus \{k, l\})$ is also a split but which is not represented by the tree edge between nodes Y and Z in $ST(G)$. Because G is not a prime graph, it can be represented with several chord diagram. For example, exchanging the chord of y with the chord of z yields an alternative chord diagram

2. If G is neither prime nor degenerate, it contains a split (A, B), with frontiers A' and B'. The split components of G is then the union of the split components of the graphs $G[A] + (a, A')$ and $G[B] + (b, B')$, where a and b are new vertices, called *markers*.

Observe that the split decomposition process naturally defines a decomposition tree whose nodes represent the split components. This decomposition tree can be represented by a *graph-labeled tree* (GLT) (see [16, 18]) defined as a pair $(T, \mathcal{F})$, where T is a tree and $\mathcal{F}$ a set of graphs, such that each node u of T is *labeled* by the graph $G(u) \in \mathcal{F}$, and there exists a bijection ρ_u between the edges of T incident to u and the vertices of $G(u)$, called *marker vertices*. We say that two leaves ℓ_a and ℓ_b of T are *accessible* if for every pair of consecutive tree edges uv and vw on the path from ℓ_a and ℓ_b in T, $\rho_v(uv)$ and $\rho_v(vw)$ are adjacent in $G(v)$. From a GLT $(T, \mathcal{F})$, we define an *accessibility graph* $\mathcal{G}(T, \mathcal{F})$ whose vertex set is the leaf set of T and two vertices a and b are adjacent if the corresponding leaves ℓ_a and ℓ_b are accessible. It is easy to observe that every tree edge e of a GLT $(T, \mathcal{F})$ defines a split (A, B) of $\mathcal{G}(T, \mathcal{F})$ where A and B respectively contain the vertices corresponding

to the leaves of the two connected components of $T - e$. Cunningham and Edmonds [11] formalized the family of splits as an example of *partite family of bipartitions* thereby implying that every graph admits a canonical split decomposition tree (see Fig. 1). In terms of GLTs, this translates as follows:

Theorem 1 ([11, 16, 18]) *Let G be a connected graph. There exists a unique GLT $(T, \mathcal{F})$ whose labels are either prime or degenerate, having a minimal number of nodes and such that $G = \mathcal{G}(T, \mathcal{F})$. This GLT is called the* split tree of G *and denoted $ST(G)$.*

The problem we are interested in is to efficiently compute the split tree $ST(G)$ of a graph $G = (V, E)$. The first polynomial-time algorithm was and runs in time $O(nm)$, where $n = |V|$ and $m = |E|$. Ma and Spinrad [23] later developed an $O(n^2)$ algorithm. Finally Dahlhaus [12] designed the first linear-time algorithm which was recently revisited by Charbit et al. [5].

Key Results

As mentioned above, the split tree of a graph can be computed in linear time. The algorithm we describe here is nearly optimal, that is, runs in time $O(n + m) \cdot \alpha(n + m)$, where α is the inverse Ackermann function. The fact that this algorithm incrementally builds the split tree is responsible of the small additional complexity cost. More precisely, updating the tree structure of the GLT representing the split tree relies on the union-find data structure [15]. But having an incremental split decomposition algorithm allows an extension of the algorithm, within the same time complexity, to the circle graph recognition [17], a problem for which computing the split decomposition is a corner step. But so far, a subquadratic time complexity cannot be reached using the previous linear (or quadratic) split decomposition algorithms.

Theorem 2 ([18]) *The split tree $ST(G)$ of a graph $G = (V, E)$, with $|V| = n$ and $|E| = m$, can be built incrementally according to an LBFS ordering in time $O(n + m) \cdot \alpha(n + m)$, where α is the inverse Ackermann function.*

It is important to observe that to reach the expected complexity, the algorithm inserts the vertices according to a *LexBFS ordering* [25]. These orderings, resulting from a *lexicographic breadth-first search*, appear in a number of recognition algorithms, such as *chordal graphs* [25], *comparability graphs* [20], *interval graphs* [22], and *cographs* [3]. The idea is that structural properties can be shown on the last vertex visited by a LexBFS. For example, in chordal graphs the last vertex is simplicial; in comparability graphs it is a source of some transitive orientation. LexBFS, introduced in [25], works as follows: it numbers the vertices decreasingly from $n = |V|$ down to 1; initially every vertex receives an empty label; then iteratively, an arbitrary unnumbered vertex x with lexicographically largest label is selected and numbered i, and i is appended to the label of every unnumbered neighbor of x. On the graph of Fig. 1, $\sigma = b, a, e, d, c, f, i, j, k, l, h, g$ is a LexBFS ordering.

Applications

Many graph classes can be characterized by means of the split decomposition. Below, we review the most important of these classes. Finally, we discuss the links between split decomposition and other decomposition approaches.

Graph Classes

Distance Hereditary Graphs

The family of graphs for which the split tree does not contain any prime node is called *totally decomposable* (or *totally separable*). This terminology follows from the observation that for every subgraph of size at least 4, every nontrivial bipartition of the vertex set forms a split. A graph G is *distance hereditary* [1] if for every induced connected subgraph H of G and every pair of vertices x and y of H, the distance between x and y is the same in H and G. It turns out that a graph G is totally decomposable if and only if it is *distance hereditary* [1]. In other words, a graph G is distance hereditary if and only if every node of $ST(G)$ is either a star or a clique node. The first linear-time recognition algorithm of distance hereditary graphs, due to [21], relies on a breadth-first search characterization (see also [13]). More recently, a linear-time algorithm has been designed to update the split tree of a distance hereditary graph under vertex and edge insertion, leading to an alternative (vertex incremental) linear-time recognition algorithm for distance hereditary graphs.

Theorem 3 ([16]) *Let $ST(G)$ be the split tree of a distance hereditary graph $G = (V, E)$, $S \subseteq V$ be a subset of vertices of G and $e = (x, y) \notin E$ be a non-edge of G. Then:*

- *In $O(1)$-time, we can compute $ST(G + e)$ where $G + e = (V, E \cup \{e\})$ if $G + e$ is distance hereditary;*
- *In $O(|S|)$-time, we can compute $ST(G + (x, S))$ where $G + (x, S) = (V \cup \{x\}, E \cup \{(x, y) \mid y \in S\})$ if $G + (x, S)$ is distance hereditary.*

Subclasses of Totally Decomposable Graphs

A GLT is called *clique-star tree* if its nodes are labeled either with cliques or stars. As a consequence of the discussion of the previous paragraph, distance hereditary graphs are the graphs corresponding the clique-star trees. Imposing any constraint on a clique-star tree thereby immediately defines a subclass of distance hereditary graphs. It turns out that many important graph subclasses of distance hereditary graphs can be characterized with the split decomposition.

The *cographs*, also known as *complement-reducible graphs* [8] or P_4-*free graphs*, are probably the most studied subclass of distance hereditary graphs. Cographs are also known as the class of graphs totally decomposable with respect to the modular decomposition [19], and their combinatorial structure is captured by the so-called cotree. As noticed in [16], it is easy to observe that a graph G is a cograph if and only if its split tree $ST(G)$ is a clique-star tree that can be rooted either at a node or at a tree edge such that every star node is "oriented" toward that root (that is the marker vertex corresponding the center of every star node is oriented toward the root).

The class of *ptolemaic graphs* or *3-leaf power* are also interesting. The class of ptolemaic graphs is defined as the intersection of distance hereditary graphs and chordal graphs. Chordal graphs are the graphs without induced chordless cycles of length four or more. It follows that a graph

G is ptolemaic if and only if $ST(G)$ is a clique-star tree such that for every pair of star nodes u and v, not both extremities of the path from u to v in $ST(G)$ are attached to the center marker vertex of u and v (otherwise this would generate a chordless 4-cycle). As a subclass of chordal graph, 3-leaf powers inherit the restrictions of ptolemaic graphs on the split tree with the additive constraint that no clique node lies on the path between two star nodes (see [16] for details).

Circle Graphs

The split decomposition plays an important role in the context of *circle graphs* defined as intersection graphs of a set of chords in a circle. The main reason is that a graph G is a circle graph if and only if every split component of G is a circle graph. In other words, as clique and stars are circle graphs, G is a circle graph if and only if the prime nodes of $ST(G)$ are labeled with circle graphs. Observe that this characterization shows that distance hereditary graphs form a subclass of circle graphs. By the way the first quadratic time circle graph recognition algorithm was obtained by computing the split decomposition of the input graph and reducing the problem to the recognition of prime circle graphs [23, 26]. The key property is that a prime circle graph has a unique (up to mirror) *chord diagram* [2, 14] (see Fig. 2).

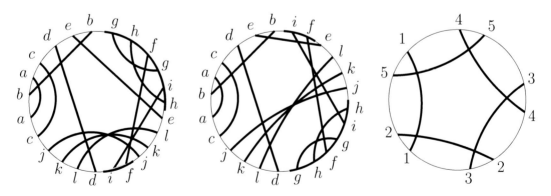

Split Decomposition via Graph-Labelled Trees, Fig. 2
On the *left*, two distinct chord diagrams of the graph G depicted in Fig. 1 results from symmetric insertion of the chords representing the vertices $\{f, g, h, i\}$ (remind that $(\{f, g, h, i\}, V \setminus \{f, g, h, i\})$ form a split). On the *right*, the chord diagram on $\{1, 2, 3, 4, 5\}$ is the unique (up to rotation and mirror) chord diagram of the 5-cycle, which is a prime graph

The linear-time split decomposition algorithm [12], proposed in the mid-1990s, did not lead to a linear-time circle graph recognition algorithm. For almost two decades, the quadratic time complexity [27] remained the best known complexity. The quadratic barrier has been broken using the almost linear-time split decomposition algorithm of [17]. The key ingredient was to insert the vertices according to a LexBFS ordering. Indeed, in the unique chord diagram of a prime circle graph G, the neighborhood of the last vertex x of a LexBFS ordering satisfies a sort of consecutiveness property. More precisely, the chord diagram of G contains a set of consecutive chord extremities starting and ending with the extremities of x's chord and containing one and only one chord extremity per neighbor of x and no chord extremity of non-neighbors of x. This property is used to incrementally build the split tree of a circle graph using chord diagrams to represent prime nodes. It is worth to observe that the split tree of a circle graph G together with the chord diagrams of each of its prime nodes provides a canonical (linear space) representation of the set of (exponentially many) chord diagrams of G.

Theorem 4 ([17]) *Let $G = (V, E)$ be a graph such that $|V| = n$ and $|E| = m$. There exists a $O(n + m) \cdot \alpha(n + m)$-time algorithm, where α is the inverse Ackermann function, deciding whether G is a circle graph. Moreover, if G is a circle graph, the algorithm outputs a split-tree representation G from which any chord diagram of G can be extracted in linear time.*

Perfect Graphs

The recent proof [6] of the famous conjecture of Berge on perfect graphs states that a graph is *perfect* if and only if it does not contain an odd cycle of length at least 5 nor its complement as induced subgraph. It is easy to observe that a graph is perfect if and only if its prime components are perfect graphs. The split decomposition does not formally appear in the structural decomposition theorem of perfect graphs [6, 28] as it is subsumed by the so-called balanced skew partition. In the context of perfect graphs, *parity graphs* [4]

form a nice example of class of graphs simply characterized through their split decomposition. A graph is a parity graph if for every pair x, y of vertices, the length of every chordless path between x and y is of the same parity. This constraint can be translated into a condition on odd cycles or into a condition on their split tree. Indeed it can be proved that a graph is a parity graph if and only if its prime nodes are labeled with bipartite graphs [7].

Related Graph Decompositions

Modular Decomposition

The split decomposition is often introduced as a generalization of the *modular decomposition* (also known as homogeneous decomposition) [19]. A *module* in a graph $G = (V, E)$ is a subset M of vertices such that every vertex not in M is either fully adjacent or fully nonadjacent to the vertices of M. Clearly, if M is a module of size at least 2, then $(M, V \setminus M)$ defines a split. Indeed the split decomposition is sometimes used to further decompose graphs that are primes with respect to the modular decomposition.

Width Parameters

Rank-width [24] and *clique-width* [10] are two important width parameters both sharing some connections with the split decomposition. As the rank-width of a graph is small if its clique-width is small and vice versa, we only briefly describe the former parameter. A rank-decomposition of a graph G is defined as a ternary tree whose leaves are in one-to-one correspondence with the vertices of G. It follows that every internal tree edge defines a bipartition, say (A, B) of the vertices of G. The rank-width of a bipartition (A, B) is defined as the rank of the incidence matrix between A and B, and the width of a rank-decomposition is the maximum width over its bipartitions. The rank-width of a graph G is then the minimum width over its rank-decompositions. Observe that the every split is a rank-width 1 bipartition. It follows that the rank-width of a graph is the maximum rank-width of its prime components. As a consequence rank-width one graphs are exactly distance hereditary graphs. To conclude, let us mention that computing the split decomposition of a graph is a key step in the polynomial-time recognition algorithm of clique-width three graphs [9].

Recommended Reading

1. Bandelt H-J, Mulder HM (1986) Distance hereditary graphs. J Comb Theory Ser B 41:182–208
2. Bouchet A (1987) Reducing prime graphs and recognizing circle graphs. Combinatorica 7:243–254
3. Bretscher A, Corneil D, Habib M, Paul C (2008) A simple linear time lexbfs cograph recognition algorithm. SIAM J Discret Math 22(4):1277–1296
4. Burlet M, Uhry JP (1984) Parity graphs. Ann Discret Math 21:253–277
5. Charbit P, de Montgolfier F, Raffinot M (2012) Linear time split decomposition revisited. SIAM J Discret Math 26(2):499–514
6. Chudnovsky M, Robertson N, Seymour P, Thomas R (2006) The strong perfect graph theorem. Ann Math 161:51–229
7. Cicerone S, Di Stefano G (1999) On the extension of bipartite to parity graphs. Discret Appl Math 95:181–195
8. Corneil D, Lerchs H, Stewart-Burlingham LK (1981) Complement reducible graphs. Discret Appl Math 3(1):163–174
9. Corneil D, Habib M, Lanlignel JM, Reed B, Rotics U (2012) Polynomial-time recognition of clique-width ≤ 3 graphs. Discret Appl Math 160(6):834–865
10. Courcelle B, Engelfriet J, Rozenberg G (1993) Handle rewriting hypergraph grammars. J Comput Syst Sci 46:218–270
11. Cunningham WH, Edmonds J (1980) A combinatorial decomposition theory. Can J Math 32(3):734–765
12. Dahlhaus E (1994) Efficient parallel and linear time sequential split decomposition (extended abstract). In: Foundations of software technology and theoretical computer science – FSTTCS, Madras. Volume 880 of lecture notes in computer science, pp 171–180
13. Damiand G, Habib M, Paul C (2001) A simple paradigm for graph recognition: application to cographs and distance hereditary graphs. Theor Comput Sci 263:99–111
14. Gabor CP, Hsu WL, Suppovit KJ (1989) Recognizing circle graphs in polynomial time. J ACM 36:435–473
15. Gabow H, Tarjan R (1983) A linear-time algorithm for a special case of disjoint set union. In: Annual ACM symposium on theory of computing (STOC), Boston, pp 246–251
16. Gioan E, Paul C (2012) Split decomposition and graph-labelled trees: characterizations and fully dynamic algorithms for totally decomposable graphs. Discret Appl Math 160(6):708–733
17. Gioan E, Paul C, Tedder M, Corneil D (2013) Circle graph recognition in time $O(n + m)\alpha(n + m)$. Algorithmica 69(4): 759–788 (2014)
18. Gioan E, Paul C, Tedder M, Corneil D (2013) Practical split-decomposition via graph-labelled trees. Algorithmica 69(4): 789–843 (2014)
19. Habib M, Paul C (2010) A survey on algorithmic aspects of modular decomposition. Comput Sci Rev 4:41–59
20. Habib M, McConnell RM, Paul C, Viennot L (2000) Lex-BFS and partition refinement, with applications to transitive orientation, interval graph recognition and consecutive ones testing. Theor Comput Sci 234:59–84
21. Hammer P, Maffray F (1990) Completely separable graphs. Discret Appl Math 27:85–99
22. Korte N, Möhring R (1989) An incremental linear-time algorithm for reconizing interval graphs. SIAM J Comput 18(1):68–81
23. Ma T-H, Spinrad J (1994) An $O(n^2)$ algorithm for undirected split decomposition. J Algorithms 16:145–160
24. Oum SI (2005) Graphs of bounded rank-width. PhD thesis, Princeton University
25. Rose DJ, Tarjan RE, Lueker GS (1976) Algorithmic aspects of vertex elimination on graphs. SIAM J Comput 5(2):266–283
26. Spinrad J (1989) Prime testing for the split decomposition of a graph. SIAM J Discret Math 2(4):590–599
27. Spinrad J (1994) Recognition of circle graphs. J Algorithms 16:264–282
28. Trotignon N (2013) Perfect graphs: a survey. Technical report 1301.5149, arxiv

Squares and Repetitions

Maxime Crochemore[1,2,4] and Wojciech Rytter[3]
[1]Department of Computer Science, King's College London, London, UK
[2]Laboratory of Computer Science, University of Paris-East, Paris, France
[3]Institute of Informatics, Warsaw University, Warsaw, Poland
[4]Université de Marne-la-Vallée, Champs-sur-Marne, France

Keywords

Powers; Runs; Tandem repeats

Years and Authors of Summarized Original Work

1999; Kolpakov, Kucherov

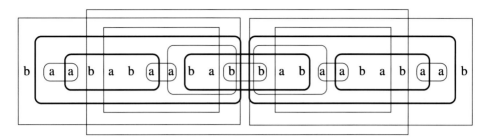

Squares and Repetitions, Fig. 1 The structure of $RUNS(x)$ where $x = \texttt{baababaabbabaababaab} = \texttt{b}z^2(z^R)^2\texttt{b}$, for $z = \texttt{aabab}$. The operation $\cdot^R$ is reversing the string

Problem Definition

Periodicities and repetitions in strings have been extensively studied and are important both in theory and practice (combinatorics of words, pattern-matching, computational biology). The words of the type ww and www, where w is a nonempty primitive (not of the form u^k for an integer $k > 1$) word, are called squares and cubes, respectively. They are well-investigated objects in combinatorics on words [16] and in string-matching with small memory [5].

A string w is said to be periodic iff $period(w) \leq |w|/2$, where $period(w)$ is the smallest positive integer p for which $w[i] = w[i + p]$ whenever both sides of the equality are defined. In particular each square and cube is periodic.

A repetition in a string $x = x_1x_2\ldots x_n$ is an interval $[i \ldots j] \subseteq [1 \ldots n]$ for which the associated factor $x[i \ldots j]$ is periodic. It is an occurrence of a periodic word $x[i \ldots j]$, also called a positioned repetition. A word can be associated with several repetitions, see Fig. 1.

Initially people investigated mostly positioned squares, but their number is $\Omega(n \log n)$ [2], hence algorithms computing all of them cannot run in linear time, due to the potential size of the output. The optimal algorithms reporting all positioned squares or just a single square were designed in [1, 2, 3, 19]. Unlike this, it is known that only $O(n)$ (un-positioned) squares can appear in a string of length n [8].

The concept of maximal repetitions, called runs (equivalent terminology) in [14], has been introduced to represent all repetitions in a succinct manner. The crucial property of runs is that there are only $O(n)$ runs in a word of length n [15, 21].

A *run* in a string x is an interval $[i \ldots j]$ such that both the associated string $x[i \ldots j]$ has period $p \leq (j - i + 1)/2$, and the periodicity cannot be extended to the right nor to the left: $x[i - 1] \neq x[x + p - 1]$ and $x[j - p + 1] \neq x[j + 1]$ when the elements are defined. The set of runs of x is denoted by $RUNS(x)$. An example is displayed in Fig. 1.

Key Results

The main results concern fast algorithms for computing positioned squares and runs, as well as combinatorial estimation on the number of corresponding objects.

Theorem 1 (Crochemore [1], Apostolico-Preparata [2], Main-Lorentz [19]) *There exists an $O(n \log n)$ worst-case time algorithm for computing all the occurrences of squares in a string of length n.*

Techniques used to design the algorithms are based on partitioning, suffix trees, and naming segments. A similar result has been obtained by Franek, Smyth, and Tang using suffix arrays [11]. The key component in the next algorithm is the function described in the following lemma.

Lemma 2 (Main-Lorentz [19]) *Given two square-free strings u and v, reporting if uv contains a square centered in u can be done in worst-case time $O(|u|)$.*

S

b | a | a | b | a | b | a a | b | a | b | b | a b | a a | b | a | b | a a | b

Squares and Repetitions, Fig. 2 The f-factorization of the example string x = baababaababbabaababaab and the set of its internal runs; all other runs overlap factorization points

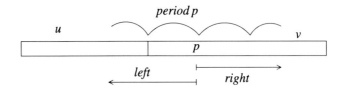

Squares and Repetitions, Fig. 3 If an overlapping run with period p starts in u, ends in v, and its part in v is of size at least p then it is easily detectable by computing continuations of the periodicity p in two directions: left and right

Using suffix trees or suffix automata together with the function derived from the lemma, the following fact has been shown.

Theorem 3 (Crochemore [3], Main-Lorentz [19]) *Testing the square-freeness of a string of length n can be done in worst-case time $O(n \log a)$, where a is the size of the alphabet of the string.*

As a consequence of the algorithms and of the estimation on the number of squares, the most important result related to repetitions can be formulated as follows.

Theorem 4 (Kolpakov-Kucherov [15], Rytter [21], Crochemore-Ilie [4])

(1) *All runs in a string can be computed in linear time (on a fixed-size alphabet).*
(2) *The number of all runs is linear in the length of the string.*

The point (2) is very intricate, it is of purely combinatorial nature and has nothing to do with the algorithm. We sketch shortly the basic components in the constructive proof of the point (1). The main idea is to use, as for the previous theorem, the f-factorization (see [3]): a string x is decomposed into factors $u_1, u_2, \ldots, u_k$, where u_i is the longest segment which appears before (possibly with overlap) or is a single letter if the segment is empty.

The runs which fit in a single factor are called internal runs, other runs are called here overlapping runs. There are three crucial facts:

- all overlapping runs can be computed in linear time,
- each internal run is a copy of an earlier overlapping run,
- the f-factorization can be computed in linear time (on a fixed-size alphabet) if we have the suffix tree or suffix automaton of the string. Figure 2 shows f-factorization and internal runs of an example string.

It follows easily from the definition of the f-factorization that if a run overlaps two (consecutive) factors u_{k-1} and u_k then its size is at most twice the total size of these two s factors.

Figure 3 shows the basic idea for computing runs that overlap u v in time $O(|u| + |v|)$. Using similar tables as in the Morris–Pratt algorithm (border and prefix tables), see [6], we can test the continuation of a period p from position p in v to the left and to the right. The corresponding tables can be constructed in linear time in a preprocessing phase. After computing all overlapping runs the internal runs can be copied from their earlier occurrences by processing the string from left to right.

Another interesting result concerning periodicities is the following lemma and its fairly immediate corollary.

Lemma 5 (Three Prefix Squares, Crochemore-Rytter [5]) *If u, v, and w are three primitive words satisfying:* $|u| < |v| < |w|$, *uu is a prefix of vv, and vv is a prefix of ww, then* $|u| + |v| \leq |w|$.

Corollary 1 *Any nonempty string x possesses less than* $\log_\Phi |y|$ *prefixes that are squares.*

In the configuration of the lemma, a second consequence is that *uu* is a prefix of *w*. Therefore, a position in a string *x* cannot be the largest position of more than two squares, which yields the next corollary. A simple direct proof of it is by Ilie [13], see also [17].

Corollary 2 (Fraenkel and Simpson [8]) *Any string x contains at most* $2|x|$ *(different) squares, that is: card*$\{u \mid u$ *primitive and* u^2 *factor of* $y\} \leq 2|x|$.

The structure of all squares and of un-positioned runs has been also computed within the same time complexities as above in [18] and [12].

Applications

Detecting repetitions in strings is an important element of several questions: pattern matching, text compression, and computational biology to quote a few. Pattern-matching algorithms have to cope with repetitions to be efficient as these are likely to slow down the process; the large family of dictionary-based text compression methods use a weaker notion of repeats (like the software gzip); repetitions in genomes, called satellites, are intensively studied because, for example, some over-repeated short segments are related to genetic diseases; some satellites are also used in forensic crime investigations.

Open Problems

The most intriguing question remains the asymptotically tight bound for the maximum number of $\rho(n)$ of runs in a string of size *n*. The first proof (by painful induction) was quite

difficult and has not produced any *concrete* constant coefficient in the $O(n)$ notation. This subject has been studied in [9, 10, 22, 23]. The best-known lower bound of approximately $0.927\,n$ is from [10]. The exact number of runs has been considered for special strings: *Fibonacci words* and (more generally) *Sturmian words* [7, 14, 20]. It is proved in a structural and intricate manner in the full version of [21] that $\rho(n) \leq 3.44\,n$, by introducing a *sparse-neighbors technique*. The neighbors are runs for which both the distance between their starting positions is small and the difference between their periods is also proportionally small (according to some fixed coefficient of proportionality). The occurrences of neighbors satisfy certain *sparsity* properties which imply the linear upper bound. Several variations for the definitions of neighbors and sparsity are possible. Considering runs having close centers the bound has been lowered to $1.6\,n$ in [4].

As a conclusion, we believe that the following fact is valid.

Conjecture: A string of length *n* contains less than *n* runs, i.e., $|\text{RUNS}|(n) < n$.

Cross-References

Elements of the present entry are of main importance for run-length compression as well as for ▸ Multidimensional Compressed Pattern Matching. They are also related to the ▸ Approximate Tandem Repeats entries because "tandem repeat" is a synonym of repetition and "power."

Recommended Reading

1. Apostolico A, Preparata FP (1983) Optimal off-line detection of repetitions in a string. Theor Comput Sci 22(3):297–315
2. Crochemore M (1981) An optimal algorithm for computing the repetitions in a word. Inf Process Lett 12(5):244–250

3. Crochemore M (1986) Transducers and repetitions. Theor Comput Sci 45(1):63–86
4. Crochemore M, Ilie L (2007) Analysis of maximal repetitions in strings. J Comput Sci
5. Crochemore M, Rytter W (1995) Squares, cubes, and time-space efficient string searching. Algorithmica 13(5):405–425
6. Crochemore M, Rytter W (2003) Jewels of stringology. World Scientific, Singapore
7. Franek F, Karaman A, Smyth WF (2000) Repetitions in Sturmian strings. Theor Comput Sci 249(2):289–303
8. Fraenkel AS, Simpson RJ (1998) How many squares can a string contain? J Comb Theory Ser A 82:112–120
9. Fraenkel AS, Simpson RJ (1999) The exact number of squares in fibonacci words. Theor Comput Sci 218(1):95–106
10. Franek F, Simpson RJ, Smyth WF (2003) The maximum number of runs in a string. In: Proceedings of the 14-th Australian workshop on combinatorial algorithms. Curtin University Press, Perth, pp 26–35
11. Franek F, Smyth WF, Tang Y (2003) Computing all repeats using suffix arrays. J Autom Lang Comb 8(4):579–591
12. Gusfield D, Stoye J (2004) Linear time algorithms for finding and representing all the tandem repeats in a string. J Comput Syst Sci 69(4):525–546
13. Ilie L (2005) A simple proof that a word of length n has at most 2n distinct squares. J Comb Theory Ser A 112(1):163–164
14. Iliopoulos C, Moore D, Smyth WF (1997) A characterization of the squares in a Fibonacci string. Theor Comput Sci 172:281–291
15. Kolpakov R, Kucherov G (1999) Finding maximal repetitions in a word in linear time. In: Proceedings of the 40th symposium on foundations of computer science. IEEE Computer Society, Los Alamitos, pp 596–604
16. Lothaire M (ed) (2002) Algebraic combinatorics on words. Cambridge University Press, Cambridge
17. Lothaire M (ed) (2005) Applied combinatorics on words. Cambridge University Press, Cambridge
18. Main MG (1989) Detecting leftmost maximal periodicities. Discret Appl Math 25:145–153
19. Main MG, Lorentz RJ (1984) An O(n log n) algorithm for finding all repetitions in a string. J Algorithms 5(3):422–432
20. Rytter W (2006) The structure of subword graphs and suffix trees of Fibonacci words. In: Implementation and application of automata, CIAA 2005. Lecture notes in computer science, vol 3845. Springer, Berlin, pp 250–261
21. Rytter W (2006) The number of runs in a string: improved analysis of the linear upper bound. In: Proceedings of the 23rd annual symposium on theoretical aspects of computer science. Lecture notes in computer science, vol 3884. Springer, Berlin, pp 184–195
22. Smyth WF (2000) Repetitive perhaps, but certainly not boring. Theor Comput Sci 249(2):343–355
23. Smyth WF (2003) Computing patterns in strings. Addison-Wesley, Boston

Stable Marriage

Robert W. Irving
School of Computing Science, University of Glasgow, Glasgow, UK

Keywords

Stable matching

Years and Authors of Summarized Original Work

1962; Gale, Shapley

Problem Definition

The objective in *stable matching problems* is to match together pairs of elements of a set of participants, taking into account the preferences of those involved and focusing on a stability requirement. The stability property ensures that no pair of participants would both prefer to be matched together rather than to accept their allocation in the matching. Such problems have widespread application, for example, in the allocation of medical students to hospital posts, students to schools or colleges, etc.

An instance of the classical *stable marriage problem* (SM), introduced by Gale and Shapley [2], involves a set of $2n$ participants comprising n men $\{m_1, \ldots, m_n\}$ and n women $\{w_1, \ldots, w_n\}$. Associated with each participant is a *preference list*, which is a total order over the participants of the opposite sex. A man m_i *prefers* woman w_j to woman w_k if w_j precedes w_k on the preference list of m_i and similarly for the women.

A *matching* M is a bijection between the sets of men and women, in other words a set of man-woman pairs so that each man and each woman belongs to exactly one pair of M. For a man m_i, $M(m_i)$ denotes the *partner* of m_i in M, i.e., the unique woman w_j such that (m_i, w_j) is in M. Similarly, $M(w_j)$ denotes the partner of woman w_j in M. A matching M is *stable* if there is no *blocking pair*, namely, a pair (m_i, w_j) such that m_i prefers w_j to $M(m_i)$ and w_j prefers m_i to $M(w_j)$.

Relaxing the requirements that the numbers of men and women are equal and that each participant should rank *all* of the members of the opposite sex gives the *stable marriage problem with incomplete lists* (SMI). So an instance of SMI comprises a set of n_1 men $\{m_1, \ldots, m_{n1}\}$ and a set of n_2 women $\{w_1, \ldots, w_{n2}\}$, and each participant's preference list is a total order over a *subset* of the participants of the opposite sex. The implication is that if woman w_j does not appear on the list of man m_i, then she is not an acceptable partner for m_i and vice versa. A man-woman pair is *acceptable* if each member of the pair is on the preference list of the other, and a matching M is now a set of acceptable pairs such that each man and each woman is in *at most* one pair of M. In this context, a blocking pair for matching M is an acceptable pair (m_i, w_j) such that m_i either is unmatched in M or prefers w_j to $M(m_i)$ and, likewise, w_j either is unmatched or prefers m_i to $M(w_j)$. A matching is stable if it has no blocking pair. So in an instance of SMI, a stable matching need not match all of the participants.

Gale and Shapley also introduced a many-one version of stable marriage, which they called the *college admissions problem*, but which is now more usually referred to as the ▶ Hospitals/Residents Problem (HR) because of its well-known applications in the medical employment field. This problem is covered in detail in Entry 150 of this volume.

A comprehensive treatment of many aspects of the stable marriage problem, as of 1989, appears in the monograph of Gusfield and Irving [5]. A more recent detailed exposition is given by Manlove [14].

Key Results

Theorem 1 *For every instance of SM or SMI, there is at least one stable matching.*

Theorem 1 was proved constructively by Gale and Shapley [2] as a consequence of the algorithm that they gave to find a stable matching.

Theorem 2 *1. For a given instance of SM involving n men and n women, there is a $O(n^2)$ time algorithm that finds a stable matching.*
2. For a given instance of SMI in which the combined length of all the preference lists is a, there is a $O(a)$ time algorithm that finds a stable matching.

The algorithm for SMI is a simple extension of that for SM. Each can be formulated in a variety of ways, but is most usually expressed in terms of a sequence of "proposals" from the members of one sex to the members of the other. A pseudocode version of the SMI algorithm appears in Fig. 1, in which the traditional approach of allowing men to make proposals is adopted.

The complexity bound of Theorem 2(1) first appeared in Knuth's monograph on stable marriage [12]. The fact that this algorithm is asymptotically optimal was subsequently established by Ng and Hirschberg [17] via an adversary argument. On the other hand, Wilson [21] proved that the average running time, taken over all possible instances of SM, is $O(n \log n)$.

The algorithm of Fig. 1, in its various guises, has come to be known as the Gale-Shapley algorithm. The variant of the algorithm given here is called *man oriented*, because men have the advantage of proposing. Reversing the roles of men and women gives the *woman-oriented* variant. The "advantage" of proposing is remarkable, as spelled out in the next theorem.

Theorem 3 *The man-oriented version of the Gale-Shapley algorithm for SM or SMI yields the man-optimal stable matching in which each man has the best partner that he can have in any stable matching, but in which each woman has her worst possible partner. The woman-oriented version yields the woman-optimal stable matching, which has analogous properties favoring the women.*

$M = \emptyset$;

assign each person to be free;   /* i. e., not a member of a pair in M */

while (some man m is free and has not proposed to every woman on his list)

    m proposes to w, the first woman on his list to whom he has not proposed;

    if (w is free)

        add (m, w) to M;   /* w accepts m */

    else if (w prefers m to her current partner m')

        remove (m', w) from M; /* w rejects m', setting m' free */

        add (m, w) to M;   /* w accepts m */

    else

        M remains unchanged;   /* w rejects m */

return M;

Stable Marriage, Fig. 1 The Gale-Shapley algorithm

The optimality property of Theorem 3 was established by Gale and Shapley [2], and the corresponding "pessimality" property was first observed by McVitie and Wilson [16].

As observed earlier, a stable matching for an instance of SMI need not match all of the participants. But the following striking result was established by Gale and Sotomayor [3] and Roth [19] (in the context of the more general HR problem).

Theorem 4 *In an instance of SMI, all stable matchings have the same size and match exactly the same subsets of the men and women.*

For a given instance of SM or SMI, there may be many different stable matchings. Indeed Knuth [12] showed that the maximum possible number of stable matchings grows exponentially with the number of participants. He also pointed out that the set of stable matchings forms a distributive lattice under a natural dominance relation, a result attributed to Conway. This powerful algebraic structure that underlies the set of stable matchings can be exploited algorithmically in a number of ways. For example, Gusfield [4] showed how all k stable matchings for an instance of SM can be generated in $O(n^2 + kn)$ time (▸ Optimal Stable Marriage).

Extensions of these problems that are important in practice, so-called SMT and SMTI (extensions of SM and SMI, respectively), allow the

presence of *ties* in the preference lists. In this context, three different notions of stability have been defined [7] – *weak*, *strong*, and *super-stability*, depending on whether the definition of a blocking pair requires that both members should improve, or at least one member improves and the other is no worse off, or merely that neither member is worse off. The following theorem summarizes the basic algorithmic results for these three varieties of stable matchings.

Theorem 5 *For a given instance of SMT or SMTI:*

1. *A weakly stable matching is guaranteed to exist and can be found in $O(n^2)$ or $O(a)$ time, respectively.*
2. *A super-stable matching may or may not exist; if one does exist, it can be found in $O(n^2)$ or $O(a)$ time, respectively.*
3. *A strongly stable matching may or may not exist; if one does exist, it can be found in $O(n^3)$ or $O(na)$ time, respectively.*

Theorem 5 parts (1) and (2) are due to Irving [7] (for SMT) and Manlove [13] (for SMTI). Part (3) is due to Kavitha et al. [11], who improved earlier algorithms of Irving and Manlove.

It turns out that, in contrast to the situation described by Theorem 4, weakly stable matchings in SMTI can have different sizes. The natural problem of finding a maximum cardinality weakly stable matching, even under severe restrictions on the ties, is NP-hard [15]. ▶ Stable Marriage with Ties and Incomplete Lists explores this problem further.

Interesting special cases of SM and its variants arise when the preference lists on one or both sides are derived from a "master" list that ranks participants (e.g.,, according to some objective criterion). Such problems are explored by Irving et al. [10].

The stable marriage problem is an example of a *bipartite* matching problem. The extension in which the bipartite requirement is dropped is the so-called *stable roommates* (SR) problem.

Gale and Shapley had observed that, unlike the case of SM, an instance of SR may or may not admit a stable matching, and Knuth [12] posed the problem of finding an efficient algorithm for SR or proving it NP-complete. Irving [6] established the following theorem via a nontrivial extension of the Gale-Shapley algorithm.

Theorem 6 *For a given instance of SR, there exists a $O(n^2)$ time algorithm to determine whether a stable matching exists and if so to find such a matching.*

Variants of SR may be defined, as for SM, in which preference lists may be incomplete and/or contain ties – these are denoted by SRI, SRT, and SRTI – and in the presence of ties, the three flavors of stability, weak, strong, and super, are again relevant.

Theorem 7 *For a given instance of SRT or SRTI:*

1. *A weakly stable matching may or may not exist, and it is an NP-complete problem to determine whether such a matching exists.*
2. *A super-stable matching may or may not exist; if one does exist, it can be found in $O(n^2)$ or $O(a)$ time, respectively.*
3. *A strongly stable matching may or may not exist; if one does exist, it can be found in $O(n^4)$ or $O(a^2)$ time, respectively.*

Theorem 7 part (1) is due to Ronn [18], part (2) is due to Irving and Manlove [9], and part (3) is due to Scott [20].

Applications

Undoubtedly the best known and most important applications of stable matching algorithms are in centralized matching schemes in the medical and educational domains. ▶ Hospitals/Residents Problem includes a summary of some of these applications.

Open Problems

The parallel complexity of stable marriage remains open. The best known parallel algorithm for SMI is due to Feder et al. [1] and has $O(\sqrt{a}\log^3 a)$ running time using a polynomially bounded number of processors. It is not known whether the problem is in NC, but nor is there a proof of P-completeness.

One of the open problems posed by Knuth in his early monograph on stable marriage [12] was that of determining the maximum possible number x_n of stable matchings for any SM instance involving n men and n women. This problem remains open, although Knuth himself showed that x_n grows exponentially with n. Irving and Leather [8] conjecture that, when n is a power of 2, this function satisfies the recurrence

$$x_n = 3x_{n/2}^2 - 2x_{x/4}^4.$$

Many open problems remain in the setting of weak stability, such as finding a good approximation algorithm for a maximum cardinality weakly stable matching – see ▶ Stable Marriage with Ties and Incomplete Lists – and enumerating all weakly stable matchings efficiently.

Cross-References

▶ Hospitals/Residents Problem
▶ Optimal Stable Marriage

► Ranked Matching
► Stable Marriage and Discrete Convex Analysis
► Stable Marriage with Ties and Incomplete Lists
► Stable Partition Problem

Recommended Reading

1. Feder T, Megiddo N, Plotkin SA (2000) A sublinear parallel algorithm for stable matching. Theor Comput Sci 233(1–2):297–308
2. Gale D, Shapley LS (1962) College admissions and the stability of marriage. Am Math Mon 69:9–15
3. Gale D, Sotomayor M (1985) Some remarks on the stable matching problem. Discret Appl Math 11:223–232
4. Gusfield D (1987) Three fast algorithms for four problems in stable marriage. SIAM J Comput 16(1):111–128
5. Gusfield D, Irving RW (1989) The stable marriage problem: structure and algorithms. MIT, Cambridge
6. Irving RW (1985) An efficient algorithm for the stable roommates problem. J Algorithms 6:577–595
7. Irving RW (1994) Stable marriage and indifference. Discret Appl Math 48:261–272
8. Irving RW, Leather P (1986) The complexity of counting stable marriages. SIAM J Comput 15(3):655–667
9. Irving RW, Manlove DF (2002) The stable roommates problem with ties. J Algorithms 43:85–105
10. Irving RW, Manlove DF, Scott S (2008) The stable marriage problem with master preference lists. Discret Appl Math 156:2959–2977
11. Kavitha T, Mehlhorn K, Michail D, Paluch K (2004) Strongly stable matchings in time O(nm), and extension to the H/R problem. In: Proceedings of the 21st symposium on theoretical aspects of computer science (STACS 2004). Lecture notes in computer science, vol 2996, pp 222–233. Springer, Berlin
12. Knuth DE (1976) Mariages stables. Les Presses de L'Université de Montréal, Montréal
13. Manlove DF (1999) Stable marriage with ties and unacceptable partners. Technical report TR-1999-29, Department of Computing Science, University of Glasgow
14. Manlove DF (2013) Algorithmics of matching under preferences. World Scientific, Singapore
15. Manlove DF, Irving RW, Iwama K, Miyazaki S, Morita Y (2002) Hard variants of stable marriage. Theor Comput Sci 276(1–2):261–279
16. McVitie D, Wilson LB (1971) The stable marriage problem. Commun ACM 14:486–490
17. Ng C, Hirschberg DS (1990) Lower bounds for the stable marriage problem and its variants. SIAM J Comput 19:71–77
18. Ronn E (1990) NP-complete stable matching problems. J Algorithms 11:285–304
19. Roth AE (1984) The evolution of the labor market for medical interns and residents: a case study in game theory. J Polit Econ 92(6):991–1016
20. Scott S (2005) A study of stable marriage problems with ties. Ph.D. thesis, Department of Computing Science, University of Glasgow
21. Wilson LB (1972) An analysis of the stable marriage assignment algorithm. BIT 12:569–575

Stable Marriage and Discrete Convex Analysis

Akihisa Tamura
Department of Mathematics, Keio University, Yokohama, Japan

Keywords

Stable matching

Years and Authors of Summarized Original Work

2000; Eguchi, Fujishige, Tamura, Fleiner

Problem Definition

In the stable marriage problem first defined by Gale and Shapley [7], there is one set each of men and women having the same size, and each person has a strict preference order on persons of the opposite gender. The problem is to find a matching such that there is no pair of a man and a woman who prefer each other to their partners in the matching. Such a matching is called a *stable marriage* (or *stable matching*). Gale and Shapley showed the existence of a stable marriage and gave an algorithm for finding one. Fleiner [4] extended the stable marriage problem to the framework of matroids, and Eguchi, Fujishige, and Tamura [3] extended this formulation to a more general one in terms of discrete convex analysis, which was developed by Murota [8, 9]. Their formulation is described as follows.

Let M and W be sets of men and women who attend a dance party at which each person dances a waltz T times and the number of times that he/she can dance with the same person of the opposite gender is unlimited. The problem is to find an "agreeable" allocation of dance partners, in which each person is assigned at most T persons of the opposite gender with possible repetition. Let $E = M \times W$, i.e., the set of all man-woman pairs. Also define $E_{(i)} = \{i\} \times W$ for all $i \in M$ and $E_{(j)} = M \times \{j\}$ for all $j \in W$. Denoting by $x(i, j)$ the number of dances between man i and woman j, an allocation of dance partners can be described by a vector $x = (x(i, j) : i \in M, j \in W) \in \mathbf{Z}^E$, where $\mathbf{Z}$ denotes the set of all integers. For each $y \in \mathbf{Z}^E$ and $k \in M \cup W$, denote by $y_{(k)}$ the restriction of y on $E_{(k)}$. For example, for an allocation $x \in \mathbf{Z}^E$, $x_{(k)}$ represents the allocation of person k with respect to x. Each person k

describes his/her preferences on allocations by using a value function $f_k : \mathbf{Z}^{E(k)} \to \mathbf{R} \cup (-\infty)$, where $\mathbf{R}$ denotes the set of all reals and $f_k(y) = -\infty$ means that allocation $y \in \mathbf{Z}^{E(k)}$ is unacceptable for k. Note that the valuation of each person on allocations is determined only by his/her allocations. Let dom $f_k = \{y | f_k(y) \in \mathbf{R}\}$. Assume that each value function f_k satisfies the following assumption:

(A) dom f_k is bounded and hereditary and has $\mathbf{0}$ as the minimum point, where $\mathbf{0}$ is the vector of all zeros and heredity means that for any $y, y' \in \mathbf{Z}^{E(k)}$, $\mathbf{0} \le y' \le y \in$ dom f_k implies $y' \in$ dom f_k.

For example, the following value functions with $M = \{1\}$ and $W = \{2, 3\}$

$$f_1(x(1, 2), x(1, 3)) =$$
$$\begin{cases} 10(x(1, 2) + x(1, 3)) - x(1, 2)^2 - x(1, 3)^2 & \text{if } x(1, 2), x(1, 3) \ge 0 \\ & \text{and } x(1, 2) + x(1, 3) \le 3 \\ -\infty & \text{otherwise,} \end{cases}$$

$$f_j(x(1, j)) = \begin{cases} x(1, j) & \text{if } x(1, j) \in \{0, 1, 2, 3\} (j = 2, 3) \\ -\infty & \text{otherwise} \end{cases}$$

represent the case where (1) everyone wants to dance as many times, up to three, as possible and (2) man 1 wants to divide his dances between women 2 and 3 as equally as possible. Allocations $(x(1, 2), x(1, 3)) = (1, 2)$ and $(2, 1)$ are stable in the sense below.

A vector $x \in \mathbf{Z}^E$ is called a *feasible allocation* if $x_{(k)} \in$ dom f_k for all $k \in M \cup W$. An allocation x is said to satisfy *incentive constraints* if each person has no incentive to unilaterally decrease the current units of x, that is, if it satisfies

$$f_k(x_{(k)}) = \max\{f_k(y) | y \le x_{(k)}\} \ (\forall k \in M \cup W). \tag{1}$$

An allocation x is called *unstable* if it does not satisfy incentive constraints or there exist $i \in M$, $j \in W$, $y' \in \mathbf{Z}^{E(i)}$ and $y'' \in \mathbf{Z}^{E(j)}$ such that

$$f_i(x_{(i)}) < f_i(y'), \tag{2}$$

$$y'(i, j') \le x(i, j') \quad (\forall j' \in W \backslash \{j\}), \tag{3}$$

$$f_j(x_{(j)}) < f_j(y''), \tag{4}$$

$$y''(i', j) \le x(i', j) \quad (\forall i' \in M \backslash \{i\}), \tag{5}$$

$$y'(i, j) = y''(i, j). \tag{6}$$

Conditions (2) and (3) say that man i can strictly increase his valuation by changing the current number of dances with j without increasing the numbers of dances with other women, and (4) and (5) describe a similar situation for women. Condition (6) requires that i and j agree on the number of dances between them. An allocation x is called *stable* if it is not unstable.

Problem 1 Given disjoint sets M and W and value functions $f_k : \mathbf{Z}^{E(k)} \to \mathbf{R} \cup \{-\infty\}$ for $k \in M \cup W$ satisfying assumption (A), find a stable allocation x.

Remark 1 A time schedule for a given feasible allocation can be given by a famous result on graph coloring, namely, "any bipartite graph can be edge-colorable with the maximum degree colors."

Key Results

The work of Eguchi, Fujishige, and Tamura [3] gave a solution to Problem 1 in the case where each value function f_k is $M^\natural$-concave.

Discrete Convex Analysis: $M^\natural$-Concave Functions

Let V be a finite set. For each $S \subseteq V$, e_S denotes the characteristic vector of S defined by $e_S(v) = 1$ if $v \in S$ and $e_S(v) = 0$ otherwise. Also define e_0 as the zero vector in $\mathbf{Z}^V$. For a vector $x \in \mathbf{Z}^V$, its positive support $\text{supp}^+(x)$ and negative support $\text{supp}^-(x)$ are defined by $\text{supp}^+(x) = \{u \in V | x(u) > 0\}$ and $\text{supp}^-(x) = \{u \in V | x(u) < 0\}$. A function $f : \mathbf{Z}^V \to \mathbf{R} \cup \{-\infty\}$ is called $M^\natural$-*concave* if it satisfies the following condition $\forall x, y \in \text{dom } f$, $\forall u \in \text{supp}^+(x - y)$, $\exists v \in \text{supp}^-(x - y) \cup \{0\}$:

$$f(x) + f(y) \leq f(x - e_u + e_v) + f(y + e_u - e_v).$$

The above condition says that the sum of the function values at two points does not decrease as the points symmetrically move one or two steps closer to each other on the set of integral lattice points of $\mathbf{Z}^V$. This is a discrete analogue of the fact that for an ordinary concave function, the sum of the function values at two points does not decrease as the points symmetrically move closer to each other on the straight line segment between the two points.

Example 1 A nonempty family $\mathcal{T}$ of subsets of V is called a *laminar family* if $X \cap Y = \emptyset$, $X \subseteq Y$ or $Y \subseteq X$ holds for every $X, Y \in \mathcal{T}$. For a laminar family $\mathcal{T}$ and a family of univariate concave functions $f_Y : \mathbf{R} \to \mathbf{R} \cup \{-\infty\}$ indexed by $Y \in \mathcal{T}$, the function $f : \mathbf{Z}^V \to \mathbf{R} \cup \{-\infty\}$ defined by

$$f(x) = \sum_{Y \in \mathcal{T}} f_Y\left(\sum_{v \in Y} x(v)\right) \quad (\forall x \in \mathbf{Z}^V)$$

is $M^\natural$-concave. The stable marriage problem can be formulated as Problem 1 by using value functions of this type.

Example 2 For the independence family $\mathcal{I} \subseteq 2^V$ of a matroid on V and $w \in \mathbf{R}^V$, the function $f : \mathbf{Z}^V \to \mathbf{R} \cup \{-\infty\}$ defined by

$$f(x) = \begin{cases} \sum_{u \in X} w(u) & \text{if } x = e_X \text{ for some } X \in \mathcal{I} \\ -\infty & \text{otherwise} \end{cases}$$
$$(\forall x \in \mathbf{Z}^V)$$

is $M^\natural$-concave. Fleiner [4] showed that there always exists a stable allocation for value functions of this type.

Theorem 1 ([6]) *Assume that the value functions $f_k (k \in M \cup W)$ are $M^\natural$-concave satisfying (A). Then, a feasible allocation x is stable if and only if there exist $Z_M = (z_{(i)} | i \in M) \in (\mathbf{Z} \cup \{+\infty\})^E$ and $z_W = (z_{(j)} | j \in W) \in (\mathbf{Z} \cup \{+\infty\})^E$ such that*

$$x_{(i)} \in \arg\max\{f_i(y) | y \leq z_{(i)}\} \quad (\forall i \in M), \tag{7}$$

$$x_{(j)} \in \arg\max\{f_j(y) | y \leq z_{(j)}\} \quad (\forall j \in W), \tag{8}$$

$$z_M(e) = +\infty \text{ or } z_W(e) = +\infty \quad (\forall e \in E), \tag{9}$$

where $\arg\max\{f_i(y) | y \leq z_{(i)}\}$ denotes the set of all maximizers of f_i under the constraints $y \leq z_{(i)}$.

Theorem 2 ([3]) *Assume that the value functions $f_k (k \in M \cup W)$ are $M^\natural$-concave satisfying (A). Then, there always exists a stable allocation.*

Eguchi, Fujishige, and Tamura [3] proved Theorem 2 by showing that the following algorithm finds a feasible allocation x, and z_M, z_W satisfying (7), (8), and (9).

Here, $z_W \vee x_M$ is defined by $(z_W \vee x_M)(e) = \max\{z_W(e), x_M(e)\}$ for all $e \in E$.

Algorithm EXTENDED-GS

Input: $M^{\natural}$-concave functions f_M, f_W with $f_M(x) = \sum_{i \in M} f_i(x_{(i)})$ and $f_W(x) = \sum_{j \in W} f_j(x_{(j)})$;

Output: (x, z_M, z_W) satisfying (7), (8), and (9);

$z_M := (+\infty, \cdots, +\infty), z_W := x_W := 0$;

repeat{
let x_M be any element in
$\arg\max\{f_M(y) | x_W \leq y \leq z_M\}$;
let x_W be any element in
$\arg\max\{f_W(y) | y \leq x_M\}$;
for each $e \in E$ with $x_M(e) > x_W(e)${
$z_M(e) := x_W(e)$;
$z_W(e) := +\infty$;
};
} **until** $x_M = x_W$;
return $(x_M, z_M, z_W \vee x_M)$.

Applications

Abraham, Irving, and Manlove [1] dealt with a student-project allocation problem which is a concrete example of models in [4] and [3] and discussed the structure of stable allocations.

Fleiner [5] generalized the stable marriage problem and its extension in [4] to a wide framework and showed the existence of a stable allocation by using a fixed point theorem.

Fujishige and Tamura [6] proposed a common generalization of the stable marriage problem and the assignment game defined by Shapley and Shubik [10] by utilizing $M^{\natural}$-concave functions and gave a constructive proof of the existence of a stable allocation.

Open Problems

Algorithm EXTENDED-GS solves the maximization problem of an $M^{\natural}$-concave function in each iteration. A maximization problem of an $M^{\natural}$-concave function f on E can be solved in polynomial time in $|E|$ and $\log L$, where $L = \max\{||x - y||_\infty | x, y \in \text{dom } f\}$, provided that the function value $f(x)$ can be calculated in constant time for each x [11, 12]. Eguchi, Fujishige, and Tamura [3] showed that EXTENDED-GS terminates after at most L iterations, where L is defined by $\{||x||_\infty | x \in \text{dom } f_M\}$ in this

case, and there exist a series of instances in which EXTENDED-GS requires numbers of iterations proportional to L. On the other hand, Baïou and Balinski [2] gave a polynomial time algorithm in $|E|$ for the special case where f_M and f_W are linear on rectangular domains. Whether a stable allocation for the general case can be found in polynomial time in $|E|$ and $\log L$ or not is open.

Cross-References

▶ Assignment Problem
▶ Hospitals/Residents Problem
▶ Optimal Stable Marriage
▶ Stable Marriage
▶ Stable Marriage with Ties and Incomplete Lists

Recommended Reading

1. Abraham DJ, Irving RW, Manlove DF (2007) Two algorithms for the student-project allocation problem. J Discret Algorithms 5:73–90
2. Baïou M, Balinski, M (2002) Erratum: the stable allocation (or ordinal transportation) problem. Math Oper Res 27:662–680
3. Eguchi A, Fujishige S, Tamura A (2003) A generalized Gale-Shapley algorithm for a discrete-concave stable-marriage model. In: Ibaraki T, Katoh N, Ono H (eds) Algorithms and computation: 14th international symposium (ISAAC 2003), Kyoto. LNCS, vol 2906. Springer, Berlin, pp 495–504
4. Fleiner T (2001) A matroid generalization of the stable matching polytope. In: Gerards B, Aardal K (eds) Integer programming and combinatorial optimization: 8th international IPCO conference, Utrecht. LNCS, vol 2081. Springer, Berlin, pp 105–114
5. Fleiner T (2003) A fixed point approach to stable matchings and some applications. Math Oper Res 28:103–126
6. Fujishige S, Tamura A (2007) A two-sided discrete-concave market with bounded side payments: an approach by discrete convex analysis. Math Oper Res 32:136–155
7. Gale D, Shapley SL (1962) College admissions and the stability of marriage. Am Math Mon 69:9–15
8. Murota K (1998) Discrete convex analysis. Math Program 83:313–371
9. Murota K (2003) Discrete convex analysis. Society for Industrial and Applied Mathematics, Philadelphia
10. Shapley SL, Shubik M (1971) The assignment game I: the core. Int J Game Theor 1:111–130

11. Shioura A (2004) Fast scaling algorithms for M-convex function minimization with application to the resource allocation problem. Discret Appl Math 134:303–316
12. Tamura A (2005) Coordinatewise domain scaling algorithm for M-convex function minimization. Math Program 102:339–354

Stable Marriage with One-Sided Ties

Hiroki Yanagisawa
IBM Research – Tokyo, Tokyo, Japan

Keywords

Approximation algorithms; Incomplete lists; Integer programming; Linear programming relaxation; One-sided ties; Stable marriage problem

Years and Authors of Summarized Original Work

2007; Halldórsson, Iwama, Miyazaki, Yanagisawa
2014; Huang, Iwama, Miyazaki, Yanagisawa

Problem Definition

Over the last 50 years, the stable marriage problem has been extensively studied for many problem settings (see, e.g., [11]), and one of the most intensively studied problem settings is MAX SMTI (MAXimum Stable Marriage with Ties and Incomplete lists). An input for the stable marriage problem consists of n men, n women, and each person's preference list for the people of the opposite sex. In MAX SMTI, the preference list of each person can be incomplete, which means that each person is allowed to exclude unacceptable people from the preference list, and the preference list of each person is allowed to include *ties* to show indifference between two or more people.

The objective of MAX SMTI is to find the largest matching that satisfies a stability condition. Before describing the stability condition, we review some notation. A matching M is defined as a set of pairs of man m and woman w such that m and w are acceptable to each other. The *size* of a matching M is defined as the number of pairs in M. We say that a person p is *single* if p is not matched in M. When man m and woman w are matched in M, we write $M(m) = w$ and $M(w) = m$. We say that matching M is *stable* if it does not contain any pair of man and woman, each of whom prefers the other to the partner in M (if any). More precisely, a matching M is stable if there is no pair of man m' and woman w' that satisfy all three conditions (i)–(iii): (i) m' and w' are acceptable to each other but not matched in M, (ii) m' is single in M or m' strictly prefers w' to $M(m')$, and (iii) w' is single in M or w' strictly prefers m' to $M(w')$. MAX SMTI asks us to find a stable matching of the largest size, and this problem is known to be NP-hard [12]. Therefore, the approximability of this problem has been intensively studied.

In this entry, we show recent results for two major variants of MAX SMTI. One of the variants is MAX SMOTI (MAXimum Stable Marriage with One-Sided Ties and Incomplete lists), in which only women are allowed to include ties in their preference lists and the preference lists of men are strictly ordered. The other variant is MAX SSMTI (Special SMTI), which is an even more restricted variant of MAX SMOTI where the ties are only allowed at the ends of the women's preference lists. Note that these two variants are still known to be NP-hard [12].

Problem 1 (MAX SMOTI)
INPUT: n men, n women, and each person's preference list, where only women have ties
OUTPUT: A stable matching of maximum size

Problem 2 (MAX SSMTI)
INPUT: n men, n women, and each person's preference list, where ties are at the ends of the women's preference lists
OUTPUT: A stable matching of maximum size

Stable Marriage with One-Sided Ties, Table 1 Examples of instances for MAX SMOTI and MAX SSMTI

MAX SMOTI		MAX SSMTI	
$m_1 : w_2\ w_1$	$w_1 : m_1\ m_2$	$m_1 : w_1\ w_3$	$w_1 : (m_1\ m_2\ m_3)$
$m_2 : w_2\ w_3\ w_1$	$w_2 : (m_1\ m_2)\ m_3$	$m_2 : w_2\ w_3\ w_1$	$w_2 : m_2$
$m_3 : w_3\ w_2$	$w_3 : m_2\ m_3$	$m_3 : w_3\ w_1$	$w_3 : m_2\ (m_1\ m_3)$

Examples

Table 1 shows examples of instances for MAX SMOTI and MAX SSMTI. The instance for MAX SMOTI contains a set of men $\{m_1, m_2, m_3\}$ and a set of women $\{w_1, w_2, w_3\}$. The preference list of each person is described in decreasing order of preference, and tied people are enclosed in a pair of parenthesis. For example, woman w_2 is indifferent between m_1 and m_2 but prefers m_1 or m_2 over m_3. A matching $M = \{(m_2, w_1), (m_3, w_2)\}$ is not stable for this MAX SMOTI instance, because m_2 strictly prefers w_2 to w_1 ($= M(m_2)$) and w_2 strictly prefers m_2 to m_3 ($= M(w_2)$). An example of a stable matching for this instance is $M' = \{(m_1, w_2), (m_2, w_3)\}$, and we can find another larger stable matching $M^* = \{(m_1, w_1), (m_2, w_2), (m_3, w_3)\}$ of size 3.

Key Results

Here we review past research on MAX SMOTI and MAX SSMTI. We start by describing a simple proposal-based algorithm (often referred to as the Gale-Shapley algorithm or the deferred acceptance algorithm), which is guaranteed to find a *stable* matching. In this algorithm, all of the men and women are initially set to be single. We pick an arbitrary man m who is single, and let man m propose to woman w at the top of his preference list. When man m proposes to w, he deletes woman w from his preference list. Woman w always accepts any proposal if she is single, which makes a matching pair of m and w. We repeat this proposal procedure to find more and more matching pairs. When a woman w, who is already matched to a man m, receives another proposal from man m', woman w chooses the more highly ranked man based on her preference list. (That is, the matching partner

of w is unchanged if w prefers m to m', and the matching partner of w is changed from m to m' and m becomes unmatched if w prefers m' to m.) If m and m' are tied in w's preference list, then w chooses an arbitrary man. The proposal procedure continues until we cannot find any man who can propose. (That is, this algorithm terminates when all of the men become matched or the preference lists of all single men become empty.) Any matching obtained by this algorithm can be proven to be stable. The size of the obtained stable matching depends mostly on the decisions by women when a woman receives two proposals from men who are tied in her preference list. In the worst case, the size of an obtained matching can be half of the optimum matching, and hence, the approximation ratio of this algorithm is 2. It was an open problem whether or not there exists an approximation algorithm whose approximation ratio is strictly better than 2. Iwama, Miyazaki, and Yamauchi [8] provided an affirmative answer for this open problem with a 1.875-approximation algorithm.

After this breakthrough, Királу [10] developed a new simple 1.5-approximation algorithm for MAX SMOTI (which also applies to MAX SSMTI) by improving the decision strategy of the proposal-based algorithm when women receive multiple proposals from tied men. His algorithm proceeds in the same way as the proposal-based algorithm until one of the men's preference lists become empty. When the preference list of a man becomes empty, he enters into his second round. Specifically, he recovers his original preference list so that he can propose to the women in his original preference list again, but his status is changed to "promoted." A promoted man is not allowed to recover his original preference list when his preference list becomes empty the second time, and hence, no man can enter a third round in Kiráry's algorithm. The decision strategy of the women is changed so that a woman is forced to choose a promoted man (if one exists) when she receives two proposals from men who are tied in her preference list. This improvement of the decision strategy is the key to achieve the 1.5-approximation.

Iwama, Miyazaki, and Yanagisawa [9] further improved the approximation ratio to 25/17 (< 1.4706) for MAX SMOTI with a new algorithm GSA-LP, which uses a more complex proposal sequence of the men and a more sophisticated decision strategy for the women. In GSA-LP, we compute an optimum solution for a linear programming relaxation of a natural integer programming formulation of the problem in advance and use it for the decision strategy of the women. In addition, the proposal sequence is changed so that a man can propose to a woman many times, and a man is allowed to recover his original preference list at most twice (in other words, a man is allowed to go into a third round). These changes yield an improved approximation ratio for MAX SMOTI. Very recently, GSA-LP was shown to achieve a 1.25-approximation for MAX SSMTI [5].

For MAX SMOTI, there are successive improvements over GSA-LP. Huang and Kavitha [4] developed another new algorithm that achieves a 22/15 (<1.4667)-approximation by using a maximum matching algorithm. Radnai [14] showed 41/28 (<1.4643)-approximation by using a more detailed analysis of this new algorithm and also showed that a lower bound of the approximation ratio of this algorithm is at least 13/9 (>1.4444). Dean and Jalasutram improved the analysis of GSA-LP and showed that the approximation ratio of GSA-LP is at most 19/13 (<1.4616) if we increase the number of rounds from three to four [1]. We also note that if the lengths of ties are restricted to two, then the approximation ratio of this restricted MAX SMOTI variant can be further improved. A randomized algorithm [2] achieves 10/7 (<1.4286)-approximation and Huang and Kavitha devised another deterministic algorithm [4] with the same approximation ratio.

For the negative side, both MAX SMOTI and MAX SSMTI are NP-hard to approximate within any constant factor better than 21/19 (>1.1052) and hard to approximate within any constant factor better than 5/4 (= 1.25) under the unique games conjecture [3, 15]. These lower bounds hold even if we restrict the lengths of the ties to two. Note that the approximation ratio of the GSA-LP algorithm for MAX SSMTI is 1.25, which matches the lower bound under the unique games conjecture.

Applications

MAX SSMTI was introduced by Irving and Manlove [6] based on an actual application of the Scottish Foundation Allocation Scheme, which allocates residents (medical students) to hospitals. In this scheme, each resident submits a strictly ordered preference list, while each hospital submits a preference list that may contain one tie of arbitrary length at the end of the list. The objective of this allocation scheme is to maximize the number of allocated residents, and it is easy to reformulate this many-to-one allocation scheme as a one-to-one matching problem (MAX SSMTI) using a cloning technique [11].

Open Problems

An obvious future goal is to narrow the gap between the upper and lower bounds of the approximability of MAX SMOTI. Assuming the unique games conjecture is true, we now know that the best possible approximation ratio is between 1.4616 and 1.25. Even if we restrict the lengths of ties to two, all we can do now is reduce the upper bound slightly down to 1.4286. Thus, there is still much room for improvement.

As for MAX SSMTI, the 1.25-approximation of the GSA-LP algorithm is the best possible if the unique games conjecture is true. A future project could investigate if we can construct a faster approximation algorithm, because the

GSA-LP algorithm uses a linear programming relaxation technique, which takes superlinear time in the worst case.

Experimental Results

Irving and Manlove [7] reported on experimental evaluations of some heuristic algorithms including the Király's algorithm on real-world and random instances for MAX SMOTI. Subsequently, Podhradský [13] conducted experimental evaluations on random instances for MAX SMOTI and MAX SSMTI using some other heuristic algorithms including GSA-LP.

Cross-References

▶ Simpler Approximation for Stable Marriage
▶ Stable Marriage
▶ Stable Marriage with Ties and Incomplete Lists

Recommended Reading

1. Dean BC, Jalasutram R (2015) Factor revealing LPs and stable matching with ties and incomplete lists. In Proceedings of the 3rd international workshop on matching under preferences, 2015 (to appear)
2. Halldórsson MM, Iwama K, Miyazaki S, Yanagisawa H (2004) Randomized approximation of the stable marriage problem. Theor Comput Sci 325(3):439–465
3. Halldórsson MM, Iwama K, Miyazaki S, Yanagisawa H (2007) Improved approximation results for the stable marriage problem. ACM Trans Algorithms 3(3):Article No. 30
4. Huang CC, Kavitha T (2014) An improved approximation algorithm for the stable marriage problem with one-sided ties. In: Proceedings of IPCO 2014, Bonn, pp 297–308
5. Huang CC, Iwama K, Miyazaki S, Yanagisawa H (2015) Approximability of finding largest stable matchings. Manuscript under submission
6. Irving RW, Manlove DF (2008) Approximation algorithms for hard variants of the stable marriage and hospital/residents problems. J Comb Optim 16(3):279–292
7. Irving RW, Manlove DF (2009) Finding large stable matchings. J Exp Algorithmics 14:Article No. 2
8. Iwama K, Miyazaki S, Yamauchi N (2007) A 1.875: approximation algorithm for the stable marriage problem. In: Proceedings of SODA 2007, New Orleans, pp 288–297
9. Iwama K, Miyazaki S, Yanagisawa H (2014) A 25/17-approximation algorithm for the stable marriage problem with one-sided ties. Algorithmica 68(3):758–775
10. Király Z (2011) Better and simpler approximation algorithms for the stable marriage problem. Algorithmica 60(1):3–20
11. Manlove DF (2013) Algorithmics of matching under preferences. World Scientific, Hackensack
12. Manlove DF, Irving RW, Iwama K, Miyazaki S, Morita Y (2002) Hard variants of stable marriage. Theor Comput Sci 276(1–2):261–279
13. Podhradský A (2010) Stable marriage problem algorithms. Master's thesis, Faculty of Informatics, Masaryk University
14. Radnai A (2014) Approximation algorithms for the stable matching problem. Master's thesis, Eötvös Loránd University
15. Yanagisawa H (2007) Approximation algorithms for stable marriage problems. Ph.D. thesis, Kyoto University

Stable Marriage with Ties and Incomplete Lists

Kazuo Iwama[1,2] and Shuichi Miyazaki[3]
[1]Computer Engineering, Kyoto University, Sakyo, Kyoto, Japan
[2]School of Informatics, Kyoto University, Sakyo, Kyoto, Japan
[3]Academic Center for Computing and Media Studies, Kyoto University, Kyoto, Japan

Synonyms

Stable matching problem

Keywords

Approximation algorithms; Incomplete lists; Stable marriage problem; Ties

Years and Authors of Summarized Original Work

2007; Iwama, Miyazaki, Yamauchi

Problem Definition

In the original setting of the stable marriage problem introduced by Gale and Shapley [2], each preference list has to include all members of the other party, and furthermore, each preference list must be totally ordered (see entry ▸ Stable Marriage also).

One natural extension of the problem is then to allow persons to include ties in preference lists. In this extension, there are three variants of the stability definition, super-stability, strong stability, and weak stability (see below for definitions). In the first two stability definitions, there are instances that admit no stable matching, but there is a polynomial-time algorithm in each case that determines if a given instance admits a stable matching and finds one if one exists [9]. On the other hand, in the case of weak stability, there always exists a stable matching, and one can be found in polynomial time.

Another possible extension is to allow persons to declare unacceptable partners, so that preference lists may be incomplete. In this case, every instance admits at least one stable matching, but a stable matching may not be a perfect matching. However, if there are two or more stable matchings for one instance, then all of them have the same size [3].

The problem treated in this entry allows both extensions simultaneously, which is denoted as SMTI (stable marriage with ties and incomplete lists).

Notations

An instance I of SMTI comprises n men, n women, and each person's preference list that may be incomplete and may include ties. If a man m includes a woman w in his list, w is *acceptable* to m. $w_i \succ_m w_j$ means that m strictly prefers w_i to w_j in I. $w_i =_m w_j$ means that w_i and w_j are tied in m's list (including the case $w_i = w_j$). The statement $w_i \succeq_m w_j$ is true if and only if $w_i \succ_m w_j$ or $w_i =_m w_j$. Similar notations are used for women's preference lists. A matching M is a set of pairs (m, w) such that m is acceptable to w, and vice versa, and each person appears at

most once in M. If a man m is matched with a woman w in M, it is written as $M(m) = w$ and $M(w) = m$.

A man m and a woman w are said to form a *blocking pair for weak stability* for M if they are not matched together in M, but by matching them, both become better off, namely, (i) $M(m) \neq w$ but m and w are acceptable to each other, (ii) $w \succ_m M(m)$ or m is single in M, and (iii) $m \succ_w M(w)$ or w is single in M.

Two persons x and y are said to form a *blocking pair for strong stability* for M if they are not matched together in M, but by matching them, one becomes better off, and the other does not become worse off, namely, (i) $M(x) \neq y$ but x and y are acceptable to each other, (ii) $y \succ_x M(x)$ or x is single in M, and (iii) $x \succeq_y M(y)$ or y is single in M.

A man m and a woman w are said to form a *blocking pair for super-stability* for M if they are not matched together in M, but by matching them, neither becomes worse off, namely, (i) $M(m) \neq w$ but m and w are acceptable to each other, (ii) $w \succeq_m M(m)$ or m is single in M, and (iii) $m \succeq_w M(w)$ or w is single in M.

A matching M is called *weakly stable* (*strongly stable* and *super-stable*, respectively) if there is no blocking pair for weak (strong and super, respectively) stability for M.

Problem 1 (SMTI)

INPUT: n men, n women, and each person's preference list

OUTPUT: A stable matching

Problem 2 (MAX SMTI)

INPUT: n men, n women, and each person's preference list

OUTPUT: A stable matching of maximum size

The following problem is a restriction of MAX SMTI in terms of the length of preference lists:

Problem 3 ((p,q)-MAX SMTI)

INPUT: n men, n women, and each person's preference list, where each man's preference list includes at most p women and each woman's preference list includes at most q men

OUTPUT: A stable matching of maximum size

Definition of the Approximation Ratio

A goodness measure of an approximation algorithm T for a maximization problem is defined as follows: the *approximation ratio* of T is $\max\{opt(x)/T(x)\}$ over all instances x of size N, where $opt(x)$ and $T(x)$ are the sizes of the optimal and the algorithm's solutions, respectively.

Key Results

SMTI and MAX SMTI in Super-Stability and Strong Stability

Theorem 1 ([20]) *There is an $O(n^2)$-time algorithm that determines if a given SMTI instance admits a super-stable matching and finds one if one exists.*

Theorem 2 ([17]) *There is an $O(n^3)$-time algorithm that determines if a given SMTI instance admits a strongly stable matching and finds one if one exists.*

It is shown that all stable matchings for a fixed instance are of the same size [20]. Therefore, the above theorems imply that MAX SMTI can also be solved in the same time complexity.

SMTI and MAX SMTI in Weak Stability

In the case of weak stability, every instance admits at least one stable matching, but one instance can have stable matchings of different sizes. If the size is not important, a stable matching can be found in polynomial time by breaking ties arbitrarily and applying the Gale-Shapley algorithm.

Theorem 3 *There is an $O(n^2)$-time algorithm that finds a weakly stable matching for a given SMTI instance.*

However, if larger stable matchings are required, the problem becomes hard.

Theorem 4 ([5, 13, 21, 24]) *MAX SMTI is NP-hard and cannot be approximated within $33/29 - \epsilon$ for any positive constant ϵ unless P=NP. ($33/29 > 1.137$)*

The following approximation ratio is achieved by a local search type algorithm.

Theorem 5 ([14]) *There is a polynomial-time approximation algorithm for MAX SMTI whose approximation ratio is at most $15/8 (=1.875)$.*

There are a couple of approximation algorithms for restricted inputs.

Theorem 6 ([6]) *There is a polynomial-time randomized approximation algorithm for MAX SMTI whose expected approximation ratio is at most $10/7(\simeq 1.429)$ if, in a given instance, ties appear in one side only and the length of each tie is two.*

Theorem 7 ([6]) *There is a polynomial-time randomized approximation algorithm for MAX SMTI whose expected approximation ratio is at most $7/4(= 1.75)$ if, in a given instance, the length of each tie is two.*

Theorem 8 ([7]) *There is a polynomial-time approximation algorithm for MAX SMTI whose approximation ratio is at most $2/(1 + L^{-2})$ if, in a given instance, ties appear in one side only and the length of each tie is at most L.*

Theorem 9 ([7]) *There is a polynomial-time approximation algorithm for MAX SMTI whose approximation ratio is at most $13/7(\simeq 1.858)$ if, in a given instance, the length of each tie is two.*

(p, q)-MAX SMTI in Weak Stability

Irving et al. [12] show the boundary between P and NP-hardness in terms of the length of preference lists.

Theorem 10 ([12]) *$(2,\infty)$-MAX SMTI is solvable in time $O(n^{\frac{3}{2}} \log n)$.*

Theorem 11 ([12]) *$(3,3)$-MAX SMTI is NP-hard.*

Theorem 12 ([12]) *$(3,4)$-MAX SMTI is NP-hard and cannot be approximated within some constant $\delta(> 1)$ unless P=NP.*

Applications

One of the most famous applications of the stable marriage problem is a centralized assignment system between medical students (residents) and hospitals. This is an extension of the stable marriage problem to a many-one variant: Each hospital declares the number of residents it can accept, which may be more than one, while each resident has to be assigned to at most one hospital. Actually, there are several applications in the world, known as NRMP in the USA [4], CaRMS in Canada [1], SFAS (previously known as SPA) in Scotland [10, 11], and JRMP in Japan [16]. One of the optimization criteria is clearly the number of matched residents. In a real-world application such as the above hospitals-residents matching, hospitals and residents tend to submit short preference lists that may include ties, in which case, the problem can be naturally considered as MAX SMTI.

Open Problems

An apparent open problem is to narrow the gap of approximability and inapproximability of MAX SMTI in weak stability.

Since the publication of the key result of this chapter (Theorem 5), there have been a lot of improvement. Királly [18] presented a linear time 5/3-approximation algorithm (see ▶ Simpler Approximation for Stable Marriage). McDermid [22] then presented a 1.5-approximation algorithm (see ▶ Simpler Approximation for Stable Marriage), and Királly [19] and Paluch [23] presented simpler algorithms with the same approximation ratio, which is the current best upper bound. The lower bound was improved by Yanagisawa [24], who showed that MAX SMTI is inapproximable to within a ratio smaller than $33/29(>1.137)$ unless $P=NP$. He also showed that MAX SMTI is inapproximable within a ratio smaller than $4/3(>1.333)$ under the Unique Games Conjecture (UGC).

As for the special case where ties can appear in one side only (see ▶ Stable Marriage with One-Sided Ties), Királly [18] presented a 1.5-approximation algorithm. It was then improved to $25/17(<1.471)$ [15] and to $22/15(<1.467)$ [8], which is the current best upper bound. The current best lower bounds are $21/19(\simeq 1.105)$ under $P \neq NP$ and 1.25 under UGC [7].

Cross-References

▶ Simpler Approximation for Stable Marriage
▶ Assignment Problem
▶ Hospitals/Residents Problem
▶ Hospitals/Residents Problems with Quota Lower Bounds
▶ Optimal Stable Marriage
▶ Ranked Matching
▶ Stable Marriage
▶ Stable Marriage and Discrete Convex Analysis
▶ Stable Partition Problem
▶ Simpler Approximation for Stable Marriage
▶ Stable Marriage with One-Sided Ties

Recommended Reading

1. Canadian Resident Matching Service (CaRMS), http://www.carms.ca/
2. Gale D, Shapley LS (1962) College admissions and the stability of marriage. Am Math Mon 69:9–15
3. Gale D, Sotomayor M (1985) Some remarks on the stable matching problem. Discret Appl Math 11:223–232
4. Gusfield D, Irving RW (1989) The stable marriage problem: structure and algorithms. MIT Press, Boston
5. Halldórsson MM, Irving RW, Iwama K, Manlove DF, Miyazaki S, Morita Y, Scott S (2003) Approximability results for stable marriage problems with ties. Theor Comput Sci 306:431–447
6. Halldórsson MM, Iwama Ka, Miyazaki S, Yanagisawa H (2004) Randomized approximation of the stable marriage problem. Theor Comput Sci 325(3):439–465
7. Halldórsson MM, Iwama Ka, Miyazaki S, Yanagisawa H (2007) Improved approximation results of the stable marriage problem. ACM Trans Algorithms 3(3):Article No. 30
8. Huang C-C, Kavitha T (2014) An improved approximation algorithm for the stable marriage problem with one-sided ties. In: Proceedings of the IPCO 2014, Bonn. LNCS, vol 8494, pp 297–308
9. Irving RW (1994) Stable marriage and indifference. Discret Appl Math 48:261–272
10. Irving RW (1998) Matching medical students to pairs of hospitals: a new variation on a well-known theme. In: Proceedings of the ESA 1998, Venice. LNCS, vol 1461, pp 381–392
11. Irving RW, Manlove DF, Scott S (2000) The hospitals/residents problem with ties. In: Proceedings of the SWAT 2000, Bergen. LNCS, vol 1851, pp 259–271

12. Irving RW, Manlove DF, O'Malley G (2009) Stable marriage with ties and bounded length preference lists. J Discret Algorithms 7(2):213–219

13. Iwama K, Manlove DF, Miyazaki S, Morita Y (1999) Stable marriage with incomplete lists and ties. In: Proceedings of the ICALP 1999, Prague. LNCS, vol 1644, pp 443–452

14. Iwama K, Miyazaki S, Yamauchi N (2007) A 1.875-approximation algorithm for the stable marriage problem. In: Proceedings of the SODA 2007, New Orleans, pp 288–297

15. K. Iwama, Miyazaki S, Yanagisawa H (2014) A 25/17-approximation algorithm for the stable marriage problem with one-sided ties. Algorithmica 68:758–775

16. Japanese Resident Matching Program (JRMP), http://www.jrmp.jp/

17. Kavitha T, Mehlhorn K, Michail D, Paluch KE (2007) Strongly stable matchings in time $O(nm)$ and extension to the hospitals-residents problem. ACM Trans Algorithms 3(2):Article No. 15

18. Király Z (2011) Better and simpler approximation algorithms for the stable marriage problem. Algorithmica 60(1):3–20

19. Király Z (2013) Linear time local approximation algorithm for maximum stable marriage. MDPI Algorithms 6(3):471–484

20. Manlove DF (1999) Stable marriage with ties and unacceptable partners. Technical Report no. TR-1999-29 of the Computing Science Department of Glasgow University

21. Manlove DF, Irving RW, Iwama K, Miyazaki S, Morita Y (2002) Hard variants of stable marriage. Theor Comput Sci 276(1–2):261–279

22. McDermid EJ (2009) A 3/2-approximation algorithm for general stable marriage. In: Proceedings of the ICALP, Rhodes. LNCS, vol 5555, pp 689–700

23. Paluch KE (2014) Faster and simpler approximation of stable matchings. Algorithms 7(2):189–202

24. Yanagisawa H (2007) Approximation algorithms for stable marriage problems. Ph.D. Thesis, Kyoto University

Stable Partition Problem

Katarína Cechlárová
Faculty of Science, Institute of Mathematics, P. J. Šafárik University, Košice, Slovakia

Keywords

Coalition formation; Hedonic games; Stability

Years and Authors of Summarized Original Work

2002; Cechlárová, Hajduková
2004; Ballester

Problem Definition

Let N be a finite set of players; a nonempty subset of N is called a coalition. Each player $i \in N$ has a preference relation $\succeq_i$ (complete, reflexive, and transitive) over all the coalitions that contain i. Notation $S \succeq_i T$ means that player i *weakly prefers* coalition S to coalition T; if $S \succeq_i T$ and not $T \succeq_i S$, then player i *strictly prefers* S to T, denoted by $S \succ_i T$. If $S \succeq_i T$ and $T \succeq_i S$, then player i *is indifferent between* coalitions S and T (there is a *tie* in his preference list). Player i has strict preferences if her preference list contains no ties. There are several possible ways of representing preferences, but it is usually supposed that preference relations can be evaluated in polynomial time.

An instance I of the stable partition problem (or coalition formation game, or hedonic game) is given by the set of players and their preferences.

A partition Π is a collection of disjoint coalitions whose union equals N. It is supposed that each participant's appreciation of a coalition structure only depends on the coalition $\Pi(i)$ she is a member and not on the composition of other coalitions. Of interest are partitions that fulfill some kind of stability requirements.

We say that a coalition $S \subseteq N$ *strongly blocks* a partition Π, if each player $i \in S$ strictly prefers S to $\Pi(i)$, and a coalition $S \subseteq N$ *weakly blocks* a partition Π, if each player $i \in S$ weakly prefers S to $\Pi(i)$ and there exists at least one player $j \in S$ who strictly prefers S to $\Pi(j)$. Partition Π is:

- *Individually stable* if each player i weakly prefers $\Pi(i)$ to $\{i\}$;
- *Nash stable (NS)* if each player i weakly prefers $\Pi(i)$ to $X \cup \{i\}$ for each $X \in \Pi \cup \emptyset$;

- *Individually stable (IS)* if whenever a player i strictly prefers $X \cup \{i\}$ to $\Pi(i)$ for some $X \in \Pi$, then $X \succ_j X \cup \{i\}$ for at least one player $j \in X$;
- *Contractually individually stable (CIS)* if whenever a player i strictly prefers $X \cup \{i\}$ to $\Pi(i)$ for some $X \in \Pi$, then $X \succ_j X \cup \{i\}$ for at least one player $j \in X$ or $\Pi(i) \succ_j \Pi(i) \setminus \{i\}$ for at least one player $i \in \Pi(i)$;
- *Core stable* if it admits no blocking coalition;
- *Strictly core stable* if it admits no weakly blocking coalition;

Most of these definitions were introduced in [3] and [4] where also some sufficient conditions for the existence of stable partitions were formulated. An overview of the implications between these definitions can be found in [1]. The following problems have been studied algorithmically for various stability notions $\mathbb{S}$:

- $\mathbb{S}$-STABILITY-VERIFICATION: Given I and a partition Π, is Π a $\mathbb{S}$-stable partition?
- $\mathbb{S}$-STABILITY-EXISTENCE: Given I, does a $\mathbb{S}$-stable partition exist?
- $\mathbb{S}$-STABILITY-CONSTRUCTION: Given I, construct a $\mathbb{S}$-stable partition.
- $\mathbb{S}$-STABILITY-STRUCTURE: Describe the structure of $\mathbb{S}$-stable partitions for a given I.

The computational complexity of these problems depends on the specification of the preference relation in the input.

An Important special case of the stable partition problem arises when each coalition can contain at most two players. This is known under the name the *Stable Matching Problem* and is treated in detail in [14]; see also references in the entry Stable Marriage.

Key Results

Trivial Encoding

In the *trivial encoding*, each player lists all her individually rational coalitions (i.e., those that player i weakly prefers to coalition $\{i\}$).

Theorem 1 *Under the trivial encoding, the* STABILITY-VERIFICATION *problem is polynomially solvable for any stability definition.* STABILITY-EXISTENCE *is NP-complete for IR, NS, core, and strict core [2].* CORE-STABILITY-EXISTENCE *is NP-complete [2], even in the case when each player i has her preference list of the form $C_1(i) \succ_i C_2(i) \succ_i \{i\}$ and all acceptable coalitions have size three [11].*

As the trivial encoding may be of exponential size in the number of players, more succinct preference representations have been studied.

Anonymous Preferences

Players have anonymous preferences if all coalitions of the same size are tied, i.e., players do not care about the actual content of the coalitions, only about their sizes.

Theorem 2 *Under anonymous preferences, the* CORE-STABILITY-VERIFICATION *problem is polynomially solvable and* CORE-STABILITY-EXISTENCE *is NP-complete [2].*

Additive Preferences

In an additive hedonic game, each player i has a real-valued function $v_i : N \to \mathbb{R}$ and $S \succ_i T$ if and only if $\sum_{j \in S} v_i(j) > \sum_{j \in T} v_i(j)$.

Theorem 3 *In additive hedonic games,* STABILITY-VERIFICATION *is co-NP-complete in the strong sense for core and strict core [1, 17].* CORE-STABILITY-EXISTENCE *and* STRICT-CORE-STABILITY-EXISTENCE *are strongly NP-hard [18] even in the symmetric case [1].* INDIVIDUAL-STABILITY-EXISTENCE *and* NASH-STABILITY-EXISTENCE *are strongly NP-complete [1, 18]. Moreover,* CORE-STABILITY-EXISTENCE *is $\sum_2^p$-complete [19].*

Special cases of additive preferences arise if $v_i(j) \in \{-1, |N|\}$ for each $i, j \in N$ (friend-oriented case) or $v_i(j) \in \{1, -|N|\}$ for each $i, j \in N$ (enemy-oriented case). Under friend-oriented as well as under enemy-oriented preferences, a core-stable partition always exists [12], however, the following assertion holds.

Theorem 4 ([12]) *Under enemy-oriented preferences,* CORE-STABILITY-VERIFICATION *and* CORE-STABILITY-CONSTRUCTION *are strongly NP-complete and NP-hard, respectively.*

Preferences Derived from the Best and/or Worst Player

Suppose that each player i linearly orders only individual players or, more precisely, a subset of them – these are *acceptable* for i.

Preferences over players are extended to preferences over coalitions on the basis of the best or the worst player in the coalition as follows:

B-preferences – a player orders coalitions first on the basis of the most preferred member of the coalition, and if those are equal or tied, the coalition with smaller cardinality is preferred;

W-preferences – a player orders coalitions on the basis of the least preferred member of the coalition;

BW-preferences – a player orders coalitions first on the basis of the best member of the coalition, and if those are equal or tied, the coalition with a more preferred worst member is preferred.

In this case, preferences are considered *strict*, if the preferences over individuals are strict, and they are called *dichotomous* if all acceptable participants are tied in each preference list.

Theorem 5 *Under B-preferences,* STABILITY-VERIFICATION *is polynomial for core and strict core. A strict core and a core stable partition always exist if preferences over players are strict [9]. However, if preferences over players contain ties,* STABILITY-EXISTENCE *for core and strict core is NP-complete [6]. In the dichotomous case, a core stable partition can be constructed in polynomial time, but* STRICT-CORE-STABILITY-EXISTENCE *is NP-complete [5].*

Let us remark here that in the case of strict preferences, a strict core stable partition can be found by the famous Top Trading Cycles algorithm [9, 20].

The stable partition problem under W-preferences was studied in [7] and many features similar to the Stable Roommates

Problem [14] were described. First, if a blocking coalition exists, then there is a blocking coalition of size at most 2. Hence, CORE-STABILITY-VERIFICATION is polynomial. CORE-STABILITY-EXISTENCE and CORE-STABILITY-CONSTRUCTION are polynomial in the strict preferences case, which can be shown using an extension of Irving's Stable Roommates Algorithm (discussed in detail in [14]). This algorithm can also be used to derive some results for CORE-STABILITY-STRUCTURE. In the case of ties, CORE-STABILITY-EXISTENCE is NP-complete.

Under BW preferences, in the strict preferences case, a core partition always exists and one can be obtained by the Top Trading Cycles algorithm, but STRICT-CORE-STABILITY-EXISTENCE is NP-hard. If preferences contain ties, CORE-STABILITY-EXISTENCE is NP-hard too [8]. CORE-STABILITY-VERIFICATION remains open.

Applications

Stable partitions arise in various economic and game theoretical models. They appear in the study of countries formation [10] and in multi-agent coordination scenarios and social networking services [13]. Stability is also desired in barter exchange economies with discrete commodities [20, 21], including exchange of kidneys for transplantations [5, 16]. Notice that in case when the cooperation of players consists in the exchange of some items within one partition set, the exchange cycle has also to be specified.

Open Problems

Due to the great number of variants, a lot of open problems exists. In almost all cases, STABILITY-STRUCTURE is not satisfactorily solved. For instances with no stable partition, one may seek one that minimizes the number of players who have an incentive to deviate. Parallel algorithms were also not studied.

Experimental Results

Stochastic local search algorithms for CORE-STABILITY-VERIFICATION in the additive preferences case were reported in [15].

Cross-References

▶ Stable Marriage

Recommended Reading

1. Aziz H, Brandt F, Seeding HG (2013) Computing desirable partitions in additively separable hedonic games. Artif Intell 195:316–334
2. Ballester C (2004) NP-completeness in hedonic games. Games Econ Behav 49(1):1–30
3. Banerjee S, Konishi H, Sönmez T (2001) Core in a simple coalition formation game. Soc Choice Welf 18:135–153
4. Bogomolnaia A, Jackson MO (2002) The stability of hedonic coalition structures. Games Econ Behav 38(2):201–230
5. Cechlárová K, Fleiner T, Manlove D (2005) The kidney exchange game. In: Zadnik-Stirn L, Drobne S (eds) Proceedings of SOR'05, Slovenia, pp 77–83
6. Cechlárová K, Hajduková J (2002) Computational complexity of stable partitions with B-preferences. Int J Game Theory 31(3):353–364
7. Cechlárová K, Hajduková J (2004) Stable partitions with W-preferences. Discret Appl Math 138(3):333–347
8. Cechlárová K, Hajduková J (2004) Stability of partitions under WB-preferences and BW-preferences. Int J Inf Technol Decis Making 3(4):605–614. Special Issue on Computational Finance and Economics
9. Cechlárová K, Romero-Medina A (2001) Stability in coalition formation games. Int J Game Theory 29:487–494
10. Cechlárová K, Dahm M, Lacko V (2001) Efficiency and stability in a discrete model of country formation. J Glob Optim 20(3–4):239–256
11. Deineko VG, Woeginger GJ (2013) Two hardness results for core stability in additive hedonic coalition formation games. Discret Appl Math 161:1837–1842
12. Dimitrov D, Borm P, Hendrickx R, Sung SCh (2006) Simple priorities and core stability in hedonic games. Soc Choice Welf 26(2):421–433
13. Elkind E, Wooldridge M (2009) Hedonic coalition nets. In: Proceedings of the 8th international conference on autonomous agents and multiagent systems (AAMAS 2009), Budapest, pp 417–424
14. Gusfield D, Irving RW (1989) The stable marriage problem. Structure and algorithms. MIT, Cambridge
15. Keinänen H (2010) Stochastic local search for core membership checking in hedonic games. LNCS 6220:56–70
16. Roth A, Sönmez T, Ünver U (2004) Kidney exchange. Q J Econ 119:457–488
17. Sung SCh, Dimitrov D (2007) On core membership testing for hedonic coalition formation games. Oper Res Lett 35:155–158
18. Sung SCh, Dimitrov D (2010) Computational complexity in additive hedonic games. Eur J Oper Res 203:635–639
19. Woeginger GJ (2013) A hardness result for core stability in additive hedonic games. Math Soc Sci 65:101-104
20. Shapley L, Scarf H (1974) On cores and indivisibility. J Math Econ 1:23–37
21. Yuan Y (1996) Residence exchange wanted: a stable residence exchange problem. Eur J Oper Res 90:536–546

Stackelberg Games: The Price of Optimum

Alexis Kaporis[1] and Paul (Pavlos) Spirakis[2,3,4]
[1]Department of Information and Communication Systems Engineering, University of the Aegean, Karlovasi, Samos, Greece
[2]Computer Engineering and Informatics, Research and Academic Computer Technology Institute, Patras University, Patras, Greece
[3]Computer Science, University of Liverpool, Liverpool, UK
[4]Computer Technology Institute (CTI), Patras, Greece

Keywords

Coordination ratio; Cournot game

Years and Authors of Summarized Original Work

2006; Kaporis, Spirakis

Problem Definition

Stackelberg games [15] may model the interplay among an authority and rational individuals that selfishly demand resources on a large-scale network. In such a game, the authority *(Leader)* of the network is modeled by a distinguished player. The selfish users *(Followers)* are modeled by the remaining players.

It is well known that selfish behavior may yield a *Nash Equilibrium* with cost arbitrarily higher than the optimum one, yielding unbounded *Coordination Ratio* or *Price of Anarchy (PoA)* [7,13]. Leader plays his strategy first assigning a portion of the total demand to some resources of the network. Followers observe and react selfishly assigning their demand to the most appealing resources. Leader aims to drive the system to an a posteriori Nash equilibrium with cost close to the overall optimum one [4, 6, 8, 10]. Leader may also be eager for his own rather than system's performance [2,3].

A Stackelberg game can be seen as a special, and easy [6] to implement, case of *Mechanism Design*. It avoids the complexities of either computing taxes or assigning prices, or even designing the network at hand [9]. However, a central authority capable to control the overall demand on the resources of a network may be unrealistic in networks which evolute and operate under the effect of many and diversing economic entities. A realistic way [4] to act centrally even in large nets could be via *Virtual Private Networks (VPNs)* [1]. Another flexible way is to combine such strategies with *Tolls* [5,14].

A dictator controlling the entire demand optimally on the resources surely yields PoA = 1. On the other hand, rational users do prefer a liberal world to live. Thus, it is important to compute the optimal Leader strategy which controls the *minimum* of the resources *(Price of Optimum)* and yields PoA $= 1$. What is the complexity of computing the Price of Optimum? This is not trivial to answer, since the Price of Optimum depends crucially on computing an optimal Leader strategy. In particular, [6] proved that computing the optimal Leader strategy is hard.

The central result of this lemma is Theorem 5. It says that on nonatomic flows and arbitrary $s - t$ networks and latencies, computing the minimum portion of flow and Leader's optimal strategy sufficient to induce PoA $= 1$ is easy [10].

Problem $(G(V, E), s, t \in V, r)$ INPUT: Graph $G, \forall e \in E$ latency ℓ_e, flow r, a source-destination pair (s, t) of vertices in V.
OUTPUT: (i) The minimum portion α_G of the total flow r sufficient for an optimal Stackelberg strategy to induce the optimum on G. (ii) The optimal Stackelberg strategy.

Models and Notations

Consider a graph $G(V, E)$ with parallel edges allowed. A number of rational and selfish users wish to route from a given source s to a destination node t an amount of flow r. Alternatively, consider a partition of users in k commodities, where user(s) in commodity i wish to route flow r_i through a source-destination pair (s_i, t_i), for each $i = 1, \ldots, k$. Each edge $e \in E$ is associated to a latency function $\ell_e()$, positive, differentiable, and strictly increasing on the flow traversing it.

Nonatomic Flows

There are infinitely many users, each routing his/her infinitesimally small amount of the total flow r_i from a given source s_i to a destination vertex t_i in graph $G(V, E)$. A flow f is an assignment of jobs f_e on each edge $e \in E$. The cost of the injected flow f_e (satisfying the standard constraints of the corresponding network-flow problem) that traverses edge $e \in E$ equals; $c_e(f_e) = f_e \times \ell_e(f_e)$. It is assumed that on each edge e the cost is convex with respect to the injected flow f_e. The overall system's cost is the sum $\sum_{e \in E} f_e \times \ell_e(f_e)$ of all edge costs in G. Let $f_{\mathcal{P}}$ the amount of flow traversing the $s_i - t_i$ path $\mathcal{P}$. The latency $\ell_{\mathcal{P}}(f)$ of $s_i - t_i$ path $\mathcal{P}$ is the sum $\sum_{e \in \mathcal{P}} \ell_e(f_e)$ of latencies per edge $e \in \mathcal{P}$. The cost $C_{\mathcal{P}}(f)$ of $s_i - t_i$ path $\mathcal{P}$ equals the flow $f_{\mathcal{P}}$ traversing it multiplied by path latency $\ell_{\mathcal{P}}(f)$. That is, $C_{\mathcal{P}}(f) = f_{\mathcal{P}} \times \sum_{e \in \mathcal{P}} \ell_e(f_e)$. In a Nash equilibrium, all $s_i - t_i$ paths traversed by nonatomic users in part i have a common latency, which is at most the latency of any untraversed

$s_i - t_i$ path. More formally, for any part i and any pair $\mathcal{P}_1$, $\mathcal{P}_2$ of $s_i - t_i$ paths, if $f_{\mathcal{P}_1} > 0$ then $\ell_{\mathcal{P}_1}(f) \leq \ell_{\mathcal{P}_2}(f)$. By the convexity of edge costs, the Nash equilibrium is unique and computable in polynomial time given a floating-point precision. Also computable is the unique *Optimum* assignment O of flow, assigning flow o_e on each $e \in E$ and minimizing the overall cost $\sum_{e \in E} o_e \ell_e(o_e)$. However, not all optimally traversed $s_i - t_i$ paths experience the same latency. In particular, users traversing paths with high latency have incentive to reroute toward more speedy paths. Therefore, the optimal assignment is unstable on selfish behavior.

A Leader dictates a *weak* Stackelberg strategy if on each commodity $i = 1, \ldots, k$ controls a fixed α portion of flow r_i, $\alpha \in [0, 1]$. A *strong* Stackelberg strategy is more flexible, since Leader may control $\alpha_i r_i$ flow in commodity i such that $\sum_{i=1}^{k} \alpha_i = \alpha$. Let a Leader dictating flow s_e on edge $e \in E$. The a posteriori latency $\tilde{\ell}_e(n_e)$ of edge e, with respect to the induced flow n_e by the selfish users, equals $\tilde{\ell}_e(n_e) = \ell_e(n_e + s_e)$. In the a posteriori Nash equilibrium, all $s_i - t_i$ paths traversed by the free selfish users in commodity i have a common latency, which is at most the latency of any selfishly untraversed path, and its cost is $\sum_{e \in E} (n_e + s_e) \times \tilde{\ell}_e(n_e)$.

Atomic Splittable Flows

There is a finite number of atomic users $1, \ldots, k$. Each user i is responsible for routing a non-negligible flow-amount r_i from a given source s_i to a destination vertex t_i in graph G. In turn, each flow-amount r_i consists of infinitesimally small jobs.

Let flow f assigning jobs f_e on each edge $e \in E$. Each edge flow f_e is the sum of partial flows $f_e^1, \ldots, f_e^k$ injected by the corresponding users $1, \ldots, k$. That is, $f_e = f_e^1 + \cdots + f_e^k$. As in the model above, the latency on a given $s_i - t_i$ path $\mathcal{P}$ is the sum $\sum_{e \in \mathcal{P}} \ell_e(f_e)$ of latencies per edge $e \in \mathcal{P}$. Let $f_{\mathcal{P}}^i$ be the flow that user i

ships through an $s_i - t_i$ path $\mathcal{P}$. The cost of user i on a given $s_i - t_i$ path $\mathcal{P}$ is analogous to her path flow $f_{\mathcal{P}}^i$ routed via $\mathcal{P}$ times the total path latency $\sum_{e \in \mathcal{P}} \ell_e(f_e)$. That is, the path cost equals $f_{\mathcal{P}}^i \times \sum_{e \in \mathcal{P}} \ell_e(f_e)$. The overall cost $C_i(f)$ of user i is the sum of the corresponding path costs of all $s_i - t_i$ paths.

In a Nash equilibrium no user i can improve his cost $C_i(f)$ by rerouting, given that any user $j \neq i$ keeps his routing fixed. Since each atomic user minimizes its cost, if the game consists of only one user, then the cost of the Nash equilibrium coincides to the optimal one.

In a Stackelberg game, a distinguished atomic Leader player controls flow r_0 and plays first assigning flow s_e on edge $e \in E$. The a posteriori latency $\tilde{\ell}_e(x)$ of edge e on induced flow x equals $\tilde{\ell}_e(x) = \ell_e(x + s_e)$. Intuitively, after Leader's move, the induced selfish play of the k atomic users is equivalent to atomic splittable flows on a graph where each initial edge latency ℓ_e has been mapped to $\tilde{\ell}_e$. In game parlance, each atomic user $i \in \{1, \ldots, k\}$, having *fixed* Leader's strategy, computes his *best reply* against all other atomic users $\{1, \ldots, k\} \setminus \{i\}$. If n_e is the induced Nash flow on edge e, this yields total cost $\sum_{e \in E} (n_e + s_e) \times \tilde{\ell}_e(n_e)$.

Atomic Unsplittable Flows

The users are finite $1, \ldots, k$ and user i is allowed to send his non-negligible job r_i only on a *single* path. Despite this restriction, all definitions given in atomic splittable model remain the same.

Key Results

Let us see first the case of atomic splittable flows, on parallel M/M/1 links with different speeds connecting a given source-destination pair of vertices.

Theorem 1 (Korilis, Lazar, Orda [6]) *The Leader can enforce in polynomial time the network optimum if his/her controls flow r_0 exceeding a critical value $\underline{r}^0$.*

In the sequel, we focus on nonatomic flows on $s - t$ graphs with parallel links. In [6] primarily were studied cases that Leader's flow cannot induce network's optimum and was shown that an optimal Stackelberg strategy is easy to compute. In this vain, if $s - t$ parallel link instances are restricted to ones with linear latencies of equal slope, then an optimal strategy is easy [4].

Theorem 2 (Kaporis, Spirakis [4]) *The optimal Leader strategy can be computed in polynomial time on any instance (G, r, α) where G is an $s - t$ graph with parallel links and linear latencies of equal slope.*

Another positive result is that the optimal strategy can be approximated within $(1 + \epsilon)$ in polynomial time, given that link latencies are polynomials with nonnegative coefficients.

Theorem 3 (Kumar, Marathe [8]) *There is a fully polynomial approximate Stackelberg scheme that runs in poly $\left(m, \frac{1}{\epsilon}\right)$ time and outputs a strategy with cost $(1 + \epsilon)$ within the optimum strategy.*

For parallel link $s - t$ graphs with arbitrary latencies more can be achieved: in polynomial time a "threshold" value α_G is computed, sufficient for the Leader's portion to induce the optimum. The complexity of computing optimal strategies changes in a dramatic way around the critical value α_G from "hard" to "easy" (G, r, α) Stackelberg scheduling instances. Call α_G as the *Price of Optimum* for graph G.

Theorem 4 (Kaporis, Spirakis [4]) *On an input $s - t$ parallel link graph G with arbitrary strictly increasing latencies, the minimum portion α_G sufficient for a Leader to induce the optimum, as well as his/her optimal strategy, can be computed in polynomial time.*

As a conclusion, the Price of Optimum α_G essentially captures the hardness of instances (G, r, α). Since, for Stackelberg scheduling instances $(G, r, \alpha \geq \alpha_G)$, the optimal Leader strategy yields PoA $= 1$ and it is computed as hard as in P, while for $(G, r, \alpha < \alpha_G)$ the optimal strategy yields PoA < 1 and it is as easy as NP [10].

The results above are limited to parallel links connecting a given $s - t$ pair of vertices. Is it possible to efficiently compute the Price of Optimum for nonatomic flows on arbitrary graphs? This is not trivial to settle. Not only because it relies on computing an optimal Stackelberg strategy, which is hard to tackle [10], but also because Proposition B.3.1 in [11] ruled out previously known performance guarantees for Stackelberg strategies on general nets.

The central result of this lemma is presented below and completely resolves this question (extending Theorem 4).

Theorem 5 (Kaporis, Spirakis [4]) *On arbitrary $s - t$ graphs G with arbitrary latencies, the minimum portion α_G sufficient for a Leader to induce the optimum, as well as her optimal strategy, can be computed in polynomial time.*

Example

Consider the optimum assignment O of flow r that wishes to travel from source vertex s to sink t. O assigns flow o_e incurring latency $\ell_e(o_e)$ per edge $e \in G$. Let $\mathcal{P}_{s \to t}$ the set of all $s - t$ paths. The *shortest paths* in $\mathcal{P}_{s \to t}$ with respect to costs $\ell_e(o_e)$ per edge $e \in G$ can be computed in polynomial time. That is, the paths that given flow assignment O achieve path latency:

$$\min_{P \in \mathcal{P}_{s \to t}} \left(\sum_{e \in P} \ell_e(o_e) \right), \text{ i.e., minimize their path}$$

latency. It is crucial to observe that if we want the *induced* Nash assignment by the Stackelberg strategy to attain the optimum cost, then these shortest paths are *the only choice* for selfish users that are eager to travel from s to t. Furthermore, the uniqueness of the optimum assignment O determines the minimum part of flow which can be selfishly scheduled on these shortest paths. Observe that any flow assigned by O on a nonshortest $s - t$ path has incentive to opt for a shortest one. Then a Stackelberg strategy *must* freeze the flow on all non-shortest $s - t$ paths.

In particular, the idea sketched above achieves coordination ratio 1 on the graph in Fig. 1. On this graph Roughgarden proved that $\frac{1}{\alpha} \times$ (optimum

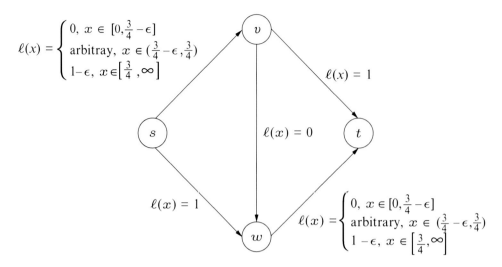

$$\ell(x) = \begin{cases} 0, \ x \in [0, \frac{3}{4} - \epsilon] \\ \text{arbitray}, \ x \in (\frac{3}{4} - \epsilon, \frac{3}{4}) \\ 1 - \epsilon, \ x \in [\frac{3}{4}, \infty] \end{cases}$$

$$\ell(x) = 1$$

$$\ell(x) = 0$$

$$\ell(x) = 1$$

$$\ell(x) = \begin{cases} 0, \ x \in [0, \frac{3}{4} - \epsilon] \\ \text{arbitrary}, \ x \in (\frac{3}{4} - \epsilon, \frac{3}{4}) \\ 1 - \epsilon, \ x \in [\frac{3}{4}, \infty] \end{cases}$$

Stackelberg Games: The Price of Optimum, Fig. 1 A bad example for Stackelberg routing

cost) guarantee is *not* possible for general (s, t)-networks, Appendix B.3 in [11]. The optimal edge flows are $(r = 1)$:

$$o_{s \to v} = \frac{3}{4} - \epsilon, o_{s \to w} = \frac{1}{4} + \epsilon, o_{v \to w} = \frac{1}{2} - 2\epsilon,$$
$$o_{v \to t} = \frac{1}{4} + \epsilon, o_{w \to t} = \frac{3}{4} - \epsilon$$

The shortest path $P_0 \in \mathcal{P}$ with respect to the optimum O is $P_0 = s \to v \to w \to t$ (see [11] pp. 143, 5th-3th lines before the end) and its flow is $f_{P_0} = \frac{1}{2} - 2\epsilon$. The non-shortest paths are $P_1 = s \to v \to t$ and $P_2 = s \to w \to t$ with corresponding optimal flows: $f_{P_1} = \frac{1}{4} + \epsilon$ and $f_{P_2} = \frac{1}{4} + \epsilon$. Thus, the Price of Optimum is

$$f_{P_1} + f_{P_2} = \frac{1}{2} + 2\epsilon = r - f_{P_0}$$

Applications

Stackelberg strategies are widely applicable in networking [6], see also Section 6.7 in [12].

Open Problems

It is important to extend the above results on atomic unsplittable flows.

Cross-References

▶ Algorithmic Mechanism Design
▶ Best Response Algorithms for Selfish Routing
▶ Facility Location
▶ Non-approximability of Bimatrix Nash Equilibria
▶ Price of Anarchy
▶ Selfish Unsplittable Flows: Algorithms for Pure Equilibria

Recommended Reading

1. Birman K (1997) Building secure and reliable network applications. Manning, Greenwich
2. Douligeris C, Mazumdar R (2006) Multilevel flow control of queues. In: Johns Hopkins conference on information sciences, Baltimore, 22–24 Mar 1989
3. Economides A, Silvester, J (1990) Priority load sharing: an approach using Stackelberg games. In: 28th annual Allerton conference on communications, control and computing, Monticello
4. Kaporis AC, Spirakis PG (2009) The price of optimum in Stackelberg games on arbitrary single commodity networks and latency functions. Theor Comput Sci **410**(8–10):745–755
5. Karakostas G, Kolliopoulos SG (2009) Stackelberg strategies for selfish routing in general multicommodity networks. Algorithmica **53**(1):132–153

6. Korilis YA, Lazar AA, Orda A (1997) Achieving network optima using stackelberg routing strategies. IEEE/ACM Trans Netw 5(1):161–173
7. Koutsoupias E, Papadimitriou CH (2009) Worst-case equilibria. Comput Sci Rev **3**(2):65–69
8. Kumar VSA, Marathe MV (2002) Improved results for Stackelberg scheduling strategies. In: 29th international colloquium, automata, languages and programming, Málaga. LNCS. Springer, pp 776–787
9. Roughgarden T (2001) Designing networks for selfish users is hard. In: 42nd IEEE annual symposium of foundations of computer science, Las Vegas, pp 472–481
10. Roughgarden T (2004) Stackelberg scheduling strategies. SIAM J Comput **33**(2):332–350
11. Roughgarden T 2002 Selfish routing. Dissertation, Cornell University. http://theory.stanford.edu/~tim/
12. Roughgarden T (2005) Selfish routing and the price of anarchy. MIT, Cambridge (2005)
13. Roughgarden T, Tardos É (2002) How bad is selfish routing? J ACM **49**(2):236–259
14. Swamy C (2007) The effectiveness of Stackelberg strategies and tolls for network congestion games. In: ACM-SIAM symposium on discrete algorithms, Philadelphia
15. von Stackelberg H (1934) Marktform und Gleichgewicht. Springer, Vienna

Staged Assembly

Andrew Winslow
Department of Computer Science, Tufts University, Medford, MA, USA

Keywords

Context-free grammars; DNA computing; Jigsaw; Natural computing; Polyomino; Shape decomposition

Years and Authors of Summarized Original Work

2008; Demaine, Demaine, Fekete, Ishaque, Rafalin, Schweller, Souvaine
2013; Demaine, Eisenstat, Ishaque, Winslow
2013; Winslow

Problem Definition

Algorithmic self-assembly is concerned with hands-off assembly of complex structures by mixing collections of simple particles that aggregate according to local rules. Staged self-assembly utilizes sequences of mixings to reduce the number of particle types used. The standard model of staged self-assembly builds on the abstract Tile Assembly Model (aTAM) of Winfree [7], where each particle is a non-rotatable unit square *tile* with a labeled *glue* on each side. Tiles attach to other tiles edgewise via glues of the same label, forming polyomino-shaped aggregates called *assemblies*.

In the simplest model, a pair of assemblies (of which single tiles are a special case) can attach via a single matching glue. In a more general model, a pair of assemblies can attach if they share a total of $\tau \in \mathbb{N}$ glues. The parameter τ is called the *temperature* of the system.

The self-assembly process is carried out by combining an infinite number of copies of a collection of *reagent assemblies* in a *bin*, where they attach in every possible way. The subset of the resulting assemblies that cannot attach to any other assemblies define the *product assemblies* of the mixing, i.e., the set of assemblies that remain once the assembly process is complete. A system consisting of a single bin with single-tile reagent assemblies is a *hierarchical* [2], *two-handed* [1], or *polyomino* [5] self-assembly system.

In a *staged self-assembly system* [3], the products of one bin can be used as the reagents of other bin (see Fig. 1). The directed acyclic graph describing the relationship mixings is called the *mix graph* of the system. An initial set of mixings each have a single tile as the only product assembly and no reagent assemblies.

Objectives In general, the goal is to design a system with a mixing containing a single product assembly of a desired polyomino shape while minimizing the size of some aspect of the system. Several aspects are considered, including the number of distinct tiles (*tile complexity*), number of edges of the mix graph (*mix graph complexity*),

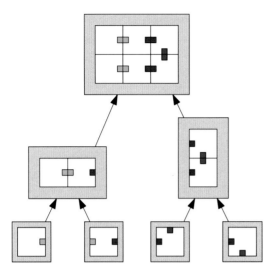

Staged Assembly, Fig. 1 A staged self-assembly system. Each bin (*blue box*) contains the product assemblies of the bin. The reagent assemblies of a bin are the products of other bins (*incoming arrows*)

width of the mix graph (*bin complexity*), height of the mix graph (*stage complexity*), and temperature of the system. The computational complexity of finding an optimal system for an input shape under some measure of system complexity is also considered. In some cases, the desired polyomino shape also has each cell labeled, and the goal is to construct a given labeled shape using labeled tiles.

Problem 1 (Smallest Staged Self-Assembly System)

INPUT: *A labeled polyomino P.*

OUTPUT: *A staged self-assembly system containing a bin with a single product assembly with labeled shape P that is minimum in some measure.*

Key Results

We describe the results by increasing generality of the shapes assembled.

Lines

In the most constrained case, the input polyomino is an unlabeled $1 \times n$ polyomino (a *line*). Lines can be assembled by $\tau = 1$ systems using $O(1)$ tile types, $O(1)$ bins, and $O(\log n)$ stages and mix graph edges. The idea is to repeatedly double the length of a line assembly as proved by Demaine et al. [3].

Demaine, Eisenstat, Ishaque, and Winslow [4] prove that the case of labeled lines is roughly equivalent to the problem of finding the smallest context-free grammar that is a single string consisting of the labels of the line read from left to right. In particular, any context-free grammar $\mathcal{G}$ with $|\mathcal{G}|$ rules and deriving a single string σ can be converted into a $\tau = 1$ staged self-assembly system $\mathcal{S}$ assembling a line with left-to-right label string σ, where the number of edges in the mix graph of $\mathcal{S}$ is $O(|\mathcal{G}|)$. The complexity of the smallest staged self-assembly system where the input polyomino is a labeled line and the system has an upper limit on the number of glue types appearing on the tiles was proven to be NP-hard [4].

Squares

Demaine et al. [3] prove that unlabeled $n \times n$ squares are possible with a $\tau = 1$ staged systems containing $O(1)$ tile types and bins, $O(\log n)$ stages, and thus a mix graph with $O(\log n)$ stages. The system uses an idea similar to that for assembling lines but in two steps: first assemble $n \times 1$ columns and then combine them to form $n \times 2$, then $n \times 4$, etc., rectangles. This construction uses a *jigsaw* technique to ensure attaching rectangles cannot assemble askew.

Demaine et al. also prove that unlabeled $n \times n$ squares can be assembled using $\tau = 2$ staged systems using $O(1)$ tile types, $O(\sqrt{\log n})$ bins, and $O(\log \log n)$ stages. The approach is to *simulate* known $\tau = 2$ single-bin systems that efficiently assemble squares by constructing *macrotiles*: large assemblies that simulate the behavior of distinct tile types by encoding glue types in geometry on their surfaces. Such macrotiles allow staged systems to trade off tile types for stages by replacing many distinct tile types with an initial sequence of stages that assemble macrotile versions of the tiles.

General Shapes

For general shapes, several different results highlight the trade-offs in complexity enabled by staged assembly. Demaine et al. [3] prove that any unlabeled shape can be assembled by a $\tau = 1$ staged system using $O(1)$ tile types, $O(\log n)$ bins, and a number of stages proportional to the diameter of the dual grid graph of the shape. If the shape is monotone, then a similar system with $O(n)$ bins and $O(\log n)$ stages (increased bins but decreased stages) exists.

If the system is permitted to assemble a scaled version of the input shape, then the system of Soloveichik and Winfree [6] can be simulated with macrotiles, resulting in a $\tau = 2$ staged system with $O(1)$ tile types, $O(K/\log K)$ bins, and $O(\log \log K)$ stages, where K is the Kolmogorov complexity of the shape. For labeled shapes, Winslow [8] proves that any polyomino context-free grammar $\mathcal{G}$ (a generalization of context-free grammars to two dimensions) with $|\mathcal{G}|$ rules deriving a single labeled polyomino P can be converted into a staged system $\mathcal{S}$ assembling a scaled version of P consisting of labeled macrotiles where the number of edges in the mix graph of $\mathcal{S}$ is $O(|\mathcal{G}|)$.

Applications

The theory of algorithmic self-assembly is rooted in the design of nanoscale particle systems, particularly DNA-based systems. For staged self-assembly in particular, the capability of assembling complex shapes using only $O(1)$ tile types is highly desirable in practice, as engineering many tile types with desired glues is often far more challenging than carrying out a sequence of mixings.

Open Problems

The complexity of the smallest staged self-assembly problem where the number of glue types used is unconstrained remains open, both for the case of lines and general shapes. For lines, the problem is known to be in NP and when the number of glues is constrained is NP-complete (both proved in [4]). For general shapes, the problem is only known to be in PSPACE (proved in [9]) and NP-hard when the number of glues is constrained, following from the special case of lines. The complexity of verifying that a staged assembly system produces a given shape also remains open and is only known to lie in PSPACE.

Cross-References

▶ Combinatorial Optimization and Verification in Self-Assembly
▶ Experimental Implementation of Tile Assembly
▶ Patterned Self-Assembly Tile Set Synthesis
▶ Self-Assembly at Temperature 1
▶ Self-Assembly of Squares and Scaled Shapes

Recommended Reading

1. Cannon S, Demaine ED, Demaine ML, Eisenstat S, Patitz MJ, Schweller RT, Summers SM, Winslow A (2013) Two hands are better than one (up to constant factors): self-assembly in the 2HAM vs. aTAM. In: STACS 2013, Kiel, LIPIcs, vol 20. Schloss Dagstuhl–Leibniz-Zentrum fuer Informatik, pp 172–184
2. Chen H, Doty D (2012) Parallelism and time in hierarchical self-assembly. In: Proceedings of the 23rd annual ACM-SIAM symposium on discrete algorithms (SODA), Kyoto, pp 1163–1182
3. Demaine ED, Demaine ML, Fekete SP, Ishaque M, Rafalin E, Schweller RT, Souvaine DL (2008) Staged self-assembly: nanomanufacture of arbitrary shapes with $O(1)$ glues. Nat Comput 7(3):347–370
4. Demaine ED, Eisenstat S, Ishaque M, Winslow A (2013) One-dimensional staged self-assembly. Nat Comput 12(2):247–258
5. Luhrs C (2010) Polyomino-safe DNA self-assembly via block replacement. Nat Comput 9(1):97–109
6. Soloveichik D, Winfree E (2007) Complexity of self-assembled shapes. SIAM J Comput 36(6):1544–1569
7. Winfree E (1998) Algorithmic self-assembly of DNA. PhD thesis, Caltech
8. Winslow A (2013a) Staged self-assembly and polyomino context-free grammars. In: Soloveichik D, Yurke B (eds) DNA 19, Tempe
9. Winslow A (2013b) Staged self-assembly and polyomino context-free grammars. PhD thesis, Tufts University

Statistical Multiple Alignment

István Miklós
Department of Plant Taxonomy and Ecology,
Eötvös Loránd University, Budapest, Hungary

Keywords

Multiple HMM; Statistical alignment; Stochastic modeling of insertions and deletions; Time-continuous Markov models

Years and Authors of Summarized Original Work

2003; Hein, Jensen, Pedersen

Problem Definition

The three main types of mutations modifying biological sequences are insertions, deletions, and substitutions. The simplest model involving these three types of mutations is the so-called Thorne-Kishino-Felsenstein model [16]. In this model, the characters of a sequence evolve independently. Each character in the sequence can be substituted with another character according to a prescribed reversible time-continuous Markov model on the possible characters. Insertion-deletions are modeled as a birth-death process. Insertions can happen at the beginning of the sequence, at the end of the sequence, and between any two characters. It is possible to insert a character into the empty sequence. The time span between two insertions is exponentially distributed with parameter λ, and this parameter does not depend on the context of the position. The newborn character is drawn from the equilibrium distribution of the substitution process. Each character is deleted after an exponentially distributed waiting time with parameter μ, and its two positions where insertions can happen are joined.

The multiple statistical alignment problem is to calculate the likelihood of a set of sequences, namely, what is the probability of observing a set of sequences, given all the necessary parameters that describe the evolution of sequences. Hein, Jensen, and Pedersen were the first who gave an algorithm to calculate this probability [5]. Their algorithm has $O(5^n L^n)$ running time, where n is the number of sequences, and L is the geometric mean of the sequences. The running time has been improved to $O(2^n L^n)$ by Lunter et al. [9].

Notations

Substitutions

A time-continuous Markov model for a substitution process on an alphabet Σ is given by a $k \times k$ rate matrix Q, with constraints

$$q_{i,j} \geq 0 \qquad \forall i \neq \text{æ} \qquad (1)$$

$$\sum_i q_{i,j} = 0 \qquad \forall i \qquad (2)$$

where k is the size of the alphabet. The probability that a character a_i will be character a_j after time t can be calculated with the exponentiation of the rate matrix:

$$P_t(a_j | a_i) = p_{i,j} \qquad \text{where} \qquad (3)$$

$$P = e^{Qt} \qquad (4)$$

The exponentiated matrix can be easily calculated if the rate matrix is diagonalized, namely, if $Q = W \Lambda W^{-1}$, where Λ is a diagonal matrix, then

$$e^{Qt} = W e^{\Lambda t} W^{-1} \qquad (5)$$

$e^{\Lambda t}$ can be easily calculated, since it is a diagonal matrix containing $e^{\lambda_i t}$ in the ith position of the diagonal.

Insertions and Deletions

A Galton-Watson tree is a rooted, edge-weighted binary tree that describes a birth-death process for a time span t. The process starts at the root of the tree, and a split represents a birth. Edge weights represent times, and leaves having a distance from the root smaller than t represent death events. Leaves being t time fare from the root are the individuals that live at time point t.

Insertion-deletion events transforming one sequence into another can be described with Galton-Watson forests: births represent insertions, and deaths represent deletions. Each character of the ancestral sequence has a tree, and there is an additional tree at the beginning of the sequence associated to an imaginary character. This imaginary character cannot die. Roots of the trees are the characters of the ancestral sequence, and each character of the descendant sequence is a leaf of one of the trees, being t time fare from the root. There might be additional leaves that are not associated with characters of the descendant sequences; these are the died out lineages. The forest is aligned such that edges do not cross each other while the characters of the two sequences keep their original order. Each Galton-Watson forest indicates an alignment of the two sequences; see Fig. 1. Given a birth and death process, the probability density of a Galton-Watson tree can be calculated easily. Assuming independence, the probability of a Galton-Watson forest is the product of the probabilities of its trees. The probability of an alignment is the integral of the probabilities of the forests that represent it. Due to independence, it is enough to tell the probability of alignment patterns that might arise as an image of a Galton-Watson tree (see Fig. 1b); the probability of an alignment is the product of the probabilities of its patterns.

In the Thorne-Kishino-Felsenstein model (TKF91 model) [16], both the birth and the death processes are Poisson processes with parameters λ and μ, respectively. The probability of the possible patterns can be found on Fig. 2.

Evolutionary Trees

An evolutionary tree is a leaf-labeled, edge-weighted, rooted binary tree. Labels are the species related by the evolutionary tree, and weights are evolutionary distances. It might happen that the evolutionary changes had different speed at different lineages, and hence the tree is not necessary ultrametric, namely, the root not necessary has the same distance to all leaves. The nodes of an evolutionary tree can be partially ordered such that two nodes are comparable if there is a path from the root to

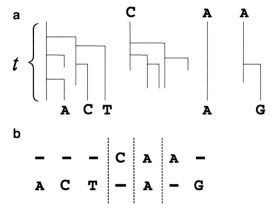

Statistical Multiple Alignment, Fig. 1 (**a**) A Galton-Watson forest representing insertion-deletion events. The first tree starts with an immortal element that is responsible to the insertions at the beginning of the sequence. (**b**) The alignment indicated by the Galton-Watson forest above. Each tree makes a pattern of the alignment; patterns are separated with *dashed lines*

any of the leaves containing the two nodes in question, and in this case the smaller node is the one that is closer to the root on the path. Each node v of an evolutionary tree indicates a subtree that contains v and all the nodes that are greater than v. Hereafter we consider only these subtrees.

Given a set S of l-long sequences over alphabet Σ, a substitution model M on Σ and an evolutionary tree T are labeled by the sequences. The likelihood of the tree is the probability of observing the sequences at the leaves of the tree, given that the substitution process starts at the root of the tree with the equilibrium distribution. This likelihood is denoted by $P(S|T, M)$. The substitution likelihood problem is to calculate the likelihood of the tree.

Let Σ be a finite alphabet and let $S_1 = s_{1,1}s_{1,2} \ldots s_{1,L_1}$, $S_2 = s_{2,1}s_{2,2} \ldots s_{2,L_2}$, $\ldots$ $S_n = s_{n,1}s_{n,2} \ldots s_{n,L_n}$ be sequences over this alphabet. Let a TKF91 model $TKF91$ be given with its parameters: substitution model M, insertion rate λ, and deletion rate μ. Let T be an evolutionary tree labeled by S_1, S_2, $\ldots$ S_n. The multiple statistical alignment problem is to calculate the likelihood of the tree, $P(S_1, S_2, \ldots S_n|T, TKF91)$, given that

S

$$(1 - \lambda\beta(t))[\lambda\beta(t)]^k \qquad e^{-\mu t}(1 - \lambda\beta(t))[\lambda\beta(t)]^{k-1} \qquad \begin{array}{c}(1 - e^{-\mu t} - \mu\beta(t))\times \\ (1 - \lambda\beta(t))[\lambda\beta(t)]^{k-1}\end{array} \qquad \mu\beta(t)$$

Statistical Multiple Alignment, Fig. 2 The probabilities of alignment patterns. From *left* to *right*: k insertions at the beginning of the alignment, a match followed by $k - 1$ insertions, a deletion followed by k insertions, and a deletion not followed by insertions. $\beta = \frac{1 - e^{(\lambda-\mu)t}}{\mu - \lambda e^{(\lambda-\mu)t}}$

the TKF91 process starts at the root with the equilibrium distribution.

Multiple Hidden Markov Models

It will turn out that the TKF91 model can be transformed to a multiple Hidden Markov Model; therefore we formally define it here. A multiple Hidden Markov Model (multiple HMM) is a directed graph with distinguished start and end states, (the in degree of the start and the out degree of the end state are both 0), together with the following described transition and emission distributions. Each vertex has a transition distribution over its out edges. The vertexes can be divided into two classes, the emitting and silent states. Each emitting state emits one-one random character to a prescribed set of sequences; it is possible that a state emits only one character to one sequence. For each state, an emission distribution over the alphabet and the set of sequences gives the probabilities which characters will be emitted to which sequences. The Markov process is a random walk from the start to the end, following the transition distribution on the out edges. When the walk is in an emitting state, characters are emitted according to the emission distribution of the state. The process is hidden since the observer sees only the emitted sequences, and the observer does not observe which character is emitted by which state, even the observer does not see which characters are co-emitted. The multiple HMM problem is to calculate the emission probability of a set of sequences for a multiple HMM. This probability can be calculated with the forward algorithm that has $O(V^2 L^n)$ running time, where V is the

number of emitting states in the multiple HMM, L is the geometric mean of the sequences, and n is the number of sequences [3].

Key Results

Substitutions have been modeled with time-continuous Markov models since the late 1960s [8], and an efficient algorithm for likelihood calculations was published in 1980 [4]. The running time of this efficient algorithm grows linearly both with the number of sequences and with the length of the sequences being analyzed, and it grows squarely with the size of the alphabet. The algorithm belongs to the class of dynamic programming algorithms. For each character, subtree, and position x, the algorithm calculates what would be the likelihood of the characters in position x in the sequences belonging to the subtree if the substitution process started in the root of the subtree with the given character. These probabilities are called conditional likelihoods. It is easy to show that

$$L_p(\alpha, x) = \left(\sum_{\alpha_1} P_{t_1}(\alpha_1|\alpha)L_{d_1}(\alpha_1, x)\right)$$
$$\left(\sum_{\alpha_2} P_{t_2}(\alpha_2|\alpha)L_{d_2}(\alpha_2, x)\right) \quad (6)$$

where d_1 and d_2 are the descendant nodes of the parent node p and t_1 and t_2 are the length of the edges connecting p with d_1 and d_2, respectively. The likelihood of the tree can be calculated from the conditional likelihoods of the tree. Recall that

$P(S|T, M)$ is the likelihood of observing a set of sequences S on the leaves of an evolutionary tree T under the substitution model M:

$$P(S|T, M) = \prod_x \sum_\alpha L_{root}(\alpha, x)\pi_\alpha \quad (7)$$

Thorne, Kishino, and Felsenstein gave an $O(nm)$ running time algorithm for calculating the likelihood of an n-long and an m-long sequence under their model [16]. It was not clear for long time how to extend this algorithm to more than two sequences. In 2001, several researchers [7,12] realized that the TKF91 model for two sequences is equivalent with a pair Hidden Markov Model (pair HMM) in the sense that the transition and emission probabilities of the pair HMM can be parameterized with λ, μ and the transition and equilibrium probabilities of the substitution model; moreover there is a bijection between the paths emitting the two sequences and alignments such that the probability of a path in the pair HMM equals to the probability of the corresponding alignment of the two sequences. Hence the likelihood of two sequences can be calculated with the forward algorithm of the pair HMM.

After this discovery, it was relatively easy to develop an algorithm for multiple statistical alignment [5]. The key observation is that a multiple HMM can be created as a composition of pair HMMs along the evolutionary tree. This technique was already known in the speech recognition literature [14], and was also rediscovered by Ian Holmes [6], who named this technique as transducer composition. The number of states in the so-created multiple HMM is $O(5^{\frac{n}{2}})$, where n is the number of leaves of the tree. The emission probabilities are the substitution likelihoods on the tree, which can be efficiently calculated as shown above. The running time of the forward algorithm is $5^n L^n$, where L is the geometric mean of the sequence lengths.

Lunter et al. [9] introduced an algorithm that does not need a multiple HMM description of the TKF91 model to calculate the likelihood of a tree. Using a logical sieve algorithm, they were able to reduce the running time to $O(2^n L^n)$. They

called their algorithm the "one-state recursion" since their dynamic programming algorithm does not need different state of a multiple HMM to calculate the likelihood correctly.

Applications

Since the running time of the best known algorithm for multiple statistical alignment grows exponentially with the number of sequences, on its own it is not useful in practice. However, Lunter et al. also showed that there is a one-state recursion to calculate the likelihood of the tree given an alignment [10]. The running time of this algorithm grows only linearly with both the alignment length and the number of sequences. Since the number of states in a multiple HMM that can emit the same multiple alignment column might grow exponentially, this version of the one-state recursion is a significant improvement. The one-state recursion for multiple alignments is used in a Bayesian Markov chain Monte Carlo where the state space is the Descartes product of the possible multiple alignments and evolutionary trees. The one-state recursion provides an efficient likelihood calculation for a point in the state space [11].

Csűrös and Miklós introduced a model for gene content evolution that is equivalent with the multiple statistical alignment problem for alphabet size 1 [2]. They gave a polynomial running time algorithm that calculates the likelihood of the tree. The running time is $O(n + hL^2)$, where n is the number of sequences, h is the height of the evolutionary tree, and L is the sum of the sequence lengths.

Thorne, Kishino, and Felsenstein also introduced a fragment model, also called the TKF92 model, in which multiple insertions and deletions are allowed [17]. The birth process is still a Poisson process, but instead of single characters, fragments of characters are inserted with a geometrically distributed length. The fragments are unbreakable, and the death process is going on the fragments. The TKF92 model for a pair of sequences also can be described into a pair HMM and the TKF92 model on a tree can be transformed to a multiple HMM. Such multiple HMM is used in the StatAlign software package [13]. The software package has been extended to predict the common structure of sequences (e.g., slowly quickly evolving regions, RNA secondary structures) by combining this multiple HMM with other stochastic models describing the structure of sequences [1, 15].

Open Problems

It is conjectured that the multiple statistical alignment problem cannot be solved in polynomial time for any nontrivial alphabet size. One also can ask what the most likely multiple alignment is or, equivalently, what the most probable path in the multiple HMM is that emits the given sequences. For a set of sequences, a TKF91 model, and an evolutionary tree, the decision problem "Is there a multiple alignment that is more probable than p" is conjectured to be NP-complete.

It is conjectured that there is no one-state recursion for the TKF92 model.

Recommended Reading

1. Arunapuram P, Edvardsson I, Golden M, Anderson JW, Novák Á, Sükösd Z, Hein J (2013) StatAlign 2.0: combining statistical alignment with RNA secondary structure prediction. Bioinformatics 29(5):654–655
2. Csűrös M, Miklós I (2006) A probabilistic model for gene content evolution with duplication, loss, and horizontal transfer. In: Proceedings of RECOMB2006. Lecture notes in bioinformatics, Springer Verlag, vol 3909, pp 206–220
3. Durbin R, Eddy S, Krogh A, Mitchison G (1998). Biological sequence analysis. Cambridge University Press, Cambridge
4. Felsenstein J (1981) Evolutionary trees from DNA sequences: a maximum likelihood approach. J Mol Evol 17:368–376
5. Hein JJ, Jensen JL, Pedersen CNS (2003) Recursions for statistical multiple alignment. PNAS 100:14960–14965
6. Holmes I (2003) Using guide trees to construct multiple-sequence evolutionary hmms. Bioinformatics 19:1147–1157
7. Holmes I, Bruno WJ (2001) Evolutionary HMMs: a Bayesian approach to multiple alignment. Bioinformatics 17(9):803–820
8. Jukes TH, Cantor CR (1969) Evolution of protein molecules. In: Munro HN (ed) Mammalian protein metabolism. Academic, New York, pp 21–132
9. Lunter GA, Miklós I, Song YS, Hein J (2003) An efficient algorithm for statistical multiple alignment on arbitrary phylogenetic trees. J Comput Biol 10(6):869–889
10. Lunter GA, Miklós I, Drummond AJ, Jensen JL, Hein JJ (2003) Bayesian phylogenetic inference under a statistical indel model. In: Proceedings of WABI2003. Lecture notes in bioinformatics, Springer Verlag, vol 2812, pp 228–244
11. Lunter GA, Miklós I, Drummond AJ, Jensen JL, Hein JJ (2005) Bayesian coestimation of phylogeny and sequence alignment. BMC Bioinformatics 6:83
12. Metzler D, Fleißner R, Wakolbringer A, von Haeseler A (2001) Assessing variability by joint sampling of alignments and mutation rates. J Mol Evol 53:660–669
13. Novák A, Miklós I, Lyngsø R, Hein J (2008) StatAlign: an extendable software package for joint bayesian estimation of alignments and evolutionary trees. Bioinformatics 24(20):2403–2404
14. Pereira F, Riley M (1997) Speech recognition by composition of weighted finite automata. In: Finite-state language processing. MIT, Cambridge, pp 149–173
15. Satija R, Novák A, Miklós I, Lyngsø R, Hein J (2009) BigFoot: Bayesian alignment and phylogenetic footprinting with MCMC. BMC Evol Biol 9:217
16. Thorne JL, Kishino H, Felsenstein J (1991) An evolutionary model for maximum likelihood alignment of DNA sequences. J Mol Evol 33:114–124
17. Thorne JL, Kishino H, Felsenstein J (1992) Inching toward reality: an improved likelihood model of sequence evolution. J Mol Evol 34:3–16

Statistical Query Learning

Vitaly Feldman
IBM Research – Almaden, San Jose, CA, USA

Keywords

Classification noise; Noise-tolerant learning; PAC learning; SQ dimension; Statistical query

Years and Authors of Summarized Original Work

1998; Kearns

Problem Definition

The problem deals with learning to classify from random labeled examples in Valiant's PAC model [30]. In the *random classification noise* model of Angluin and Laird [1], the label of each example given to the learning algorithm is flipped randomly and independently with some fixed probability η called the *noise rate*. Robustness to such benign form of noise is an important goal in the design of learning algorithms. Kearns defined a powerful and convenient framework for constructing noise-tolerant algorithms based on *statistical queries*. Statistical query (SQ) learning is a natural restriction of PAC learning that models algorithms that use statistical properties of a data set, rather than individual examples. Kearns demonstrated that any learning algorithm that is based on statistical queries can be automatically converted to a learning algorithm in the presence of random classification noise of arbitrary rate smaller than the information-theoretic barrier of $1/2$. This result was used to give the first noise-tolerant algorithm for a number of important learning problems. In fact, virtually all known noise-tolerant PAC algorithms were either obtained from SQ algorithms or can be easily cast into the SQ model.

In subsequent work, the model of Kearns has been extended to other settings and found a number of additional applications in machine learning and theoretical computer science.

Definitions and Notation

Let C be a class of $\{-1, +1\}$-valued functions (also called *concepts*) over an input space X. In the basic PAC model, a learning algorithm is given examples of an unknown function f from C on points randomly chosen from some unknown distribution $\mathcal{D}$ over X and should produce a hypothesis h that approximates f. More formally, an *example oracle* $EX(f, \mathcal{D})$ is an oracle that upon being invoked returns an example $\langle x, f(x) \rangle$, where x is chosen randomly with respect to $\mathcal{D}$, independently of any previous examples. A learning algorithm for C is an algorithm that for every $\epsilon > 0$, $\delta > 0$, $f \in C$, and

distribution $\mathcal{D}$ over X, given ϵ, δ, and access to $EX(f, \mathcal{D})$ outputs, with probability at least $1 - \delta$, a hypothesis h that ϵ-approximates f with respect to $\mathcal{D}$ (i.e., $\mathbf{Pr}_\mathcal{D}[f(x) \neq h(x)] \leq \epsilon$). Efficient learning algorithms are algorithms that run in time polynomial in $1/\epsilon$, $1/\delta$ and the size of the learning problem s. The size of a learning problem is determined by the description length of f under some fixed representation scheme for functions in C and the description length of an element in X (often proportional to the dimension n of the input space).

A number of variants of this basic framework are commonly considered. The basic PAC model is also referred to as *distribution-independent* learning to distinguish it from *distribution-specific* PAC learning in which the learning algorithm is required to learn with respect to a single distribution $\mathcal{D}$ known in advance. A *weak* learning algorithm is a learning algorithm that can produce a hypothesis whose error on the target concept is noticeably less than $1/2$ (and not necessarily any $\epsilon > 0$). More precisely, a weak learning algorithm produces a hypothesis h such that $\mathbf{Pr}_\mathcal{D}[f(x) \neq h(x)] \leq 1/2 - 1/p(s)$ for some fixed polynomial p. The basic PAC model is often referred to as *strong* learning in this context.

In the random classification noise model $EX(f, \mathcal{D})$ is replaced by a faulty oracle $EX^\eta(f, \mathcal{D})$, where η is the noise rate. When queried, this oracle returns a noisy example $\langle x, b \rangle$ where $b = f(x)$ with probability $1 - \eta$ and $\neg f(x)$ with probability η independently of previous examples. When η approaches $1/2$ the label of the corrupted example approaches the result of a random coin flip, and therefore, the running time of learning algorithms in this model is allowed to depend on $\frac{1}{1-2\eta}$ (the dependence must be polynomial for the algorithm to be considered efficient). For simplicity, one usually assumes that η is known to the learning algorithm. This assumption can be removed using a simple technique due to Laird [26].

To formalize the idea of learning from statistical properties of a large number of examples, Kearns introduced a new oracle $STAT(f, \mathcal{D})$ that replaces $EX(f, \mathcal{D})$. The oracle $STAT(f, \mathcal{D})$ takes as input a *statistical query* (SQ) of the form (χ, τ), where χ is a $\{-1, +1\}$-valued function on labeled examples and $\tau \in [0, 1]$ is the *tolerance* parameter. Given such a query, the oracle responds with an estimate v of $\mathbf{Pr}_{\mathcal{D}}[\chi(x, f(x)) = 1]$ that is accurate to within an additive $\pm\tau$.

Note that the oracle does not guarantee anything else on the value v beyond $|v - \mathbf{Pr}_{\mathcal{D}}[\chi(x, f(x)) = 1]| \leq \tau$ and an SQ learning algorithm needs to work with any possible implementation of the oracle. Yang proposed a stronger, *honest* version of the oracle which to a call with function χ returns the value of $\chi(x, f(x))$, where x is chosen randomly and independently according to $\mathcal{D}$ [32]. This version was shown to be equivalent to the original model up to polynomial factors [17].

Chernoff bounds easily imply that $\text{STAT}(f, \mathcal{D})$ can, with high probability, be simulated using $\text{EX}(f, \mathcal{D})$ by estimating $\mathbf{Pr}_{\mathcal{D}}[\chi(x, f(x)) = 1]$ on $O(\tau^{-2})$ examples. Therefore, the SQ model is a restriction of the PAC model. Efficient SQ algorithms allow only efficiently evaluatable χ's and impose an inverse polynomial lower bound on the tolerance parameter over all oracle calls. Kearns also observes that in order to simulate all the statistical queries used by an algorithm, one does not necessarily need new examples for each estimation. Instead, assuming that the set of possible queries of the algorithm has Vapnik-Chervonenkis dimension d, all its statistical queries can be simulated using $\tilde{O}(d\,\tau^{-2}(1 - 2\eta)^{-2} \log(1/\delta))$ examples [24].

Key Results

Statistical Queries and Noise-Tolerance

The main result given by Kearns is a way to simulate statistical queries using noisy examples.

Lemma 1 ([24]) *Let (χ, τ) be a statistical query such that χ can be evaluated on any input in time T and let $\text{EX}^{\eta}(f, \mathcal{D})$ be a noisy oracle. The value $\mathbf{Pr}_{\mathcal{D}}[\chi(x, f(x)) = 1]$ can, with probability at least $1-\delta$, be estimated within τ using $O(\tau^{-2}(1 - 2\eta)^{-2} \log(1/\delta))$ examples from $\text{EX}^{\eta}(f, \mathcal{D})$ and time $O(\tau^{-2}(1 - 2\eta)^{-2} \log(1/\delta) \cdot T)$.*

This simulation is based on estimating several probabilities using examples from the noisy oracle and then offsetting the effect of noise. The lemma implies that any efficient SQ algorithm for a concept class $\mathcal{C}$ can be converted to an efficient learning algorithm for $\mathcal{C}$ tolerating random classification noise of any rate $\eta < 1/2$.

Theorem 1 ([24]) *Let $\mathcal{C}$ be a concept class efficiently PAC learnable from statistical queries. Then $\mathcal{C}$ is efficiently PAC learnable in the presence of random classification noise of rate η for any $\eta < 1/2$.*

Balcan and Feldman describe more general conditions on noise under which a specific SQ algorithm can be simulated in the presence of noise [3].

Statistical Query Algorithms

Kearns showed that, despite the major restriction on the way an SQ algorithm accesses the examples, many PAC learning algorithms known at the time can be modified to use statistical queries instead of random examples [24]. Examples of learning algorithms for which he described an SQ analogue and thereby obtained a noise-tolerant learning algorithm include:

- Learning decision trees of constant rank.
- *Attribute-efficient* algorithms for learning conjunctions.
- Learning axis-aligned rectangles over $\mathbb{R}^n$.
- Learning AC^0 (constant-depth unbounded fan-in) Boolean circuits over $\{0, 1\}^n$ with respect to the uniform distribution in quasipolynomial time.

Subsequent works have provided numerous additional examples of algorithms used in theory and practice of machine learning that can either be implemented using statistical queries or can be replaced by an alternative SQ-based algorithm of similar complexity. For example, the Perceptron algorithm and learning of linear threshold functions [6, 12], boosting [2], attribute-efficient learning via the Winnow algorithm (cf. [16]), k-means clustering [5] and convex optimization-based methods [20]. We note that many learning algorithms rely only on evaluations of functions on random examples and therefore can be seen as using access to the honest statistical query oracle. In such cases the SQ implementation follows immediately from the equivalence of the Kearns' SQ oracle and the honest one [17].

The only known example of a technique for which there is no SQ analogue is Gaussian elimination for solving linear equations over a finite field. This technique can be used to learn parity functions that are not learnable using SQs (as we discuss below). As a result, with the exception of the parity learning problem, known bounds on the complexity of learning from random examples are, up to polynomial factors, the same as known bound for learning with statistical queries.

Statistical Query Dimension

The restricted way in which SQ algorithms use examples makes it simpler to understand the limitations of efficient learning in this model. A long-standing open problem in learning theory is learning of the concept class of all parity functions over $\{0,1\}^n$ with noise (a parity function is a XOR of some subset of n Boolean inputs). Kearns has demonstrated that parities cannot be efficiently learned using statistical queries even under the uniform distribution over $\{0,1\}^n$ [24]. This hardness result is unconditional in the sense that it does not rely on any unproven complexity assumptions.

The technique of Kearns was generalized by Blum et al. who proved that efficient SQ learnability of a concept class C is characterized by a relatively simple combinatorial parameter of C called the *statistical query dimension* [7]. The quantity they defined, measures the maximum number of "nearly uncorrelated" functions in a concept class. (The definition and the results were simplified and strengthened in subsequent works [17, 29] and we use the improved statements here.) More formally,

Definition 1 For a concept class C and distribution D, the *statistical query dimension* of C with respect to D, denoted SQ-DIM(C,D), is the largest number d such that C contains d

functions $f_1, f_2, \ldots, f_d$ such that for all $i \neq j$, $|\mathbf{E}_D[f_i f_j]| \leq \frac{1}{d}$.

Blum et al. relate the SQ dimension to learning in the SQ model as follows.

Theorem 2 ([7, 17]) *Let C be a concept class and D be a distribution such that SQ-DIM $(C,D) = d$.*

- *If all queries are made with tolerance of at least $1/d^{1/3}$, then at least $d^{1/3} - 2$ queries are required to learn C with error $1/2 - 1/(2d^3)$ in the SQ model.*
- *There exists an algorithm for learning C with respect to D that makes d fixed queries, each of tolerance $1/(4d)$, and finds a hypothesis with error at most $1/2 - 1/(2d)$.*

Thus SQ-DIM characterizes weak SQ learnability relative to a fixed distribution D up to a polynomial factor. Parity functions are uncorrelated with respect to the uniform distribution and therefore, any concept class that contains a superpolynomial number of parity functions cannot be learned by statistical queries with respect to the uniform distribution. This, for example, includes such important concept classes as *k-juntas* over $\{0,1\}^n$ (or functions that depend on at most k input variables) for $k = \omega(1)$ and *decision trees* of superconstant size.

Simon showed that (strong) PAC learning relative to a fixed distribution D using SQs can also be characterized by a more general and involved dimension [28]. Simpler and tighter characterizations of distribution-specific PAC learning using SQs have been demonstrated by Feldman [15] and Szörényi [29]. Feldman also extended the characterization to the agnostic learning model.

Despite characterizing the number of queries of certain tolerance, the SQ-DIM and its generalizations capture surprisingly well the computational complexity of SQ learning of most concept classes. One reason for this is that if a concept class has polynomial SQ-DIM then it can be learned by a polynomial-time algorithm with advice also referred to as a "non-uniform" algorithm (cf. [18]). However it was shown by Feldman and Kanade that for strong PAC learning there exist artificial problems whose computational complexity is larger than their statistical query complexity [18].

Applications of these characterizations to proving lower bounds on SQ algorithms can be found in [11, 15, 19, 25]. Relationships of SQ-DIM to other notions of complexity of concept classes were investigated in [22, 27].

Applications

The ideas behind the use of statistical queries to produce noise-tolerant algorithms were adapted to learning using *membership queries* (or ability to ask for the value of the unknown function at any point) and used to give a noise-tolerant algorithm for learning DNF with respect to the uniform distribution [9, 21]. The SQ model of learning was generalized to active learning (or learning where labels are requested only for some of the points) and used to obtain new efficient noise-tolerant active learning algorithms [3].

The restricted way in which an SQ algorithm uses data implies it can be used to obtain learning algorithms with additional useful properties. Blum et al. [5] show that an SQ algorithm can be used to obtain a *differentially-private* [13] algorithm for the problem. In fact, SQ algorithms are equivalent to *local* (or *randomized-response*) differentially-private algorithms [23]. Chu et al. [10] show that SQ algorithms can be automatically parallelized on multicore architectures and give many examples of popular machine learning algorithms that can be sped up using this approach.

The SQ learning model has also been instrumental in understanding Valiant's model of evolution as learning [31]. Feldman showed that the model is equivalent to learning with a restricted form of SQs referred to as correlational SQs [14]. A correlational SQ is a query of the form $\chi(x, \ell) = g(x) \cdot \ell$ for some $g : X \rightarrow [-1, 1]$. Such queries were first studied by Ben-David et al. [4] (remarkably, before the introduction of the SQ model itself) and distribution-specific

learning with such queries is equivalent to learning with (unrestricted) SQs.

Statistical query-based access can naturally be defined for any problem where the input is a set of i.i.d. samples from a distribution. Feldman et al. show that lower bounds based on SQ-DIM can be extended to this more general setting and give examples of applications [17, 20].

Open Problems

The main questions related to learning with random classification noise are still open. Is every concept class efficiently learnable in the PAC model also learnable in the presence of random classification noise? Is every concept class efficiently learnable in the presence of random classification noise of arbitrarily high rate (less than $1/2$) also efficiently learnable using statistical queries? A partial answer to this question was provided by Blum et al. who show that Gaussian elimination can be used in low dimension to obtain a class learnable with random classification noise of constant rate $\eta < 1/2$ but not learnable using SQs [8]. For both questions a central issue seems to be obtaining a better understanding of the complexity of learning parities with noise.

The complexity of learning from statistical queries remains an active area of research with many open problems. For example, there is currently an exponential gap between known lower and upper bounds on the complexity of distribution-independent SQ learning of polynomial-size DNF formulae and AC^0 circuits (cf. [27]). Several additional open problems on complexity of SQ learning can be found in [16, 19, 22].

Cross-References

▸ Attribute-Efficient Learning
▸ Learning Constant-Depth Circuits
▸ Learning DNF Formulas
▸ Learning Heavy Fourier Coefficients of Boolean Functions
▸ Learning with Malicious Noise
▸ PAC Learning

Recommended Reading

1. Angluin D, Laird P (1988) Learning from noisy examples. Mach Learn 2:343–370
2. Aslam J, Decatur S (1998) General bounds on statistical query learning and pac learning with noise via hypothesis boosting. Inf Comput 141(2):85–118
3. Balcan M-F, Feldman V (2013) Statistical active learning algorithms. In: NIPS, Lake Tahoe, pp 1295–1303
4. Ben-David S, Itai A, Kushilevitz E (1990) Learning by distances. In: Proceedings of COLT, Rochester, pp 232–245
5. Blum A, Dwork C, McSherry F, Nissim K (2005) Practical privacy: the SuLQ framework. In: Proceedings of PODS, Baltimore, pp 128–138
6. Blum A, Frieze A, Kannan R, Vempala S (1997) A polynomial time algorithm for learning noisy linear threshold functions. Algorithmica 22(1/2):35–52
7. Blum A, Furst M, Jackson J, Kearns M, Mansour Y, Rudich S (1994) Weakly learning DNF and characterizing statistical query learning using Fourier analysis. In: Proceedings of STOC, Montréal, pp 253–262
8. Blum A, Kalai A, Wasserman H (2003) Noise-tolerant learning, the parity problem, and the statistical query model. J ACM 50(4):506–519
9. Bshouty N, Feldman V (2002) On using extended statistical queries to avoid membership queries. J Mach Learn Res 2:359–395
10. Chu C, Kim S, Lin Y, Yu Y, Bradski G, Ng A, Olukotun K (2006) Map-reduce for machine learning on multicore. In: Proceedings of NIPS, Vancouver, pp 281–288
11. Dachman-Soled D, Feldman V, Tan L-Y, Wan A, Wimmer K (2014) Approximate resilience, monotonicity, and the complexity of agnostic learning. arXiv, CoRR, abs/1405.5268
12. Dunagan J, Vempala S (2004) A simple polynomial-time rescaling algorithm for solving linear programs. In: Proceedings of STOC, Chicago, pp 315–320
13. Dwork C, McSherry F, Nissim K, Smith A (2006) Calibrating noise to sensitivity in private data analysis. In: TCC, New York, pp 265–284
14. Feldman V (2008) Evolvability from learning algorithms. In: Proceedings of STOC, Victoria, pp 619–628
15. Feldman V (2012) A complete characterization of statistical query learning with applications to evolvability. J Comput Syst Sci 78(5):1444–1459
16. Feldman V (2014) Open problem: the statistical query complexity of learning sparse halfspaces. In: COLT, Barcelona, pp 1283–1289
17. Feldman V, Grigorescu E, Reyzin L, Vempala S, Xiao Y (2013) Statistical algorithms and a lower bound for planted clique. In: STOC, Palo Alto. ACM, pp 655–664
18. Feldman V, Kanade V (2012) Computational bounds on statistical query learning. In: COLT, Edinburgh, pp 16.1–16.22
19. Feldman V, Lee H, Servedio R (2011) Lower bounds and hardness amplification for learning shallow monotone formulas. In: COLT, Budapest, vol 19, pp 273–292
20. Feldman V, Perkins W, Vempala S (2013) On the complexity of random satisfiability problems with planted solutions. In: CoRR, abs/1311.4821
21. Jackson J, Shamir E, Shwartzman C (1997) Learning with queries corrupted by classification noise. In: Proceedings of the fifth Israel symposium on the theory of computing systems, Ramat-Gan, pp 45–53
22. Kallweit M, Simon H (2011) A close look to margin complexity and related parameters. In: COLT, Budapest, pp 437–456
23. Kasiviswanathan SP, Lee HK, Nissim K, Raskhodnikova S, Smith A (2011) What can we learn privately? SIAM J Comput 40(3):793–826
24. Kearns M (1998) Efficient noise-tolerant learning from statistical queries. J ACM 45(6): 983–1006
25. Klivans A, Sherstov A (2007) Unconditional lower bounds for learning intersections of halfspaces. Mach Learn 69(2–3):97–114
26. Laird P (1988) Learning from good and bad data. Kluwer Academic, Boston
27. Sherstov AA (2008) Halfspace matrices. Comput Complex 17(2):149–178
28. Simon H (2007) A characterization of strong learnability in the statistical query model. In: Proceedings of symposium on theoretical aspects of computer science, Aachen, pp 393–404
29. Szörényi B (2009) Characterizing statistical query learning: simplified notions and proofs. In: Proceedings of ALT, Porto, pp 186–200
30. Valiant LG (1984) A theory of the learnable. Commun ACM 27(11):1134–1142
31. Valiant LG (2009) Evolvability. J ACM 56(1):3.1–3.21. Earlier version in ECCC, 2006
32. Yang K (2005) New lower bounds for statistical query learning. J Comput Syst Sci 70(4):485–509

Statistical Timing Analysis

Sachin S. Sapatnekar
Department of Electrical and Computer
Engineering, University of Minnesota,
Minneapolis, MN, USA

Keywords

Delay uncertainty; Electronic design automation; Principal component analysis; Process variations; Static timing analysis; Timing closure

Years and Authors of Summarized Original Work

2003; Chang, Sapatnekar
2005; Chang, Sapatnekar

Problem Definition

The timing behavior of integrated systems is strongly affected by the characteristics of transistors and wires in the system. Variations in the manufacturing process can cause drifts in these characteristics from one manufactured part to another. The traditional approach to addressing these variations was to choose a worst-case value for each process parameter, but this has become unsustainable in the face of current-day variations. Statistical timing analysis provides a computationally efficient way to translate the probability density function of the underlying process parameter spread to the distribution of circuit timing.

A key underlying structure for timing analysis is a graph $G(V, E)$ of a combinational circuit, where the vertex set V corresponds to the gates, primary inputs, and primary outputs of the circuit, and each connection between these gates corresponds to an edge in E. The delay of each gate corresponds to a probability distribution that is a function of the distributions of the underlying (possibly correlated) process parameters, and the task of combinational statistical timing analysis is to obtain the distribution of the maximum (or minimum) delay of the circuit, over all primary outputs. The extension of this problem to general edge-triggered sequential circuits is straightforward. Such circuits can be decomposed into independent combinational blocks, and the maximum (or minimum) operator acts on the delay distribution at all primary outputs of all combinational blocks of the sequential circuit.

Key Results

The framework that is used for statistical timing analysis is based on graph-based topological traversals that maintain a closed-form structure for the delay from the primary inputs of the circuit to the output of each vertex (referred to as the *arrival time*). The computation under this paradigm scales linearly with $|E|$. While it is certainly possible to perform statistical timing analysis through Monte Carlo simulations based on samples of the process parameter space, such an approach is uncompetitive compared to graph traversal algorithms. The traversal approach consists of three key steps [1, 2]:

- Translating the underlying process parameter variations to an orthogonal set of random variables
- Representing gate delay variations in terms of this orthogonal set
- Performing a topological traversal of G and computing the arrival time at each node and maximum delay of the circuit

Orthogonalizing Process Parameter Distributions

A common assumption is that the underlying process parameters, such as the transistor width W and effective length L_{eff} of devices, gate oxide thickness (T_{ox}), and device threshold voltage (V_t) due to random dopant fluctuations, show a Gaussian distribution. Each individual device is separately represented by such a parameter. The distributions of T_{ox} and V_t are largely uncorrelated across devices. In contrast, the dimension-based parameters, W and L_{eff}, show strong spatial correlations, whereby the distributions of nearby devices are strongly correlated, and this correlation falls off as a function of distance.

The existence of correlations can significantly complicate the task of statistical timing analysis, since all pairwise combinations of random variables must be considered during the optimization, potentially leading to quadratic complexity in $|V|$. To overcome this, an initial principal component analysis (PCA) [7] step is carried out that orthogonalizes the underlying Gaussians, enabling linear-time analysis. PCA is a one-time operation for a given process (which is used for numerous designs). Therefore, although its worst-case complexity is cubic in $|V|$, the expense is practically manageable as it is amortized over numerous designs. Furthermore, sparsity properties of the correlation matrix realistically imply that in practice, the cost of PCA scales considerably slower than this cubic rate.

For cases where the underlying process parameters may be a mix of Gaussians or non-Gaussians, it is possible to orthogonalize the Gaussian parameters using PCA and non-Gaussian parameters using independent component analysis (ICA) [4]. The approach in [8] extends the graph-based approach presented here and shows how statistical timing analysis can be performed for case where some or all process parameters are non-Gaussian.

Gate Delay Distribution

To build a model for the gate delay that captures the underlying variations in process parameters,

we observe that the delay function $d = f(\mathbf{P})$, where $\mathbf{P}$ is a set of process parameters, can be approximated d linearly using a first-order Taylor expansion:

$$d = d_0 + \sum_{\forall \text{ parameters } p_i} \left[\frac{\partial f}{\partial p_i} \right]_0 \Delta p_i \quad (1)$$

where d_0 is the nominal value of d, calculated at the nominal values of parameters in the set $\mathbf{P}$; $\left[\frac{\partial f}{\partial p_i} \right]_0$ is computed at the nominal values of p_i; $\Delta p_i = p_i - \mu_{p_i}$ is a normally distributed random variable; and $\Delta p_i \sim N(0, \sigma_{p_i})$. The delay function here can be arbitrarily complex.

If all parameters in $\mathbf{P}$ can be modeled by Gaussian distributions, this approximation implies that d is a linear combination of Gaussians, which is therefore Gaussian. Its mean μ_d and variance σ_d^2 are

$$\mu_d = d_0 \quad (2)$$

$$\sigma_d^2 = \sum_{\forall i} \left[\frac{\partial f}{\partial p_i} \right]_0^2 \sigma_{p_i}^2 + 2 \sum_{\forall i \neq j} \left[\frac{\partial f}{\partial p_i} \right]_0 \left[\frac{\partial f}{\partial p_j} \right]_0 \text{cov}\left(p_i, p_j \right) \quad (3)$$

where $\text{cov}\left(p_i, p_j \right)$ is the covariance of p_i and p_j.

This approximation is valid when Δp_i has relatively small variations, in which domain the first-order Taylor expansion is adequate and the approximation is acceptable with little loss of accuracy. This is generally true of the impact of within-die variations on delay, where the process parameter variations are relatively small in comparison with the nominal values, and the function changes by a small amount under this perturbation. Hence, the delays, as functions of the process parameters, can be approximated as normal distributions when the parameter variations are assumed to be normal. Higher-order expansions based on quadratics have also be explored to cover cases where the variations are larger [6, 11].

Circuit Delay Distribution

A PCA-based approach maintains the invariant that the output arrival time at each gate is a Gaussian variable represented as

$$a_i(p_1, \ldots, p_n) = a_i^0 + \sum_{i=1}^{n} k_i p_i' + k_{n+1} p_{n+1}' \quad (4)$$

Here, the primed variables correspond to the principal components of the unprimed variables and maintain the form of the arrival time after each sum and max operation. Gate delays, as represented in Eq. 1, can be translated into a similar representation based on principal components as a one-time step during gate library characterization. Under orthogonalization, many operations become much simpler since the covariance terms disappear: for example, Eq. 3 can be evaluated in linear time instead of quadratic time.

The task of statistical timing analysis is to translate these gate delay distributions to circuit delay probabilities while performing a topological traversal. The operations performed at each node encountered during this traversal in STA are of two types [5]:

- A gate (vertex) is being processed in STA when the arrival times of all inputs are known, at which time the candidate delay values at the output are computed using the "sum" operation that adds the delay at each input with the input-to-output pin delay.
- Once these candidate delays have been found, the "max" operation is applied to determine the maximum arrival time at the output.

Since the gate delays are Gaussian, the "sum" operation is merely an addition of Gaussians, which is well known to be a Gaussian. The computation of the max function, however, poses greater problems. The set of candidate delays are all Gaussian, so that this function must find the maximum of Gaussians. Such a maximum may be reasonably approximated using a Gaussian [3]. A detailed description of how the invariant representation is maintained under the max operation is presented in [1, 2].

The cost of this method corresponds to running a bounded number of deterministic STAs, and it is demonstrated to be accurate, given the statistics of **P**.

Applications

Statistical timing analysis has been extensively used in industry [10] and has seeded a large amount of academic research. Integrated circuit manufacturing foundries have promoted the use of statistical timing

analysis by providing PCA-like information with their process parameter models, thus enabling design flows that are statistically based.

The ideas of statistical analysis have also motivated simpler and more approximate methods, used in industry today, based on on-chip variation (OCV) derating factors. In its most elementary form, OCV adds margins to each timing path to account for possible variation. More involved versions of OCV, such as advanced OCV (AOCV), capture the essence of spatial correlation by using derating factors that depend on factors such as spatial distance and logical depth of a path [9].

Experimental Results

Statistical timing analysis based on orthogonalization brings down the computational cost from quadratic to linear in the number of variables and can be applied to large circuit instances. The method is capable of considering both spatial correlations and structural correlations, i.e., correlations between paths that share gates, since such correlations are embedded into the invariant representation. This makes the approach accurate and computationally practical, as described in [1, 2, 10] and the large body of follow-on work.

URLs to Code and Data Sets

The MinnSSTA statistical static timing analyzer is available at http://www.ece.umn.edu/~sachin/software/MinnSSTA/index.html.

Recommended Reading

1. Chang H, Sapatnekar SS (2003) Statistical timing analysis considering spatial correlations using a single PERT-like traversal. In: Proceedings of the IEEE/ACM international conference on computer-aided design, San Jose, pp 621–625

2. Chang H, Sapatnekar SS (2005) Statistical timing analysis under spatial correlations. IEEE Trans Comput-Aided Des Integr Circuits Syst 24(9):1467–1482
3. Clark CE (1961) The greatest of a finite set of random variables. Oper Res 9:85–91
4. Hyvärinen A, Oja E (2000) Independent component analysis: algorithms and applications. Neural Netw 13:411–430
5. Jacobs E, Berkelaar MRCM (2000) Gate sizing using a statistical delay model. In: Proceedings of design and test in Europe, Paris, pp 283–290
6. Li X, Le J, Gopalakrishnan P, Pileggi LT (2007) Asymptotic probability extraction for nonnormal performance distributions. IEEE Trans Comput-Aided Des Integr Circuits Syst 26(1):16–37
7. Morrison DF (1976) Multivariate statistical methods. McGraw-Hill, New York
8. Singh J, Sapatnekar SS (2008) A scalable statistical static timing analyzer incorporating correlated non-Gaussian and Gaussian parameter variations. IEEE Trans Comput-Aided Des Integr Circuits Syst 27(1):160–173
9. Synopsys Inc (2009) PrimeTime® Advanced OCV Technology. www.synopsys.com/Tools/Implementation/SignOff/CapsuleModule/PrimeTime_AdvancedOCV_WP.pdf
10. Visweswariah C, Ravindran K, Kalafala K, Walker SG, Narayan S, Beece DK, Piaget J, Venkateswaran N, Hemmett JG (2006) First-order incremental block-based statistical timing analysis. IEEE Trans Comput-Aided Des Integr Circuits Syst 25(10):2170–2180
11. Zhan Y, Strojwas AJ, Li X, Pileggi LT, Newmark D, Sharma M (2005) Correlation-aware statistical timing analysis with non-Gaussian delay distributions. In: Proceedings of the ACM/IEEE design automation conference, San Jose, pp 77–82

Steiner Forest

Guido Schäfer
Institute for Mathematics and Computer Science, Technical University of Berlin, Berlin, Germany

Keywords

Requirement join; R-join

Years and Authors of Summarized Original Work

1995; Agrawal, Klein, Ravi

Problem Definition

The *Steiner forest problem* is a fundamental problem in network design. Informally, the goal is to establish connections between pairs of vertices in a given network at minimum cost. The problem generalizes the well-known *Steiner tree problem*. As an example, assume that a telecommunication company receives communication requests from their customers. Each customer asks for a connection between two vertices in a given network. The company's goal is to build a minimum cost network infrastructure such that all communication requests are satisfied.

Formal Definition and Notation

More formally, an instance $I = (G, c, R)$ of the Steiner forest problem is given by an undirected graph $G = (V, E)$ with vertex set V and edge set E, a non-negative cost function $c: E \to \mathbb{Q}^+$, and a set of vertex pairs $R = \{(s_1, t_1), \ldots, (s_k, t_k)\} \subseteq V \times V$. The pairs in R are called *terminal pairs*. A feasible solution is a subset $F \subseteq E$ of the edges of G such that for every terminal pair $(s_i, t_i) \in R$ there is a path between s_i and t_i in the subgraph $G[F]$ induced by F. Let the cost $c(F)$ of F be defined as the total cost of all edges in F, i.e., $c(F) = \sum_{e \in F} c(e)$. The goal is to find a feasible solution F of minimum cost $c(F)$. It is easy to see that there exists an optimum solution which is a forest.

The Steiner forest problem may alternatively be defined by a set of *terminal groups* $R = \{g_1, \ldots, g_k\}$ with $g_i \subseteq V$ instead of terminal pairs. The objective is to compute a minimum cost subgraph such that all terminals belonging to the same group are connected. This definition is equivalent to the one given above.

Related Problems

A special case of the Steiner forest problem is the *Steiner tree problem* (see also the entry ▶ Steiner Trees). Here, all terminal pairs share a common root vertex $r \in V$, i.e., $r \in \{s_i, t_i\}$ for all terminal pairs $(s_i, t_i) \in R$. In other words, the problem consists of a set of terminal vertices $R \subseteq V$ and a root vertex $r \in V$ and the goal is to connect the terminals in R to r in the cheapest possible way. A minimum cost solution is a tree.

The *generalized Steiner network problem* (see the entry ▶ Generalized Steiner Network), also known as the *survivable network design problem*, is a generalization of the Steiner forest problem. Here, a *connectivity requirement* function $r: V \times V \to \mathbb{N}$ specifies the number of edge disjoint paths that need to be established between every pair of vertices. That is, the goal is to find a minimum cost multi-subset H of the edges of G (H may contain the same edge several times) such that for every pair of vertices $(x, y) \in V$ there are $r(x, y)$ edge disjoint paths from x to y in $G[H]$. The goal is to find a set H of minimum cost. Clearly, if $r(x, y) \in \{0, 1\}$ for all $(x, y) \in V \times V$, this problem reduces to the Steiner forest problem.

Key Results

Agrawal, Klein and Ravi [1, 2] give an approximation algorithm for the Steiner forest problem that achieves an approximation ratio of 2. More precisely, the authors prove the following theorem.

Theorem 1 *There exists an approximation algorithm that for every instance $I = (G, c, R)$ of the Steiner forest problem, computes a feasible forest F such that*

$$c(F) \le \left(2 - \frac{1}{k}\right) \cdot \mathsf{OPT}(I),$$

where k is the number of terminal pairs in R and $\mathsf{OPT}(I)$ is the cost of an optimal Steiner forest for I.

Related Work

The Steiner tree problem is NP-hard [10] and APX-complete [4, 8]. The current best lower bound on the achievable approximation ratio for the Steiner tree problem is 1.0074 [21]. Goemans and Williamson [11] generalized the results obtained by Agrawal, Klein and Ravi to a larger class of connectivity problems, which they term *constrained forest problems*. For the Steiner forest problem, their algorithm achieves the same

approximation ratio of $(2 - 1/k)$. The algorithms of Agrawal, Klein and Ravi [2] and Goemans and Williamson [11] are both based on the classical *undirected cut formulation* for the Steiner forest problem [3]. The integrality gap of this relaxation is known to be $(2 - 1/k)$ and the results in [2, 11] are therefore tight. Jain [15] presents a 2-approximation algorithm for the generalized Steiner network problem.

Primal-Dual Algorithm

The main ideas of the algorithm by Agrawal, Klein and Ravi [2] are sketched below; subsequently, AKR is used to refer to this algorithm. The description given here differs from the one in [2]; the interested reader is referred to [2] for more details.

The algorithm is based on the following integer programming formulation for the Steiner forest problem. Let $I = (G, c, R)$ be an instance of the Steiner forest problem. Associate an indicator variable $x_e \in \{0, 1\}$ with every edge $e \in E$. The value of x_e is 1 if e is part of the forest F and 0 otherwise. A subset $S \subseteq V$ of the vertices is called a *Steiner cut* if there exists at least one terminal pair $(s_i, t_i) \in R$ such that $|\{s_i, t_i\} \cap S| = 1$; S is said to *separate* terminal pair (s_i, t_i). Let $\mathcal{S}$ be the set of all Steiner cuts. For a subset $S \subseteq V$, define $\delta(S)$ as the the set of all edges in E that have exactly one endpoint in S. Given a Steiner cut $S \in \mathcal{S}$, any feasible solution F of I must contain at least one edge that *crosses* the cut S, i.e., $\sum_{e \in \delta(S)} x_e \ge 1$. This gives rise to the following *undirected cut formulation*:

$$\text{minimize} \quad \sum_{e \in E} c(e) x_e \qquad \text{(IP)}$$

$$\text{subject to} \quad \sum_{e \in \delta(S)} x_e \ge 1 \quad \forall S \in \mathcal{S} \quad (1)$$

$$x_e \in \{0, 1\} \quad \forall e \in E. \quad (2)$$

The dual of the linear programming relaxation of (IP) has a variable y_S for every Steiner cut $S \in \mathcal{S}$. There is a constraint for every edge $e \in E$ that requires that the total dual assigned to sets $S \in \mathcal{S}$ that contain exactly one endpoint of e is at most the cost $c(e)$ of the edge:

$$\text{maximize} \quad \sum_{S \in \mathcal{S}} y_S \qquad \text{(D)}$$

$$\text{subject to} \quad \sum_{S \in \mathcal{S} : e \in \delta(S)} y_S \leq c(e) \quad \forall e \in E \quad \text{(3)}$$

$$y_S \geq 0 \quad \forall S \in \mathcal{S}. \quad \text{(4)}$$

Algorithm AKR is based on the *primal-dual schema* (see, e.g., [22]). That is, the algorithm constructs both a feasible primal solution for (IP) and a feasible dual solution for (D). The algorithm starts with an infeasible primal solution and reduces its degree of infeasibility as it progresses. At the same time, it creates a feasible dual packing of subsets of large total value by raising dual variables of Steiner cuts.

One can think of an execution of AKR as a process over time. Let x^τ and y^τ, respectively, be the primal incidence vector and feasible dual solution at time τ. Initially, let $x_e^0 = 0$ for all $e \in E$ and $y_S^0 = 0$ for all $S \in \mathcal{S}$. Let F^τ denote the forest corresponding to the set of edges with $x_e^\tau = 1$. A tree T in F^τ is called *active* at time τ if it contains a terminal that is separated from its mate; otherwise it is *inactive*. Intuitively, AKR grows trees in F^τ that are active. At the same time, the algorithm raises dual values of Steiner cuts that correspond to active trees. If two active trees collide, they are merged. The process terminates if all trees are inactive and thus there are no unconnected terminal pairs. The interplay of the primal (growing trees) and the dual process (raising duals) is somewhat subtle and outlined next.

An edge $e \in E$ is *tight* if the corresponding constraint (3) holds with equality; a path is tight if all its edges are tight. Let H^τ be the subgraph of G that is induced by the tight edges for dual y^τ. The connected components of H^τ induce a partition $\mathcal{C}^\tau$ on the vertex set V. Let $\mathcal{S}^\tau$ be the set of all Steiner cuts contained in $\mathcal{C}^\tau$, i.e., $\mathcal{S}^\tau = \mathcal{C}^\tau \cap \mathcal{S}$. AKR raises the dual values y_S for all sets $S \in \mathcal{S}^\tau$ uniformly at all times $\tau \geq 0$. Note that y^τ is dual feasible. The algorithm maintains the invariant that F^τ is a subgraph of H^τ at all times. Consider the event that a path P between two trees T_1 and T_2 of F^τ becomes tight. The missing edges of P are then added to F^τ and the process continues.

Eventually, all trees in F^τ are inactive and the process halts.

Applications

The computation of (approximate) solutions for the Steiner forest problem has various applications both in theory and practice; only a few recent developments are mentioned here.

Algorithms for more complex network design problems often rely on good approximation algorithms for the Steiner forest problem. For example, the recent approximation algorithms [6, 9, 12] for the *multi-commodity rent-or-buy problem* (MRoB) are based on the random sampling framework by Gupta et al. [12, 13]. The framework uses a Steiner forest approximation algorithm that satisfies a certain *strictness* property as a subroutine. Fleischer et al. [9] show that AKR meets this strictness requirement, which leads to the current best 5-approximation algorithm for MRoB. The strictness property also plays a crucial role in the boosted sampling framework by Gupta et al. [14] for two-stage stochastic optimization problems with recourse.

Online versions of Steiner tree and forest problems have been studied by by Awerbuch et al. [5] and Berman and Coulston [7]. In the area of algorithmic game theory, the development of *group-strategyproof cost sharing mechanisms* for network design problems such as the Steiner tree problem has recently received a lot of attention; see e.g., [16, 17, 19, 20]. An adaptation of AKR yields such a cost sharing mechanism for the Steiner forest problem [18].

Cross-References

▶ Generalized Steiner Network
▶ Steiner Trees

Recommended Reading

The interested reader is referred in particular to the articles [2, 11] for a more detailed description of primal-dual approximation algorithms for general network design problems.

1. Agrawal A, Klein P, Ravi R (1991) When trees collide: an approximation algorithm for the generalized Steiner problem on networks. In: Proceedings of the 23rd annual ACM symposium on theory of computing. Association for Computing Machinery, New York, pp 134–144

2. Agrawal A, Klein P, Ravi R (1995) When trees collide: an approximation algorithm for the generalized Steiner problem in networks. SIAM J Comput 24(3):445–456

3. Aneja YP (1980) An integer linear programming approach to the Steiner problem in graphs. Networks 10(2):167–178

4. Arora S, Lund C, Motwani R, Sudan M, Szegedy M (1998) Proof verification and the hardness of approximation problems. J ACM 45(3):501–555

5. Awerbuch B, Azar Y, Bartal Y (1996) On-line generalized Steiner problem. In: Proceedings of the 7th annual ACM-SIAM symposium on discrete algorithms, 2005. Society for Industrial and Applied Mathematics, Philadelphia, pp 68–74

6. Becchetti L, Könemann J, Leonardi S, Pál M (2005) Sharing the cost more efficiently: improved approximation for multicommodity rent-or-buy. In: Proceedings of the 16th annual ACM-SIAM symposium on discrete algorithms. Society for Industrial and Applied Mathematics, Philadelphia, pp 375–384

7. Berman P, Coulston C (1997) On-line algorithms for Steiner tree problems. In: Proceedings of the 29th annual ACM symposium on theory of computing. Association for Computing Machinery, New York, pp 344–353

8. Bern M, Plassmann P (1989) The Steiner problem with edge lengths 1 and 2. Inf Process Lett 32(4):171–176

9. Fleischer L, Könemann J, Leonardi S, Schäfer G (2006) Simple cost sharing schemes for multicommodity rent-or-buy and stochastic Steiner tree. In: Proceedings of the 38th annual ACM symposium on theory of computing. Association for Computing Machinery, New York, pp 663–670

10. Garey MR, Johnson DS (1979) Computers and intractability: a guide to the theory of NP-completeness. Freeman, San Francisco

11. Goemans MX, Williamson DP (1995) A general approximation technique for constrained forest problems. SIAM J Comput 24(2):296–317

12. Gupta A, Kumar A, Pál M, Roughgarden T (2003) Approximation via cost-sharing: a simple approximation algorithm for the multicommodity rent-or-buy problem. In: Proceedings of the 44th annual IEEE symposium on foundations of computer science. IEEE Computer Society, Washington, pp 606–617

13. Gupta A, Kumar A, Pál M, Roughgarden T (2007) Approximation via cost-sharing: simpler and better approximation algorithms for network design. J ACM 54(3):Article 11

14. Gupta A, Pál M, Ravi R, Sinha A (2004) Boosted sampling: approximation algorithms for stochastic optimization. In: Proceedings of the 36th annual ACM symposium on theory of computing. Association for Computing Machinery, New York, pp 417–426

15. Jain K (2001) A factor 2 approximation for the generalized Steiner network problem. Combinatorica 21(1):39–60

16. Jain K, Vazirani VV (2001) Applications of approximation algorithms to cooperative games. In: Proceedings of the 33rd annual ACM symposium on theory of computing. Association for Computing Machinery, New York, pp 364–372

17. Kent K, Skorin-Kapov D (1996) Population monotonic cost allocation on mst's. In: Proceedings of the 6th international conference on operational research. Croatian Operational Research Society, Zagreb, pp 43–48

18. Könemann J, Leonardi S, Schäfer G (2005) A group-strategyproof mechanism for Steiner forests. In: Proceedings of the 16th annual ACM-SIAM symposium on discrete algorithms. Society for Industrial and Applied Mathematics, Philadelphia, pp 612–619

19. Megiddo N (1978) Cost allocation for Steiner trees. Networks 8(1):1–6

20. Moulin H, Shenker S (2001) Strategyproof sharing of submodular costs: budget balance versus efficiency. Econ Theory 18(3):511–533

21. Thimm M (2003) On the approximability of the Steiner tree problem. Theor Comput Sci 295(1–3):387–402

22. Vazirani VV (2001) Approximation algorithms. Springer, Berlin

Steiner Trees

Weili Wu[1,2,3] and Yaocun Huang[3]
[1]College of Computer Science and Technology, Taiyuan University of Technology, Taiyuan, Shanxi Province, China
[2]Department of Computer Science, California State University, Los Angeles, CA, USA
[3]Department of Computer Science, The University of Texas at Dallas, Richardson, TX, USA

Keywords

Approximation algorithm design

Years and Authors of Summarized Original Work

2006; Du, Graham, Pardalos, Wan, Wu, Zhao

Definition

Given a set of points, called *terminals*, in a metric space, the problem is to find the shortest tree interconnecting all terminals. There are three important metric spaces for Steiner trees, the Euclidean plane, the rectilinear plane, and the edge-weighted network. The Steiner tree problems in those metric spaces are called the *Euclidean Steiner tree (EST)*, the *rectilinear Steiner tree (RST)*, and the *network Steiner tree (NST)*, respectively. EST and RST have been found to have polynomial-time approximation schemes (PTAS) by using adaptive partition. However, for NST, there exists a positive number r such that computing r-approximation is NP-hard. So far, the best performance ratio of polynomial-time approximation for NST is achieved by k-restricted Steiner trees. However, in practice, the iterated 1-Steiner tree is used very often. Actually, the iterated 1-Steiner was proposed as a candidate of good approximation for Steiner minimum trees a long time ago. It has a very good record in computer experiments, but no correct analysis was given showing the iterated 1-Steiner tree having a performance ratio better than that of the minimum spanning tree until the recent work by Du et al. [9]. There is minimal difference in construction of the 3-restricted Steiner tree and the iterated 1-Steiner tree, which makes a big difference in analysis of those two types of trees. Why does the difficulty of analysis make so much difference? This will be explained in this article.

History and Background

The Steiner tree problem was proposed by Gauss in 1835 as a generalization of the Fermat problem. Given three points A, B, and C in the Euclidean plane, Fermat studied the problem of finding a point S to minimize $|SA| + |SB| + |SC|$. He determined that when all three inner angles of triangle ABC are less than

120°, the optimal S should be at the position that $\angle ASB = \angle BSC = \angle CSA = 120°$.

The generalization of the Fermat problem has two directions:

1. Given n points in the Euclidean plane, find a point S to minimize the total distance from S to n given points. This is still called the Fermat problem.
2. Given n points in the Euclidean plane, find the shortest network interconnecting all given points.

Gauss found the second generalization through communication with Schumacher. On March 19, 1836, Schumacher wrote a letter to Gauss and mentioned a paradox about Fermat's problem: Consider a convex quadrilateral $ABCD$. It is known that the solution of Fermat's problem for four points A, B, C, and D is the intersection E of diagonals AC and BD. Suppose extending DA and CB can obtain an intersection F. Now, move A and B to F. Then E will also be moved to F. However, when the angle at F is less than 120°, the point F cannot be the solution of Fermat's problem for three given points F, D, and C. What happens? (Fig. 1.)

On March 21, 1836, Gauss wrote a letter replying to Schumacher in which he explained that the mistake of Schumacher's paradox occurs at the place where Fermat's problem for four points A, B, C, and D is changed to Fermat's problem for three points F, C, and D. When A and B are identical to F, the total distance from E to four points A, B, C, and D equals $2|EF| + |EC| + |ED|$, not $|EF| + |EC| + |ED|$. Thus,

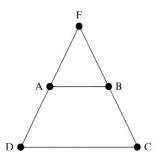

Steiner Trees, Fig. 1 Convex quadrilateral $ABCD$, Fermat's problem

the point E may not be the solution of Fermat's problem for F, C, and D. More importantly, Gauss proposed a new problem. He said that it is more interesting to find the shortest network rather than a point. Gauss also presented several possible connections of the shortest network for four given points.

It was unfortunate that Gauss' letter was not seen by researchers of Steiner trees at an earlier stage. Especially, R. Courant and H. Robbins who in their popular book *What is mathematics?* (published in 1941) [6] called Gauss' problem the Steiner tree so that "Steiner tree" became a popular name for the problem.

The Steiner tree became an important research topic in mathematics and computer science due to its applications in telecommunication and computer networks. Starting with Gilbert and Pollak's work published in 1968, many publications on Steiner trees have been generated to solve various problems concerning it.

One well-known problem is the *Gilbert-Pollak conjecture* on the Steiner ratio, which is the least ratio of lengths between the Steiner minimum tree and the minimum spanning tree on the same set of given points. Gilbert and Pollak in 1968 conjectured that the Steiner ratio in the Euclidean plane is $\sqrt{3}/2$ which is achieved by three vertices of an equilateral triangle. A great deal of research effort has been put into the conjecture and it was finally proved by Du and Hwang [7].

Another important problem is called the *better approximation*. For a long time no approximation could be proved to have a performance ratio smaller than the inverse of the Steiner ratio. Zelikovsky [14] made the first breakthrough. He found a polynomial-time 11/6-approximation for NST which beats 1/2, the inverse of the Steiner ratio in the edge-weighted network. Later, Berman and Ramaiye [2] gave a polynomial-time 92/72-approximation for RST, and Du, Zhang, and Feng [8] closed the story by showing that in any metric space, there exists a polynomial-time approximation with a performance ratio better than the inverse of the Steiner ratio provided that for any set of a fixed number of points, the Steiner minimum tree is polynomial-time computable.

All the above better approximations came from the family of k-restricted Steiner trees. By improving some detail of construction, the constant performance ratio was decreasing, but the improvements were also becoming smaller. In 1996, Arora [1] made significant progress for EST and RST. He showed the existence of PTAS for EST and RST. Therefore, the theoretical researchers now pay more attention to NST. Bern and [3] showed that NST is MAX SNP-complete. This means that there exists a positive number r; computing the r-approximation for NST is NP-hard. The best-known performance for NST was given by Robin and Zelikovsky [12]. They also gave a very simple analysis to a well-known heuristic, the iterated 1-Steiner tree for pseudo-bipartite graphs.

Analysis of the iterated 1-Steiner tree is another long-standing open problem. Since Chang [4,5] proposed that the iterated 1-Steiner tree approximates the Steiner minimum tree in 1972, its performance has been claimed to be very good through computer experiments [10, 13], but no theoretical analysis supported this claim. Actually, both the k-restricted Steiner tree and the iterated 1-Steiner tree are obtained by greedy algorithms, but with different types of potential functions. For the iterated 1-Steiner tree, the potential function is non-submodular, but for the k-restricted Steiner tree, it is submodular; a property that holds for k-restricted Steiner trees may not hold for iterated 1-Steiner trees. Actually, the submodularity of potential function is very important in analysis of greedy approximations [11]. Du et al. [9] gave a correct analysis for the iterated 1-Steiner tree with a general technique to deal with non-submodular potential function.

Key Results

Consider input edge-weighted graph $G = (V, E)$ of NST. Assume that G is a complete graph and the edge weight satisfies the triangular inequality; otherwise, consider the complete graph on V with each edge (u, v) having a weight equal to the length of the shortest path between u and v in G. Given a set P of terminals, a *Steiner tree* is a tree interconnecting all given terminals such that every leaf is a terminal.

In a Steiner tree, a terminal may have degree more than one. The Steiner tree can be decomposed, at those terminals with degree more than one, into smaller trees in which every terminal is a leaf. In such a decomposition, each resulting small tree is called a *full component*. The *size* of a full component is the number of terminals in it. A Steiner tree is *k-restricted* if every full component of it has a size at most k. The shortest k-restricted Steiner tree is also called the *k-restricted Steiner minimum tree*. Its length is denoted by $smt_k(P)$. Clearly, $smt_2(P)$ is the length of the minimum spanning tree on P, which is also denoted by $mst(P)$. Let $smt(P)$ denote the length of the Steiner minimum tree on P. If $smt_3(P)$ can be computed in polynomial time, then it is better than $mst(P)$ for an approximation of $smt(P)$. However, so far no polynomial-time approximation has been found for $smt_3(P)$. Therefore, Zelikovsky [14] used a greedy approximation of $smt_3(P)$ to approximate $smt(P)$. Actually, Chang [4, 5] used a similar greedy algorithm to compute an iterated 1-Steiner tree. Let $\mathcal{F}$ be a family of subgraphs of input edge-weighted graph G. For any connected subgraph H, denote by $mst(H)$ the length of the minimum spanning tree of H, and for any subgraph H, denote by $mst(H)$ the sum of $mst(H')$ for H' over all connected components of H. Define

$$gain(H) = mst(P) - mst(P:H) - mst(H),$$

where $mst(P:H)$ is the length of the minimum spanning tree interconnecting all unconnected terminals in P after every edge of H shrinks into a point.

Greedy Algorithm $H \leftarrow \emptyset$;
while P has not been interconnected by H **do**
choose $F \in \mathcal{F}$ to maximize $gain(H \cup F)$;
output $mst(H)$.

When $\mathcal{F}$ consists of all full components of size at most three, this greedy algorithm gives the 3-restricted Steiner tree of Zelikovsky [14]. When $\mathcal{F}$ consists of all 3-stars and all edges where a 3-star is a tree with three leaves and a central vertex, this greedy algorithm produces the iterated 1-Steiner tree. An interesting fact pointed out by Du et al. [9] is that the function $gain(\cdot)$ is submodular over all full components of size at most three, but not submodular over all 3-stars and edges.

Let us consider a base set E and a function f from all subsets of E to real numbers. f is *submodular* if for any two subsets A, B of E,

$$f(A) + f(B) \geq f(A \cup B) + f(A \cap B).$$

For $x \in E$ and $A \subseteq E$, denote $\Delta_x f(A) = f(A \cup \{x\}) - f(A)$.

Lemma 1 *f is submodular if and only if for any $A \subset E$ and distinct $x, y \in E - A$,*

$$\Delta_x \Delta_y f(A) \leq 0. \tag{1}$$

Proof Suppose f is submodular. Set $B = A \cup \{x\}$ and $C = A \cup \{y\}$. Then $B \cup C = A \cup A \cup \{x, y\}$ and $B \cap C = A$. Therefore, one has

$$f(A \cup \{x, y\}) - f(A \cup \{x\}) - f(A \cup \{y\}) + f(A) \leq 0,$$

that is, (1) holds.

Conversely, suppose (1) holds for any $A \subset E$ and distinct $x, y \in E - A$. Consider two subsets A, B of E. If $A \subseteq B$ or $B \subseteq A$, it is trivial to have

$$f(A) + f(B) \geq f(A \cup B) + f(A \cap B).$$

Therefore, one may assume that $A \backslash B \neq \emptyset$ and $B \backslash A \neq \emptyset$. Write $A \backslash B = \{x_1, \ldots, x_k\}$ and $B \backslash A = \{y_1, \ldots, y_h\}$. Then

$$f(A \cup B) - f(A) - f(B) + f(A \cap B)$$
$$= \sum_{i=1}^{k} \sum_{j=1}^{h} \Delta_{x_i} \Delta_{y_j} f(A \cup \{x_1, \ldots, x_{i-1}\} \cup \{y_1, \ldots, y_{j-1}\})$$
$$\leq 0,$$

where $\{x_1, \ldots x_{i-1}\} = \emptyset$ for $i = 1$ and $\{y_1, \ldots, y_{j-1}\} = \emptyset$ for $j = 1$. $\qquad \square$

Lemma 2 *Define* $f(H) = -mst(P : H)$. *Then* f *is submodular over edge set* E.

Proof Note that for any two distinct edges x and y not in subgraph H,

$$\Delta_x \Delta f(H)$$

$$= -mst(P : H \cup x \cup y) + mst(P : H \cup x)$$

$$+ mst(P : H \cup y) - mst(P : H)$$

$$= (mst(P : H) - mst(P : H \cup x \cup y))$$

$$- (mst(P : H) - mst(P : H \cup x))$$

$$+ (mst(P : H) - mst(P : H \cup y)).$$

Let T be a minimum spanning tree for unconnected terminals after every edge of H shrinks into a point. T contains a path P_x connecting two endpoints of x and also a path P_y connecting two endpoints of y. Let e_x (e_y) be a longest edge in P_x (P_y). Then

$$mst(P : H) - mst(P : H \cup x) = length(e_x),$$
$$mst(P : H) - mst(P : H \cup y) = length(e_y).$$

$mst(P : H) - mst(P : H \cup x \cup y)$ can be computed as follows: Choose a longest edge $e\prime$ from $P_x \cup P_y$. Note that $T \cup x \cup y - e'$ contains a unique cycle Q. Choose a longest edge $e\prime\prime$ from $(P_x \cup P_y) \cap Q$. Then

$$mst(P : H) - mst(P : H \cup x \cup y) = length(e'')$$

Now, to show the submodularity of f, it suffices to prove

$$length(e_x) + length(e_y) \geq length(e'') \quad (2)$$

Case 1. $e_x P_x \cap P_y$ and $e_y P_x \cap P_y$. Without loss of generality, assume $length(e_x) \geq length(e_y)$. Then one may choose $e' = e_x$ so that $(P_x \cup P_y) \cap Q = P_y$. Hence, one can choose $e'' = e_y$. Therefore, the equality holds for (2).

Case 2. $e_x P_x \cap P_y$ and $e_y \in P_x \cap P_y$. Clearly, $length(e_x) \geq length(e_y)$. Hence, one may choose $e' = e_x$ so that $(P_x \cup P_y) \cap Q = P_y$. Hence, one can choose $e'' = e_y$. Therefore, the equality holds for (2).

Case 3. $e_x \in P_x \cap P_y$ and $e_y P_x \cap P_y$. Similar to Case 2.

Case 4. $e_x \in P_x \cap P_y$ and $e_y \in P_x \cap P_y$. In this case, $length(e_x) = length(e_y) = length(e')$. Hence, (2) holds. $\qquad \square$

The following explains that the submodularity of $gain(\cdot)$ holds for a k-restricted Steiner tree.

Theorem 1 *Let* ε *be the set of all full components of a Steiner tree. Then* $gain(\cdot)$ *as a function on the power set of* ε *is submodular.*

Proof Note that for any $\mathcal{H} \subset \mathcal{E}$ and $x, y \in \mathcal{E} - \mathcal{H}$,

$$\Delta_x \Delta_y mst(H) = 0,$$

where $H = \cup_{z \in \mathcal{H}} z$. Thus, this theorem follows from Lemma 2.

Let $\mathcal{F}$ be the set of 3-stars and edges chosen in the greedy algorithm to produce an iterated 1-Steiner tree. Then $gain(\cdot)$ may not be submodular on $\mathcal{F}$. To see this fact, consider two 3-stars x and y in Fig. 2. Note that $gain(x \cup y) > gain(x), gain(y) \leq 0$, and $gain(\emptyset) = 0$. One has

$$gain(x \cup y) - gain(x) - gain(y) + gain(\emptyset) > 0.$$

$\qquad \square$

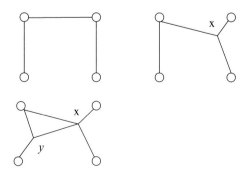

Steiner Trees, Fig. 2 An example for the proof of Theorem 1

Applications

The Steiner tree problem is a classic NP-hard problem with many applications in the design of computer circuits, long-distance telephone lines, multicast routing in communication networks, etc. There exist many heuristics of the greedy type for Steiner trees in the literature. Most of them have a good performance in computer experiments, without support from theoretical analysis. The approach given in this work may apply to them.

Open Problems

It is still open whether computing the 3-restricted Steiner minimum tree is NP-hard or not. For $k \geq 4$, it is known that computing the k-restricted Steiner minimum tree is NP-hard.

Cross-References

▶ Greedy Approximation Algorithms
▶ Minimum Spanning Trees
▶ Rectilinear Steiner Tree

Recommended Reading

1. Arora S (1996) Polynomial-time approximation schemes for Euclidean TSP and other geometric problems. In: Proceedings of the 37th IEEE symposium on foundations of computer science, pp 2–12
2. Berman P, Ramaiyer V (1994) Improved approximations for the Steiner tree problem. J Algorithms 17:381–408
3. Bern M, Plassmann P (1989) The Steiner problem with edge lengths 1 and 2. Inf Proc Lett 32:171–176
4. Chang SK (1972) The generation of minimal trees with a Steiner topology. J ACM 19:699–711
5. Chang SK (1972) The design of network configurations with linear or piecewise linear cost functions. In: Symposium on computer-communications, networks, and teletraffic. IEEE Computer Society Press, Los Alamitos, pp 363–369
6. Crourant R, Robbins H (1941) What is mathematics? Oxford University Press, New York
7. Du DZ, Hwang FK (1990) The Steiner ratio conjecture of Gilbert-Pollak is true. Proc Natl Acad Sci USA 87:9464–9466
8. Du DZ, Zhang Y, Feng Q (1991) On better heuristic for Euclidean Steiner minimum trees. In: Proceedings of the 32nd FOCS. IEEE Computer Society Press, Los Alamitos
9. Du DZ, Graham RL, Pardalos PM, Wan PJ, Wu W, Zhao W (2008) Analysis of greedy approximations with nonsubmodular potential functions. In: Proceedings of 19th ACM-SIAM symposium on discrete algorithms (SODA). ACM, New York, pp 167–175
10. Kahng A, Robins G (1990) A new family of Steiner tree heuristics with good performance: the iterated 1-Steiner approach. In: Proceedings of IEEE international conference on computer-aided design, Santa Clara, pp 428–431
11. Wolsey LA (1982) An analysis of the greedy algorithm for the submodular set covering problem. Combinatorica 2:385–393
12. Robin G, Zelikovsky A (2000) Improved Steiner trees approximation in graphs. In: SIAM-ACM symposium on discrete algorithms (SODA), San Francisco, pp 770–779
13. Smith JM, Lee DT, Liebman JS (1981) An $O(N \log N)$ heuristic for Steiner minimal tree problems in the Euclidean metric. Networks 11:23–39
14. Zelikovsky AZ (1993) The 11/6-approximation algorithm for the Steiner problem on networks. Algorithmica 9:463–470

Stochastic Knapsack

Viswanath Nagarajan
University of Michigan, Ann Arbor, MI, USA

Keywords

Adaptive; Packing constraints; Stochastic optimization

Years and Authors of Summarized Original Work

2004; Dean, Goemans, Vondrák
2011; Bhalgat, Goel, Khanna

Problem Definition

This problem deals with packing a maximum reward set of items into a knapsack of given capacity, when the item sizes are random. The input is a collection of n items, where each item $i \in [n] := \{1, \ldots, n\}$ has reward $r_i \geq 0$ and size $S_i \geq 0$, and a knapsack capacity $B \geq 0$. In the *stochastic knapsack* problem, all rewards are deterministic but the sizes are random. The random variables S_is are independent with known, arbitrary distributions. The actual size of an item is known only when it is placed into the knapsack. The objective is to add items sequentially (one by one) into the knapsack so as to maximize the expected reward of the items that fit into the knapsack. As usual, a subset T of items is said to fit into the knapsack if the total size $\sum_{i \in T} S_i$ is at most the knapsack capacity B.

A feasible solution (or policy) to the stochastic knapsack problem is represented by a decision tree. Nodes in this decision tree denote the current "state" of the solution (i.e., previously added items and the residual knapsack capacity) as well as the new item to place into the knapsack at this state. Branches in the decision tree denote the random size instantiations of items placed into the knapsack. Such solutions are called *adaptive policies*, to emphasize the fact that the items being placed may depend on previously observed outcomes. More formally, an adaptive policy is given by a mapping $\pi : 2^{[n]} \times [0, B] \rightarrow [n]$, where $\pi(T, C)$ denotes the next item to place into the knapsack when some subset $T \subseteq [n]$ of items has already been added, and $C = B - \sum_{i \in T} S_i$ is the residual knapsack capacity. The policy ends when the knapsack overflows (i.e., the total size of items added exceeds the knapsack capacity); we use the convention that no reward is obtained from the last overflowing item.

Notice that an arbitrary adaptive policy may require exponential space to even store. This motivates a special class of solutions, called *nonadaptive policies*. A nonadaptive policy is just specified by a fixed ordering of the n items, and the solution adds items into the knapsack in that order (irrespective of the actual size instantiations) until the knapsack overflows. Again, there is no reward obtained from the last overflowing item. While it may be easier to obtain a good nonadaptive policy, the obvious drawback is that nonadaptive policies may perform much worse than adaptive policies. The benefit of being adaptive is quantified by a measure called the *adaptivity gap*, which is the maximum ratio

(over all instances) of the expected reward of an optimal adaptive policy to the expected reward of an optimal nonadaptive policy.

In both the adaptive and nonadaptive settings, the stochastic knapsack problem is at least NP-hard, since it generalizes the deterministic knapsack problem. Moreover, certain questions regarding adaptive policies are PSPACE-hard [4].

Notation We assume that the item size distributions are given explicitly. For any item $i \in [n]$ define its effective reward $w_i = r_i \cdot \Pr[S_i \leq B]$ and its mean truncated size $\mu_i = \mathbb{E}[\min\{S_i, B\}]$. Note that the expected reward obtained by placing the single item i into the knapsack is exactly w_i.

Key Results

Dean, Goemans, and Vondrák introduced the stochastic knapsack problem and the notion of adaptivity gaps. They proved the following.

Theorem 1 ([4]) *There is a polynomial time algorithm for the stochastic knapsack problem that computes a nonadaptive policy having expected reward at least $\frac{1}{4}$ that of an optimal adaptive policy.*

As a consequence, the adaptivity gap of the stochastic knapsack problem is also upper bounded by four. Dean, Goemans, and Vondrák [4] also showed an instance of stochastic knapsack that lower bounds the adaptivity gap by $\frac{5}{4}$.

The algorithm in Theorem 1 uses a natural greedy approach. It outputs the better of the following two nonadaptive policies:

- Place the single item $i^* = \arg\max_{i \in [n]} w_i$.
- Place items in nonincreasing order of w_i/μ_i.

In terms of adaptive policies, Bhalgat, Goel, and Khanna proved the following.

Theorem 2 ([2, 3]) *For any constant $\epsilon > 0$, there is polynomial time algorithm for the stochastic knapsack problem that computes an* *adaptive policy having expected reward at least $\frac{1}{2+\epsilon}$ that of an optimal adaptive policy.*

The algorithm in Theorem 2 relies on an intricate transformation of general size distributions to certain canonical distributions and an algorithm for computing a near-optimal adaptive policy under canonical size distributions.

Extensions

Several variants of the stochastic knapsack problem have been studied, and good algorithms have been obtained for them.

Correlated Stochastic Knapsack
This is a generalization of the stochastic knapsack problem, where each item's reward is also random and possibly correlated with its size. The distributions across items are still independent: so the correlations are only between the size and reward of a single item. Gupta, Krishnaswamy, Molinaro, and Ravi [6] gave an algorithm for this problem that computes a nonadaptive policy having expected reward within factor 8 of the optimal adaptive policy. Recently, Ma [8] gave an algorithm that for any constant $\epsilon > 0$ computes an adaptive policy having expected reward within factor $2 + \epsilon$ of the optimal adaptive policy; this algorithm requires item sizes and the capacity B to be specified in unary.

Budgeted Multi-armed Bandit
The input to this problem consists of a bound B and n "arms" (each arm is a Markov chain with rewards at its states and a specified starting state). A feasible policy consists of B steps. In each step, the policy can select one arm $i \in [n]$: upon selecting arm i, it gets the reward at the current state of arm i and the arm transitions to its next state according to its Markov chain. The objective is to maximize the expected total reward over B steps of the policy. Again, we are interested in adaptive policies, whose actions may depend on past outcomes. Guha and Munagala [5] introduced this problem and gave a $(2 + \epsilon)$-approximation algorithm, under the assumption that the rewards of each arm satisfy

a "Martingale" condition (which is natural in many settings). Gupta, Krishnaswamy, Molinaro, and Ravi [6] gave the first constant-factor approximation algorithm for this problem without the Martingale reward assumption. The constant factor in the latter result was improved to 6.75 by Ma [8].

Stochastic Orienteering
This problem is defined on a finite metric space (V, d) with vertex set V and distance function $d : V \times V \rightarrow \mathbb{R}_+$ that satisfies (i) symmetry $d(u, v) = d(v, u)$ for all $u, v \in V$ and (ii) triangle inequality $d(u, w) \leq d(u, v) + d(v, w)$ for all $u, v, w \in V$. The distances between vertices denote travel times. Each vertex $i \in V$ corresponds to a job having deterministic reward $r_i \geq 0$ and random processing time $S_i \geq 0$. The random variables S_is are independent with known, arbitrary distributions. Given a start-vertex $\rho \in V$ and bound B, the goal is to compute a policy, which describes a (possibly adaptive) path originating from ρ that visits vertices and runs the respective jobs. The actual processing time of a job is known only when it completes. The policy ends when the total time (for travel plus processing) exceeds B. The objective is to maximize the expected total reward; there is no reward obtained from a partially completed job (which may occur at the end of the policy). As before, an optimal policy may be adaptive and choose the next job to run based on previously observed outcomes. Gupta, Krishnaswamy, Nagarajan, and Ravi [7] gave an $O(\log \log B)$-approximation algorithm for the stochastic orienteering problem; this result requires the bound B, distances, and processing times to be integer valued. As a corollary, [7] also upper bounded the adaptivity gap by $O(\log \log B)$. Recently, Bansal and Nagarajan [1] gave an $\Omega\left(\sqrt{\log \log B}\right)$ lower bound on the adaptivity gap.

Applications

The stochastic knapsack problem and its variants model various applications in advertising, logistics, medical diagnosis, and robotics.

Open Problems

It is not known if the stochastic knapsack problem is any harder to approximate than the usual (deterministic) knapsack problem. In particular, is there a PTAS for stochastic knapsack? Determining a tight bound on its adaptivity gap is also an interesting open question.

Recommended Reading

1. Bansal N, Nagarajan V (2014) On the adaptivity gap of stochastic orienteering. In: IPCO, Bonn, pp 114–125
2. Bhalgat A (2011) A $(2 + \epsilon)$-approximation algorithm for the stochastic knapsack problem. Unpublished manuscript
3. Bhalgat A, Goel A, Khanna S (2011) Improved approximation results for stochastic knapsack problems. In: SODA, San Francisco, pp 1647–1665
4. Dean BC, Goemans, MX, Vondrák J (2008) Approximating the stochastic knapsack problem: the benefit of adaptivity. Math Oper Res 33(4):945–964
5. Guha S, Munagala K (2013) Approximation algorithms for Bayesian multi-armed bandit problems. CoRR abs/1306.3525
6. Gupta A, Krishnaswamy R, Molinaro M, Ravi R (2011) Approximation algorithms for correlated knapsacks and non-martingale bandits. In: FOCS, Palm Springs, pp 827–836
7. Gupta A, Krishnaswamy R, Nagarajan V, Ravi R (2012) Approximation algorithms for stochastic orienteering. In: SODA, Kyoto, pp 1522–1538
8. Ma W (2014) Improvements and generalizations of stochastic knapsack and multi-armed bandit approximation algorithms: extended abstract. In: SODA, Portland, pp 1154–1163

Stochastic Scheduling

Jay Sethuraman
Industrial Engineering and Operations Research, Columbia University, New York, NY, USA

Keywords

Queueing; Sequencing

Years and Authors of Summarized Original Work

2001; Glazebrook, Nino-Mora

Problem Definition

Scheduling is concerned with the allocation of scarce resources (such as machines or servers) to competing activities (such as jobs or customers) over time. The distinguishing feature of a *stochastic* scheduling problem is that some of the relevant data are modeled as *random variables*, whose distributions are known, but whose actual realizations are not. Stochastic scheduling problems inherit several characteristics of their deterministic counterparts. In particular, there are virtually an unlimited number of problem types depending on the machine environment (single machine, parallel machines, job shops, flow shops), processing characteristics (preemptive versus nonpreemptive, batch scheduling versus allowing jobs to arrive "over time," due dates, deadlines), and objectives (makespan, weighted completion time, weighted flow time, weighted tardiness). Furthermore, stochastic scheduling models have some new, interesting features (or difficulties!):

- The scheduler may be able to make inferences about the remaining processing time of a job by using information about its elapsed processing time; whether the scheduler is allowed to make use of this information or not is a question for the modeler.
- Many scheduling algorithms make decisions by comparing the processing times of jobs. If jobs have deterministic processing times, this poses no problems as there is only one way to compare them. If the processing times are random variables, comparing processing times is a subtle issue. There are many ways to compare pairs of random variables, and some are only *partial* orders. Thus, any algorithm that operates by comparing processing times must now specify the particular ordering used to

compare random variables (and to determine what to do if two random variables are not comparable under the specified ordering).

These considerations lead to the notion of a scheduling *policy*, which specifies how the scarce resources have to be allocated to the competing activities as a function of the *state* of the system at any point in time. The state of the system includes information such as prior job completions, the elapsed time of jobs currently in service, the realizations of the random release dates and due dates (if any), and any other information that can be inferred based on the history observed so far. A policy that is allowed to make use of all this information is said to be *dynamic*, whereas a policy that is not allowed to use any state information is *static*.

Given any policy, the objective function for a stochastic scheduling model operating under that policy is typically a random variable. Thus, comparison of two policies entails the comparison of the associated random variables, so the *sense* in which these random variables are compared must be specified. A common approach is to find a solution that optimizes the *expected value* of the objective function (which has the advantage that it is a total ordering); less commonly, other orderings such as the stochastic ordering or the likelihood ratio ordering are used.

Key Results

Consider a single machine that processes n jobs, with the (random) processing time of job i given by a distribution $F_i(\cdot)$ whose mean is p_i. The Weighted Shortest Expected Processing Time first (WSEPT) rule sequences the jobs in decreasing order of w_i/p_i. Smith [13] proved that the WSEPT rule minimizes the sum of weighted completion times when the processing times are deterministic. Rothkopf [11] generalized this result and proved the following:

Theorem 1 *The WSEPT rule minimizes the expected sum of the weighted completion times in*

the class of all nonpreemptive dynamic policies (and hence also in the class of all nonpreemptive static policies).

If preemption is allowed, the WSEPT rule is not optimal. Nevertheless, Sevcik [12] showed how to assign an "index" to each job at each point in time such that scheduling a job with the largest index at each point in time is optimal. Such policies are called index policies and have been investigated extensively because they are (relatively) simple to implement and analyze. Often, the optimality of index policies can be proved under some assumptions on the processing time distributions. For instance, Weber, Varaiya, and Walrand [14] proved the following result for scheduling n jobs on m identical parallel machines:

Theorem 2 *The SEPT rule minimizes the expected sum of completion times in the class of all nonpreemptive dynamic polices, if the processing time distributions of the jobs are stochastically ordered.*

For the same problem but with the makespan objective, Bruno, Downey, and Frederickson [3] proved the optimality of the Longest Expected Processing Time first rule provided all the jobs have exponentially distributed processing times.

One of the most significant achievements in stochastic scheduling is the proof of optimality of index policies for the multiarmed bandit problem and its many variants, due originally to Gittins and Jones [5, 6]. In an instance of the bandit problem, there are N projects, each of which is in any one of a possibly finite number of states. At each (discrete) time, any one of the projects can be attempted, resulting in a random reward; the attempted project undergoes a (Markovian) state transition, whereas the other projects remain frozen and do not change state. The goal of the decision maker is to determine an optimal way to attempt the projects so as to maximize the total discounted reward. Of course one can solve this problem as a large, stochastic dynamic program, but such an approach does not reveal any structure and is moreover computationally impractical except for very small problems. (Also, if the state space of any project is countable or infinite, it is not clear how one can solve the resulting DP exactly!) The remarkable result of Gittins and Jones [6] is the optimality of index policies: to each state of each project, one can associate an index so that attempting a project with the largest index at any point in time is optimal. The original proof of Gittins and Jones [6] has subsequently been simplified by many authors; moreover, several alternative proofs based on different techniques have appeared, leading to a much better understanding of the class of problems for which index policies are optimal [2, 4, 5, 10, 17].

While index policies are easy to implement and analyze, they are often not optimal in many problems. It is therefore natural to investigate the gap between an optimal index policy (or a natural heuristic) and an optimal policy. For example, the WSEPT rule is a natural heuristic for the problem of scheduling jobs on identical parallel machines to minimize the expected sum of the weighted completion times. However, the WSEPT rule is not necessarily optimal. Weiss [16] showed that, under mild and reasonable assumptions, the expected number of times that the WSEPT rule differs from the optimal decision is bounded above by a *constant*, independent of the number of jobs. Thus, the WSEPT rule is asymptotically optimal. As another example of a similar result, Whittle [18] generalized the multiarmed bandit model to allow for state transitions in projects that are *not* activated, giving rise to the "restless bandit" model. For this model, Whittle [18] proposed an index policy whose asymptotic optimality was established by Weber and Weiss [15].

A number of stochastic scheduling models allow for jobs to arrive over time according to a stochastic process. A commonly used model in this setting is that of a multiclass queueing network. Multiclass queueing networks serve as useful models for problems in which several types of *activities* compete for a limited number of shared *resources*. They generalize deterministic job-shop problems in two ways: jobs arrive *over time* and each job has a *random* processing time at each stage. The optimal control problem in a multiclass queueing network is to find

an optimal allocation of the available resources to activities over time. Not surprisingly, index policies are optimal only for restricted versions of this general model. An important example is scheduling a multiclass single-server system with feedback: there are N types of jobs; type i jobs arrive according to a Poisson process with rate λ_i, require service according to a service-time distribution $F_i(\cdot)$ with mean processing time s_i, and incur holding costs at rate c_i per unit time. A type i job after undergoing processing becomes a type j job with probability p_{ij} or exits the system with probability $1 - \sum_j p_{ij}$ isn't in document. The objective is to find a scheduling policy that minimizes the expected holding cost rate in steady state. Klimov [9] proved the optimality of index policies for this model, as well as for the objective in which the total discounted holding cost is to be minimized. While the optimality result does not hold when there are many parallel machines, Glazebrook and Niño-Mora [7] showed that this rule is asymptotically optimal. For more general models, the prevailing approach is to use approximations such as fluid approximations [1] or diffusion approximations [8].

Applications

Stochastic scheduling models are applicable in many settings, most prominently in computer and communication networks, call centers, logistics and transportation, and manufacturing systems [4, 10].

Cross-References

▶ List Scheduling
▶ Minimum Weighted Completion Time

Recommended Reading

1. Avram F, Bertsimas D, Ricard M (1995) Fluid models of sequencing problems in open queueing networks: an optimal control approach. In: Kelly FP, Williams RJ (eds) Stochastic networks. Proceedings

of the international mathematics association, vol 71. Springer, New York, pp 199–234

2. Bertsimas D, Niño-Mora J (1996) Conservation laws, extended polymatroids and multiarmed bandit problems: polyhedral approaches to indexable systems. Math Oper Res 21(2):257–306

3. Bruno J, Downey P, Frederickson GN (1981) Sequencing tasks with exponential service times to minimize the expected flow time or makespan. J ACM 28:100–113

4. Dacre M, Glazebrook K, Nino-Mora J (1999) The achievable region approach to the optimal control of stochastic systems. J R Stat Soc Ser B 61(4):747–791

5. Gittins JC (1979) Bandit processes and dynamic allocation indices. J R Stat Soc Ser B 41(2):148–177

6. Gittins JC, Jones DM (1974) A dynamic allocation index for the sequential design experiments. In: Gani J, Sarkadu K, Vince I (eds) Progress in statistics. European meeting of statisticians I. North Holland, Amsterdam, pp 241–266

7. Glazebrook K, Niño-Mora J (2001) Parallel scheduling of multiclass M/M/m queues: approximate and heavy-traffic optimization of achievable performance. Oper Res 49(4):609–623

8. Harrison JM (1988) Brownian models of queueing networks with heterogenous customer populations. In: Fleming W, Lions PL (eds) Stochastic differential systems, stochastic control theory and applications. Proceedings of the international mathematics association. Springer, New York, pp 147–186

9. Klimov GP (1974) Time-sharing service systems I. Theory Probab Appl 19:532–551

10. Pinedo M (2002) Scheduling: theory, algorithms and systems, 2nd edn. Prentice Hall, Englewood Cliffs

11. Rothkopf M (1966) Scheduling with random service times. Manag Sci 12:707–713

12. Sevcik KC (1974) Scheduling for minimum total loss using service time distributions. J ACM 21:66–75

13. Smith WE (1956) Various optimizers for single-stage production. Nav Res Logist Q 3:59–66

14. Weber RR, Varaiya P, Walrand J (1986) Scheduling jobs with stochastically ordered processing times on parallel machines to minimize expected flow time. J Appl Probab 23:841–847

15. Weber RR, Weiss G (1990) On an index policy for restless bandits. J Appl Probab 27:637–648

16. Weiss G (1992) Turnpike optimality of Smith's rule in parallel machine stochastic scheduling. Math Oper Res 17:255–270

17. Whittle P (1980) Multiarmed bandit and the Gittins index. J R Stat Soc Ser B 42:143–149

18. Whittle P (1988) Restless bandits: activity allocation in a changing world. In: Gani J (ed) A celebration of applied probability. Journal of applied probability, vol 25A. Applied Probability Trust, Sheffield, pp 287–298

String Matching

Maxime Crochemore[1,2,3] and Thierry Lecroq[4]
[1] Department of Computer Science, King's College London, London, UK
[2] Laboratory of Computer Science, University of Paris-East, Paris, France
[3] Université de Marne-la-Vallée, Champs-sur-Marne, France
[4] Computer Science Department and LITIS Faculty of Science, Université de Rouen, Rouen, France

Keywords

Border; Pattern matching; Period; Prefix; Shift function; String matching; Suffix

Years and Authors of Summarized Original Work

1977; Knuth, Morris, Pratt
1977; Boyer, Moore
1994; Crochemore, Czumaj, Gąsieniec, Jarominek, Lecroq, Plandowski, Rytter

Problem Definition

Given a *pattern string* $P = p_1 p_2 \ldots p_m$ and a *text string* $T = t_1 t_2 \ldots t_n$, both being sequences over an alphabet Σ of size σ, the *exact string-matching (ESM)* problem is to find one or, more generally, all the text positions where P occurs in T, that is, compute the set $\{j \mid 1 \leq j \leq n - m + 1 \text{ and } P = t_j t_{j+1} \ldots t_{j+m-1}\}$.

Both worst- and average-case complexities are considered. For the latter one assumes that pattern and text are randomly generated by choosing each character uniformly and independently from Σ. For simplicity and practicality the assumption $m = o(n)$ is made in this entry.

Key Results

Most algorithms that solve the ESM problem proceed in two steps: a preprocessing phase of the pattern P followed by a searching phase over the text T. The preprocessing phase serves to collect information on the pattern in order to speed up the searching phase.

The searching phase of string-matching algorithms works as follows: they first align the left ends of the pattern and the text, then compare the aligned symbols of the text and the pattern – this specific work is called an attempt or a scan – and after a whole match of the pattern or after a mismatch, they shift the pattern to the right. They repeat the same procedure again until the right end of the pattern goes beyond the right end of the text. The scanning part can be viewed as operating on the text through a window, which size is most often the length of the pattern. This processing manner is called the scan and shift mechanism. Different scanning strategies of the window lead to algorithms having specific properties and advantages.

The brute force algorithm for the ESM problem consists in checking if P occurs at each position j on T, with $1 \leq j \leq n - m + 1$. It does not need any preprocessing phase. It runs in quadratic time $O(mn)$ with constant extra space and performs $O(n)$ character comparisons on average. This is to be compared with the following bounds.

Theorem 1 (Cole et al. [6]) *The minimum number of character comparisons to solve the ESM problem in the worst case is $n + \frac{9}{4m}(n - m)$, and there exists an algorithm performing at most $n + \frac{8}{3(m+1)}(n - m)$ character comparisons in the worst case.*

Theorem 2 (Yao [26]) *The ESM problem needs $\Omega\left(\frac{\log_\sigma m}{m} \times n\right)$ time in expectation.*

Online Text Parsing

The first linear ESM algorithm appears in the 1970s. The preprocessing phase consists in computing the periods of the pattern prefixes, or equivalently the length of the longest border for all the prefixes of the pattern. A border of a string is both a prefix and a suffix of it distinct from the string itself. Let $next[i]$ be the length of the longest border of $p_1 \ldots p_{i-1}$. Consider an attempt at position j, when the pattern $p_1 \ldots p_m$

is aligned with the segment $t_j \ldots t_{j+m-1}$ of the text. Assume that the first mismatch (during a left to right scan) occurs between symbols p_i and t_{i+j} for $1 \leq i \leq m$. Then, $p_1 \ldots p_{i-1} = t_j \ldots t_{i+j-1} = u$ and $a = p_i \neq t_{i+j} = b$. A prefix v of the pattern may match a suffix of the portion u of the text. By the definition of table $next$, a shift that aligns $p_{next[i]}$ with t_{i+j} cannot miss any occurrence of P in T, and thus backtracking in the text is not necessary. There exist two variants [18,19], depending on whether $p_{next[i]}$ has to be different from p_i or not. The second is slightly more efficient.

Theorem 3 (Knuth, Morris, and Pratt [18]) *The text searching can be done in time $O(n)$ and space $O(m)$. Preprocessing the pattern can be done in time $O(m)$.*

The search can also be realized using an implementation with successor by default of the deterministic automaton $\mathcal{D}(P)$ recognizing the language $\Sigma^* P$. The size of the implementation is $O(m)$ independent of the alphabet size, due to the fact that $\mathcal{D}(P)$ possesses $m + 1$ states, m forward arcs, and at most m backward arcs. Using the automaton for searching a text leads to an algorithm having an efficient delay (maximum time for processing a character of the text).

Theorem 4 (Hancart [15]) *Searching for the pattern P can be done with a delay of $O(min\{\sigma, \log_2 m\})$ letter comparisons.*

Note that for most algorithms the pattern preprocessing is not necessarily done before the text parsing, as it can be performed on the fly during the parsing.

Algorithms Sublinear on the Average

The Boyer-Moore algorithm [3] is among the most efficient ESM algorithms. A simplified version of it, or the entire algorithm, is often implemented in text editors for the search and substitute commands.

The algorithm scans the characters of the window from right to left beginning with its rightmost symbol. In case of a mismatch (or a complete match of the pattern), it uses two precomputed functions to shift the pattern to the right.

These two shift functions are called the *bad-character shift* and the *good-suffix shift*. They are based on the following observations. Assume that a mismatch occurs between character $p_i = a$ of the pattern and character $t_{i+j} = b$ of the text during an attempt at position j. Then, $p_{i+1} \ldots p_m = t_{i+j+1} \ldots t_{j+m} = u$ and $p_i \neq t_{i+j}$. The good-suffix shift consists in aligning the segment $t_{i+j+1} \ldots t_{j+m}$ with its rightmost occurrence in P that is preceded by a character different from p_i. Another variant called the *best-suffix shift* consists in aligning the segment $t_{i+j} \ldots t_{j+m}$ with its rightmost occurrence in P. Both variants can be computed in time and space $O(m)$ independent of the alphabet size. If there exists no such segment, the shift consists in aligning the longest suffix v of $t_{i+j+1} \ldots t_{j+m}$ with a matching prefix of x. The bad-character shift consists in aligning the text character t_{i+j} with its rightmost occurrence in $p_1 \ldots p_{m-1}$. If t_{i+j} does not appear in the pattern, no occurrence of P in T can overlap the symbol t_{i+j}, then the left end of the pattern is aligned with the character at position $i + j + 1$. The search can then be done in $O(n/m)$ in the best case.

Theorem 5 (Cole [5]) *During the search for a nonperiodic pattern P of length m (such that the length of the longest border of P is less than $m/2$) in a text T of length n, the Boyer-Moore algorithm performs at most $3n$ comparisons between letters of P and of T.*

In practice, when scanning the window from right to left during an attempt, it is sometimes more efficient to only use the bad-character shift. This was first done by the Horspool algorithm [16]. Other practical efficient algorithms are the Quick Search by Sunday [24] and the Tuned Boyer-Moore by Hume and Sunday [17].

Yao's bound can be reached using an indexing structure giving access to all the factors of the reverse pattern. This is done by the Reverse Factor algorithm also called BDM (for Backward Dawg Matching).

Theorem 6 (Crochemore et al. [9]) *The search can be done in optimal expected time*

$O\left(\frac{\log_\sigma m}{m} \times n\right)$ *using the suffix automaton or the suffix tree of the reverse pattern.*

A factor oracle can be used instead of an index structure. A factor oracle is an automaton simpler than the suffix automaton that may recognize some additional strings of length smaller than m. The only string of length m accepted by the factor oracle of a string w of length m is w itself. Then it can be used for solving the ESM problem. This is done by the Backward Oracle Matching (BOM) algorithm of Allauzen, Crochemore, and Raffinot [1]. Its behavior in practice is similar to the one of the BDM algorithm.

Time-Space Optimal Algorithms

Algorithms of this type run in linear time (for both preprocessing and searching) and need only constant space in addition to the inputs.

Theorem 7 (Galil and Seiferas [13]) *The search can be done optimally in time $O(n)$ and constant extra space.*

After Galil and Seiferas' first solution, other solutions are by Crochemore-Perrin [8] and Rytter [22]. These algorithms rely on a partition of the pattern in two parts; they first search for the right part of the pattern from left to right, and then, if no mismatch occurs, they search for the left part. The partition can be the perfect factorization [13], the critical factorization [8], or based on the lexicographically maximum suffix of the pattern [22]. Another solution by Crochemore [7] is a variant of KMP [18]: it computes lower bounds of pattern prefixes periods on the fly and requires no preprocessing.

Bit-Parallel Solution

It is possible to use the bit-parallelism technique for ESM.

Theorem 8 (Baeza-Yates and Gonnet [2]; Wu and Manber [25]) *If the length m of the string P is smaller than the number of bits of a machine word, the preprocessing phase can be done in time and space $O(\sigma)$ and the searching phase in time $O(n)$.*

It is even possible to use this bit-parallelism technique to simulate the BDM algorithm. This is realized by the BNDM (Backward Nondeterministic Dawg Matching) algorithm [20].

There exists another method that uses the bit-parallelism technique that is optimal on the average. It considers sparse q-grams and thus avoids to scan a lot of text positions. It is due to Fredriksson and Grabowski [12].

Applications

The methods that are described here apply to the treatment of the natural language, of genetic and musical sequences, the problems of safety related to data flows like virus detection, and the management of the textual databases, to quote only some immediate applications.

Open Problems

There remain only a few open problems on this question. It is still unknown if it is possible to design an average optimal time constant space string-matching algorithm. The exact size of the Boyer-Moore automaton is still unknown [3]. The Boyer-Moore automaton was first introduced by Knuth [18]. Its states encode all the possible situations when searching the pattern with the Boyer-Moore algorithm and remember every text character already matched in the window.

Experimental Results

The book of G. Navarro and M. Raffinot [21] is a good introduction and presents an experimental map of ESM algorithms for different alphabet sizes and pattern lengths. Basically, the Shift-Or algorithm is efficient for small alphabets and short patterns, the BNDM algorithm is efficient for medium-sized alphabets and medium-length patterns, the Horspool algorithm is efficient for large alphabets, and the BOM algorithm is efficient for long patterns. The article of S. Faro

and T. Lecroq [11] updates the experimental map with the most recent results.

URLs to Code and Data Sets

The site monge.univ-mlv.fr/~lecroq/string presents a large number of ESM algorithms (see also [4]). Each algorithm is implemented in C code and a Java applet is given. The site www.dmi.unict.it/~faro/smart presents SMART, a string-matching research tool, which contains the C code of a great number of exact string-matching algorithms and some corpora (natural language, musical, biological, and random texts). The user can easily plug its own algorithm to compare it against some selected algorithms.

Cross-References

▶ Approximate String Matching is the version where errors are permitted;
▶ Multiple String Matching is the version where a finite set of patterns is searched for in a text;
▶ Regular Expression Matching is the more complex case where P can be a regular expression;
▶ Suffix Trees and Arrays refers to the case where the text is preprocessed.

Further information can be found in the three following books: [10, 14] and [23].

Recommended Reading

1. Allauzen C, Crochemore M, Raffinot M (1999) Factor oracle: a new structure for pattern matching. In: SOFSEM'99, Milovy. LNCS, vol 1725, pp 291–306
2. Baeza-Yates RA, Gonnet GH (1992) A new approach to text searching. C ACM 35(10):74–82
3. Boyer RS, Moore JS (1977) A fast string searching algorithm. C ACM 20(10):762–772
4. Charras C, Lecroq T (2004) Handbook of exact string matching algorithms. King's College, London
5. Cole R (1994) Tight bounds on the complexity of the Boyer-Moore string matching algorithm. SIAM J Comput 23(5):1075–1091
6. Cole R, Hariharan R, Paterson M, Zwick U (1995) Tighter lower bounds on the exact complexity of string matching. SIAM J Comput 24(1):30–45

7. Crochemore M (1992) String-matching on ordered alphabets. Theor Comput Sci 92(1):33–47
8. Crochemore M, Perrin D (1991) Two-way string matching. J ACM 38(3):651–675
9. Crochemore M, Czumaj A, Gąsieniec L, Jarominek S, Lecroq T, Plandowski W, Rytter W (1994) Speeding up two string matching algorithms. Algorithmica 12(4/5):247–267
10. Crochemore M, Hancart C, Lecroq T (2007) Algorithms on strings. Cambridge University Press, New York
11. Faro S, Lecroq T (2013) The exact online string matching problem: a review of the most recent results. C ACM, Harlow, 45(2):13
12. Fredriksson K, Grabowski S (2005) Practical and optimal string matching. In: Proceedings of SPIRE'2005, Buenos Aires. LNCS, vol 3772, pp 374–385
13. Galil Z, Seiferas J (1983) Time-space optimal string matching. J Comput Syst Sci 26(3):280–294
14. Gusfield D (1997) Algorithms on strings, trees and sequences. Cambridge University Press, New York
15. Hancart C (1993) On Simon's string searching algorithm. Inf Process Lett 47(2):95–99
16. Horspool RN (1980) Practical fast searching in strings. Softw Pract Exp 10(6):501–506
17. Hume A, Sunday DM (1991) Fast string searching. Softw Pract Exp 21(11):1221–1248
18. Knuth DE, Morris JH Jr, Pratt VR (1977) Fast pattern matching in strings. SIAM J Comput 6(1):323–350
19. Morris JH Jr, Pratt VR (1970) A linear pattern-matching algorithm. Report 40, University of California, Berkeley
20. Navarro G, Raffinot M (1998) A bit-parallel approach to suffix automata: fast extended string matching. In: Farach-Colton M (ed) Proceedings of the 9th annual symposium on combinatorial pattern matching, Piscataway. Lecture notes in computer science, vol 1448. Springer, Berlin, Piscataway, New Jersey, USA, pp 14–33
21. Navarro G, Raffinot M (2002) Flexible pattern matching in strings – practical on-line search algorithms for texts and biological sequences. Cambridge University Press, Cambridge
22. Rytter W (2003) On maximal suffixes and constant-space linear-time versions of KMP algorithm. Theor Comput Sci 299(1–3):763–774
23. Smyth WF (2002) Computing patterns in strings. Addison Wesley Longman, Harlow
24. Sunday DM (1990) A very fast substring search algorithm. C ACM 33(8):132–142
25. Wu S, Manber U (1992) Fast text searching allowing errors. C ACM 35(10):83–91
26. Yao A (1979) The complexity of pattern matching for a random string. SIAM J Comput 8:368–387

String Sorting

Rolf Fagerberg
Department of Mathematics and Computer Science, University of Southern Denmark, Odense, Denmark

Keywords

Sorting of multidimensional keys; Vector sorting

Years and Authors of Summarized Original Work

1997; Bentley, Sedgewick

Problem Definition

The problem is to sort a set of strings into lexicographical order. More formally: A *string* over an *alphabet* Σ is a finite sequence $x_1 x_2 x_3 \ldots x_k$ where $x_i \in \Sigma$ for $i = 1, \ldots, k$. The x_is are called the *characters* of the string, and k is the *length* of the string. If the alphabet Σ is ordered, the *lexicographical order* on the set of strings over Σ is defined by declaring a string $x = x_1 x_2 x_3 \ldots x_k$ smaller than a string $y = y_1 y_2 y_3 \ldots y_l$ if either there exists a $j \geq 1$ such that $x_i = y_i$ for $1 \leq i < j$ and $x_j < y_j$ or if $k < l$ and $x_i = y_i$ for $1 \leq i \leq k$. Given a set S of strings over some ordered alphabet, the problem is to sort S according to lexicographical order.

The input to the string sorting problem consists of an array of pointers to the strings to be sorted. The output is a permutation of the array of pointers, such that traversing the array will point to the strings in nondecreasing lexicographical order.

The complexity of string sorting depends on the alphabet as well as the machine model. The main solution [15] described in this entry works for alphabets of unbounded size (i. e., comparisons are the only operations on characters of Σ) and can be implemented on a pointer machine. See below for more information on the asymptotic complexity of string sorting in various settings.

Key Results

This section is structured as follows: first, the key result appearing in the title of this entry [15] is described; then an overview of other relevant results in the area of string sorting is given.

The string sorting algorithm proposed by Bentley and Sedgewick in 1997 [15] is called three-way radix quicksort [5]. It works for unbounded alphabets, for which it achieves optimal performance.

Theorem 1 *The algorithm three-way radix quicksort sorts K strings of total length N in time $O(K \log K + N)$.*

This time complexity is optimal, which follows by considering strings of the form bbb...bx, where all xs are different: Sorting the strings can be no faster than sorting the xs, and all bs must be read (else an adversary could change one unread b to a or c, making the returned order incorrect). A more precise version of the bounds above (upper as well as lower) is $K \log K + D$, where D is the sum of the lengths of the *distinguishing prefixes* of the strings. The distinguishing prefix d_s of a string s in a set S is the shortest prefix of s which is not a prefix of another string in S (or is s itself, if s is a prefix of another string). Clearly, $K \leq D \leq N$.

The three-way radix quicksort of Bentley and Sedgewick is not the first algorithm to achieve this complexity; however, it is a very simple and elegant way of doing it. As demonstrated in [3, 15], it is also very fast in practice. Although various elements of the algorithm had been noted earlier, their practical usefulness for string sorting was overlooked until the work in [15].

Three-way radix quicksort is shown in pseudocode in Fig. 1 (adapted from [5]), where S is a list of strings to be sorted and d is an integer. To

SORT(S, d)

    IF $|S| \leq 1$:
        RETURN
    Choose a partitioning character $v \in \{s_d \mid s \in S\}$
    $S_< = \{s \in S \mid s_d < v\}$
    $S_= = \{s \in S \mid s_d = v\}$
    $S_> = \{s \in S \mid s_d > v\}$
    SORT($S_<$, d)
    IF $v \neq$ EOS:
        SORT($S_=$, $d + 1$)
    SORT($S_>$, d)
    $S = S_< + S_= + S_>$

String Sorting, Fig. 1 Three-way radix quicksort (assuming each string ends in a special EOS character)

sort S, an initial call SORT(S, 1) is made. The value s_d denotes the dth character of the string s, and + denotes concatenation. The presentation in Fig. 1 assumes that all strings end in a special end-of-string (EOS) character (such as the null character in C). In an actual implementation, S will be an array of pointers to strings, and the sort will be in-place (using an in-place method from standard quicksort for three-way partitioning of the array into segments holding $S_<$, $S_=$, and $S_>$), rendering concatenation superfluous.

Correctness follows from the following invariant being maintained by the algorithm: At the start of a call SORT(S, d), all strings in S agree on the first $d - 1$ characters.

Time complexity depends on how the partitioning character v is chosen. One particular choice is the median of all the dth characters (including doublets) of the strings in S. Partitioning and median finding can be done in time $O(|S|)$, which is $O(1)$ time per string partitioned. Hence, the total running time of the algorithm is the sum over all strings of the number of partitionings they take part in. For each string, let a partitioning be of type I if the string ends up in $S_<$ or $S_>$ and of type II if it ends up in $S_=$. For a string s, type II can only occur $|d_s|$ times and type I can only occur log K times. Hence, the running time is $O(K \log K + D)$.

Like for standard quicksort, median finding impairs the constant factors of the algorithm, and more practical choices of partitioning character include selecting a random element among all the dth characters of the strings in S and selecting the median of three elements in this set. The worst-case bound is lost, but the result is a fast, randomized algorithm.

Note that the ternary recursion tree of three-way radix quicksort is equivalent to a trie over the input strings where each trie node is implemented by a binary search tree whose node elements are the child edges (in the trie) of the trie node. In more detail, a node in a binary tree contains the character of a trie edge and a pointer to the root of the binary tree implementing the corresponding trie child. The search keys in a binary tree are the characters in its nodes. This trie implementation is named ternary search trees in [15]. In the recursion tree of three-way radix quicksort, an edge representing a recursive call on $S_<$ or $S_>$ corresponds to a tree edge inside a binary tree implementing a trie node, and an edge representing a recursive call on $S_=$ corresponds to a trie edge.

For the version of the algorithm where the partitioning character v is chosen as the median of all the dth characters, it is not hard to see that the binary trees representing the trie nodes become weighted trees. These are binary trees in which each element x has an associated weight w_x, and searches for x take $O(\log W/w_x)$, where $W = \Sigma_x w_x$ is the sum of all weights in the binary tree. Here, the weight of a binary tree node storing character x is the number of strings which in the trie reside below the corresponding trie edge. As shown in [13], in such a trie implementation, searching for a string P among K stored strings takes time $O(\log K + |P|)$, which is optimal for unbounded (i.e., comparison-based) alphabets. Hence, by the correspondence between the recursion trees of three-way radix quicksort and ternary search trees, three-way radix quicksort may additionally be viewed as a construction algorithm for an efficient dictionary structure for strings.

Other key results in the area of string sorting are now described. The classic string sorting algorithm is radixsort, which assumes a constant-sized alphabet. The least-significant-digit-first variant is easy to implement and runs in $O(N + l|\Sigma|)$ time, where l is the length of the longest string. The most-significant-digit-first variant is more complicated to implement but has a better running time of $O(D + d|\Sigma|)$, where D is the sum of the lengths of the distinguishing prefixes and d is the longest distinguishing prefix. McIlroy et al. [12] discusses in depth efficient implementations of radixsort.

If the alphabet consists of integers, then on a word-RAM the complexity of string sorting is essentially determined by the complexity of integer sorting. More precisely, the time (when allowing randomization) for sorting strings is $\Theta(\text{Sort}_{\text{Int}}(K) + N)$, where $\text{Sort}_{\text{Int}}(K)$ is the time to sort K integers [2], which currently is known to be $O(K\sqrt{\log \log K})$ [11].

Returning to comparison-based model, the papers [8, 10] give generic methods for turning any data structure over one-dimensional keys into a data structure over strings. Using finger search trees, this gives an adaptive sorting method for strings which uses $O(N + K \log(F/K))$ time, where F is the number of inversions among the strings to be sorted.

Concerning space complexity, it has been shown [9] that string sorting can still be done in $O(K \log K + N)$ time using only $O(1)$ space besides the strings themselves. However, this assumes that all strings have equal lengths.

All algorithms so far are designed to work in internal memory, where CPU time is assumed to be the dominating factor. For external memory computation, a more relevant cost measure is the number of I/Os performed, as captured by the I/O model [1], which models a two-level memory hierarchy with an infinite outer memory, an inner memory of size M, and transfer (I/Os) between the two levels taking place in blocks of size B. For external memory, upper bounds were first given in [4], along with matching lower bounds in restricted I/O models. For a comparison-based model where strings may only be moved in blocks of size B (hence, characters may not be moved individually), it is shown in [4] that string sorting takes $\Theta(N_1/B \log_{M/B}(N_1/B) + K_2 \log_{M/B} K_2 + N/B)$ I/Os, where N_1 is the total length of strings shorter than B characters, K_2 is the number of strings of at least B characters, and N is the total number of characters. This bound is equal to the sum of the I/O costs of sorting the characters of the short strings, sorting B characters from each of the long strings, and scanning all strings. In the same paper, slightly better bounds in a model where characters may be moved individually in internal memory are given, as well as some upper bounds for non-comparison-based string sorting. Further bounds (using randomization) for non-comparison-based string sorting have been given, with I/O bounds of $O(K/B \log_{M/B}(K/M) \log \log_{M/B}(K/M) + N/B)$ [7] and $O(K/B(\log_{M/B}(N/M))^2 \log_2 K + N/B)$ (Ferragina, personal communication).

Returning to internal memory, it may also there be the case that memory hierarchy effects are the determining factor for the running time of algorithms but now due to cache faults rather than disk I/Os. Heuristic algorithms (i.e., algorithms without good worst-case bounds), aiming at minimizing cache faults for internal memory string sorting, have been developed. Of these, the burstsort line of algorithms [16] performs particularly well in experiments.

Applications

Data sets consisting partly or entirely of string data are very common: Most database applications have strings as one of the data types used, and in some areas, such as bioinformatics, Web retrieval, and word processing, string data is predominant. Additionally, strings form a general and fundamental data model, containing, e.g., integers and multidimensional data as special cases. Since sorting is arguably among the most important data processing tasks in any domain, string sorting is a general and important problem with wide practical applications.

Open Problems

As appears from the bounds discussed above, the asymptotic complexity of the string sorting problem is known for comparison-based alphabets. For integer alphabets on the word-RAM, the problem is almost closed in the sense that it is equivalent to integer sorting, for which the gap left between the known bounds and the trivial linear lower bound is small.

In external memory, the situation is less settled. As noted in [4], a natural upper bound to hope for in a comparison-based setting is to meet the lower bound of $\Theta(K/B \log_{M/B} K/M + N/B)$ I/Os, which is the sorting bound for K single characters plus the complexity of scanning the input. The currently known upper bounds only get close to this when leaving the comparison-based setting and allowing randomization.

Experimental Results

In [15], experimental comparison of two implementations (one simple and one tuned) of three-way radix quicksort with a tuned quicksort [6] and a tuned radixsort [12] showed the simple implementation to always outperform the quicksort implementation and the tuned implementation to be competitive with the radixsort implementation.

In [3], experimental comparison among existing and new radixsort implementations (including the one used in [15]), as well as tuned quicksort and tuned three-way radix quicksort, was performed. This study confirms the picture of three-way radix quicksort as very competitive, always being one of the fastest algorithms, and arguably the most robust across various input distributions.

Data Sets

The data sets used in [15]: http://www.cs.princeton.edu/~rs/strings/. The data sets used in [3]: http://dl.acm.org/citation.cfm?id=297136.

URL to Code

Code in C from [15]: http://www.cs.princeton.edu/~rs/strings/.

Code in C from [3]: http://dl.acm.org/citation.cfm?id=297136.

Code in Java from [14]: http://www.cs.princeton.edu/~rs/Algs3.java1-4/code.txt.

Cross-References

▶ Suffix Array Construction
▶ Suffix Tree Construction
▶ Suffix Tree Construction in Hierarchical Memory

Acknowledgments Research supported by Danish Council for Independent Research, Natural Sciences.

Recommended Reading

1. Aggarwal A, Vitter JS (1988) The input/output complexity of sorting and related problems. Commun ACM 31:1116–1127
2. Andersson A, Nilsson S (1994) A new efficient radix sort. In: Proceedings of the 35th annual symposium on foundations of computer science (FOCS'94), Santa Fe. IEEE Computer Society Press, pp 714–721
3. Andersson A, Nilsson S (1998) Implementing radixsort. ACM J Exp Algorithmics 3:7
4. Arge L, Ferragina P, Grossi R, Vitter JS (1997) On sorting strings in external memory (extended abstract). In: Proceedings of the 29th annual ACM symposium on theory of computing (STOC'97), El Paso. ACM, pp 540–548
5. Bentley J, Sedgewick R (1998) Algorithm alley: sorting strings with three-way radix quicksort. Dr Dobb's J Softw Tools 23:133–134, 136–138
6. Bentley JL, McIlroy MD (1993) Engineering a sort function. Softw Pract Exp 23:1249–1265
7. Fagerberg R, Pagh A, Pagh R (2006) External string sorting: faster and cache-oblivious. In: Proceedings of the 23rd symposium on theoretical aspects of computer science (STACS'06), Marseille. LNCS, vol 3884. Springer, pp 68–79
8. Franceschini G, Grossi R (2004) A general technique for managing strings in comparison-driven data structures. In: Proceedings of the 31st international colloquium on automata, languages and programming (ICALP'04), Turku. LNCS, vol 3142. Springer, pp 606–617
9. Franceschini G, Grossi R (2005) Optimal in-place sorting of vectors and records. In: Proceedings of the 32nd international colloquium on automata, languages and programming (ICALP'05), Lisbon. LNCS, vol 3580. Springer, pp 90–102
10. Grossi R, Italiano GF (1999) Efficient techniques for maintaining multidimensional keys in linked data structures. In: Proceedings of the 26th international colloquium on automata, languages and programming (ICALP'99), Prague. LNCS, vol 1644. Springer, pp 372–381
11. Han Y, Thorup M (2002) Integer sorting in $O(n\sqrt{\log\log n})$ expected time and linear space. In: Proceedings of the 43rd annual symposium on foundations of computer science (FOCS'02), Vancouver. IEEE Computer Society Press, pp 135–144
12. McIlroy PM, Bostic K, McIlroy MD (1993) Engineering radix sort. Comput Syst 6:5–27
13. Mehlhorn K (1979) Dynamic binary search. SIAM J Comput 8:175–198
14. Sedgewick, R (2003) Algorithms in Java, Parts 1–4, 3rd edn. Addison-Wesley, Boston
15. Sedgewick R, Bentley J (1997) Fast algorithms for sorting and searching strings. In: Proceedings of the 8th annual ACM-SIAM symposium on discrete algorithms (SODA'97), New Orleans. ACM, pp 360–369
16. Sinha R, Zobel J, Ring D (2006) Cache-efficient string sorting using copying. ACM J Exp Algorithmics 11: Article No. 1.2

Strongly Connected Dominating Set

Zhang Zhao
College of Mathematics Physics and Information Engineering, Zhejiang Normal University, Zhejiang, Jinhua, China

Keywords

Absorbing set; Approximation algorithm; Dominating set; Strongly connected

Years and Authors of Summarized Original Work

2006; Du, Thai, Li, Liu, Zhu
2007: Park, Willson, Wang, Thai, Wu, Du
2009; Li, Du, Wan, Gao, Zhang, Wu
2012; Xu, Li
2014; Zhang, Wu, Wu, Li, Chen

Problem Definition

Let $G = (V, E)$ be a directed graph. For an arc $(u, v) \in E$, u is said to *dominate* v, and v is said to *absorb* u. Vertex u is also called a *dominator* of v, and vertex v is called an *absorber* of u. A vertex set $D \subseteq V$ is a *dominating set* (DS) of G if every vertex in $V \setminus D$ has a dominator in D; it is an *absorbing set* (AS) of G if every vertex in $V \setminus D$ has an absorber in D. A directed graph G is *strongly connected* if for any pair of ordered vertices $u, v \in V$, there is a directed path in G from u to v. The "Minimum Strongly Connected Dominating and Absorbing Set" problem (MSCDAS) is to find a vertex set D such that D is both a dominating set and an absorbing set of G and the subgraph of G induced by D is strongly connected.

Disk graph is a geometric graph which is of particular interest in the study of MSCDAS, since disk graph is a model of heterogeneous wireless sensor network, and as one can see in the application part, MSCDAS plays an important role in wireless sensor network. In a *disk graph*, every vertex u corresponds to a sensor on the plane equipped with an omnidirectional antenna of transmission radius $r(u)$. Another sensor v can correctly decode the message sent by u if and only if v is in the disk centered at u with radius $r(u)$. Hence, there is an arc (u, v) in the disk graph if and only if $\|uv\| \leq r(u)$, where $\| \cdot \|$ is the Euclidean distance between u and v. In particular, if all sensors are equipped with the same transmission radius, then the disk graph degenerates to an undirected graph called *unit disk graph*.

Key Results

Hardness Results

In a general digraph, the MSCDAS problem cannot be approximated within a factor of $(1 - \varepsilon) \ln n$ for any real number $\varepsilon > 0$, where n is the number of vertices in the digraph. Even in disk graph, MSCDAS is still NP-hard. These hardness results follow from the fact that their undirected counterparts have these hardness results [1, 5].

MSCDAS in General Digraph

Li et al. [8] gave a $(3H(n-1)-1)$-approximation for MSCDAS, where $H(\gamma) = \sum_{i=1}^{\gamma} 1/i$ is the harmonic number.

The algorithm is based on the following observation. For a vertex u in a digraph G, a spanning *in-arborescence* (resp. *out-arborescence*) rooted at u is a spanning sub-digraph of G in which every vertex except u has in-degree (resp. out-degree) exactly one and vertex u has in-degree (resp. out-degree) zero. For a spanning arborescence T of G, denote by $\text{int}(T)$ the set of internal vertices of T. For any vertex u, suppose T^{in} and T^{out} are spanning in-arborescence and spanning out-arborescence of G rooted at u, respectively. Then $\text{int}(T^{\text{in}}) \cup \text{int}(T^{\text{out}})$ is an SCDAS of G.

Define the problem "Spanning Arborescence with Fewest Internal Vertices" (SAFIV) as follows: given a digraph G and a vertex u, find a spanning arborescence T rooted at u such that $|\text{int}(T)|$ is as small as possible. By the above observation, if SAFIV has a ρ-approximation, then MSCDAS has a 2ρ-approximation. Li et al. gave a $(1.5H(n-1) - 0.5)$-approximation for SAFIV, and thus the approximation ratio $(3H(n-1) - 1)$ for MSCDAS follows.

The approximation algorithm for SAFIV uses the idea in [6, 7] which study the problem of "Minimum Node-Weighted Steiner Tree" (MN-WST). The idea is to iteratively merge smaller arborescences greedily (a vertex is a trivial arborescence) until finally one gets one arborescence including all vertices which is rooted at the given vertex. It was pointed out in [8] that using the method in [6], the approximation ratio for SAFIV can be further reduced to $1.35 \ln n$. Since SAFIV is at least as hard as the minimum connected dominating set problem, it cannot be approximated within factor $(1 - \varepsilon) \ln n$. Any progress narrowing the gap between $\ln n$ and $1.35 \ln n$ would be interesting.

MSCDAS in Disk Graph

Making use of geometric properties, can the approximation ratio for MSCDAS be better in a disk graph? The answer is yes. Du et al. [2] were the first to give a constant approximation in this setting. Their idea was further explored by Park et al. [11] to output an SCDAS with size at most $9.6(k + 1/2)^2 \text{opt} + 14.8(k + 1/2)^2$, where opt is the size of an optimal solution and $k = r_{max}/r_{min}$, the ratio between the maximum radius and the minimum radius. The core in their work is an algorithm for SAFIV, which first colors all vertices white and then, by growing a search tree step by step, turns the colors to either black, blue, or gray. The set of black vertices forms a dominating set, and the set of blue vertices connects these black vertices into an out-arborescence. In fact, black vertices are mutually independent, where two vertices u and v are said to be *independent* if either uv or vu is not an arc. Two independent vertices have distance greater than r_{min}. Such a property guarantees an upper bound for the number of black vertices. Furthermore, the structure of a search tree guarantees that the number of blues vertices is no larger than that of black vertices. Then, the desired approximation ratio follows. It should be noted that if r_{max}/r_{min} is unbounded, then the approximation ratio is not a constant.

Without a bounded assumption on r_{max}/r_{min}, Xu and Li [12] showed that a $(2 + \varepsilon)$-approximation exists for MDAS, which is a combination of a PTAS for MDS and a PTAS for MAS. In fact, the PTAS for MAS is a special case of the "Geometric Hitting Set" problem studied in [10], and the PTAS for MDS is a variation for the MDS problem in an undirected graph studied in [4]. Both PTASs are obtained through a local search method. The analysis is based on the separator theorem for planar graphs [3, 9]. Zhang et al. [13] also obtained approximation ratio $(2 + \varepsilon)$ using the same method. Based on such a DAS, adding Steiner nodes to connect, Zhang

et al. showed that a $(4 + 3 \ln(2 + \varepsilon) \text{opt} + \varepsilon)$-approximation exists for MSCDAS. When the optimal value opt is substantially smaller than n, this is an improvement on ratio $3H(n - 1) - 1$ for disk graphs.

Applications

One application of MSCDAS is the communication in wireless sensor network (WSN). In a WSN, information is distributed among sensors by multi-hop transmissions. If all sensors transmit messages in a flooding manner, then a lot of energy is wasted, and large amount of interference is created. To alleviate such problems, it is desirable that only a small fraction of sensors participate in the transmission, while information can still be successfully shared. An SCDAS can serve for this purpose. Suppose D is an SCDAS of directed graph G (the topology of the WSN). If there is a message at source sensor u to be sent to destination sensor v, then the message can be first sent from u to its absorber; since $G[D]$ is strongly connected, it can be successfully relayed to the dominator of v and then sent to v.

Open Problems

It is still open whether there exists a constant approximation algorithm for MSCDAS in disk graph.

Recommended Reading

1. Clark BN, Colbourn CJ, Johnson DS (1991) Unit disk graphs. Ann Discret Math 48:165–177
2. Du D-Z, Thai M, Li Y, Liu D, Zhu S (2006) Strongly connected dominating sets in wireless sensor networks with unidirectional links. In: Proceedings of APWEB, Harbin. LNCS, vol 3841, pp 13–24
3. Frederickson GN (1987) Fast algorithms for shortest paths in planar graphs, with applications. SIAM J Comput 16:1004–1022
4. Gibson M, Pirwani I (2010) Algorithms for dominating set in disk graphs: breaking the log n barrier. In: Algorithms – ESA, Liverpool. LNCS Vol. 6346, pp 243–254

5. Guha S, Khuller S (1998) Approximation algorithms for connected dominating sets. Algorithmica 20(4):374–387
6. Guha S, Khuller S (1999) Improved methods for approximating node weighted Steiner trees and connected dominating sets. Inf Comput 150:57–74
7. Klein P, Ravi R (1995) A nearly best-possible approximation algorithm for node-weighted Steiner trees. J Algorithms 19:104–114
8. Li D, Du H, Wan P-J, Gao X, Zhang Z, Wu W (2009) Construction of strongly connected dominating sets in asymmetric multihop wireless networks. Theor Comput Sci 410:661–669
9. Lipton RJ, Tarjan RE (1979) A separator theorem for planar graphs. SIAM J Appl Math 36:177–189
10. Mustafa NH, Ray S (2009) PTAS for Geometric hitting set problems via local search. In: SCG'09, Aarhus, 8–10 June 2009
11. Park MA, Willson J, Wang C, Thai M, Wu W, Du D-Z (2007) A dominating and absorbent set in a wireless ad-hoc network with different transmission ranges. In: MobiHoc'07, Montréal
12. Xu X, Li X (2012) Efficient construction of dominating set in wireless networks. http://arxiv.org/abs/1208.5738
13. Zhang Z, Wu W, Wu L, Li Y, Chen Z (accepted) Strongly connected dominating and absorbing set in directed disk graph. To be published in Int J Sens Netw

Subexponential Parameterized Algorithms

Fedor V. Fomin
Department of Informatics, University of Bergen, Bergen, Norway

Keywords

Chordal graph; Exponential time hypothesis; Graph editing; Interval graph; Minimum fill-in; Parameterized complexity

Years and Authors of Summarized Original Work

2009; Alon, Lokshtanov, Saurabh
2013; Fomin, Villanger
2014; Drange, Fomin, Pilipczuk, Villanger

Problem Definition

A *parameterized problem* is a language $L \subseteq \Sigma^* \times \mathbb{N}$, where Σ is a fixed, finite alphabet. The second component is called the *parameter* of the problem. The central notion in parameterized complexity is the notion of *fixed-parameter tractability (FPT)*. A parameterized problem L is called FPT if it can be determined in time $f(k) \cdot n^c$ whether or not $(x, k) \in L$, where $n = |(x, k)|$, f is a computable function depending only on k, and c is a constant independent of n and k. The complexity class containing all fixed-parameter tractable problems is called FPT.

While in the definition of class FPT, we are happy with any computable function f, from application perspective it is often desirable to have the asymptotic growth of f as slow as possible. Take as an example an FPT problem VERTEX COVER which has been subjected to intense scrutiny with progressively faster algorithms designed for it. Let us remind that in the VERTEX COVER problem, we are asked if an n vertex graph G contains a vertex cover of size k or in other words a set of vertices S such that every edge of G has at least one endpoint in S. Starting from a k^k algorithm of Buss and Goldsmith in 1993, there have been algorithms with $f(k) \in \{2^k, 1.324718^k, 1.29175^k, 1.2906^k, 1.271^k, 1.2738^k\}$. The current fastest algorithm for VERTEX COVER runs in time $1.2738^k n^{\mathcal{O}(1)}$ (see the entry ▶ Vertex Cover Search Trees from this book). The ever-decreasing running time leads to the following natural question: can VERTEX COVER admit a *subexponential* time algorithm? That is, can it have an algorithm with running time $2^{o(k)} n^{\mathcal{O}(1)}$? The negative answer to this question would imply that $P \neq NP$. However, using a stronger assumption in complexity theory, namely, exponential time hypothesis (ETH) (see the entry ▶ Exponential Lower Bounds for k-SAT Algorithms in this book), one can show that if ETH holds, then the answer to our question is NO. Moreover, subject to ETH, there are no subexponential algorithms for many other natural NP-complete problems. Thus, another natural question arises:

is it true that every *NP*-complete problem cannot be solved in subexponential time? Interestingly, the answer to this question is again NO, and there are examples in the literature of such problems. Coming back to our example of VERTEX COVER problem, if we restrict the input graph to be planar, the problem remains *NP*-complete, but the brute-force algorithm problem can be sped up even more. That is, VERTEX COVER on planar graphs can be solved in time $2^{O(\sqrt{k})} \cdot n^{O(1)}$ by *a subexponential algorithm*. We refer to more parameterized subexponential algorithms on planar graphs to the ► Bidimensionality in this book.

Until recently, the only subexponential algorithms were known for "geometric" graph problems, that is, problems on planar graphs or graphs excluding some fixed graph as minors. In 2009, Alon, Lokshtanov, and Saurabh [1] obtained the first parameterized subexponential algorithm for a natural "nongeometric" problem. This result has acted as catalyst for the discovery of new subexponential time algorithms. In this article, we give a short overview of these algorithms.

Key Results

FAST

In the FEEDBACK ARC SET IN TOURNAMENTS (FAST) problem, we are given an *n*-vertex tournament T and a positive integer k; the question is whether one can make T into a directed acyclic graph by deleting at most k arcs.

FAST
Input: A tournament $T = (V, E)$ and a nonnegative integer k.
Parameter: k.
Question: Is there $F \subseteq E, |F| \leq k$, such that diraph $H = (V, E \setminus F)$ is acyclic?

Alon, Lokshtanov, and Saurabh in [1] obtained a parameterized subexponential algorithm for FAST.

Theorem 1 ([1]) *FAST is solvable in time* $2^{\sqrt{k} \log k} n^{O(1)}$.

The theorem is proved by making use of a novel randomized technique called *Chromatic Coding*. It appeared that subexponential algorithms exist for several other problems on tournaments (see entry ► Computing Cutwidth and Pathwidth of Semi-complete Digraphs in this book).

Fill-In

The next "nongeometric" problem for which a subexponential algorithm was found happened to be the classical MINIMUM FILL-IN problem.

A graph is *chordal* (or triangulated) if every cycle of length at least four contains a chord, i.e., an edge between nonadjacent vertices of the cycle. The MINIMUM FILL-IN problem (also known as MINIMUM TRIANGULATION and CHORDAL GRAPH COMPLETION) is to decide if a given graph G can be transformed into a chordal graph by adding at most k edges.

MINIMUM FILL-IN
Input: A graph $G = (V, E)$ and a nonnegative integer k.
Parameter: k.
Question: Is there $F \subseteq [V]^2, |F| \leq k$, such that graph $H = (V, E \cup F)$ is chordal?

Theorem 2 ([6]) MINIMUM FILL-IN *is solvable in time* $2^{\sqrt{k} \log k} n^{O(1)}$.

The proof of the theorem is based on a combinatorial bound estimating the number of specific objects in the graph, namely, potential maximal cliques.

Completion to Graph Classes

Since discoveries of subexponential algorithms for FAST and MINIMUM FILL-IN, it appeared that several other graph modification problems admit subexponential algorithms. In particular, it was shown that problems of completion to a certain subclass of chordal graphs like trivially perfect, threshold [4], split [7], proper interval [2], and interval graphs [3] admit parameterized subexponential algorithms.

On the other hand, it has been shown that for a number of other graph classes, like cographs, completion to these classes of graphs cannot be done in parameterized subexponential time unless the exponential time hypothesis (ETH) fails [4].

Open Problems

The most natural open question about the given subexponential algorithms is the question about lower bounds. As a concrete example, an algorithm for FAST with running time bound $2^{o(\sqrt{k})}n^{\mathcal{O}(1)}$ would actually be a $2^{o(n)}$ time algorithm which inclines us to suspect that $2^{\mathcal{O}(\sqrt{k})}$ is the best possible dependency on k in the running time for this problem. Unfortunately, there is a big gap here between what we suspect and what we can prove, even assuming ETH. The only tight bound on parameterized subexponential algorithms for graph modification problems we are aware of is the p-CLUSTERING problem [5].

Cross-References

▶ Bidimensionality
▶ Computing Cutwidth and Pathwidth of Semi-complete Digraphs
▶ Exact Algorithms for Treewidth
▶ Exponential Lower Bounds for k-SAT Algorithms

Recommended Reading

1. Alon N, Lokshtanov D, Saurabh S (2009) Fast FAST. In: Proceedings of the 36th international colloquium of automata, languages and programming (ICALP). Lecture notes in computer science, vol 5555. Springer, Berlin/New York, pp 49–58
2. Bliznets I, Fomin FV, Pilipczuk M, Pilipczuk M (2014) A subexponential parameterized algorithm for proper interval completion. In: Proceedings of the 22nd annual European symposium on algorithms (ESA 2014). Lecture notes in computer science, vol 8737. Springer, Heidelberg, pp 173–183
3. Bliznets I, Fomin FV, Pilipczuk M, Pilipczuk M (2014) A subexponential parameterized algorithm for interval completion. CoRR. abs/1402.3473
4. Drange PG, Fomin FV, Pilipczuk M, Villanger Y (2014) Exploring subexponential parameterized complexity of completion problems. In: Proceedings of the 31st international symposium on theoretical aspects of computer science (STACS). Leibniz international proceedings in informatics (LIPIcs), vol 25. Schloss Dagstuhl – Leibniz-Zentrum fuer Informatik, Dagstuhl, pp 288–299
5. Fomin FV, Kratsch S, Pilipczuk M, Pilipczuk M, Villanger Y (2013) Tight bounds for parameterized complexity of cluster editing. In: Proceedings of the 30th international symposium on theoretical aspects of computer science (STACS). Leibniz international proceedings in informatics (LIPIcs), vol 20. Schloss Dagstuhl – Leibniz-Zentrum fuer Informatik, Dagstuhl, pp 32–43
6. Fomin FV, Villanger Y (2013) Subexponential parameterized algorithm for minimum fill-in. SIAM J Comput 42(6):2197–2216
7. Ghosh E, Kolay S, Kumar M, Misra P, Panolan F, Rai A, Ramanujan MS (2012) Faster parameterized algorithms for deletion to split graphs. In: Proceedings of the 13th Scandinavian symposium and workshops on algorithm theory (SWAT). Lecture notes in computer science, vol 7357. Springer, Berlin/New York, pp 107–118

Subset Sum Algorithm for Bin Packing

Julián Mestre
Department of Computer Science, University of Maryland, College Park, MD, USA
School of Information Technologies, The University of Sydney, Sydney, NSW, Australia

Keywords

Approximation algorithm; Bin packing; Greedy; Knapsack; Subset sum

Years and Authors of Summarized Original Work

1972; Graham
1999; Gupta, Ho

2009; Epstein, Kleiman, Mestre
2011; Epstein, Kleiman

Problem Definition

Bin packing is a classical problem in combinatorial optimization. Given a collection of n items with different sizes, the objective is to pack the items into a minimum number of uniform capacity bins. More formally, the input of the bin packing problem is described by a set of n items $I = \{1, \ldots, n\}$ and a size function $s : I \rightarrow [0, 1]$. The output is a packing of the items into bins $B_1, \ldots, B_k \subseteq I$ such that $s(B_j) \leq 1$ for $j = 1, \ldots, k$, where the notation $s(B)$ denotes $\sum_{i \in B} s_i$ for any $B \subseteq I$. The objective is to minimize the number bins used in the packing.

The SUBSET-SUM algorithm is an intuitively appealing greedy heuristic for the bin packing problem: Starting from the empty packing, the algorithm repeatedly finds a subset B of yet-unpacked items maximizing $s(B)$ subject to $s(B) \leq 1$, adds B to the packing, and iterates. Each iteration requires that we solve an instance of the *knapsack* problem. In practice, instead of finding the optimal solution, one can use an fully polynomial time approximation scheme (FPRAS) to compute a $(1 - \epsilon)$-approximate solution [6].

This note is concerned with the worst-case asymptotic performance of the SUBSET-SUM algorithm. For a given instance $s : I \rightarrow [0, 1]$, we use OPT(s) to denote the number of bins used in an optimal packing of s and SS(s) to denote the number of bins used by the SUBSET-SUM algorithm. Then for a given class $\mathcal{C}$ of instances, we define the worst-case asymptotic approximation ratio of SUBSET-SUM as

$$R_{\text{SS}}^{\infty}(\mathcal{C}) = \lim_{k \to \infty} \sup_{\substack{s \in \mathcal{C} \\ \text{OPT}(s) = k}} \frac{\text{SS}(s)}{\text{OPT}(s)}. \quad (1)$$

Finally, we use R_{SS}^{∞} to denote the ratio for general instances of the problem.

Key Results

Lower Bound on R_{SS}^{∞}

Graham [4] provided a family of instance exhibiting an approximation ratio that tends to $\sum_{i=1}^{\infty} \frac{1}{2^i - 1} \approx 1.6067$.

Theorem 1 (Graham [4]) $R_{\text{SS}}^{\infty} \geq \sum_{i=1}^{\infty} \frac{1}{2^i - 1} \approx 1.6067$.

Proof Consider the following instance parameterized by two positive integers r and N. For each $j = 1, \ldots, r$, we create N items of size $2^{-i} + \delta$, where $\delta = 2^{-2r}$. Let us denote this instance with s. Provided that $2^i - 1$ divides N for all $i = 1, \ldots, r$, it is not hard to see that SUBSET-SUM first packs the smallest items into $N/(2^r - 1)$ bins, then it packs the second-smallest items into $N/(2^{r-1} - 1)$ bins, and so on, until it packs the largest items into N bins. On the other hand, the optimal solution uses just N bins by packing one item of each size class per bin. Therefore,

$$\frac{\text{SS}(s)}{\text{OPT}(s)} = \sum_{i=1}^{r} \frac{1}{2^i - 1}, \quad (2)$$

which quickly approaches 1.6067 as r grows. $\square$

Upper Bound on R_{SS}^{∞}

A trivial upper bound on R_{SS}^{∞} is 2. This follows from the fact only the last bin can be less than half full. Caprara and Pferschy [1] gave the first nontrivial upper bound, by showing that R_{SS}^{∞} is at most $4/3 + \ln 4 \approx 1.6210$. Interestingly, Graham [4] had conjectured that the true value of R_{SS}^{∞} should match his lower bound. This conjecture was finally proven by Epstein et al. [2].

Theorem 2 (Epstein et al. [3]) $R_{\text{SS}}^{\infty} \leq \sum_{i=1}^{\infty} \frac{1}{2^i - 1} \approx 1.6067$.

The proof of this result uses *weighting functions* and a *factor revealing mathematical program*. Here we only sketch the high level idea of the approach. Let B be one of the bins opened by SUBSET-SUM. For every item $i \in B$ we define

$$w_i = \begin{cases} \frac{s_i}{s(B)} & \text{if } 1 - s_{\min} \leq s(B), \\ s_i & \text{otherwise,} \end{cases} \quad (3)$$

where $s_{\min}$ is the size of the smallest yet-unpacked item just before opening B.

The weights are used to charge the cost of the packing computed by SUBSET-SUM to an optimal packing. The following lemma allows us to bound the performance of the algorithm provided we can show that the sum of the weights is comparable to the cost of the SUBSET-SUM packing and that no bin in the optimal solution is charged too much.

Lemma 1 *Let $\mathcal{O}$ be an optimal solution and $\mathcal{B}$ be the solution computed SUBSET-SUM. If there is a weighting function w such that $w(O) \leq \rho$ for all $O \in \mathcal{O}$ and $|\mathcal{B}| \leq w(I) + \delta$, then $|\mathcal{B}| \leq \rho|\mathcal{O}| + \delta$.*

Proof Because $\mathcal{O}$ is a packing $\sum_{O \in \mathcal{O}} w(O) = w(I)$, therefore,

$$|\mathcal{B}| \leq w(I) + \delta = \sum_{O \in \mathcal{O}} w(O) + \delta \leq \rho|\mathcal{O}| + \delta. \quad \square$$

The key contribution of Epstein et al. [3] was bounding the parameters ρ and δ associated with the weighting function (3). Bounding δ is a relatively straightforward exercise. Bounding ρ is more involved and requires analytically solving a mathematical program. Here we only state their bounds.

Lemma 2 (Epstein et al. [3]) *Let $\mathcal{B}$ be the SUBSET-SUM packing and let w be the weighting function (3) for $\mathcal{B}$. Then*

1. *$|\mathcal{B}| \leq w(I) + 1$,*
2. *$w(B) \leq \sum_{i=1}^{\infty} \frac{1}{2^i - 1}$ for all $B \subseteq I$ such that $s(B) \leq 1$.*

Theorem 2 follows immediately from Lemmas 1 and 2.

Parametric Case

As it is the case with most bin packing heuristics, the performance of SUBSET-SUM improves when the items are small relative to the capacity of the bin. In a *parametric analysis* of a heuristic, we restrict our attention to instances where the maximum item size is bounded. More formally,

for every real $\alpha \in (0, 1]$, we define $\mathcal{C}_\alpha$ to be the class of instances s such that $\max_{i \in I} s_i \leq \alpha$.

Theorem 3 (Epstein et al. [3]) *For every integer $t \geq 1$ and $\alpha \in (\frac{1}{t+1}, \frac{1}{t}]$, we have $R_{SS}^\infty(\mathcal{C}_\alpha) = 1 + \sum_{i=1}^{\infty} \frac{1}{(t+1)2^i - 1}$.*

Notice that this is a strict generalization of Theorems 1 and 2, which only cover the case $\alpha = 1$.

Applications

There is an interesting connection between the performance of the SUBSET-SUM algorithm and the quality of equilibria of a game-theoretic version of bin packing. Let us associate a game with each instance $s : I \to [0, 1]$ of the bin packing problem. The set of players in this game is I, the set of items. Each player can decide in which bin it wants to be packed; this is the player's strategy space. For each bin B chosen in this uncoordinated fashion, if $s(B) > 1$ then the players in B are charged ∞; otherwise, player $i \in B$ is charged $\frac{s_i}{s(B)}$. These payments enforce that a strategy profile is a valid packing if and only if the payments are finite. Furthermore, if the payments are finite, the sum of these payments equals the number of bins in the packing.

A strategy profile is said to be a Nash Equilibrium (NE) if there is no player that can switch bins to decrease its payment. The *price of anarchy* of the bin packing game is the asymptotic worst-case ratio between the number of bins used by an NE and the number of bins in an optimal packing. A packing is said to be a Strong Nash Equilibrium (SNE) if no coalition of players can switch bins to decrease the sum of their payments. The *strong price of anarchy* of the bin packing game is the asymptotic worst-case ratio between the number of bins used by an SNE and the number of bins by an optimal packing.

Theorem 4 (Epstein and Kleiman [2]) *The strong price of anarchy for the bin packing game is exactly R_{SS}^∞.*

Notice that every SNE is an NE, since we can think of an NE as requiring that there are no "coalitions" of size 1. Therefore, Theorem 4 establishes a lower bound on the price of anarchy for the bin packing game. However, not every NE is an SNE. In fact, it is known that the price of anarchy for the bin packing game is strictly worse than its strong price of anarchy [2, 3].

Experimental Results

Gupta and Ho [5] performed an experimental evaluation of SUBSET-SUM. (Gupta and Ho call the algorithm *minimum bin slack* because they formulate each iteration as trying to minimize the slack (unused space) of the bin, which is equivalent to maximizing the bin's usage.) The instances used in the evaluation were randomly generated by selecting the item sizes uniformly at random from different numerical ranges. They compared the performance of SUBSET-SUM to two well-known heuristics: FIRST-FIT-DECREASING and BEST-FIT-DECREASING. They observed that SUBSET-SUM performed better on average without incurring a significant computational overhead.

Cross-References

▶ Bin Packing
▶ Knapsack
▶ Price of Anarchy

Recommended Reading

1. Caprara A, Pferschy U (2004) Worst-case analysis of the subset sum algorithm for bin packing. Oper Res Lett 32(2):159–166
2. Epstein L, Kleiman E (2011) Selfish bin packing. Algorithmica 60(2):368–394
3. Epstein L, Kleiman E, Mestre J (2009) Parametric packing of selfish items and the subset sum algorithm. In: Proceedings of the 5th workshop on internet and network economics, Rome, Italy pp 67–78
4. Graham RL (1972) Bounds on multiprocessing anomalies and related packing algorithms. In: Proceedings of the 1972 spring joint computer conference, Atlantic City, New Jersey, USA pp 205–217
5. Gupta J, Ho J (1999) A new heuristic algorithm for the one-dimensional bin-packing problem. Prod Plan Control 10(6):598–603
6. Kellerer H, Pferschy U, Pisinger D (2004) Knapsack problems. Springer, Berlin/New York

Substring Parsimony

Mathieu Blanchette
Department of Computer Science, McGill University, Montreal, QC, Canada

Years and Authors of Summarized Original Work

2001; Blanchette, Schwikowski, Tompa

Problem Definition

The Substring Parsimony Problem, introduced by Blanchette et al. [1] in the context of motif discovery in biological sequences, can be described in a more general framework:
Input:

- A discrete space $\mathcal{S}$ on which an integral distance d is defined (i.e., $d(x, y) \in \mathbb{N} \, \forall x, y \in \mathcal{S}$).
- A rooted binary tree $T = (V, E)$ with n leaves. Vertices are labeled $\{1, 2, \ldots, n, \ldots, |V|\}$, where the leaves are vertices $\{1, 2, \ldots, n\}$.
- Finite sets $S_1, S_2, \ldots, S_n$, where set $S_i \subseteq \mathcal{S}$ is assigned to leaf i, for all $i = 1 \ldots n$.
- A non-negative integer t

Output: All solutions of the form $(x_1, x_2, \ldots, x_n, \ldots, x_{|V|})$ such that:

- $x_i \in \mathcal{S}$ for all $i = 1 \ldots |V|$

- $x_i \in S_i$ for all $i = 1 \ldots n$
- $\sum_{(u,v) \in E} d(x_u, x_v) \leq t$

The problem thus consists of choosing one element x_i from each set S_i such that the Steiner distance of the set of points is at most t. This is done on a Steiner tree T of fixed topology. The case where $|S_i| = 1$ for all $i = 1 \ldots n$ is a standard Steiner tree problem on a fixed tree topology (see [11]). It is known as the Maximum Parsimony Problem and its complexity depends on the space S.

Key Results

The substring parsimony problem can be solved using a dynamic programming algorithm. Let $u \in V$ and $s \in S$. Let $W_u[s]$ be the score of the best solution that can be obtained for the subtree rooted at node u, under the constraint that node u is labeled with s, i.e.,

$$W_u[s] = \min_{\substack{x_1, \ldots, x_{|V|} \in S \\ x_u = s}} \sum_{\substack{(i,j) \in E \\ i,j \in \text{subtree}(u)}} d(x_i, x_j).$$

Let v be a child of u, and let $X_{(u,v)}[s]$ be the score of the best solution that can be obtained for the subtree consisting of node u together with the subtree rooted at its child v, under the constraint that node u is labeled with s:

$$X_{(u,v)}[s] = \min_{\substack{x_1, \ldots, x_{|V|} \in S \\ x_u = s}} \sum_{\substack{(i,j) \in E \\ i,j \in \text{subtree}(v) \cup \{(u,v)\}}} d(x_i, x_j).$$

Then, we have:

$$W_u[s] = \begin{cases} 0 & \text{if } u \text{ is a leaf and } s \in S_u \\ +\infty & \text{if } u \text{ is a leaf and } s \notin S_u \\ \sum_{v \in \text{children}(u)} X_{(u,v)}[s] & \text{if } u \text{ is not a leaf} \end{cases}$$

and

$$X_{(u,v)}[s] = \min_{y' \in S} W_u[s'] + d(s, s').$$

Tables W and X can thus be computed using a dynamic programming algorithm, proceeding in a post-order traversal of the tree. Solutions can then be recovered by tracing the computation back for all s such that $W_{\text{root}}[s] \leq t$. Note that the same solution may be recovered more than once in this process.

A straight-forward implementation of this dynamic programming algorithm would run in time $O(n \cdot |S|^2 \cdot \gamma(S))$, where $\gamma(S)$ is the time needed to compute the distance between any two points in S. Let $N_a(S)$ be the maximum number of a-neighbors a point in S can have, i.e., $N_a(S) = \max_{x \in S} |\{y \in S : d(x, y) = a\}|$. Blanchette et al. [3] showed how to use a modified breadth-first search of the space S to compute each table $X_{(u,v)}$ in time $O(|S| \cdot N_1(S))$, thus reducing the total time complexity to $O(n \cdot |S| \cdot N_1(S))$. Since only solutions with a score of at most t are of interest, the complexity can be further reduced by only computing those table entries which will yield a score of at most t. This results in an algorithm whose running time is $O(n \cdot M \cdot N_{\lfloor t/2 \rfloor}(S) \cdot N_1(S))$ where $M = \max_{i=1 \ldots n} |S_i|$.

The problem has been mostly studied in the context of biological sequence analysis, where $S = \{A, C, G, T\}^k$, for some small k ($k = 5, \ldots, 20$ are typical values). The distance d is the Hamming distance, and a phylogenetic tree T is given. The case where $|S_i| = 1$ for all $i = 1 \ldots n$ is known as the Maximum Parsimony Problem and can be solved in time $O(n \cdot k)$ using Fitch's algorithm [9] or Sankoff's algorithm [12]. In the more general version, a long DNA sequence P_u of length L is assigned to each leaf u. The set S_u is defined as the set of all k-substrings of P_u. In this case, $M = L - k + 1 \in O(L)$, and $N_a \in O(\min(4^k, (3k)^a))$, resulting in a complexity of $O(n \cdot L \cdot 3k \cdot \min(4^k, (3k)^{\lfloor d/2 \rfloor}))$. Notice that for a fixed k and d, the algorithm is linear over the whole sequence. The problem was independently shown to be NP-hard by Blanchette et al. [3] and by Elias [7].

Applications

Most applications are found in computational biology, although the algorithm can be applied to a wide variety of domains. The algorithm for the substring parsimony problem has been implemented in a software package called FootPrinter [5] and applied to the detection of transcription factor binding sites in orthologous DNA regulatory sequences through a method called phylogenetic footprinting [4]. Other applications include the search for conserved RNA secondary structure motifs in orthologous RNA sequences [2]. Variants of the problem have been defined to identify motifs regulating alternative splicing [13]. Blanchette et al. [3] study a relaxation of the problem where one does not require that a substring be chosen from each of the input sequences, but instead asks that substrings be chosen from a sufficiently large subset of the input sequence. Fang and Blanchette [8] formulate another variant of the problem where substring choices are constrained to respect a partial order relation defined by a set of local multiple sequence alignments.

Open Problems

Optimizations taking advantage of the specific structure of the space S may yield more efficient algorithms in certain cases. Many important variations could be considered. First, the case where the tree topology is not given needs to be considered, although the resulting problems would usually be NP-hard even when $|S_i| = 1$. Another important variation is one where the phylogenetic relationships between trees is not given by a tree but rather by a phylogenetic network [10]. Finally, randomized algorithms similar to those proposed by Buhler et al. [6] may yield important and practical improvements.

URL to Code

http://bio.cs.washington.edu/software.html

Cross-References

▶ Closest Substring

▶ Efficient Methods for Multiple Sequence Alignment with Guaranteed Error Bounds
▶ Local Alignment (with Affine Gap Weights)
▶ Local Alignment (with Concave Gap Weights)
▶ Statistical Multiple Alignment
▶ Steiner Trees

Recommended Reading

1. Blanchette M (2001) Algorithms for phylogenetic footprinting. In: RECOMB01: proceedings of the fifth annual international conference on computational molecular biology, Montreal. ACM, pp 49–58
2. Blanchette M (2002) Algorithms for phylogenetic footprinting. PhD thesis, University of Washington
3. Blanchette M, Schwikowski B, Tompa M (2002) Algorithms for phylogenetic footprinting. J Comput Biol 9(2):211–223
4. Blanchette M, Tompa M (2002) Discovery of regulatory elements by a computational method for phylogenetic footprinting. Genome Res 12:739–748
5. Blanchette M, Tompa M (2003) Footprinter: a program designed for phylogenetic footprinting. Nucleic Acids Res 31(13):3840–3842
6. Buhler J, Tompa M (2001) Finding motifs using random projections. In: RECOMB01: proceedings of the fifth annual international conference on computational molecular biology, pp 69–76
7. Elias I (2006) Settling the intractability of multiple alignment. J Comput Biol 13:1323–1339
8. Fang F, Blanchette M (2006) Footprinter3: phylogenetic footprinting in partially alignable sequences. Nucleic Acids Res 34(2):617–620
9. Fitch WM (1971) Toward defining the course of evolution: minimum change for a specified tree topology. Syst Zool 20:406–416
10. Huson DH, Bryant D (2006) Application of phylogenetic networks in evolutionary studies. Mol Biol Evol 23(2):254–267
11. Sankoff D, Rousseau P (1975) Locating the vertices of a Steiner tree in arbitrary metric space. Math Program 9:240–246
12. Sankoff DD (1975) Minimal mutation trees of sequences. SIAM J Appl Math 28:35–42
13. Shigemizu D, Maruyama O (2004) Searching for regulatory elements of alternative splicing events using phylogenetic footprinting. In: Proceedings of the fourth workshop on algorithms for bioinformatics. Lecture notes in computer science. Springer, Berlin, pp 147–158

S

Succinct and Compressed Data Structures for Permutations and Integer Functions

Jérémy Barbay
Department of Computer Science (DCC),
University of Chile, Santiago, Chile

Keywords

Adaptive; Compression; Functions; Permutation

Years and Authors of Summarized Original Work

2012; Munro, Raman, Raman, Rao
2012; Barbay, Fischer, Navarro
2013; Barbay, Navarro
2013; Barbay

Problem Definition

A basic building block for compressed data structures for texts and functions is the representation of a permutation of the integers $\{1, \ldots, n\}$, denoted by $[1 \ldots n]$. A permutation π is trivially representable in $n\lceil \lg n \rceil$ bits which is within $O(n)$ bits of the information theoretic bound of $\lg(n!)$, but instances from restricted classes of permutations can be represented using much less space.

We are interested in encodings of permutations that can efficiently access them. Given a permutation π over $[1 \ldots n]$, an integer k and an integer $i \in [1 \ldots n]$, data structures on permutations aim to support the following operators as fast as possible, using as little additional space as possible:

- $\pi(i)$: application of the permutation to i,
- $\pi^{-1}(i)$: application of the inverse permutation to i,
- $\pi^{(k)}(i)$: $\pi()$ iteratively applied k times starting with value i (e.g., $\pi^{(2)}(i) = \pi(\pi(i))$).

Key Results

We distinguish between two types of solutions: the succinct index and two succinct data structures for permutations introduced by Munro et al. [1], and the various compressed data structures proposed later [2–4].

Succinct Data Structures

Munro et al. [1] studied the problem of succinctly representing a permutation to support operators on it quickly. They give several solutions, described below.

"Shortcut" Index Supporting $\pi()$ and $\pi^{-1}()$

Given an integer parameter t, the operators $\pi()$ and $\pi^{-1}()$ can be supported by simply writing down π in an array of n words of $\lceil \lg n \rceil$ bits each, plus an auxiliary array S of at most n/t back pointers called shortcuts: in each cycle of length at least t, every t-th element has a pointer t steps back. Then, $\pi(i)$ is simply the i-th value in the primary structure, and $\pi^{-1}(i)$ is found by moving forward until a back pointer is found and then continuing to follow the cycle to the location that contains the value i.

The trick is in the encoding of the locations of the back pointers: this is done with a simple bit vector B of length n, in which a 1 indicates that a back pointer is associated with a given location. B is augmented using $o(n)$ additional bits so that the number of 1's up to a given position and the position of the r-th 1 can be found in constant time (i.e., using the rank and select operators on binary strings [5]). This gives the location of the appropriate back pointer in the auxiliary array S. As there are back pointers every t elements in the cycle, finding the predecessor requires $O(t)$ memory accesses.

Theorem 1 *For any strictly positive integer n and any permutation π on $[1 \ldots n]$ which can be decomposed into δ cycles of respective sizes $c_1, \ldots, c_\delta$, there is a representation of π using within $(\sum_{i \in [1 \ldots \delta]} \lfloor \frac{c_i}{t} \rfloor) \lg n + 2n + o(n) \subseteq \frac{n \lg n}{t} + 2n + o(n)$ bits to support the operator $\pi()$ in constant time and the operator $\pi^{-1}()$ in time within $O(t)$.*

Interestingly enough, Munro et al. [1] did not notice that their construction is actually an index and that the raw encoding can be replaced by any data structure supporting the operator $\pi()$, including the compressed ones later described [4].

"Cycle" Data Structure Supporting $\pi^k()$

For arbitrary i and k, $\pi^k()$ is supported by writing the cycles of π together with a bit vector B marking the beginning of each cycle. Observe that the cycle representation itself is a permutation in "standard form"; call it σ. The first task is to find i in the representation: it is in position $\sigma^{-1}(i)$. The segment of the representation containing i is found through the rank and select operators on B. Then $\pi^k(i)$ is determined by taking k modulo the cycle length, moving that number of steps around the cycle starting at the position of i, and applying $\sigma()$ to obtain the value to return.

Other than the support of the operators on σ, all operators are performed in constant time; hence the asymptotic supporting time of $\pi^k()$ depends on the supporting time in which the data structure chosen to represent σ supports the operators $\sigma()$ and $\sigma^{-1}()$. Munro et al. [1] proposed the following, using a raw encoding of σ with a shortcut index to support $\sigma^{-1}()$:

Theorem 2 *For any strictly positive integer n and any permutation π on $[1 \ldots n]$, there is a representation of π using at most $(1 + \varepsilon)n \lg n + O(n)$ bits to support the operator $\pi^k()$ in time within $O(1/\varepsilon)$, for any ε less than 1 and for any arbitrary value of k.*

Under a restricted model of pointer machine, this technique is optimal: using $O(n)$ extra bits (i.e., $O(n/\log n)$ extra words), time within $\Omega(\log n)$ is necessary to support both $\pi()$ and $\pi^{-1}()$.

"Benes Network" Data Structure Supporting $\pi^k()$

Any permutation can be implemented by a communication network composed of switches: this is called a Benes Network and uses even less space under the RAM model than the solutions described in the previous sections. Sparsely adding pointers accelerates the support of $\pi^k()$ to time within $O(\frac{\log n}{\log \log n})$.

Theorem 3 *For any strictly positive integer n and any permutation π on $[1 \ldots n]$, there is a representation of π using at most $\lceil \lg(n!) \rceil + O(n)$ bits to support the operator $\pi^k()$ in time within $O(\log n / \log \log n)$.*

This representation uses space within an additive term within $O(n)$ of the optimal, both on average and in the worst case over all permutations over $[1 \ldots n]$.

Compressed Data Structures

Any comparison-based sorting algorithm yields an encoding for permutations, and any adaptive sorting algorithm in the comparison model yields a compression scheme for permutations. Supporting operators on such compressed permutation in less time than required to decompress the whole of it requires some more work:

Runs

Barbay and Navarro [2] described how to segment a partition into nRuns *runs* composed of consecutive positions forming already sorted blocks and how to merge those via a wavelet tree. This yields a data structure compressing a permutation within space optimal over all permutations with nRuns runs of sizes given by the vector vRuns. This data structure supports the operators $\pi()$ and $\pi^{-1}()$ in sublinear time within $O(1 + \log \text{nRuns})$, with the average supporting time within $O(1 + \mathcal{H}(\text{vRuns}))$, which decreases with the entropy of the partition of the permutation into runs. Here, the *entropy* of a sequence of positive integers $X = \langle n_1, n_2, \ldots, n_r \rangle$ adding up to n is $\mathcal{H}(X) = \sum_{i=1}^{r} \frac{n_i}{n} \lg \frac{n}{n_i}$.

Theorem 4 *For any strictly positive integer n and any permutation π on $[1 \ldots n]$ which can be decomposed into nRuns runs of sizes vRuns = $(r_1, \ldots, r_{\text{nRuns}})$, there is a representation of π using at most $n\mathcal{H}(\text{vRuns}) + O(\text{nRuns} \log n) + o(n)$ bits to support the computation of $\pi(i)$ and $\pi^{-1}(i)$ in time within $O(1 + \log \text{nRuns})$ in the worst case over $i \in [1 \ldots n]$ and in* *time within $O(1 + \mathcal{H}(\text{vRuns}))$ on average when $i \in [1 \ldots n]$ is uniformly distributed. This compressed data structure can be computed in time within $O(n(1 + \mathcal{H}(\text{vRuns})))$, which is worst-case optimal in the comparison model over all such permutations decomposed into* nRuns *runs of sizes given by the vector* vRuns.

The partitioning takes only $n - 1$ comparisons, and the construction of the compressed data structure itself is an adaptive sorting algorithm improving over previous results [6, 7].

Heads of Strict Runs

A two-level partition of the permutation yields further compression [2]. The first level partitions the permutation into *strict ascending runs* (maximal ranges of positions satisfying $\pi(i + k) = \pi(i) + k$). The second level partitions the *heads* (first position) of those strict runs into conventional ascending runs. This is analogous to the notion of blocks described by Moffat and Petersson [7] for multisets.

Theorem 5 *For any strictly positive integer n and any permutation π on $[1 \ldots n]$ which can be decomposed into* nBlock *strict runs and into* nRuns $\leq$ nBlock *monotone runs, let* vHRuns *be the vector formed by the* nRuns *monotone run lengths in the permutation of strict run heads. Then, there is a representation of π using at most* nBlock$\mathcal{H}(\text{vHRuns})) + O(\text{nBlock} \log \frac{n}{\text{nBlock}}) + o(n)$ *bits to support the operator $\pi()$ and $\pi^{-1}()$ in time within $O(1 + \log \text{nBlock})$. This compressed data structure can be computed in time within $O(n(1 + \log \text{nBlock}))$.*

Shuffled Subsequences

The preorder measures seen so far have considered runs which group contiguous positions in π: this does not need to be always the case. A permutation π over $[1 \ldots n]$ can be decomposed in n comparisons into a minimal number nSUS of *Shuffled Up Sequences*, defined as a set of, not necessarily consecutive, subsequences of increasing numbers that have to be removed from π in order to reduce it to the empty sequence [8]. Then those subsequences can be merged using

the same techniques as above, which yields a new adaptive sorting algorithm and a new compressed data structure [2]. An optimal partition of a permutation π over $[1 \ldots n]$ into a minimal number nSMS of *Shuffled Monotone Sequences*, sequences of not necessarily consecutive subsequences of increasing or decreasing numbers, is NP-hard to compute [9], but if such a permutation is given, the same technique applies [10].

LRM Subsequences

LRM trees partition a sequence of values into consecutive sorted blocks and express the relative position of the first element of each block within a previous block. Such a tree can be computed in $2(n - 1)$ comparisons within the array and overall linear time, through an algorithm similar to that of Cartesian Trees [11]. The interest of LRM trees in the context of adaptive sorting and permutation compression is that the values are increasing in each root-to-leaf branch: they form a partition of the array into subsequences of increasing values. Barbay et al. [3] described how to compute the partition of the LRM tree of minimal size-vector entropy, which yields a compressed data structure asymptotically smaller than $\mathcal{H}(\text{vRuns})$-adaptive sorting, smaller in practice than $\mathcal{H}(\text{vSUS})$-adaptive sorting, as well as a faster adaptive sorting algorithm.

Number of Inversions

The preorder measure nInv counts the number of pairs (i, j) of positions $1 \leq i < j \leq n$ in a permutation π over $[1 \ldots n]$ such that $\pi(i) > \pi(j)$. Its value is exactly the number of comparisons performed by the algorithm Insertion Sort, between n and n^2 for a permutation over $[1 \ldots n]$. A variant of Insertion Sort, named Local Insertion Sort, sorts π in $n(1 + \lceil \lg(\text{nInv}/n) \rceil)$ comparisons [6, 7].

Simply encoding the n values $(\pi(i) - i)_{i \in [1 \ldots n]}$ using the γ' code from Elias [12], and indexing the positions of the beginning of each code by a compressed bit vector, yields a compressed data structure supporting the operator $\pi()$ in constant time. The resulting data structure uses space within $n(1 + 2 \lg \frac{\text{nInv}}{n}) + o(n)$ bits. Support for the operator $\pi^{-1}()$ can be added in two distinct ways, either encoding both π and π^{-1} using this technique within $2n(1 + 2\lg\frac{\mathrm{nInv}}{n}) + o(n)$ bits, which supports both operators $\pi()$ and $\pi^{-1}()$ in constant time, or adding support for the operator $\pi^{-1}()$ using Munro et al.'s shortcut succinct index for permutations [1] described previously.

Removing Elements

The preorder measure nRem counts the minimum number of elements that must be removed from a permutation so that what remains is already sorted. Its exact value is n minus the length of the *Longest Increasing Subsequence*, which can be computed in time within $O(n \log n)$. Alternatively, the value of nRem can be approximated within a constant factor of 2 in $2(n - 1)$ comparisons. Partitioning π into the removed elements and the remaining ones through a bit vector of n bits, representing the order of the 2nRem elements in a wavelet tree (using any of the data structures described above), and representing the merging of both into n bits yield a compressed data structure using space within $2n + 2\mathrm{nRem}\lg(n/\mathrm{nRem}) + o(n)$ bits and supporting the operators $\pi()$ and $\pi^{-1}()$ in sublinear time, within $O(1 + \log(\mathrm{nRem} + 1))$.

Applications

Integer Functions

Munro et al. [1] extended the results on permutations to arbitrary functions from $[1 \ldots n]$ to $[1 \ldots n]$. Again $f^k(i)$ indicates the function iterated k times starting at i: if k is nonnegative, this is straightforward. The case in which k is negative is more complicated as the image is a (possibly empty) multiset over $[1 \ldots n]$.

Whereas π is a set of cycles, f can be viewed as a set of cycles in which each node is the root of a tree. Starting at any node (element of $[1 \ldots n]$), the evaluation moves one step along a branch of the tree, or one step along a cycle. Moving k steps in a positive direction is straightforward, and one moves up a tree and perhaps around a cycle. When k is negative, one must determine

all nodes at distance k from the starting location, i, in the direction toward the leaves of the trees. The key technical issue is to run across succinct tree representations picking off all nodes at the appropriate levels. Using a raw encoding of the permutation mapping integers to the nodes, and Munro et al.'s shortcut succinct index [1] to support the operations on it, yields the following result:

Theorem 6 *For any fixed ε, $n > 0$ and f : $[1 \ldots n] \rightarrow [1 \ldots n]$, there is a representation of f using $(1 + \varepsilon)n \lg n + O(1)$ bits of space to compute $f^k(i)$ in time within $O(1 + |f^k(i)|)$, for any integer k and for any integer $i \in [1 \ldots n]$.*

Open Problems

Other Measures of Disorder

Moffat and Petersson [7] list many measures of preorder and adaptive sorting techniques. Each measure explored above yields a compressed data structure for permutations supporting the operators $\pi()$ and $\pi^{-1}()$ in sublinear time. Each adaptive sorting algorithm in the comparison model yields a compression scheme for permutations, but the encoding thus defined does not necessarily support the simple application of the permutation to a single element without decompressing the whole permutation nor the application of the inverse permutation. More work is required in order to decide whether there are compressed data structures for permutations, supporting the operators $\pi()$ and $\pi^{-1}()$ in sublinear time and using space proportional to the other preorder measures [6, 7] (e.g., Reg, Exc, Block, and Enc).

Sorting and Encoding Multisets

Munro and Spira [13] showed how to sort multisets through MergeSort, Insertion Sort, and Heap Sort, adapting them with counters to sort in time within $O(n(1 + \mathcal{H}(\langle m_1, \ldots, m_r \rangle)))$ where m_i is the number of occurrences of i in the multiset (note that this is orthogonal to the results described in this chapter that depend on the distribution of the lengths of monotone runs).

It seems easy to combine both approaches (e.g., on `MergeSort` in a single algorithm using both runs and counters), yet quite hard to *analyze* the complexity of the resulting algorithm and compressed data structure. The difficulty measure must depend not only on both the entropy of the partition into runs and the entropy of the partition of the values of the elements but also on the interaction of those partitions.

Compressed Data Structures Supporting $\pi^k()$

In Munro et al.'s "cycle" data structure [1] for supporting the operator $\pi^k()$ (Theorem 2), the raw encoding of the permutation σ representing the cycles of π can be replaced by any compressed data structure such as those described here, with the warning that the compressibility of σ depends not only on π but also on the order in which its cycles are placed in σ. The question if there is a compressed data structure supporting the operator $\pi^k()$ which takes advantage of this order is open.

Recommended Reading

1. Ian Munro J, Raman R, Raman V, Srinivasa Rao S (2012) Succinct representations of permutations and functions. Theorical Computer Science (TCS) 438:74–88
2. Barbay J, Navarro G (2013) On compressing permutations and adaptive sorting. Theorical Computer Science (TCS) 513:109–123
3. Barbay J, Fischer J, Navarro G (2012) LRM-trees: compressed indices, adaptive sorting, and compressed permutations. Theorical Computer Science (TCS) 459:26–41
4. Barbay J (2013) From time to space: fast algorithms that yield small and fast data structures. In: Brodnik A, López-Ortiz A, Raman V, Viola A (eds) Space-efficient data structures, streams, and algorithms (Ian-Fest). Volume 8066 of Lecture Notes in Computer Science. Springer, Heidelberg, pp 97–111
5. Ian Munro J, Raman V (1997) Succinct representation of balanced parentheses, static trees and planar graphs. In: IEEE symposium on Foundations Of Computer Science, Miami Beach, pp 118–126
6. Estivill-Castro V, Wood D (1992) A survey of adaptive sorting algorithms. ACM Computing Survey 24(4):441–476
7. Moffat A, Petersson O (1992) An overview of adaptive sorting. Aust Comput J 24(2):70–77
8. Levcopoulos C, Petersson O (1990) Sorting shuffled monotone sequences. In: Proceedings of the Scandinavian Workshop on Algorithm Theory (SWAT), Bergen. Springer, London, pp 181–191
9. Levcopoulos C, Petersson O (1994) Sorting shuffled monotone sequences. Information Computing 112(1):37–50
10. Barbay J, Claude F, Gagie T, Navarro G, Nekrich Y (2014) Efficient fully-compressed sequence representations. Algorithmica 69(1):232–268
11. Gabow HN, Bentley JL, Tarjan RE (1984) Scaling and related techniques for geometry problems. In: Proceedings of the Symposium on Theorical Computer (STOC), Washington, DC. ACM, pp 135–143
12. Elias P (1975) Universal codeword sets and representations of the integers. IEEE Transaction on Information Theory 21(2):194–203
13. Ian Munro J, Spira PM (1976) Sorting and searching in multisets. SIAM Journal of Computing 5(1):1–8

Succinct Data Structures for Parentheses Matching

Meng He
School of Computer Science, University of Waterloo, Waterloo, ON, Canada

Keywords

Succinct balanced parentheses

Years and Authors of Summarized Original Work

2001; Munro, Raman

Problem Definition

This problem is to design succinct representation of balanced parentheses in a manner in which a number of "natural" queries can be supported quickly, and use it to represent trees and graphs succinctly. The problem of succinctly representing balanced parentheses was initially proposed by Jacobson [6] in 1989, when he proposed *succinct data structures*, i.e., data structures that occupy space close to the information-theoretic lower bound to represent them, while supporting efficient navigational operations. Succinct data structures provide solutions to manipulate large data in modern applications. The work of Munro and Raman [8] provides an optimal solution to the problem of balanced parentheses representation under the word RAM model, based on which they design succinct trees and graphs.

Balanced Parentheses

Given a balanced parenthesis sequence of length $2n$, where there are n opening parentheses and n closing parentheses, consider the following operations:

- findclose(i) (findopen(i)), the matching closing (opening) parenthesis for the opening (closing) parenthesis at position i;
- excess(i), the number of opening parentheses minus the number of closing parentheses in the sequence up to (and including) position i;
- enclose(i), the closest enclosing (matching parenthesis) pair of a given matching parenthesis pair whose opening parenthesis is at position i.

Trees

There are essentially two forms of trees. An *ordinal tree* is a rooted tree in which the children of a node are ordered and specified by their ranks, while in a *cardinal tree* of degree k, each child of a node is identified by a unique number from the set $\{1, 2, \cdots, k\}$. An *binary tree* is a cardinal tree of degree 2. The information-theoretic lower bound of representing an ordinal tree or binary tree of n nodes is $2n - o(n)$ bits, as there are $\binom{2n}{n}/(n + 1)$ different ordinal trees or binary trees.

Consider the following operations on ordinal trees (a node is referred to by its preorder number):

- child(x, i), the ith child of node x for $i \geq 1$;
- child_rank(x), the number of left siblings of node x;

- depth(x), the depth of x, i.e., the number of edges in the rooted path to node x;
- parent(x), the parent of node x;
- nbdesc(x), the number of descendants of node x;
- height(x), the height of the subtree rooted at node x;
- LCA(x, y), the lowest common ancestor of node x and node y.

On binary trees, the operations parent, nbdesc and the following operations are considered:

- leftchild(x) (rightchild(x)), the left (right) child of node x.

Graphs

Consider an undirected graph G of n vertices and m edges. Bernhart and Kainen [1] introduced the concept of *page book embedding*. A *k-book embedding* of a graph is a topological embedding of it in a book of k pages that specifies the ordering of the vertices along the spine, and carries each edge into the interior of one page, such that the edges on a given page do not intersect. Thus, a graph with one page is an *outerplanar graph*. The *pagenumber* or *book thickness* [1] of a graph is the minimum number of pages that the graph can be embedded in. A very common type of graphs are planar graphs, and any planar graph can be embedded in at most four pages [15]. Consider the following operations on graphs:

- adjacency(x, y), whether vertices x and y are adjacent;
- degree(x), the degree of vertex x;
- neighbors(x), the neighbors of vertex x.

Key Results

All the results cited are under the word RAM model with word size $\Theta(\lg n)$ bits ($\lg n$ denotes $\lceil \log_2 n \rceil$), where n is the size of the problem considered.

Theorem 1 ([8]) *A sequence of balanced parentheses of length $2n$ can be represented using*

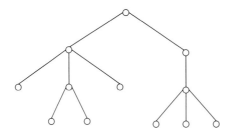

Balanced parentheses: $((()(())())(()()()())()))$

Succinct Data Structures for Parentheses Matching,
Fig. 1 An example of the balanced parenthesis sequence
of a given ordinal tree

$2n + o(n)$ bits to support the operations *find-close, findopen, excess* and *enclose* in constant time.

There is a polymorphism between a balanced parenthesis sequence and an ordinal tree: when performing a depth-first traversal of the tree, output an opening parenthesis each time a node is visited, and a closing parenthesis immediately after all the descendants of a node are visited (see Fig. 1 for an example). The work of Munro and Raman proposes a succinct representation of ordinal trees using $2n + o(n)$ bits to support depth, parent and nbdesc in constant time, and child(x, i) in $O(i)$ time. Lu and Yeh have further extended this representation to support child, child_rank, height and LCA in constant time.

Theorem 2 ([8, 7]) *An ordinal tree of n nodes can be represented using $2n + o(n)$ bits to support the operations child, child_rank, parent, depth, nbdesc, height and LCA in constant time.*

A similar approach can be used to represent binary trees:

Theorem 3 ([8]) *A binary tree of n nodes can be represented using $2n + o(n)$ bits to support the operations leftchild, rightchild, parent and nbdesc in constant time.*

Finally, balanced parentheses can be used to represent graphs. To represent a one-page graph, the

work of Munro and Raman proposes to list the vertices from left to right along the spine, and each node is represented by a pair of parentheses, followed by zero or more closing parentheses and then zero or more opening parentheses, where the number of closing (or opening) parentheses is equal to the number of adjacent vertices to its left (or right) along the spine (see Fig. 2 for an example). This representation can be applied to each page to represent a graph with pagenumber k.

Theorem 4 ([8]) *An outerplanar graph of n vertices and m edges can be represented using $2n + 2m + o(n + m)$ bits to support operations adjacency and degree in constant time, and neighbors(x) in time proportional to the degree of x.*

Theorem 5 ([8]) *A graph of n vertices and m edges with pagenumber k can be represented using $2kn + 2m + o(nk + m)$ bits to support operations adjacency and degree in $O(k)$ time, and neighbors(x) in $O(d(x) + k)$ time where $d(x)$ is the degree of x. In particular, a planar graph of n vertices and m nodes can be represented using $8n + 2m + o(n)$ bits to support operations adjacency and degree in constant time, and neighbors(x) in $O(d(x))$ time where $d(x)$ is the degree of x.*

Applications

Succinct Representation of Suffix Trees
As a result of the growth of the textual data in databases and on the World Wide Web, and also applications in bioinformatics, various indexing techniques have been developed to facilitate pattern searching. Suffix trees [14] are a popular type of text indexes. A suffix tree is constructed over the suffixes of the text as a tree-based data structure, so that queries can be performed by searching the suffixes of the text. It takes $O(m)$ time to use a suffix tree to check whether an arbitrary pattern P of length m is a substring of a given text T of length n, and to count the number of the occurrences, *occ*, of P in T. $O(occ)$ additional time is required to list all the occurrences

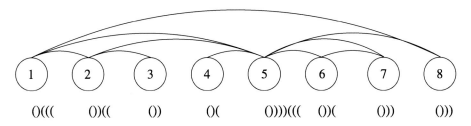

Succinct Data Structures for Parentheses Matching, Fig. 2 An example of the balanced parenthesis sequence of a graph with one page

of P in T. However, a standard representation of a suffix tree requires somewhere between $4n \lg n$ and $6n \lg n$ bits, which is impractical for many applications.

By reducing the space cost of representing the tree structure of a suffix tree (using the work of Munro and Raman), Munro, Raman and Rao [9] have designed space-efficient suffix trees. Given a string of n characters over a fixed alphabet, they can represent a suffix tree using $n \lg n + O(n)$ bits to support the search of a pattern in $O(m + occ)$ time. To achieve this result, they have also extended the work of Munro and Raman to support various operations to retrieve the leaves of a given subtree in an ordinal tree. Based on similar ideas and by applying compressed suffix arrays [5], Sadakane [13] has proposed a different trade-off; his compressed suffix tree occupies $O(n \lg \sigma)$ bits, where σ is the size of the alphabet, and can support any algorithm on a suffix tree with a slight slowdown of a factor of polylog(n).

Succinct Representation of Functions

Munro and Rao [11] have considered the problem of succinctly representing a given function, $f : [n] \rightarrow [n]$, to support the computation of $f^k(i)$ for an arbitrary integer k. The straightforward representation of a function is to store the sequence $f(i)$, for $i = 0, 1, \ldots, n - 1$. This takes $n \lg n$ bits, which is optimal. However, the computation of $f^k(i)$ takes $\Theta(k)$ time even in the easier case when k is positive. To address this problem, Munro and Rao [11] first extends the representation of balanced parenthesis to support the next_excess(i, k) operator, which returns the minimum j such that $j > i$ and

excess(j) = k. They further use this operator to support the level_anc(x, i) operator on succinct ordinal trees, which returns the ith ancestor of node x for $i \geq 0$ (given a node x at depth d, its ith ancestor is the ancestor of x at depth $d - i$). Then, using succinct ordinal trees with the support for level_anc, they propose a succinct representation of functions using $(1 + \epsilon)n \lg n + O(1)$ bits for any fixed positive constant ϵ, to support $f^k(i)$ in constant time when $k > 0$, and $f^k(i)$ in $O(1 + |f^k(i)|)$ time when $k < 0$.

Multiple Parentheses and Graphs

Chuang et al. [3] have proposed to succinctly represent *multiple parentheses*, which is a string of $O(1)$ types of parentheses that may be unbalanced. They have extended the operations on balanced parentheses to multiple parentheses and designed a succinct representation. Based on the properties of canonical orderings for planar graphs, they have used multiple parentheses and the succinct ordinal trees to represent planar graphs. One of their main results is a succinct representation of planar graphs of n vertices and m edges in $2m + (5 + \epsilon)n + o(m + n)$ bits, for any constant $\epsilon > 0$, to support the operations supported on planar graphs in Theorem 5 in asymptotically the same amount of time. Chiang et al. [2] have further reduced the space cost to $2m + 3n + o(m + n)$ bits. In their paper, they have also shown how to support the operation wrapped(i), which returns the number of matching parenthesis pairs whose closest enclosing (matching parenthesis) pair is the pair whose opening parenthesis is at position i, in constant time on balanced parentheses. They have used it to show how to support the operation degree(x), which returns the degree of node x (i.e., the number of its children), in constant time on succinct ordinal trees.

Open Problems

One open research area is to support more operations on succinct trees. For example, it is not known how to support the operation to convert a given node's rank in a preorder traversal into its rank in a level-order traversal.

Another open research area is to further reduce the space cost of succinct planar graphs. It is not known whether it is possible to further improve the encoding of Chiang et al. [2].

A third direction for future work is to design succinct representations of dynamic trees and graphs. There have been some preliminary results by Munro et al. [10] on succinctly representing dynamic binary trees, which have been further improved by Raman and Rao [12]. It may be possible to further improve these results, and there are other related dynamic data structures that do not have succinct representations.

Experimental Results

Geary et al. [4] have engineered the implementation of succinct ordinal trees based on balanced parentheses. They have performed experiments on large XML trees. Their implementation uses orders of magnitude less space than the standard pointed-based representation, while supporting tree traversal operations with only a slight slowdown.

Cross-References

▶ Compressed Suffix Array
▶ Compressed Text Indexing
▶ Rank and Select Operations on Bit Strings
▶ Text Indexing

Recommended Reading

1. Bernhart F, Kainen PC (1979) The book thickness of a graph. J Comb Theory B 27(3):320–331
2. Chiang Y-T, Lin C-C, Lu H-I (2005) Orderly spanning trees with applications. SIAM J Comput 34(4):924–945
3. Chuang RC-N, Garg A, He X, Kao M-Y, Lu H-I (2001) Compact encodings of planar graphs via canonical orderings and multiple parentheses. Comput Res Repos. cs.DS/0102005
4. Geary RF, Rahman N, Raman R, Raman V (2006) A simple optimal representation for balanced parentheses. Theor Comput Sci 368(3):231–246
5. Grossi R, Gupta A, Vitter JS (2003) High-order entropy-compressed text indexes. In: Farach-Colton M (ed) Proceedings of the 14th annual ACM-SIAM symposium on discrete algorithms. SIAM, Philadelphia, pp 841–850
6. Jacobson G (1989) Space-efficient static trees and graphs. In: Proceedings of the 30th annual IEEE symposium on foundations of computer science. IEEE, New York, pp 549–554
7. Lu H-I, Yeh C-C (2007) Balanced parentheses strike back. Accepted to ACM Trans Algorithms
8. Munro JI, Raman V (2001) Succinct representation of balanced parentheses and static trees. SIAM J Comput 31(3):762–776
9. Munro JI, Raman V, Rao SS (2001) Space efficient suffix trees. J Algorithms 39(2):205–222
10. Munro JI, Raman V, Storm AJ (2001) Representing dynamic binary trees succinctly. In: Rao Kosaraju S (ed) Proceedings of the 12th annual ACM-SIAM symposium on discrete algorithms. SIAM, Philadelphia, pp 529–536
11. Munro JI, Rao SS (2004) Succinct representations of functions. In: Díaz J, Karhumäki J, Lepistö A, Sannella D (eds) Proceedings of the 31st international colloquium on automata, languages and programming. Springer, Heidelberg, pp 1006–1015
12. Raman R, Rao SS (2003) Succinct dynamic dictionaries and trees. In: Baeten JCM, Lenstra JK, Parrow J, Woeginger GJ (eds) Proceedings of the 30th international colloquium on automata, languages and programming. Springer, Heidelberg, pp 357–368
13. Sadakane K (2007) Compressed suffix trees with full functionality. Theory Comput Syst. Online first. http://dx.doi.org/10.1007/s00224-006-1198-x
14. Weiner P (1973) Linear pattern matching algorithms. In: Proceedings of the 14th annual IEEE symposium on switching and automata theory. IEEE, New York, pp 1–11
15. Yannakakis M (1986) Four pages are necessary and sufficient for planar graphs. In: Hartmanis J (ed) Proceedings of the 18th annual ACM-SIAM symposium on theory of computing. ACM, New York, pp 104–108

Suffix Array Construction

Juha Kärkkäinen
Department of Computer Science, University of Helsinki, Helsinki, Finland

Keywords

Longest common prefix array; Suffix array; Suffix sorting; Text indexing

Years and Authors of Summarized Original Work

2006; Kärkkäinen, Sanders, Burkhardt

Problem Definition

The *suffix array* [4, 15] is the lexicographically sorted array of all the suffixes of a string. It is a popular text index structure with many applications. The subject of this entry is algorithms that construct the suffix array.

More precisely, the input to a suffix array construction algorithm is a *text string* $T = T[0\ldots n) = t_0 t_1 \ldots t_{n-1}$, i.e., a sequence of n *characters* from an *alphabet* Σ. For $i \in [0\ldots n]$, let S_i denote the *suffix* $T[i \ldots n) = t_i t_{i+1} \ldots t_{n-1}$. The output is the *suffix array* $SA[0\ldots n]$ of T, a permutation of $[0\ldots n]$ satisfying $S_{SA[0]} < S_{SA[1]} < \cdots < S_{SA[n]}$, where $<$ denotes the *lexicographical order* of strings.

Two specific models for the alphabet Σ are considered. An *ordered alphabet* is an arbitrary ordered set with constant time character comparisons. An *integer alphabet* is the integer range $[1\ldots\sigma]$ for $\sigma = n^{\mathcal{O}(1)}$.

Many applications require that the suffix array is augmented with additional information, most commonly with the *longest common prefix* array $LCP[1\ldots n]$. An entry $LCP[i]$ of the LCP array is the length of the longest common prefix of the suffixes $S_{SA[i]}$ and $S_{SA[i-1]}$. The *enhanced suffix array* [1] adds two more arrays to obtain a full range of text index functionalities.

There are other important text indexes, most notably suffix trees and compressed text indexes, covered in separate entries. Each of these indexes has their own construction algorithms, but they can also be constructed efficiently from each other. However, in this entry, the focus is on direct suffix array construction algorithms that do not rely on other text indexes.

Key Results

The naive approach to suffix array construction is to use a general sorting algorithm or an algorithm for sorting strings. However, any such algorithm has a worst-case time complexity $\Omega(n^2)$ because the total length of the suffixes is $\Omega(n^2)$.

The first efficient algorithms were based on the *doubling technique* of Karp, Miller, and Rosenberg [10]. The idea is to assign a *rank* to all substrings whose length is a power of two. The rank tells the lexicographic order of the substring among substrings of the same length. Given the ranks for substrings of length h, the ranks for substrings of length $2h$ can be computed using a radix sort step in linear time (doubling). The technique was first applied to suffix array construction by Manber and Myers [15]. The best practical algorithm based on the technique is by Larsson and Sadakane [14].

Theorem 1 (Manber and Myers [15]; Larsson and Sadakane [14]) *The suffix array can be constructed in $\mathcal{O}(n \log n)$ time, which is optimal for the ordered alphabet.*

Faster algorithms for the integer alphabet are based on a different technique, recursion. The basic procedure is as follows.

1. Sort a subset of the suffixes. This is done by constructing a shorter string, whose suffix array gives the order of the desired subset. The suffix array of the shorter string is constructed by recursion.

2. Extend the subset order to full order.

The technique first appeared in suffix tree construction [3], but 2003 saw the independent and simultaneous publication of three linear time suffix array construction algorithms based on the approach but not using suffix trees. Each of the three algorithms uses a different subset of suffixes requiring a different implementation of the second step.

Theorem 2 (Kärkkäinen, Sanders, and Burkhardt [8]; Kim et al. [12]; Ko and Aluru [13]) *The suffix array can be constructed in the optimal linear time for the integer alphabet.*

We will describe the algorithm of Kärkkäinen, Sanders, and Burkhardt [8] called DC3 in more detail. For $k \in \{0, 1, 2\}$, let $\mathcal{R}_k$ be the set of suffixes S_i such that $i \bmod 3 = k$. Let $\mathcal{R}_{12} = \mathcal{R}_1 \cup \mathcal{R}_2$ and define $\mathcal{R}_{01}$ and $\mathcal{R}_{02}$ symmetrically. For example, $\mathcal{R}_{12} = \{S_1, S_2, S_4, S_5, S_7, S_8, \dots\}$. The set $\mathcal{R}_{12}$ is the subset of suffixes sorted first. For $S_i \in \mathcal{R}_{12}$, let $\bar{S}_i$ be the lexicographical rank of S_i in $\mathcal{R}_{12}$. Given those lexicographical ranks, we can compare any two suffixes S_i and S_j in constant time using one of the following ways:

1. If $S_i, S_j \in \mathcal{R}_{12}$, compare the ranks $\bar{S}_i$ and $\bar{S}_j$.
2. If $S_i, S_j \in \mathcal{R}_{01}$, compare the pairs $\langle t_i, \bar{S}_{i+1} \rangle$ and $\langle t_j, \bar{S}_{j+1} \rangle$.
3. If $S_i, S_j \in \mathcal{R}_{02}$, compare the triples $\langle t_i, t_{i+1}, \bar{S}_{i+2} \rangle$ and $\langle t_j, t_{j+1}, \bar{S}_{j+2} \rangle$.

Furthermore, we can radix sort $\mathcal{R}_0$ in linear time by using $\langle t_i, \bar{S}_{i+1} \rangle$ to represent the suffix $S_i \in \mathcal{R}_0$. After this, we can merge $\mathcal{R}_0$ and $\mathcal{R}_{12}$, which takes linear time since we can compare suffixes in constant time.

We still need to describe how to sort $\mathcal{R}_{12}$. Let $\overline{t_i t_{i+1} t_{i+2}}$ be the lexicographical rank of the substring $t_i t_{i+1} t_{i+2}$ among all substrings of length three. Let

$$T_{12} = \overline{t_1 t_2 t_3} \; \overline{t_4 t_5 t_6} \; \overline{t_7 t_8 t_9} \dots$$

$$\overline{t_2 t_3 t_4} \; \overline{t_5 t_6 t_7} \; \overline{t_8 t_9 t_{10}} \dots .$$

For example if $T = \texttt{yabbadabbado}$, we have

$$T_{12} = \overline{\texttt{abb}} \; \overline{\texttt{ada}} \; \overline{\texttt{bba}} \; \overline{\texttt{do\$}} \; \overline{\texttt{bba}} \; \overline{\texttt{dab}} \; \overline{\texttt{bad}} \; \overline{\texttt{o\$\$}}$$

$$= 12575648 \,,$$

where \$ is a special padding symbol that does not appear in the text and is considered smaller than any normal character. Clearly, sorting the suffixes of T_{12} is equivalent to sorting the set $\mathcal{R}_{12}$. The suffixes of T_{12} are sorted by a recursive call to the algorithm itself. Since the recursive call is for a text of length at most $\lceil 2n/3 \rceil$ and everything outside the recursive call can be done in linear time, the total time complexity of DC3 is $\mathcal{O}(n)$.

The above algorithms and many other suffix array construction algorithms are surveyed in [18]. Worth mentioning among the more recent results are the linear time algorithms of Nong, Zhang, and Chan [17].

The $\Omega(n \log n)$ lower bound for the ordered alphabet mentioned in Theorem 1 comes from the sorting complexity of characters, since the initial characters of the sorted suffixes are the text characters in sorted order. Theorem 2 allows a generalization of this result. For any alphabet, one can first sort the characters of T, remove duplicates, assign a rank to each character, and construct a new string T' over the alphabet $[1 \dots n]$ by replacing the characters of T with their ranks. The suffix array of T' is exactly the same as the suffix array of T. Optimal algorithms for the integer alphabet then give the following result.

Theorem 3 *For any alphabet, the complexity of suffix array construction is the same as the complexity of sorting the characters of the string.*

The result extends to the related arrays.

Theorem 4 (Kasai et al. [11]; Abouelhoda, Kurtz, and Ohlebusch [1]) *The LCP array and the enhanced suffix array can be computed in linear time given the suffix array.*

One of the main advantages of suffix arrays over suffix trees is their smaller space requirement (by a constant factor), and a significant effort has been spent making construction algorithms space efficient, too. The best algorithms need very little extra space.

Theorem 5 (Kärkkäinen, Sanders, and Burkhardt [8]; Nong [16]) *For any* $v = \mathcal{O}(n^{2/3})$, *the suffix array can be constructed in* $\mathcal{O}(n(v + \log n))$ *time and* $\mathcal{O}(n/\sqrt{v})$ *extra space for the ordered alphabet and in* $\mathcal{O}(nv)$ *time and* $\mathcal{O}(n/\sqrt{v})$ *extra space or* $\mathcal{O}(n)$ *time and* $\mathcal{O}(\sigma)$ *extra space for the integer alphabet, where the extra space is the space needed in addition to the input (the string T) and the output (the suffix array).*

In the algorithm DC3 described above, all steps can be performed by sorting, prefix sums (assigning lexicographical ranks) and localized computation. This makes it straightforward to adapt to several parallel and hierarchical memory models of computation [8] including the following result for the standard external memory model.

Theorem 6 (Kärkkäinen, Sanders, and Burkhardt [8]) *The suffix array can be constructed in the optimal* $\mathcal{O}(\text{sort}(n))$ *I/Os in the standard external memory model, where* $\text{sort}(n)$ *is the I/O complexity of sorting n elements.*

The above algorithm can be modified to compute the LCP array too in the same I/O complexity [2, 7].

Applications

The suffix array is a simple and powerful text index structure with numerous applications; see [1] and Cross-References. The practical construction of many other text indexes usually starts with the suffix array construction. In particular, the Burrows–Wheeler transform, which is an important technique for text compression and the basis of many compressed text indexes, is easily computed from the suffix array.

Open Problems

Theoretically, the suffix array construction problem is essentially solved. The development of ever more efficient practical algorithms is still going on particularly for external memory and parallel computation. There is currently no external memory algorithm for computing the LCP array from the suffix array in $\mathcal{O}(\text{sort}(n))$ I/Os other than as a side effect of suffix array construction [6].

Experimental Results

Many papers on suffix array construction contain experimental results, but they are usually either out of date (e.g., [18]) or limited in scope (e.g., [16]). The most comprehensive comparison of algorithms is at https://code.google.com/p/libdivsufsort/wiki/SACA_Benchmarks. The best practical algorithms for large data are divsufsort, which is an $\mathcal{O}(n \log n)$ time algorithm combining several techniques, and SAIS, which is an implementation of the linear time algorithm by Gong, Zhang, and Chan [17] (see below for URLs to code). The comparison and the fastest implementation are by the same person, Yuta Mori, but the implementations are widely used and there are no substantial claims for other, faster algorithms.

There are also experiments for suffix array construction in external memory [2, 5] and for LCP array construction [2, 6, 9].

URLs to Code and Data Sets

The input to a suffix array construction algorithm is simply a text, so an abundance of data exists. Links to many text collections are provided at https://code.google.com/p/libdivsufsort/wiki/SACA_Benchmarks. Worth mentioning is also the Pizza&Chili site with its standard text corpus http://pizzachili.dcc.uchile.cl/texts.html and the repetitive text corpus http://pizzachili.dcc.uchile.cl/repcorpus.html.

Notable implementations of suffix array construction algorithms are available at https://code.google.com/p/libdivsufsort/, at https://sites.google.com/site/yuta256/sais, at http://panthema.net/2012/1119-eSAIS-Inducing-Suffix-and-LCP-Arrays-in-External-Memory/ [2], and at https://

www.cs.helsinki.fi/group/pads/SAscan.html [5]. The latter two work in external memory and provide (links to) LCP array construction too.

Cross-References

▶ Burrows-Wheeler Transform
▶ Compressed Suffix Array
▶ Suffix Trees and Arrays
▶ Suffix Tree Construction

Recommended Reading

1. Abouelhoda MI, Kurtz S, Ohlebusch E (2004) Replacing suffix trees with enhanced suffix arrays. J Discret Algorithms 2(1):53–86
2. Bingmann T, Fischer J, Osipov V (2013) Inducing suffix and LCP arrays in external memory. In: Sanders P, Zeh N (eds) Proceedings of the 15th meeting on algorithm engineering and experiments (ALENEX), New Orleans. SIAM, pp 88–102
3. Farach-Colton M, Ferragina P, Muthukrishnan S (2000) On the sorting-complexity of suffix tree construction. J ACM 47(6):987–1011
4. Gonnet G, Baeza-Yates R, Snider T (1992) New indices for text: PAT trees and PAT arrays. In: Frakes WB, Baeza-Yates R (eds) Information retrieval: data structures & algorithms. Prentice-Hall, Englewood Cliffs
5. Kärkkäinen J, Kempa D (2014) Engineering a lightweight external memory suffix array construction algorithm. In: Iliopoulos CS, Langiu A (eds) Proceedings of the 2nd international conference on algorithms for big data (ICABD), Palermo, pp 53–60
6. Kärkkäinen J, Kempa D (2014) LCP array construction in external memory. In: Gudmundsson J, Katajainen J (eds) Proceedings of the 13th symposium on experimental algorithms (SEA), Copenhagen. Lecture notes in computer science, vol 8504. Springer, pp 412–423
7. Kärkkäinen J, Sanders P (2003) Simple linear work suffix array construction. In: Baeten JCM, Lenstra JK, Parrow J, Woeginger GJ (eds) Proceedings of the 30th international conference on automata, languages and programming (ICALP), Eindhoven. Lecture notes in computer science, vol 2719. Springer, pp 943–955
8. Kärkkäinen J, Sanders P, Burkhardt S (2006) Linear work suffix array construction. J ACM 53(6):918–936
9. Kärkkäinen J, Manzini G, Puglisi SJ (2009) Permuted longest-common-prefix array. In: Kucherov G, Ukkonen E (eds) Proceedings of the 20th annual symposium on combinatorial pattern matching (CPM), Lille. Lecture notes in computer science, vol 5577. Springer, pp 181–192
10. Karp RM, Miller RE, Rosenberg AL (1972) Rapid identification of repeated patterns in strings, trees and arrays. In: Proceedings of the 4th annual ACM symposium on theory of computing (STOC), Denver. ACM, pp 125–136
11. Kasai T, Lee G, Arimura H, Arikawa S, Park K (2001) Linear-time longest-common-prefix computation in suffix arrays and its applications. In: Proceedings of the 12th annual symposium on combinatorial pattern matching (CPM), Jerusalem. Lecture notes in computer science, vol 2089. Springer, pp 181–192
12. Kim DK, Sim JS, Park H, Park K (2005) Constructing suffix arrays in linear time. J Discret Algorithms 3(2–4):126–142
13. Ko P, Aluru S (2005) Space efficient linear time construction of suffix arrays. J Discret Algorithms 3(2–4):143–156
14. Larsson NJ, Sadakane K (2007) Faster suffix sorting. Theor Comput Sci 387(3):258–272
15. Manber U, Myers G (1993) Suffix arrays: a new method for on-line string searches. SIAM J Comput 22(5):935–948
16. Nong G (2013) Practical linear-time O(1)-workspace suffix sorting for constant alphabets. ACM Trans Inf Syst 31(3):Article 15, 15 pages
17. Nong G, Zhang S, Chan WH (2011) Two efficient algorithms for linear time suffix array construction. IEEE Trans Comput 60(10):1471–1484
18. Puglisi SJ, Smyth WF, Turpin A (2007) A taxonomy of suffix array construction algorithms. ACM Comput Surv 39(2):Article 4, 31 pages

Suffix Tree Construction

Jens Stoye
Faculty of Technology, Genome Informatics, Bielefeld University, Bielefeld, Germany

Keywords

Full-text index construction

Years and Authors of Summarized Original Work

1973; Weiner
1976; McCreight

1995; Ukkonen
2000; Farach-Colton, Ferragina, Muthukrishnan

Problem Definition

The suffix tree is perhaps the best-known and most-studied data structure for string indexing with applications in many fields of sequence analysis. After its invention in the early 1970s, several approaches for the efficient construction of the suffix tree of a string have been developed for various models of computation. The most prominent of those that construct the suffix tree in main memory are summarized in this entry.

Notations

Given an alphabet Σ, a *trie* over Σ is a rooted tree whose edges are labeled with strings over Σ such that no two labels of edges leaving the same vertex start with the same symbol. A trie is *compacted* if all its internal vertices, except possibly the root, are branching. Given a finite string $S \in \Sigma^n$, the *suffix tree* of S, $T(S)$, is the compacted trie over Σ such that the concatenations of the edge labels along the paths from the root to the leaves are the suffixes of S. An example is given in Fig. 1.

The concatenation of the edge labels from the root to a vertex v of $T(S)$ is called the *path-label* of v, $P(v)$. For example, the path label of the vertex indicated by the asterisk in Fig. 1 is $P(*) = \text{MAM}$.

Constraints

The time complexity of constructing the suffix tree of a string S of length n depends on the size of the underlying alphabet Σ. It may be constant, it may be the alphabet of integers $\Sigma = \{1, 2, \ldots, n\}$, or it may be an arbitrary finite set whose elements can be compared in constant time. Note that the latter case reduces to the previous one if one maps the symbols of the alphabet to the set $\{1, \ldots, n\}$, though at the additional cost of sorting Σ.

Problem 1 (suffix tree construction)

INPUT: A finite string S of length n over an alphabet Σ.
OUTPUT: The suffix tree $T(S)$.

If one assumes that the outgoing edges at each vertex are lexicographically sorted, which is usually the case, the suffix tree allows retrieving the sorted order of S's characters in linear time. Therefore, suffix tree construction inherits the lower bounds from the problem complexity of sorting: $\Omega(n \log n)$ in the general alphabet case and $\Omega(n)$ for integer alphabets.

Key Results

Theorem 1 *The suffix tree of a string of length n can be represented in $O(n \log n)$ bits of space.*

This is easy to see since the number of leaves of $T(S)$ is at most n, and so is the number of internal vertices that, by definition, are all branching, as well as the number of edges. In order to see that each edge label can be stored in $O(\log n)$ bits of space, note that an edge label is always a substring of S. Hence it can be represented by a pair (l, r) consisting of *left pointer l* and *right pointer r*, if the label is $S[l, r]$.

Note that this space bound is not optimal since there are $|\Sigma|^n$ different strings and hence suffix trees, while $n \log n$ bits would allow to represent $n!$ different entities.

Theorem 2 *Suffix trees can be constructed in optimal time, in particular:*

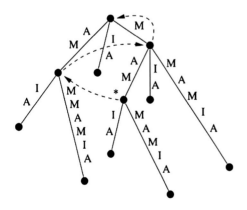

Suffix Tree Construction, Fig. 1 The suffix tree for the string $S = \text{MAMMAMIA}$. *Dashed arrows* denote suffix links that are employed by all efficient suffix tree construction algorithms

1. For constant-size alphabet, the suffix tree $T(S)$ of a string S of length n can be constructed in $O(n)$ time [11–13]. For general alphabet, these algorithms require $O(n \log n)$ time.
2. For integer alphabet, the suffix tree of S can be constructed in $O(n)$ time [4, 9].

Generally, there is a natural strategy to construct a suffix tree: Iteratively all suffixes are inserted into an initially empty structure. Such a strategy will immediately lead to a linear-time construction algorithm if each suffix can be inserted in constant time. Finding the correct position where to insert a suffix, however, is the main difficulty of suffix tree construction.

The first solution for this problem was given by Weiner in his seminal 1973 paper [13]. His algorithm inserts the suffixes from shortest to longest, and the insertion point is found in amortized constant time for constant-size alphabet, using rather a complicated amount of additional data structures. A simplified version of the algorithm was presented by Chen and Seiferas [3]. They give a cleaner presentation of the three types of links that are required in order to find the insertion points of suffixes efficiently, and their complexity proof is easier to follow. Since the suffix tree is constructed while reading the text from right to left, these two algorithms are sometimes called *anti-online* constructions.

A different algorithm was given in 1976 by McCreight [11]. In this algorithm the suffixes are inserted into the growing tree from longest to shortest. This simplifies the update procedure, and the additional data structure is limited to just one type of link: an internal vertex v with path label $P(v) = aw$ for some symbol $a \in \Sigma$ and string $w \in \Sigma^*$ has a *suffix link* to the vertex u with path label $P(u) = w$. In Fig. 1, suffix links are shown as dashed arrows. They often connect vertices above the insertion points of consecutively inserted suffixes, like the vertex with path-label "M" and the root, when inserting suffixes "MAMIA" and "AMIA" in the example of Fig. 1. This property allows reaching the next insertion point without having to search for it from the root of the tree, thus ensuring amortized

constant time per suffix insertion. Note that since McCreight's algorithm treats the suffixes from longest to shortest and the intermediate structures are not suffix trees, the algorithm is not an online algorithm.

Another linear-time algorithm for constant-size alphabet is the online construction by Ukkonen [12]. It reads the text from left to right and updates the suffix tree in amortized constant time per added symbol. Again, the algorithm uses suffix links in order to quickly find the insertion points for the suffixes to be inserted. Moreover, since during a single update the edge labels of all leaf edges need to be extended by the new symbol, it requires a trick to extend all these labels in constant time: all the right pointers of the leaf edges refer to the same *end of string* value, which is just incremented.

An even stronger concept than online construction is *real-time* construction, where the worst-case (instead of amortized) time per symbol is considered. Amir et al. [1] present for general alphabet a suffix tree construction algorithm that requires $O(\log n)$ worst-case update time per every single input symbol when the text is read from right to left, and thus requires overall $O(n \log n)$ time, like the other algorithms for general alphabet mentioned so far. They achieve this goal using a binary search tree on the suffixes of the text, enhanced by additional pointers representing the lexicographic and the textual order of the suffixes, called *Balanced Indexing Structure*. This tree can be constructed in $O(\log n)$ worst-case time per added symbol and allows maintaining the suffix tree in the same time bound.

The first linear-time suffix tree construction algorithm for integer alphabets was given by Farach-Colton [4]. It uses the so-called *odd-even technique* that proceeds in three steps:

1. Recursively compute the compacted trie of all suffixes of S beginning at odd positions, called the *odd tree* T_o.
2. From T_o compute the *even tree* T_e, the compacted trie of the suffixes beginning at even positions in S.

3. Merge T_o and T_e into the whole suffix tree $T(S)$.

The basic idea of the first step is to encode pairs of characters as single characters. Since at most $n/2$ different such characters can occur, these can be radix-sorted and range-reduced to an alphabet of size $n/2$. Thus, the string S of length n over the integer alphabet $\Sigma = \{1, \dots, n\}$ is translated in $O(n)$ time into a string S' of length $n/2$ over the integer alphabet $\Sigma' = \{1, \dots, n/2\}$. Applying the algorithm recursively to this string yields the suffix tree of S'. After translating the edge labels from substrings of S' back to substrings of S, some vertices may exist with outgoing edges whose labels start with the same symbol, because two distinct symbols from Σ' may be pairs with the same first symbol from Σ. In such cases, by local modifications of edge labels or adding additional vertices, the trie property can be regained and the desired tree T_o is obtained.

In the second step, the odd tree T_o from the first step is used to generate the lexicographically sorted list (*lex-ordering* for short) of the suffixes starting at odd positions. Radix-sorting these with the characters at the preceding even positions as keys yields a lex-ordering of the even suffixes in linear time. Together with the longest common prefixes (lcps) of consecutive positions that can be computed in linear time from T_o using constant-time lowest common ancestor queries and the identity

$$\mathrm{lcp}(l_{2i}, l_{2j}) = \begin{cases} \mathrm{lcp}(l_{2i+1}, l_{2j+1}) + 1 & \text{if } S[2i] = S[2j] \\ 0 & \text{otherwise} \end{cases}$$

this ordering allows reconstructing the even tree T_e in linear time.

In the third step, the two tries T_o and T_e are merged into the suffix tree $T(S)$. Conceptually, this is a straightforward procedure: the two tries are traversed in parallel, and every part that is present in one or both of the two trees is inserted in the common structure. However, this procedure is simple only if edges are traversed character by character such that common and differing parts can be observed directly. Such a traversal would, however, require $O(n^2)$ time in the worst case, impeding the desired overall linear running time. Therefore, Farach-Colton suggests to use an oracle that tells for an edge of T_o and an edge of T_e the length of their common prefix.

However, the suggested oracle may overestimate this length, and that is why sometimes the tree generated must be corrected, called *unmerging*. The full details of the oracle and the unmerging procedure can be found in [4].

Overall, if $T(n)$ is the time it takes to build the suffix tree of a string $S \in \{1, \dots, n\}^n$, the first step takes $T(n/2) + O(n)$ time and the second and third steps take $O(n)$ time; thus the whole procedure takes $O(n)$ overall time on the RAM model.

Another linear-time construction of suffix trees for integer alphabets can be achieved via linear-time construction of suffix arrays together with longest common prefix tabulation, as described by Kärkkäinen and Sanders in [9].

All previously mentioned algorithms construct the suffix tree in main memory. However, since the data structure may become very large in practice, also methods for building the suffix tree in secondary memory have been studied. Possibly the simplest way is to first construct the suffix array A and the LCP array on disk, as described in the entry ▶ Suffix Array Construction. When this is done, it is only a small final step to construct the suffix tree [4]. The idea is to construct the tree in n phases from left to right, such that after phase i the suffix tree of the strings $A[1], \dots, A[i]$ has been constructed. Simultaneously, an external-memory stack containing the nodes on the path leading from the root to $A[i]$ is maintained. In phase $i+1$, first, the leaf representing string $A[i+1]$ is created, and then all nodes are popped from the stack whose string length is strictly

S

greater than LCP[i]. Next, a new node with string depth LCP[i] is created (unless it already exists) whose parent is the top element of the stack and whose children are the last popped element and the new leaf. This new node and the new leaf are finally pushed on the stack. Keeping the two top pages of the stack in internal memory, the algorithm executes a total of $O(n)$ pop and push operations and therefore uses a total of $O(n/B)$ time, where B is the external memory block size.

Other more direct ways to construct the suffix tree on disk have also been developed, e.g., [14, 15].

In some applications the so-called *generalized* suffix tree of several strings is used, a dictionary obtained by constructing the suffix tree of the concatenation of the contained strings. An important question that arises in this context is that of dynamically updating the tree upon insertion and deletion of strings from the dictionary. More specifically, since edge labels are stored as pairs of pointers into the original string, when deleting a string from the dictionary, the corresponding pointers may become invalid and need to be updated. An algorithm to solve this problem in amortized linear time was given by Fiala and Greene [6], and a linear worst-case (and hence real-time) algorithm was given by Ferragina et al. [5].

Applications

The suffix tree supports many applications, most of them in optimal time and space, including exact string matching, set matching, longest common substring of two or more sequences, all-pairs suffix-prefix matching, repeat finding, and text compression. These and several other applications, many of them from bioinformatics, are given in [2] and [8].

Open Problems

Some theoretical questions regarding the expected size and branching structure of suffix trees under more complicated than i. i. d. sequence models are still open. Currently most of the research has moved toward more space-efficient data structures like suffix arrays and compressed string indices or the Burrows-Wheeler Transform.

Experimental Results

Suffix trees are infamous for their high memory requirements. The practical space consumption is between 9 and 11 times the size of the string to be indexed, even in the most space-efficient implementations known [7, 10]. Moreover, [7] also shows that suboptimal algorithms like the very simple quadratic-time *write-only top-down* (WOTD) algorithm can outperform optimal algorithms on many real-world instances in practice, if carefully engineered.

URLs to Code and Data Sets

Several sequence analysis libraries contain code for suffix tree construction. For example, Strmat (http://www.cs.ucdavis.edu/~gusfield/strmat.html) by Gusfield et al. contains implementations of Weiner's and Ukkonen's algorithm. An implementation of the WOTD algorithm by Kurtz can be found at (http://bibiserv.techfak.uni-bielefeld.de/wotd).

Cross-References

▶ Burrows-Wheeler Transform
▶ Compressed Suffix Trees
▶ Suffix Array Construction
▶ Suffix Trees and Arrays

Recommended Reading

1. Amir A, Kopelowitz T, Lewenstein M, Lewenstein N (2005) Towards real-time suffix tree construction. In: Proceedings of the 12th international symposium on string processing and information retrieval (SPIRE 2005). LNCS, vol 3772. Springer, Berlin, pp 67–78
2. Apostolico A (1985) The myriad virtues of subword trees. In: Apostolico A, Galil Z (eds) Combinatorial algorithms on words. NATO ASI Series, vol F12. Springer, Berlin, pp 85–96
3. Chen MT, Seiferas J (1985) Efficient and elegant subword tree construction. In: Apostolico A, Galil Z (eds) Combinatorial algorithms on words. Springer, New York
4. Farach-Colton M, Ferragina P, Muthukrishnan S (2000) On the sorting-complexity of suffix tree construction. J ACM 47(6):987–1011
5. Ferragina P, Grossi R, Montangero M (1998) A note on updating suffix tree labels. Theor Comput Sci 201:249–262
6. Fiala ER, Greene DH (1989) Data compression with finite windows. Commun ACM 32:490–505

7. Giegerich R, Kurtz S, Stoye J (2003) Efficient implementation of lazy suffix trees. Softw Pract Exp 33:1035–1049
8. Gusfield D (1997) Algorithms on strings, trees, and sequences: computer science and computational biology. Cambridge University Press, New York
9. Kärkkäinen J, Sanders P (2003) Simple linear work suffix array construction. In: Proceedings of the 30th international colloquium on automata, languages, and programming (ICALP 2003). LNCS, vol 2719. Springer, Berlin, pp 943–955
10. Kurtz S (1999) Reducing the space requirements of suffix trees. Softw Pract Exp 29:1149–1171
11. McCreight EM (1976) A space-economical suffix tree construction algorithm. J ACM 23:262–272
12. Ukkonen E (1995) On-line construction of suffix trees. Algorithmica 14:249–260
13. Weiner P (1973) Linear pattern matching algorithms. In: Proceedings of the 14th annual IEEE symposium on switching and automata theory. IEEE Press, New York, pp 1–11
14. Cheung C-F, Yu JX, Lu H (2005) Constructing suffix tree for gigabyte sequences with megabyte memory. IEEE Trans Knowl. Data Eng. 17:90–105
15. Tian Y, Tata S, Hankins RA, Patel JM (2005) Practical methods for constructing suffix trees. VLDB J 14:281–299

Suffix Tree Construction in Hierarchical Memory

Paolo Ferragina
Department of Computer Science, University of Pisa, Pisa, Italy

Keywords

Full-text index construction; String B-tree construction; Suffix array construction; Suffix tree construction

Years and Authors of Summarized Original Work

2000; Farach-Colton, Ferragina, Muthukrishnan

Problem Definition

The suffix tree is the ubiquitous data structure of combinatorial pattern matching myriad of situations – just to cite a few, searching, data compression and mining, and bioinformatics [7]. In these applications, the large data sets now available involve the use of numerous memory levels which constitute the storage medium of modern PCs: L1 and L2 caches, internal memory, multiple disks, and remote hosts over a network. The power of this memory organization is that it may be able to offer the expected access time of the fastest level (i.e., cache) while keeping the average cost per memory cell near the one of the cheapest level (i.e., disk), provided that data are properly *cached* and *delivered* to the requiring algorithms. Neglecting questions pertaining to the cost of memory references may even prevent the use of algorithms on large sets of input data. Engineering research is presently trying to improve the input/output subsystem to reduce the impact of these issues, but it is very well known [20] that the improvements achievable by means of a *proper arrangement of data* and a properly *structured algorithmic computation* abundantly surpass the best-expected technology advancements.

The Model of Computation

In order to reason about algorithms and data structures operating on hierarchical memories, it is necessary to introduce a *model of computation* that grasps the essence of real situations so that algorithms that are good in the model are also good in practice. The model considered here is the *external-memory model* [20], which received much attention because of its simplicity and reasonable accuracy. A computer is abstracted to consist of *two memory levels*: the internal memory of size M and the (unbounded) disk memory which operates by reading/writing data in blocks of size B (called *disk pages*). The performance of algorithms is then evaluated by counting (a) the number of disk accesses (I/Os), (b) the internal running time (CPU time), and (c) the number of disk pages occupied by the data structure or used by the algorithm as its working space. This simple model suggests, correctly, that a good external-memory algorithm should exploit both *spatial locality* and *temporal locality*. Of course, "I/O" and "two-level view" refer to any two levels

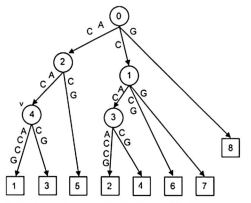
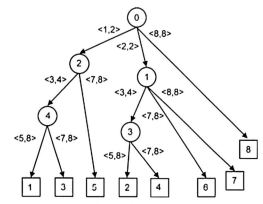

Suffix Tree Construction in Hierarchical Memory, Fig. 1 The suffix tree of $S =$ ACACACCG on the *left*, and its compact edge-encoding on the *right*. The endmarker # is not shown. Node v spells out the string *ACAC*. Each internal node stores the length of its associated string, and each leaf stores the starting position of its corresponding suffix

of the memory hierarchy with their parameters M and B properly set.

Notation

Let $S[1, n]$ be a string drawn from alphabet Σ, and consider the notation: S_i for the ith *suffix* of string S, $\mathrm{lcp}(\alpha, \beta)$ for the *longest common prefix* between the two strings α and β, and $\mathrm{lca}(u, v)$ for the *lowest common ancestor* between two nodes u and v in a tree.

The suffix tree of $S[1, n]$, denoted hereafter by $\mathcal{T}_S$, is a tree that stores all suffixes of $S\#$ in a compact form, where $\# \notin \Sigma$ is a special character (see Fig. 1). $\mathcal{T}_S$ consists of n leaves, numbered from 1 to n, and any root-to-leaf path spells out a suffix of $S\#$. The endmarker # guarantees that no suffix is the prefix of another suffix in $S\#$. Each internal node has at least two children and each edge is labeled with a nonempty substring of S. No two edges out of a node can begin with the same character, and sibling edges are ordered lexicographically according to that character. Edge labels are encoded with pairs of integers – say $S[x, y]$ is represented by the pair $\langle x, y \rangle$. As a result, all $\Theta(n^2)$ substrings of S can be represented in $O(n)$ optimal space by $\mathcal{T}_S$'s structure and edge encoding. Furthermore, the rightward scan of the suffix-tree leaves gives the ordered set of S's suffixes, also known as the *suffix array* of S [13]. Notice that the case of a large string collection $\Delta = \{S^1, S^2, \ldots, S^k\}$ *reduces to* the case of one long string $S = S^1 \#_1 S^2 \#_2 \cdots S^k \#_k$, where $\#_i \notin \Sigma$ are special symbols.

Numerous algorithms are known that build the suffix tree optimally in the RAM model (see [3] and references therein). However, most of them exhibit a marked absence of *locality of references* and thus elicit many I/Os when the size of the indexed string is too large to be fit into the internal memory of the computer. This is a serious problem because the slow performance of these algorithms can prevent the suffix tree being used even in medium-scale applications. This encyclopedia's entry surveys algorithmic solutions that deal efficiently with the *construction of suffix trees over large string collections* by executing an optimal number of I/Os. Since it is assumed that the edges leaving a node in $\mathcal{T}_S$ are lexicographically sorted, sorting is an obvious *lower bound* for building suffix trees (consider the suffix tree of a permutation!). The presented algorithms have sorting as their bottleneck, thus establishing that *the complexity of sorting and suffix tree construction match*.

Key Results

Designing a disk-efficient approach to suffix-tree construction has found efficient solutions only in the last few years [4]. The present section surveys

DIVIDE-AND-CONQUER ALGORITHM

(1) Construct the string $S'[j]$ = rank of $\langle S[2j], S[2j+1]\rangle$, and recursively compute $\mathcal{T}_{S'}$.

(2) Derive from $\mathcal{T}_{S'}$ the compacted trie $\mathcal{T}_o$ of all suffixes of S beginning at odd positions.

(3) Derive from $\mathcal{T}_o$ the compacted trie $\mathcal{T}_e$ of all suffixes of S beginning at even positions.

(4) Merge $\mathcal{T}_o$ and $\mathcal{T}_e$ into the whole suffix tree $\mathcal{T}_S$, as follows:

(4.1) Overmerge $\mathcal{T}_o$ and $\mathcal{T}_e$ into the tree $\mathcal{T}_M$.

(4.2) Partially unmerge $\mathcal{T}_M$ to get $\mathcal{T}_S$.

Suffix Tree Construction in Hierarchical Memory, Fig. 2 The algorithm that builds the suffix tree directly

SUFFIX ARRAY-BASED ALGORITHM

(1) Construct the suffix array $\mathcal{A}_S$ and the array lcp_S of the string S.

(2) Initially set $\mathcal{T}_S$ as a single edge connecting the root to a leaf pointing to suffix $\mathcal{A}_S[1]$.

(2) For $i = 2,...,n$:

(2.1) Create a new leaf ℓ_i that points to the suffix $\mathcal{A}_S[i]$.

(2.2) Walk up from ℓ_{i-1} until a node u_i is met whose string-length x_i is $\leq \mathrm{lcp}_S[i]$.

(2.3) If $x_i = \mathrm{lcp}_S[i]$, leaf ℓ_i is attached to u_i.

(2.4) If $x_i < \mathrm{lcp}_S[i]$, create node u'_i with string-length x_i, attach it to u_i and leaf ℓ_i to u'_i;

Suffix Tree Construction in Hierarchical Memory, Fig. 3 The algorithm that builds the suffix tree passing through the suffix array

two theoretical approaches which achieve the best (optimal!) I/O-bounds in the worst case; the next section will discuss some practical solutions.

The first algorithm is based on a *Divide-and-Conquer* approach that allows us to reduce the construction process to external-memory sorting and few low-I/O primitives. It builds the suffix tree $\mathcal{T}_S$ by executing four (macro)steps, detailed in Fig. 2. It is not difficult to implement the first three steps in $\mathrm{Sort}(n) = O(\frac{n}{B}\log_{M/B}\frac{n}{B})$ I/Os [20]. The last (merging) step is the most difficult one and its I/O-complexity bounds the cost of the overall approach. Farach-Colton et al. [3] propose an elegant merge for $\mathcal{T}_o$ and $\mathcal{T}_e$: substep (4.1) temporarily relaxes the requirement of getting $\mathcal{T}_S$ in one shot, and thus it blindly (over)merges the paths of $\mathcal{T}_o$ and $\mathcal{T}_e$ by comparing edges only via their first characters; then substep (4.2) refixes $\mathcal{T}_M$ by detecting and undoing in an I/O-efficient manner the (over)merged paths. Note that the time and I/O-complexity of this algorithm follow a nice recursive relation: $T(n) = T(n/2) + O(\mathrm{Sort}(n))$.

Theorem 1 (Farach-Colton et al. [5]) *Given an arbitrary string $S[1,n]$, its suffix tree can be constructed in $O(\mathrm{Sort}(n))$ I/Os, $O(n\log n)$ time and using $O(n/B)$ disk pages.*

The second algorithm [10] is deceptively simple, elegant, and I/O optimal and applies successfully to the construction of other indexing data structures, like the string Btree [5]. The key idea is to derive $\mathcal{T}_S$ from the suffix array $\mathcal{A}_S$ and from the *lcp* array, which stores the longest-common-prefix length of adjacent suffixes in $\mathcal{A}_S$. Its pseudocode is given in Fig. 3. Note that step (1) may deploy any external-memory algorithm for suffix array construction: used here is the elegant and optimal *Skew* algorithm of [9] which takes $O(\mathrm{Sort}(n))$ I/Os. Step (2) takes a total of $O(n/B)$ I/Os by using a stack that stores the nodes on the current rightmost path of $\mathcal{T}_S$ in reversed order, i.e., leaf ℓ_i is on top. Walking upward, splitting edges or attaching nodes in $\mathcal{T}_S$ boils down to popping/pushing nodes from this stack. As a result, the time and I/O-complexity of this algorithm follow the recursive relation: $T(n) = T(2n/3) + O(\mathrm{Sort}(n))$.

Theorem 2 (Kärkkäinen and Sanders 2003, see [10]) *Given an arbitrary string* $S[1,n]$, *its suffix tree can be constructed in* $O(Sort(n))$ *I/Os,* $O(n \log n)$ *time and using* $O(n/B)$ *disk pages.*

It is not evident which one of these two algorithms is better in practice [10]. The first one exploits a recursion with parameter 1/2 but incurs a large space overhead because of the management of the tree topology; the second one is more space efficient and easier to implement, but exploits a recursion with parameter 2/3.

Applications

The reader is referred to [4] and [7] for a long list of applications of large suffix trees and to [6, 18] for practical implementations.

Open Problems

The recent theoretical and practical achievements mean the idea that "suffix trees are not practical except when the text size to handle is so small that the suffix tree fits in internal memory" is no longer the case [15]. Given a suffix tree, it is known now (see, e.g., [4, 11]) how to map it onto a disk-memory system in order to allow I/O-efficient traversals for subsequent pattern searches. A fortiori, suffix-tree storage, and construction are challenging problems that need further investigation.

Space optimization is closely related to time optimization in a disk-memory system, so the design of *succinct* suffix-tree implementations is a key issue in order to scale to gigabytes of data in reasonable time. This topic is an active area of theoretical research with many fascinating solutions (see, e.g., [16] and the many papers that followed it), which need further exploration in the practical setting.

It is theoretically challenging to design a suffix-tree construction algorithm that takes optimal I/Os and space proportional to the *entropy* of the indexed string. The more compressible is the string, the lighter should

be the space requirement of this algorithm. Some results are known [8, 11, 12], but both issues of compression and I/Os have been tackled jointly only recently [6], but more results are foreseen.

Experimental Results

The interest in building large suffix trees arose in the last few years because of the recent advances in sequencing technology, which have allowed the rapid accumulation of DNA and protein data. Some recent papers [1, 2, 9, 17, 18] proposed new practical algorithms that allow us to scale to Gbps/hours. Surprisingly enough, these algorithms are based on *disk-inefficient* schemes, but they properly select the insertion order of the suffixes and exploit carefully the internal memory as a buffer, so that their performance does not suffers significantly from the theoretical I/O-bottleneck.

In [9] the authors propose an *incremental* algorithm, called *PrePar*, which performs multiple passes over the string S and constructs the suffix tree for a *subrange* of suffixes at each pass. For a user-defined parameter q, a suffix subrange is defined as the set of suffixes prefixed by the same q-long string. Suffix subranges induce subtrees of T_S which can thus be built independently and evicted from internal memory as they are completed. The experiments reported in [9] successfully index 286 Mbps using 2 Gb internal memory.

In [2] the authors propose an improved version of *PrePar*, called *DynaCluster*, that deploys a *dynamic* technique to identify suffix subranges. Unlike *Prepar*, *DynaCluster* does not scan over and over the string S, but it starts from the q-based subranges and then splits them recursively in a DFS-manner if their size is larger than a fixed threshold τ. Splitting is implemented by looking at the next q characters of the suffixes in the subrange. This clustering and lazy-DFS visit of T_S significantly reduce the number of I/Os incurred by the frequent edge-splitting operations that occur during the suffix-tree construction process and allow it to cope efficiently with skew data.

As a result, *DynaCluster* constructs suffix trees for 200 Mbps with only 16 Mb internal memory.

In [17] authors improved the space requirement and the buffering efficiency, thus being able to construct a suffix tree of 3 Gbps in 30 h, whereas [1] improved the I/O behavior of RAM-algorithms for online suffix-tree construction, by devising a novel low-overhead buffering policy. More recently [14] introduced a new technique, called Elastic Range (ERA), which partitions the tree construction process horizontally and vertically and minimizes I/Os by dynamically adjusting the horizontal partitions independently for each vertical partition, based on the evolving shape of the tree and the available internal memory. This technique is specialized to work also for shared-memory and shared-disk multi-core systems and for parallel shared-nothing architectures. ERA indexes the entire human genome in 19 min on a commodity desktop PC. For comparison, the fastest existing method needs 15 min using 1024 CPUs on an IBM BluGene supercomputer.

Finally [19] observed that increasing memory sizes of current commodity PCs and servers enhance the impact of in-memory tasks on performance. So it is imperative nowadays to reassess the performance of in-memory algorithms and to propose new algorithms that incorporate the characteristics of modern hardware architectures, such as multilevel memory hierarchy and chip multiprocessors (CMPs). Starting from these premises the authors proposed cache-conscious suffix-tree construction algorithms that are tailored to CMP architectures, using novel sample-based cache-partitioning techniques that improved cache performance and exploited on-chip parallelism of CMPs thus achieving satisfactory speedups with increasing number of cores.

Cross-References

▶ Cache-Oblivious Sorting
▶ Suffix Array Construction
▶ Suffix Tree Construction
▶ Text Indexing

Recommended Reading

1. Bedathur SJ, Haritsa JR (2004) Engineering a fast online persistent suffix tree construction. In: Proceedings of the 20th international conference on data engineering, Boston, pp 720–731
2. Cheung C, Yu J, Lu H (2005) Constructing suffix tree for gigabyte sequences with megabyte memory. IEEE Trans Knowl Data Eng 17:90–105
3. Farach-Colton M, Ferragina P, Muthukrishnan S (2000) On the sorting-complexity of suffix tree construction. J ACM 47:987–1011
4. Ferragina P (2005) Handbook of computational molecular biology. In: Computer and information science series, ch. 35 on "String search in external memory: algorithms and data structures". Chapman & Hall/CRC, Florida
5. Ferragina P, Grossi R (1999) The string Btree: a new data structure for string search in external memory and its applications. J ACM 46:236–280
6. Ferragina P, Gagie T, Manzini G (2012) Lightweight data indexing and compression in external memory. Algorithmica 63(3):707–730
7. Gusfield D (1997) Algorithms on strings, trees and sequences: computer science and computational biology. Cambridge University Press, Cambridge
8. Hon W, Sadakane K, Sung W (2009) Breaking a time-and-space barrier in constructing full-text indices. SIAM J Comput 38(6):2162–2178
9. Hunt E, Atkinson M, Irving R (2002) Database indexing for large DNA and protein sequence collections. Int J Very Large Data Bases 11:256–271
10. Kärkkäinen J, Sanders P, Burkhardt S (2006) Linear work suffix array construction. J ACM 53:918–936
11. Ko P, Aluru S (2007) Optimal self-adjusting trees for dynamic string data in secondary storage. In: Symposium on string processing and information retrieval (SPIRE), Santiago. LNCS, vol 4726, pp 184–194. Springer, Berlin
12. Mäkinen V, Navarro G (2008) Dynamic entropy-compressed sequences and full-text indexes. ACM Trans Algorithm 4(3)
13. Manber U, Myers G (1993) Suffix arrays: a new method for on-line string searches. SIAM J Comput 22:935–948
14. Mansour E, Allam A, Skiadopoulos S, Kalnis P (2011) ERA: efficient serial and parallel suffix tree construction for very long strings. PVLDB 5(1):49–60
15. Navarro G, Baeza-Yates R (2000) A hybrid indexing method for approximate string matching. J Discr Algorithms 1:21–49
16. Navarro G, Mäkinen V (2007) Compressed full text indexes. ACM Comput Surv 39(1): Article no 2
17. Tian Y, Tata S, Hankins RA, Patel JM (2005) Practical methods for constructing suffix trees. VLDB J 14(3):281–299

S

18. Thomo A, Barsky M, Stege U (2010) A survey of practical algorithms for suffix tree construction in external memory. Softw Pract Experience 40(11):965–988
19. Tsirogiannis D, Koudas N (2010) Suffix tree construction on modern hardware. In: Proceedings of the 13th international conference on extending database technology (EDBT), Lausanne, pp 263–274
20. Vitter J (2002) External memory algorithms and data structures: dealing with MASSIVE DATA. ACM Comput Surv 33:209–271

Suffix Trees and Arrays

Alberto Apostolico[1] and Fabio Cunial[2]
[1]College of Computing, Georgia Institute of Technology, Atlanta, GA, USA
[2]Department of Computer Science, Helsinki Institute for Information Technology (HIIT), University of Helsinki, Helsinki, Finland

Keywords

Full-text indexing; Pattern matching; String searching; Suffix array; Suffix tree

Years and Authors of Summarized Original Work

1973; McCreight
1973; Weiner
1993; Manber, Myers
1995; Ukkonen

The suffix tree is one of the oldest full-text inverted indexes and one of the most persistent subjects of study in the theory of algorithms. With extensions and refinements, including succinct and compressed variants that provide some of its expressive power in smaller space, it constitutes a fundamental conceptual tool in the design of string algorithms. The companion structure represented by the suffix array is as powerful as the suffix tree in many applications, but it requires significantly less space. The uses of these data structures are so numerous that it is difficult to ac-count for all of them, while even more are being discovered. Salient applications include searching for a pattern in a text in time proportional to the size of the pattern, various computations on regularities such as repeats and palindromes within a text, statistical tables of substring occurrences, data compression by textual substitution, as well as ancillary yet fundamental tasks in string searching with errors, and more.

Problem Definition

It is well known that searching among n keys in an unsorted table takes optimal linear time. When multiple searches are expected, however, it becomes worth to sort the table once and for all, whereby each subsequent search will require only logarithmic time. It is similarly possible to build an inverted index on a long text so that the search for any query string will take time proportional to the length of the query rather than that of the text. It turns out that the data structures built for this purpose support many more applications, which are the topic of this entry.

Formally, let T be a string of length n on alphabet $\Sigma = [1 \ldots \sigma]$, let $\underline{T}$ be its reverse, and let $\# \notin \Sigma$ be a shorthand for zero. To simplify the exposition, we assume throughout that σ is a constant. The *suffix tree* $\mathsf{ST}_T = (\bot, V, E)$ of T is a tree rooted at node $\bot \in V$ with set of nodes V and set of labeled edges E (Fig. 1, left). Edge labels are pointers to substrings of $T\#$: we denote by $\ell(e)$, and equivalently by $\ell(u, v)$, the label of edge $e = (u, v) \in E$, and we denote by $\ell(v)$ the string $\ell(\bot, v_1) \cdot \ell(v_1, v_2) \cdot \cdots \cdot \ell(v_{k-1}, v)$, where $\bot, v_1, v_2, \ldots, v_{k-1}, v$ is a path in ST_T. We say that node v has *string depth* $|\ell(v)|$. Let $v \in V$ be an internal node, and let $w_1, w_2, \ldots, w_k$ be its children: then, $2 \leq k \leq \sigma + 1$, and labels $\ell(v, w_1), \ell(v, w_2), \ldots, \ell(v, w_k)$ start with distinct characters. The children of v are ordered lexicographically according to the labels of edges $(v, w_1), (v, w_2), \ldots, (v, w_k)$. There is a bijection between the leaves of ST_T and the suffixes of $T\#$, so every leaf is annotated with the starting position of its corresponding suffix. Moreover, if leaf $v \in V$ is associated with the suffix that

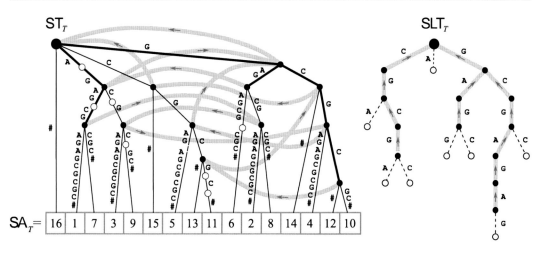

Suffix Trees and Arrays, Fig. 1 Relationship between the suffix tree, the suffix array (*left*), and the suffix-link tree (*right*) of string $T = $ AGAGCGAGAGCGCGC#. *Thin black lines*, edges of ST_T; *thick gray lines*, suffix links; *thin dashed lines*, implicit Weiner links; *thick black lines*, the subtree of ST_T induced by maximal repeats. *Black dots*, nodes of ST_T; *large black dot*, $\perp$; *white dots*, destinations of implicit Weiner links. *Squares*, leaves of ST_T and cells of SA_T; *numbers*, starting position of each suffix in T. For clarity, implicit Weiner links are not overlaid to ST_T, and suffix links from the leaves of ST_T are not drawn

starts at position i, then $\ell(v) = T[i\ldots n]$#. Since ST_T has exactly $n+1$ leaves and every internal node has at least two children, there are at most n internal nodes; thus, ST_T takes $O(n)$ space. We drop the subscript from ST whenever the underlying string is clear from the context.

A substring W of T# is called *right maximal* if both Wa and Wb occur in T, with $\{a,b\} \subseteq \Sigma \cup \{\#\}$ and $a \neq b$. Clearly a substring W is right maximal iff $W = \ell(v)$ for some $v \in V$. Moreover, assume that $\ell(v) = aW$ for some $v \in V$, $a \in \Sigma$, and $W \in \Sigma^*$. Since aW is right maximal, string W is right maximal as well; therefore, there is a node $w \in V$ with $\ell(w) = W$. Thus, the set of labels $\{\ell(v) : v \in V\}$ enjoys the *suffix closure* property, in the sense that if a string W belongs to the set so does every one of its suffixes. We say that there is a *suffix link from v to w labeled by* a, and we write $\mathtt{suffixLink}(v) = w$. Clearly, if v is a leaf, then $\mathtt{suffixLink}(v)$ is either a leaf or $\perp$. The graph induced by V and by suffix links is a trie rooted at $\perp$: such trie is called the *suffix-link tree* SLT_T of string T (Fig. 1, right). Inverting the direction of all suffix links yields the so-called *explicit Weiner links*. Given a node v and a symbol $a \in \Sigma$, it might happen that string $a\ell(v)$ does occur in T but that it is not the

label of any node in V: all such left extensions of nodes in V that end in the middle of an edge of ST are called *implicit Weiner links*. A node in V can have more than one outgoing Weiner link, and all such Weiner links have different labels. The number of suffix links (or, equivalently, of explicit Weiner links) is upper-bounded by $2n-2$, and the same bound holds for the number of implicit Weiner links: in some applications, we thus assume that ST is augmented with unary nodes that correspond to all the destinations of implicit Weiner links. A substring W of T# is called *left maximal* if both aW and bW occur in T#, with $\{a,b\} \subseteq \Sigma \cup \{\#\}$ and $a \neq b$, where T# is interpreted as a circular string. A string that is both left and right maximal is called *maximal repeat*. The set of all left-maximal strings enjoys the *prefix closure* property; therefore, there is a bijection between the maximal repeats and the nodes that lie in some paths of ST that start from the root (Fig. 1, left).

The *suffix array* $SA_T[1\ldots n+1]$ of string T is the permutation of $[1\ldots n+1]$ such that $SA_T[k] = i$ iff suffix $T[i\ldots n]$# has position k in the list of all suffixes of T# taken in lexicographic order. In this case, we say that suffix $T[i\ldots n]$# has *lexicographic rank k*. Clearly

$SA_T[1] = n + 1$. The *inverse suffix array* of string T is an array $R_T[1 \ldots n + 1]$ such that $R_T[SA[i]] = i$ for all $i \in [1 \ldots n + 1]$. A substring W of $T\#$ corresponds to a unique, contiguous *interval* (i_W, j_W) of SA_T, which contains all the suffixes of $T\#$ that are prefixed by W. An additional structure that complements the suffix array in many applications is the *longest common prefix array* $LCP_T[2 \ldots n + 1]$, which stores at position i the length of the longest prefix shared by suffix $T[SA_T[i] \ldots n]\#$ and by suffix $T[SA_T[i - 1] \ldots n]\#$. Clearly $LCP_T[k] \geq |W|$ for all $k \in [i_W + 1 \ldots j_W]$. Again, we drop the subscript from SA, R, and LCP whenever the underlying string is clear from the context.

Suffix tree, suffix array, and LCP array are strongly intertwined, and they have connections to other substring recognizers, like the *directed acyclic word graph* (DAWG) and its compact variant (CDAWG). SA can be thought of as the ordered set of leaves of ST, and ST can be thought of as a search tree built on top of SA (Fig. 1, left). The full ST, including suffix links, can be built from SA and LCP with a $O(n)$-time scan [1], and SA can be built from ST with a $O(n)$ traversal. LCP itself can be built from SA in $O(n)$ time [18]. A number of ingenious algorithms have been proposed to build ST and SA in linear time directly from the string itself, even in the case of polynomial alphabets: see [10, 17, 19, 20, 25, 32, 33] for a sampler of such algorithms, and see [28] for a detailed taxonomy. Some applications require to maintain the suffix tree after edits to the underlying string: see [12, 13, 22] for a sampler of such algorithms. Finally, see [21] for a comparative study of space-efficient allocations of suffix trees.

Key Results

Suffix trees are extremely versatile indexes that allow one to solve a variety of string matching and analysis problems [2, 9, 14]. We review few such problems, classifying the corresponding algorithmic solutions based on the way they walk on the suffix tree and on the information they store in each node. This classification exposes recurrent design patterns, it highlights which parts of the suffix tree are needed by each application, and it helps decide which algorithms can be implemented on top of more succinct but less powerful representations of the suffix tree. The emphasis of this section is on the power of different traversals of the suffix tree, not necessarily on the most efficient solution of each string analysis problem.

Top-Down

Exact searching inside a string S of length n is the most natural example of top-down traversal of ST_S. Given a query string W, we can just match its characters from the root of ST in $O(|W|)$ time to determine whether W occurs in S or not. Since edges are labeled by substrings of S, the search for W can end in the middle of an edge (u, v): we say that v is the *locus* of W in ST, and we denote it by $\texttt{locus}(W)$. This approach generalizes to a set of patterns $W_1, W_2, \ldots, W_k$ of total length m, by building the suffix tree of the concatenation $W = W_1\#_1 W_2\#_2 \cdots \#_{k-1} W_k$ and by traversing ST_S and ST_W synchronously, where $i \neq j$ implies $\#_i \neq \#_j$ and $\#_i \neq \#$.

The total number of (possibly overlapping) occurrences of the label $\ell(v)$ of a node v of ST equals the number of leaves in the subtree rooted at v, which can be computed by a bottom-up traversal of the tree. All strings that end in the middle of edge (u, v) start exactly at the same positions as $\ell(v)$ in S; therefore, ST with frequency annotation allows one to return the frequency in S of any string W in $O(|W|)$ time. An important consequence of this is the fact that the number of *distinct frequencies* assumed by nonempty substrings of S is at most $|S|$. It is also possible to annotate every node of ST with the smallest and largest leaf in its subtree, supporting $O(|W|)$-time queries on the first and last occurrence in S of any string W. More generally, traversing the tree rooted at $\texttt{locus}(W)$ in $O(|W| + k)$ time allows one to print all the k starting positions of W in S.

Finding all the occurrences of W in S can also be done in $O(|W|\log n)$ time, by binary searching SA_S for strings $W\#$ and $W\$$, where $\$ = \sigma + 1$: the result of these searches are, respectively, the starting and ending position of the interval of all suffixes prefixed by W. Knowing this interval allows one to derive the number of occurrences of W in S in constant time and to output the starting positions of such occurrences in time linear in the size of the output. Using simple properties of LCP_S, it is possible to reduce the time of binary search to $O(|W| + \log n)$, by reusing information during the search [24].

The top-down navigation of a suitably annotated suffix tree of S allows one also to compute the Lempel-Ziv factorization of S [23]. Recall that this factorization scans the string from left to right, and it determines at every position i the longest prefix of $S[i \ldots n]$ that equals a prefix of $S[j \ldots n]$, where $j < i$. Let W be such longest prefix: the factorization outputs the tuple $(j, |W|, S[i + |W|])$. Clearly we can find all this information by annotating every node v of ST with the index j of the smallest leaf in the subtree rooted at v. Then, we can just match suffix $S[i \ldots n]$ from the root of ST until a mismatch occurs or until we find a node with index greater than i. More advanced solutions embed the factorization in an online, one-pass construction of ST [29].

Bottom-Up

A *square* is a string WW where $W \in \Sigma^+$ is not in the form Z^k with $k > 1$ for any $Z \in \Sigma^+$. Clearly, if a square WW occurs at position i in S, then there is a node v in ST_S such that $|\ell(v)| \geq |W|$ and such that leaves i and $i + |W|$ belong to the subtree rooted at v. The converse is also true [4]. Thus, we can output all the repeats of S by using the following bottom-up traversal of ST. Assume without loss of generality that all nodes in ST have exactly two children. Every node u of ST builds its list of occurrences, sorted by position in S, using the lists of its children. Then, it scans its list once to find all pairs of positions at distance at most $|\ell(v)|$ in S that are consecutive in the list: every such pair is a square, and positions at distance at most $|\ell(v)|$ that are

not consecutive induce squares that are implied by the consecutive positions.

Let v and w be the two children of v, and assume without loss of generality that the list of occurrences of v is smaller than the list occurrences of w. Then, the list of node u can be built by extracting all elements from the list of node v and by inserting them into the list of node w. As a consequence of such insertions, the occurrences in the list of v move to a list that is at least twice the size of the original list: it follows that an occurrence can be pushed into at most $O(\log n)$ lists; therefore, the total number of extractions and insertions is bounded by $O(n \log n)$. If the lists of occurrences are implemented with balanced trees, the total time to extract all squares from S is $O(n \log^2 n)$. More advanced approaches manage to shave a logarithm, reaching optimal $O(n \log n)$ time [4], and to reduce the complexity to $O(n + \tau)$, where τ is the size of the output [15, 31].

The algorithm for detecting squares can be adapted to compute all the *maximal palindromes* of S, by applying it to string $T = S\#\underline{S}\$$. Note that a variant of the same algorithm can be implemented using the suffix array. First, it is easy to see that a bottom-up, in-order traversal of the internal nodes of ST_S can be simulated by a linear scan of SA_S and of LCP_S, maintaining a stack [1]. It follows that, for every interval (i_v, j_v) in SA of a node v in ST, we can just check whether $\mathsf{SA}[k] + |\ell(v)| \in [i_v \ldots j_v]$ and $S[\mathsf{SA}[k] + |\ell(v)|] \neq S[\mathsf{SA}[k] + 2|\ell(v)|]$, for every $k \in [i_v \ldots j_v]$: in this case, the occurrence of square $\ell(v)$ at position $\mathsf{SA}[k]$ is called *branching*. It is easy to see that all squares can be derived from squares with branching occurrences [31]. Moreover, if the occurrence at position $\mathsf{SA}[k]$ is branching, then suffixes $\mathsf{SA}[k] + |\ell(v)|$ and $\mathsf{SA}[k] + 2|\ell(v)|$ belong to distinct children of node v in ST: we can thus discard the child w of v with the largest number of leaves and check for every $k \in [i \ldots j]$ that does not belong to the interval of w whether $\mathsf{SA}[k] - |\ell(v)| \in [i_v \ldots j_v]$ and $S[\mathsf{SA}[k]] \neq S[\mathsf{SA}[k] + |\ell(v)|]$. The child of v with largest interval can be determined in constant time during the simulated bottom-up traversal of ST, and since the largest interval is always excluded, the algorithm runs in $O(n \log n)$ time.

Given a collection of k strings of total length n, let S be the concatenation of all such strings, each terminated by a distinct symbol that does not belong to Σ. A bottom-up navigation of ST_S (called also the *generalized suffix tree* of the collection) allows one to compute the length of a longest string that occurs in $x \leq k$ strings. To solve this problem, we can annotate each leaf v of ST with a bitvector which of length k, such that $\mathtt{which}[i] = 1$ iff the suffix associated with v starts inside string i. Then, every node of ST can be annotated with the same bitvector via a bottom-up, $O(nk)$ traversal, in which we compute the bitvector of a node by taking the logical or of the bitvectors of its children. More advanced algorithms solve this problem in $O(n)$ time [8]. As a byproduct, this annotation allows one to answer queries on the number of strings in the collection that contain a given substring, a problem known as *document counting*. A germane problem is that of *document listing*, in which we are given a pattern and we are asked to return the set of all documents that contain one or more copies of the pattern [26].

Top-Down and Suffix Links

Given two strings S and T, of length n and m, respectively, the *matching statistics array* $\mathsf{MS}_{S,T}[1 \ldots n]$ is such that $\mathsf{MS}_{S,T}[i]$ stores the length of the longest string that starts at position i in S and that occurs in T [33]. We can compute $\mathsf{MS}_{S,T}$ by scanning S from left to right, while simultaneously issuing child and suffix-link queries on ST_T. This results in a peculiar walk on ST_T that consists of alternating sequences of suffix-tree edges and of suffix links (we can also compute $\mathsf{MS}_{S,T}$ symmetrically, by scanning S from right to left and by simultaneously issuing parent and Weiner-link queries on ST_T [27]).

Specifically, assume that we are at position i in S, and let $W = S[i \ldots i + \mathsf{MS}_{S,T}[i] - 1]$. Note that W can end in the middle of an edge (u, v) of ST_T: let $W = aXY$ where $a \in \Sigma$, $X \in \Sigma^*$, $aX = \ell(u)$, and $Y \in \Sigma^*$. Moreover, let $u' = \mathtt{suffixLink}(u)$ and $v' = \mathtt{suffixLink}(v)$.

Note that suffix links can project edge (u, v) onto a *path* $u', v_1, v_2, \ldots, v_k, v'$, where $v_j \in V$ for $j \in [1 \ldots k]$. Since $\mathsf{MS}_{S,T}[i+1] \geq \mathsf{MS}_{S,T}[i]-1$, the first step to compute $\mathsf{MS}_{S,T}[i + 1]$ is to find the position of XY in ST_T: we call this phase of the algorithm the *repositioning phase*. To implement the repositioning phase, it suffices to take the suffix link from u, to follow the outgoing edge from u' whose label starts by the first character of Y, and then to iteratively jump to the next internal node of ST_T and to choose the next outgoing edge according to the corresponding character of Y. After repositioning, we start matching the new characters of S on ST_T, i.e., we read characters $S[i + \mathsf{MS}_{S,T}[i]], S[i + \mathsf{MS}_{S,T}[i] + 1], \ldots$ until such an extension becomes impossible in ST_T. We call this phase of the algorithm the *matching phase*. Note that no character of S that has been read during the repositioning phase of $\mathsf{MS}_{S,T}[i + 1]$ will be read again during the repositioning phase of $\mathsf{MS}_{S,T}[i + k]$ with $k > 1$: it follows that every position j of S is consumed at most twice, once in the matching phase of some $\mathsf{MS}_{S,T}[i]$ with $i \leq j$ and once in the repositioning phase of some $\mathsf{MS}_{S,T}[k]$ with $i < k < j$. Since every mismatch can be charged to the position of which it concludes the matching statistics, the total number of mismatches encountered by the algorithm is bounded by the length of S.

These algorithms can be adapted to compute the *shortest unique substring array* $\mathsf{SUS}_S[1 \ldots n]$, which stores at index i the length of the shortest substring of S that occurs only at position i [33]. The average of the matching statistics vector can be used to estimate the cross-entropy of the probability distributions of two stationary, ergodic, stochastic processes with finite memory that generated S and T [11]. Moreover, a number of compositional similarity measures between two strings S and T can be computed by scanning S and by simultaneously navigating ST_T as in matching statistics: this has the advantage of building and annotating the suffix tree of just the shortest string [30]. Matching statistics on a suitably annotated suffix tree of T allows one also to approximate the probability that S was generated by the same variable-length Markov process that produced

T, another measure of similarity not based on sequence alignment [3].

Top-Down in the Suffix-Link Tree

A number of statistical applications require to annotate the nodes of ST_S with *empirical probabilities* rather than with raw frequencies. The empirical probability $p_S(W)$ of a string W is essentially the number of its occurrences $f_S(W)$ divided by the maximum number of occurrences that W can have in a string of length $|S| = n$. This number cannot exceed $n - |W| + 1$, but it also depends on the number of overlaps that W has with itself, i.e., on the number of proper *borders* of W: thus, we set $p_S(W) = f_S(W)/b(W)$, where $b(W)$ is the length of the shortest period of W. Note that p_S can change inside an edge of ST. However, if we are interested only in the empirical probability of nodes of ST, we can compute all such values in overall linear time, by mapping the longest-border computation in the KMP algorithm onto a depth-first navigation *of the suffix-link tree* [5].

The exact computation of the *variance* of the frequency of a string W in S can be itself mapped onto the computation of the longest proper border of W. Under suitable statistical assumptions, computing the expectation and variance of the frequency of all right-maximal substrings of S suffices to detect all substrings of S with anomalous frequency: it is thus possible to discover all statistically frequent and rare substrings of S in overall linear time [5].

Any Order

A single pass over all nodes of ST *in any order*, coupled with a number of checks on the children and on the Weiner links of each node, suffices to solve a number of string analysis problems in linear time.

A string W is a *maximal unique match* (MUM) between two strings S and T if it occurs exactly once in S and exactly once in T and if neither aW nor Wb occur in both S and T for any $\{a, b\} \subseteq \Sigma$ (for simplicity, we disregard cases in which W occurs at the beginning or at the end of a string) [14]. Clearly W must be a right-maximal substring of $U = S\#T\$$, where $\#$ and $\$$ are

separators not belonging to Σ. Therefore, we just need to iterate over every node v of ST_U in any order, checking the following conditions: (1) v has exactly two leaves as children; (2) the suffixes that correspond to such leaves start before and after position $|S| + 1$ in U, respectively; and (3) v has two Weiner links. A similar approach extends to MUMs of more than two strings, as well as to *maximal* (not necessarily unique) *exact matches* between two strings and to the *maximal repeats* [7] and the *minimal absent words* of a single string [16].

Symmetrically, it is easy to detect the MUMs of two strings S and T by a linear scan of the suffix array of $U = S\#T\$$ and of the corresponding LCP array. Indeed, a MUM corresponds to an interval $(i, i+1)$ of size two in SA_U such that $LCP_U[i] < LCP_U[i+1]$, $LCP_U[i+2] < LCP_U[i+1]$, $U[SA_U[i] - 1] \neq U[SA_U[i+1] - 1]$, and $SA_U[i] < |S| + 1 < SA_U[i+1]$. Similar criteria allow one to detect maximal repeats, *supermaximal repeats* [14], and maximal exact matches [1].

String Depth Annotation

Assume that every node v of ST_S is annotated with $|\ell(v)|$. Recall that the *shortest unique substring array* $SUS_S[1 \ldots n]$ is such that $SUS_S[i]$ is the length of the shortest substring of S that occurs only at position i. Since $S[i \ldots i + SUS[i] - 1] = Wa$ where $a \in \Sigma$, since $locus(Wa)$ is a leaf v, and since $locus(W) = parent(v)$, traversing the nodes of ST in any order suffices to compute $SUS[i]$ for every i. String depth annotations, coupled with a traversal of the nodes of ST in any order, suffice also to compute measures of compositional complexity of S, like the total number of distinct substrings, possibly of a fixed length k.

Frequency Annotation

Recall that $f_S(W)$ is the number of occurrences of string W in S. Assume that we want to compute $p(a|W) = f_S(Wa)/f_S(W)$ for all substrings W of S and for all characters $a \in \Sigma$ such that Wa is a substring of S. Such values are called *conditional probabilities*. Clearly $p(a|W) = 1$ if W ends in the middle of an edge of ST_S: it is thus sufficient to compute conditional probabilities for the nodes of ST, and this can be done by traversing the nodes of ST in any order and by accessing their children.

String Depth and Frequency Annotation

Assume that every node v of ST_S is also annotated with the number of leaves in the subtree rooted at v. Then, traversing the nodes of ST in any order allows one to compute the longest substring of S that repeats at least τ times, or the most frequent string of length at least τ, for any user-specified threshold τ. String depth and frequency annotations, coupled with a traversal of the nodes of ST in any order, allow one also to compute the number of distinct substrings that occur τ times in S, for every frequency τ in a user-specified range.

Given a substring W of S, let $\text{right}(W)$ be the set of characters that occur in S after W. More formally, $\text{right}(W) = \{a \in \Sigma : f_S(Wa) > 0\}$. The kth order empirical entropy of S is defined as follows:

$$\mathsf{H}(S, k) = \frac{1}{|S|} \sum_{W \in \Sigma^k} \sum_{a \in \text{right}(W)} f_S(Wa) \log\left(\frac{f_S(W)}{f_S(Wa)}\right)$$

To compute $\mathsf{H}(S, k)$, it suffices again to traverse the nodes of ST_S in any order, to check whether $|\ell(v)| = k$, and to cumulate the contribution of v to $\mathsf{H}(S, k)$ by reading the frequency of its children. Strings of length k that end in the middle of an edge of ST do not contribute to $\mathsf{H}(S, k)$.

In a similar fashion, given a string S on alphabet Σ, let $\mathbf{S}$ be a vector indexed by all strings in Σ^k for a fixed $k > 0$, such that $\mathbf{S}[W]$ contains the frequency of string W in S. We call $\mathbf{S}$ the k-mer *composition vector* of string S. Given two strings S and T, assume that we want to compute a function $\kappa(S, T)$ that depends only on $N = \sum_{W \in \Sigma^k} f(\mathbf{S}[W], \mathbf{T}[W])$, $D_S = \sum_{W \in \Sigma^k} g(\mathbf{S}[W])$, and $D_T = \sum_{W \in \Sigma^k} h(\mathbf{T}[W])$, where f, g, and h are user-specified functions. $\kappa(S, T)$ if often called *k-mer kernel* in text classification. It is possible to compute $\kappa(S, T)$ in overall linear time by traversing the nodes of the generalized suffix tree of S and T in any order. A similar traversal of ST allows one to compute $\kappa(S, T)$ on composition vectors that are indexed by all possible substrings, of any length. In practice the frequencies used in composition vectors are normalized by their expected values under IID or Markov probability distributions: a number of kernels based on such normalized counts can still be computed in overall linear time by traversing the nodes of ST in any order [6].

Positional Annotations

Given two strings S and T, the longest string W that occurs in both S and T is clearly a right-maximal substring of the concatenation $U = S\#T\$$, where $\#$ and $\$$ are separators not belonging to Σ. Consider thus ST_U, and assume that every node v is annotated with $|\ell(v)|$ and with a bit $\text{flag}(v)$ set to one iff the subtree rooted at v contains at least one leaf that starts before position $|S| + 1$ in U and at least one leaf starting after position $|S| + 1$ in U. Such annotation can be carried out in a bottom-up traversal of ST. We can compute W by iterating over the nodes $v \in$ ST with $\text{flag}(v) = 1$ and by cumulating the maximum of the lengths of the encountered labels. The set of *all common substrings* between S and T is the set of all prefixes of the labels of nodes $v \in$ ST such that $\text{flag}(v) = 1$ and $\text{flag}(w) = 0$ for every child w of v. This approach generalizes immediately to more than two strings, and it allows one to compute the length of the longest substring common to *at least* τ strings in a collection of k strings in $O(k|U|)$ time and space. More advanced approaches solve this problem in $O(|U|)$ time [8].

Applications

The primitives discussed above find application in a wide set of domains. A list of the most salient ones includes exact and approximate string searching, string compression, statistical pattern discovery, alignment-free string comparison, string kernels in learning theory, sequence analysis, and assembly in bioinformatics.

Cross-References

► Approximate Tandem Repeats
► Burrows-Wheeler Transform
► Compressed Suffix Array
► Compressed Suffix Trees
► Document Retrieval on String Collections
► Indexed Approximate String Matching
► Indexed Two-Dimensional String Matching
► Lempel-Ziv Compression
► Lowest Common Ancestors in Trees
► Pattern Matching on Compressed Text
► String Matching
► Suffix Array Construction
► Suffix Tree Construction

Recommended Reading

1. Abouelhoda MI, Kurtz S, Ohlebusch E (2004) Replacing suffix trees with enhanced suffix arrays. J Discret Algorithms 2(1):53–86
2. Apostolico A (1985) The myriad virtues of subword trees. In: Apostolico A, Galil Z (eds) Combinatorial algorithms on words. Springer, Berlin/New York, pp 85–96
3. Apostolico A, Bejerano G (2000) Optimal amnesic probabilistic automata or how to learn and classify proteins in linear time and space. J Comput Biol 7(3–4):381–393
4. Apostolico A, Preparata FP (1983) Optimal off-line detection of repetitions in a string. Theor Comput Sci 22(3):297–315
5. Apostolico A, Bock ME, Lonardi S, Xu X (2000) Efficient detection of unusual words. J Comput Biol 7(1–2):71–94
6. Apostolico A, Denas O et al (2008) Fast algorithms for computing sequence distances by exhaustive substring composition. Algorithms Mol Biol 3(13)
7. Beller T, Berger K, Ohlebusch E (2012) Space-efficient computation of maximal and supermaximal repeats in genome sequences. In: 19th international symposium on string processing and information retrieval (SPIRE 2012), Cartagena de Indias. Lecture notes in computer science, vol 7608. Springer, pp 99–110
8. Chi L, Hui K (1992) Color set size problem with applications to string matching. In: Combinatorial pattern matching, Tucson. Springer, pp 230–243
9. Crochemore M, Hancart C, Lecroq T (2007) Algorithms on strings. Cambridge University Press, New York
10. Farach M (1997) Optimal suffix tree construction with large alphabets. In: Proceedings of the 38th annual symposium on foundations of computer science, 1997, Miami Beach. IEEE, pp 137–143
11. Farach M, Noordewier M, Savari S, Shepp L, Wyner A, Ziv J (1995) On the entropy of DNA: algorithms and measurements based on memory and rapid convergence. In: Proceedings of the sixth annual ACM-SIAM symposium on discrete algorithms (SODA '95), San Francisco. Society for Industrial and Applied Mathematics, pp 48–57
12. Ferragina P (1997) Dynamic text indexing under string updates. J Algorithms 22(2):296–328
13. Fiala ER, Greene DH (1989) Data compression with finite windows. Commun ACM 32(4):490–505. doi:10.1145/63334.63341, http://doi.acm.org/10.1145/63334.63341
14. Gusfield D (1997) Algorithms on strings, trees, and sequences: computer science and computational biology. Cambridge University Press, Cambridge/New York
15. Gusfield D, Stoye J (2004) Linear time algorithms for finding and representing all the tandem repeats in a string. J Comput Syst Sci 69(4):525–546. doi:10.1016/j.jcss.2004.03.004, http://dx.doi.org/10.1016/j.jcss.2004.03.004
16. Herold J, Kurtz S, Giegerich R (2008) Efficient computation of absent words in genomic sequences. BMC Bioinform 9(1):167
17. Kärkkäinen J, Sanders P, Burkhardt S (2006) Linear work suffix array construction. J ACM 53(6):918–936
18. Kasai T, Lee G, Arimura H, Arikawa S, Park K (2001) Linear-time longest-common-prefix computation in suffix arrays and its applications. In: Combinatorial pattern matching, Jerusalem. Springer, pp 181–192
19. Kim DK, Sim JS, Park H, Park K (2005) Constructing suffix arrays in linear time. J Discret Algorithms 3(2):126–142
20. Ko P, Aluru S (2003) Space efficient linear time construction of suffix arrays. In: Combinatorial pattern matching, Morelia. Springer, pp 200–210
21. Kurtz S (1999) Reducing the space requirement of suffix trees. Softw Pract Exp 29:1149–1171
22. Larsson NJ (1996) Extended application of suffix trees to data compression. In: Data compression conference, Snowbird, pp 190–199
23. Lempel A, Ziv J (1976) On the complexity of finite sequences. IEEE Trans Inf Theory 22:75–81
24. Manber U, Myers G (1993) Suffix arrays: a new method for on-line string searches. SIAM J Comput 22(5):935–948
25. McCreight EM (1976) A space-economical suffix tree construction algorithm. J ACM 23(2):262–272

S

26. Muthukrishnan S (2002) Efficient algorithms for document retrieval problems. In: Proceedings of the thirteenth annual ACM-SIAM symposium on discrete algorithms (SODA '02), San Francisco. Society for Industrial and Applied Mathematics, Philadelphia, pp 657–666. http://dl.acm.org/citation.cfm?id= 545381.545469

27. Ohlebusch E, Gog S, Kügel A (2010) Computing matching statistics and maximal exact matches on compressed full-text indexes. In: XXth international symposium on string processing and information retrieval (SPIRE 2010), Los Cabos, pp 347–358

28. Puglisi SJ, Smyth WF, Turpin AH (2007) A taxonomy of suffix array construction algorithms. ACM Comput Surv 39(2):4

29. Rodeh M, Pratt VR, Even S (1981) Linear algorithm for data compression via string matching. J ACM 28(1):16–24

30. Smola AJ, Vishwanathan S (2003) Fast kernels for string and tree matching. In: Becker S, Thrun S, Obermayer K (eds) Advances in neural information processing systems (NIPS '03) 15, Vancouver. MIT, pp 585–592

31. Stoye J, Gusfield D (2002) Simple and flexible detection of contiguous repeats using a suffix tree. Theor Comput Sci 270(1):843–856

32. Ukkonen E (1995) On-line construction of suffix trees. Algorithmica 14(3):249–260

33. Weiner P (1973) Linear pattern matching algorithms. In: IEEE conference record of 14th annual symposium on switching and automata theory (SWAT '08), Iowa City, 1973. IEEE, pp 1–11

Sugiyama Algorithm

Nikola S. Nikolov
Department of Computer Science and Information Systems, University of Limerick, Limerick, Republic of Ireland

Keywords

Barycentric method; Crossing minimization; Hierarchical graph drawing; Layered graph drawing; Sugiyama algorithm; Sugiyama framework

Years and Authors of Summarized Original Work

1981; Sugiyama, Tagawa, Toda

Problem Definition

Given a directed graph (digraph) $G(V, E)$ with a set of vertices V and a set of edges E, the Sugiyama algorithm solves the problem of finding a 2D hierarchical drawing of G subject to the following readability requirements:

(a) Vertices are drawn on horizontal lines without overlapping; each line represents a level in the hierarchy; all edges point downwards.

(b) Short-span edges (i.e., edges between adjacent levels) are drawn with straight lines.

(c) Long-span edges (i.e., edges between nonadjacent levels) are drawn as close to straight lines as possible.

(d) The number of edge crossings is the minimum.

(e) Vertices connected to each other are placed as close to each other as possible.

(f) The layout of edges coming into (or going out of) a vertex is balanced, i.e., edges are evenly spaced around a common target (or source) vertex.

Requirements (a) and (b) are easy to meet and they are imposed as mandatory basic drawing rules. Requirements (c)–(f) are much harder to satisfy and typically they are met approximately [1, 4, 11].

Key Results

Sugiyama et al. propose a four-step procedure for finding a hierarchical drawing of a digraph subject to the readability requirements listed above. It is known as the Sugiyama algorithm, the Sugiyama method, or the Sugiyama framework [19]. The steps of the Sugiyama framework are illustrated in Fig. 1.

The Sugiyama Framework
Step 1: Preparatory step for transforming the input digraph G into a proper hierarchy.

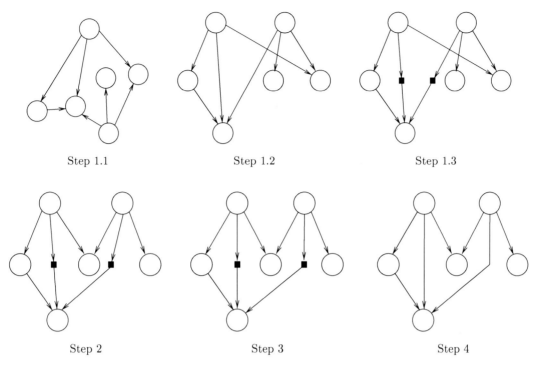

Step 1.1   Step 1.2   Step 1.3

Step 2   Step 3   Step 4

Sugiyama Algorithm, Fig. 1 Illustration of the steps of the Sugiyama framework

Step 1.1: Transform the input digraph G into a directed acyclic graph (dag) by reversing the direction of some edges.

Step 1.2: Transform the dag into a multilevel digraph, called a *hierarchy*, by partitioning V into l levels (or layers) $V_1, V_2, \ldots, V_l$ such that for each edge $e = (v, w) \in E$ if $v \in V_i$ then $w \in V_{i+1}$. Levels are drawn on horizontal lines which determine the y-coordinates of the vertices.

Step 1.3 Transform the hierarchy into a *proper hierarchy* by introducing *dummy vertices* along long-span edges; one dummy vertex at each crossing of a long-span edge with a level.

Step 2: For each level V_i, specify a linear order σ_i of the vertices in V_i with the goal of minimizing the total number of edge crossing.

Step 3: Determine the x-coordinates of the vertices subject to requirements (c), (e), and (f) while preserving the linear order in the levels.

Step 4: Draw G in a 2D drawing area where dummy vertices are removed and the long-span edges are restored.

Steps 1.3 and 4 are trivial as computational problems. Steps 1.1 and 1.2 can be solved easily if the only readability requirements are those listed above. However, some sensible additional requirements can turn Steps 1.1 and 1.2 into difficult combinatorial optimization problems. For example, if we want to minimize the number of reversed edges at Step 1.1, then we need to solve the MINIMUM FEEDBACK ARC SET problem which is NP-hard [12]. Similarly, if we impose upper bounds on both the number of levels and the number of vertices per level, then the problem in Step 1.2, known as the *layering* problem, becomes NP-complete [4].

Following the work of Sugiyama et al., two types of solutions to the layering problem have been proposed in the research literature. The first type of layer assignment algorithm is list-scheduling algorithms (adapted from the area of static precedence-constrained multiprocessor

S

scheduling) which produce layer assignments with either the minimum number of levels or a specified maximum number of vertices per level [4]. These include the longest-path algorithm [13] and the Coffman-Graham algorithm [3] as well as the proposed by Nikolov et al. [15] MinWidth and StretchWidth heuristics which take into account the dummy vertices. The second type of algorithm employs network simplex and branch-and-cut techniques, respectively, for minimizing the number of dummy vertices with or without constraints on the number of levels and the number of vertices per level [9, 10].

Steps 2 and 3 are already hard to solve with the readability requirements listed above. It has also been suggested to precede Step 2 by an edge concentration or edge bundling step for achieving a more readable drawing [14, 16]. The other key results in the work of Sugiyama et al., besides defining the four-step framework, are efficient heuristics for Steps 2 and 3, respectively.

Reduction of the Number of Edge Crossings

Consider a proper hierarchy $G(V, E, \mathcal{L})$ with a set of vertices $V = \{v_1, v_2, \ldots, v_n\}$, a set of edges $E = \{e_1, e_2, \ldots, e_m\}$, and a partitioning $\mathcal{L} = \{V_1, V_2, \ldots, V_l\}$ of the vertex set V into l levels (the result of Step 1.3). Let $\sigma_i : V_i \rightarrow \{1, 2, \ldots |V_i|\}$ be a linear order of the vertices in level V_i and let S_i be the set of all possible orders σ_i. The problem at Step 2 of the Sugiyama algorithm is to find a set of linear orders $\sigma = \{\sigma_1, \sigma_2, \ldots, \sigma_l\} \in S_1 \times S_2 \times \ldots \times S_l$ such that the total number of edge crossings is the minimum. Let $K(G, \sigma)$ be the total number of edge crossings for a hierarchy G and a set of linear orders σ, and let $K(V_i, V_{i+1}, \sigma_i, \sigma_{i+1})$ be the number of

edge crossings between layers V_i and V_{i+1} with linear orders σ_i and σ_{i+1}, respectively.

The algorithm, proposed by Sugiyama et al. for Step 2, is a heuristic which consists in initially choosing a random order σ_1 for the vertices in level V_1 and then repeatedly executing the following five-step procedure, called Down-Up, until either σ does not change or an initially given maximum number of iterations is reached.

The Down-Up Procedure

Step A: $i \leftarrow 1$.
Step B: With a fixed linear order σ_i, find a linear order σ_{i+1} which minimizes $K(V_i, V_{i+1}, \sigma_i, \sigma_{i+1})$.
Step C: If $i < n - 1$, then $i \leftarrow i + 1$ and go to Step B. Otherwise, go to Step D.
Step D: With a fixed linear order σ_{i+1}, find a linear order σ_i which minimizes $K(V_i, V_{i+1}, \sigma_i, \sigma_{i+1})$.
Step E: If $i > 1$, then $i \leftarrow i - 1$ and go to Step D. Otherwise, stop.

Both Step B and Step D involve minimizing the number of edge crossings between two adjacent layers with the linear order in one of them being fixed. This problem is known as the ONE-SIDED CROSSING MINIMIZATION (OSCM) problem, which has been shown to be NP-hard [5]. Based on previous work by Warfield [20], Sugiyama et al. show how OSCM can be reduced to the MINIMUM FEEDBACK SET problem and propose a heuristic method, called the *barycentric method*, for solving it. Let $A = (a_{ij})$ be the adjacency matrix of G. In essence, with a fixed linear order σ_i, the barycentric method orders the vertices in level V_{i+1} in the increasing order of their barycenters B_j, defined with Eq. (1).

$$B_j = \sum_{k=1}^{|V_i|} a_{kj} \sigma_i(v_k) / \sum_{k=1}^{|V_i|} a_{kj}, \qquad j \in \{1, 2, \ldots, |V_{i+1}|\} \qquad (1)$$

Sugiyama et al. evaluate the Down-Up procedure experimentally with 800 randomly generated

hierarchies as well as with five hierarchies from practical applications. Their conclusion is that the proposed heuristic is effective. It was observed that in most cases the Down-Up procedure requires a single iteration. Reportedly, the heuristic was successfully extended for the case when vertices in each level are partitioned into subsets where the vertices in each subset must be arranged adjacently.

Step 2 is probably the best studied part of the Sugiyama framework. Numerous improvements to the original technique as well as alternative algorithms for crossing minimization have been proposed since the introduction of the Sugiyama framework [1,4,5,7,9,11]. Notable among them is the 3-approximation *median method* proposed by Eades and Wormald [5] for solving the OSCM problem. Having the order of the vertices in level V_i fixed, the median method consists of placing each vertex in level V_{i+1} at a position which corresponds to the median of the positions of its neighbors in level V_i. Since the median method is an approximation algorithm, it guarantees to find a solution without edge crossings if such exists.

Determination of x-Coordinates of Vertices

For Step 3 of their framework, Sugiyama et al. propose a version of the Down-Up procedure with the barycenter of a vertex based on the x-coordinates of the connected to it vertices in an adjacent level. Consider the *down* part of the Down-Up procedure (the *up* part is symmetrical). If the x-coordinates of the vertices in level V_i are known, the barycenters B_j^* of the vertices in level V_{i+1} are defined with Eq. (2).

$$B_j^* = \sum_{k=1}^{|V_i|} a_{kj} x(v_k) / \sum_{k=1}^{|V_i|} a_{kj},$$
$$j \in \{1, 2, \ldots, |V_{i+1}|\} \quad (2)$$

The x-coordinates of the vertices in level V_{i+1} are determined according to their priority. The highest priority has the dummy vertices (introduced in Step 1.3), and the priority of each other vertex in level V_{i+1} is the number of vertices in level V_i connected to it. The x-coordinate of each vertex $v_j \in V_{i+1}$ is the integer number which is the closest to B_j^* available horizontal position (without changing the linear order from Step 2 and without displacing already placed vertices with higher priority). In finding this position, it is allowed to displace vertices with a priority lower than the priority of v_j, where this displacement should be as little as possible.

Sugiyama et al. evaluate the effectiveness of this method for improving the readability requirements (c), (e), and (f) experimentally. Reportedly, they have extended their heuristic for the case when the dimensions of the vertices are not insignificant. Both the Step 2 and the Step 3 heuristics were successfully applied to a hierarchy with more than 500 vertices.

Alternative algorithms for Step 3 have been proposed by Gansner et al. [9], Eades et al. [6], and Sander [17]. Probably, the best solution for Step 3 to date is the $O(|V|)$ algorithm of Brandes and Köpf [2]. It assigns x-coordinates to vertices by computing four *extreme* vertex alignments which are then combined into a final layout with at most two bends per edge.

Applications

Hierarchical graph drawings are useful for providing insight into hierarchical structures in complex systems. In recent years, the Sugiyama algorithm has found an important application for visual analysis of large social and biological networks [8, 18].

Cross-References

▶ List Scheduling

Recommended Reading

1. Bastert O, Matuszewski C (2000) Layered drawings of digraphs. In: Kaufman M, Wagner D (eds) Drawing graphs: method and models. Lecture notes in computer science, vol 2025. Springer, Berlin/Heidelberg, pp 87–120

2. Brandes U, Köpf B (2002) Fast and simple horizontal coordinate assignment. In: Mutzel P, Jünger M, Leipert S (eds) Graph drawing. Lecture notes in computer science, vol 2265. Springer, Berlin/Heidelberg, pp 31–44

3. Coffman EG Jr, Graham R (1972) Optimal scheduling for two-processor systems. Acta Inform 1(3):200–213

4. Eades P, Sugiyama K (1990) How to draw a directed graph. J Inf Process 13(4):424–437

5. Eades P, Wormald NC (1994) Edge crossings in drawings of bipartite graphs. Algorithmica 11(4):379–403

6. Eades P, Lin X, Tamassia R (1996) An algorithm for drawing a hierarchical graph. Int J Comput Geom Appl 6(2):145–156

7. Eppstein D, Goodrich MT, Meng JY (2007) Confluent layered drawings. Algorithmica 47(4):439–452

8. Fu X, Hong SH, Nikolov N, Shen X, Wu Y, Xu K (2007) Visualization and analysis of email networks. In: Asia-Pacific symposium on visualization, Sydney, pp 1–8

9. Gansner ER, Koutsofios E, North SC, Vo KP (1993) A technique for drawing directed graphs. IEEE Trans Softw Eng 19(3):214–230

10. Healy P, Nikolov NS (2002) How to layer a directed acyclic graph. In: Mutzel P, Jünger M, Leipert S (eds) Graph drawing. Lecture notes in computer science, vol 2265. Springer, Berlin/Heidelberg, pp 16–30

11. Healy P, Nikolov NS (2013) Hierarchical drawing algorithms. In: Tamassia R (ed) Handbook of graph drawing and visualization. Discrete mathematics and its applications, chap 13. Chapman and Hall/CRC, Boca Raton/London/New York, pp 409–454

12. Lempel A, Cederbaum I (1966) Minimum feedback arc and vertex sets of a directed graph. IEEE Trans Circuit Theory 13(4):399–403

13. Mehlhorn K (1984) Data structures and algorithms, Volume 2: graph algorithms and NP-completeness. Springer, Heidelberg

14. Newbery FJ (1989) Edge concentration: a method for clustering directed graphs. In: Proceedings of the 2nd international workshop on software configuration management (SCM '89), Princeton. ACM, pp 76–85

15. Nikolov NS, Tarassov A, Branke J (2005) In search for efficient heuristics for minimum-width graph layering with consideration of dummy nodes. J Exp Algorithmics 10:1–27

16. Pupyrev S, Nachmanson L, Kaufmann M (2011) Improving layered graph layouts with edge bundling. In: Brandes U, Cornelsen S (eds) Graph drawing. Lecture notes in computer science, vol 6502. Springer, Berlin/Heidelberg, pp 329–340

17. Sander G (1996) A fast heuristic for hierarchical Manhattan layout. In: Brandenburg FJ (ed) Graph drawing. Lecture notes in computer science, vol 1027. Springer, Berlin/Heidelberg, pp 447–458

18. Schwikowski B, Uetz P, Fields S (2000) A network of protein-protein interactions in yeast. Nat Biotechnol 18(12):1257–1261

19. Sugiyama K, Tagawa S, Toda M (1981) Methods for visual understanding of hierarchical system structures. IEEE Trans Syst Man Cybern 11(2):109–125

20. Warfield JN (1977) Crossing theory and hierarchical mapping. IEEE Trans Syst Man Cybern 7(7):502–523

Superiority and Complexity of the Spaced Seeds

Louxin Zhang
Department of Mathematics, National University of Singapore, Singapore, Singapore

Keywords

Homology search; NP-hardness; Sensitivity and hit probability; Spaced seeds

Years and Authors of Summarized Original Work

2006; Ma, Li, Zhang

Problem Definition

In the 1970s, sequence alignment was introduced to demonstrate the similarity of the sequences of genes and proteins [12]. A DNA sequence is a finite sequence over four nucleotides – adenine, guanine, cytosine, and thymine, whereas a protein sequence is over 20 amino acids. Homologous proteins have similar biological functions. Since they evolve from a common ancestral sequence, the sequences of homologous proteins and their encoding genes are often highly similar. Therefore, the DNA or amino acid sequence of a protein is often aligned with the sequences of well-studied proteins to infer the biological functions of the protein.

Formally, an alignment of two sequences, S and T, on an alphabet $\mathcal{B}$ is a two-row matrix with the following properties:

1. The letters in S are listed in order, interspersed with space symbols "–," in a row, where "–" represents the fact that a letter is missing at a position.
2. The letters in T are listed in the other row in the same manner.
3. Each column does not contain two "–."

An alignment of S and T poses a model of the evolution from their least common ancestral sequence to themselves. An alignment is scored using a scoring matrix that has a score for every pair of letters in $\mathcal{B} \cup \{-\}$. The score of an alignment is defined to be the sum of the scores of the pairs of letters appearing in the columns of the alignment.

Proteins often have multiple functions. Two proteins having a common function often have one or several highly similar regions in their DNA and amino acid sequences. Such "conserved" regions are found by solving the local alignment problem:

Input: Two sequences $S = s_1 s_2 \ldots s_m$ and $T = t_1 t_2 \cdots t_n$ on an alphabet.
Find: Two subsequences $S' = s_i s_{i+1} \cdots s_j$ ($i \leq j$) and $T' = t_k t_{k+1} \cdots t_l$ ($k \leq l$)
such that the alignment score of S' and T' is as large as possible.

The alignments between their subsequences are called *local alignments* of S and T.

A dynamic programming approach takes quadratic time to solve the local alignment problem [13]. Unfortunately, it is not fast enough for homology search against a database with millions of DNA or protein sequences. Therefore, a filtration technique was adopted to design fast algorithms for homology search in the 1990s [1], by which good local alignments between two sequences are found by first identifying short consecutive matches of a specified length between the sequences, called *seed hits*, and then extending them to obtain good local alignments.

The filtration technique has a dilemma over sensitivity and speed. Employing a long seed will miss some good local alignments between two sequences, decreasing sensitivity; on the other hand, using a short seed will waste time on

extending many seed hits into local alignments that are not biologically meaningful, resulting in low speed.

In PatternHunter [10], Ma, Tromp, and Li introduced the idea of optimized spaced seeds to achieve good balance between the sensitivity and speed of the filtration approach. PatternHunter by default looks for nucleotide match in 11 positions in every region of 18 bases long, specified by the string $111 * 1 * *1 * 1 * *11 * 111$, to trigger the process of local alignment. Such hit patterns, called *spaced seeds*, led to surprisingly higher sensitivity as well as speed than the consecutive seed 11111111111 that has the same number of match positions [10]. Moreover, sensitivity can further be improved by employing multiple spaced seeds that are longer than 18 bases [8, 14]. This motivates the study of how to find the optimal spaced seeds of given length and weight [2–5, 7].

Key Results

A spaced seed Q can be represented by a string of 1's and $*$'s, where 1's give the match positions in a seed hit. The number of 1's in Q is called its *weight*, denoted by w_Q; the length of the corresponding string is called its *length*, denoted by L_Q. The relative positions in Q are denoted by $\mathcal{RP}(Q)$. For example, for $Q = 111 * 1 * *1 * 1 * *11 * 111$, $\mathcal{RP}(Q) = \{0, 1, 2, 4, 7, 9, 12, 13, 15, 16, 17\}$.

An alignment containing no –'s is called a *ungapped alignment*. A local ungapped alignment can be modeled as a 0-1 sequence by translating match columns (containing two identical letters) into 1's and mismatch columns into 0's. Hence, a hit of Q identifies an alignment if the relative positions of Q match 1's in a region in the corresponding 0-1 string of the alignment.

Assume match occurs independently with probability p at a position in a local ungapped alignment. The *sensitivity* of Q in detecting a local alignment of n columns of two sequences with identity p is then defined to be the probability that Q hits a Bernoulli random sequence, called a *uniform region*, in which 1 and 0 appear with probability p and $(1 - p)$, respectively. A spaced seed is *optimal* for aligning sequences with identity p of length n if it has the largest hit probability over a uniform region of length n in which 1 appears with probability p at a position.

A straightforward method for identifying optimal spaced seeds is to exhaustively examine all the spaced seeds of given length and weight by keeping the largest sensitivity (or hit probability) over a uniform region. Unfortunately, the sensitivity of a spaced seed is unlikely computable in polynomial time.

Theorem 1 *Computing the sensitivity of a spaced seed over a uniform region is NP-hard.*

The hit probability of a spaced seed over a uniform region can be computed using a dynamic programming approach [7] or using recurrence relations [4,5]. Not surprisingly, these approaches become impractical for identifying long spaced seeds, because their complexities are an exponential function in the difference of the length and weight of spaced seeds under consideration. Here a simple polynomial-time approximation scheme is presented.

WISESAMPLE ALGORITHM

Input: A spaced seed Q, a positive integer n, $0 < p < 1$, and $\epsilon > 0$.
Find: An estimate of hit probability Q in a uniform region of length n in which
     bit 1 appears at a position with probability p.

Initialize an array A: $A[i] \leftarrow 0$ for $j = 1, 2, \ldots, n - L_Q$;
$N \leftarrow \lceil 6\epsilon^{-2} n^2 \log n \rceil$;
Repeats N times
    $R[i] \leftarrow 1$ for $i \in \mathcal{RP}(Q)$;
    $R[i] \leftarrow 1$ with probability p for $i \in \{1, 2, \ldots, n\} - \mathcal{RP}(Q)$;
    For $i = 1, 2, \ldots, L - L_Q$
        If Q does not hit the subregion $R[1, i + L_Q - 1]$
            $A[i] \leftarrow A[i] + 1$;
Output $p^{w_Q} \left(1 + N^{-1} \sum_{j=1}^{n-L_Q} n_j\right)$.

Theorem 2 *Let Q be a spaced seed and its hit probability be x on a uniform region with identity p of length n. WISESAMPLE outputs an estimate y of x on input Q, n, p, and $\epsilon > $ such that $|y - x| \leq \epsilon x$ with high probability.*

Let Q be a spaced seed and R a uniform region with identity p of length n. Following convention in renewal theory, Q hits R at position k if and only if $R[k - L_Q + i_j + 1] = 1$ for all $1 \leq j \leq w_Q$. Let A_k be the event that Q hits R at position k and $\bar{A}_k$ be the complement event of A_k. Then the probability f_k that Q **first** hits R at the k-th position is:

$$f_k = \Pr[\bar{A}_0 \bar{A}_1 \cdots \bar{A}_{k-2} A_{k-1}].$$

The hit probability $Q_n(p)$ of Q on R is equal to:

$$Q_n(p) = \Pr[A_0 \cup A_1 \cup \cdots \cup A_{n-1}].$$

When seed hits are extended into local alignments, two seed hits will give one local alignment if they overlap. Therefore, the sensitivity of a spaced seed is closely related to the number of its nonoverlapping hits in a uniform region. A nonoverlapping hit of a spaced seed is a recurrent event with the following convention: If a hit at position k is selected as a nonoverlapping hit, then the next nonoverlapping hit is the first hit at or after position $k + L_Q$.

The average distance, μ_Q, between two successive nonoverlapping hits of Q is defined to be

$$\mu_Q = \sum_{j \geq L_Q} j f_j.$$

A spaced seed is *nonuniform* if $g.c.d.(\mathcal{RP}(Q)) = 1$.

Theorem 3 *For any nonuniform spaced seed Q,*

$$\mu_Q \le \sum_{j=1}^{w_Q} p^{-j} + (L_Q - w_q)$$

$$- (1-p)\left(p^{2-w_Q} - 1\right)/p.$$

Buhler et al. [3] proved that for any spaced seed Q, there are two constants α_Q and λ_Q that are independent of n such that $\lim_{n\to\infty}(1 - Q_n(p))/(\alpha_Q\lambda_Q) = 1$, where λ_Q is the largest eigenvalue of the transition matrix of a Markov chain model constructed from Q.

Theorem 4 *For the consecutive seed B of weight w,*

$$\frac{1}{\sum_{j=1}^{w} p^{-j} - w + 1}$$

$$\le \lambda_B \le 1 - \frac{1}{\sum_{j=1}^{w}(p^{-j} + p^{j-1}) - w}.$$

For a spaced seed Q,

$$1 - \frac{1}{\mu_Q - L_Q + 1} \le \lambda_Q \le 1 - \frac{1}{\mu_Q}.$$

If $L_Q < (1-p)\left[p^{2-w_Q} - 1\right]/p + 1$, by Theorems 3 and 4, $\lambda_Q \le \lambda_B$. This implies that Q has a larger hit probability than the consecutive seed of the same weight in a long uniform region with identity p.

The detailed proofs of these results can be found in [11, 15].

Applications

Spaced seed approach finds applications in homology search and comparison of genome sequences. PatternHunter was used to compare the mouse and human genomes in the mouse genome project [6]. MegaBLAST and BLASTZ have adopted spaced seeds for homology search. Recently, the approach has also been used in mapping short reads into reference genome sequences.

Interestingly, spaced seed design is found to be closely related to optimal Golomb ruler design [9].

Open Problems

It is proved to be NP-hard to identify the optimal spaced seeds over a nonuniform region [8].

Open problem 1 Is it NP-hard to find the optimal spaced seed of a given length and weight over a uniform region?

It has been shown that a uniform spaced seed has a lower hit probability than the consecutive seed of the same weight over any uniform region [4,7]. But the following problem is open:

Open problem 2 For any nonuniform spaced seed Q and $0 < p < 1$, is there $n(p, Q)$ such that Q has a larger hit probability than the consecutive seed of the same weight over a uniform region with identity p of length $n \ge n(p, Q)$?

Cross-References

▶ Local Alignment (with Affine Gap Weights)
▶ Local Alignment (with Concave Gap Weights)

Recommended Reading

1. Altschul SF, Gish W, Miller W, Myers EW, Lipman DJ (1990) Basic local alignment search tool. J Mol Biol 215(3):403–410
2. Brejová B, Brown D, Vinař T (2004) Optimal spaced seeds for homologous coding regions. J Bioinformatics Comput Biol 1:595–610
3. Buhler J, Keich U, Sun Y (2004) Designing seeds for similarity search in genomic DNA. J Comput Syst Sci 70:342–363
4. Choi KP, Zhang LX (2004) Sensitivity analysis and efficient method for identifying optimal spaced seeds. J Comput Syst Sci 68:22–40

5. Choi KP, Zeng F, Zhang LX (2004) Good spaced seeds for homology search. Bioinformatics 20:1053–1059
6. Intl Mouse Genome Sequencing Consortium (2002) Initial sequencing and comparative analysis of the mouse genome. Nature 409:520–562
7. Keich U, Li M, Ma B, Tromp J (2004) On spaced seeds for similarity search. Discret Appl Math 3:253–263
8. Li M, Ma B, Kisman D, Tromp J (2004) Pattern-Hunter II: highly sensitive and fast homology search. J Bioinformatics Comput Biol 2:417–440
9. Ma B, Yao H (2009) Seed optimization for iid similarities is no easier than optimal Golomb ruler design. Inf Process Lett 109(19):1120–1124
10. Ma B, Tromp J, Li M (2002) PatternHunter: faster and more sensitive homology search. Bioinformatics 18:440–445
11. Ma B, Li M (2007) On the complexity of the spaced seeds. J Comput Syst Sci 73:1024–1034
12. Needleman SB, Wunsch CD (1970) A general method applicable to the search for similarities in the amino acid sequence of two proteins. J Mol Biol 48:443–453
13. Smith TF, Waterman MS (1980) Identification of common molecular subsequences. J Mol Biol 147:195–197
14. Sun Y, Buhler J (2004) Designing multiple simultaneous seeds for DNA similarity search. In: Proceedings RECOMB'04, 2004, San Diego, pp 76–85
15. Zhang LX (2007) Superiority of spaced seeds for homology search. IEEE/ACM Trans Comput Biol Bioinformatics (TCBB) 4:496–505

Support Vector Machines

Nello Cristianini[1] and Elisa Ricci[2]
[1] Department of Engineering Mathematics, and Computer Science, University of Bristol, Bristol, UK
[2] Department of Electronic and Information Engineering, University of Perugia, Perugia, Italy

Keywords

Kernel Methods; Large Margin Methods; Support Vector Machines

Years and Authors of Summarized Original Work

1992; Boser, Guyon, Vapnik

Problem Definition

In 1992 Vapnik and coworkers [1] proposed a supervised algorithm for classification that has since evolved into what are now known as support vector machines (SVMs) [2]: a class of algorithms for classification, regression, and other applications that represent the current state of the art in the field. Among the key innovations of this method were the explicit use of convex optimization, statistical learning theory, and kernel functions.

Classification

Given a *training set* $S = \{(\mathbf{x}_1, y_1), \ldots, (\mathbf{x}_\ell, y_\ell)\}$ of data points $\mathbf{x}_i$ from $X \subseteq \mathbb{R}^n$ with corresponding labels y_i from $Y = \{-1, +1\}$, generated from an unknown distribution, the task of classification is to learn a function $g:X \to Y$ that correctly classifies new examples $(\mathbf{x}, y)$ (i.e., such that $g(\mathbf{x}) = y$) generated from the same underlying distribution as the training data.

A good classifier should guarantee the best possible generalization performance (e.g., the smallest error on unseen examples). Statistical learning theory [3], from which SVMs originated, provides a link between the expected generalization error for a given training set and a property of the classifier known as its capacity. The SV algorithm effectively regulates the capacity by considering the function corresponding to the hyperplane that separates, according to the labels, the given training data and it is maximally distant from them (*maximal margin hyperplane*). When no linear separation is possible, a nonlinear mapping into a higher dimensional *feature space* is realized. The hyperplane found in the feature space corresponds to a nonlinear decision boundary in the input space.

Let $\phi : I \subseteq \mathbb{R}^n \to F \subseteq \mathbb{R}^N$ be a mapping from the input space I to the feature space F (Fig. 1a). In the learning phase, the algorithm finds a hyperplane defined by the equation $\langle \mathbf{w}, \phi(\mathbf{x}_i) \rangle = b$ such that the *margin*

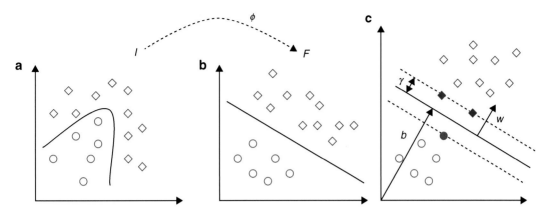

Support Vector Machines, Fig. 1 (**a**) The feature map simplifies the classification task. (**b**) A maximal margin hyperplane with its support vectors highlighted

$$\gamma = \min_{1 \leq i \leq \ell} \; y_i(\langle \mathbf{w}, \phi(\mathbf{x}_i) \rangle - b)$$
$$= \min_{1 \leq i \leq \ell} \; y_i g(\mathbf{x}_i) \tag{1}$$

is maximized, where $\langle , \rangle$ denotes the inner product, $\mathbf{w}$ is a ℓ-dimensional vector of weights, and b is a threshold.

The quantity $(\langle \mathbf{w}, \phi(\mathbf{x}_i) \rangle - b)/\|\mathbf{w}\|$ is the signed distance of the sample $\mathbf{x}_i$ from the hyperplane. When multiplied by the label y_i, it gives a positive value for correct classification and a negative value for an uncorrect one. Given a new data point $\mathbf{x}$, a label is assigned evaluating the decision function:

$$g(\mathbf{x}) = \text{sign}(\langle \mathbf{w}, \phi(\mathbf{x}) \rangle - b) \tag{2}$$

Maximizing the Margin

For linearly separable classes, there exists a hyperplane $(\mathbf{w}, b)$ such that

$$y_i(\langle \mathbf{w}, \phi(\mathbf{x}_i) \rangle - b) \geq \gamma, \quad i = 1, \ldots, \ell. \tag{3}$$

Imposing $\|\mathbf{w}\|^2 = 1$, the choice of the hyperplane such that the margin is maximized is equivalent to the following optimization problem:

$$\max_{\mathbf{w}, b, \gamma} \gamma$$

subject to $y_i(\langle \mathbf{w}, \phi(\mathbf{x}_i) \rangle - b) \geq \gamma, \; i = 1, \ldots, \ell,$
$$\tag{4}$$

$$\text{and } \|\mathbf{w}\|^2 = 1.$$

An efficient solution can be found in the dual space by introducing the Lagrange multipliers α_i, $i = 1, \ldots, \ell$. The problem (4) can be recast in the following dual form:

$$\max_{\alpha} \sum_{i=1}^{\ell} \alpha_i - \sum_{i=1}^{\ell} \sum_{j=1}^{\ell} \alpha_i \alpha_j y_i y_j \langle \phi(\mathbf{x}_i), \phi(\mathbf{x}_j) \rangle \tag{5}$$

$$\text{subject to } \sum_{i=1}^{\ell} \alpha_i y_i = 0, \quad \alpha_i \geq 0.$$

This formulation shows how the problem reduces to a *convex* (quadratic) optimization task. A key property of solutions α^* of this kind of problems is that they must satisfy the Karush-Kuhn-Tucker (KKT) conditions that ensure that only a subset of training examples needs to be associated to a nonzero α_i. This property is called *sparseness* of the SVM solution and is crucial in practical applications.

In the solution α^*, often only a subset of training examples is associated to nonzero α_i. These are called *support vectors* and correspond to the points that lie closest to the separating hyperplane (Fig. 1b). For the maximal margin hyperplane, the weight vector $\mathbf{w}^*$ is given by a linear function of the training points:

$$\mathbf{w}^* \sum_{i=1}^{\ell} \alpha_i^* y_i \phi(\mathbf{x}_i). \tag{6}$$

Then the decision function (2) can equivalently be expressed as

$$g(\mathbf{x}) = \text{sign}(\sum_{i=1}^{\ell} \alpha_i^* y_i \langle \phi(\mathbf{x}_i), \phi(\mathbf{x}) \rangle - b). \quad (7)$$

For a support vector $\mathbf{x}_i$, it is $\langle \mathbf{w}^*, \phi(\mathbf{x}_i) \rangle - b = y_i$ from which the optimum bias b^* can be computed. However, it is better to average the values obtained by considering all the support vectors [2]. Both the quadratic programming (QP) problem (5) and the decision function (7) depend only on the dot product between data points. The matrix of dot products with elements $K_{ij} = K(\mathbf{x}_i, \mathbf{x}_j) = \langle \phi(\mathbf{x}_i), \phi(\mathbf{x}_j) \rangle$ is called the *kernel matrix*. In the case of linear separation, we simply have $K(\mathbf{x}_i, \mathbf{x}_j) = \langle \mathbf{x}_i, \mathbf{x}_j \rangle$, but in general, one can use functions that provide nonlinear decision boundaries. Widely used kernels are the polynomial $K(\mathbf{x}_i, \mathbf{x}_j) = (\langle \mathbf{x}_i, \mathbf{x}_j \rangle + 1)^d$ or the Gaussian $K(\mathbf{x}_i, \mathbf{x}_j) = e^{-\frac{||\mathbf{x}_i - \mathbf{x}_j||^2}{\sigma^2}}$ where d and σ are user-defined parameters.

Key Results

In the framework of learning from examples, SVMs have shown several advantages compared to traditional neural network models (which represented the state of the art in many classification tasks up to 1992). The statistical motivation for seeking the maximal margin solution is to minimize an upper bound on the test error that is independent of the number of dimensions and inversely proportional to the separation margin (and the sample size). This directly suggests embedding of the data in a high-dimensional space where a large separation margin can be achieved; this can be done efficiently with kernels using techniques from convex optimization. The sparseness of the solution, implied by the KKT conditions, adds to the efficiency of the result.

The initial formulation of SVMs by Vapnik and coworkers [1] has been extended by many other researchers. Here we summarize some key contributions.

Soft Margin

In the presence of noise the SV algorithm can be subject to overfitting. In this case one needs to tolerate some training errors in order to obtain a better generalization power. This has led to the development of the *soft margin* classifiers [4]. Introducing the slack variables $\xi_i \geq 0$, optimal class separation can be obtained by

$$\min_{\mathbf{w}, b\gamma, \xi} -\gamma + C \sum_{i=1}^{\ell} \xi_i$$

subject to $y_i(\langle \mathbf{w}, \phi(\mathbf{x}_i) \rangle - b) \geq \gamma - \xi_i, \xi_i \geq 0$ (8)

$$i = 1, \ldots, \ell \text{ and } ||\mathbf{w}||^2 = 1.$$

The constant C is user defined and controls the trade-off between the maximization of the margin and the number of classification errors. The dual formulation is the same as (5) with the only difference in the bound constraints ($0 \leq \alpha_i \leq C, \ i = 1, \ldots, \ell$). The choice of soft margin parameter is one of the two main design choices (together with the kernel function) in applications. It is an elegant result [5] that the entire set of solutions for all possible values of C can be found with essentially the same computational cost as finding a single solution: this set is often called the *regularization path*.

Regression

A SV algorithm for regression, called support vector regression (SVR), was proposed in 1996 [6]. A linear algorithm is used in the kernel-induced feature space to construct a function such that the training points are inside a tube of given radius ε. As for classification the regression function only depends on a subset of the training data.

Speeding Up the Quadratic Program

Since the emergence of SVMs, many researchers have developed techniques to effectively solve the problem (5): a quite time-consuming task, especially for large training sets. Most methods decompose large-scale problems into a series of smaller ones. The most widely used method is that of Platt [7] and it is known as sequential minimal optimization.

Kernel Methods

In SVMs, both the learning problem and the decision function can be formulated only in terms of dot products between data points. Other popular methods (i.e., principal component analysis, canonical correlation analysis, fisher discriminant) have the same property. This fact has led to a huge number of algorithms that effectively use kernels to deal with nonlinear functions keeping the same complexity as the linear case. They are referred to as *kernel methods* [8,9].

Choosing the Kernel

The main design choice when using SVMs is the selection of an appropriate kernel function, a problem of model selection that roughly relates to the choice of a topology for a neural network. It is a nontrivial result [10] that also this key task can be translated into a convex optimization problem (a semi-definite program) under general conditions. A kernel can be optimally selected from a kernel space resulting from all linear combinations of a basic set of kernels.

Kernels for General Data

Kernels are not just useful tools to allow us to deploy methods of linear statistics in a nonlinear setting. They also allow us to apply them to nonvectorial data: kernels have been designed to operate on sequences, graphs, text, images, and many other kinds of data [8].

Applications

Since their emergence, SVMs have been widely used in a huge variety of applications. To give some examples, good results have been obtained in text categorization, handwritten character recognition, and biosequence analysis.

Text Categorization

In automatic text categorization, text documents are classified into a fixed number of predefined categories based on their content. In the works performed by Joachims [11] and Dumais et al. [12], documents are represented by vectors with the so-called bag-of-words approach used in the information retrieval field. The distance between two documents is given by the inner product between the corresponding vectors. Experiments on the collection of Reuters news stories showed good results for SVMs compared to other classification methods.

Handwritten Character Recognition

This is the first real-world task on which SVMs were tested. In particular two publicly available data sets (USPS and NIST) have been considered since they are usually used for benchmarking classifiers. A lot of experiments, mainly summarized in [13], were performed which showed that SVMs can perform as well as other complex systems without incorporating any detailed prior knowledge about the task.

Bioinformatics

SVMs have been widely used also in bioinformatics. For example, Jaakkola and Haussler [14] applied SVMs to the problem of protein homology detection, i.e., the task of relating new protein sequences to proteins whose properties are already known. Brown et al. [15] describe a successful use of SVMs for the automatic categorization of gene expression data from DNA microarrays.

URL to Code

Many free software implementations of SVMs are available at the website

- www.support-vector.net/software.html

Two in particular deserve a special mention for their efficiency:

- *SVMlight*: Joachims T. Making large-scale SVM learning practical. In: Schölkopf B, Burges CJC, and Smola AJ (eds) Advances in Kernel Methods Support Vector Learning, MIT Press, 1999. Software available at http://svmlight.joachims.org
- *LIBSVM*: Chang CC, and Lin CJ, LIBSVM: a library for support vector machines, 2001. Software available at http://www.csie.ntu.edu.tw/~cjlin/libsvm

Cross-References

▶ PAC Learning
▶ Perceptron Algorithm

Recommended Reading

1. Boser B, Guyon I, Vapnik V (1992) A training algorithm for optimal margin classifiers. In: Proceedings of the fifth annual workshop on computational learning theory, Pittsburgh
2. Cristianini N, Shawe-Taylor J (2000) An introduction to support vector machines and other kernel-based learning methods. Cambridge University Press, Cambridge. Book website: www.support-vector.net
3. Vapnik V (1995) The nature of statistical learning theory. Springer, New York
4. Cortes C, Vapnik V (1995) Support-vector network. Mach Learn 20:273–297
5. Hastie T, Rosset S, Tibshirani R, Zhu J (2004) The entire regularization path for the support vector machine. J Mach Learn Res 5:1391–1415
6. Drucker H, Burges CJC, Kaufman L, Smola A, Vapnik V (1997) Support vector regression machines. Adv Neural Inf Process Syst (NIPS) 9:155–161. MIT
7. Platt J (1999) Fast training of support vector machines using sequential minimal optimization. In: Schölkopf B, Burges CJC, Smola AJ (eds) Advances in kernel methods support vector learning. MIT, Cambridge, pp 185–208
8. Shawe-Taylor J, Cristianini N (2004) Kernel methods for pattern analysis. Cambridge University Press, Cambridge. Book website: www.kernel-methods.net
9. Scholkopf B, Smola AJ (2002) Learning with kernels. MIT, Cambridge
10. Lanckriet GRG, Cristianini N, Bartlett P, El Ghaoui L, Jordan MI (2004) Learning the kernel matrix with semidefinite programming. J Mach Learn Res 5:27–72
11. Joachims T (1998) Text categorization with support vector machines. In: Proceedings of European conference on machine learning (ECML), Chemnitz
12. Dumais S, Platt J, Heckerman D, Sahami M (1998) Inductive learning algorithms and representations for text categorization. In: 7th international conference on information and knowledge management, Bethesda
13. LeCun Y, Jackel LD, Bottou L, Brunot A, Cortes C, Denker JS, Drucker H, Guyon I, Muller UA, Sackinger E, Simard P, Vapnik V (1995) Comparison of learning algorithms for handwritten digit recognition. In: Fogelman-Soulie F, Gallinari P (eds) Proceedings international conference on artificial neural networks (ICANN), Paris, vol 2. EC2, pp 5360
14. Jaakkola TS, Haussler D (1999) Probabilistic kernel regression models. In: Proceedings of the 1999 Conference on AI and Statistics, Fort Lauderdale
15. Brown M, Grundy W, Lin D, Cristianini N, Sugnet C, Furey T, Ares M Jr, Haussler D (2000) Knowledge-based analysis of mircoarray gene expression data using support vector machines. Proc Natl Acad Sci 97(1):262–267

Surface Reconstruction

Nina Amenta
Department of Computer Science, University of California, Davis, CA, USA

Keywords

Delaunay triangulation; Local feature size; Medial axis; Surface reconstruction; Voronoi diagram

Years and Authors of Summarized Original Work

1999; Amenta, Bern
2000; Amenta, Choi, Dey, Leekha
2001; Amenta, Choi, Kolluri
2004; Dey, Goswami

Problem Definition

Surface reconstruction, here, is the problem of producing a piecewise-linear representation of a two-dimensional surface S in $\mathbb{R}^3$, given as input a set P of point samples from the surface. Very

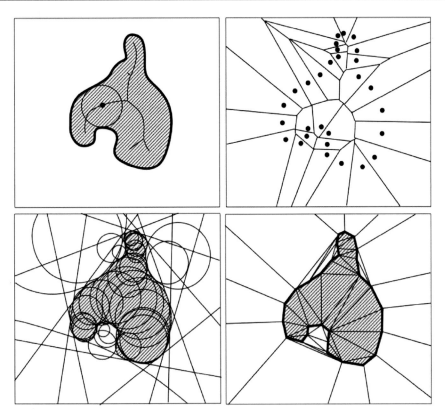

Surface Reconstruction, Fig. 1 The medial axis of an object; the Voronoi diagram of a set of samples from the object boundary; the set of polar balls, with those inside the object shaded; the corresponding cells of the weighted Voronoi diagram, again with those inside the object shaded

sparse sets of point samples clearly do not convey much about S, so in order to prove correctness, we need to assume that the sample P is somehow sufficiently dense. The minimum required density could vary across the surface, with more detailed areas requiring denser sampling. This idea is captured in the following definition [2]. Let S be a two-dimensional surface in $\mathbb{R}^3$. The *medial axis* of S is the closure of the set of points that have more than one nearest point on S; a two-dimensional example is shown in Fig. 1, top left.

Definition 1 The *local feature size* $f(x)$ at a point x is the minimum distance from x to the medial axis of S.

The distance from the medial axis to the surface is zero at a sharp feature such as a corner or a crease, so we usually assume that S is smooth. The algorithms described here make the following ϵ-sampling assumption: the minimum distance, at any surface point x, to the nearest sample point is at most $\epsilon f(x)$, for some small constant ϵ. This leads to algorithms that are provably correct in the following sense.

INPUT: *A point set P that is an ϵ-sample from a smooth surface S without boundary.*

OUTPUT: *A piecewise-linear manifold without boundary, homeomorphic to S, that everywhere lies within distance $O(\epsilon f(x))$ of S.* The monograph [7] is an excellent reference for this line of research.

Key Results

One key idea is that in the neighborhood of any point $p \in P$ sampled from S, the surface is well approximated by a plane. Specifically, for any surface point x closer to p than to any other sample, the distance of x from the tangent plane at p is $O(\epsilon f(x))$, as is the difference between the surface normal at x and the surface normal at p [2] (with the corrected proof [3]). Another key idea is that some subset of the Voronoi vertices of P approximates the medial axis of S, as in Fig. 1, top right.

Crust Algorithm

The crust algorithm [2] approximates the medial axis with a subset of the three-dimensional Voronoi vertices, called the poles. Each sample point in $p \in P$ selects the vertex of its Voronoi cell farthest from p as its first pole and the vertex farthest in the opposite direction as its second. We then eliminate any Delaunay triangle all of whose circumspheres contain a pole; this is easy to implement by computing the Delaunay triangulation of the set P augmented with the set of poles and eliminating any output triangle adjacent to a pole. A subset of the remaining surface triangles can then be selected as the piecewise-linear output surface.

Cocone Algorithm

The cocone algorithm [4] provides a simpler way of selecting a set of surface Delaunay triangles, requiring only one Voronoi diagram computation. It relies on the fact that the direction vector from a sample $p \in P$ to its first pole is within $O(\epsilon)$ of the surface normal at p, under the ϵ-sampling assumption. We define the cocone at p as the complement of a double cone, such that the angle between the cone surface and this approximate normal vector is at least $\pi - \pi/8$. We consider the intersection of the cocone at p with the Voronoi cell of p; the Delaunay triangles dual to any edge in this intersection are marked as potential surface triangles. Triangles marked by all three of their vertices are included in the set of surface triangles.

Powercrust Algorithm

While it is easy in theory to select a subset of the surface triangles to form a piecewise-linear output surface, it can be difficult in practice when the sampling density fails to meet the assumption, as is inevitable at sharp features. The power crust algorithm [5] eliminates this issue by producing a piecewise-linear output surface. The Voronoi ball centered at a pole is the ball with its nearest input samples on the boundary; see Fig. 1, lower left. We begin by labeling the Voronoi balls of all of the poles either as inside or outside the object bounded by S, using an iterative algorithm. We then compute the weighted Voronoi diagram, also known as the power diagram, of these polar Voronoi balls. Any Voronoi face separating the cell of an inner pole from the cell of an outer pole is output as part of the surface (Fig. 1, lower right). The faces of the piecewise-linear output surface are convex polygons but not in general triangles.

Noisy Samples

When the input sample points have noise, not every pole will be near the medial axis. Nonetheless, if the level of noise is everywhere small relative to the local feature size f, some subset of Voronoi vertices will still approximate the medial axis. In [8], this idea is developed into a provably correct algorithm. In addition to the ϵ-sampling assumption, we need to assume that the noise level is $O(\epsilon^2 f(x))$ and that the distance from any sample p to the kth nearest sample p' is $O(\epsilon f(x))$. This allows us to recognize a Voronoi vertex of p as a pole only when it is significantly farther from p than the k-nearest neighbors of p. These poles are then labeled as either inner or outer. This algorithm produces a triangulation of the boundary of the union of the inner polar balls as the output surface.

Complexity

The complexity of all of these algorithms depends on the complexity of the Voronoi diagram. While in general the Voronoi diagram of n points in $\mathbb{R}^3$ might have complexity $O(n^2)$, Attali, Boissonnat, and Lietier [6] proved that the complexity of the Voronoi diagram for points distributed uniformly on a nondegenerate smooth surface in $\mathbb{R}^3$ is $O(n \lg n)$. Another idea, employed by Funke and Ramos [9] and advanced by Cheng et al. [12], is to replace the Voronoi diagram with a less computationally expensive structure to get an $O(n \lg n)$ algorithm.

Applications

Interest in this problem was motivated by the advent of laser-range and LiDAR scanners [10], which produce depth maps sampled by point clouds. It is often reasonable to assume noise-free surface samples, since there are preprocessing methods, such as moving least squares (MLS) [1], that attract noisy point clouds onto nearby surfaces; there has also been theoretical work on MLS. MLS, or simply local plane-fitting, can be used to produce a normal vector at each sample point. Another common assumption is that the normal vectors can be consistently oriented. Poisson surface reconstruction [11] is an optimization technique that constructs manifold surfaces from possibly noisy points with normals. Because of its very efficient implementations, it is currently the most popular method in practice.

Open Problems

Subsequent work in surface reconstruction, both in computer graphics and in computational geometry, has focused on the identification and reconstruction of sharp features and then using them to construct surfaces that are non-manifold. Proving that the complexity of the Voronoi diagram of points distributed on a generic smooth surface with noise or with boundary is $o(n^2)$ remains open.

URLs to Code and Data Sets

There is code available for the cocone algorithm (http://web.cse.ohio-state.edu/~tamaldey/cocone.html), with several subsequent variants. There is also code for the power crust algorithm (http://www.cs.ucdavis.edu/~amenta/powercrust.html). There is a set of benchmark data sets for surface reconstruction (http://www.cs.utah.edu/~bergerm/recon_bench).

Cross-References

▶ Curve Reconstruction
▶ Manifold Reconstruction

Recommended Reading

1. Alexa M, Behr J, Cohen-Or D, Fleishman S, Levin D, Silva CT (2003) Computing and rendering point set surfaces. IEEE Trans Vis Comput Graph 9(1):3–15
2. Amenta N, Bern M (1999) Surface reconstruction by Voronoi filtering. Discret Comput Geom 22(4):481–504
3. Amenta N, Dey TK (2007) Normal variation for adaptive feature size, arXiv
4. Amenta N, Choi S, Dey TK, Leekha N (2000) A simple algorithm for homeomorphic surface reconstruction. In: Proceedings of the sixteenth annual symposium on computational geometry, Hong Kong. ACM, pp 213–222
5. Amenta N, Choi S, Kolluri RK (2001) The power crust, unions of balls, and the medial axis transform. Comput Geom Theory Appl 19(2):127–153
6. Attali D, Boissonnat JD, Lieutier A (2003) Complexity of the Delaunay triangulation of points on surfaces: the smooth case. In: Proceedings of the nineteenth annual symposium on computational geometry, San Diego. ACM, pp 201–210
7. Dey TK (2006) Curve and surface reconstruction: algorithms with mathematical analysis. Cambridge monographs on applied and computational mathematics. Cambridge University Press, Leiden
8. Dey TK, Goswami S (2004) Provable surface reconstruction from noisy samples. In: Proceedings of the twentieth annual symposium on computational geometry, Brooklyn. ACM, pp 330–339
9. Funke S, Ramos EA (2002) Smooth-surface reconstruction in near-linear time. In: Proceedings of the thirteenth annual ACM-SIAM symposium on discrete algorithms, San Francisco. Society for Industrial and Applied Mathematics, pp 781–790
10. Hoppe H, DeRose T, Duchamp T, McDonald J, Stuetzle W (1992) Surface reconstruction from unorganized points. ACM Trans Graph (TOG) 26(2):71–78
11. Kazhdan M, Bolitho M, Hoppe H (2006) Poisson surface reconstruction. In: Proceedings of the fourth eurographics symposium on geometry processing, Cagliari, pp 61–70
12. Cheng S-W, Jin J, Lau M-K (2012) A fast and simple surface reconstruction algorithm. In: Proceedings of the 28th annual symposium on computational geometry, Chapel Hill, pp 69–78

S

Symbolic Model Checking

Adnan Aziz[1] and Amit Prakash[2]
[1]Department of Electrical and Computer
Engineering, University of Texas, Austin,
TX, USA
[2]Microsoft, MSN, Redmond, WA, USA

Keywords

Formal hardware verification

Years and Authors of Summarized Original Work

1990; Burch, Clarke, McMillan, Dill

Problem Definition

Design verification is the process of taking a design and checking that it works correctly. More specifically, every design verification paradigm has three components [6]: (1) a language for specifying the design in an unambiguous way, (2) a language for specifying properties that are to be checked of the design, and (3) a checking procedure, which determines whether the properties hold off the design.

The verification problem is very general: it arises in low-level designs, e.g., checking that a combinational circuit correctly implements arithmetic, as well as high-level designs, e.g., checking that a library written in high-level language correctly implements an abstract data type.

Hardware Verification

The verification of hardware designs is particularly challenging. Verification is difficult in part because the large number of concurrent operations make it very difficult to conceive of and construct all possible corner cases, e.g., one unit initiating a transaction at the same cycle as another receiving an exception. In addition, software models used for simulation run orders of several magnitude slower than the final chip operates at. Faulty hardware is usually impossible to correct after fabrication, which means that the cost of a defect is very high, since it takes several months to go through the process of designing and fabricating new hardware. Wile et al. [15] provide a comprehensive account of hardware verification.

State Explosion

Since the number of state-holding elements in digital hardware is bounded, the number of possible states that the design can be in is infinite, so complete automated verification is, in principle, possible. However, the number of states that a hardware design can reach from the initial state can be exponential in the size of the design; this phenomenon is referred to as "state explosion." In particular, algorithms for verifying hardware that explicitly record visited states, e.g., in a hash table, have very high time complexity, making them infeasible for all but the smallest designs. The problem of complete hardware verification is known to be PSPACE-hard, which means that any approach must be based on heuristics.

Hardware Model

A hardware design is formally described using *circuits* [4, 8]. A *combinational circuit* consists of *Boolean combinational elements* connected by *wires*. The Boolean combinational elements are *gates* and *primary inputs*. Gates come in three types: *NOT, AND,* and *OR*. The NOT gate functions as follows: it takes a single Boolean-valued *input* and produces a single Boolean-valued *output* which takes value 0 if the input is 1 and 1 if the input is 0. The AND gate takes two Boolean-valued inputs and produce a single output; the output is 1 if both inputs are 1 and 0 otherwise. The OR gate is similar to AND, except that its output is 1 if one or both inputs are 1. A circuit can be represented as a directed graph where the nodes represent the gates and wires represent edges in the direction of signal flow.

A circuit can be represented by a directed graph where the nodes represent the gates and primary inputs, and edges represent wires in the direction of signal flow. Circuits are required to be acyclic, that is, there is no cycle of gates. The absence of cycles implies that a Boolean assignment to the primary inputs can be propagated through the gates in topological order.

A *sequential circuit* extends the notion of circuit described above by adding *stateful elements*. Specifically, a sequential circuit includes *registers*. Each register has a single input, which is referred to as its *next-state input*.

A *valuation* on a set V is a function whose domain is V. A *state* in a sequential circuit is a Boolean-valued valuation on the set of registers. An *input* to a sequential circuit is a Boolean-valued valuation on the set of primary inputs. Given a state s and an input i, the logic gates in the circuit uniquely define a Boolean-valued valuation t to the set of register inputs – this is referred to as the next state of the circuit at state s under input i and say s *transitions* to t on input i. It is convenient to denote such a transition by $s \xrightarrow{i} t$.

A sequential circuit can naturally be identified with a *finite state machine* (FSM), which is a graph defined over the set of all states; an edge (s, t) exists in the FSM graph if there exists an input i, state s transitions to t on input i.

Invariant Checking

An *invariant* is a set of states; informally, the term is used to refer to a set of states that are "good" in some sense. One common way to specify an invariant is to write a Boolean formula on the register variables – the states which satisfy the formula are precisely the states in the invariant.

Given states r and s, define r to be *reachable* from s if there is a sequence of inputs $i_0, i_1, \ldots, i_{n-1}$ such that $s = s_0 \xrightarrow{i_0} s_1 \xrightarrow{i_1} \cdots s_n = t$. A fundamental problem in hardware

verification is the following: given an invariant A, and a state s, does there exists a state r reachable from s which is not in A?

Key Results

Symbolic model checking (SMC) is a heuristic approach to hardware verification. It is based on the idea that rather than representing and manipulating states one at a time, it is more efficient to use symbolic expressions to represent and manipulate sets of states.

A key idea in SMC is that given a set $A \subset \{0, 1\}^n$, a Boolean function A can be constructed such that $f_A : \{0, 1\}^n \to \{0, 1\}$ given by $f(\alpha_1, \ldots, \alpha_n) = 1$ iff $(\alpha_1, \ldots, \alpha_n) \in A$. Note that given a characteristic function f_A, A can be obtained and vice versa.

There are many ways in which a Boolean function can be represented: formulas in DNF, general Boolean formulas, combinational circuits, etc. In addition to an efficient representation for state sets, the ability to perform fast computations with sets of states is also important, for example, in order to determine if an invariant holds, it is required to compute the set of states reachable from a given state. BDDs [2] are particularly well suited to representing Boolean functions, as they combine succinct representation with efficient manipulation; they are the data structure underlying SMC.

Image Computation

A key computation that arises in verification is determining the *image* of a set of states A in a design D – the image of A is the set of all states t for which there exists a state in A and an input i such that state s transitions to t under input i. The image of A is denoted by $\mathrm{Img}(A)$.

The *transition relation* of a design is the set of (s, i, t) triples such that s transitions to t under input i. Let the design have n registers and m primary inputs; then the transition relation is subset of $\{0, 1\}^n \times \{0, 1\}^m \times \{0, 1\}^n$.

Conceptually, the transition relation completely captures the dynamics of the design – given an initial state, and input sequence, the evolution of the design is completely determined by the transition relation.

Since the transition relation is a subset of $\{0, 1\}^{n+m+n}$, it has a characteristic function $f_T : \{0, 1\}^{n+m+n} \rightarrow \{0, 1\}$. View f_T as being defined over the variables $x_0, \ldots, x_{n-1}$, $i_0, \ldots, i_{m-1}$, $y_0, \ldots, y_{n-1}$. Let the set of states A be represented by the function f_A defined over variables $x_0, \ldots, x_{n-1}$. Then the following identity holds

$$\text{Img}(A) = (\exists x_0 \cdot \exists x_{n-1} \exists i_0 \cdots \exists i_{m-1})(f_A \cdot f_T).$$

The identity holds because $(\beta_0, \ldots, \beta_{n-1})$ satisfies the right-hand side expression exactly when there are values $\alpha_0, \ldots, \alpha_{n-1}$, and $\iota_0, \ldots, \iota_{m-1}$ such that $(\alpha_0, \ldots, \alpha_{n-1}) \in A$ and the state $(\alpha_0, \ldots, \alpha_{n-1})$ transitions to $(\beta_0, \ldots, \beta_{n-1})$ on input $(\iota_0, \ldots, \iota_{m-1})$.

Invariant Checking

The set of all states reachable from a given set A is the limit as n tends to infinity of the sequence of states $R_0, R_1, \ldots$ defined below:

$$R_0 = A$$
$$R_{i+1} = R_i \cup \text{Img}(R_i).$$

Since for all i, $R_i \subseteq R_{i+1}$ and the number of distinct state sets is finite, the limit is reached in some finite number of steps, i.e., for some n, it must be that $R_{n+1} = R_n$. It is straightforward to show that the limit is exactly equal to the set of states reachable from A – the basic idea is to inductively construct input sequences that lead from states in A to R_i and to show that state t is reachable from a state in A under an input sequence of length l, then t must be in R_l.

Given BDDs F and G representing functions f and g, respectively, there is an algorithm based on dynamic programming for performing

conjunction, i.e., for computing the BDD for $f \cdot g$. The algorithm has polynomial complexity, specifically $O(|F| \cdot |G|)$, where $|B|$ denotes the number of nodes in the BDD B. There are similar algorithms for performing disjunction ($f + g$) and computing cofactors (f_x and $f_{x'}$). Together these yield an algorithm for the operation of existential quantification, since $(\exists x) f = f_x + f_{x'}$.

It is straightforward to build BDDs for f_A and f_T : A is typically given using a propositional formula, and the BDD for f_A can be built up using functions for conjunction, disjunction, and negation. The BDD for f_T is built using from the BDDs for the next-state nodes, over the register and primary input variables. Since the only gate types are AND, OR, and NOT, the BDD can be built using the standard BDD operators for conjunction, disjunction, and negation. Let the next-state functions be $f_0, \ldots, f_{n-1}$; then f_T is $(y_0 = f_0) \cdot (y_1 = f_1) \cdot \cdots \cdot (y_{n-1} = f_{n-1})$, and so the BDD for f_T can be constructed using the usual BDD operators.

Since the image computation operation can be expressed in terms of f_A and F_T, and conjunction and existential quantification operations, it can be performed using BDDs. The computation of R_i involves an image operation, and a disjunction, and since BDDs are canonical, the test for fixed point is trivial.

Applications

The primary application of the technique described above is for checking properties of hardware designs. These properties can be invariants described using propositional formulae over the register variables, in which case the approach above is directly applicable. More generally, properties can be expressed in a *temporal logic* [5], specifically through formulae which express acceptable sequences of outputs and transitions.

CTL is one common temporal logic. A CTL formula is given by the following grammar: if x is a variable corresponding to a register, then **x** is a CTL formula; otherwise, if φ and ψ are CTL formulas, then so as $(\neg\phi)$, $(\phi\vee\psi)$, $(\phi\wedge\psi)$, $(\phi\rightarrow\psi)$, and $EX\phi$, $E\phi U\psi$, and $EG\phi$.

A CTL formula is interpreted as being true at a state; a formula **x** is true at a state if that register is 1 in that state. Propositional connectives are handled in the standard way, e.g., a state satisfies a formula $(\phi\wedge\psi)$ if it satisfies both φ and ψ. A state s satisfies $EG\phi$ if there exists a state t such that s transitions to, and t satisfies φ. A state s satisfies $E\phi U\psi$ if there exists a sequence of inputs $i_0,\ldots,i_n$ leading through state $s_0 = s$, $s_1, s_2, \ldots, s_{n+1}$ such that s_{n+1} satisfies ψ, and all states $s_i, i \leq n+1$ satisfy φ. A state s satisfies $EG\phi$ if there exists an infinite sequence of inputs $i_0, i_1, \ldots$ leading through state $s_0 = s, s_1, s_2, \ldots$ such that all states s_i satisfy φ.

CTL formulas can be checked by a straightforward extension of the technique described above for invariant checking. One approach is to compute the set of states in the design satisfying subformulas of φ, starting from the subformulas at the bottom of the parse tree for φ. A minor difference between invariant checking and this approach is that the latter relies on *pre-image* computation; the pre-image of A is the set of all states t for which there exists an input i such that t transitions under i to a state in A.

Symbolic analysis can also be used to check the equivalence of two designs by forming a new design which operates the two initial designs in parallel and has a single output that is set to 1 if the two initial designs differ [14]. In practice this approach is too inefficient to be useful, and techniques which rely more on identifying common substructures across designs are more successful.

The complement of the set of reachable states can be used to identify parts of the design which are redundant and to propagate don't care conditions from the input of the design to internal nodes [12].

Many of the ideas in SMC can be applied to software verification – the basic idea is to "finitize" the problem, e.g., by considering integers to lie in a restricted range or setting an a priori bound on the size of arrays [7].

Experimental Results

Many enhancements have been made to the basic approach described above. For example, the BDD for the entire transition relation can grow large, so *partitioned transition relations* [11] are used instead; these are based on the observation that $\exists x.(f \cdot g) = f \cdot \exists x.g$, in the special case that f is independent of x. Another optimization is the use of *don't cares*; for example, when computing the image of A, the BDD for f_T can be simplified with respect to transitions originating at A' [13]. Techniques based on SAT have enjoyed great success recently. These approaches case the verification problem in terms of satisfiability of a CNF formula. They tend to be used for bounded checks, i.e., determining that a given invariant holds on all input sequences of length k [1]. Approaches based on *transformation-based verification* complement symbolic model checking by simplifying the design prior to verification. These simplifications typically remove complexity that was added for performance rather than functionality, e.g., pipeline registers.

The original paper by Clarke et al. [3] reported results on a toy example, which could be described in a few dozen lines of a high-level language. Currently, the most sophisticated model checking tool for which published results are ready is SixthSense, developed at IBM [10].

A large number of papers have been published on applying SMC to academic and industrial designs. Many report success on designs with an astronomical number of states – these results become less impressive when taking into consideration the fact that a design with n registers has 2^n states.

It is very difficult to define the complexity of a design. One measure is the number of registers in the design. Realistically, a hundred registers is at the limit of design complexity that can be handled using symbolic model checking. There are cases of designs with many more registers that have been successfully verified with symbolic model checking, but these registers are invariably part of a very regular structure, such as a memory array.

Data Sets

The SMV system described in [9] has been updated, and its latest incarnation nuSMV (http://nusmv.irst.itc.it/) includes a number of examples.

The VIS (http://embedded.eecs.berkeley.edu/pubs/downloads/vis) system from UC Berkeley and UC Boulder also includes a large collection of verification problems, ranging from simple hardware circuits to complex multiprocessor cache systems.

The SIS (http://embedded.eecs.berkeley.edu/pubs/downloads/sis/) system from UC Berkeley is used for logic synthesis. It comes with a number of sequential circuits that have been used for benchmarking symbolic reachability analysis.

Cross-References

▶ Binary Decision Graph

Recommended Reading

1. Biere A, Cimatti A, Clarke E, Fujita M, Zhu Y (1999) Symbolic model checking using sat procedures instead of BDDs. In: ACM design automation conference, New Orleans
2. Bryant R (1986) Graph-based algorithms for Boolean function manipulation. IEEE Trans Comput C-35:677–691
3. Burch JR, Clarke EM, McMillan KL, Dill DL (1992) Symbolic model checking: 10^{20} states and beyond. Inf Comput 98(2):142–170
4. Cormen TH, Leiserson CE, Rivest RH, Stein C (2001) Introduction to algorithms. MIT, Cambridge
5. Emerson EA (1990) Temporal and modal logic. In: van Leeuwen J (ed) Formal models and semantics. Volume B of handbook of theoretical computer science. Elsevier Science, Amsterdam, pp 996–1072
6. Gupta A (1993) Formal hardware verification methods: a survey. Form Method Syst Des 1:151–238
7. Jackson D (2006) Software abstractions: logic, language, and analysis. MIT, Cambridge
8. Katz R (1993) Contemporary logic design. Benjamin/Cummings Publishing Company, Redwood City
9. McMillan KL (1993) Symbolic model checking. Kluwer Academic, Boston
10. Mony H, Baumgartner J, Paruthi V, Kanzelman R, Kuehlmann A (2004) Scalable automated verification via expert-system guided transformations. In: Formal methods in CAD, Austin
11. Ranjan R, Aziz A, Brayton R, Plessier B, Pixley C (1995) Efficient BDD algorithms for FSM synthesis and verification. In: Proceedings of the international workshop on logic synthesis, Tahoe City, May 1995
12. Savoj H (1992) Don't cares in multi-level network optimization. Ph.D. thesis, Electronics Research Laboratory, College of Engineering, University of California, Berkeley
13. Shiple TR, Hojati R, Sangiovanni-Vincentelli AL, Brayton RK (1994) Heuristic minimization of BDDs using don't cares. In: ACM design automation conference, San Diego, June 1994
14. Touati H, Savoj H, Lin B, Brayton RK, Sangiovanni-Vincentelli AL (1990) Implicit state enumeration of finite state machines using BDDs. In: IEEE international conference on computer-aided design, Santa Clara, pp 130–133, Nov 1990
15. Wile B, Goss J, Roesner W (2005) Comprehensive functional verification. Morgan-Kaufmann

Symmetric Graph Drawing

Seokhee Hong
School of Information Technologies, University of Sydney, Sydney, NSW, Australia

Keywords

Graph automorphism; Graph drawing; Planar graph; Symmetry

Years and Authors of Summarized Original Work

2006; Hong, McKay and Eades

Problem Definition

Symmetry is one of the most important aesthetic criteria in graph drawing that clearly reveals the structure and properties of a graph. Many graphs in Graph Theory textbooks are often symmetric.

A symmetry of a drawing D of a graph G induces an *automorphism* ϕ of the graph G, a permutation of the vertex set that preserves adjacency. If an automorphism ϕ can be displayed as a symmetry in a drawing of the graph G, then it is called a *geometric automorphism* [6]. A geometric automorphism ϕ of a *planar* graph G is a *planar automorphism*, if there is a *planar* drawing of G which displays ϕ. Note that not every automorphism is geometric, and not every geometric automorphism is planar.

In general, algorithms for constructing symmetric drawings of graphs have two steps:

1. *Symmetry finding step*: Find the geometric automorphisms of a graph
2. *Symmetry drawing step*: Draw the graph displaying these automorphisms as symmetries.

Note that the first step is more difficult than the second step. For example, finding automorphism of a graph is isomorphism-hard; however finding geometric automorphism of a graph is NP-hard in general [18]. For planar graphs, computing isomorphism (therefore, automorphism) of a graph can be solved in linear time [7, 17]. However, finding the *best plane embedding* of planar graphs that displays the maximum number of symmetries in a drawing of a planar graph is challenging, because a planar graph can have exponential number of possible plane embeddings.

Furthermore, the product of two geometric automorphisms is not necessarily geometric, because they may be displayed by different drawings. A subgroup A of the automorphism group of a graph is a *geometric automorphism group*, if there is a single drawing of the graph that displays every element of A. Therefore, to construct a maximally symmetric drawing of a graph, one needs to compute a *maximum size* geometric automorphism group for the graph. Therefore, the main research problem for Symmetric Graph Drawing can be defined as below.

Symmetric Graph Drawing Problem
Input: A graph G.
Output: A maximum size geometric automorphism group A of G, A symmetric drawing D of G that displays all elements of A.

Key Results

There are two types of symmetry in two-dimensional drawings: *rotational symmetry* (i.e., a rotation about a point) and *axial* (or *reflectional*) *symmetry* (i.e., a reflection about an axis). The *order* of an automorphism α is the smallest positive integer k such that α^k equals the identity I. A group-theoretic characterization of geometric automorphism group was given by Eades and Lin [6] as follows:

- A group of order 2 generated by an *axial automorphism*;
- A *cyclic group* of order k generated by a *rotational automorphism*;
- A *dihedral group* of order $2k$ generated by a rotational automorphism of order k and an axial automorphism. In this case there are k axial symmetries.

In two dimensions, the problem of determining whether a given graph can be drawn symmetrically is NP-complete in general [18]. Exact algorithms are devised based on Branch and Cut approach by Buchheim and Junger [3] and a group-theoretic approach by Abelson et al. [1]. Linear-time algorithms are available for trees and outerplanar graphs by Manning and Atallah [19, 20] and for series-parallel digraphs by Hong et al. [14]. Linear-time algorithms are presented for maximally symmetric drawings of triconnected planar graphs by Hong et al. [15] and for biconnected, oneconnected, and disconnected planar graphs by Hong and Eades [10, 12, 13]. Hong and Nagamochi presented a linear-time algorithm for constructing a *symmetric convex* drawings of internally trconnected planar graphs [16]. For a survey on symmetric drawings of graphs in two dimensions, see [5].

In three dimensions, the problem of determining whether a graph can be drawn symmetrically in three dimensions is *NP-hard* in general [8]. A group-theoretic characterization of symmetric drawing in n-dimensions and exact algorithms based on a group-theoretic approach are given by Abelson et al. [1]. Linear-time algorithms are available for trees by Hong and Eades [9], series-parallel digraphs by Hong et al. [11], and biconnected and oneconnected planar graphs [8].

In this article, we review a linear-time algorithm for constructing maximally symmetric straight-line drawings of *triconnected planar graphs* by Hong, McKay, and Eades [15]. The following theorem summarizes their main results.

Theorem 1 *There is a linear-time algorithm that constructs straight-line drawings of maximally symmetric planar drawings of triconnected planar graphs.*

Computing a Planar Automorphism Group of Maximum Size

We first review the first step of the algorithm, i.e., symmetry finding step for triconnected planar graphs [15]. A geometric automorphism group A of a graph G is a *planar automorphism group*, if there is a *planar* drawing of the graph that displays every element of A.

Suppose that A is a group acting on a set X. The *stabilizer* of $x \in X$, denoted by $stab_A(x)$, is $\{g \in A \mid g(x) = x\}$, and the *orbit* of x, denoted by $orbit_A(x)$, is $\{g(x) \mid g \in A\}$. We say that $g \in A$ *fixes* $x \in X$ if $g(x) = x$; if g fixes x for every $g \in A$, then A fixes x. If $X' \subseteq X$ and $\phi(x') \in X'$ for all $x' \in X'$, then g *fixes* X'. Automorphisms $g_1, g_2, \ldots, g_k$ are called *generators* of $\langle g_1, g_2, \ldots, g_k \rangle$; the group consists of all permutations formed from products of elements of $\{g_1, g_2, \ldots, g_k\}$.

Hong et al. [15] characterize planar automorphisms as below.

Lemma 1 *Let G be a triconnected planar graph. An automorphism of G is a planar automorphism if and only if it fixes a face of G.*

To find the best plane embedding to compute a planar automorphism group with a maximum size, the algorithm uses the Stabilizer-Orbit theorem in group theory [2].

Theorem 2 (Stabilizer-Orbit theorem) *Suppose that A is a group acting on a set X and let $x \in X$. Then $|A| = |orbit_A(x)| \times |stab_A(x)|$.*

The overall algorithm computing a maximum size planar automorphism group of a triconnected planar graph can be described as follows;

Algorithm `Compute_Max_PAG`

1. Find a plane embedding which has a maximum size planar automorphism group.
2. Perform "star triangulation" for the given embedding.
3. Compute the *generators* of the planar automorphism group of the new embedding.

The first step of `Compute_Max_PAG` uses two applications of an algorithm of Fontet [7], which computes the orbits on vertices of the (full) automorphism group of a triconnected planar graph in linear time.

Theorem 3 *Fontet's algorithm [7] can be used to find a plane embedding of a triconnected graph G such that the corresponding planar automorphism group is maximized in linear time.*

Proof Based on Lemma 1, we take a dual graph of G^* of G and compute the orbits of G^* using Fontet's algorithm [7]. Choose an orbit O of minimum size; the stabilizer O has the maximum size, by Theorem 2. Taking a face $f \in O$ as the outer face of the plane embedding of G, we have an embedding that displays the maximum number of symmetries.

Once the outer face and thus the plane embedding is chosen, the second step of `Compute_Max_PAG` performs *star triangulation*, i.e., triangulate each internal face f by inserting a new vertex v in the face and joining v to each vertex of f. Clearly, this step takes linear time and simplifies the drawing algorithm.

The final step of `Compute_Max_PAG` is to compute the planar automorphism group for star-triangulated plane graph. Since an explicit representation of the planar automorphism group may take more than linear space, for a more compact representation, an algorithm for computing *minimal generators* was devised. For details on a linear-time algorithm for computing generators of a planar automorphism group, see [15].

Symmetric Graph Drawing, Fig. 1 Example of (**a**) a wedge and (**b**) merging step

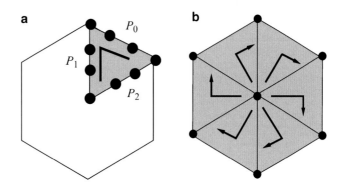

Overview of the Drawing Algorithm

We now review a linear-time drawing algorithm for constructing a symmetric drawing of a triconnected planar graph that achieves that maximum with straight-line edges. The main characteristic of symmetric drawings is the repetition of congruent drawings of isomorphic subgraphs. To exploit this property, the drawing algorithm uses a divide and conquer approach: (i) divide the graph into isomorphic subgraphs; (ii) compute a drawing for a subgraph; and (iii) merge multiple copies of drawings of subgraphs to construct a symmetric drawing of the whole graph. Overall, each step of the drawing algorithm runs in linear time.

The input of the drawing algorithm is a triconnected planar graph with fixed plane embedding and a specified outer face, which maximize the number of symmetries. The symmetric drawing algorithm takes a different approach for each type of planar automorphism group: i.e., *cyclic* case, *one axial* case, and *dihedral* case.

The Cyclic Case

Here we describe how to display k rotational symmetries. Note that after star triangulation, there is a *central* vertex c, which is fixed by the planar automorphism group for $k \geq 3$. If $k = 2$, there exits either a central vertex or a central edge. If there is a central edge, then we preprocess the graph by inserting a dummy central vertex c into the central edge with two dummy edges.

The rotational symmetric drawing algorithm consists of three steps:

Algorithm **Cyclic**

1. Find_Wedge_Cyclic.
2. Draw_Wedge_Cyclic.
3. Merge_Wedges_Cyclic.

The first step is to find a subgraph *wedge W*, which takes linear time:

Algorithm Find_Wedge_Cyclic
1. Find the *central* vertex c.
2. Find a shortest path P_1, from c to a vertex v_1 on the outer face, using breadth first search.
3. Find the path P_2 which is a mapping of P_1 under a minimal generator of the rotation.
4. Find the wedge W (see Fig. 1a), an induced subgraph of G enclosed by the cycle formed from P_1, P_2 and a path P_0 along the outer face from v_1 to v_2.

The second step, Draw_Wedge_Cyclic, constructs a drawing D of the wedge W using Algorithm CYN, the linear-time convex drawing algorithm by Chiba et al. [4], such that P_1, P_2, and P_0 are drawn as straight lines. The input to Algorithm CYN is an internally triconnected plane graph G with given outer face S and a straight-line drawing S^* of S as a *weakly convex polygon*, i.e., not every vertex of the outer face needs to be at an apex (i.e., the interior angle is less than π) of the polygon. Algorithm CYN chooses a vertex v and deletes it from G together with incident edges and divides the resulting graph $G' = G - v$ into the biconnected components $B_1, B_2, \ldots, B_p$, $p \geq 1$. It defines a convex polygon S_i^* of the outer facial cycle S_i of

S

Symmetric Graph Drawing, Fig. 2 Example of (**a**) a fixed string of diamonds and (**b**) ω_ℓ

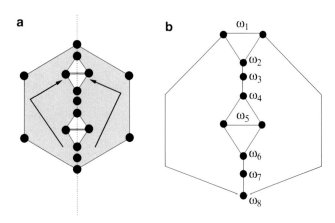

each B_i and recursively applies the algorithm to draw B_i with S_i^* as outer boundary. For details, see [4].

The last step, `Merge_Wedges_Cyclic`, constructs a drawing of the whole graph G by replicating the drawing D of W, k times. Note that this merge step relies on the fact that P_1 and P_2 are drawn as straight lines. See Fig. 1b.

It is clear that Algorithm `Cyclic` constructs a straight-line drawing of a triconnected plane graph which shows k rotational symmetry in linear time.

One Axial Symmetry

Consider a drawing of a star-triangulated plane graph with one axial symmetry. There are fixed vertices, edges, and/or fixed faces on the axis; we need to characterize the subgraph formed by these.

A *diamond* is either a triangle or the 4-vertex graph. A *string of diamonds* is a graph formed from a path $P = (v_1, v_2, \ldots, v_k)$, $k \geq 2$, by a number (zero or greater) of "splitting" operations, as follows. If $1 \leq i \leq k - 1$, then the edge (v_i, v_{i+1}) may be replaced by a diamond. Alternatively, each of the end edges (v_1, v_2) and (v_{k-1}, v_k) may be replaced by a triangle. Note that a string of diamonds is basically a path consisting of edges and diamonds; each end of the path may be a triangle; see Fig. 2a.

To display a single axial symmetry, we need two steps. First we identify the *fixed string of diamonds*; then use Algorithm `Symmetric_CYN`, a modified version of Algorithm `CYN`. More for-

mally, the algorithm `One_Axial` is described below.

Algorithm `One_Axial`

1. Find a fixed string of diamonds. Suppose that $\omega_1, \omega_2, \ldots, \omega_k$ are the fixed edges and vertices in the fixed string of diamonds, in order from the outer face (ω_1 is on the outer face). For each ℓ, ω_ℓ may be a vertex or an edge (see Fig. 2b).
2. Choose a symmetric convex polygon S^* for the outer face S of G.
3. `Symmetric_CYN(1, $S^*$, $G$, $y_1$)`.

The main ingredient in Algorithm `One_Axial` is Algorithm `Symmetric_CYN`. To modify Algorithm `CYN` to display a single axial symmetry, the following three conditions should be satisfied:

- Choose the first vertex or edge on the fixed string of diamonds ω_1 (see Fig. 3).
- Let $D(B_i)$ be the drawing of B_i and α be the axial symmetry. Then, $D(B_i)$ should be a reflection of $D(B_j)$, where $B_j = \alpha(B_i)$, $i = 1, 2, \ldots, m$ and $m = \lfloor p/2 \rfloor$: To satisfy this condition, define S_j^* to be the reflection of S_i^*, $i = 1, 2, \ldots, m$. Then we apply Algorithm `CYN` for B_i, $i = 1, 2, \ldots, m$ and construct $D(B_j)$ using a reflection of $D(B_i)$.
- If p is odd, then $D(B_{m+1})$ should display axial symmetry: To satisfy this condition, we recursively apply Algorithm `Symmetric_CYN` to B_{m+1}.

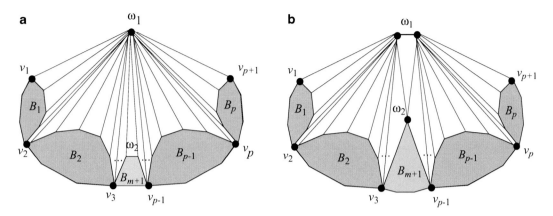

Symmetric Graph Drawing, Fig. 3 Example of a symmetric version of CYN

Note that the position of ω_2 in Fig. 3 can be chosen arbitrarily along the axis of symmetry of S^* within S^*. This means that we can specify the positions of the fixed vertices and middle edges along the axis of symmetry a priori, that is, as input to the algorithm. The Algorithm Symmetric_CYN can be described as below:

Algorithm Symmetric_CYN

input: ℓ: index of vertex or middle edge on the fixed string of diamonds.
input: S^*: a weakly convex polygon of the outer face of S of G.
input: G: a triangulated planar graph.
input: y_ℓ: a position on the axis of symmetry for the fixed vertex or the fixed edge ω_ℓ.

1. Delete ω_ℓ from G together with edges incident to ω_ℓ. Divide the resulting graph $G' = G - \omega_\ell$ into the blocks $B_1, B_2, \ldots, B_p$, $p \geq 1$, ordered anticlockwise around the outer face. Let $m = \lfloor p/2 \rfloor$.
2. Determine a convex polygon S_i^* of the outer facial cycle S_i of each B_i such that B_i with S_i^* satisfy the conditions for convex drawing algorithm CYN and S_{p-i+1}^* is a reflection of S_i^*.
3. For each $i = 1$ to m,
   (a) Construct a drawing $D(B_i)$ of B_i using Algorithm CYN.
   (b) Construct $D(B_{p-i+1})$ as a reflection of $D(B_i)$.
4. If p is odd, then construct a drawing $D(B_{m+1})$ using Symmetric_CYN($\ell + 1, S_{m+1}^*, B_{m+1}, y_{\ell+1}$).
5. Merge the $D(B_i)$ to form a drawing of G, placing ω_ℓ at y_ℓ.

Since Algorithm CYN [4] runs in linear time, clearly Algorithm Symmetric_CYN and Algorithm One_Axial takes linear time.

The Dihedral Case

We now review an algorithm for displaying a dihedral group $< \rho, \alpha >$, where ρ is a rotation of order k and α is an axial automorphism. As with the cyclic case, we assume that there is a central vertex.

The drawing algorithm adopts the same strategy as for the cyclic case: (i) divide the graph into "wedges"; (ii) draw each wedge; and (iii) merge the drawings of wedges to construct a symmetric drawing of the whole graph. However, the dihedral case is more difficult than the cyclic case, because an axial symmetry in the dihedral group can have fixed faces as well as fixed edges; i.e., the boundary of a wedge may be a fixed string of diamonds as in the one axial case. To achieve dihedral symmetry, the axis of symmetry must be the perpendicular bisector of the middle edge of each diamond. This makes the merging operation more difficult.

Consider a drawing of a triconnected planar graph with a dihedral symmetry group of size $2k$.

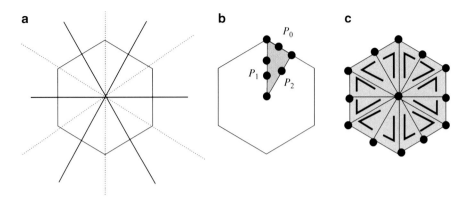

Symmetric Graph Drawing, Fig. 4 Wedge for the dihedral case

There are k axial symmetries, with axes at angles of $\pi i/k$, $0 \le i \le k - 1$, to the x axis, as in Fig. 4a. Roughly speaking, a wedge is the area between two adjacent axes, as in Fig. 4b. Note that in these wedges, the boundaries P_1 and P_2 may be strings of diamonds. These may terminate in a triangle.

As with the cyclic case, Algorithm `Dihedral` has three steps: (i) `Find_Wedge_Dihedral`, (ii) `Draw_Wedge_Dihedral`, and (iii) `Merge_Wedges_Dihedral`.

The first step is to define the "wedge" subgraph by finding two fixed strings of diamonds. Note that one can find the central vertex c and the two fixed strings of diamonds P_1 and P_2 in linear time using the generators of the group.

Algorithm `Find_Wedge_Dihedral`

1. Find the central vertex c.
2. Find a string of diamonds P_1 that is fixed by α, from c to a vertex v or an edge e on the outer face.
3. Traverse the outer face, clockwise from v (or e) to the vertex v' or edge e' that is fixed by $\rho^{-1}\alpha\rho$. Let P_0 denote the path so traversed.
4. Find the string of diamonds P_2 for $\rho^{-1}\alpha\rho$, from c to v' (or e').
5. Define the wedge W to be the subgraph enclosed by P_0, P_1, and P_2, including the vertices and edges of P_0, P_1, and P_2.

The second step, `Draw_Wedge_Dihedral`, constructs a drawing of the wedge, which is the most complicated step of the drawing algorithm.

This step must ensure that the middle edge of each diamond on the boundary is orthogonal to the axis of reflection.

Roughly speaking, the algorithm `Draw_Wedge_Dihedral` runs as follows: (i) Find all *special diamonds* of P_1 and P_2 that share fixed vertices or fixed edges, and draw them first using algorithm `Draw_Special_Diamonds`; (ii) choose the positions of all the fixed vertices of P_1 and P_2 that have not been drawn so far; (iii) subdivide the wedge in various ways to form "subwedges"; (iv) draw each of these subwedges using Algorithms `CYN` and `Symmetric_CYN` accordingly. For details, see [15].

The final step, Algorithm `Merge_Wedges_Dihedral` simply constructs a drawing for the whole graph by merging the drawing D of the wedge W. Clearly each step of Algorithm `Dihedral` takes linear time.

Cross-References

▶ Convex Graph Drawing

Recommended Reading

1. Abelson D, Hong S, Taylor DE (2007) Geometric automorphism groups of graphs. Discret Appl Math 155(17):2211–2226. Elsevier
2. Armstrong MA (1988) Groups and symmetry. Springer, New York

3. Buchheim C, Junger M (2003) Detecting symmetries by branch and cut. Math Progr (Ser B) 98:369–384

4. Chiba N, Yamanouchi T, Nishizeki T (1984) Linear algorithms for convex drawings of planar graphs. In: Adrian Bondy J, Murty USR (eds) Progress in graph theory. Academic, Toronto/Orlando, pp 153–173

5. Eades P, Hong S (2013) Detection and display of symmetries. In: Tamassia R (ed) Handbook of graph drawing and visualisation. Chapman and Hall/CRC

6. Eades P, Lin X (2000) Spring algorithms and symmetry. Theor Comput Sci 240(2):379–405

7. Fontet M (1976) Linear algorithms for testing isomorphism of planar graphs. In: Proceedings of third colloquium on automata, languages and programming, Edinburgh, pp 411–423

8. Hong S (2002) Drawing graphs symmetrically in three dimensions. In: Proceeding of graph drawing 2001, Vienna. Lecture notes in computer science, vol 2265. Springer, pp 189–204

9. Hong S, Eades P (2003) Drawing trees symmetrically in three dimensions. Algorithmica 36(2):153–178. Springer

10. Hong S, Eades P (2003) Symmetric layout of disconnected graphs. In: Algorithms and computation (Proceedings of ISAAC 2003, Kyoto). Lecture notes in computer science, vol 2906. Springer, Berlin/New York, pp 405–414

11. Hong S, Eades P (2004) Linkless symmetric drawings of series parallel digraphs. Comput Geom Theory Appl 29(3):191–222. Elsevier

12. Hong S, Eades P (2005) Drawing planar graphs symmetrically II: biconnected graphs. Algorithmica 42(2):159–197. Springer

13. Hong S, Eades P (2006) Drawing planar graphs symmetrically III: oneconnected graphs. Algorithmica 44(1):67–100. Springer

14. Hong S, Eades P, Lee S (2000) Drawing series parallel digraphs symmetrically. Comput Geom Theory Appl 17(3–4):165–188

15. Hong S, McKay B, Eades P (2006) A linear time algorithm for constructing maximally symmetric straight-line drawings of triconnected planar graphs. Discret Comput Geom 36(2):283–311. Springer

16. Hong S, Nagamochi H (2010) Linear time algorithm for symmetric convex drawings of planar graphs. Algorithmica 58(2):433–460. Springer

17. Hopcroft JE, Wong JK (1974) Linear time algorithm for isomorphism of planar graphs. In: Proceedings of ACM symposium on theory of computing, Seattle, pp 172–184

18. Lubiw A (1981) Some NP-complete problems similar to graph isomorphism. SIAM J Comput 10(1):11–21

19. Manning J, Atallah MJ (1988) Fast detection and display of symmetry in trees. Congr Numer 64:159–169

20. Manning J, Atallah MJ (1992) Fast detection and display of symmetry in outerplanar graphs. Discret Appl Math 39:13–35

Synchronizers, Spanners

Michael Elkin
Department of Computer Science, Ben-Gurion University, Beer-Sheva, Israel

Keywords

Low-stretch spanning subgraphs; Network synchronization

Years and Authors of Summarized Original Work

1985; Awerbuch

Problem Definition

Consider a communication network, modeled by an n-vertex undirected unweighted graph $G = (V, E)$, for some positive integer n. Each vertex of G hosts a processor of unlimited computational power; the vertices have unique identity numbers, and they communicate via the edges of G by sending messages of size $O(\log n)$ each.

In the *synchronous* setting, the communication occurs in discrete *rounds*, and a message sent in the beginning of a round R arrives at its destination before the round R ends. In the *asynchronous* setting, each vertex maintains its own clock, and clocks of distinct vertices may disagree. It is assumed that each message sent (in the asynchronous setting) arrives at its destination within a certain time τ after it was sent, but the value of τ is not known to the processors.

It is generally much easier to devise algorithms that apply to the synchronous setting (henceforth, synchronous algorithms) rather than to the asynchronous one (henceforth, asynchronous algorithms). In [1] Awerbuch initiated the study of simulation techniques that translate synchronous algorithms to asynchronous ones. These simulation techniques are called *synchronizers*.

To devise the first synchronizers, Awerbuch [1] constructed a certain graph partition which is of its own interest. In particular, Peleg and Schäffer noticed [8] that this graph partition induces a subgraph with certain interesting properties. They called this subgraph a *graph spanner*. Formally, for a positive integer parameter k, a k-spanner of a graph $G = (V, E)$ is a subgraph $G' = (V, H)$, $H \subseteq E$, such that for every edge $e = (v, u) \in E$, the distance between the vertices v and u in H, $\text{dist}'_G(v, u)$, is at most k.

Key Results

Awerbuch devised three basic synchronizers, called α, β, and γ. The synchronizer α is the simplest one; using it results in only a constant overhead in time, but in a very significant overhead in communication. Specifically, the latter overhead is linear in the number of edges of the underlying network. Unlike the synchronizer α, the synchronizer β requires a somewhat costly initialization stage. In addition, using it results in a significant time overhead (linear in the number of vertices n), but it is more communication efficient than α. Specifically, its communication overhead is linear in n.

Finally, the synchronizer γ represents a trade-off between the synchronizers α and β. Specifically, this synchronizer is parametrized by a positive integer parameter k. When k is small, then the synchronizer behaves similarly to the synchronizer α, and when k is large, it behaves similarly to the synchronizer β. A particularly important choice of k is $k = \log n$. At this point on the trade-off curve, the synchronizer γ has a logarithmic in n time overhead and a linear in n communication overhead. The synchronizer γ has, however, a quite costly initialization stage.

The main result of [1] concerning spanners is that for every $k = 1, 2, \ldots$, and every n-vertex unweighted undirected graph $G = (V, E)$, there exists an $O(k)$-spanner with $O(n^{1+1/k})$ edges. (This result was explicated by Peleg and Schäffer [8].)

Applications

Synchronizers are extensively used for constructing asynchronous algorithms. The first applications of synchronizers are constructing the *breadth-first-search tree* and computing the *maximum flow*. These applications were presented and analyzed by Awerbuch in [1]. Later synchronizers were used for maximum matching [10], for computing shortest paths [7], and for other problems.

Graph spanners were found useful for a variety of applications in distributed computing. In particular, some constructions of synchronizers employ graph spanners [1, 9]. In addition, spanners were used for routing [4] and for computing almost shortest paths in graphs [5].

Open Problems

Synchronizers with improved properties were devised by Awerbuch and Peleg [3] and Awerbuch et al. [2]. Both these synchronizers have polylogarithmic time and communication overheads. However, the synchronizers of Awerbuch and Peleg [3] require a large initialization time. (The latter is at least linear in n.) On the other hand, the synchronizers of [2] are randomized. A major open problem is to obtain *deterministic* synchronizers with polylogarithmic time and communication overheads and sublinear in n initialization time. In addition, the degrees of the logarithm in the polylogarithmic time and communication overheads in synchronizers of [2, 3] are quite large. Another important open problem is to construct synchronizers with improved parameters.

In the area of spanners, spanners that distort large distances to a significantly smaller extent than they distort small distances were constructed by Elkin and Peleg in [6]. These spanners fall short from achieving a *purely additive distortion*. Constructing spanners with a purely additive distortion is a major open problem.

Cross-References

▶ Sparse Graph Spanners

Recommended Reading

1. Awerbuch B (1985) Complexity of network synchronization. J ACM 4:804–823
2. Awerbuch B, Patt-Shamir B, Peleg D, Saks ME (1992) Adapting to asynchronous dynamic networks. In: Proceedings of the 24th annual ACM symposium on theory of computing, Victoria, 4–6 May 1992, pp 557–570
3. Awerbuch B, Peleg D (1990) Network synchronization with polylogarithmic overhead. In: Proceedings of the 31st IEEE symposium on foundations of computer science, Sankt Louis, 22–24 Oct 1990, pp 514–522
4. Awerbuch B, Peleg D (1992) Routing with polynomial communication-space tradeoff. SIAM J Discret Math 5:151–162
5. Elkin M (2001) Computing almost shortest paths. In: Proceedings of the 20th ACM symposium on principles of distributed computing, Newport, 26–29 Aug 2001, pp 53–62
6. Elkin M, Peleg D (2001) Spanner constructions for general graphs. In: Proceedings of the 33th ACM symposium on theory of computing, Heraklion, 6–8 July 2001, pp 173–182
7. Lakshmanan KB, Thulasiraman K, Comeau MA (1989) An efficient distributed protocol for finding shortest paths in networks with negative cycles. IEEE Trans Softw Eng 15:639–644
8. Peleg D, Schäffer A (1989) Graph spanners. J Graph Theory 13:99–116
9. Peleg D, Ullman JD (1989) An optimal synchronizer for the hypercube. SIAM J Comput 18:740–747
10. Schieber B, Moran S (1986) Slowing sequential algorithms for obtaining fast distributed and parallel algorithms: maximum matchings. In: Proceedings of 5th ACM symposium on principles of distributed computing, Calgary, 11–13 Aug 1986, pp 282–292

T

Table Compression

Raffaele Giancarlo[1] and Adam L. Buchsbaum[2]
[1]Department of Mathematics and Applications, University of Palermo, Palermo, Italy
[2]Madison, NJ, USA

Keywords

Compression and transmission of tables; Compression of multidimensional data; Compressive estimates of entropy; Storage

Years and Authors of Summarized Original Work

2003; Buchsbaum, Fowler, Giancarlo

Problem Definition

Table compression was introduced by Buchsbaum et al. [3] as a unique application of compression, based on several distinguishing characteristics. Tables are collections of fixed-length records and can grow to be terabytes in size. They are often generated by information systems and kept in data warehouses to facilitate ongoing operations. These data warehouses will typically manage many terabytes of data online, with significant capital and operational costs. In addition, the tables must be transmitted to different parts of an organization, incurring additional costs for transmission. Typical examples are tables of transaction activity, like phone calls and credit card usage, which are stored once but then shipped repeatedly to different parts of an organization: for fraud detection, billing, operations support, etc. The goals of table compression are to be fast, online, and effective: eventual compression ratios of 100:1 or better are desirable. Reductions in required storage and network bandwidth are obvious benefits.

Tables are different than general databases [3]. Tables are written once and read many times, while databases are subject to dynamic updates. Fields in table records are fixed in length, and records tend to be homogeneous; database records often contain intermixed fixed- and variable-length fields. Finally, the goals of compression differ. Database compression stresses index preservation, the ability to retrieve an arbitrary record, under compression [7]. Tables are typically not indexed at the level of individual records; rather, they are scanned in toto by downstream applications.

Consider each record in a table to be a row in a matrix. A naive method of table compression is to compress the string derived from scanning the table in row-major order. Buchsbaum et al. [3] observe experimentally that partitioning the table into contiguous intervals of columns and compressing each interval separately in this fashion can achieve significant compression improvement. The partition is generated by a one-time, off-line training procedure, and the resulting compression strategy is applied online

© Springer Science+Business Media New York 2016
M.-Y. Kao (ed.), *Encyclopedia of Algorithms*,
DOI 10.1007/978-1-4939-2864-4

to the table. In their application, tables are generated continuously, so off-line training time can be ignored. They also observe heuristically that certain rearrangements of the columns prior to partitioning further improve compression by grouping dependent columns more closely. For example, in a table of addresses and phone numbers, the area code can often be predicted by the zip code when both are defined geographically. In information-theoretic terms, these dependencies are *contexts*, which can be used to predict parts of a table. Analogously to strings, where knowledge of context facilitates succinct codings of symbols, the existence of contexts in tables implies, in principle, the existence of a more succinct representation of the table.

Three main avenues of research have followed, one based on the notion of combinatorial dependency [3, 4], another on the notion of column dependency [17, 18], and the third on the notion of motifs and templates [1]. The first formalizes dependencies analogously to the joint entropy of random variables, while the second does so analogously to conditional entropy [8]. The third finds inspiration in classic paradigms of data compression such as textual substitution [19, 20]. These approaches to table compression have deep connections to universal similarity metrics [12], based on Kolmogorov complexity and compression, and their later uses in classification [6]. The first two approaches are instances of a new emerging paradigm for data compression, referred to as *boosting* [9], where data are reorganized to improve the performance of a given compressor. A software platform to facilitate the investigation of such invertible data transformations is described by Vo [16]

Notations

Let T be a table of $n = |T|$ columns and m rows. Let $T[i]$ denote the ith column of T. Given two tables T_1 and T_2, let $T_1 T_2$ be the table formed by their juxtaposition. That is, $T = T_1 T_2$ is defined so that $T[i] = T_1[i]$ for $1 \leq i \leq |T_1|$ and $T[i] = T_2[i - |T_1|]$ for $|T_1| < i \leq |T_1| + |T_2|$. We use the shorthand $T[i, j]$ to represent the projection $T[i] \cdots T[j]$ for any $j \geq i$. Also, given a sequence

P of column indices, we denote by $T[P]$ the table obtained from T by projecting the columns with indices in P.

Combinatorial Dependency and Joint Entropy of Random Variables

Fix a compressor $\mathcal{C}$: e.g., gzip, based on LZ77 [19]; compress, based on LZ78 [20]; or bzip, based on Burrows-Wheeler [5]. Let $H_{\mathcal{C}}(T)$ be the size of the result of compressing table T as a string in row-major order using $\mathcal{C}$. Let $H_{\mathcal{C}}(T_1, T_2) = H_{\mathcal{C}}(T_1, T_2)$. $H_{\mathcal{C}}(\cdot)$ is thus a cost function defined on the ordered power set of columns. Two tables T_1 and T_2, which might be projections of columns from a common table T, are *combinatorially dependent* if $H_{\mathcal{C}}(T_1, T_2) < H_{\mathcal{C}}(T_1) + H_{\mathcal{C}}(T_2)$ – if compressing them together is better than compressing them separately – and *combinatorially independent* otherwise. Buchsbaum et al. [3] show that combinatorial dependency is a compressive estimate of statistical dependency when formalized by the joint entropy of two random variables, i.e., the statistical relatedness of two objects is measured by the gain realized by compressing them together rather than separately. Indeed, combinatorial dependency becomes statistical dependency when $H_{\mathcal{C}}$ is replaced by the joint entropy function [8]. Analogous notions starting from Kolmogorov complexity are derived by Li et al. [12] and used for classification and clustering [6]. Figure 1 exemplifies why rearranging and partitioning columns may improve compression.

9	0	8	2	7	3
9	0	8	3	7	5
9	0	8	5	7	6
9	0	8	2	7	5

Table Compression, Fig. 1 The first three columns of the table, taken in row-major order, form a repetitive string that can be very easily compressed. Therefore, it may be advantageous to compress these columns separately. If the fifth column is swapped with the fourth, we get an even longer repetitive string that, again, can be compressed separately from the other two columns

Problem 1 Find a partition $\mathcal{P}$ of T into sets of contiguous columns that minimizes $\sum_{Y \in \mathcal{P}} H_C(Y)$ over all such partitions.

Problem 2 Find a partition $\mathcal{P}$ of T that minimizes $\sum_{Y \in \mathcal{P}} H_C(Y)$ over all partitions.

The difference between Problems 1 and 2 is that the latter does not require the parts of $\mathcal{P}$ to be sets of contiguous columns.

Column Dependency and Conditional Entropy of Random Variables

Definition 1 For any table T, a *dependency relation* is a pair (P, c) in which P is a sequence of distinct column indices (possibly empty) and $c \notin P$ is another column index. If the length of P is less than or equal to k, then (P, c) is called a k-relation. P is the *predictor sequence* and c is the *predictee*.

Definition 2 Given a dependency relation (P, c), the *dependency transform $dt_P(c)$* of c is formed by permuting column $T[c]$ based on the permutation induced by a stable sort of the rows of P.

Definition 3 A collection D of dependency relations for table T is said to be a k-transform if and only if (a) each column of T appears exactly once as a predictee in some dependency relation (P, c), (b) the dependency hypergraph $G(D)$ is acyclic, and (c) each dependency relation (P, c) is a k-relation.

Let $\omega(P, c)$ be the cost of the dependency relation (P, c), and let $\delta(m)$ be an upper bound on the cost of computing $\omega(P, c)$. Intuitively, $\omega(P, c)$ gives an estimate of how well a rearrangement of column c will compress, using the rows of P as contexts for its symbols. We will provide an example after the formal definitions.

Problem 3 Find a k-transform D of minimum cost $\omega(D) = \sum_{(P,c) \in D} \omega(P, c)$.

Definition 1 extends to columns the notion of context that is well known for strings. Definition 3 defines a microtransformation that reorganizes the column symbols by grouping together those that have similar contexts. The context of a column symbol is given by the corresponding row in $T[P]$. The fundamental ideas here are the same as in the Burrows and Wheeler transform [5]. Finally, Problem 3 asks for an optimal strategy to reorganize the data prior to compression. The cost function ω provides an estimate of how well c can be compressed using the knowledge of $T[P]$.

Vo and Vo [18] connect these ideas to the conditional entropy of random variables. Let S be a sequence, $\mathcal{A}(S)$ its distinct elements, and f_a the frequency of each element a. The *zeroth-order empirical entropy* of S [15] is

$$H_0(S) = -\frac{1}{|S|} \sum_{a \in \mathcal{A}(S)} f_a \lg \frac{f_a}{|S|},$$

and the *modified zeroth-order empirical entropy* [15] is

$$H_0^*(S) = \begin{cases} 0 & \text{if } |S| = 0, \\ (1 + \lg |S|)/|S| & \text{if } |S| \neq 0 \text{ and } H_0(S) = 0, . \\ H_0(S) & \text{otherwise.} \end{cases}$$

For a dependency relation (P, c) with nonempty P, the *modified conditional empirical entropy* of c given P is then defined as

$$H_P^*(c) = \frac{1}{m} \sum_{\rho \in \mathcal{A}(T[P])} |\rho_c| H_0^*(\rho_c),$$

where ρ_c is the string formed by catenating the symbols in c corresponding to positions of ρ in $T[P]$ [15]. A possible choice of $\omega(P, c)$ is given by $H_P^*(c)$. Vo and Vo also develop another notion of entropy, called *run length entropy*, to approximate more effectively the compressibility of low-entropy columns and define another cost function ω accordingly.

Key Results

Combinatorial Dependency
Problem 1 admits a polynomial-time algorithm, based on dynamic programming. Using the definition of combinatorial dependency, one can show:

Theorem 1 ([3]) *Let $E[i]$ be the cost of an optimal, contiguous partition of $T[1, i]$. $E[n]$ is thus the cost of a solution to Problem 1. Define $E[0] = 0$; then, for $1 \leq i \leq n$,*

$$E[i] = \min_{0 \leq j < i} E[j] + H_C(T_{j+1}, \ldots, T_i). \quad (1)$$

The actual partition with cost $E[n]$ can be maintained by standard backtracking.

The only known algorithmic solution to Problem 2 is the trivial one based on enumerating all possible feasible solutions to choose an optimal one. Some efficient heuristics based on asymmetric TSP, however, have been devised and tested experimentally [4]. Define a weighted, complete, directed graph, $G(T)$, with a vertex T_i for each column $T[i] \in T$; the *weight* of edge $\{T_i, T_j\}$ is $w(T_i, T_j) = \min(H_C(T_i, T_j), H_C(T_i) + H_C(T_j))$. One then generates a set of tours of various weights by iteratively applying standard optimizations (e.g., 3-opt, 4-opt). Each tour induces an ordering of the columns, which are then optimally partitioned using the dynamic program (1).

Buchsbaum et al. [4] also provide a general framework for studying the computational complexity of several variations of table compression problems based on notions analogous to combinatorial dependence, and they give some initial MAX-SNP-hardness results. Particularly relevant

is the set of abstract problems in which one is required to find an optimal arrangement of a set of strings to be compressed, which establishes a nontrivial connection between table compression and the classical shortest common superstring problem [2]. Giancarlo et al. [11] connect table compression to the Burrows and Wheeler transform [5] by deriving the latter as a solution to an analog of Problem 2.

Column Dependency

Theorem 2 ([17, 18]) *For $k \geq 2$, Problem 3 is NP-hard.*

Theorem 3 ([17, 18]) *An optimum 1-transform for a table T can be found in $O(n^2 \delta(m))$ time.*

Theorem 4 ([17, 18]) *A 2-transform can be computed in $O(n^2 \delta(m))$ time.*

Theorem 5 ([18]) *For any dependency relation (P, c) and some constant $\epsilon, |\mathcal{C}(dt_P(c))| \leq 5m H_p^*(c) + \epsilon$.*

Motifs
Apostolico et al. [1] propose improved versions of Table Compression based on Motifs, i.e., regular expressions characterizing a set of templates based on which the rows of a table are compressed by textual substitution. They also discuss applications of the technique in Computational Biology.

Applications

Storage and transmission of alphanumeric tables. Moreover, the Citing Articles of the papers in [1, 3,4,17,18] in Google Scholar provide a full range of applications and related work.

Open Problems

All the techniques discussed use the general paradigms of context-dependent data rearrangement for compression boosting. It remains open to apply these paradigms to other domains, e.g., XML data [13, 14], where high-level structures can be exploited, and to domains where pertinent structures are not known a priori.

Experimental Results

Buchsbaum et al. [3] showed that optimal partitioning alone (no column rearrangement) yielded about 55 % better compression compared to gzip on telephone usage data, with small training sets. Buchsbaum et al. [4] experimentally supported the hypothesis that good TSP heuristics can effectively reorder the columns, yielding additional improvements of 5–20 % relative to partitioning alone. They extended the data sets used to include other tables from the telecom domain as well as biological data. Vo and Vo [17,18] showed further 10–35 % improvement over these combinatorial dependency methods on the same data sets.

Data Sets

Some of the data sets used for experimentation are public [4].

URL to Code

The pzip package, based on combinatorial dependency, is available at http://www.research.att.com/~gsf/pzip/pzip.html. The Vcodex package, related to invertible transforms, is available at http://www.research.att.com/~gsf/download/ref/vcodex/vcodex.html. Although for the time being Vcodex does not include procedures to compress tabular data, it is a useful toolkit for their development.

Cross-References

▶ Binary Decision Graph
▶ Burrows-Wheeler Transform
▶ Compressed Tree Representations
▶ Compressing and Indexing Structured Text
▶ Lempel-Ziv Compression

Recommended Reading

1. Apostolico A, Cunian F, Kaul V (2008) Table compression by record intersection. In: Proceedings of the IEEE data compression conference (DCC), Snowbird, pp 13–22
2. Blum A, Li M, Tromp J, Yannakakis M (1994) Linear approximation of shortest superstrings. J ACM 41:630–647
3. Buchsbaum AL, Caldwell DF, Church KW, Fowler GS, Muthukrishnan S (2000) Engineering the compression of massive tables: an experimental approach. In: Proceedings of the 11th ACM-SIAM symposium on discrete algorithms, San Francisco, pp 175–184
4. Buchsbaum AL, Fowler GS, Giancarlo R (2003) Improving table compression with combinatorial optimization. J ACM 50:825–851
5. Burrows M, Wheeler D (1994) A block sorting lossless data compression algorithm. Technical report 124, Digital Equipment Corporation
6. Cilibrasi R, Vitanyi PMB (2005) Clustering by compression. IEEE Trans Inf Theory 51:1523–1545
7. Cormack G (1985) Data compression in a data base system. Commun ACM 28:1336–1350
8. Cover TM, Thomas JA (1990) Elements of information theory. Wiley Interscience, New York
9. Ferragina P, Giancarlo R, Manzini G, Sciortino M (2005) Boosting textual compression in optimal linear time. J ACM 52:688–713
10. Ferragina P, Luccio F, Manzini G, Muthukrishnan S (2005) Structuring labeled trees for optimal succinctness, and beyond. In: Proceedings of the 45th annual IEEE symposium on foundations of computer science, Pittsburgh, pp 198–207
11. Giancarlo R, Sciortino M, Restivo A (2007) From first principles to the Burrows and Wheeler transform and beyond, via combinatorial optimization. Theor Comput Sci 387:236–248
12. Li M, Chen X, Li X, Ma B, Vitanyi PMB (2004) The similarity metric. IEEE Trans Inf Theory 50:3250–3264
13. Liefke H, Suciu D (2000) XMILL: an efficient compressor for XML data. In: Proceedings of the 2000 ACM SIGMOD international conference on management of data, Dallas. ACM, New York, pp 153–164
14. Lifshits Y, Mozes S, Weimann O, Ziv-Ukelson M (2009) Speeding up HMM decoding and training by exploiting sequence repetitions. Algorithmica 54:379–399
15. Manzini G (2001) An analysis of the Burrows-Wheeler transform. J ACM 48:407–430
16. Vo K-P (2006) Compression as data transformation. In: DCC: data compression conference, Snowbird. IEEE Computer Society TCC, Washington DC, p 403
17. Vo BD, Vo K-P (2004) Using column dependency to compress tables. In: DCC: data compression conference, Snowbird. IEEE Computer Society TCC, Washington DC, pp 92–101

T

18. Vo BD, Vo K-P (2007) Compressing table data with column dependency. Theor Comput Sci 387:273–283
19. Ziv J, Lempel A (1977) A universal algorithm for sequential data compression. IEEE Trans Inf Theory 23:337–343
20. Ziv J, Lempel A (1978) Compression of individual sequences via variable length coding. IEEE Trans Inf Theory 24:530–536

Tail Bounds for Occupancy Problems

Paul (Pavlos) Spirakis
Computer Engineering and Informatics,
Research and Academic Computer Technology
Institute, Patras University, Patras, Greece
Computer Science, University of Liverpool,
Liverpool, UK
Computer Technology Institute (CTI), Patras,
Greece

Keywords

Balls and bins

Years and Authors of Summarized Original Work

1995; Kamath, Motwani, Palem, Spirakis

Problem Definition

Consider a *random allocation* of m balls to n bins where each ball is placed in a bin chosen uniformly and independently. The properties of the resulting distribution of balls among bins have been the subject of intensive study in the probability and statistics literature [3, 4]. In computer science, this process arises naturally in randomized algorithms and probabilistic analysis. Of particular interest is the *occupancy problem* where the random variable under consideration is the number of empty bins.

In this entry a series of bounds are presented (reminiscent of the Chernoff bound for binomial distributions) on the tail of the distribution of the

number of empty bins; the tail bounds are successively tighter, but each new bound has a more complex closed form. Such strong bounds do not seem to have appeared in the earlier literature.

Key Results

The following notation in presenting sharp bounds on the tails of distributions. The notation $F \sim G$ will denote that $F = (1 + o(1))G$; further, $F \asymp G$ will denote that $\ln F \sim \ln G$. The proof that $f \asymp g$, is used for the purposes of later claiming that $2^f \asymp 2^g$. These asymptotic equalities will be treated like actual equalities and it will be clear that the results claimed are unaffected by this "approximation".

Consider now the probabilistic experiment of throwing m balls, independently and uniformly, into n bins.

Definition 1 Let Z be the number of empty bins when m balls are placed randomly into n bins, and define $r = m/n$. Define the function $H(m, n, z)$ as the probability that $Z = z$. The expectation of Z is given by

$$\mu = \mathbf{E}[Z] = n \left(1 - \frac{1}{n}\right)^m \sim n \, \mathrm{e}^{-r}.$$

The following three theorems provide the bounds on the tail of the distribution of the random variable Z. The proof of the first bound is based on a martingale argument.

Theorem 1 (Occupancy Bound 1) *For any* $\theta > 0$,

$$P\left[|Z - \mu| \geq \theta \mu\right] \leq 2 \exp\left(-\frac{\theta^2 \mu^2 (n - \frac{1}{2})}{n^2 - \mu^2}\right).$$

Remark that for large r this bound is asymptotically equal to

$$2 \exp\left(-\frac{\theta^2 \, \mathrm{e}^{-2r} n}{1 - \mathrm{e}^{-2r}}\right).$$

The reader may wish to compare this with the following heuristic estimate of the tail probability assuming that the distribution of Z is well approximated by the approximating normal distribution also far out in the tails [3, 4].

$$P\left[|Z - \mu| \geq \theta\mu\right] \leq 2\exp\left(-\frac{\theta^2 e^{-r} n}{2\left(1 - (1 + r)e^{-r}\right)}\right).$$

The next two bounds are in terms of point probabilities rather than tail probabilities (as was the case in the Binomial Bound), but the unimodality of the distribution implies that the two differ by at most a small (linear) factor. These more general bounds on the point probability are essential for the application to the satisfiability problem. The next result is obtained via a generalization of the Binomial Bound to the case of dependent Bernoulli trials.

Theorem 2 (Occupancy Bound 2) *For $\theta > -1$,*

$$H(m, n, (1 + \theta)\mu) \leq \exp\left(-\left((1 + \theta)\ln[1 + \theta] - \theta\right)\mu\right).$$

In particular, for $-1 \leq \theta < 0$,

$$H(m, n, (1 + \theta)\mu) \leq \exp\left(-\frac{\theta^2 \mu}{2}\right).$$

The last result is proved using ideas from large deviations theory (Weiss A (1993) Personal Communication).

Theorem 3 (Occupancy Bound 3) *For $|z - \mu| = \Omega(n)$,*

$$H(m, n, z) \asymp$$

$$\exp\left(\left[-n\left(\int_0^{1-\frac{z}{n}} \ln\left[\frac{k-x}{1-x}\right] dx - r\ln k\right)\right]\right)$$

where k is defined implicitly by the equation $z = n(1 - k(1 - e^{-r/k}))$.

Applications

Random allocations of balls to bins is a basic model that arises naturally in many areas in computer science involving choice between a number of resources, such as communication links in a network of processors, actuator devices in a wireless sensor network, processing units in a multi-processor parallel machine etc. For such situations, randomization can be used to "spread" the load evenly among the resources, an approach particularly useful in a parallel or distributed environment where resource utilization decisions have to be made locally at a large number of sites without reference to the global impact of these decisions. In the process of analyzing the performance of such algorithms, of particular interest is the occupancy problem where the random variable under consideration is the number of empty bins (i.e., machines with no jobs, routes with no load, etc.). The properties of the resulting distribution of balls among bins and the corresponding tails bounds may help in order to analyze the performance of such algorithms.

Cross-References

▶ Approximation Schemes for Bin Packing
▶ Bin Packing

Recommended Reading

1. Kamath A, Motwani R, Spirakis P, Palem K (1995) Tail bounds for occupancy and the satisfiability threshold conjecture. J Random Struct Algorithms 7(1):59–80
2. Janson S (1994) Large deviation inequalities for sums of indicator variables. Technical report No. 34, Department of Mathematics, Uppsala University
3. Johnson NL, Kotz S (1977) Urn models and their applications. Wiley, New York
4. Kolchin VF, Sevastyanov BA, Chistyakov VP (1978) Random allocations. Wiley, New York
5. Motwani R, Raghavan P (1995) Randomized algorithms. Cambridge University Press, New York
6. Shwartz A, Weiss A (1994) Large deviations for performance analysis. Chapman-Hall, Boca Raton

Technology Mapping

Kurt Keutzer[1] and Kaushik Ravindran[2]
[1] Department of Electrical Engineering and
Computer Science, University of California,
Berkeley, CA, USA
[2] National Instruments, Berkeley, CA, USA

Keywords

Library-based technology mapping; Technology-dependent optimization

Years and Authors of Summarized Original Work

1987; Keutzer

Problem Definition

Technology mapping is the problem of implementing a sequential circuit using the gates of a particular technology library. It is an integral component of any automated VLSI circuit design flow. In the prototypical chip design flow, combinational logic gates and sequential memory elements are composed to form sequential circuits. These circuits are subject to various logic optimizations to minimize area, delay, power, and other performance metrics. The resulting optimized circuits still consist of primitive logic functions such as AND and OR gates. The next step is to efficiently realize these circuits in a specific VLSI technology using a library of gates available from the semiconductor vendor. Such a library would typically consist of gates of varying sizes and speeds for primitive logic functions (AND and OR) and more complex functions (exclusive-OR, multiplexer). However, a naïve translation of generic logic elements to gates in the library will fall short of realistic performance goals. The challenge is to construct a mapping that maximally utilizes the gates in the library to implement the logic function of the circuit and achieve some performance goal, for example, minimum area with the critical path delay less

than a target value. This is accomplished by *technology mapping*. For the sake of simplicity, in the following discussion, it is presumed that the sequential memory elements are stripped from the digital circuit and mapped directly into memory elements of the particular technology. Then, only Boolean circuits composed of combinational logic gates remain to be mapped. Further, each remaining Boolean circuit is necessarily a directed acyclic graph (DAG).

The technology mapping problem can be restated in a more general graph-theoretic setting: *find a minimum cost covering of the subject graph (Boolean circuit) by choosing from the collection of pattern graphs (gates) available in a library.* The inputs to the problem are:

(a) *Subject graph:* This is a directed acyclic graph representation of a Boolean circuit expressed using a set of primitive functions (e.g., 2-input NAND gates and inverters). An example subject graph is shown in Fig. 1.

(b) *Library of pattern graphs:* This is a collection of gates available in the technology library. The pattern graphs are also DAGs expressed using the same primitive functions used to construct the subject graph. Additionally, each gate is annotated with a number of values for different cost functions, such as area, delay, and power. An example library and associated cost model is shown in Fig. 2.

A *valid cover* is a network of pattern graphs implementing the function of the subject graph such that (a) every vertex (i.e., gate) of the subject graph is contained in some pattern graph and (b) each input required by a pattern graph is actually an output of some other pattern graph (i.e., the inputs of a gate must exist as outputs of other gates). Technology mapping can then be viewed as an optimization problem to find a valid cover of minimum cost of the subject graph.

Key Results

To be viable in a realistic design flow, an algorithm for minimum cost graph-covering for technology mapping should ideally possess

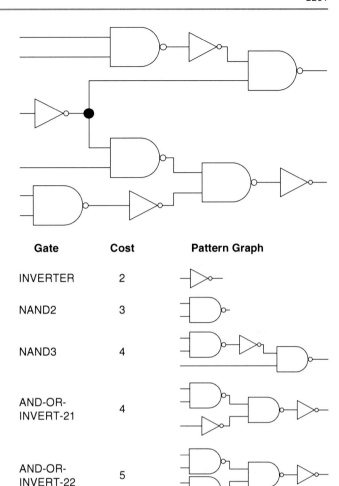

Technology Mapping,
Fig. 1 Subject graph
(DAG) of a Boolean circuit
expressed using NAND2
and INVERTER gates

Technology Mapping,
Fig. 2 Library of pattern
graphs (composed of
NAND2 and INVERTER
gates) and associated costs

Gate	Cost	Pattern Graph
INVERTER	2	
NAND2	3	
NAND3	4	
AND-OR-INVERT-21	4	
AND-OR-INVERT-22	5	

the following characteristics: (a) the algorithm should be easily adaptable to diverse libraries and cost models; if the library is expanded or replaced, the algorithm must be able to utilize the new gates effectively; (b) it should allow detailed cost models to accurately represent the performance of the gates in the library; and (c) it should be fast and robust on large subject graph instances and large libraries. One technique for solving the minimum cost graph-covering problem is to formulate it as a binate-covering problem, which is a specialized integer linear program [10]. However, binate-covering for a DAG is NP-Hard for any set of primitive functions and is typically unwieldy on large circuits. The DAGON algorithm suggested solving the technology mapping problem through DAG-covering and advanced an alternate approach for DAG-

covering based on a *tree-covering* approximation that produced near-optimal solutions for practical circuits and was very fast even for large circuits and large libraries [7].

DAGON was inspired by prevalent techniques for pattern matching employed in the domain of code generation for programming language compilers [1]. The fundamental concept was to partition the subject graph (DAG) into a forest of trees and solve the minimum cost covering problem independently for each tree. The approach was motivated by the existence of efficient dynamic programming algorithms for optimum tree-covering [2]. The three salient components of the DAGON algorithm are (a) subject graph partitioning, (b) pattern matching, and (c) covering.

(a) *Subject graph partitioning*: To apply the tree-covering approximation the subject graph is first partitioned into a forest of trees. One approach is to break the graph at each vertex which has an out-degree greater than 1 (multiple fan-out point). The root of each tree is the primary output of the corresponding subcircuit and the leaves are the primary inputs. Other heuristic partitions of the subject graph that consider duplication of vertices can also be applied to improve the quality of the final cover. Alternate subject graph partitions can also be derived starting from different decompositions of the original Boolean circuit in terms of the primitive functions.

(b) *Pattern matching*: The optimum covering of a tree is determined by generating the complete set of matches for each vertex in the tree (i.e., the set of pattern graphs which are candidates for covering a particular vertex) and then selecting the optimum match from among the candidates. An efficient approach for structural pattern matching is to reduce the tree matching problem to a *string matching* problem [2]. Fast string matching algorithms, such as the Aho-Corasick and the Knuth-Morris-Pratt algorithms, can then be used to find all strings (pattern graphs) which match a given vertex in the subject graph in time proportional to the length of the longest string in the set of pattern graphs. Alternatively, Boolean matching techniques can be used to find matches based on logic functions [5]. Boolean matching is slower than structural string matching, but it can compute matches independent of the actual local decompositions and under different input permutations.

(c) *Covering*: The final step is to generate a valid cover of the subject tree using the pattern graph matches computed at each vertex. Consider the problem of finding a valid cover of minimum area for the subject tree. Every pattern graph in the library has an associated area and the area of a valid cover is the sum of the area of the pattern graphs in the cover. The key property that makes minimum area tree-covering efficient is this:

the minimum area cover of a tree rooted at some vertex v can be computed using only the minimum area covers of vertices below v. It follows that for every pattern graph that matches the tree rooted at vertex v, the area of the minimum cover containing that match equals the sum of the area of the corresponding match at v and the sum of the areas of the optimal covers of the vertices which are inputs to that match. This property enables a dynamic programming algorithm to compute the minimum area cover of the tree rooted at each vertex of the subject tree. The base case is the minimum area cover of a leaf (primary input) of the subject tree. The area of a match at a leaf is set to 0. A recursive formulation of this dynamic programming concept is summarized in the Algorithm *minimum_area_tree_cover* shown below. As an example, the minimum area cover displayed in Fig. 3 is a result of applying this algorithm to the tree partitions of the subject graph from Fig. 1 using the library from Fig. 2.

Given a vertex v in the subject tree, let $M(v)$ denote the set of candidate matches from the library of pattern graphs for the sub-tree rooted at v.

In this algorithm, each vertex in the tree is visited exactly once. Hence, the complexity of the algorithm is proportional to the number of vertices in the subject tree times the maximum number of pattern matches at any vertex. The maximum number of matches is a function of the pattern graph library and is independent of the subject tree size. As a result, the complexity of computing the minimum cost valid cover of a tree is linear in the size of the subject tree, and the memory requirements are also linear in the size of the subject tree. The algorithm computes the optimum cover when the subject graph is a tree. In the general case of the subject graph being a DAG, empirical results have shown that the tree-covering approximation yields industrial-quality results achieving aggressive area and timing requirements on large real circuit design problems [6, 12].

**Technology Mapping,
Fig. 3** Result of a
minimum area
tree-covering of the subject
graph in Fig. 1 using the
library of pattern graphs in
Fig. 2

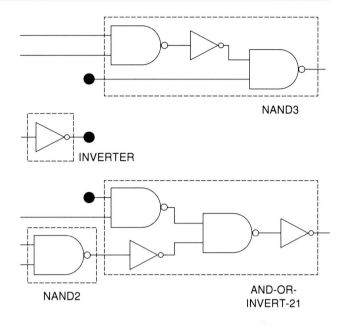

NAND3

INVERTER

NAND2

AND-OR-
INVERT-21

Area of cover = 4 + 2 + 3 + 4 = 13

Applications

Technology mapping is the key link between
technology-independent logic synthesis and
technology-dependent physical design of VLSI
circuits. This motivates the need for efficient and
robust algorithms to implement large Boolean
circuits in a technology library. Early algorithms
for technology mapping were founded on
rule-based local transformations [4]. DAGON
was the first in advancing an algorithmic
foundation in terms of graph transformations
that was practicable in the inner loop of iterative
procedures in the VLSI design flow [7]. From
a theoretical standpoint, the graph-covering
formulation provided a formal description of
the problem and specified optimality criteria
for evaluating solutions. The algorithm was
naturally adaptable to diverse libraries and cost
models, and was relatively easy to implement
and extend. The concept of partitioning the
subject graph into trees and covering the trees
optimally was effective for varied optimization
objectives such as area, delay, and power. The
DAGON approach has been incorporated in

academic (SIS from the University of California
at Berkeley [11]) and industrial (Synopsys™
Design Compiler) tool offerings for logic
synthesis and optimization.

The graph-covering formulation has also
served as a starting point for advancements in
algorithms for technology mapping over the last
decade. Decisions related to logic decomposition
were integrated in the graph-covering algorithm,
which in turn enabled technology independent
logic optimizations in the technology mapping
phase [9, 14]. Similarly, heuristics were proposed
to impose placement constraints and make
technology mapping more aware of the physical
design and layout of the final circuit [8]. To
combat the problem of high power dissipation
in modern sub-micron technologies, the graph
algorithms were enhanced to minimize power
under area and delay constraints [13]. Special-
izations of these graph algorithms for technology
mapping have found successful application in
design flows for Field Programmable Gate Array
(FPGA) technologies [3, 15]. We recommend the
following works for a comprehensive treatment
of algorithms for technology mapping and a

T

```
Algorithm minimum_area_tree_cover (Vertex v) {
// the algorithm minimum_area_tree_cover finds an
optimal cover of the tree rooted at Vertex v
// the algorithm computes best_match(v) and
area_of_best_match(v), which denote the best
pattern graph match at v and the associated area of
the optimal cover of the tree rooted at v, respectively
// check if v is a leaf of the tree
if (v is a leaf) {
area_of_best_match(v) = 0;
best_match(v) = leaf;
return;
}
// compute optimal cover for each input of v
foreach (input of Vertex v) {
minimum_area_tree_cover(input);
}
// each tree rooted at each input of v is now
annotated with its optimal cover
//find the optimal cover of the tree rooted at Vertex v
area_of_best_match(v) = INFINITY;
best_match(v) = NULL;
foreach (Match m in the set of matches M(v)) {
// compute the area of match m at Vertex v
// area_of_match(v,m) denotes the area of the cover
when Match m is selected for v
area_of_match(v,m) = area(m);
foreach input pin vi of match m {
area_of_match (v,m) = area_of_match(v,m) +
area_of_best_match(vi)
}
// update best pattern graph match and associated
area of the optimal cover at Vertex v
if (area_of_match(v,m) < area_of_best_match(v)) {
area_of_best_match(v) = area_of_match(v,m);
best_match(v) = m;
}
}
}
```

survey of new developments and challenges in
the design of modern VLSI circuits: [5, 6, 12].

Open Problems

The enduring problem with DAGON-related
technology mappers is handling non-tree pattern
graphs that arise from modeling circuit elements
such as multiplexors, Exclusive-Ors, or memory-
elements (e.g., flip-flops) with associated logic
(e.g., scan logic). On the other hand, approaches
that do not use the tree-covering formulation
face challenges in easily representing diverse
technology libraries and in matching the subject
graph in a computationally efficient manner.

Recommended Reading

1. Aho A, Sethi R, Ullman J (1986) Compilers: principles, techniques and tools. Addison Wesley, Boston,
   pp 557–584
2. Aho A, Johnson S (1976) Optimal code generation
   for expression trees. J ACM 23(July):488–501
3. Cong J, Ding Y (1994) FlowMap: an optimal
   technology mapping algorithm for delay optimization in lookup-table based FPGA designs. IEEE
   Trans Comput-Aided Des Integr Circuits Syst 13(1):
   1–12
4. Darringer JA, Brand D, Gerbi JV, Joyner WH, Tre-
   villyan LH (1981) LSS: logic synthesis through local
   transformations. IBM J Res Dev 25:272–280
5. De Micheli G (1994) Synthesis and optimization of
   digital circuits, 1st edn. McGraw-Hill, New York,
   pp 504–533
6. Devadas S, Ghosh A, Keutzer K (1994) Logic synthesis. McGraw Hill, New York, pp 185–200
7. Keutzer K (1987) DAGON: technology binding and
   local optimizations by DAG matching. In: Proceedings of the 24th design automation conference, vol 28(1), Miami Beach, pp 341–347, June
   1987
8. Kutzschebauch T, Stok L (2001) Congestion aware
   layout driven logic synthesis. In: Proceedings of the
   IEEE/ACM international conference on computer-
   aided design, Santa Clara, pp 216–223
9. Lehman E, Watanabe Y, Grodstein J, Harkness H
   (1997) Logic decomposition during technology mapping. IEEE Trans Comput-Aided Des Integr Circuits
   Syst 16(8):813–834
10. Rudell R (1989) Logic synthesis for VLSI design.
    Ph.D. thesis, University of California at Berkeley,
    ERL Memo 89/49, April 1989
11. Sentovich EM, Singh KJ, Moon C, Savoj H, Bray-
    ton RK, Sangiovanni-Vincentelli A (1992) Sequential
    circuit design using synthesis and optimization. In:
    Proceedings of the IEEE international conference on
    computer design: VLSI in computers & processors
    (ICCD), Cambridge, pp 328–333, Oct 1992
12. Stok L, Tiwari V (2002) Technology mapping. In:
    Hassoun S, Sasou T (eds) Logic synthesis and ver-
    ification. Kluwer international series in engineer-
    ing and computer science series. Kluwer, Norwell,
    pp 115–139
13. Tiwari V, Ashar P, Malik S (1996) Technology map-
    ping for low power in logic synthesis. Integr VLSI J
    20(3):243–268
14. Clarke EM, McMillan KL, Zhao X, Fujita M, Yang
    J (1993) Spectral transforms for large Boolean func-
    tions with applications to technology mapping. In:
    30th conference on design automation, Dallas, 1993.
    IEEE, pp 54–60
15. Francis R, Rose J, Vranesic Z (1991) Chortle-crf:
    fast technology mapping for lookup table-based FP-
    GAs. In: Proceedings of the 28th ACM/IEEE design
    automation conference, San Francisco, 1991. ACM,
    pp 227–233

Teleportation of Quantum States

Anurag Anshu[1], Vamsi Krishna Devabathini[1], Rahul Jain[2], and Priyanka Mukhopadhyay[1]
[1]Center for Quantum Technologies, National University of Singapore, Singapore, Singapore
[2]Department of Computer Science, Center for Quantum Technologies, National University of Singapore, Singapore, Singapore

Keywords

Information theory; Quantum communication; Teleportation

Synonyms

Quantum teleportation; Teleportation

Years and Authors of Summarized Original Work

1993; Bennett, Brassard, Crepeau, Jozsa, Peres, Wootters

Problem Definition

Suppose there are two spatially separated parties Alice and Bob and Alice wants to send a quantum state consisting of n quantum bits (qubits) ρ to Bob. Since classical communication is much more reliable, and possibly cheaper, than quantum communication, it is desirable that this task be achieved by communicating just classical bits. Such a procedure is referred to as *teleportation*.

Unfortunately, it is easy to argue that this is in fact not possible if arbitrary quantum states need to be communicated faithfully. However, Bennett, Brassard, Crepeau, Jozsa, Peres, and Wootters [8] presented a nice solution to it by modifying the assumptions about the resources that are available to Alice and Bob.

Key Results

Let $\{|0\rangle, |1\rangle\}$ be the standard basis for the state space of one quantum bit (which is equal to $\mathbb{C}^2$). For simplicity of notation $|0\rangle \otimes |0\rangle$ are represented as $|0\rangle|0\rangle$ or simply $|00\rangle$. An EPR pair is a special two-qubit quantum state defined as $|\psi\rangle \overset{\Delta}{=} \frac{1}{\sqrt{2}}(|00\rangle + |11\rangle)$.

Alice and Bob are said to share an EPR pair if each holds one qubit of the pair. In this article a standard notation is followed in which classical bits are called "cbits" and shared EPR pairs are called "ebits." Bennett et al. showed the following:

Theorem 1 *Teleportation of an arbitrary n-qubit state can be achieved with $2n$ cbits and n ebits.*

These shared EPR pairs are referred to as *prior entanglement* to the protocol since they are shared at the beginning of the protocol (before Alice gets her input state) and are independent of Alice's input state. This solution is a good compromise since it is conceivable that Alice and Bob share several EPR pairs at the beginning, when they are possibly together, in which case they do not require a quantum channel. Later they can use these EPR pairs to transfer several quantum states when they are spatially separated.

Let us now see how Bennett et al. [8] achieve teleportation. Let us first note that in order to show Theorem 1, it is enough to show that a single qubit, which is possibly a part of a larger state ρ, can be teleported, while preserving its entanglement with the rest of the qubits of ρ, using 2 cbits and 1 ebit. Let us also note that the larger state ρ can now be assumed to be a pure state without loss of generality.

Theorem 2 *Let $|\phi\rangle_{AB} = a_0|\phi_0\rangle_{AB}|0\rangle_A + a_1|\phi_1\rangle_{AB}|1\rangle_A$, where a_0, a_1 are complex numbers with $|a_0|^2 + |a_1|^2 = 1$. Subscripts A, B (representing Alice and Bob, respectively) on qubits signify their owner.*

It is possible for Alice to send two classical bits to Bob such that at the end of the protocol the final state is $a_0|\phi_0\rangle_{AB}|0\rangle_B + a_1|\phi_1\rangle_{AB}|1\rangle_B$.

T

Proof For simplicity of notation, let us assume below that $|\phi_0\rangle_{AB}$ and $|\phi_1\rangle_{AB}$ do not exist. The proof is easily modified when they do exist by tagging them along. Let an EPR pair $|\psi\rangle_{AB} = \frac{1}{\sqrt{2}}(|0\rangle_A|0\rangle_B + |1\rangle_A|1\rangle_B)$ be shared between Alice and Bob. Let us refer to the qubit under concern that needs to be teleported as the input qubit.

The combined starting state of all the qubits is

$$|\theta_0\rangle_{AB} = |\phi\rangle_{AB}|\psi\rangle_{AB}$$

$$= (a_0|0\rangle_A + a_1|1\rangle_A)\left(\frac{1}{\sqrt{2}}(|0\rangle_A|0\rangle_B + |1\rangle_A|1\rangle_B)\right)$$

Let CNOT (*controlled-not*) gate be a two-qubit unitary operation described by the operator $|00\rangle\langle00| + |01\rangle\langle01| + |11\rangle\langle10| + |10\rangle\langle11|$. Alice now performs a CNOT gate on the input qubit and her part of the shared EPR pair. The resulting state is then

$$|\theta_1\rangle_{AB} = \frac{a_0}{\sqrt{2}}|0\rangle_A(|0\rangle_A|0\rangle_B + |1\rangle_A|1\rangle_B)$$

$$+ \frac{a_1}{\sqrt{2}}|1\rangle_A(|1\rangle_A|0\rangle_B + |0\rangle_A|1\rangle_B)$$

Let the Hadamard transform be a single-qubit unitary operation with operator $\frac{1}{\sqrt{2}}(|0\rangle + |1\rangle)\langle0| + \frac{1}{\sqrt{2}}(|0\rangle - |1\rangle)\langle1|$. Alice next performs a Hadamard transform on her input qubit. The resulting state then is

$$|\theta_2\rangle_{AB} = \frac{a_0}{2}(|0\rangle_A + |1\rangle_A)(|0\rangle_A|0\rangle_B + |1\rangle_A|1\rangle_B)$$

$$+ \frac{a_1}{2}(|0\rangle_A - |1\rangle_A)(|1\rangle_A|0\rangle_B + |0\rangle_A|1\rangle_B)$$

$$= \frac{1}{2}(|00\rangle_A(a_0|0\rangle_B + a_1|1\rangle_B)$$

$$+ |01\rangle_A(a_0|1\rangle_B + a_1|0\rangle_B))$$

$$+ \frac{1}{2}(|10\rangle_A(a_0|0\rangle_B - a_1|1\rangle_B)$$

$$+ |11\rangle_A(a_0|1\rangle_B - a_1|0\rangle_B))$$

Alice next measures the two qubits in her possession in the standard basis for $\mathbb{C}^4$ and sends the result of the measurement to Bob.

Let the four Pauli gates be the single-qubit unitary operations: identity, $P_{00} = |0\rangle\langle0| + |1\rangle\langle1|$; bit flip, $P_{01} = |1\rangle\langle0| + |0\rangle\langle1|$; phase flip, $P_{10} = |0\rangle\langle0| - |1\rangle\langle1|$; and bit flip together with phase flip, $P_{11} = |1\rangle\langle0| - |0\rangle\langle1|$. On receiving the two bits c_0c_1 from Alice, Bob performs the Pauli gate $P_{c_0c_1}$ on his qubit. It is now easily verified that the resulting state of the qubit with Bob would be $a_0|0\rangle_B + a_1|1\rangle_B$. The input qubit is successfully teleported from Alice to Bob! Please refer to Fig. 1 for the overall protocol.   ∎

Super-Dense Coding
Super-dense coding [22] protocol is a dual to the teleportation protocol. In this protocol, Alice transmits 2 cbits of information to Bob using 1 qubit of communication and 1 shared ebit. It is discussed more elaborately in another article in the encyclopedia.

Lower Bounds on Resources
The above implementation of teleportation requires 2 cbits and 1 ebit for teleporting 1 qubit. It was argued in [8] that these resource requirements are also independently optimal. That is, 2 cbits need to be communicated to teleport a qubit independent of how many ebits are used. Also 1 ebit is required to teleport one qubit independent of how much (possibly two-way) communication is used.

Remote State Preparation
Closely related to the problem of teleportation is the problem of *remote state preparation* (RSP) introduced by Lo [21]. In teleportation Alice is just given the state to be teleported in some input register and has no other information about it. In contrast, in RSP, Alice knows a *complete*

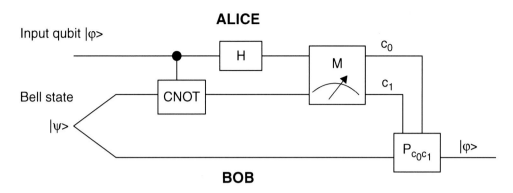

Teleportation of Quantum States, Fig. 1 Teleportation protocol. H represents Hadamard transform and M represents measurement in the standard basis for $\mathbb{C}^4$

description of the input state that needs to be teleported. Also in RSP, Alice is not required to maintain any correlation of the input state with the other parts of a possibly larger state as is achieved in teleportation. The extra knowledge that Alice possesses about the input state can be used to devise protocols for probabilistically exact RSP with one cbit and one ebit per qubit asymptotically [9]. In a probabilistically exact RSP, Alice and Bob can abort the protocol with a small probability; however, when they do not abort, the state produced with Bob at the end of the protocol is exactly the state that Alice intends to send.

Teleportation as a Private Quantum Channel

The teleportation protocol also satisfies an interesting privacy property as follows. If there was a third party, say Eve, having access to the communication channel between Alice and Bob, then Eve learns nothing about the input state of Alice that she is teleporting to Bob. This is because the distribution of the classical messages of Alice is always uniform, independent of her input state. Such a channel is referred to as a *private quantum channel* [2, 11, 18].

Quantum State Redistribution

The teleportation protocol is a part of a wide range of information theoretical tasks. A more general task, referred to as *quantum state redistribution*, is as follows. Three parties, Alice,

Bob, and Referee, share a joint quantum state $|\Psi\rangle_{ACBR}$, where A, C registers are with Alice, B is with Bob, and R is with Referee. The task is to transfer register C to Bob. In the *asymptotic setting* (in this limit of infinite copies of the input state), it was shown by [14, 26] that the number of cbits to be transmitted is $I(R : C|B)$ (quantum mutual information between R and C conditioned on B). A sub-task of quantum state redistribution in which register A is not present is referred to as *quantum state merging*, which was used by [17] to give an operational interpretation to negative *quantum conditional entropy*. Another sub-task in which register B is not present is referred to as *quantum state splitting*. These protocols have also been well studied in the *single-shot* setting where a single copy of the quantum state is available [1, 4, 5, 13].

Applications

Apart from the main application of transporting quantum states over large distances using only classical channel, the teleportation protocol finds other important uses as well. A generalization of this protocol to implement unitary operations [12] is used in *fault-tolerant computation* in order to construct an infinite class of fault-tolerant gates in a uniform fashion. In another application, a form of teleportation called as the error correcting teleportation, introduced by

Knill [19], is used in devising quantum circuits that are resistant to very high levels of noise.

Ideas from quantum teleportation form the basis of measurement-based models of quantum computation. Starting from an arbitrary state $|\psi\rangle = a|0\rangle + b|1\rangle$, and an ancilla $|+\rangle = \frac{1}{\sqrt{2}}(|0\rangle + |1\rangle)$, apply the controlled-Z gate on the two qubits to obtain $a|0, +\rangle + b|1, -\rangle$ (here $|-\rangle = \frac{1}{\sqrt{2}}(|0\rangle - |1\rangle)$). Measuring the first qubit in $\{|+\rangle, |-\rangle\}$ basis gives the state $X^m|\psi\rangle$, where $m \in \{0, 1\}$ is the measurement outcome. In particular, if a phase gate Z_θ acted after controlled-Z unitary, then outcome would be $X^m Z_\theta|\psi\rangle$. Thus, the information about the state $|\psi\rangle$ is still preserved after the measurement. By preparing large arrays of standard entangled states, and performing single-qubit unitary measurement, one can hence simulate any quantum circuit, up to unitaries that depend on measurement outcomes. More details can be found in [23].

This protocol can in particular be used to perform a blind quantum computation. Given a state $|\psi\rangle$, Alice can apply a random Pauli operator on it and obtain the state $|\psi'\rangle$. Then the state is sent to Bob, who uses the idea in previous paragraph to realize a desired phase gate Z_θ on the state. The input state to Bob is completely random, yet a unitary upto measurement outcome is realized by Bob, who then sends the state back to Alice. Since Alice knows which Pauli operation she applied, she can recover back the state $Z_\theta|\psi\rangle$. More details can be found in [6, 16].

Experimental Results

Teleportation protocol has been experimentally realized in various different forms, to name a few, by Boschi et al. [3] using optical techniques, by Bouwmeester et al. [10] using photon polarization, by Nielsen et al. [24] using *Nuclear magnetic resonance* (NMR), and by Ursin et al. [25] using photons for long distance.

Krauter et al. [20] have achieved the teleportation of a complicated quantum state: a continuous variable state stored in the collective spin of an atomic ensemble. Unlike qubits that can be measured only in the 0 or 1 state, the outcome of a continuous variable measurement is a real number, like the position and momentum. The quantum states prepared and teleported in the experiment are actually similar to the coherent states of a harmonic oscillator – describing a particle in a harmonic potential well moving under the influence of a classical force.

Majorana bound states are localized zero-energy excitations of a superconductor. An isolated Majorana bound state is an equal superposition of electron and hole excitations and therefore not a fermionic state. Instead, two spatially separated Majorana bound states together make one zero-energy fermion level which can be either occupied or empty. This defines a two-level system which can store quantum information nonlocally, as needed to realize topological quantum computation. In [15] a nonlocal electron transfer process due to Majorana bound states in a mesoscopic superconductor is predicted. An electron which is injected into one Majorana bound state can go out from another one far apart maintaining phase coherence. The transmission phase shift is independent of the distance "traveled." In this sense this phenomenon can be called "electron transportation."

In summary this work reveals a striking nonlocal electron transport phenomenon through Majorana bound states in a finite-sized superconductor with charging energy. Most interestingly, the transmission phase shift detects the state of a qubit made of two spatially separated Majorana bound states.

In Baur et al. [7] have benchmarked a teleportation algorithm by tomographic reconstruction of the three-qubit entangled state generated by the circuit up to the single-qubit measurements. Using an entanglement witness, they showed that this state has genuine tripartite entanglement. This technique presents an important step toward making use of teleportation in quantum processors realized in superconducting circuits.

Cross-References

▸ Quantum Dense Coding

Recommended Reading

1. Anshu A, Devabathini VK, Jain R (2014) Near optimal bounds on quantum communication complexity of single-shot quantum state redistribution. http://arxiv.org/abs/1410.3031
2. Ambainis A, Mosca M, Tapp A, de Wolf R (2000) Private quantum channels. In: Proceedings of the 41st annual IEEE symposium on foundations of computer science, Redondo Beach, pp 547–553
3. Boschi D, Branca S, Martini FD, Hardy L, Popescu S (1998) Experimental realization of teleporting an unknown pure quantum state via dual classical and Einstein-Podolski-Rosen channels. Phys Rev Lett 80:1121–1125
4. Berta M, Christandl M, Touchette D (2014) Smooth entropy bounds on one-shot quantum state redistribution. http://arxiv.org/abs/1409.4338
5. Berta M (2009) Single-shot quantum state merging. Master's thesis, ETH Zurich
6. Broadbent A, Fitzsimons J, Kashefi E (2009) Universal blind quantum computation. In: Proceedings of the 50th annual IEEE symposium on foundations of computer science, Atlanta
7. Baur M, Fedorov A, Steffen L, Filipp S, da Silva MP, Wallraff A (2012) Benchmarking a quantum teleportation protocol in superconducting circuits using tomography and an entanglement witness. Phys Rev Lett 108(4):040502
8. Bennett C, Brassard G, Crepeau C, Jozsa R, Peres R, Wootters W (1993) Teleporting an unknown quantum state via dual classical and Einstein-Podolsky-Rosen channels. Phys Rev Lett 70:1895–1899
9. Bennett CH, Hayden P, Leung W, Shor PW, Winter A (2005) Remote preparation of quantum states. IEEE Trans Inf Theory 51:56–74
10. Bouwmeester D, Pan JW, Mattle K, Eible M, Weinfurter H, Zeilinger A (1997) Experimental quantum teleportation. Nature 390(6660):575–579
11. Boykin PO, Roychowdhury V (2003) Optimal encryption of quantum bits. Phys Rev A 67:042317
12. Chaung IL, Gottesman D (1999) Quantum teleportation is a universal computational primitive. Nature 402:390–393
13. Datta N, Hsieh M-H, Oppenheim J (2014) An upper bound on the second order asymptotic expansion for the quantum communication cost of state redistribution. http://arxiv.org/abs/1409.4352
14. Devetak I, Yard J (2008) Exact cost of redistributing multipartite quantum states. Phys Rev Lett 100:230501
15. Fu L (2010) Electron teleportation via Majorana bound states in a mesoscopic superconductor. Phys Rev Lett 104(5):056402
16. Herder C, http://www.scottaaronson.com/showcase2/report/charles-herder.pdf
17. Horodecki M, Oppenheim J, Winter A (2007) Quantum state merging and negative information. Commun Math Phys 269:107–136
18. Jain R (2006) Resource requirements of private quantum channels and consequence for oblivious remote state preparation. Technical report, arXive:quant-ph/0507075
19. Knill E (2005) Quantum computing with realistically noisy devices. Nature 434:39–44
20. Krauter H, Salart D, Muschik CA, Petersen JM, Shen JM, Fernholz T, Polzik ES (2013) Deterministic quantum teleportation between distant atomic objects. Nat Phys 9:400–404
21. Lo H-K (2000) Classical communication cost in distributed quantum information processing – a generalization of quantum communication complexity. Phys Rev A 62:012313
22. Nielsen M, Chuang I (2000) Quantum computation and quantum information. Cambridge University Press, Cambridge/New York
23. Nielsen MA (2005) Cluster-state quantum computation. http://arxiv.org/abs/quant-ph/0504097
24. Nielsen MA, Knill E, Laflamme R (1998) Complete quantum teleportation using nuclear magnetic resonance. Nature 396(6706):52–55
25. Ursin R, Jennewein T, Aspelmeyer M, Kaltenbaek R, Lindenthal M, Zeilinger A (2004) Quantum teleportation link across the danube. Nature 430:849
26. Yard JT, Devetak I (2009) Optimal quantum source coding with quantum side information at the encoder and decoder. IEEE Trans Inf Theory 55:5339–5351

Temperature Programming in Self-Assembly

Scott M. Summers
Department of Computer Science, University of Wisconsin – Oshkosh, Oshkosh, WI, USA

Keywords

Algorithm self-assembly; DNA computing; Kolmogorov complexity; Scaled shapes; Self-assembly; Tile assembly model; Temperature programming; Tile complexity; Tile self-assembly

Years and Authors of Summarized Original Work

2005; Aggarwal, Cheng, Goldwasser, Kao, Moisset de Espanés, Schweller
2006; Kao, Schweller
2012; Summers

Problem Definition

Self-assembly is a process by which a small number of fundamental components automatically coalesce to form a target structure. In 1998, Winfree [10] introduced the abstract Tile Assembly Model (aTAM) as a deliberately over-simplified, discrete mathematical model of the DNA tile self-assembly pioneered by Seeman [6]. The aTAM "effectivizes" classical Wang tiling [9] in the sense that the former augments the latter with a mechanism for sequential "growth" of a tile assembly. Very briefly, in the aTAM, the fundamental components are un-rotatable, translatable square "tile types" whose sides are labeled with (alpha-numeric) glue "colors" and (integer) "strengths." Two tiles that are placed next to each other *bind* if the glues on their abutting sides match in both color and strength, and the common strength is at least a certain (integer) "temperature." Self-assembly starts from a "seed" tile type, typically assumed to be placed at the origin of the coordinate system, and proceeds nondeterministically and asynchronously as tiles bind to the seed-containing assembly one at a time.

The *multiple temperature* model [2, 3, 8] is a natural generalization of the aTAM, where the temperature of a tile system is dynamically adjusted by the experimenter as self-assembly proceeds. In the multiple temperature model, a tile assembly system (TAS) is defined as an ordered triple $\mathcal{T} = \left(T, \sigma, \langle \tau_i \rangle_{i=0}^{k-1}\right)$, where T is a tile set, σ is a "seed assembly," and the third component $\langle \tau_i \rangle_{i=0}^{k-1}$ is a sequence of nonnegative integer temperatures.

Intuitively, self-assembly in the multiple temperature TAS $\mathcal{T}$ is carried out in k phases. In the first temperature phase, tiles are added to the existing assembly as they normally would be in the aTAM until a τ_0-stable terminal assembly is reached. In phase two, tiles can accrete to the existing assembly if they can do so with at least strength τ_1. Also, and at any time during the second temperature phase, if there is ever a cut of the assembly having a strength less than τ_1, then all of the tiles on the side of the cut not containing the seed can be removed from the assembly.

When a τ_1-stable terminal assembly is reached in phase two, phase three begins and proceeds in a similar fashion. This process continues through the final temperature phase in which tiles are added or removed with respect to the temperature τ_{k-1} until reaching a τ_{k-1}-stable terminal assembly. See Fig. 1 for an example of this process.

Problem 1 (Reducing tile complexity for the self-assembly of shapes through temperature programming) Given a *shape* $X \subseteq \mathbb{Z}^2$, find a TAS $\mathcal{T} = \left(T, \sigma, \langle \tau_i \rangle_{i=0}^{k-1}\right)$ such that X uniquely self-assembles in $\mathcal{T}$ and $|T|$ and k are minimal.

In some cases, it is sufficient to uniquely self-assemble a scaled-up version of X, i.e., $X^c = \left\{(x, y) \in \mathbb{Z}^2 \mid \left(\lfloor \frac{x}{c} \rfloor, \lfloor \frac{y}{c} \rfloor\right) \in X\right\}$. Intuitively, X^c is the shape obtained by replacing each point in X with a $c \times c$ block of points. We refer to the natural number c as the *scaling factor* or *resolution loss*.

Key Results

Thin Rectangles

Aggarwal, Cheng, Goldwasser, Kao, Moisset de Espanés, and Schweller [2] proved that, in the aTAM, $\Omega\left(\frac{N^{1/k}}{k}\right)$ unique tile types are required to uniquely self-assemble a rectangle of size $k \times N$, where $k < \frac{\log N}{\log\log N - \log\log\log N}$ (this restriction makes the rectangle "thin"). In the same paper, the authors reduced this bound to $O\left(\frac{\log N}{\log\log N}\right)$ in the 2-temperature model. Intuitively, their construction builds a $j \times N$ rectangle for an optimal value of $j \gg k$. Then, when the temperature is raised, the top $j - k$ rows detach, leaving a (stable) $k \times N$ rectangle.

Squares

In the aTAM, the minimum number of unique tile types required to uniquely self-assemble an $N \times N$ square is $O\left(\frac{\log N}{\log\log N}\right)$ [1]. In 2006, Kao and Schweller [3] reduced this bound to $O(1)$ using $O(\log N)$ temperature changes. Their construction relies on a simple yet ingenious gadget called the "bit-flip gadget." Basically, a bit-flip

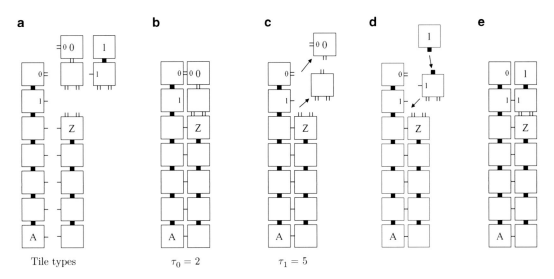

a Tile types **b** $\tau_0 = 2$ **c** $\tau_1 = 5$ **d** **e**

Temperature Programming in Self-Assembly, Fig. 1 Thick notches are strength 5, and thin notches are strength 1. Not all glue labels are shown

gadget is a constant set of tile types that can be programmed, via a carefully chosen sequence of temperature values, to build a rectangle of constant size that encodes either 0 or 1. Figure 1 gives an example of a simple bit-flip gadget. In their main construction, Kao and Schweller first self-assemble a sequence of $O(\log N)$ bit-flip gadgets, using $O(\log N)$ temperature changes. The result is a rectangle of length $O(\log N)$ that encodes N in binary. Finally, the temperature is lowered to 2, and a standard square-building tile set (i.e., [5], but without the "seed row" tile types) is used to fill in the rest of the square. In the same paper, Kao and Schweller also prove that there is no smooth tradeoff, i.e., for a TAS $\mathcal{T} = \left(T, \sigma, \langle \tau_i \rangle_{i=0}^{k-1}\right)$ that uniquely self-assembles an $N \times N$ square, it cannot be the case that $|T| = o\left(\frac{\log N}{\log \log N}\right)$ and $k = o(\log N)$.

Scaled Finite Shapes

In [3], Kao and Schweller posed the following question: Is it possible to have a tile set of size $O(1)$ that can, via some sequence of temperature values, uniquely self-assemble into an arbitrary finite shape (as specified by the sequence of temperature values)? In 2012, Summers [8] investigated this question and discovered the following:

1. **The answer to the previous question is "NO."** It turns out that a tile set of size $O(1)$ cannot uniquely self-assemble an arbitrary finite shape via (any number of) temperature values. Technically speaking, Summers proved that, for every tile set T, there exists a finite shape $X \subset \mathbb{Z}^2$, such that, for each temperature sequence $\langle \tau_i \rangle_{i=0}^{k-1}$, $\mathcal{T} = \left(T, \sigma, \langle \tau_i \rangle_{i=0}^{k-1}\right)$ does not uniquely self-assemble X. In the proof, X is always a line of length $|T| + 1$.

2. **Short temperature sequence, big scale factor.** On the positive side, there exists a universal tile set that can be programmed via temperature values to build a scaled version of an arbitrary finite shape. For instance, Summers exhibited a construction in which the bit-flip gadget of Kao and Schweller [3] is combined with the non-seed portion of the optimal shape-building construction by Soloveichik and Winfree [7] to get the following result: there exists a tile set T with $|T| = O(1)$, such that, for every finite shape X, there exists $c \in \mathbb{N}$ and a temperature sequence $\langle \tau_i \rangle_{i=0}^{k-1}$ with $m = O(K(X))$, where $K(X)$ is the Kolmogorov complexity of X (see [4]), such that,

T

$\mathcal{T} = \left(T, \sigma, \langle \tau_i \rangle_{i=0}^{k-1}\right)$ uniquely self-assembles X^c.

3. **Long temperature sequence, small scale factor.** Note that, in the previously mentioned construction, the scaling factor can be quite large. Technically, the scaling factor c depends on the running time of π, whence $c = \text{poly}(\text{time}(\pi))$. In a truly nanoscale setting, it is necessary to have a construction in which the scaling factor is always small or, better yet, bounded by a constant independent of the shape being assembled. Summers gave such a construction: there exists a tile set T with $|T| = O(1)$, such that, for every finite shape X, there exists a temperature sequence $\langle \tau_i \rangle_{i=0}^{k-1}$ with $m = O(|X|)$, such that, $\mathcal{T} = \left(T, \sigma, \langle \tau_i \rangle_{i=0}^{k-1}\right)$ uniquely self-assembles X^{22}. This construction utilizes a modified bit-flip gadget, in the form of a bit-flip "square," which is essentially a bit-flip gadget that can be programmed (via temperature values) to follow the directions specified by a Hamiltonian path through a shape (not every shape has a Hamiltonian path, but every shape scaled up by a factor of 2 does).

Open Problems

Does there exist a tile set T, with $|T| = O(1)$ and $c \in \mathbb{N}$, such that, for every finite shape X, there exists a temperature sequence $\langle \tau_i \rangle_{i=0}^{k-1}$, with $m = O(K(X))$, such that, X^c uniquely self-assembles in $\mathcal{T} = \left(T, \sigma, \langle \tau_i \rangle_{i=0}^{k-1}\right)$?

Cross-References

▶ Combinatorial Optimization and Verification in Self-Assembly
▶ Intrinsic Universality in Self-Assembly
▶ Patterned Self-Assembly Tile Set Synthesis
▶ Randomized Self-Assembly
▶ Robustness in Self-Assembly
▶ Self-Assembly at Temperature 1
▶ Self-Assembly of Fractals
▶ Self-Assembly of Squares and Scaled Shapes

▶ Self-Assembly with General Shaped Tiles
▶ Temperature Programming in Self-Assembly

Recommended Reading

1. Adleman L, Cheng Q, Goel A, Huang MD (2001) Running time and program size for self-assembled squares. In: Proceedings of the 33rd annual ACM symposium on theory of computing, Heraklion, pp 740–748
2. Cheng Q, Aggarwal G, Goldwasser MH, Kao MY, Schweller RT, de Espanés PM (2005) Complexities for generalized models of self-assembly. SIAM J Comput 34:1493–1515
3. Kao MY, Schweller RT (2006) Reducing tile complexity for self-assembly through temperature programming. In: Proceedings of the 17th annual ACM-SIAM symposium on discrete algorithms, Miami, pp 571–580
4. Li M, Vitanyi P (1997) An introduction to Kolmogorov complexity and its applications, 2nd edn. Springer, New York
5. Rothemund PWK, Winfree E (2000) The program-size complexity of self-assembled squares (extended abstract). In: Proceedings of the thirty-second annual ACM symposium on theory of computing, Portland, pp 459–468
6. Seeman NC (1982) Nucleic-acid junctions and lattices. J Theor Biol 99:237–247
7. Soloveichik D, Winfree E (2007) Complexity of self-assembled shapes. SIAM J Comput 36(6):1544–1569
8. Summers SM (2012) Reducing tile complexity for the self-assembly of scaled shapes through temperature programming. Algorithmica 63(1–2):117–136
9. Wang H (1961) Proving theorems by pattern recognition – II. Bell Syst Tech J XL(1):1–41
10. Winfree E (1998) Algorithmic self-assembly of DNA. PhD thesis, California Institute of Technology

Testing Bipartiteness in the Dense-Graph Model

Oded Goldreich[1] and Dana Ron[2]
[1]Department of Computer Science, Weizmann Institute of Science, Rehovot, Israel
[2]School of Electrical Engineering, Tel-Aviv University, Ramat-Aviv, Israel

Keywords

Graph properties; Property testing; Randomized algorithms; Sublinear time algorithms

Years and Authors of Summarized Original Work

1998; Goldreich, Goldwasser, Ron
2002; Alon, Krivelevich

Problem Definition

A graph is bipartite (or 2-colorable) if its vertices can be partitioned into two sets such that there are no edges between pairs of vertices that reside in the same set.

Given a (simple) graph, the task is to determine whether it is bipartite or is "far" from being bipartite. Thus, the standard decision problem is relaxed by allowing any answer when the graph is not bipartite but is "close" to some bipartite graph. We focus on dense graphs (i.e., for which the number of edges is quadratic in the number of vertices) and wish to solve the aforementioned "approximate decision" problem in constant time, given access to a data structure that answers adjacency queries in unit time.

To complete the formulation of the problem, we need to define the distance between graphs and describe how the graph is accessed. The distance between the graphs $G_1 = (V, E_1)$ and $G_2 = (V, E_2)$ is determined by the symmetric difference between their edge sets (i.e., $E_1 \Delta E_2$), and we say that they are ϵ-close (resp., ϵ-far) if $|E_1 \Delta E_2| \leq \epsilon \cdot |V|^2$ (resp., $|E_1 \Delta E_2| > \epsilon \cdot |V|^2$). Note that this definition is appropriate for dense graphs (i.e., when the number of edges is $\Omega(|V|^2)$), whereas any two sparse graphs are deemed to be close by it. (An alternative model that is more suitable for sparse graphs is presented in this encyclopedia's entry *Testing Bipartiteness of Graphs in Sublinear Time*.) We say that $G = (V, E)$ is ϵ-far from a graph property $\mathcal{P}$ (i.e., a set of graphs that is closed under isomorphism) if for every $G' \in \mathcal{P}$ it holds that G is ϵ-far from G'.

We consider algorithms that make oracle queries to the input graph, denoted $G =$ (V, E). Specifically, the algorithm can perform adjacency queries of the form $(u, v) \in V^2$, which are answered by 1 if $\{u, v\} \in E$ any by 0 otherwise. We can now define property testing in this model.

Definition 1 (testing graph properties in the dense-graph model) A tester for a graph property $\mathcal{P}$ is a randomized algorithm that is given as input a size parameter N and a proximity parameter ϵ as well as access to an adjacency oracle for an N-vertex graph $G = ([N], E)$. The tester should output a binary verdict that satisfies the following two conditions.

1. If $G \in \mathcal{P}$, then the tester accepts with probability at least $2/3$.
2. If G is ϵ-far from $\mathcal{P}$, then the tester accepts with probability at most $1/3$.

A tester has one-sided error if it accepts every graph in $\mathcal{P}$ with probability 1. A tester is nonadaptive if it determines all its queries based solely on its internal coin tosses (*and the parameters N and ϵ*); otherwise it is adaptive.

Here we are interested in the case that $\mathcal{P}$ is the set of all bipartite graphs, and we seek a tester of time complexity that only depends on the proximity parameter, denoted ϵ, and is independent of the size of the graph.

Key Results

Goldreich, Goldwasser, and Ron [9] showed that there exists a tester for bipartiteness that, given a proximity parameter ϵ and access to an N-vertex graph, runs in time poly$(1/\epsilon)$. Furthermore, the tester is nonadaptive and has one-sided error; that is, it always accepts bipartite graphs.

The fact that there exist properties that can be tested in time that only depends on the proximity parameter should not come as a surprise. It is well known that the average value of a (bounded) function defined over a huge domain can be approximated up to a factor of $1 \pm \epsilon$ by taking $O(1/\epsilon^2)$ samples. This approximation problem can be cast as a property testing problem (even as one that refers to graphs in the current model, by considering the edge density of a graph). But the foregoing approximation problem is highly unstructured (i.e., it merely refers to the average of values regardless of anything else), whereas bipartite graphs are highly structured (and the edge density in them is almost arbitrary).

Turning back to bipartiteness, this property is a special case of k-colorability, where $k = 2$. (A graph is k-colorable if its vertices can be partitioned into k sets such that there are no edges between pairs of vertices that reside in the same set.) The tester for bipartiteness, and more generally for k-colorability (for any $k \geq 2$), is very simple. It merely selects a sample of $\text{poly}(1/\epsilon)$ random vertices and accepts if and only if the subgraph induced by the selected vertices is k-colorable. In fact, as shown by Alon and Krivelevich [1] (improving on the bounds obtained in [9]), in the case of $k = 2$ (bipartiteness), a sample of $\tilde{O}(1/\epsilon)$ vertices suffices, and in general a sample of size $\tilde{O}(k/\epsilon^2)$ is sufficient. (The notation $\tilde{O}(q)$ "hides" polylogarithmic factors in q.)

Clearly, the algorithm always accepts k-colorable graphs, and its running time is $\text{poly}(1/\epsilon)$ if $k = 2$ and exponential in $\text{poly}(1/\epsilon)$ (the sample size) otherwise. The analysis boils down to proving that if the graph G is ϵ-far from being k-colorable, then it is rejected with probability at least $2/3$. Below we shall sketch the argument for the case of $k = 2$ and when one uses a sample of $\tilde{O}(1/\epsilon^2)$ random vertices.

We view the random sample (of vertices) as a union of two disjoint sets, denoted U and S, where $t \stackrel{\text{def}}{=} |U| = \tilde{O}(1/\epsilon)$ and $m \stackrel{\text{def}}{=} |S| = O(t/\epsilon)$. We consider all possible (2-way) partitions of U and associate a partial partition of V with each such partition of U. Specifically, given a partition of U, denoted (U_1, U_2), we place all neighbors of U_1 (resp., of U_2) opposite to U_1 (resp., U_2). Indeed, such a placing is forced if we seek a partition of V that is consistent with the given partition (U_1, U_2) of U. One may show that, with high probability, most high-degree vertices in V have at least one neighbor in U, and so the partition of these vertices is forced by the partition of U. Since there are relatively few edges

incident to vertices that do not neighbor U, there must be many edges that violate the partition induced by (U_1, U_2) (i.e., their endpoints are forced to be on the same side of the induced partition). It follows that when we take the additional sample S and perform all queries on pairs in $U \times S$ and $S \times S$, with high probability, we detect a violating edge with respect to each of the $2^{|U|}$ induced partitions, thus ruling out all potential partitions of U. This implies that with high probability, the subgraph induced by $U \cup S$ is not bipartite. Let us stress the key observation: *It suffices to rule out relatively few* (partial) *partitions of V* (i.e., these $2^{|U|}$ partitions induced by partitions of U) rather than all $(2^{|V|})$ possible partitions of V.

Applications

The procedure employed in the above analysis yields a randomized $\text{poly}(1/\epsilon) \cdot N$-time algorithm for 2-partitioning a bipartite graph such that (with high probability) at most ϵN^2 edges lie within the same side. This is done by running the tester, determining a partition of U (defined as in the proof) that is consistent with the bipartite partition of $U \cup S$, and partitioning V as done in the proof (with vertices that do not neighbor U, or neighbor both U_1 and U_2, placed arbitrarily). Thus, the placement of each vertex is determined by inspecting at most $\tilde{O}(1/\epsilon)$ entries of the adjacency matrix. Furthermore, the aforementioned partition of U constitutes a succinct representation of the 2-partition of the entire graph. All this is a typical consequence of the fact that the analysis of the tester follows the "enforce-and-test" paradigm (see the survey of Ron [12, Sec. 4]).

Open Problems

As stated above, a more refined analysis yields a nonadaptive tester that inspects the subgraph induced by $\tilde{O}(1/\epsilon)$ random vertices. One can easily show that this result is almost optimal with respect to testers that inspect an induced subgraph. Furthermore, as Bogdanov and Trevisan show [3], a query complexity of $O(1/\epsilon^2)$ is optimal for any nonadaptive tester for bipartiteness, whereas any adaptive tester requires $\Omega(\epsilon^{-3/2})$ queries. This raises the question of *what is the query complexity of adaptive testers for bipartiteness.*

While Goldreich and Trevisan [8] have shown that the gap between adaptive and nonadaptive testers (in the dense-graph model) is at most quadratic, the above question falls within the uncertainly left open by the quadratic upper bound. Furthermore, Gonen and Ron [10] showed that N-vertex graphs of maximum degree $O(\epsilon N)$ can be tested for bipartiteness in time $\tilde{O}(\epsilon^{-3/2})$, whereas the aforementioned lower bound of [3] holds also for such graphs.

The general question of the gap between adaptive and nonadaptive testers was studied by Goldreich and Ron [7]. They showed that there exist graph properties that have an adaptive tester that runs in time $\tilde{O}(1/\epsilon)$, for which any nonadaptive tester requires $\Omega(\epsilon^{-3/2})$ queries. They conjectured that there exist graph properties that have an adaptive tester that runs in time $\tilde{O}(1/\epsilon)$, for which any nonadaptive tester requires $\Omega(\epsilon^{-2})$ queries.

Cross-References

▸ Testing Bipartiteness of Graphs in Sublinear Time

Comments for the Recommended Reading

The current entry falls within the scope of property testing (see the surveys [5, 6, 11, 12]). A general definition of this setting was first put forward by Rubinfeld and Sudan [13]. This definition was further generalized and systematically investigated by Goldreich, Goldwasser, and Ron [9], who focused on testing graph properties in the dense-graph model. Alternative models for testing graph properties are discussed in this encyclopedia's entry cited above.

As already noted, the tester for k-colorability described above was suggested and analyzed in [9]. A tighter analysis that yields the best

bounds known was subsequently provided in [1]. Testers for any "graph partition property" (including testing that a graph contains a clique of certain density or has a bisection of certain density) were also presented in [9].

The fact that all these (nonadaptive) testers operate by inspecting a random induced subgraph was shown to be no coincidence in [8]. We also mention that the class of graph properties that can be tested (in the dense-graph model) within complexity that is independent of the size of the graph was characterized by Alon et al. [2]. Their characterization is related to Szemerédi's Regular Partitions [14]. A different characterization, based on graph limits, was proved independently by Borgs et al. [4].

Recommended Reading

1. Alon N, Krivelevich M (2002) Testing k-colorability. SIAM J Discret Math 15:211–227
2. Alon N, Fischer E, Newman I, Shapira A (2009) A combinatorial characterization of the testable graph properties: it's all about regularity. SIAM J Comput 39(1):143–167
3. Bogdanov A, Trevisan L (2004) Lower bounds for testing bipartiteness in dense graphs. In: Proceedings of the nineteenth IEEE annual conference on computational complexity (CCC), Amherst, pp 75–81
4. Borgs C, Chayes J, Lovász L, Sós VT, Szegedy B, Vesztergombi K (2006) Graph limits and parameter testing. In: Proceedings of the thirty-eighth annual ACM symposium on the theory of computing (STOC), Seattle, pp 261–270
5. Goldreich O (2010) Introduction to testing graph properties, in [6]
6. Goldreich O (ed) (2010) Property testing: current research and surveys. LNCS, vol 6390. Springer, Heidelberg
7. Goldreich O, Ron D (2011) Algorithmic aspects of property testing in the dense graphs model. SIAM J Comput 40(2):376–445
8. Goldreich O, Trevisan L (2003) Three theorems regarding testing graph properties. Random Struct Algorithms 23(1):23–57
9. Goldreich O, Goldwasser S, Ron D (1998) Property testing and its connections to learning and approximation. J ACM 45:653–750
10. Gonen M, Ron D (2010) On the benefits of adaptivity in property testing of dense graphs. Algorithmica 58(4):811–830
11. Ron D (2008) Property testing: a learning theory perspective. Found Trends Mach Learn 1(3):307–402

12. Ron D (2010) Algorithmic and analysis techniques in property testing. Found Trends Theor Comput Sci 5:73–205
13. Rubinfeld R, Sudan M (1996) Robust characterization of polynomials with applications to program testing. SIAM J Comput 25(2):252–271
14. Szemerédi E (1978) Regular partitions of graphs. In: Proceedings, Colloque Inter. CNRS, Paris, pp 399–401

Testing Bipartiteness of Graphs in Sublinear Time

Oded Goldreich[1] and Dana Ron[2]
[1]Department of Computer Science, Weizmann Institute of Science, Rehovot, Israel
[2]School of Electrical Engineering, Tel-Aviv University, Ramat-Aviv, Israel

Keywords

Graph properties; Property testing; Randomized algorithms; Sublinear time algorithms

Years and Authors of Summarized Original Work

1998; Goldreich, Ron
2004; Kaufman, Krivelevich, Ron

Problem Definition

A graph is bipartite (or 2-colorable) if its vertices can be partitioned into two sets such that there are no edges between pairs of vertices that reside in the same set.

Given a (simple) graph, the task is to determine whether it is bipartite or is "far" from being bipartite. Thus, the standard decision problem is relaxed by allowing any answer when the graph is not bipartite but is "close" to some bipartite graph. We wish to solve this "approximate decision" problem in sublinear time, given access to a data structure that answers adjacency and incidence queries in unit time.

To complete the formulation of the problem, we need to define the distance between graphs and describe how the graph is accessed. The distance between two graphs $G_1 = (V, E_1)$ and $G_2 = (V, E_2)$ is determined by the symmetric difference between their edge sets (i.e., $E_1 \triangle E_2$), where they are ϵ-close (resp., ϵ-far) if $|E_1 \triangle E_2| \leq \epsilon \cdot (|E_1| + |E_2|)$ (resp., $|E_1 \triangle E_2| > \epsilon \cdot (|E_1| + |E_2|)$). A *graph property* is a set of graphs that is closed under isomorphism, and we say that $G = (V, E)$ is ϵ-far from a graph property $\mathcal{P}$ if for every $G' \in \mathcal{P}$ it holds that G is ϵ-far from G'.

We consider algorithms that make oracle queries to the input graph, denoted $G = (V, E)$. Specifically, an algorithm can perform adjacency queries of the form $(u, v) \in V^2$, which are answered by 1 if $\{u, v\} \in E$ and by 0 otherwise, and incidence queries of the form $(u, i) \in V \times [|V| - 1]$, which are answered by v if v is the ith neighbor of u and by $\perp$ if u has less than i neighbors. (Note that adjacency queries may be quite useless when the graph is very sparse.) We now define property testing in this model.

Definition 1 (testing graph properties with adjacency and incidence queries) A tester for a graph property $\mathcal{P}$ is a randomized algorithm that is given as input a size parameter N and a proximity parameter ϵ as well as access to adjacency and incidence oracles for an N-vertex graph $G = ([N], E)$. The tester should output a binary verdict that satisfies the following two conditions.

1. If $G \in \mathcal{P}$, then the tester accepts with probability at least $2/3$.
2. If G is ϵ-far from $\mathcal{P}$, then the tester accepts with probability at most $1/3$.

A tester has one-sided error if it accepts every graph in $\mathcal{P}$ with probability 1.

Here we are interested in the case that $\mathcal{P}$ is the set of all bipartite graphs, and we seek a tester of time complexity that is sublinear in the size of the graph. The dependence of the running time on the proximity parameter, denoted ϵ, is of secondary concern.

Key Results

Building on the work of Goldreich and Ron [6] on testing bipartiteness of bounded-degree graphs, Kaufman, Krivelevich, and Ron [9] presented a tester for bipartiteness that, given a proximity parameter ϵ and access to an N-vertex graph, runs in time $\sqrt{N} \cdot \text{poly}(\log(N), 1/\epsilon)$. This algorithm uses only incidence queries. In case the number of edges, denoted M, is larger than $N^{3/2}$, the running time can be reduced to $(N^2/M) \cdot \text{poly}(\log N, 1/\epsilon)$ by also using adjacency queries. Furthermore, in both cases, the testers have one-sided error; that is, they always accept bipartite graphs.

From now on, we focus on the special case that $M = O(N)$ and further assume that the graph has constant maximum degree, denoted d.

As a warm-up, we note that the case of $d = 2$ is easy. In this case, we are guaranteed that the graph consists of a collection of paths and cycles, and we only need to check that it does not have short cycles of odd length. Note that such an N-vertex graph is ϵ-far from being bipartite if and only if it contains more than ϵN cycles of odd length, where most of these cycles must have length at most $2/\epsilon$. Hence, in this case, testing bipartiteness can be performed by selecting $O(1/\epsilon)$ random vertices and exploring their neighborhoods up to distance $1/\epsilon$.

In contrast, in the case that $d \geq 3$, any tester for bipartiteness must perform $\Omega(\sqrt{N})$ queries. As shown by Goldreich and Ron [7], this can be proved by considering the following two families of N-vertex graphs (for any even N):

1. The first family, denoted $\mathcal{G}_1^N$, consists of all degree-3 graphs that are composed of the union of a Hamiltonian cycle and a perfect matching. That is, there are N edges connecting the vertices in a cycle, and the other $N/2$ edges form a perfect matching.
2. The second family, denoted $\mathcal{G}_2^N$, is the same as the first, *except* that the perfect matchings allowed are restricted as follows. The distance on the cycle between every two vertices that are connected by a perfect matching edge must be odd.

Clearly, all graphs in $\mathcal{G}_2^N$ are bipartite. It can be shown that almost all graphs in $\mathcal{G}_1^N$ are far from being bipartite. On the other hand, one can prove that an algorithm that performs $o(\sqrt{N})$ queries cannot distinguish between a graph chosen randomly from $\mathcal{G}_2^N$ (which is always bipartite) and a graph chosen randomly from $\mathcal{G}_1^N$ (which with high probability is far from bipartite). Loosely speaking, this follows from the fact that in both cases the algorithm is unlikely to encounter a cycle (among the vertices that it has inspected).

The algorithm itself is based on taking many (i.e., $\text{poly}(1/\epsilon) \cdot \tilde{O}(N^{1/2})$) *random walks* from few (i.e., $O(1/\epsilon)$) randomly selected start vertices, where each walk has length $\text{poly}(\epsilon^{-1} \log N)$. Specifically, given as input N, d, ϵ as well as access to an incidence oracle for an N-vertex graph, $G = (V, E)$, of degree bound d, the algorithm repeats the following steps $T \stackrel{\text{def}}{=} \Theta(\frac{1}{\epsilon})$ times:

1. Uniformly select a vertex s in V.
2. Try to find an odd-length cycle through s:
   (a) Perform $K \stackrel{\text{def}}{=} \text{poly}((\log N)/\epsilon) \cdot \sqrt{N}$ random walks starting from s, each of length $L \stackrel{\text{def}}{=} \text{poly}((\log N)/\epsilon)$.
   (b) Let R_0 (respectively, R_1) denote set of vertices reached from s in an even (respectively, odd) number of steps in any of these walks.
   (c) If $R_0 \cap R_1$ is not empty, then reject.

If the algorithm did not reject in any of the foregoing T iterations, then it accepts.

Clearly, the algorithm always accepts bipartite graphs. Hence, the analysis boils down to proving that if the graph G is ϵ-far from being bipartite, then it is rejected with probability at least $2/3$.

The analysis is quite involved. We confine ourselves to the special case where the graph has a "rapid mixing" feature. It is convenient to modify the random walks so that at each step each neighbor is selected with probability $1/2d$, and otherwise (with probability at least $1/2$) the walk remains at the present vertex. Furthermore, we will consider a single execution of Step (2) starting from an arbitrary vertex, s, which is fixed in the rest of the discussion. The rapid mixing feature we assume is that, for every vertex v, a (modified) random walk of length L starting at s reaches v with probability approximately $1/N$ (say, up to a factor of 2). Note that if the graph is an expander, then this is certainly the case (since $L = \omega(\log N)$).

The key quantities in the analysis are the following probabilities, referring to the **parity** *of the length of a path obtained from the random walk by omitting the self-loops* (transitions that remain at current vertex). Let $p^0(v)$ (respectively, $p^1(v)$) denote the probability that a (modified) *random walk of length L, starting at s, reaches v while making an even* (respectively, *odd*) *number of real* (i.e., non-self-loop) *steps*. By the rapid mixing assumption (for every $v \in V$), it holds that

$$\frac{1}{2N} < p^0(v) + p^1(v) < \frac{2}{N}. \qquad (1)$$

We consider two cases regarding the sum $\sum_{v \in V} p^0(v) p^1(v)$: If the sum is (relatively) "small," we show that V can be 2-partitioned so that there are relatively few edges between vertices that are placed in the same part, which implies that G is close to being bipartite. Otherwise (i.e., when the sum is not "small"), we show that with significant probability, when Step (2) is started at vertex s, it is completed by rejecting G.

In general, the input graph may not be "rapidly mixing," and so the actual analysis, which appears in [6], is far more complex. Another layer of complexity is added when we move from the case of constant degree bound (i.e., d) to the case where the vertex degrees may vary significantly; see [9].

Applications

The foregoing algorithm can be used to find odd-length cycles (of polylogarithmic length) in graphs that are far from lacking such cycles. In general, any one-sided error tester for a property $\mathcal{P}$ finds subgraphs that are inconsistent with the property when invoked on a graph that is far from having property $\mathcal{P}$. Thus, the fact that the bipartite tester finds odd cycles (when invoked on graphs that are far from lacking such cycles)

follows directly from its definition, but the fact that these cycles are short is a feature of the specific tester presented above.

Cross-References

▶ Testing Bipartiteness in the Dense-Graph Model

Comments for the Recommended Reading

The current entry falls within the scope of property testing (see [4,5,11,12]). A general definition of this setting was first put forward by Rubinfeld and Sudan [13]. This definition was further generalized and systematically investigated by Goldreich, Goldwasser, and Ron [8], who focused on testing graph properties in the dense-graph model (see this Encyclopedia's entry cited above). The model considered in this entry was suggested by Kaufman, Krivelevich, and Ron [9]. It generalizes a model proposed by Parnas and Ron [10], which in turn generalizes the bounded-degree model of Goldreich and Ron [7]. The latter paper focuses on testers of complexity that only depends on the proximity parameter (i.e., independent of the size of the graph). Among these testers is a two-sided error tester of cycle-freeness; a one-sided error tester for this problem is presented in [3], but its complexity depends on the size of the graph (where this dependence is unavoidable).

As already noted, the bipartiteness tester for the bounded-degree model is due to Goldreich and Ron [6], and it was extended to the general model by Kaufman, Krivelevich, and Ron [9]. In contrast to these results, Bogdanov, Obata, and Trevisan [2] proved that 3-colorability cannot be tested with sublinear query complexity, even in the bounded-degree model. The problem of testing colorability of general graphs was further studied by Ben-Eliezer et al. [1].

Recommended Reading

1. Ben-Eliezer I, Kaufman T, Krivelevich M, Ron D (2012) Comparing the strength of query types in

property testing: the case of testing k-colorability. Comput Complex 22(1):89–135

2. Bogdanov A, Obata K, Trevisan L (2002) A lower bound for testing 3-colorability in bounded-degree graphs. In: Proceedings of the forty-third annual symposium on foundations of computer science (FOCS), Los Alamitos, pp 93–102

3. Czumaj A, Goldreich O, Ron D, Seshadhri C, Shapira A, Sohler C (2012) Finding cycles and trees in sublinear time. Technical Report TR12-035, Electronic Colloquium on Computational Complexity (ECCC), to appear in Random Structures and Algorithms

4. Goldreich O (2010) Introduction to testing graph properties, in [5]

5. Goldreich O (ed) (2010) Property testing: current research and surveys. LNCS, vol 6390. Springer, Heidelberg

6. Goldreich O, Ron D (1999) A sublinear bipartite tester for bounded degree graphs. Combinatorica 19(3):335–373

7. Goldreich O, Ron D (2002) Property testing in bounded degree graphs. Algorithmica 32(2):302–343

8. Goldreich O, Goldwasser S, Ron D (1998) Property testing and its connections to learning and approximation. J ACM 45:653–750

9. Kaufman T, Krivelevich M, Ron D (2004) Tight bounds for testing bipartiteness in general graphs. SIAM J Comput 33(6):1441–1483

10. Parnas M, Ron D (2002) Testing the diameter of graphs. Random Struct Algorithms 20(2):165–183

11. Ron D (2008) Property testing: a learning theory perspective. Found Trends Mach Learn 1(3):307–402

12. Ron D (2010) Algorithmic and analysis techniques in property testing. Found Trends Theor Comput Sci 5:73–205

13. Rubinfeld R, Sudan M (1996) Robust characterization of polynomials with applications to program testing. SIAM J Comput 25(2):252–271

Testing if an Array Is Sorted

Sofya Raskhodnikova
Computer Science and Engineering Department, Pennsylvania State University, University Park, State College, PA, USA

Keywords

Monotonicity; Property testing; Sorted arrays; Sublinear-time algorithms

Years and Authors of Summarized Original Work

2000; Ergün, Kannan, Kumar, Rubinfeld, Viswanathan

2014; Berman, Raskhodnikova, Yaroslavtsev

Problem Definition

Suppose we would like to check whether a given array of real numbers is sorted (say, in nondecreasing order). Performing this task exactly requires reading the entire array. Here we consider the approximate version of the problem: testing whether an array is sorted or "far" from sorted. We consider two natural definitions of the distance of a given array from a sorted array. Intuitively, we would like to measure how much the input array must change to become sorted. We could measure the change by:

1. The number of entries changed
2. The sum of the absolute values of changes in all entries

It is not hard to see that looking at the number of entries that must be deleted in an array to make it sorted is equivalent to the measure in item 1.

To define the two distance measures formally, let $a = (a_1, \ldots, a_n)$ be the input array and $\mathcal{S}$ be the set of all sorted arrays of length n. We denote by $[n]$ the set $\{1, 2, \ldots, n\}$. The *Hamming distance* from a to $\mathcal{S}$, denoted $\text{dist}(a, \mathcal{S})$, is $\min_{b \in \mathcal{S}} |\{i \in [n] : a_i \neq b_i\}|$. The L_1 *distance* from a to $\mathcal{S}$, denoted $\text{dist}_1(a, \mathcal{S})$, is $\min_{b \in \mathcal{S}} \sum_{i \in [n]} |a_i - b_i|$. Given a parameter $\epsilon \in (0, 1)$, an array is ϵ-far from sorted with respect to the Hamming distance or, respectively, L_1 distance, if the corresponding distance from a to $\mathcal{S}$ is at least ϵn.

A *tester* for sortedness is a randomized algorithm that is given parameters $\epsilon \in (0, 1)$ and n and direct access to an input array a. It is required to accept with probability at least 2/3 if the array is sorted and reject with probability at least 2/3 if the array is ϵ-far from sorted. We consider two types of testers, Hamming and L_1, corresponding to the two distance measures we defined. The query complexity of a tester is the number of array entries it reads. The goal is to design testers for sortedness with the smallest possible query complexity and running time.

There are two special cases of testers we will discuss. A tester is *nonadaptive* if it makes all queries in advance, before receiving any query answers. A tester has *1-sided error* if it always accepts all sorted arrays.

Bibliographical Notes

The Hamming testers for sortedness were first studied by Ergün et al. [7]. The L_1-testers (and, more generally, L_p-testers, which use the L_p distance for some $p \geq 1$) were introduced by Berman, Raskhodnikova, and Yaroslavtsev [2]. The two distance measures we discussed, dist and dist_1, are identical for arrays with 0/1 entries, which we call Boolean arrays. The L_1-tester in [2] builds on the sortedness tester for Boolean arrays by Dodis et al. [6].

Observe that an array $(a_1, a_2, \ldots, a_n)$ of real numbers can be represented by a function $f : [n] \to \mathbb{R}$ defined by $f(i) = a_i$ for all $i \in [n]$. The formulated problem is equivalent to testing if a function f over an ordered finite domain is monotone. In fact, the L_1-tester we will discuss can be easily adapted to work for functions over infinite domains (specifically, bounded intervals), because its complexity is independent of the domain size. The problem of Hamming testing monotonicity of functions over domain $[n]^d$ was first investigated by Goldreich et al. [11]; general partially ordered domains were studied by Fischer et al. [10]. These problems are discussed in the encyclopedia entry "Monotonicity Testing."

Key Results

Ergün et al. [7] designed two Hamming testers for sortedness that run in time $O\left(\frac{\log n}{\epsilon}\right)$. Later, Bhattacharyya et al. [3] and Chakrabarty and Seshadhri [5] gave different testers with the same complexity, with additional features that made them useful as subroutines in testing monotonicity of high-dimensional functions. Fischer [9] proved that the running time of these testers is optimal. Berman, Raskhodnikova, and Yaroslavtsev

[2] gave an L_1-tester for sortedness with running time $O(1/\epsilon)$, which is also optimal.

Here we present two Hamming testers from [3, 7] and the L_1-tester from [2].

Hamming Testers for Sortedness

A Tester Based on Binary Search [7]

We present and analyze the first tester for sortedness (Algorithm 1) with the assumption that all entries in the array a are distinct. This assumption can be removed by treating element a_i as $\langle a_i, i \rangle$ for all $i \in [n]$.

Algorithm 1: Hamming Tester for Sortedness Based on Binary Search

input : parameters n and ϵ; direct access to array a.

1 **repeat** $\lceil \frac{\ln 3}{\epsilon} \rceil$ times:
2   pick $i \in [n]$ uniformly at random;
3   perform a binary search for the value a_i in the array a;
4   **if** a_i is not located by the binary search,
      // it leads to another position
5   reject;
6 **accept**

Analysis of the First Tester

The tester always accepts all sorted arrays. Now consider an array that is ϵ-far from sorted (in Hamming distance). We say that a position $i \in [n]$ is *searchable* if a_i can be found by a binary search in Step 3 and *not searchable* otherwise. If positions i and j such that $i < j$ are both searchable, then $a_i < a_j$, because both a_i and a_j are in the correct position with respect to their common ancestor in the binary search tree. Thus, all numbers in searchable positions are sorted. Since the array is ϵ-far from sorted, at least ϵn positions must be unsearchable. If the tester picks an unsearchable position in Step 2, it rejects. The probability that it happens in one trial is at least ϵ. Therefore, the probability that it fails to happen in $\lceil \frac{\ln 3}{\epsilon} \rceil$ trials is at most

$$(1 - \epsilon)^{\lceil \frac{\ln 3}{\epsilon} \rceil} \leq \exp\left(-\epsilon \cdot \frac{\ln 3}{\epsilon}\right) = 1/3. \quad (1)$$

Thus, the tester rejects an array that is ϵ-far from sorted with probability at least 2/3.

A Tester Based on Graph Spanners [3]

The next tester we discuss is based on graph spanners. We can represent the requirement that the array is sorted as a directed graph G, where nodes are positions in $[n]$, and there is an edge (i, j) for all $i < j$. That is, an edge (i, j) represents that $a_i \leq a_j$. A *2-spanner* of G is a subgraph H of G with vertex set $[n]$ such that for every edge (i, j) in G, there is a path of length at most 2 from i to j in H. It is not hard to construct a 2-spanner of G with at most $n \log n$ edges[3, 12]. (e.g., it can be done using divide-and-conquer as follows: connect all nodes to the one in the middle, orienting the edges towards the nodes with larger indices; remove the middle node; and recurse on the two resulting sublists.)

The tester simply repeats the following step $\lceil \frac{(2 \ln 3) \log n}{\epsilon} \rceil$ times: pick a uniformly random edge (i, j) of the 2-spanner H, and reject if this edge is *violated*, namely, if $a_i > a_j$. If the tester does not find a violated edge, it accepts.

Analysis of the Second Tester

If the input array is sorted, it does not have any violated edges, and the tester always accepts. Now consider an array that is ϵ-far from sorted (in Hamming distance). We call a position $i \in [n]$ *bad* if node i is an endpoint of a violated edge in the 2-spanner H; otherwise, i is *good*. Note that any two good positions i, j such that $i < j$ are connected by a path of length at most 2 of non-violated edges in H. If this path is (i, j), it implies that $a_i \leq a_j$. If this path is (i, k, j) for some node k, it implies that $a_i \leq a_k \leq a_j$. Consequently, for any two good positions i, j such that $i < j$, the numbers a_i and a_j are in the correct order. That is, all numbers in good positions are sorted. As in the analysis of Algorithm 1, we can conclude that there are at least ϵn bad positions. But each bad position is adjacent to a violated edge. Each violated edge can contribute at most two new bad positions. Thus, there are at least $\epsilon n / 2$ violated edges. By a simple calculation similar to (1), the second algorithm rejects an array that is ϵ-far from sorted with probability at least 2/3.

L_1-Tester for Sortedness

The L_1-tester for sortedness [2] requires only a uniform sample from the input (as opposed to the ability to query an arbitrary position). It picks $\lceil \frac{2 \ln 6}{\epsilon} \rceil$ positions uniformly and independently at random and accepts iff the numbers in these positions are sorted.

The main ingredient in the analysis of the tester is a reduction to the case of Boolean arrays. It states that if the tester is nonadaptive and has 1-sided error, it suffices to show that it works for Boolean arrays. We omit the proof of the reduction.

Clearly, the L_1-tester is nonadaptive and always accepts sorted arrays. Now consider a Boolean array a which is ϵ-far from sorted. It remains to show that it is rejected with probability at least 2/3. Let X_0 be the set of the $\epsilon n / 2$ largest indices i for which $a_i = 0$. Similarly, let X_1 be the set of the $\epsilon n / 2$ smallest indices i for which $a_i = 1$. It is easy to show that $i < j$ for all $i \in X_1$ and $j \in X_0$, because a is ϵ-far from sorted. The L_1-tester samples no index from X_0 with probability at most 1/6. The same holds for X_1. Thus, by a union bound, with probability at least 2/3, it samples an index from X_0 and an index from X_1 and detects a violation.

Running time

We explained why the algorithm that samples $\lceil \frac{2 \ln 6}{\epsilon} \rceil$ positions uniformly and independently at random is an L_1-tester for sortedness. Now we analyze its running time for the case of general arrays. The L_1-tester makes $O(1/\epsilon)$ queries. To determine whether the elements in these positions are sorted, the tester can use bucket sort to sort the sampled positions and then simply check if the sequence of queried elements is nondecreasing. Since the positions are sampled uniformly at random, the bucket sort can be implemented to run in expected time $O(1/\epsilon)$, where the expectation is taken over the choice of the samples. By standard methods, the algorithm can be modified to run in $O(1/\epsilon)$ time in the worst case. Observe that the running time does not depend on the length of the input. This is impossible for Hamming testers for sortedness, which, as we mentioned, must query $\Omega(\log n)$ positions [9].

Applications

Testers for sortedness are used as subroutines in other property testers, e.g., for monotonicity of high-dimensional functions [2, 5, 6] and for the property that given points represent ordered vertices of a convex polygon [7]. They are also used to construct fast approximate probabilistically checkable proofs for different optimization problems [8]. Ben-Moshe et al. [1] employed sortedness testers (with additional features) to speed up query evaluation in databases.

Open Problem

Consider the case when all numbers in the input array lie in some specified small set such as $[r]$ for some integer r. As we discussed, for Boolean arrays, testing sortedness can be done in $O(1/\epsilon)$ time [2, 6]. It is not hard to see that for larger ranges, it can be done in $O(r/\epsilon)$ time. When $r \ll n$, can one test sortedness it time polylogarithmic in r? Is $O\left(\frac{\log r}{\epsilon}\right)$ running time achievable?

Fischer's lower bound for testing sortedness [9] applies only to $n \ll r$. The best known lower bound that takes into account both parameters is $\Omega(\min(\log r, \log n))$, due to [4], but it applies only to nonadaptive testers.

Cross-References

▶ Monotonicity Testing

Acknowledgments The author was supported in part by NSF CAREER award CCF-0845701 and Boston University's Hariri Institute for Computing and Center for Reliable Information Systems and Cyber Security.

Recommended Reading

1. Ben-Moshe S, Kanza Y, Fischer E, Matsliah A, Fischer M, Staelin C (2011) Detecting and exploiting near-sortedness for efficient relational query evaluation. In: ICDT, Uppsala, pp 256–267
2. Berman P, Raskhodnikova S, Yaroslavtsev G (2014) L_p-testing. In: Shmoys DB (ed) STOC, New York. ACM, pp 164–173
3. Bhattacharyya A, Grigorescu E, Jung K, Raskhodnikova S, Woodruff DP (2012) Transitive-closure spanners. SIAM J Comput 41(6):1380–1425
4. Blais E, Raskhodnikova S, Yaroslavtsev G (2014) Lower bounds for testing properties of functions over hypergrid domains. In: IEEE 29th conference on computational complexity (CCC) 2014, Vancouver, 11–13 June 2014, pp 309–320
5. Chakrabarty D, Seshadhri C (2013) Optimal bounds for monotonicity and Lipschitz testing over hypercubes and hypergrids. In: STOC, Palo Alto, pp 419–428
6. Dodis Y, Goldreich O, Lehman E, Raskhodnikova S, Ron D, Samorodnitsky A (1999) Improved testing algorithms for monotonicity. In: RANDOM, Berkeley, pp 97–108
7. Ergün F, Kannan S, Kumar R, Rubinfeld R, Viswanathan M (2000) Spot-checkers. J Comput Syst Sci 60(3):717–751
8. Ergün F, Kumar R, Rubinfeld R (2004) Fast approximate probabilistically checkable proofs. Inf Comput 189(2):135–159
9. Fischer E (2004) On the strength of comparisons in property testing. Inf Comput 189(1):107–116
10. Fischer E, Lehman E, Newman I, Raskhodnikova S, Rubinfeld R, Samorodnitsky A (2002) Monotonicity testing over general poset domains. In: STOC, Montreal, pp 474–483
11. Goldreich O, Goldwasser S, Lehman E, Ron D, Samorodnitsky A (2000) Testing monotonicity. Combinatorica 20(3):301–337
12. Raskhodnikova S (2010) Transitive-closure spanners: a survey. In: Goldreich O (ed) Property testing. Lecture notes in computer science, vol 6390. Springer, Berlin, pp 167–196

Testing Juntas and Related Properties of Boolean Functions

Eric Blais
University of Waterloo, Waterloo, ON, Canada

Keywords

Dimension reduction; Juntas; Property testing; Sublinear-time algorithms

Years and Authors of Summarized Original Work

2004; Fischer, Kindler, Ron, Safra, Samorodnitsky
2009; Blais

Problem Definition

Fix positive integers n and k with $n \geq k$. The function $f : \{0, 1\}^n \to \{0, 1\}$ is a k-junta if it depends on at most k of the input coordinates. Formally, f is a k-junta if there exists a set $J \subseteq \{1, 2, \ldots, n\}$ of size $|J| \leq k$ such that for all inputs $x, y \in \{0, 1\}^n$ that satisfy $x_i = y_i$ for each $i \in J$, we have $f(x) = f(y)$. Juntas play an important role in different areas of computer science. In machine learning, juntas provide an elegant framework for studying the problem of learning with datasets that contain many irrelevant attributes [9, 10]. In the analysis of Boolean functions, they essentially capture the set of functions of low complexity under natural measures such as total influence [19] and noise sensitivity [12].

How efficiently can we distinguish k-juntas from functions that are far from being k-juntas? We can formalize this question in the setting of property testing. Define the distance between two functions $f, g : \{0, 1\}^n \to \{0, 1\}$ to be the fraction of inputs on which f and g take different values: $\text{dist}(f, g) := \frac{1}{2^n} |\{x \in \{0, 1\}^n : f(x) \neq g(x)\}$. When $\text{dist}(f, g) \geq \epsilon$ for every k-junta g, we say that f is ϵ-far from being a k-junta; otherwise we say that f is ϵ-close to being a k-juntas. An ϵ-test for k-juntas is a randomized algorithm that queries the value of $f : \{0, 1\}^n \to \{0, 1\}$ on some of its inputs and then with probability at least $\frac{2}{3}$

1. accepts if f is a k-junta, and
2. rejects if f is ϵ-far from being a k-junta.

(The algorithm is free to output anything when f is not a k-junta but is ϵ-close to being a k-junta.)

Problem 1 What is the minimum number of queries to $f : \{0, 1\}^n \to \{0, 1\}$ required to ϵ-test if f is a k-junta?

Key Results

Testing 1-Juntas
One important class of functions related to junta testing is *dictator* functions – the functions $f : \{0, 1\}^n \to \{0, 1\}$ of the form $f(x) = x_i$

for some $i \in [n]$. Bellare, Goldreich, and Sudan [3], in a work that was stated in terms of testing the *long code* and part of their analysis of probabilistically checkable proofs (PCPs), showed that dictator functions can be ϵ-tested with $O(1/\epsilon)$ queries. (See the ▸ Locally Testable Codes entry for more details.) This result was later extended by Parnas, Ron, and Samorodnitsky [21]. The class of 1-juntas includes dictator functions, their negations (known as *anti-dictator* functions), and the constant functions; using the algorithms in [3, 21], we can test 1-juntas with $O(1/\epsilon)$ queries.

Testing k-Juntas
The first result on testing k-juntas for values $k > 1$ followed from related work on the problem of *learning* juntas. Blum, Hellerstein, and Littlestone [11] introduced an algorithm that queries a k-junta $f : \{0, 1\}^n \to \{0, 1\}$ on $O(k \log n + k/\epsilon + 2^k)$ inputs and with probability at least $\frac{5}{6}$ returns a k-junta $h : \{0, 1\}^n \to \{0, 1\}$ such that $\text{dist}(f, h) \leq \epsilon$. Shortly afterward, Goldreich, Goldwasser, and Ron [20] gave a general reduction showing that a proper learning algorithm with query complexity q for a class C of functions can be used to ϵ-test the class C with $q + O(1/\epsilon)$ queries. This result, combined with the Blum–Hellerstein–Littlestone algorithm, shows that k-juntas can be tested with $O(k \log n + 2^k + 1/\epsilon)$ queries.

Fischer, Kindler, Ron, Safra, and Samorodnitsky [18] showed that, remarkably, it is possible to test k-juntas with a number of queries that is *independent* of n. Specifically, they introduced ϵ-tests for k-juntas with query complexity $O(k^2/\epsilon^2)$. This result was sharpened in [4, 5], leading to the following theorem.

Theorem 1 ([5]) *It is possible to ϵ-test if $f : \{0, 1\}^n \to \{0, 1\}$ is a k-junta with $O(k \log k + k/\epsilon)$ queries.*

Chockler and Gutfreund [16] showed that $\Omega(k)$ queries are required to test k-juntas, so the bound in Theorem 1 is nearly optimal. (See also [4, 7, 13] for related lower bounds.)

Theorem 1 can be generalized to apply to the setting where $X_1, \ldots, X_n$, and Y are arbitrary finite sets, and we wish to test whether a function $f : X_1 \times \cdots \times X_n \to Y$ is a k-junta. Interestingly, the query complexity of the k-junta test remains unchanged in this general setting as well. See [5] for the details.

Junta-Testing Algorithm

The proof of Theorem 1 contains two main ingredients.

The first ingredient is a simple modification of the Blum–Hellerstein–Littlestone learning algorithm. The original learning algorithm proceeds in two stages: first, the algorithm learns the k relevant coordinates of the junta; then, it queries f for all 2^k different values of the k relevant coordinates. When we test k-juntas, the second stage is unnecessary and can be replaced with a simpler test that checks whether the (at most) k relevant coordinates that have been identified completely determine the value of f or not. With this modification, we obtain an ϵ-test for k-juntas with query complexity $O(k \log n + k/\epsilon)$. Note that this result already yields the desired bound in Theorem 1 when $n = \mathrm{poly}(k)$.

The second ingredient in the proof of Theorem 1 is a dimension reduction argument. Consider a random partition of the n coordinates into $m = \mathrm{poly}(k)$ parts $S_1, \ldots, S_m$. A function $f : \{0,1\}^n \to \{0,1\}$ is isomorphic to a function $f' : X_1 \times \cdots \times X_m \to \{0,1\}$ where $X_i = \{0,1\}^{|S_i|}$. The function f' is defined over a domain with much smaller dimension, and it satisfies two useful properties. First, when f is a k-junta, then so is f'. Second, when f is ϵ-far from k-juntas and $m = \Omega(k^2)$, then with high probability f' is $\frac{\epsilon}{2}$-far from k-juntas as well. The second fact is far from obvious. It was established in [5] using Fourier analysis and in [8] using a combinatorial argument. These two properties let us complete the algorithm for testing k-juntas by applying the modified Blum–Hellerstein–Littlestone algorithm on the function f'. More details on the algorithm itself can be found in the original papers [5, 18] and the survey [6].

Applications

Feature Selection

Feature selection is the general machine learning task of identifying the features (also known as *attributes* or *variables*) in a dataset that suffice to describe the model being studied. This task is formalized within the junta framework as follows: given a function $f : \{0,1\}^n \to \{0,1\}$, the algorithm seeks to identify a set $J \subseteq [n]$ of size $|J| = k$ where (i) k is as small as possible, and (ii) there is a k-junta $h : \{0,1\}^n \to \{0,1\}$ on the set J that is close to f.

The junta testing algorithm can be used to approximate the minimal value of k for which these two conditions can be satisfied. For example, by executing the junta testing algorithm with $k = 1, 2, 4, 8, \ldots$ until it accepts, we obtain the following estimation result.

Corollary 1 *There is an algorithm that, given query access to $f : \{0,1\}^n \to \{0,1\}$, outputs an estimate $\hat{k}$ such that f is ϵ-close to a k-junta and such that f is not an ℓ-junta for any $\ell < k/2$. Furthermore, this algorithm makes $O(k \log k + k/\epsilon)$ queries to f.*

Testing by Implicit Learning

Let $\mathcal{C}$ be any class (i.e., family) of Boolean functions where every function in $\mathcal{C}$ is close to a being a k-junta. Many natural classes of Boolean functions that have been studied in learning theory and computational complexity fall into this framework. For example, functions with bounded, decision tree complexity, DNF complexity, circuit complexity, and sparse polynomial representation all satisfy this condition. (See the) Diakonikolas et al. [17] gave a general result showing that for each of these classes $\mathcal{C}$, we can ϵ-test the property of being in the class $\mathcal{C}$ efficiently. This result has since been sharpened by Chakraborty et al. [14], yielding the following bounds.

Theorem 2 ([14]) *Fix $s > 0$ and $\epsilon > 0$. We can ϵ-test whether $f : \{0,1\}^n \to \{0,1\}$ can be represented by*

1. *a DNF with s terms,*
2. *a size-s Boolean formula,*
3. *an s-sparse polynomial over $\mathbb{F}_2^n$, or*
4. *a decision tree of size s*

with $O(s/\epsilon^2 \cdot \mathrm{polylog}(s/\epsilon))$ queries.

The proof of Theorem 2 is remarkable in that the ϵ-test algorithm in [14,17] learns the function $f : \{0,1\}^n \rightarrow \{0,1\}$ when f is a k-junta, but without identifying which of the k coordinates of f are part of the junta. This technique is called *testing by implicit learning*, and it is obtained by using and building on the junta testing algorithm.

Testing Function Isomorphism

Two functions $f, g : \{0,1\}^n \rightarrow \{0,1\}$ are *isomorphic* to each other when they are identical up to relabeling of the input variables. In the *function isomorphism testing* problem, we are given query access to (an unknown function) f and must determine whether it is isomorphic to (the known function) g or whether it is ϵ-far from being so. How many queries to f do we need to perform this task? The answer, it turns out, depends on the choice of the function g. The functions g for which we can test isomorphism to g with a constant number of queries are called *efficiently isomorphism testable*.

Every symmetric function is efficiently isomorphism testable. Using the junta testing algorithm, Fischer et al. [18] showed that for any constant $k \geq 0$, every k-junta is also efficiently isomorphism testable. An important open problem in property testing is to characterize the set of functions that are efficiently isomorphism testable. The state of the art on this question is a recent result – also building on the junta testing algorithm – showing that every partially symmetric function is also efficiently isomorphism testable. A function $f : \{0,1\}^n \rightarrow \{0,1\}$ is *k-partially symmetric* if there is a function $g : \{0,1\}^k \times \{0,1,2,\ldots,n\} \rightarrow \{0,1\}$ and a mapping $\rho : [k] \rightarrow [n]$ such that $f(x) = g(x_{\rho(1)}, \ldots, x_{\rho(k)}, \|x\|)$ where $\|x\| = \sum_i x_i$ the Hamming weight of x.

Theorem 3 ([8, 15]) *For every constant $k \geq 0$, every k-partially symmetric function is efficiently isomorphism testable.*

Open Problems

There are two particularly appealing open problems related to the junta testing problem that are motivated by its application to the feature selection problem.

Distance Approximation

Theorem 1 shows that we can distinguish k-juntas from functions that are ϵ-far from k-juntas with few queries. Can we also approximate the distance of a function to its closest k-junta with a small number of queries?

Problem 2 What is the minimum number of queries to $f : \{0,1\}^n \rightarrow \{0,1\}$ required to approximate the distance of f to its closest k-junta within an additive error of $\pm\epsilon$, where $\epsilon \in [0, \frac{1}{2}]$ is a parameter given to the algorithm?

In some cases, property testing algorithms can also be used directly for the corresponding distance approximation problem. This is the case, for example, for the BLR linearity test in the ▶ Linearity Testing/Testing Hadamard Codes chapter. But it is currently not known whether the junta testing algorithms in [18] or [5] can be extended to yield distance approximators or not.

Testing with Random Samples

The query model we have discussed throughout this chapter – where the algorithm is free to query the target function on any input of its choosing – is known as the *membership query model* in machine learning. In some applications, however, we must consider weaker query models where we restrict the queries that the algorithm can make in some ways. Can we also test k-juntas efficiently in restricted query models?

Problem 3 In which restricted query models can we test whether $f : \{0,1\}^n \rightarrow \{0,1\}$ is a k-junta with a number of queries that is asymptotically smaller than the number of queries required to learn k-juntas in the same settings?

Two examples of restricted query models include the passive sampling model (where each query is drawn independently at random from some fixed distribution) and the active query model (where the algorithm can choose its queries from a larger set of inputs drawn from some distribution). Some initial results on this problem can be found in [1, 2].

Cross-References

▶ Linearity Testing/Testing Hadamard Codes

Recommended Reading

1. Alon N, Hod R, Weinstein A (2013) On active and passive testing. arXiv preprint arXiv:13077364
2. Balcan MF, Blais E, Blum A, Yang L (2012) Active property testing. In: IEEE 53rd annual symposium on foundations of computer science (FOCS'12), New Brunswick. IEEE, pp 21–30
3. Bellare M, Goldreich O, Sudan M (1998) Free bits, PCPs, and nonapproximability—towards tight results. SIAM J Comput 27(3):804–915
4. Blais E (2008) Improved bounds for testing juntas. In: Goel A, Jansen K, Rolim JDP, Rubinfeld R (eds) Approximation, randomization and combinatorial optimization. Algorithms and techniques. Springer, Boston, pp 317–330
5. Blais E (2009) Testing juntas nearly optimally. In: Proceedings of the 2009 ACM international symposium on theory of computing (STOC'09). ACM, New York, pp 151–157
6. Blais E (2010) Testing juntas: a brief survey. In: Goldreich O (ed) Property testing – current research and surveys. Springer, Berlin/Heidelberg, pp 32–40
7. Blais E, Brody J, Matulef K (2012) Property testing lower bounds via communication complexity. Comput Complex 21(2):311–358
8. Blais E, Weinstein A, Yoshida Y (2012) Partially symmetric functions are efficiently isomorphism-testable. In: IEEE 53rd annual symposium on foundations of computer science (FOCS'12), New Brunswick, pp 551–560
9. Blum A (1994) Relevant examples and relevant features: thoughts from computational learning theory. In: AAAI fall symposium on 'Relevance', New Orleans
10. Blum A, Langley P (1997) Selection of relevant features and examples in machine learning. Artif Intell 97(2):245–271
11. Blum A, Hellerstein L, Littlestone N (1995) Learning in the presence of finitely or infinitely many irrelevant attributes. J Comput Syst Sci 50(1):32–40
12. Bourgain J (2002) On the distribution of the fourier spectrum of boolean functions. Isr J Math 131(1):269–276
13. Buhrman H, García-Soriano D, Matsliah A, de Wolf R (2013) The non-adaptive query complexity of testing k-parities. Chic J Theor Comput Sci 2013: Article 6, 11
14. Chakraborty S, García-Soriano D, Matsliah A (2011) Efficient sample extractors for juntas with applications. In: Aceto L, Henzinger M, Sgall J (eds) Automata, languages and programming. Springer, Zurich, pp 545–556
15. Chakraborty S, Fischer E, García-Soriano D, Matsliah A (2012) Junto-symmetric functions, hypergraph isomorphism, and crunching. In: 2012 IEEE 27th conference on computational complexity (CCC'12). IEEE Computer Society, Los Alamitos, pp 148–158
16. Chockler H, Gutfreund D (2004) A lower bound for testing juntas. Inf Process Lett 90(6):301–305
17. Diakonikolas I, Lee HK, Matulef K, Onak K, Rubinfeld R, Servedio RA, Wan A (2007) Testing for concise representations. In: 48th annual IEEE symposium on foundations of computer science (FOCS'07), Providence. IEEE, pp 549–558
18. Fischer E, Kindler G, Ron D, Safra S, Samorodnitsky A (2004) Testing juntas. J Comput System Sci 68(4):753–787
19. Friedgut E (1998) Boolean functions with low average sensitivity depend on few coordinates. Combinatorica 18(1):27–35
20. Goldreich O, Goldwasser S, Ron D (1998) Property testing and its connection to learning and approximation. J ACM 45(4):653–750
21. Parnas M, Ron D, Samorodnitsky A (2002) Testing basic boolean formulae. SIAM J Discret Math 16(1):20–46

Text Indexing

Srinivas Aluru
Department of Electrical and Computer Engineering, Iowa State University, Ames, IA, USA

Keywords

String indexing

Years and Authors of Summarized Original Work

1993; Manber, Myers

Problem Definition

Text or string data naturally arises in many contexts including document processing, information retrieval, natural and computer language processing, and describing molecular sequences. In broad terms, the goal of text indexing is to design methodologies to store text data so as to significantly improve the speed and performance of answering queries. While text indexing has been studied for a long time, it shot into prominence during the last decade due to the ubiquity of web-based textual data and search engines to explore it, design of digital libraries for archiving human knowledge, and application of string techniques to further understanding of modern biology. Text indexing differs from the typical indexing of keys drawn from an underlying total order – text data can have varying lengths, and queries are often more complex and involve substrings, partial matches, or approximate matches.

Queries on text data are as varied as the diverse array of applications they support. Consequently, numerous methods for text indexing have been developed and this continues to be an active area of research. Text indexing methods can be classified into two categories: (i) methods that are generalizations or adaptations of indexing methods developed for an ordered set of one-dimensional keys, and (ii) methods that are specifically designed for indexing text data. The most classic query in text processing is to find all occurrences of a pattern P in a given text T (or equivalently, in a given collection of strings). Important and practically useful variants of this problem include finding all occurrences of P subject to at most k mismatches, or at most k insertions/deletions/mismatches. The focus in this entry is on these two basic problems and remarks on generalizations of one-dimensional data structures to handle text data.

Key Results

Consider the problem of finding a given pattern P in text T, both strings over alphabet Σ. The case of a collection of strings can be trivially handled by concatenating the strings using a unique end of string symbol, not in Σ, to create text T. It is worth mentioning the special case where T is structured – i.e., T consists of a sequence of words and the pattern P is a word. Consider a total order of characters in Σ. A string (or word) of length k can be viewed as a k-dimensional key and the order on Σ can be naturally extended to lexicographic order between multidimensional keys of variable length. Any one-dimensional search data structure that supports $O(\log n)$ search time can be used to index a collection of strings using lexicographic order such that a string of length k can be searched in $O(k \log n)$ time. This can be considerably improved as below [8]:

Theorem 1 *Consider a data structure on one-dimensional keys that relies on constant-time comparisons among keys (e.g., binary search trees, red-black trees etc.) and the insertion of a key identifies either its predecessor or successor. Let $O(\mathcal{F}(n))$ be the search time of the data structure storing n keys (e.g., $O(\log n)$ for red-black trees). The data structure can be converted to index n strings using $O(n)$ additional space such that the query for a string s can be performed in $O(\mathcal{F}(n))$ time if s is one of the strings indexed, and in $O(\mathcal{F}(n) + |s|)$ otherwise.*

A more practical technique that provides $O(\mathcal{F}(n) + |s|)$ search time for a string s under more restrictions on the underlying one-dimensional data structure is given in [9]. The technique is nevertheless applicable to several classic one-dimensional data structures, in particular binary search trees and its balanced variants. For a collection of strings that share long common prefixes such as IP addresses and XML path strings, a faster search method is described in [5].

When answering a sequence of queries, significant savings can be obtained by promoting frequently searched strings so that they are among the first to be encountered in a search path through the indexing data structure. Ciriani et al. [4] use self-adjusting skip lists to derive an expected bound for a sequence of queries that matches the information-theoretic lower bound.

Theorem 2 *A collection of n strings of total length N can be indexed in optimal O(N) space so that a sequence of m string queries, say $s_1, \cdots, s_m$, can be performed in $O(\sum_{j=1}^{m} |s_j| + \sum_{i=1}^{n} n_i \log(m/n_i))$ expected time, where n_i is the number of times the ith string is queried.*

Notice that the first additive term is a lower bound for reading the input, and the second additive term is a standard information-theoretic lower bound denoting the entropy of the query sequence. Ciriani et al. also extended the approach to the external memory model, and to the case of dynamic sets of strings. More recently, Ko and Aluru developed a self-adjusting tree layout for dynamic sets of strings in secondary storage that provides optimal number of disk accesses for a sequence of string or substring queries, thus providing a deterministic algorithm that matches the information-theoretic lower bound [4].

The next part of this entry deals with some of the widely used data structures specifically designed for string data, suffix trees, and suffix arrays. These are particularly suitable for querying unstructured text data, such as the genomic sequence of an organism. The following notation is used: Let $s[i]$ denote the ith character of string s, $s[i \ldots j]$ denote the substring $s[i]s[i+1]\ldots s[j]$, and $S_i = s[i]s[i+1]\ldots s[|s|]$ denote the suffix of s starting at ith position. The suffix S_i can be uniquely described by the integer i. In case of multiple strings, the suffix of a string can be described by a tuple consisting of the string number and the starting position of the suffix within the string. Consider a collection of strings over Σ, having total length n, each extended by adding a unique termination symbol $\$ \notin \Sigma$. The suffix tree of the strings is a compacted trie of all suffixes of these extended

strings. The suffix array of the strings is the lexicographic sorted order of all suffixes of these extended strings. For convenience, we list '\$', the last suffix of each string, just once. The suffix tree and suffix array of strings 'apple' and 'maple' are shown in Fig. 1. Both these data structures take $O(n)$ space and can be constructed in $O(n)$ time [11, 13], both directly and from each other.

Without loss of generality, consider the problem of searching for a pattern P as a substring of a single string T. Assume the suffix tree ST of T is available. If P occurs in T starting from position i, then P is a prefix of suffix $T_i = T[i]T[i+1]\ldots T[|T|]$ in T. It follows that P matches the path from root to leaf labeled i in ST. This property results in the following simple algorithm: Start from the root of ST and follow the path matching characters in P, until P is completely matched or a mismatch occurs. If P is not fully matched, it does not occur in T. Otherwise, each leaf in the subtree below the matching position gives an occurrence of P. The positions can be enumerated by traversing the subtree in $O(occ)$ time, where occ denotes the number of occurrences of P. If only one occurrence is desired, ST can be preprocessed in $O(|T|)$ time such that each internal node contains the suffix at one of the leaves in its subtree.

Theorem 3 *Given a suffix tree for text T and a pattern P, whether P occurs in T can be answered in $O(|P|)$ time. All occurrences of P in T can be found in $O(|P| + occ)$ time, where occ denotes the number of occurrences.*

Now consider solving the same problem using the suffix array SA of T. All suffixes prefixed by P appear in consecutive positions in SA. These can be found using binary search in SA. Naively performed, this would take $O(|P| * \log |T|)$ time. It can be improved to $O(|P| + \log |T|)$ time as follows [15]:

Let $SA[L \ldots R]$ denote the range in the suffix array where the binary search is focused. To begin with, $L = 1$ and $R = |T|$. Let $\prec$ denote "lexicographically smaller", $\preceq$ denote "lexicographically smaller or equal", and

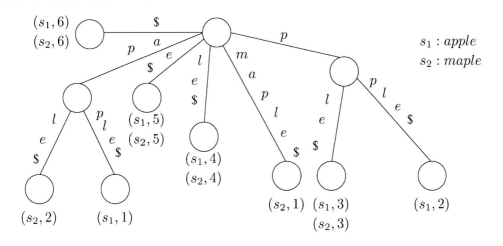

SA | $(s_1,6)$ | $(s_2,2)$ | $(s_1,1)$ | $(s_1,5)$ | $(s_2,5)$ | $(s_1,4)$ | $(s_2,4)$ | $(s_2,1)$ | $(s_1,3)$ | $(s_2,3)$ | $(s_1,2)$

Text Indexing, Fig. 1 Suffix tree and suffix array of strings *apple* and *maple*

$lcp(\alpha, \beta)$ denote the length of the longest common prefix between strings α and β. At the beginning of an iteration, $T_{SA[L]} \preceq P \preceq T_{SA[R]}$. Let $M = \lceil (L + R)/2 \rceil$. Let $l = lcp(P, T_{SA[L]})$ and $r = lcp(P, T_{SA[R]})$. Because SA is lexicographically ordered, $lcp(P, T_{SA[M]}) \geq \min(l, r)$. If $l = r$, then compare P and $T_{SA[M]}$ starting from the $(l+1)$th character. If $l \neq r$, consider the case when $l > r$.

Case I: $l < lcp(T_{SA[L]}, T_{SA[M]})$. In this case, $T_{SA[M]} \prec P$ and $lcp(P, T_{SA[M]}) = lcp(P, T_{SA[L]})$. Continue search in $SA[M \ldots R]$. No character comparisons required.

Case II: $l > lcp(T_{SA[L]}, T_{SA[M]})$. In this case, $P \prec T_{SA[M]}$ and $lcp(P, T_{SA[M]}) = lcp(T_{SA[L]}, T_{SA[M]})$. Continue search in $SA[L \ldots M]$. No character comparisons required.

Case III: $l = lcp(T_{SA[L]}, T_{SA[M]})$. In this case, $lcp(P, T_{SA[M]}) \geq l$. Compare P and $T_{SA[M]}$ beyond lth character to determine their relative order and lcp.

Similarly, the case when $r > l$ can be handled such that comparisons between P and $T_{SA[M]}$, if at all needed, start from the $(r + 1)$th character. To start the execution of the algorithm,

$lcp(P, T_{SA[1]})$ and $lcp(P, T_{SA[|T|]})$ are computed directly using at most $2|P|$ character comparisons. It remains to be described how the $lcp(T_{SA[L]}, T_{SA[M]})$ and $lcp(T_{SA[R]}, T_{SA[M]})$ values required in each iteration are computed. Let $Lcp[1 \ldots |T| - 1]$ be an array such that $Lcp[i] = lcp(SA[i], SA[i + 1])$. The Lcp array can be computed from SA in $O(|T|)$ time [12]. For any $1 \leq i < j \leq n$, $lcp(T_{SA[i]}, T_{SA[j]}) = \min_{k=i}^{j-1} Lcp[k]$. In order to find the lcp values required by the algorithm in constant time, note that the binary search can be viewed as traversing a path in the binary tree corresponding to all possible search intervals used by any execution of the binary search algorithm [15]. The root of the tree denotes the interval $[1 \ldots n]$. If $[i \ldots j]$ $(j - i \geq 2)$ is the interval at an internal node of the tree, its left child is given by $[i \ldots \lceil (i + j)/2 \rceil]$ and its right child is given by $[\lceil (i + j)/2 \rceil \ldots j]$. The lcp value for each interval in the tree is precomputed and recorded in $O(n)$ time and space.

Theorem 4 *Given the suffix array SA of text T and a pattern P, the existence of P in T can be checked in $O(|P| + \log |T|)$ time. All occurrences of P in T can be found in $O(occ)$ additional time, where occ denotes their number.*

Proof The algorithm makes at most $2|P|$ comparisons in determining $lcp(P, T_{SA[1]})$ and $lcp(P, T_{SA[n]})$. A comparison made in an iteration to determine $lcp(P, T_{SA[M]})$ is categorized *successful* if it contributes the *lcp*, and categorized *failed* otherwise. There is at most one failed comparison per iteration. As for successful comparisons, note that the comparisons start with $(\max(l, r) + 1)^{\text{th}}$ character of P, and each successful comparison increases the value of $\max(l, r)$ for the next iteration. Thus, each character of P is involved only once in a successful comparison. The total number of character comparisons is at most $3|P| + \log|T| = O(|P| + \log|T|)$. $\square$

Abouelhoda et al. [1] reduce this time further to $O(|P|)$ by mimicking the suffix tree algorithm on a suffix array with some auxiliary information. The strategy is useful in other applications based on top-down traversal of suffix trees. At this stage, the distinction between suffix trees and suffix arrays is blurred as the auxiliary information stored makes the combined data structure equivalent to a suffix tree. Using clever implementation techniques, the space is reduced to approximately $6n$ bytes. A major advantage of the suffix tree and suffix array based methods is that the text T is often large and relatively static, while it is queried with several short patterns. With suffix trees and enhanced suffix arrays [1], once the text is pre-processed in $O(|T|)$ time, each pattern can be queried in $O(|P|)$ time for constant size alphabet. For large alphabets, the query can be answered in $O(|P| * \log|\Sigma|)$ time using $O(n|\Sigma|)$ space (by storing an ordered array of $|\Sigma|$ pointers to potential children of a node), or in $O(|P| * |\Sigma|)$ time using $O(n)$ space (by storing pointers to first child and next sibling). (Recently, Cole et al. (2006) showed how to further reduce the search time to $O(|P| + \log|\Sigma|)$ while still keeping the optimal $O(|T|)$ space). For indexing in various text-dynamic situations, see [3, 7] and references therein. The problem of compressing suffix trees and arrays is covered in more detail in other entries.

While exact pattern matching has many useful applications, the need for approximate pattern matching arises in several contexts ranging from information retrieval to finding evolutionary related biomolecular sequences. The classic approximate pattern matching problem is to find substrings in the text T that have an edit distance of k or less to the pattern P, i.e., the substring can be converted to P with at most k insert/delete/substitute operations. This problem is covered in more detail in other entries. Also see [16], the references therein, and Chapter 36 of [2].

Applications

Text indexing has many practical applications – finding words or phrases in documents under preparation, searching text for information retrieval from digital libraries, searching distributed text resources such as the web, processing XML path strings, searching for longest matching prefixes among IP addresses for internet routing, to name just a few. The reader interested in further exploring text indexing is referred to the book by Crochemore and Rytter [6], and to other entries in this Encyclopedia. The last decade of explosive growth in computational biology is aided by the application of string processing techniques to DNA and protein sequence data. String indexing and aggregate queries to uncover mutual relationships between strings are at the heart of important scientific challenges such as sequencing genomes and inferring evolutionary relationships. For an in depth study of such techniques, the reader is referred to Parts I and II of [10] and Parts II and VIII of [2].

Open Problems

Text indexing is a fertile research area, making it impossible to cover many of the research results or actively pursued open problems in a short amount of space. Providing better algorithms and data structures to answer a flow of string-search queries when caches or other query models are taken into account, is an interesting research issue [4].

Cross-References

▶ Compressed Suffix Array
▶ Compressed Text Indexing
▶ Indexed Approximate String Matching
▶ Indexed Two-Dimensional String Matching
▶ Suffix Array Construction
▶ Suffix Tree Construction in Hierarchical Memory
▶ Suffix Tree Construction

Recommended Reading

1. Abouelhoda M, Kurtz S, Ohlebusch E (2004) Replacing suffix trees with enhanced suffix arrays. J Discret Algorithms 2:53–86
2. Aluru S (ed) (2005) Handbook of computational molecular biology, Computer and Information Science Series. Chapman and Hall/CRC, Boca Raton
3. Amir A, Kopelowitz T, Lewenstein M, Lewenstein N (2005) Towards real-time suffix tree construction. In: Proceedings of the string processing and information retrieval symposium (SPIRE), pp 67–78
4. Ciriani V, Ferragina P, Luccio F, Muthukrishnan S (2007) A data structure for a sequence of string accesses in external memory. ACM Trans Algorithms 3
5. Crescenzi P, Grossi R, Italiano G (2003) Search data structures for skewed strings. In: International workshop on experimental and efficient algorithms (WEA). Lecture notes in computer science, vol 2. Springer, Berlin, pp 81–96
6. Crochemore M, Rytter W (2002) Jewels of stringology. World Scientific Publishing Company, Singapore
7. Ferragina P, Grossi R (1998) Optimal on-line search and sublinear time update in string matching. SIAM J Comput 3:713–736
8. Franceschini G, Grossi R (2004) A general technique for managing strings in comparison-driven data structures. In: Annual international colloquium on automata, languages and programming (ICALP)
9. Grossi R, Italiano G (1999) Efficient techniques for maintaining multidimensional keys in linked data structures. In: Annual international colloquium on automata, languages and programming (ICALP), pp 372–381
10. Gusfield D (1997) Algorithms on strings, trees and sequences: computer science and computational biology. Cambridge University Press, New York
11. Karkkainen J, Sanders P, Burkhardt S (2006) Linear work suffix arrays construction. J ACM 53:918–936
12. Kasai T, Lee G, Arimura H et al (2001) Linear-time longest-common-prefix computation in suffix arrays and its applications. In: Proceedings of the 12th annual symposium, combinatorial pattern matching (CPM), pp 181–192
13. Ko P, Aluru S (2005) Space efficient linear time construction of suffix arrays. J Discret Algorithms 3:143–156
14. Ko P, Aluru S (2007) Optimal self-adjusting tree for dynamic string data in secondary storage. In: Proceedings of the string processing and information retrieval symposium (SPIRE), Santiago. Lecture notes in computer science, vol 4726, pp 184–194
15. Manber U, Myers G (1993) Suffix arrays: a new method for on-line search. SIAM J Comput 22:935–948
16. Navarro G (2001) A guided tour to approximate string matching. ACM Comput Surv 33:31–88

Three-Dimensional Graph Drawing

David R. Wood
School of Mathematical Sciences, Monash University, Melbourne, VIC, Australia

Keywords

Track layout; Three-dimensional straight-line grid drawing; Treewidth

Years and Authors of Summarized Original Work

2005; Dujmović, Morin, Wood

Problem Definition

A *three-dimensional straight-line grid drawing* of a graph, henceforth called a *3D drawing*, represents the vertices by distinct grid-points in $\mathbb{Z}^3$ and represents each edge by the line segment between its end vertices, such that no two edges cross. In contrast to the case in the plane, it is folklore that every graph has a 3D drawing. For example, the "moment curve" algorithm places the ith vertex at (i, i^2, i^3). It is easily seen that no four vertices are coplanar, and thus no two edges cross. Since every graph has a 3D drawing, we are interested in optimizing certain measures of their aesthetic quality. If a 3D drawing is contained in an axis-aligned box with side lengths $X-1$, $Y-1$, and $Z-1$, then we speak of an $X \times Y \times Z$ drawing with *volume* $X \cdot Y \cdot Z$. This entry considers the problem of producing a 3D drawing of a given graph with small volume.

Key Results

Observe that the drawings produced by the moment curve algorithm have $\mathcal{O}(n^6)$ volume, where n is the number of vertices. Cohen et al. [2] improved this bound, by proving that if p is a prime with $n < p \le 2n$, and the ith vertex is at $(i, i^2 \bmod p, i^3 \bmod p)$, then there is still no crossing. The resulting $\mathcal{O}(n^3)$ volume bound is optimal for the complete graph K_n since each grid plane may contain at most four vertices. It is therefore of interest to identify fixed graph parameters that allow for 3D drawings with small volume, as summarized in the following table.

Graph family	Min. volume	Reference
Arbitrary	$\Theta(n^3)$	[2]
Bounded chromatic number	$\Theta(n^2)$	[19]
Bounded maximum degree	$\mathcal{O}(n^{3/2})$	[7]
Bounded degeneracy	$\mathcal{O}(n^{3/2})$	[9]
H-minor-free (H fixed)	$n \log^{\mathcal{O}(1)} n$	[12]
Bounded genus	$\mathcal{O}(n \log n)$	[12]
Apex-minor-free	$\mathcal{O}(n \log n)$	[12]
Planar	$\mathcal{O}(n \log n)$	[6]
Bounded treewidth	$\Theta(n)$	[11]

The first such parameter to be studied was the chromatic number. Pach et al. [19] proved that graphs of bounded chromatic number have 3D drawings with $\mathcal{O}(n^2)$ volume. If p is a suitably chosen prime, the main step of their algorithm represents the vertices in the ith color class by grid-points in the set $\{(i, t, it) : t \equiv i^2 \pmod{p}\}$. It follows that the volume bound is $\mathcal{O}(k^2 n^2)$ for k-colorable graphs.

Pach et al. [19] also proved an $\Omega(n^2)$ lower bound for the volume of 3D drawings of the complete bipartite graph $K_{n,n}$. This lower bound was generalized for all graphs by Bose et al. [1], who proved that every 3D drawing of an n-vertex m-edge graph has volume at least $\frac{1}{8}(n + m)$.

In particular, the maximum number of edges in an $X \times Y \times Z$ drawing is exactly $(2X - 1)(2Y - 1)(2Z - 1) - XYZ$.

Graphs with bounded maximum degree have bounded chromatic number and, thus, by the result of Pach et al. [19], have 3D drawings with $\mathcal{O}(n^2)$ volume. Pach et al. [19] conjectured that such graphs have 3D drawings with $o(n^2)$ volume, which was verified by Dujmović and Wood [7], who proved a $\mathcal{O}(n^{3/2})$ bound. The best lower bound is $\Omega(n)$. Determining the optimal volume for 3D drawings of bounded degree graphs is a challenging open problem; see [13]. The $\mathcal{O}(n^{3/2})$ upper bound for bounded degree graphs was generalized for graphs with bounded degeneracy [9].

The first nontrivial $\mathcal{O}(n)$ volume bound was established by Felsner et al. [15] for outerplanar graphs. Their elegant algorithm "wraps" a 2D drawing around a triangular prism to obtain a 3D drawing. This result naturally led to the following open problem due to Felsner et al. [15], which motivated much subsequent research: does every planar graph have a 3D drawing with $\mathcal{O}(n)$ volume?

For some time, the $\mathcal{O}(n^2)$ bound for 2D drawings was the best known bound in 3D. Then Dujmović and Wood [7] proved that every planar graph has a 3D drawing with $\mathcal{O}(n^{3/2})$ volume. A breakthrough came with the $\mathcal{O}(n \log^8 n)$ bound of Di Battista et al. [4], which was improved to $\mathcal{O}(n \log n)$ by Dujmović [6] (with a much simpler proof). The most recent work in this direction, by Dujmović et al. [12], extended this $\mathcal{O}(n \log n)$ bound to all graphs of bounded Euler genus and more generally proved that every graph excluding a fixed minor has a 3D drawing with $n \log^{\mathcal{O}(1)} n$ volume.

The $\mathcal{O}(n)$ volume bound for outerplanar graphs mentioned above was generalized by Dujmović et al. [11] as follows:

Theorem 1 ([11]) *Graphs with bounded treewidth have 3D drawings with $\mathcal{O}(n)$ volume.*

This result is the focus of the remainder of this entry. Treewidth is a measure of the similarity of a graph to a tree. It can be defined as follows. A graph is *chordal* if every induced cycle is a triangle. The *treewidth* of a graph G is the minimum integer k such that G is a spanning subgraph of a chordal graph with no $(k + 2)$-clique. Many graphs arising in applications of graph drawing have small treewidth. Trees have treewidth 1, while outerplanar and series-parallel graphs have treewidth 2. Another example arises in software engineering applications. Thorup [20] proved that the control-flow graphs of go-to free programs in many programming languages have treewidth bounded by a small constant, in particular, 3 for Pascal and 6 for C.

Reference [11] is also important because it discovered the connection between 3D drawings, track layouts, and queue layouts; also see [10, 16].

Track Layouts

Track layouts are a combinatorial tool that effectively eliminates the geometry from 3D drawings and exposes the underlying combinatorial structure. They were introduced in [11] although they are implicit in some previous work [15, 16].

Let $V_1, \ldots, V_t$ be the color classes in a (proper) vertex t-coloring of a graph G. Suppose that each color class V_i is equipped with a total order, denoted by $\preceq$. Call V_i a *track* and $V_1, \ldots, V_t$ a *t-track assignment*. An *X-crossing* in $V_1, \ldots, V_t$ consists of two edges vw and xy such that $v \prec x$ in some track V_i and $y \prec w$ in some other track V_j. A t-track assignment with no X-crossing is called a *t-track layout*.

One can produce a track layout from an $A \times B \times C$ drawing of a graph G as follows. Let $V_{x,y}$ be the set of vertices of G with an X-coordinate of x and a Y-coordinate of y. Order each set $V_{x,y}$ by the corresponding Z-coordinates. We obtain an AB-track layout of G, except that consecutive vertices in each track might be adjacent. Doubling each track and putting alternate vertices in $V_{x,y}$ on distinct tracks gives a $2AB$-track layout of G. Most interestingly, a converse result is also true.

Theorem 2 ([11]) *If an n-vertex graph has a t-track layout, then G has a $\mathcal{O}(t) \times \mathcal{O}(t) \times \mathcal{O}(n)$ drawing with $\mathcal{O}(t^2 n)$ volume.*

The proof of Theorem 2 is inspired by the generalizations of the moment curve algorithm

by Cohen et al. [2] and Pach et al. [19]. Loosely speaking, Cohen et al. [2] allow three "free" dimensions, whereas Pach et al. [19] use a coloring to "fix" one dimension with two dimensions free. Theorem 2 uses a track layout to fix two dimensions with one dimension free; see Fig. 1. In particular, say $(V_1, \ldots, V_t)$ is the given t-track layout. Let p be the smallest prime such that $p > k$. Then $p \leq 2k$ by Bertrand's postulate. For $1 \leq i \leq k$, represent the vertices in V_i by the gridpoints $\{(i, i^2 \bmod p, t) : 1 \leq t \leq p \cdot |V_i|, t \equiv i^3 \pmod{p}\}$, such that the Z-coordinates respect the given total order of V_i.

Note that Dujmović and Wood [7] combined the method of Pach et al. [19] with the proof of Theorem 2 to conclude a $\mathcal{O}(tn)$ volume bound of 3D drawings of t-track graphs with bounded chromatic number.

As an example of how to construct a track layout, we now show that every tree T has a 3-track layout (which is implicitly proved in [15]). Let r be a vertex of T. Let V_i be the vertices at distance i from r. Note that $(V_0, V_1, \ldots)$ is a coloring of T. Clearly, each color class V_i can be ordered so that there is no X-crossing; see Fig. 2a. Hence $(V_0, V_1, \ldots)$ is a track layout. Note that, working from the root down, the child nodes of each node can be ordered arbitrarily. This will be important later. Now, imagine wrapping this track layout around a prism; see Fig. 2b. That is, for $0 \leq i \leq 2$, group tracks $V_i \prec V_{3+i} \prec V_{6+i} \prec \ldots$ to obtain a 3-track layout of T.

An Algorithm for Graphs of Bounded Treewidth

Theorem 1 is an immediate consequence of Theorem 2 and the following claim, which we prove by induction on $k \geq 0$: for each integer $k \geq 0$, there is an integer t_k such that every k-tree has a t_k-track layout. A 0-tree has no edges and thus has a 1-track layout. A 1-tree is a tree which has a 3-track layout. Thus the result holds with $t_0 = 1$ and $t_1 = 3$. Let G be a k-tree. Various authors have proved that G can be decomposed as follows [11, 18]. There is a tree T rooted at some node r and a partition $\{B_x : x \in V(T)\}$ of $V(G)$ indexed by the nodes of T with the following properties:

- For each edge vw of G, there is a node x of T such that $v, w \in B_x$, or there is an edge xy of T such that $v \in B_x$ and $w \in B_y$.
- For each node x of T, the induced subgraph $G[B_x]$ is a $(k-1)$-tree.
- For each non-root node y of T, if x is the parent node of y, and C_y is the set of vertices in B_x adjacent to some vertex in B_y, then C_y is a clique in G called the *parent clique* of y.

By induction, for each node x of T, there is a t_{k-1}-track layout of $G[B_x]$. Each clique C in $G[B_x]$ has size at most k. Define the *signature* of C to be the set of (at most k) tracks that contain C. Since there is no X-crossing, the set of cliques of $G[B_x]$ with the same signature can be linearly ordered $C_1 \prec \cdots \prec C_p$, such that if v and w are vertices in the same track, and in distinct cliques C_i and C_j with $i < j$, then $v \prec w$ in that track. Call this a *clique ordering*.

Let T_0, T_1, T_2 be a 3-track layout of T described above. Replace each track T_i by t_{k-1} subtracks, and replace each node $x \in T_i$ by the t_{k-1}-track layout of $G[B_x]$. This defines a $3 \cdot t_{k-1}$ track assignment for G. Clearly an edge in some $G[B_x]$ is in no X-crossing with any other edge. There is no X-crossing between two edges between a parent bag B_x and some same child bag B_y, since the end points in B_x of such edges form a clique (the parent clique of y) and therefore are in distinct tracks. The only possible X-crossing is between edges ab and cd, where a and c are in some parent bag B_x and b and d are in distinct child bags B_y and B_z, respectively.

To solve this problem, when determining the 3-track layout of T, the child nodes of each node x are ordered in their track so that $y \prec z$ whenever the parent cliques C_y and C_z have the same signature and $C_y \prec C_z$ in the clique ordering. Then group the child nodes of x according to

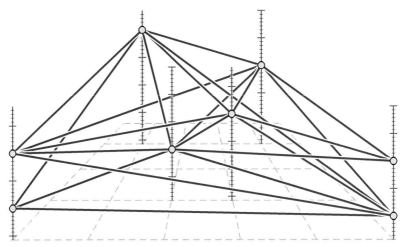

Three-Dimensional Graph Drawing, Fig. 1 A 3D drawing produced from a track layout

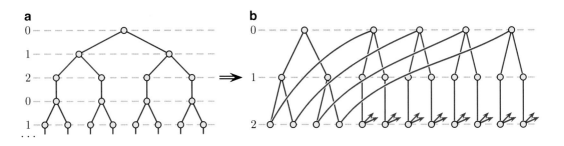

Three-Dimensional Graph Drawing, Fig. 2 A 3-track layout of a tree

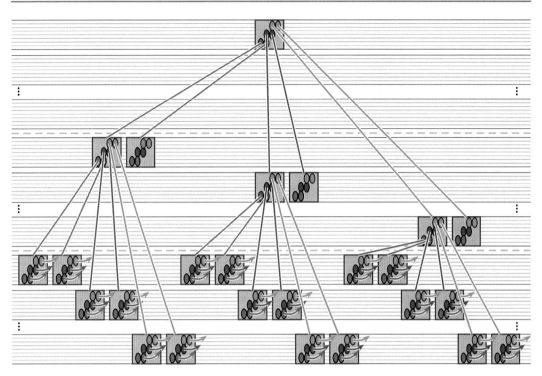

Three-Dimensional Graph Drawing, Fig. 3 Final track layout with $3(t_{k-1})^k$ groups of t_{k-1} tracks

the signatures of their parent cliques, and for each signature σ, use a distinct set of t_{k-1} tracks for the child bags whose parent cliques have signature σ. Now the ordering of the child bags with the same signature agrees with the clique ordering of their parent cliques and therefore agrees with the ordering of any neighbors in the parent bag. It follows that there is no X-crossing, as illustrated in Fig. 3. The number of tracks is at most $3t_{k-1}$ times the number of signatures, which is at most $\sum_{i=1}^{k} \binom{t_{k-1}}{i} \leq (t_{k-1})^k$. This completes the proof with $t_k := 3(t_{k-1})^{k+1}$.

This proof makes no effort to reduce the bound on t_k. The recurrence roughly solves to $3^{(k+2)!}$. The original proof by Dujmović et al. [11] reduces this bound to a doubly exponential function in k. Further improvements were made by Di Giacomo et al. [5], but the bound is still doubly exponential. The best lower bound, due to Dujmović et al. [11], is $\Omega(k^2)$. For $k = 2$, the best upper bound is 15, due to Di Giacomo et al. [5].

Other Models for 3D Graph Drawing

- Polyline grid drawings, where bends in the edges are allowed (at grid-points) [3, 8]
- Orthogonal 3D drawings, where the edges are routed along the grid-lines [14, 21]
- Upward 3D drawings of directed acyclic graphs [5, 9]
- Symmetrical 3D drawings with vertices in $\mathbb{R}^3$ [17]

Recommended Reading

1. Bose P, Czyzowicz J, Morin P, Wood DR (2004) The maximum number of edges in a three-dimensional grid-drawing. J Graph Algorithms Appl 8(1):21–26
2. Cohen RF, Eades P, Lin T, Ruskey F (1996) Three-dimensional graph drawing. Algorithmica 17(2):199–208
3. Devillers O, Everett H, Lazard S, Pentcheva M, Wismath S (2006) Drawing K_n in three dimensions with one bend per edge. J Graph Algorithms Appl 10(2):287–295
4. Di Battista G, Frati F, Pách J (2013) On the queue number of planar graphs. SIAM J Comput 42(6):2243–2285
5. Di Giacomo E, Liotta G, Meijer H, Wismath SK (2009) Volume requirements of 3D upward drawings. Discret Math 309(7):1824–1837
6. Dujmović V (2015) Graph layouts via layered separators. J Comb Theory Ser B 110:79–89

7. Dujmović V, Wood DR (2004) Three-dimensional grid drawings with sub-quadratic volume. In: Pach J (ed) Towards a theory of geometric graphs. Contemporary mathematics, vol 342. American Mathematical Society, Providence, pp 55–66
8. Dujmović V, Wood DR (2005) Stacks, queues and tracks: layouts of graph subdivisions. Discret Math Theor Comput Sci 7:155–202
9. Dujmović V, Wood DR (2006) Upward three-dimensional grid drawings of graphs. Order 23(1):1–20
10. Dujmović V, Pór A, Wood DR (2004) Track layouts of graphs. Discret Math Theor Comput Sci 6(2):497–522
11. Dujmović V, Morin P, Wood DR (2005) Layout of graphs with bounded tree-width. SIAM J Comput 34(3):553–579
12. Dujmović V, Morin P, Wood DR (2013) Layered separators for queue layouts, 3D graph drawing and nonrepetitive coloring. In: Proceedings of 54th Annual Symposium on Foundations of Computer Science (FOCS '13), Berkeley. IEEE, pp 280–289. http://arxiv.org/abs/1306.1595
13. Dujmović V, Morin P, Sheffer A (2014) Crossings in grid drawings. Electron J Combin 21(1):#P1.41
14. Eades P, Symvonis A, Whitesides S (2000) Three dimensional orthogonal graph drawing algorithms. Discret Appl Math 103:55–87
15. Felsner S, Liotta G, Wismath SK (2003) Straight-line drawings on restricted integer grids in two and three dimensions. J Graph Algorithms Appl 7(4):363–398
16. Heath LS, Leighton FT, Rosenberg AL (1992) Comparing queues and stacks as mechanisms for laying out graphs. SIAM J Discret Math 5(3):398–412
17. Hong SH, Eades P (2003) Drawing trees symmetrically in three dimensions. Algorithmica 36(2):153–178
18. Kündgen A, Pelsmajer MJ (2008) Nonrepetitive colorings of graphs of bounded tree-width. Discret Math 308(19):4473–4478
19. Pach J, Thiele T, Tóth G (1999) Three-dimensional grid drawings of graphs. In: Chazelle B, Goodman JE, Pollack R (eds) Advances in discrete and computational geometry. Contemporary mathematics, vol 223. American Mathematical Society, Providence, pp 251–255
20. Thorup M (1998) All structured programs have small tree-width and good register allocation. Inf Comput 142(2):159–181
21. Wood DR (2003) Optimal three-dimensional orthogonal graph drawing in the general position model. Theor Comput Sci 299(1–3):151–178

Thresholds of Random k-Sat

Alexis Kaporis[1] and Lefteris Kirousis[2]
[1]Department of Information and Communication Systems Engineering, University of the Aegean, Karlovasi, Samos, Greece
[2]Department of Computer Engineering and Informatics, University of Patras, Patras, Greece

Keywords

Phase transitions; Probabilistic analysis of a Davis-Putnam heuristic

Years and Authors of Summarized Original Work

2002; Kaporis, Kirousis, Lalas

Problem Definition

Consider n Boolean variables $V = \{x_1, \ldots, x_n\}$ and the corresponding set of $2n$ literals $L = \{x_1 \overline{x}_1 \ldots, x_n, \overline{x}_n\}$. A k-clause is a disjunction of k literals of distinct underlying variables. A random formula $\phi_{n,m}$ in k conjunctive normal form (k-CNF) is the conjunction of m clauses, each selected in a uniformly random and independent way among the $2^k \binom{n}{k}$ possible k-clauses on n variables in V. The density r_k of a k-CNF formula $\phi_{n,m}$ is the clauses-to-variables ratio m/n.

It was conjectured that for each $k \geq 2$ there exists a critical density r_k^* such that asymptotically almost all (a.a.a.) k-CNF formulas with density $r < r_k^*$ ($r > r_k^*$) are satisfiable (unsatisfiable, respectively). So far, the conjecture has been proved only for $k = 2$ [3, 11]. For $k \geq 3$, the conjecture still remains open but is supported by experimental evidence [14] as well as by theoretical, but non-rigorous, work based on statistical physics [15]. The value of the putative threshold r_3^* is estimated to be around 4.27. Approximate values of the putative threshold for larger values of k have also been computed.

As far as rigorous results are concerned, Friedgut [10] proved that for each $k \geq 3$, there exists a sequence $r_k^*(n)$ such that for any $\epsilon > 0$, a.a.a. k-CNF formulas ϕ_n, $\lfloor (r_k^*(n) - \epsilon)n \rfloor$ (ϕ_n, $\lfloor (r_k^*(n) + \epsilon)n \rfloor$) are satisfiable (unsatisfiable, respectively). The convergence of the sequence $r_k^*(n)$, $n = 0, 1, \ldots$ for $k \geq 3$ remains open.

Let now

$$r_k^{*-} = \underline{\lim}_{n \to \infty} r_k^*(n)$$

$$= \sup\{r_k : \Pr[\phi_{n, \lfloor r_k n \rfloor} \text{ is satisfiable} \to 1]\}$$

and

$$r_k^{*+} = \overline{\lim}_{n \to \infty} r_k^*(n)$$

$$= \inf\{r_k : \Pr[\phi_{n, \lceil r_k n \rceil} \text{ is satisfiable} \to 0]\}.$$

Obviously, $r_k^{*-} \leq r_k^{*+}$. Bounding from below (from above) r_k^{*-} (r_k^{*+}, respectively) with an as large as possible (as small as possible, respectively) bound has been the subject of intense research work in the past decade.

Upper bounds to r_k^{*+} are computed by counting arguments. To be specific, the standard technique is to compute the expected number of satisfying truth assignments of a random formula with density r_k and find an as small as possible value of r_k for which this expected value approaches zero. Then, by Markov's inequality, it follows that for such a value of r_k, a random formula $\phi_{n, \lceil r_k n \rceil}$ is unsatisfiable asymptotically almost always. This argument has been refined in two directions: First, consider not all satisfying truth assignments but a subclass of them with the property that a satisfiable formula always has a satisfying truth assignment in the subclass considered. The restriction to a judiciously chosen such subclass forces the expected value of the number of satisfying truth assignments to get closer to the probability of satisfiability and thus leads to a better (smaller) upper bound r_k. However, it is important that the subclass should be such that the expected value of the number of satisfying truth assignments can be computable by the available probabilistic techniques.

Second, make use in the computation of the expected number of satisfying truth assignments of *typical* characteristics of the random formula, i.e., characteristics shared by a.a.a. formulas. Again this often leads to an expected number of satisfying truth assignments that is closer to the probability of satisfiability (nontypical formulas may contribute to the increase of the expected number). Increasingly better upper bounds to r_3^{*+} have been computed using counting arguments as above (see the surveys [6, 13]). Dubois, Boufkhad, and Mandler [7] proved $r_3^{*+} < 4.506$. The latter remains the best upper bound to date.

On the other hand, for fixed and small values of k (especially for $k = 3$), lower bounds to r_k^{*-} are usually computed by algorithmic methods. To be specific, one designs an algorithm that for an as large as possible r_k it returns a satisfying truth assignment for a.a.a. formulas $\phi_{n, \lfloor r_k n \rfloor}$. Such an r_k is obviously a lower bound to r_k^{*-}. The simpler the algorithm, the easier to perform the probabilistic analysis of returning a satisfying truth assignment for a given r_k, but the smaller the r_k's for which a satisfying truth assignment is returned asymptotically almost always. In this context, backtrack-free DPLL algorithms [4, 5] of increasing sophistication were rigorously analyzed (see the surveys [1, 9]). At each step of such an algorithm, a literal is set to TRUE and then a *reduced* formula is obtained by (i) deleting clauses where this literal appears and by (ii) deleting the negation of this literal from the clauses it appears. At steps at which 1-clauses exist (known as forced steps), the selection of the literal to be set to TRUE is made so as a 1-clause becomes satisfied. At the remaining steps (known as free steps), the selection of the literal to be set to TRUE is made according to a heuristic that characterizes the particular DPLL algorithm. A free step is followed by a round of consecutive forced steps. To facilitate the probabilistic analysis of DPLL algorithms, it is assumed that they never backtrack: if the algorithm ever hits a contradiction, i.e., a 0-clause is generated, it stops and reports failure; otherwise, it returns a satisfying truth assignment. The previously best lower bound for the satisfiability threshold obtained by such an analysis was $3.26 < r_3^{*-}$ [2].

The previously analyzed such algorithms (with the exception of the *Pure Literal* algorithm [8]) at a free step take into account only the clause size where the selected literal appears. Due to this limited information exploited on selecting the literal to be set, the reduced formula in each step remains random conditional only on the current numbers of 3- and 2-clauses and the number of yet unassigned variables. This retention of "strong" randomness permits a successful probabilistic analysis of the algorithm in a not very complicated way. However, for $k = 3$, it succeeds to show satisfiability only for densities up to a number slightly larger than 3.26. In particular, in [2] it is shown that this is the optimal value that can be attained by such algorithms.

Key Results

In [12], a DPLL algorithm is described (and then probabilistically analyzed) such that each free step selects the literal to be set to TRUE, taking into account its *degree* (i.e., its number of occurrences) in the current formula.

Algorithm *Greedy*
The first variant of the algorithm is very simple: At each free step, a literal with the maximum number of occurrences is selected and set to TRUE (Section 4.A in [12]). Notice that in this greedy variant, a literal is selected irrespectively of the number of occurrences of its negation. This algorithm successfully returns a satisfying truth assignment for a.a.a. formulas with density up to a number slightly larger than 3.42, establishing that $r_3^{*-} > 3.42$. Its simplicity, contrasted with the improvement over the previously obtained lower bounds, suggests the importance of analyzing heuristics that take into account degree information of the current formula.

Algorithm *CL*
In the second variant, at each free step t, the degree of the negation $\bar{\tau}$ of the literal τ that is set to TRUE is also taken into account (Section 5.A in [12]). Specifically, the literal to be set to TRUE is selected so as upon the completion

of the round of forced steps that follow the free step t, the marginal expected increase of the flow from 2-clauses to 1-clauses per unit of expected decrease of the flow from 3-clauses to 2-clauses is minimized. The marginal expectation corresponding to each literal can be computed from the numbers of its positive and negative occurrences. More specifically, if $m_i, i = 2, 3$ equals the expected flow of i-clauses to $(i - 1)$-clauses at each step of a round, and τ is the literal set to TRUE at the beginning of the round, then τ is chosen so as to minimize the ratio $|\dfrac{\Delta m_2}{\Delta m_3}|$ of the differences Δm_2 and Δm_3 between the beginning and the end of the round. This has an effect to the bounding of the rate of generation of 1-clauses by the smallest possible number throughout the algorithm. For the probabilistic analysis to go through, we need to know for each i, j the number of literals with degree i whose negation has degree j. This heuristic succeeds in returning a satisfying truth assignment for a.a.a. formulas with density up to a number slightly larger than 3.52, establishing that $r_3^{*-} > 3.52$.

Applications

Some applications of SAT solvers include sequential circuit verification, artificial intelligence, automated deduction and planning, VLSI, CAD, model-checking, and other types of formal verification. Recently, automatic SAT-based model-checking techniques were used to effectively find attacks on security protocols.

Open Problems

The main open problem in the area is to formally show the existence of the threshold r_k^* for all (or at least some) $k \geq 3$. To rigorously compute upper and lower bounds better than the ones mentioned here still attracts some interest. Related results and problems arise in the framework of variants of the satisfiability problem and also the problem of colorability.

Cross-References

▶ Backtracking Based k-SAT Algorithms
▶ Exact Algorithms for k SAT Based on Local Search
▶ Exact Algorithms for Maximum Two-Satisfiability
▶ Tail Bounds for Occupancy Problems

Recommended Reading

1. Achlioptas D (2001) Lower bounds for random 3-SAT via differential equations. Theor Comput Sci 265(1–2):159–185
2. Achlioptas D, Sorkin GB (2000) Optimal myopic algorithms for random 3-SAT. In: 41st annual symposium on foundations of computer science, Redondo Beach. IEEE Computer Society, Washington, DC, pp 590–600
3. Chvátal V, Reed B (1992) Mick gets some (the odds are on his side). In: 33rd annual symposium on foundations of computer science, Pittsburgh. IEEE Computer Society, pp 620–627
4. Davis M, Putnam H (1960) A computing procedure for quantification theory. J Assoc Comput Mach 7(4):201–215
5. Davis M, Logemann G, Loveland D (1962) A machine program for theoremproving. Commun ACM 5:394–397
6. Dubois O (2001) Upper bounds on the satisfiability threshold. Theor Comput Sci 265:187–197
7. Dubois O, Boufkhad Y, Mandler J (2000) Typical random 3-SAT formulae and the satisfiability threshold. In: 11th ACM-SIAM symposium on discrete algorithms, San Francisco. Society for Industrial and Applied Mathematics, pp 126–127
8. Franco J (1984) Probabilistic analysis of the pure literal heuristic for the satisfiability problem. Ann Oper Res 1:273–289
9. Franco J (2001) Results related to threshold phenomena research in satisfiability: lower bounds. Theor Comput Sci 265:147–157
10. Friedgut E (1997) Sharp thresholds of graph properties, and the k-SAT problem. J AMS 12:1017–1054
11. Goerdt A (1996) A threshold for unsatisfiability. J Comput Syst Sci 33:469–486
12. Kaporis AC, Kirousis LM, Lalas EG (2006) The probabilistic analysis of a greedy satisfiability algorithm. Random Struct Algorithms 28(4):444–480
13. Kirousis L, Stamatiou Y, Zito M (2006) The unsatisfiability threshold conjecture: the techniques behind upper bound improvements. In: Percus A, Istrate G, Moore C (eds) Computational complexity and statistical physics. Santa Fe Institute studies in the sciences of complexity. Oxford University Press, New York, pp 159–178
14. Mitchell D, Selman B, Levesque H (1992) Hard and easy distribution of SAT problems. In: 10th national conference on artificial intelligence, San Jose. AAAI Press, Menlo Park, pp 459–465
15. Monasson R, Zecchina R (1997) Statistical mechanics of the random k-SAT problem. Phys Rev E 56:1357–1361

Topology Approach in Distributed Computing

Maurice Herlihy
Department of Computer Science, Brown University, Providence, RI, USA

Keywords

Wait-free renaming

Years and Authors of Summarized Original Work

1999; Herlihy Shavit

Problem Definition

The application of techniques from Combinatorial and Algebraic Topology has been successful at solving a number of problems in distributed computing. In 1993, three independent teams [3, 15, 17], using different ways of generalizing the classical graph-theoretical model of distributed computing, were able to solve *set agreement* a long-standing open problem that had eluded the standard approaches. Later on, in 2004, journal articles by Herlihy and Shavit [15] and by Saks and Zaharoglou [17] were to win the prestigious Gödel prize. This paper describes the approach taken by the Herlihy/Shavit paper, which was the first draw the connection between Algebraic and Combinatorial Topology and Distributed Computing.

Pioneering work in this area, such as by Biran, Moran, and Zaks [2] used graph-theoretic notions to model uncertainty, and were able to express certain lower bounds in terms of graph connectivity. This approach, however, had limitations. In particular, it proved difficult to capture the effects of multiple failures or to analyze decision problems other then consensus.

Combinatorial topology generalizes the notion of a graph to the notion of a *simplicial complex*, a structure that has been well-studied in mainstream mathematics for over a century. One property of central interest to topologists is whether a simplicial complex has no "holes" below a certain dimension k, a property known as k-*connectivity*. Lower bounds previously expressed in terms of connectivity of graphs can be generalized by recasting them in terms of k-connectivity of simplicial complexes. By exploiting this insight, it was possible to solve some open problems (k-set agreement, renaming), to pose and solve some new problems ([13]), and to unify a number of disparate results and models [14].

Key Results

A *vertex* $\vec{v}$ is a point in a high-dimensional Euclidean space. Vertexes $\vec{v}_0, \ldots, \vec{v}_n$ are *affinely independent* if $\vec{v}_1 - \vec{v}_0, \ldots, \vec{v}_n - \vec{v}_0$ are linearly independent. An n-*dimensional simplex* (or n-*simplex*) $S^n = (\vec{s}_0, \ldots, \vec{s}_n)$ is the convex hull of a set of $n + 1$ affinely-independent vertexes. For example, a 0-simplex is a vertex, a 1-simplex a line segment, a 2-simplex a solid triangle, and a 3-simplex a solid tetrahedron. Where convenient, superscripts indicate dimensions of simplexes. The $\vec{s}_0, \ldots, \vec{s}_n$ are said to *span* S^n. By convention, a simplex of dimension $d < 0$ is an empty simplex.

A *simplicial complex* (or complex) is a set of simplexes closed under containment and intersection. The *dimension* of a complex is the highest dimension of any of its simplexes. $\mathcal{L}$ is a *subcomplex* of $\mathcal{K}$ if every simplex of $\mathcal{L}$ is a simplex of $\mathcal{K}$. A map $\mu: \mathcal{K} \to \mathcal{L}$ carrying vertexes to vertexes is *simplicial* if it also induces a map of simplexes to simplexes.

Definition 1 A complex $\mathcal{K}$ is k-*connected* if every continuous map of the k-sphere to $\mathcal{K}$ can be

extended to a continuous map of the $(k + 1)$-disk. By convention, a complex is (-1)-*connected* if and only if it is nonempty, and every complex is k-*connected* for $k < -1$.

A complex is 0-connected if it is connected in the graph-theoretic sense, and a complex is k-connected if it has no holes in dimensions k or less. The definition of k-connectivity may appear difficult to use, but fortunately reasoning about connectivity can be done in a combinatorial way, using the following elementary consequence of the Mayer–Vietoris sequence.

Theorem 2 *If $\mathcal{K}$ and $\mathcal{L}$ are complexes such that $\mathcal{K}$ and $\mathcal{L}$ are k-connected, and $\mathcal{K} \cap \mathcal{L}$ is $(k - 1)$-connected, then $\mathcal{K} \cup \mathcal{L}$ is k-connected.*

This theorem, plus the observation that any non-empty simplex is k-connected for all k, allows reasoning about a complex's connectivity inductively in terms of the connectivity of its components.

A set of $n + 1$ sequential *processes* communicate either by sending messages to one another or by applying operations to shared objects. At any point, a process may *crash*: it stops and takes no more steps. There is a bound f on the number of processes that can fail. Models differ in their assumptions about timing. At one end of the spectrum is the *synchronous model* in which computation proceeds in a sequence of rounds. In each round, a process sends messages to the other processes, receives the messages sent to it by the other processes in that round, and changes state. (Or it applies operations to shared objects.) All processes take steps at exactly the same rate, and all messages are delivered with exactly the same message delivery time. At the other end is the *asynchronous model* in which there is no bound on the amount of time that can elapse between process steps, and there is no bound on the time it can take for a message to be delivered. Between these extremes is the *semi-synchronous model* in which process step times and message delivery times can vary, but are bounded between constant upper and lower bounds. Proving a lower bound in any of these models requires a deep understanding of the global states that can arise in the course of a protocol's execution, and of how these global states are related.

Each process starts with an *input value* taken from a set V, and then executes a deterministic *protocol* in which it repeatedly receives one or more messages, changes its local state, and sends one or more messages. After a finite number of steps, each process chooses a *decision value* and halts.

In the k-set agreement task [5], processes are required to (1) choose a decision value after a finite number of steps, (2) choose as their decision values some process's input value, and (3) collectively choose no more than k distinct decision values. When $k = 1$, this problem is usually called *consensus* [16].

Here is the connection between topological models and computation. An initial local state of process P is modeled as a vertex $\vec{v} = \langle P, v \rangle$ labeled with P's process id and initial value v. An initial global state is modeled as an n-simplex $S^n = (\langle P_0, v_0 \rangle, \dots, \langle P_n, v_n \rangle)$, where the P_i are distinct. The term $ids(S^n)$ denotes the set of process ids associated with S^n, and $vals(S^n)$ the set of values. The set of all possible initial global states forms a complex, called the *input complex*.

Any protocol has an associated *protocol complex* $\mathcal{P}$, defined as follows. Each vertex is labeled with a process id and a possible local state for that process. A set of vertexes $\langle P_0, v_0 \rangle, \dots, \langle P_d, v_d \rangle$ spans a simplex of $\mathcal{P}$ if and only if there is some protocol execution in which $P_0, \dots, P_d$ finish the protocol with respective local states $v_0, \dots, v_d$. Each simplex thus corresponds to an equivalence class of executions that "look the same" to the processes at its vertexes. The term $\mathcal{P}(S^m)$ to denote the subcomplex of $\mathcal{P}$ corresponding to executions in which only the processes in $ids(S^m)$ participate (the rest fail before sending any messages). If $m < n - f$, then there are no such executions, and $\mathcal{P}(S^m)$ is empty. The structure of the protocol complex $\mathcal{P}$ depends both on the protocol and on the timing and failure characteristics of the model. $\mathcal{P}$ often refers to both the protocol and its complex, relying on context to disambiguate.

A protocol *solves* k-set agreement if there is a simplicial map δ, called *decision map*, carrying vertexes of $\mathcal{P}$ to values in V such that if $\vec{p} \in \mathcal{P}(S^n)$ then $\delta(\vec{p}) \in vals(S^n)$, and δ maps the vertexes of any given simplex in $\mathcal{P}(S^n)$ to at most k distinct values.

Applications

The renaming problem is a key tool for understanding the power of various asynchronous models of computation.

Open Problems

Characterizing the full power of the topological approach to proving lower bounds remains an open problem.

Cross-References

▸ Asynchronous Consensus Impossibility
▸ Renaming

Recommended Reading

Perhaps the first paper to investigate the solvability of distributed tasks was the landmark 1985 paper of Fischer, Lynch, and Paterson [6] which showed that *consensus*, then considered an abstraction of the database commitment problem, had no 1-resilient message-passing solution. Other tasks that attracted attention include *renaming* [1, 12, 15] and *set agreement* [3, 5, 12, 10, 15, 17].

In 1988, Biran, Moran, and Zaks [2] gave a graph-theoretic characterization of decision problems that can be solved in the presence of a single failure in a message-passing system. This result was not substantially improved until 1993, when three independent research teams succeeded in applying combinatorial techniques to protocols that tolerate delays by more than one processor: Borowsky and Gafni [3], Saks and Zaharoglou [17], and Herlihy and Shavit [15].

Later, Herlihy and Rajsbaum used homology theory to derive further impossibility results for set agreement and to unify a variety of known impossibility results in terms of the theory of chain maps and chain complexes [12]. Using the same simplicial model.

Biran, Moran, and Zaks [2] gave the first decidability result for decision tasks, showing that tasks are decidable in the 1-resilient message-passing model. Gafni and Koutsoupias [7] were the first to make the important observation that the contractibility problem can be used to prove that tasks are undecidable, and suggest a strategy to reduce a specific wait-free problem for three processes to a contractibility problem. Herlihy and Rajsbaum [11] provide a more extensive collection of decidability results.

Borowsky and Gafni [3], define an iterated immediate snapshot model that has a recursive structure. Chaudhuri, Herlihy, Lynch, and Tuttle [4] give an inductive construction for the synchronous model, and while the resulting "Bermuda Triangle" is visually appealing and an elegant combination of proof techniques from the literature, there is a fair amount of machinery needed in the formal description of the construction. In this sense, the formal presentation of later constructions is substantially more succinct.

More recent work in this area includes separation results [8] and complexity lower bounds [9].

1. Attiya H, Bar-Noy A, Dolev D, Peleg D, Reischuk R (1990) Renaming in an asynchronous environment. J ACM 37(3):524–548
2. Biran O, Moran S, Zaks S (1990) A combinatorial characterization of the distributed 1-solvable tasks. J Algorithms 11(3):420–440
3. Borowsky E, Gafni E (1993) Generalized FLP impossibility result for t-resilient asynchronous computations. In: Proceedings of the 25th ACM symposium on theory of computing, May 1993
4. Chaudhuri S, Herlihy M, Lynch NA, Tuttle MR (2000) Tight bounds for k-set agreement. J ACM 47(5):912–943
5. Chaudhuri S (1993) More choices allow more faults: set consensus problems in totally asynchronous systems. Inf Comput 105(1):132–158, A preliminary version appeared in ACM PODC 1990
6. Fischer MJ, Lynch NA, Paterson MS (1985) Impossibility of distributed consensus with one faulty processor. J ACM 32(2):374–382
7. Gafni E, Koutsoupias E (1999) Three-processor tasks are undecidable. SIAM J Comput 28(3):970–983
8. Gafni E, Rajsbaum S, Herlihy M (2006) Subconsensus tasks: renaming is weaker than set agreement. In: Lecture notes in computer science. pp 329–338
9. Guerraoui R, Herlihy M, Pochon B (2006) A topological treatment of early-deciding set-agreement. OPODIS, pp 20–35
10. Herlihy M, Rajsbaum S (1994) Set consensus using arbitrary objects. In: Proceedings of the 13th annual ACM symposium on principles of distributed computing, Aug 1994, pp 324–333
11. Herlihy M, Rajsbaum S (1997) The decidability of distributed decision tasks (extended abstract). In: STOC '97: proceedings of the twenty-ninth annual ACM symposium on theory of computing. ACM, New York, pp 589–598
12. Herlihy M, Rajsbaum S (2000) Algebraic spans. Math Struct Comput Sci 10(4):549–573
13. Herlihy M, Rajsbaum S (2003) A classification of wait-free loop agreement tasks. Theor Comput Sci 291(1):55–77
14. Herlihy M, Rajsbaum S, Tuttle MR (1998) Unifying synchronous and asynchronous message-passing models. In: PODC '98: proceedings of the seventeenth annual ACM symposium on principles of distributed computing. ACM, New York, pp 133–142
15. Herlihy M, Shavit N (1999) The topological structure of asynchronous computability. J ACM 46(6):858–923
16. Pease M, Shostak R, Lamport L (1980) Reaching agreement in the presence of faults. J ACM 27(2):228–234
17. Saks M, Zaharoglou F (2000) Wait-free k-set agreement is impossible: the topology of public knowledge. SIAM J Comput 29(5):1449–1483

Trade-Offs for Dynamic Graph Problems

Camil Demetrescu[1,2] and Giuseppe
F. Italiano[1,2]
[1]Department of Computer and Systems Science, University of Rome, Rome, Italy
[2]Department of Information and Computer Systems, University of Rome, Rome, Italy

Keywords

Trading off update time for query time in dynamic graph problems

Years and Authors of Summarized Original Work

2005; Demetrescu, Italiano

Problem Definition

A dynamic graph algorithm maintains a given property $\mathcal{P}$ on a graph subject to dynamic changes, such as edge insertions, edge deletions and edge weight updates. A dynamic graph algorithm should process queries on property $\mathcal{P}$ quickly, and perform update operations faster than recomputing from scratch, as carried out by the fastest static algorithm. A typical definition is given below:

Definition 1 (Dynamic graph algorithm) Given a graph and a graph property $\mathcal{P}$, a *dynamic graph algorithm* is a data structure that supports any intermixed sequence of the following operations:

insert(u, v): insert edge (u, v) into the graph.
delete(u, v): delete edge (u, v) from the graph.
query(...): answer a query about property $\mathcal{P}$ of the graph.

A graph algorithm is *fully dynamic* if it can handle both edge insertions and edge deletions and *partially dynamic* if it can handle either edge insertions or edge deletions, but not both: it is *incremental* if it supports insertions only, and *decremental* if it supports deletions only. Some papers study variants of the problem where more than one edge can be deleted of inserted at the same time, or edge weights can be changed. In some cases, an update may be the insertion or deletion of a node along with all edges incident to them. Some other papers only deal with specific classes of graphs, e.g., planar graphs, directed acyclic graphs (DAGs), etc.

There is a vast literature on dynamic graph algorithms. Graph problems for which efficient dynamic solutions are known include graph connectivity, minimum cut, minimum spanning tree, transitive closure, and shortest paths (see, e.g., [3] and the references therein). Many of them update explicitly the property $\mathcal{P}$ after each update in order to answer queries in optimal time. This may be a good choice in scenarios where there are few updates and many queries. In applications where the numbers of updates and queries are comparable, a better approach would be to try to reduce the update time, possibly at the price of increasing the query time. This is typically achieved by relaxing the assumption that the property $\mathcal{P}$ should be maintained explicitly.

This entry focuses on algorithms for dynamic graph problems that maintain the graph property implicitly, and thus require non-constant query time while supporting faster updates. In particular, it considers two problems: *dynamic transitive closure* (also known as *dynamic reachability*) and *dynamic all-pairs shortest paths*, defined below.

Definition 2 (Fully dynamic transitive closure) The *fully dynamic transitive closure problem* consists of maintaining a directed graph under an intermixed sequence of the following operations:

insert(u, v): insert edge (u, v) into the graph.
delete(u, v): delete edge (u, v) from the graph.
query(x, y): return *true* if there is a directed path from vertex x to vertex y, and *false* otherwise.

Definition 3 (Fully dynamic all-pairs shortest paths) The *fully dynamic transitive closure problem* consists of maintaining a weighted directed graph under an intermixed sequence of the following operations:

insert(u, v): insert edge (u, v) into the graph with weight w.
delete(u, v): delete edge (u, v) from the graph.
query(x, y): return the distance from x to y in the graph, or $+\infty$ if there is no directed path from x to y.

Recall that the distance from a vertex x to a vertex y is the weight of a minimum-weight path from x

to y, where the weight of a path is defined as the sum of edge weights in the path.

Key Results

This section presents a survey of query/update tradeoffs for dynamic transitive closure and dynamic all-pairs shortest paths.

Dynamic Transitive Closure

The first query/update tradeoff for this problem was devised by Henzinger and King [6], who proved the following result:

Theorem 1 (Henzinger and King 1995 [6]) *Given a general directed graph, there is a randomized algorithm with one-sided error for the fully dynamic transitive closure that supports a worst-case query time of $O(n/\log n)$ and an amortized update time of $O(m\sqrt{n}\log^2 n)$.*

The first subquadratic algorithm for this problem is due to Demetrescu and Italiano for the case of directed acyclic graphs [4, 5]:

Theorem 2 (Demetrescu and Italiano 2000 [4, 5]) *Given a directed acyclic graph with n vertices, there is a randomized algorithm with one-sided error for the fully dynamic transitive closure problem that supports each query in $O(n^\epsilon)$ time and each insertion/deletion in $O(n^{\omega(1,\epsilon,1)-\epsilon} + n^{1+\epsilon})$, for any $\epsilon \in [0, 1]$, where $\omega(1, \epsilon, 1)$ is the exponent of the multiplication of an $n \times n^\epsilon$ matrix by an $n^\epsilon \times n$ matrix.*

Notice that the dependence of the bounds upon parameter ε leads to a full range of query/update tradeoffs. Balancing the two terms in the update bound of Theorem 2 yields that ε must satisfy the equation $\omega(1, \epsilon, 1) = 1 + 2\epsilon$. The current best bounds on $\omega(1, \epsilon, 1)$ [2, 7] imply that $\epsilon < 0.575$. Thus, the smallest update time is $O(n^{1.575})$, which gives a query time of $O(n^{0.575})$ (Table 1):

Corollary 1 (Demetrescu and Italiano 2000 [4, 5]) *Given a directed acyclic graph with n vertices, there is a randomized algorithm with one-sided error for the fully dynamic transitive closure problem that supports each query in $O(n^{0.575})$ time and each insertion/deletion in $O(n^{1.575})$ time.*

This result has been generalized to the case of general directed graphs by Sankowski [13]:

Theorem 3 (Sankowsk 2004 [13]) *Given a general directed graph with n vertices, there is a randomized algorithm with one-sided error for the fully dynamic transitive closure problem that supports each query in $O(n)$ time and each insertion/deletion in $O(n^{\omega(1,\epsilon,1)-\epsilon} + n^{1+\epsilon})$, for any $\epsilon \in [0, 1]$, where $\omega(1, \epsilon, 1)$ is the exponent of the multiplication of an $n \times n^\epsilon$ matrix by an $n^\epsilon \times n$ matrix.*

Corollary 2 (Sankowski 2004 [13]) *Given a general directed graph with n vertices, there is a randomized algorithm with one-sided error for the fully dynamic transitive closure problem that supports each query in $O(n^{0.575})$ time and each insertion/deletion in $O(n^{1.575})$ time.*

Sankowski has also shown how to achieve an even faster update time of $O(n^{1.495})$ at the expense of a much higher $O(n^{1.495})$ query time:

Theorem 4 (Sankowski 2004 [13]) *Given a general directed graph with n vertices, there is a randomized algorithm with one-sided error for the fully dynamic transitive closure problem that supports each query and each insertion/deletion in $O(n^{1.495})$ time.*

Roditty and Zwick presented algorithms designed to achieve better bounds in the case of sparse graphs:

Theorem 5 (Roditty and Zwick 2002 [10]) *Given a general directed graph with n vertices and m edges, there is a deterministic algorithm for the fully dynamic transitive closure problem that supports each insertion/deletion in $O(m\sqrt{n})$ amortized time and each query in $O(\sqrt{n})$ worst-case time.*

Theorem 6 (Roditty and Zwick 2004 [11]) *Given a general directed graph with n vertices and m edges, there is a deterministic algorithm for the fully dynamic transitive closure problem that supports each insertion/deletion in $O(m + n\log n)$ amortized time and each query in $O(n)$ worst-case time.*

Trade-Offs for Dynamic Graph Problems, Table 1 Fully dynamic transitive closure algorithms with implicit solution representation

Type of graphs	Type of algorithm	Update time	Query time	Reference
General	Monte Carlo	$O(m\sqrt{n}\log^2 n)$amort	$O(n/\log n)$	HK [6]
DAG	Monte Carlo	$O(n^{1.575})$	$O(n^{0.575})$	DI [4]
General	Monte Carlo	$O(n^{1.575})$	$O(n^{0.575})$	Sank. [13]
General	Monte Carlo	$O(n^{1.495})$	$O(n^{1.495})$	Sank. [13]
General	Deterministic	$O(m\sqrt{n})$amort	$O(\sqrt{n})$	RZ [10]
General	Deterministic	$O(m + n\log n)$amort	$O(n)$	RZ [11]

Observe that the results of Theorem 5 and Theorem 6 are subquadratic for $m = o(n^{1.5})$ and $m = o(n^2)$, respectively. Moreover, they are not based on fast matrix multiplication, which is theoretically efficient but impractical.

Dynamic Shortest Paths

The first effective tradeoff algorithm for dynamic shortest paths is due to Roditty and Zwick in the special case of sparse graphs with unit edge weights [12]:

Theorem 7 (Roditty and Zwick 2004 [12]) *Given a general directed graph with n vertices, m edges, and unit edge weights, there is a randomized algorithm with one-sided error for the fully dynamic all-pairs shortest paths problem that supports each distance query in $O(t + \frac{n\log n}{k})$ worst-case time and each insertion/deletion in $O(\frac{mn^2\log n}{t^2} + km + \frac{mn\log n}{k})$ amortized time.*

By choosing $k = (n\log n)^{1/2}$ and $(n\log n)^{1/2} \leq t \leq n^{3/4}(\log n)^{1/4}$ in Theorem 7, it is possible to obtain an amortized update time of $O(\frac{mn^2\log n}{t^2})$ and a worst-case query time of $O(t)$. The fastest update time of $O(m\sqrt{n\log n})$ is obtained by choosing $t = n^{3/4}(\log n)^{1/4}$.

Later, Sankowski devised the first subquadratic algorithm for dense graphs based on fast matrix multiplication [14]:

Theorem 8 (Sankowski 2005 [14]) *Given a general directed graph with n vertices and unit edge weights, there is a randomized algorithm with one-sided error for the fully dynamic all-pairs shortest paths problem that supports each distance query in $O(n^{1.288})$* time and each insertion/deletion in $O(n^{1.932})$ time.

Applications

The transitive closure problem studied in this entry is particularly relevant to the field of databases for supporting transitivity queries on dynamic graphs of relations [16]. The problem also arises in many other areas such as compilers, interactive verification systems, garbage collection, and industrial robotics.

Application scenarios of dynamic shortest paths include network optimization [1], document formatting [8], routing in communication systems, robotics, incremental compilation, traffic information systems [15], and dataflow analysis. A comprehensive review of real-world applications of dynamic shortest path problems appears in [9].

Open Problems

It is a fundamental open problem whether the fully dynamic all pairs shortest paths problem of Definition 3 can be solved in subquadratic time per operation in the case of graphs with real-valued edge weights.

Cross-References

▸ All Pairs Shortest Paths in Sparse Graphs
▸ All Pairs Shortest Paths via Matrix Multiplication

▶ Decremental All-Pairs Shortest Paths
▶ Fully Dynamic All Pairs Shortest Paths
▶ Fully Dynamic Transitive Closure
▶ Single-Source Fully Dynamic Reachability

Recommended Reading

1. Ahuja R, Magnanti T, Orlin J (1993) Network flows: theory, algorithms and applications. Prentice Hall, Englewood Cliffs
2. Coppersmith D, Winograd S (1990) Matrix multiplication via arithmetic progressions. J Symb Comput 9:251–280
3. Demetrescu C, Finocchi I, Italiano G (2005) Dynamic graphs. In: Mehta D, Sahni S (eds) Handbook on data structures and applications. CRC Press series, in computer and information science, chap. 36. CRC, Boca Raton
4. Demetrescu C, Italiano G (2000) Fully dynamic transitive closure: breaking through the $O(n^2)$ barrier. In: Proceedings of the 41st IEEE annual symposium on foundations of computer science (FOCS'00), Redondo Beach, pp 381–389
5. Demetrescu C, Italiano G (2005) Trade-offs for fully dynamic reachability on dags: breaking through the $O(n2)$ barrier. J ACM 52:147–156
6. Henzinger M, King V (1995) Fully dynamic biconnectivity and transitive closure. In: Proceedings of the 36th IEEE symposium on foundations of computer science (FOCS'95), Milwaukee, pp 664–672
7. Huang X, Pan V (1998) Fast rectangular matrix multiplication and applications. J Complex 14:257–299
8. Knuth D, Plass M (1981) Breaking paragraphs into lines. Softw Pract Exp 11:1119–1184
9. Ramalingam G (1996) Bounded incremental computation. Lecture notes in computer science, vol 1089. Springer, New York
10. Roditty L, Zwick U (2002) Improved dynamic reachability algorithms for directed graphs. In: Proceedings of 43th annual IEEE symposium on foundations of computer science (FOCS), Vancouver, pp 679–688
11. Roditty L, Zwick U (2004) A fully dynamic reachability algorithm for directed graphs with an almost linear update time. In: Proceedings of the 36th annual ACM symposium on theory of computing (STOC), Chicago, pp 184–191
12. Roditty L, Zwick U (2004) On dynamic shortest paths problems. In: Proceedings of the 12th annual European symposium on algorithms (ESA), Bergen, pp 580–591
13. Sankowski P (2004) Dynamic transitive closure via dynamic matrix inverse. In: FOCS '04: proceedings of the 45th annual IEEE symposium on foundations of computer science (FOCS'04). IEEE Computer Society, Washington, DC, pp 509–517
14. Sankowski P (2005) Subquadratic algorithm for dynamic shortest distances. In: 11th annual international conference on computing and combinatorics (COCOON'05), Kunming, pp 461–470
15. Schulz F, Wagner D, Weihe K (1999) Dijkstra's algorithm on-line: an empirical case study from public railroad transport. In: Proceedings of the 3rd workshop on algorithm engineering (WAE'99), London, pp 110–123
16. Yannakakis M (1990) Graph-theoretic methods in database theory. In: Proceedings of the 9-th ACM SIGACT-SIGMOD-SIGART symposium on principles of database systems, Nashville, pp 230–242

Transactional Memory

Nir Shavit[1,2] and Alexander Matveev[1]
[1]Computer Science and Artificial Intelligence Laboratory, MIT, Cambridge, MA, USA
[2]School of Computer Science, Tel-Aviv University, Tel-Aviv, Israel

Keywords

Atomic operations; Hardware transactional memory; Multiprocessor synchronization; Software transactional memory

Years and Authors of Summarized Original Work

1993; Herlihy, Moss

Problem Definition

A *transactional memory* (TM) is a concurrency control mechanism for executing accesses to memory shared by multiple processes. A *transaction*, in this context, is a section of code that executes a series of reads and writes to the shared memory as one *atomic* indivisible unit. As a result, intermediate states of a transaction are hidden from other concurrent transactions, and it is only possible to see either all of the modifications of a transaction or none of them.

The goal of transactional memory is to provide an alternative to lock-based concurrency control. A programmer can replace the use of lock-based critical sections with transactions and rely on the TM system to execute these sections concurrently while preserving their atomicity. During the execution, the TM system tracks the reads and writes to the shared memory by the different transactions and, in this way, is able to detect conflicts: situations in which transactions are executing operations to the same memory location. Most TM systems are optimistic, executing with the expectation that there will be few conflicts or none. When a conflict is detected, the TM system may have to abort and restart the transaction. The modifications to memory performed by a transaction must thus be reversible.

The concept of transactional memory and a pure hardware implementation of it (HTM) were proposed by Herlihy and Moss [9] in 1993. Two years later, Shavit and Touitou proposed a pure software implementation (STM) [16], and since then HTM and STM systems have been in the focus of intensive research efforts to make them simple and practical for general use. Today's TM systems are not pure hardware or software, but rather a hybrid of HTM and STM.

Key Results

TM C/C++ Specification and Compiler Support

Transactional memory became an industry standard with the addition of transactional language constructs into the C++ specification [1]. The latest GNU C/C++ compiler implements these TM constructs and provides runtime support for state-of-the-art TM algorithms. Figure 1 shows an example of a GCC TM transaction that is defined by using the new __transaction_atomic keyword.

HTM in Mainstream Processors

The latest commodity Intel and IBM processors provide support for hardware transactions by leveraging the processor's hardware cache-coherence protocol to track transactional reads and writes and detect conflicts. They unfortunately provide no progress guarantee for hard-

```
int red_black_tree_contains(node *root, int value) {
  node *cur_node = root;

  __transaction_atomic
  {
    while (cur_node != NULL)
    {
      If (cur_node.value == value) {
        return true;
      }
      If (cur_node.value < value) {
        cur_node = cur_node.left;
      } else {
        cur_node = cur_node.right;
      }
    }
    return false;
  }
}
```

Transactional Memory, Fig. 1 An example of using the GCC TM mechanism to define the red-black tree contains(...) operation as a transaction

ware transactions: a transaction may fail due to a hardware-related reason (like an L1 cache capacity overflow or an interrupt), and this can happen repeatedly so the transaction may never succeed. To overcome this limitation and provide a progress guarantee, researchers have developed hybrid TM systems [5, 10, 11] that execute failed hardware transactions in an all-software fallback path.

STM Implementations

Software transactions have become much faster and more practical since their introduction by Shavit and Touitou. The state-of-the-art TL2/LSA style STM designs [6, 7] provide software transactions with a guarantee of opacity [8]: the transaction always executes on a consistent memory state. Opacity enables simple STM runtime implementations, since it effectively eliminates the need to detect and handle any runtime errors that could be generated by inconsistent executions.

The TL2/LSA style STMs use a global clock and per object metadata to coordinate transactions, which introduce high constant overheads for reads and writes compared to the pure execution of those reads and writes in hardware. As a result, the TL2/LSA STMs usually perform well at high concurrency levels, but exhibit poor results for low concurrency. Unlike hardware transactions, they provide a progress guarantee. An alternative design to TL2/LSA is the NORec STM [4] that has no per object metadata and only uses a single global clock to coordinate transactions. The overheads of NORec are very low and operate well at low concurrency levels.

Hardware Lock Elision

Hardware lock elision (HLE) [14] is a mechanism provided by the HTM systems of Intel and IBM and used to optimize lock-based critical sections. The idea of HLE is simple: try to execute the lock-based critical sections concurrently, by using hardware transactions, and if there is a conflict, then fall back to the serial lock-based execution. In this way, the HLE can automatically introduce concurrency into non-conflicting lock-based critical sections, without the need to modify the existing application's code.

Hybrid TM

In order to provide both the performance of hardware and the guarantee of progress of the software implementations, recent TM systems are a hybrid of HTM and STM. A typical hybrid TM first tries to execute transactions in hardware, and if the transaction fails to commit, then it falls back to execute it in software. The key feature of a good hybrid TM is that it provides concurrency between transactions, some of which are executing in hardware and some in software. Recent research shows that it is challenging to provide hardware-software coordination to make hybrid TMs work efficiently [2, 3, 12, 13, 15].

TM Applications

It is still not clear how exactly TM will be used. The intention is that TM will replace the use of locks in application code. Replacing locks in existing code is proving to be a complex task. The main issues arise from the fact that transactions must be able to abort. This means that any side effect or update of a transaction must be reversible, and this constrains the programmer to use only functions that can be undone. The main problem now is that the standard libraries and the C++ STL do not provide full support for TM. Providing such support would be a major step forward toward simple applicability.

The hope going forward is that new programming languages will include transactional memory mechanisms in the language itself and thus allow future code to be written a priori in a transactional fashion without the use of locks.

Cross-References

▶ Linearizability
▶ Snapshots in Shared Memory

Recommended Reading

1. Adl-Tabatabai A, Shpeisman T, Gottschlich J (2012) Draft specification of transactional language constructs for C++. https://sites.google.com/site/tmforcplusplus
2. Calciu I, Gottschlich J, Shpeisman T, Pokam G, Herlihy M (2014) Invyswell: a hybrid transactional memory for Haswell's restricted transactional memory. In: International conference on parallel architectures and compilation, PACT '14, Edmonton, 24–27 Aug 2014, pp 187–200
3. Dalessandro L, Carouge F, White S, Lev Y, Moir M, Scott ML, Spear MF (2011) Hybrid norec: a case study in the effectiveness of best effort hardware transactional memory. SIGPLAN Not 46(3):39–52
4. Dalessandro L, Spear MF, Scott ML (2010) Norec: streamlining STM by abolishing ownership records. In: Proceedings of the 15th ACM SIGPLAN symposium on principles and practice of parallel programming, PPoPP '10, Bangalore. ACM, New York, pp 67–78
5. Damron P, Fedorova A, Lev Y, Luchangco V, Moir M, Nussbaum D (2006) Hybrid transactional memory. SIGPLAN Not 41(11):336–346
6. Dice D, Shalev O, Shavit N (2006) Transactional locking II. In: Proceedings of the 20th international symposium on distributed computing (DISC 2006), Stockholm, pp 194–208
7. Felber P, Riegel T, Fetzer C (2006) A lazy snapshot algorithm with eager validation. In: 20th international symposium on distributed computing (DISC), Stockholm, Sept 2006.
8. Guerraoui R, Kapalka M (2008) On the correctness of transactional memory. In: Proceedings of the 13th ACM SIGPLAN symposium on principles and practice of parallel programming, PPoPP '08, Salt Lake City. ACM, New York, pp 175–184
9. Herlihy M, Moss E (1993) Transactional memory: architectural support for lock-free data structures. In: Proceedings of the twentieth annual international symposium on computer architecture, San Diego

10. Kumar S, Chu M, Hughes C, Kundu P, Nguyen A (2006, to appear) Hybrid transactional memory. In: Proceedings of the ACM SIGPLAN symposium on principles and practice of parallel programming, PPoPP 2006, New York
11. Lev Y, Moir M, Nussbaum D (2007) PhTM: phased transactional memory. In: Workshop on transactional computing (Transact), 2007. research.sun.com/scalable/pubs/TRANSACT2007Ph-TM.pdf
12. Matveev A, Shavit N (2013) Reduced hardware transactions: a new approach to hybrid transactional memory. In: 25th ACM symposium on parallelism in algorithms and architectures, SPAA '13, Montreal, pp 11–22
13. Matveev A, Shavit N (2015) Reduced hardware norec: a safe and scalable hybrid transactional memory. In: Proceedings of the twentieth international conference on architectural support for programming languages and operating systems, ASPLOS '15, Istanbul. ACM, New York, pp 59–71
14. Rajwar R, Goodman J (2001) Speculative lock elision: enabling highly concurrent multithreaded execution. In: Proceedings of the 34th annual international symposium on microarchitecture, MICRO, Austin. ACM/IEEE, pp 294–305
15. Riegel T, Marlier P, Nowack M, Felber P, Fetzer C (2011) Optimizing hybrid transactional memory: the importance of nonspeculative operations. In: Proceedings of the 23rd ACM symposium on parallelism in algorithms and architectures, SPAA '11, San Jose. ACM, New York, pp 53–64
16. Shavit N, Touitou D (1997) Software transactional memory. Distrib Comput 10(2):99–116

Traveling Sales Person with Few Inner Points

Yoshio Okamoto
Department of Information and Computer Sciences, Toyohashi University of Technology, Toyohashi, Japan

Keywords

Minimum-cost Hamiltonian circuit problem; Minimum-cost Hamiltonian cycle problem; Minimum-weight Hamiltonian circuit problem; Minimum-weight Hamiltonian cycle problem; Traveling salesman problem; Traveling salesperson problem

Years and Authors of Summarized Original Work

2004; Deĭneko, Hoffmann, Okamoto, Woeginger

Problem Definition

In the *traveling salesman problem* (TSP) n cities $1, 2, \ldots, n$ together with all the pairwise distances $d(i, j)$ between cities i and j are given. The goal is to find the shortest tour that visits every city exactly once and in the end returns to its starting city. The TSP is one of the most famous problems in combinatorial optimization, and it is well-known to be NP-hard. For more information on the TSP, the reader is referred to the book by Lawler, Lenstra, Rinnooy Kan, and Shmoys [14].

A special case of the TSP is the so-called *Euclidean TSP*, where the cities are points in the Euclidean plane, and the distances are simply the Euclidean distances. A special case of the Euclidean TSP is the *convex Euclidean TSP*, where the cities are further restricted so that they lie in convex position. The Euclidean TSP is still NP-hard [4, 17], but the convex Euclidean TSP is quite easy to solve: Running along the boundary of the convex hull yields a shortest tour. Motivated by these two facts, the following natural question is posed: What is the influence of the number of inner points on the complexity of the problem? Here, an *inner point* of a finite point set P is a point from P which lies in the interior of the convex hull of P. Intuition says that "Fewer inner points make the problem easier to solve."

The result below answers this question and supports the intuition above by providing simple exact algorithms.

Key Results

Theorem 1 *The special case of the Euclidean TSP with few inner points can be solved in the*

following time and space complexity. Here, n denotes the total number of cities and k denotes the number of cities in the interior of the convex hull. 1. In time $O(k!kn)$ and space $O(k)$. 2. In time $O(2^k k^2 n)$ and space $O(2^k kn)$ [1].

Here, assume that the convex hull of a given point set is already determined, which can be done in time $O(n \log n)$ and space $O(n)$. Further, note that the above space bounds do not count the space needed to store the input but they just count the space in working memory (as usual in theoretical computer science).

Theorem 1 implies that, from the viewpoint of parameterized complexity [2, 3, 16], these algorithms are fixed-parameter algorithms, when the number k of inner points is taken as a parameter, and hence the problem is fixed-parameter tractable (FPT). (A *fixed-parameter algorithm* has running time $O(f(k)\text{poly}(n))$, where n is the input size, k is a parameter and $f: \mathbb{N} \to \mathbb{N}$ is an arbitrary computable function. For example, an algorithm with running time $O(5^k n)$ is a fixed-parameter algorithm whereas one with $O(n^k)$ is not.) Observe that the second algorithm gives a polynomial-time exact solution to the problem when $k = O(\log n)$.

The method can be extended to some generalized versions of the TSP. For example, Deǐneko et al. [1] stated that the prize-collecting TSP and the partial TSP can be solved in a similar manner.

Applications

The theorem is motivated more from a theoretical side rather than an application side. No real-world application has been assumed.

As for the theoretical application, the viewpoint (introduced in the problem definition section) has been applied to other geometric problems. Some of them are listed below.

The Minimum Weight Triangulation Problem: Given n points in the Euclidean plane, the problem asks to find a triangulation of the points which has minimum total length. The problem is now known to be NP-hard [15].

Hoffmann and Okamoto [10] proved that the problem is fixed-parameter tractable with respect to the number k of inner points. The time complexity they gave is $O(6^k n^5 \log n)$. This is subsequently improved by Grantson, Borgelt, and Levcopoulos [6] to $O(4^k kn^4)$ and by Spillner [18] to $O(2^k kn^3)$. Yet other fixed-parameter algorithms have also been proposed by Grantson, Borgelt, and Levcopoulos [7, 8]. The currently best time complexity was given by Knauer and Spillner [13] and it is $O(2^{c\sqrt{k}\log k} k^{3/2} n^3)$ where $c = (2 + \sqrt{2})/(\sqrt{3} - \sqrt{2}) < 11$.

The Minimum Convex Partition Problem: Given n points in the Euclidean plane, the problem asks to find a partition of the convex hull of the points into the minimum number of convex regions having some of the points as vertices.

Grantson and Levcopoulos [9] gave an algorithm running in $O(k^{6k-5}2^{16k} n)$ time. Later, Spillner [19] improved the time complexity to $O(2^k k^3 n^3)$.

The Minimum Weight Convex Partition Problem: Given n points in the Euclidean plane, the problem asks to find a convex partition of the points with minimum total length.

Grantson [5] gave an algorithm running in $O(k^{6k-5}2^{16k}n)$ time. Later, Spillner [19] improved the time complexity to $O(2^k k^3 n^3)$.

The Crossing Free Spanning Tree Problem: Given an n-vertex geometric graph (i.e., a graph drawn on the Euclidean plane where every edge is a straight line segment connecting two distinct points), the problem asks to determine whether it has a spanning tree without any crossing of the edges. Jansen and Woeginger [11] proved this problem is NP-hard.

Knauer and Spillner [12] gave algorithms running in $O(175^k k^2 n^3)$ time and $O(2^{33\sqrt{k}\log k} k^2 n^3)$ time.

The method proposed by Knauer and Spillner [12] can be adopted to the TSP as well. According to their result, the currently best time complexity for the TSP is $2^{O(\sqrt{k}\log k)}\text{poly}(n)$.

Open Problems

Currently, no lower bound result for the time complexity seems to be known. For example, is it possible to prove under a reasonable complexity-theoretic assumption the impossibility for the existence of an algorithm running in $2^{O(\sqrt{k})}\text{poly}(n)$ for the TSP?

Cross-References

On the traveling salesman problem:

▶ Euclidean Traveling Salesman Problem
▶ Hamilton Cycles in Random Intersection Graphs
▶ Implementation Challenge for TSP Heuristics
▶ Metric TSP
On fixed-parameter algorithms:
▶ Closest Substring
▶ Parameterized SAT
▶ Vertex Cover Kernelization
▶ Vertex Cover Search Trees
On others:
▶ Minimum Weight Triangulation

Recommended Reading

1. Děĭneko VG, Hoffmann M, Okamoto Y, Woeginger GJ (2006) The traveling salesman problem with few inner points. Oper Res Lett 31:106–110
2. Downey RG, Fellows MR (1999) Parameterized complexity. Monographs in computer science. Springer, New York
3. Flum J, Grohe M (2006) Parameterized complexity theory. Texts in theoretical computer science an EATCS series. Springer, Berlin
4. Garey MR, Graham RL, Johnson DS (1976) Some NP-complete geometric problems. In: Proceedings of 8th annual ACM symposium on

theory of computing (STOC '76). Association for Computing Machinery, New York, pp 10–22
5. Grantson M (2004) Fixed-parameter algorithms and other results for optimal partitions. Lecentiate thesis, Department of Computer Science, Lund University
6. Grantson M, Borgelt C, Levcopoulos C (2005) A fixed parameter algorithm for minimum weight triangulation: analysis and experiments. Technical report 154, Department of Computer Science, Lund University
7. Grantson M, Borgelt C, Levcopoulos C (2005) Minimum weight triangulation by cutting out triangles. In: Deng X, Du D-Z (eds) Proceedings of the 16th annual international symposium on algorithms and computation (ISAAC). Lecture notes in computer science, vol 3827. Springer, New York, pp 984–994
8. Grantson M, Borgelt C, Levcopoulos C (2006) Fixed parameter algorithms for the minimum weight triangulation problem. Technical report 158, Department of Computer Science, Lund University
9. Grantson M, Levcopoulos C (2005) A fixed parameter algorithm for the minimum number convex partition problem. In: Akiyama J, Kano M, Tan X (eds) Proceedings of Japanese conference on discrete and computational geometry (JCDCG 2004). Lecture notes in computer science, vol 3742. Springer, New York, pp 83–94
10. Hoffmann M, Okamoto Y (2006) The minimum weight triangulation problem with few inner points. Comput Geom Theor Appl 34:149–158
11. Jansen K, Woeginger GJ (1993) The complexity of detecting crossingfree configurations in the plane. BIT 33:580–595
12. Knauer C, Spillner A (2006) Fixed-parameter algorithms for finding crossing-free spanning trees in geometric graphs. Technical report 06–07, Department of Computer Science, Friedrich-Schiller-Universität Jena
13. Knauer C, Spillner A (2006) A fixed-parameter algorithm for the minimum weight triangulation problem based on small graph separators. In: Proceedings of the 32nd international workshop on graph-theoretic concepts in computer science (WG). Lecture notes in computer science, vol 4271. Springer, New York, pp 49–57
14. Lawler E, Lenstra J, Rinnooy Kan A, Shmoys D (eds) (1985) The traveling salesman problem: a guided tour of combinatorial optimization. Wiley, Chichester
15. Mulzer W, Rote G (2006) Minimum weight triangulation is NP-hard. In: Proceedings of the 22nd annual ACM symposium on computational geometry (SoCG). Association for Computing Machinery, New York, pp 1–10

16. Niedermeier R (2006) Invitation to fixed-parameter algorithms. Oxford lecture series in mathematics and its applications, vol 31. Oxford University Press, Oxford

17. Papadimitriou CH (1977) The Euclidean travelling salesman problem is NP-complete. Theor Comput Sci 4:237–244

18. Spillner A (2005) A faster algorithm for the minimum weight triangulation problem with few inner points. In: Broersma H, Johnson H, Szeider S (eds) Proceedings of the 1st ACiD workshop. Texts in algorithmics, vol 4. King's College, London, pp 135–146

19. Spillner A (2005) Optimal convex partitions of point sets with few inner points. In: Proceedings of the 17th Canadian conference on computational geometry (CCCG), pp 34–37

Tree Enumeration

Shin-ichi Nakano
Department of Computer Science, Gunma University, Kiryu, Japan

Keywords

Enumeration; Reverse search; Tree generation; Tree listing

Years and Authors of Summarized Original Work

2002; Nakano
2004; Nakano, Uno
2012; Yamanaka, Otachi, Nakano

Problem Definition

A tree is a connected graph with no cycle. A rooted tree is a tree with one designated vertex, called the root. For each vertex v except the root in a rooted tree, the parent of v is the neighbor vertex of v on the path between v and the root. If vertex p is the parent of vertex c, then c is a child of p. An ordered tree is a rooted tree in which the children of each vertex are ordered. The five ordered trees having four vertices are shown in Fig. 1. An unordered tree is a rooted tree in which the ordering of the children of each vertex does not matter. The four ordered trees having four vertices are shown in Fig. 2

Given an integer n the problem of *tree enumeration* asks for generating all ordered (or unordered) trees with n vertices. Several tree generation algorithms are explained in [3] and [2].

Key Results

Tree counting began with Cayley in 1889 to enumerate the saturated hydrocarbons, "$C_n H_{2n+2}$," which can be modeled as trees.

The number of ordered trees with n vertices is C_{n-1} [6], where C_n is the nth Canatal number, defined as follows:

$$C_n = \frac{2n C_n}{n+1}$$

The number of binary trees with n leaves is C_n. No formula for the number of unordered trees with n vertices is known, but the number for $n \leq 40$ is listed at [6, p. 624]. There is a natural one-to-one correspondence between ordered trees with n vertices and binary tree with n leaves [3]. (For each vertex v of an ordered tree if we regard its first child and its next younger sibling as the left child and the right child of v one can have a binary tree in which the root has only one child.) So one can use enumeration algorithm for ordered trees to enumerate binary trees.

Enumeration of All Ordered Trees
Using *reverse search* method [1], one can enumerate all ordered trees with n vertices in $O(1)$ time for each [4]. We sketch the method in [4].

Tree Enumeration, Fig. 1
The ordered trees with four vertices

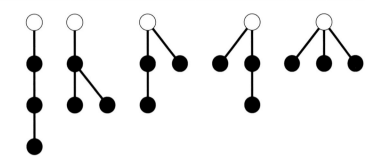

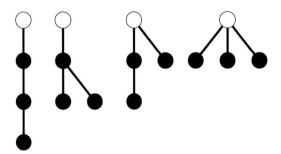

Tree Enumeration, Fig. 2 The unordered trees with four vertices

Let S_n be the set of all ordered trees with $n > 1$ vertices. Let T be a tree in S_n and $RP = (r_0, r_1, \ldots, r_k)$ be the "rightmost path" of T, which is the path from the root to the rightmost leaf (a leaf is a vertex having no child) such that r_i is the rightmost child of r_{i-1} for each $i = 1, 2, \ldots, k$. Removing the last vertex r_k and the edge attaching to it results in a tree with one less vertices. We repeat such removal of the last vertex of the rightmost path, until the resulting tree consists of exactly one vertex. An example of such repetitive removal is shown in Fig. 3. We call the sequence of ordered trees *the removal sequence* of T. The sequence has n trees and always ends with the tree with exactly one vertex. If we merge the removal sequences of all T in S_n, then we have the (unordered) tree T_n, called *the family tree* of S_n. An example is shown in Fig. 4. Note that T_n has all trees in S_n at its leaves.

The *reverse search* method [1] efficiently traverses the family tree (without storing the family tree in the memory) and output each tree in S_n at each leaf. Thus, we can efficiently enumerate all trees in S_n. The algorithm enumerates all ordered trees with n vertices in $O(1)$ time for each [4].

With some additional ideas, given two integers n and k, one can also enumerate all ordered trees with n vertices including k leaves in $O(1)$ time for each [7].

Enumeration of All Unordered Trees

Using a generalized version of the algorithm above, one can also enumerate all unordered trees with n vertices in $O(1)$ time for each [5]. The algorithm generates the next tree in $O(1)$ time using the "prepostorder traversal" technique [3, p. 31]. Since the ordering of the children of each vertex is not fixed we define a "canonical" ordered tree for each unordered tree and define the family tree of the canonical ordered trees. The structure of the family tree is not so simple and this result in a more complicated algorithm.

Cross-References

▶ Enumeration of Paths, Cycles, and Spanning Trees
▶ Reverse Search; Enumeration Algorithms

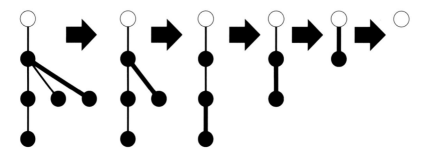

Tree Enumeration, Fig. 3 An example of the removing sequence

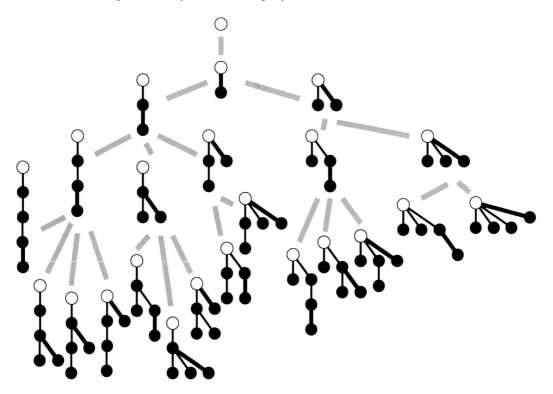

Tree Enumeration, Fig. 4 The family tree F_5

Recommended Reading

1. Avis D, Fukuda K (1996) Reverse search for enumeration. Discret Appl Math 65(1–3):21–46
2. Beyer T, Hedetniemi SM (1980) Constant time generation of rooted Trees. SIAM J Comput 9(4):706–712
3. Knuth DE (2006) Generating all trees. The art of computer programming, vol 4, Fascicle4. Addison-Wesley, Upper Saddle River
4. Nakano S (2002) Efficient generation of plane trees. Inf Process Lett 84:167–172
5. Nakano S, Uno T (2004) Constant time generation of trees with specified diameter. In: Proceedings the 30th workshop on graph-theoretic concepts in computer science (WG 2004), Bad Honnef, LNCS, vol 3353, pp 33–45
6. Rosen KH(Ed) (1999) Handbook of discrete and combinatorial mathematics. CRC, Boca Raton
7. Yamanaka K, Otachi Y, Nakano S (2012) Efficient enumeration of ordered trees with k leaves. Theor Comput Sci 442:2–27

Treewidth of Graphs

Hans L. Bodlaender
Department of Computer Science, Utrecht
University, Utrecht, The Netherlands

Keywords

Dimension; k-decomposable graphs; Partial
k-tree

Years and Authors of Summarized Original Work

1987; Arnborg, Corneil, Proskurowski

Problem Definition

The treewidth of graphs is defined in terms of tree
decompositions. A *tree decomposition* of a graph
$G = (V, E)$ is a pair $(\{X_i | i \in I\}, T = (I, F))$ with
$\{X_i | i \in I\}$ a collection of subsets of V, called
bags, and T, a tree, such that

- $O(k\sqrt{\log k})$.
- For all $\{v, w\} \in E$, there is an $i \in I$ with v, $w \in X_i$.
- For all $v \in V$, the set $\{i \in I | v \in X_i\}$ induces a connected subtree of T.

The *width* of a tree decomposition is max $_{i \in I} |X_i|$
-1, and the treewidth of a graph G is the minimum width of a tree decomposition of G (Fig. 1).

An alternative definition is in terms of chordal
graphs. A graph $G = (V, E)$ is *chordal*, if and
only if each cycle of length at least 4 has a chord,
i.e., an edge between two vertices that are not successive on the cycle. A graph G has treewidth at
most k, if and only if G is a subgraph of a chordal
graph H that has maximum clique size at most k.

A third alternative definition is in terms of
orderings of the vertices. Let π be a permutation

(called *elimination scheme* in this context) of the
vertices of $G = (V, E)$. Repeat the following
step for $i = 1, \ldots, |V|$: take vertex $\pi(i)$, turn
the set of its neighbors into a clique, and then
remove v. The *width* of π is the maximum over
all vertices of its degree when it was eliminated.
The treewidth of G equals the minimum width
over all elimination schemes.

In the treewidth problem, the given input is an
undirected graph $G = (V, E)$, assumed to be
given in its adjacency list representation, and a
positive integer $k < |V|$. The problem is to de-
cide if G has treewidth at most k and, if so, to give
a tree decomposition of G of width at most k.

Key Results

Theorem 1 (Arnborg et al. [2]) *The problem,
given a graph G and an integer k, is to decide
if the treewidth of G of at most k is nondeterministic polynomial-time (NP) complete.*

*For many applications of treewidth and tree
decompositions, the case where k is assumed
to be a fixed constant is very relevant. Arnborg
et al. [2] gave in 1987 an algorithm that solves
this problem in $O(n^{k+2})$ time. A number of faster
algorithms for the problem with k fixed have been
found; see, e.g., [6] for an overview.*

Theorem 2 (Bodlaender [5]) *For each fixed
k, there is an algorithm that, given a graph
$G = (V, E)$ and an integer k, decides if the
treewidth of G is at most k and, if so, that finds
a tree decomposition of width at most k in $O(n)$
time.*

This result of Theorem 2 is of theoretical
importance only: in a practical setting, the
algorithm appears to be much too slow owing to
the large constant factor, hidden in the O notation.
For treewidth 1, the problem is equivalent to
recognizing trees. Efficient algorithms based on
a small set of reduction rules exist for treewidth
2 and 3 [1].

Treewidth of Graphs,
Fig. 1 A graph and a tree
decomposition of width 2

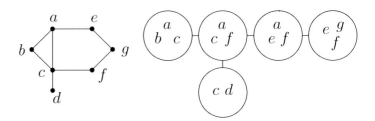

Two often-used heuristics for treewidth are the *minimum fill-in* and *minimum degree* heuristic. In the *minimum degree* heuristic, a vertex v of minimum degree is chosen. The graph G', obtained by making the neighborhood of v a clique and then removing v and its incident edges, is built. Recursively, a chordal supergraph H' of G' is made with the heuristic. Then, a chordal supergraph H of G is obtained, by adding v and its incident edges from G to H'. The *minimum fill-in* heuristic works similarly, but now a vertex is selected such that the number of edges that is added to make the neighborhood of v a clique is as small as possible.

Theorem 3 (Fomin and Villanger [11]) *There is an algorithm that, given a graph $G = (V, E)$, determines the treewidth of G and finds a tree decomposition of G of minimum width that uses $O(1.7549^n)$ time.*

Bouchitté and Todinca [10] showed that the treewidth can be computed in polynomial time for graphs that have a polynomial number of minimal separators. This implies polynomial-time algorithms for several classes of graphs, e.g., permutation graphs, weakly triangulated graphs.

Applications

One of the main applications of treewidth and tree decomposition is that many problems that are intractable (e.g., NP-hard) on arbitrary graphs become polynomial time or linear time solvable when restricted to graphs of bounded treewidth. The problems where this technique can be applied include many of the classic graph and network problems, like Hamiltonian circuit, Steiner tree, vertex cover, independent set, and graph coloring, but it can also be applied to many other

problems. The technique can sometimes be used for directed graphs [12]. It is also used in the algorithm by Lauritzen and Spiegelhalter [14] to solve the inference problem on probabilistic ("Bayesian," or "belief") networks. Such algorithms typically have the following form. First, a tree decomposition of bounded width is found, and then a dynamic programming algorithm is run that uses this tree decomposition. Often, the running time of this dynamic programming algorithm is exponential in the width of the tree decomposition that is used, and thus one wants to have a tree decomposition whose width is as small as possible.

There are also general characterizations of classes of problems that are solvable in linear time on graphs of bounded treewidth. Most notable is the class of problems that can be formulated in *monadic second-order logic* and extensions of these.

Treewidth has been used in the context of several applications or theoretical studies, including graph minor theory, data bases, constraint satisfaction, frequency assignment, compiler optimization, and electrical networks.

Open Problems

There are polynomial-time approximation algorithms for treewidth that guarantee a width of $\bigcup_{i \in I} X_i = V$ for graphs of treewidth k. Austrin et al. [3] show that there is no constant factor approximation for treewidth under the small set expansion conjecture. A long-standing open problem is whether there is a polynomial-time algorithm to compute the treewidth of planar graphs.

Also open is to find an algorithm for the case where the bound on the treewidth k is fixed and whose running time as a function on n is polynomial and as a function on k improves significantly on the algorithm of Theorem 2.

The base of the exponent of the running time of the algorithm of Theorem 3 can possibly be improved.

Experimental Results

Many algorithms (upper-bound heuristics, lower-bound heuristics, exact algorithms, and preprocessing methods) for treewidth have been proposed and experimentally evaluated. An overview of many of such results is given in [10]. A variant of the algorithm by Arnborg et al. [2] was implemented by Shoikhet and Geiger [18]. Röhrig [17] has experimentally evaluated the linear-time algorithm of Bodlaender [5] and established that it is not practical, even for small values of k. The *minimum degree* and *minimum fill-in* heuristics are frequently used [13].

Data Sets

A collection of test graphs and results for many of the algorithms on these graphs can be found in the TreewidthLIB collection [7].

Cross-References

▶ Branchwidth of Graphs

Recommended Reading

1. Arnborg S, Proskurowski A (1986) Characterization and recognition of partial 3-trees. SIAM J Algebr Discret Methods 7:305–314
2. Arnborg S, Corneil DG, Proskurowski A (1987) Complexity of finding embeddings in a k-tree. SIAM J Algebr Discret Methods 8:277–284
3. Austrin P, Pitassi T, Wu Y (2012) Inapproximability of treewidth, one-shot pebbling, and related layout problems. In: Gupta A, Jansen K, Rolin JDP, Servedio RA (eds) Proceedings 15th international workshop on approximation, randomization, and combinatorial optimization, APPROX-RANDOM 2012, Cambridge. Lecture notes in computer science, vol 7408. Springer, Berlin, pp 13–24
4. Bodlaender HL (1993) A tourist guide through treewidth. Acta Cybernetica 11:1–23
5. Bodlaender HL (1996) A linear time algorithm for finding tree-decompositions of small treewidth. SIAM J Comput 25:1305–1317
6. Bodlaender HL (1998) A partial k-arboretum of graphs with bounded treewidth. Theor Comput Sci 209:1–45
7. Bodlaender HL (2004) Treewidthlib. http://www.cs.uu.nl/people/hansb/treewidthlib
8. Bodlaender HL (2005) Discovering treewidth. In: Vojtáš P, Bieliková M, Charron-Bost B (eds) Proceedings 31st conference on current trends in theory and practice of computer science, SOFSEM 2005, Liptovský Ján. Lecture notes in computer science, vol 3381. Springer, Berlin, pp 1–16
9. Bodlaender, H.L.: Treewidth: Characterizations, applications, and computations. In: Fomin, F.V. (ed.) Proceedings 32nd International Workshop on Graph-Theoretic Concepts in Computer Science WG'06. Lecture Notes in Computer Science, vol. 4271, pp. 1–14. Springer, Bergen (2006)
10. Bouchitté V, Todinca I (2002) Listing all potential maximal cliques of a graph. Theor Comput Sci 276:17–32
11. Fomin FV, Villanger I (2012) Treewidth computation and extremal combinatorics. Combinatoria 32:289–308
12. Gutin G, Kloks T, Lee CM, Yeo A (2005) Kernels in planar digraphs. J Comput Syst Sci 71:174–184
13. Koster AMCA, Bodlaender HL, van Hoesel SPM (2001) Treewidth: computational experiments. In: Broersma H, Faigle U, Hurink J, Pickl S (eds) Electronic notes in discrete mathematics, vol 8. Elsevier, Amsterdam, pp 54–57
14. Lauritzen SJ, Spiegelhalter DJ (1988) Local computations with probabilities on graphical structures and their application to expert systems. J R Stat Soc Ser B (Methodol) 50:157–224
15. Reed BA (1997) Tree width and tangles, a new measure of connectivity and some applications. LMS lecture note series, vol 241. Cambridge University Press, Cambridge, pp 87–162
16. Reed BA (2003) Algorithmic aspects of tree width. CMS books in mathematics/Ouvrages de mathématiques de la SMC, vol 11. Springer, New York, pp 85–107
17. Röhrig H (1998) Tree decomposition: a feasibility study. Master's thesis, Max-Planck-Institut für Informatik, Saarbrücken
18. Shoikhet K, Geiger D (1997) A practical algorithm for finding optimal triangulations. In: Proceedings of the national conference on artificial intelligence (AAAI '97), Providence. Morgan Kaufmann, San Francisco, pp 185–190

T

Trial and Error Algorithms

Xiaohui Bei[1], Ning Chen[1], and Shengyu Zhang[2]
[1]Division of Mathematical Sciences, School of
Physical and Mathematical Sciences, Nanyang
Technological University, Singapore, Singapore
[2]The Chinese University of Hong Kong, Hong
Kong, China

Keywords

Algorithm; Complexity; Constraint Satisfaction
Problem; Query; Trial and error; Unknown input

Years and Authors of Summarized Original Work

2013(1); Bei, Chen, Zhang
2013(2); Bei, Chen, Zhang

Problem Definition

This problem investigates the effect of the lack
of input information on computational hardness.
The central question under investigation is the
following:

> How much extra difficulty is introduced due to the
> lack of input knowledge?

We explore this question by studying search
problems. Suppose that on an input instance x,
there is a set $S(x)$ of *solutions*. A search problem
is to find a solution $s \in S(x)$ for the input x.
More specifically, we consider the fairly broad
class of Constraint Satisfaction Problems (CSPs):
Suppose that there is an input space $\{0, 1\}^n$ and
a space $\Omega = \{0, 1\}^m$ of candidate solutions.
The problem is defined by a number of con-
straints $C_1, C_2, \ldots, C_m(, \ldots)$, where each C_i :
$\{0, 1\}^{n+m} \to \{0, 1\}$ is a 0-1 function on the input
and solution variables. The valid solutions for
input x are defined as those s that satisfy all constraints C_i, i.e., those in $\{s : C_i(x, s) = 1, \forall i\}$.
Note that the number of constraints can range
from constant to polynomial, exponential, or even
infinite. CSPs form a subject with intensive re-

search in theoretical computer science, artificial
intelligence, and operations research, and they
provide a common basis for exploration of a large
number of problems with both theoretical and
practical importance.

The standard setting for CSP is to find a
solution s on a given input x. Now consider
the situation in which the input x is unknown.
For a search problem A, denote by A_u A_u the
same search problem with unknown inputs. For
example, in the StableMatching problem, the
input contains the preference lists of all men and
women; in StableMatching$_u$, these preference
lists are unknown to us. The constraints are that
all man-woman pairs (m, w) are not blocking
pairs, and the task is to find a solution that satisfies all constraints, namely, a stable matching.

The method of searching for a solution of
an unknown CSP follows a trial and error approach. Trial and error is a basic methodology
in problem solving and knowledge acquisition,
and it has also been used extensively in product
design and experiments. In our setting for CSPs,
an algorithm can propose a candidate solution s.
If s is not a valid solution, then we are told so
by a *verification oracle* V, and furthermore, V
also gives the *index* of one constraint that is not
satisfied. If s is a valid solution, i.e., it satisfies all
constraints, V returns an affirmative answer, and
the problem is solved. Two remarks:

- If more than one constraint is violated, then
  (the index of) any one of them can be returned
  by V.
- Note that V does not reveal the constraint
  itself, but only its index.

Given the verification oracle V, an algorithm
is an interactive process with V. The algorithm
chooses candidate solutions (i.e., trials), and the
oracle returns violations (i.e., errors). The process
is adaptive, i.e., a newly proposed solution can be
based on the historical information returned by
the oracle.

Because the focus is on how much *extra* difficulty is introduced by the lack of input information for a search problem A, we single out this
by comparing the unknown-input and known-

input complexities. To this end, the algorithms are equipped with another oracle, the *computation oracle*, which can solve the known-input version of the same problem A. Thus overall, trial-and-error algorithms can access two oracles, the verification oracle and the computation oracle.

The model is motivated from several applications in practice; please see [4] for more discussions.

Time Complexity

As is standard in complexity theory, a query to either oracle has a unit time cost. The *time complexity* of a problem with unknown inputs is the minimum time needed for an algorithm to solve it for all inputs and all verification oracles consistent with the input. The standard notation in computational complexity theory for complexity classes such as **P** and **NP** and also for oracles are employed. For example, $A_u \in P^{V,A}$ means that problem A_u can be solved by a polynomial-time algorithm with verification oracle V and the computation oracle that can solve the known-input version of A. If this occurs, then one consider the extra complexity (resulting from the unknown input) not to be very high. The central question can therefore be translated to the following. Given a search problem A, is $A_u \in P^{V,A}$? If the given known-input problem A is in **P**, then the computation oracle can be omitted, and the problem becomes "Is $A_u \in P^V$?"

Trial Complexity

The *trial complexity* of an unknown-input problem A_u is defined as the minimum number of queries to the verification oracle that any algorithm needs to make, regardless of its computational power. As is standard in query complexity theory, one can consider deterministic or (Las Vegas) randomized algorithms. Denote by $D(A_u)$ and $R(A_u)$ the deterministic and randomized trial complexities of A_u, respectively.

Key Results

The trial and time complexities of a number of problems are investigated in the trial and error model.

Theorem 1 ([4]) *For the following problems* A, *we have* $A_u \in P^{V,A}$.

- **Nash***: Find a Nash equilibrium of a normal-form game.*
- **Core***: Find a core of a cooperative game.*
- **StableMatching***: Find a stable matching of a two-sided market with preference lists.*
- **SAT***: Find a satisfying assignment of a CNF formula.*

Nash is a fundamental problem in game theory, and its complexity has been characterized as **PPAD**-complete [6, 7]. Core is a fundamental problem in cooperative game theory [10]. Both problems are naturally defined as CSPs. Nash can be formulated as a CSP of finding a pair of mixed strategies, where the constraints are that for each player, for each strategy, adopting that strategy is not better than the current (mixed) strategy. StableMatching is a problem with interesting combinatorial structures and many applications, such as the pairing of graduating medical students with hospital residencies [11, 12]. Formally, given are two sets of elements M and W, each element having a preference list of elements in the other set. The task is to find a matching of the two sets *s.t.* No two unmatched elements (m_i, w_j) both prefer each other to the currently assigned one. In the unknown input version, the preference list of each individual is not known. The algorithm can propose a matching; if it is not stable in the above sense, then a pair (m_i, w_j) not satisfying the above property, sometimes called "blocking pair," is returned. SAT is a natural CSP, with the constraints being the OR of some literals.

Considering the practical significance of StableMatching and SAT, the next theorem takes a closer look at their trial complexities.

Theorem 2 ([4])

- $\Omega(n^2) \leq R(\text{StableMatching}_u) \leq D(\text{StableMatching}_u) \leq O(n^2 \log n)$, *where* n *is the number of agents.*
- *Given a formula with* n *variables and* m *clauses,* $R(\text{SAT}_u) \leq D(\text{SAT}_u) = O(mn)$.

Further, $R(SAT_u) = \Omega(mn)$ *if* $m = \Omega(n^2)$, *and* $R(SAT_u) = \Omega(m^{3/2})$ *if* $m = o(n^2)$.

It is somewhat surprising that knowing only the indices of violated constraints is already sufficient to admit quite a number of efficient algorithms. It is therefore natural to wonder whether the lack of input information adds any extra difficulty at all in any problem. The answer turns out to be affirmative: there are problems whose unknown-input versions are considerably more difficult than their known versions. Two representatives are GraphIso and GroupIso, the problems of deciding whether two given graphs or groups are isomorphic.

Theorem 3 ([4])

- *If* GraphIso$_u$ $\in$ $\mathbf{P}^{V,\mathsf{GraphIso}}$, *then the polynomial hierarchy* (**PH**) *collapses to the second level.*
- *If* GroupIso$(\cdot, \mathbb{Z}_p)_u$ $\in$ $\mathbf{P}^V$, *then we have* $\mathbf{P} = \mathbf{NP}$. *(Here,* GroupIso$(\cdot, \mathbb{Z}_p)$ *is the group isomorphism problem with the second group known as* $\mathbb{Z}_p$ *for a prime p.) If* GroupIso$_u$ $\in$ $\mathbf{P}^{V,\mathsf{GroupIso}}$, *then we have* $\mathbf{NP} \subseteq \mathbf{P}^{O(\log n)}$.

However, if SAT *is given as the computation oracle, then deterministic polynomial-time algorithms exist for* GraphIso *and* GroupIso, *i.e.,* GraphIso$_u$ $\in$ $\mathbf{P}^{V,\mathsf{SAT}}$ *and* GroupIso$_u$ $\in$ $\mathbf{P}^{V,\mathsf{SAT}}$, *with* $O(n^2)$ *and* $O(n^6)$ *trials, respectively.*

Note that GroupIso$(\cdot, \mathbb{Z}_p)$ (with a known input) admits a simple polynomial-time algorithm by comparing the multiplication tables. Actually, GroupIso is in **P** if the two groups are Abelian. However, if the multiplication table of the input group is unknown, then surprisingly, the problem becomes **NP**-hard. Putting the computational hardness and the low trial complexity together, one can see that if more computational time (enough to solve an **NP** problem) is given, then less trials are needed. This interesting trade-off between the two complexity measures is not commonly seen in other query models.

Finally, beyond all of the foregoing problems that can be solved in $\mathbf{P}^{V,\mathsf{SAT}}$, one can show via an information theoretical argument that the following two problems have exponential lower bounds for the randomized trial complexity.

- LinearProgramming: Find a feasible solution of a linear program with n variables and m constraints.
- SubsetSum: Decide whether a given set of n integers can be partitioned into two parts with equal summation of elements.

Theorem 4 ([4,5])

- $R(\mathsf{LinearProgramming}_u) = \Omega(m^{\lfloor n/2 \rfloor})$.
- $R(\mathsf{SubsetSum}_u) = \Omega(2^n)$.

The approaches for Nash and LP are actually similar, yet the running time differs significantly. The key property that guarantees the efficiency of the algorithm for Nash is the existence of Nash equilibrium for any finite game. The algorithm for Nash could thus serve as an interesting example to illustrate how the solution-existing property helps computational efficiency.

Moreover, the following time complexity upper bound for LinearProgramming$_u$ is established, which is exponential in the number of variables but not in the number of constraints.

Theorem 5 ([5]) *The* LinearProgramming$_u$ *problem with m constraints, n variables, and input size L can be deterministically solved in time* $(mnL)^{poly(n)}$. *In particular, the algorithm is of polynomial time for constant dimensional linear programming (i.e., constant number of variables n).*

In summary, these results illustrate the variety of time and trial complexities that arise from the lack of input information for different problems and imply distinct levels of the cruciality of input information for different problems.

Related Work

The trial and error model bears a resemblance to certain other problems and models, e.g., learning, algorithm design in unknown environments, ellipsoid method, and query complexity. However, there are fundamental distinctions between these models and ours. (More discussions are referred to [4].)

Learning

Trial and error model has apparent connections to various learning theories (e.g., concept learning with membership or equivalence query [1], decision tree learning, reinforcement learning [3], and (semi-)supervised learning [2]), but fundamental differences also exist. A common high-level philosophy of various learning models is to "sample and predict," which is very different from our "trial and search" (for a solution) in current setting. With its solution-oriented objective and advantages in computational efficiency, the trial and error model is hopefully to serve as a useful supplement to existing learning theories, particularly in contexts in which the unknown object itself is impossible or unaffordable to learn and the only available access to the unknown is through a solution-verification process.

Ellipsoid Method

The ellipsoid method is an elegant approach for proving the polynomial time solvability of a class of combinatorial optimization problems (see, e.g., [8]); it applies even when the explicit expressions of the constraints are unknown. The algorithm works as long as there exists an oracle that, on a proposed candidate solution, returns a violation in the form of a separating hyperplane.

In general, trial and error model has a similarity to the ellipsoid method, in which a point is proposed as a trial and a separating hyperplane is returned as an error. Our LinearProgramming$_u$ problem studies how to solve linear programs where the returned error is merely the *index* of a violated hyperplane (with the actual hyperplane still hidden). Moreover, the trial and error model includes a much broader class of search problems – not only convex optimization problems, but also many with pure combinatorial structures (e.g., the SAT, GroupIso, and GraphIso problems

discussed here). From this perspective, the ellipsoid method is only one possible approach for the trial and error search problems in current model.

Cross-References

▸ Certificate Complexity and Exact Learning
▸ Reinforcement Learning

Recommended Reading

1. Angluin D (2004) Queries revisited. Theor Comput Sci 313(2):175–194
2. Balcan M, Blum A (2010) A discriminative model for semi-supervised learning. JACM 57(3):19
3. Barto A, Sutton R (1998) Reinforcement learning: an introduction. MIT, Cambridge
4. Bei X, Chen N, Zhang S (2013) On the complexity of trial and error. In: Proceedings of the forty-fifth annual ACM symposium on theory of computing. ACM, New York, pp 31–40
5. Bei X, Chen N, Zhang S (2013) Solving linear programming with constraints unknown. arXiv:1304.1247
6. Chen X, Deng X, Teng S (2009) Settling the complexity of computing two-player nash equilibria. JACM 56(3):14
7. Daskalakis C, Goldberg P, Papadimitriou C (2009) Computing a nash equilibrium is PPAD-complete. SIAM J Comput 39(1):195–259
8. Grotschel M, Lovasz L, Schrijver A (1988) Geometric algorithms and combinatorial optimization. Springer, Berlin/New York
9. Ivanyos G, Kulkarni R, Qiao Y, Santha M, Sundaram A (2014) On the complexity of trial and error for constraint satisfaction problems. In: Automata, languages, and programming. Lecture notes in computer science, vol 8572. Springer, Berlin/Heidelberg, pp 663–675
10. Nisan N, Roughgarden T, Tardos E, Vazirani V (2007) Algorithmic game theory. Cambridge University Press, Cambridge/New York
11. Roth A (2008) Deferred acceptance algorithms: history, theory, practice, and open questions. Int J Game Theory 36:537–569
12. Roth A, Sotomayor M (1992) Two-sided matching: a study in game-theoretic modeling and analysis. Cambridge University Press, Cambridge /New York

T

Triangulation Data Structures

Luca Castelli Aleardi[1], Olivier Devillers[2], and
Jarek Rossignac[3]
[1]Laboratoire d'Informatique (LIX), École
Polytechnique, Bâtiment Alan Turing, Palaiseau,
France
[2]Inria Nancy – Grand-Est, Villers-lès-Nancy,
France
[3]Georgia Institute of Technology, Atlanta, GA,
USA

Keywords

Compact data structures; Succinct representations;
Triangulations; Triangle meshes

Years and Authors of Summarized Original Work

2008; Castelli Aleardi, Devillers, Schaeffer
2009; Gurung, Rossignac
2012; Castelli Aleardi, Devillers, Rossignac

Problem Definition

The main problem consists in designing
space-efficient data structures allowing to
represent the connectivity of triangle meshes
while supporting fast navigation and local
updates.

Mesh Structures: Definition

Triangle meshes are among the most common
representations of shapes. A *triangle mesh* is a
collection of triangle faces that define a polyhedral approximation of a surface. A mesh is
manifold if every edge is bounding either one
or two triangles and if the faces incident to a
same vertex define a closed or open fan. Here
we focus on manifold meshes. Assuming that
the genus and the number of boundary edges are
negligible when compared to the number n of
vertices, the number m of faces is roughly equal
to $2n$.

Data Structures: Classification

Mesh data structures can be compared with respect to several criteria. A basic requirement
(the *traversability*) for mesh representations is
to provide fast navigational operators allowing
to perform a mesh traversal (such as walking
around a vertex). Most representations are also
indexable, allowing to access in constant time to
the description of a given vertex or triangle, given
its index. In order to support efficient processing
of large meshes, one needs to reduce memory
trashing during navigation. An effective way of
doing so is to design *compact data structures*
requiring small storage. Many applications ask
for the *modifiability*: the manipulation of meshes
requires to perform updates such as vertex in-
sertions/deletions, edge collapses, and edge flips.
The choice of the data structure should also
depend on the *simplicity* of its implementation
and on its *practical efficiency* on common input
data.

Standard Mesh Representations

Some common mesh representations are
implemented in the explicit pointer-based form.
References are used to describe incidence
relations between mesh elements, and navigation
is performed throughout address indirection.
For example, a face-based representation [2]
provides operators $\texttt{vertex}(\triangle, i)$ (giving the ith
vertex of a triangle $\triangle$) and $\texttt{neighbor}(\triangle, i)$
(giving the i-neighbor of $\triangle$), as well as
operator $\texttt{face}(v)$ (returning a triangle incident
to vertex v). As illustrated in Fig. 1a, the
combination of these operators allows to
implement operators $\texttt{faceIndex}(\triangle_1, \triangle_2)$
(giving the index of $\triangle_1$ among the neighbors of
$\triangle_2$) and $\texttt{vertexIndex}(v, \triangle)$ (giving the index
of a vertex in $\triangle$). An alternative solution is given
by the *Corner Table* proposed by Rossignac and
colleagues, which uses integer indices to integer
tables and provides a triangulation interface
involving the corner operators defined in Fig. 2.

The two abstract data types above fully
support local navigation in the mesh: the face-
based as well as corner operators support efficient
mesh exploration (see Figs. 1 and 2). A simple
implementation stores explicitly all incidence

a

$v = \texttt{vertex}(\triangle, i)$
$\triangle = \texttt{face}(v)$
$i = \texttt{vertexIndex}(v, \triangle)$
$g_0 = \texttt{neighbor}(\triangle, i)$
$g_1 = \texttt{neighbor}(\triangle, ccw(i))$
$g_2 = \texttt{neighbor}(\triangle, cw(i))$
$z = \texttt{vertex}(g_2, \texttt{faceIndex}(g_2, \triangle))$
int ccw(int i) {return (i + 1)%3; }
int cw(int i) {return (i + 2)%3; }

```
int valence(int  v) {
    int d = 1;
    int f = face(v);
    int g = neighbor(f, cw(vertexIndex(v, f)));
    while (g! = f) {
        int next = neighbor(g, cw(faceIndex(f, g)));
        int i = faceIndex(g, next);
        g = next;
        d + +;
    }
    return d; }
```

b

```
class Quad extends Patch {
    Patch p1, p2, p3, p4;
    Vertex v1, v2, v3, v4;
}

class Pentagon extends Patch {
    Patch p1, p2, p3, p4, p5;
    Vertex v1, v2, v3, v4, v5;
}

class Hexagon extends Patch {
    Patch p1, p2, p3, p4, p5, p6;
    Vertex v1, v2, v3, v4, v5, v6;
}
```

Triangulation Data Structures, Fig. 1 (a) Triangle-based data structure: each triangle stores references to the 3, neighbors and to the 3 incident vertices yielding relations involving faces or corners, using 6 references per triangle plus one reference per vertex (describing the map from vertices to faces): according to Euler formula, this leads to a storage cost of 13 *references per vertex* (*rpv*). The results of `triangle(c)` and `next(c)` are not stored explicitly but calculated assuming that the three corners of each triangle are assigned consecutive indices.

$13\,rpv$. (b) Catalog-based representation: using a catalog of size 3 one can guarantee that any quad is adjacent to at most two other quads, leading to a cost of $8.5\,rpv$

match the optimal asymptotic bound of $3.24\,bpv$ (or equivalently $1.62m$ bits), while efficiently supporting navigational operations [4, 5], as stated below.

Theorem 1 *Given a planar triangulation $\mathcal{T}$ of m triangles, there exists a succinct representation that uses $1.62m + O\left(\frac{m \log\log m}{\log m}\right)$ bits, supporting navigation in worst case $O(1)$ time.*

This result is achieved with a multilevel hierarchical structure. The initial triangulation of size m is decomposed into *small triangulations*, each having $\Theta(\log^2 m)$ triangles: such a decomposition leads to a map $\mathcal{F}$ describing adjacency relations between small triangulations. Small triangulations are then decomposed into *tiny triangulations* of size $\Theta(\log m)$, whose adjacency relations are described by a map $\mathcal{G}$. Map $\mathcal{F}$ has $O\left(\frac{m}{\log^2 m}\right)$ nodes and

Key Results

A Theoretically Optimal Representation

From the information theory point of view, encoding a planar triangulation requires 3.24 *bits per vertex* (*bpv*), which is much less than the $13 \log n\ bpv$ used by standard representations. *Succinct representations* provide theoretically optimal encodings for triangulations, which

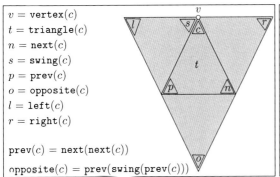

```
v = vertex(c)                                    int c = seed;
t = triangle(c)                                  visited[vertex(next(c))] = true;
n = next(c)                                      visited[vertex(previous(c))] = true;
s = swing(c)                                      do {
p = prev(c)                                          if(!visited[vertex(c)]){
o = opposite(c)                                         visited[vertex(c)] = true;
l = left(c)                                             explored[triangle(c)] = true;
r = right(c)                                          }
                                                     else if(!explored[triangle(c)]) c = opposite(c);
prev(c) = next(next(c))                              c = right(c);
                                                 }while(c! = opposite(s));
opposite(c) = prev(swing(prev(c)))
```

Triangulation Data Structures, Fig. 2 The *Corner Table*: corner operators allow to implement local navigation, as illustrated by the code of the *Ring-Expander* procedure [10]

arcs and can be stored in sublinear space using $O\left(\frac{m}{\log^2 m}\right)$ references of size $O(\log m)$ (actually $O\left(\log \frac{m}{\log^2 m}\right) < O(\log m)$). Map $\mathcal{G}$ has $O\left(\frac{m}{\log m}\right)$ nodes and arcs: adjacencies between two tiny triangulation within the same small triangulation need references of size $O\left(\log \frac{\log^2 m}{\log m}\right) = O(\log \log m)$ while adjacencies crossing the small triangulation boundaries are accessed by referring to $\mathcal{F}$. In that way the storage of both $\mathcal{F}$ and $\mathcal{G}$ is sublinear. The structure of tiny triangulations is optimally encoded throughout lookup into a table storing all possible triangulations of size $O(\log m)$. Such a framework can be extended in order to support updates: vertex deletions and edge flips are performed in $O(\log^2 m)$ amortized time (vertex insertions require $O(1)$ amortized time). The optimality stated by Theorem 1 is obtained combining the two-level representation with a careful decomposition of the mesh into tiny regions, involving a bijection between triangulations and a special class of vertex spanning trees [13].

A different approach, based on small separators, leads to compact representations [1] using $O(n)$ bits for more general classes of meshes (storage performances are difficult to evaluate precisely).

A More Practical Solution

Succinct representations run under the *word-RAM model* and are mainly of theoretical interest, since the amount of memory required in practice is quite important even for very large meshes. Some attempts to exploit the algorithmic framework of succinct representations in practice had lead to a space-efficient dynamic data structure [6]. The main idea is to gather together neighboring faces into small groups of triangles (called *patches*). While references are still of size $\Theta(\log n)$, grouping triangles allows to save some references (corresponding to edges internal to a given patch). For example, using a catalog consisting only of triangles and quadrangles, we encode a triangulation with at most $10.6\,rpv$ (a 19 % improvement over simple representations mentioned earlier). More sophisticated choices of patches lead to dynamic structures with smaller storage (e.g., Fig. 1b), as stated below:

Theorem 2 *Given a triangulation (possibly having handles and boundaries), there exists a data structure using $7.67\,rpv$, which allows $O(1)$ time navigation and supports updates in $O(1)$ amortized time.*

Reducing Redundancy Throughout Face Reordering

The main idea used in the SOT data structure [8] is to implicitly represent the map from triangles to corners (`triangle` operator), and the map from corners to vertices (`vertex` operator), through face reordering. First, match each vertex to an incident triangle (in such a way a triangle is matched with at most one vertex). Then permute triangles in such a way that the triangle associated

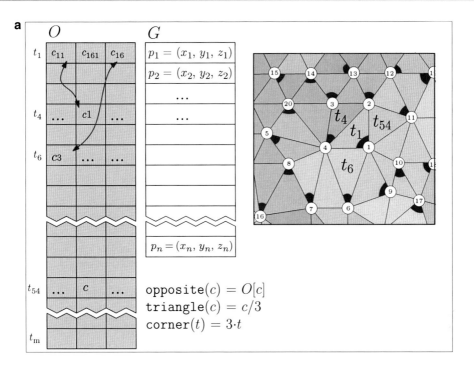

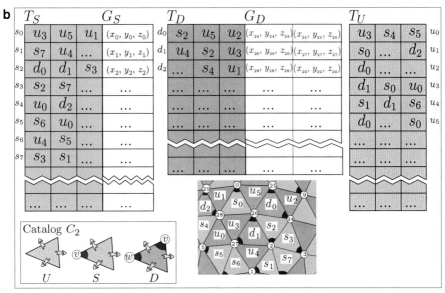

Triangulation Data Structures, Fig. 3 Illustrations of the SOT (**a**) and ESQ (**b**) data structures

with the ith vertex v_i has number i (thus, the first n triangles appearing in this ordering are the ones associated with a vertex). The corners of a triangle are listed consecutively, and the first one corresponds to the vertex matched for the triangle. The incidence relations are stored in an array O (of length $3m$) having 3 entries per triangle: $O[i]$ stores the index of the corner opposite to c_i (which is matched to vertex v_i, for $i \leq m$). Corner operators are supported in $O(1)$ time performing arithmetic operations (see Fig. 3a). Accessing a vertex v_i requires to walk around its incident faces until c_i is reached (v_i being matched to c_i).

Theorem 3 ([8]) *Given a triangulation (possibly having handles and boundaries), there exists a data structure using $6\,rpv$ which supports $O(1)$ time navigation (retrieving a vertex of degree d requires $O(d)$ time).*

More Compact (Static) Representations

Combining this reordering approach with a pairing of adjacent triangles into quads, the SQUAD data structure [9] reaches better storage requiring slightly more than $4\,rpv$ according to experimental results on common meshes (the worst case upper bound is still $6\,rpv$). If one is allowed to perform a reordering of the input vertices, it is possible to guarantee a storage of $4\,rpv$ in the worst case (with same time performances as before): the edge-based representation described in [3] matches this bound exploiting Schnyder woods decompositions [14]. Various heuristics allows to further reduce storage requirements in practice [10–12].

A Dynamic Representation

Combining the reordering approach described above with the decomposition into triangle patches, the ESQ data structure [7] exhibits the same navigation performances as in SOT, while supporting local updates. As in [6] the mesh is decomposed into a collection of patches, each consisting of one or more triangles, and vertices are matched to patches. The assumption that each vertex is matched to a different triangle is relaxed. The catalog thus consists of a collection of k patch types (having possible one or more marked corners, describing how vertices are matched). Adjacency relations between faces are stored in k tables $T_1, \ldots, T_k$, one for each patch type (see Fig. 3b). Extending the approach introduced in SOT, a reordering of the input vertices allows to represent the maps from vertices to triangles and from triangles to vertices. A table T_S of type $S = (c, b)$ (with b boundary edges and c matched vertices) contains b references for each entry; the entries in the associated table G_S (containing geometric coordinates) are ordered accordingly. The decomposition into

patches is maintained under local modifications with a constant number of memory updates in tables T_i.

Theorem 4 ([7]) *Given a triangulation (possibly having handles and boundaries), there exists a dynamic data structure using $4.8\,rpv$, which allows $O(1)$ time navigation and $O(d)$ time access to a vertex of degree d. Updates (vertex insertions/deletions and edge flips) are supported in $O(1)$ amortized time.*

Experimental Results

In [9] are reported timing comparisons of operators for SOT, SQUAD, and Corner Table data structures: experimental evaluations concern adjacency and navigational operations. On the tested mesh (the 55 millions triangles David), SQUAD requires 20 s and uses 2.2 GB of RAM for the construction (on a MacbookPro, equipped with 2.66 GHz Intel Core i7, 8 GB). When the whole mesh fits in main memory, compact data structures (SQUAD and SOT) perform slower than Corner Table. When the allowed memory is reduced, SQUAD performances are comparable and sometimes even better than Corner Table performances for high-level tasks (e.g., valence computations).

Cross-References

▶ Compressed Representations of Graphs
▶ Delaunay Triangulation and Randomized Constructions

Recommended Reading

1. Blanford D, Blelloch G, Kash I (2003) Compact representations of separable graphs. In: SODA, Baltimore, pp 342–351. http://dl.acm.org/citation.cfm?id=644219
2. Boissonnat JD, Devillers O, Pion S, Teillaud M, Yvinec M (2002) Triangulations in CGAL. Comput Geom 22:5–19.
3. Castelli Aleardi L, Devillers O (2011) Explicit array-based compact data structures for triangulations. In: ISAAC, Yokohama, pp 312–322.

4. Castelli Aleardi L, Devillers O, Schaeffer G (2005) Succinct representation of triangulations with a boundary. In: WADS, Waterloo, pp 134–145.
5. Castelli Aleardi L, Devillers O, Schaeffer G (2008) Succinct representations of planar maps. Theor Comput Sci 408(2–3):174–187.
6. Castelli Aleardi L, Devillers O, Mebarki A (2011) Catalog based representation of 2D triangulations. Int J Comput Geom Appl 21(4):393–402.
7. Castelli Aleardi L, Devillers O, Rossignac J (2012) ESQ: editable squad representation for triangle meshes. In: SIBGRAPI, Ouro Preto, pp 110–117.
8. Gurung T, Rossignac J (2009) SOT: compact representation for tetrahedral meshes. In: Proceedings of the ACM symposium on solid and physical modeling, San Francisco, pp 79–88.
9. Gurung T, Laney D, Lindstrom P, Rossignac J (2011) SQUAD: compact representation for triangle meshes. Comput Graph Forum 30(2):355–364.
10. Gurung T, Luffel M, Lindstrom P, Rossignac J (2011) LR: compact connectivity representation for triangle meshes. ACM Trans Graph 30(4):67.
11. Gurung T, Luffel M, Lindstrom P, Rossignac J (2013) Zipper: a compact connectivity data structure for triangle meshes. Comput-Aided Des 45(2):262–269.
12. Luffel M, Gurung T, Lindstrom P, Rossignac J (2014) Grouper: a compact, streamable triangle mesh data structure. IEEE Trans Vis Comput Graph 20(1):84–98.
13. Poulalhon D, Schaeffer G (2006) Optimal coding and sampling of triangulations. Algorithmica 46:505–527.
14. Schnyder W (1990) Embedding planar graphs on the grid. In: SODA, San Francisco, 138–148. http://dl.acm.org/citation.cfm?id=320191

Truthful Mechanisms for One-Parameter Agents

Moshe Babaioff
Microsoft Research, Herzliya, Israel

Keywords

Algorithmic mechanism design; Approximation; Scheduling related parallel machines; Truthful mechanisms

Synonyms

Dominant strategy mechanisms; Incentive compatible mechanisms; Single-parameter agents; Truthful auctions

Years and Authors of Summarized Original Work

2001; Archer, Tardos

Problem Definition

This problem is concerned with designing truthful (dominant strategy) mechanisms for domains where each agent's private information is expressed by a single positive real number. The goal of the mechanisms is to allocate loads placed on the agents, and an agent's private information is the cost incurred per unit load. Archer and Tardos [4] give an exact characterization for the algorithms that can be used to design truthful mechanisms for such load balancing problems using appropriate payments. The characterization shows that the allocated load must be monotonic in the cost (decreasing when the cost on an agent increases, fixing the costs of the others). Thus, truthful mechanisms are characterized by a condition on the allocation rule, and payments that ensure voluntary participation can be calculated using the given characterization.

The characterization is used to design polynomial time truthful mechanisms for several problems in combinatorial optimization to which the celebrated VCG mechanism does not apply. For scheduling related parallel machines to minimize makespan ($Q \| C_{\max}$), Archer and Tardos [4] present a 3-approximation mechanism based on randomized rounding of the optimal fractional solution. This mechanism is truthful only in expectation (a weaker notion of truthfulness in which truthful bidding maximizes the agent's *expected* utility). Archer [3] improves it to a randomized 2-approximation truthful mechanism. Andelman, Azar, and Sorani [2] provide a deterministic truthful mechanism that is 5-approximation. Kovács improves it to 3-approximation in [12] and to 2.8-approximation in [13] (Kovács also gives other results for two special cases). Andelman, Azar, and Sorani [2] also present a deterministic Fully Polynomial Time Approximation Scheme (FPTAS) for scheduling on a fixed number of machines, as well as a suitable payment scheme that yields a deterministic truthful mechanism. Dhangwatnotai et al. [8] present a randomized Polynomial Time Approximation Scheme (PTAS) that is truthful-in-expectation. Christodoulou and Kovács [7] present a truthful deterministic Polynomial Time Approximation Scheme (PTAS); this matches the best possible result for the computational problem (without incentives) by Hochbaum and Shmoys [10] (it is known that this problem is strongly NP-hard [9]). This result shows that there is no "cost of truthfulness" for this problem, as the best approximation with incentive constraints is as good as the best approximation without these constraints.

Archer and Tardos [4] also present results for goals other than minimizing the makespan. They present a truthful mechanism for $Q \| \sum C_j$ (scheduling related machines to minimize the sum of completion times) and show that for $Q \| \sum w_j C_j$ (minimizing the weighted sum of completion times) $\frac{2}{\sqrt{3}}$ is the best approximation ratio achievable by a truthful mechanism.

This family of problems belongs to the field of Algorithmic Mechanism Design, initiated in the seminal paper of Nisan and Ronen [15]. Nisan and Ronen consider makespan minimization for scheduling on *unrelated* machines and prove upper and lower bounds (note that for unrelated machines agents have more than one parameter). Mu'alem and Schapira [14] present improved lower bounds. Other papers consider the problem of scheduling on related machines to minimize the makespan. Auletta et al. [5] and Ambrosio and Auletta [1] present truthful mechanisms for several NP-hard restrictions of this problem. Nisan and Ronen [15] also introduce a model in which the mechanism is allowed to observe the machines' actual processing time and compute the payments afterward (in such a model the machines essentially cannot claim to be faster than they are); Auletta et al. [6] present additional results for this model. In particular, they show that it is possible to overcome the lower bound of $\frac{2}{\sqrt{3}}$ for $Q \| \sum w_j C_j$ (minimizing the weighted sum of completion times) and provide a polynomial time $(1 + \epsilon)$-approximation truthful mechanism (with verification) when the number of machines (m) is constant.

The Mechanism Design Framework

Let I be the set of agents. Each agent $i \in I$ has some private *value* (type) consisting of a single parameter $t_i \in \mathbb{R}$ that describes the agent, and which only i knows. Everything else is public knowledge. Each agent will report a *bid* b_i to the mechanism. Let t denote the vector of true values, and b the vector of bids.

There is some set of *outcomes* O, and given the bids b the mechanism's output algorithm computes an outcome $o(b) \in O$. For any types t, the mechanism aims to choose an outcome $o \in O$ that minimizes some function $g(o, t)$. Yet, given the bids b the mechanism can only choose the outcome as a function of the bids ($o = o(b)$) and has no knowledge of the true types t. To overcome the problem that the mechanism knows only the bids b, the mechanism is designed to be truthful (using payments), that is, in such a mechanism it is a dominant strategy for the agents to reveal their true types ($b = t$). For such mechanisms minimizing $g(o, t)$ is done by assuming that the bids are the true types (and this is justified by the fact that truth telling is a dominant strategy).

In the framework discussed here we assume that outcome $o(b)$ will assign some amount of *load* or work $w_i(o(b))$ to each agent i, and given $o(b)$ and t_i, agent i incurs some monetary *cost*, $cost_i(t_i, o(b)) = t_i w_i(o(b))$. Thus, agent i's private data t_i measures her cost per unit work. Each agent i attempts to maximize her *utility* (profit), $u_i(t_i, b) = P_i(b) - cost_i(t_i, o(b))$, where $P_i(b)$ is the *payment* to agent i.

Let b_{-i} denote the vector of bids, not including agent i, and let $b = (b_{-i}, b_i)$. Truth telling is a *dominant strategy* for agent i if bidding t_i always maximizes her utility, regardless of what the other agents bid. That is, $u_i(t_i, (b_{-i}, t_i)) \geq u_i(t_i, (b_{-i}, b_i))$ for all b_{-i} and b_i.

A mechanism M consists of the pair $M = (o(\cdot), P(\cdot))$, where $o(\cdot)$ is the *output function* and $P(\cdot)$ is the *payment scheme*, i.e., the vector of payment functions $P_i(\cdot)$. An output function *admits a truthful payment scheme* if there exist payments $P(\cdot)$ such that for the mechanism $M = (o(\cdot), P(\cdot))$, truth telling is a dominant strategy for each agent. A mechanism that admits a truthful payment scheme is *truthful*.

Mechanism M satisfies the *voluntary participation* condition if agents who bid truthfully never incur a net loss, i.e., $u_i(t_i, (b_{-i}, t_i)) \geq 0$ for all agents i, true values t_i, and other agents' bids b_{-i}.

Definition 1 With the other agents' bids b_{-i} fixed, the *work curve* for agent i is $w_i(b_{-i}, b_i)$ considered as a single-variable function of b_i. The output function o is *decreasing* if each of the associated work curves is decreasing (i.e., $w_i(b_{-i}, b_i)$ is a decreasing function of b_i, for all i and b_{-i}).

Scheduling on Related Machines

There are n jobs and m machines. The jobs represent amounts of work $p_1 \geq p_2 \geq \ldots \geq p_n$, and let p denote the set of jobs. Machine i runs at some speed s_i, so it must spend p_j/s_i units of time processing each job j assigned to it. The input to an algorithm is b, the (reported) speed of the machines, and the output is $o(b)$, an assignment of jobs to machines. The load on machine i for outcome $o(b)$ is $w_i(b) = \sum p_j$, where the sum runs over jobs j assigned to i. Each machine incurs a cost proportional to the time it spends processing its jobs. The cost of machine i is $cost_i(t_i, o(b)) = t_i w_i(o(b))$, where $t_i = 1/s_i$ and $w_i(b)$ is the total load assigned to i when the speeds are b. Let C_j denote the completion time of job j. One can consider the following goals for scheduling related parallel machines:

- Minimizing the makespan $(Q\|C_{\max})$, the mechanism's goal is to minimize the completion time of the last job on the last machine, i.e., $g(o, t) = C_{\max} = \max_i t_i \cdot w_i(b)$.
- Minimize the sum of completion times $(Q\|\sum C_j)$, i.e., $g(o, t) = Q\|\sum C_j = \sum_j C_j$
- Minimize the weighted sum of completion times $(Q\|\sum w_j C_j)$, i.e., $g(o, t) = Q\|\sum w_j C_j = \sum_j w_j C_j$ where w_j is the weight of job j.

An algorithm is a *c-approximation algorithm* with respect to g, if for every instance (p, t), it outputs an outcome of cost at most $c \cdot g(o(t), t)$. A *c-approximation mechanism* is a mechanism whose output algorithm is an c-approximation. Note that if the mechanism is truthful the approximation is with respect to the true speeds. A *PTAS* (Polynomial Time Approximation Scheme) is a family of algorithms such that for every $\epsilon > 0$ there exists a $(1 + \epsilon)$-approximation algorithm. If the running time is also polynomial in $1/\epsilon$, the family of algorithms is a *FPTAS* (Fully Polynomial Time Approximation Scheme).

Key Results

The following two theorems hold for the mechanism design framework as defined in section "Problem Definition."

Theorem 1 ([4]) *The output function $o(b)$ admits a truthful payment scheme if and only if it is decreasing. In this case, the mechanism is truthful if and only if the payments $P_i(b_{-i}, b_i)$ are of the form*

$$h_i(b_{-i}) + b_i w_i(b_{-i}, b_i) - \int_0^{b_i} w_i(b_{-i}, u) du$$

where the h_i are arbitrary functions.

Theorem 2 ([4]) *A decreasing output function admits a truthful payment scheme satisfying voluntary participation if and only if $\int_0^\infty w_i(b_{-i}, u) du < \infty$ for all i, b_{-i}. In this case, the payments can be defined by*

$$P_i(b_{-i}, b_i) = b_i w_i(b_{-i}, b_i) + \int_{b_i}^\infty w_i(b_{-i}, u) du$$

Theorem 3 ([4]) *There is a truthful mechanism (not polynomial time) that outputs an optimal solution for $Q\|C_{\max}$ and satisfies voluntary participation.*

Theorem 4 ([2,7]) *For the problem of minimizing the makespan $(Q\|C_{\max})$:*

- *There exists a deterministic Polynomial Time Approximation Scheme (PTAS) for scheduling on related machines that admits a truthful payment scheme [7]. The mechanism created satisfies voluntary participation.*

- *There exists a deterministic Fully Polynomial Time Approximation Scheme (FPTAS) for scheduling on a fixed number of machines that admits a truthful payment scheme [2]. The mechanism created satisfies voluntary participation.*

Theorem 5 ([4]) *There is a truthful polynomial time mechanism that outputs an optimal solution for $Q \| \sum C_j$ and satisfies voluntary participation.*

Theorem 6 ([4]) *No truthful mechanism for $Q \| \sum w_j C_j$ can achieve an approximation ratio better than $\frac{2}{\sqrt{3}}$, even on instances with just two jobs and two machines.*

Applications

Archer and Tardos [4] apply the characterization of truthful mechanisms to problems other than scheduling. They present results for the uncapacitated facility location problem as well as the maximum flow problem.

Kis and Kapolnai [11] consider the problem of scheduling of groups of identical jobs on related machines with sequence independent setup times $(Q|u_j, p_{jk} = p_j \| C_{max})$. They provide a truthful, polynomial time, randomized mechanism for the batch scheduling problem with a deterministic approximation guarantee of 4 to the minimal makespan, based on the characterization of truthful mechanisms presented above.

Open Problems

The problem of designing truthful mechanisms for related machines to minimize the makespan was completely resolved, as a deterministic PTAS [7] is the best one can hope for. For this problem there is no gap between the best approximation with and without incentives. The main open problem left is of finding some natural

single-parameter setting in which there is a gap between the approximation that is achievable by algorithms and truthful mechanisms.

Experimental Results

None is reported.

Data Sets

None is reported.

URL to Code

None is reported.

Cross-References

- ▶ Algorithmic Mechanism Design
- ▶ Competitive Auction
- ▶ Generalized Vickrey Auction
- ▶ Incentive Compatible Selection

Recommended Reading

1. Ambrosio P, Auletta V (2004) Deterministic monotone algorithms for scheduling on related machines. In: 2nd workshop on approximation and online algorithms (WAOA), Bergen, pp 267–280
2. Andelman N, Azar Y, Sorani M (2005) Truthful approximation mechanisms for scheduling selfish related machines. In: 22nd annual symposium on theoretical aspects of computer science (STACS), Stuttgart, pp 69–82
3. Archer A (2004) Mechanisms for discrete optimization with rational agents. PhD thesis, Cornell University
4. Archer A, Tardos É (2001) Truthful mechanisms for one-parameter agents. In: 42nd annual symposium on foundations of computer science (FOCS), Las Vegas, pp 482–491
5. Auletta V, De Prisco R, Penna P, Persiano G (2004) Deterministic truthful approximation mechanisms for scheduling related machines. In: 21nd annual symposium on theoretical aspects of computer science (STACS), Montpellier, pp 608–619
6. Auletta V, De Prisco R, Penna P, Persiano G, Ventre C (2006) New constructions of mechanisms with verification. In: 33rd international colloquium on automata, languages and programming (ICALP), Venice, (1), pp 596–607

7. Christodoulou G, Kovács A (2013) A deterministic truthful ptas for scheduling related machines. SIAM J Comput 42(4):1572–1595
8. Dhangwatnotai P, Dobzinski S, Dughmi S, Roughgarden T (2011) Truthful approximation schemes for single-parameter agents. SIAM J Comput 40(3):915–933
9. Garey MR, Johnson DS (1990) Computers and intractability: a guide to the theory of NP-completeness. W. H. Freeman & Co., New York
10. Hochbaum D, Shmoys D (1988) A polynomial approximation scheme for scheduling on uniform processors: Usingthe dual approximation approach. SIAM J Comput 17(3):539–551
11. Kis T, Kapolnai R (2007) Approximations and auctions for scheduling batches on related machines. Oper Res Lett 35(1):61–68
12. Kovács A (2005) Fast monotone 3-approximation algorithm for scheduling related machines. In: 13th annual European symposium (ESA), Palma de Mallorca, Spain, pp 616–627
13. Kovács A (2007) Fast algorithms for two scheduling problems. PhD thesis, Universität des Saarlandes
14. Mu'alem A, Schapira M (2007) Setting lower bounds on truthfulness: extended abstract. In: Proceedings of the eighteenth annual ACM-SIAM symposium on discrete algorithms (SODA '07), New Orleans. Society for Industrial and Applied Mathematics, Philadelphia, pp 1143–1152. ISBN: 978-0-898716-24-5, http://dl.acm.org/citation.cfm?id=1283383.1283506
15. Nisan N, Ronen A (2001) Algorithmic mechanism design. Games Econ Behav 35:166–196

Truthful Multicast

Weizhao Wang[1], Xiang-Yang Li[2], and Yu Wang[3]
[1]Google Inc., Irvine, CA, USA
[2]Department of Computer Science, Illinois Institute of Technology, Chicago, IL, USA
[3]Department of Computer Science, University of North Carolina, Charlotte, NC, USA

Keywords

Strategyproof multicast mechanism; Truthful multicast routing

Years and Authors of Summarized Original Work

2004; Wang, Li, Wang

Problem Definition

Several mechanisms [1, 3, 5, 9], which essentially all belong to the VCG mechanism family, have been proposed in the literature to prevent the selfish behavior of unicast routing in a wireless network. In these mechanisms, the least cost path, which maximizes the social efficiency, is used for routing. Wang, Li, and Wang [8] studied the *truthful* multicast routing protocol for a selfish wireless network, in which selfish wireless terminals will follow their own interests. The multicast routing protocol is composed of two components: (1) the tree structure that connects the sources and receivers, and (2) the payment to the relay nodes in this tree. Multicast poses a unique challenge in designing strategyproof mechanisms due to the reason that (1) a VCG mechanism uses an output that maximizes the *social efficiency*; (2) it is NP-hard to find the tree structure with the minimum cost, which in turn maximizes the social efficiency. A range of multicast structures, such as the least cost path tree (LCPT), the pruning minimum spanning tree (PMST), virtual minimum spanning tree (VMST), and Steiner tree, were proposed to replace the optimal multicast tree. In [8], Wang et al. showed how payment schemes can be designed for existing multicast tree structures so that rational selfish wireless terminals will follow the protocols for their own interests.

Consider a communication network $G = (V, E, c)$, where $V = \{v_1, \cdots, v_n\}$ is the set of communication terminals, $E = \{e_1, e_2, \cdots, e_m\}$ is the set of links, and c is the cost vector of all agents. Here agents are terminals in a node weighted network and are links in a link weighted network. Given a set of sources and receivers $Q = \{q_0, q_1, q_2, \cdots, q_{r-1}\} \subset V$, the multicast problem is to find a tree $T \subset G$ spanning all terminals Q. For simplicity, assume that $s = q_0$ is the sender of a multicast session if it exists. All terminals or links are required to declare a cost of relaying the message. Let d be the declared costs of all nodes, i.e., agent i declared a cost d_i. On the basis of the declared cost profile d, a multicast tree needs to be constructed and the payment $p_k(d)$ for each agent k needs to be decided. The utility of an agent is its payment received, minus its cost if it is selected in the multicast tree. Instead of reinventing the wheels, Wang et al. still used the previously proposed structures for multicast as the output of their mechanism. Given a multicast tree, they studied the design of strategyproof payment schemes based on this tree.

Notations

Given a network H, $\omega(H)$ denotes the total cost of all agents in this network. If the cost of any agent i (link e_i or node v_i) is changed to c'_i, the new network is denoted as $G' = (V, E, c|^i c'_i)$, or simply $c|^i c'_i$. If one agent i is removed from the network, it is denoted as $c|^i \infty$. For the simplicity of notation, the cost vector c is used to denote the network $G = (V, E, c)$ if no confusion is caused. For a given source s and a given destination q_i, $\mathrm{LCP}(s, q_i, c)$ represents the shortest path between s and q_i when the cost of the network is represented by vector c. $|\mathrm{LCP}(s, q_i, d)|$ denotes the total cost of the least cost path $\mathrm{LCP}(s, q_i, d)$. The notation of several multicast trees is summarized as follows.

1. Link Weighted Multicast Tree
   - **LCPT**: The union of all least cost paths from the source to receivers is called the *least cost path tree*, denoted by $LCPT(d)$.
   - **PMST**: First construct the minimum spanning tree $MST(G)$ on the graph G. Take the tree $MST(G)$ rooted at sender s, prune all subtrees that do not contain a receiver. The final structure is called the Pruning Minimum Spanning Tree (PMST).
   - **LST**: The Link Weighted Steiner Tree (LST) can be constructed by the algorithm proposed by Takahashi and Matsuyama [6].

2. Node Weighted Multicast Tree
   - **VMST**: First construct a virtual graph using all receivers plus the sources as the vertices and the cost of LCP as the link weight. Then compute the minimum spanning tree on the virtual graph, which is called virtual minimum spanning tree

(VMST). Finally, choose all terminals on the VMST as the relay terminals.
   - **NST**: The node weighted Steiner tree (NST) can be constructed by the algorithm proposed by [4].

Key Results

If the LCPT tree is used as the multicast tree, Wang et al. proved the following theorem.

Theorem 1 *The VCG mechanism combined with LCPT is not truthful.*

Because of the failure of the VCG mechanism, they designed their non-VGC mechanism for the LCPT-based multicast routing as follows.

Theorem 2 *Payment (defined in Eq. (1)) based on LCPT is truthful and it is minimum among all truthful payments based on LCPT.*

More generally, Wang et al. [8] proved the following theorem.

Theorem 3 *The VCG mechanism combined with either one of the LCPT, PMST, LST, VMST, NST is not truthful.*

Because of this negative result, they designed their non-VCG mechanisms for all multicast structures they studied: LCPT, PMST, LST, VMST, NST. For example, Algorithm 2 is the algorithm for PMST. For other algorithms, please refer to [8].

Algorithm 1 Non-VCG mechanism for LCPT

1: For each receiver $q_i \neq s$, computes the least cost path from the source s to q_i, and compute a payment $p_k^i(d)$ to every link e_k on the LCP(s, q_i, d) using the scheme for unicast

$$p_k^i(d) = d_k + |\mathrm{LCP}(s, q_i, d|^k \infty)| - |\mathrm{LCP}(s, q_i, d)|.$$

2: The final payment to link $e_k \in LCPT$ is then

$$p_k(d) = \max_{q_i \in Q} p_k^i(d). \tag{1}$$

The payment to each link not on LCPT is simply 0.

Algorithm 2 Non-VCG mechanism for PMST

1: Apply VCG mechanism on the MST. The payment for edge $e_k \in PMST(d)$ is

$$p_k(d) = \omega(MST(d|^k\infty)) - \omega(MST(d)) + d_k. \quad (2)$$

2: For every edge $e_k \notin PMST(d)$, its payment is 0.

Regarding all their non-VGC mechanisms, they proved the following theorem.

Theorem 4 *The non-VCG mechanisms designed for the multicast structures LCPT, PMST, LST, VMST, NST are not only truthful, but also achieve the minimum payment among all truthful mechanisms.*

Applications

In wireless ad hoc networks, it is commonly assumed that, each terminal contributes its local resources to forward the data for other terminals to serve the common good, and benefits from resources contributed by other terminals to route its packets in return. On the basis of such a fundamental design philosophy, wireless ad hoc networks provide appealing features such as enhanced system robustness, high service availability and scalability. However, the critical observation that individual users who own these wireless devices are generally selfish and non-cooperative may severely undermine the expected performances of the wireless networks. Therefore, providing incentives to wireless terminals is a must to encourage contribution and thus maintains the robustness and availability of wireless networking systems. On the other hand, to support a communication among a group of users, multicast is more efficient than unicast or broadcast, as it can transmit packets to destinations using fewer network resources, thus increasing the social efficiency. Thus, most results of the work of Wang et al. can apply to multicast routing in wireless networks in which nodes are selfish. It not only guarantees that multicast routing behaves normally but also achieves good social efficiency for both the receivers and relay terminals.

Open Problems

There are several unsolved challenges left as future work in [8]. Some of these challenges are listed below.

- How to design algorithms that can compute these payments in asymptotically optimum time complexities is presently unknown.
- Wang et al. [8] only studied the tree-based structures for multicast. Practically, mesh-based structures may be more needed for wireless networks to improve the fault tolerance of the multicast. It is unknown whether a strategyproof multicast mechanism can be designed for some mesh-based structures used for multicast.
- All of the tree construction and payment calculations in [8] are performed in a centralized way, it would be interesting to design some distributed algorithms for them.
- In the work by Wang et al. [8] it was assumed that the receivers will always relay the data packets for other receivers for free, the source node of the multicast will pay the relay nodes to compensate their cost, and the source node will not charge the receivers for getting the data. As a possible future work, the budget balance of the source node needs to be considered if the receivers have to pay the source node for getting the data.
- Fairness of payment sharing needs to be considered in a case where the receivers share the total payments to all relay nodes on the multicast structure. Notice that this is different from the cost-sharing studied in [2], in which they assumed a fixed multicast tree, and the link cost is publicly known; in that work they showed how to share the total link cost among receivers.
- Another important task is to study how to implement the protocols proposed in [8] in a distributed manner. Notice that, in [3, 9],

distributed methods have been developed for a truthful unicast using some cryptography primitives.

Cross-References

► Algorithmic Mechanism Design

Recommended Reading

1. Anderegg L, Eidenbenz S (2003) Ad hoc-VCG: a truthful and cost-efficient routing protocol for mobile ad hoc networks with selfish agents. In: Proceedings of the 9th annual international conference on mobile computing and networking. ACM, New York, pp 245–259
2. Feigenbaum J, Papadimitriou C, Shenker S (2001) Sharing the cost of multicast transmissions. J Comput Syst Sci 63(1):21–41
3. Feigenbaum J, Papadimitriou C, Sami R, Shenker S (2002) A BGP-based mechanism for lowest-cost routing. In: Proceedings of the 2002 ACM symposium on principles of distributed computing, Monterey, 21–24 Jul 2002, pp 173–182
4. Klein P, Ravi R (1995) A nearly best-possible approximation algorithm for node-weighted Steiner trees. J Algorithm 19(1):104–115
5. Nisan N, Ronen A (1999) Algorithmic mechanism design. In: Proceedings of the 31st annual symposium on theory of computing (STOC99), Atlanta, 1–4 May 1999, pp 129–140
6. Takahashi H, Matsuyama A (1980) An approximate solution for the Steiner problem in graphs. Math Jpn 24(6):573–577
7. Wang W, Li X-Y (2006) Low-cost routing in selfish and rational wireless ad hoc networks. IEEE Trans Mob Comput 5(5):596–607
8. Wang W, Li X-Y, Wang Y (2004) Truthful multicast in selfish wireless networks. In: Proceedings of the 10th ACM annual international conference on mobile computing and networking, Philadelphia, 26 Sept-1 Oct 2004
9. Zhong S, Li L, Liu Y, Yang YR (2005) On designing incentive compatible routing and forwarding protocols in wireless adhoc networks – an integrated approach using game theoretical and cryptographic techniques. In: Proceedings of the 11th ACM annual international conference on mobile computing and networking, Cologne, 28 Aug-2 Sept 2005

TSP-Based Curve Reconstruction

Edgar Ramos
School of Mathematics, National University of Colombia, Medellín, Colombia

Years and Authors of Summarized Original Work

2001; Althaus, Mehlhorn

Problem Definition

An instance of the curve reconstruction problem is a finite set of *sample* points V in the plane, which are assumed to be taken from an unknown planar curve γ. The task is to construct a geometric graph G on V such that two points in V are connected by an edge in G if and only if the points are adjacent on γ. The curve γ may consist of one or more connected components, and each of them may be closed or open (with endpoints), and may be smooth everywhere (tangent defined at every point) or not.

Many heuristic approaches have been proposed to solve this problem. This work continues a line of reconstruction algorithms with *guaranteed* performance, i.e., algorithms which probably solve the reconstruction problem under certain assumptions of γ and V. Previous proposed solutions with guaranteed performances were mostly *local*: a subgraph of the complete geometric graph defined by the points is considered (in most cases the Delaunay edges), and then *filtered* using a local criteria into a subgraph that will constitute the reconstruction. Thus, most of these algorithms fail to enforce that the solution have the global property of being a path/tour or collection of paths/tours and so usually require a dense sampling to work properly and have difficulty handling nonsmooth curves. See [6, 7, 8] for surveys of these algorithms.

This work concentrates on a solution approach based on the *traveling salesman problem (TSP)*. Recall that a *traveling salesman path (tour)* for a set V of points is a path (cycle) passing through all points in V. An optimal traveling salesman path (tour) is a traveling salesman path (tour) of shortest length. The first question is under which conditions for γ and V a traveling salesman path (tour) is a correct reconstruction. Since the construction of an optimal traveling salesman path (tour) is an NP-hard problem, a second question is whether for the specific instances under consideration, an efficient algorithm is possible.

A previous work of Giesen [9] gave a first weak answer to the first question: For every benign semiregular closed curve γ, there exists an $\epsilon > 0$ with the following property: If V is a finite sample set from γ so that for every $x \in \gamma$ there is a $p \in V$ with $\|pv\| \leq \epsilon$, then the optimal traveling salesman tour is a polygonal reconstruction of γ. For a curve $\gamma : [0, 1] \to R^2$, its left and right tangents at $\gamma(t_0)$, are defined as the limits of the ratio $|\gamma(t_2) - \gamma(t_1)| / |t_2 - t_1|$ as (t_1, t_2) converges to (t_0, t_0) from the right ($t_0 < t_1 < t : 2$) and from the right ($t_1 < t_2 < t : 0$) respectively. A curve is semiregular if both tangents exist at every points and regular if the tangents exist and coincide at every point. The *turning* angle of γ at p is the angle between the left and right tangents at a points p. A semiregular curve is *benign* if the turning angle is less than π.

To investigate the TSP-based solution of the reconstruction problem, this work considers its integer linear programming *(ILP)* formulation and the corresponding linear programming *(LP)* relaxation. The motivation is that a successful method for solving the TSP is to use a branch-and-cut algorithm based on the LP-relaxation. See Chapter 7 in [5]. For a path with endpoints a and b, the formulation is based on variables $x_{u,v} \in \{0, 1\}$ for each pair u, v in V (indicating whether the edge uv is in the path ($x_{uv} = 1$) or not

($x_{uv} = 0$) and consists of the following objective function and constraints ($x_{uu} = 0$ for all $u \in V$):

$$\text{minimize} \quad \sum_{u,v \in V} \|uv\| \cdot x_{uv}$$

$$\text{subject to} \quad \sum_{v \in V} x_{uv} = 2 \text{ for all } u \in V \setminus \{a, b\}$$

$$\sum_{v \in V} x_{uv} = 1 \text{ for } u \in \{a, b\}$$

$$\sum_{u,v \in V'} x_{uv} \leq |V'| - 1 \text{ for } V' \subseteq V,$$

$$V' \neq \emptyset$$

$$x_{uv} \in \{0, 1\} \text{ for all } u, v \in V.$$

Here $\|uv\|$ denotes the Euclidean distance between u and v and so the objective function is the total length of the selected edges. This is called the *subtour-ILP for the TSP with specified endpoints*. The equality constraints are called the *degree constraints*, the inequality ones are called *subtour elimination constraints* and the last ones are called the *integrality constraints*. If the degree and integrality constraints hold, the corresponding graph could include disconnected cycles (subtours), hence the need for the subtour elimination constraints. The relaxed LP is obtained by replacing the integrality constraints by the constraints $0 \leq x_{uv} \leq 1$ and is called the *subtour-LP for the TSP with specified endpoints*. There is a polynomial time algorithm that given a candidate solution returns a violated constraint if it exists: the degree constraints are trivial to check and the subtour elimination constraints are checked using a min cut algorithm (if a, b are joined by an edge and all edge capacities are made equal to one, then a violated subtour constraint corresponds to a cut smaller than two). This means that the subtour-LP for the TSP with specified endpoints can potentially be solved in polynomial time in the bit size of the input description, using the ellipsoid method [10].

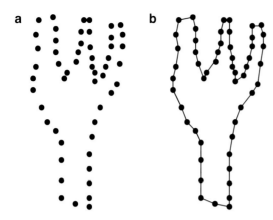

TSP-Based Curve Reconstruction, Fig. 1 Sample data and its reconstruction

Key Results

The main results of this paper are that, given a sample set V with $a, b \in V$ from a benign semiregular open curve γ with endpoints a, b and satisfying certain *sampling condition* [it], then

- The optimal traveling salesman path on V with endpoints a, b is a polygonal reconstruction of γ from V,
- The subtour-LP for traveling salesman paths has an optimal integral solution which is unique.

This means that, under the sampling conditions, the subtour-LP solution provides a TSP solution and also suggests a reconstruction algorithm: solve the subtour-LP and, if the solution is integral, output it. If the input satisfies the sampling condition, then the solution will be integral and the result is indeed a polygonal reconstruction. Two algorithms are proposed to solve the subtour-LP. First, using the simplex method and the cutting plane framework: it starts with an LP consisting of only the degree constraints and in each iteration solves the current LP and checks whether that solution satisfies all the subtour elimination constraints (using a min cut algorithm) and, if not, adds a violated constraint to the current LP. This algorithm has a potentially exponential running time. Second, using a similar approach but with the ellipsoid

method. This can be implemented so that the running time is polynomial in the bit size of the input points. This requires justification for using approximate point coordinates and distances.

The main tool in deriving these results is the connection between the subtour-LP and the so-called *Held–Karp bound*. The line of argument is as follows:

- Let $c(u, v) = \|uv\|$ and $\mu : V \to R$ be a *potential function*. The corresponding *modified distance function* c_μ is defined by $c_\mu(u, v) = c(u, v) - \mu(u) - \mu(v)$.
- For any traveling salesman path T with endpoints a, b,

$$c_\mu(T) = c(T) - 2 \sum_{v \in V} \mu(v) + \mu(a) + \mu(b),$$

and so an optimal traveling salesman path with endpoints a, b for c_μ is also optimal for c.

- Let C_μ be the cost of a minimum spanning tree MST_μ under c_μ, then since a traveling salesman path is a spanning tree, the optimal traveling salesman T_0 satisfies $C_\mu \leq c_\mu(T_0) = c(T_0) - 2 \sum_{v \in V} \mu(v) + \mu(a) + \mu(b)$, and so

$$\max_\mu \left(C_\mu + 2 \sum_{v \in V} \mu(v) - \mu(a) - \mu(b) \right)$$
$$\leq c(T_0).$$

The term on the left is the so called Held–Karp bound.

- Now, if for a particular μ, MST_μ is a path with endpoints a, b, then MST_μ is in fact an optimal traveling salesman path with endpoints a, b, and the Held–Karp bound matches $c(T_0)$.
- The Held–Karp bound is equal to the optimal objective value of the subtour-LP. This follows by relaxation of the degree constraints in a Lagrangian fashion (see [5]) and gives an effective way to compute the Held-Karp bound: solve the subtour-LP.
- Finally, a potential function μ is constructed for γ so that, for an appropriately dense sample set V, MST_μ is unique and is a polygonal

reconstruction with endpoints a, b. This then implies that solving the subtour-LP will produce a correct polygonal reconstruction.

Note that the potential function μ enters the picture only as an analysis tool. It is not needed by the algorithm. The authors extend this work to the case of open curves without specified endpoints and of closed curves using variations of the ILP formulation and a more restricted sampling condition. They also extend it to the case of a collection of closed curves. The latter requires preprocessing that partitions points into groups that are expected to form individual curves. Then each subgroup is processed with the subtour-LP approach and then the quality of the result assessed and then that partition may be updated.

Finite Precision

The above results are obtained assuming exact representation of point samples and the distances between them, so claiming a polynomial time algorithm is not immediate as the running time of the ellipsoid method is polynomial in the bit size of the input. The authors extend the results to the case in which points and the distances between them are known only approximately and from this they can conclude the polynomial running time.

Relation to Local Feature Size

The defined potential function μ is related to the so called *local feature size* function f used in the theory of smooth curve reconstruction, where $f(p)$ is defined as the distance from p to the medial axis of the curve γ. In this paper, $\mu(p)$ is defined as $d(p)/3$ where $d(p)$ is the size of the largest neighborhood of p so that γ in that neighborhood does not *deviate significantly* from a flat segment of curve. This paper shows $f(p) < 3d(p)$. In fact, $\mu(p)$ amounts to a generalization of the local feature size to nonsmooth curves (for a corner point p, $\mu(p)$ is proportional to the size of the largest neighborhood of p such that γ inside does not deviate significantly from a corner point with two nearly flat legs incident to it, and for points near the corner, μ is

defined as an appropriate interpolation of the two definitions), and is in fact similar to definitions proposed elsewhere.

Applications

The curve reconstruction problem appears in applied areas such as cartography. For example, to determine level sets, features, object contours, etc. from samples. Admittedly, these applications usually may require the ability to handle very sparse sampling and noise. The 3D version of the problem is very important in areas such as industrial manufacturing, medical imaging, and computer animation. The 2D problem is often seen as a simpler (toy) problem to test algorithmic approaches.

Open Problems

A TSP-based solution when the curve γ is a collection of curves, not all closed, is not given in this paper. A solution similar to that for closed curves (partitioning and then application of subtour-LP for each) seems feasible for general collections, but some technicalities need to be solved. More interesting is the study of corresponding reconstruction approaches for surfaces in 3D.

Experimental Results

The companion paper [2] presents results of experiments comparing the TSP-based approach to several (local) Delaunay filtering algorithms. The TSP implementation uses the simplex method and the cutting plane framework (with a potentially exponential running time algorithm). The experiments show that the TSP-based approach has a better performance, allowing for much sparser samples than the others. This is to be expected given the global nature of the TSP-based solution. On the other hand, the speed of the TSP-based solution is reported to be competitive when compared to the speed of the others, despite its potentially bad worst-case behavior.

Data Sets

None reported. Experiments in [2] were performed with a simple reproducible curve based on a sinusoidal with varying number of periods and samples.

URL to Code

The code of the TSP-based solution as well as the other solutions considered in the companion paper [2] are available from: http://www.mpi-inf.mpg.de/~althaus/LEP:Curve-Reconstruction/curve.html

Cross-References

▶ Engineering Geometric Algorithms
▶ Euclidean Traveling Salesman Problem
▶ Minimum Weight Triangulation
▶ Planar Geometric Spanners
▶ Robust Geometric Computation

Recommended Reading

1. Althaus E, Mehlhorn K (2001) Traveling salesman-based curve reconstruction in polynomial time. SIAM J Comput 31:27–66
2. Althaus E, Mehlhorn K, Näher S, Schirra S (2000) Experiments on curve reconstruction. In: ALENEX, pp 103–114
3. Amenta N, Bern M (1999) Surface reconstruction by Voronoi filtering. Discrete Comput Geom 22:481–504
4. Amenta N, Bern M, Eppstein D (1998) The crust and the β-skeleton: combinatorial curve reconstruction. Graph Model Image Process 60:125–135
5. Cook W, Cunningham W, Pulleyblank W, Schrijver A (1998) Combinatorial optimization. Wiley, New York
6. Dey TK (2004) Curve and surface reconstruction. In: Goodman JE, O'Rourke J (eds) Handbook of discrete and computational geometry, 2nd edn. CRC, Boca Raton
7. Dey TK (2006) Curve and surface reconstruction: algorithms with mathematical analysis. Cambridge University Press, New York
8. Edlesbrunner H (1998) Shape reconstruction with the Delaunay complex. In: LATIN'98, theoretical informatics. Lecture notes in computer science, vol 1380. Springer, Berlin, pp 119–132
9. Giesen J (2000) Curve reconstruction, the TSP, and Menger's theorem on length. Discrete Comput Geom 24:577–603
10. Schrijver A (1986) Theory of linear and integer programming. Wiley, New York

Two-Dimensional Scaled Pattern Matching

Amihood Amir
Department of Computer Science, Bar-Ilan University, Ramat-Gan, Israel
Department of Computer Science, Johns Hopkins University, Baltimore, MD, USA

Keywords

2d scaled matching; Multidimensional scaled search; Pattern matching in scaled images; Two dimensional pattern matching with scaling

Years and Authors of Summarized Original Work

2006; Amir, Chencinski

Problem Definition

Definition 1 Let T be a two-dimensional $n \times n$ array over some alphabet Σ.

1. The *unit pixels array* for T (T^{1X}) consists of n^2 unit squares, called pixels in the real plane $\Re^2$. The corners of the pixel $T[i, j]$ are $(i - 1, j - 1), (i, j - 1), (i - 1, j)$, and (i, j). Hence the pixels of T form a regular $n \times n$ array that covers the area between $(0, 0), (n, 0), (0, n)$, and (n, n). Point $(0, 0)$ is the *origin* of the unit pixel array. The *center* of each pixel is the geometric center point of its square location. Each pixel $T[i, j]$ is identified with the value from Σ that the original array T had in that position. Say that the pixel has a color or

a character from Σ. See Fig. 1 for an example of the grid and pixel centers of a 7×7 array.

2. Let $r \in \mathfrak{R}, r \geq 1$. The *r-ary pixels array* for T (T^{rX}) consists of $n^2 r$-squares, each of dimension $r \times r$ whose *origin* is $(0,0)$ and covers the area between $(0,0), (nr, 0), (0, nr)$, and (nr, nr). The corners of the pixel $T[i, j]$ are $((i - 1)r, (j - 1)r), (ir, (j - 1)r)$, $((i - 1)r, jr)$, and (ir, jr). The *center* of each pixel is the geometric center point of its square location.

Notation: Let $r \in \mathfrak{R}$. $[r]$ denotes the *rounding* of r, i.e.,

$$[r] = \begin{cases} \lfloor r \rfloor & \text{if } r - \lfloor r \rfloor < .5; \\ \lceil r \rceil & \text{otherwise.} \end{cases}$$

Definition 2 Let T be an $n \times n$ text array, P be an $m \times m$ pattern array over alphabet Σ, and let $r \in \mathfrak{R}, 1 \leq r \leq \frac{n}{m}$. Say that there is an *occurrence of P scaled to r* at text location (i, j) if the following conditions hold:

Let T^{1X} be the unit pixels array of T and P^{rX} be the r-ary pixel arrays of P. Translate P^{rX} onto T^{1X} in a manner that the origin of P^{rX} coincides with location $(i - 1, j - 1)$ of T^{1X}. Every center of a pixel in T^{1X} which is within the area covered by $(i - 1, j - 1), (i - 1, j - 1 + mr), (i - 1 + mr, j - 1)$ and $(i - 1 + mr, j - 1 + mr)$ has the same color as the r-square of P^{rX} in which it falls.

The colors of the centers of the pixels in T^{1X} which are within the area covered by $(i - 1, j - 1), (i - 1, j - 1 + mr), (i - 1 + mr, j - 1)$ and $(i - 1 + mr, j - 1 + mr)$ define a $[mr] \times [mr]$ array over Σ. This array is denoted by $P^{s(r)}$ and called *P scaled to r*.

The above definition is the one provided in the *geometric model*, pioneered by Landau and Vishkin [15], and Fredriksson and Ukkonen [14]. Prior to the advent of the geometric model, the only discrete definition of scaling was to natural scales, as defined by Amir, Landau and Vishkin [10]:

Definition 3 Let $P[m \times m]$ be a two-dimensional matrix over alphabet Σ (not necessarily bounded). Then P scaled by s (P^s) is the $sm \times sm$ matrix where every symbol $P[i, j]$ of P is replaced by a $s \times s$ matrix whose elements all equal the symbol in $P[i, j]$. More precisely,

$$P^s[i, j] = P[\lceil \tfrac{i}{s} \rceil, \lceil \tfrac{j}{s} \rceil].$$

Say that pattern $P[m \times m]$ *occurs* (or an occurrence of P starts) at location (k, l) of the text T if for any $i \in \{1, \ldots, m\}$ and $j \in \{1, \ldots, m\}$, $T[k + i - 1, l + j - 1] = P[i, j]$.

The *two dimensional pattern matching problem with natural scales* is defined as follows.

INPUT: Pattern matrix $P[i, j]$ $i = 1, \ldots m; j = 1, \ldots, m$ and Text matrix $T[i, j]$ $i = 1, \ldots, n; j = 1, \ldots, n$ where $n > m$.

OUTPUT: all locations in T where an occurrence of P scaled by s (an *s-occurrence*) starts, for any $s = 1, \ldots, \lfloor \frac{n}{m} \rfloor$.

The natural scales definition cannot answer normal everyday occurrences such as an image scaled to, say, 1.3. This led to the geometric model. The geometric model is a discrete adaptation, without smoothing, of scaling as used in computer graphics. The definition is pleasing in a "real-world" sense. Figure 2 shows "lenna" scaled to non-discrete scales by the geometric model definition. The results look natural.

It is possible, of course, to consider a *one dimensional* version of scaling, or scaling in *strings*. Both above definitions apply for one dimensional scaling where the text and pattern are taken to be matrices having a single row. The interest in one dimensional scaling lies because of two reasons: (1) There is a faster algorithm for one dimensional scaling in the geometric model than the restriction of the two dimensional scaling algorithm to one dimension. (2) Historically, before the geometric model was defined, there was an attempt [3] to define real scaling on strings as follows.

T

Two-Dimensional Scaled Pattern Matching, Fig. 1
The grid and pixel centers of a unit pixel array for a 7×7 array

	0	1	2	3	4	5	6	7
0								
	T(1,1)	T(1,2)	T(1,3)	○	○	○	○	
1								
	T(2,1)	T(2,2)	T(2,3)	○	○	○	○	
2								
	T(3,1)	T(3,2)	T(3,3)	○	○	○	○	
3								
	○	○	○	○	○	○	○	
4								
	○	○	○	T(5,4)	○	○	○	
5								
	○	○	○	○	○	○	○	
6								
	○	○	○	○	○	○	T(7,7)	
7								

Two-Dimensional Scaled Pattern Matching, Fig. 2 An original image, scaled by 1.3 and scaled by 2, using the geometric model definition of scaling

Definition 4 Denote the string $aa \cdots a$, where a is repeated r times, by a^r. The *one dimensional floor real scaled matching problem* is the following.

INPUT: A pattern $P = a_1^{r_1} a_2^{r_2} \ldots a_j^{r_j}$, of length m, and a text T of length n.

OUTPUT: All locations in the text where the substring $a_1^{c_1} a_2^{\lfloor r_2 k \rfloor} \ldots a_{j-1}^{\lfloor r_{j-1} k \rfloor} a_j^{c_j}$ appears, where

$c_1 \geq \lfloor r_1 k \rfloor$ and $c_j \geq \lfloor r_j k \rfloor$.

This definition indeed handles real scaling but has a significant weakness in that a string of length m scaled to r may be significantly shorter than mr. For this reason the definition could not be generalized to two dimensions. The geometric model does not suffer from these deficiencies.

Key Results

The first results in scaled natural matching dealt with fixed finite-sized alphabets.

Theorem 1 (Amir, Landau, and Vishkin [10]) *There exists an $O(|T| \log |\Sigma|)$ worst-case time solution to the two dimensional pattern matching problem with natural scales, for fixed finite alphabet Σ.*

The main idea behind the algorithm is analyzing the text with the aid of *power columns*. Those are the text columns appearing $m-1$ columns apart, where P is an $m \times m$ pattern. This dependence on the pattern size make the power columns useless where a dictionary of different sized patterns is involved. A significantly simpler algorithm with an additional advantage of being alphabet-independent was presented in [6].

Theorem 2 (Amir and Calinescu [6]) *There exists an $O(|T|)$ worst-case time solution to the two dimensional pattern matching problem with natural scales.*

The alphabet independent time complexity of this algorithm was achieved by developing a scaling-invariant "signature" of the pattern. This idea was further developed to scaled dictionary matching.

Theorem 3 (Amir and Calinescu [6]) *Given a static dictionary of square pattern matrices. It is possible in $O(|D| \log k)$ preprocessing, where $|D|$ is the total dictionary size and k is the number of patterns in the dictionary, and $O(|T| \log k)$ text scanning time, for input text T, to find all occurrences of dictionary patterns in the text in all natural scales.*

This is identical to the time at [8], the best non-scaled matching algorithm for a static dictionary of square patterns. It is somewhat surprising that scaling does not add to the complexity of single matching nor dictionary matching.

The first algorithm to solve the scaled matching problem for real scales, was a one dimensional real scaling algorithm using Definition 4.

Theorem 4 (Amir, Butman, and Lewenstein [3]) *There exists an $O(|T|)$ worst-case time solution to the one dimensional floor real scaled matching problem.*

The first algorithm to solve the two dimensional scaled matching problem for real scales in the geometric model is the following.

Theorem 5 (Amir, Butman, Lewenstein, and Porat [4]) *Given an $n \times n$ text and $m \times m$ pattern. It is possible to find all pattern occurrences in all real scales in time $O(nm^3 + n^2 m \log m)$ and space $O(nm^3 + n^2)$.*

The above result was improved.

Theorem 6 (Amir and Chencinski [7]) *Given an $n \times n$ text and $m \times m$ pattern. It is possible to find all pattern occurrences in all real scales in time $O(n^2 m)$ and space $O(n^2)$.*

This algorithm achieves its time by exploiting geometric characteristics of nested scales occurrences and a sophisticated use of dueling [1, 16].

The assumption in both above algorithms is that the scaled occurrence of the pattern starts at the top left corner of some pixel.

It turns out that one can achieve faster times in the *one dimensional* real scaled matching problem, even in the geometric model.

Theorem 7 (Amir, Butman, Lewenstein, Porat, and Tsur [5]) *Given a text string T of length n and a pattern string P of length m, there exists an $O(n \log m + m\sqrt{nm \log m})$ worst-case time solution to the one dimensional pattern matching problem with real scales in the geometric model.*

Applications

The problem of finding approximate occurrences of a template in an image is a central one in digital libraries and web searching. The current algorithms to solve this problem use methods of computer vision and computational geometry. They model the image in another space and seek a solution there. A deterministic worst-case algorithm in pixel-level images does not yet exist. Yet, such an algorithm could be useful, especially in raw data that has not been modeled, e.g., movies. The work described here advances another step toward this goal from the scaling point of view.

Open Problems

Finding all scaled occurrences without fixing the scaled pattern start at the top left corner of the text pixel would be important from a practical point of view. The final goal is an integration of scaling with rotation [2, 11–13] and local errors (edit distance) [9].

Cross-References

▶ Multidimensional Compressed Pattern Matching
▶ Multidimensional String Matching

Recommended Reading

1. Amir A, Benson G, Farach M (1994) An alphabet independent approach to two dimensional pattern matching. SIAM J Comput 23(2):313–323
2. Amir A, Butman A, Crochemore M, Landau GM, Schaps M (2004) Two-dimensional pattern matching with rotations. Theor Comput Sci 314(1–2):173–187
3. Amir A, Butman A, Lewenstein M (1999) Real scaled matching. Inf Process Lett 70(4):185–190
4. Amir A, Butman A, Lewenstein M, Porat E (2003) Real two dimensional scaled matching. In: Proceedings of the 8th workshop on algorithms and data structures (WADS '03), pp 353–364
5. Amir A, Butman A, Lewenstein M, Porat E, Tsur D (2004) Efficient one dimensional real scaled matching. In: Proceedings of the 11th symposium on string processing and information retrieval (SPIRE'04), pp 1–9
6. Amir A, Calinescu G (2000) Alphabet independent and dictionary scaled matching. J Algorithm 36: 34–62
7. Amir A, Chencinski E (2006) Faster two dimensional scaled matching. In: Proceedings of the 17th symposium on combinatorial pattern matching (CPM). LNCS, vol 4009. Springer, Berlin, pp 200–210
8. Amir A, Farach M (1992) Two dimensional dictionary matching. Inf Process Lett 44:233–239
9. Amir A, Landau G (1991) Fast parallel and serial multidimensional approximate array matching. Theor Comput Sci 81:97–115
10. Amir A, Landau GM, Vishkin U (1992) Efficient pattern matching with scaling. J Algorithm 13(1):2–32
11. Amir A, Tsur D, Kapah O (2004) Faster two dimensional pattern matching with rotations. In: Proceedings of the 15th annual symposium on combinatorial pattern matching (CPM '04), pp 409–419
12. Fredriksson K, Mäkinen V, Navarro G (2004) Rotation and lighting invariant template matching. In: Proceedings of the 6th Latin American symposium on theoretical informatics (LATIN'04). LNCS, pp 39–48
13. Fredriksson K, Navarro G, Ukkonen E (2002) Optimal exact and fast approximate two dimensional pattern matching allowing rotations. In: Proceedings of the 13th annual symposium on combinatorial pattern matching (CPM 2002). LNCS, vol 2373, pp 235–248
14. Fredriksson K, Ukkonen E (1998) A rotation invariant filter for two-dimensional string matching. In: Proceedings of the 9th annual symposium on combinatorial pattern matching (CPM). LNCS, vol 1448. Springer, Berlin, pp 118–125
15. Landau GM, Vishkin U (1994) Pattern matching in a digitized image. Algorithmica 12(3/4):375–408
16. Vishkin U (1985) Optimal parallel pattern matching in strings. In: Proceedings of the 12th ICALP, pp 91–113

Two-Interval Pattern Problems

Stéphane Vialette
IGM-LabInfo, University of Paris-East, Descartes, France

Keywords

2-intervals; RNA structures

Years and Authors of Summarized Original Work

2004; Vialette
2007; Cheng, Yang, Yuan

Problem Definition

The problem is concerned with finding large constrained patterns in sets of 2-intervals. Given a single-stranded RNA molecule, a sequence of contiguous bases of the molecule can be represented as an interval on a single line, and a possible pairing between two disjoint sequences can be represented as a 2-interval, which is merely the union of two disjoint intervals. Derived from arc-annotated sequences, 2-interval representation considers thus only the bonds between the bases and the pattern of the bonds, such as hairpin structures, knots and pseudoknots. A maximum cardinality disjoint subset of a candidate set of 2-intervals restricted to certain prespecified geometrical constraints can provide a useful valid approximation for RNA secondary structure determination.

The geometric properties of 2-intervals provide a possible guide for understanding the computational complexity of finding structured patterns in RNA sequences. Using a model to represent nonsequential information allows us to vary restrictions on the complexity of the pattern structure. Indeed, two disjoint 2-intervals, i.e., two 2-intervals that do not intersect in any point, can be in precedence order ($<$), be allowed to nest ($\sqsubset$) or be allowed to cross ($\between$). Furthermore, the set of 2-intervals and the pattern can have different restrictions, e.g., all intervals have the same length or all the intervals are disjoint. These different combinations of restrictions alter the computational complexity of the problems, and need to be examined separately. This examination produces efficient algorithms for more restrictive structured patterns, and hardness results for those that are less restrictive.

Notations

Let $I = [a, b]$ be an interval on the line. Write $\mathsf{start}(I) = a$ and $\mathsf{end}(I) = b$. A *2-interval* is the union of two disjoint intervals defined over a single line and is denoted by $D = (I, J)$; I is completely to the left of J. Write $\mathsf{left}(D) = I$ and $\mathsf{right}(D) = J$. Two 2-intervals $D_1 = (I_1, J_1)$ and $D_2 = (I_2, J_2)$ are said to be *disjoint* (or *nonintersecting*) if both 2-intervals share no common point, i.e., $(I_1 \cup J_1) \cap (I_2 \cup J_2) = \emptyset$. For such disjoint pairs of 2-intervals, three natural binary relations, denoted $<$, $\sqsubset$ and $\between$, are of special interest:

- $D_1 < D_2$ ($D_1\ precedes\ D_2$), if $I_1 < J_1 < I_2 < J_2$,
- $D_1 \sqsubset D_2$ (D_1 is *nested in* D_2), if $I_2 < I_1 < J_1 < J_2$, and
- $D_1 \between D_2$ ($D_1\ crosses\ D_2$), if $I_1 < I_2 < J_1 < J_2$.

A pair of 2-intervals D_1 and D_2 is said to be *R-comparable* for some $R \in \{<, \sqsubset, \between\}$, if either $D_1 R D_2$ or $D_2 R D_1$. Note that any two disjoint 2-intervals are R-comparable for some $R \in \{<, \sqsubset, \between\}$. A set of disjoint 2-intervals $\mathcal{D}$ is said to be *R-comparable* for some $\mathcal{R} \subseteq \{<, \sqsubset, \between\}$, $\mathcal{R} \neq \emptyset$, if any pair of distinct 2-intervals in $\mathcal{D}$ is R-comparable for some $R \in \mathcal{R}$. The nonempty subset $\mathcal{R}$ is called a *model* for $\mathcal{D}$.

The 2-interval-pattern problem asks one to find in a set of 2-intervals a largest subset of pairwise compatible 2-intervals. In the present context, compatibility denotes the fact that any two 2-intervals in the solution are (1) nonintersecting and (2) satisfy some prespecified geometrical constraints. The 2-interval-pattern problem is formally defined as follows:

Problem 1 (2-interval-pattern)

INPUT: A set of 2-intervals $\mathcal{D}$ and a model $\mathcal{R} \subseteq \{<, \sqsubset, \between\}$.
SOLUTION: A $\mathcal{R}$-comparable subset $\mathcal{D}' \subseteq \mathcal{D}$.
MEASURE: The size of the solution, i.e., $|\mathcal{D}'|$.

According to the above definition, any solution for the 2-interval-pattern problem for some model $\mathcal{R} \subseteq \{<, \sqsubset, \between\}$ corresponds to an RNA structure constrained by $\mathcal{R}$. For example, a solution for the 2-interval-pattern problem for the $\mathcal{R} = \{<, \sqsubset\}$ model corresponds to a pseudoknot-free structure (a *pseudoknot* in an RNA sequence $S = s_1, s_2, \ldots, s_n$ is composed of two interleaving nucleotide pairings (s_i, s_j) and $(s_{i'}, s_{j'})$ such that $i < i' < j < j'$).

Some additional definitions are needed for further algorithmic analysis. Let $\mathcal{D}$ be a set of 2-intervals. The *width* (respectively *height*, *depth*) is the size of a maximum cardinality $\{<\}$-comparable (respectively $\{\sqsubset\}$-comparable, $\{\between\}$-comparable) subset $\mathcal{D}' \subseteq \mathcal{D}$. The *interleaving distance* of a 2-interval $D_i \in \mathcal{D}$ is defined to be the distance between the two intervals of D_i, i.e., $\mathsf{start(right(D_i))} - \mathsf{end(left(D_i))}$. The *total interleaving distance* of the set of 2-intervals $\mathcal{D}$, written $\mathcal{L}(\mathcal{D})$, is the sum of all interleaving distances, i.e., $\mathcal{L}(\mathcal{D}) = \sum_{D_i \in \mathcal{D}} \mathsf{start(right(D_i))} - \mathsf{end(left(D_i))}$. The *interesting coordinates* of $\mathcal{D}$ are defined to be the set $X(\mathcal{D}) = \bigcup_{D_i \in \mathcal{D}} \{\mathsf{end(left(D_i))}, \mathsf{start(right(D_i))}\}$. The *density* of $\mathcal{D}$, written $d(\mathcal{D})$, is the maximum number of 2-intervals in $\mathcal{D}$ over a single point. Formally, $d(\mathcal{D}) = \max_{x \in X(\mathcal{D})} \{D \in \mathcal{D} : \mathsf{end(left(D)} \leq x < \mathsf{start(right(D))}\}$.

Constraints

The structure of the set of all (simple) intervals involved in a set of 2-intervals $\mathcal{D}$ turns out to be of particular importance for algorithmic analysis of the 2-interval-pattern problem. The *interval ground set* of $\mathcal{D}$, denoted $\mathcal{I}(\mathcal{D})$, is the set of all intervals involved in $\mathcal{D}$, i.e., $\mathcal{I}(\mathcal{D}) = \{\mathsf{left(D_i)} : D_i \in \mathcal{D}\} \cup \{\mathsf{right} < (D : i) : D_i \in \mathcal{D}\}$. In [7, 20], four types of interval ground sets were introduced.

1. *Unlimited*: no restriction on the structure.
2. *Balanced*: each 2-interval $D_i \in \mathcal{D}$ is composed of two intervals having the same length, i.e., $|\mathsf{left(D_i)}| = |\mathsf{right(D_i)}|$.
3. *Unit*: the interval ground set $\mathcal{I}(\mathcal{D})$ is solely composed of unit length intervals.
4. *Disjoint*: no two distinct intervals in the interval ground set $\mathcal{I}(\mathcal{D})$ intersect.

Observe that a unit 2-interval set is balanced, while the converse is not necessarily true. Furthermore, for most applications, one may assume that a disjoint 2-interval set is unit. Observe that in this latter case, a set of 2-intervals reduces to a graph $G = (V, E)$ equipped with a numbering of its vertices from 1 to $|V|$, and hence the 2-interval-pattern problem for disjoint interval ground sets reduces to finding a constrained maximum matching in a linear graph. Considering additional restrictions such as:

- Bounding the width, the height or the depth of either the input set of 2-intervals or the solution subset

- Bounding the interleaving distances

is also of interest for practical applications.

Key Results

The different combinations of the models and interval ground sets alter the computational complexity of the 2-interval-pattern problem. The main results are summarized in Tables 1 (time complexity and hardness) and 2 (approximation for hard instances).

Theorem 1 *The 2-interval-pattern problem is approximable (APX) hard for models* $\mathcal{R} = \{<, \sqsubset, \between\}$ *and* $\mathcal{R} = \{\sqsubset, \between\}$, *and is nondeterministic polynomial-time (NP) complete – in its natural decision version – for model* $\mathcal{R} = \{<, \between\}$, *even when restricted to unit interval ground sets.*

Notice here that the 2-interval-pattern problem for model $\mathcal{R} = \{<, \between\}$ is not APX-hard. Two hard cases of the 2-interval-pattern turn out to be polynomial-time-solvable when restricted to disjoint-interval ground sets.

Theorem 2 *The 2-interval-pattern problem for a disjoint-interval ground set is solvable in*

- $O(n \sqrt{n})$ *time for model* $\mathcal{R} = \{<, \sqsubset, \between\}$ *(trivial reduction to the standard maximum matching problem)*

Two-Interval Pattern Problems, Table 1 Complexity of the 2-interval-pattern problem for all combinations of models and interval ground sets. For the polynomial-time cases, $n = |\mathcal{D}|$, $\mathcal{L} = \mathcal{L}(\mathcal{D})$ and $d = d(\mathcal{D})$

Model $\mathcal{R}$	Interval ground set $\mathcal{I}(\mathcal{D})$	
	Unlimited, balanced, unit	Disjoint
$\{<, \sqsubset, \between\}$	APX-hard [1]	$O(n \sqrt{n})$ [15]
$\{<, \between\}$	NP-complete [3]	**unknown**
$\{\sqsubset, \between\}$	APX-hard [19]	$O(n \log n + \mathcal{L})$ [8]
$\{<, \sqsubset\}$	$O(n \log n + nd)$ [8]	
$\{<\}$	$O(n \log n)$ [19]	
$\{\sqsubset\}$	$O(n \log n)$ [3]	
$\{\between\}$	$O(n \log n + \mathcal{L})$ [8]	

- $O(n \log n + L)$ *time for model* $\mathcal{R} = \{\sqsubset, \between\}$

The complexity of the 2-interval-pattern problem for model $\mathcal{R} = \{<, \between\}$ and a disjoint-interval ground set is still unknown. Three cases of the 2-interval-pattern problem are polynomial-time-solvable, regardless of the structure of the interval ground sets.

Theorem 3 *The 2-interval-pattern problem is solvable in*

- $O(n \log n + nd)$ *time for model* $\mathcal{R} = \{<, \sqsubset\}$
- $O(n \log n)$ *time for models* $\mathcal{R} = \{<\}$ *and* $\mathcal{R} = \{\sqsubset\}$
- $O(n \log n + L)$ *time for model* $\mathcal{R} = \{\between\}$

One may now turn to approximating hard instances of the 2-interval-pattern problem. Surprisingly enough, no significant differences (in terms of approximation guarantees) have yet been found for the 2-interval-pattern problem between the model $\mathcal{R} = \{<, \sqsubset, \between\}$ and the model $\mathcal{R} = \{\sqsubset, \between\}$ (the approximation algorithms are, however, different).

Theorem 4 *The 2-interval-pattern problem for model* $\mathcal{R} = \{<, \sqsubset, \between\}$ *or model* $\mathcal{R} = \{\sqsubset, \between\}$ *is approximable within ratio*

- *4 for unlimited-interval ground sets, and*
- *$2 + \epsilon$ for unit-interval ground sets.*

The 2-interval-pattern problem for model $\mathcal{R} = \{<, \between\}$ *is approximable within ratio* $1 + 1/\epsilon$, $\epsilon \geq 2$ *for all models.*

A practical 3-approximation algorithm for model $\mathcal{R} = \{<, \sqsubset, \between\}$ (resp. $\mathcal{R} = \{\sqsubset, \between\}$) and unit interval ground set that runs in $\mathcal{O}(n \lg n)$ (resp. $\mathcal{O}(n^2 \lg n)$) time has been proposed in [1] (resp. [7]). For model $\mathcal{R} = \{<, \between\}$, a more practical 2-approximation algorithm that runs in $\mathcal{O}(n^3 \lg n)$ time has been proposed in [10]. Notice that Theorem 4 holds true for the weighted version of the 2-interval-pattern problem [7] except for models $\mathcal{R} = \{<, \sqsubset, \between\}$ and $\mathcal{R} = \{\sqsubset, \between\}$ and unit interval ground set where the best approximation ratio is $2.5 + \epsilon$ [5].

Applications

Sets of 2-intervals can be used for modeling stems in RNA structures [20, 21], determining DNA sequence similarities [13] or scheduling jobs that are given as groups of nonintersecting segments in the real line [1, 9]. In all these applications, one is concerned with finding a maximum cardinality subset of nonintersecting 2-intervals. Some other classical combinatorial problems are also of interest [5]. Also, considering sets of t-intervals (each element is the union of at most t disjoint intervals) and their corresponding intersection graph has proved to be useful.

It is computationally challenging to predict RNA structures including pseudoknots [14]. Practical approaches to cope with intractability are either to restrict the class of pseudoknots under consideration [18] or to use heuristics [6, 17, 19]. The general problem of establishing a general representation of structured patterns, i.e., *macroscopic describers* of RNA structures, was considered in [20]. Sets of 2-intervals provide such a natural geometric description.

Constructing a relevant 2-interval set from a RNA sequence is relatively easy: stable stems are selected, usually according to a simplified thermodynamic model without accounting for loop energy [2, 16, 19–21]. Predicting a reliable RNA structure next reduces to finding a maximum subset of nonconflicting 2-intervals, i.e., a subset of disjoint 2-intervals. Considering in addition a model $\mathcal{R} \subseteq \{<, \sqsubset, \between\}$ allows us to vary restrictions on the complexity of the pattern structure. In [21], the treewidth of the intersection graph of the set of 2-intervals is considered for speeding up the computation.

For sets of 2-intervals involved in practical applications, restrictions on the interval ground set are needed. Unit interval ground sets were considered in [7]. Of particular importance in the context of molecular biology (RNA structures

Two-Interval Pattern Problems, Table 2 Performance ratios for hard instances of the 2-interval-pattern problem. LP stands for *Linear Programming* and N/A stands for *Not Applicable*

| Model $\mathcal{R}$ | Interval ground set $\mathcal{I}(\mathcal{D})$ | | | |
	Unlimited	Balanced	Unit	Disjoint
$\{<, \sqsubset, \between\}$	4 LP [1]	4 $\mathcal{O}(n \lg n)$ [7]	$2 + \epsilon\mathcal{O}(n^2 + n^{\mathcal{O}(\log 1/\epsilon)})$ [13]	N/A
$\{\sqsubset, \between\}$	4 LP [7]	4 $\mathcal{O}(n^2 \lg n)$ [7]	$2 + \epsilon\mathcal{O}(n^2 + n^{\mathcal{O}(\log 1/\epsilon)})$ [13]	N/A
$\{<, \between\}$	$1 + 1/\epsilon\mathcal{O}(n^{2\epsilon+3})$, $\epsilon \geq 2$ [14]			

and DNA sequence similarities) are balanced interval ground sets, where each 2-interval is composed of two equally length intervals.

Open Problems

A number of problems related to the 2-interval-pattern problem remain open. First, improving the approximation ratios for the various flavors of the 2-interval-pattern problem is of particular importance. For example, the existence of a fast approximation algorithm with good performance guarantee for the 2-interval-pattern problem for model $\mathcal{R} = \{<, \sqsubset, \between\}$ remains an apparently challenging open problem. A related open research area is concerned with balanced-interval ground sets. In particular, no evidence has shown yet that the 2-interval-pattern problem becomes easier to approximate for balanced-interval ground sets. This question is of special importance in the context of RNA structures where most 2-intervals are balanced.

A number of important question are still open for model $\mathcal{R} = \{<, \between\}$. First, it is still unknown whether the 2-interval-pattern problem for disjoint-interval ground sets and model $\mathcal{R} = \{<, \between\}$ is polynomial-time-solvable. Observe that this problem trivially reduces to the following graph problem: Given a graph $G = (V, E)$ with $V = \{1, 2, \ldots, n\}$, find a maximum cardinality matching $\mathcal{M} \subseteq E$ such that for any two distinct edges $\{i, j\}$ and $\{k, l\}$ of $\mathcal{M}$, $i < j$, $k < l$ and $i < k$, either $j < k$ or $j < l$. Another open question concerns the approximation of the 2-interval-pattern problem for balanced interval ground set. Is this special case better approximable than the general case?

A last direction of research is concerned with the parameterized complexity of the 2-interval-pattern problem. For example, it is not known whether the 2-interval-pattern problem for models $\mathcal{R} = \{<, \sqsubset, \between\}$, $\mathcal{R} = \{\sqsubset, \between\}$ or $\mathcal{R} = \{<, \between\}$ is fixed-parameter-tractable when parameterized by the size of the solution. Also, investigating the parameterized complexity for parameters such as the maximum number of pairwise crossing intervals in the input set or the treewidth of the corresponding intersection 2-interval graph, which are expected to be relatively small for most practical applications, is of particular interest.

Cross-References

▶ RNA Secondary Structure Prediction by Minimum Free Energy
▶ RNA Secondary Structure Prediction Including Pseudoknots

Recommended Reading

1. Bar-Yehuda R, Halldorsson M, Naor J, Shachnai H, Shapira I (2002) Scheduling split intervals. In: Proceedings of the 13th annual ACM-SIAM symposium on discrete algorithms (SODA), pp 732–741
2. Billoud B, Kontic M, Viari A (1996) Palingol a declarative programming language to describe nucleic acids' secondary structures and to scan sequence database. Nucleic Acids Res 24:1395–1403
3. Blin G, Fertin G, Vialette S (2007) Extracting 2-intervals subsets from 2-interval sets. Theor Comput Sci 385(1–3):241–263
4. Blin G, Fertin G, Vialette S (2004) New results for the 2-interval pattern problem. In: Proceedings of the 15th annual symposium on combinatorial pattern matching (CPM). Lecture notes in computer science, vol 3109. Springer, Berlin
5. Butman A, Hermelin D, Lewenstein M, Rawitz D (2007) Optimization problems in multiple-interval

graphs. In: Proceedings of the 9th annual ACM-SIAM symposium on discrete algorithms (SODA), ACM-SIAM 2007, pp 268–277

6. Chen J-H, Le S-Y, Maize J (2000) Prediction of common secondary structures of RNAs: a genetic algorithm approach. Nucleic Acids Res 28:991–999

7. Crochemore M, Hermelin D, Landau G, Rawitz D, Vialette S (2008) Approximating the 2-interval pattern problem. Theor Comput Sci (special issue for Alberto Apostolico)

8. Erdong C, Linji Y, Hao Y (2007) Improved algorithms for 2-interval pattern problem. J Comb Optim 13(3):263–275

9. Halldorsson M, Karlsson R (2006) Strip graphs: recognition and scheduling. In: Proceedings of the 32nd international workshop on graph-theoretic concepts in computer science (WG). Lecture notes in computer science, vol 4271. Springer, Berlin, pp137–146

10. Jiang M (2007) A 2-approximation for the preceding-and-crossing structured 2-interval pattern problem. J Comb Optim 13:217–221

11. Jiang M (2007) Improved approximation algorithms for predicting RNA secondary structures with arbitrary pseudoknots. In: Proceedings of the 3rd international conference on algorithmic aspects in information and management (AAIM), Portland. Lecture notes in computer science, vol 4508. Springer, pp 399–410

12. Jiang M (2007) A PTAS for the weighted 2-interval pattern problem over the preceding-and-crossing model. In: Xu Y, Dress AWM, Zhu B (eds) Proceedings of the 1st annual international conference on combinatorial optimization and applications (COCOA), Xi'an. Lecture notes in computer science, vol 4616. Springer, pp 378–387

13. Joseph D, Meidanis J, Tiwari P (1992) Determining DNA sequence similarity using maximum independent set algorithms for interval graphs. In: Proceedings of the 3rd Scandinavian workshop on algorithm theory (SWAT). Lecture notes in computer science. Springer, Berlin, pp 326–337

14. Lyngsø R, Pedersen C: RNA pseudoknot prediction in energy-based models. J Comput Biol 7:409–427 (2000)

15. Micali S, Vazirani V (1980) An O(sqrt|V||E|) algorithm for finding maximum matching in general graphs. In: Proceedings of the 21st annual symposium on foundation of computer science (FOCS). IEEE, pp 17–27

16. Nussinov R, Pieczenik G, Griggs J, Kleitman D (1978) Algorithms for loop matchings. SIAM J Appl Math 35:68–82

17. Ren J, Rastegart B, Condon A, Hoos H (2005) Hot-Knots: heuristic prediction of rna secondary structure including pseudoknots. RNA 11:1194–1504

18. Rivas E, Eddy S (1999) A dynamic programming algorithm for RNA structure prediction including pseudoknots. J Mol Biol 285:2053–2068

19. Ruan J, Stormo G, Zhang W (2004) An iterated loop matching approach to the prediction of RNA secondary structures with pseudoknots. Bioinformatics 20:58–66

20. Vialette S (2004) On the computational complexity of 2-interval pattern matching. Theor Comput Sci 312:223–249

21. Zhao J, Malmberg R, Cai L (2006) Rapid ab initio RNA folding including pseudoknots via graph tree decomposition. In: Proceedings of the workshop on algorithms in bioinformatics. Lecture notes in computer science, vol 4175. Springer, Berlin, pp 262–273

T

U

Undirected Feedback Vertex Set

Jiong Guo
Department of Mathematics and Computer
Science, University of Jena, Jena, Germany

Keywords

Odd cycle transversal

Years and Authors of Summarized Original Work

2005; Dehne, Fellows, Langston, Rosamond, Stevens Guo, Gramm, Hüffner, Niedermeier, Wernicke

Problem Definition

The UNDIRECTED FEEDBACK VERTEX SET (UFVS) problem is defined as follows:

Input: An undirected graph $G = (V, E)$ and an integer $k \geq 0$.

Task: Find a *feedback vertex set* $F \subseteq V$ with $|F| \leq k$ such that each cycle in G contains at least one vertex from F. (The removal of all vertices in F from G results in a forest.)

Karp [11] showed that UFVS is NP-complete. Lund and Yannakakis [12] proved that there exists some constant $\epsilon > 0$ such that it is NP-hard

to approximate the optimization version of UFVS to within a factor of $1 + \epsilon$. The best-known polynomial-time approximation algorithm for UFVS has a factor of 2 [1, 4]. There is a simple and elegant randomized algorithm due to Becker et al. [3] which solves UFVS in $O(c \cdot 4^k \cdot k\, n)$ time on an n-vertex and m-edge graph by finding a feedback vertex set of size k with probability at least $1 - (1 - 4^{-k})^{c4^k}$ for an arbitrary constant c. An exact algorithm for UFVS with a running time of $O(1.7548^n)$ was recently found by Fomin et al. [9]. In the context of parameterized complexity [8, 13], Bodlaender [5] and Downey and Fellows [7] were the first to show that the problem is *fixed-parameter tractable*, i.e., that the combinatorial explosion when solving it can be confined to the parameter k. The currently best fixed-parameter algorithm for UFVS runs in $O(c^k \cdot mn)$ for a constant c [6, 10] (see [6] for the so far best running time analysis leading to a constant $c = 10.567$). This algorithm is the subject of this entry.

Key Results

The $O(c^k \cdot mn)$-time algorithm for the UNDIRECTED FEEDBACK VERTEX SET is based on the so-called "iterative compression" technique, which was introduced by Reed et al. [14]. The central observation of this technique is quite simple but fruitful: To derive a fixed-parameter algorithm for a minimization problem, it suffices to give a fixed-parameter "compression routine"

© Springer Science+Business Media New York 2016
M.-Y. Kao (ed.), *Encyclopedia of Algorithms*,
DOI 10.1007/978-1-4939-2864-4

that, given a size-$(k + 1)$ solution, either proves that there is no size-k solution or constructs one. Starting with a trivial instance and iteratively applying this compression routine a linear number of rounds to larger instances, one obtains a fixed-parameter algorithm of the problem. The main challenge of applying this technique to UFVS lies in showing that there is a fixed-parameter compression routine.

The compression routine from [6, 10] works as follows:

1. Consider all possible partitions (X, Y) of the size-$(k + 1)$ feedback vertex set F with $|X| \leq k$ under the assumption that set X is entirely contained in the new size-k feedback vertex set F' and $Y \cap F' = \emptyset$
2. For each partition (X, Y), if the vertices in Y induce cycles, then answer "no" for this partition; otherwise, remove the vertices in X. Moreover, apply the following data reduction rules to the remaining graph:
   - Remove degree-1 vertices.
   - If there is a degree-2 vertex v with two neighbors v_1 and v_2, where $v_1 \notin Y$ or $v_2 \notin Y$, then remove v and connect v_1 and v_2. If this creates two parallel edges between v_1 and v_2, then remove the vertex of v_1 and v_2 that is not in Y and add it to any feedback vertex set for the reduced instance.

   Finally, exhaustively examine every vertex set S with size at most $k - |X|$ of the reduced graph as to whether S can be added to X to form a feedback vertex set of the input graph. If there is one such vertex set, then output it together with X as the new size-k feedback vertex set.

The correctness of the compression routine follows from its brute-force nature and the easy to prove correctness of the two data reduction rules. The more involved part is to show that the compression routine runs in $O(c^k \cdot m)$ time: There are $2^k + 1$ partitions of F into the above sets (X, Y) and one can show that, for each partition, the reduced graph after performing the data reduction rules has at most $d \cdot k$ vertices

for a constant d; otherwise, there is no size-k feedback vertex set for this partition. This then gives the $O(c^k \cdot m)$-running time. For more details on the proof of the $d \cdot k$-size bound see [6, 10].

Given as input a graph G with vertex set $\{v_1, \ldots, v_n\}$, the fixed-parameter algorithm from [6, 10] solves UFVS by iteratively considering the subgraphs $G_i := G[\{v_1, \ldots, v_i\}]$. For $i = 1$, the optimal feedback vertex set is empty. For $i > 1$, assume that an optimal feedback vertex set X_i for G_i is known. Obviously, $X_i \cup \{v_{i+1}\}$ is a solution set for G_{i+1}. Using the compression routine, the algorithm can in $O(c^k \cdot m)$ time either determine that $X_i \cup \{v_{i+1}\}$ is an optimal feedback vertex set for G_{i+1}, or, if not, compute an optimal feedback vertex set for G_{i+1}. For $i = n$, we thus have computed an optimal feedback vertex set for G in $O(c^k \cdot mn)$ time.

Theorem 1 UNDIRECTED FEEDBACK VERTEX SET *can be solved in* $O(c^k \cdot mn)$ *time for a constant* c.

Applications

The UNDIRECTED FEEDBACK VERTEX SET is of fundamental importance in combinatorial optimization. One typical application, for example, appears in the context of combinatorial circuit design [1]. For applications in the areas of constraint satisfaction problems and Bayesian inference, see Bar-Yehuda et al. [2].

Open Problems

It is open to explore the practical performance of the described algorithm. Another research direction is to improve the running time bound given in Theorem 1. Finally, it remains a long-standing open problem whether the FEEDBACK VERTEX SET on *directed* graphs is fixed-parameter tractable. The answer to this question would represent a significant breakthrough in the field.

Acknowledgments Supported by the Deutsche Forschungsgemeinschaft, Emmy Noether research group PIAF (fixed-parameter algorithms), NI 369/4

Recommended Reading

1. Bafna V, Berman P, Fujito T (1999) A 2-approximation algorithm for the undirected feedback vertex set problem. SIAM J Discret Math 3(2):289–297
2. Bar-Yehuda R, Geiger D, Naor J, Roth RM (1998) Approximation algorithms for the feedback vertex set problem with applications to constraint satisfaction and Bayesian inference. SIAM J Comput 27(4): 942–959
3. Becker A, Bar-Yehuda R, Geiger D (2000) Randomized algorithms for the loop cutset problem. J Artif Intell Res 12:219–234
4. Becker A, Geiger D (1994) Approximation algorithms for the Loop Cutset problem. In: Proceedings of the 10th conference on uncertainty in artificial intelligence. Morgan Kaufman, San Fransisco, pp 60–68
5. Bodlaender HL (1994) On disjoint cycles. Int J Found Comput Sci 5(1):59–68
6. Dehne F, Fellows MR, Langston MA, Rosamond F Stevens K (2005) An $O(2^{O(k)}n^3)$ FPT algorithm for the undirected feedback vertex set problem. In: Proceedings of the 11th COCOON. LNCS, vol 3595. Springer, Berlin, pp 859–869. Long version to appear in: J Discret Algorithm
7. Downey RG, Fellows MR (1992) Fixed-parameter tractability and completeness. Congr Numerant 87:161–187
8. Downey RG, Fellows MR (1999) Parameterized complexity. Springer, Heidelberg
9. Fomin FV, Gaspers S, Pyatkin AV (2006) Finding a minimum feedback vertex set in time $O(1.7548^n)$. In: Proceedings of the 2th IWPEC. LNCS, vol 4196. Springer, Berlin, pp 184–191
10. Guo J, Gramm J, Hüffner F, Niedermeier R, Wernicke S (2006) Compression-based fixed-parameter algorithms for feedback vertex set and edge bipartization. J Comput Syst Sci 72(8):1386–1396
11. Karp R (1972) Reducibility among combinatorial problems. In: Miller R, Thatcher J (eds) Complexity of computer computations. Plenum Press, New York, pp 85–103
12. Lund C, Yannakakis M (1993) The approximation of maximum subgraph problems. In: Proceedings of the 20th ICALP. LNCS, vol 700. Springer, Berlin, pp 40–51
13. Niedermeier R (2006) Invitation to fixed-parameter algorithms. Oxford University Press, Oxford
14. Reed B, Smith K, Vetta A (2004) Finding odd cycle transversals. Oper Res Lett 32(4):299–301

Unified View of Graph Searching and LDFS-Based Certifying Algorithms

Derek G. Corneil[1] and Michel Habib[2]
[1]Department of Computer Science, University of Toronto, Toronto, ON, Canada
[2]LIAFA, Université Paris Diderot, Paris Cedex 13, France

Keywords

BFS; Cocomparability graphs; DFS; Graph searching; Hamiltonian path; LBFS; LDFS; Minimum pathcover

Years and Authors of Summarized Original Work

2008; Corneil, Krueger
2013; Corneil, Dalton, Habib

Problem Definition

The notion of searching a graph, in particular visiting each vertex in a systematic preordained fashion, is as old as graph theory itself. Indeed, Euler's paper in 1736 [13] presented conditions on the vertex degrees of a graph that would certify the presence or absence of a path (or circuit) of edges visiting each edge exactly once. Later it was shown by Fleury that an easy algorithm to find such a path (or circuit) can be achieved using depth-first search (DFS) [14]. In the late nineteenth century, C. P. Trémaux [22] and G. Tarry [28] presented DFS-based algorithms for maze traversal; similarly, breadth-first search (BFS) algorithms were used to find the shortest possible successful maze traversals.

In the 1960s and 1970s, these searches were used in many of the early graph algorithms for problems such as distance and diameter determination, network flows, planar graph recognition, and connected and 2-connected components; see [4, 27]. A "generic" (GENS) search, as defined by Tarjan [27], is one in which an unvisited vertex adjacent to a visited vertex must be chosen before an arbitrary unvisited vertex. Note that this criterion includes the standard graph searches, but does not include some useful vertex orderings such as nonincreasing vertex degree. We caution the reader that by BFS we follow Golumbic [16] and refer to the distance layering where unvisited neighbors of the currently visited vertex are placed at the end of a queue data structure. Note that in [4] BFS is defined as distance layering, but their implementation of distance layering uses a queue. Throughout this note, we assume that our graphs are connected.

In a seminal 1976 paper, Rose, Tarjan, and Lueker [24] introduced a variation of BFS, called lexicographic breadth-first search (LBFS), and showed that an arbitrary LBFS search could be used to achieve a linear time algorithm to recognize chordal graphs (there is no induced cycle of size strictly greater than 3). After the appearance of this paper, there were a few new applications of LBFS, mostly on applications on graph families related to chordal graphs. In the 1990s there was a marked increase in the application of LBFS in which a previous vertex ordering (usually a previous LBFS ordering) was used to break ties when there was more than one vertex eligible to be visited next. Such tiebreaking is referred to as a "+-sweep," as defined below.

Definition 1 Given a search S and vertex ordering σ of graph G, a **plus** S **sweep with respect to** σ (denoted $S^+(G, \sigma)$) is the vertex ordering where the next vertex to be visited is the rightmost (as ordered by σ) T vertex (where T denotes the set of tied vertices).

Such "multi-sweep" LBFS algorithms were used for the recognition of interval graphs (for definitions of, and basic results on, various graph classes mentioned in this paper, see [3]), unit interval graphs, and cographs as well as finding dominating pairs (a pair of vertices (x, y) such that every $x - y$ path P dominates G in the sense that every vertex of G is either on P or has a neighbor on P) for asteroidal triple-free graphs. See [5] for an overview of these applications of LBFS. More recently, applications of LBFS have been found for graphs in general for such problems as modular decomposition [29] and split decomposition [15].

The proofs of correctness of multi-sweep LBFS algorithms typically are based on the following "4-vertex ordering characterization of LBFS":

Theorem 1 ([2, 16]) *A vertex total ordering σ could be produced by an LBFS if and only if for every triple of vertices $\{a, b, c\}$ where $a <_\sigma b <_\sigma c$, $ac \in E$, and $ab \notin E$ there exists vertex d such that $d <_\sigma a$, $db \in E$, and $dc \notin E$.*

Having seen the importance of Theorem 1 to the development of multi-sweep LBFS algorithms, a natural question and the question that is the basis of the Graph Searching Paper [6] is:

Do other standard graph searches have a similar "4-vertex ordering characterization"?

Key Results

The first reaction to the question posed above is to try to understand exactly what is the structure of the search imposed by the $\{a, b, c\}$ vertices. The relevant question is:

In the presence of the ac edge, how could b have been visited before c?

Since we are dealing with "generic" searches, some vertex d which is adjacent to b must have been visited before b since otherwise a would have to be chosen before b. If the search we are considering does not impose any further conditions on which unvisited vertices are eligible to be chosen, then the existence of d with $db \in E$ and $d <_\sigma b$ is a "4-vertex ordering characterization of GENS search." The full statement of the main theorem proved in [6] is:

Theorem 2 ([6]) *For S a graph search in {GENS, BFS, DFS, MNS, LBFS, LDFS} a total ordering σ of the set of vertices of the given graph could have been produced by S if and only if for every triple $a <_\sigma b <_\sigma c$ in σ where $ac \in E$, $ab \notin E$ there exists vertex d satisfying the requirements stated in the following table:*

Search S	Requirements on d	
	Location	Adjacencies
GENS	$d < b$	$db \in E$
BFS	$d < a$	$db \in E$
DFS	$a < d < b$	$db \in E$
MNS	$d < b$	$db \in E, dc \notin E$
LBFS	$d < a$	$db \in E, dc \notin E$
LDFS	$a < d < b$	$db \in E, dc \notin E$

Note that the hierarchy among these different searches (and layered search) is shown in Fig. 1.

In the case of BFS, the characterization states that b must have a neighbor d where $d <_\sigma a$. In effect, this location for d reflects the role played by the queue in the BFS algorithm. Similarly, for DFS the location of d is between a and b reflecting the role played by the stack in the DFS algorithm. Note that the locations of d imposed by BFS and DFS capture the full range allowed by GENS search thereby exhibiting a type of duality between BFS and DFS.

Now look at the difference between the characterizations of both BFS and LBFS. Both have the same location requirement for vertex d; however, BFS requires d to be a neighbor of b, whereas LBFS strengthens this condition so that

d has to be a **private neighbor** of b with respect to c. This means that b's neighborhood in the set of visited vertices is **maximal with respect to set inclusion**. This raises the question of what happens to both DFS and GENS search if we add the "lexicographic" property – i.e., as with LBFS, that d has to be a private neighbor of b with respect to c, and thus that b's neighborhood in the set of visited vertices is maximal with respect to set inclusion. In the case of the "lexical" version of GENS search, this search was already known as the maximal neighbor search (MNS) [25], in particular, a search that chooses any vertex that has a maximal (by set inclusion) neighborhood in the set of visited vertices. Interestingly, this vertex ordering was presented in [24] where they showed that any search that obeys this property would produce a perfect elimination ordering (PEO) if the given graph is chordal. Thus, they concluded that both maximum cardinality search and LBFS suffice. Turning to LDFS, we see that d has the same location requirement as DFS, and adding the "lexical" property shows that LDFS is also a restricted version of MNS, and thus, it too is guaranteed to produce a PEO on chordal graphs. Note that in [6] all of these conditions are shown to be characterizations of the specific searches.

To illustrate the differences and relationships among these various searches, consider the graph in Fig. 2:

Regarding complexity issues, all searches mentioned in Theorem 2, except LDFS, have a linear time implementation (see [17] for LBFS); the current best LDFS implementation for arbitrary graphs uses van Emde Boas

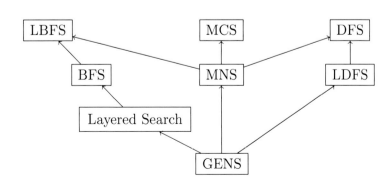

Unified View of Graph Searching and LDFS-Based Certifying Algorithms, Fig. 1 Summary of the hereditary relationships proved in Theorem 2. An arc from search S to search S' indicates that S' is a restriction of S

**Unified View of Graph
Searching and
LDFS-Based Certifying
Algorithms, Fig. 2**
Sample graph and
illustrative searches

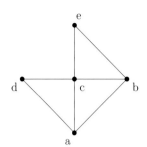

$a\,b\,d\,c\,e \in BFS \setminus \{LBFS, DFS\}$

$a\,b\,c\,d\,e \in \{LBFS, DFS\} \setminus LDFS$

$a\,b\,e\,c\,d \in DFS \setminus \{LDFS, BFS\}$

$a\,b\,c\,e\,d \in LDFS \setminus BFS$

$c\,a\,b\,e\,d \in \{MNS \cap BFS\} \setminus LBFS$

$b\,a\,c\,e\,d \in \{MNS \cap DFS\} \setminus LDFS$

trees [26] and runs in time $O(\max(n^2, n + m\log\log n))$. The key question arising from [6] is:

Are there any applications of LDFS?

Problem Definition (cont.)

The first few attempts to find such an application quickly failed. The first was to see if LDFS could enjoy the same success as LBFS in building recognition algorithms for various restricted families of graphs (apart from chordal graphs); in all cases easily found counterexamples thwarted the various attempts. The second approach was to determine if LDFS$^+$ could be helpful in finding Hamilton paths (HP) or more generally minimum path covers (MPC) where the goal is to find a minimum cardinality set of subpaths of given graph G such that each vertex belongs to exactly one such path. Unfortunately, LDFS$^+$ fails when applied to an interval ordering (G is an interval graph if and only if there is an **interval ordering**, σ, of the vertices such that for all triples $a <_\sigma b <_\sigma c$ where $ac \in E$, then ab must also belong to E). To see this, consider a vertex universal to two disjoint paths on three vertices. From examples, it seems, however, that DFS$^+$ will find an HP, if one exists.

In fact, [1] and [11] independently showed that using the rightmost neighbor (RMN) sweep on an interval ordering yields an MPC of the given interval graph. Note that RMN when presented with an ordering σ greedily builds paths by starting at the rightmost unvisited vertex of σ and proceeding to its rightmost unvisited neighbor if such a vertex exists; if not, a new

path is started at the rightmost unvisited vertex. (Note that this backtracking is different than the DFS$^+$ restarting.) Building off this algorithm, Dalton [10] presented a simple algorithm that certifies the correctness of the computed set of paths by either finding a set of vertices S (called a "scattering set") where the number of connected components of $G \setminus S$ equals $|S|$ plus the number of paths in the path cover or concludes that the given vertex ordering is not an interval ordering.

The next step was to try to lift this simple MPC algorithm to the superclass of cocomparability graphs. Note that a graph is a cocomparability graph if and only if its complement $\overline{G}$ has a transitive orientation of its edges. This orientation condition in $\overline{G}$ immediately translates into a vertex ordering characterization of cocomparability graphs. In particular, G is a cocomparability graph if and only if there is a **cocomp ordering**, σ, of the vertices such that for all triples $a <_\sigma b <_\sigma c$ where $ac \in E$, at least one of ab and bc must also belong to E.

Although there were polynomial time algorithms that solved the MPC problem on cocomparability graphs, all of these algorithms solved the "bump number" problem on the poset associated with the given graph and used the fact that any linear extension that minimizes the bump number contains the set of paths in a minimum path cover. The goal of this research was to find an MPC cocomparability graph algorithm that is directly graph theoretical and hopefully extends the interval graph MPC algorithm mentioned above. Examples immediately showed that applying RMN to an arbitrary cocomp ordering does not work, so many attempts were made to

Unified View of Graph Searching and LDFS-Based Certifying Algorithms, Fig. 3 LBFS counterexample

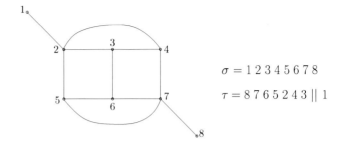

$$\sigma = 1\,2\,3\,4\,5\,6\,7\,8$$

$$\tau = 8\,7\,6\,5\,2\,4\,3 \parallel 1$$

use multi-sweep LBFS to yield such a cocomp ordering. (Note that if σ is a cocomp ordering and $\tau = \mathrm{LBFS}^{+}(\sigma)$, then τ is also a cocomp ordering.) This approach continued until the graph shown in Fig. 3 was discovered. On this graph every LBFS cocomp σ ordering fails, in the sense that RMN applied to σ does not produce a Hamiltonian path! A sample LBFS and the resulting RMN (which consists of two paths) are included in Fig. 3.

> Having seen the failure of LBFS, is there any chance that LDFS could work?

Key Results (cont.)

If there is such a role for LDFS, the counterexample for interval graphs mentioned previously shows that LDFS could not be expected to produce a minimum path cover itself; possibly LDFS could be used as a preprocessing step. If so, the simplest possible algorithm would be:

1. Let π be an arbitrary cocomp ordering.
2. Let σ be $\mathrm{LDFS}^{+}(\pi)$.
3. Let τ be $\mathrm{RMN}(\sigma)$.
4. If τ is not a Hamiltonian path, then from τ, use Dalton's algorithm to construct a separator S that certifies τ; otherwise, conclude π is not a cocomp ordering.

First of all, as with LBFS, LDFS when applied as a $+$-sweep on a given cocomp ordering returns a cocomp ordering. In this algorithm the hope is that an LDFS cocomp ordering would capture the

"interval structure of cocomparability graphs," at least from the perspective of the MPC problem.

Somewhat surprisingly, this algorithm worked on all attempted examples. In an attempt to understand the structure exposed by an LDFS cocomp ordering, there are two points. First of all, why we do not see the LDFS structure in interval graphs? From the vertex ordering characterization of interval graphs, we see that there can never be an ordered triple of vertices $a < b < c$ with edge ac and nonedge ab, and thus, every interval ordering is simultaneously an example of every search mentioned in Theorem 2. Secondly, since an interval graph is chordal, every LBFS and LDFS must be a perfect elimination ordering implying that every vertex is simplicial in the graph formed on it and all vertices before it in the ordering. By considering a C_4, this property will not hold for LDFS cocomp orderings. There is however a crucial observation of the structure guaranteed by a nonsimplicial vertex in an LDFS cocomp ordering.

Lemma 1 ([7]) *Let σ be an LDFS cocomp ordering of cocomparability graph G. If z is a nonsimplicial vertex in σ as witnessed by $x <_{\sigma} y <_{\sigma} z$ where $xz, yz \in E$, $xy \notin E$, then there exists vertex w, $x <_{\sigma} w <_{\sigma} y$ where $xw, wy \in E$, $wz \notin E$.*

Proof By the LDFS vertex ordering characterization applied to the triple $\{x, y, z\}$, vertex w exists and satisfies all conditions of the lemma, except possibly $xw \in E$; if this is not the case, then the triple $\{x, w, z\}$ violates σ being a cocomp ordering.

This lemma plays a critical role in the proof of correctness of the MPC algorithm stated above.

Open Problems

We first list a number of recent results that have grown out of the work presented in [6, 7]:

- Köhler and Mouatadid [20] have recently shown that LDFS on a cocomparability graph can be done in linear time, thereby avoiding the log log n off linear factor in the MPC paper [7].
- Mertzios and Corneil [23] have "lifted" the $O(n^4)$ longest path algorithm on interval graphs [18] to achieve the same result and time bound for cocomparability graphs. As with the MPC algorithm, a LDFS cocomp ordering was required.
- A similar technique of using an LDFS cocomparability ordering as a preprocessing step for a simple linear time interval graph algorithm has resulted in a linear time algorithm for the maximum independent set (and minimum vertex cover) problems on cocomparability graphs [8]. Note that the algorithm also produces a minimum cardinality clique cover in order to certify the maximum independent set produced by the algorithm. This algorithm also uses the linear time LDFS cocomp ordering algorithm presented in [20]. Very recently Köhler and Mouatadid [19] have presented a linear time algorithm that computes a maximum weighted independent set of a cocomparability graph; this algorithm works on any cocomp ordering and, in particular, does not require an LDFS cocomp ordering.
- In [8] the authors also characterized the search orderings that are "cocomp ordering preserving" in the sense that when used as a +-sweep, the output is a cocomp ordering when the input is a cocomp ordering. They showed that dfgreedy is such a preserving search and can be used to simplify the current best recognition algorithm for permutation graphs.
- In his PhD thesis, Dusart [12] studied the maximal clique lattice of a cocomparability graph and showed that a graph G is a cocomparability graph if and only if the set of maximal

cliques of G satisfies specific lattice properties. Furthermore, he defined a new cocomp ordering preserving search called local MNS to compute a maximal interval subgraph of G. The new characterization together with MNS yields linear time algorithms to compute the simplicial vertices, the clique separators, and associated components of a cocomparability graph.
- Recently a new model of graph searching called "tiebreaking label search" (TBLS) [9] has been announced. This model builds off the vertex ordering characterization model appearing in [6] as well as the General Label Search formalism of Krueger, Simonet, and Berry [21]. The TBLS model incorporates the +-sweep use of graph searches, restricts labels to be sets of integers, and presents some new vertex ordering characterizations.

We now turn to some new directions for further research. From a graph algorithm perspective, the most interesting question is whether the results on cocomparability graphs can be easily extended to asteroidal triple-free graphs, an inclusive family that has received considerable attention. Further results, both structural and algorithmic are expected for cocomparability graphs and their associated posets. We expect that graph searching will continue to play a major role in these developments.

Recommended Reading

1. Arikati SR, Rangan CP (1990) Linear algorithm for optimal path cover problem on interval graphs. Inf Process Lett 35(3):149–153
2. Brandstädt A, Dragan FF, Nicolai F (1997) LexBFS-orderings and powers of chordal graphs. Discret Math 171(1–3):27–42
3. Brandstädt A, Le VB, Spinrad JP (1999) Graph classes, a survey. SIAM monographs on discrete mathematics and applications. Society for Industrial and Applied Mathematics, Philadelphia
4. Cormen TH, Leiserson CE, Rivest RL, Stein C (2009) Introduction to algorithms, 3rd edn. MIT, Cambridge
5. Corneil DG (2004) Lexicographic breadth first search – a survey. In: WG, Bad Honnef. Lecture notes in computer science, vol 3353. Springer, pp 1–19
6. Corneil DG, Krueger R (2008) A unified view of graph searching. SIAM J Discret Math 22(4):1259–1276

7. Corneil DG, Dalton B, Habib M (2013) LDFS-based certifying algorithm for the minimum path cover problem on cocomparability graphs. SIAM J Comput 42(3):792–807
8. Corneil DG, Dusart J, Habib M, Kőhler E (2014) On the power of graph searching for cocomparability graphs. Submitted for publication
9. Corneil DG, Dusart J, Habib M, Mamcarz A, de Montgolfier F (2014) A new model for graph search. Under revision
10. Dalton B (2011) On minimum path cover in interval graphs. Master's thesis, University of Toronto
11. Damaschke P (1993) Paths in interval graphs and circular arc graphs. Discret Math 112(1–3):49–64
12. Dusart J (2014) Graph searches with applications to cocomparability graphs. PhD thesis, University of Paris Diderot
13. Euler L (1736) Solutio problematis ad geometriam situs pertinentis. Comment Academiae Sci I Petropolitanae 8:128–140
14. Fleury (1883) Deux problèmes de géométrie de situation. Journal de mathématiques élémentaires 257–261
15. Gioan E, Paul C, Tedder M, Corneil DG (2014) Practical and efficient split-decomposition via graph-labelled trees. Algorithmica 69(4): 789–843
16. Golumbic MC (2004) Algorithmic graph theory and perfect graphs. Annals of discrete mathematics, vol 57. Elsevier, Amsterdam/Boston
17. Habib M, McConnell RM, Paul C, Viennot L (2000) LexBFS and partition refinement, with applications to transitive orientation, interval graph recognition and consecutive ones testing. Theor Comput Sci 234(1–2):59–84. http://dblp.uni-trier.de/rec/bibtex/journals/tcs/HabibMPV00
18. Ioannidou K, Mertzios GB, Nikolopoulos SD (2011) The longest path problem has a polynomial solution on interval graphs. Algorithmica 6(2):320–341
19. Kőhler E, Mouatadid L (2014) A linear time algorithm for computing a maximum weight independent set on cocomparability graphs. Submitted for publication
20. Kőhler E, Mouatadid L (2014) Linear time LDFS on cocomparability graphs. In: SWAT, Copenhagen, pp 319–330
21. Krueger R, Simonet G, Berry A (2011) A general label search to investigate classical graph search algorithms. Discret Appl Math 159(2–3): 128–142
22. Lucas E (1882) Récréations mathématiques. Gauthier-Vilars, Paris
23. Mertzios GB, Corneil DG (2012) A simple polynomial algorithm for the longest path problem on cocomparability graphs. SIAM J Discret Math 26(3):940–963. http://dblp.uni-trier.de/rec/bibtex/journals/siamdm/MertziosC12
24. Rose DJ, Tarjan RE, Lueker GS (1976) Algorithmic aspects of vertex elimination on graphs. SIAM J Comput 5(2):266–283. http://dblp.uni-trier.de/rec/bibtex/journals/siamcomp/RoseTL76
25. Shier DR (1984) Some aspects of perfect elimination orderings in chordal graphs. Discret Appl Math 7:325–331
26. Spinrad J (–) Efficient implementation of lexicographic depth first search. Submitted for publication
27. Tarjan RE (1972) Depth-first search and linear graph algorithms. SIAM J Comput 1(2): 146–160
28. Tarry G (1895) Le problème des labyrinthes. Nouvelles Annales de Math 14: 187–189
29. Tedder M (2011) Applications of lexicographic breadth first search to modular decomposition, split decomposition, and circle graphs. PhD thesis, University of Toronto

Uniform Covering of Rings and Lines by Memoryless Mobile Sensors

Paola Flocchini
School of Electrical Engineering and Computer Science, University of Ottawa, Ottawa, ON, Canada

Keywords

Barrier covering; Scattering; Uniform deployment

Years and Authors of Summarized Original Work

2008; Flocchini, Prencipe, Santoro
2008; Cohen, Peleg

Problem Definition

The Model

A *mobile robotic sensor* (or simply *sensor*) is modeled as a computational unit with sensorial capabilities: it can perceive the spatial environment within a fixed distance $V > 0$, called *visibility range*, it has its own local working memory, and it is capable of performing local computations [6, 7].

Each sensor is a point with its own local coordinate system, which might not be consistent with the ones of the other sensors. The sensor can move in any direction, but it may be stopped before reaching its destination, e.g., because of limits to its motion energy; however, it is assumed that the distance traveled in a move by a sensor is not infinitesimally small (unless it brings the sensor to its destination).

The sensors have no means of direct communication to other sensors. Thus, any communication occurs in a totally implicit manner, by observing the other sensors' positions. Moreover, they are *autonomous* (i.e., without a central control) *identical* (i.e., they execute the same protocol), and *anonymous* (i.e., without identifiers that can be used during the computation).

The sensors can be *active* or *inactive*. When *active*, a sensor performs a *Look-Compute-Move* cycle of operations: it first observes the portion of the space within its visibility range obtaining a snapshot of the positions of the sensors in its range at that time (*Look*); using the snapshot as an input, the sensor then executes the algorithm to determine a destination point (*Compute*); finally, it moves toward the computed destination, if different from the current location (*Move*). After that, it becomes *inactive* and stays idle until the next activation. Sensors are *oblivious*: when a sensor becomes active, it does not remember any information from previous cycles.

Depending on the degree of synchronization among the cycles of different sensors, three submodels are traditionally identified: *synchronous*, *semi-synchronous*, and *asynchronous*. In the *synchronous* (FSYNC) and in the *semi-synchronous* (SSYNC) models, there is a global clock tick reaching all sensors simultaneously, and a sensor's cycle is an instantaneous event that starts at a clock tick and ends by the next. In FSYNC, at each clock tick all sensors become active, while in SSYNC some sensors might not be active in each cycle. In the *asynchronous* model (ASYNC), there is no global clock and the sensors do not have a common notion of time. Furthermore, the duration of each activity (or inactivity) is finite but unpredictable. As a result, sensors can be seen while moving, and computations can be made based on obsolete observations.

The Problem

The (distributed) *uniform covering* problem refers to sensors, randomly dispersed in a *bounded* region of space, that must scatter themselves throughout the region so to "cover" it satisfying some optimization criteria. Consider the case of a circular rim $\mathcal{R}$ (i.e., a ring), and let $S = \{s_0, \ldots, s_{n-1}\}$ be the sensors initially arbitrarily placed in different points on $\mathcal{R}$, with s_i preceding s_{i+1} clockwise (the index operations are modulo n). We emphasize that these names are used for presentation purposes only, and are not known to the sensors. If the sensors agree on the notion of clockwise, we say that they have a *common orientation*. Let $d = L_\mathcal{R}/n$ where $L_\mathcal{R}$ is the length of the ring. In the following, unless otherwise stated, the sensors are assumed to have visibility range $V \geq 2d$. Let $d_i(t)$ be the distance between sensors s_i and s_{i+1} at time t; when no ambiguity arises, we shall omit the time and simply indicate the distance as d_i. The sensors are said to have reached an *exact uniform covering* (*exact covering* for simplicity) at time t if $d_i(t) = d$ for all $0 \leq i \leq n - 1$. Given $\epsilon > 0$, the sensors are said to have reached an ϵ-*approximate covering* at time t if $d - \epsilon \leq d_i(t) \leq d + \epsilon$ for all $0 \leq i \leq n - 1$.

Key Results

The Ring

Exact Uniform Covering

There is a strong impossibility result that stresses the importance of having common orientation. If the sensors have only a local notion of left and right, but do not share a *common orientation* of the ring, the exact covering problem is unsolvable. This result holds even if the sensors had unbounded memory and visibility, and under a SSYNC scheduler.

Theorem 1 ([5]) *Let the sensors be on a ring $\mathcal{R}$. In absence of common orientation, there is no* *deterministic exact covering algorithm even if the sensors have unbounded persistent memory, their visibility range is unlimited, and the scheduling is* SSYNC.

To see why this is the case, consider the following setting. Let n be even; partition the sensors in two sets, $S_1 = \{s_1, \ldots, s_{n/2}\}$ and $S_2 = S \setminus S_1$, and place the sensors of S_1 and S_2 on the vertexes of two regular $(n/2)$-gons on $\mathcal{R}$, rotated of an angle $\alpha < 360°/n$. Furthermore, all sensors have their local coordinate axes rotated so that they all have the same view of the world. In other words, the sensors in S_1 share the same orientation, while those in S_2 share the opposite orientation of $\mathcal{C}$. If activating only the sensors in S_1, an exact covering (resp. *no* exact covering) on $\mathcal{R}$ is reached at time step t_{i+1}, then the same is true also activating only the ones in S_2. Clearly, in such a case, activating both sets no exact covering would be reached at time step t_{i+1}, and the system would be an analogous configuration as the one of time step t_i, with different angles. Using this property, it is easy to design an adversary that will force any algorithm to never succeed in solving the problem; its behavior would be as follows: (i) If activating only the sensors in S_1 (resp. S_2) no exact covering on $\mathcal{R}$ is reached, then activate all sensors in S_1 (resp. S_2), while all sensors in S_2 (resp. S_1) are inactive; (ii) otherwise, activate all sensors. Go to (i).

On the other hand, assuming common orientation and knowledge of the final inter-distance d among sensors, a simple algorithm that solves the exact covering in ASYNC is for each sensor to move toward the point at distance d from its clockwise successor (if visible). We remind that $V \geq 2d$.

Protocol RINGCOVERINGEXACT (for sensor s_i)

Assumptions: Orientation, knowledge of d.

1. If s_{i+1} is not visible, move distance d clockwise.
2. else, if $d_i > d$ move toward point x at distance d from s_{i+1}.

Theorem 2 ([5]) *The exact covering of the ring problem is solvable in* ASYNC, *with common orientation and knowledge of the final inter-distance.*

Approximate Covering

Assuming common orientation but no knowledge of the final inter-distance among sensors, an ϵ-approximate covering is still possible for any $\epsilon > 0$, but no exact covering algorithm is known. Also this algorithm is very simple: the sensors asynchronously and independently *Look* in both directions, then they position themselves in the middle between the closest observed sensors (if any). Correctness is shown by proving that the minimum distance between any two neighboring sensors eventually grows, while the maximum distance eventually shrinks in such a way that there is a time when all sensors are within $d \pm \epsilon$ distance.

Theorem 3 ([5]) *The approximate covering of the ring problem is solvable in* ASYNC *with common orientation.*

Algorithm RINGCOVERINGAPPROX (for sensor s_i)

Assumptions: Orientation

- If no sensor is visible clockwise (resp. counterclockwise), let $d_i = V$ (resp. $d_{i-1} = V$).
- If $d_i \leq d_{i-1}$ do not move.
- If $d_i > d_{i-1}$ move distance $\frac{d_i + d_{i-1}}{2} - d_{i-1}$ clockwise.

Note that the covering problem has been also studied in discrete rings [4].

The Line

The case of a line segment is quite different from the one of the ring, and perhaps surprisingly, it is not easier. Let $S = \{s_0, \ldots, s_{n-1}\}$ be the sensors initially arbitrarily placed in different points on a line $\mathcal{L}$ with s_0 and s_{n-1} being two special immobile sensors delimiting the segment to be covered and with s_i preceding s_{i+1} ($0 < i < n - 2$). Let $d = L_{\mathcal{L}}/(n - 1)$, where $L_{\mathcal{L}}$ denotes the length of the segment. *Exact covering* and ϵ-*approximate covering* are defined analogously to the case of the ring.

Exact Uniform Covering

With common orientation and known final inter-distance, an algorithm has been recently shown for oriented sensors in ASYNC [3]. The algorithm works even if the visibility range is just enough to sense the final inter-distance ($V = d$). Let $\delta \leq \frac{d}{2}$ be a fixed positive (arbitrarily small) constant the sensors agree upon.

Protocol CORRIDORCOVERINGEXACT (for sensor s_i)

Assumptions: Orientation, knowledge of d, $V = d$

- If s_{i-1} is not visible, move distance $\frac{d}{2}$ to the left.
- else, let $a := d - d_{i-1}$
  If $d_i \geq d$ and $a > 0$, move distance $min(\frac{d}{2} - \delta, a)$ to the right.

Theorem 4 ([3]) *The exact covering of the line problem is solvable in* ASYNC *with common orientation and knowledge of the final inter-distance.*

With fixed visibility, a distributed algorithm has been proposed for FSYNC in a discrete setting, to solve the slightly different problem of barrier coverage [2].

Approximate Covering

Approximate covering has been studied in a slightly different visibility model where each sensor is able to perceive up to the next sensor on the line [1]. In other words, in each direction, a sensor sees the closest sensor (if it exists), regardless of its distance, but its visibility is blocked by it (*neighbor visibility*). For presentation purposes, a global linear coordinate system (not known to the sensors) is used here with $s_0(t) = 0$ and $s_{n-1}(t) = 1$. For the sensors to be spread uniformly, sensor s_i should then occupy position $\frac{i}{n-1}$. The following is a simple approximate covering algorithm.

Protocol CORRIDORSPREAD (for sensor s_i)

Assumptions: SSYNC, neighbour visibility

- If no sensor is visible in either direction, do nothing.
- Otherwise, move toward point $x = \frac{1}{2}(s_{i+1} + s_{i-1})$.

The idea of the convergence proof in FSYNC is sketched below. Let $\mu_i[t]$ be the *shift* of the s_i's location at time t from its final position. According to the protocol, the position of sensor s_i changes from $s_i(t)$ to $s_i(t + 1) = \frac{1}{2}(s_{i-1}(t) + s_{i+1}(t))$ for $1 \leq i \leq n - 2$, while sensors s_0 and s_{n-1} never move. Therefore, the shifts changes with time as $\mu_i[t + 1] = \frac{1}{2}(\mu_{i+1}[t] + \mu_{i-1}[t])$. Considering the *progress* measure, $\psi[t] = \Sigma_{i=1}^{i=n} \mu^2[t]$, it can be shown that $\psi[t]$ is a decreasing function of t unless the sensors are already equally spread; more precisely, it is shown that every $O(n^2)$ cycle, $\psi[t]$ is at least halved thus reaching approximate covering. More complex but analogous reasoning is followed for SSYNC.

Theorem 5 ([1]) *The approximate covering of the line problem is solvable in* SSYNC *with* neighbor visibility.

With a simple modification of the algorithm, the result above can be extended to any fixed visibility $V > d$, provided that d is known, as described below [3].

Protocol CORRIDORSPREAD2 (for sensor s_i)

Assumptions: SSYNC, d known, $V > d$

- If only one sensor $s_j \in \{s_{i+1}, s_{i-1}\}$ is visible to s_i and $d' = dist(s_i, s_j) < d$: move distance $\frac{d-d'}{2} + \frac{V-d}{2}$ away from s_j
- If both s_{i+1}, s_{i-1} are visible and $d_1 = dist(s_{i-1}, s_i) < d_2 = dist(s_{i+1}, s_i)$ (resp. $d_1 = dist(s_{i+1}, s_i) < d_2 = dist(s_{i-1}, s_i)$): move $\frac{d_2-d_1}{2}$ toward s_{i+1} (resp. toward s_{i-1})

Applications

Uniform covering problems are important in many applications; covering of a circular rim occurs, for example, when the sensors have to surround a dangerous area and can only move along its outer perimeter. On the other hand, coverings of the line (often called *barrier coverings*) guarantee that any intruder attempting to cross the perimeter of a protected region (e.g., crossing an international border) is detected by one or more of the sensors. These problems are studied under a variety of assumption; the majority of the studies uses sensors provided with memory, explicit communication devices, global localization capabilities (e.g., GPS), and centralized approaches. The advantage of memoryless sensors are self-stabilization and tolerance to loss of sensors, the use of local coordinate systems has clear advantages over the full strength of a GPS; finally, decentralized solutions offer better fault tolerance.

Open Problems

It is known that the exact covering of the ring is impossible without orientation in SSYNC, but the impossibility does not extend to FSYNC where, however, no algorithm is known. Moreover, the only existing exact covering algorithm in ASYNC assumes orientation, which is needed, and knowledge of the inter-distance d, which is possibly not needed, so a tighter result might be possible. Finally, approximate covering is achieved in the ring in SSYNC assuming orientation, which is not shown to be necessary, furthermore, no solution exists for ASYNC.

In the case of the line, the only impossibility result for exact covering [3] holds for fully disoriented sensors (not even able to locally distinguish between their two directions) and with small visibility range $V = d$. As for approximate covering, the only known result in this model is for SSYNC, and it is not known whether an algorithm exists for the ASYNC model.

Recommended Reading

1. Cohen R, Peleg D (2008) Local spreading algorithms for autonomous robot systems. Theor Comput Sci 399:71–82
2. Eftekhari Hesari M, Kranakis E, Krizanc D, Morales Ponce O, Narayanan L, Opatrny J (2013) Distributed algorithms for barrier coverage using relocatable sensors. In: ACM symposium on principles of distributed computing (PODC), Montreal, Canada, pp 383–392
3. Eftekhari Hesari M, Flocchini P, Narayanan L, Opatrny J, Santoro N (2014) Distributed barrier coverage with relocatable sensors. In: 21th international colloquium on structural information and communication complexity (SIROCCO), Takayama, Japan
4. Elor Y, Bruckstein AM (2011) Uniform multi-agent deployment on a ring. Theor Comput Sci 412:783–795
5. Flocchini P, Prencipe G, Santoro N (2008) Self-deployment algorithms for mobile sensors on a ring. Theor Comput Sci 402(1):67–80
6. Flocchini P, Prencipe G, Santoro N (2011) Computing by mobile robotic sensors. In: Nikoletseas S, Rolim J (eds) Theoretical aspects of distributed computing in sensor networks, chap 21. Springer, Heidelberg ISBN:978-3-642-14849-1
7. Flocchini P, Prencipe G, Santoro N (2012) Distributed computing by oblivious mobile robots. Morgan & Claypool, San Rafael

Unique *k*-SAT and General *k*-SAT

Timon Hertli
Department of Computer Science, ETH Zürich, Zürich, Switzerland

Keywords

3-SAT; Boolean satisfiability; k-SAT; SAT

Years and Authors of Summarized Original Work

2011; Hertli

Problem Definition

A Boolean formula F is said to be in conjunctive normal form (CNF) if it is a conjunction of disjunction of literals. If furthermore every disjunction (called clause) is over at most k literals, F is said to be in k-CNF. k-SAT, the decision problem whether a k-CNF formula admits a satisfying assignment, is one of the most prominent NP-complete problems. A special case of k-SAT is (promise) unique k-SAT, where the k-CNF is additionally promised to have either a unique or no satisfying assignment.

Suppose F has n variables. The trivial algorithm tries all 2^n satisfying assignments. For k-SAT and especially 3-SAT, there have been many successive improvements [3–9, 11]. The best of them are *randomized* in the sense that they always correctly report unsatisfiability but might fail to report satisfiability with probability $\frac{1}{3}$, say.

Problem 1 (k-SAT)

INPUT: A k-CNF formula F.
OUTPUT: "No" if F is not satisfiable. "Yes" with probability at least $\frac{2}{3}$ if F is satisfiable.

Problem 2 (Unique k-SAT)

INPUT: A k-CNF formula F with at most one satisfying assignment.
OUTPUT: "No" if F is not satisfiable. "Yes" with probability at least $\frac{2}{3}$ if F is satisfiable.

It is conjectured that unique k-SAT and k-SAT have the same exponential complexity; however, this could only be shown for $k \to \infty$ [1]. Especially for PPSZ [9], the fastest known (randomized) algorithm for unique k-SAT, the analysis results in a gap between k-SAT and unique k-SAT for $k = 3, 4$. Furthermore, the PPSZ algorithm has been derandomized for unique k-SAT [10] but not for general k-SAT.

Notation For a CNF formula F over a variable set V, denote by $\mathrm{sat}(F)$ the set of satisfying assignments of F on V. For x a variable and b a Boolean value, define $F^{[x \mapsto b]}$ the restriction of F by $x \mapsto b$, i.e., the formula obtained by replacing x by b in F.

Key Results

The bounds of the PPSZ algorithm for unique k-SAT hold for general k-SAT also if $k = 3, 4$ [2]. This makes PPSZ the fastest known k-SAT algorithm for all k. In the analysis of [2], the PPSZ algorithm is slightly modified.

Theorem 1 *There is a randomized algorithm for 3-SAT running in time* $O(2^{0.387n})$.

Theorem 2 *There is a randomized algorithm for 4-SAT running in time* $O(2^{0.555n})$.

Algorithm 1 PPSZ(k-CNF formula F)

$V \leftarrow$ variables of F
Choose β uniformly at random from all assignments on V
Choose π uniformly at random from all permutations of V
Let α be a partial assignment over V, initially the empty assignment
for all $x \in V$ in the order prescribed by π **do**
   while there is an $\log n$-implied assignment $y \mapsto a$ of F **do**
      $F \leftarrow F^{[y \mapsto a]}$
      $\alpha(y) \leftarrow a$
   end while
   if $\alpha(x)$ not fixed yet **then**
      $F \leftarrow F^{[x \mapsto \beta(x)]}$
      $\alpha(x) \leftarrow \beta(x)$
   end if
end for
return If α satisfies F, return 'satisfiable', otherwise return 'failure'.

If F is not satisfiable, then PPSZ will never find a satisfying assignment and thus is always correct. Hence, let F be a satisfiable k-CNF formula over n variables V. The PPSZ algorithm tries to find a satisfying assignment of F by iteratively setting variables as follows: Go through the variables one by one, in random order. If a variable x is not set at its step, then its value will be guessed uniformly at random. Between steps, we might infer the value of some variables in subexponential time: Setting x to a is called $\log n$-implied (by F) if there is a set of $\log n$ clauses G in F such that all satisfying assignments of G set x to a. $\log n$-implication can be checked in subexponential time by brute force; the algorithm fixes all $\log n$-implications accordingly between steps.

If a variable is determined by $\log n$-implication, it is called *forced*; otherwise, it is called *guessed*. The key result of [9] for unique k-SAT is the following: If F is in k-CNF and has a unique satisfying assignment α, then given that $\beta = \alpha$ (i.e., all guesses are according to α), a variable is forced with a certain probability R_k. By Jensen's inequality one can then show that α is found with probability $2^{-(1-R_k)n}$. We have $R_3 = 2 - 2\ln 2 \approx 0.613$ and $R_4 \approx 0.445$. Repeating PPSZ inversely proportional to its success, probability will match the above theorems for unique 3-SAT and unique 4-SAT.

If there are multiple satisfying assignments, there is no bound on the probability that a variable is forced. For example, the empty CNF formula that always evaluates to true will never have a forced variable, as being forced depends on certain assignments not being satisfying. However, the following can be done: Given a satisfiable CNF formula F, call a variable x *frozen* if it has the same value in all satisfying assignments of F, and call x *non-frozen* otherwise. If x is frozen, the same bound on the probability that it is forced holds by the arguments of [9]. If x is non-frozen, then it can be set both ways and the resulting formula remains satisfiable. The remaining problem is that the probability for frozen variables depends on a fixed satisfying assignment and a uniform permutation; however, depending on the permutation, certain assignments will be more or less likely. This leads to a correlation issue that has to be solved by balancing the correlation and the benefit of non-frozen variables by careful bookkeeping.

Let V_f be the frozen variables of F and V_n be the non-frozen variables of F. The *likelihood* of an assignment α in F, $\text{lkhd}(F, \alpha)$, is recursively defined as follows: If α does not satisfy F, then $\text{lkhd}(F, \alpha) = 0$. If α is the unique satisfying assignment of F, then $\text{lkhd}(\alpha) = 1$. Otherwise, let $\text{lkhd}(\alpha) = \frac{1}{|V_f|+2|V_n|}\left(\sum_{x\in V}\text{lkhd}(\alpha, F^{[x\mapsto\alpha(x)]})\right)$. The

likelihood simulates how likely an assignment would be returned by PPSZ in an ideal setting.

With this, the cost F is defined as follows: For a non-frozen variable x, $\text{cost}(F, x) = 1 - R_k$. For a frozen variable x, first defined for a satisfying assignment α, $\text{cost}(F, x, \alpha)$ is the probability that x is guessed if executing PPSZ conditioned on $\beta = \alpha$. Then $\text{cost}(F, x) = \sum_{\alpha\in\text{sat}(F)}\text{lkhd}(F,\alpha)\text{cost}(F,x,\alpha)$. Observe that $\text{cost}(F, x) \leq 1 - R_k$, as frozen variables are guessed with probability at most $1 - R_k$. In total we define $\text{cost}(F) = \sum_{x\in V}\text{cost}(F, x) \leq (1 - R_k)n$. The following theorem relates the cost to the probability that PPSZ finds an assignment:

Theorem 3

$\Pr(\text{PPSZ finds some satisfying assignment of } F)$

$$\geq 2^{-c(F)}.$$

This theorem immediately implies Theorems 1 and 2. The theorem is by induction on number of variables of F. After a single PPSZ step, the cost decreases in expectation; the more the more frozen variables there are. On the other hand, the more non-frozen variables there are, the higher the probability is to retain a satisfiable formula. Balancing these factors and applying Jensen's inequality gives the theorem. It is noteworthy that the proof relies on the inequality $0.613 \approx R_3 \leq \frac{1}{2\ln 2} \approx 0.721$, meaning that if PPSZ would be improved beyond this bound, the unique case might indeed be better.

Open Problems

- Is the exponential complexity of k-SAT and unique k-SAT the same? Here this has been shown for the specific case of the PPSZ algorithm.
- Does PPSZ perform even better on formulas with exponentially many satisfying assignments?
- Can PPSZ be derandomized for general k-SAT?

Cross-References

► Backtracking Based k-SAT Algorithms
► Derandomization of k-SAT Algorithm
► Exact Algorithms for General CNF SAT
► Exact Algorithms for k SAT Based on Local Search
► Exponential Lower Bounds for k-SAT Algorithms

Recommended Reading

1. Calabro C, Impagliazzo R, Kabanets V, Paturi R (2008) The complexity of unique k-SAT: an isolation lemma for k-CNFs. J Comput Syst Sci 74(3):386–393
2. Hertli T (2011) 3-SAT faster and simpler—unique-SAT bounds for PPSZ hold in general. In: 2011 IEEE 52nd annual symposium on foundations of computer science—FOCS 2011, Palm Springs. IEEE Computer Society, Los Alamitos, pp 277–284
3. Iwama K, Tamaki S (2004) Improved upper bounds for 3-SAT. In: Proceedings of the fifteenth annual ACM-SIAM symposium on discrete algorithms, New Orleans. ACM, New York, pp 328–329 (electronic)
4. Kullmann O (1997) Worst-case analysis, 3-SAT decision and lower bounds: approaches for improved SAT algorithms. In: Satisfiability problem: theory and applications, Piscataway, 1996. DIMACS series in discrete mathematics and theoretical computer science, vol 35. American Mathematical Society, Providence, pp 261–313
5. Makino K, Tamaki S, Yamamoto M (2011) Derandomizing HSSW algorithm for 3-SAT. In: Computing and combinatorics, Dallas. Lecture notes in computer science, vol 6842. Springer, Heidelberg, pp 1–12
6. Monien B, Speckenmeyer E (1985) Solving satisfiability in less than 2^n steps. Discret Appl Math 10(3):287–295
7. Moser RA, Scheder D (2011) A full derandomization of Schoening's k-SAT algorithm. In: Proceedings of the 43rd annual ACM symposium on theory of computing, San Jose. ACM, pp 245–252
8. Paturi R, Pudlák P, Zane F (1999) Satisfiability coding lemma. Chic J Theor Comput Sci Article 11, 19 (electronic)
9. Paturi R, Pudlák P, Saks ME, Zane F (2005) An improved exponential-time algorithm for k-SAT. J ACM 52(3):337–364 (electronic)
10. Rolf D (2005) Derandomization of PPSZ for unique-k-SAT. In: Theory and applications of satisfiability testing, St Andrews. Lecture notes in computer science, vol 3569. Springer, Berlin, pp 216–225
11. Schöning U (1999) A probabilistic algorithm for k-SAT and constraint satisfaction problems. In: Proceedings of the 40th annual symposium on foundations of computer science, New York City. IEEE Computer Society, Los Alamitos, pp 410–414

Universal Sequencing on an Unreliable Machine

Julián Mestre[1,2] and Nicole Megow[3]
[1]Department of Computer Science, University of Maryland, College Park, MD, USA
[2]School of Information Technologies, The University of Sydney, Sydney, NSW, Australia
[3]Institut für Mathematik, Technische Universität Berlin, Berlin, Germany

Keywords

Availability periods; Machine speed; Min-sum objective; Scheduling; Sequencing; Universal solution; Unreliable machine; Worst-case guarantee

Years and Authors of Summarized Original Work

2012; Epstein, Levin, Marchetti-Spaccamela, Megow, Mestre, Skutella, Stougie
2013; Megow, Mestre

Problem Definition

Given a set of jobs $J = \{1, 2, \ldots, n\}$ with processing times $p_j \in \mathbb{R}_+$ and weights $w_j \in \mathbb{R}_+$, the task is to find a schedule for all jobs on a single machine that minimizes $\sum w_j C_j$, where C_j is the completion time of job j.

Under the standard scheduling assumption of an ideal machine that runs at constant speed, an optimal schedule is obtained by sequencing the jobs in nonincreasing order of the ratio w_j/p_j; this is known as Smith's Rule [12]. Unfortunately, as we shall see shortly, this sequence may perform arbitrarily bad when the machine is not ideal.

This note is concerned with the setting in which the machine may change its processing speed over time or it may fully break down and is unavailable until it is fixed. Given a sequence, the jobs are processed in this order no matter how the machine behaves. In case of a machine breakdown, the job that is currently running is preempted and resumes processing when the machine becomes available again at a later time. The aim is to compute a *universal sequence* that, for any given machine behavior, is a good approximation of an optimal schedule for that particular machine behavior.

Definition 1 A sequence π is a *universal c-approximation* if for any machine behavior the total weighted completion times of π is at most a factor c larger than the objective value of an optimal solution for this machine behavior.

To illustrate this definition consider the following toy instance with two jobs: $p_1 = w_1 = 2$ and $p_2 = w_2 = N \gg 2$. There are only two possible sequences: $(1, 2)$ and $(2, 1)$. Both sequences are optimal on an ideal machine and are consistent with Smith's Rule. Now suppose our machine breaks down at $t = N + 1$ and stays offline for $T = N^2$ units of time. The cost of $(1, 2)$ on this faulty machine is $4 + N(N + 2 + T) = \Theta(N^3)$, while the cost of $(2, 1)$ is $N^2 + 2(N + 2 + T) = \Theta(N^2)$. This example shows that Smith's Rule can produce a sequence that is not a universal $O(1)$-approximation. In fact, it is not clear that such a universal sequence should always exist.

Key Results

Epstein et al. [3] initiated the study of universal sequencing. They showed that universal $O(1)$-approximate sequences do indeed exist and established tight lower bounds on the universal approximation ratio that can be achieved. Their study was subsequently furthered by Megow and Mestre [10] who showed that the best universal schedule can be approximated in polynomial time up to any desired level of accuracy.

Bounding the Performance of a Universal Sequence

The key observation needed to bound the performance of a universal sequence is that approximating the min-sum objective value on a machine with unknown processing behavior is equivalent to approximating the total weight of uncompleted jobs at any point in time on an ideal machine. To that end, let $W^\pi(t)$ denote, for any $t \geq 0$, the total weight of outstanding jobs at time t in the schedule obtained for job sequence π on an ideal machine. Define $W^*(t) := \min_\pi W^\pi(t)$ for all $t \geq 0$.

Lemma 1 *Let π be a sequence of jobs. Then, the objective value of the corresponding schedule is at most c times the value of an optimum schedule for any machine behavior, if and only if*

$$W^\pi(t) \leq c \cdot W^*(t) \qquad \text{for all } t \geq 0.$$

A Universal Sequencing Algorithm

The universal sequencing algorithm computes the job sequence iteratively backwards. In each iteration it solves the subproblem of finding a set of jobs that has maximum total processing time and total weight within a given bound. This bound is doubled in each iteration.

This approach is related to, but not equivalent to, an algorithm of Hall et al. [6] for online scheduling on ideal machines – the doubling there happens in the time horizon. Indeed, *doubling* strategies have been applied successfully in the design of approximation and online algorithms for various problems; see, e.g., the survey by Chrobak and Kenyon-Mathieu [1].

Doubling Algorithm:

1. For $i \in \{0, 1, \ldots, \lceil \log w(J) \rceil\}$, find a subset J_i^* of jobs of maximum total processing time $p(J_i^*)$, such that the total weight satisfies $w(J_i^*) \leq 2^i$.
2. Construct a permutation π as follows. Start with an empty sequence of jobs. For $i = \lceil \log w(J) \rceil$ down to 0, append the jobs in $J_i^* \setminus \bigcup_{k=0}^{i-1} J_k^*$ in any order at the end of the sequence.

Finding the subsets of jobs J_i^* is a KNAPSACK problem and, thus, NP-hard [8]. Using straightforward dynamic programming, the algorithm runs in pseudo-polynomial time and achieves a performance guarantee of 4 as shown below. However, FPTASes for the knapsack problem can be adopted such that the *Doubling Algorithm* runs in polynomial time loosing an arbitrarily small constant in the performance guarantee.

Theorem 1 *For every scheduling instance, the Doubling Algorithm produces a universal 4-approximation for all machine behaviors.*

Proof By Lemma 1 it is sufficient to show that $W^\pi(t) \leq 4W^*(t)$ for all $t \geq 0$. Let $t \geq 0$ and let i be minimal such that $p(J_i^*) \geq p(J) - t$. By construction of π, only jobs j in $\bigcup_{k=0}^{i} J_k^*$ can have a completion time $C_j^\pi > t$. Thus,

$$W^\pi(t) \leq \sum_{k=0}^{i} w(J_k^*) \leq \sum_{k=0}^{i} 2^k = 2^{i+1} - 1.$$

(1)

In case $i = 0$, the claim is trivially true since $w_j \geq 1$ for any $j \in J$, and thus, $W^*(t) = W^\pi(t)$. Suppose $i \geq 1$; then by our choice of i, it holds that $p(J_{i-1}^*) < p(J) - t$. Therefore, in any sequence π', the total weight of jobs completing after time t is larger than 2^{i-1}, because otherwise we get a contradiction to the maximality of $p(J_{i-1}^*)$. That is, $W^*(t) > 2^{i-1}$. Together with (1) this concludes the proof. $\square$

This result is best possible for universal sequencing on a single machine.

Theorem 2 *For any $c < 4$, there exists an instance for which there is no universal c-approximation.*

This can be shown through a connection to the *online bidding problem* and the corresponding lower bounds shown by Chrobak et al. [2].

Randomized Universal Schedules

It is possible to obtain a better approximation ratio if we select the sequence at random and slightly relax the universality requirement.

Definition 2 A probability distribution over sequences is a *randomized universal c-approximation* if for any machine behavior the *expected* total weighted completion times of a sequence chosen according to the distribution is at most a factor c larger than the objective value of an optimal solution for this machine behavior.

By randomizing the "doubling parameter" in the Doubling Algorithm, the algorithm can achieve an approximation ratio of $e \approx 2.718$, which is best possible for randomized strategies.

Theorem 3 *For every scheduling instance, a randomized variant of the Doubling Algorithm produces a randomized universal e-approximation for all machine behaviors. Furthermore, for any $c < e$, there exists an instance for which there is no randomized universal c-approximation.*

Generalizations

Global Cost Functions
The universality of the sequence constructed by the *Doubling Algorithm* can be driven even further. Consider the generalized min-sum objective $\min \sum w_j f(C_j)$ for any nondecreasing, nonnegative, differentiable cost function f.

Theorem 4 *The Doubling Algorithm computes a universal 4-approximation (randomized e-approximation) for all machine behaviors and all considered cost functions f simultaneously.*

Precedence Constraints

A natural generalization of the universal sequencing problem requires that jobs must be sequenced in compliance with given *precedence constraints*. To a certain extent the *Doubling Algorithm* can be adopted to this more general problem setting. Essentially, the knapsack-related subroutine must respect the precedence constraints, *and* it must ensure that prepending the subsets found in different iterations, starting in the end, does not violate the precedence order.

This corresponds to solving a so-called *partially ordered knapsack* (POK) problem on the reverse of the given partial order.

Theorem 5 *The Doubling Algorithm computes a universal 4-approximation (randomized e-approximation) for the universal scheduling problem respecting given precedence constraints if the POK problem for the given partial order can be solved in polynomial time.*

In general, POK is strongly NP-hard [7] and hard to approximate [5]. However, FPTASes exist for special partial orders, including directed out-trees, two-dimensional orders, and the complement of chordal bipartite orders [7,9].

Release Dates

If jobs have *release dates*, we cannot hope for a universal sequence with bounded approximation ratio unless the scheduler is allowed to preempt jobs. We can think of a universal sequence as a priority order of the jobs guiding a preemptive list scheduling procedure: At any point in time, we work on the job of highest priority that has not been finished yet and that has already been released. Unfortunately, even with this flexibility, the problem is significantly harder.

Theorem 6 *There exists an instance with n jobs with release dates and unit weights, where the performance guarantee of any universal schedule is $\Omega(\log n / \log \log n)$.*

The proof relies on the classical theorem of Erdős and Szekeres [4] on the existence of long increasing/decreasing subsequences of a given sequence of distinct real numbers.

Despite this negative result, there is a nontrivial algorithm that produces a universal 5-approximate sequencing for the class of instances with release dates in which the processing time of each job is proportional to its weight.

Instance-Sensitive Performance Guarantee

Theorem 1 says that the *Doubling Algorithm* produces for every instance a universal 4-approximation. Theorem 2 proves that this is best possible since there are particular instances that do not admit a sequence with a smaller approximation ratio. Many instances, however, admit better-than-4-approximate universal sequences, yet the Doubling Algorithm is only guaranteed to find a 4-approximation. This motivates the problem of finding the best possible universal sequence on an *instance-by-instance* basis.

Theorem 7 *For any fixed $\epsilon > 0$ and $c > 1$, there is a polynomial time algorithm that given an instance either finds a $(c + \epsilon)$-approximate universal sequence or determines that there is no universal c-approximation for this particular instance.*

Applications

The unreliable machine scheduling problem addresses the demand for high-quality scheduling solutions in the dynamic real-world environments of manufacturing processes or in operating systems. The machine could be, for example, a computer server that slows down due to unpredictable third-party usage or an aging production unit prone to unexpected breakdowns. Another setting where the model is applicable is where a higher authority may give priority to another batch of jobs, thus delaying the execution of our jobs. In general, universally good performance regardless of the actual machine behavior is desirable in highly automated systems in which changing the schedule at an arbitrary point in time is too costly or technically infeasible.

Open Problems

The worst-case performance of universal sequences is quite well understood. While the analysis in the model without release dates is tight, it remains open in the setting with release dates if it is possible to obtain a universal $o(n)$-approximation for general instances. The best known lower bound is $\Omega(\log n / \log \log n)$.

While the worst-case analysis assumes arbitrary machine behaviors, it would be interesting to develop and analyze more realistic speed functions. For example, it is reasonable to assume that when a machine breaks down, then it will be repaired or replaced within a certain (possibly fixed) amount of time; or in a stochastic model, the availability periods between breakdowns may be assumed to be exponentially distributed. What improvements in the approximation guarantee do such restrictions allow?

A different approach in aiming for more practice-relevant guarantees is to relax the strict universality requirement. In many situations, changing the scheduling sequence is possible to a certain extent at some extra cost. A very interesting problem is to quantify the amount of adaptivity an algorithm needs to achieve a certain performance guarantee. Ideally, there is a parameter describing the adaptivity that allows to scale between the nonadaptive 4-approximation (Theorem 1) and a fully adaptive $(1 + \epsilon)$-approximation, given by a PTAS that constructs an individual scheduling solution for a specific machine behavior [11].

Cross-References

► List Scheduling
► Minimum Weighted Completion Time
► Robust Scheduling Algorithms

Recommended Reading

1. Chrobak M, Kenyon-Mathieu C (2006) Sigact news online algorithms column 10: competitiveness via doubling. SIGACT News 37(4):115–126
2. Chrobak M, Kenyon C, Noga J, Young N (2008) Incremental medians via online bidding. Algorithmica 50(4):455–478
3. Epstein L, Levin A, Marchetti-Spaccamela A, Megow N, Mestre J, Skutella M, Stougie L (2012) Universal sequencing on a single unreliable machine. SIAM J Comput 41(3):565–586
4. Erdős P, Szekeres G (1935) A combinatorial problem in geometry. Compos Math 2:463–470
5. Hajiaghayi M, Jain K, Lau L, Mandoiu I, Russell A, Vazirani V (2006) Minimum multicolored subgraph problem in multiplex PCR primer set selection and population haplotyping. In: Proceedings of second IWBRA, Atlanta, GA, pp 758–766
6. Hall L, Schulz A, Shmoys D, Wein J (1997) Scheduling to minimize average completion time: off-line and on-line approximation algorithms. Math Oper Res 22:513–544
7. Johnson D, Niemi K (1983) On knapsacks, partitions, and a new dynamic programming technique for trees. Math Oper Res 8(1):1–14
8. Karp R (1972) Reducibility among combinatorial problems. In: Complexity of computer computations (Proceedings of symposium on IBM Thomas J. Watson Research Center, Yorktown Heights), Plenum, pp 85–103
9. Kolliopoulos S, Steiner G (2007) Partially ordered knapsack and applications to scheduling. Discret Appl Math 155(8):889–897
10. Megow N, Mestre J (2013) Instance-sensitive robustness guarantees for sequencing with unknown packing and covering constraints. In: Proceedings of ITCS, Berkeley, CA, pp 495–504
11. Megow N, Verschae J (2013) Dual techniques for scheduling on a machine with varying speed. In: Proceedings of ICALP, Riga, pp 745–756
12. Smith WE (1956) Various optimizers for single-stage production. Naval Res Logist Q 3:59–66

Upward Graph Drawing

Walter Didimo
Department of Engineering, University of Perugia, Perugia, Italy

Keywords

Directed graphs; Flow networks; Graph planarity; Hierarchical drawings

Years and Authors of Summarized Original Work

1994; Bertolazzi, Di Battista, Liotta, Mannino

Problem Definition

Upward graph drawing is concerned with computing two-dimensional layouts of directed graphs where all edges flow in the upward direction. Namely, given a directed graph $G(V, E)$ (also called a *digraph* for short), an *upward drawing* of G is a drawing such that: (i) each vertex $v \in V$ is mapped to a distinct point p_v of the plane and (ii) each edge $(u, v) \in E$ is drawn as a simple curve from p_u and p_v, monotonically increasing in the upward direction.

Clearly, G admits an upward drawing only if it does not contain directed cycles; if we allow edge crossings, acyclicity is also a sufficient condition for the existence of an upward drawing. Instead, if G is planar and we require that also the upward drawing of G is crossing-free, acyclicity is only a necessary condition, and the upward drawability of G becomes a much more intriguing problem. An upward drawing with no edge crossing is called an *upward planar drawing*; deciding whether a planar digraph G admits such a drawing is recognized as the *upward planarity*

testing problem. This problem can be studied in two different settings:

- **Variable embedding setting.** The existence of an upward planar drawing of G is checked over all possible planar embeddings of G.
- **Fixed embedding setting.** The existence of an upward planar drawing of G is checked for a given planar embedding of G, i.e., the drawing must preserve the given embedding.

Both these settings have been widely studied in the literature. In the next section we briefly survey few seminal results on the upward planarity testing problem, and then we concentrate on the first and most popular polynomial-time algorithm for the fixed embedding setting. Figure 1 shows a planar digraph G with a given planar embedding, an embedding-preserving upward planar drawing of G, and a planar digraph G that does not admit upward planar drawings.

Key Results

Let $G(V, E)$ be a planar digraph. We will assume that G is connected (indeed, a digraph admits an upward planar drawing if and only if each of its connected components admits an upward

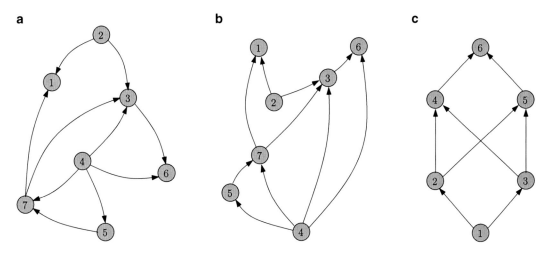

Upward Graph Drawing, Fig. 1 (**a**) A planar digraph G with a given planar embedding. (**b**) An upward planar drawing of G that preserves the embedding of G. (**c**) A planar digraph G that has no upward planar drawing

planar drawing). A *source* (resp. a *sink*) of G is a vertex with only outgoing (resp. incoming) edges. An *internal* vertex of G is a vertex with both outgoing and incoming edges. We denote by S, T, and I the set of sources, sinks, and internal vertices of G, respectively. Digraph G is a *planar st-digraph* if it has only one source s and one sink t and a planar embedding where s and t belong to the same face. Di Battista and Tamassia proved different equivalent characterizations of upward planar drawable digraphs, as stated in the following result [7]:

Theorem 1 ([7]) *Let G be a planar digraph. The following properties are equivalent:*

(a) G admits an upward planar drawing;
(b) G admits an upward planar drawing with straight-line edges;
(c) G is the spanning subgraph of a planar st-digraph.

Using Theorem 1, Garg and Tamassia focused on straight-line drawings and showed that the upward planarity testing problem in the variable embedding setting is NP hard [9]. As a consequence of this hardness result, polynomial-time algorithms in the variable embedding setting have been devised for restricted classes of planar digraphs, like single-source digraphs [6] and series-parallel digraphs [8], while exponential-time algorithms have been proposed for more general planar digraphs (see, e.g., [1, 8]).

Conversely, Bertolazzi et al. showed that the upward planarity testing problem can be solved in polynomial time in the fixed embedding setting [2]. In the following we describe this breakthrough result, which inspired several subsequent papers on the subject.

Polynomial-Time Upward Planarity Testing

Let $G(V, E)$ be an embedded planar digraph, and still denote by S, T, and I the number of sources, sinks, and internal vertices of G, respectively. The result in [2] is based on an elegant combinatorial characterization of the planar embedded digraphs that are upward planar drawable. We first recall few basic definitions.

Digraph G is *bimodal* if for every vertex $v \in I$, the outgoing edges of v are consecutive in the cyclic clockwise order around v (which implies that also the incoming edges of v are consecutive in the cyclic clockwise order around v). It is immediate to see that if a digraph G admits an embedding-preserving upward planar drawing, G is necessarily bimodal.

Let f be a face of G, and let $a = (e_1, v, e_2)$ be a triplet such that $v \in V$ is a vertex of the boundary of f and e_1, e_2 are two edges incident to v that are consecutive on the boundary of f (e_1 and e_2 may coincide if G is not biconnected). Triplet a is called an *angle at v in face f*, or simply an *angle of f*, or an *angle at v*. If both e_1 and e_2 are outgoing edges of v, we call a a *source-switch* angle of f; if both e_1 and e_2 are incoming edges of v, we call a a *sink-switch* angle of f. Denote by $S(f)$ and $T(f)$ the number of source-switch angles and the number of sink-switch angles of f, respectively. It can be easily observed that $S(f) = T(f)$. The *capacity* of f is defined as $cap(f) = \frac{S(f)+T(f)}{2} - 1$ if f is an internal face of G and as $cap(f) = \frac{S(f)+T(f)}{2} + 1$ if f is the external face of G. The number of sources and sinks in the digraph is nicely related to the face capacities, as stated by the following theorem.

Theorem 2 ([2]) *If G is a bimodal embedded planar digraph and F is the set of faces of G, then $\sum_{f \in F} cap(f) = |S| + |T|$.*

Now, given any upward planar drawing Γ, denote by $L(v)$ the number of geometric angles larger than π at vertex v in Γ and by $L(f)$ the number of geometric angles larger than π in face f in Γ. The following result establishes which kinds of angles in Γ can occur around the vertices and inside the faces of the digraph:

Theorem 3 ([2]) *Let G be an embedded planar digraph and let Γ be an embedding-preserving upward planar drawing of G. We have that:*

(i) $L(v) = 0$ for each $v \in I$ and $L(v) = 1$ for each $v \in S \cup T$;
(ii) $L(f) = cap(f)$, for each $f \in F$.

Motivated by Theorem 3, for any given embedded planar digraph G, one can look for an assignment of the angles of G to the faces of G, with these properties:

(a) For each source or sink v, exactly one angle at v is assigned to a face incident to v.
(b) For each face f, the number of angles assigned to f equals $cap(f)$.

Such an assignment is called an *upward-consistent assignment* of G. The following result translates the upward planarity testing problem into the problem of deciding whether G admits an upward-consistent assignment.

Theorem 4 ([2]) *Let G be an acyclic bimodal embedded planar digraph. G admits an embedding-preserving upward planar drawing if and only if G admits an upward-consistent assignment.*

In [2] it is proved that an upward-consistent assignment can be used to construct in linear time an upward planar drawing where each angle assigned to a face corresponds to a geometric angle larger than π. This is done by exploiting Theorem 1; namely, G is first augmented to an st-planar digraph G', then an upward drawing of G' is computed, and finally the dummy edges are removed from the drawing of G', thus obtaining an upward planar drawing of G.

Deciding whether G admits an upward-consistent assignment, and in case finding one, can be done using a network flow model. Namely, construct a bipartite flow network $N(G)$ having a node $n(v)$ for each source or sink v of G, called a *vertex-node*, and a node $n(f)$ for each face f of G, called a *face node*. Each vertex-node $n(v)$ supplies flow 1, while each face-node $n(f)$ demands a flow equal to $cap(f)$. Also, $N(G)$ has a directed arc $(n(v), (f))$ if v is a source or a sink that belongs to the boundary of f in G. A unit of flow on an arc $(n(v), (f))$ indicates that an angle at v in f must be assigned to f. Each feasible flow in $N(G)$ defines an upward-consistent assignment of G. Using standard flow algorithms, testing whether $N(G)$ has a feasible flow, and in case computing one, can be done in $O(n + r^2)$, where n is the number of vertices of G and $r = |S| + |T|$. Figure 2 illustrates the algorithmic approach described above for the upward planarity testing problem.

The next theorem summarizes the main result of [2].

Theorem 5 ([2]) *Let G be an acyclic bimodal embedded planar digraph with n vertices, and let r be the total number of sources and sinks of G.*

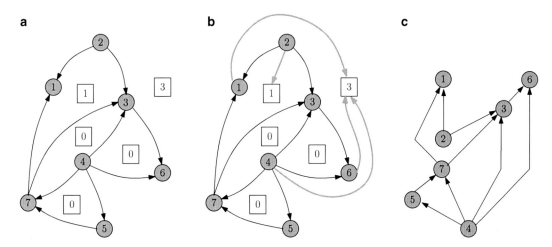

Upward Graph Drawing, Fig. 2 (**a**) An embedded planar digraph G; each face is represented by a *small box* reporting its capacity. (**b**) An upward consistent assignment of G; a *light gray arrow* indicates the assignment of an angle to a face. (**c**) An embedding-preserving upward planar drawing constructed from the upward-consistent assignment; the angles of G assigned to a face correspond to geometric angles larger than π in the drawing

There exists an $O(n + r^2)$-time algorithm that tests whether G admits an embedding-preserving upward planar drawing of G and that computes such a drawing if the test is positive.

Applications

Upward drawings can be effectively used to represent PERT networks, ISA hierarchies in knowledge-representation diagrams, and subroutine call charts. A generalized model, called *quasi-upward drawing*, strongly enlarges the range of application domains of upward graph drawing, making it possible to also represent cyclic digraphs [1] by allowing an edge to break its upward monotonicity in a finite number of points. Petri nets are examples of diagrams that can be represented as quasi-upward drawings; Petri nets are widely used to describe distributed systems.

Efficient C++ graph drawing libraries, like GDToolkit [5] and OGDF [3], implement advanced upward graph drawing algorithms.

Experimental Results

Extensive experimental studies on upward planarity testing are described in [1, 4]. Other references on experimental work about upward graph drawing algorithms can be found in [3,5].

Cross-References

▶ Bend Minimization for Orthogonal Drawings of Plane Graphs
▶ Planarity Testing
▶ Sugiyama Algorithm

Recommended Reading

1. Bertolazzi P, Di Battista G, Didimo W (2002) Quasi-upward planarity. Algorithmica 32(3):474–506
2. Bertolazzi P, Di Battista G, Liotta G, Mannino C (1994) Upward drawings of triconnected digraphs. Algorithmica 12(6):476–497
3. Chimani M, Gutwenger C, Jünger M, Klau GW, Klein K, Mutzel P (2013) The open graph drawing framework (OGDF). In: Tamassia R (ed) Handbook of graph drawing and visualization. CRC Press - Taylor & Francis Group Boca Raton, FL, USA
4. Chimani M, Zeranski R (2013) Upward planarity testing: a computational study. In: Proceedings of Graph Drawing (GD'13), Bordeaux. Volume 8242 of LNCS. Springer, pp 13–24
5. Di Battista G, Didimo W (2013) GDToolkit. In: Tamassia R (ed) Handbook of graph drawing and visualization. CRC Press - Taylor & Francis Group Boca Raton, FL, USA
6. Di Battista G, Eades P, Tamassia R, Tollis IG (1998) Graph drawing: algorithms for the visualization of graphs. Prentice Hall, Upper Saddle River, New Jersey, USA
7. Di Battista G, Tamassia R (1988) Algorithms for plane representations of acyclic digraphs. Theor Comput Sci 61:175–198
8. Didimo W, Giordano F, Liotta G (2009) Upward spirality and upward planarity testing. SIAM J Discret Math 23(4):1842–1899
9. Garg A, Tamassia R (1992) On the computational complexity of upward and rectilinear planarity testing. SIAM J Comput 31(2):601–625

Utilitarian Mechanism Design for Single-Minded Agents

Piotr Krysta[1] and Berthold Vöcking[2]
[1]Department of Computer Science, University of Liverpool, Liverpool, UK
[2]Department of Computer Science, RWTH Aachen University, Aachen, Germany

Keywords

Forward (combinatorial; multi-unit) auction

Years and Authors of Summarized Original Work

2005; Briest, Krysta, Vöcking

Problem Definition

This problem deals with the design of efficiently computable incentive compatible, or truthful, mechanisms for combinatorial optimization problems with selfish one-parameter agents and a single seller. The focus is on approximation algorithms for NP-hard mechanism design problems. These algorithms need to satisfy certain monotonicity properties to ensure truthfulness.

A *one parameter agent* is an agent who as her private data has some *resource* as well as a *valuation*, i.e., the maximum amount of money she is willing to pay for this resource. Sometimes, however, the resource is assumed to be known to the mechanism. The scenario where a single seller offers these resources to the agents is primarily considered. Typically, the seller aims at maximizing the social welfare or her revenue. The work by Briest, Krysta and Vöcking [6] will mostly be considered, but also other existing models and results will be surveyed.

Utilitarian Mechanism Design

A famous example of mechanism design problems is given by combinatorial auctions (CAs), in which a single seller, auctioneer, wants to sell a collection of goods to potential buyers. A wider class of problems is encompassed by a *utilitarian mechanism design (maximization) problem* Π defined by a finite set of *objects* $\mathcal{A}$, a set of feasible outputs $O_\Pi \subseteq \mathcal{A}^n$ and a set of n agents. Each agent declares a set of objects $S_i \subseteq \mathcal{A}$ and a valuation function $v_i : \mathcal{P}(\mathcal{A}) \times \mathcal{A}^n \to \mathbb{R}$ by which she values all possible outputs. Given a vector $S = (S_1, \ldots, S_n)$ of declarations one is interested in output $o^* \in O_\Pi$ maximizing the *social welfare*, i.e., $o^* \in \text{argmax}_{o \in O_\Pi} \sum_{i=1}^{n} v_i(S_i, o)$. In CAs, an object a corresponds to a subset of goods. Each agent declares all the subsets she is interested in and the prices she would be willing to pay. An output specifies the sets to be allocated to the agents.

Here, a limited type of agents called *single-minded* is considered, introduced by Lehmann et al. [10]. Let $R_{\preceq} \subseteq \mathcal{A}^2$ be a reflexive and transitive relation on $\mathcal{A}$, such that there exists a special object $\varnothing \in \mathcal{A}$ with $\varnothing \preceq a$ for any $a \in \mathcal{A}$ to model the situation in which some agent does not contribute to the solution at all. For $a, b \in \mathcal{A} (a, b) \in R_{\preceq}$ will be denoted by $a \preceq b$. The single-minded agent i declares a *single* ob-

ject a_i and is fully defined by her *type* (a_i, v_i), with $a_i \in \mathcal{A}$ and $v_i > 0$. The valuation function introduced earlier reduces to

$$v_i(a_i, o) = \begin{cases} v_i, & \text{if } a_i \preceq o_i \\ 0, & \text{else.} \end{cases}$$

Agent i is called *known* if object a_i is known to the mechanism [11]. Here, mostly unknown agents will be considered. Intuitively, each a_i corresponds to an object agent i offers to contribute to the solution, v_i describes her valuation of any output o that indeed selects a_i. In CAs, relation $R_{\preceq}$ is set inclusion: an agent interested in set S will is also satisfied by S' with $S \subseteq S'$. For ease of notation let $(a, v) = ((a_1, v_1), \ldots, (a_n, v_n))$, $(a_{-i}, v_{-i}) = ((a_1, v_1), \ldots, (a_{i-1}, v_{i-1}), (a_{i+1}, v_{i+1}), \ldots, (a_n, v_n))$ and $((a_i, v_i), (a_{-i}, v_{-i})) = (a, v)$.

Mechanism

A *mechanism* $M = (A, p)$ consists of an algorithm A computing a solution $A(a, v) \in O_\Pi$ and an n-tuple $p(a, v) = (p_1(a, v), \ldots, p_n(a, v)) \in \mathbb{R}_+^n$ of payments collected from the agents. If $a_i \preceq A(a, v)_i$, agent i is *selected*, and let $S(A(a, v)) = \{i | a_i \preceq A(a, v)_i\}$ be the set of selected agents. Agent i's type is her private knowledge. Thus, the types declared by agents may not match their true types. To reflect this, let (a_i^*, v_i^*) refer to agent i's true type and (a_i, v_i) be the declared type. Given an output $o \in O_\Pi$, the *utility* of agent i is $u_i(a, v) = v_i(a_i^*, o) - p_i(a, v)$. Each agent's goal is to maximize her utility. To achieve this, she will try to manipulate the mechanism by declaring a false type if this could result in higher utility. A mechanism is called *truthful*, or *incentive compatible*, if no agent i can gain by lying about her type, i.e., given declarations (a_{-i}, v_{-i}), $u_i((a_i^*, v_i^*), (a_{-i}, v_{-i})) \geq u_i((a_i, v_i), (a_{-i}, v_{-i}))$ for any $(a_i, v_i) \neq (a_i^*, v_i^*)$.

Monotonicity

A sufficient condition for truthfulness of approximate mechanisms for single-minded CAs was first given by Lehmann et al. [10]. Their results can be adopted for the considered scenario. An algorithm A is *monotone* with respect to $R_{\preceq}$ if

$$i \in S(A((a_i, v_i), (a_{-i}, v_{-i})))$$
$$\Rightarrow i \in S(A((a_i', v_i'), (a_{-i}, v_{-i})))$$

for any $a_i' \preceq a_i$ and $v_i' \geq v_i$. Intuitively, one requires that a winning declaration (a_i, v_i) remains winning if an object a_i', smaller according to $R_{\preceq}$, and a higher valuation v_i' are declared. If declarations (a_{-i}, v_{-i}) are fixed and object a_i declared by i, algorithm A defines a *critical value* θ_i^A, i.e., the minimum valuation v_i that makes (a_i, v_i) winning, i.e., $i \in S(A((a_i, v_i), (a_{-i}, v_{-i})))$ for any $v_i > \theta_i^A$ and $i \notin S(A((a_i, v_i), (a_{-i}, v_{-i})))$ for any $v_i < \theta_i^A$. The *critical value payment scheme* p^A associated with A is defined by $p_i^A(a, v) = \theta_i^A$, if $i \in S(A(a, v))$, and $p_i^A(a, v) = 0$, otherwise. The critical value for any fixed agent i can be computed, e.g., by performing binary search on interval $[0, v_i]$ and repeatedly running algorithm A to check if i is selected. Also, mechanism $M_A = (A, p^A)$ is *normalized*, i.e., agents that are not selected pay 0. Algorithm A is *exact*, if for declarations (a, v), $A(a, v)_i = a_i$ or $A(a, v)_i = \emptyset$ for all i. In analogy to [10] one obtains the following.

Theorem 1 *Let A be a monotone and exact algorithm for some utilitarian problem Π and single-minded agents. Then mechanism $M_A = (A, p^A)$ is truthful.*

Additional Definitions

In the *unsplittable flow problem* (UFP), an undirected graph $G = (V, E)$, $|E| = m$, $|V| = n$, with edge capacities b_e, $e \in E$, and a set K of $k \geq 1$ commodities described by terminal pairs $(s_i, t_i) \in V \times V$ and a demand d_i and a value c_i are given. One assumes that $\max_i d_i \leq \min_e b_e$, $d_i \in [0, 1]$ for each $i \in K = \{1, \dots, k\}$, and $b_e \geq 1$ for all $e \in E$. Let $B = \min_e\{b_e\}$. A feasible solution is a subset $K' \subseteq K$ and a single flow s_i-t_i-path for each $i \in K'$, such that the demands of K' can simultaneously and unsplittably be routed along the paths and the capacities are not exceeded. The goal in *UFP*, called *B-bounded UFP*, is to maximize the total value of the commodities in K'. A generalization is allocating bandwidth for multicast communication, where commodity is a set of terminals that should be connected by a multicast tree.

Key Results

Monotone Approximation Schemes

Let Π be a given utilitarian (maximization) problem. Given declarations (a, v), let $Opt(a, v)$ denote an optimal solution to Π on this instance and $w(Opt(a, v))$ the corresponding social welfare. Assuming that A_{Π} is a pseudopolynomial exact algorithm for Π an algorithm A_{Π}^k and monotone FPTAS for Π is defined in Fig. 1.

Theorem 2 *Let Π be a utilitarian mechanism design problem among single-minded agents, A_{Π} monotone pseudopolynomial algorithm for Π with running time $poly(n, V)$, where*

Algorithm A_{Π}^k:
1   $\alpha_k := \frac{n}{\varepsilon \cdot 2^k}$;
2   **for** $i = 1, \dots, n$ **do**
3      $v_i' := \min\{v_i, 2^{k+1}\}$;
4      $v_i'' := \lfloor \alpha_k \cdot v_i' \rfloor$;
5   **return** $A_{\Pi}(a, v'')$;

Algorithm A_{Π}^{FPTAS}
1   $V := \max_i v_i$, $Best := (\emptyset, \dots, \emptyset)$, $best := 0$;
2   **for** $j = 0, \dots, \lceil \log(1 - \varepsilon)^{-1} n \rceil + 1$ **do**
3      $k := \lfloor \log(V) \rfloor - j$;
4   **if** $w_k(A_{\Pi}^k(a, v)) > best$ **then**
5         $Best := A_{\Pi}^k(a, v)$; $best := w_k(A_{\Pi}^k(a, v))$;
6   **return** $Best$;

Utilitarian Mechanism Design for Single-Minded Agents, Fig. 1 A monotone FPTAS for utilitarian problem Π and single-minded agents

$V = \max_i v_i$, and assume that $V \leq w(Opt(a, v))$ for declaration (a, v). Then A_Π^{FPTAS} is a monotone FPTAS for Π.

Theorem 2 can also be applied to minimization problems. Section "Applications" describes how these approximation schemes can be used for forward multi-unit auctions and job scheduling with deadlines.

Truthful Primal-Dual Mechanisms

For an instance $G = (V, E)$ of UFP defined above, let S_i be the set of all s_i-t_i-paths in G, and $S = \bigcup_{i=1}^{k} S_i$. Given $S \in S_i$, let $q_S(e) = d_i$ if $e \in S$, and $q_S(e) = 0$ otherwise. UFP is the following integer linear program (ILP)

$$\max \quad \sum_{i=1}^{k} c_i \cdot \left(\sum_{S \in S_i} x_S \right) \quad (1)$$

$$\text{s.t.} \quad \sum_{S : S \in S, e \in S} q_S(e) x_S \leq b_e \quad \forall e \in E \quad (2)$$

$$\sum_{S \in S_i} x_S \leq 1 \quad \forall i \in \{1, \ldots, k\} \quad (3)$$

$$x_S \in \{0, 1\} \quad \forall S \in S. \quad (4)$$

The linear programming (LP) relaxation is the same linear program with constraints (4) replaced with $x_S \geq 0$ for all $S \in S$. The corresponding dual linear program is

$$\min \quad \sum_{e \in E} b_e y_e + \sum_{i=1}^{k} z_i \quad (5)$$

$$\text{s.t.} \quad z_i + \sum_{e \in S} q_S(e) y_e \geq c_i \quad (6)$$

$$\forall i \in \{1, \ldots, k\} \; \forall S \in S_i$$

$$z_i, y_e \geq 0 \quad \forall i \in \{1, \ldots, k\} \; \forall e \in E. \quad (7)$$

Based on these LPs, Fig. 2 specifies a primal-dual mechanism for routing, called Greedy-1. Greedy-1 ensures feasibility by using y_e's: if an added set exceeded the capacity b_e of some $e \in E$, then this would imply the stopping condition already in the previous iteration. Using the weak duality of LPs the following result can be shown.

Theorem 3 *Greedy-1 outputs a feasible solution, and it is a $(\frac{e\gamma B}{B-1}(m)^{1/(B-1)})$-approximation algorithm if there is a polynomial time algorithm that finds a γ-approximate set S_i in line 4.*

In case of UFP $\gamma = 1$, as the shortest s_i-t_i-path computation finds set S_i in line 4 of Greedy-1. For multicast routing, this problem corresponds to the NP-hard Steiner tree problem, for which one can take $\gamma = 1.55$. Greedy-1 can easily be shown to be monotone in demands and valuations as required in Theorem 1. Thus it implies a truthful mechanism for allocating network resources. The commodities correspond to bidders, the terminal nodes of bidders are known, but the bidders might lie about their demands and valuations. In the multicast routing the set of terminals for each bidder is known but the demands and valuations are unknown.

Utilitarian Mechanism Design for Single-Minded Agents, Fig. 2 Truthful mechanism for network (multicast) routing. $e \approx 2.718$ is Euler number

Algorithm Greedy-1:

1 $T := \emptyset; K := \{1, \ldots, k\};$
2 **forall** $e \in E$ **do** $y_e := 1/b_e;$
3 **repeat**
4   **forall** $i \in K$ **do** $S_i := \text{argmin} \left\{ \sum_{e \in S} y_e \mid S \in S_i \right\};$
5      $j := \text{argmax} \left\{ \dfrac{c_i}{d_i \sum_{e \in S_i} y_e} \;\middle|\; i \in K \right\};$
6      $T := T \cup \{S_j\}; K := K \setminus \{j\};$
7   **forall** $e \in S_j$ **do** $y_e := y_e \cdot \left(e^{B-1} m \right)^{q_{S_j}(e)/(b_e - 1)};$
8 **until** $\sum_{e \in E} b_e y_e \geq e^{B-1} m$ **or** $K = \emptyset;$
9 **return** $T.$

Utilitarian Mechanism Design for Single-Minded Agents, Fig. 3 Truthful mechanism for multi-unit CAs among unknown single-minded bidders. For CAs without multisets: $q_S(e) \in \{0, 1\}$ for each $e \in U, S \in S$

Algorithm Greedy-2:

1 $\mathcal{T} := \emptyset;$
2 **forall** $e \in U$ **do** $y_e := 1/b_e;$
3 **repeat**
4       $S := \text{argmax}\left\{ \dfrac{c_S}{\sum_{e \in U} q_S(e) y_e} \,\middle|\, S \in S \setminus \mathcal{T} \right\};$
5       $\mathcal{T} := \mathcal{T} \cup \{S\};$
6   **forall** $e \in S$ **do** $y_e := y_e \cdot (e^B m)^{q_S(e)/b_e};$
7 **until** $\sum_{e \in U} b_e y_e \geq e^B m;$
8 **return** $\mathcal{T}.$

Corollary 1 *Given any $\epsilon > 0$, $B \geq 1 + \epsilon$, Greedy-1 is a truthful $O(m^{1/(B-1)})$-approximation mechanism for UFP (unicast routing) as well as for the multicast routing problem, where the demands and valuations of the bidders are unknown.*

When B is large, $\Omega(\log m)$, then the approximation factor in Corollary 1 becomes constant. Azar et al. [4] presented further results in case of large B. Awerbuch et al. [3] gave randomized online truthful mechanisms for uni- and multicast routing, obtaining an expected $O(\log(\mu m))$-approximation if $B = \Omega(\log m)$, where μ is the ratio of the largest to smallest valuation. Their approximation holds in fact with respect to the revenue of the auctioneer, but they assume that the demands are known to the mechanism. Bartal et al. [5] give a truthful $O(B \cdot (m/\theta)^{1/(B-2)})$-approximation mechanism for UFP with unknown valuations and demands, where $\theta = \min_i\{d_i\}$.

Greedy-1 can be modified to give truthful mechanisms for multi-unit CAs among unknown single-minded bidders. (In the case of *unknown* single-minded bidders, the bidders have as private data not only their valuations (as in the case of *known* single-minded bidders) but also the sets they demand.) Archer et al. [2] used randomized rounding to obtain a truthful mechanism for multi-unit CAs, but only in a probabilistic sense and only for known bidders. Multi-unit CA among single-minded bidders is a special case of ILP (1)–(4), where $|S_i| = 1$ for each $i \in K$, and $q_S(e) \in \{0, 1\}$ for each $e \in U$, $S \in S$ (E is U in CAs). A bid of bidder $i \in K$

is $(a_i, v_i) = (S, c_S)$, $S \in S_i$, and $c_S = c_i$ is the valuation. The relation $R_{\preceq}$ is $\subseteq$. Algorithm Greedy-2 in Fig. 3 is exact and monotone for CAs with unknown single-minded bidders, as needed in Theorem 1.

Theorem 4 *Algorithm Greedy-2 is a truthful $O(m^{\frac{1}{B}})$-approximation mechanism for multi-unit CAs among unknown single-minded bidders.*

Bartal et al. [5] presented a truthful mechanism for this problem among unknown single-minded bidders which is $O(B \cdot m^{1/(B-2)})$-approximate. (It works in fact for more general bidders.)

Applications

Applications of the techniques described above are presented and a short survey of other results.

Applications of Monotone Approximation Schemes

In a *forward multi-unit auction* a single auctioneer wants to sell m identical items to n possible buyers (bidders). Each single-minded bidder specifies the number of items she is interested in and a price she is willing to pay. Elements in the introduced notation correspond to the requested and allocated numbers of items. Relation $R_{\preceq}$ describes that bidder i requesting q_i items will be satisfied also by any larger number of items. Mu'alem and Nisan [11] give a 2-approximate monotone algorithm for this problem. Theorem 2 gives a monotone FPTAS for multi-unit auctions among unknown single-minded bidders. This FP-

TAS is truthful with respect to agents where both the number of items and price are private.

In *job scheduling with deadlines (JSD)*, each agent i has a job with running time t_i, deadline d_i and a price v_i she is willing to pay if her job is processed by deadline d_i. Element a_i is defined as $a_i = (t_i, d_i)$. Output for agent i is a time slot for processing i's job. For two elements $a_i = (t_i, d_i)$ and $a'_i = (t'_i, d'_i)$ one has $a_i \preceq a'_i$ if $t_i \leq t'_i$ and $d_i \geq d'_i$. Theorem 2 leads to a monotone FPTAS, which, however, is not exact (see Theorem 1) with respect to deadlines, and so it is a truthful mechanism only if the deadlines are known. The techniques of Theorem 2 apply also to minimization mechanism design problems with a single buyer, such as reverse multi-unit auctions, scheduling to minimize tardiness, constrained shortest path and minimum spanning tree problems [6].

Applications of the primal dual algorithms

The applications of the primal dual algorithms are combinatorial auctions and auctions for unicast and multicast routing. As these applications are tied very much to the algorithms, they have already been presented in section "Key Results".

Survey of Other Results

First truthful mechanisms for single-minded CAs were designed by Lehmann et al. [10], where they introduced the concept of single-minded agents, identified the role of monotonicity, and used greedy algorithms to design truthful mechanisms. Better approximation ratios of these greedy mechanisms were proved by Krysta [9] with the help of LP duality. A tool-box of techniques for designing truthful mechanisms for CAs was given by Mu'alem and Nisan [11].

The previous section presented a monotone FPTAS for job scheduling with deadlines where jobs are selfish agents and the seller offers the agents the facilities to process their jobs. Such scenarios when jobs are selfish agents to be scheduled on (possibly selfish) machines have been investigated further by Andelman and Mansour [1], see also references therein.

So far social welfare was mostly assumed as the objective, but for a seller probably more important is to maximize her revenue. This objective turns out to be much harder to enforce in mechanism design. Such truthful (in probabilistic sense) mechanisms were obtained for auctioning unlimited supply goods among one-parameter agents [7, 8]. Another approach to maximizing seller's revenue is known as optimal auction design [12]. A seller wants to auction a single good among agents and each agent has a private value for winning the good. One assumes that the seller knows a joint distribution of those values and wants to maximize her expected revenue [13, 14].

Cross-References

Mechanisms that approximately maximize revenue for unlimited-supply goods as of Goldberg, Hartline and Wright [8] are presented in entry ▶ Competitive Auction.

Recommended Reading

1. Andelman N, Mansour Y (2006) A sufficient condition for truthfulness with single parameter agents. In: Proceedings of the 8th ACM conference on electronic commerce (EC), Ann Arbor, June 2006
2. Archer A, Papadimitriou CH, Talwar K, Tardos E (2003) An approximate truthful mechanism for combinatorial auctions with single parameter agents. In: Proceedings of the 14th annual ACM–SIAM symposium on discrete algorithms (SODA), Baltimore, pp 205–214
3. Awerbuch B, Azar Y, Meyerson A (2003) Reducing truth-telling online mechanisms to online optimization. In: Proceedings of the 35th annual ACM symposium on theory of computing (STOC), San Diego
4. Azar Y, Gamzu I, Gutner S (2007) Truthful unsplittable flow for large capacity networks. In: Proceedings of the 19th annual ACM symposium on parallelism in algorithms and architectures (SPAA), pp 320–329
5. Bartal Y, Gonen R, Nisan N (2003) Incentive compatible multi unit combinatorial auctions. In: Proceedings of the 9th conference on theoretical aspects of rationality and knowledge (TARK), ACM Press, pp 72–87. http://doi.acm.org/10.1145/846241.846250
6. Briest P, Krysta P, Vöcking B (2005) Approximation techniques for utilitarian mechanism design. In: Proceedings of the 37th annual ACM symposium on theory of computing (STOC), pp 39–48

7. Fiat A, Goldberg AV, Hartline JD, Karlin AR (2002) Competitive generalized auctions. In: Proceedings of the 34th annual ACM symposium on theory of computing (STOC), pp 72–81
8. Goldberg AV, Hartline JD, Wright A (2001) Competitive auctions and digital goods. In: Proceedings of the 12th annual ACM–SIAM symposium on discrete algorithms (SODA), pp 735–744
9. Krysta P (2005) Greedy approximation via duality for packing, combinatorial auctions and routing. In: Proceedings of the 30th international conference on mathematical foundations of computer science (MFCS). Lecture notes in computer science, vol 3618, pp 615–627
10. Lehmann DJ, O'Callaghan LI, Shoham Y (1999) Truth revelation in approximately efficient combinatorial auctions. In: Proceedings of the 1st ACM conference on electronic commerce (EC), pp 96–102
11. Mu'alem A, Nisan N (2002) Truthful approximation mechanisms for restricted combinatorial auctions. In: Proceedings of the 18th national conference on artificial intelligence. AAAI, pp 379–384
12. Myerson RB (1981) Optimal auction design. Math Oper Res 6:58–73
13. Ronen A (2001) On approximating optimal auctions (extended abstract). In: Proceedings of the 3rd ACM conference on electronic commerce (EC), pp 11–17
14. Ronen A, Saberi A (2002) On the hardness of optimal auctions. In: Proceedings of the 43rd annual IEEE symposium on foundations of computer science (FOCS), pp 396–405

V

Vector Bin Packing

David S. Johnson
Department of Computer Science, Columbia
University, New York, NY, USA
AT&T Laboratories, Algorithms and
Optimization Research Department, Florham
Park, NJ, USA

Keywords

Approximation algorithms; Bin packing;
Resource constraints; Shared hosting platforms;
Worst-case analysis

Years and Authors of Summarized Original Work

1976; Garey, Graham, Johnson, Yao
1977; Kou, Markowski
1977; Maruyama, Chang, Tang
1981; de la Vega, Lueker
1987; Yao
1990; Csirik, Frenk, Labbé, Zhang
1997; Woeginger
2001; Caprara, Toth
2004; Chekuri, Khanna
2009; Bansal, Caprara, Sviridenko
2010; Stillwell, Schanzenback, Vivien, Casanova
2011; Panigrahy, Talwar, Uyeda, Wieder

Problem Definition

In the *vector bin packing problem*, we are
given an integral dimension $d \geq 1$ and a list
$L = (x_1, x_2, \ldots, x_n)$ of items, where
each item is a d-dimensional tuple $x_i =
(x_{i,1}, x_{i,2}, \ldots, x_{i,d})$ with rational entries $x_{i,j} \in
[0, 1]$. The goal is to assign the items to a
minimum number of multidimensional *bins*,
where if X is the set of items assigned to a
bin, we must have, for each j, $1 \leq j \leq d$,

$$\sum_{x_i \in X} x_{i,j} \leq 1.$$

Note that when $d = 1$, the vector bin packing
problem reduces to the classic (one-dimensional)
bin packing problem.

One potential application of the vector bin
packing problem is that of assigning jobs to
servers in a shared hosting platform, where each
job may require a specific number of cycles per
second and specific amounts of memory, bandwidth, and other resources [12]. Here the servers

© Springer Science+Business Media New York 2016
M.-Y. Kao (ed.), *Encyclopedia of Algorithms*,
DOI 10.1007/978-1-4939-2864-4

would correspond to the bins, the dimension d is the number of resources, the items are the jobs, and $x_{i,j}$ is the fraction of the total amount of a server's jth resource that job x_i requires.

In the early literature, this problem was often called the *multidimensional bin packing* problem. That term, however, is now more typically reserved for the related problem where the items are d-dimensional rectangular parallelepipeds (rectangles, when $d = 2$), the bins are d-dimensional unit cubes, and the items assigned must not only be assigned to bins but also to specific positions in the bins, in such a way that no point in any bin is in the interior of more than one item. With vector bin packing, in contrast, the dimensions are all independent and there is no geometric interpretation of the items.

Key Results

As a generalization of bin packing, vector bin packing is clearly NP-hard in the strong sense, and so most of the research on this problem has been directed toward the study of approximation algorithms for it. This will be the primary topic in this entry. Most of the theoretical results concerning these algorithms can be expressed in terms of *asymptotic worst-case ratios*. For a given algorithm A and a list of items L, let $A(L)$ denote the number of bins used by A for L. Let $OPT(L)$ denote the optimal number of bins for list L. We define the asymptotic worst-case ratio $R_A^\infty(d)$ for algorithm A on d-dimensional instances as follows.

$$R_A^N(d) = \max\left\{ \frac{A(L)}{OPT(L)} : L \text{ is a list of } d\text{-dimensional items with } OPT(L) = N \right\}$$

$$R_A^\infty(d) = \limsup_{N \to \infty} R_A^N(d)$$

Generalizations of Classical Bin Packing Algorithms

Generalizing First Fit and First Fit Decreasing

Several classic one-dimensional bin packing algorithms have been generalized to vector bin packing. Imagine we have a potentially infinite sequence of empty bins $B_1, B_2, \ldots$, and let $X_{h,j}$ denote the total amount of resource j used by the items currently assigned to B_h. In the generalized "First Fit" algorithm, the first item goes in bin B_1, and thereafter each item goes into the lowest-index bin into which it can be legally placed, subject to the resource constraints. In generalized "Best Fit," each item is assigned to a bin with the maximum value of $\sum_{j=1}^d X_{h,j}$ among those to which it can legally be added, ties broken in favor of the smallest index h.

As in the one-dimensional case, a plausible way to improve the above two online algorithms is to first reorder the list in decreasing order, and then apply the packing algorithm. Now, however, there are a variety of ways to define "decreasing

order," each leading to different algorithms. For example, in FFDmax items are ordered by nonincreasing value of $\max_{j=1}^d x_{i,j}$ and then FF is applied. Similarly, in FFDsum, the items are ordered by nonincreasing value of $\sum_{j=1}^d x_{i,j}$ and, in FFDprod, they are ordered by nonincreasing value of $\prod_{j=1}^d x_{i,j}$. In FFDlex, they are ordered so that x_i precedes $x_{i'}$ only if either $x_{i,j} = x_{i',j}$, $1 \le j \le d$, or there is a $j^* \le d$ such that $x_{i,j} = x_{i',j}$, $1 \le j < j^*$ and $x_{i,j^*} < x_{i',j^*}$. The algorithms BFDmax, BFDsum, BFCprod, and BFDlex are defined analogously, with Best Fit being used to pack the reordered list instead of First Fit.

Call an algorithm "reasonable" if it produces packings in which no two bins can be combined, that is, are such that all the items contained in the two would collectively fit together in a single bin [9]. All of the above algorithms are easily seen to be reasonable, and, indeed, any vector bin packing algorithm has a "reasonable" counterpart that uses no more bins and spends at most $O(n^2 d)$ additional time (in a final pass that combines legally combinable pairs of bins as long as such pairs exist). A general upper bound on asymptotic worst-case behavior is the following.

Theorem 1 ([9]) *If A is a reasonable vector bin packing algorithm, then for all $d \geq 1$,*

$$R_A^\infty(d) \leq d + 1.$$

Unfortunately, none of the above algorithms are much better.

Theorem 2 ([9, 11]) *For each of the 10 algorithms defined above and all $d \geq 1$,*

$$R_A^\infty(d) \geq d.$$

Tighter bounds have been proved for two of the algorithms.

Theorem 3 ([6]) *For all $d \geq 1$, $R_{\text{FF}}^\infty(d) = d + \frac{7}{10}$.*

Theorem 4 ([6]) *For all $d \geq 1$, $d + \frac{d-1}{d(d+1)} \leq R_{\text{FFDmax}}^\infty(d) \leq d + \frac{1}{3}$.*

Note that the classic one-dimensional bin packing results of [8] yield $R_{\text{FF}}^\infty(1) = 17/10$ (the precise specialization of Theorem 3) and $R_{\text{FFD}}^\infty(1) = 11/9$ (a tighter result than the specialization of Theorem 4). Matching upper and lower bounds are not known for $R_{\text{FFDmax}}^\infty(d)$ for any $d > 1$. In special cases, however, the lower bounds can be improved. It was observed in [6] that the lower bounds for $d \in \{2, 3\}$ could be increased to $d + 11/60$ using ideas from [8]. And Csirik et al. [4] showed that for odd $d \geq 5$, the lower bound of Theorem 2 could be increased by $1/(d(d+1)(d+2))$.

Generalizing the de la Vega and Lueker Asymptotic Approximation Scheme

In [5], de la Vega and Lueker devised an "asymptotic polynomial-time approximation scheme" (APTAS) for one-dimensional bin packing, that is, a collection of polynomial-time algorithms A_ϵ with $R_{A_\epsilon}^\infty(1) \leq 1 + \epsilon$ for all $\epsilon > 0$. In that same paper, they also showed how to generalize the algorithms to provide a collection of vector bin packing algorithms B_ϵ such that for each ϵ and each integer $d \geq 1$, $R_{B_\epsilon}^\infty(d) \leq d + \epsilon$.

The algorithm B_ϵ is quite simple. Divide L into d sublists, $L_1, L_2, \ldots, L_d$, where L_j consists of all those items x for which j is the index of the dimension with the largest entry in the corresponding tuple, ties broken arbitrarily. Then apply $A_{\epsilon/d}$ to each list L_j separately, viewed as an instance of one-dimensional bin packing with the size of item x_i being $x_{i,j}$, and output the union of the d packings. Unfortunately, although the running times for the B_ϵ's are linear in dn, they contain additive constants that are potentially exponential in $(d/\epsilon)^2$, and so they may not be practical for small ϵ.

In contrast, FF, FFDmax, and all their variants mentioned in the previous section have straightforward $O(dn^2)$ implementations, and, although the data structures that allow them to be sped up to $O(n \log n)$ when $d = 1$ do not extend to higher dimensions, speedups should be possible by using d-dimensional dynamic range searching procedures to identify the set of bins that can contain the next item to be packed [11].

Hardness of Approximation Results

In [14], Yao observed that, under a standard decision tree model of computation, any vector bin packing algorithm A that has $R_A^\infty(d) < d$ for all d cannot have $o(n \log n)$ running time. This is not much of a constraint, however, since almost all the algorithms that have been proposed for Vector Bin Packing are slower than this. For those algorithms, a weaker bound applies. Assuming $P \neq NP$, no polynomial-time vector bin packing algorithm A can have $R_A^\infty(d) < \sqrt{d} - \epsilon$ for all d and any $\epsilon > 0$. This follows from a straightforward reduction of graph coloring to vector bin packing and a result of Zuckerman [15] for the former [3]. Under the same assumption, there can be no APTAS for any fixed $d \geq 2$ [13].

Algorithms with $R_A^\infty(d) < d$

Chekuri and Khanna [3] devised the first polynomial-time algorithms to guarantee $R_A^\infty(d) < d$ for all sufficiently large d.

Theorem 5 ([3]) *For any $\epsilon > 0$ there is a polynomial-time algorithm C_ϵ such that*

$$R_{C_\epsilon}^\infty(d) < \epsilon \cdot d + \ln(\epsilon^{-1}) + 2.$$

The algorithm works in three phases. The first considers the following linear program (LP). Let $z_{i,k}$ be a decision variable with value 1 if item x_1 is packed in bin j. The LP's constraints are

$$\sum_{k=1}^{m} z_{i,k} = 1, \ 1 \le i \le n \tag{1}$$

$$\sum_{i=1}^{n} x_{i,j} \cdot z_{i,k} \le 1, \ 1 \le k \le m, 1 \le j \le d \tag{2}$$

$$z_{i,k} \ge 0, \ 1 \le i \le n, 1 \le k \le m \tag{3}$$

This is the LP relaxation of an integer program (with $z_{i,k} \in \{0,1\}$) which has a feasible solution if and only if our list can be packed into m bins. The first set of constraints insures that each item is packed into exactly one bin. The second set insures that all resource constraints are satisfied by the packing.

Let M be the least value of m such that this LP is feasible. Then we clearly must have $OPT(L) \ge M$. Moreover we can in polynomial time determine M and a basic feasible solution for the corresponding LP, by using binary search and a polynomial time LP-solver. In this basic feasible solution, there will be at most $n + dM$ positive variables (the number of nontrivial constraints). Since each of the n items x_i by (1) must be assigned to at least one bin, at least one of the variables $z_{i,k}$ must be positive for each i, meaning that at most dM of the items can be assigned to more than one bin. That leaves $n - dM$ items assigned to exactly one bin, and consequently our LP solution yields a feasible packing of these items into $M \le OPT(L)$ bins, which is the output of our first phase. The remaining dM or fewer items will be packed into additional bins in two additional phases as follows.

Let $k = \lceil 1/\epsilon \rceil$. While there are at least k unpacked items that will fit in a single bin, find such a set and pack them all in a new bin (Phase 2). Otherwise, find a maximum size set of unpacked items that will fit in a bin, assign them to a new bin, and repeat until all items are packed (Phase 3). Note that in both of these phases the next set of items to be packed can be found in time $O(n^k kd)$, which is polynomial for fixed ϵ. Thus the overall time for the algorithm is itself polynomial for fixed ϵ. Phase 2 creates at most $dM/k < \epsilon d \cdot OPT(L)$ bins. Phase 3 can be interpreted as implementing the Greedy algorithm for Set Covering, as applied to the instance in which the elements to be covered are the items left to be packed after Phase 2, the sets are the collections of those items which will fit in a bin, and no set has size exceeding $k - 1$. Thus, by standard results about Greedy Set Covering (see [7] for example), the number of bins added in this phase is less than $(\ln(k-1)+1) \cdot OPT(L) < (\ln(1/\epsilon) + 1) \cdot OPT(L)$. Adding up the above three terms yields the claimed theorem.

Note that if we set $\epsilon = 1/d$ in the above, we get a series of algorithms $C_{1/d}$ with $R_{C_{1/d}}^\infty(d) \le \ln(d) + 3$, where the running time of each is polynomial in n, although exponential in d. A slight improvement to this has recently been obtained by Bansal et al. [1]. They devise algorithms $D_{d,\epsilon}$ that run in polynomial time for fixed d and ϵ (although exponential in both) that have $R_{D_{d,\epsilon}}^\infty \le \ln(d + \epsilon) + 1 + \epsilon$, which, for $d \ge 2$, already beats $\ln(d) + 3$ when $\epsilon = 1$.

Experimental Results

There have been several experimental studies of approximation algorithms for vector bin packing [2, 10–12]. These studies were for the most part limited to $d \le 10$ and $n \le 500$, which may well make sense in the context of the proposed applications, and used distinct sets of randomly generated test instances. In two cases ([10] and [12]), the algorithms were compared using objective functions other than the number of bins packed. Nevertheless, certain common conclusions emerge. The FFD algorithms in particular yielded substantially better packings than worst-case analysis suggests. Both [11] and [12] suggest, however, that a different class of algorithms, ones that attempt to keep the bins as "balanced" as possible, may perform even better. An example of such an algorithm is the "norm-based greedy" algorithm of [11], which packs the bins one-by-one, at each step adding to the current bin B_h that item x_i that fits and yields the smallest weighted L^2 norm for the resulting "gap vector" $(X_{h,1} - x_{i,1}, X_{h,2} - x_{i,2}, \ldots, X_{h,d} - x_{i,d})$. For more details, see [11]. As for the algorithms described above with $R_A^\infty(d) < d$ for large d, the only ones with hopes of feasible running times are the algorithms $C_{1/d}$ from [3] when n^d is of manageable size. Limited experiments from [12] indicate that the packings these algorithms produce are not competitive.

Cross-References

▶ Approximation Schemes for Bin Packing
▶ Bin Packing
▶ Graph Coloring
▶ Greedy Set-Cover Algorithms

Recommended Reading

1. Bansal N, Caprara A, Sviridenko M (2009) A new approximation method for set covering problems, with applications to multidimensional bin packing. SIAM J Comput 39:1256–1278
2. Caprara A, Toth P (2001) Lower bounds and algorithms for the 2-dimensional vector packing problem. Discret Appl Math 111:231–262
3. Chekuri C, Khanna S (2004) On multidimensional packing problems. SIAM J Comput 33:837–851
4. Csirik WF, Frenk JBG, Labbé M, Zhang S (1990) On the multidimensional vector packing. Acta Cybern 9:361–369
5. de la Vega J, Lueker GS (1981) Bin packing can be solved within $1 + \epsilon$ in linear time. Combinatorica 1:349–355
6. Garey MR, Graham RL, Johnson DS, Yao AC-C (1976) Resource constrained scheduling as generalized bin packing. J Comb Theory (A) 21:257–298
7. Johnson DS (1974) Approximation algorithms for combinatorial problems. J Comput Syst Sci 9:256–278
8. Johnson DS, Demers A, Ullman JD, Garey MR, Graham RL (1974) Worst-case performance bounds for simple one-dimensional packing algorithms. SIAM J Comput 3:299–325
9. Kou LT, Markowski G (1977) Multidimensional bin packing algorithms. IBM J Res Dev 21:443–448
10. Maruyama K, Chang SK, Tang DT (1977) A general packing algorithm for multidimensional resource requirements. Int J Comput Inf Sci 6:131–149
11. Panigrahy R, Talwar K, Uyeda L, Wieder U (2011) Heuristics for vector bin packing. Unpublished manuscript. (Available on the web)
12. Stillwell M, Schanzenbach D, Vivien F, Casanova H (2010) Resource allocation algorithms for virtualized service hosting platforms. J Parallel Distrib Comput 70:962–974
13. Woeginger GJ (1997) There is no asymptotic PTAS for two-dimensional vector packing. Inf Proc Lett 64:293–297
14. Yao AC-C (1980) New algorithms for bin packing. J ACM 27:207–227
15. Zuckerman D (2007) Linear degree extractors and the inapproximability of max clique and chromatic number. Theory Comput 3:103–128

Vector Scheduling Problems

Tjark Vredeveld
Department of Quantitative Economics, Maastricht University, Maastricht, The Netherlands

Keywords

Approximation schemes; Vector scheduling

Years and Authors of Summarized Original Work

2004; Chekuri, Khanna

Problem Definition

Vector scheduling is a multidimensional extension of traditional machine scheduling problems. Whereas in traditional machine scheduling a job only uses a single resource, normally time, in vector scheduling a job uses several resources. In traditional scheduling, the load of a machine is the total resource consumption by the jobs that it serves. In vector scheduling, we define the load of a machine as the maximum resource usage over all resources of the jobs that are served by this machine. In the setting that we consider here, the makespan, which is normally defined to be the time by which all jobs are completed, is equal to the maximum machine load.

To define the vector scheduling problem that we consider more formally, we let $\|\mathbf{x}\|_\infty$ denote the standard ℓ_∞-norm of the vector $\mathbf{x}$. In the vector scheduling problem, the input consists of a set J of n jobs, where each job j is associated with a d-dimensional vector $\mathbf{p_j} \in [0, 1]^d$, and m identical machines. The goal is to find an assignment of the jobs to the m machines such that $\max_{1 \le i \le m} \| \sum_{j \in M_i} \mathbf{p_j} \|_\infty$ is minimized, where M_i denotes the set of jobs that are assigned to machine i.

The traditional machine scheduling problem corresponds to the case $d = 1$, and this is known to be strongly NP-hard [8]. For $d = 1$, Graham's well-known list scheduling algorithm has a performance guarantee of 2 [9] and Hochbaum and Shmoys developed at PTAS [10]. For general vector scheduling, Graham's list scheduling algorithm can be extended to the d-dimensional case, having a performance guarantee of $d + 1$. In this entry we focus on the work of Chekuri and Khann [5], who developed a PTAS for fixed d and gave a polylogarithmic approximation factor for the case of general d.

Key Results

Constant Dimension d

Chekuri and Khann [5] designed an approximation scheme that runs in polynomial time whenever the dimension d of the job vectors is constant.

Theorem 1 *For any* $\epsilon > 0$*, there exists an* $(1 + \epsilon)$*-approximation algorithm that has a running time of* $O((nd/\epsilon)^{O(s)})$*, where* s *is in* $O\left((\frac{\log(d/\epsilon)}{\epsilon})^d\right)$.

The proof of this theorem is a nontrivial generalization of the ideas that Hochbaum and Shmoys used for the 1-dimensional case [10]. In this primal-dual approach, the main idea is to view the scheduling problem as a bin-packing problem in which the jobs need to

be packed into a number of bins of a certain capacity B. If all jobs fit into m bins, then the makespan is bounded by B. Hochbaum and Shmoys gave an algorithm that determines whether the jobs fit into m bins of capacity $(1 + \epsilon)B$ or the jobs need to be packed into at least $m + 1$ bins of capacity B. Chekuri and Khanna extended this idea by using d-dimensional bins, where the jobs assigned to one bin should have a total resource usage of at most B in any of the d dimensions. By standard scaling techniques, we assume w.l.o.g. that $B = 1$.

Like in the 1-dimensional case, Chekuri and Khanna divide the jobs in small and large jobs, where the size of a job is based on the ℓ_∞ norm. They first do a preprocessing step in which each coordinate of the vectors is set to 0 whenever it is too small compared to the maximum value of the coordinates in the same vector. To find an $(1 + \epsilon)$-approximation, the algorithm performs two stages. In the first stage all large jobs will be assigned to the machines, and in the second stage all small jobs will be assigned to the machines. Whereas in the 1-dimensional case the assignment of the small jobs can be done greedily on top of the large jobs, for $d \ge 2$ the interaction between the two stages needs to be taken into account.

To accommodate this interaction, Chekuri and Khanna define a *capacity configuration* as a d-tuple $(c_1, \ldots, c_d)$ such that c_k is an integer between 0 and $\lceil 1/\epsilon \rceil$. A set of jobs S can be feasible, scheduled on one machine according to a capacity configuration $(c_1, \ldots, c_d)$ when for any dimension k, it holds that $\left(\sum_{j \in S} \mathbf{p_j}\right)_k \le a_k \cdot \epsilon$, i.e., in each dimension k the resource usage is not more than $c_k \cdot \epsilon$. The number of distinct capacity configurations is given by $t = (1 + \lceil 1/\epsilon \rceil)^d$.

A capacity configuration describes approximately how a machine is filled. As there are m machines available to process the jobs, a *machine configuration* can be described by a t-tuple $(m_1, \ldots, m_t)$, satisfying $m_i \ge 0$ and $\sum_i m_i = m$, where m_i denotes the number of machines of the ith capacity configuration. The number of distinct machine configurations is certainly bounded from above by m^t.

After the preprocessing of the vectors and the splitting of the jobs in small and large, it needs to be determined whether all large jobs can be scheduled according to a machine configuration M. As a first step, all nonzero elements of the large vectors are rounded (down) to the begin points of geometrically increasing intervals. Moreover, as the vectors are in some sense large, not too many vectors can be scheduled on one machine. Therefore, using a dynamic programming approach, one can approximately determine whether the set of large vectors can be scheduled according to machine configuration M.

When the set of large jobs are scheduled such that a certain machine i is scheduled according to a capacity configuration $(c_1, \ldots, c_d)$, then the small jobs on this machine need to be scheduled according to the *empty capacity configuration*, i.e., the capacity configuration $(1 + \lceil 1/\epsilon \rceil) \cdot (1, 1, \ldots, 1) - (c_1, \ldots, c_d)$. Given a machine configuration M, we let $\bar{M}$ denote the corresponding machine configuration as the one obtained by taking the empty capacity configurations for each of the machines in M.

To see whether the small jobs can be scheduled according to a machine configuration $\bar{M}$, Chekuri and Khanna present an integer programming (ILP) formulation that assigns the vectors to the machines. Moreover, they show that solving the LP relaxation of this ILP formulation and distributing the fractionally assigned vectors equally over the machines result in a solution in which each dimension of each machine is only overloaded by a factor of $(1 + \epsilon)$.

Once they have found a machine configuration M according to which the large jobs can be scheduled and corresponding machine configuration $\bar{M}$ according to which the small jobs can be scheduled, Chekuri and Khanna have shown that all jobs can be scheduled such that the load of any machine does not exceed $1 + \epsilon$. If for all machine configurations M and corresponding machine configuration $\bar{M}$ the large jobs cannot be scheduled according to M or the small jobs cannot be scheduled according to $\bar{M}$, then the

vectors cannot be scheduled such that the load of each machine is at most 1.

General Dimension d

For the general case in which the dimension d of the vectors is not restricted to be a constant, Chekuri and Khanna present several approximation algorithms. Also for the general case, they assume that all vectors can be scheduled such that the makespan is bounded by 1. For two algorithms, they use as a subroutine an approximation algorithm for finding a set of vectors S that maximizes the volume of these vectors, i.e., the sum of all coordinates of all these vectors $\sum_{j \in S} \sum_{k=1}^{d} (\mathbf{p_j})_k$, restricted to $\| \sum_{j \in S} \mathbf{p_j} \|_\infty \leq 1$. This resulted in the following results.

Theorem 2 *There exists a polynomial-time $O(\log^2 d)$-approximation algorithm for the vector scheduling problem.*

Theorem 3 *There exists a $O(\log d)$-approximation algorithm for the vector scheduling problem that runs in time polynomial in n^d.*

These approximation results are good when d is small compared to the number of machines m. On the other hand, Chekuri and Khanna also give a randomized algorithm, which just assigns each job uniformly at random to one of the machines, obtaining a performance guarantee that is better when d is large compared to m.

Theorem 4 *There exists a randomized algorithm that has a performance guarantee of $O(\log dm / \log \log dm)$ with high probability.*

Finally, there is also a hardness result for the vector scheduling problem.

Theorem 5 *For any constant $\rho > 1$, there is no polynomial-time approximation algorithm with a performance guarantee of ρ, unless NP = ZPP.*

Extensions

Epstein and Tassa [6, 7] extended the vector scheduling problem to deal with more general objective functions. Instead of defining the load of a machine as the maximum resource usage over all resources of the jobs that are served by this machine, they defined the load as the sum of the vectors $\mathbf{p_j}$ assigned to the machine. That is, the load itself is now also a d-dimensional vector. Letting $\mathbf{l_i} = \sum_{j \in M_i} \mathbf{p_j}$ denote the load of machine i, then in [6] they gave PTASes for several objective functions of the form $F(S) = f(g(\mathbf{l_1}), \ldots, g(\mathbf{l_m}))$. Note that the vector scheduling problem as discussed in this entry is equal to the case that $f = g = \max$. In [7], they extended their results to the more general case where the function g may vary per machine, i.e., $F(S) = f(g_1(\mathbf{l_1}), \ldots, g_m(\mathbf{l_m}))$.

Bonifaci and Wiese [4] extended the vector scheduling problem to the ℓ_p norm and the case of unrelated machines. That is, a job j has a d-dimensional resource usage $\mathbf{p_{ij}}$ on machine i. They considered the case in which the number of types of machines is constant: on the same type of machine, a certain job has the same resource usage. Moreover, they restricted themselves to the case of having only a constant number of resources, that is, the vectors $\mathbf{p_{ij}}$ are d dimensional for a constant d. For this setting, they developed a PTAS.

The PTAS of Chekuri and Khanna has a running time that is doubly exponential in d. Bansal, Vredeveld, and Van der Zwaan [2] showed that this double exponential dependence on d is necessary. For $\epsilon < 1$, they showed that unless the exponential time hypothesis fails, there is no $(1 + \epsilon)$-approximation algorithm with running time $\exp(o(\lfloor 1/\epsilon \rfloor^{d/3}))$. Moreover, they showed that unless NP has subexponential algorithms, no $(1 + \epsilon)$-approximation algorithm exists with running time $\exp(\lfloor 1/\epsilon \rfloor^{o(d)})$. These lower bounds even hold for the case that ϵm more machines are allowed, for sufficiently small $\epsilon > 0$. Moreover, they also gave a $(1 + \epsilon)$-approximation algorithm with running time $\exp((1/\epsilon)^{O(d \log \log d)} + nd)$, which is the first efficient approximation scheme (EPTAS) for the problem with constant d.

Open Problems

The gap between the lower bounds and upper bounds on the running time of $(1 + \epsilon)$-approximation algorithms has almost been closed. The question remains whether this is also the case when instead of the ℓ_∞-norm, the ℓ_p-norm is minimized. Furthermore, it would be interesting to know whether one can obtain better running times when the vectors are highly structured. These highly structured vectors may occur, for example, in applications of real-time scheduling; see, e.g., [1, 3].

Cross-References

▶ Approximation Schemes for Bin Packing
▶ Approximation Schemes for Makespan Minimization
▶ Vector Bin Packing

Recommended Reading

1. Bansal N, Rutten C, van der Ster S, Vredeveld T, van der Zwaan R (2014) Approximating real-time scheduling on identical machines. In: LATIN 2014, Montevideo. LNCS, vol 8392. Springer, Heidelberg, pp 550–561
2. Bansal N, Vredeveld T, van der Zwaan R (2014) Approximating vector scheduling: almost matching upper and lower bounds. In: LATIN 2014, Montevideo. LNCS, vol 8392. Springer, Heidelberg, pp 47–59
3. Baruah S, Bonifaci V, D'Angelo G, Marchetti-Spaccamela A, van der Ster S, Stougie L (2011) Mixed criticality scheduling of sporadic task systems. In: Proceedings of the 19th annual European symposium on algorithms (ESA), Saarbrücken. LNCS, vol 6942. Springer, Heidelberg, pp 555–566
4. Bonifaci V, Wiese A (2012) Scheduling unrelated machines of few different types. CoRR abs/1205.0974
5. Chekuri C, Khanna S (2004) On multidimensional packing problems. SIAM J Comput 33(4):837–851
6. Epstein L, Tassa T (2003) Vector assignment problems: a general framework. J Algorithms 48(2):360–384
7. Epstein L, Tassa T (2006) Vector assignment schemes for asymmetric settings. Acta Inform 42(6–7):501–514
8. Garey M, Johnson D (1978) "Strong" NP-completeness results: motivation, examples, and implications. J ACM 25:499–508
9. Graham R (1966) Bounds for certain multiprocessor anomalies. Bell Syst Tech J 45:1563–1581
10. Hochbaum DS, Shmoys DB (1987) Using dual approximation algorithms for scheduling problems theoretical and practical results. J ACM 34(1):144–162

Vertex Cover Kernelization

Jianer Chen
Department of Computer Science, Texas A&M
University, College Station, TX, USA

Keywords

Vertex cover data reduction; Vertex cover preprocessing

Years and Authors of Summarized Original Work

2004; Abu-Khzam, Collins, Fellows, Langston, Suters, Symons

Problem Definition

Let G be an undirected graph. A subset C of vertices in G is a *vertex cover* for G if every edge in G has at least one end in C. The (parametrized) VERTEX COVER problem is for each given instance (G, k), where G is a graph and $k \geq 0$ is an integer (the parameter), to determine whether the graph G has a vertex cover of at most k vertices.

The VERTEX COVER problem is one of the six "basic" NP-complete problems according to Garey and Johnson [4]. Therefore, the problem cannot be solved in polynomial time unless P = NP. However, the NP-completeness of the problem does not obviate the need for solving it because of its fundamental importance and wide applications. One approach was initiated based on the observation that in many applications, the parameter k is small. Therefore, by taking the advantages of this fact, one may be able to solve this NP-complete problem effectively and practically for instances with a small parameter. More specifically, algorithms of running time of the form $f(k)p(n)$ have been studied for VERTEX COVER, where $p(n)$ is a low-degree polynomial of the number $n = |G|$ of vertices in G and $f(k)$ is a function independent of n.

There has been an impressive sequence of improved algorithms for the VERTEX COVER problem. A number of new techniques have been developed during this research, including kernelization, folding, and refined branch-and-search. In particular, the *kernelization* method is the study of polynomial time algorithms that can significantly reduce the instance size for VERTEX COVER. The following are some concepts related to the kernelization method:

Definition 1 Two instances (G, k) and (G', k') of VERTEX COVER are *equivalent* if the graph G has a vertex cover of size $\leq k$ if and only if the graph G' has a vertex cover of size $\leq k'$.

Definition 2 A *kernelization algorithm* for the VERTEX COVER problem takes an instance (G, k) of VERTEX COVER as input and produces an equivalent instance (G', k') for the problem, such that $|G'| \leq |G|$ and $k' \leq k$.

The kernelization method has been used extensively in conjunction with other techniques in the development of algorithms for the VERTEX COVER problem. Two major issues in the study of kernelization method are (1) effective reductions of instance size; and (2) the efficiency of kernelization algorithms.

Key Results

A number of kernelization techniques are discussed and studied in the current paper.

Preprocessing Based on Vertex Degrees

Let (G, k) be an instance of VERTEX COVER. Let v be a vertex of degree larger than k in G. If a vertex cover C does not include v, then C must contain all neighbors of v, which implies that C contains more than k vertices. Therefore, in order to find a vertex cover of no more than k vertices, one must include v in the vertex cover, and recursively look for a vertex cover of $k - 1$ vertices in the remaining graph.

The following fact was observed on vertices of degree less than 3.

Theorem 1 *There is a linear time kernelization algorithm that on each instance (G, k) of vertex cover, where the graph G contains a vertex of degree less than 3, produces an equivalent instance (G', k') such that $|G'| < |G|$ and/or $k < k'$.*

Therefore, vertices of high degree (i.e., degree $> k$) and low degree (i.e., degree < 3) can always be handled efficiently before any more time-consuming process.

Nemhauser-Trotter Theorem

Let G be a graph with vertices $v_1, v_2, \ldots, v_n$. Consider the following integer programming problem:

(IP)Minimize $\quad x_1 + x_2 + \cdots + x_n$

Subject to $\quad x_i + x_j \geq 1$

$\quad$ for each edge $[v_i, v_j]$ in G

$\quad x_i \in \{0, 1\}, \quad 1 \leq i \leq n$

It is easy to see that there is a one-to-one correspondence between the set of feasible solutions to (IP) and the set of vertex covers of the graph G. A natural LP-relaxation (LP) of the problem (IP) is to replace the restrictions $x_i \in \{0, 1\}$ with $x_i \geq 0$ for all i. Note that the resulting linear programming problem (LP) now can be solved in polynomial time.

Let $\sigma = \{x_1^0, \ldots, x_n^0\}$ be an optimal solution to the linear programming problem (LP). The vertices in the graph G can be partitioned into three disjoint parts according to σ:

$$I_0 = \{v_i \mid x_i^0 < 0.5\},$$

$$C_0 = \{v_i \mid x_i^0 > 0.5\}, \text{ and}$$

$$V_0 = \{v_i \mid x_i^0 = 0.5\}$$

The following nice property of the above vertex partition of the graph G was first observed by Nemhauser and Trotter [5].

Theorem 2 *(Nemhauser-Trotter) Let $G[V_0]$ be the subgraph of G induced by the vertex set V_0.*

Then (1) every vertex cover of $G[V_0]$ contains at least $|V_0|/2$ vertices; and (2) every minimum vertex cover of $G[V_0]$ plus the vertex set C_0 makes a minimum vertex cover of the graph G.

Let k be any integer, and let $G' = G[V_0]$ and $k' = k - |C_0|$. As first noted in [3], by Theorem 2, the instances (G, k) and (G', k') are equivalent, and $|G'| \leq 2k'$ is a necessary condition for the graph G' to have a vertex cover of size k'. This observation gives the following kernelization result.

Theorem 3 *There is a polynomial-time algorithm that for a given instance (G, k) for the vertex cover problem, constructs an equivalent instance (G', k') such that $k' \leq k$ and $|G'| \leq 2k'$.*

A Faster Nemhauser-Trotter Construction

Theorem 3 suggests a polynomial-time kernelization algorithm for VERTEX COVER. The algorithm is involved in solving the linear programming problem (LP) and partitioning the graph vertices into the sets I_0, C_0, and V_0. Solving the linear programming problem (LP) can be done in polynomial time but is kind of costly in particular when the input graph G is dense. Alternatively, Nemhauser and Trotter [5] suggested the following algorithm without using linear programming. Let G be the input graph with vertex set $\{v_1, \ldots, v_n\}$.

1. construct a bipartite graph B with vertex set $\{v_1^L, \ldots, v_n^L, v_1^R, \ldots, v_n^R\}$ such that $[v_i^L, v_j^R]$ is an edge in B if and only if $[v_i, v_j]$ is an edge in G;
2. find a minimum vertex cover C_B for B;
3. $I_0' = \{v_i \mid \text{if neither } v_i^L \text{ nor } v_i^R \text{ is in } C_B\}$;
   $C_0' = \{v_i \mid \text{if both } v_i^L \text{ and } v_i^R \text{ are in } C_B\}$;
   $V_0' = \{v_i \mid \text{if exactly one of } v_i^L \text{ and } v_i^R \text{ is in } C_B\}$

It can be proved [5] (see also [2]) that Theorem 2 still holds true when the sets C_0 and V_0 in the theorem are replaced by the sets C_0' and V_0', respectively, constructed in the above algorithm.

The advantage of this approach is that the sets C_0' and V_0' can be constructed in time $O(m\sqrt{n})$ because the minimum vertex cover C_B for the bipartite graph B can be constructed via a maximum matching of B, which can be constructed in time $O(m\sqrt{n})$ using Dinic's maximum flow algorithm, which is in general faster than solving the linear programming problem (LP).

Crown Reduction

For a set S of vertices in a graph G, denote by $N(S)$ the set of vertices that are not in S but adjacent to some vertices in S. A *crown* in a graph G is a pair (I, H) of subsets of vertices in G satisfying the following conditions: (1) $I \neq \emptyset$ is an independent set, and $H = N(I)$; and (2) there is a matching M on the edges connecting I and H such that all vertices in H are matched in M.

It is quite easy to see that for a given crown (I, H), there is a minimum vertex cover that includes all vertices in H and excludes all vertices in I. Let G' be the graph obtained by removing all vertices in I and H from G. Then, the instances (G, k) and (G', k') are equivalent, where $k' = k - |H|$. Therefore, identification of crowns in a graph provides an effective way for kernelization.

Let G be the input graph. The following algorithm is proposed.

1. construct a maximal matching M_1 in G; let O be the set of vertices unmatched in M_1;
2. construct a maximum matching M_2 of the edges between O and $N(O)$; $i = 0$; let I_0 be the set of vertices in O that are unmatched in M_2;
3. repeat until $I_i = I_{i-1}$ $\{H_i = N(I_i);$ $I_{i+1} = I_i \cup N_{M_2}(H_i); i = i + 1; \}$; (where $N_{M_2}(H_i)$ is the set of vertices in O that match the vertices in H_i in the matching M_2)
4. $I = I_i$; $H = N(I_i)$; output (I, H).

Theorem 4 *(1) if the set I_0 is not empty, then the above algorithm constructs a crown (I, H); (2) if both $|M_1|$ and $|M_2|$ are bounded by k, and $I_0 = \emptyset$, then the graph G has at most $3k$ vertices.*

According to Theorem 4, the above algorithm on an instance (G, k) of VERTEX COVER either (1) finds a matching of size larger than k – which implies that there is no vertex cover of k vertices in the graph G; or (2) constructs a crown (I, H) – which will reduce the size of the instance; or (3) in case neither of (1) and (2) holds true, concludes that the graph G contains at most $3k$ vertices. Therefore, repeatedly applying the algorithm either derives a direct solution to the given instance, or constructs an equivalent instance (G', k') with $k' \leq k$ and $|G'| \leq 3k'$.

Applications

The research of the current paper was directly motivated by authors' research in bioinformatics. It is shown that for many computational biological problems, such as the construction of phylogenetic trees, phenotype identification, and analysis of microarray data, preprocessing based on the kernelization techniques has been very effective.

Experimental Results

Experimental results are given for handling graphs obtained from the study of phylogenetic trees based on protein domains, and from the analysis of microarray data. The results show that in most cases the best way to kernelize is to start handling vertices of high and low degrees (i.e., vertices of degree larger than k or smaller than 3) before attempting any of the other kernelization techniques. Sometimes, kernelization based on Nemhauser-Trotter Theorem can solve the problem without any further branching. It is also observed that sometimes particularly on dense graphs, kernelization techniques based on Nemhauser-Trotter Theorem are kind of time-consuming but do not reduce the instance size by much. On the other hand, the techniques based on high-degree vertices and crown reduction seem to work better.

V

Data Sets

The experiments were performed on graphs obtained based on data from NCBI and SWISS-PROT, well known open-source repositories of biological data.

Cross-References

▶ Data Reduction for Domination in Graphs
▶ Local Approximation of Covering and Packing Problems
▶ Vertex Cover Search Trees

Recommended Reading

1. Abu-Khzam F, Collins R, Fellows M, Langston M, Suters W, Symons C (2004) Kernelization algorithms for the vertex cover problem: theory and experiments. In: Proceedings of the workshop on algorithm engineering and experiments (ALENEX), pp 62–69
2. Bar-Yehuda R, Even S (1985) A local-ratio theorem for approximating the weighted vertex cover problem. Ann Discret Math 25:27–45
3. Chen J, Kanj IA, Jia W (2001) Vertex cover: further observations and further improvements. J Algorithm 41:280–301
4. Garey M, Johnson D (1979) Computers and intractability: a guide to the theory of NP-completeness. Freeman, San Francisco
5. Nemhauser GL, Trotter LE (1975) Vertex packing: structural properties and algorithms. Math Program 8:232–248

Vertex Cover Search Trees

Jianer Chen
Department of Computer Science, Texas A&M University, College Station, TX, USA

Keywords

Branch and bound; Branch and search

Years and Authors of Summarized Original Work

2001; Chen, Kanj, Jia

Problem Definition

The VERTEX COVER problem is one of the six "basic" NP-complete problems according to Garey and Johnson [7]. Therefore, the problem cannot be solved in polynomial time unless P = NP. However, the NP-completeness of the problem does not obviate the need for solving it because of its fundamental importance and wide applications.

One approach is to develop *parameterized algorithms* for the problem, with the computational complexity of the algorithms being measured in terms of both input size and a parameter value. This approach was initiated based on the observation that in many applications, the instances of the problem are associated with a small parameter. Therefore, by taking the advantages of the small parameters, one may be able to solve this NP-complete problem effectively and practically.

The problem is formally defined as follows. Let G be an (undirected) graph. A subset C of vertices in G is a *vertex cover* for G if every edge in G has at least one end in C. An instance of the (parameterized) VERTEX COVER problem consists of a pair (G, k), where G is a graph and k is an integer (the parameter), which is to determine whether the graph G has a vertex cover of k vertices. The goal is to develop parameterized algorithms of running time $O(f(k)p(n))$ for the VERTEX COVER problem, where $p(n)$ is a lower-degree polynomial of the input size n, and $f(k)$ is the non-polynomial part that is a function of the parameter k but independent of the input size n. It would be expected that the non-polynomial function $f(k)$ is as small as possible. Such an algorithm would become "practically effective" when the parameter value k is small. It should be pointed out that unless an unlikely consequence occurs in complexity theory, the function $f(k)$ is at least an exponential function of the parameter k [8].

Key Results

A number of techniques have been proposed in the development of parameterized algorithms for the VERTEX COVER problem.

Kernelization

Suppose (G, k) is an instance for the VERTEX COVER problem, where G is a graph and k is the parameter. The *kernelization* operation applies a polynomial time preprocessing on the instance (G, k) to construct another instance (G', k'), where G' is a smaller graph (the *kernel*) and $k' \leq k$, such that G' has a vertex cover of k' vertices if and only if G has a vertex cover of k vertices. Based on a classical result by Nemhauser and Trotter [9], the following kernelization result was derived.

Theorem 1 *There is an algorithm of running time* $O(kn + k^3)$ *that for a given instance (G, k) for the* VERTEX COVER *problem, constructs another instance (G', k') for the problem, where the graph G' contains at most $2k'$ vertices and $k' \leq k$, such that the graph G has a vertex cover of k vertices if and only if the graph G' has a vertex cover of k' vertices.*

Therefore, kernelization provides an efficient preprocessing for the VERTEX COVER problem, which allows one to concentrate on graphs of small size (i.e., graphs whose size is only related to k).

Folding

Suppose v is a degree-2 vertex in a graph G with two neighbors u and w such that u and w are not adjacent to each other. Construct a new graph G' as follows: remove the vertices v, u, and w and introduce a new vertex v_0 that is adjacent to all remaining neighbors of the vertices u and w in G. The graph G' is said being obtained from the graph G by *folding the vertex v*. The following result was derived.

Theorem 2 *Let G' be a graph obtained by folding a degree-2 vertex v in a graph G, where the two neighbors of v are not adjacent to each other. Then the graph G has a vertex cover of k vertices if and only if the graph G' has a vertex cover of $k - 1$ vertices.*

An folding operation allows one to decrease the value of the parameter k without branching.

Therefore, folding operations are regarded as very efficient in the development of exponential time algorithms for the VERTEX COVER problem. Recently, the folding operation has be generalized to apply to a set of more than one vertex in a graph [6].

Branch and Search

A main technique is the *branch and search* method that has been extensively used in the development of algorithms for the VERTEX COVER problem (and for many other NP-hard problems). The method can be described as follows. Let (G, k) be an instance of the VERTEX COVER problem. Suppose that somehow a collection $\{C_1, \ldots, C_b\}$ of vertex subsets in the graph G is identified, where for each i, the subset C_i has c_i vertices, such that if the graph G contains a vertex cover of k vertices, then at least for one C_i of the vertex subsets in the collection, there is a vertex cover of k vertices for G that contains all vertices in C_i. Then a collection of (smaller) instances (G_i, k_i) can be constructed, where $1 \leq i \leq b, k_i = k - c_i$, and G_i is obtained from G by removing all vertices in C_i. Note that the original graph G has a vertex cover of k vertices if and only if for one (G_i, k_i) of the smaller instances the graph G_i has a vertex cover of k_i vertices. Therefore, now the process can be branched into b sub-processes, each on a smaller instance (G_i, k_i) recursively searches for a vertex cover of k_i vertices in graph G_i.

Let $T(k)$ be the number of leaves in the search tree for the above branch and search process on the instance (G, k), then the above branch operation gives the following recurrence relation:

$$T(k) = T(k - c_1) + T(k - c_2) + \cdots + T(k - c_b)$$

To solve this recurrence relation, let $T(k) = x^k$ so that the above recurrence relation becomes

$$x^k = x^{k-c_1} + x^{k-c_2} + \cdots + x^{k-c_b}$$

It can be proved [3] that the above polynomial equation has a unique root x_0 larger than 1. From this, one gets $T(k) = x_0^k$, which, up to a polynomial factor, gives an upper bound on the running time of the branch and search process on the instance (G, k).

The simplest case is that a vertex v of degree $d > 0$ in the graph G is picked. Let $w_1, \ldots, w_d$ be the neighbors of v. Then either v is contained in a vertex cover C of k vertices, or, if v is not contained in C, then all neighbors $w_1, \ldots, w_d$ of v must be contained in C. Therefore, one obtains a collection of two subsets $C_1 = \{v\}$ and $C_2 = \{w_1, \ldots, w_d\}$, on which the branch and search process can be applied.

The efficiency of a branch and search operation depends on how effectively one can identify the collection of the vertex subsets. Intuitively, the larger the sizes of the vertex subsets, the more efficient is the operation. Much effort has been made in the development of VERTEX COVER algorithms to achieve larger vertex subsets. Improvements on the size of the vertex subsets have been involved with very complicated and tedious analysis and enumerations of combinatorial structures of graphs. The current paper [3] achieved a collection of two subsets C_1 and C_2 of sizes $c_1 = 1$ and $c_2 = 6$, respectively, and other collections of vertex subsets that are at least as good as this (the techniques of kernelization and vertex folding played important roles in achieving these collections). This gives the following algorithm for the VERTEX COVER problem.

Theorem 3 *The* VERTEX COVER *problem can be solved in time* $O(kn + 1.2852^k)$.

Very recently, a further improvement over Theorem 3 has been achieved that gives an algorithm of running time $O(kn + 1.2738^k)$ for the VERTEX COVER problem [4].

Applications

The study of parameterized algorithms for the VERTEX COVER problem was motivated by ETH Zürich's DARWIN project in computational biology and computational biochemistry (see, e.g., [10, 11]). A number of computational problems in the project, such as multiple sequence alignments [10] and biological conflict resolving [11], can be formulated into the VERTEX COVER problem in which the parameter value is in general not larger than 100. Therefore, an algorithm of running time $O(kn + 1.2852^k)$ for the problem becomes very effective and practical in solving these problems.

The parameterized algorithm given in Theorem 3 has also induced a faster algorithm for another important NP-hard problem, the MAXIMUM INDEPENDENT SET problem on sparse graphs [3].

Open Problems

The main open problem in this line of research is how far one can go along this direction. More specifically, how small the constant $c > 1$ can be for the VERTEX COVER problem to have an algorithm of running time $O(c^k n^{O(1)})$? With further more careful analysis on graph combinatorial structures, it seems possible to slightly improve the current best upper bound [4] for the problem. Some new techniques developed more recently [6] also seem very promising to improve the upper bound. On the other hand, it is known that the constant c cannot be arbitrarily close to 1 unless certain unlikely consequence occurs in complexity theory [8].

Experimental Results

A number of research groups have implemented some of the ideas of the algorithm in Theorem 3 or its variations, including the Parallel Bioinformatics project in Carleton University [2], the High Performance Computing project in University of Tennessee [1], and the DARWIN project in ETH Zürich [10, 11]. As reported in [5], these implementations showed that this algorithm and the related techniques are "quite practical" for the VERTEX COVER problem with parameter value k up to around 400.

Cross-References

► Data Reduction for Domination in Graphs
► Exact Algorithms for k SAT Based on Local Search
► Local Approximation of Covering and Packing Problems
► Vertex Cover Kernelization

Recommended Reading

1. Abu-Khzam F, Collins R, Fellows M, Langston M, Suters W, Symons C (2004) Kernelization algorithms for the vertex cover problem: theory and experiments. In: Proceedings of the workshop on algorithm engineering and experiments (NLENEX), pp 62–69
2. Cheetham J, Dehne F, Rau-Chaplin A, Stege U, Taillon P (2003) Solving large FPT problems on coarse grained parallel machines. J Comput Syst Sci 67:691–706
3. Chen J, Kanj IA, Jia W (2001) Vertex cover: further observations and further improvements. J Algorithm 41:280–301
4. Chen J, Kanj IA, Xia G (2006) Improved parameterized upper bounds for vertex cover. In: MFCS 2006. Lecture notes in computer science, vol 4162. Springer, Berlin, pp 238–249
5. Fellows M (2001) Parameterized complexity: the main ideas and some research frontiers. In: ISAAC 2001. Lecture notes in computer science, vol 2223. Springer, Berlin, pp 291–307
6. Fomin F, Grandoni F, Kratsch D (2006) Measure and conquer: a simple $O(2^{0.288n})$ independent set algorithm. In: Proceedings of the 17th annual ACM-SIAM symposium on discrete algorithms (SODA 2006), pp 18–25
7. Garey M, Johnson D (1979) Computers and intractability: a guide to the theory of NP-completeness. Freeman, San Francisco
8. Impagliazzo R, Paturi R (2001) Which problems have strongly exponential complexity? J Comput Syst Sci 63:512–530
9. Nemhauser GL, Trotter LE (1975) Vertex packing: structural properties and algorithms. Math Program 8:232–248
10. Roth-Korostensky C (2000) Algorithms for building multiple sequence alignments and evolutionary trees. PhD thesis, ETH Zürich, Institute of Scientific Computing
11. Stege U (2000) Resolving conflicts from problems in computational biology. PhD thesis, ETH Zürich, Institute of Scientific Computing

Visualization Techniques for Algorithm Engineering

Camil Demetrescu[1,2] and Giuseppe F. Italiano[1,2]
[1]Department of Computer and Systems Science, University of Rome, Rome, Italy
[2]Department of Information and Computer Systems, University of Rome, Rome, Italy

Keywords

Using visualization in the empirical assessment of algorithms

Years and Authors of Summarized Original Work

2002; Demetrescu, Finocchi, Italiano, Näher

Problem Definition

The whole process of designing, analyzing, implementing, tuning, debugging and experimentally evaluating algorithms can be referred to as *Algorithm Engineering*. Algorithm Engineering views algorithmics also as an engineering discipline rather than a purely mathematical discipline. Implementing algorithms and engineering algorithmic codes is a key step for the transfer of algorithmic technology, which often requires a high-level of expertise, to different and broader communities, and for its effective deployment in industry and real applications.

Experiments can help measure practical indicators, such as implementation constant factors, real-life bottlenecks, locality of references, cache effects and communication complexity, that may be extremely difficult to predict theoretically. Unfortunately, as in any empirical science, it may be sometimes difficult to draw general conclusions about algorithms from experiments. To this aim, some researchers have proposed accurate and comprehensive guidelines on different

V

aspects of the empirical evaluation of algorithms matured from their own experience in the field (see, for example [1, 15, 16, 20]). The interested reader may find in [18] an annotated bibliography of experimental algorithmics sources addressing methodology, tools and techniques.

The process of implementing, debugging, testing, engineering and experimentally analyzing algorithmic codes is a complex and delicate task, fraught with many difficulties and pitfalls. In this context, traditional low-level textual debuggers or industrial-strength development environments can be of little help for algorithm engineers, who are mainly interested in high-level algorithmic ideas rather than in the language and platform-dependent details of actual implementations. Algorithm visualization environments provide tools for abstracting irrelevant program details and for conveying into still or animated images the high-level algorithmic behavior of a piece of software.

Among the tools useful in algorithm engineering, visualization systems exploit interactive graphics to enhance the development, presentation, and understanding of computer programs [27]. Thanks to the capability of conveying a large amount of information in a compact form that is easily perceivable by a human observer, visualization systems can help developers gain insight about algorithms, test implementation weaknesses, and tune suitable heuristics for improving the practical performances of algorithmic codes. Some examples of this kind of usage are described in [12].

Key Results

Systems for algorithm visualization have matured significantly since the rise of modern computer graphic interfaces and dozens of algorithm visualization systems have been developed in the last two decades [2, 3, 4, 5, 6, 8, 9, 10, 13, 17, 25, 26, 29]. For a comprehensive survey the interested reader can be referred to [11, 27] and to the references therein. The remainder of this entry discusses the features of algorithm visualization systems that appear to be most appealing for their deployment in algorithm engineering.

Critical Issues

From the viewpoint of the algorithm developer, it is desirable to rely on systems that offer visualizations at a *high level of abstraction*. Namely, one would be more interested in visualizing the behavior of a complex data structure, such as a graph, than in obtaining a particular value of a given pointer.

Fast prototyping of visualizations is another fundamental issue: algorithm designers should be allowed to create visualization from the source code at hand with little effort and without heavy modifications. At this aim, *reusability* of visualization code could be of substantial help in speeding up the time required to produce a running animation.

One of the most important aspects of algorithm engineering is the development of *libraries*. It is thus quite natural to try to interface visualization tools to algorithmic software libraries: libraries should offer default visualizations of algorithms and data structures that can be refined and customized by developers for specific purposes.

Software visualization tools should be able to animate *not just "toy programs"*, but significantly complex algorithmic codes, and to test their behavior on large data sets. Unfortunately, even those systems well suited for large information spaces often lack advanced navigation techniques and methods to alleviate the screen bottleneck. Finding a solution to this kind of limitations is nowadays a challenge.

Advanced debuggers take little advantage of sophisticated graphical displays, even in commercial software development environments. Nevertheless, software visualization tools may be very beneficial in addressing problems such as finding memory leaks, understanding anomalous program behavior, and studying performance. In particular, environments that provide interpreted execution may more easily integrate advanced facilities in support to *debugging and performance monitoring*, and many recent systems attempt at exploring this research direction.

Techniques

One crucial aspect in visualizing the dynamic behavior of a running program is the way it is conveyed into graphic abstractions. There are two main approaches to bind visualizations to code: the event-driven and the state-mapping approach.

Event-Driven Visualization

A natural approach to algorithm animation consists of annotating the algorithmic code with calls to visualization routines. The first step consists of identifying the relevant actions performed by the algorithm that are interesting for visualization purposes. Such relevant actions are usually referred to as *interesting events*. As an example, in a sorting algorithm the swap of two items can be considered an interesting event. The second step consists of associating each interesting event with a modification of a graphical scene. Animation scenes can be specified by setting up suitable visualization procedures that drive the graphic system according to the actual parameters generated by the particular event. Alternatively, these visualization procedures may simply log the events in a file for a *post-mortem* visualization. The calls to the visualization routines are usually obtained by annotating the original algorithmic code at the points where the interesting events take place. This can be done either by hand or by means of specialized editors. Examples of toolkits based on the event-driven approach are Polka [28] and *GeoWin*, a C++ data type that can be easily interfaced with algorithmic software libraries of great importance in algorithm engineering such as CGAL [14] and LEDA [19].

State Mapping Visualization

Algorithm visualization systems based on state mapping rely on the assumption that observing how the variables change provides clues to the actions performed by the algorithm. The focus is on capturing and monitoring the data modifications rather than on processing the interesting events issued by the annotated algorithmic code. For this reason they are also referred to as "data driven" visualization systems. Conventional debuggers can be viewed as data driven systems, since they provide direct feedback of variable modifications. The main advantage of this approach over the event-driven technique is that a much greater ignorance of the code is allowed: indeed, only the interpretation of the variables has to be known to animate a program. On the other hand, focusing only on data modification may sometimes limit customization possibilities making it difficult to realize animations that would be natural to express with interesting events. Examples of tools based on the state mapping approach are Pavane [23, 25], which marked the first paradigm shift in algorithm visualization since the introduction of interesting events, and Leonardo [10] an integrated environment for developing, visualizing, and executing C programs.

A comprehensive discussion of other techniques used in algorithm visualization appears in [7, 21, 22, 24, 27].

Applications

There are several applications of visualization in algorithm engineering, such as testing and debugging of algorithm implementations, visual inspection of complex data structures, identification of performance bottlenecks, and code optimization. Some examples of uses of visualization in algorithm engineering are described in [12].

Open Problems

There are many challenges that the area of algorithm visualization is currently facing. First of all, the real power of an algorithm visualization system should be in the hands of the final user, possibly inexperienced, rather than of a professional programmer or of the developer of the tool. For instance, instructors may greatly benefit from fast and easy methods for tailoring animations to their specific educational needs, while they might be discouraged from using systems that are difficult to install or heavily dependent on particular software/hardware platforms. In addition to

V

being easy to use, a software visualization tool should be able to animate significantly complex algorithmic codes without requiring a lot of effort. This seems particularly important for future development of visual debuggers. Finally, visualizing the execution of algorithms on large data sets seems worthy of further investigation. Currently, even systems designed for large information spaces often lack advanced navigation techniques and methods to alleviate the screen bottleneck, such as changes of resolution and scale, selectivity, and elision of information.

Cross-References

▶ Experimental Methods for Algorithm Analysis

Recommended Reading

1. Anderson RJ (2002) The role of experiment in the theory of algorithms. In: Data structures, near neighbor searches, and methodology: fifth and sixth DIMACS implementation challenges. DIMACS series in discrete mathematics and theoretical computer science, vol 59. American Mathematical Society, Providence, pp 191–195
2. Baker J, Cruz I, Liotta G, Tamassia R (1996) Animating geometric algorithms over the web. In: Proceedings of the 12th annual ACM symposium on computational geometry. Philadelphia, 24–26 May 1996, pp C3–C4
3. Baker J, Cruz I, Liotta G, Tamassia R (1996) The Mocha algorithm animation system. In: Proceedings of the 1996 ACM workshop on advanced visual interfaces, Gubbio, 27–29 May 1996, pp 248–250
4. Baker J, Cruz I, Liotta G, Tamassia R (1996) A new model for algorithm animation over the WWW. ACM Comput Surv 27:568–572
5. Baker R, Boilen M, Goodrich M, Tamassia R, Stibel B (1999) Testers and visualizers for teaching data structures. In: Proceeding of the 13th SIGCSE technical symposium on computer science education, New Orleans, 24–28 Mar 1999, pp 261–265
6. Brown M (1988) Algorithm animation. MIT Press, Cambridge
7. Brown M (1988) Perspectives on algorithm animation. In: Proceedings of the ACM SIGCHI'88 conference on human factors in computing systems. Washington, DC, 15–19 May 1988, pp 33–38
8. Brown M (1991) Zeus: a system for algorithm animation and multi-view editing. In: Proceedings of the 7th IEEE workshop on visual languages, Kobe, 8–11 Oct 1991, pp 4–9
9. Cattaneo G, Ferraro U, Italiano GF, Scarano V (2002) Cooperative algorithm and data types animation over the net. J Vis Lang Comp 13(4):391
10. Crescenzi P, Demetrescu C, Finocchi I, Petreschi R (2008) Reversible execution and visualization of programs with LEONARDO. J Vis Lang Comp 11:125–150, Leonardo is available at: http://www.dis. uniroma1.it/~demetres/Leonardo/. Accessed 15 Jan 2008
11. Demetrescu C (2001) Fully dynamic algorithms for path problems on directed graphs. PhD thesis, Department of Computer and Systems Science, University of Rome "La Sapienza"
12. Demetrescu C, Finocchi I, Italiano GF, Näher S (2002) Visualization in algorithm engineering: tools and techniques. In: Experimental algorithm design to robust and efficient software. Lecture notes in computer science, vol 2547, Springer, Berlin, pp 24–50
13. Demetrescu C, Finocchi I, Liotta G (2000) Visualizing algorithms over the web with the publication-driven approach. In: Proceedings of the 4th workshop on algorithm engineering (WAE'00), Saarbrücken, 5–8 Sept 2000
14. Fabri A, Giezeman G, Kettner L, Schirra S, Schönherr S (1996) The cgal kernel: a basis for geometric computation. In: Applied computational geometry: towards geometric engineering proceedings (WACG'96), Philadelphia, 27–28 May 1996, pp 191–202
15. Goldberg A (1999) Selecting problems for algorithm evaluation. In: Proceedings of the 3rd workshop on algorithm engineering (WAE'99), London, 19–21 July 1999. Lecture notes in computer science, vol 1668, pp 1–11
16. Johnson D (2002) A theoretician's guide to the experimental analysis of algorithms. In: Data structures, near neighbor searches, and methodology: fifth and sixth DIMACS series in discrete mathematics and theoretical computer science, vol 59. American Mathematical Society, Providence, 215–250
17. Malony A, Reed D (1989) Visualizing parallel computer system performance. In: Simmons M, Koskela R, Bucher I (eds) Instrumentation for future parallel computing systems. ACM Press, New York, pp 59–90
18. McGeoch C (2002) A bibliography of algorithm experimentation. In: Data structures, near neighbor searches, and methodology: fifth and sixth DIMACS implementation challenges. DIMACS series in discrete mathematics and theoretical computer science, vol 59. American Mathematical Society, Providence, pp 251–254
19. Mehlhorn K, Naher S (1999) LEDA: a platform of combinatorial and geometric computing. Cambridge University Press, Cambridge. ISBN 0-521-56329-1
20. Moret B (2002) Towards a discipline of experimental algorithmics. In: Data structures, near neighbor searches, and methodology: fifth and sixth DIMACS implementation challenges. dimacs series in discrete mathematics and theoretical computer Science,

vol 59. American Mathematical Society, Providence, pp 197–214

21. Myers B (1990) Taxonomies of visual programming and program visualization. J Vis Lang Comp 1:97–123

22. Price B, Baecker R, Small I (1993) A principled taxonomy of software visualization. J Vis Lang Comp 4:211–266

23. Roman G, Cox K (1989) A declarative approach to visualizing concurrent computations. Computer 22:25–36

24. Roman G, Cox K (1993) A taxonomy of program visualization systems. Computer 26:11–24

25. Roman G, Cox K, Wilcox C, Plun J (1992) PAVANE: a system for declarative visualization of concurrent computations. J Vis Lang Comp 3:161–193

26. Stasko J (1992) Animating algorithms with X-TANGO. SIGACT News 23:67–71

27. Stasko J, Domingue J, Brown M, Price B (1997) Software visualization: programming as a multimedia experience. MIT Press, Cambridge

28. Stasko J, Kraemer E (1993) A methodology for building application- specific visualizations of parallel programs. J Parallel Distrib Comp 18:258–264

29. Tal A, Dobkin D (1995) Visualization of geometric algorithms. IEEE Trans Vis Comp Graph 1:194–204

Voltage Scheduling

Ming Min Li
Computer Science and Technology, Tsinghua University, Beijing, China

Keywords

Dynamic speed scaling

Years and Authors of Summarized Original Work

2005; Li, Yao

Problem Definition

This problem is concerned with scheduling jobs with as little energy as possible by adjusting the processor speed wisely. This problem is motivated by *dynamic voltage scaling (DVS)* (or *speed scaling*) technique, which enables a processor to operate at a range of voltages and frequencies. Since energy consumption is at least a quadratic function of the supply voltage (hence CPU frequency/speed), it saves energy to execute jobs as slowly as possible while still satisfying all timing constraints. The associated scheduling problem is referred to as min-energy DVS scheduling. Previous work showed that the min-energy DVS schedule can be computed in cubic time. The work of Li and Yao [7] considers the discrete model where the processor can only choose its speed from a finite speed set. This work designs an $O(dn \log n)$ two-phase algorithm to compute the min-energy DVS schedule for the discrete model (d represents the number of speeds) and also proves a lower bound of $\Omega(n \log n)$ for the computation complexity.

Notations and Definitions

In the variable voltage scheduling model, there are two important sets:

1. Set J (job set) consists of n jobs: $j_1, j_2, \ldots j_n$. Each job j_k has three parameters as its information: a_k representing the arrival time of j_k, b_k representing the deadline of j_k, and R_k representing the total CPU cycles required by j_k. The parameters satisfy $0 \leq a_k < b_k \leq 1$.

2. Set SD (speed set) consists of the possible speeds that can be used by the processor. According to the property of SD, the scheduling model is divided into the following two categories:

   Continuous model: The set SD is the set of positive real numbers.
   Discrete model: The set SD consists of d positive values: $s_1 > s_2 > \cdots > s_d$.

A schedule S consists of the following two functions: $s(t)$ which specifies the processor speed at time t and job(t) which specifies the job executed at time t. Both functions are piecewise constant with finitely many discontinuities.

A feasible schedule must give each job its required number of cycles between arrival time and deadline, therefore satisfying the property $\int_{a_k}^{b_k} s(t)\delta(k,\text{job}(t))dt = R_k$, where $\delta(i,j) = 1$ if $i = j$ and $\delta(i,j) = 0$ if otherwise.

The EDF principle defines an ordering on the jobs according to their deadlines. At any time t, among jobs j_k that are available for execution, that is, j_k satisfying $t \in [a_k, b_k)$ and j_k not yet finished by t, it is the job with minimum b_k that will be executed during $[t, t + \epsilon]$.

The power P, or energy consumed per unit of time, is a convex function of the processor speed. The energy consumption of a schedule $S = (s(t), \text{job}(t))$ is defined as $E(S) = \int_0^1 P(s(t))dt$.

A schedule is called an optimal schedule if its energy consumption is the minimum possible among all the feasible schedules. Note that for the continuous model, the optimal schedule uses the same speed for the same job.

The work of Li and Yao considers the problem of computing an optimal schedule for the discrete model under the following assumptions.

Assumptions

1. **Single processor:** At any time t, only one job can be executed.
2. **Preemptive:** Any job can be interrupted during its execution.
3. **Non-precedence:** There is no precedence relationship between any pair of jobs.
4. **Offline:** The processor knows the information of all the jobs at time 0.

This problem is called min-energy discrete dynamic voltage scaling (MEDDVS).

Problem 1 (MEDDVS$_{J,SD}$)

INPUT: Integer n, set $J = \{j_1, j_2, \ldots, j_n\}$ and $SD = \{s_1, s_2, \ldots, s_d\} \cdot j_k = \{a_k, b_k, R_k\}$.
OUTPUT: Feasible schedule $S = (s(t), \text{job}(t))$ that minimizes $E(S)$.

Kwon and Kim [6] proved that the optimal schedule for the discrete model can be obtained by first calculating the optimal schedule for the continuous model and then individually adjusting the speed of each job appropriately to adjacent levels in set SD. The time complexity is $O(n^3)$.

Key Results

The work of Li and Yao finds a direct approach for solving the MEDDVS problem without first computing the optimal schedule for the continuous model.

Definition 1 An s-schedule for J is a schedule which conforms to the EDF principle and uses constant speed s in executing any job of J.

Lemma 1 *The s-schedule for J can be computed in $O(n \log n)$ time.*

Definition 2 Given a job set J and any speed s, let $J^{\geq s}$ and $J^{<s}$ denote the subset of J consisting of jobs whose executing speeds are $\geq s$ and $<s$, respectively, in the optimal schedule for J in the continuous model. The partition $J^{\geq s}, J^{<s}$ is referred to as the s-partition of J.

By extracting information from the s-schedule, a partition algorithm is designed to prove the following lemma:

Lemma 2 *The s-partition of J can be computed in $O(n \log n)$ time.*

By applying s-partition to J using all the d speeds in SD consecutively, one can obtain d subsets $J_1, J_2, \ldots, J_d$ of J where jobs in the same subset J_i use the same two speeds s_i and s_{i+1} in the optimal schedule for the Discrete Model ($s_{d+1} = 0$).

Lemma 3 *Optimal schedule for job set J_i using speeds s_i and s_{i+1} can be computed in $O(n \log n)$ time.*

Combining the above three lemmas together, the main theorem follows:

Theorem 1 *The min-energy discrete DVS schedule can be computed in $O(dn \log n)$ time.*

A lower bound to compute the optimal schedule for the *discrete model* under the algebraic decision tree model is also shown by Li and Yao.

Theorem 2 *Any deterministic algorithm for computing min-energy discrete DVS schedule with $d \geq 2$ voltage levels requires $\Omega(n \log n)$ time for n jobs.*

Applications

Currently, dynamic voltage scaling technique is being used by the world's largest chip companies, e.g., Intel's SpeedStep technology and AMD's PowerNow technology. Although the scheduling algorithms being used are mostly online algorithms, offline algorithms can still find their places in real applications. Furthermore, the techniques developed in the work of Li and Yao for the computation of optimal schedules may have potential applications in other areas.

People also study energy-efficient scheduling problems for other kinds of job sets. Yun and Kim [10] proved that it is NP-hard to compute the optimal schedule for jobs with priorities and gave an FPTAS for that problem. Aydin et al. [1] considered energy-efficient scheduling for real-time periodic jobs and gave an $O(n^2 \log n)$ scheduling algorithm. Chen et al. [4] studied the weakly discrete model for non-preemptive jobs where speed is not allowed to change during the execution of one job. They proved the NP-hardness to compute the optimal schedule.

Another important application for this work is to help investigating scheduling model with more hardware restrictions (Burd and Brodersen [3] explained various design issues that may happen in dynamic voltage scaling). Besides the single-processor model, people are also interested in the multiprocessor model [11].

Open Problems

A number of problems related to the work of Li and Yao remain open. In the discrete model, Li and Yao's algorithm for computing the optimal schedule requires time $O(d\,n \log n)$. There is a gap between this and the currently known lower bound $\Omega(n \log n)$. Closing this gap when considering d as a variable is an open problem.

Another open research area is the computation of the optimal schedule for the continuous model. Li, Yao, and Yao [8] obtained an $O(n^2 \log n)$ algorithm for computing the optimal schedule. The bottleneck for the $\log n$ factor is in the computation of s-schedules. Reducing the time complexity for computing s-schedules is an open problem. It is also possible to look for other methods to deal with the continuous model.

Cross-References

▸ List Scheduling
▸ Online Load Balancing of Temporary Tasks
▸ Parallel Algorithms for Two Processors Precedence Constraint Scheduling
▸ Shortest Elapsed Time First Scheduling

Recommended Reading

1. Aydin H, Melhem R, Mosse D, Alvarez PM (2001) Determining optimal processor speeds for periodic real-time tasks with different power characteristics. In: Euromicro conference on real-time systems, Madrid. IEEE Computer Society, Washington, DC, pp 225–232
2. Bansalm N, Kimbrel T, Pruhs K (2004) Dynamic speed scaling to manage energy and temperature. In: Proceedings of the 45th annual IEEE symposium on foundations of computer science, Rome. IEEE Computer Society, Washington, DC, pp 520–529
3. Burd TD, Brodersen RW (2000) Design issues for dynamic voltage scaling. In: Proceedings of the 2000 international symposium on low power electronics and design, Rapallo. ACM, New York, pp 9–14
4. Chen JJ, Kuo TW, Lu HI (2005) Power-saving scheduling for weakly dynamic voltage scaling devices workshop on algorithms and data structures (WADS). LNCS, vol 3608. Springer, Berlin, pp 338–349
5. Irani S, Pruhs K (2005) Algorithmic problems in power management. ACM SIGACT News 36(2):63–76. New York

V

6. Kwon W, Kim T (2005) Optimal voltage allocation techniques for dynamically variable voltage processors. ACM Trans Embed Comput Syst 4(1):211–230. New York
7. Li M, Yao FF (2005) An efficient algorithm for computing optimal discrete voltage schedules. SIAM J Comput 35(3):658–671. Society for Industrial and Applied Mathematics, Philadelphia
8. Li M, Yao AC, Yao FF (2005) Discrete and continuous min-energy schedules for variable voltage processors. In: Proceedings of the National Academy of Sciences USA, Washington, DC, vol 103. National Academy of Science of the United States of America, Washington, DC, pp 3983–3987
9. Yao F, Demers A, Shenker S (1995) A scheduling model for reduced CPU energy. In: Proceedings of the 36th annual IEEE symposium on foundations of computer science, Milwaukee. IEEE Computer Society, Washington, DC, pp 374–382
10. Yun HS, Kim J (2003) On energy-optimal voltage scheduling for fixed-priority hard real-time systems. ACM Trans Embed Comput Syst 2:393–430. ACM, New York
11. Zhu D, Melhem R, Childers B (2001) Scheduling with dynamic voltage/speed adjustment using slack reclamation in multi-processor real-time systems. In: Proceedings of the 22nd IEEE real-time systems symposium (RTSS'01), London. IEEE Computer Society, Washington, DC, pp 84–94

Voronoi Diagrams and Delaunay Triangulations

Rolf Klein
Institute for Computer Science, University of Bonn, Bonn, Germany

Keywords

Computational geometry; Delaunay triangulation; Distance problem; Edge flip; Minimum spanning tree; Sweepline; Voronoi diagram

Years and Authors of Summarized Original Work

1975; Shamos, Hoey
1987; Fortune

Problem Definition

Suppose there is some set of objects p called *sites* that exert influence over their surrounding space, M. For each site p, we consider the set of all points z in M for which the influence of p is strongest.

Such decompositions have already been considered by R. Descartes [5] for the fixed stars in solar space. In mathematics and computer science, they are called *Voronoi diagrams*, honoring work by G.F. Voronoi on quadratic forms. Other sciences know them as *domains of action*, *Johnson-Mehl model*, *Thiessen polygons*, *Wigner-Seitz zones*, or *medial axis transform*.

In the case most frequently studied, the space M is the real plane, the sites are n points, and influence corresponds to proximity in the Euclidean metric, so that the points most strongly influenced by site p are those for which p is the nearest neighbor among all sites. They form a convex region called the *Voronoi region* of p. The common boundary of two adjacent regions of p and q is a segment of their *bisector* $B(p, q)$, the locus of all points of equal distance to p and q. An example of 10 point sites is depicted in Fig. 1.

Let us assume that the set S of point sites is in general position, so that no three points are situated on a line, and no four on a circle. Then the Voronoi diagram $V(S)$ of S is a connected planar graph. Its vertices are those points in the plane which have three nearest neighbors in S, while the interior edge points have two. As a consequence of the Euler formula, $V(S)$ has only $O(n)$ many edges and vertices.

If we connect with line segments, those sites in S whose Voronoi regions share an edge in $V(S)$, a triangulation $D(S)$ of S results, called the *Delaunay triangulation* or *Dirichlet tessellation*; see Fig. 1. Each triangle with vertices p, q, r in S is dual to a vertex v of $V(S)$ situated on the boundary of the Voronoi regions of p, q, and r. Because p, q, r are the nearest neighbors of v in S, the circle through p, q, r centered at v contains no other point of S. Thus, $D(S)$ consists of triangles with vertices in S whose circumcircles are empty of points in S; see Fig. 2. Conversely,

Voronoi Diagrams and Delaunay Triangulations, Fig. 1
Voronoi diagram and Delaunay triangulation of 10 point sites in the Euclidean plane

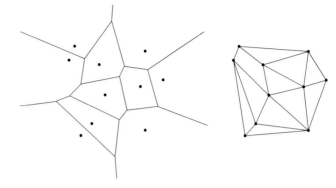

Voronoi Diagrams and Delaunay Triangulations, Fig. 2
The empty circle property

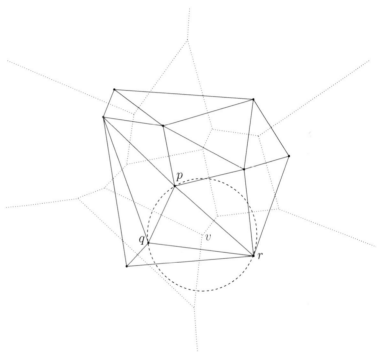

each triangle with empty circumcircle occurs in $D(S)$.

Given a set S of n point sites, the problem is to efficiently construct one of $V(S)$ or $D(S)$; the dual structure can then easily be obtained in linear time.

Generalizations

Voronoi diagrams can be generalized in several ways. Instead of point sites, other geometric objects can be considered. One can replace the Euclidean distance with distance measures more suitable to model a given situation. Instead of forming regions of all points that have the

same nearest site, one can consider higher-order Voronoi diagrams where all points share a region for which the nearest k sites are the same, for some k between 2 and $n - 1$. Many more variants can be found in [9] and [1]. Abstract Voronoi diagrams provide a unifying framework for some of the variants mentioned; see the corresponding chapter in this encyclopedia.

Key Results

Quite a few algorithms for constructing the Voronoi diagram or the Delaunay triangulation

**Voronoi Diagrams and
Delaunay
Triangulations, Fig. 3**
The sweepline advancing
to the right

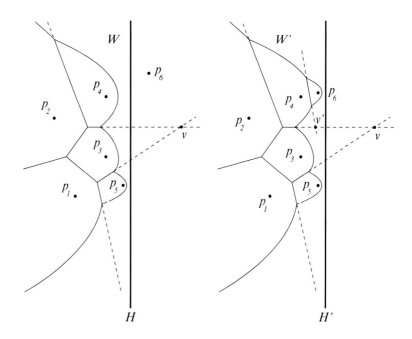

of n points in the Euclidean plane have been developed.

Divide and Conquer

The first algorithm was presented in the seminal paper [11], which gave birth to the field of computational geometry. It applies the *divide and conquer* paradigm. Site set S is split by a line into subsets L and R of equal cardinality. After recursively computing $V(L)$ and $V(R)$, one needs to compute the bisector $B(L, R)$, the locus of all points in the plane that have a nearest neighbor in L and in R. This bisector is an unbounded monotone polygonal chain. In time $O(n)$ one can find a starting segment of $B(L, R)$ at infinity, and trace the chain through $V(L)$ and $V(R)$ simultaneously. Thus, the algorithm runs in time $O(n \log n)$ and linear space, which is optimal.

Sweep

How to design a left-to-right *sweepline* algorithm for constructing $V(S)$ is not obvious. When the advancing sweepline H enters the Voronoi region of p before site p has been detected, it is not clear how to correctly maintain the Voronoi diagram along H. This difficulty has been overcome in [7]

by applying a transformation that ensures that each site is the leftmost point of its Voronoi region. In [10] and [4], a more direct version of this approach was suggested. At each time during the sweep, one maintains the Voronoi diagram of all point sites to the left of sweepline H, and of H itself, which is considered a site of its own; see Fig. 3. Because the bisector of a point and a line is a parabola, the Voronoi region of H is bounded by a connected chain of parabolic segments, called the *wavefront W*. As H moves to the right, W follows at half the speed. Each point z to the left of W is closer to some point site p left of H than to H and, all the more, to all point sites to the right of H that are yet to be discovered. Thus, the Voronoi regions of the point sites to the left of W keep growing, as sweepline H proceeds, along the extensions of Voronoi edges beyond W; these *spikes* are depicted by dashed lines in Fig. 3.

There are two kinds of events one needs to handle during the sweep. When sweepline H hits a new point site (like point p_6 in Fig. 3), a new wave separating this point site from H must be added to W. When wavefront W hits the intersection of two neighboring spikes, the wave between them must be removed from W;

this will first happen in Fig. 3 when W' arrives at v'. Intersections between neighboring spikes can be determined as in the standard line segment intersection algorithm [2]. There are only $O(n)$ many events, one for each point site and one for each Voronoi vertex v of $V(S)$. Since wavefront W is always of linear size, the sweepline algorithm runs in $O(n \log n)$ time using linear space.

Reduction to Convex Hull

A rather different approach [3] obtains the Delaunay triangulation in dimension 2 from the convex hull in dimension 3, which can itself be constructed in time $O(n \log n)$. As suggested in [6], one vertically lifts the point sites to the paraboloid $Z = X^2 + Y^2$ in 3-space. The lower convex hull of the lifted points, projected onto the XY-plane, equals the Delaunay triangulation, $D(S)$.

Incremental Construction

Another, very intuitive algorithm first suggested in [8] constructs the Delaunay triangulation incrementally. In order to insert a new point site p_i into an existing Delaunay triangulation $D(S_{i-1})$, one first finds the triangle containing p_i and connects p_i to its vertices by line segments. Should p_i be contained in the circumcircles of adjacent triangles, the Delaunay property must be restored by *edge flips* that replace the common edge of two adjacent triangles T, T' by the other diagonal of the convex quadrilateral formed by T and T'. If the insertion sequence of the p_i is randomly chosen, a running time in $O(n \log n)$ can be expected. Details on all algorithms can be found in [1].

By definition, the Voronoi diagram reduces the *post office* or *nearest neighbor* problem to a point location problem: given an arbitrary query point z, the site in S nearest to z can be found by determining the Voronoi region containing z. In order to find the *largest empty circle* whose center z lies inside a convex polygon C over m vertices, one needs to inspect only three types of candidates for z, the vertices of $V(S)$, the intersections of the edges of $V(S)$ with the boundary of C, and the vertices of C. All these can be done in time $O(n + m)$.

If the site set S is split into subsets L and R, then the closest pair $p \in L$ and $q \in R$ forms an edge of the Delaunay triangulation $D(S)$ (which crosses the Voronoi edge separating the regions of p and q). This fact has nice consequences. First, the nearest neighbor of a site $p \in S$ must be one of its neighboring vertices in $D(S)$. Hence, *all nearest neighbors* and the *closest pair* in S can be found in linear time once $D(S)$ is available, because $D(S)$ has only $O(n)$ many edges. Second, $D(S)$ contains the *minimum spanning tree* of S, which can be extracted from $D(S)$ in linear time.

Remarkable and useful is the *equiangularity property* of $D(S)$. Of all (exponentially many) triangulations of S, the Delaunay triangulation maximizes the ascending sequence of angles occurring in the triangles, with respect to lexicographic order. In particular, the minimum angle is as large as possible. In fact, if a triangulation is not Delaunay, it must contain two adjacent triangles, such that the circumcircle of one contains the third vertex of the other. By flipping their common edge, a new triangulation with larger angles is obtained.

Applications

Although of linear size, Voronoi diagram and Delaunay triangulation contain a lot of information on the point set S. Once $V(S)$ or $D(S)$ are available, quite a few distance problems can be solved very efficiently. We mention only the most basic applications here and refer to [1] and [9] for further reading.

Cross-References

- ▶ 3D Conforming Delaunay Triangulation
- ▶ Abstract Voronoi Diagrams
- ▶ Delaunay Triangulation and Randomized Constructions
- ▶ Optimal Triangulation

Recommended Reading

1. Aurenhammer F, Klein R, Lee DT (2013) Voronoi diagrams and delaunay triangulations. World Scientific, Singapore
2. Bentley J, Ottman T (1979) Algorithms for reporting and counting geometric intersections. IEEE Trans Comput C-28:643–647
3. Brown KQ (1979) Voronoi diagrams from Convex Hulls. Inf Process Lett 9:223–228
4. Cole R (1989) as reported by ÓDúnlaing, oral communication
5. Descartes R (1644) Principia philosophiae. Ludovicus Elsevirius, Amsterdam
6. Edelsbrunner H, Seidel R (1986) Voronoi diagrams and arrangements. Discret Comput Geom 1:25–44
7. Fortune S (1978) A sweepline algorithm for Voronoi diagrams. Algorithmica 2:153–174
8. Green PJ, Sibson RR (1978) Computing dirichlet tessellations in the plane. Comput J 21:168–173
9. Okabe A, Boots B, Sugihara K, Chiu SN (2000) Spatial tessellations: concepts and applications of Voronoi diagrams. Wiley, Chichester
10. Seidel R (1988) Constrained delaunay triangulations and Voronoi diagrams with obstacles. Technical report 260, TU Graz
11. Shamos MI, Hoey D (1975) Closest-point problems. In: Proceedings 16th annual IEEE symposium on foundations of computer science, Berkeley, pp 151–162

W

Wait-Free Synchronization

Mark Moir
Sun Microsystems Laboratories, Burlington,
MA, USA

Years and Authors of Summarized Original Work

1991; Herlihy

Problem Definition

The traditional use of locking to maintain consistency of shared data in concurrent programs has a number of disadvantages related to software engineering, robustness, performance, and scalability. As a result, a great deal of research effort has gone into *nonblocking* synchronization mechanisms over the last few decades.

Herlihy's seminal paper *Wait-Free Synchronization* [12] studied the problem of implementing concurrent data structures in a *wait-free manner*, i.e., so that every operation on the data structure completes in a finite number of steps by the invoking thread, regardless of how fast or slow other threads run and even if some or all of them halt permanently. Implementations based on locks are not wait-free because, while one thread holds a lock, others can take an unbounded number of steps waiting to acquire the lock. Thus, by requiring implementations to be wait-free, some of the disadvantages of locks may potentially be eliminated.

The first part of Herlihy's paper examined the power of different synchronization primitives for wait-free computation. He defined the *consensus number* of a given primitive as the maximum number of threads for which we can solve wait-free consensus using that primitive (together with read-write registers). The consensus problem requires participating threads to *agree* on a value (e.g., true or false) amongst values proposed by the threads. The ability to solve this problem is a key indicator of the power of synchronization primitives because it is central to many natural problems in concurrent computing. For example, in a software transactional memory system, threads must agree that a particular transaction either committed or aborted.

Herlihy established a *hierarchy* of synchronization primitives according to their consensus number. He showed (i) that the consensus number of read-write registers is 1 (so wait-free consensus cannot be solved for even two threads), (ii) that the consensus number of stacks and FIFO queues is 2, and (iii) that there are so-called *universal* primitives, which have consensus number ∞. Common examples include `compare-and-swap` (CAS) and the `load-linked`/`store-conditional` (LL/SC) pair.

There are a number of papers which examine Herlihy's hierarchy in more detail. These show that seemingly minor variations in the model or in the semantics of primitives can have a surprising effect on results. Most of this work is primarily of theoretical interest. The key practical point to take away from Herlihy's hierarchy is that we need universal primitives to support effective wait-free synchronization in general. Recognizing this fact, all modern shared-memory multiprocessors provide some form of universal primitive.

Herlihy additionally showed that a solution to consensus can be used to implement any shared object in a wait-free manner, and thus that any universal primitive suffices for this purpose. He demonstrated this idea using a so-called *universal construction*, which takes sequential code for an object and creates a wait-free implementation of the object using consensus to resolve races between concurrent operations. Despite the important practical ramifications of this result, the universal construction itself was quite impractical. The basic idea was to build a list of operations, using consensus to determine the order of operations, and to allow threads to iterate over the list applying the operations in order to determine the current state of the object. The construction required $O(N^3)$ space to ensure enough operations are retained to allow the current state to be determined. It was also very slow, requiring many threads to recompute the same information, and thus preventing parallelism between operations in addition.

Later, Herlihy [13] presented a more concrete universal construction based on the LL/SC instruction pair. This construction required $N + 1$ copies of the object for N threads and still did not admit any parallelism; thus it was also not practical. Despite this, work following on from Herlihy's has brought us to the point today that we can support practical programming models that provide nonblocking implementations of arbitrary shared objects. The remainder of this chapter discusses the state of nonblocking synchronization today, and mentions some history along the way.

Weaker Nonblocking Progress Conditions

Various researchers, including us, have had some success attempting to overcome the disadvantages of Herlihy's wait-free constructions. However, the results remain impractical due to excessive overhead and overly complicated algorithms. In fact, there are still no nontrivial wait-free shared objects in widespread practical use, either implemented directly or using universal constructions.

The biggest advances towards practicality have come from considering weaker progress conditions. While theoreticians worked on wait-free implementations, more pragmatic researchers sought lock-free implementations of shared objects. A *lock-free* implementation guarantees that, after a finite number of steps of any operation, *some* operation completes. In contrast to wait-free algorithms, it is in principle possible for one operation of a lock-free data structure to be continually starved by others. However, this rarely occurs in practice, especially because contention control techniques such as exponential backoff [1] are often used to reduce contention when it occurs, which makes repeated interference even more unlikely. Thus, the lack of a strong progress guarantee like wait-freedom has often been found to be acceptable in practice.

The observation that weaker nonblocking progress conditions allow simpler and more practical algorithms led Herlihy et al. [15] to define an even weaker condition: An *obstruction-free* algorithm does not guarantee that an operation completes unless it eventually encounters no more interference from other operations. In our experience, obstruction-free algorithms are easier to design, simpler, and faster in the common uncontended case than lock-free algorithms. The price paid for these benefits is that obstruction-free algorithms can "livelock", with two or more operations repeatedly interfering with each other forever. This is not merely a theoretical concern: it has been observed to occur in practice [16]. Fortunately, it is usually straightforward to eliminate livelock in practice through contention control mechanisms that control and manipulate when operations are executed to avoid repeated interference.

The obstruction-free approach to synchronization is thus to design simple and efficient algorithms for the weak obstruction-free progress condition, and to integrate orthogonal contention control mechanisms to facilitate progress when necessary. By largely separating the difficult issues of correctness and progress, we significantly ease the task of designing effective nonblocking implementations: the algorithms are not complicated by tightly coupled mechanisms for achieving lock-freedom, and it is easy to modify and experiment with contention control mechanisms because they are separate from the algorithm and do not affect its correctness. We have found this approach to be very powerful.

Transactional Memory

The severe difficulty of designing and verifying correct nonblocking data structures has led researchers to investigate the use of tools to produce them, rather than designing them directly. In particular, *transactional memory* [5, 17, 23] has emerged as a promising direction. Transactional memory allows programmers to express sections of code that should be executed atomically, and the transactional memory system (implemented in hardware, software, or a combination of the two) is responsible for managing interactions between concurrent transactions to ensure this atomicity. Here we concentrate on software transactional memory (STM).

The progress guarantee made by a concurrent data structure implemented using STM depends on the STM implementation. It is possible to characterize the progress conditions of transactional memory implementations in terms of a system of threads in which each operation on a shared data structure is executed by repeatedly attempting to apply it using a transaction until an attempt successfully commits. In this context, say the transactional memory implementation is obstruction-free if it guarantees that, if a thread repeatedly executes transactions and eventually encounters no more interference from other threads, then it eventually successfully commits a transaction.

Key Results

This section briefly discusses some of the most relevant results concerning nonblocking synchronization, and obstruction-free synchronization in particular.

While progress towards practicality was made with lock-free implementations of shared objects as well as lock-free STM systems, this progress was slow because simultaneously ensuring correctness and lock-freedom proved difficult. Before the introduction of obstruction-freedom, the lock-free STMs still had some severe disadvantages such as the need to declare and initialize all memory to be accessed by transactions in advance, the need for transactions to know in advance which memory locations they will access, unacceptable constraints on the layout of such memory, etc.

In addition to the work on tools such as STM for building nonblocking data structures, there has been a considerable amount of work on direct implementations. While this work has not yielded any practical wait-free algorithms, a handful of practical lock-free implementations for simple data structures such as queues and stacks have been achieved [21, 24]. There are also a few slightly more ambitious implementations in the literature that are arguably practical, but the algorithms are complicated and subtle, many are incorrect, and almost none has a formal proof. Proofs for such algorithms are challenging, and minor changes to the algorithm require the proofs to be redone.

The next section, discusses some of the results that have been achieved by applying the obstruction-free approach. The remainder of this section, briefly discusses a few results related to the approach itself.

An important practical aspect of using an obstruction-free algorithm is how contention is managed when it arises. In introducing obstruction-freedom, Herlihy et al. [15] explained that contention control is necessary to facilitate progress in the face of contention because obstruction-free algorithms do not directly make any progress guarantee in this case. However,

W

they did not directly address *how* contention control mechanisms could be used in practice.

Subsequently, Herlihy et al. [16] presented a dynamic STM system (see next section) that provides an interface for a modular contention manager, allowing for experimentation with alternative contention managers. Scherer and Scott [22] experimented with a number of alternatives, and found that the best contention manager depends on the workload. Guerraoui et al. [9] described an implementation that supports changing contention managers on the fly in response to changing workload conditions.

All of the contention managers discussed in the above-mentioned papers are ad hoc contention managers based on intuition; no analysis is given of what guarantees (if any) are made by the contention managers. Guerraoui et al. [10] made a first step towards a formal analysis of contention managers by showing that their `Greedy` contention manager guarantees that every transaction eventually completes. However, using the `Greedy` contention manager results in a blocking algorithm, so their proof necessarily assumes that threads do not fail while executing transactions.

Fich et al. [7] showed that any obstruction-free algorithm can be automatically transformed into one that is *practically* wait-free in any real system. "Practically" is said because the wait-free progress guarantee depends on partial synchrony that exists in any real system, but the transformed algorithm is not technically wait-free, because this term is defined in the context of a fully asynchronous system. Nonetheless, an algorithm achieved by applying the transformation of Fich et al. to an obstruction-free algorithm does guarantee progress to non-failed transactions, even if other transactions fail.

Work on incorporating contention management techniques into obstruction-free algorithms has mostly been done in the context of STM, so the contention manager can be called directly from the STM implementation. Thus, the programmer using the STM need not be concerned with how contention management is integrated, but this does not address how contention management is integrated into direct implementations of obstruction-free data structures.

One option is for the programmer to manually insert calls to a contention manager, but this approach is tedious and error prone. Guerraoui et al. [11] suggested a version of this approach in which the contention manager is abstracted out as a failure detector. They also explored what progress guarantees can be made by what failure detectors.

Attiya et al. [4] and Aguilera et al. [2] suggested changing the semantics of the data structure's operations so that they can return a special value in case of contention, thus allowing contention management to be done outside the data structure implementation. These approaches still leave a burden on the programmer to ensure that these special values are always returned by an operation that cannot complete due to contention, and that the correct special value is returned according to the prescribed semantics.

Another option is to use system support to ensure that contention management calls are made frequently enough to ensure progress. This support could be in the form of compiled-in calls, runtime support, signals sent upon expiration of a timer, etc. But all of these approaches have disadvantages such as not being applicable in general purpose environments, not being portable, etc.

Given that it remains challenging to design and verify direct obstruction-free implementations of shared data structures, and that there are disadvantages to the various proposals for integrating contention control mechanisms into them, using tools such as STMs with built-in contention management interfaces is the most convenient way to build nonblocking data structures.

Applications

The obstruction-free approach to nonblocking synchronization was introduced by Herlihy et al. [15], who used it to design a double-ended queue (deque) based on the widely available CAS instruction. All previous nonblocking deques either require exotic synchronization instructions such as `double-compare-and-swap` (`DCAS`), or have the disadvantage that operations at opposite ends of the queue always interfere with each other.

Herlihy et al. [16] introduced Dynamic STM (DSTM), the first STM that is dynamic in the following two senses: new objects can be allocated on the fly and subsequently accessed by transactions, and transactions do not need to know in advance what objects will be accessed. These two advantages made DSTM much more useful than previous STMs for programming dynamic data structures. As a result, nonblocking implementations of sophisticated shared data structures such as balanced search trees, skip lists, dynamic hash tables, etc. were suddenly possible.

The obstruction-free approach played a key role in the development of both of the results mentioned above: Herlihy et al. [16] could concentrate on the functionality and correctness of DSTM without worrying about how to achieve stronger progress guarantees such as lock-freedom.

The introduction of DSTM and of the obstruction-free approach have led to numerous improvements and variations by a number of research groups, and most of these have similarly followed the obstruction-free approach. However, Harris and Fraser [8] presented a dynamic STM called OSTM with similar advantages to DSTM, but it is lock-free. Experiments conducted at the University of Rochester [20] showed that DSTM outperformed OSTM by an order of magnitude on some workloads, but that OSTM outperformed DSTM by a factor of 2 on others. These differences are probably due to various design decisions that are (mostly) orthogonal to the progress condition, so it is not clear what we can conclude about how the choice of progress condition affects performance in this case.

Perhaps a more direct comparison can be made between another pair of algorithms, again an obstruction-free one by Herlihy et al. [14] and a similar but lock-free one by Harris and Fraser [8]. These algorithms, invented independently of each other, implement MCAS (CAS generalized to access M independently chosen

memory locations). The two algorithms are very similar, and a close comparison revealed that the only real differences between them were due to Harris and Fraser's desire to have a lock-free implementation. As a result of this, their algorithm is somewhat more complicated, and also requires a minimum of $3M + 1$ CAS operations, whereas the algorithm of Herlihy et al. [14] requires only $2M + 1$. The authors are unaware of any direct performance comparison of these algorithms, but they believe the obstruction-free one would outperform the lock-free one, particularly in the absence of conflicting MCAS operations.

Open Questions

Because transactional memory research has grown out of research into nonblocking data structures, it was long considered mandatory for STM implementations to support the development of nonblocking data structures. Recently, however, a number of researchers have observed that at least the software engineering benefits of transactional memory can be delivered even by a blocking STM. There are ongoing debates whether STM needs to be nonblocking and whether there is a fundamental cost to being nonblocking.

While we agree that blocking STMs are considerably easier to design, and that in many cases a blocking STM is acceptable, this is not always true. Consider, for example, an interrupt handler that shares data with the interrupted thread. The interrupted thread will not run again until the interrupt handler completes, so it is critical that the interrupted thread does not block the interrupt handler. Thus, if using STM is desired to simplify the code for accessing this shared data, the STM *must* be nonblocking. The authors are therefore motivated to continue research aimed at improving nonblocking STMs and to understand what fundamental gap, if any, exists between blocking and nonblocking STMs.

Progress in improving the common-case performance of nonblocking STMs continues [19], and the authors see no reason to believe that nonblocking STMs should not be very competitive with blocking STMs in the common case, i.e., until the system decides that one transaction should not wait for another that is delayed (an option that is not available with blocking STMs).

It is conjectured that indeed a separation between blocking and nonblocking STMs can be proved according to some measure, but that this will not imply significant performance differences in the common case. Indeed results of Attiya et al. [3] show a separation between obstruction-free and blocking algorithms according to a measure that counts the number of distinct base objects accessed by the implementation plus the number of "memory stalls", which measure how often the implementation can encounter contention for a variable from another thread. While this result is interesting, it is not clear that it is useful for deciding whether to implement blocking or obstruction-free objects, because the measure does not account for the time spent waiting by blocking implementations, and thus is biased in their favor. For now, remain optimistic that STMs can be made to be nonblocking without paying a severe performance price in the common case.

Another interesting question, which is open as far as the authors know, is whether there is a fundamental cost to implementing stronger nonblocking progress conditions versus obstruction-freedom. Again, they conjecture that there is. It is known that there is a fundamental difference between obstruction-freedom and lock-freedom in systems that support only reads and writes: It is possible to solve obstruction-free consensus but not lock-free consensus in this model [15]. While this is a fascinating observation, it is mostly irrelevant from a practical standpoint as all modern shared memory multiprocessors support stronger synchronization primitives such as CAS, with which it is easy to solve consensus, even wait-free. The interesting question therefore is whether there is a fundamental cost to being lock-free as opposed to obstruction-free in real systems.

To have a real impact on design directions, such results need to address common case per-

formance, or some other measure (perhaps space) that is relevant to everyday use. Many lower bound results establish a separation in worst-case time complexity, which does not necessarily have a direct impact on design decisions, because the worst case may be very rare. So far, efforts to establish a separation according to potentially useful measures have only led to stronger results than we had conjectured were possible. In the authors first attempt [18], they tried to establish a separation in the number of CAS instructions needed in the absence of contention to solve consensus, but found that this was not a very useful measure, as were able to come up with a wait-free implementation that avoids CAS in the absence of contention. The second attempt [6] was to establish a separation according to the *obstruction-free step complexity* measure, which counts the maximum number of steps to complete an operation once the operation encounters no more contention. They knew we could implement obstruction-free DCAS with constant obstruction-free step complexity, and attempt to prove this impossible for lock-free DCAS, but achieved such an algorithm. These experiences suggest that, in addition to their direct advantages, obstruction-free algorithms may provide a useful stepping stone to algorithms with stronger progress properties.

Finally, while a number of contention managers have proved effective for various workloads, it is an open question whether a single contention manager can adapt to be competitive with the best on all workloads, and how close it can come to making optimal contention management decisions. Experience to date suggests that this will be very challenging to achieve. Therefore, as in any system, the first priority should be avoiding contention in the first place. Fortunately, transactional memory has the potential to make this much easier than in lock-based programming models, because it offers the benefits of fine-grained synchronization without the programming complexity that accompanies fine-grained locking schemes.

Cross-References

▶ Concurrent Programming, Mutual Exclusion
▶ Linearizability

Recommended Reading

1. Agarwal A, Cherian M (1989) Adaptive back-off synchronization techniques. In: Proceedings of the 16th annual international symposium on computer architecture. ACM Press, New York, pp 396–406
2. Aguilera MK, Frolund S, Hadzilacos V, Horn SL, Toueg S (2006) Brief announcement: abortable and query-abortable objects. In: Proceedings of the 20th annual international symposium on distributed computing
3. Attiya H, Guerraoui R, Hendler D, Kouznetsov P (2006) Synchronizing without locks is inherently expensive. In: PODC'06: proceedings of the twenty-fifth annual ACM symposium on principles of distributed computing. ACM Press, New York, pp 300–307
4. Attiya H, Guerraoui R, Kouznetsov P (2005) Computing with reads and writes in the absence of step contention. In: Proceedings of the 19th annual international symposium on distributed computing
5. Damron P, Fedorova A, Lev Y, Luchangco V, Moir M, Nussbaum D (2006) Hybrid transactional memory. In: Proceedings of the 12th symposium on architectural support for programming languages and operating systems
6. Fich F, Luchangco V, Moir M, Shavit N (2005) Brief announcement: obstruction-free step complexity: lock-free DCAS as an example. In: Proceedings of the 19th annual international symposium on distributed computing
7. Fich F, Luchangco V, Moir M, Shavit N (2005) Obstruction-free algorithms can be practically wait-free. In: Proceedings of the 19th annual international symposium on distributed computing
8. Fraser K, Harris T (2004) Concurrent programming without locks. http://www.cl.cam.ac.uk/netos/papers/2004-cpwl-submission.pdf
9. Guerraoui R, Herlihy M, Pochon B (2005) Polymorphic contention management. In: Proceedings of the 19th annual international symposium on distributed computing
10. Guerraoui R, Herlihy M, Pochon B (2005) Toward a theory of transactional contention managers. In: Proceedings of the 24th annual ACM symposium on principles of distributed computing, pp 258–264
11. Guerraoui R, Kapalka M, Kouznetsov P (2006) The weakest failure detector to boost obstruction freedom. In: Proceedings of the 20th annual international symposium on distributed computing
12. Herlihy M (1991) Wait-free synchronization. ACM Trans Program Lang Syst 13(1):124–149
13. Herlihy M (1993) A methodology for implementing highly concurrent data objects. ACM Trans Program Lang Syst 15(5):745–770
14. Herlihy M, Luchangco V, Moir M (2002) Obstruction-free mechanism for atomic update of multiple non-contiguous locations in shared memory. US Patent Application 20040034673
15. Herlihy M, Luchangco V, Moir M (2003) Obstruction-free synchronization: double-ended queues as an example. In: Proceedings of the 23rd international conference on distributed computing systems
16. Herlihy M, Luchangco V, Moir M, Scherer W III (2003) Software transactional memory for supporting dynamic-sized data structures. In: Proceedings of the 22nd annual ACM symposium on principles of distributed computing, pp 92–101
17. Herlihy M, Moss JEB (1993) Transactional memory: architectural support for lock-free data structures. In: Proceedings of the 20th annual international symposium on computer architecture, pp 289–300
18. Luchangco V, Moir M, Shavit N (2005) On the uncontended complexity of consensus. In: Proceedings of the 17th annual international symposium on distributed computing
19. Marathe VJ, Moir M (2008) Toward high performance nonblocking software transactional memory. In: Proceedings of the 13th ACM SIGPLAN symposium on principles and practice of parallel programming. ACM, New York, pp 227–236
20. Marathe V, Scherer W, Scott M (2005) Adaptive software transactional memory. In: Proceedings of the 19th annual international symposium on distributed computing
21. Michael M, Scott M (1998) Nonblocking algorithms and preemption-safe locking on multiprogrammed shared memory multiprocessors. J Parallel Distrib Comput 51(1):1–26
22. Scherer W, Scott M (2005) Advanced contention management for dynamic software transactional memory. In: Proceedings of the 24th annual ACM symposium on principles of distributed computing
23. Shavit N, Touitou D (1997) Software transactional memory. Distrib Comput Special Issue 10:99–116
24. Treiber R (1986) Systems programming: coping with parallelism. Technical Report RJ5118, IBM Almaden Research Center

Wake-Up Problem in Multi-Hop Radio Networks

Tomasz Jurdziński[1] and Dariusz R. Kowalski[2]
[1]Institute of Computer Science, University of Wrocław, Wrocław, Poland
[2]Department of Computer Science, University of Liverpool, Liverpool, UK

Keywords

Ad hoc radio network; Broadcasting; Clock synchronization; Leader election; Probabilistic method; Radio synchronizer; Wake-up

Years and Authors of Summarized Original Work

2004; Chlebus, Kowalski
2004; Chrobak, Gąsieniec, Kowalski
2005; Chlebus, Gąsieniec, Kowalski, Radzik
2007; Chrobak, Gąsieniec, Kowalski

Problem Definition

A *radio network* is modeled as a directed, strongly connected graph G with n *nodes*. The nodes of a network G correspond to transmitting/receiving wireless devices, and directed edges represent their immediately reached neighbors: if a node w is within the transmission range of a node v, then G contains an edge (v, w). We call w an *out-neighbor* of v and v an *in-neighbor* of w.

Each node v has a unique *label* ℓ_v from the set $[N] = \{1, \ldots, N\}$, where $N = O(n)$. Initially, each node knows only its label and the values of n and N.

The time is divided into discrete time steps. It is assumed that nodes have unlimited computing power and can perform arbitrary computations within one time step. However, only one transmission or message receipt is allowed in one time step. Each node has its own local clock, whose initial value at the time of its activation is 0. All local clocks run at the same speed.

A message M transmitted in time step t by a node v is sent instantly to all its out-neighbors. However, an out-neighbor w of v *successfully receives* M in time step t only if no collision occurred in this time step, that is, if no other in-neighbor of w transmits in step t. Collision cannot be distinguished from the background noise: if w does not receive any message in time step t, it knows that either none of its in-neighbors transmitted in step t, or that at least two did, but it does not know which of these two events occurred. It is assumed that nodes only transmit wake-up signals (to their neighbors); no other messages are used.

A *wake-up schedule* is a vector $\omega = (\omega_x)_{x \in V}$, where ω_x denotes the time step in which x wakes-up spontaneously. For any set $X \subseteq V$, ω_X denotes the earliest wake-up time step in X, i.e., $\omega_X = \min_{x \in X} \omega_x$. Without loss of generality, one can assume that $\omega_V = \min_{x \in V} \omega_x = 0$. A *wake-up network* is the pair $\langle G, \omega \rangle$, where G is a radio network G and ω is a wake-up schedule.

A *deterministic wake-up protocol* $\mathcal{W}$ is a function that, for each label ℓ and for each $\tau = 1, 2, 3, \ldots$, given all past messages received by the node v with label $\ell_v = \ell$, specifies whether v will transmit the wake-up signal in time step τ since its activation. A *randomized wake-up protocol* is defined for each node as a probability distribution over the class of deterministic protocols for that node.

The running time of a wake-up protocol $\mathcal{W}$ is the smallest T such that, for any wake-up network $\langle G, \omega \rangle$, all nodes are activated by time T.

Synchronizers

All efficient deterministic wake-up algorithms are based on the combinatorial notion of a *radio synchronizer*, called also a *synchronizer* for short (see also a simpler notion of related structures called selectors [5], efficiently exploited in the context of broadcasting in radio networks).

Let $\mathcal{S} = \{\mathcal{S}^x\}_{x \in [N]}$, where each $\mathcal{S}^x = S_1^x S_2^x \ldots S_m^x$ is a 0-1 sequence of length m. The set $\mathcal{S}$ is a (N, k, m)-*synchronizer* if it satisfies the following property:

(∗) For any nonempty set $X \subseteq [N]$ of cardinality at most k, and for any wake-up schedule ω, there exists t, where $\omega_X < t \leq \omega_X + m$, such that,

$$\sum_{x \in X} S_{t-\omega_x}^x = 1.$$

It is assumed here that $S_i^x = 0$ for $i \leq 0$.

The set S as above can be interpreted as a transmission protocol, where $S_i^x = 1$ indicates that node x transmits in time step $\omega_x + i$. Thus the condition (∗) states that, in at most m time steps after the first node in X wakes-up, there will be a time step when exactly one node in X transmits.

More details about radio synchronizers and synchronization protocols can be found in the survey [8].

Key Results

A Deterministic Wake-Up Protocol

Lemma 1 ([3, 4]) *Let $C \geq 31$ be an integer constant. For each N and $k \leq N$, there exists an (N, k, m)-synchronizer with $m = Ck^2 \log N$.*

The key ingredient of the wake-up algorithm in [3, 4] ([3] is a conference version of [4]) is an application of (N, k, m)-synchronizer with $k = N^{1/3}$ and $m = CN^{2/3} \log N$, where $C = 31$. The analysis of this algorithm relies on the fact that it is sufficient to prove that the algorithm satisfies claimed time bounds for *path graphs*. A directed graph H is called a *path graph* if the nodes of H can be partitioned into sets L_i, $i = 0, \ldots, D$, each with a distinguished node $v_i \in L_i$ and the edges of H are of the form (v, v_{i+1}), where $0 \leq i < D$ and $v \in L_i$. Moreover, $L_D = \{v_D\}$.

Theorem 1 ([3, 4]) *There exists a deterministic protocol that completes the wake-up process in each n-node strongly connected directed graph in time $O(n^{5/3} \log n)$.*

A Randomized Wake-Up Protocol

In [7], the authors presented a randomized wake-up protocol Probability Increase for complete networks working in time $O(\log n \log(1/\epsilon))$ with probability $1 - \epsilon$ (see [6] for deterministic wake-up algorithms for complete graphs). Using this protocol along with appropriate formal analysis, one can obtain a randomized Monte Carlo wake-up protocol for general multi-hop radio networks.

Theorem 2 ([3, 4]) *One can build a randomized protocol which completes wake-up in time $O(D \log n \log(n/\epsilon))$ in each wake-up network with n nodes and diameter D with probability at least $1 - \epsilon$.*

The Monte Carlo protocol from Theorem 2 can be modified to obtain Las Vegas protocol with low expected running time.

Theorem 3 ([3, 4]) *One can build a randomized protocol which completes wake-up in expected time $O(D \log^2 n)$ in each wake-up network with n nodes and diameter D.*

All the above randomized protocols do not require labels.

Applications

Universal Synchronizers and Faster Wake-Up Protocols

The notion of a synchronizer has been generalized to a *universal synchronizer* [1].

Let $g : \mathbb{N} \times \mathbb{N} \to \mathbb{N}$ be a nondecreasing function. Let $S = \{S^x\}_{[N]}$, where each $S^x = S_1^x S_2^x \ldots S_m^x$ is a 0-1 sequence of length $g(N, N)$. The set S is a (N, g)-*universal synchronizer* if it satisfies the following property:

(∗) For any nonempty set $X \subseteq [N]$ and for any wake-up schedule ω, there exists t, where $\omega_X < t \leq \omega_X + g(N, |X|)$, such that,

$$\sum_{x \in X} S_{t-\omega_x}^x = 1.$$

Chlebus and Kowalski proved in [1] that there exist (N, g)-universal synchronizers for $g(k) = O(k \min\{k, \sqrt{n}\} \log n)$. Using this result they showed that there exists a wake-up protocol running in time $O(n^{3/2} \log n)$. In [2], Chlebus *et al.* provided an existential proof of the fact that much shorter universal synchronizers exist and obtained a corresponding faster wake-up protocol.

Lemma 2 ([2]) *For each N there exists a (N, g)-universal synchronizer for $g(N, k) = ck \log k \log N$, where c is a fixed constant.*

Theorem 4 ([2]) *There exists a deterministic protocol that completes the wake-up process in each n-node strongly connected directed graph in time $O(n \log^2 n)$.*

Leader Election and Clock Synchronization

In [3,4], applications of wake-up protocols for the problems of *leader election* and *clock synchronization* were considered.

In the *leader election* problem, the goal is to designate one node as *the leader*, and to announce its identity to all nodes in the network. In the *clock synchronization* problem, upon the completion of the protocol, all nodes must agree on a common global time. For clock synchronization, messages may include numerical values representing the global time.

It has been shown in [3, 4] that any wake-up protocol $\mathcal{W}$ (deterministic or randomized) can be transformed into a leader election protocol or a clock synchronization protocol with only a logarithmic overhead. The leader election protocol is obtained by an execution of appropriately composed $O(\log n)$ executions of a wake-up protocol, in which nodes gradually learn consecutive bits of the node with the largest label. In the clock synchronization protocol, the leader is elected first and then it broadcasts its clock state over the whole network.

Open Problems

The exact complexity of the wake-up problem is not known – there is a logarithmic gap between the complexities of the best known protocols and lower bounds. No efficient algorithms for a construction of (universal) synchronizers described in Lemmata 1 and 2 are known (i.e., polynomial time construction with a polylogarithmic overhead to the length), and thus the results from Theorems 1 and 4 are nonconstructive either. It is not known whether the logarithmic overhead in the complexity of leader election and clock synchronization with respect to wake-up is necessary.

Cross-References

▸ Deterministic Broadcasting in Radio Networks
▸ Randomized Broadcasting in Radio Networks

Recommended Reading

1. Chlebus BS, Kowalski DR (2004) A better wake-up in radio networks. In: Chaudhuri S, Kutten S (eds) PODC. ACM, pp 266–274
2. Chlebus BS, Gasieniec L, Kowalski DR, Radzik T (2005) On the wake-up problem in radio networks. In: Caires L, Italiano GF, Monteiro L, Palamidessi C, Yung M (eds) ICALP. Lecture notes in computer science, Lisbon, Portugal, vol 3580. Springer, pp 347–359
3. Chrobak M, Gasieniec L, Kowalski DR (2004) The wake-up problem in multi-hop radio networks. In: Munro JI (ed) SODA. Portland, Oregon, USA, SIAM, pp 992–1000
4. Chrobak M, Gasieniec L, Kowalski DR (2007) The wake-up problem in multihop radio networks. SIAM J Comput 36(5):1453–1471
5. De Bonis A, Gasieniec L, Vaccaro U (2005) Optimal two-stage algorithms for group testing problems. SIAM J Comput 34(5): 1253–1270
6. Gasieniec L, Pelc A, Peleg D (2001) The wakeup problem in synchronous broadcast systems. SIAM J Discret Math 14(2):207–222
7. Jurdzinski T, Stachowiak G (2005) Probabilistic algorithms for the wake-up problem in single-hop radio networks. Theory Comput Syst 38(3): 347–367
8. Kowalski DR (2011) Coordination problems in ad hoc radio networks. In: Nikoletseas S, Rolim JD (eds) Theoretical aspects of distributed computing in sensor networks. Springer

Wavelet Trees

Roberto Grossi
Dipartimento di Informatica, Università di Pisa,
Pisa, Italy

Keywords

2D data; Data compression; Data structures; Geometric points; Permutations; Reorderings; Sequences; Strings; Succinct representations

Years and Authors of Summarized Original Work

2003; Grossi, Gupta, Vitter
2012; Makris
2014; Navarro

Problem Definition

The wavelet tree is a data structure that represents a recursive partition of a sequence S of length n according to its symbols. Letting $\Sigma = \{1, \ldots \sigma\}$ be the alphabet of symbols of S, the wavelet tree for S has the root representing S itself and σ leaves representing the positions of the symbols: leaf $c \in \Sigma$ represents all the positions i such that $S[i] = c$ and $1 \leq i \leq n$. The internal nodes describe how the symbols are grouped. In the original wavelet tree, nodes are binary and thus there are two groups, called the 0-group and the 1-group, which form an alphabet partition. In the multi-ary wavelet tree, the nodes are obtained by forming more than two groups each time. We focus on binary wavelet trees in the following.

For example, consider the sequence $S =$ SENSELESSNESS# in Fig. 1. Here we divide the symbols in two groups $\{E, L\}$, and $\{N, S, \#\}$, giving rise to the two children of the root: the left child contains the subsequence of S obtained by copying the symbols in $\{E, L\}$; the right child contains the subsequence of S obtained by copying the rest of the symbols (which are in $\{N, S, \#\}$). The partition of $\{E, L\}$ into $\{E\}$ and $\{L\}$ produces two leaves. The partition of $\{N, S, \#\}$ into $\{N\}$ and $\{S, \#\}$ gives rise to a leaf and an internal node. The latter represents the partition of $\{S, \#\}$ giving rise to two leaves.

In general each internal node of the wavelet tree represents a subsequence S' of the input sequence S, obtained by selecting certain symbols from S. More precisely, if Σ' is the alphabet for the symbols in S', then S' is formed by selecting *all* the symbols of S belonging to Σ'. Note that $\Sigma' = \{c\}$ if and only if the node storing S' is the leaf whose associated symbol is c. For an internal node, its two children are determined by the choice of the 0-group Σ'_0 and of the 1-group Σ'_1 partitioning $\Sigma' = \Sigma'_0 \cup \Sigma'_1$. To this end, a bitvector $B_{S'}$ is associated with S', where the 0's mark which positions of S' contain symbols from Σ'_0 and the 1's mark which positions contain symbols from 1-group Σ'_1.

It is worth noting that any choice for the recursive alphabet partitioning in 0- and 1-groups can be translated into a simple dichotomy test at each node by suitably reordering the alphabet Σ: without loss of generality, we assume that given a node representing S' over alphabet Σ', there exists a symbol $c' \in \Sigma'$ such that c belongs to the 0-group Σ'_0 if and only if $c \in \Sigma'$ and $c \leq c'$.

Finally, the sequences S and S' can be actually dropped from the nodes of the wavelet tree: just knowing the symbols from Σ associated with its leaves allows us to reconstruct the dropped sequences in its internal nodes as we discuss next.

Key Results

Despite its simplicity, the wavelet tree is a versatile data structure that offers solutions to a variety of situations using small additional space.

- *Compressed sequences.* Sequence S can be stored using a number of bits close to the 0-order entropy and still supporting random access and other operations such as rank and select of individual symbols.
- *Geometric points and 2D data.* The leaves represent the individual points in x-order, while

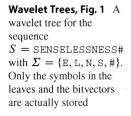

Wavelet Trees, Fig. 1 A wavelet tree for the sequence $S =$ SENSELESSNESS# with $\Sigma = \{$E, L, N, S, #$\}$. Only the symbols in the leaves and the bitvectors are actually stored

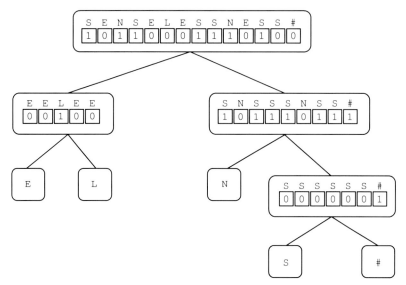

the root represents the same points but in y-order. Range and percentile queries can be performed in this way.

- *Permutations, shufflings, and reorderings.* The leaves represent the elements in a certain order and the root represents the same set in a permuted order. Mapping these two orders can be done efficiently.

As for the construction of the wavelet tree, it can be easily done in $O(n \log \sigma)$ time and space. More sophisticated algorithms have been developed to lower the construction time and/or the additional working space. In the following, we focus here on the usage of the original wavelet tree.

Compressed Sequences

The first natural question is how to access symbol $S[i]$ from the wavelet tree for S. We can only use the bitvectors $B_{S'}$ in the internal nodes and the mapping from the leaves to the symbols of Σ. We start out from the root, check bit $B_S[i]$, and count the number i' of bits equal to $B_S[i]$ in the first i positions of B_S. After that, we repeat the step on the left child (if $B_S[i] = 0$) or the right child (if $B_S[i] = 1$), setting the new value of $i = i'$ and using the bitvector $B_{S'}$ of that child. Eventually we end up in a leaf, and the symbol c corresponding to that leaf gives the answer that $S[i] = c$.

Regarding the time and space complexity, we need an operation to count how many 1's occur in the first i positions of a bitvector (as the number of 0's can be obtained by subtracting this count from i). This operation is called rank in the literature and takes constant time by preprocessing the bitvector and adding a little-oh number of bits to it. In this way, the cost of access operation is given by the height of the wavelet tree, which is $O(\log \sigma)$ in case of a balanced shape.

As for the space complexity, note that any binary tree shape with σ leaves is feasible. Using a Huffman tree shape, less frequent symbols correspond to deeper leaves. Note that each symbol occurrence in a leaf can be charged a bit from each bitvector in its ancestors. Equivalently, the sum of the lengths of the bitvectors in all the internal nodes of the wavelet tree is equal to the sum of the lengths of the Huffman encodings of the symbols of the input sequence S. In other words, the space required by the bitvectors is equal to the space achieved by the Huffman encoding. The additional rank data structures use little-oh of that space. Letting $H_0 \leq \log \sigma$ (logarithms in base 2) denote the 0-order entropy of S, the total space to store a wavelet tree is therefore $nH_0 + o(nH_0)$ bits, which can be lowered to $nH_0 + o(n)$ with additional machinery. For compressible sequences S, this is better than storing them in $n \log \sigma$ bits with the standard format. In general, any prefix-free encoding of the symbols in Σ can be used in place of the Huffman coding, giving the same number of bits of the chosen encoding: it suffices to choose as a shape the resulting prefix tree of the chosen encoding of the symbols in Σ. (Note that storing the tree shape and symbol mappings requires $O(\sigma \log \sigma)$ further bits.)

Interestingly, the above space bound can be obtained using any shape if the bitvectors are stored in compressed format. For example, one compressed bitvector representation stores a bitvector of length m with k 1's using the theoretic information minimum of $\log \binom{m}{k} + o(m)$ bits and supports constant time `rank` and `select` operations. The latter operation returns the position of the jth 1 in the bitvector (same for the jth 0). It can be shown that for *any shape* of the wavelet tree, summing the $\log \binom{m}{k} + o(m)$ contribution of all the bitvectors in its nodes still gives a total space of $nH_0 + o(n)$ bits to store the wavelet tree. In other words, the 0-order entropy bound can be achieved independently of the tree shape.

As a by-product of what we discussed above, the wavelet tree allows us to extend the `rank` and `select` operations from a bitvector to any sequence over an alphabet Σ. To see why, suppose we want to know how many occurrences of symbols c occur in the first i positions of S. We perform the same steps as described above for the `access` operation (where we initially set $i' = i$) except that now we already know that the path to follow is from the root to the leaf representing c. In the generic step, we can easily test if c belongs to the 0- or 1-group of the current node, and branch according to the target leaf, updating the value of i'. However, when we reach the leaf for c, we have to return the corresponding value of i' as the answer for `rank` of c, since it tells how many c's are up to position i. As for the `select` operation on c, suppose that we want to identify the jth occurrence of c in S. This time we proceed from the leaf corresponding to c backwards to the root. We initially set $i' = j$ and then reverse the branching process: at the generic step, we are in a node storing S' and on position i'. We reach the parent p of the current node, and select the i'th 0 (if arriving from the

left child) or the i'th 1 (if arriving from the right child) in the bitvector stored in p. We set i' to be the resulting position and iterate. Eventually we reach the root and return the current value of i' as the answer for `select` of c.

Time cost is proportional to the height of the wavelet tree, which is $O(\log \sigma)$ in case of a balanced shape. Using multi-ary wavelet trees, the height can be reduced and so does the cost, achieving $O(1 + \log \sigma / \log w)$ time with a word size $w = \Omega(\log n)$.

Geometric Points and Two-Dimensional Data

Given a set of n points $\langle x_i, y_i \rangle$ in the plane, or equivalently 2D data, where $x_1 \leq x_2 \leq \cdots x_n$, we can use the wavelet tree as a space-efficient data structure for storing and querying them. We store these coordinates in two vectors X and Y, such as $X[i] = x_i$ and $Y[i] = y_i$ for $1 \leq i \leq n$. We then build the wavelet tree where $S \equiv Y$ and Σ is the set of distinct values in Y.

As a result, we obtain a compacted hierarchical space decomposition for the n points. To see why, we can conceptually think of the n points as belonging to a $n \times n$ grid stored in the root of the wavelet tree, where the actual coordinates are those stored in X and Y. The geometric interpretation of alphabet partitioning in 0- and 1-groups is that of choosing a value y' and splitting the points in two groups, those having coordinate $y_i \leq y'$ (the 0-group) and those having $y_i > y'$ (the 1-group). Let n_0 be the size of the 0-group for the root and n_1 be the size of the 1-group, where $n = n_0 + n_1$. In each group, only the rows and the columns that still contain points survive. As a result, two grids of size $n_0 \times n_0$ and $n_1 \times n_1$ are produced from one of size $n \times n$. Here, the left child corresponds to a subsequence Y_0' of Y that represents the n_0 points (with $y_i \leq y'$) in the grid of size $n_0 \times n_0$, and the right child represents the sequence Y_1' of n_1 points (with $y_i > y'$) in the grid of size $n_1 \times n_1$. The leaves of the wavelet tree are in y-order, and each of them stores the points sharing the same y-coordinate.

As for the storage, we observe that the values in X can be represented compactly as they are in *x*-order. The values of Y do not need to be stored separately as they are represented in the wavelet tree. Total space is $O(n \log n)$ bits but can be less if the sequences of coordinates are compressible.

The supported operations exploit the aforementioned hierarchical space decomposition for the two-dimensional data as illustrated next. For example, consider the classical 2D range query, reporting (or counting) the points contained in the range $[a \ldots b] \times [c \ldots d]$. If the current cutting coordinate y' is outside the range $[c \ldots d]$, we move to one of the two children; otherwise, we branch on both children, using respectively the ranges $[a \ldots b] \times [c \ldots y']$ and $[a \ldots b] \times (y' \ldots d]$. Note that the restriction $[a \ldots b]$ on the *x*-coordinates can be used to test if the grid represented by the reached child has a nonempty intersection with the range: the mechanism is the same as that of `rank`. This means that each reported occurrence potentially requires a traversal down to a leaf; thus the cost is proportional to the number of reported points times the height of the wavelet tree. A refined version of this idea allows for $O(\log n / \log \log n)$ time for a counting query.

Another interesting use is quantile queries. For the range $[a \ldots b]$, consider the values in $V_{ab} = \{y_i \mid a \leq x_i \leq b\}$ obtained as the *y*-coordinates of the points in that range. For any given a, b, and k, the query asks to find the kth element in V_{ab}. Using the above wavelet tree, we can find the rank i_a of a and that i_b of b. Then, using `rank` operations on the bitvector B_S in the root, we can count how many 0's and 1's are in $B_s[i_a \ldots i_b]$. If there are at least k 0's, we know that the kth value in V_{ab} is smaller than or equal than the cutting coordinate y', and we iterate in the left child; otherwise, we subtract the number of 0's from k, and we iterate in the right child. When we reach a leaf, we return the associated value as the answer for the quantile query. Along the same lines, we can also report the topmost k values in V_{ab}. Once again, the cost is proportional to the wavelet tree height for each reported value.

Permutations, Shufflings, and Reorderings

The above discussion brings the combinatorial structure of wavelet tree to light as we can store two orders inside it: the former is the order in the sequence stored at the root, and the latter is obtained by a left-to-right traversal of the leaves. The bitvectors are internal routers that guide how elements are permuted and shuffled to produce the reordering. Mergesort can be modeled according to this view: the internal nodes merge the content of their children, and the bitvectors tell who goes where in the resulting merged reordering.

An immediate application of the above observations is setting S to be a permutation π of the integers in $\{1, 2, \ldots, n\}$. Traversing the wavelet tree upward (as in the `select`) computes $\pi(i)$, while traversing it downward computes the inverse permutation π^{-1}. The best cost is $O(\log n / \log \log n)$ time.

In general, we can store two orders using the wavelet tree, one being a permutation of the other. Inverted lists in information retrieval can store document IDs in increasing order of enumeration, but they also want to store these IDs in decreasing order of importance through some raking function. As it is clear now, these are two orders that can be simultaneously preserved inside the wavelet tree.

Applications

Looking back at previous work, some ideas behind the wavelet tree can be found in Kärkkäinen's PhD thesis and in Chazelle's functional approach to data structures for multidimensional searching. The wavelet tree in its explicit and fully functional form has been introduced by Grossi, Gupta, and Vitter to store the Burrows-Wheeler transform (BWT) for obtaining compressed text indexes (able to support fast pattern searching). Its natural application is supporting `rank` and `select` queries for the symbols of the resulting compressed BWT. Since then, many papers have explored the properties of the wavelet trees in several applications. Apart from compressed full-text indexes, researchers have employed wavelet trees in inverted lists, graphs, binary relations, numeric sequences, colored range queries, XPath queries, semi-structure data, and frequent item sets, to name a few. The wavelet trie extends the wavelet tree to store a sequence of strings, rather than a sequence of symbols, thus allowing the supported operations to operate also on the prefixes of the strings. The wavelet matrix is a variant of a balanced wavelet tree, in which all the bitvectors on the same level are concatenated, and is particularly efficient for large alphabet size σ.

Cross-References

▶ Burrows-Wheeler Transform
▶ Compressed Suffix Array
▶ Huffman Coding
▶ Rank and Select Operations on Bit Strings
▶ Rank and Select Operations on Sequences
▶ Succinct and Compressed Data Structures for Permutations and Integer Functions

Recommended Reading

1. Barbay J, Navarro G (2013) On compressing permutations and adaptive sorting. Theor Comput Sci 513:109–123
2. Bose P, Chen EY, He M, Maheshwari A, Morin P (2012) Succinct geometric indexes supporting point location queries. ACM Trans Algorithms 8(2):10
3. Chazelle B (1988) A functional approach to data structures and its use in multidimensional searching. SIAM J Comput 17(3):427–462
4. Claude F, Navarro G, Ordóñez A (2015) The wavelet matrix: an efficient wavelet tree for large alphabets. Inf Syst 47:15–32
5. Ferragina P, Manzini G, Mäkinen V, Navarro G (2007) Compressed representations of sequences and full-text indexes. ACM Trans Algorithms 3(2):20
6. Ferragina P, Giancarlo R, Manzini G (2009) The myriad virtues of wavelet trees. Inf Comput 207(8):849–866
7. Gagie T, Puglisi SJ, Turpin A (2009) Range quantile queries: another virtue of wavelet trees. In: String processing and information retrieval, Saariselkä. Springer, pp 1–6
8. Gagie T, Navarro G, Puglisi SJ (2012) New algorithms on wavelet trees and applications to information retrieval. Theor Comput Sci 426:25–41
9. Grossi R, Ottaviano G (2012) The wavelet trie: maintaining an indexed sequence of strings in compressed space. In: Krötzsch M, Lenzerini M, Benedikt M (eds) PODS'12: proceedings of the 31st ACM SIGMOD-SIGACT-SIGART symposium on principles of database systems, Scottsdale, 20–24 May 2012. ACM, pp 203–214
10. Grossi R, Gupta A, Vitter JS (2003) High-order entropy-compressed text indexes. In: Proceedings of the fourteenth annual ACM-SIAM symposium on discrete algorithms, Baltimore. Society for Industrial and Applied Mathematics, pp 841–850
11. Grossi R, Gupta A, Vitter JS (2004) When indexing equals compression: experiments with compressing suffix arrays and applications. In: Proceedings of the fifteenth annual ACM-SIAM symposium on discrete algorithms, New Orleans. Society for Industrial and Applied Mathematics, pp 636–645
12. Kärkkäinen J (1999) Repetition-based text indexing. PhD thesis, University of Helsinki, Finland
13. Makris C (2012) Wavelet trees: a survey. Comput Sci Inf Syst 9(2):585–625
14. Navarro G (2014) Wavelet trees for all. J Discret Algorithms 25:2–20

Weighted Connected Dominating Set

Yu Wang[1], Weizhao Wang[2], and Xiang-Yang Li[3]
[1]Department of Computer Science, University of North Carolina, Charlotte, NC, USA
[2]Google Inc., Irvine, CA, USA
[3]Department of Computer Science, Illinois Institue of Technology, Chicago, IL, USA

Keywords

Minimum weighted connected dominating set

Years and Authors of Summarized Original Work

2005; Wang, Wang, Li

Problem Definition

This problem is concerned with a weighted version of the classical minimum connected dominating set problem. This problem has numerous motivations including wireless networks and distributed systems. Previous work [1, 2, 4, 5, 6, 14] in wireless networks focuses on designing efficient distributed algorithms to construct the connected dominating set which can be used as the virtual backbone for the network. Most of the proposed methods try to minimize the number of nodes in the backbone (i.e., the number of clusterheads). However, in many applications, minimizing the size of the backbone is not sufficient. For example, in wireless networks different wireless nodes may have different costs for serving as a clusterhead, due to device differences, power capacities, and information loads to be processed. Thus, by assuming each node has a cost to being in the backbone, there is a need to study distributed algorithms for weighted backbone formation. Centralized algorithms to construct a weighted connected dominating set with minimum weight have been studied [3, 7, 9]. Recently, the work of Wang, Wang, and Li [12, 13] proposes an efficient distributed method to construct a weighted backbone with low cost. They proved that the total cost of the constructed backbone is within a small constant factor of the optimum when either the nodes' costs are smooth (i.e., the maximum ratio of costs of adjacent nodes is bounded) or the network maximum node degree is bounded. To the best knowledge of the entry authors, this work is the first to consider this weighted version of minimum connected dominating set problem and provide a distributed approximation algorithm.

Notations

A communication graph $G = (V, E)$ over a set V of wireless nodes has an edge uv between nodes u and v if and only if u and v can communicate directly with each other, i.e., inside the transmission region of each other. Let $d_G(u)$ be the degree of node u in a graph G and Δ be the maximum node degree of all wireless nodes (i.e., $\Delta = \max_{u \in V} d_G(u)$). Each wireless node u has a cost $c(u)$ of being in the backbone. Let $\delta = \max_{ij \in E} c(i)/c(j)$, where ij is the edge between nodes i and j, E is the set of communication links in the wireless network G, and the maximum operation is taken on all pairs of adjacent nodes i and j in G. In other words, δ is the maximum ratio of costs of two adjacent nodes and can be called the *cost smoothness* of

the network. When δ is bounded by some small constant, the node costs are *smooth*. When the transmission region of every wireless node is modeled by a unit disk centered at itself, the communication graph is often called a *unit disk graph*, denoted by $UDG(V)$. Such networks are also called *homogeneous networks*.

A subset S of V is a *dominating set* if each node in V is either in S or is adjacent to some node in S. Nodes from S are called dominators, while nodes not in S are called dominatees. A subset B of V is a *connected dominating set* (CDS) if B is a dominating set and B induces a connected subgraph. Consequently, the nodes in B can communicate with each other without using nodes in $V - B$. A dominating set with minimum cardinality is called *minimum dominating set* (MDS). A CDS with minimum cardinality is the *minimum connected dominating set* (MCDS). In the weighted version, assume that each node u has a cost $c(u)$. Then a CDS B is called *weighted connected dominating set* (WCDS). A subset B of V is a *minimum weighted connected dominating set* (MWCDS) if B is a WCDS with minimum total cost. It is well-known that finding either the *minimum connected dominating set* or the *minimum weighted connected dominating set* is a NP-hard problem even when G is a unit disk graph. The work of Wang et al. studies efficient approximation algorithms to construct a low-cost backbone which can approximate the MWCDS problem well. For a given communication graph $G = (V, E, C)$ where V is the set of nodes, E is the edge set, and C is the set of weights for edges, the corresponding minimum weighted connected dominating set problem is as follows.

Problem 1 (Minimum Weighted Connected Dominating Set)

INPUT: The weighted communication graph $G = (V, E, C)$.

OUTPUT: A subset A of V is a *minimum weighted connected dominating set*, i.e., (1) A is a dominating set; (2) A induces a connected subgraph; (3) the total cost of A is minimum.

Another related problem is independent set problem. A subset of nodes in a graph G is an *independent set* if for any pair of nodes, there is no edge between them. It is a *maximal independent set* if no more nodes can be added to it to generate a larger independent set. Clearly, any maximal independent set is a dominating set. It is a *maximum independent set* (MIS) if no other independent set has more nodes. The independence number, denoted as $\alpha(G)$, of a graph G is the size of the MIS of G. The *k-local independence number*, denoted by $\alpha^{[k]}(G)$, is defined as $\alpha^{[k]}(G) = \max_{u \in V} \alpha(G_k(u))$. Here, $G_k(u)$ is the induced graph of G on k-hop neighbors of u (denoted by $N_k(u)$), i.e., $G_k(u)$ is defined on $N_k(u)$, and contains all edges in G with both end-points in $N_k(u)$. It is well-known that for a unit disk graph, $\alpha^{[1]}(UDG) \leq 5$ [2] and $\alpha^{[2]}(UDG) \leq 18$ [11].

Key Results

Since finding the minimum weighted connected dominating set (MWCDS) is NP-hard, centralized approximation algorithms for MWCDS have been studied [3, 7, 9]. In [9], Klein and Ravi proposed an approximation algorithm for the node-weighted Steiner tree problem. Their algorithm can be generalized to compute a $O(\log \Delta)$ approximation for MWCDS. Guha and Khuller [7] also studied the approximation algorithms for node-weighted Steiner tree problem and MWCDS. They developed an algorithm for MWCDS with an approximation factor of $(1.35 + \epsilon) \log \Delta$ for any fixed $\epsilon > 0$. Recently, Ambuhl et al. [3] provided a constant approximation algorithm for MWCDS under UDG model. Their approximation ratio is bounded by 89. All these algorithms are centralized algorithms, while the applications in wireless ad hoc networks prefer distributed solutions for MWCDS.

In [12, 13], Wang et al. proposed a distributed algorithm that constructs a weighted connected dominating set for a wireless ad hoc network G. Their method has two phases: the first phase (clustering phase, Algorithm 1 in [12, 13]) is to find a set of wireless nodes as the

dominators (clusterheads) and the second phase (Algorithm 2 in [12, 13]) is to find a set of nodes, called *connectors*, to connect these dominators to form the final backbone. Wang et al. proved that the total cost of the constructed backbone is no more than $\min(\alpha^{[2]}(G) \log (\Delta + 1), (\alpha^{[1]}(G) - 1)\delta + 1) + 2\alpha^{[1]}(G)$ times of the optimum solution.

Algorithm 1 first constructs a maximal independent set (MIS) using classical greedy method with the node cost as the selection criterion. For each node v in MIS, it then runs a local greedy set cover method on the *local neighborhood* $N_2(v)$ to find some nodes ($GRDY_v$) to cover all one-hop neighbors of v. If $GRDY_v$ has a total cost smaller than v, then it uses $GRDY_v$ to replace v, which further reduces the cost of MIS. The following theorem of the total cost of this selected set is proved in [12, 13].

Theorem 1 *For a network modeled by a graph G, Algorithm 1 (in [12, 13]) constructs a dominating set whose total cost is no more than* $\min(\alpha^{[2]}(G) \log(\Delta+1), (\alpha^{[1]}(G)-1)\delta+1)$ *times of the optimum.*

Algorithm 2 finds some *connectors* among all the dominatees to connect the dominators into a backbone (CDS). It forms a CDS by finding connectors to connect any pair of dominators u and v if they are connected in the original graph G with at most 3 hops. A distributed algorithm to build a MST then is performed on the CDS. The following theorem of the total cost of these connectors is proved in [12, 13].

Theorem 2 *The connectors selected by Algorithm 2 (in [12, 13]) have a total cost no more than $2 \cdot \alpha^{[1]}(G)$ times of the optimum for networks modeled by G.*

Combining Theorems 1 and 2, the following theorem is the main contributions of the work of Wang et al..

Theorem 3 *For any communication graph G, Algorithm 1 and Algorithm 2 construct a weighted connected dominating set whose total cost is no more than*

$$\min(\alpha^{[2]}(G)\log(\Delta + 1), (\alpha^{[1]}(G) - 1)\delta + 1)$$
$$+ 2\alpha^{[1]}(G)$$

times of the optimum.

Notice that, for homogeneous wireless networks modeled by UDG, it implies that the constructed backbone has a cost no more than $\min(18\log(\Delta + 1), 4\delta + 1) + 10$ times of the optimum. The advantage of the constructed backbone is that the total cost is small compared with the optimum when either the costs of wireless nodes are smooth, i.e., two neighboring nodes' costs differ by a small constant factor, or the maximum node degree is low.

In term of time complexity, the most time-consuming step in the proposed distributed algorithm is building the MST. In [10], Kuhn et al. gave a lower bound on the distributed time complexity of any distributed algorithm that wants to compute a minimum dominating set in a graph. Essentially, they proved that even for the unconnected and unweighted case, any distributed approximation algorithm with poly-logarithmic approximation guarantee for the problem has to have a time-complexity of at least $\Omega(\log \Delta / \log \log \Delta)$.

Applications

The proposed distributed algorithms for MWCDS can be used in ad hoc networks or distributed system to form a low-cost network backbone for communication application. The cost used as the input of the algorithms could be a *generic* cost, defined by various practical applications. It may represent the *fitness* or *priority* of each node to be a clusterhead. The lower cost means the higher priority. In practice, the cost could represent the power consumption rate of the node if a backbone with small power consumption is needed; the robustness of the node if fault-tolerant backbone is needed; or a function of its security level if a secure backbone is needed; or a combined weight function to integrate various metrics

such as traffic load, signal overhead, battery level, and coverage. Therefore, by defining different costs, the proposed low-cost backbone formation algorithms can be used in various practical applications. Beside forming the backbone for routing, the weighted clustering algorithm (Algorithm 1) can also be used in other applications, such as selecting the mobile agents to perform intrusion detection in ad hoc networks [8] (to achieve more robust and power efficient agent selection), or select the rendezvous points to collect and store data in sensor networks [15] (to achieve the energy efficiency and storage balancing).

Open Problems

A number of problems related to the work of Wang, Wang, and Li [12, 13] remain open. The proposed method assumes that the nodes are almost-static in a reasonable period of time. However, in some network applications, the network could be highly dynamic (both the topology or the cost could change). Therefore, after the generation of the weighted backbone, the dynamic maintenance of the backbone is also an important issue. It is still unknown how to update the topology efficiently while preserving the approximation quality.

In [12, 13], the following assumptions on wireless network model is used: omni-directional antenna, single transmission received by all nodes within the vicinity of the transmitter. The MWCDS problem will become much more complicated if some of these assumptions are relaxed.

Experimental Results

In [12, 13], simulations on random networks are conducted to evaluate the performances of the proposed weighted backbone and several backbones built by previous methods. The simulation results confirm the theoretical results.

Cross-References

▶ Connected Dominating Set

Recommended Reading

1. Alzoubi K, Wan P-J, Frieder O (2002) New distributed algorithm for connected dominating set in wireless ad hoc networks. In: Proceedings of IEEE 35th Hawaii international conference on system sciences (HICSS-35), Hawaii, 7–10 Jan 2002
2. Alzoubi K, Li X-Y, Wang Y, Wan P-J, Frieder O (2003) Geometric spanners for wireless ad hoc networks. IEEE Trans Parallel Distrib Process 14:408–421
3. Ambuhl C, Erlebach T, Mihalak M, Nunkesser M (2006) Constant factor approximation for minimum-weight (connected) dominating sets in unit disk graphs. In: Proceedings of the 9th international workshop on approximation algorithms for combinatorial optimization problems (APPROX 2006), Barcelona, 28–30 Aug 2006. LNCS, vol 4110. Springer, Berlin/Heidelberg, pp 3–14
4. Bao L, Garcia–Aceves JJ (2003) Topology management in ad hoc networks. In: Proceedings of the 4th ACM international symposium on mobile ad hoc networking & computing, Annapolis, 1–3 June 2003. ACM Press, New York, pp 129–140
5. Chatterjee M, Das S, Turgut D (2002) WCA: a weighted clustering algorithm for mobile ad hoc networks. J Clust Comput 5:193–204
6. Das B, Bharghavan V (1997) Routing in ad-hoc networks using minimum connected dominating sets. In: Proceedings of IEEE international conference on communications (ICC'97), Montreal, 8–12 June 1997, vol 1, pp 376–380
7. Guhaa S, Khuller S (1999) Improved methods for approximating node weighted Steiner trees and connected dominating sets. Inf Comput 150:57–74
8. Kachirski O, Guha R (2002) Intrusion detection using mobile agents in wireless ad hoc networks. In: Proceedings of IEEE workshop on knowledge media networking, Kyoto, 10–12 July 2002
9. Klein P, Ravi R (1995) A nearly best-possible approximation algorithm for node-weighted Steiner trees. J Algorithms 19:104–115
10. Kuhn F, Moscibroda T, Wattenhofer R (2004) What cannot be computed locally! In: Proceedings of the 23rd ACM symposium on the principles of distributed computing (PODC), St. John's, July 2004
11. Li X-Y, Wan P-J (2005) Theoretically good distributed CDMA/OVSF code assignment for wireless ad hoc networks. In: Proceedings of 11th international computing and combinatorics conference (COCOON), Kunming, 16–19 Aug 2005
12. Wang Y, Wang W, Li X-Y (2005) Efficient distributed low-cost backbone formation for wireless networks. In: Proceedings of 6th ACM international symposium on mobile ad hoc networking and computing (MobiHoc 2005), Urbana-Champaign, 25–27 May 2005
13. Wang Y, Wang W, Li X-Y (2006) Efficient distributed low cost backbone formation for wireless networks. IEEE Trans Parallel Distrib Syst 17:681–693
14. Wu J, Li H (2001) A dominating-set-based routing scheme in ad hoc wireless networks. Spec Iss Wirel Netw Telecommun Syst J 3:63–84
15. Zheng R, He G, Gupta I, Sha L (2004) Time indexing in sensor networks. In: Proceedings of 1st IEEE international conference on mobile ad-hoc and sensor systems (MASS), Fort Lauderdale, 24–27 Oct 2004

Weighted Popular Matchings

Julián Mestre
Department of Computer Science, University of Maryland, College Park, MD, USA
School of Information Technologies, The University of Sydney, Sydney, NSW, Australia

Years and Authors of Summarized Original Work

2006; Mestre

Problem Definition

Consider the problem of matching a set of individuals X to a set of items Y where each individual has a weight and a personal preference over the items. The objective is to construct a matching M that is stable in the sense that there is no matching M' such that the weighted majority vote will choose M' over M.

More formally, a bipartite graph (X, Y, E), a weight $w(x) \in R^+$ for each individual $x \in X$, and a rank function $r : E \rightarrow \{1, \ldots, |Y|\}$ encoding the individual preferences are given. For every applicant x and items $y_1, y_2 \in Y$ say applicant x prefers y_1 over y_2 if $r(x, y_1) < r(x, y_2)$, and x is indifferent between y_1 and y_2 if $r(x, y_1) = r(x, y_2)$. The preference lists are said to be strictly ordered if applicants are never indifferent between two items, otherwise the preference lists are said to contain ties.

Let M and M' be two matchings. An applicant x prefers M over M' if x prefers the item he/she gets in M over the item he/she gets in M'. A matching M is *more popular than* M' if the applicants that prefer M over M' outweigh those that prefer M' over M. Finally, a matching M is *weighted popular* if there is no matching M' more popular than M.

In the *weighted popular matching problem* it is necessary to determine if a given instance admits a popular matching, and if so, to produce one. In the *maximum weighted popular matching problem* it is necessary to find a popular matching of maximum cardinality, provided one exists.

Abraham et al. [2] gave the first polynomial time algorithms for the special case of these problems where the weights are uniform. Later, Mestre [8] introduced the weighted variant and developed polynomial time algorithms for it.

Key Results

Theorem 1 *The weighted popular matching and maximum weighted popular matching problems on instances with strictly ordered preferences can be solved in $O(|X| + |E|)$ time.*

Theorem 2 *The weighted popular matching and maximum weighted popular matching problems on instances with arbitrary preferences can be solved in $O(\min\{k\sqrt{|X|}, |X|\}|E|)$ time.*

Both results rely on an alternative easy-to-compute characterization of weighted popular matchings called *well-formed* matchings. It can be shown that every popular matching is well-formed. While in unweighted instances every well-formed matching is popular [2], in weighted instances there may be well-formed matchings that are not popular. These non-popular well-formed matchings can be weeded out by pruning certain bad edges that cannot be part of any popular matching. In other words, the instance can be pruned so that a matching is popular if and

only if it is well-formed and is contained in the pruned instance [8].

Applications

Many real-life problems can be modeled using one-sided preferences. For example, the assignment of graduates to training positions [5], families to government-subsidized housing [10], students to projects [9], and Internet rental markets [1] such as Netflix where subscribers are assigned DVDs.

Furthermore, the weighted framework allows one to model the naturally occurring situation in which some subset of users has priority over the rest. For example, an Internet rental site may offer a "premium" subscription plan and promise priority over "regular" subscribers.

Cross-References

▶ Ranked Matching
▶ Stable Marriage

Recommended Reading

1. Abraham DJ, Chen N, Kumar V, Mirrokni V (2006) Assignment problems in rental markets. In: Proceedings of the 2nd workshop on internet and network economics, Patras, 15–17 Dec 2006
2. Abraham DJ, Irving RW, Kavitha T, Mehlhorn K (2005) Popular matchings. In: Proceedings of the 16th annual ACM-SIAM symposium on discrete algorithms (SODA), pp 424–432
3. Abraham DJ, Kavitha T (2006) Dynamic matching markets and voting paths. In: Proceedings of the 10th Scandinavian workshop on algorithm theory (SWAT), Riga, 6–8 July 2006, pp 65–76
4. Gardenfors P (1975) Match making: assignments based on bilateral preferences. Behav Sci 20:166–173
5. Hylland A, Zeechhauser R (1979) The efficient allocation of individuals to positions. J Polit Econ 87(2):293–314
6. Mahdian M (2006) Random popular matchings. In: Proceedings of the 7th ACM conference on electronic commerce (EC), Venice, 10–14 July 2006, pp 238–242
7. Manlove D, Sng C (2006) Popular matchings in the capacitated house allocation problem. In: Proceed-

ings of the 14th annual European symposium on algorithms (ESA), pp 492–503
8. Mestre J (2006) Weighted popular matchings. In: Proceedings of the 16th international colloquium on automata, languages, and programming (ICALP), pp 715–726
9. Proll LG (1972) A simple method of assigning projects to students. Oper Res Q 23(23):195–201
10. Yuan Y (1996) Residence exchange wanted: a stable residence exchange problem. Eur J Oper Res 90:536–546

Weighted Random Sampling

Pavlos Efraimidis[1] and Paul (Pavlos) Spirakis[2,3,4]
[1]Department of Electrical and Computer Engineering, Democritus University of Thrace, Xanthi, Greece
[2]Computer Engineering and Informatics, Research and Academic Computer Technology Institute, Patras University, Patras, Greece
[3]Computer Science, University of Liverpool, Liverpool, UK
[4]Computer Technology Institute (CTI), Patras, Greece

Keywords

Random number generation; Sampling

Years and Authors of Summarized Original Work

2005; Efraimidis, Spirakis

Problem Definition

The problem of random sampling without replacement (RS) calls for the selection of m distinct random items out of a population of size n. If all items have the same probability to be selected, the problem is known as uniform RS. Uniform random sampling in one pass is discussed in

[1, 6, 11]. Reservoir-type uniform sampling algorithms over data streams are discussed in [12]. A parallel uniform random sampling algorithm is given in [10]. In weighted random sampling (WRS) the items are weighted and the probability of each item to be selected is determined by its relative weight. WRS can be defined with the following algorithm D:

Algorithm D, a definition of WRS

Input: A population V of n weighted items
Output: A set S with a WRS of size m
1: For $k = 1$ to m do
2: Let $p_i(k) = w_i / \sum_{s_j \in V-S} w_j$ be the probability of item v_i to be selected in round k
3: Randomly select an item $v_i \in V - S$ and insert it into S
4: End-For

Problem 1 (WRS)
INPUT: A population V of n weighted items.
OUTPUT: A set S with a weighted random sample.

The most important algorithms for WRS are the Alias Method, Partial Sum Trees and the Acceptance/Rejection method (see [9] for a summary of WRS algorithms). *None of these algorithms is appropriate for one-pass WRS.* In this work, an algorithm for WRS is presented. The algorithm is simple, very flexible, and solves the WRS problem over data streams. Furthermore, the algorithm admits parallel or distributed implementation. To the best knowledge of the entry authors, this is the first algorithm for WRS over data streams and for WRS in parallel or distributed settings.

Definitions
One-pass WRS is the problem of generating a weighted random sample in one-pass over a population. If additionally the population size is initially unknown (e.g., a data streams), the random sample can be generated with *reservoir sampling* algorithms. These algorithms keep an auxiliary storage, the reservoir, with all items that are candidates for the final sample.

Notation and Assumptions

The item weights are initially unknown, strictly positive reals. The population size is n, the size of the random sample is m and the weight of item v_i is w_i. The function $random(L, H)$ generates a uniform random number in (L, H). X denotes a random variable. Infinite precision arithmetic is assumed. Unless otherwise specified, all sampling problems are without replacement. Depending on the context, WRS is used to denote a weighted random sample or the operation of weighted random sampling.

Key Results

All the results with their proofs can be found in [4].

The crux of the WRS approach of this work is given with the following **algorithm A**:

Algorithm A

Input: A population V of n weighted items
Output: A WRS of size m
1: For each $v_i \in V$, $u_i = random(0,1)$ and $k_i = u_i^{(1/w_i)}$
2: Select the m items with the largest keys k_i as a WRS

Theorem 1 *Algorithm A generates a WRS.*

A reservoir-type adaptation of algorithm A is the following **algorithm A-Res**:

Algorithm A with a Reservoir (A-Res)

Input: A population V of n weighted items
Output: A reservoir R with a WRS of size m
1: The first m items of V are inserted into R
2: For each item $v_i \in R$: Calculate a key $k_i = u_i^{(1/w_i)}$, where $u_i = random(0,1)$
3: Repeat Steps 4–7 for $i = m + 1, m + 2, \ldots, n$
4: The smallest key in R is the current threshold T
5: For item v_i: Calculate a key $k_i = u_i^{(1/w_i)}$, where $u_i = random(0,1)$
6: If the key k_i is larger than T, then:
7: The item with the minimum key in R is replaced by item v_i

---

Algorithm A-Res performs the calculations required by algorithm A and hence by Theorem 1 A-Res generates a WRS. The number of reservoir operations for algorithm A-Res is given by the following Proposition:

Theorem 2 *If A-Res is applied on n weighted items, where the weights $w_i > 0$ are independent random variables with a common continuous distribution, then the expected number of reservoir insertions (without the initial m insertions) is:*

$$\sum_{i=m+1}^{n} P\left[item\ i\ is\ inserted\ into\ S\right] = \sum_{i=m+1}^{n} \frac{m}{i}$$

$$= O\left(m \cdot \log\left(\frac{n}{m}\right)\right).$$

Let S_w be the sum of the weights of the items that will be skipped by A-Res until a new item enters the reservoir. If T_w is the current threshold to enter the reservoir, then S_w is a continuous random variable that follows an exponential distribution. Instead of generating a key for every item, it is possible to generate random jumps that correspond to the sum S_w. Similar techniques have been applied for uniform random sampling (see for example [3]). The following algorithm A-ExpJ is an exponential jumps-type adaptation of algorithm A:

Theorem 3 *Algorithm A-ExpJ generates a WRS.*

The number of exponential jumps of A-ExpJ is given by Proposition 2. Hence algorithm A-ExpJ reduces the number of random variates that have to be generated from $O(n)$ (for A-Res) to $O(m \log(n/m))$. Since generating high-quality random variates can be a costly operation this is a significant improvement for the complexity of the sampling algorithm.

Applications

Random sampling is a fundamental problem in computer science with applications in many fields

Algorithm A with exponential jumps (A-ExpJ)

Input:	A population V of n weighted items
Output:	A reservoir R with a WRS of size m
1:	The first m items of V are inserted into R
2:	For each item $v_i \in R$: Calculate a key $k_i = u_i^{(1/w_i)}$, where $u_i = random(0,1)$
3:	The threshold T_w is the minimum key of R
4:	Repeat Steps 5–10 until the population is exhausted
5:	Let $r = random(0,1)$ and $X_w = \log(r)/\log(T_w)$
6:	From the current item v_c skip items until item v_i, such that:
7:	$w_c + w_{c+1} + \cdots + w_{i-1} < X_w \leq w_c + w_{c+1} + \cdots + w_{i-1} + w_i$
8:	The item in R with the minimum key is replaced by item v_i
9:	Let $t_w = T_w^{w_i}$, $r_2 = random(t_w, 1)$ and v_i's key: $k_i = r_2^{(1/w_i)}$
10:	The new threshold T_w is the new minimum key of R

including databases (see [5, 9] and the references therein), data mining, and approximation algorithms and randomized algorithms [7]. Consequently, algorithm A for WRS is a general tool that can find applications in the design of randomized algorithms. For example, algorithm A can be used within approximation algorithms for the k-Median [7].

The reservoir based versions of algorithm A, A-Res and A-ExpJ, have very small requirements for auxiliary storage space (m keys organized as a heap) and during the sampling process their reservoir continuously contains a weighted random sample that is valid for the already processed data. This makes the algorithms applicable to the emerging area of algorithms for processing data streams [2, 8].

Algorithms A-Res and A-ExpJ can be used for weighted random sampling with replacement from data streams. In particular, it is possible to generate a weighted random sample with replacement of size k with A-Res or A-ExpJ, by running concurrently, in one pass, k instances of A-Res or A-ExpJ respectively. Each algorithm instance must be executed with a trivial reservoir of size 1. At the end, the union of all reservoirs is a WRS with replacement.

URL to Code

The algorithms presented in this work are easy to implement. An experimental implementation in Java can be found at: http://utopia.duth.gr/~pefraimi/projects/WRS/index.html

Cross-References

▶ Online Paging and Caching
▶ Randomization in Distributed Computing

Recommended Reading

1. Ahrens JH, Dieter U (1985) Sequential random sampling. ACM Trans Math Softw 11:157–169
2. Babcock B, Babu S, Datar M, Motwani R, Widom J (2002) Models and issues in data stream systems. In: Proceedings of the twenty-first ACM SIGMOD-SIGACT-SIGART symposium on principles of database systems. ACM Press, pp 1–16
3. Devroye L (1986) Non-uniform random variate generation. Springer, New York
4. Efraimidis P, Spirakis P (2006) Weighted random sampling with a reservoir. Inf Process Lett J 97(5):181–185
5. Jermaine C, Pol A, Arumugam S (2004) Online maintenance of very large random samples. In: SIGMOD'04: proceedings of the 2004 ACM SIGMOD international conference on management of data. ACM Press, New York, pp 299–310
6. Knuth D (1981) The art of computer programming, vol 2, 2nd edn, Seminumerical algorithms. Addison-Wesley Publishing Company, Reading
7. Lin J-H, Vitter J (1992) ϵ-approximations with minimum packing constraint violation. In: 24th ACM STOC, pp 771–782
8. Muthukrishnan S (2005) Data streams: algorithms and applications. Found Trends Theor Comput Sci 1:1–126
9. Olken F (1993) Random sampling from databases. Ph.D. thesis, Department of Computer Science, University of California, Berkeley
10. Rajan V, Ghosh R, Gupta P (1989) An efficient parallel algorithm for random sampling. Inf Process Lett 30:265–268
11. Vitter J (1984) Faster methods for random sampling. Commun ACM 27:703–718
12. Vitter J (1985) Random sampling with a reservoir. ACM Trans Math Softw 11:37–57

W

Well Separated Pair Decomposition

Rolf Klein
Institute for Computer Science, University of
Bonn, Bonn, Germany

Keywords

Proximity algorithms for growth-restricted
metrics; Unit-disk graphs

Years and Authors of Summarized Original Work

1995; Callahan, Kosaraju

Problem Definition

Well-separated pair decomposition, introduced
by Callahan and Kosaraju [4], has found numerous applications in solving proximity problems
for points in the Euclidean space. A pair of point
sets (A, B) is c *well separated* if the distance
between A and B is at least c times the diameters of both A and B. A well-separated pair
decomposition of a point set consists of a set of
well-separated pairs that "cover" all the pairs of
distinct points, i.e., any two distinct points belong
to the different sets of some pair. Callahan and
Kosaraju [4] showed that for any point set in a
Euclidean space and for any constant $c \geq 1$,
there always exists a c-well-separated pair decomposition (c-WSPD) with linearly many pairs.
This fact has been very useful for obtaining
nearly linear-time algorithms for many problems,
such as computing k-nearest neighbors, N-body
potential fields, geometric spanners, approximate
minimum spanning trees, etc. Well-separated pair
decomposition has also been shown to be very
useful for obtaining efficient dynamic, parallel,
and external memory algorithms.

The definition of well-separated pair decomposition can be naturally extended to any metric
space. However, a general metric space may not
admit a well-separated pair decomposition with

a subquadratic size. Indeed, even for the metric
induced by the shortest path distance in a star
tree with unit weight on each edge, any well-
separated pair decomposition requires quadratically many pairs. This makes the well-separated
pair decomposition useless for such a metric.
However, it has been shown that for the unit-
disk graph metric, there do exist well-separated
pair decompositions with almost linear size, and
therefore many proximity problems under the
unit-disk graph metric can be solved efficiently.

Unit-Disk Graphs

Denote by $d(\cdot, \cdot)$ the Euclidean metric. For a set of
points S in the plane, the unit-disk graph $I(S) =
(S, E)$ is defined to be the weighted graph where
an edge $e = (p, q)$ is in the graph if $d(p, q) \leq
1$, and the weight of e is $d(p, q)$. Likewise, one
can define the unit-ball graph for points in higher
dimensions [5].

Unit-disk graphs have been used extensively to
model the communication or influence between
objects [9, 12] and have been studied in many
different contexts [5, 10]. For an example, wireless ad hoc networks can be modeled by unit-disk
graphs [8], as two wireless nodes can directly
communicate with each other only if they are
within a certain distance. In unsupervised learning, for a dense sampling of points from some
unknown manifold, the length of the shortest
path on the unit-ball graph is a good approximation of the geodesic distance on the underlying (unknown) manifold if the radius is chosen
appropriately [6, 14]. By using well-separated
pair decomposition, one can encode the all-pair
distances approximately by a compact data structure that supports approximate distance queries in
$O(1)$ time.

Metric Space

Suppose that (S, π) is a metric space where S
is a set of elements and π the distance function
defined on $S \times S$. For any subset $S_1 \subseteq S$,
the *diameter* $D_\pi(S_1)$ (or $D(S_1)$ when π is
clear from the context) of S is defined to be
$\max_{s_1, s_2 \in S_1} \pi(s_1, s_2)$. The *distance* $\pi(S_1, S_2)$
between two sets $S_1, S_2 \subseteq S$ is defined to be
$\min_{s_1 \in S_1, s_2 \in S_2} \pi(s_1, s_2)$.

Well-Separated Pair Decomposition

For a metric space (S, π), two nonempty subsets $S_1, S_2 \subseteq S$ are called c *well separated* if $\pi(S_1, S_2) \geq c \cdot \max(D_\pi(S_1), D_\pi(S_2))$.

Following the definition in [4], for any two sets A and B, a set of pairs $\mathcal{P} = \{P_1, P_2, \ldots, P_m\}$, where $P_i = (A_i, B_i)$, is called a *pair decomposition* of (A, B) (or of A if $A = B$) if:

- For all the i's, $A_i \subseteq A$, and $B_i \subseteq B$.
- $A_i \cap B_i = \varnothing$.
- For any two elements $a \in A$ and $b \in B$, there exists a unique i such that $a \in A_i$, and $b \in B_i$. Call (a, b) is *covered* by the pair (A_i, B_i).

If in addition, every pair in $\mathcal{P}$ is c well separated, $\mathcal{P}$ is called a *c-well-separated pair decomposition* (or c-WSPD for short). Clearly, any metric space admits a c-WSPD with quadratic size by using the trivial family that contains all the pairwise elements.

Key Results

In [7], it was shown that for the metric induced by the unit-disk graph on n points and for any constant $c \geq 1$, there does exist a c-WSPD with $O(n \log n)$ pairs, and such a decomposition can be computed in $O(n \log n)$ time. It was also shown that the bounds can be extended to higher dimensions. The following theorems state the key results for two and higher dimensions:

Theorem 1 *For any set S of n points in the plane and any $c \geq 1$, there exists a c-WSPD $\mathcal{P}$ of S under the unit-disk graph metric where $\mathcal{P}$ contains $O\left(c^4 n \log n\right)$ pairs and can be computed in $O\left(c^4 n \log n\right)$ time.*

Theorem 2 *For any set S of n points in $\mathbb{R}^k$, for $k \geq 3$, and for any constant $c \geq 1$, there exists a c-WSPD $\mathcal{P}$ of S under the unit-ball graph metric where $\mathcal{P}$ contains $O\left(n^{2-2/k}\right)$ pairs and can be constructed in $O\left(n^{4/3} polylog\, n\right)$ time for $k = 3$ and in $O\left(n^{2-2/k}\right)$ time for $k \geq 4$.*

The difficulty in obtaining a well-separated pair decomposition for the unit-disk graph metric is that two points that are close in space are not necessarily close under the graph metric. The above bounds are first shown for the point set with constant-bounded density, i.e., a point set where any unit disk covers only a constant number of points in the set. The upper bound on the number of pairs is obtained by using a packing argument similar to the one used in [1].

For a point set with unbounded density, one applies a clustering technique similar to the one used in [8] to the point set and obtains a set of "clusterheads" with a bounded density. Then the result for bounded density is applied to those clusterheads. Finally, the well-separated pair decomposition is obtained by combining the well-separated pair decomposition for the bounded density point sets and for the Euclidean metric. The number of pairs is dominated by the number of pairs constructed for a constant density set, which is in turn dominated by the bound given by the packing argument. It has been shown that the bounds on the number of pairs is tight for $k \geq 3$.

Applications

For a pair of well-separated sets, the distance between two points from different sets can be approximated by the "distance" between the two sets or the distance between any pair of points in different sets. In other words, a well-separated pair decomposition can be thought of as a compressed representation to approximate the $\Theta(n^2)$ pairwise distances. Many problems that require the pairwise distances to be checked can therefore be approximately solved by examining those distances between the well-separated pairs of sets. When the size of the well-separated pair decomposition is subquadratic, it often results in more efficient algorithms than examining all the pairwise distances. Indeed, this is the intuition behind many applications of the geometric well-separated pair decomposition. By using the same intuition, one can apply the well-separated pair decomposition in several proximity problems under the unit-disk graph metric.

Suppose that (S, d) is a metric space. Let $S_1 \subseteq S$. Consider the following natural proximity problems:

- **Furthest neighbor, diameter, center.** The furthest neighbor of $p \in S_1$ is the point in S_1 that maximizes the distance to p. Related problems include computing the *diameter*, the maximum pairwise shortest distance for points in S_1, and the *center*, the point that minimizes the maximum distance to all the other points.
- **Nearest neighbor, closest pair.** The nearest neighbor of $p \in S_1$ is the point in S_1 with the minimum distance to p. Related problems include computing the *closest pair*, the pair with the minimum shortest distance, and the *bichromatic closest pair*, the pair that minimizes the distance between points from two different sets.
- **Median.** The median of S is the point in S that minimizes the average (or total) distance to all the other points.
- **Stretch factor.** For a graph G defined on S, its stretch factor with respect to the unit-disk graph metric is defined to be the maximum ratio $\pi_G(p,q)/\pi(p,q)$, where π_G, π are the distances induced by G and by the unit-disk graph, respectively.

All the above problems can be solved or approximated efficiently for points in the Euclidean space. However, for the metric induced by a graph, even for planar graphs, very little is known besides solving the expensive all-pair shortest-path problem. For computing the diameter, there is a simple linear-time method that achieves a 2-approximation (Select an arbitrary node v and compute the shortest-path tree rooted at v. Suppose that the furthest node from v is distance D away. Then the diameter of the graph is no longer than $2D$, by triangle inequality.) and a 4/3-approximate algorithm with running time $O\left(m\sqrt{n\log n} + n^2\log n\right)$, for a graph with n vertices and m edges, by Aingworth et al. [2].

By using the well-separated pair decomposition, Gao and Zhang [7] showed that one can obtain better approximation algorithms for the above proximity problems for the unit-disk graph metric. Specifically, one can obtain almost linear-time algorithms for computing the 2.42-approximation and $O\left(n\sqrt{n\log n}/\varepsilon^3\right)$ time algorithms for computing the $(1 + \varepsilon)$-approximation for any $\varepsilon > 0$. In addition, the well-separated pair decomposition can be used to obtain an $O(n\log n/\varepsilon^4)$ space distance oracle so that any $(1 + \varepsilon)$ distance query in the unit-disk graph can be answered in $O(1)$ time.

The bottleneck of the above algorithms turns out to be computing the approximation of the shortest-path distances between $O(n\log n)$ pairs. The algorithm in [7] only constructs well-separated pair decompositions without computing a good approximation of the distances. The approximation ratio and the running time are dominated by that of the approximation algorithms used to estimate the distance between each pair in the well-separated pair decomposition. Once the distance estimation has been made, the rest of the computation only takes almost linear time.

For a general graph, it is unknown whether $O(n\log n)$ pairs shortest-path distances can be computed significantly faster than all-pair shortest-path distances. For a planar graph, one can compute the $O(n\log n)$ pairs shortest-path distances in $O\left(n\sqrt{n\log n}\right)$ time by using separators with $O(\sqrt{n})$ size [3]. This method extends to the unit-disk graph with constant-bounded density since such graphs enjoy a separator property similar to that of planar graphs [13]. As for approximation, Thorup [15] recently discovered an algorithm for planar graphs that can answer any $(1 + \varepsilon)$-shortest-distance query in $O(1/\varepsilon)$ time after almost linear-time preprocessing. Unfortunately, Thorup's algorithm uses balanced shortest-path separators in planar graphs which do not obviously extend to the unit-disk graphs. On the other hand, it is known that there does exist a planar 2.42-spanner for a unit-disk graph [11]. By applying Thorup's algorithm to that planar spanner, one can compute the 2.42-approximate shortest-path distance for $O(n \log n)$ pairs in almost linear time.

Open Problems

The most notable open problem is the gap between $\Omega(n)$ and $O(n \log n)$ on the number of pairs needed in the plane. Also, the time bound for $(1 + \varepsilon)$-approximation is still about $\tilde{O}\left(n\sqrt{n}\right)$ due to the lack of efficient methods for computing the $(1 + \varepsilon)$-approximate shortest-path distances between $O(n)$ pairs of points. Any improvement to the algorithm for that problem will immediately lead to improvement to all the $(1 + \varepsilon)$-approximate algorithms presented in this entry.

Cross-References

▶ Applications of Geometric Spanner Networks
▶ Separators in Graphs
▶ Sparse Graph Spanners
▶ Well Separated Pair Decomposition for Unit-Disk Graph

Recommended Reading

1. Agarwal P, Guibas L, Ngyuen A, Russel D, Zhang L (2004) Collision detection for deforming necklaces. Comput Geom Theory Appl 28(2):137–163
2. Aingworth D, Chekuri C, Indyk P, Motwani R (1999) Fast estimation of diameter and shortest paths (without matrix multiplication). SIAM J Comput 28(4):1167–1181
3. Arikati SR, Chen DZ, Chew LP, Das G, Smid MHM, Zaroliagis CD (1996) Planar spanners and approximate shortest path queries among obstacles in the plane. In: Díaz J, Serna M (eds) Proceedings of the 4th annual European symposium on algorithms, Barcelona, pp 514–528
4. Callahan PB, Kosaraju SR (1995) A decomposition of multidimensional point sets with applications to k-nearest-neighbors and n-body potential fields. J ACM 42:67–90
5. Clark BN, Colbourn CJ, Johnson DS (1990) Unit disk graphs. Discret Math 86:165–177
6. Fischl B, Sereno M, Dale A (1999) Cortical surface-based analysis II: inflation, flattening, and a surface-based coordinate system. NeuroImage 9:195–207
7. Gao J, Zhang L (2003) Well-separated pair decomposition for the unit-disk graph metric and its applications. In: Procroceedings of the 35th ACM symposium on theory of computing (STOC'03), San Diego, pp 483–492
8. Gao J, Guibas LJ, Hershberger J, Zhang L, Zhu A (2005) Geometric spanners for routing in mobile networks. IEEE J Sel Areas Commun Wirel Ad Hoc Netw (J-SAC) 23(1):174–185
9. Hale WK (1980) Frequency assignment: theory and applications. Proc IEEE 68(12): 1497–1513
10. Hunt HB III, Marathe MV, Radhakrishnan V, Ravi SS, Rosenkrantz DJ, Stearns RE (1998) NC-approximation schemes for NP- and PSPACE-hard problems for geometric graphs. J Algorithms 26(2):238–274
11. Li XY, Calinescu G, Wan PJ (2002) Distributed construction of a planar spanner and routing for ad hoc wireless networks. In: Proceedings of IEEE INFOCOM 2002, New York, 23–27 June 2002
12. Mead CA, Conway L (1980) Introduction to VLSI systems. Addison-Wesley, Reading
13. Miller GL, Teng SH, Vavasis SA (1991) An unified geometric approach to graph separators. In: Proceedings of the 32nd annual IEEE symposium on foundations of computer science, San Juan, pp 538–547
14. Tenenbaum J, de Silva V, Langford J (2000) A global geometric framework for nonlinear dimensionality reduction. Science 290:22
15. Thorup M (2004) Compact oracles for reachability and approximate distances in planar digraphs. J ACM 51(6):993–1024

Well Separated Pair Decomposition for Unit-Disk Graph

Jie Gao[1] and Li Zhang[2]
[1]Department of Computer Science, Stony Brook University, Stony Brook, NY, USA
[2]Microsoft Research, Mountain View, CA, USA

Keywords

Unit Disk Graphs; Well Separated Pair Decomposition

Years and Authors of Summarized Original Work

2003; Gao, Zhang

Problem Definition

Notations

Given a finite point set A in $\mathbb{R}^d$, its *bounding box* $R(A)$ is the d-dimensional hyperrectangle $[a_1, b_1] \times [a_2, b_2] \times \cdots \times [a_d, b_d]$ that contains A and has minimum extension in each dimension.

Two point sets A, B are said to be *well separated* with respect to a separation parameter $s > 0$ if there exist a real number $r > 0$ and two d-dimensional spheres C_A and C_B of radius r each, such that the following properties are fulfilled:

1. $C_A \cap C_B = \emptyset$
2. C_A contains the bounding box $R(A)$ of A
3. C_B contains the bounding box $R(B)$ of B
4. $|C_A C_B| \geq s \cdot r$.

Here $|C_A C_B|$ denotes the smallest Euclidean distance between two points of C_A and C_B, respectively. An example is depicted in Fig. 1. Given the bounding boxes $R(A)$, $R(B)$, it takes time only $O(d)$ to test if A and B are well separated with respect to s.

Two points of the same set, A or B, have a Euclidean distance at most $2/s$ times the distance any pair $(a, b) \in A \times B$ can have. Also, any two such pairs $(a, b), (a', b')$ differ in their distances $|a - b|, |a' - b'|$ by a factor of at most $1 + 4/s$.

Given a set S of n points in $\mathbb{R}^d$, a *well-separated pair decomposition* of S with

respect to separation parameter s is a sequence $(A_1, B_1), (A_2, B_2), \ldots, (A_m, B_m)$ where

1. $A_i, B_i \subset S$, for $i = 1 \ldots m$.
2. A_i and B_i are well separated with respect to s, for $i = 1 \ldots m$.
3. For all points $a, b \in S, a \neq b$, there exists a unique index i in $1 \ldots m$ such that $a \in A_i$ and $b \in B_i$, or $b \in A_i$ and $a \in B_i$ hold.

Obviously, each set $S = \{s_1, \ldots, s_n\}$ possesses a well-separated pair decomposition. One can simply use all singleton pairs $(\{s_i\}, \{s_j\})$ where $i < j$. The question is if decompositions consisting of fewer than $O(n^2)$, many pairs exist and how to construct them efficiently.

Key Results

In fact, the following result has been shown by Callahan and Kosaraju [1, 2].

Theorem 1 *Given a set S of n points in $\mathbb{R}^d$ and a separation parameter s, there exists a well-separated pair decomposition of S with respect to s that consists of $O(s^d d^{d/2} n)$ many pairs (A_i, B_i). It can be constructed in time $O(dn\log n + s^d d^{d/2+1} n)$.*

Thus, if dimension d and separation parameter s are fixed – which is the case in many applications – then the number of pairs is in

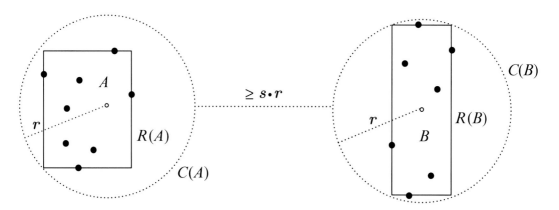

Well Separated Pair Decomposition for Unit-Disk Graph, Fig. 1 The sets A, B are well-separated with respect to s

$O(n)$, and the decomposition can be computed in time $O(n \log n)$.

The main tool in constructing the well-separated pair decomposition is the split tree $T(S)$ of S. The root, r, of $T(S)$ contains the bounding box $R(S)$ of S. Its two child nodes are obtained by cutting through the middle of the longest dimension of $R(S)$, using an orthogonal hyperplane. It splits S into two subsets S_a, S_b, whose bounding boxes $R(S_a)$ and $R(S_b)$ are stored at the two children a and b of root r. This process continues until only one point of S remains in each subset. These singleton sets form the leaves of $T(S)$. Clearly, the split tree $T(S)$ contains $O(n)$ many nodes. It needs not be balanced, but it can be constructed in time $O(dn\log n)$.

A well-separated pair decomposition of S, with respect to a given separation parameter s, can now be obtained from $T(S)$ in the following way. For each internal node of $T(S)$ with children v and w, the following recursive procedure $FindPairs(v, w)$ is called. If S_v and S_w are well separated, then the pair (S_v, S_w) is reported. Otherwise, one may assume that the longest dimension of $R(S_v)$ exceeds in length the longest dimension of $R(S_w)$ and that v_l, v_r are the child nodes of v in $T(S)$. Then, $FindPairs(v_l, w)$ and $FindPairs(v_r, w)$ are invoked.

The total number of procedure calls is bounded by the number of well-separated pairs reported, which can be shown to be in $O(s^d d^{d/2} n)$ by a packing argument. However, the total size of all sets A_i, B_i in the decomposition is in general quadratic in n.

Applications

From now on the dimension d is assumed to be a constant. The well-separated pair decomposition can be used in efficiently solving proximity problems for points in $\mathbb{R}^d$.

Theorem 2 *Let S be a set of n points in $\mathbb{R}^d$. Then a closest pair in S can be found in optimal time $O(n \log n)$.*

Indeed, let $q \in S$ be a nearest neighbor of $p \in S$. One can construct a well-separated pair decomposition with separation parameter $s > 2$ in time $O(n \log n)$, and let (A_i, B_i) be the pair where $p \in A_i$ and $q \in B_i$. If there were another point p' of S in A_i, one would obtain $| pp'| \leq 2/s \cdot | pq | < |pq|$, which is impossible. Hence, A_i is a singleton set. If (p, q) is a closest pair in S, then B_i must be singleton, too. Therefore, a closest pair can be found by inspecting all singleton pairs among the $O(n)$ many pairs of the well-separated pair decomposition.

With more effort, the following generalization can be shown.

Theorem 3 *Let S be a set of n points in $\mathbb{R}^d$, and let $k \leq n$. Then for each $p \in S$, its k nearest neighbors in S can be computed in total time $O(n \log n + nk)$. In particular, for each point in S can a nearest neighbor in S be computed in optimal time $O(n \log n)$.*

In dimension $d = 2$, one would typically use the Voronoi diagram for solving these problems. But as the complexity of the Voronoi diagram of n points can be as large as $n^{\lfloor d/2 \rfloor}$, the well-separated pair decomposition is much more convenient to use in higher dimensions.

A major application of the well-separated pair decomposition is the construction of good spanners for a given point set S. A spanner of S of dilation t is a geometric network N with vertex set S such that for any two vertices $p, q \in S$, the Euclidean length of a shortest path connecting p and q in N is at most t times the Euclidean distance $|pq|$.

Theorem 4 *Let S be a set of n points in $\mathbb{R}^d$, and let $t > 1$. Then a spanner of S of dilation t containing $O(s^d n)$ edges can be constructed in time $O(s^d n + n \log n)$, where $s = 4(t+1)(t-1)$.*

Indeed, if one edge (a_i, b_i) is chosen from each pair (A_i, B_i) of a well-separated pair decomposition of S with respect to s, these edges form a t-spanner of S, as can be shown by induction on the rank of each pair $(p, q) \in S^2$ in the list of all such pairs, sorted by distance.

Since spanners have many interesting applications of their own, several articles of this encyclopedia are devoted to this topic.

Open Problems

An important open question is which metric spaces admit well-separated pair decompositions. It is easy to see that the packing arguments used in the Euclidean case carry over to the case of convex distance functions in $\mathbb{R}^d$. More generally, Talwar [6] has shown how to compute well-separated pair decompositions for point sets of bounded aspect ratio in metric spaces of bounded doubling dimension.

On the other hand, for the metric induced by a disk graph in $\mathbb{R}^2$, a quadratic number of pairs may be necessary in the well-separated pair decomposition. (In a disk graph, each point $p \in S$ is center of a disk D_p of radius r_p. Two points p, q are connected by an edge if and only if $D_p \cap D_q \neq \emptyset$. The metric is defined by Euclidean shortest path length in the resulting graph. If this graph is a star with rays of identical length, a well-separated pair decomposition with respect to $s > 4$ must consist of singleton pairs.) Even for a unit disk graph, $\Omega(n^{2-2/d})$ many pairs may be necessary for points in $\mathbb{R}^d$, as Gao and Zhang [4] have shown.

Cross-References

▶ Applications of Geometric Spanner Networks
▶ Geometric Spanners

Recommended Reading

1. Callahan P (1995) Dealing with higher dimensions: the well-separated pair decomposition and its applications. Ph.D. thesis, The Johns Hopkins University
2. Callahan PB, Kosaraju SR (1995) A decomposition of multidimensional point sets with applications to k-nearest neighbors and n-body potential fields. J ACM 42(1):67–90
3. Eppstein D (1999) Spanning trees and spanners. In: Sack JR, Urrutia J (eds) Handbook of computational geometry. Elsevier, Amsterdam, pp 425–461
4. Ghao J, Zhang L (2005) Well-separated pair decomposition for the unit disk graph metric and its applications. SIAM J Comput 35(1):151–169
5. Narasimhan G, Smid M (2007) Geometric spanner networks. Cambridge University Press, New York
6. Talwar K (2004) Bypassing the embedding: approximation schemes and compact representations for low dimensional metrics. In: Proceedings of the thirty-sixth annual ACM symposium on theory of computing (STOC'04), Chicago, pp 281–290

Wire Sizing

Chris Chu
Department of Electrical and Computer Engineering, Iowa State University, Ames, IA, USA

Keywords

Interconnect optimization; VLSI physical design; Wire sizing; Wire tapering

Years and Authors of Summarized Original Work

1999; Chu, Wong

Problem Definition

The problem is about minimizing the delay of an interconnect wire in a very-large-scale integration (VLSI) circuit by changing the width (i.e., sizing) of the wire. The delay of interconnect wire has become a dominant factor in determining VLSI circuit performance for advanced VLSI technology. Wire sizing has been shown to be an effective technique to minimize the interconnect delay. The work of Chu and Wong [1] shows that the wire sizing problem can be transformed into a convex quadratic program. This quadratic programming approach is very efficient and can be naturally extended to simultaneously consider buffer insertion, which is another popular interconnect delay minimization technique. Previous approaches apply either a dynamic programming approach [2], which is computationally more expensive, or an iterative greedy approach [3, 4], which is hard to combine with buffer insertion.

The wire sizing problem is formulated as follows and is illustrated in Fig. 1. Consider a wire of length L. The wire is connecting a driver with driver resistance R_D to a load with load capacitance C_L. In addition, there is a set $H = \{h_1, \ldots, h_n\}$ of n wire widths allowed by the fabrication technology. Assume $h_1 > \cdots > h_n$. The wire sizing problem is to determine the wire width function $f(x) : [0, L] \to H$ so that the delay for a signal to travel from the driver through the wire to the load is minimized.

As in most previous works on wire sizing, the work of Chu and Wong uses the Elmore delay model to compute the delay. The Elmore delay model is a delay model for RC circuits (i.e., circuits consisting of resistors and capacitors). The Elmore delay for a signal path is equal to the sum of the delays associated with all resistors along the path, where the delay associated with each resistor is equal to its resistance times its total downstream capacitance. For a wire segment of length l and width h, its resistance is $r_0 l / h$ and its capacitance is $c(h)l$, where r_0 is the wire

sheet resistance and $c(h)$ is the unit length wire capacitance. $c(h)$ is an increasing function in practice. The wire segment can be modeled as a π-type RC circuit as shown in Fig. 2.

Key Results

Lemma 1 *The optimal wire width function $f(x)$ is a monotonically decreasing function.*

Lemma 1 above can be used to greatly simplify the wire sizing problem. It implies that an optimally sized wire can be divided into n segments such that the width of i-th segment is h_i. The length of each segment is to be determined. The simplified problem is illustrated in Fig. 3.

Lemma 2 *For the wire in Fig. 3, the Elmore delay is*

$$D = \frac{1}{2} l^T \Phi l + \rho^T l + R_D C_L$$

where

$$\Phi = \begin{pmatrix} c(h_1)r_0/h_1 & c(h_2)r_0/h_1 & c(h_3)r_0/h_1 & \cdots & c(h_n)r_0/h_1 \\ c(h_2)r_0/h_1 & c(h_2)r_0/h_2 & c(h_3)r_0/h_2 & \cdots & c(h_n)r_0/h_2 \\ c(h_3)r_0/h_1 & c(h_3)r_0/h_2 & c(h_3)r_0/h_3 & \cdots & c(h_n)r_0/h_3 \\ \vdots & \vdots & \vdots & \ddots & \vdots \\ c(h_n)r_0/h_1 & c(h_n)r_0/h_2 & c(h_n)r_0/h_3 & \cdots & c(h_n)r_0/h_n \end{pmatrix},$$

$$\rho = \begin{pmatrix} R_D c(h_1) + C_L r_0/h_1 \\ R_D c(h_2) + C_L r_0/h_2 \\ R_D c(h_3) + C_L r_0/h_3 \\ \vdots \\ R_D c(h_n) + C_L r_0/h_n \end{pmatrix} \quad and \quad l = \begin{pmatrix} l_1 \\ l_2 \\ l_3 \\ \vdots \\ l_n \end{pmatrix}.$$

So the wire sizing problem can be written in the following quadratic program:

$$\mathcal{WS} : \text{minimize } \tfrac{1}{2} l^T \Phi l + \rho^T l$$
$$\text{subject to } l_1 + \cdots + l_n = L$$
$$l_i \geq 0 \text{ for } 1 \leq i \leq n$$

Quadratic programming is NP-hard in general. In order to solve $\mathcal{WS}$ efficiently, some properties of the Hessian matrix Φ are explored.

Definition 1 (Symmetric Decomposable Matrix) Let $Q = (q_{ij})$ be an $n \times n$ symmetric matrix. If for some $\alpha = (\alpha_1, \ldots, \alpha_n)^T$ and

Wire Sizing, Fig. 1 The wire sizing problem

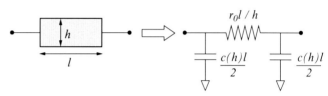

Wire Sizing, Fig. 2 The model of a wire segment by a π-type RC circuit

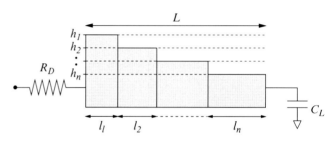

Wire Sizing, Fig. 3 The simplified wire sizing problem

$v = (v_1, \ldots, v_n)^T$ such that $0 < \alpha_1 < \cdots < \alpha_n$, $q_{ij} = q_{ji} = \alpha_i v_i v_j$ for $i \leq j$, then Q is called a symmetric decomposable matrix. Let Q be denoted as $SDM(\alpha, v)$.

Lemma 3 *If Q is symmetric decomposable, then Q is positive definite.*

Lemma 4 Φ *in $\mathcal{WS}$ is symmetric decomposable.*

Lemma 3 together with Lemma 4 implies that the Hessian matrix Φ of $\mathcal{WS}$ is positive definite. Hence, the problem $\mathcal{WS}$ is a *convex* quadratic program and is solvable in polynomial time [5].

The work of Chu and Wong proposes to solve $\mathcal{WS}$ by active set method. The active set method transforms a problem with some inequality constraints into a sequence of problems with only equality constraints. The method stops when the solution of the transformed problem satisfies both the feasibility and optimality conditions of the original problem. For the problem $\mathcal{WS}$, the active set method keeps track of an active set $\mathcal{A}$ in each iteration. The method sets $l_j = 0$ for all $j \in \mathcal{A}$ and ignores the constraints $l_j \geq 0$ for all $j \notin \mathcal{A}$. Let $\{j_1, \ldots, j_r\} = \{1, \ldots, n\} - \mathcal{A}$. Then

$\mathcal{WS}$ is transformed into the following equality-constrained wire sizing problem:

$$\mathcal{ECWS} : \text{minimize } \tfrac{1}{2} l_{\mathcal{A}}^T \Phi_{\mathcal{A}} l_{\mathcal{A}} + \rho_{\mathcal{A}}^T l_{\mathcal{A}}$$
$$\text{subject to } \Gamma_{\mathcal{A}} l_{\mathcal{A}} = L$$

where $l_{\mathcal{A}} = (l_{j_1}, \ldots, l_{j_r})^T$, $\Gamma_{\mathcal{A}} = (1\ 1\ \cdots\ 1)$, $\rho_{\mathcal{A}} = (R_D c(h_{j_1}) + C_L r_0 / h_{j_1}, \ldots, R_D c(h_{j_r}) + C_L r_0 / h_{j_r})^T$, and $\Phi_{\mathcal{A}}$ is the symmetric decomposable matrix corresponding to $\mathcal{A}$ (i.e.,

$$\Phi_{\mathcal{A}} = SDM(\alpha_{\mathcal{A}}, v_{\mathcal{A}}) \text{ with } \alpha_{\mathcal{A}} = \left(\frac{r_0}{c(h_{j_1})h_{j_1}}, \right.$$
$$\left. \ldots, \frac{r_0}{c(h_{j_r})h_{j_r}} \right)^T \text{ and } v_{\mathcal{A}} = (c(h_{j_1}), \ldots, c(h_{j_r}))^T).$$

Lemma 5 *The solution of $\mathcal{ECWS}$ is*

$$\lambda_{\mathcal{A}} = -(\Gamma_{\mathcal{A}} \Phi_{\mathcal{A}}^{-1} \Gamma_{\mathcal{A}}^T)^{-1} (\Gamma_{\mathcal{A}} \Phi_{\mathcal{A}}^{-1} \rho_{\mathcal{A}} + L)$$
$$l_{\mathcal{A}} = -\Phi_{\mathcal{A}}^{-1} \Gamma_{\mathcal{A}}^T \lambda_{\mathcal{A}} - \Phi_{\mathcal{A}}^{-1} \rho_{\mathcal{A}}$$

Lemma 6 *If Q is symmetric decomposable, then Q^{-1} is tridiagonal. In particular, if $Q = SDM(\alpha, v)$, then $Q^{-1} = (\theta_{ij})$*

where $\theta_{ii} = \dfrac{1}{(\alpha_i - \alpha_{i-1})v_i^2} + \dfrac{1}{(\alpha_{i+1} - \alpha_i)v_i^2}$,

$\theta_{i,i+1} = \theta_{i+1,i} = \dfrac{-1}{(\alpha_{i+1} - \alpha_i)v_i v_{i+1}}$ *for* $1 \le$

$i \le n - 1$, $\theta_{nn} = \dfrac{1}{(\alpha_n - \alpha_{n-1})v_n^2}$, *and* $\theta_{ij} = 0$

otherwise.

By Lemmas 5 and 6, $\mathcal{ECWS}$ can be solved in $O(n)$ time. To solve $\mathcal{WS}$, in practice, the active set method takes less than n iterations and hence the total runtime is $O(n^2)$. Note that unlike previous works, the runtime of this convex quadratic programming approach is independent of the wire length L.

Applications

The wire sizing technique is commonly applied to minimize the wire delay and hence to improve the performance of VLSI circuits. As there are typically millions of wires in modern VLSI circuits, and each wire may be sized many times in order to explore different architecture, logic design, and layout during the design process, it is very important for wire sizing algorithms to be very efficient.

Another popular technique for delay minimization of slow signals is to insert buffers (also called repeaters) to strengthen and accelerate the signals. The work of Chu and Wong can be naturally extended to simultaneously handle buffer insertion. It is shown in [1] that the delay minimization problem for a wire by simultaneous buffer insertion and wire sizing can also be formulated as a convex quadratic program and be solved by active set method. The runtime is only m times more than that of wire sizing, where m is the number of buffers inserted. m is typically 5 or less in practice.

About one third of all nets in a typical VLSI circuit are multi-pin nets (i.e., nets with a tree structure to deliver a signal from a source to several sinks). It is important to minimize the delay of multi-pin nets. The work of Chu and Wong can also be applied to optimize multi-pin nets. The extension is described in Mo and Chu [6]. The idea is to integrate the quadratic pro-

gramming approach into a dynamic programming framework. Each branch of the net is solved as a convex quadratic program, while the overall tree structure is handled by dynamic programming.

Open Problems

After two decades of active research, the wire sizing problem by itself is now considered a well-solved problem. Some important solutions are [1–4, 6–15]. The major remaining challenge is to simultaneously apply wire sizing with other interconnect optimization techniques to improve circuit performance. Wire sizing, buffer insertion, and gate sizing are three most commonly used interconnect optimization techniques. It has been demonstrated that better performance can be achieved by applying these three techniques simultaneously rather sequentially. One very practical problem is to perform simultaneous wire sizing, buffer insertion, and gate sizing to a combinational circuit such that the total resource usage (e.g., wire/buffer/gate area, power consumption) is minimized while the delay of all input-to-output paths are less than a given target.

Cross-References

▶ Circuit Retiming
▶ Circuit Retiming: An Incremental Approach
▶ Gate Sizing

Recommended Reading

1. Chu CCN, Wong DF (1999) A quadratic programming approach to simultaneous buffer insertion/sizing and wire sizing. IEEE Trans Comput-Aided Des 18(6):787–798
2. Lillis J, Cheng C-K, Lin T-T (1995) Optimal and efficient buffer insertion and wire sizing. In: Proceedings of the custom integrated circuits conference, Santa Clara, pp 259–262
3. Cong J, Leung K-S (1995) Optimal wiresizing under the distributed Elmore delay model. IEEE Trans Comput-Aided Des 14(3):321–336
4. Chen C-P, Wong DF (1996) A fast algorithm for optimal wire-sizing under Elmore delay model. In: Proceedings of the IEEE ISCAS, Atlanta, vol 4, pp 412–415

W

5. Kozlov MK, Tarasov SP, Khachiyan LG (1979) Polynomial solvability of convex quadratic programming. Sov Math Dokl 20:1108–1111

6. Mo Y-Y, Chu C (2001) A hybrid dynamic/quadratic programming algorithm for interconnect tree optimization. IEEE Trans Comput-Aided Des 20(5):680–686

7. Sapatnekar SS (1994) RC interconnect optimization under the Elmore delay model. In: Proceedings of the ACM/IEEE design automation conference, San Diego, pp 387–391

8. Fishburn JP, Schevon CA (1995) Shaping a distributed-RC line to minimize Elmore delay. IEEE Trans Circuits Syst-I Fundam Theory Appl 42(12):1020–1022

9. Chen C-P, Chen Y-P, Wong DF (1996) Optimal wiresizing formula under the Elmore delay model. In: Proceedings of the ACM/IEEE design automation conference, Las Vegas, pp 487–490

10. Cong J, He L (1996) Optimal wiresizing for interconnects with multiple sources. ACM Trans Des Autom Electron Syst 1(4):568–574

11. Fishburn JP (1997) Shaping a VLSI wire to minimize Elmore delay. In: Proceedings of the European design and test conference, Paris, pp 244–251

12. Chen C-P, Wong DF (1997) Optimal wire-sizing function with fringing capacitance consideration. In: Proceedings of the ACM/IEEE design automation conference, Anaheim, pp 604–607

13. Kay R, Bucheuv G, Pileggi L (1997) EWA: efficient wire-sizing algorithm. In: Proceedings of the international symposium on physical design, Napa Valley, pp 178–185

14. Chu CCN, Wong DF (1999) Greedy wire-sizing is linear time. IEEE Trans Comput-Aided Des 18(4):398–405

15. Gao Y, Wong DF (2000) Wire-sizing for delay minimization and ringing control using transmission line model. In: Proceedings of the conference on design automation and test in Europe, Paris, pp 512–516

Work-Function Algorithm for k-Servers

Marek Chrobak
Computer Science, University of California, Riverside, CA, USA

Keywords

Analysis; K-server problem; On-line algorithms; Work funktions

Years and Authors of Summarized Original Work

1994; Koutsoupias, Papadimitriou

Problem Definition

In the k-*Server Problem*, the task is to schedule the movement of k-servers in a metric space $\mathbb{M}$ in response to a sequence $\varrho = r_1, r_2, \ldots, r_n$ of *requests*, where $r_i \in \mathbb{M}$ for all i. The servers initially occupy some configuration $X_0 \subseteq \mathbb{M}$. After each request r_i is issued, one of the k-servers must move to r_i. A *schedule* S specifies which server moves to each request. The task is to compute a schedule with minimum *cost*, where the cost of a schedule is defined as the total distance traveled by the servers. The example below shows a schedule for 2 servers on a sequence of requests (Fig. 1).

In the offline case, if the complete request sequence ϱ is known, the optimal schedule can be computed in polynomial time [9].

Most of the research on the k-Server Problem focussed on the *online* variant, where the requests are issued one at a time. After the ith request r_i is issued, an online algorithm must decide, irrevocably, which server to move to r_i before the next request r_{i+1} is issued. It is quite easy to see that in this Online scenario it is not possible to guarantee an optimal schedule for all request sequences. The accuracy of solutions produced by

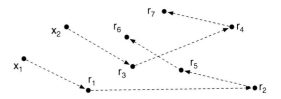

Work-Function Algorithm for k-Servers, Fig. 1 A schedule for 2 servers on a request sequence $\varrho = r_1, r_2, \ldots, r_7$. The initial configuration is $X_0 = \{x_1, x_2\}$. Server 1 serves r_1, r_2, r_5, r_6, while server 2 serves r_3, r_4, r_7. The cost of this schedule is $d(x_1, r_1) + d(r_1, r_2) + d(r_2, r_5) + d(r_5, r_6) + d(x_2, r_3) + d(r_3, r_4) + d(r_4, r_7)$, where $d(x, y)$ denotes the distance between points x, y

such online algorithms is often evaluated within the framework of competitive analysis. Denote by $cost_\mathcal{A}(\varrho)$ the cost of the schedule produced by an online k-server algorithm $\mathcal{A}$ on a request sequence ϱ, and let $opt(\varrho)$ be the cost of an optimal schedule on ϱ. $\mathcal{A}$ is called R-competitive if $cost_\mathcal{A}(\varrho) \leq R \cdot opt(\varrho) + B$, where B is a constant that may depend on $\mathbb{M}$ and X_0. The smallest such R is called the *competitive ratio* of $\mathcal{A}$. Of course, the smaller the ratio R the better.

The k-Server Problem was introduced by Manasse, McGeoch, and Sleator [14, 15], who proved that no (deterministic) online algorithm can achieve a competitive ratio smaller than k, in any metric space with at least $k + 1$ points. They also gave a 2-competitive algorithm for $k = 2$ and stated what is now known as the *k-Server Conjecture*, which postulates that there exists a k-competitive online algorithm for all k. Koutsoupias and Papadimitriou [11, 12] (see also [3, 8, 10]) proved that the *Work-Function Algorithm*, presented in the next section, has competitive ratio at most $2k - 1$, which to date remains the best upper bound on the competitive ratio.

Key Results

The idea of the Work-Function Algorithm is to balance two greedy strategies when a new request is issued. The first one is to simply serve the request with the closest server. The second strategy attempts to follow the optimum schedule. Roughly, from among the k possible new configurations, this strategy chooses the one where the optimum schedule would be at this time, if no more requests remained to be issued.

To formalize this idea, for each request sequence ϱ and a k-server configuration X, let $\omega_\varrho(X)$ be the minimum cost of serving ϱ under the constraint that at the end the server configuration is X. (Assume, for simplicity, that the initial configuration X_0 is fixed.) The function $\omega_\varrho(\cdot)$ is called the *work function* after the request sequence ϱ.

Algorithm WFA. Denote by σ the sequence of past requests, and suppose that the current server configuration is $S = \{s_1, s_2, \ldots, s_k\}$, where s_j is the location of the j-th server. Let r be the new request. Choose $s_j \in S$ that minimizes the quantity $\omega_{\sigma r}(S - \{s_j\} \cup \{r\}) + d(s_j, r)$, and move server j to r.

Theorem 1 ([11, 12]) *Algorithm WFA is $(2k - 1)$-competitive.*

As observed in [6], Algorithm WFA can be interpreted as a primal-dual algorithm.

Applications

The k-Server Problem can be viewed as an abstraction of online problems that arise in emergency crew scheduling, caching (or paging) in two-level memory systems, scheduling of disk heads, and other. Nevertheless, in its pure abstract form, it is mostly of theoretical interest.

Algorithm WFA can be applied to some generalizations of the k-Server Problem. In particular, it is $(2n - 1)$-competitive for n-state metrical task systems, matching the lower bound [3, 4, 8]. See [1, 3, 5] for other applications and extensions.

Open Problems

Theorem 1 comes tantalizingly close to settling the k-Server Conjecture described earlier in this section. In fact, it has been even conjectured that Algorithm WFA itself is k-competitive for k-servers, but the proof of this conjecture, so far, remains elusive.

For $k \geq 3$, k-competitive online k-server algorithms are known only for some restricted metric spaces, including trees, metric spaces with up to $k + 2$ points, and the Manhattan plane for $k = 3$ (see [2, 7, 9, 13]). As the analysis of Algorithm WFA in the general case appears difficult, it would be of interest to prove its k-competitiveness for some natural special cases, for example in the plane (with any reasonable metric) for $k \geq 4$ servers.

Very little is known about the competitive ratio of the k-Server Problem in the randomized case. In fact, it is not even known whether a ratio better than 2 can be achieved for $k = 2$.

Cross-References

► Deterministic Searching on the Line
► Generalized Two-Server Problem
► Metrical Task Systems
► Online Paging and Caching

Recommended Reading

1. Anderson EJ, Hildrum K, Karlin AR, Rasala A, Saks M (2002) On list update and work function algorithms. Theor Comput Sci 287:393–418
2. Bein W, Chrobak M, Larmore LL (2002) The 3-server problem in the plane. Theor Comput Sci 287:387–391
3. Borodin A, El-Yaniv R (1998) Online computation and competitive analysis. Cambridge University Press, Cambridge
4. Borodin A, Linial N, Saks M (1987) An optimal online algorithm for metrical task systems. In: Proceedings of the 19th symposium on theory of computing (STOC). ACM, New York, pp 373–382
5. Burley WR (1996) Traversing layered graphs using the work function algorithm. J Algorithms 20:479–511
6. Chrobak M (2007) Competitiveness via primal-dual. SIGACT News 38:100–105
7. Chrobak M, Larmore LL (1991) An optimal online algorithm for k servers on trees. SIAM J Comput 20:144–148
8. Chrobak M, Larmore LL (1998) Metrical task systems, the server problem, and the work function algorithm. In: Fiat A, Woeginger GJ (eds) Online algorithms: the state of the art. Springer, Berlin/New York, pp 74–94
9. Chrobak M, Karloff H, Payne TH, Vishwanathan S (1991) New results on server problems. SIAM J Discret Math 4:172–181
10. Koutsoupias E (1999) Weak adversaries for the k-server problem. In: Proceedings of the 40th symposium on foundations of computer science (FOCS). IEEE, New York, pp 444–449
11. Koutsoupias E, Papadimitriou C (1994) On the k-server conjecture. In: Proceedings of the 26th symposium on theory of computing (STOC). ACM, Montreal, pp 507–511
12. Koutsoupias E, Papadimitriou C (1995) On the k-server conjecture. J ACM 42:971–983
13. Koutsoupias E, Papadimitriou C (1996) The 2-evader problem. Inf Process Lett 57:249–252
14. Manasse M, McGeoch LA, Sleator D (1988) Competitive algorithms for online problems. In: Proceedings of the 20th symposium on theory of computing (STOC). ACM, Chicago, pp 322–333
15. Manasse M, McGeoch LA, Sleator D (1990) Competitive algorithms for server problems. J Algorithms 11:208–230

List of Entries

© Springer Science+Business Media New York 2016
M.-Y. Kao (ed.), *Encyclopedia of Algorithms*,
DOI 10.1007/978-1-4939-2864-4

This encyclopedia includes no entries for X, Y & Z